Homayoon Beigi

Fundamentals of
Speaker Recognition

 Springer

Homayoon Beigi
Recognition Technologies, Inc.
Yorktown Heights, NY, USA
beigi@recotechnologies.com

ISBN 978-0-387-77591-3   e-ISBN 978-0-387-77592-0
DOI 10.1007/978-0-387-77592-0
Springer New York Dordrecht Heidelberg London

Library of Congress Control Number: 2011941119

Printed on acid-free paper

Springer is part of Springer Science+Business Media (www.springer.com)

I dedicate this book to my inspiring and supportive wife, Pargol, my dear and wonderful son, Ara, my motivating father, the memory of my dedicated mother, and to my nurturing grandmother!

Preface

When I was being interviewed at the handwriting recognition group of IBM T.J. Watson Research Center in December of 1990, one of the interviewers asked me why, being a mechanical engineer, I was applying for a position in that group. Well, he was an electrical engineer and somehow was under the impression that handwriting recognition was an electrical engineering field! My response was that I had done research on Kinematics, Dynamics, Control, Signal Processing, Optimization, Neural Network Learning theory and lossless image compression during the past 7 years while I was in graduate school. I asked him what background he thought would have been more relevant to do research in handwriting recognition.

Anyhow, I joined the on-line handwriting recognition group which worked side-by-side with the speech recognition group. Later, I transferred to the speech recognition group and worked on speaker recognition. Aside from the immediate front-end processing, on-line handwriting recognition, signature verification, speech recognition and speaker recognition have a lot in common. During the 10 years at IBM I also worked on many complementary problems such as phonetics, statistical learning theory, language modeling, information theoretic research, etc. This continued with further work on real-time large-scale optimization, interactive voice response systems, standardization and more detailed speaker recognition research at Recognition Technologies, Inc. to the present date, not to mention the many years of code optimization, integer arithmetic, software architecture and alike within the past 25 years.

The reason for sharing this story with the reader is to point out the extreme multidisciplinary nature of the topic of speaker recognition. In fact, every one of the fields which I mentioned above, was quite necessary for attaining a deep understanding of the subject. This was the prime motivation which lead me to the writing of this book. As far as I know, this is the first textbook (reference book) on the subject which tries to deal with every aspect of the field, as much as possible. I have personally designed and implemented (coded) two full-featured speaker recognition systems and in the process have had to deal with many different aspects of the subject from theory to practice.

One problem with which many researchers are faced, when dealing with highly multi-disciplinary subjects such as speaker recognition, is the scattered information in all the relevant fields. Usually, most treatments of the subject try to use hand-waving to get the reader through all the different aspects of the subject. In their treatment, most survey papers, throw a plethora of references at the reader so that he/she would follow up on each of the many leads – which is usually impractical. This causes a half-baked understanding of the subject and its details which will be carried over from master to apprentice, leaving the field crippled at times.

In the above description, while qualifying this book, I used the word textbook, but I also parenthetically referred to it as a reference book. Well, originally when I was asked to write it, we had a textbook approach in mind. However, as I delved deeper into the attempt of presenting all the necessary material, based on the motivation which was stated earlier, the coverage of the different subjects grew quickly from an intended 300 page textbook to nearly 900 pages which probably qualifies as a reference book. In fact, most of the book may be used as reference material for many related subjects.

In my many years of teaching different courses at Columbia University, such as Speech Recognition, Signal Recognition, and Digital Control, I have noticed the following. Since today's technologies are built layer-upon-layer on top of existing basic technologies, the amount of underlying knowledge necessary for understanding the topics at the tips of these theoretical hierarchies has grown exponentially. This makes it quite hard for a researcher in a multi-disciplinary topic to grasp the intricacies of the underlying theory. Often, to deal with the lack of time, necessary for an in-depth understanding of the underlying theory, it is either skipped or left to the pursuance of the students, of their own volition.

In this book, I have tried to cover as much detail as possible and to keep most of the necessary information self-contained and rigorous. Although, you will see many references presented at the end of each chapter and finally as a collection in a full bibliography, the references are only meant for the avid reader to follow up into the nitty-gritty details upon interest. Most of the high-level details are stated in the 26 chapters which make up this book.

To be able to present the details, and yet have a smooth narrative in the main text, a large amount of the detailed material is included in the last 4 chapters of the book, categorized as *Background Material*. These chapters start with the coverage of some necessary *linear algebra* and related mathematical bases followed by a very detailed chapter on *integral transforms*. Since integral transforms are central to the *signal processing* end of the subject, and they heavily rely on an intimate knowledge of *complex variable theory*, Chapter 24 tries to build that foundation for the reader. Moreover, the essence of theoretical subjects such as *neural networks* and *support vector machines* is the field of *numerical optimization* which has been covered in some detail in Chapter 25. The last chapter covers details on *standards*, related to the speaker recognition field. This is a practical aspect which is usually left out in most textbooks and, in my opinion, should be given much more attention.

The main narrative of the book has *three major parts*:
Part I covers the *introductory and basic theory* of the subject including *anatomy, signal representation, phonetics, signal processing and feature extraction, probability theory, information theory, metrics and distortion measures, Bastian learning theory, parameter estimation and leaning, clustering, parameter transformation, hidden Markov modeling, neural networks,* and *support vector machines*.

The second part, *advanced theory*, covers subjects which deal more directly with speaker recognition. These topics are *speaker modeling, speaker recognition implementation,* and *signal enhancement and compensation*.

Part III, *practice*, discusses topics specifically related to the implementation of speaker recognition or related issues. These are *representation of results, time-lapse effects, adaptation techniques,* and finally, *overall design issues*.

Every effort has been made to deliver the contents of the book in a hierarchical fashion. In other words, think of writing an efficient program in a class-based programming language where the main program is simply a few lines. The main program, in this, case would be the chapters in *Part III (Practice)*. The classes that are instantiated within the main program, mainly come from *Part II (advanced theory)* and they in-turn include more specialized classes from *Part I (basic theory)*. Part II and Part I classes make calls to methods in *Part IV (background material)*.

Yorktown Heights, New York, August 2011   *Homayoon Beigi*

Acknowledgments

I would like to convey my greatest appreciation for the invaluable reviews and persistent moral support of my dearest friends and colleagues Dr. Sassan Pejhan and Dr. Fereydoun Maali, as well as my lovely and supportive wife, Pargol, throughout this enormous task!

I am also most grateful to my sister, Mandis, for her kind and generous help with some of the references and to Dr. Judith Markowitz for our valuable discussions on standards.

Furthermore, I would like to acknowledge the tremendous effort of the staff at Springer for making this book possible.

Contents

Acronyms and Abbreviations

ADPCM	Adaptive Differential Pulse Code Modulation
AEP	Asymptotic Equipartition Property
AGN	Automatic Gain Normalization
AHC	Agglomorative Hierarchical Clustering
ANSI	American National Standards Institute
API	Application Programming Interface
ASR	Automatic Speech Recognition
BFGS	Broyden-Fletcher-Goldfarb-Shanno
BIC	Bayesian Information Criterion
BioAPI	Biometric Application Programming Interface
CBEFF	Common Biometric Exchange Formats Framework
CDMA	Code Division Multiple Access
CELP	Code Excited Linear Prediction
CHN	Cepstral Histogram Normalization
CMA	Constant Modulus Algorithm
CMN	Cepstral Mean Normalization
CMS	Cepstral Mean Subtraction
CMVN	Cepstral Mean and variance Normalization
CNG	Comfort Noise Generation
CoDec	Coder/Decoder
CS-ACELP	Conjugate Structure Algebraic Code Excited Linear Prediction
dB	deci Bel (decibel)
DC	Direct Current
DCF	Detection Cost Function
DCT	Discrete Cosine Transform
DET	Detection Error Trade-Off
DFP	Davidon-Fletcher-Powell
DHC	Divisive Hierarchical Clustering
DPCM	Differential Pulse Code Modulation
DTMF	Dual Tone Multi-Frequency

EER	Equal-Error Rate
e.g.	exempli gratia (for example)
EIH	Ensemble Interval Histogram
ELRA	European Language Resources Association
EM	Expectation Maximization
EMD	Empirical Mode Decomposition
EMMA	Extensible Multimodal Annotation
ETSI	European Telecommunications Standards Institute
FA	Factor Analysis
FAR	False Acceptance Rate
FBI	Federal Bureau of Investigation
FFT	Fast Fourier Transform
FRR	False Rejection Rate
FTP	File Transfer Protocol
GLR	General Likelihood Ratio
GMM	Gaussian Mixture Model(s)
GrXML	Grammar eXtensible Markup Language
GSM	Groupe Spécial Mobile *or* Global System for Mobile Communications
GSM-EFR	GSM Enhanced Full Rate
HE-AAC	High Efficiency Advanced Audio Coding
HEQ	Histogram Equalization
HME	Hierarchical Mixtures of Experts
HMM	Hidden Markov Model(s)
H-Norm	Handset Normalization
HTER	Half Total Error Rate
HTTP	HyperText Transfer Protocol
Hz	Hertz
IBM	International Business Machines
ID	Identity; Identification
iDEN	Integrated Digital Enhanced Network
i.e.	id est (that is)
IEC	International Electrotechnical Commission
IETF	Internet Engineering Task Force
IFG	Inferior Frontal Gyrus (of the Brain)
i.i.d.	Independent and Identically Distributed (Description of a type of Random Variable)
IMF	Intrinsic Mode Function
INCITS	InterNational Committee for Information Technology Standards
ISO	International Organization for Standardization
ISV	Independent Software Vendor
ITU	International Telecommunications Union
ITU-T	ITU Telecommunication Standardization Sector
JFA	Joint Factor Analysis
JTC	Joint ISO/IEC Technical Committee
IVR	Interactive Voice Response

KLT	Karhunen-Loève Transformation
LBG	Linde-Buzo-Gray
LFA	Latent Factor Analysis
kHz	kilo-Hertz
LDC	Linguistic Data Consortium
LAR	Log Area Ratio
LLN	Law of Large Numbers
LLR	Log-Likelihood Ratio
LPC	Linear Predictive Coding, also, Linear Predictive Coefficients
LPCM	Linear Pulse Code Modulation
MAP	Maximum A-Posteriori
MFCC	Mel Frequency Cepstral Coefficients
MFDWC	Mel Frequency Discrete Wavelet Coefficients
MIT-LL	Massachusetts Institute of Technology's Lincoln Laboratories
MLE	Maximum Likelihood Estimation or Maximum Likelihood Estimate
MLLR	Maximum Likelihood Linear Regression
MMIE	Maximum Mutual Information Estimation
MPEG	Moving Picture Experts Group
MRCP	Media Resource Control Protocol
NAP	Nuisance Attribute Projection
N.B.	Nota Bene (Note Well) – Note that
NIST	National Institute of Standards and Technology
NLSML	Natural Language Semantics Markup Language
NLU	Natural Language Understanding
OGI	Oregon Graduate Institute
PAM	Pulse Amplitude Modulation (Sampler)
PARCOR	Partial Correlation
PCA	Principal Component Analysis
PCM	Pulse Code Modulation
PCMA	A-Law Pulse Code Modulation
PCMU	μ-Law Pulse Code Modulation
PDC	Personal Digital Cellular
ppm	Parts per Million
pRAM	Probabilistic Random Access Memory
PSTN	Public Switched Telephone Network
PWM	Pulse Width Modulation (Sampler)
PWPAM	Pulse Width Pulse Amplitude Modulation (Sampler)
QCELP	Qualcomm Code Excited Linear Prediction
Q.E.D.	Quod Erat Demonstradum (That which was to be Demostrated)
QOS	Quality of Service
rad.	radians
RASTA	RelAtive SpecTrAl
RBF	Radial Basis Function
RFC	Request for Comments

RIFF	Resource Interchange File Format
RNN	Recurrent Neural Network
ROC	Receiver Operator Characteristic
RTP	Real-time Transport Protocol
SAFE	Standard Audio Format Encapsulation
SC	Subcommittee
SI	Systèm International
SIMM	Sequential Interacting Multiple Models
SIP	Session Initiation Protocol
SIV	Speaker Identification and Verification
SLLN	Strong Law of Large Numbers
SPHERE	SPeech HEader REsources
SPI	Service Provider Interface
SRAPI	Speech Recognition Application Programming Interface
SSML	Speech Synthetic Markup Language
SVAPI	Speaker Verification Application Programming Interface
SVM	Support Vector Machine(s)
TCP	Transmission Control Protocol
TD-SCDMA	Time Division Synchronous Code Division Multiple Access
TLS	Transport Layer Security
TDMA	Time Division Multiple Access
TDNN	Time-Delay Neural Network
T-Norm	Test Normalization
TTS	Text To Speech
U8	Unsigned 8-bit Storage
U16	Unsigned 16-bit Storage
U32	Unsigned 32-bit Storage
U64	Unsigned 64-bit Storage
UDP	User Datagram Protocol
VAD	Voice Activity Detection
VAR	Value Added Reseller
VB	Variational Bayesian Technique
VBWG	Voice Browser Working Group
VoiceXML	Voice eXtensible Markup Language
VoIP	Voice Over Internet Protocol
VQ	Vector Quantization
W3C	World Wide Web Consortium
WG	Workgroup
WCDMA	Wideband Code Division Multiple Access
WCDMA HSPA	Wideband Code Division Multiple Access High Speed Packet Access
WLLN	Weak Law of Large Numbers
XML	eXtensible Markup Language

Nomenclature

In this book, lower-case bold letters are used to denote vectors and upper-case bold letters are used for matrices. For set, measure, and probability theory, as much as possible, special style guidelines have been used such that the letter X when written as $\mathscr{X}$ signifies a set and when written as $\mathfrak{X}$ is a class of (sub)sets. The following is a list of symbols used in the text:

$\{\varnothing\}$	Empty Set
$\overline{(\alpha+i\beta)}$	Complex Conjugate of $(\alpha+i\beta)$ equal to $(\alpha-i\beta)$
$\lvert.\rvert$	Determinant of .
$(\mathbf{a})_{[i]}$	i^{th} element of vector $\mathbf{a}$.
$(\mathbf{A})_{[i][j]}$	Element in row i and column j of matrix $\mathbf{A}$.
$(\mathbf{A})_{[i]}$	Column i of matrix $\mathbf{A}$.
$*$	Convolution, e.g., $g*h$.
$\circ$	Correlation (Cross-Correlation), e.g., $g\circ h$, $g\circ g$.
$\tilde{\cdot}$	Estimate of $\cdot$
$\wedge$	Logical And
$\vee$	Logical Or
$\longmapsto$	Maps to, e.g. $\mathscr{R}^N \mapsto \mathscr{R}^M$
$\longleftrightarrow$	Mutual Mapping (used for signal/transform pairs, e.g. $h(t) \leftrightarrow H(s)$).
$\therefore$	Therefore
$\overset{R}{\equiv}$	Equivalent with respect to equivalence relation R.
$\sim$	Distributed According to $\cdots$ (a Distribution).
$\preceq$	$\mathbf{a} \preceq \mathbf{b}$ is read, $\mathbf{a}$ *precedes* $\mathbf{b}$ – i.e. in an ordered set of vectors.
$\prec$	$\mathbf{a} \prec \mathbf{b}$ is read, $\mathbf{a}$ *strictly precedes* $\mathbf{b}$ – i.e. in an ordered set of vectors.
$\succeq$	$\mathbf{a} \succeq \mathbf{b}$ is read, $\mathbf{a}$ *succeeds* $\mathbf{b}$ – i.e. in an ordered set of vectors.
$\succ$	$\mathbf{a} \succ \mathbf{b}$ is read, $\mathbf{a}$ *strictly succeeds* $\mathbf{b}$ – i.e. in an ordered set of vectors.
$\bar{x}$	Mean (Expected Value) of x
$\mathscr{A}$	A generic set.
$\mathscr{A}^{\complement}$	Complement of set $\mathscr{A}$.
$\mathscr{A}\backslash\mathscr{B}$	The difference between $\mathscr{A}$ and $\mathscr{B}$.

$\mathbf{A}$	Jacobian matrix of optimization constraints with respect to $\mathbf{x}$
$\mathscr{B}$	A generic set.
$\mathscr{B}_c$	Center Frequency of a Critical Band
$\mathscr{B}_w$	Bandwidth of a Critical Band
$\mathbb{C}$	Set of Complex Numbers
$\mathscr{C}$	Cost Function
$\mathscr{C}^n$	n-dimensional Complex Space
D	Dimension of the feature vector
Δ	Step Change
$\mathscr{D}$	Domain of a Function
$\Upsilon_{\mathscr{A}}(x)$	Characteristic function of $\mathscr{A} \in \mathfrak{X}$ for random variable X
$\mathscr{D}_F(. \leftrightarrow .)$	f-Divergence
$\mathscr{D}_J(. \leftrightarrow .)$	Jeffreys Divergence
$\mathscr{D}_{KL}(. \rightarrow .)$	Kullback-Leibler Divergence
$d_E(.,.)$	Euclidean Distance
$d_{WE}(.,.)$	Weighted Euclidean Distance
$d_H(.,.)$	Hamming Distance
$d_{He}(.,.)$	Hellinger's Distance
$d_M(.,.)$	Mahalanobis Distance
$\nabla_{\mathbf{x}}E$	Gradient of E with respect to $\mathbf{x}$
$E(.)$	Objective Function of Optimization
$\mathscr{E}\{\cdot\}$	Expectation of $\cdot$
e	Euler's Constant $(2.7182818284\ldots)$
$\mathbf{e}_n$	Error vector
$\bar{\mathbf{e}}_N$	N-dimensional vector of all ones, i.e. $\bar{\mathbf{c}}: \mathscr{R}^1 \mapsto \mathscr{R}^N$ such that, $(\bar{\mathbf{e}}_N)_{[n]} = 1 \ for all \ n = \{1, 2, \cdots, N\}$
$\hat{\mathbf{e}}_k$	Unit vector whose k^{th} element is 1 and all other elements are 0
$exp\{\cdot\}$	Exponential function $(e^{\{\cdot\}})$
ϕ	Sample Space of the Parameter Vector, $\boldsymbol{\varphi}$
$\boldsymbol{\varphi}_\gamma$	Parameter Vector for the cluster γ
$\boldsymbol{\Phi}$	Matrix of parameter vectors
F_s	Spectral Flatness
$\mathscr{F}\{\cdot\}$	Fourier Transform of $\cdot$
$\mathscr{F}^{-1}\{\cdot\}$	Inverse Fourier Transform of $\cdot$
$\mathfrak{F}$	A Field
$\mathscr{I}_F(\boldsymbol{\varphi}|\mathbf{x})$	Fisher Information matrix for parameter vector $\boldsymbol{\varphi}$ given $\mathbf{x}$
f	Frequency measured in Hertz $(\frac{cycles}{s})$
f_c	Nyquist Critical Frequency measured in Hertz $(\frac{cycles}{s})$
f_s	Sampling Frequency measured in Hertz $(\frac{cycles}{s})$
Γ	Number of clusters – mostly Gaussian clusters
γ	Cluster index – mostly for Gaussian clusters
$\boldsymbol{\gamma}_{n_c}$	Column n_c of Jacobian matrix ($\mathbf{J}$) of optimization constraints
$\mathbf{G}$	Hessian Matrix

$\mathbf{g}$	Gradient Vector
$\mathscr{H}(p)$	Entropy
$\mathscr{H}(p\|q)$	Conditional Entropy
$\mathscr{H}(p,q)$	Joint Entropy
$\mathscr{H}(p \to q)$	Cross Entropy
$\mathbf{H}$	Inverse Hessian Matrix
$\mathscr{H}$	Hilbert Space
$\mathfrak{H}$	Borel Field of the Borel Sets in Hilbert Space
$\mathscr{H}_p$	Pre-Hilbert Space
$\mathfrak{H}_p$	Borel Field of the Borel Sets in Pre-Hilbert Space
H_0	Null Hypothesis
H_1	Alternative Hypothesis
$H(f)$	Fourier Transform of the signal $h(t)$
$H(s)$	Laplace Transform of the signal $h(t)$
$H(s)$	Any Generic Function of a Complex Variable
$H(\omega)$	Fourier Transform of the signal $h(t)$ in Terms of the Angular Frequency ω
H_{kl}	Discrete Fourier Transform of the sampled signal h_{nl} in frame l for the linear frequency index k
$\breve{H}_{ml}$	Mel-scale Discrete Fourier Transform of the sampled signal h_{nl} in frame l for the Mel frequency index m
$h(t)$	A Continuous Function of Time or a Continuous Signal
$\hbar(p)$	Differential Entropy (Continuous Entropy)
$\hbar(p \to q)$	Differential Cross Entropy (Continuous Cross Entropy)
I_0	Standard Intensity Threshold for Hearing
I	Intensity of Sound
I_r	Relative Intensity of Sound
$\mathscr{I}$	Information
$\mathscr{I}(X;Y)$	Mutual Information between Random Variables X and Y
$\mathscr{I}_J(X;Y)$	Jeffrey's Mutual Information between Random Variables X and Y
$\mathbb{I}$	Set of Imaginary Numbers
$\mathbf{I}$	Identity Matrix
$\mathscr{I}m$	The Imaginary part of variable $\{s : s \in \mathbb{C}\}$
$\mathbf{I}_N$	N-dimensional Identity Matrix
i	The Imaginary Number $(\sqrt{-1})$
iff	If and Only If ($\iff$)
inf	Infimum
$\mathscr{K}(t,s)$	Kernel Function of t and s used in Integral Transforms
$\mathbf{\Lambda}$	Diagonal matrix of Eigenvalues
λ	Lebesgue Measure
$\tilde{\lambda}$	Wavelength
$\bar{\lambda}$	Forgetting Factor

$\hat{\mathcal{R}}$	Eigenvalue
λ	Lagrange Multiplier
L	Total number of frames
$\mathscr{L}(\boldsymbol{\varphi}|\mathbf{x})$	Likelihood of $\boldsymbol{\varphi}$ given $\mathbf{x}$
$\mathscr{L}\{\cdot\}$	Laplace Transform of $\cdot$
$\mathscr{L}^{-1}\{\cdot\}$	Inverse Laplace Transform of $\cdot$
$\mathfrak{L}_p$	Class of extended real valued p-integrable functions
l	Frame Index
$\ell(\boldsymbol{\varphi}|\mathbf{x})$	Log-Likelihood of $\boldsymbol{\varphi}$ given $\mathbf{x}$
$ln(\cdot)$	Napierian Logarithm, Natural Logarithm, or Hyperbolic Logarithm ($log_e(\cdot)$)
$log(\cdot)$	Common Logarithm ($log_{10}(\cdot)$)
$\boldsymbol{\mu}$	Mean Vector
$\hat{\boldsymbol{\mu}}$	Sample mean vector, as a shortcut for $\overline{X}|_N$
$\hat{\boldsymbol{\mu}}_\gamma$	Sample mean vector for cluster γ
M	Number of Models, number of critical bands
M	Number of samples in a partition of the Welch PSD computation
M	Dimension of the parameter vector
$\mathscr{M}$	Matrix of the weights for mapping the linear frequency to the Mel scale critical filter bank frequencies
$\mathscr{N}(\boldsymbol{\mu},\boldsymbol{\Sigma})$	Gaussian or Normal Distribution with mean $\boldsymbol{\mu}$ and Variance-Covariance $\boldsymbol{\Sigma}$
N	Window size
N	Number of samples
N	Number of hypotheses
n	Sample index which is not necessarily time aligned – see t for time aligned sample index
N_γ	Number of samples associated with cluster γ
N_s	Number of samples associated with state s
$\mathbb{N}$	The set of Natural Numbers
O	Observation random variable
$\mathscr{O}$	Observation sample space
$\mathscr{O}$	Bachmann-Landau asymptotic notation – Big-O notation
$\mathfrak{O}$	Borel Fields of the Borel Sets of sample space $\mathscr{O}$
o	An observation sample
ϖ	Pulsewidth of Pulse Amplitude Modulation Sampler
$\varpi(o|s)$	Penalty (loss) associated with decision o conditioned on state s
$\varpi(o|x)$	Conditional Risk in Bayesian Decision Theory
$\wp$	Pitch
$\boldsymbol{\Pi}$	Penalty matrix in Bayesian Decision Theory.
P	Probability
P	Pressure Differential
P_0	Pressure Threshold
$\mathscr{P}$	Total Power

$\mathscr{P}_d$	Power Spectral Density
$\mathscr{P}_d^\circ$	Power Spectral Density in Angular Frequency
p	Probability Distribution
$\wp$	Training patten index for a Neural Network
q	Probability Distribution
$\mathbb{R}$	Set of Real Numbers
R	Redundancy
$\mathscr{R}(h)$	Range of Function h – Set of values which function h may take on
$\mathscr{R}e(s)$	The Real part of variable $\{s : s \in \mathbb{C}\}$
$\mathscr{R}^n$	n-dimensional Euclidean Space
$\boldsymbol{\Sigma}$	Covariance (Variance-Covariance) Matrix
$\hat{\boldsymbol{\Sigma}}$	Biased Sample Covariance (Variance-Covariance) Matrix
$\tilde{\boldsymbol{\Sigma}}$	Unbiased Sample Covariance (Variance-Covariance) Matrix
$\hat{\boldsymbol{\Sigma}}_\gamma$	Biased Sample Covariance Matrix for cluster γ
Ⓢ	Number of States
S	State Random variable
$\mathscr{S}$	State sample space
$\mathbb{S}$	State Borel Field of the Borel Sets of sample space $\mathscr{S}$
$\mathbf{S}\vert_N$	Second Order Sum $(\sum_{i=1}^N \mathbf{x}_i\mathbf{x}_i^T)$
s	A sample of the state random variable
$\mathbf{s}\vert_N$	First Order Sum $(\sum_{i=1}^N \mathbf{x}_i)$
sup	Supremum
$\varsigma(\boldsymbol{\varphi}\vert\mathbf{x})$	Score Statistic (Fisher Score) for parameters vector $\boldsymbol{\varphi}$ given $\mathbf{x}$
T	Total Number of Samples, and sometimes the Sampling Period
t	Sample index in time
T_c	Nyquist Critical Sampling Period
T_s	Sampling Period
$\hat{\mathbf{u}}$	Unit Vector
ω	Angular Frequency measured in $\frac{rad.}{s}$
ω_c	Nyquist Critical Angular Frequency measured in $\frac{rad.}{s}$
ω_s	Angular Sampling Frequency measured in $\frac{rad.}{s}$
W_N	The Twiddle Factor used for expressing DFT ($e^{i\frac{2\pi}{N}}$)
W_N^{kn}	$W_N^{(k\times n)}$
Ξ	Seconds of shift in feature computation
$\mathfrak{X}$	Borel Field (the smallest σ-field) of the Borel Sets of Sample Space, $\mathscr{X}$
$\mathscr{X}$	Sample Space
$\mathbf{x}$	Feature Vector
$\mathscr{Z}\{\cdot\}$	z Transform of ·
$\mathscr{Z}^{-1}\{\cdot\}$	Inverse z Transform of ·
$\mathbb{Z}$	The Set of Integers
$\mathbf{z}_k$	Direction of the Inverse Hessian Update in Optimization

List of Figures

List of Tables

List of Definitions

List of Theorems, Lemmas, and Laws

List of Properties

List of Examples

List of Problems

Part I
Basic Theory

Chapter 1
Introduction

What I know is not what others knew; what they shall know is
not what I know! They wrote what they knew and I write what I
know; in hopes that they will write what they shall know!
Homayoon Beigi – July 31, 2009

1.1 Definition and History

Speaker recognition, sometimes referred to as *speaker biometrics*, includes *identification*, *verification (authentication)*, *classification*, and by extension, *segmentation*, *tracking* and *detection* of speakers. It is a generic term used for any procedure which involves knowledge of the identity of a person based on his/her voice.

In addressing the act of *speaker recognition* many different terms have been coined, some of which have caused great confusion. *Speech recognition* research has been around for a long time and, naturally, there is some confusion in the public between *speech* and *speaker* recognition. One term that has added to this confusion is *voice recognition*.

The term *voice recognition* has been used in some circles to double for *speaker recognition*. Although it is conceptually a correct name for the subject, it is recommended that we steer away from using this term. *Voice recognition* [37, 46, 48], in the past, has been mistakenly applied to *speech recognition* and these terms have become synonymous for a long time. In a speech recognition application, it is not the voice of the individual which is being recognized, but the contents of his/her speech. Alas, the term has been around and has had the wrong association for too long.

Other than the aforementioned, there have been a myriad of different terminologies used to refer to this subject. These include, *voice biometrics* [74], *speech biometrics* [8, 43], *biometric speaker identification* [16, 35], *talker identification* [1, 11], *talker clustering* [25], *voice identification* [70], *voiceprint identification* [36], and so on. With the exception of the term *speech biometrics* which also introduces the addition of a speech knowledge-base to speaker recognition task, the rest do not present any additional information.

Part of the problem is that there has been no standard reference for the subject. In fact, this is the first text book addressing *automatic speaker recognition*. Of course, there have been other texts on the subject such as Nolan's, *The Phonetic bases of Speaker Recognition* [49] and Tosi's, *Voice Identification: Theory and Legal Applications* [70]. These books are quite valuable, but have had a completely different viewpoint. They have deeply delved into the phonetic and psychological aspects of speaker recognition and have discussed it in its forensic and legal applications in so far as human experts can tell speakers apart. Yet, no complete treatment of the *automatic speaker recognition* class of problems has been produced in textbook form until now. It should be mentioned that although there has been no textbook, the author estimates that there are in excess of 3500 research papers, to date, on the subject. The earliest known papers on speaker recognition were published in the 1950s. [54, 63] In the course of writing this book (about 3 years), more than 2400 publications were reviewed, some in more detail than others.

To avoid any further confusion, the author proposes standard usage of the most popular and concise terms for the subject in addressing this discipline. These terms are *speaker recognition* for the whole class of problems and *speaker identification*, *speaker verification*, *speaker classification*, *speaker segmentation*, *speaker tracking*, and *speaker detection* for the specific branches of the discipline. Of course there are other combinations of speaker recognition ideas with other knowledge sources such as *speaker diarization* and *speech biometrics*.

A speaker recognition system first tries to model the vocal tract characteristics of a person. This may be a mathematical model of the physiological system producing the human speech [45, 24] or simply a statistical model with similar output characteristics as the human vocal tract. [8] Once a model is established and has been associated with an individual, new instances of speech may be assessed to determine the likelihood of them having been generated by the model of interest in contrast with other observed models. This is the underlying methodology for all speaker recognition applications.

In 2006, in the movie, *Mission Impossible III*, Tom Cruise claims the identity of Philip Seymour Hoffman by putting on a mask of his face as it is customary in all *Mission Impossible* programs. However, this time, he forces the person being impersonated to read an excerpt (similar to the enrollment in speaker recognition) and uploads the audio to a remote notebook computer which builds a model of the person's voice. The model parameters are in-turn transmitted to a device on Tom Cruise's neck, located over his trachea. This device adaptively modifies his vocal characteristics to mimic the voice of Mr. Hoffman. In this scenario, the objective is to *spoof* the most familiar speaker recognition engine, namely the human perception. Of course, this idea is not new to the movie industry. In the 1971 James Bond film, "Diamonds are Forever," too, Blofeld who is Sean Connery's nemesis uses a cassette tape which includes the resonance information (formants) for the voice of Mr. Whyte to modify his vocal characteristics to those of the space program administrator.

As for the importance of speaker recognition, it is noteworthy that *speaker identity* is the only biometric which may be easily tested (identified or verified) remotely through the existing infrastructure, namely the telephone network. This makes speaker recognition quite valuable and unrivaled in many real-world applications. It needs not be mentioned that with the growing number of cellular (mobile) telephones and their ever-growing complexity, speaker recognition will become more popular in the future.

1.2 Speaker Recognition Branches

The speaker recognition discipline has many branches which are either directly or indirectly related. In general, it manifests itself in 6 different ways. The author categorizes these branches into two different groups, *Simple* and *Compound*. *Simple* speaker recognition branches are those which are self-contained. On the other hand, *Compound* branches are those which utilize one or more of the simple manifestations possibly with added techniques. The *Simple* branches of speaker recognition are *speaker verification*, *speaker identification*, and *speaker classification*. By the above definition, the *Compound* branches of speaker recognition are *speaker segmentation*, *speaker detection*, and *speaker tracking*. Currently, speaker verification (speaker authentication) is the most popular branch due to its importance in security and access control and the fact that it is an easier problem to handle than the first runner up, *speaker identification*. The reason for the difficulty in handling speaker identification will be made apparent later in Sections 1.2.2 and 17.3.

1.2.1 Speaker Verification (Speaker Authentication)

In a generic speaker verification application, the person being verified (known as the test speaker), identifies himself/herself, usually by non-speech methods (e.g., a username, an identification number, et cetera). By non-speech, we are strictly talking about content-based methods; such information may still be delivered using the speech medium, but the speech *carrier signal* is not directly used for identification, in the, so called, non-speech methods. The provided ID is used to retrieve the model for that person from a database. This model is called the *target speaker model*.[1] Then, the speech signal of the test speaker is compared against the target speaker model to verify the test speaker. Of course, comparison against the target speaker's

---

[1] In some circles this is referred to as the *reference model*, but *target speaker model* is used here.

model is not enough.

There is always a need for contrast when making a comparison. Take the evolution of the monotheistic religions in human history. The first known monotheistic religion, by some accounts, Zoroastrianism which developed in Iran derived its concepts from the older religions of Indo-European. In its initial developments, darkness was attributed to Ahriman and light was attributed to Ahura-Mazda (God). Initially, these two forces were almost equally powerful. Even later, when the role of Ahura-Mazda became much more important, hence the creation of a monotheistic religion, darkness was still deemed necessary to be able to contrast light and goodness. This ideology found itself in following monotheistic religions such as Judaism, Christianity and Islam. The devil was always a philosophical necessity to give followers of these religions an appreciation for the good forces. This stems from the need for contrasting poles in order to assess the quantitative closeness of something to one pole. Namely, the definition of something being good, needs to include how bad it is not.

In analogy, imagine trying to assess the brightness of an object. It would be hard to come up with a measure without having an opposite sense, which would be darkness. We can have a model for brightness and we can compare to it, but we will not be able to make any quantitative judgment of the amount of light without having a model for darkness (or zero light). The same is true for speaker verification (or any other verification system). To be able to get a quantitative assessment of the likeness of the test speaker to the target speaker we would have to know how the test speaker is unlike other speakers. This is partly due to the fuzzy nature of speech. It is impossible for two instances of speech to be identical due to many reasons including the content of speech, the nature of speech (low information content being transmitted by a high capacity signal), and many other reasons.

To properly assess the closeness of the test speaker to a target speaker, there are several approaches. The Two major approaches, in the literature, deal with the said contrast by introducing one or more competing models. The first method uses a *Background Model* or a *Universal Background Model*.[57] This is usually a model based on data from a large population. The idea behind it is that, if the test speaker is closer to the average population than the target speaker, then he/she is most likely not the target speaker.

The second method uses a, so called, *cohort model*.[8] The members of the cohort of the target speaker are speakers who sound similar to the target speaker. The philosophy behind this approach is that if the test speaker happens to be closer to the target speaker compared to the cohort, then most likely the test speaker is the same as the target speaker. In this method, there is no need to involve the rest of the population. The comparison is done between the target speaker and his/her cohort.

As we have seen, the speaker verification process involves a small number of comparisons (generally two); so as the population grows, the amount of computation needed for the recognition stays constant. This is in part responsible for its popularity among vendors – namely, it is an easier problem to solve. Of course, this should not be interpreted as the problem being generally easy. Again, as it was stated, there is a relative degree to everything. It is easier than speaker identification, but it certainly has its own share of problems which make it quite challenging.

1.2.2 Speaker Identification (Closed-Set and Open-Set)

There are two different types of speaker identification, *closed-set* and *open-set*. Closed-set identification is the simpler of the two problems. In closed-set identification, the audio of the test speaker is compared against all the available speaker models and the speaker ID of the model with the closest match is returned.[2] Note that in closed-set identification, the ID of one of the speakers in the database will always be closest to the audio of the test speaker; there is no rejection scheme.

One may imagine a case where the test speaker is a 5-year old child and where all the speakers in the database are adult males. Still, the child will match against one of the adult male speakers in the database. Therefore, closed-set identification is not very practical. Of course, like anything else, closed-set identification also has its own applications. An example would be a software program which would identify the audio of a speaker so that the interaction environment may be customized for that individual. In this case, there is no great loss by making a mistake. In fact, some match needs to be returned just to be able to pick a customization profile. If the speaker does not exist in the database, then there is generally no difference in what profile is used, unless profiles hold personal information in which case rejection or diversion to a different profile will become necessary.

Open-set identification may be seen as a combination of closed-set identification and speaker verification. For example, a closed-set identification may be conducted and the resulting ID may be used to run a speaker verification session. If the test speaker matches the target speaker, based on the ID returned from the closed-set identification, then the ID is accepted and it is passed back as the true ID of the test speaker. On the other hand, if the verification fails, the speaker may be rejected all-together with no valid identification result. An open-set identification problem is therefore at least as complex as a speaker verification task (the limiting case being when there is only one speaker in the database) and most of the time it is more complex. In fact, another way of looking at verification is as a special case of open-set identification in which there is only one speaker in the list. Also, the complexity gen-

---

[2] In practice, usually, the top best matching candidates are returned in a ranked list with corresponding confidence or likelihood scores.

erally increases linearly with the number of speakers enrolled in the database, since, theoretically, the test speaker should be compared against all the speaker models in the database.[3]

1.2.3 Speaker and Event Classification

The goal of classification is a bit more vague. It is the general label for any technique that pools similar audio signals into individual bins. Some examples of the many classification scenarios are *gender classification*, *age classification*, and *event classification*. Gender classification, as is apparent from its name, tries to separate male speakers and female speakers. More advanced versions also distinguish children and place them into a separate bin; classifying male and female is not so simple in children since their vocal characteristics are quite similar before the onset of puberty.

As it will be made more clear in section 17.5, classification may use slightly different sets of features from those used in verification and identification. For instance, vowels and fricatives have much more information regarding the gender of the speaker since they carry a lot more information about the fundamental frequency of the vocal tract and its higher harmonics. These harmonic variations stem from the variations in the vocal tract lengths [17] and shapes among adult males and females, and children. For example, the fundamental frequencies of the vocal tracts of males, females, and children lie around 130 Hz, 220 Hz, and 265 Hz respectively.[53] Pitch has therefore been, quite popularly, used to determine the gender of speakers.

To classify people into different age groups, also, specialized features have been studied. Some such features are *jitter and shimmer* which are defined based on pitch variations.[47] Spectral envelopes have also been used for performing such classification.[32] Of course, the classic features used in verification and identification are still used with good results.[52, 65]

Similar to the above examples of gender and age classification, it is more of an art to come up with the proper features when looking for specific features which would be able to classify audio events such as blasts, gun shots, music, screams, whistles, horns, etc. For this reason, there is no cookbook method which can be used to classify such events, giving classification the vagueness of which we spoke at the beginning of this section.

---

[3] In practice, this may be avoided by tolerating some accuracy degradation.[7]

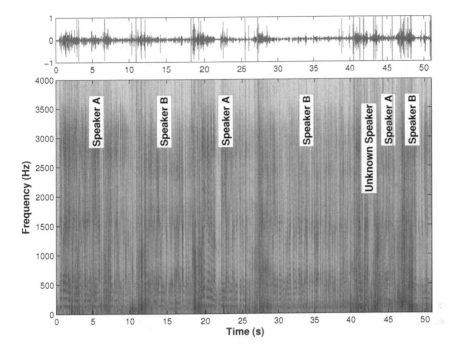

Fig. 1.1: Open-Set Segmentation Results for a Conference Call
Courtesy of Recognition Technologies, Inc.

1.2.4 Speaker Segmentation

Automatic segmentation of an audio stream into parts containing the speech of distinct speakers, music, noise, and different background conditions has many applications. This type of segmentation is elementary to the practical considerations of speaker recognition as well as speech and other audio-related recognition systems. Different specialized recognizers may be used for recognition of distinct categories of audio in a stream. An example of such tasks is audio transcription, like the *ARPA HUB4* evaluation task consisting of automatic transcription of radio broadcast news shows from the *Market Place* program.[3, 4]

A typical radio broadcast news contains speech and non-speech signals from a large variety of sources like clean speech, band-limited speech such as telephony sources, music segments, speech over music, speech over ambient noise, speech over speech, etc. The segmentation challenge is to be able to separate the speech produced by different speakers from each other. It is also desirable to separate, music and other non-speech segments.

It is worth noting that most speech recognizers will break down if they are presented with music instead of speech. Therefore, it is important to separate the music from recognizable speech. In addition, one may wish to remove all the music and only store the speech in an archiving scenario to save space.

Another example is the ever-growing tele-conferencing application. An array of conference calling systems have been established which allow telephone conversations among multiple speakers. Usually, a host makes an appointment for a conference call and notifies attendees to call a telephone number and to join the conference using a special access code. There is an increasing interest from the involved parties to obtain transcripts (minutes) of these conversations. In order to fully transcribe the conversations, it is necessary to know the speaker of each statement. If an enrolled model exists for each speaker, then prior to identifying the active speaker, the audio of that speaker should be segmented and separated from adjoining speakers.

Here, we consider speaker segmentation a type of speaker recognition since the process of segmenting audio is quite similar to other speaker recognition techniques. Normally, statistical models of the local characteristics of two adjoining segments of audio are created. Based on the difference between the underlying model parameters and features that are appropriate for modeling speaker characteristics, an assessment of the similarity of the two segments is made. If the segments are deemed sufficiently dissimilar, a segmentation mark is realized at this point of transition.

Once the basic segmentation points are identified, it is useful to classify the data into segments associated with known speakers. Even if the speaker identities are unknown, by knowing the number of speakers who have participated in the conversation, one may classify the speakers with some common label for identical speakers. As it will be seen in Section 17.4, knowledge of the number of speakers in the conversation is quite helpful and hard to estimate. The underlying difficulty in the estimation of the number of speakers is linked to the "contrast" argument which was made earlier, in Section 1.2.1.

We qualify the operation of tagging the speakers as *speaker classification*, since it performs a classification of the speech associated with an unknown number of unknown speakers – see Sections 1.2.3 and 17.5. If the speaker identities are known, then the sub-problem becomes an identification problem – see Sections 1.2.2 and 17.3. If all the speakers in a conversation have been enrolled in the system, then it may not be necessary to know the total number of speakers in the conversation in contrast with the case where the speakers are unknown and would have to be tagged by the system as speaker A, speaker B, etc.

Note that the segmentation problem has two stages: a basic stage which entails the elementary segmentation of the audio into small pieces with uniform production properties and a more advanced stage which labels these segments and sometimes merges similar adjoining segments. Some references in the literature only consider the initial stage as segmentation, but since as it shall be seen later, there will be feedback between these two stages, they are usually inseparable. Here, it is preferred to include the whole process under the auspices of one category, called *speaker segmentation*.

1.2.5 Speaker Detection

Speaker detection is the act of detecting one or more specific speaker in a stream of audio. Therefore, the underlying theory encompasses segmentation as well as identification and/or verification of speakers. Enrollment data is usually necessary. The choice of speaker identification, verification, or to use them both is mostly dependent on the problem formulation. For example, if there are many speakers in a stream of audio and it is known that there will always be a speaker from the database speaking at any time, with no possibility of extraneous data such as music, then closed-set identification may be applied to the results of the speaker segmentation to identify the speaker for each segment.

A more complex problem would be one in which there are speakers outside the known set of speakers or there may be music or other types of audio in the stream. In this case, if the list of speakers to detect is not large, then a verification session may be conducted on each segment for every one of the members of the list. On the other hand if the list is large, then an identification may be conducted and the result of the identification may be used as the claimed ID of a subsequent verification. If the identified speaker is verified and if it is a member of the list of speakers to detect, then the result is returned. One can imagine many different possible scenarios in which a combination of speaker identification and verification may be used in conjunction with the results of segmentation.

1.2.6 Speaker Tracking

Speaker Tracking is somewhat similar to speaker detection with the subtle difference that one or more of the speakers are tracked across the stream. In this case, one may envision conditions where no enrollment data is available, but not only is the audio segmented into single speaker segments, but the segments are also tagged with labels signifying the individual speakers in the stream. If enrollment data is available, then the speaker labels for the segments may be adjusted to reflect the true speakers of those segments from the enrollment database. However, in the general sense, it may not be necessary to have specific labels portraying real speakers. In most cases, as long as the different speakers in the stream are identified with general labels such as alphanumeric tags, then the goal is achieved. The most important application of tracking is the tagging of speakers in a conversation such as a telephone conference call.

1.3 Speaker Recognition Modalities

Theoretically speaker recognition may be implemented using different modalities which are tied to the use of linguistics, context, and other means. However, in the practical sense these modalities are only relevant for speaker verification. In this branch, based on the requirements of the application, different sources of information may be mixed in with the acoustic information present from the vocal tract. Of course, identification and other branches may also be able to use some extra information for improving performance, but the following modalities are most relevant to speaker verification. Although, they are appended with the phrase, "speaker recognition," to show generality.

1.3.1 Text-Dependent Speaker Recognition

In the 1992 film, *Sneakers*, Robert Redford plays the role of an expert who is hired to find vulnerabilities in security systems. When the plot plays out, he tries to access a secure laboratory which is protected with many means including a *text-dependent* speaker verification system. To be able to access the lab, the following was to be spoken: "Hi. My name is Werner Brandes. My voice is my passport. Verify me." He pulls it off by having expected this system and preparing for it.

Since sophisticated digital recording was not common-place at that time, he sends out a woman with the lab owner, Werner Brandes. She is wearing a recording device and tries to ask Mr. Brandes different questions in the course of the evening for which he would have to say the words in the expected prompt in a scattered fashion. The audio tape is then spliced to create the expected prompt. This could be achieved much more easily these days with the existence of small and high quality digital recording devices and digital editing techniques.

This, so called, *liveness challenge*, is the very reason why the *text-dependent* modality is flawed and should not be used in any serious application. So, why do people still work on this type of recognizer? The answer is, because of its relative high accuracy. As we shall see in the next few sections, the liveness issue is still a problem with other speaker recognition modalities, but it is remedied in simple fashion in those modalities. Also, Section 1.3.3 shows that it is possible to view the *text-prompted* modality as an extension of *text-dependent* recognition. The *text-prompted* modality may be used to remedy the liveness problem of *text-dependent* systems.

The text-dependent modality only applies to the speaker verification branch. Most other branches cannot be used with specific phrases since they happen in a more passive manner where a recognizer listens to an utterance and makes a decision.

1.3.2 Text-Independent Speaker Recognition

Text-independent speaker recognition is the most versatile of the modalities. It is also the only viable modality which may be used in all branches of speaker recognition. There are different degrees of text-independence. Some recognizers are completely text- and language-independent. There are engines which are somewhat language-dependent. Some are completely language-dependent, but text-independent to a degree. Chapter 17 will discuss these different possibilities in more detail.

A purely text-independent and language-independent system only relies on the vocal tract characteristics of the speaker and makes no assumption about the context of the speech. One of the most important problems plaguing text-independent systems is the possibility of a poor coverage of the part of speech. Take an enrollment utterance for example. Under the auspices of a text-independent process, generally, there is no constraint on the enrollment text. Also, a common goal of all recognizers is to try to minimize the length of enrollment and test segments.

As enrollment and test data lengths are reduced, the possibility of a common coverage of the phonetic space is reduced. Therefore, parts of the test utterance may never have been seen at the enrollment time. So, it is plausible that the phones in the enrollment utterance of a non-target speaker and the test utterance of the target speaker have more in common, acoustically, than the enrollment utterance of the target speaker and the test segment for that speaker. This commonality may contribute toward a mis-recognition. To account for this problem, most text-independent speaker recognition engines require longer enrollment and test utterances to be able to achieve acceptable accuracies.

Text-independent speaker recognition also suffers from the liveness assessment problem described earlier (see Section 1.3.1). In this case, one does not even need to have specific words spoken by the individual to be able to spoof the system. Any type of high quality recording of the individual's voice will be enough. Sections 1.3.3 and 1.3.4 show different solutions to the liveness problem.

1.3.3 Text-Prompted Speaker Recognition

Text-prompted speaker recognition, as it may be apparent from its name, prompts the speaker to say a specific phrase at the time of testing. It was mainly developed to combat spoofing from impostors. If the speaker is not anticipating the text of the prompt, he/she will not be able to prepare for fooling the system. Take the example of section 1.3.1. If the system, being utilized, had used a text-prompted engine, the system would not have been easily fooled.

There are generally two main approaches to the design of a *text-prompted* system. The first method would modify a *text-dependent* system to generate somewhat random phrases for its prompts. These systems will randomly generate a phrase and then build the *text-dependent* language model for that prompt. The response will therefore have to match the vocal characteristics of the target speaker as well as the context of the prompted phrase. This process will be discussed in more detail in Chapter 17.

One of the main advantages of doing *text-dependent* recognition is the sufficience of shorter enrollment texts. However, to be able to perform *text-prompted* recognition through the text-dependent approach, more enrollment data has to be collected to cover the most common phones and phone sequences. Section 17.2.1 sheds more light onto this process.

The second approach would also generate a random prompt, but will use the combination of a *text-independent* speaker recognition engine and a speech recognizer to perform recognition. To recognize an individual, the vocal characteristics of the individual would have to be matched. Also, the recognized text coming from the speech recognizer has to match the expectation of the prompted phrase.

Note that *text-prompted* recognition only makes sense for speaker verification. Most other branches of speaker recognition cannot use specific prompts. This is also the case for *text-dependent* recognition. It is true that one can dream up special cases where other branches of speaker recognition may be used with dictated prompts, but it will not be true in general.

1.3.4 Knowledge-Based Speaker Recognition

A knowledge-based speaker recognition system is usually a combination of a *text-independent* or *text-prompted* speaker recognition system with a speech recognizer and sometimes a natural language understanding engine or more. It is somewhat related to the basic *text-prompted* modality with the difference that there is another abstraction layer in the design. This layer uses knowledge, supplied by the speaker, to test for liveness.

Consider the very familiar security check used by many financial institutions in which they ask for responses to questions that only the caller should know. This has advanced through the years from the old "Mother's Maiden Name," "Last Four Digits of a Social Security Number," and so on to the more versatile systems which require the client to come up with questions, the answers to which are only apparent to him/her. There have also been newer incarnations, especially on Internet-Based systems, where at the time of enrollment, the system asks the user to pick an image. Later, at the verification time, the user is asked to divulge what he/she had chosen at the time of enrollment. Since the choice would have to be limited to a few images, the systems would allow impostors in, with the probability of at best $\frac{1}{Number of choices}$, even if the impostor would pick an image randomly.

Therefore, such systems are usually used to strengthen other authentication processes. An example of such *fusion* is to use the image-based authentication in place of the submit button in another type of authentication. However, for personal use, I have found it hard to remember what image I had used at the time of enrollment, especially with systems which are seldom used.

Similar ideas may be used in conjunction with speaker recognition so that the vocal tract characteristics are used in addition to the knowledge from the individual. This is sometimes seen as a *fusion* of two different information sources, but its main objective is to handle the liveness issue described earlier.

In this scenario, the speaker has to provide his/her voice as well as some knowledge-base to the system so that he/she may be challenged with proper questions assessing his/her liveness. This scenario is true for a speaker verification system, however, knowledge based systems may also be relevant for other branches of speaker recognition, but in a slightly different way. An example is the use of the content of speech as well as the vocal characteristics to come up with a match in a speaker indexing problem. Imagine hours upon hours of audio from a newswire or a similar source. The objective may be to find a specific speaker (speaker detection), but with additional constraints.

For instance, the user may be searching for a certain speaker (speaker detection) when that speaker is talking about a specific topic. This is an example of knowledge-based speaker recognition, where there is no need for the enrollment within the context, since it follows the search topic as recognized by a speech recognition engine, with possible natural language understanding (NLU) capabilities. Also, the speaker voice model may be an excerpt in the same stream, chosen by the user.[72]

1.4 Applications

There is truly no limit to the applications of speaker recognition. If audio is involved, one or more of the speaker recognition branches may be used. However, in terms of deployment, speaker recognition is in its early stages of infancy. This is partly due to unfamiliarity of the general public with the subject and its existence, partly because of the limited development in the field. Also, there are some deterrents that feed the skeptics, the most important of which are channel mismatch and quality issues. These topics have been discussed in detail in Sections 22.4 and 22.12 respectively.

Some of the major applications of speaker recognition have been discussed in the following few sections. This list is by no means complete and it is not in any specific order. These examples have been chosen in an effort to try and cover some of the most popular applications. Also, some attention has been paid to covering examples for the different branches of speaker recognition.

1.4.1 Financial Applications

It is hard to lump everything having to do with financial institutions in one category. There are so many different applications that basically span all the branches of speaker recognition. However, here, we will try to cover some of the more popular applications which have direct relevance to the experiences of all of us as users of financial services.

Most of us may have been in the position of contacting a financial institution for questions regarding our accounts. These may be credit-card accounts or simply standard bank accounts. Since financial data is sensitive and should only be accessed by the owners of the accounts, there are usually a number of procedures which are used by financial companies to establish the identity of the individual (on the telephone or in person).

At the present, most institutions provide fully automated account information, accessible through the telephone. They usually require your account number and a pin number to establish your identity. Then full access is granted to the account which could be detrimental if the wrong person gains access. Pin numbers have also been limited to 4 digits by most financial institutions to be compatible with an international standard. Many of these institutions also disallow the use of 0 or 1 at the beginning and the end of the pin number, considerably reducing the number of permutations. Add to this the fact that most people use easy-to-remember numbers such as birthdays or important dates in their immediate family and you have a recipe for a simple breach of security.

Also, when speaking to a customer support representative, no pin number is necessary. The customer is asked for the account number in addition to some very simple questions, the answers to which are quite easily obtainable. Examples of these questions are "Mother's Maiden Name" which is public knowledge for most people, "Favorite Color" which happens to be blue for over 40% of the population, "Last Four Digits of Social Security Number" which is also something quite accessible to the persistent impostor, etc. Some variations may exist to make these questions somewhat harder to answer. However, to retain the loyalty of their clients, these institutions cannot make the questions too hard to answer.

An important security breach which is hardly considered these days is the possibility of sniffing DTMF sequences by tapping into the telephone line of an individual while pin-based authentications are performed. This is quite simple and does not require much skill. The tapping may be done close to the source (close to the user) or in more serious cases close to the institution performing the authentication. Once the DTMF information is recorded, it may readily be catalogued and used by impostors with dire consequences. Another potential security breach is the readily available personal information being sold by special sites on the Internet for typical $19.95 prices. These are some of the reasons for the enormous number of identity thefts being reported.

Speaker verification is a great match for this type of access problem. It may be used in an automatic fashion. It requires an enrollment which may be performed once a rigorous authentication is conducted by an agent. It can also be used in conjunction or in lieu of existing security protocols. In some cases such as any customer-agent interaction, verification may be done in a passive manner, namely, the verification engine can listen in on the conversation between the agent and the customer. Since the agent is an employee of the institution, his/her voice may be pre-enrolled into the system so that it would be possible to separate the audio of the customer using speaker segmentation followed by verification. In most cases, separate-channel recording will even alleviate the separation problem. Once the customer's audio is isolated, it may be verified as the conversation between the agent and the customer is in progress. The agent may obtain feedback on the possibility of fraud on his/her computer screen which allows for further scrutiny of the caller without the caller even being aware.

Another application is in passive fraud control. Instead of running speaker verification on the valid users, one may hold a list of known fraudsters' voices. In most cases, a limited number of professional fraudsters call frequently to try and fool the security system. An identification engine can try to alert the security agents of these attacks by listening in on all established communication channels. This can also run in parallel with the verification process described earlier. Imagine a fraudster who has successfully been able to assume a valid client's identity. This fraudster may have also gone through the enrollment process and enrolled himself/herself as the true client. Having this parallel fraud monitor can alert the agents in the institution that there is a likelihood of fraud.

These and other applications could prove to be priceless for financial institutions where fraud could be quite costly. Also, the same argument applies to many other similar institutions with the same security and access requirements, such as health institutions which are required by law to keep the health status and personal information of their customers extremely confidential. There are U.S. government mandates trying to limit the requests by agents for personal information such as social security information, in an effort to reduce the chance of misappropriation of such information.

1.4.2 Forensic and Legal Applications

Speech has a unique standing due to its non-intrusive nature. It may be collected without the speaker's knowledge or may even be processed as a biometric after it has been collected for other purposes. This makes it a prime candidate for forensic and legal applications which deal with passive recognition of the speakers or non-cooperative users. Passive recognition involves tasks in which the application does not generally dictate the flow and type of data being processed. For example, even if speaker verification is used for some specific needs in Forensics, it cannot generally be text-prompted or text-dependent. It will have to be mainly utilized on whatever audio is available and most of the time it is done without the knowledge of the speaker, in which case even the knowledge-based modality does not apply in the strict sense. It is true that information about textual contents of the audio may also be used, in addition to the vocal tract information, but it will be done in the most independent sense.

There are other biometrics such as fingerprint and DNA recognition that also allow some degree of passivity in terms of data collection. That is why they have been successfully used in forensics and legal applications. However, they are not as convenient as speech which may be collected, intercepted, and transmitted much more effectively with the existing infrastructure.

Forensic applications rely on a few different modalities of speaker recognition. Based on the above discussion, speaker identification [10, 19, 21, 27, 51] is the main modality of interest in these applications. For example, it may be used to identify an individual against a list of suspects. In addition, *speaker segmentation* and *classification* can be quite useful. Segmentation is needed to separate the audio of the target individual in a stream of audio consisting of several sources of speech and other types of audio. Classification can help in categorizing the segmented audio. Also, it may be used to look for anomalies such as abrupt events (gun shots, blasts, screaming, et cetera).

1.4.3 Access Control (Security) Applications

Access control is another place where speech may be utilized as a very effective biometric. Entering secure locations is only a small part of the scope of speaker biometrics. In that domain they compete head-to-head with most other biometrics and possess pros and cons like any other – see Section 1.5. Where speaker recognition truly excels with respect to other biometrics is in remote access control in which the user is not physically at the location where access should take place. Take, for example, the case of accessing sensitive data through a telephony network or a conventional computer network. With any biometric the biometric sensor has to be located where the user is. This, for speech, is a microphone which is readily available in many devices we use. Most other biometric sensors have had no other use in the past and therefore will have to be exclusively utilized for the biometric at hand. In addition, the telephony network is so well distributed that it would be hard to imagine an application that does not have access, at least, to a telephone.

Most access control cases would utilize the speaker verification branch of this biometric. Sections 1.3 and 17.2 describe speaker verification and its modalities in detail.

1.4.4 Audio and Video Indexing (Diarization) Applications

Indexing is a major application of speaker recognition which involves many of its branches, in addition to other technologies such as speech recognition. It has also, quite successfully, been fused with other biometrics such as face recognition.[73] It requires segmentation of the audio stream, detection and/or tracking of speakers, sometimes with the added burden of identifying these speakers. Generally speaking, detection or tracking may have localized scope over a limited set of speakers, such as the distinct speakers which may be identified in a stream of audio. After performing such localized tasks, sometimes a global identification may be required. In some cases, though, the speaker list, used for detection and tracking, is not limited to a

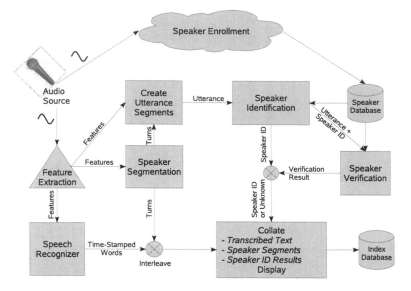

Fig. 1.2: Diagram of a Full Speaker Diarization System including Transcription for the Generation of an Indexing Database to be Used for Text+ID searches

small set. Figure 1.2 shows the diagram for a full *diarization* system, including the transcription of the audio. The results are used to build an indexing database which allows for doing searches based on the text, the speakers or the combination of these two sources.

1.4.5 Surveillance Applications

Surveillance applications (lawful intercept) are really very similar to forensic applications discussed in Section 1.4.2. All surveillance applications, by definition, have to be conducted in a passive manner as discussed in the Forensics section. Unfortunately, they can sometimes be misappropriated by some governments and private organizations due to their relative ease of implementation. There have been many controversies on this type of intercept especially in the last few years. However, if done lawfully, they could be implemented with great efficiency.

An obvious case is one where a system would be searching on telephone networks for certain perpetrators which have been identified by the legal process and need to be found at large. Of course, speaker segmentation would be essential in any such application. Also, identification would have to be used to achieve the final goal. Essentially, the subtle difference between forensic and surveillance applications is that the former deals with identification while the latter requires the compound branch of speaker recognition, *speaker tracking.*

Another method in which speaker recognition can help in the field of surveillance is at a capacity very similar to that of speaker indexing (Section 1.4.4). Imagine having to transcribe the speech of a target speaker. On the other hand, transcribing the audio of other individuals, not allowed by the legal intercept rules, may not be acceptable. Speaker recognition can concentrate the transcription effort (which is also usually quite costly) to the speech of the target individual.

1.4.6 Teleconferencing Applications

Teleconferencing can also benefit from different branches of speaker recognition. It used to be the case that teleconferencing was limited to large corporations which had the infrastructure to perform such meetings. With the increasing number of free and paid sites, many more teleconferencing sessions are taking place daily. Many of these sites are looking for ways to improve their services in hope of being more competitive in this growing market. Many are looking for speaker diarization capabilities to add as a service. The application discussed in Section 1.4.4 may be used for providing this value-added service. In addition, companies may implement their own diarization systems to make sure that the minutes of the meetings are made available to the individuals on the conference call as well as the companies' archives for future reference.

1.4.7 Proctorless Oral Testing

Figure 1.3 shows a distant learning application of speaker recognition. This is used for performing proctorless oral language proficiency testing. These tests take place on a telephone network. The candidate is usually in a different location from the tester. There is also a set of second tier raters who offer supplementary opinions about the rating of the candidate. In one such application, the candidate is matched by the testing office to a tester for the specific language of interest. Most of the time the tester and the candidate are not even in the same country.

The date of the test is scheduled at the time of matching the tester and the candidate. In addition, the candidate is asked to speak into the Interactive Voice Response (IVR) system which is enabled by speaker recognition technology to be enrolled in the speaker recognition system. The speaker recognition system will then enroll the candidate and save the resulting speaker model for future recognition sessions. Once

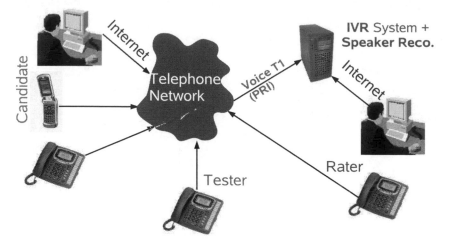

Fig. 1.3: Proctorless Oral Language Proficiency Testing
Courtesy of Recognition Technologies, Inc.

it is time for the candidate to call in for performing the oral exam, he/she calls the IVR system and enters a test code which acts as the key into the database holding the candidate's test details. The candidate is first asked to say something so that a verification process may be conducted on his/her voice. The ID of the candidate is known from the test code entered earlier, so verification may be performed on the audio.

If the candidate is verified, he/she is connected to the tester and the oral examination takes place. In the process of taking the oral examination, the speaker recognition system which is listening in on the conversation between the candidate and the tester keeps doing further verifications. Since the tester is known, the audio of the candidate may be segmented and isolated by the recognition engine from the conversation, to be verified. This eliminates the need for a proctor to be present with the candidate at the time of the examination which reduces the cost of the test. The conversation is also recorded and made available for the second tier rater(s) to be evaluated further.

This application makes use of speaker verification and segmentation. It also does speaker tracking to make sure the candidate has not handed the phone to another person in the middle of the test.

1.4.8 Other Applications

The applications of speaker recognition are not limited to those described here. There are countless other applications which are either known or will be made apparent as more advancement is made in this subject and more people are made aware of its existence. This is one of the many goals of this textbook, namely to promote awareness of the speaker recognition discipline so that it may be applied in new fields.

1.5 Comparison to Other Biometrics

In 1997, Bruce Feirstein touched upon an array of biometrics including *speaker verification* in the screenplay of the James Bond movie *Tomorrow Never Dies*, acted by Pierce Brosnan as James Bond (007). It is certain that speaker recognition is not alone in the biometric arena. Also, it is conceded that there is no single way to solve any problem. In fact there is a place for every kind of biometric to be used and as we will discuss later, it is most of the time beneficial to combine a few techniques to achieve better performance. All of us use a combination of biometric measures in our daily lives to make decisions about the identity of those around us. It is important to be well informed about the strengths and weaknesses of all that is available to us to be able to make a decision about the combination of systems we would need to utilize for any specific application. Although this book concentrates on speaker recognition, in this section, we attempt to review the most popular biometrics and try and compare them with speaker recognition whenever possible.

Many have attempted to categorize the different types of biometrics into two different categories, *Behavioral* and *Physiological*. The problem with this kind of categorization is that just like any other clustering, things are not always so clear cut. Many biometrics have elements from both categories. In fact, the specific treatment of some biometrics may place them in any of the two categories or the combination of the two.

Take, for example, the subject matter of this book. *Speaker recognition* could be construed as a behavioral biometric if the recognition system concentrates on the transitions of the audio and the manner of speaking. In contrast, it may be considered as a physiological biometric based on the characteristics of the vocal tract. In fact, most *text-independent* treatments of the subject view it as a physiological biometric. *Text-dependent* systems, in addition to the physiological information, also use many behavioral tips to make their assessments.

Other, so called, behavioral biometrics also include considerable amounts of physiological information. Behavior is something that may be consciously adapted, however, many of the characteristics of our voice, signature (handwriting), gait (style of walking), keystrokes, and so on may only be changed in a limited fashion and in accordance with our physiology. In general, there are certain biometrics which are purely physiological. *DNA, fingerprint, palm, iris, retina, thermogram,* and *vein* are some such biometrics.

Of course, there will always be more inventive biometric techniques some of which will not be so practical, such as the *lip identification* system, as featured by *Eric Horsted*, the writer of "A Taste of Freedom[4]," an episode of the animated television series, *Futurama*. Although it may be possible to do that, having to kiss a glass scanner may not be so sanitary! In some cases, although the sensors for a specific biometric may be harmless, public perception will dictate its success. An example is *retina* imaging which happens to be very accurate and predates *iris* imaging, but since the sensors shine a laser onto the retina, the public has been quite resistant toward accepting this biometric.

With the exception of a few, most biometrics are not usable by every member of the population since they rely on body parts or features which may be lacking, defective or disabled in part of the population. For speaker recognition to be useful, the person has to be able to speak. In most cases, hearing as a feedback measure is also necessary. It is hard to find any separate census information regarding the hearing impaired and mute persons. There have been numbers available since 1850 for the deaf and mute as a whole. Based on the US Census Bureau results, the percentage of the population who was both deaf and mute was 0.04% in 1850 [29], 0.07% in 1880 [29] and 0.4% in 2005 [71]. It is hard to compare these numbers since the number in 2005 has two categories of hearing impaired individuals. The number quoted here is those over the age of 15 who were able to hear any conversation at all. There is another number also reported for those who can hear conversations partially. We are not considering those individuals. However, most of the growth in the percentage may possibly be attributed to more proper census practices. The 2005 number is presumed to be more accurate.

1.5.1 Deoxyribonucleic Acid (DNA)

The idea behind *DNA recognition* is to start with a target sequence of the 4 nucleotides which make up the coding of a *DNA strand*, namely, *A, C, G,* and *T*. At the time of recognition, one or more samples of a DNA or its fragments are first replicated using *polymerase chain reaction (PCR)*. Table 1.1 shows the 4 *DNA nucleotides* and the corresponding industrial *triphosphates* used in the replication

---

[4] Episode 59, fourth season. The series was created by Matt Groening and developed by him and David X. Cohen. this episode originally aired on December 22, 2002. [76]

process.[23] This will replicate the original sample, increasing the number by a few orders of magnitude. Then, an *hybridization process* is used to compare the replicated sequence with the sample sequence.

Abbreviation	Short Name	Chemical Formula	Triphosphate	Triphosphate Formula
A	Adenine	$C_5H_5N_5$	dATP[a]	$C_{10}H_{13}N_5O_{12}P_3Na_3$
C	Cytosine	$C_4H_5N_3O$	dCTP[b]	$C_9H_{13}N_3O_{13}P_3Na_3$
G	Guanine	$C_5H_5N_5O$	dGTP[c]	$C_{10}H_{13}N_5O_{13}P_3Na_3$
T	Thymine	$C_5H_6N_2O_2$	dTTP[d]	$C_{10}H_{14}N_2O_{14}P_3Na_3$

[a] dATP: Deoxyadenosine Triphosphate

[b] dCTP: Deoxycytidine Triphosphate

[c] Deoxyguanosine Triphosphate

[d] Thymidine Triphosphate

Table 1.1: DNA Nucleotides

Currently, the *DNA recognition* procedure is at its infancy. It could potentially be very accurate once the system matures. At the present, chips are being developed to aid in the hybridization process of the recognition. However, with current technology, only *single strands of DNA (ssDNA)* of *pathogenic bacteria* can be recognized using electronic technology.[28, 66] Still the *PCR* and *hybridization processes* are done separately and manual intervention is necessary. Work is being done to create a single chip capable of doing the whole recognition process.[66]

Fortunately, *DNA* seems to be one of those biometrics which is available for every human being. It makes *DNA* a powerful biometric, but there are certainly some great disadvantages to this biometric. It will still be a while until human DNA strands can be automatically recognized in a practical time interval. Although some work has been done on different aspects, including the classification problem [33]. Even when automatic DNA recognition matures, there are serious other limitations to DNA recognition. One problem is that people are not comfortable with giving up their DNA. Part of it may be stored while being replicated by the *PCR* process. This could seriously jeopardize a person's security and could possibly be misused if it came into the possession of the wrong people.

1.5.2 Ear

The folds in the *pinna* (the *auricle* of the ear) and the shape of the *ear canal* are different among individuals. These differences are quite pronounced and are easily realizable by a visual inspection of the outer ear. The Ear has recently been used for establishing the identity of individuals through different approaches. We lump everything related to the ear together, in this section. To date, two separate branches of *ear recognition* have been studied. Aside from the *visual* approach, there is also an *acoustic* method for ear recognition.

The first branch of techniques uses images of the ear for recognition of the individual. This problem has several phases. Generally, a side image of the face is taken and the ear is segmented out. Then, depending on the algorithm, several processes may happen. To achieve invariance, most researchers use some flavor of the *Principal Component Analysis* (PCA) method – see Section 12.1.[22, 50] To handle rotational invariance, techniques such as conversion to polar coordinates and the usage of the polar coordinate version of the Fourier Transform to obtain *generic Fourier descriptor* (GFD) features have been successful.[22]

Some use a general 2-dimensional image of the ear.[80, 79, 50] Others use the more expensive apparatus of 3-dimensional scanning to obtain more detailed information about the contours.[18, 15] The 3-d scans usually require an additional 2-d reference image for color information. Therefore, the 3-d systems do not seem very practical. Some have used multiple view 2-d images to alleviate the expense and complexity associated with the 3-d systems.[81, 42] These systems do become more complex in the definition of the amount of rotation between different views and the possibility of repeating the same conditions with a practical apparatus.

To improve the accuracy of image-based ear recognition, many have fused this biometric with face recognition results to obtain a *multimodal biometric*.[31, 75, 78] Results of methods with these combinations and the best of breed seem to be in the order of about 2.5% error-rate for identification. The largest population seen in the 43 reviewed references was only in the order of 400 individuals. To date, no large-population study has been seen.

There are some major problems associated with image-based ear recognition approaches. Changes due to *illumination variations* plague this technique in a similar manner as in any other image-based biometric recognition system. These techniques usually have to work hard to attain rotation invariance, with some degree of success. The ear may be covered wholly or partially by hair, especially for individuals with long hair. This will create a gender bias since in general women have longer hair. Finally, there are issues with automatic segmentation of the image to extract the ear from the side-view image of the head. This problem is magnified with *illumination variations*.

The second ear recognition approach uses the acoustic properties of the pinna to establish the identity of an individual. In this approach, a small speaker device and a microphone, both point into the ear canal. The speaker sends out a wave (1.5-kHz

- 22-kHz) into the ear canal at an angle. Once the wave goes through the canal and reflects back from the ear drum and the wall of the canal, the microphone picks up the reflection wave. The wave, manipulated by this reflection, is related to the transfer function which is made up of the transfer functions of the speaker device, the *pinna*, the *ear canal* and the *microphone*. This transfer function is estimated based on the input and the reflected output.[2]

The phase difference between the emitted wave and the received wave is quite sensitive and is present even among within-class samples. To avoid these within-class variations and at the expense of losing of some biometric information, Reference [2] only uses the amplitude of the spectrum of the wave and throws away the phase information.

Unfortunately there are not that many researchers working on this branch of ear recognition and no test on large populations seems to be available for this method. The tests in [2] have only been done on $17 - 31$ subjects with best results of about $1.5\% - 7\%$ equal error rate (EER) – see Section 19.1.1. Therefore, this method does seem to show some promise, but is mostly inconclusive. An interesting point in [2] is that several small earphone and cellular (mobile) phone installations were custom-made for the research. The earphones had much better performance than the cellular phones, as it would be expected, intuitively.

1.5.3 Face

Automatic face recognition has received quite a bit of attention in the last decade mostly due to the availability of the many video cameras in public locations for security purposes. Although, there has been active research in this field for more than 3 decades.[14] There have also been a handful of books written on the subject, in recent years.[41, 82] Face recognition may be achieved in two major forms, cooperative and passive.

The term *cooperative* is used to describe systems which basically work on mug-shots of individuals. These are systems installed in airports, as well as systems for cataloging criminals. The airport versions are usually used in conjunction with other biometrics such as fingerprint and iris recognition. Normally, a frontal profile is captured using a digital camera and it is compared against a database of pictures. The lighting conditions are controlled in these systems and mostly take place at the desk of an official with a fixed apparatus. This type of recognition can be most effective, since at the time of capturing the photograph the officer will make sure that the target is not wearing any blocking attire such as glasses or a hat. Systems of this kind use an array of different types of features from geometric features and templates [14] to analogs of force field transforms [31]. Features are normally based on the locations, shapes and distances of the different components of the face. A popular treatment uses *principal component analysis* (PCA) and goes by the name of *Eigenfaces*. This method uses normalization techniques to transform different frontal snapshots of faces into the same lighting condition as well as size and pixel distribution. Then *principal component analysis* is used to parametrize the faces.[38, 62]

The more challenging form is the passive one. In this method, usually a video camera constantly surveys an area and the individual is not necessarily cooperative in the data acquisition. In this case, the angle of the camera, the attire, lighting conditions, style of walking (the way the face is pointed) and many other variables dictate the quality of the outcome. Of all video-style face recognition applications, analyzing video from security cameras proves most challenging. Most modern cities such as New York and London have thousands of video cameras installed in public areas for surveillance. However, without an automatic face recognition system, the thousands of hours of video which are captured per day are not so useful. Another problem is the high bandwidth required for capturing video compared to audio. A speaker recognition system with segmentation and event classification/detection may be used to greatly reduce the work needed for searching these video streams. This type of speaker recognition may even be applied on-site at the location of the camera to selectively record high resolution video, based on islands of high audio activity. In lower audio activity situations, lower frame rates may be utilized to reduce the amount of storage and processing needed. There are, however, some legal challenges with the public recording of audio. Audio-related regulations are not always treated in the same way as video regulations.[56, 34]

There are simpler cases where the video is obtained in studio quality and under controlled lighting conditions. Broadcast news is one such example.[73] This type of video-face recognition really belongs with the cooperative kind even if the individual in the footage is not aware of face recognition being utilized on the video.

1.5.4 Fingerprint and Palm

Live scanning of fingerprints has been made possible in the last few decades. The minimum resolution required for scanning a finger is about 300 dots-per-inch (dpi), although the minimum required by the United States Federal Bureau of Investigation (FBI) is 500 dpi. There are many kinds of fingerprint scanners in the market, including optical, solid-state, and ultrasound.[44] These scanners started being mass produced years ago and are quite inexpensive. Some solid-state scanners are very small and use a sliding apparatus – the finger is slid on the scanner and the image is reconstructed from the data stream. Many notebook computers have these senors built-in. This is a testament to the popularity of fingerprint recognition.

The fingerprint pattern is classically considered to include ridges, valleys, singularities (deltas, loops, and cores), and minutiae (local discontinuities and special features). Using these patterns, fingerprints are classified into classes starting from five major classes of patterns which encompass most prints (*left loop, right loop, whirl, arch,* and *tented arch*).[44] Once the main class categories are identified, the minutiae of the test prints are usually matched against those of the target templates on file.

Liveness is one of the major issues with fingerprint recognition. Fingerprints are left behind when an individual touches any hard surface. In the same way that forensic techniques have been collecting finger printers for more than a century, the prints of an individual may easily be lifted and with today's advanced latex molding techniques, a replica of the target person's finger may be created out of latex or similar materials. Imagine a latex fingertip, which may be worn over anyone's finger, that has been molded from the target individual's fingerprint. To be able to access anything that the target individual is allowed, all the impostor has to do is to wear the latex replica and pass the fingerprint recognition test. These latex impressions can be made so thin so that they are not easily visible to the naked eye.

Another problem is that close to 2% of the population do not have usable fingerprints. This is mostly due to damage caused by years of manual labor. Depending on the application, this percentage could be much higher. An example is the use of fingerprints for setting off dynamites for construction purposes. Most construction foremen get to that position after having years of hands-on experience as laborers. The percentage of construction workers that have non-usable prints is quite high. For this population, the use of fingerprint recognition would simply not work. Therefore, the dynamite detonation security systems would have to use other types of biometrics.

The whole palm may also be used in pretty much the same way as a fingerprint is used to identify individuals. The patterns on the palm of the hand possess a uniqueness based on the random generation of the patterns while the hands is formed. Recognition algorithms try to match the pattern of the ridges and shapes of the different zones of the hand to a database. Palm recognition suffers from most of the same problems as stated for fingerprints, including the liveness issue and the fact that palms could be somewhat damaged due to problems such as engagement in manual labor. New fingerprint sensors with light that penetrates the skin have been proposed to alleviate the population with damaged fingerprints. These techniques resemble the techniques of Section 1.5.9 far more than those discussed in this section.

1.5.5 Hand and Finger Geometry

Hand and finger geometry techniques usually use the back of the hand. An example is the technique used by [60] and [39] which is based on a special apparatus with a surface containing pegs designed to keep the fingers and the hand in a specific configuration. Envision a hand placed flat on a surface with the fingers kept wide apart. The amount of distance between the fingers is dictated by the pivot pegs on the specialized apparatus. Then, the shape of the fingers in that standard orientation is photographed and studied to utilize the different sizes of fingers and knuckle locations, and other features for establishing a unique identity. The geometry may of course be obtained by taking a photograph of the top of the hand once it is kept in the constrained position [60, 39] or just to scan the palm for the same information [59]. This is quite limited, due to the specialized apparatus which is certainly not very portable and not designed for everyday remote recognition of individuals the way speaker recognition can be utilized.

There are also techniques that concentrate on the fingers and do simple imaging of the upper surface of the hand to establish the locations of the knuckles. They use the lines in the knuckles just to come up with the locations of the knuckles and then the finger geometry is used to identify the individual.[40] It is unclear how unique these distances are since no large population study seems to be available like it is for some other biometrics.

1.5.6 Iris

After Fingerprint recognition, Iris has received the most attention in the biometrics field. It is partly because of the fact that based on large studies involving millions of different people, the uniqueness of the iris pattern has been established. In 2001, Reference [20] studied 2.3 million pairs of irides and based on the information in the patterns of the examined images, it concluded that the chance of two identical irides extrapolates to 1 in 7-billion. The same paper discusses a smaller study based on comparing the left and right eyes of 324 individuals and shows that the left eye and the right eye have the same *cross entropy* (see Definition 7.17) as the eyes of different people. This suggests that the iris pattern enjoys an epigenetic randomness which makes it an ideal candidate for a biometric measure.

As seen with other biometrics, iris recognition also has its downfall. One of the most serious hurdles is the problem of obtaining usable images from the eye. Illumination changes the shape of the iris as the pupil size changes. Also, non-cooperative subjects can pose problems in obtaining usable images at the proper angle. In addition, for a good image, the eye has to be still and should be close to the camera.

Another issue is that to reduce reflectivity from the cornea, an infrared leaning light should be used. The lighting conditions are paramount in the success of iris recognition. Again, these issues make the simplicity of obtaining audio samples for speaker recognition stand out in competition.

Note that the same issue as mentioned for most other biometrics regarding defective or missing body parts is also true for the iris. There is no known statistic on this issue, but there are certainly a number of people with damaged irides, either congenitally or through an accident or an illness. Another problem is the security of keeping around data which has been obtained directly from body parts. Speech samples do not have direct consequences on the person's identity and by changing the request for the audio different types of samples may be requested. Most iris recognition systems can easily be fooled by using a high quality image of the face of the person, so identity theft could become a real issue. Of course like any other problem, these also have solutions such as trying to purposefully change the pupil size of the individual while obtaining the images to test for liveness. However, it adds to the complexity of the problem and the problems stemming from different shapes of the iris, discussed earlier.

1.5.7 Retina

The idea for retinal scanning for biometric identification was first introduced by a New York Mount Sinai Hospital ophthalmologist in 1935.[68, 64] It was used to catalog criminals. It was shown that the tissue and vein patterns of the retina are almost unique to each individual, even more than the iris pattern, due to the larger surface area.

Realizing retinal patterns, has been achieved using several different signal processing techniques, mainly through the usage of Fourier and Mellin transforms.[13, 67] Most of the time, the color information in the image is discarded and the gray-scale image is used to conduct the pattern analysis.[13] There does not seem to be any real technical challenge with the image capture and the pattern recognition. However, retinal recognition has never been accepted for many other reasons, although it has been around longer than most other biometrics.

The failure of being widely accepted is mostly due to the invasive nature of the imaging as well as the difficulty of obtaining an image. Traditionally, a near infrared light was shone into the subject's eye to obtain the retinal image. This worries many users for the possible risk of damage to their retina. Also, the retina does not remain unchanged through the years. Illnesses such as *diabetes* and *glaucoma* can change the patterns of the retina. Furthermore, there are degenerative diseases such as *retinitis pigmentosa* and other retinal dystrophies which can change this pattern. In addition, the image may be distorted by advanced *astigmatism* and *cataracts*. All of these problems attribute to its lack of popularity.

The special optical systems used for retinal image acquisition have been quite expensive in the past. To remedy this issue, there are new techniques for using normal cameras at the expense of some accuracy.[13]

1.5.8 Thermography

Thermographic biometrics utilize the distribution of thermal energy on the skin (generally of the face). As seen in Section 1.5.3, face recognition is plagued by the illumination curse. Namely, the within class variability due to illumination discrepancies could easily exceed those across different classes. The main motivation behind using thermographic imaging is to utilize light at a wavelength which is not abundantly available in normal lighting conditions. Therefore, variation in the lighting would not cause as many problems. This is an advantage of thermal imaging, however, like any other biometric, there are disadvantages.

Thermographic imaging inherits all other disadvantages pointed out for face recognition in Section 1.5.3 with the exception of the remedied illumination issue. However, as a trade-off which offsets the illumination solution, the cost of the photography increases substantially. The wavelength of the infrared light used in these operations is in the range of $8\mu m - 12\mu m$.[61] It cannot be detected using normal Charge-Coupled Device (CCD) cameras. For this purpose, costly *Microbolometer* technology has to be used which adds to the cost significantly. Prices of such cameras at the time of writing this book were in excess of $US\$26,000.00$!

1.5.9 Vein

The vein pattern (mostly of the hands) is used to identify individuals. It works by using near-infrared light which penetrates through the skin, but gets absorbed by the iron-rich red blood cells in the veins. The result is a pattern in which the veins look like dark lines and the rest of the light is reflected off the subcutaneous tissues of the hand. The vein pattern seems to be quite unique to the individual and great results are seen in the identification results.

Like any other biometric, vein recognition has many advantages as well as disadvantages. One of the advantages of the vein pattern is that it does not change readily, so it seems to be less susceptible to time lapse effects (see Chapter 20). Also, everyone has veins in contrast with fingerprints which are not present in about 2% of the population. Of course, we are considering people with limbs in this argument. Missing limbs would affect both fingerprint and vein recognition in the same way. That is more analogous to the mute population for speaker recognition.

One of the disadvantages of this technique is its need for specialized, expensive and typically large camera scanners. This makes it unsuitable for deployment in mobile applications such as in cellular (mobile) phones and notebook computer applications. A second problem which makes it hard for it to compete with speaker recognition is that the technology may seem somewhat invasive to the layperson. Although there is no indication that the light used in the process is harmful, it is not as well understood by the general public as talking.

Another important disadvantage, which may be addressed in time, is the unavailability of large independent studies for this biometric. Most results have been published by vendors and this makes them biased. Also, it seems like there is no standards development for this biometric at the time of writing of this book.

1.5.10 Gait

Some researchers have tried to recognize people based on their style of walking. The length of a person's *stride* and his/her *cadence* are somewhat behavioral, but they also possess some physiological aspects. They are affected by the person's height, weight, and gender among other factors.[9] Cadence is a function of the periodicity of the walk and by knowing the distance the person travels, his/her stride length can be estimated.

The sensor for gait recognition is usually a camera which has to first try to decipher the elements of walking from a video. Of course, the same issues as seen in other image related biometrics (lighting and angle consistencies) come up again. In addition there are specific problems related to video techniques in contrast with still techniques such as speed and direction of movement and their effects on recognition.

There have been very limited studies with this biometric, with small number of trials. Results seem to show some tangible correlation, but are not good enough to be used in the same sense as the more popular biometrics listed here. Reference [9] shows identification rates close to 40% for a handful of individuals in the database. Also, the verification equal error rates (EER) are higher than 10%.

Furthermore, it seems like the behavioral aspect of gait is also influenced by the company which is kept. For example, when a group of people walk together they affect one another in the style of walking for that session. This short-term behavioral influence reduces the usability of this technique for serious recognition activities.

1.5.11 Handwriting

Handwriting may also be used in a similar fashion as speech. In fact it is quite similar in the sense that it also possesses behavioral (learned style of writing) and physiological (motor control) aspects.[6, 30] Although at a first glance it may seem like handwriting is more behavioral, it is quite apparent that we are constrained by our motor control in how we move a pen – if we are to do it in a natural way.[6, 30] This is very similar to the natural characteristics of our vocal tract versus our ability to somewhat affect the resulting audio which is uttered.

Signature verification greatly resembles speaker verification. Also, in the same way that the speech of a person may be used to assess his/her identity, handwriting may also be used. Of course, it is a bit vague to simply talk about handwriting. The handwriting we normally talk about is the trace of a pen on paper. However, what we are discussing mostly in this section is not limited to this trace. To truly be able to use handwriting for recognition purposes, a digitizer tablet should be used to capture not only the trace, but also the dynamics of the pen on the surface of the tablet. The tip of the pen is digitized at a fixed sampling rate. The resulting points will provide the location and speed of the pen motion. Sometimes a pressure sensitive pen is used to be able to use the pressure differentials as well. An offline assessment of handwriting loses most essential information about the physiological aspects of handwriting, hence losing most of the biometric capacity. This is not as pronounced when the objective is simply to read the content of a handwritten text. That problem is more related to speech recognition and may show relative success without the existence of the online features – this is the optical character recognition (OCR) problem. Although, in general, the online features of handwriting always help in both scenarios.

Therefore, since an online tablet seems to be a necessary device in using handwriting as a biometric, it becomes much harder to implement – if only for the infrastructural problems. Of course, if the will is present, it may be used in a wider sense. Take the ever-growing number of signature capture devices at department stores, your mail courier, etc. Unfortunately, at this point, these devices have been designed to perform a very simple capture with very little performance. Relaxed legal definitions of signature have not pushed these devices to their full potentials. You may have noticed that the signatures captured on most of these devices generate many spurious points and have a great miss rate which result in illegible traces of the signatures. High quality tablets do exist which allow for high quality capture of the online signature. If the topic receives more attention, the better quality tablets may be placed in service. Until that time, it is no competition for speaker recognition and even then it will only tend to capture a specialized part of the market. Handwriting recognition, due to its independence from speech is a great candidate for fusion and multimodal use with speaker recognition.

1.5.12 Keystroke

Much in the same way that handwriting has its physiological angle, typing is governed by our motor control. The speed of transition between keys is not only behavioral, but is limited by the physiology of our fingers and hand movements. Also, the behavioral part will present a bias toward words we have typed more often in the past. If the text is original (of our own cognition), then another aspect becomes essential and that is the frequency of word usage and their relative placement. In other words, every one of us has his/her own personal built-in language model with word frequencies and personalized n-grams [26, 5].

Most software applications do not register the speed of transition between keys, although it is not so hard to accomplish. In a training stage, the statistics of the key transitions and the speed of transition may be learned by an enrollment system which may create a model in the same way as a speaker model is generated for speaker recognition. The model may then be used to recognize the typist at the time of testing.

1.5.13 Multimodal

Section 1.4.4 described an indexing application which can operate on the audio track of a video stream as well as plain audio. Imagine the example of using the audio track from a video clip. For each segment whose boundary is realized based on a speaker change, it is possible to produce a sorted list of speaker labels – sorted by a likelihood score. This list may be combined with other biometrics such as face recognition for a video stream as well as textual information either in the form of results from a natural language understanding system or raw and in the form of a search list. Such combinations have been shown to portray promising results.[73, 12, 77]

Figures 1.4 and 1.5 show the two parallel audio and video indexing processes applied on a video stream. Results are combined and indexed for future searches on the text, speaker by voice and speaker by face. The results of the face and voice may be combined for a more seamless search where the search query will ask for a combination of a text string being spoken by a certain speaker. The speaker tag-

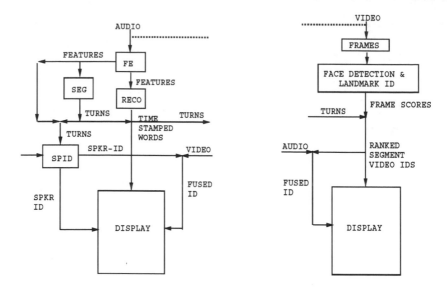

Fig. 1.4: Indexing Based on Audio   **Fig. 1.5:** Indexing Based on Video

ging includes results from matching the audio and the face of the target individual. Table 1.2 shows the ranking results based on Audio, Video and Fusion scores for a sample clip. In this case, the ID of the correct individual, OH, is returned by the fused system.[73]

Rank	Audio		Video		Fused	
1	UM	1.000	JW	1.000	OH	0.990
2	OH	0.997	OH	0.988	AK	0.966
3	UF	0.989	AK	0.961	GB	0.941
4	AK	0.988	SF	0.939	JM	0.936
5	JM	0.986	GB	0.932	JW	0.808
6	GB	0.980	JM	0.925	SF	0.759

Table 1.2: Sample Audio/Video and Fusion Results For Multimodal Speaker Recognition [73]

Since the basis for speaker recognition is the vocal tract characteristics of individuals and their uniqueness, it makes sense to combine this source of information with other presumably unrelated sources of biometric information. In fact speaker recognition may be combined with any of the biometrics listed here to improve results. Another example of such a multimodal system is the fusion of *speaker recognition* with *fingerprint recognition* [69].

Also, there have been studies which use microphone arrays in fixed settings such as conference rooms to identify the active speaker by the position of the speaker in the room. It works by comparing the strength of the signal being inputted into the different elements of the array. This information is then combined with more classical speaker segmentation and identification techniques to come up with a better performance.[58]

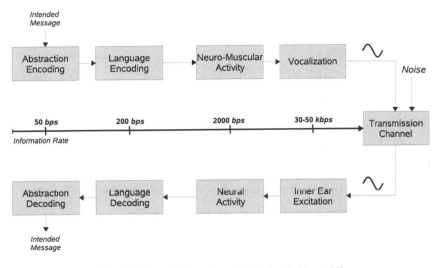

Fig. 1.6: Speech Generation Model after Rabiner [55]

1.5.14 Summary of Speaker Biometric Characteristics

Telephones and microphones have been around for a long time and have prepared the public for their acceptance. Usage of a biometric system based on microphones seems to be much better tolerated than newer systems using unfamiliar means. The negative aspect of this existing infrastructure is the presence of somewhat antiquated systems in use which degrade the quality of speaker recognition systems. Examples are band-limited analog networks which are still in use in many countries. This is true with all established infrastructures, but the positive aspects of the existence of the infrastructure outweigh the difficulties with legacy networks.

Major advantages of using speech as a biometric are its the non-intrusive nature (especially based on the modern culture revolving around the telephone) and the possibility of remote processing, again based on the telephone and the Internet infrastructures.

Notable disadvantages on the other hand are its variability, channel effects, and background noise susceptibilities. The variability aspect may be due to illnesses such as nasal congestion, laryngitis, behavioral variations, and variations due to lack of coverage. Let us discuss the lack of coverage in some more detail. In most biometrics, the samples are quite repeatable and one or at most a handful of samples would be enough for acceptable recognition. In general, the speech signal is a high capacity signal conveying a small amount of information – See Figure 1.6. In certain cases, long samples of speech may be obtained with a small coverage of all possible speech sequences. This means that the data seen in training and enrollment stages does not necessarily possess sufficient coverage of the data being seen at the recognition time.

The trouble with channel effects, mentioned above, may be present in different forms such as noise on the channel, channel variability, and compression effects. The noise present on a channel is usually well modulated into the speech signal with almost impossible separation. The characteristics of the noise can manifest themselves by modifying the properties of the speaker characteristics observed in the signal. This is also true about the channel characteristics. Since telephony networks are quite complex with unpredictable paths, these characteristics may change with each instance of communication. Also, due to the small amount of information present in the high-capacity speech signal, most of the time very aggressive compression schemes are utilized, which still allow intelligibility of the content of speech, but may modify the speaker characteristics significantly by eliminating some of the dynamics of the signal.

References

1. Abu-El-Quran, A., Gammal, J., Goubran, R., Chan, A.a.: Talker Identification Using Reverberation Sensing System. In: Sensors, 2007 IEEE, pp. 970–973 (2007)
2. Akkermans, A., Kevenaar, T., Schobben, D.a.: Acoustic ear recognition for person identification. In: Automatic Identification Advanced Technologies, 2005. Fourth IEEE Workshop on, pp. 219–223 (2005)
3. ARPA: Proceedings of the DARPA Speech Recognition Workshop (1996)
4. ARPA: Proceedings of the DARPA Speech Recognition Workshop (1997)
5. Beigi, H.S.: Character Prediction for On-line Handwriting Recognition. In: Proceedings of the Canadian Conference on Electrical and Computer Engineering, vol. II, pp. TM10.3.1–TM10.3.4 (1992)
6. Beigi, H.S.: Pre-Processing the Dynamics of On-Line Handwriting Data, Feature Extraction and Recognition. In: A. Downton, S. Impedovo (eds.) Progress in Handwriting Recognition, pp. 191–198. World Scientific Publishers, New Jersey (1997). ISBN: 981-02-3084-2
7. Beigi, H.S., Maes, S.H., Chaudhari, U.V., Sorensen, J.S.: A Hierarchical Approach to Large-Scale Speaker Recognition. In: EuroSpeech 1999, vol. 5, pp. 2203–2206 (1999)

8. Beigi, H.S.M., Maes, S.H., Chaudhari, U.V., Sorensen, J.S.: IBM Model-Based and Frame-By-Frame Speaker Recognition. In: Speaker Recognition and its Commercial and Forensic Appications (1998)

9. BenAbdelkader, C., Cutler, R., Davis, L.a.: Stride and cadence as a biometric in automatic person identification and verification. In: Automatic Face and Gesture Recognition, 2002. Proceedings. Fifth IEEE International Conference on, pp. 372–377 (2002)

10. Bengherabi, M., Tounsi, B., Bessalah, H., Harizi, F.: Forensic Identification Reporting Using A GMM Based Speaker Recognition System Dedicated to Algerian Arabic Dialect Speakers. In: 3rd International Conference on Information and Communication Technologies: From Theory to Applications (ICTTA 2008), pp. 1–5 (2008)

11. Bennani, Y., Gallinari, P.a.: A modular connectionist architecture for text-independent talker identification. In: Neural Networks, 1991., IJCNN-91-Seattle International Joint Conference on, vol. ii, pp. 857–860 (1991)

12. Besson, P., Popovici, V., Vesin, J.M., Thiran, J.P., Kunt, M.a.: Extraction of Audio Features Specific to Speech Production for Multimodal Speaker Detection. Multimedia, IEEE Transactions on 10(1), 63–73 (2008)

13. Borgen, H., Bours, P., Wolthusen, S.: Visible-Spectrum Biometric Retina Recognition. In: International Conference on Intelligent Information Hiding and Multimedia Signal Processing (IIHMSP2008), pp. 1056–1062 (2008)

14. Brunelli, R., Poggio, T.: Face Recognition: Features versus Templates. IEEE Transactions on Pattern Analysis and Machine Intelligence 15(10), 1042–1052 (1993)

15. Cadavid, S., Abdel-Mottaleb, M.a.: Human Identification based on 3D Ear Models. In: Biometrics: Theory, Applications, and Systems, 2007. BTAS 2007. First IEEE International Conference on, pp. 1–6 (2007)

16. Cetingul, H., Yemez, Y., Erzin, E., Tekalp, A.a.: Discriminative lip-motion features for biometric speaker identification. In: Image Processing, 2004. ICIP '04. 2004 International Conference on, vol. 3, pp. 2023–2026 (2004)

17. Chau, C.K., Lai, C.S., Shi, B.E.: Feature vs. Model Based Vocal Tract Length Normalization for a Speech Recognition-Based Interactive Toy. In: Active Media Technology, Lecture Notes in Computer Science, pp. 134–143. Springer, Berlin/Heidelberg (2001). ISBN: 978-3-540-43035-3

18. Chen, H., Bhanu, B.a.: Human Ear Recognition in 3D. Pattern Analysis and Machine Intelligence, IEEE Transactions on 29(4), 718–737 (2007)

19. Chunrong, X., Jianhuan, Z., and, L.F.: A Dynamic Feature Extraction Based on Wavelet Transforms for Speaker Recognition. In: Electronic Measurement and Instruments, 2007. ICEMI '07. 8th International Conference on, pp. 1–595–1–598 (2007)

20. Daugman, J., Downing, C.: Epigenetic Randomness, Complexity and Singularity of Human Iris Patterns. Biological Sciences 268(1477), 1737–1740 (2001)

21. Drygajlo, A.a.: Forensic Automatic Speaker Recognition [Exploratory DSP]. Signal Processing Magazine, IEEE 24(2), 132–135 (2007)

22. Fabate, A., Nappi, M., Riccio, D., Ricciardi, S.a.: Ear Recognition by means of a Rotation Invariant Descriptor. In: Pattern Recognition, 2006. ICPR 2006. 18th International Conference on, vol. 4, pp. 437–440 (2006)

23. Fermentas Nucleotides Catalog. Website (2009). URL http://www.fermentas.com/catalog/nucleotides

24. Flanagan, J.L.: Speech Analysis, Synthesis and Perception, 2nd edn. Springer-Verlag, New York (1972). ISBN: 0-387-05561-4

25. Foote, J., Silverman, H.a.: A model distance measure for talker clustering and identification. In: Acoustics, Speech, and Signal Processing, 1994. ICASSP-94., 1994 IEEE International Conference on, vol. i, pp. I/317–I/320 (1994)

26. Fujisaki, T., Beigi, H., Tappert, C., Ukelson, M., Wolf, C.: Online Recognition of Unconstrained Handprinting: A Stroke-based System and Its Evaluation. In: S. Impedovo, J. Simon (eds.) From Pixels to Features III: Frontiers in Handwriting, pp. 297–312. North Holland, Amsterdam (1992). ISBN: 0-444-89665-1

27. Gonzalez-Rodriguez, J., Fierrez-Aguilar, J., Ortega-Garcia, J.a.: Forensic identification reporting using automatic speaker recognition systems. In: Acoustics, Speech, and Signal Processing, 2003. Proceedings. (ICASSP '03). 2003 IEEE International Conference on, vol. 2, pp. II–93–6 (2003)
28. Hairer, G., Vellekoop, M., Mansfeld, M., Nohammer, C.: Biochip for DNA Amplification and Label-free DNA Detection. In: IEEE Conference on Sensors, pp. 724–727 (2007)
29. Hodges, N.D.C.: Census of the Defective Classes. Science **VIII** (1889). URL http://www.census.gov
30. Hollerbach, J.M.: An Oscillation Theory of Handwriting. MIT Press (1980). PhD Thesis
31. Hurley, D., Nixon, M., Carter, J.a.: A new force field transform for ear and face recognition. In: Image Processing, 2000. Proceedings. 2000 International Conference on, vol. 1, pp. 25–28 (2000)
32. J., A.: Effect of age and gender on LP smoothed spectral envelope. In: The IEEE Odyssey Speaker and Language Recognition Workshop, pp. 1–4 (2006)
33. Jaakkola, T., Haussler, D.: Exploiting Generative Models in Discriminative Classifiers. In: Advances in Neural Information Processing Systems, vol. 11, pp. 487–493. MIT Press (1998)
34. Kablenet: No Snooping on the Public. World Wide Web (2007). URL http://www.theregister.co.uk/2007/08/03/cctv_audio_recording_consultation
35. Kanak, A., Erzin, E., Yemez, Y., Tekalp, A.a.: Joint audio-video processing for biometric speaker identification. In: Acoustics, Speech, and Signal Processing, 2003. Proceedings. (ICASSP '03). 2003 IEEE International Conference on, vol. 2, pp. II–377–80 (2003)
36. Kersta, L.G.: Voiceprint Identification. Nature **196**, 1253–1257 (1962)
37. Kijima, Y., Nara, Y., Kobayashi, A., Kimura, S.a.: Speaker adaptation in large-vocabulary voice recognition. In: Acoustics, Speech, and Signal Processing, IEEE International Conference on ICASSP '84., vol. 9, pp. 405–408 (1984)
38. Kirby, M., Sirovich, L.: Application of the Karhunen-Loeve Procedure for the Characterization of Human Faces. IEEE Transactions on Pattern Analysis and Machine Intelligence **12**(1), 103–108 (1990)
39. Kukula, E., Elliott, S.: Implementation of hand geometry: an analysis of user perspectives and system performance. IEEE Aerospace and Electronic Systems Magazine **21**(3), 3–9 (2006)
40. Kumar, A., Ravikanth, C.: Personal Authentication Using Finger Knuckle Surface. IEEE Transactions on Information Forensics and Security **4**(1), 1–13 (2009)
41. Li, S.Z., Jain, A.K. (eds.): Handbook of Face Recognition. Springer, New York (2005). ISBN: 978-0-387-40595-7
42. Liu, H., and, J.Y.: Multi-view Ear Shape Feature Extraction and Reconstruction. In: Signal-Image Technologies and Internet-Based System, 2007. SITIS '07. Third International IEEE Conference on, pp. 652–658 (2007)
43. Maes, S.H., Beigi, H.S.: Open SESAME! Speech, Password or Key to Secure Your Door? In: Asian Conference on Computer Vision (1998)
44. Maltoni, D.: A Tutorial on Fingerprint Recognition. In: Advanced Studies in Biometrics, *Lecture Notes in Computer Science*, vol. 3161, pp. 43–68. Springer, Berlin/Heidelberg (2005). ISBN: 978-3-540-26204-6
45. Miller, R.L.: Nature of the Vocal Cord Wave. Journal of the Acoustical Society of America **31**, 667–677 (1959)
46. Morito, M., Yamada, K., Fujisawa, A., Takeuchi, M.a.: A single-chip speaker independent voice recognition system. In: Acoustics, Speech, and Signal Processing, IEEE International Conference on ICASSP '86., vol. 11, pp. 377–380 (1986)
47. Naini, A.S., Homayounpour, M.M.: Speaker age interval and sex identification based on Jitters, Shimmers and Mean MFCC using supervised and unsupervised discriminative classification methods. In: The 8th International Conference on Signal Processing, vol. 1 (2006)
48. Nava, P., Taylor, J.a.: Speaker independent voice recognition with a fuzzy neural network. In: Fuzzy Systems, 1996., Proceedings of the Fifth IEEE International Conference on, vol. 3, pp. 2049–2052 (1996)
49. Nolan, F.: The Phonetic Bases of Speaker Recognition. Cambridge University Press, New York (1983). ISBN: 0-521-24486-2

50. Nosrati, M.S., Faez, K., Faradji, F.a.: Using 2D wavelet and principal component analysis for personal identification based On 2D ear structure. In: Intelligent and Advanced Systems, 2007. ICIAS 2007. International Conference on, pp. 616–620 (2007)

51. Ortega-Garcia, J., Cruz-Llanas, S., Gonzalez-Rodriguez, J.a.: Speech variability in automatic speaker recognition systems for forensic purposes. In: Security Technology, 1999. Proceedings. IEEE 33rd Annual 1999 International Carnahan Conference on, pp. 327–331 (1999)

52. Parris, E., Carey, M.a.: Language independent gender identification. In: Acoustics, Speech, and Signal Processing, 1996. ICASSP-96. Conference Proceedings., 1996 IEEE International Conference on, vol. 2, pp. 685–688 (1996)

53. Peterson, G., Barney, H.L.: Control Methods Used in a Study of the Vowels. The Journal of the Acoustical Society of America (JASA) 24(2), 175–185 (1952)

54. Pollack, I., Pickett, J.M., Sumby, W.: On the Identification of Speakers by Voice. Journal of the Acoustical Society of America 26, 403–406 (1954)

55. Rabiner, L., Juang, B.H.: Fundamentals of Speech Recognition. Prentice Hall Signal Processing Series. PTR Prentice Hall, New Jersey (1990). ISBN: 0-130-15157-2

56. RCFP: Can We Tape? World Wide Web (2008). URL http://www.rcfp.org/taping

57. Reynolds, D.A., Quatieri, T.F., , Dunn, R.B.: Speaker Verification Using Adapted Gaussian Mixture Models. Digital Signal Processing 10, 19–41 (2000)

58. Rozgic, V., Busso, C., Georgiou, P., Narayanan, S.a.: Multimodal Meeting Monitoring: Improvements on Speaker Tracking and Segmentation through a Modified Mixture Particle Filter. In: Multimedia Signal Processing, 2007. MMSP 2007. IEEE 9th Workshop on, pp. 60–65 (2007)

59. Saeed, K., Werdoni, M.: A New Approach for hand-palm recognition. In: Enhanced Methods in Computer Security, Biometric and Artificial Interlligence Systems, Lecture Notes in Computer Science, pp. 185–194. Springer, London (2005). ISBN: 1-4020-7776-9

60. Sanchez-Reillo, R., Sanchez-Avila, C., Gonzalez-Marcos, A.: Biometric identification through hand geometry measurements. IEEE Transactions on Pattern Analysis and Machine Intelligence 22(10), 1168–1171 (2000)

61. Selinger, A., Socolinsky, D.A.: Appearance-Based Facial Recognition Using Visible and Thermal Imagery: A Comparative Study. Computer Vision and Image Understanding 91(1–2), 72–114 (2003)

62. Sharkas, M., Elenien, M.A.: Eigenfaces vs. fisherfaces vs. ICA for face recognition; a comparative study. In: IEEE 9th International Conference on Signal Procesing (ICSP2008), pp. 914–919 (2008)

63. Shearme, J.N., Holmes, J.N.: An Experiment Concerning the Recognition of Voices. Language and Speech 2, 123–131 (1959)

64. Simon, C., Goldstein, I.: Retinal Method of Identification. New York State Journal of Medicine 15 (1936)

65. Slomka, S., Sridharan, S.a.: Automatic gender identification optimised for language independence. In: TENCON '97. IEEE Region 10 Annual Conference. Speech and Image Technologies for Computing and Telecommunications'., Proceedings of IEEE, vol. 1, pp. 145–148 (1997)

66. Stagni, C., Guiducci, C., Benini, L., Ricco, B., Carrara, S., Paulus, C., Schienle, M., Thewes, R.: A Fully Electronic Label-Free DNA Sensor Chip. IEEE Sensors Journal 7(4), 577–585 (2007)

67. Tabatabaee, H., Fard, A., Jafariani, H.: A Novel Human Identifier System using Retinal Image and Fuzzy Clustering. In: International Conference on Intelligent Information Hiding and Multimedia Signal Processing (IIHMSP2008), vol. 1, pp. 1031–1036 (2006)

68. Eye Prints. Time Magazine (1935)

69. Toh, K.A., and, W.Y.Y.: Fingerprint and speaker verification decisions fusion using a functional link network. Systems, Man, and Cybernetics, Part C: Applications and Reviews, IEEE Transactions on 35(3), 357–370 (2005)

70. Tosi, O.I.: Voice Identification: Theory and Legal Applications. University Park Press, Baltimore (1979). ISBN: 978-0-839-11294-5

71. Disability Census Results for 2005. World Wide Web (2005). URL http://www.census.gov
72. Viswanathan, M., Beigi, H., Tritschler, A., Maali, F.a.: Information access using speech, speaker and face recognition. In: Multimedia and Expo, 2000. ICME 2000. 2000 IEEE International Conference on, vol. 1, pp. 493–496 (2000)
73. Viswanathan, M., Beigi, H.S., Maali, F.: Information Access Using Speech, Speaker and Face Recognition. In: IEEE International Conference on Multimedia and Expo (ICME2000) (2000)
74. Voice Biometrics. Meeting (2008). URL http://www.voicebiocon.com
75. Wang, Y., Mu, Z.C., Liu, K., and, J.F.: Multimodal recognition based on pose transformation of ear and face images. In: Wavelet Analysis and Pattern Recognition, 2007. ICWAPR '07. International Conference on, vol. 3, pp. 1350–1355 (2007)
76. Wikipedia: A   Taste   of   Freedom.   Website.   URL http://en.wikipedia.org/wiki/A_Taste_Of_Freedom
77. Xiong, Z., Chen, Y., Wang, R., Huang, T.a.: A real time automatic access control system based on face and eye corners detection, face recognition and speaker identification. In: Multimedia and Expo, 2003. ICME '03. Proceedings. 2003 International Conference on, vol. 3, pp. III–233–6 (2003)
78. Xu, X., and, Z.M.: Feature Fusion Method Based on KCCA for Ear and Profile Face Based Multimodal Recognition. In: Automation and Logistics, 2007 IEEE International Conference on, pp. 620–623 (2007)
79. Yuan, L., Mu, Z.C., and, X.N.X.: Multimodal recognition based on face and ear. In: Wavelet Analysis and Pattern Recognition, 2007. ICWAPR '07. International Conference on, vol. 3, pp. 1203–1207 (2007)
80. Zhang, H.J., Mu, Z.C., Qu, W., Liu, L.M., and, C.Y.Z.: A novel approach for ear recognition based on ICA and RBF network. In: Machine Learning and Cybernetics, 2005. Proceedings of 2005 International Conference on, vol. 7, pp. 4511–4515 (2005)
81. Zhang, Z., and, H.L.: Multi-view ear recognition based on B-Spline pose manifold construction. In: Intelligent Control and Automation, 2008. WCICA 2008. 7th World Congress on, pp. 2416–2421 (2008)
82. Zhou, S.K., Chellappa, R., Zhao, W.: Unconstrained Face Recognition, *International Series on Biometrics*, vol. 5. Springer, New York (2008). ISBN: 978-0-387-26407-3

Chapter 2
The Anatomy of Speech

The shadow of a sound, a voice without a mouth, and words without a tongue.

Horace Smith (Paul Chatfield)
The Tin Trumpet [6]: Echo, 1836

To achieve an understanding of human speech production, first, one should study the anatomy of the vocal system (the speech signal production machinery). It is fair to say that one should grasp the process of speech production, before attempting to model a system that would understand it. Once this mechanism is better understood, we may try to create systems that recognize its distinguishing characteristics and nuances, thus recognizing the individual speaker.

Furthermore, since evolution has a way of fine tuning our anatomy, the vocal and auditory systems have evolved to work in unison. An understanding of the auditory system can help us do a better job at picking up characteristics of our voice. In addition, evolution has already built in a recognition capability into our auditory perception which should be studied before we attempt to create our own automatic speaker recognition system. This natural recognition system has been designed to allow us to recognize the voices of our parents at infancy and it generally develops even before we start to make sense of their speech content. Reference [2] shows that infants start recognizing human voices from other sounds between the ages of 4 and 7 months, way in advance of when they start to understand speech.

Aside from the actual speech production and sensory organs, the nervous system, in charge of producing and deciphering the signals, plays a major role. As we shall see later, different parts of the brain, although quite intertwined and in permanent communication among themselves, specialize in handling specific tasks. Low level audio processing is done in both hemispheres in the auditory cortex. Parts of the left hemisphere specialize in the production and understanding of languages and their mirror images in the right hemisphere of the brain are involved in understanding and production of musical characteristics including pitch, tempo, rhythm, and speakers' voices.[1]

In the following two sections, we will study the mechanical parts of speech production and perception, namely the *vocal system* and the *auditory system*. Then, we will continue with the discussion of language and music production and understanding mechanisms in the human nervous system.

2.1 The Human Vocal System

Figure 2.1 shows a sagittal section of the *nose, mouth, pharynx* and *larynx*. This is the upper portion and the most significant part of the speech production machinery in the human. The only thing missing in this schematic is the pressure production system or the lungs. In fact the lungs do not directly play any role in the audio production other than providing the pressure difference which is necessary to produce speech.

2.1.1 Trachea and Larynx

Starting our way up from the bottom of Figure 2.1, we see a cross section of the *trachea* followed by the *laryngeal* section which includes the *vocal folds* (vocal chords). Figures 2.2 and 2.3 show a more detailed sagittal and coronal view of the upper part of the *trachea* as well as the *larynx*. Figure 2.4 displays the top view of the interior of the *larynx* as it may be seen using a laryngoscope. The opening displays the pathway to the *trachea* from the top. The *folds* are surrounded by cartilage from two sides and are controlled by muscles which can fully close and open them.

2.1.2 Vocal Folds (Vocal Chords)

The opening in the *vocal folds* is in the shape of a triangle with a slight downward slope from the front to the back. Figure 2.4 gives more detail. The front slopes up in the form of the *epiglottal wall*. The back is surrounded by two types of cartilage,

---

[1] The major language functions actually lie in the *dominant hemisphere* of the brain. For the majority of people (over 95% of right-handed people [22] and about 60% of left-handed people), this is the left hemisphere. In the rest of the population, the right hemisphere is dominant. For the sake of simplicity, in this chapter, we speak of the more common case which is the brain structure of an individual with a dominant left brain hemisphere.

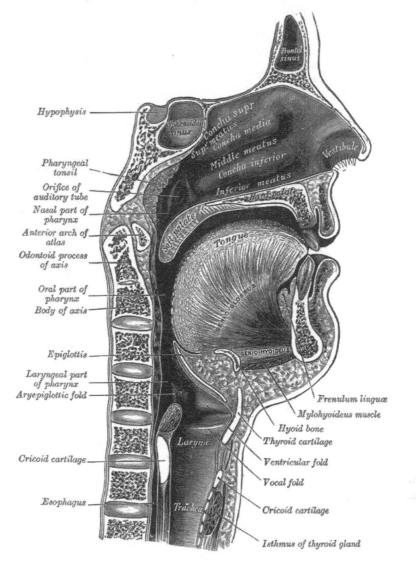

Fig. 2.1: Sagittal section of Nose, Mouth, Pharynx, and Larynx; *Source: Gray's Anatomy [13]*

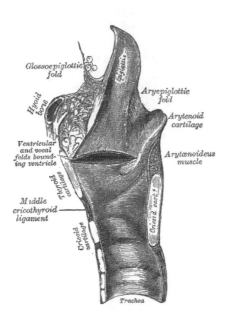

Fig. 2.2: Sagittal Section of Larynx and Upper Part of Trachea; *Source: Gray's Anatomy [13]*

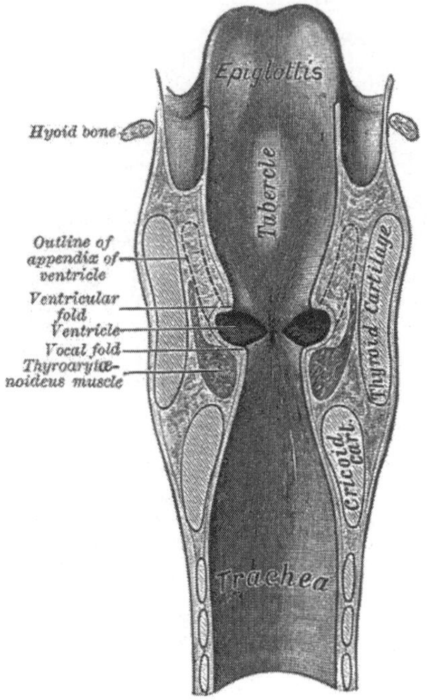

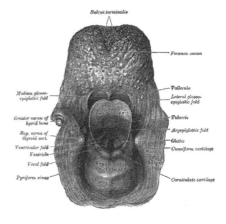

Fig. 2.3: Coronal Section of Larynx and Upper Part of Trachea; *Source: Gray's Anatomy [13]*

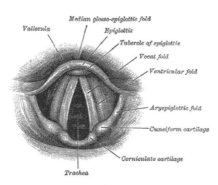

Fig. 2.4: Laryngoscopic View of the interior Larynx; *Source: Gray's Anatomy [13]*

Fig. 2.5: The Entrance to the Larynx, Viewed from Behind; *Source: Gray's Anatomy [13]*

the *corniculate cartilage* in the back and the *cuneiform cartilages* toward the sides.

The area immediately above the *vocal folds* is called the *glottis* and is the starting point of the control units for articulation. The *glottis* is the transitional area between the *larynx* and the *pharynx*. As we mentioned earlier, going up toward the front of the tract, one reaches the *epiglottis* and parallel to it toward the back side of the tract, we go through the *laryngeal part of the pharynx* – see Figure 2.1.

The more closed the v-shaped opening of the *vocal folds* becomes, the more tense the muscles of the folds would be. When fully opened, the muscles are completely relaxed and provide the least resistance to the air which flows from the *trachea* to the upper part of the *larynx* and the *pharynx*. Depending on the amount of tension in the muscles of the *vocal folds* and the difference in the air pressure between the *trachea* and the *pharynx*, either puffs of air or voiced sounds are produced.

2.1.3 Pharynx

Working its way up, after the air has passed through the *vocal folds* and almost immediately has exited the *larynx*, the *pharynx* starts which has an irregular shape with many sections. This is essentially the beginning of what is called the *vocal tract* since all articulation control starts here. Depending on the amount of air going through the *vocal folds* and their state of tension, vibrating laminar or turbulent air [10] will be going through the different sections of the *pharynx* to produce different kinds of sounds.

The next interesting spot, as the air travels upward in the vocal tract, is the *epiglottis* which leads to the back of the *tongue*. The *tongue* is a rounded muscular mass which at this point in the journey defines a moving frontal boundary for the *oral part* of the *pharynx*. Moving up the *oral part of the pharynx*, the air may travel into two different cavities. The first is along the surface of the *tongue* which is bounded on the top by the *palate*. This is the *oral cavity* which eventually leads out to the ambiance through the opening in the lips.

Another path, the air may take, is up through the *nasal cavity* or the, so called, *nasal part* of the *pharynx* – see Figure 2.1. The air then hits low pressure at the exit of the nasal cavity, or the *vestibule*.

2.1.4 Soft Palate and the Nasal System

The *soft palate* starts with the *uvula* which hangs in the back of the *soft palate* and ends at the boney part of palate called the *hard palate*. The *uvula* and the *soft palate* (*velum*) are primarily responsible for the diversion of the air into the *nasal cavity* or the *oral cavity*. In fact, *Laver* [18] categorizes the *soft palate* and the *nasal cavity* as the *nasal system*. The *nasal cavity* is essentially a sound box which mostly dissipates the energy in speech by releasing the air into ambient pressure through the *vestibule*.

2.1.5 Hard Palate

If the air passage to the *nasal cavity* is blocked using the shape of the *tongue* and the position of the *uvula* and the *soft palate*, it will flow along the surface of the *tongue* into the *oral cavity*, constrained on top by the *hard palate*. As the *hard palate* extends toward the upper teeth, a ridge appears called the *alveolar ridge* – rub the tip of your tongue on the *hard palate* starting from the roots of the top front teeth to the back.

2.1.6 Oral Cavity Exit

At the exit point, the air, traveling up the vocal tract and into the oral cavity, meets the two final hurdles which may shape the articulation. These are the upper and lower *teeth* and the *lips*. As it will be seen later, they play important roles in articulation.

2.2 The Human Auditory System

The complete auditory system includes the whole ear assembly, the *vestibulocochlear nerve bundle* (auditory nerve bundle) and the *auditory cortex* of the brain. We shall quickly examine the different parts of this system which are responsible for registering the audio signal in our brain. Later, we shall examine complementary parts of the brain which are responsible for interpreting these signals into parts of language and speaker identity.

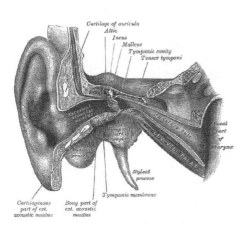
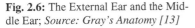

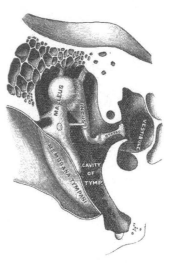

Fig. 2.6: The External Ear and the Middle Ear; *Source: Gray's Anatomy [13]*

Fig. 2.7: The Middle Ear; *Source: Gray's Anatomy [13]*

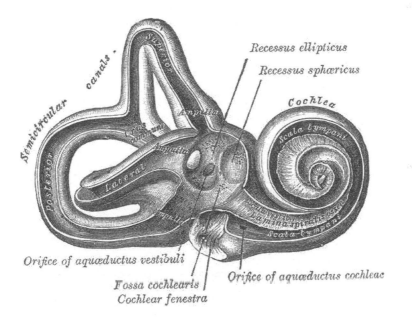

Fig. 2.8: The Inner Ear; *Source: Gray's Anatomy [13]*

2.2.1 The Ear

The auditory system includes a mechanical system and a nervous system. In this section, we mainly concentrate on the function of the ear. Later, we shall see the nervous-system associated with hearing. The mechanical part of hearing (the *ear*) is made up of three sections. The first section, *external ear*, is the combination of cartilages in the *auricula* and the *external acoustic meatus* (the *ear canal*) – see Figure 2.6.

Second, the *middle ear* includes the *tympanic membrane* (*ear drum*) and everything to its right, shown in Figure 2.6. This is mainly the membrane itself and the three special bones which transfer the motion of the *tympanic membrane* induced by the sound waves going through and being amplified by the *external ear*. These bones are in turn called, *malleus* (hammer), *incus* (anvil), and *stapes* (stirrup) – see Figure 2.7. The vibrations are transmitted from the *tympanic membrane* to the *malleus*, from the *malleus* to the *incus*, and from the *incus* to the *stapes*. From there these vibrations are transferred to the inner ear through the *cochlear fenestra ovalis* (oval window of the cochlea) – see Figure 2.8.

Finally, the *inner ear* (Figure 2.8) is made up of the *cochlea*, a snail-like cavity, and three semicircular canals called the *superior ampulla*, the *anterior ampulla*, and the *posterior ampulla*. The cochlea and the three semicircular canals are filled with an incompressible fluid [10] which is excited by the motion of the *stapes* bone at the extreme entrance of the *cochlea* called the *cochlear fenestra ovalis*. The motion of the *stapes* induces pressure waves in the fluid of the *inner ear* which in turn excites the thousands of hairs (*cilia*) inside the spiral of the cochlea (the *scala tympani*). The *cilia* are arranged in four rows, one row lines the inner side of the spiral (points on the inside of the spiral closest to the center of curvature). The *cilia* in this row are connected to the auditory nerve bundle and transmit the motion signal to the brain for cognition. The other three rows in the outer extreme of the scala receive feedback from the brain which allows for pre-amplification of the motion of the fluid – see Section 2.3.6.1.

The spiral shape of the *scala tympani* provides a semi-logarithmic cognitive ability of sound which is important in the development of the speaker models and features. The end point of it is called the *helicotrema* [25]. Starting at the *cochlear fenestra ovalis*, high-pitch audio is realized. As the sound travels toward the *helicotrema*, the high-pitched components are suppressed and only the lower pitched components survive such that, close to the helicotrema, only the lowest pitch is realized.

Section 5.1.1 describes measures taken to match this semi-logarithmic (combination of linear and logarithmic) cognition using special warping of the frequency scale used in feature analysis. In addition, the sizes of the hairs also achieve a logarithmic cognition of the amplitude as well as the said frequency warping.

Once the cilia are excited, the signal they generate is carried through the auditory nerve bundle to the auditory cortex located in the right and left hemispheres of the brain – see Figures 2.20, 2.21, and 2.22. In the process of traveling from the cochlea to the auditory cortex, the auditory signal passes through several relay stations in the brain stem and the thalamus. In the next few sections, we will study the nervous system involved in speech production and perception, including the path that the auditory signal takes to reach the brain.

2.3 The Nervous System and the Brain

Considering that the ear, in all its complexity, is only a transducer and that we really hear with our brains, there has been very limited coverage of the role of the brain in the speech and speaker recognition literature. This is partly due to the relatively recent developments in brain sciences attributed to modern imaging techniques. On the other hand, as engineers, we may be more biased toward electromechanically more apparent systems. In fact, until very recently, the electrical activity of the brain was not so readily measurable.

This may be the reason why some old cultures such as ancient Egyptians and the Greeks downplayed the importance of the brain and placed all their emphasis on the heart since it actually seemed to do something apparent. *Aristotle* believed that the blood turned into vapor and rose up to the brain (at the top of the body) and then the brain acted as a condenser and turned it back to fluid and returned it to the lower part of the body.[1]

Here, we will try to study speech generation and perception, both from a content perception point of view and the identity recognition aspect. The nervous system is a hierarchical control system with motor-control, feedback, relay stations, and decision centers at different points of the hierarchy. A special cell called the *neuron* has evolved to handle the transmission and decision needs of this hierarchical system. Here, we will start from the anatomy of a neuron and work toward the higher levels of the hierarchy. We will only concern ourselves with parts of the nervous system which are directly involved in the processing, understanding, and production of speech and other forms of audio signals.

2.3.1 Neurons – Elementary Building Blocks

Neurons are the elementary building blocks of the brain and the rest of the nervous system. There are in the order of 100 billion neurons in the human brain alone. Their number peaks in the first few years of life. Every cell is developed in its predetermined relative position to give the brain its functional structure. This structure is coded into our genome. All the neurons are instructed to make a predetermined set of connections at the time of the development of the brain. The genes specify the predefined blueprint for building this network. In the course of a life-time, the connections which are used become strengthened and the ones which are not used become weakened. This is related to the concept of re-enforcement learning which is used in artificial learning algorithms and which is what goes on in our brain throughout our lives.

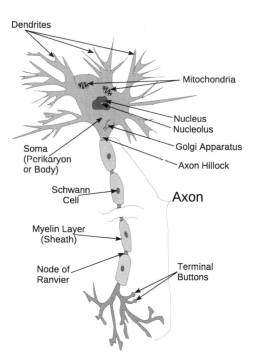

Fig. 2.9: A Typical Neuron

Neurons generally do not reproduce or replicate after the nervous system development is completed. This means that they can live over 100 years and their number will decrease after our first few birthdays. For most of our lives, each person loses an average of 2×10^5 neurons per day. This translates to about a 10% reduction in the number of cells in our life-time.

A neuron is made up of different parts – see Figure 2.9. The main part of a neuron is its *soma* or *perikaryon* (body of the neuron) which contains the *nucleus*. The nucleus contains the genes of the neuron. The job of the genes is to encode proteins, hormones, enzymes and structural proteins which are produced and extruded into the *cellular cytoplasm*.

In addition to the nucleus and the cellular cytoplasm, the soma includes several other mechanisms involved in sustaining the neuron. Figure 2.9 shows some of these units. In the Figure, the *mitochondria* produce ATP^2 to provide the neuron with its much needed chemical energy. Within the nucleus, the *nucleolus* produces *ribosomes* which help introduce genetic information into proteins and to build them from basic amino acids. These ribosomes are stored into *Nissl granules* in the soma and are used for protein synthesis by the cell. The *Golgi apparatuses* were discovered by Golgi [12]. They are structures which aid in the packaging of proteins and *neurotransmitters* and introduce them into *vesicles* which are the transport mechanisms for delivering these materials to different parts of the cell, most importantly the *axon*.

During the development stage of the brain, as a neuron is created, it extends an *axon* and many *dendrites*. Most neurons seem to be polarized in their communication. They receive electrical signals through their dendrites which are usually connected to the axons of other neurons and will transmit a signal through their axon to the dendrites of another neuron. The axon keeps growing until it makes a connection with another neuron's dendrites. This is the most prevalent connection style. Although, there are cases when the axon may bond with another axon or with the body of another neuron.

The elements of the axon that finally make the connection with the dendrites of the receiving neuron are called *Terminal Buttons*. The dendrites contain elements called *receptors* which are at the receiving end of the electrical signal being transmitted. The connection itself is called a *synapse*. At a synapse, the terminal buttons of the axon are slightly removed from the receptors of the dendrites of the receiving neurons. This space (known as the *synaptic cleft*) is used to carry chemical messages in the form of *Serotonin* or other neurotransmitters across from the terminal buttons of the axons to the receptors of the dendrites. The two sides of the synapse are called *presynaptic* and *postsynaptic* nerve endings. Depending on the excitation level of the transmitting neuron, different levels of neurotransmitters are released at the synapse, governing the electrical excitation of the dendrites and hence that of the receiving neuron.

---

2 Adenosine-5-Triphosphate is a nucleotide which is responsible for intercellular energy transfer. It is related to the dATP nucleotide since it is basically a dATP bonded with a simple sugar called *ribose* which is a 5-carbon sugar (*pentose*) – see Section 1.5.1.

The *soma* is in the order of about $10^{-5}m$ in diameter, but the axon is quite thin and long. Length of some tactile axons can reach several feet from the sensor locations to the spinal cord. In fact, the cytoplasm embedded in the axon could be up to an order of 10^5 times more than that embedded in the *soma*. This shows the relative volume of an axon with respect to the body of the neuron. However, the nucleus of the neuron has to provide the energy and oxygen needed for the cytoplasm of the whole neuron which includes the relatively huge volume of the axon. This makes neurons the most energy absorbing cells in the body.[15]

There are several different types of neurons in the body. They may have no axon, or sometimes up to two. Some conduct one-way communication, some do two. As far as our topic is concerned, we are interested in the more abundant and relevant neurons which have a single axon and communicate in a single direction – from the dendrites to the axon. *Camillo Golgi* [12], the recipient of the Nobel prize in medicine in 1906, categorized neurons into two types. The first type, also known as *Type I* neurons have a very long axon. The second type of neurons have very short axons or none at all. These are called *Type II* neurons. In *Type I* neurons, the axon is inhabited by small cells called *Schwann cells* which produce a fatty substance called *Myelin* surrounding the axon. The *Myelin layer (sheath)* acts as an electrical insulator which increases the electrical conductivity, therefore increasing the efficiency of the axon. The length of the axon is not completely covered by Myelin. The Myelin is extruded around the Schwann cells, so the axon looks somewhat like a string of sausages. The small portions of the axon in between adjacent Schwann cells, not covered in Myelin, are called *Ranvier nodes*. These nodes allow some electrical current to escape from the sides of the axon.

2.3.2 The Brain

The central nervous system has two major parts, the *spinal cord* and the *brain* (see Figure 2.10). The brain is further divided into three major sections, the *forebrain* (*prosencephalon*), the *midbrain* (*mesencephalon*), and the *hindbrain* (*rhombencephalon*). The *forebrain* contains the *cerebrum* (*telencephalon*) and the *diencephalon*. The *cerebrum* is at the top of the control hierarchy. The *diencephalon* includes the *Thalamus*, the *Hypothalamus*, the *Epithalamus* and the *Ventriculus Tertius* (Third Ventricle). The *midbrain* (*mesencephalon*) is mainly a neural fiber bundle designed for communications between the cerebrum and the spinal cord. The *hindbrain* (*rhombencephalon*) starts from the top of the spinal cord with the *medulla oblongata*, and follows up to the *pons* and finally to the *cerebellum* (the little brain). *Midbrain* and *hindbrain* together constitute the *brain stem*.[13]

All parts of the brain, with the exception of part of the cerebrum, are mostly made of *white matter* and generally handle low-level functions and communications in the

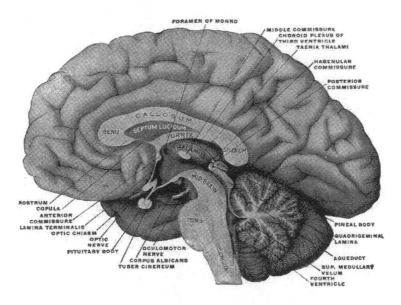

Fig. 2.10: Sagittal Section of the Human Brain (*Source: Gray's Anatomy [13]*)

brain. *White matter* is called by that name since it is mostly made up of neurons with long axons designed for transmission of messages. The fatty Myelin cover around the axons, gives white matter its white appearance. In general, neurons with longer axons have myelinated axons and those with shorter axons do not.[19]

The *cerebrum* is the most interesting part of brain, so far as this topic is concerned. It, also, has many parts, the most important of which is the *cerebral cortex*. The *cerebral cortex* is the gray mantel of tissue on the surface of the cerebrum. It is sometimes called *gray matter* because of its gray shade due to the high density of short (unmyelinated) neurons [19] which are packed together to provide the higher functional capabilities of the brain. It is really made up of the outer layer which only constitutes a thickness of up to 4 *mm*. However, to fit this, nearly 2 square meter sheet, into the limited space of the scull, it is folded into hills (*gyri*) and valleys (*sulci*). Therefore, despite its thinness, it is the largest part of the human brain.

The *cerebral cortex* (gray matter) is involved in handling most of the higher functions such as thinking, language production and understanding, vision, hearing, etc. The inner part of the cerebrum, is made up of white matter and is responsible for lower level functions. Its neurons are generally much longer and therefore have myelinated axons [19]. The cerebral cortex (gray matter) was developed much later in the evolutionary path.

As it was briefly mentioned, the cerebral cortex is a sheet of cells which has been folded to fit within the confines of the scull. *David van Essen*'s lab in Washington University has developed a software called *Caret* [8] which takes a *Magnetic Resonance Image* (MRI) of the cerebral cortex (Figure 2.11) and inflates the folded cerebral cortex into a balloon-like surface in which the sulci are depicted in darker shades and the gyri in light (Figure 2.12). This demonstrates the fact that the cerebral cortex is actually made up of a sheet of neurons. This structure is then flattened into a sheet of cells. Several other algorithms have also been proposed for unfolding of the brain. For example, see [14, 7].

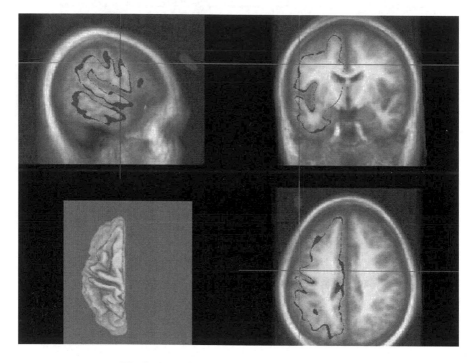

Fig. 2.11: MRI of the Left Hemisphere of the Brain

The Cerebral Cortex is divided into 4 *lobes*. Figure 2.14 shows these divisions on the left hemisphere of the human brain. The right hemisphere is basically a mirror image of the left hemisphere. The lobes are separated by major fissures.

The *Occipital lobe*, located at the back of the cerebral cortex, is mostly in charge of the visual function. The *Temporal lobe*, located at the bottom, performs higher processing of visual information, handles memory, processes sound and its relation to linguistic and musical representation. The *Parietal lobe*, located on top, combines information from different senses including visual, auditory and tactile senses. It for-

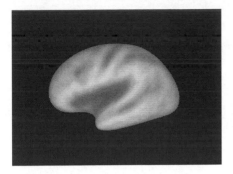

Fig. 2.12: Left Cerebral Cortex (Inflated)

Fig. 2.13: Left Cerebral Cortex (Flattened)

mats motor control to orient the body. It also handles spatial orientation. Finally, the *Frontal lobe*, located in the front as suggested by its name, is the most highly evolved part of the cerebral cortex in humans. It is most directly responsible for decisions and actions, including motor control, and may be thought of as the central controller.

In our study of audio processing and production, the lateral sulcus separating the *Temporal* lobe from the *Frontal* and *Parietal* lobes plays an important role. This fissure is called the *Sylvian fissure* (or simply the *lateral sulcus*) [13, 5].

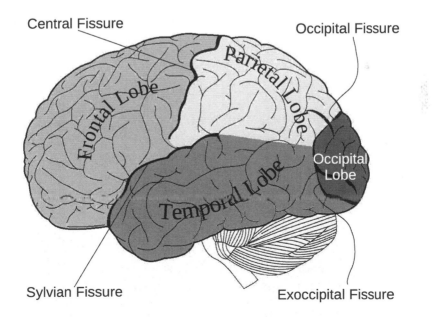

Fig. 2.14: Left Hemisphere of the Human Brain (*Modified from: Gray's Anatomy [13]*)

2.3.2.1 Brodmann Areas

In 1905 and 1909 [4], A German neuropathologist, *Korbinian Brodmann* (1868-1918), published three maps of the brain of the primates which divided the lateral and medial parts of the cerebral cortex into 50 areas based on the cell structure, neuron type, layers and tissue characteristics of these areas.[3] Some areas are split into smaller divisions and marked using a letter suffix in certain primates. [28] He numbered these areas from 1 to 52 with an exceptionally named area, 8*a*.[4] These areas are known as *Brodmann's cytoarchitectonic areas* or simply as *Brodmann areas*. Figure 2.15 shows the centers of the Brodmann areas in the lateral left cerebral cortex. The missing numbers in this figure are areas which are only part of the medial cerebral cortex and hence are not listed in the figure. Different *Brodmann areas* have distinct shapes and sizes, however, since we were not able to use color in the printing of this book, no shading is attempted. For a good color representation of the area boundaries see [5]. *Brodmann areas* are patches on this flattened sheet of cells which show localization of the functionality of the brain (Figure 2.13).[7]

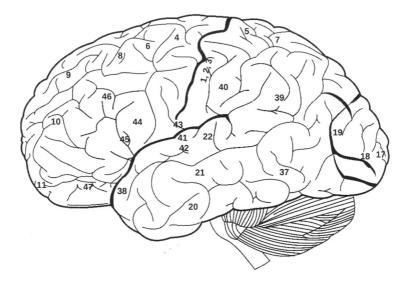

Fig. 2.15: Centers of the Lateral Brodmann Areas

---

[3] This study is known as *Cytoarchitectonics*.

[4] All together there are 50 different Brodmann areas which have been numbered from 1 to 52. Two of these areas (14 and 15) are only present in non-human primates. Also, areas $49 - 51$ have not been assigned. Area 8*a* is the only addition which is distinguished from its counterpart sharing the same number, area 8. Area 8*a* was only mentioned in the 1905 Brodmann map of the brain. It was not mentioned in the 1909 map.

2.3.3 Function Localization in the Brain

It is very important to note that most basic functions have been shown to be handled by specific parts of the brain specializing in those functions. *Franz Joseph Gall* (1758-1828) first hypothesized this function localization in the brain. Gall believed that every mental function had to be biological and created and performed by the brain. Namely, there were no external (spiritual) forces at work. Most importantly, he postulated that different mental functions could be localized to different regions of the cerebral cortex. In fact he stated that the cerebral cortex is a collection of different organs, each of which specializes in a different function.[16]

To localize these functions, he utilized the symmetry of the two hemispheres of the brain and studied the four lobes as the basic divisions. At the time of Gall, main brain operations had been categorized into 40 different functions. Since the 4 lobes were not divisive enough for these 40, seemingly distinct functions, there had to be more than 4 specialized areas involved.[16]

2.3.3.1 Broca's Area

Gall started observing the shapes of people's sculls to see if he could deduce some of these divisions based on specific characteristics. In his limited sample, he established that more intellectual people had larger foreheads, so the frontal lobe had to be related to intelligence.[21, 16] Although the frontal lobe is important in the matters of higher function and complex thoughts, the French neuropathologist, *Pierre-Paul Broca* (1824-1880) correctly deduced that Gall was wrong and that it is not the outer shape of the scull that governs the function of the brain. Although, he did agree that the brain has localized functionality. *Broca* started examining the brain itself to make his assessments about the functionality of its different areas. In 1861, Broca stated, "I had thought that if there were ever a phrenological science, it would be the phrenology of convolutions [in the cortex], and not the phrenology of bumps [on the head]."[21]

Due to its high intellectual requirements, as compared with other animals, language is a good function to study in an effort to localize higher functions of the brain. From the onset, language disorders were studied to assess the localization of function in the human brain. This was done by finding correlations between language disorders and problems in different areas of the brains of the individuals with these disorders. The first documented person to study such disorders was *Jean-Baptiste Bouillaud*, who in 1825, started publishing papers in support of Gall's theory by noting that damage to the frontal lobe caused speech impairments.[21]

However, Broca, in 1865, realized that language impairment (*aphasia*) was localized to the frontal lobe at the left hemisphere and did not involve the right hemisphere of the brain. This area is known as *Broca's area* of the brain which is responsible for the construction of linguistic content.[21]

Broca first noticed this in one of his patients by the name of *Leborgne*. Leborgne could not produce intelligible linguistic phrases. However, he could understand spoken language very well. He could even hum or whistle a tune, but could not produce any linguistic output either in the form of speech or in written form.

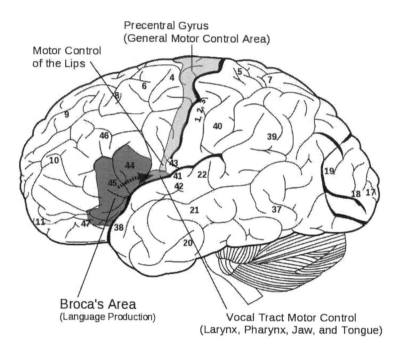

Fig. 2.16: Areas of Speech Production in the Human Brain

When Leborgne died, Broca performed an autopsy on his brain, only to find that a portion of his brain, about midway on the left side of the frontal lobe, just above the *Sylvian fissure*, was damaged. This area is coincident with Brodmann areas 44 and 45 (Figure 2.15) located at the *Inferior Frontal Gyrus* (IFG). Apparently, Leborgne had a syphilitic lesion in that region. Broca associated that area with the production of language. To this date, this portion of the left hemisphere is known as *Broca's area* – see Figure 2.16.

Later, Broca started looking for patients with similar language production impediment (aphasias). He found several cases and realized that in all of them, roughly the same area in the left hemisphere was damaged. This type of language production impairment was named *Leborgne aphasia* after Broca's patient.

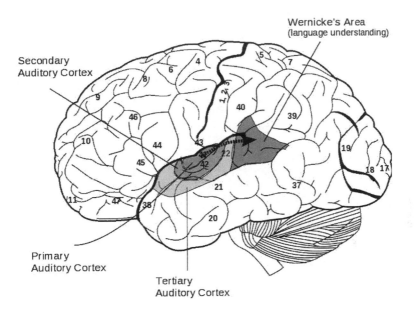

Fig. 2.17: Areas of Speech Understanding in the Human Brain

It is important to note that language production is one of the many tasks being performed by Broca's area. Other functions include, but are not limited to, motor imagery, object manipulation, motor planning, and understanding.[22]

2.3.3.2 Wernicke's Area

This was the first proof of the localization of functions in the human brain. Later, others started looking into other parts of the brain and associated them with different functions. One such person was a German neuroscientist by the name of *Karl Wernicke* (1848-1905).[21, 16] Wernicke had a patient with a language impediment manifesting itself in the opposite way compared to Leborgne's condition. This patient was able to articulate linguistic phrases very well, but he could not grasp any linguistic input. He could hear and understand words. He could also create words and sentences, but he could not understand phrases and sentences. In 1876, When his patient died, Wernicke performed an autopsy and found that he had a lesion at the top of the temporal lobe, just below the Sylvian Fissure – see Figure 2.17. Considering Broca's area and the newly found *Wernicke's area* [9], Wernicke decided that complex functions which included more than one basic function (such as language) must be handled within a few areas which are interconnected to allow for interaction among these areas.

2.3.4 Specializations of the Hemispheres of the Brain

Although most of the Brodmann areas are common to both hemispheres of the brain, some areas may have different functionalities in the left and right hemispheres. Starting with Broca, the different hemispheres were studied to better understand the special roles of the two hemispheres. The most prevalent differences are involved in higher functions such as the processing of language and musical discourse. The, so called, *Wada test* has been used in the past to evaluate functionalities of the different sides of the brain. Since the blood supplies to the different hemispheres are almost independent of one another, *Juhn Atsushi Wada*, a Japanese-Canadian neurologist, in 1949, proposed to anesthetize the different hemispheres of the brain, one at a time, to assess the usage of each side in activities such as language production, language perception and memory.[29]

More than ten years later, with the help of his colleague, *Theodore Rasmussen*, a safe delivery of a barbiturate through injection was developed and some results were achieved in larger clinical trials.[30, 23] This method became quite common as a preamble to surgical alleviation of epilepsy in order to be able to avoid any damage to the language sections of the individuals undergoing surgery. Since the barbiturate would anesthetize the hemisphere of interest while the patient was still alert, different tasks could be asked of the patient to see the level of disability due to the nonoperational hemisphere. Results of these tests showed that in most cases the left hemisphere is responsible for most major language functions including generation and perception.[23] Of course, in later studies, some variabilities were reported including gender relations with regards to inter-hemispheric reorganization of speech.[26]

About 20 years ago, a revolution in brain imaging techniques was begun. Today, with the existence of advanced imaging methods, there is no longer any need for invasive tests such as the *Wada test*.

fMRI (*functional Magnetic Resonance Imaging*), for example, shows the increase in the blood flow and oxygen due to the increased oxygen demand in active areas of the brain. In fact, as the *hemoglobin* loses its oxygen, it becomes more *ferromag-*

netic, thus increasing the magnetic resonance attenuation, creating a more active signal. Conversely, oxygenated hemoglobin loses its ferromagnetic signature and produces less attenuation in the magnetic resonance signal. Therefore, the less active resonance signal correlates with more active, hence oxygenated, parts of the brain.

This type of imaging facilitates the understanding of different functional areas of the brain. Examples are *language production* and *perception, face recognition, speaker recognition, music discourse*, etc.

Imaging technology has shed some light into complex questions about language production and understanding. Some recent imaging studies of the brain, related to language generation and understanding, have shown support for the fact that after puberty, it is harder to learn new languages. If one is bilingual and learns two languages at the same time, then the same areas (*Broca* and *Wernicke*) are responsible for both languages. However, if one learns one language followed by the second language, the first language will reside in the standard areas, but the new language will extend the Broca and the Wernicke areas to the sides of these areas, utilizing new areas for language processing.[20, 31]

Imaging has shown that tonality, rhythm, tempo, intonation and stress are deciphered mostly in the right hemisphere of the brain. Although, Broca's area (in the left hemisphere) has been shown to be also quite active during estimation of timing and production of rhythmic content.[3] Their production happens in the same Brodmann areas (44 and 45) as Broca's area, but mostly in its right hemispheric mirror image. Similarly, formant structures and musical discourse are processed at the mirror image of Wernicke's area in the right hemisphere [17]. Also, for languages that utilize pitch, the musical areas of the right hemisphere are used in combination with the traditional language areas of the left hemisphere. For example, it was shown using fMRI studies that lexical analysis of the *Mandarin dialect* of the *Chinese* language (see Section 4.3.1.1) involves parts of areas 22, 42 and 45 in the right hemisphere.[31] Also, it was shown that American speakers who learned Mandarin as a second language, underwent cortical reorganization to achieve the pitch-related lexical analysis associate with Mandarin.[31]

fMRI has also shown that human activities, which may be normally processed by different parts of the brain, are automatically processed using the language areas, when these activities are used for communication. An example is *sign language* which is used by deaf and mute individuals. Normally, *gestures* are recognized using the right hemisphere. However, in those individuals who use sign language for communication, the signs are processed in the *Wernicke area*. Sign production is also processed in *Broca's area*. Both of these areas are in the left hemisphere! This is also true for languages such as the *Spanish Silbo* language of the people of *La Gomera* in the *Canary islands* (see Section 4.1.8) who use different *whistle intona-*

tions for communication. fMRI has shown that they also process whistles in their language areas in the left hemisphere, whereas most people process whistles in the right hemisphere with other *tonal content*.[5]

Speaker recognition, is mostly conducted in the right hemisphere along with gender classification and pitch recognition.[17] This suggests that it uses cues from tonality, rhythm, intonation and stress in the speech of the individual. In automatic speaker recognition, we use very different features. In fact, in general, we use the same features that are used for speech recognition. To mimic the human-style of speaker recognition, one should use supra-segmental information and spectral analysis based on pitch variations which is usually processed by the right hemisphere of the brain. This is an argument for introducing such features to conventional speaker recognition techniques which developed out of basic speech recognition methods[5]. However, because our brains use such information for recognizing speakers, we can be easily fooled by an impostor who tries to mimic a target speaker's pitch. This type of perception is why impostors usually try to match the target speaker's pitch. As we will see later in this book, automatic speaker recognition techniques, discussed here, are not as sensitive to pitch values and cannot be fooled as easily using conventional impostor techniques.

At this point, we review the role of the nervous system in speech and music production and the auditory systems. This includes a quick overview of the relevant parts of the left and right hemispheres of the brain and their roles in the production and understanding of sounds. With regard to speech, individual parts of the system are responsible for the different blocks represented in Figure 1.6. Figure 2.18 is a replica of Figure 1.6, in which boundaries of the mechanical and nervous parts of speech production and perception have been marked.

2.3.5 Audio Production

As it was briefly discussed in the last section, linguistic abstraction and grammar are produced in Broca's area in the left hemisphere – see Figure 2.16. The figure also shows the motor control areas in charge of movements in the vocal tract and the lips. The larynx, the pharynx, the jaw, and the tongue are all handled by the same area. The lips are controlled by an adjacent section at the bottom tip of the motor control strip in the frontal lobe. These areas are adjacent to Broca's area. The language code produced by Broca's area is therefore fed to these motor areas to cause speech

---

[5] It is not all that unusual to use speech recognition features to conduct speaker recognition. Reference [17], in its conclusion, suggests that there may be a significant overlap between speech and speaker recognition in the human brain.

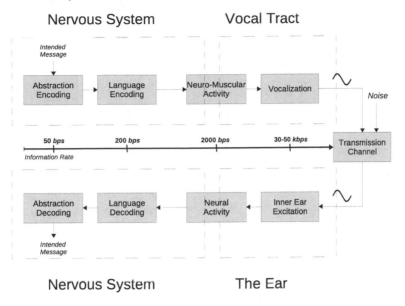

Fig. 2.18: Speech Generation and Perception – Adapted From Figure 1.6

output by the vocal tract, as discussed in Section 2.1.

The motor control area is just above the *auditory cortex* (see Section 2.3.6.1) in the frontal lobe. It is situated at the lower part of the precentral gyrus adjacent to the central fissure – see Figure 2.16. The motor control area is needed to interact with Broca's area to produce speech and writing. In fact, the lower part of the motor control region is responsible for the control of the vocal tract. It is also known to be related to the *FOXP2 gene* which has been shown to be responsible for the exceptional speech motor control in humans. Mutations in the FOXP2 gene have shown to create speech production impairments.[27]

The right hemisphere is more or less symmetric with the left. Broca's area is the combination of Brodmann areas 44 and 45. The *homologue* of Broca's area in the right hemisphere is basically the combination of areas 44 and 45 in the right hemisphere. Area 44 of the left hemisphere is slightly larger than the one in the right. Area 45 in both hemispheres is the same size. It is also interesting to note that the size of area 44 may vary by up to ten-fold from one individual to the next.[22]

Broca's area and its homologue portray significant differences in their neural anatomy (*cytoarchitecture*).[22]

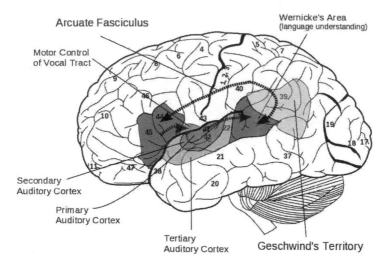

Fig. 2.19: Language Production and Understanding Regions in the Brain (*Basic Figure was adopted from Gray's Anatomy [13]*)

2.3.6 Auditory Perception

In all mammals, audio is processed by a section below the *Sylvian fissure*. This portion of the brain is known as the *auditory cortex*. As discussed earlier, Wernicke's area is primarily responsible for decoding language – both spoken and written. Its *homologue* in the right hemisphere is also credited for the deciphering and understanding of musical features such as *tempo*, *rhythm*, *intonation*, and *stress*.

Figure 2.19 shows Wernicke's area and its relation to the auditory cortex, Broca's area and the connective fibers called *Arcuate Fasciculus* and *Geschwind's Territory*, in charge of interconnectivities between these regions to provide the means for higher-level language dynamics. The following sections describe the different regions related to language and music understanding and their relation to discourse areas in the two hemispheres.

2.3.6.1 The Auditory Cortex

The nervous system includes an auditory nerve bundle (*vestibulocochlear nerve*) which goes from the *cochlea* to the *medial geniculate nuclei* at the rear of the *Thalamus* (situated at the anatomical core of the brain – see Figure 2.21) which is linked to *auditory cerebral cortex* (just below the Sylvian fissure in both hemispheres). The *auditory cortex* neighbors areas of the brain involved in perceiving speech (in the left hemisphere) and music (in the right hemisphere).[6]

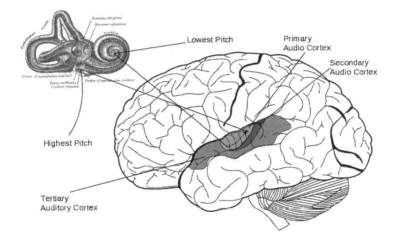

Fig. 2.20: Auditory Mapping of the Brain and the Cochlea (*Basic figures were adopted from Gray's Anatomy [13]*)

In Figure 2.20, the three parts of the auditory cortex are highlighted. These are known as the *primary*, the *secondary*, and the *tertiary* audio cortices.

The *primary auditory cortex* which is coincident with *Brodmann area* 41 is at the receiving end of the relayed signals from the cochlea.[7] Note that the neurons in the primary auditory cortex are organized *tonotopically*.[8] This means that the pitch sensitivity of the neurons in this area is arranged to map the sensitivity of the cilia along the length of the cochlea. Figure 2.20 shows the mapping between the cilia in the different parts of the cochlea with the different sections of the primary auditory cortex. Therefore, different regions of the primary auditory cortex are trained to be excited by specific ranges of frequencies. Pitch-sensitive mapping is preserved at every stage of the neural transmission from the cochlea to the primary auditory cortex – see Figure 2.21. There are also feedback pathways [24] going from the primary auditory cortex back to the *cochlear nuclei*. These pathways are shown in

---

[6] It is important to note that although we will be speaking about certain areas being responsible for certain functions, it does not mean that they are alone in their involvement. It solely suggests that the noted areas are also involved. The reader should be mindful of this loose expression of the involvement of an area in any function.

[7] See Section 2.3.2.1

[8] Arranged according to tone

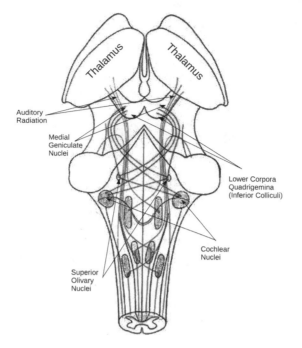

Fig. 2.21: The Auditory Neural Pathway – Relay Path toward the Auditory Cortex

Figure 2.22 and were briefly discussed in Section 2.2.1.

Coincident with *Brodmann area* 42, the *secondary auditory cortex* surrounds the primary cortex toward the bottom of the superior temporal gyrus – see Figure 2.20. Although, present in both hemispheres, the secondary auditory cortex in the left hemisphere is known to map the individual sounds from the primary cortex into phonetic elements. In the right hemisphere, on the other hand, the secondary auditory cortex extracts the harmonic, melodic and rhythmic patterns of the basic sounds which are realized by the primary auditory cortex. This is the beginning of the specialization of the different hemispheres in linguistic and musical functions.

The *tertiary auditory cortex* surrounds the secondary cortex, hence the primary cortex, to the bottom of the superior temporal gyrus. It is mostly coincident with *Brodmann area* 22 [11] which is basically most of the rest of the superior temporal gyrus. This cortex, in the left hemisphere, is responsible for mapping the phonetic patterns deciphered by the secondary auditory cortex into lexical semantics. In the right hemisphere, it is responsible with extracting the musical discourse of the heard audio signal by processing the output of the secondary auditory cortex.

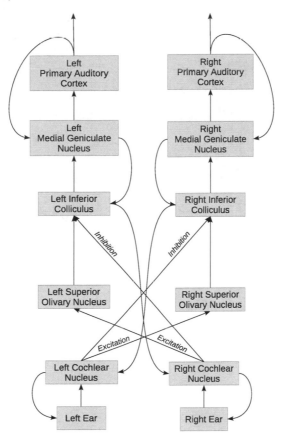

Fig. 2.22: Speech Signal Transmission between the Ears and the Auditory Cortex – See Figure 2.21 for the connection into the lower portion indicated by clouds

2.3.6.2 Speech and Music Perception – Language and Music Understanding

The role of the auditory cortex is to process the audio signal and transform it into higher level representations so that *Wernicke's area* in the left hemisphere may use the information to understand and decipher language. The same is true for the understanding of the intended musical discourse being processed by the homologue of Wernicke's area in the right hemisphere. The *visual cortex* is in the *Occipital lobe* in the back of the brain. This allows for sensory information from reading to be shared with Wernicke's area for understanding the read text. Wernicke, realized that the visual cortex and the auditory cortex are connected to Wernicke's area. Also, he identified a set of neural fibers connecting Broca's area and Wernicke's area through the inner part of the cortex, called *Arcuate Fasciculus*. This is a bi-directional neural pathway which enables a feedback mechanism between the language production (Broca) area and the language understanding (Wernicke) area. A new lesion that interrupts the Arcuate Fasciculus is called *Aconduction Aphasia*. It has been noted that these connections are far stronger in the dominant (left) hemisphere of the brain than in the right.[22]

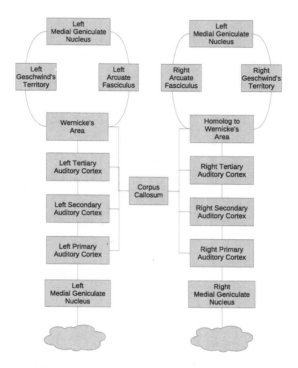

Fig. 2.23: The connectivity and relation among the audio cortices and audio perception areas in the two hemispheres of the cerebral cortex

Figure 2.23 shows the hierarchical connectivity among the different areas of the auditory cortices in the left and right hemispheres and their connectivity to the language understanding (Wernicke) and Music Understanding (homologue of Wernicke in the right hemisphere). The two auditory cortices are connected through the *corpus callosum* [13]. Figure 2.24 shows this body which is in charge of communication between the two hemispheres. Disconnections in the *corpus callosum* have shown many different aphasias.

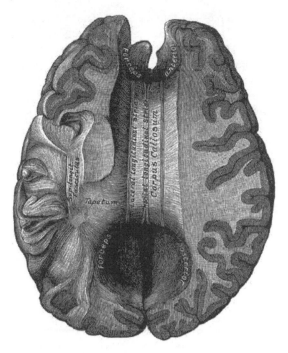

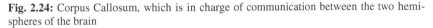

Fig. 2.24: Corpus Callosum, which is in charge of communication between the two hemispheres of the brain

2.3.7 Speaker Recognition

It is important to note that speaker recognition in humans is not simply based on the apparent pitch content of the individual speaker's utterances. It is a complex combination of many different features, some of which are quite simple to understand and seem very apparent, and others which are not well understood. The complex interactions between the two hemispheres of the brain in performing different tasks make matters much more complicated. An evidence of this complicated process is in the importance of the right hemisphere in contextual relations [3] and prosody [22] while using idiosyncratic differences in speech content through the involvement of the left hemisphere utilizing the connectivity of the two hemispheres through the *corpus callosum*. Add to this complex structure, the feedback mechanism of the auditory cortex to the cilia in the cochlea for an adaptive sensitivity of different frequencies. Also, take into account the fact that some aspects such as rhythm and timing are performed by both hemispheres with special dominance depending on linguistic (left hemisphere) and musical (right hemisphere) content. Far more complex interactions are added through the activities via the *Arcuate Fasciculus* and the *Geschwind's Territory* and many other connective networks. The final decision in recognizing a speaker is then based on a multitude of different facets, some of which may include as complex a technique as specific experiences with psychological evaluations developed over the course of our interactions with target individuals.

The automated techniques discussed in this book and all the research done in the field of speaker recognition only scratch the surface of this complex process. The material discussed here is at best a building block in the process of performing such complex functionality. In this chapter we tried to give an overview of all the different parts involved in performing the speaker recognition task by humans. It is important to note that a machine-based recognition mechanism need not be identical to human methods. In fact, it will be shown that the methods used here are far from the human speaker recognition methods. Knowing about the human approach will allow us to enhance the basic techniques used in this book and generally in most research and practical systems in use.

References

1. Aristotle: On Sleep and Sleeplessness. HTTP (350 BCE). URL http://classics.mit.edu/Aristotle/sleep.html. Translated by J. Beare
2. Belin, P., Grosbras, M.H.: Before Speech: Cerebral Voice Processing in Infants. Neuron **65**(6), 733–735 (2010)
3. Bookheimer, S.: Functional MRI of Language: New Approaches to Understanding the Cortical Organization of Semantic Processing. Annual Reviews of Neuroscience **25**, 151–158 (2002)
4. Brodmann, K.: Vergleichende Lokalisationslehre der Grosshirnrinde in ihren Prinzipien dargestellt auf Grund des Zellenbaues. J. A. Barth, Leipzig (1909)
5. Carter, R., Aldridge, S., martyn page, steve parker: The Human Brain Book. Dorling Kindersley Ltd., New York (2009)
6. Chatfield, P.: The Tin Trumpet. Volume I. Whittaker & Co., London (1836). P. 183
7. Essen, D.C.V.: Surface-Based Approaches to Spatial Localization and Registration in Primate Cerebral Cortex. Nueroimage **23**(Supplement 1), S97–S107 (2004)
8. Essen, D.C.V., Drury, H.A., Dickson, J., Harwell, J., Hanlon, D., Anderson, C.H.: An Integrated Software Suite for Surface-based Analyses of Cerebral Cortex. Journal of American Medical Informatics Association **8**(5), 443–459 (2001)
9. Forehand, C.J.: Integrative Functions of the Nervous System. In: R. Rhoades, D.R. Bell (eds.) Medical Physiology: Principles of Clinical Medicine, 3rd edn. Lippincott Williams and Wilkins, Baltimore, MD (2009). ISBN: 0-7817-6852-8
10. Fox, R.W.: Introduction to Fluid Mechanics, 2nd edn. Addison-Wesley Publishing Company, New york (1978). ISBN: 0-417-01909-7
11. Gaab, N.: The Auditory Cortex: Perception, Memory, Plasticity and the Influence of Musicianship. the University of Zürich, Zürich, Switzerland (2004). PhD Thesis
12. Golgi, C.: The neuron doctrine - theory and facts. Lecture (1906). URL http://nobelprize.org/nobel_prizes/medicine/laureates/1906/golgi-lecture.pdf
13. Gray, H.: Anatomy of the Human Body, 20th edn. LEA and FEBIGER, Philadelphia (1918). URL http://www.Bartleby.com. Online version, New York (2000)
14. James, D.B.: A Method of Unfolding the Cerebral Cortex or Any Other Folded Surface. Kybernetes **27**(8), 959–961 (1998)
15. Kandel, E., Rose, C.: Charlie Rose Series on the Brain. Television (2009). URL http://www.charlierose.com
16. Kandel, E.R.: Making Your Mind, Molecules, Motion and Memory. Lecture (2008)

17. Lattner, S., Meyer, M.E., Fiederici, A.D.: Voice Perception: Sex, Pitch and the Right Hemisphere. Human Brain Mapping **24**(1), 11 20 (2005)
18. Laver, J.: Principles of Phonetics. Cambridge Press, New York (1994). ISBN: 0-521-45031-4
19. Lee, K.H., Chung, K., Chung, J.M., Coggeshall, R.E.: Correlation of Cell Body Size, Axon Size, and Signal Conduction Velocity for Invidually Labelled Dorsal Root Ganglion Cells in the Cat. The Journal of Comparative Neurology **243**(3), 335–346 (1986)
20. Marian, V., Shildkrot, Y., Blumenfeld, H.K., Kaushanskaya, M., Faroqi-Shah, Y., Hirsch, J.: Cortical Activation During Word Processing in Late Bilinguals: Similarities and Differences as Revealed by Functional Magnetic Resonance Imaging. Journal of Clinical and Experimental Neuropsychology **29**(3), 247–265 (2007)
21. Marshall, J.C., Fink, G.R.: Cerebral Localization, Then and Now. NeuroImage **20**(Supplement 1), S2–S7 (2003). Convergence and Divergence of Lesion Studies and Functional Imaging of Cognition
22. Nishitani, N., Schürmann, M., Amunts, K., Hari, R.: Brocas Region: From Action to Language. Physiology **20**(1), 60–69 (2005)
23. Rasmussen, T.B., Milner, B.: The role of early left-brain injury in determining lateralization of cerebral speech functions. Annals of the New York Academy of Science **299**, 355–369 (1977)
24. Schofield, B.R., Coomes, D.L.: Pathways from Auditory Cortex to the Cochlear Nucleus in Guinea Pigs. Hearing Research **216–217**, 81–89 (2006)
25. Steinberg, J.C.: Position of Stimulatioin in Cochlea by Pure Tones. Journal of the Acoustical Society of America **8**(3), 176–180 (1937)
26. Strauss, E., Wada, J.A., Goldwater, B.: Sex differences in interhemispheric reorganization of speech. Neuropsychologia **30**, 353–359 (1992)
27. Vargha-Khadem, F., Gadian, D.G., Copp, A., Mishkin, M.: FOXP2 and the Neuroanatomy of Speech and Language. Nature Reviews. Neuroscience **6**(2), 131–137 (2005)
28. Vogt, B.A., Pandya, D.N., Rosene, D.L.: Cingulate Cortex of the Rhesus Monkey: I. Cytoarchitecture and Thalamic Afferents. The Journal of Comparative Neurology **262**(2), 271–289 (1987). Online version published, Oct. 9, 2004
29. Wada, J.A.: A New Method for the Determination of the Side of Cerebral Speech Dominance. A Preliminary Report of the Intra-Cartoid Injection of Sodium Amytal in Man. Igaju to Seibutsugaki **14**, 221–222 (1949)
30. Wada, J.A., Rasmussen, T.B.: Intracarotid Injection of Sodium Amytal for the Lateralization of Cerebral Speech Dominance: Experimental and Clinical Observations. Journal of Neurosurgery **17**, 266–282 (1960)
31. Wang, Y., Sereno, J.A., Jongman, A., Hirsch, J.: fMRI Evidence for Cortical Modification During Learning of Mandarin Lexical Tones. Journal of Cognitive Neuroscience **15**(7), 1019–1027 (2003)

Chapter 3
Signal Representation of Speech

Nothing can please many, and please long,
but just representations of general nature.

Samuel Johnson
Preface to the Plays of William Shakespeare, 1765

The main focus of this chapter is the signal representation of speech. Hence, before going any further we should define the concept of a signal.

Definition 3.1 (Signal). *A signal is an observed measurement of a physical phenomenon. It generally describes an observation of a higher level physical phenomenon in correlation with lower level measurement concepts such as time or space.*

In mathematical terms, definition 3.1 may be written as,

Definition 3.2 (Signal). *A signal, $h(\xi_1, \cdots, \xi_n)$ is a function that maps any point in the generalized coordinate system defined by $(\xi_1, \cdots, \xi_n)$ in its domain $\{(\xi_1, \cdots, \xi_n) : (\xi_1, \cdots, \xi_n) \in \mathscr{D}(h)\}$ to a point in its range $\{h(\xi_1, \cdots, \xi_n) : h(\xi_1, \cdots, \xi_n) \in \mathscr{R}(h)\}$ to describe the relation of any physical phenomenon to its domain – i.e. $\{h(\xi_1, \cdots, \xi_n) : (\xi_1, \cdots, \xi_n) \in \mathscr{D}(h) \mapsto h \in \mathscr{R}(h)\}$.*

Note that the domain $\mathscr{D}(h)$, and the range $\mathscr{R}(h)$ are *sets* in the strict sense of the word and may take on continuous values in an interval or be a set of limited number of values. There is no restriction on what kind of a *set* may be used.

Therefore, a signal is a mapping of a point in the low level bases such as time or space into the higher level measurement. Since space and time are continuous, all physical phenomena observed in their presence must be related to those bases in a continuous fashion. So, it is fair to say that most natural interactions happen in an analog (continuous) domain.

At this point, let us customize Definition 3.2 to one that relates to the speech signal, namely a *time-dependent signal*.

Definition 3.3 (Time-Dependent Signal). *A time-dependent signal, $h(t)$ is a function that maps any instance of time t in its domain $\{t : t \in \mathscr{D}(h)\}$ to a point in its range $\{h(t) : h(t) \in \mathscr{R}(h)\}$. i.e. $\{h(t) : t \in \mathscr{D}(h) \mapsto h \in \mathscr{R}(h)\}$ in order to describe a physical phenomenon.*

The speech signal is an observed measurement, done with respect to the passing of time as defined by Definition 3.3. It may be viewed as the mapping of time into the strength of the speech waves at any given instance of time. It is important to note that this value is not solely related to any single wave with a specific frequency. This is an important note to understand when we discuss the sampling theorem in Section 3.1.1.

To simplify the processing of continuous signals, the infinite set of possible values the independent continuum may take on in a finite interval $[a, b]$, may be reduced to a finite set through another mapping process called *sampling*. This action is called discretization and the newly defined signal, capable of mapping this finite set of points to a higher level measurement, is called a discrete signal. Of course, in general, one may also impose a similar restriction to the range of the mapping and reduce it to a finite set, but that is not necessary.

Now, let us examine a normal speech signal in more detail. A speech signal changes shape as the vocal tract state is changed – see Chapter 4. As you will see in chapter 4, the human vocal system is quite dynamic and, in fact, it is designed to change the form of the signal as a function of time. According to[3], an average *phone* (see Definition 4.2) lasts about 80ms. Even within the utterance of each phone, we know that there are several transitions that happen, changing the characteristics of the signal along the way. This categorizes the speech production system into a nonlinear system with constantly changing parameters.

Therefore, the speech signal is known as a *non-stationary* signal.

Definition 3.4 (Stationary and Non-Stationary Signals). *A stationary signal is a signal whose statistical parameters do not change over time. These are parameters which may be used to describe the signal in statistical terms such as intensity, variance, etc. For a non-stationary signal, the statistical parameters vary over time.*

All non-periodic signals are by definition non-stationary. Non-stationary signals may be periodic for a finite interval, but taken over a longer duration, if the periodicity ceases or changes to a different shape over the same period, the signal is still non-stationary. The speech signal is a good example of a non-stationary signal.

In this book we only concern ourselves with *discrete* signal processing techniques and algorithms which make use of sampled audio. Although it is generally possible to perform speaker recognition based on the analog audio signal, it is outside the scope of this book. It is well understood that analog processing is quite complicated and usually involves the design of elaborate circuitry to do simple recognition tasks without much flexibility. The powerful digital computing apparatuses available to us and the rich theoretical advancements in discrete information and system theories (starting in the early 20th century and growing exponentially to date) really leave us no choice but to adopt the discrete approach. In fact without discrete processing, speaker recognition, along with many similar disciplines would not have been too practical.

Here, we consciously steer away from using the term "digital" such as digital signal processing, digital to analog conversion, etc. It is mostly due to the misconceptions that it creates. Digital has a binary connotation, but we are really dealing with discrete systems which may be represented and manipulated using digital representations, although the theory relies mostly on the discretization concepts. Also, in the process of digitization, the imposition holds that both the domain and the range of the signal should be discretized and possibly mapped to a binary representation, whereas, most of the theory of discrete signal processing only assumes the discretization of the domain of the signal.

3.1 Sampling The Audio

Since speaker recognition is basically a passive process and only observes the audio signal to make a decision, it is considered to use *signal processing* techniques in contrast with active systems such as *control systems* which contribute to the dynamics of the system in which they are involved.[1] Therefore, we are only concerned with a sampling process at the beginning and once the signal is in a sampled state, the algorithms are independent of the analog world. Of course in some other speech related disciplines this is not the case. For example, take Speech Synthesis which has to deal with the conversion of a sampled signal to an analog signal. Therefore it needs to deal with data reconstruction techniques such as hold devices [10].

The natural starting point is to sample the analog audio signal to be processed later. There are several possible ways to sample a signal, namely, *periodic, cyclicrate, multirate, random,* and *pulse width modulated* sampling.[10] In speech processing we usually use periodic sampling in which the sampling frequency (rate of sampling) is fixed. Although, it is conceivable that speech-related applications would deal with low-activity signals and that they may use variable sampling techniques. Many lossy compression techniques utilize variable sampling. Examples are MP3, HE-AAC, and OGG Vorbis coded signals – see Section 26.1. For the sake of simplicity of operation, usually speaker recognition systems convert these representations to those using a periodic sampling rate such as Pulse Code Modulation (PCM) – see Section 26.1.1.

Narrowing down the possibilities to the periodic sampling of the signal, there is a fundamental question: *what sampling period (or sampling frequency) should we be using?* To answer this very important question, let us examine a fundamental theo-

---

[1] Of course the actual speech production system is an active system which we have modeled as a control system in Chapter 5. But here we are only concerned with the observation end of the system. In Chapter 5 this distinction will become more clear.

·rem in information theory and signal processing called, *the Sampling Theorem.*

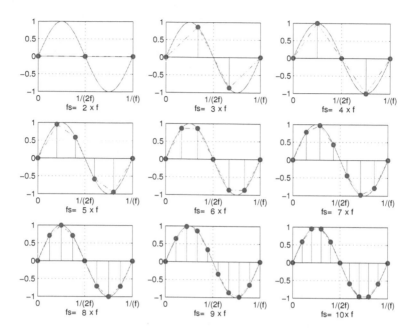

Fig. 3.1: Sampling of a Simple Sine Signal at Different Sampling Rates; f = Signal Frequency f_s = Sampling Frequency – The Sampling Rate starts at $2f$ and goes up to $10f$

3.1.1 The Sampling Theorem

The origin of the sampling theorem, one of the most crucial and basic theorems in engineering, has been quite controversial! Although, it has even been attributed to, as far back as, Cauchy [1], more evidently it was used in Interpolation Theory by E. T. Whittaker [17] in 1915. It was known to Nyquist [12] and J. M. Whittaker [18, 19]. In the Soviet Union it is attributed to Kotelnikov through his conference proceedings publication [9] in 1933. In communication theory, in its basic form, it is attributed to Shannon from a classic article he published in 1949 [14]. Shannon's statement of the theorem is the most popular one in the literature, although there are further extensions which were made to it later, to make it more descriptive. For a good overview of the sampling theorem see the tutorial by Jerri [7] and the corresponding errata [8].

The following is the statement of the sampling theorem per Shannon [14] – with some modification to his original nomenclature for uniformity with the rest of the material in this book,

Theorem 3.1 (Sampling Theorem). *If a function $h(t)$ contains no frequencies higher than f_c cycles per second, it is completely determined by giving its ordinates at a series of points spaced $\frac{1}{2f_c}$ seconds apart.*

Proof.
We defer the proof of the theorem to the end of this section, so that the statement of the proof follows more smoothly. $\square$

This theorem is so fundamental to signal processing and communication that as Shannon states in his paper [14], "this is a fact which is common knowledge in the communication art." In fact without subscribing to this recipe in practice, it will very quickly be realized that there is something wrong with the signal, the side effects of which will be discussed in detail. Indeed, the formal statement and the theoretical backing which follows is what makes this theorem quite useful.

The statement that the function has no frequencies higher than f_c Hz is equivalent to saying that the function is *bandlimited* to frequency f_c (bandwidth) from the top, or in mathematical terms, $H(f) = 0 \ \forall |f| \ge f_c$, where $H(f)$ is the spectral representation of $h(t)$ – see Chapter 24. f_c is known as the *Nyquist Critical Frequency* and it sets the limits of what is more widely known as the signal *bandwidth*. Therefore, the sampling rate of the signal must be $f_s \ge 2f_c$. There is another terminology used in the literature; *Nyquist Rate*, is equivalent to the lower limit of the sampling rate, namely, $2f_c$.

Recall the statement that was made earlier in this chapter while describing the speech signal. It was said that the value of a signal at any point in time is not solely related to any single wave with a specific frequency. Imagine such a point at any time t and consider its value. If at that moment there were n different sources each creating a specific tone (frequency), then the value of the signal recorded at the moment would be the sum of the values of the individual waves. In general, this notion extends to the fact that at any point in time, the value of the signal is the sum of the values of an infinite number of waves with all possible frequencies. Think about the opposite scenario. Consider an amplitude associated with a single frequency; then this amplitude is a sum of all the signal strengths at all moments of time having the same frequency. This is the idea behind a Fourier series expansion, discussed in quite a bit of detail in Chapter 24.

As we mentioned earlier, *Nyquist* [12] knew the basic idea of the statement in Theorem 3.1 to the extent that he prescribed that a signal must be sampled at a rate, f_s which is at least $2 \times f_c$. So, what makes Shannon's statement of the theorem so special?! It is that part of the theorem which deals with the reconstruction of the signal. Let us take a detailed look at the reconstruction of a bandlimited signal from its points. The statement of Shannon's theorem (Theorem 3.1) says that the signal is completely determined from its ordinates at a series of points which are $\frac{1}{2f_c}$ seconds apart.

E. T. Whittaker [17] noted that if any finite number of sampled points are given, there will be a family of an infinite number of *co-tabular functions* that will pass through all the sampled points. He stated that one of these functions will have the lowest harmonic constituents and that it is represented as an infinite series involving the samples weighed by the *sinc* function (see Figure 3.2). This is the, so called, *cardinal function* of the signal. He approached the problem from a function interpolation perspective. *J. M. Whittaker* [18, 19], *Kotelnikov* [9] and *Shannon* [14], later presented exactly the same cardinal series representation of the signal. *Shannon's* paper [14] studied the reconstruction in some detail. Equation 3.1, is the common reconstruction given by all the above researchers. For this reason, the theorem is known in most circles as the WKS Theorem (Whittaker-Kotelnikov-Shannon Theorem).

$$h(t) = \sum_{n=-\infty}^{\infty} h_n \frac{\sin(\omega_c t - n\pi)}{\omega_c t - n\pi} \qquad (3.1)$$

where,

$$h_n \overset{\Delta}{=} h(\frac{n}{f_s})$$
$$= h(\frac{n}{2f_c}) \qquad (3.2)$$

In general, the sampling rate f_s should be larger than $2f_c$, however, in Shannon's statement of the theorem, he uses the lower limit of the sampling frequency to show the reconstruction. In practice, one has to be careful, since it is quite important to know the value of f_c. For example, if you happen to sample a pure sinusoid at its zero-crossings and exactly at the Nyquist Rate ($2f_c$), all the samples will be zero. Of course, since we are interested in the cardinal function, as defined above, Shannon's theorem will still be valid, but $h(t)$ could theoretically become zero. A small perturbation of the sample will, of course, alleviate this problem.

Proof. − The WKS Sampling Theorem
To prove the validity of Equation 3.1, consider the definition of the *complex Fourier transform* of a function given by Equation pair 24.390 and 24.391. In the process of this derivation and for the rest of this chapter, we shall be using the following different relations,

$\omega_c \overset{\Delta}{=} 2\pi f_c$ (Nyquist Critical Angular Frequency)

$T_c \overset{\Delta}{=} \frac{1}{f_c}$ (Nyquist Critical Period)

$\omega_s \overset{\Delta}{=} 2\pi f_s$ (Sampling Angular Frequency)   (3.3)

$T_s \overset{\Delta}{=} \frac{1}{f_s}$ (Sampling Period)

Also, we have re-stated Equations 24.390 and 24.391 for the convenience of the reader (Equations 3.4 and 3.5).

$$H(\omega) = \int_{-\infty}^{\infty} h(t) e^{-i(\omega t)} dt \tag{3.4}$$

and,

$$h(t) = \frac{1}{2\pi} \int_{-\infty}^{\infty} H(\omega) e^{i(\omega t)} d\omega \tag{3.5}$$

Based on the statement of Theorem 3.1, $h(t)$ contains no frequencies higher than f_c. Using the relationship between frequency and angular frequency, Equation 3.3, it means,

$$H(\omega) = 0 \ \forall \begin{cases} \omega < -\omega_c \\ \omega > \omega_c \end{cases} \tag{3.6}$$

Therefore, Equation 3.5 may be written as follows,

$$h(t) = \frac{1}{2\pi} \int_{-\omega_c}^{\omega_c} H(\omega) e^{i(\omega t)} d\omega \tag{3.7}$$

The statement of the theorem says that the samples will be $\frac{1}{2f_c}$ apart. In other words,

$$f_s = 2f_c \tag{3.8}$$

So, each time sample, t, would be given by,

$$\begin{aligned} t &= \frac{n}{f_s} \\ &= \frac{n}{2f_c} \\ &= \frac{n}{2\frac{\omega_c}{2\pi}} \\ &= \frac{n\pi}{\omega_c} \end{aligned} \tag{3.9}$$

Plugging t from Equation 3.9 into Equation 3.7, we get,

$$h_n \overset{\Delta}{=} h\left(\frac{n\pi}{\omega_c}\right) = \frac{1}{2\pi}\int_{-\omega_c}^{\omega_c} H(\omega)e^{i\omega\frac{n\pi}{\omega_c}}\,d\omega \tag{3.10}$$

Now, let us write the *complex Fourier expansion* of $H(\omega)$ using the equation pair 24.325 and 24.326 substituting $t = -\omega$ and $T = -\omega_c$,

$$h(\omega) \approx \sum_{n=-\infty}^{\infty} c_n e^{-i\left(\frac{n\pi\omega}{\omega_c}\right)} \tag{3.11}$$

where,

$$c_n = \frac{1}{2\omega_c}\int_{-\omega_c}^{\omega_c} h(\omega)e^{i\left(\frac{n\pi\omega}{\omega_c}\right)}\,d\omega \tag{3.12}$$

Comparing Equations 3.10 and 3.12, we see that they would become identical if we set,

$$c_n = \frac{\pi}{\omega_c}h_n \tag{3.13}$$

Using Relation 3.13 in Equation 3.11 and changing the approximation to an equal sign for convenience,

$$H(\omega) = \sum_{n=-\infty}^{\infty} \frac{\pi}{\omega_c}h_n e^{-in\pi\left(\frac{\omega}{\omega_c}\right)} \tag{3.14}$$

Using the inverse Fourier Transform, Equation 3.7, with $H(\omega)$ substituted from Equation 3.14,

$$\begin{aligned}
h(t) &= \frac{1}{2\pi}\int_{-\omega_c}^{\omega_c} H(\omega)e^{i(\omega t)}\,d\omega \\
&= \frac{1}{2\pi}\int_{-\omega_c}^{\omega_c} \sum_{n=-\infty}^{\infty} \frac{\pi}{\omega_c}h_n e^{-in\pi\left(\frac{\omega}{\omega_c}\right)} e^{i(\omega t)}\,d\omega \\
&= \frac{1}{2\omega_c}\sum_{n=-\infty}^{\infty} h_n \int_{-\omega_c}^{\omega_c} e^{i\omega\left(t-\frac{n\pi}{\omega_c}\right)}\,d\omega \\
&= \frac{1}{2\omega_c}\sum_{n=-\infty}^{\infty} h_n \left.\frac{e^{i\omega\left(t-\frac{n\pi}{\omega_c}\right)}}{i\left(t-\frac{n\pi}{\omega_c}\right)}\right|_{-\omega_c}^{\omega_c} \\
&= \frac{1}{2\omega_c}\sum_{n=-\infty}^{\infty} h_n \left[\frac{e^{i(\omega_c t-n\pi)} - e^{-i(\omega_c t-n\pi)}}{i(\omega_c t - n\pi)}\right] \tag{3.15}
\end{aligned}$$

Using Euler's identities (see Property 24.5) in Equation 3.15,

$$h(t) = \frac{1}{2\omega_c} \sum_{n=-\infty}^{\infty} h_n \left[\frac{2\sin(\omega_c t - n\pi)}{t - \frac{n\pi}{\omega_c}} \right]$$

$$= \sum_{n=-\infty}^{\infty} h_n \frac{\sin(\omega_c t - n\pi)}{\omega_c t - n\pi} \qquad (3.16)$$

$$= \sum_{n=-\infty}^{\infty} h_n \text{sinc}(\omega_c t - n\pi) \qquad (3.17)$$

Equation 3.17 proves the statement of Shannon's Theorem (WKS Theorem). Figure 3.2 shows the cardinal function (sinc function) which was used in the expansion of the signal $h(t)$ in Equation 3.17. As it was mentioned earlier, Whittaker [17] presented this expansion for the purpose of doing the lowest constituent harmonic interpolation in 1915. Slight variations of this equation were later presented in chronological order in papers by Whittaker [18, 19], Kotelnikov [9], and Shannon [14]. □

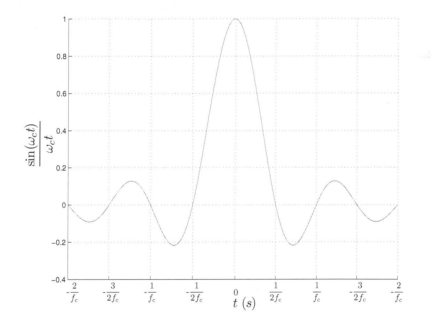

Fig. 3.2: *sinc* function which is known as the cardinal function of the signal – f_c is the Nyquist Critical Frequency and ω_c is the corresponding Nyquist Angular Frequency ($\omega_c = 2\pi f_c$)

3.1.2 Convergence Criteria for the Sampling Theorem

$h(t)$ in the sampling theorem should be piecewise continuous (have a finite number of discontinuities in a finite interval) which is generally the case with audio signals unless there are problems with the data acquisition apparatus in which case, the signal would be unintelligible and would be filled with pops and clicks. Still, theoretically, if the order of magnitude of the number of pops and clicks is less than true sampled points in the audio signal, the Fourier representation will converge. See Chapter 24 for more information about the convergence of the Fourier Series.

3.1.3 Extensions of the Sampling Theorem

Shannon [14] goes on with certain other observations relating to the sampling theorem. He states that if we consider $f_s = 2f_c$, for a time period, τ seconds, then we will have $2\tau f_c$ samples and that these samples do not have to be obtained using periodic sampling. This means that the time between every two consecutive samples need not be fixed. However, if the $2\tau f_c$ samples are scattered within the time period τ, the reconstruction will become more complicated. Also, if there are dense clusters of samples in some regions within the period τ, the accuracy of sampling should be higher in order to be able to reconstruct the original signal.

Another observation which was later expanded and proven by Fogel [5] is that if the first, second, and higher derivatives of the signal are also observed, the signal may be reconstructed with a reduced number of sample points, essentially reducing the sampling rate each time a higher derivative is added to the sample pool. Fogel's statement of the extended sampling theorem is a bit vague and has been slightly modified, here, for the sake of clarity. Also, the nomenclature has been changed to match the one used in this book.

Theorem 3.2 (Extended Sampling Theorem – Fogel). *If a function $h(t)$ contains no frequencies higher than f_c cycles per second, it is determined by giving M function and/or derivative values at each of a series of points extending throughout the time domain with the sampling interval of $T = \frac{M}{2f_c}$ being the period between instantaneous observations.*

Fogel [5] provides the proof of this theorem in the appendix of the paper. However, the proof is stronger than the statement of the theorem given in Theorem 3.2. To make this theorem a bit more general and to convey what the proof of Reference [5] really conveys, the following statement of the theorem is proposed here,

Theorem 3.3 (Extended Sampling Theorem – Proposed here). *If a function $h(t)$ contains no frequencies higher than f_c cycles per second, and it is sampled for a*

period of τ seconds, then only $2\tau f_c$ number of values made up of function values or derivatives of any order at any point of the interval are sufficient for reconstructing the original signal.

Fogel [5] proves Theorem 3.3 and presents the essence of it in the form of examples, but does not actually state it in the universal form which is stated here. Jerri [7, 8] describes, in detail, the Generalized Sampling Theorem. Weiss [16] extended the WKS Sampling Theorem to use integral transforms with *kernels* (see Definitions 24.54 and 24.55) which are solutions to the *Sturm-Liouville problem* (see Section 24.6) in place of the exponential family kernels. This extension was later completed by Kramer to cover kernels which are solutions to general n^{th}-order differential equations. He demonstrated the case of a *Bessel function* for the Kernel. This Generalized Sampling Theorem is known as the WKSK (*Whittaker-Kotelnikov-Shannon-Kramer*) Sampling Theorem. The WKSK Generalized Sampling Theorem is interesting to pursue, for the avid reader, but is outside the scope of this text.

Another set of extensions due to *Papoulis* [13] deal with issues such as non-uniform sampling. Again, due to the many subjects that should be covered in this book, we refer the reader to the source and to *Jerri's* coverage of these extensions [7, 8] as well as [2].

3.2 Quantization and Amplitude Errors

Quantization acts on the amplitude of the signal. As we stated in the introduction of this chapter, we deal with the discretization of the signal. It was also pointed out that the range of the mapping (amplitude of the signal) may be continuous and comprise an infinite set. However, in practice, to be able to use a digital device such as a computer, we have to quantize the possible values of the signal amplitude. These quantized values are the values associated with the samples of the signal. Each sample point can take on one such quantized value.

When speaking of an audio signal, one normally specifies the sampling rate as well as the quantization level of the signal. The quantization level is normally given in terms of the number of bits of the range of the signal. So, when a signal is a 16-bit signal, the values may vary from -32768 to 32767. Figure 3.3 shows a signal which has been quantized to 11 levels. A small number of quantization levels is used so that the quantization error may be more visually apparent.

The choice of the quantization level could make a huge difference in the storage requirements of a signal. Note that as the number of bits used for representing the samples increases, the number gets multiplied by the total number of samples. A simple minded approach to quantization is to use a linear mapping of the amplitude

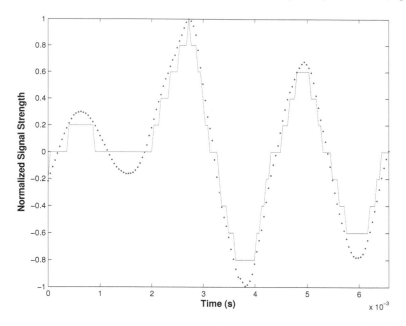

Fig. 3.3: Portion of a speech waveform sampled at $f_s = 22050$ Hz – Solid line shows the signal quantized into 11 levels and the dots show original signal

of the signal. This will use the same number of bits to represent small values in the amplitude, as it would for large values. However, human perception is such that small differences between large values of the amplitude are not perceived as much as small changes in small values. Because of this logarithmic perception model, when we try to conserve storage, a logarithmic mapping of the amplitude is used. Some examples of these logarithmic maps are μ-law and a-law algorithms for amplitude mapping which have been introduced by the ITU-T G.711 recommendations [6] and are included in PCMU and PCMA formats – see Chapter 26 for more information.

Aside from the quantization error which may be present in a signal, the amplitude representation may suffer from other errors such as additive noise or uncertainty in the sampling. The uncertainty is mostly hardware and process related and could take on many different forms all of which are apparatus-dependent. The additive noise is a hard problem with which one may have to deal at some point in the process of speaker recognition. It is especially pronounced in low quality microphones, analog systems and finally ambient noise which accounts for most of the trouble. Different kinds of filters may be designed to deal with the noise. Unfortunately, all types of filters will affect the signal as well as the noise and will somewhat color or modify the original signal. We will be talking more about noise contamination in later chapters.

3.3 The Speech Waveform

Given the different flavors of the sampling theorem, as stated in Sections 3.1.1 and 3.1.3, we may sample the output of an analog microphone to produce the, so called, speech signal. Using amplitude quantization along with sampling in the frequency domain, we will have a representation of the speech signal called the speech waveform. Chapter 26 covers the many different standards available for storing and transmitting such a signal. Figure 3.4 is a plot of such waveform sampled at 22050 Hz.

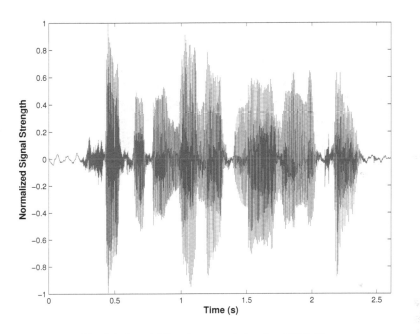

Fig. 3.4: Speech Waveform sampled at $f_s = 22050$ Hz

3.4 The Spectrogram

A spectrogram is a three dimensional representation of the spectral content of the speech signal. It represents the power of the different spectral components of each instance of speech. See Figure 3.5. The time domain is represented by the horizontal axis and the vertical access shows the frequency. The extent of the frequency is from 0-Hz, corresponding to the DC level of the signal, to f_c (Nyquist Critical Frequency) which is the highest frequency component of the signal. For the example of Figure 3.5, the sampling frequency is 22050Hz (22kHz). Therefore, $f_c = 11025$Hz

(11kHz).[2]

For an instance of time, the energy level for each frequency component is designated by a shade. In the graph shown in Figure 3.5, the higher the energy is, the darker the marking would be. So, white spots represent 0 energy in that frequency and completely black spots represent the highest energy for that frequency at the designated instance of time. In chapter 5, spectral techniques for computing the energy level are presented in detail.

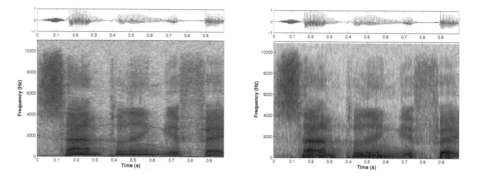

Fig. 3.5: Narrowband spectrogram using ~ 23 ms widows (43Hz Band)   **Fig. 3.6:** Wideband spectrogram using ~ 6 ms widows (172Hz Band)

Figures 3.5 and 3.6 show a narrowband and a *wideband spectrogram* of speech, sampled at 22050Hz (22kHz). For the narrowband spectrogram, the bandwidth of the sliding window which was used for the computation was 512 samples wide which works out to about 43-Hz. Notice the horizontal lines at different frequency levels along the timeline. These are the spectral harmonics corresponding to pitch during voiced speech such as vowels. The wideband spectrogram of Figure 3.6 was generated using 128 sample windows which come out to nearly 172 Hz for each each band. The wideband spectrogram is characterized by the vertical lines for different time instances, along the frequency axis. These are the spectral envelopes of individual periods of the speech waveform. The difference between the two is due to the differences in averaging effects.

To give you a better idea of the shapes of spectrograms for different types of speech, the spectrograms of numbers from 0 to 9 have been plotted in Figures 3.7 through 3.16. In chapter 4, we will discuss the features of spectrograms and some

---

[2] We are using f_c loosely in this section. As we shall see later, this frequency ($\frac{f_s}{2}$) is really the *folding frequency* (f_f) and not the *Nyquist critical frequency* (f_c). Here we have made the shady assumption that the sampling frequency has been picked to be $2f_c$ when in reality, f_c for speech signals is much higher. This will become more clear toward the end of the chapter. For now, we make this incorrect assumption since we have not yet defined *the folding frequency*.)

heuristics to be able to recognize different parts of speech from the spectrogram.

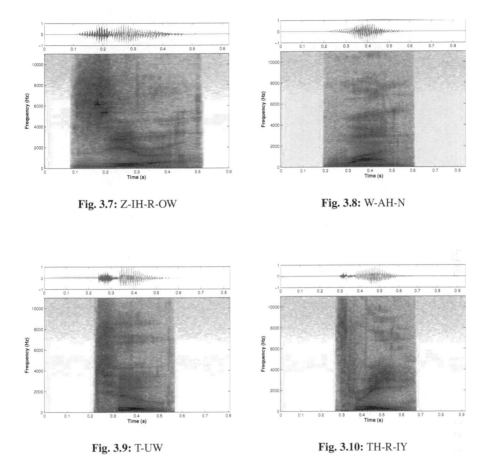

Fig. 3.7: Z-IH-R-OW   **Fig. 3.8:** W-AH-N

Fig. 3.9: T-UW   **Fig. 3.10:** TH-R-IY

3.5 Formant Representation

The formants of speech are resonant regions within the spectrogram – see Figure 3.17. The vocal tract is changing shape so that the resonance is changing. The definition and estimation of formant locations is a difficult task. In general, the vocal tract length is inversely proportional to the height of the format in the frequency range of a speaker. This means that the longer the vocal tract length (for example in adult males), the lower the format. As the vocal tract length is shortened (for example in female speakers and children), the formant locations move up higher in the frequency domain. Figures 3.18 and 3.19 demonstrate this effect. They are identical

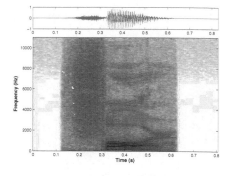

Fig. 3.11: F-OW-R

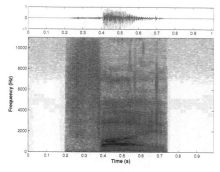

Fig. 3.12: F-AY-V

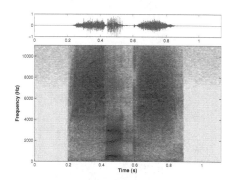

Fig. 3.13: S-IH-K-S

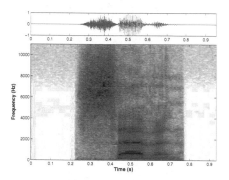

Fig. 3.14: S-EH-V-AX-N

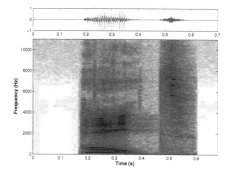

Fig. 3.15: EY-T

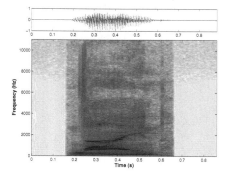

Fig. 3.16: N-AY-N

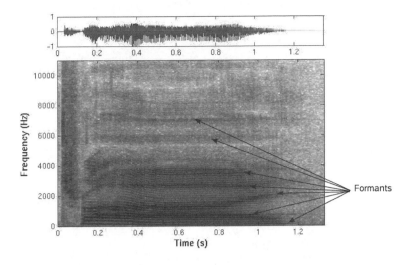

Fig. 3.17: Formants shown for an elongated utterance of the word [try] – see Figure 4.29 for an explanation.

phrases spoken by a 2 year old boy and a 44 year old man. It is quite apparent that the formants for the adult male are much lower than those of the young boy. The variability from male to female speakers is on the average about 20% and is much more pronounced in case of speech generated in an open vocal tract configuration (when the vocal chords are open). [4] Chapter 4 discusses this difference in much more detail for the fundamental frequency. Formants will be revisited in more detail in the rest of the book.

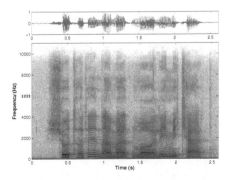

Fig. 3.18: Adult male (44 years old)

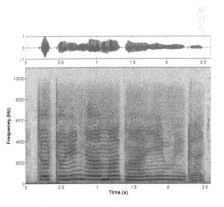

Fig. 3.19: Male child (2 years old)

3.6 Practical Sampling and Associated Errors

There are three major types of samplers which may be used for sampling time-dependent signals such as the speech signal: Pulse Amplitude Modulation (PAM), Pulse Width Modulation (PWM), and Pulse Width Pulse Amplitude Modulation (PWPAM) samplers. The way PAM sampler works is that it performs an amplitude modulation of a carrier signal, $p(t)$ using the information in the original signal, $h(t)$. $p(t)$ is essentially a pulse train with each pulse lasting for a period of ϖ seconds [10]. The time period between two consequent time pulses (from the beginning of one pulse to the beginning of the next pulse) is the sampling period of the signal, T_s seconds. T_s is the reciprocal of the sampling frequency, f_s, i.e.,

$$T_s = \frac{1}{f_s} \tag{3.18}$$

$$= \frac{2\pi}{\omega_s} \tag{3.19}$$

Figure 3.20 shows this process. The top part of the Figure shows the original signal $h(t)$ represented in a waveform against time. The middle part of the figure shows the pulse train, $p(t)$. Each pulse has an amplitude of 1 and a width of ϖ seconds. When the two signals, $h(t)$ and $p(t)$ are multiplied, the resulting signal will be the sampled signal given by $h^*_{\varpi}(t)$, shown at the bottom of the figure. The bottom figure, the original signal, $h(t)$ is also laid over $h^*_{\varpi}(t)$, in a dotted line for reference. This sampling technique is basically the method used for doing Linear Pulse Code Modulation (PCM) defined in the ITU-T G.711 recommendations.[6] There are also other implementations of this PAM sampler where instead of the samples being proportional to the instances of the amplitude of the original signal, $h(t)$, they are a nonlinear function of those values. In speech, these nonlinear functions are usually logarithmic in nature due to the logarithmic perception of our auditory system. Examples of such logarithmic mappings are the PCMU (μ-law Pulse Code Modulation) and PCMA (a-law Pulse Code modulation) coding of the signal. In these cases, the signal processing is usually carried out in the linear domain, as we will be discussing here, and the mapping is considered a secondary operation which happened for the sole purpose of storage. The intermediate representations are mostly carried out in the proportional mapping which will be discussed here in detail.

The second technique, Pulse Width Modulation, is represented in Figure 3.21. At the top of the figure, the same sample signal as in Figure 3.20 is plotted in its waveform. The bottom of Figure 3.21 shows the pulse width modulated samples. The sampling period is still T_s which is measured from the beginning of one sample to the beginning of the next (marked in the figure). The amplitude of the sampled signal is always 1 in the PWM sampler. The width of the samples is a function of the instantaneous value of $h(t)$ at onset of each sample. As in the case of the PAM sampler, this function may also be nonlinear (logarithmic or other), but that would similarly be used for storage purposes and for the sake of simplicity most operations would assume a linear function. In most cases, this function is a linearly

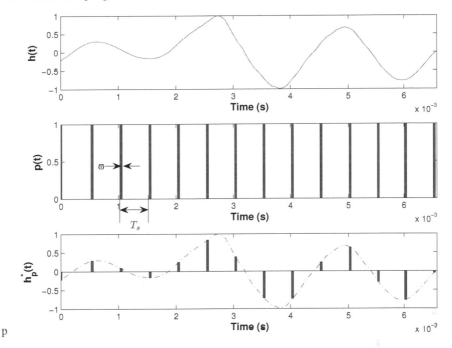

Fig. 3.20: Uniform Rate Pulse Amplitude Modulation Sampler. top: Waveform plot of a section of a speech signal. middle: Pulse Train p(t) at $T_s = 5 \times 10^{-4}$s (2kHz) and $\varpi = \frac{T_s}{10}$ bottom: Pulse Amplitude Modulated samples overlaid with the original signal for reference.

proportional function to the instantaneous values of $h(t)$. Here, we have used the following relation for computing the pulse width,

$$\varpi(t) \stackrel{\Delta}{=} \frac{h(t)}{\left| \max_t h(t) \right|} T_s \tag{3.20}$$

The PWM sampler [10] has been used in some audio technologies such as the, so called, switching amplifiers. However, for the purposes of speaker recognition and in the spirit of economy, we will not treat this sampling technique in any more detail.

The third type of sampler, Pulse Amplitude Pulse Width Modulation sampler, modifies the amplitude and width of each pulse. It is a sophisticated sampling technique which is outside the scope of this text. For the rest of this section, we will be concentrating on the PAM sampling technique which will be used for theoretical development of the rest of this textbook. Furthermore, we only concern ourselves with proportional PAM sampling which is conducted with a uniform rate.

Figure 3.22 and Equation 3.21 show the block diagram and equation representation of the PAM sampler.

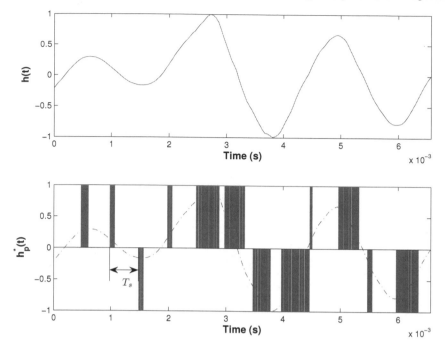

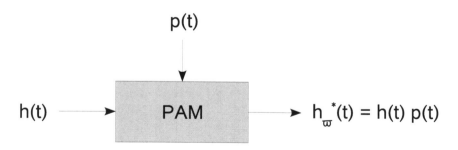

Fig. 3.21: Pulse Width Modulation Sampler. top: Waveform plot of a section of a speech signal. bottom: Pulse Width Modulated samples overlaid with the original signal for reference.

$$h_{\varpi}^{*}(t) = h(t)p(t) \tag{3.21}$$

Fig. 3.22: Pulse Amplitude Modulation Sampler Block Diagram (after [10])

$h(t)$ is the original analog signal, $p(t)$ is a train of pulses with unit amplitude and width of ϖ seconds which are T_s seconds apart (from the beginning of one pulse to the beginning of the next one). $h_{\varpi}^{*}(t)$ is the output of the sampler. Each pulse of $p(t)$ may be written as the difference between two unit step (Heaviside) functions ($u(t)$), ϖ seconds apart, located at nT_s for the nth pulse. In mathematical form,

$$p(t) = \sum_{n=-\infty}^{\infty} \left[u(t - nT_s) - u(t - nT_s - \varpi) \right] \tag{3.22}$$

where $\varpi < T_s$ and the unit step or Heaviside function is defined as,

$$u(t) \triangleq \begin{cases} 1 \ \forall \ t \geq 0 \\ 0 \ \forall \ t < 0 \end{cases} \tag{3.23}$$

$p(t)$ has sharp edges and is basically a square wave. Therefore, its Fourier Transform has many harmonic components, making the resulting output of the samples a harmonic generator. Chapter 24 shows that the *complex Fourier series* is defined for periodic functions. $p(t)$ is a *periodic function* with period T_s and within the fundamental period, it is defined as,

$$p(t) = \begin{cases} 1 \ \forall \ 0 \leq t \leq \varpi \\ 0 \ \forall \ \varpi < t < T_s \end{cases} \tag{3.24}$$

Take the definition of the *complex Fourier series*, Equations 24.325 and 24.326. Also note that the pulse train, $p(t)$ has the period $[0, T_s]$, in contrast with $[-T, T]$ used in the definition. Therefore, the *complex Fourier series* expansion of $p(t)$ becomes,

$$p(t) \approx \sum_{-\infty}^{\infty} c_n e^{i \frac{2n\pi}{T_s} t} \tag{3.25}$$

where,

$$c_n = \frac{1}{T_s} \int_0^{T_s} p(t) e^{-i \frac{2n\pi}{T_s} t} dt \tag{3.26}$$

Using Equation 3.24, we may rewrite Equation 3.26 as follows,

$$\begin{aligned} c_n &= \frac{1}{T_s} \left[\int_0^{\varpi} 1 \cdot e^{-i \frac{2n\pi}{T_s} t} dt + \int_{\varpi}^{T_s} 0 \cdot e^{-i \frac{2n\pi}{T_s} t} dt \right] \\ &= \frac{1}{T_s} \int_0^{\varpi} e^{-i \frac{2n\pi}{T_s} t} dt \\ &- \frac{1}{T_s} \int_0^{\varpi} e^{-in\omega_s t} dt \end{aligned} \tag{3.27}$$

Performing the integration in Equation 3.27, we have,

$$\begin{aligned} c_n &= \frac{1}{T_s} \frac{1}{-in\omega_s} e^{-in\omega_s t} \Big|_0^{\varpi} \\ &= \frac{1}{T_s(-in\omega_s)} \left[e^{-in\omega_s \varpi} - 1 \right] \\ &= -\frac{\varpi}{T_s} \frac{1}{in\omega_s \varpi} \left[\cos(n\omega_s \varpi) - i\sin(n\omega_s \varpi) - 1 \right] \end{aligned} \tag{3.28}$$

Note the following double angle relations,

$$\sin(2\theta) = 2\sin(\theta)\cos(\theta) \tag{3.29}$$

$$\cos(2\theta) = \cos^2(\theta) - \sin^2(\theta) \tag{3.30}$$
$$= 1 - 2\sin^2(\theta) \tag{3.31}$$

Using Equations 3.29 and 3.31, we may rewrite Equation 3.28 as follows,

$$c_n = -\frac{\varpi}{T_s}\frac{1}{in\omega_s\varpi}$$

$$\left[\cancel{1} - 2\sin^2\left(\frac{n\omega_s\varpi}{2}\right) - i2\sin\left(\frac{n\omega_s\varpi}{2}\right)i\cos\left(\frac{n\omega_s\varpi}{2}\right)\cancel{-1}\right] \tag{3.32}$$

$$= -\frac{\varpi}{T_s}\frac{1}{in\omega_s\varpi}\left[-2\sin\left(\frac{n\omega_s\varpi}{2}\right)\left(\sin\left(\frac{n\omega_s\varpi}{2}\right)+i\cos\left(\frac{n\omega_s\varpi}{2}\right)\right)\right] \tag{3.33}$$

$$= \frac{\varpi}{T_s}\frac{\sin\left(\frac{n\omega_s\varpi}{2}\right)}{\left(\frac{n\omega_s\varpi}{2}\right)}\left[\frac{i\sin\left(\frac{n\omega_s\varpi}{2}\right)-\cos\left(\frac{n\omega_s\varpi}{2}\right)}{-1}\right] \tag{3.34}$$

$$= \frac{\varpi}{T_s}\mathrm{sinc}\left(\frac{n\omega_s\varpi}{2}\right)\left[e^{-i\frac{n\omega_s\varpi}{2}}\right] \tag{3.35}$$

If we plug Equation 3.35 into Equation 3.25, we will have the following expression for $p(t)$,

$$p(t) \approx \frac{\varpi}{T_s}\sum_{n=-\infty}^{\infty}\mathrm{sinc}\left(\frac{n\omega_s\varpi}{2}\right)e^{-i\frac{n\omega_s\varpi}{2}}e^{in\omega_s t} \tag{3.36}$$

$$= \frac{\varpi}{T_s}\sum_{n=-\infty}^{\infty}\mathrm{sinc}\left(\frac{n\omega_s\varpi}{2}\right)e^{in\omega_s\left(t-\frac{\varpi}{2}\right)} \tag{3.37}$$

Using Equation 3.37 in 3.21,

$$h_{\varpi}^*(t) = \sum_{n=-\infty}^{\infty}c_n h(t)e^{in\omega_s t} \tag{3.38}$$

$$= \frac{\varpi}{T_s}\sum_{n=-\infty}^{\infty}h(t)\mathrm{sinc}\left(\frac{n\omega_s\varpi}{2}\right)e^{in\omega_s\left(t-\frac{\varpi}{2}\right)} \tag{3.39}$$

Based on the definition of the *complex Fourier transform* given by Equations 24.390 and 24.391, let us write the transform of $h_{\varpi}^*(t)$,

$$H_{\varpi}^*(\omega) = \int_{-\infty}^{\infty}h_{\varpi}^*(t)e^{-i\omega t}\,dt$$

$$= \int_{-\infty}^{\infty}h(t)p(t)e^{-i\omega t}\,dt$$

$$= \int_{-\infty}^{\infty}h(t)\left(\sum_{n=-\infty}^{\infty}c_n e^{in\omega_s t}\right)e^{-i\omega t}\,dt$$

$$= \sum_{n=-\infty}^{\infty}c_n\int_{-\infty}^{\infty}h(t)e^{in\omega_s t}e^{-i\omega t}\,dt \tag{3.40}$$

Using the shifting theorem for the Fourier Transform (see Section 24.9.4),

$$H_{\bar{\omega}}^*(\omega) = \sum_{n=-\infty}^{\infty} H(\omega - n\omega_s) \qquad (3.41)$$

where $H(\omega)$ is the *complex Fourier transform* of $h(t)$.

Since the range of n is $[-\infty, \infty]$,

$$H_{\bar{\omega}}^*(\omega) = \sum_{n=-\infty}^{\infty} H(\omega + n\omega_s) \qquad (3.42)$$

This shows that the sampling process has produced higher harmonics at $n \neq 0$. If we only consider the part of Equation 3.35, where $n = 0$, we have,

$$c_0 = \frac{\bar{\omega}}{T_s} \qquad (3.43)$$

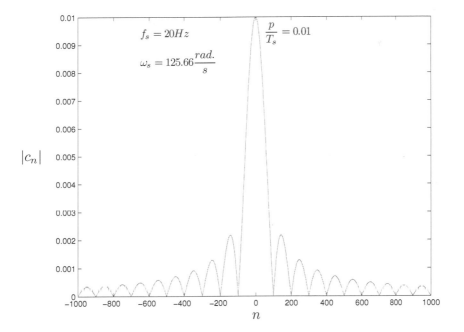

Fig. 3.23: Magnitude of the *complex Fourier series coefficients* of a uniform-rate fixed pulsewidth sampler

Therefore, the part of the Fourier Transform of the output evaluated for $n = 0$ is,

$$H_{\varpi}^*|_{n=0}(\omega) = c_0 H(\omega)$$

$$= \frac{\varpi}{T_s} H(\omega) \tag{3.44}$$

which shows that the sampled output has an amplitude which is scaled by $\frac{\varpi}{T_s}$ from the Fourier Transform of the signal itself. Also, the magnitudes of the complex Fourier coefficients are,

$$|c_n| = \left| \text{sinc}\left(\frac{n\omega_s \varpi}{2} \right) \right| \left| e^{-i\frac{n\omega_s \varpi}{2}} \right|$$

$$= \left| \text{sinc}\left(\frac{n\omega_s \varpi}{2} \right) \right| \tag{3.45}$$

Figure 3.23 shows a plot of $|c_n|$.

Equation 3.45 enables us to compute an upper bound for the magnitude of the sampled output as follows,

$$|H_{\varpi}^*(\omega)| = \left| \sum_{n=-\infty}^{\infty} c_n H(\omega + n\omega_s) \right|$$

$$\leq \sum_{n=-\infty}^{\infty} |c_n| |H(\omega + n\omega_s)| \tag{3.46}$$

3.6.1 Ideal Sampler

Consider the finite pulse width of the sampler given in Equation 3.22 and make it small such that it approaches 0. Then, the pulse train, $p(t)$ will change to the impulse train, $I(t)$ as follows,

$$I(t) = \sum_{n=-\infty}^{\infty} \delta(t - nT_s) \tag{3.47}$$

Therefore, the output of the sampler will be,

$$h^*(t) = h(t)I(t)$$

$$= h(t) \sum_{n=-\infty}^{\infty} \delta(t - nT_s) \tag{3.48}$$

Taking the *complex Fourier transform* of $h^*(t)$,

$$H^*(\omega) = \int_{-\infty}^{\infty} h^*(t) e^{-i\omega t} dt$$

$$= \int_{-\infty}^{\infty} h(t) \sum_{n=-\infty}^{\infty} \delta(t - nT_s) e^{-i\omega t} dt$$

$$= \frac{1}{T_s} \sum_{n=-\infty}^{\infty} H(\omega - n\omega_s)$$

$$= \frac{1}{T_s} \sum_{n=-\infty}^{\infty} H(\omega + n\omega_s) \tag{3.49}$$

Sampling when ϖ has a finite width, the amplitude of the output spectrum, $|H_{\varpi}^*(\omega)|$ decreases as ω increases (see Figure 3.23). However, based on Equation 3.49, the amplitude of all the harmonics of the spectrum of the sampled signal are equal, i.e., $\frac{1}{T_s}$.

Note that for the case when the sampling angular frequency is greater than twice the *Nyquist critical angular frequency* ($\omega_s > 2\omega_c$), it means that applying a low pass filter to the sampled data with a cut-off angular frequency of ω_c will produce the original signal. However, if $\omega_s < 2\omega_c$, then, there will be some distortion present in the filtered signal from the leakage of the overlapping side-bands of the sampled spectrum, $H^*(\omega)$. This overlap is called aliasing and the output signal will portray a different frequency from the original signal. This output frequency is called the *alias frequency*. The limit, $\frac{\omega_s}{2}$ is called the *folding frequency*. When there is overlapping of the higher harmonics with the fundamental portion of the spectrum, the effect is called folding.

Figure 3.24 shows the reflection of two poles marked by $\times$ on the Laplace plane. The poles are folded to the higher frequencies in $n\omega_s$ intervals. This manifests itself as a term of $\pm 2n\pi i$ added to the polar representation in the z plane.

There are several other errors related to sampling. These are aliasing, truncation error, jitter, and finally partial loss of information. Let's take a look at these errors and try to analyze them in such a way that we would minimize their effect.

3.6.2 Aliasing

Figure 3.25 shows a short segment of speech which starts with high frequency components. The sampling rate used in this case was 22kHz. Most of the activity in the first 0.15 seconds of the speech is between 4kHz and 10kHz. Since the sampling rate is higher than twice the highest component of the signal, we see this activity. In Figure 3.26, the data has simply been sampled at $\frac{1}{4}$ of the sampling frequency of Figure 3.25. Notice the high frequency component present in the waveform repre-

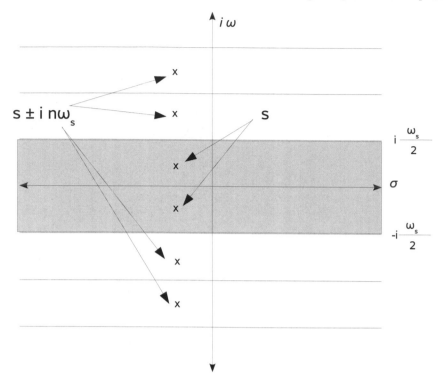

Fig. 3.24: Reflections in the Laplace plane due to folding of the Laplace Transform of the output of an ideal sampler – x marks a set of poles which are also folded to the higher frequencies

sentation of Figure 3.26. This high frequency component shows up as leakage into the low frequency region of the spectrogram. Although the 22kHz sampling of the signal did not show much energy in the $0 - 4$kHz band, in Figure 3.26, this band does shows some high energy. This leakage is due to the aliasing effect described at the end of the previous section.

To remedy the aliasing problem, the data should be passed through a low-pass filter (as discussed earlier). The low-pass filter ensures that the high frequency components of the signal are removed before sampling takes effect. This filter is sometimes called an *anti-aliasing* filter. The cut-off frequency of this filter should be less than or equal to the Nyquist critical frequency of the new sampler. Therefore, in this case, since the new sampling rate is 5512Hz, the cut-off frequency of the anti-aliasing low-pass filter should be at most 2756Hz.

Figure 3.27 shows the result of sampling with a rate of 5512Hz after passing the data through a low-pass filter with the cut-off frequency of 2756Hz. Notice that the high frequency component is no longer present in the waveform representation of the signal in Figure 3.27. Also, there is no high-energy component in the $0 - 3\text{kHz}$ band.

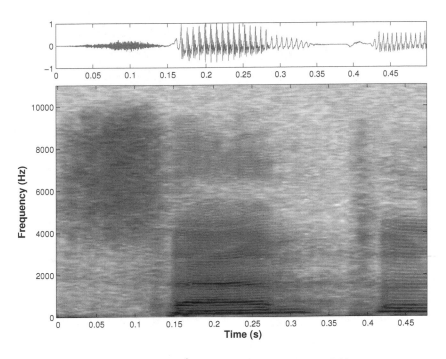

Fig. 3.25: The first $\frac{1}{2}$ second of the signal in Figure 3.28

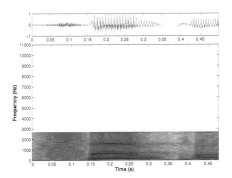

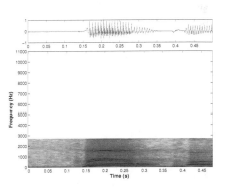

Fig. 3.26: Original signal was subsampled by a factor of 4 with no filtering done on the signal

Fig. 3.27: The original signal was subsampled by a factor of 4 after being passed through a low-pass filter

3.6.3 Truncation Error

Consider the reconstruction equation of the sampling theorem, Equation 3.17. In practice, it will not be feasible to compute the whole infinite series. One method is to truncate the summation to go from $-N$ to N instead of going from $-\infty$ to ∞. In doing this, we will be computing $2N+1$ terms of the summation resulting in the following approximation to the signal,

$$h(t) \approx \sum_{n=-N}^{N} h_n \mathrm{sinc}(\omega_c t - n\pi) \tag{3.50}$$

subtracting Equation 3.50 from Equation 3.17 we obtain the expression for the truncation error, E_t,

$$E_t = \sum_{n=-\infty}^{\infty} h_n \mathrm{sinc}(\omega_c t - n\pi) - \sum_{n=-N}^{N} h_n \mathrm{sinc}(\omega_c t - n\pi) \tag{3.51}$$

There have been many different upper bounds given in the literature.[7] Here, we include a general bound for a finite time interval, $-\tau \leq t \leq \tau$.[15] Reference [7] reviews many other bounds which may have been developed for special cases of the signal at hand. For a general bound, consider the power of the band-limited signal, $\mathscr{P}$,

$$\mathscr{P} \triangleq \int_{-\omega_c}^{\omega_c} |H(\omega)|^2 d\omega \tag{3.52}$$

The truncation Error, E, is a function of t, bounded by the following expression,

$$|E_t(t)| \leq \frac{\sqrt{2}}{\pi} \mathscr{P} \left| \sin(\frac{\pi t}{\Delta t}) \right| \sqrt{\frac{\tau \Delta t}{\tau^2 - t^2}} \tag{3.53}$$

where the signal is being considered in the finite domain of $-\tau \leq t \leq \tau$ and $\Delta t < \frac{1}{f_c}$. Basically, we can be assured that $|E_t(t)|$ will always be smaller than the larger value produced by the right hand side of Equation 3.53 depending on what is used for Δt given the restrictions.

Also, in case of the WKS Sampling theorem, [11] presented a least upper bound for $E_t(t)$ for bandlimited signals with bounds on their energy. They reached the important conclusion that a few percent of oversampling the signal produces significant improvement in error reduction.

Truncation error is usually not very significant for speaker recognition, since we seldom reconstruct the sampled signal. However, we briefly covered it here, since the indirect consequences of the results from reducing the reconstruction errors help in choosing the correct sampling strategy.

3.6.4 Jitter

There are two types of *jitter* with which we are faced, in the processing of speech. Here, we call them *micro-jitter* and *macro-jitter*. Recall that the WKS sampling theorem deals with a period sampling rate. It means that the sampling period must be fixed for the theory to hold. There are extensions to the sampling theorem such as those due to Papoulis [13] which deal with non-uniform sampling rates (see Section 3.1.3). However, the theory being used throughout this book and with most practical systems, relies on uniform sampling (periodic sampling).

Given this restriction, whenever there is the slightest discrepancy in the timing of the samples, the resulting error in the signal reconstruction (signal representation) is called *jitter*. We call this type of jitter, *micro-jitter*, based on the fact that it is at the lowest possible level, namely the sample level. Equation 3.54 represents the micro-jitter error (E_j) of a sample:

$$E_j(n) = h(T_s - \tau_j(n)) \tag{3.54}$$

where $T_s \triangleq \frac{1}{f_s}$ and $|\tau_j(n)| \ll T_s$. $\tau_j(n)$ is the jitter in the sampling time for sample n.

The second kind of jitter happens in network-based telephony or other types of audio systems (such as video conferencing). In this case, the timing difference is not at the sample level, but it is more on the packet level. Most network-based audio transmission protocols are asynchronous and do not guarantee that the packets would arrive at the same time. When this happens, it is natural for the packets that arrive earlier than their predecessors to be held until a later time for presentation. This is done at the remote end (receiving end) in the capacity of a buffer, called a *jitter buffer*. Jitter buffers will add more delay to the overall delay of the system since they hold on to packets for later retrieval. The larger they are, the more the packets may be delayed, but also there is less of a chance of a packet loss.

You may have experienced something similar to this when automobile CD players had first entered the market. Expensive memory meant that only a few seconds of audio could be buffered for jitter mitigation. Some may recall that the larger the buffer, the more expensive the CD players were. This is a very similar idea, so that when the car hit a bump and the CD reader lost its place for a brief moment, the buffer would use the information it had read earlier and played it back so that there was no interruption in the audio. If the buffers were too small, large puddles would still cause an interruption in the playback of audio.

Since this type of jitter relates to delays and timing differences at a macro level, namely at the packet level, we call it a *macro-jitter* to distinguish it from the *micro-jitter*, introduced earlier.

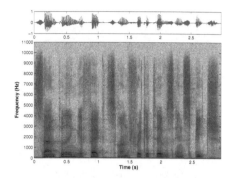

Fig. 3.28: "Sampling Effects on Fricatives in Speech" (Sampling Rate: 22 kHz)

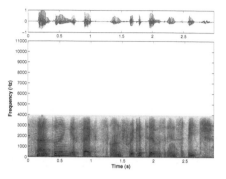

Fig. 3.29: "Sampling Effects on Fricatives in Speech" (Sampling Rate: 8 kHz)

3.6.5 Loss of Information

It is worthwhile noting that the speech signal is not band-limited, at least to the extent that we usually sample it. Typical sampling rates used in speech processing systems go from about 8-kHz upwards, but they seldom go past 22-kHz per channel for speaker recognition tasks. The actual speech signal has a much wider band. Fricatives can produce frequencies upward of 30-kHz and our ears are usually capable of hearing signals to those frequency levels. This means that we would have to sample each channel of speech at 60+ kHz to maintain most of the information.

Figures 3.28 and 3.29 show the spectrogram of an utterance with some fricatives sampled at 22-kHz and 8-kHz respectively. Notice the significant amount of information which is lost in the higher harmonics of the audio. Therefore, even if we try to avoid all the types of errors listed in the previous sections, we usually lose a significant amount of information to limitations of the sampling rate. This problem will probably be corrected as time progresses, since the computational power, cost of memory, and quality of the telecommunication infrastructure keep improving, making it possible to use higher sampling rates in the speech acquisition apparatuses. Once we can comfortably sample speech at 96kHz and transmit it with ease, the problems reported in this section will no longer be significant.

References

1. Black, H.S.: Modulation Theory. van Nostrand Publishers, New York (1953)
2. Cheung, K.F.: A Multidimensional Extension of Papoulis' Generalized Sampling Expansion with Application in Minimum Density Sampling. In: R.J.M. II (ed.) Advanced Topics in Shannon Sampling and Interpolation Theory. Springer Verlag, New York (1993)
3. Deng, L., O'Shaughnessy, D.: Speech Processing, A Dynamic and Optimization-Oriented Approach. Marcel Dekker, Inc., New york (2003). ISBN: 0-824-74040-8
4. Edie, E., Gish, H.: A Parametric Approach to Vocal Tract Length Normalization. In: IEEE International Conference on Acoustics, Speech and Signal Processing (ICASSP 1996), vol. 1, pp. 346–348 (1996)
5. Fogel, L.: A Note on the Sampling Theorem. The Institute of Radio Engineers Transactions on Information Theory $1(1)$, 47–48 (1955)
6. ITU-T: G.711: Pulse Code Modulation (PCM) of Voice Frequencies. ITU-T Recommendation (1988). URL http://www.itu.int/rec/T-REC-G.711-198811-I/en
7. Jerri, A.J.: The Shannon Sampling Theorem – Its Various Extensions and Applications: A Tutorial Review. Proceedings of the IEEE $65(11)$, 1565–1596 (1977)
8. Jerri, A.J.: Correction to The Shannon Sampling Theorem – Its Various Extensions and Applications: A Tutorial Review. Proceedings of the IEEE $67(4)$, 695–695 (1979)
9. Kotel'nikov, V.A.: On the Transmission apacity of Ether and Wire in Electrocommunications. In: Izd. Red. Upr. Svyazi RKKA (The First All-Union Conference on Questions of Communications) (1933). English Translation by C.C. Bissell and V. E. Katsnelson
10. Kuo, B.C.: Digital Control Systems, 2nd edn. Oxford University Press, New York (1992). ISBN: 0-195-12064-7
11. Mendelovicz, E., Sherman, J.W.: Truncation Error Bounds for Signal Sampling. In: 9th Annual Asilomar Conference on Circuits Systems and Computers, p. 16 (1975)
12. Nyquist, H.: Certain Topics in Telegraph Transmission Theory. Transactions of the American Institute of Electrical Engineers (AIEE) 47, 617–644 (1928). Reprint in Proceedings of the IEEE (2002), Vol. 90, No. 2, pp. 280–305
13. Papoulis, A.: Signal Analysis. McGraw Hill, New York (1977)
14. Shannon, C.E.: Communication in the Presence of Noise. Proceedings of the Institute of Radio Engineers $37(1)$, 10–21 (1949). Reprint available at: Proceedings of the IEEE, Vol. 86, No. 2, Feb. 1998
15. Tsybakov, B.S., Iakoviev, V.P.: On the Accuracy of Restoring a Function with a Finite Number of Terms of Kotel'nikov Series. Radio Engineering and Electronic Physics $4(3)$, 274–275 (1959)
16. Weiss: Sampling Theorems Associated with Sturm-Liouville Systems. Bulletin of the Mathmatical Society 63, 242 (1957)
17. Whittaker, E.T.: On the Functions which are Represented by the Expansion of Interpolating Theory. Proceedings of the Royal Society of Edinburgh 35, 181–194 (1915)
18. Whittaker, J.M.: The Fourier Theory of the Carndinal Functions. Proceedings of the Mathematical Society of Edinburgh 1, 169–176 (1929)
19. Whittaker, J.M.. Interpolatory Function Theory. No. 33 in Cambridge Tracts in Mathematics and Mathematical Physics. Cambridge University Press, Cambridge, England (1935)

Chapter 4
Phonetics and Phonology

Language is a form of human reason, and has its reasons which are unknown to man.

Claude Lévi-Strauss
The Savage Mind (Chapter 9), 1962

According to Summer Institute of Linguistics (SIL International) [20], the linguistic hierarchy from one of the leaves to the top is as follows, *Phonetics, Phonology, Morphology, Syntax, Semantics*, and *Pragmatics*. In Chapter 2, we reviewed the anatomy of the human speech production and perception. In this chapter we will start by exploring the range and limitations imposed by the speech production system, so called *phonetics*. Then, we will follow to a higher level in the hierarchy by studying how sounds are organized and used in human languages, so called *Phonology*, along with the rest of the hierarchy which we will call *linguistics* as a whole. In the last part of this chapter, we will pay specific attention to *suprasegmental*[1] flow of human speech called *prosody*. This is to give the reader a basic understanding of the types of sounds produced by the vocal tract. Of course, as with many of the other topics covered in this book, we will only scratch the surface and will concentrate on portions of the discipline that are more directly relevant to the speaker recognition task.

4.1 Phonetics

This section is concerned with the phonetic aspects of speech. In choosing the areas of coverage, we have had speaker recognition in mind, although this section is quite useful for other speech related disciplines such as speech recognition and speech synthesis. Let us begin with the definition of *Phonetics*.

Definition 4.1 (Phonetics). *The term Phonetics refers to the study of sounds which are produced by the human vocal system regardless of their associated languages.*

---

[1] Suprasegmental features of speech are those which surpass the segment boundaries – phone and phoneme boundaries.

We will start by defining the elementary segments in speech. The following definitions have subtle differences which depend on the perspective of interest.

Definition 4.2 (phone). *Phones are elementary sounds that occur in a language.*

In another perspective, phones may be viewed as the smallest segments of speech in a linguistic viewpoint with distinct vocal patterns.

Definition 4.3 (Phoneme). *Phonemes are semantically significant sounds that occur in a language.*

Another perspective would consider phonemes as the smallest conceptual segments of speech in a linguistic viewpoint which may encompass several vocal variations with the same objective.

Definition 4.4 (Allophone). *Allophones are different phones which convey the same phonemic information.*

To distinguish these three concepts, take the English word, *pop*. There are two *p*s in the word with the first *p* being aspirated (followed by the release of a puff of air). We write this as p^h in a phonetic representation. The second *p* is not aspirated, so it is just written as *p* in phonetic representation. However, they are both represented as *p* in a phonemic representation. Thus, the phonetic transcription of the word is $p^h op$. In English the aspiration does not change the meaning of *p*, so p^h and *p*, although different *phones*, are considered to be instances of the same *phoneme*. In some languages such as most Indian languages (e.g., Sindhi), interchanging p^h and *p* changes the meaning of the word. Thus, they are considered as separate phonemes in these languages.

In general, any sound contains three main components. The type of sound which is resonant and periodic can be seen in a spectrogram through the apparent activity of the different formants. Most of these periodic sounds are a product of the vibration of the vocal folds by tightening, but keeping them slightly open so that air may pass through the small opening and generate the resonance associated with the mode of vibration of the folds combined with the resonance chamber of the vocal tract above them. Some other period generators also exist in the vocal tract, so periodic sounds may be generated by other means than the vocal folds as well. Another major category of sounds is composed of those which are produced due to turbulent air flow and contains a full spectrum of different vibrations. A third category is related to impulsive sounds. These are categorized by a very brief burst of energy which quickly drops off. Please note that these categories apply to all the sounds being generated by the vocal system and should not be confused with the phonation categories which will be described later.

According to *Laver* [9], there are four major elements of speech production, *initiation*, *phonation*, *articulation*, and *coordination*. All these elements contribute to the creation of speech.

4.1.1 Initiation

Initiation is a function of the airstream mechanism and the direction of airflow. The airstream may be either *pulmonic*, *glottalic*, or *velaric*. The *pulmonic* airstream is initiated from the lungs. When the glottis is closed, the *glottalic* airstream is initiated by the vertical motion of the larynx. This produces voiceless sounds. It can also vibrate, in which case it produces voiced sounds through the alternating pressure differential in the vocal tract above it. When the tongue initiates the air pressure differential in an air-filled cavity, the airstream is called *velaric*. For any of these three initiation airstreams, the air may move outward, in which case it is called *egressive*, or inward, which is called *ingressive*.

A glottalic egressive sound is called an *ejective* sound and in contrast, an ingressive glottalic sound is called, *implosive*. Velaric ingressive sounds are called *clicks*. Clicks may be combined with voiced pulmonic egressive sounds to create *voiced clicks* – see Section 4.1.9.1. Voiced implosives are a product of the combination of the pulmonic egressive voiced sounds and glottalic ingressives.

4.1.2 Phonation

Phonation deals with the acoustic energy generated by the vocal folds at the larynx. The different kinds of phonation are *unvoiced*, *voiced*, and *whisper*.

Unvoiced phonation may be either in the form of *nil phonation* which corresponds to zero energy or *breath phonation* which is based on relaxed vocal folds passing a turbulent air stream.

Majority of voiced sounds are generated through *normal voiced* phonation which is basically when the vocal folds are vibrating at a periodic rate and generate certain resonance in the upper chamber of the vocal tract. Another category of voice phonation is called *laryngealization (creaky voice)*. It is when the arytenoid cartilages fix the posterior portion of the vocal folds, only allowing the anterior part of the vocal folds to vibrate. Yet another type voiced phonation is a falsetto which is basically the un-natural creation of a high pitched voice by tightening the basic shape of the vocal folds to achieve a false high pitch.

Whispered phonation happens when the speaker acts like generating a voiced phonation with the exception that the vocal folds are made more relaxed so that a greater flow of air can pass through them, generating more of turbulent airstream compared to a voiced resonance. However, the vocal folds are not relaxed enough to generate an unvoiced phonation.

4.1.3 Articulation

Articulation deals with three different notions. The first is the *place of articulation* which is the location where the vocal tract has the most constriction. Secondly, the *degree of stricture* is of interest. That is the amount of closure and the proximity of it within the location of articulation. Thirdly, the aspect of articulation which is a collection of higher level concepts such as factors related to conformation, topology and transition. The following are some resulting phonations based on the manner and degree of the stricture.

1. *Stops* – Maximum closure at some point in the vocal tract. Refer to Chapter 2. There are two exits in the vocal tract. Depending on where the maximum stricture takes place, the following two types of stops are possible, *Oral* and *Nasal*. These may be further categorized based on the routing of the air into *central* and *lateral* stops.)

   a. *Oral*
   b. Nasal

2. *Fricatives* – This happens when the stricture is such that it is slightly open so that the air stream is partially obstructed and a turbulent flow develops making many high frequency components in the signal. Most fricatives have a frequency content of more than 4kHz. Examples are /s/, /f/, and /ʃ/.

   a. *Normal* – These are the more nominal fricatives which although they have quite a good coverage of higher frequencies, still possess more of turbulent nature with higher air flow so that the pitch does not surpass the limits of audible range. An Example is /f/ in English.
   b. *Sibilant* – These are fricatives with exceptionally high pitched sounds such as /s/ and /ʃ/ in English.

3. *Resonants* – Resonant flows are those which are produced by the passing of the air stream through a tight opening producing vocal harmonics. These are categorized into centrally and laterally resonant sounds.

   a. *Central Resonants* – These are sounds for which the air flow passes through the central part of the vocal tract and the resonance stricture. Most sounds, especially in English are of this kind.

      i. *Vocoids (Syllabic)* – These are central resonant vocoids (non-contoids) which mostly include vowels and some glides. Vowels may have a stable medial phase of resonance (see *Laver* [9]) in which case they are known as monophthongs, or they may have a transitional medial phase in which case they are called diphthongs. Triphthongs are also seen in English such as in the word, [flower]. Triphthongs are similar to diphthongs, except there are two transitions and three stable regions in the transition.

      ii. *Approximants (Nonsyllabic)* – nonsyllabic central resonance phonations such as initial sounds in the English words [you] /ju/ and [want] /wɑnt/.

  b. *Lateral Resonant Contoids* – These are lateral resonant sounds which means that they are generated by the diversion of the air stream to a lateral part of the vocal tract for stricture. /l/ in [leaf] /lif/ is a lateral resonant. In this example, although the front of the mouth is constricted by the tip of the tongue, the air flows out of the sides of tongue creating the *gliding* sound in /l/.

4. *Affricates* – This is a compound stricture which is made up of a stop followed by a homorganically generated fricative[2]. An example is the word [Tsunami]. The t^s is an affricate which is made up of the stop /t/ followed by the short fricative /s/. Other examples are the /ǰ/ and /č/ sounds in English words such as [John] and [Charles].

5. *Trills (Rolls)* – This is a low frequency motion of one articulator with respect to another where there is a slight flapping effect at the close proximity points. Examples are the way the Scottish pronounce words starting with r such as the word [royal] or the pronunciation of the long [r] sound in Spanish.

6. Flaps and Taps – A sound created by a quick collision of one articulator against another. A slow collision in passing, would be called a *flap* and a quick almost impulsive collision would be a *tap*. Examples are different usages of /t/ in English and Indian languages.

4.1.4 Coordination

Coordination is the temporal and collaborative nature of articulatory organs in unison to produce an advanced sound. The concept of coordination is interconnected with articulation and cannot be separated. It generally involves neighboring articulators due to physiological restrictions, but it may be associated with farther articulators such as the combination of velar ingressives and pulmonic voiced egressives shown in Table 4.1 for Nama.

---

[2] This means that the fricative is generated at the same location as the point of the stop.

As we have seen, there are many overlapping possibilities for classifying phones, depending on the initiation, phonation, articulation and coordination. Rather than the prolific subcategorization of each of the noted segments which will produce many overlapped segments of parts of speech, we prefer to categorize the sounds based on a combination of articulatory classes first relying on the location of articulation, followed by the type of coordination and then phonation and degrees of stricture. To start, we will cover vowels, which are the most important elements used in conducting speaker recognition due to their resonant nature, using the phonation categorization. Afterwards, consonants are split into two major groups of pulmonic and non-pulmonic (utilizing the initiation categories with some modifications). Within the pulmonic category of consonants which is the largest one, subcategories are based on coordination and degree of stricture (a subclass of articulation).

In the following few sections, we will try to produce the phonetic transcription of the sounds of interest using the *International Phonetic Alphabet (IPA)* [8], whenever we discuss phones. Of course because of technical reasons, at some instances, we were forced to use similar representations. An effort has been made to give comprehensive examples in English or other popular languages so that the reader would be able to grasp the subtleties without having the need to jump back and forth between references.

4.1.5 Vowels

Voiced and voiceless vowels exist. Here we will discuss voiced vowels. Voiceless vowels exist in some languages like *Amerindian languages* [3] such as Comanche and Tlingit (Alaska). There are other old languages with voiceless vocoids. See Section 4.2.2 for a more detailed discussion.

The Most important part of the anatomy used for generating different vowels is the tongue. The basic oscillations come from the partially opened vocal folds. The tongue manipulates higher harmonics to develop different vowels. In addition, the amount of roundedness of the opening of the lips is a factor in determining the sounds of vowels. The tongue contributes in two ways. The height of the body of the tongue is one and its position relative to the front and back of the month is another. The combination of the tongue shape and position, with the roundedness of the opening of the lips changes the first and second formants (second and third harmonics).

[18] conducted a series of experiments on 10 common vowels in the English language. 33 men, 28 women and 15 children (a total of 76 speakers) were asked to

---

[3] Languages spoken by native inhabitants of the Americas.

say 10 words (two times each) and their utterances were recorded. The words were designed to examine the 10 vowels in context of an *h* to the left and a *d* to the right, namely, hid, hɪd, hɛd, hæd, hɑd, həd, hʊd, hud, hʌd, and hɝd. [8] includes a CD-ROM with the pronunciation of these vowels.

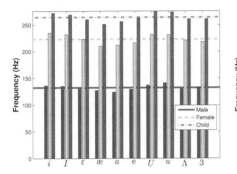

Fig. 4.1: Fundamental Frequencies for Men, Women and Children while uttering 10 common vowels in the English Language – Data From [18]

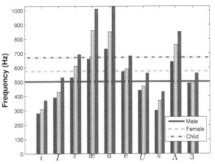

Fig. 4.2: Formant 1 Frequencies for Men, Women and Children while uttering 10 common vowels in the English Language – Data From [18]

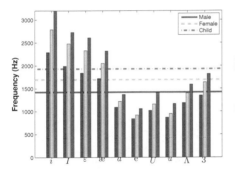

Fig. 4.3: Formant 2 Frequencies for Men, Women and Children while uttering 10 common vowels in the English Language – Data From [18]

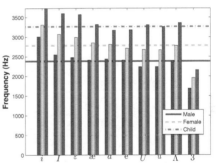

Fig. 4.4: Formant 3 Frequencies for Men, Women and Children while uttering 10 common vowels in the English Language – Data From [18]

Figure 4.1 shows that there is little deviation in the fundamental frequency for an individual. Consider a man, a woman and a child as representatives of the total population. Note that the fundamental frequency (formant 0) does not change much for different vowels. This is the fundamental frequency of the vocal tract based on a normal opening of the vocal folds when one is producing a vowel. However, formants 1 and 2 do vary considerably depending on which vowel is being uttered (see

Figures 4.2 and 4.3). Formant 3 does not change significantly for different vowels (see Figure 4.4). This is the basis for the, so called, formant triangle which defines different vowel locations as a two-dimensional function of formants 1 and 2 (Figure 4.8). Of course, it is really a trapezoid and not a triangle, but it is called a triangle due to legacy. Front ($\bar{i}$ – in) Mid ($\bar{a}$ – father) and Back (o – obey).

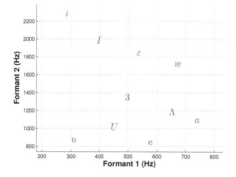

Fig. 4.5: Position of the 10 most common vowels in the English Language as a function of formants 1 and 2 – Average Male Speaker

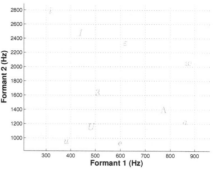

Fig. 4.6: Position of the 10 most common vowels in the English Language as a function of formants 1 and 2 – Average Female Speaker

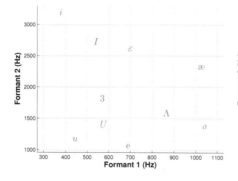

Fig. 4.7: Position of the 10 most common vowels in the English Language as a function of formants 1 and 2 – Average Child Speaker

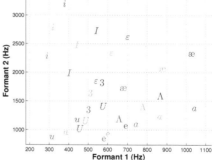

Fig. 4.8: Position of the 10 most common vowels in the English Language as a function of formants 1 and 2 – Male, Female and Child

It is notable that children's vocal tract lengths are different from those of adults and the relative locations and sizes of the articulators, the larynx, the shape of the vocal tract create completely different resonant chambers. For this reason, the articulation method of children for generating the same vowels as adults, differ substantially. As they grow, however, they constantly relearn the articulation techniques to achieve similar outcomes. These changes are tantamount to shifts in the formant triangle (trapezoid) as the children readjust for the new manner of articulation.

Vowels are not so important in speech recognition, but are of utter importance in speaker recognition. The reason is that the goal of speech recognition is to consolidate all speakers into one model and to be able to understand the contents of what is being said. On the other hand, speaker recognition's aim is to differentiate speakers and generally is uninterested in the content of speech. Of course, this statement is only true to a certain degree – when we focus on the primary task of speaker recognition. In the case of text-dependent speaker recognition, the content becomes somewhat relevant, yet still secondary, compared to the primary goal of modeling the speaker's vocal system. Note that this discrepancy stems from the fact that vowels contain most of the periodic parts of speech. Hence, vowels possess more information about the resonance of the vocal tract, namely the fundamental frequency and the formants.

- Vowels are very easy to recognize since they are all voiced and spectrally very different.
- Formants may be easily used for recognizing vowels.
- Vowel triangle – plot of F1 (200Hz - 800Hz) vs F2 (800Hz-2400Hz) and you will see a triangle which is called the vowel triangle.

4.1.6 Pulmonic Consonants

There are two major types of consonants, *pulmonic* and *non-pulmonic*. Pulmonic consonants get their energy from the air stream that is generated in the lungs. Non-pulmonic consonants, on the other hand, usually get their total energy from the activities in the mouth itself. Most consonants are pulmonic. We will discuss non-pulmonic consonants briefly in Section 4.1.9. The air stream that is produced in the lungs may be used to produce different sounds depending on the location of the sound production in the vocal tract and the different parts of the vocal tracts which are used for the production.

Pulmonic consonants exist in a different variety. Some are voiced, which means that there is some periodic vibration present in the production of the sound, triggering formant activity. Some may be unvoiced, which means that the air stream is somewhat turbulent and the vibrations are not as periodic as the voiced counterpart.

4.1.6.1 Glottal Sounds

This part, in combination with instantaneous changes in the state of the vocal folds, closure and opening, is responsible for generating glottal sounds which exist in some languages such as in Arabic, Hebrew, and Gimi (a language from central Papua-New Guinea).[8] Of course, depending on the dialect, even English possesses certain glottalized sounds especially in some local accents in Britain in which *plosives*, $/t/$, $/p/$, and $/k/$ are glottalized.[9] Also, the hard /h/ sound which comes after certain vowels may be categorized as a glottal sound.

4.1.6.2 Epiglottal and Pharyngeal Sounds

Epiglottal and Pharyngeal sounds involve pulling the root of the tongue (epiglottis) back toward the back wall of the pharynx.[8] It is physically impossible to produce a nasal epiglottal or pharyngeal sound since it works using a constriction very low in the vocal tract and that will stop the flow of air to the nasal passage. Pharyngeal fricatives exist in Semitic languages such as Arabic and Hebrew. Examples of Arabic words with these sounds are the word for "beloved," /ħabib/ and the word for "powerful," /ʕziz/.

4.1.6.3 Uvular Sounds

Uvular sounds are /q/, /G/, nasal version of them and /N/. They do not exist in most English dialects. They are quite prominent in Eskimo and other Amerindian languages, Semitic languages, and some Indo-European languages such as Persian and French. An example in French is the sound of /r/ in the word, [Paris] pronounced in the French language, which is a trilled uvular sound. Example of /N/ is the word [ni.hon] which in Japanese means Japan. Another example is the Quechua[4], [q'aʎu], meaning "tomato sauce."[8] An example in Persian is the word for frog, [qurbāqe].

4.1.6.4 Velar Sounds

In producing Velar sounds, the back of the tongue touches the soft palate. Examples are /k/, /g/, /η/ and /x/. Take the English word, *sing* for instance. The "ng" produces a velar consonant, /η/. /η/ is a Velar nasal. /k/ and /g/ are Velar stops. /x/ is a Velar fricative. It has been lost in modern English, but it does exist in Dutch and German.

---

[4] An American Indian language spoken in Bolivia, Chile and Peru

An example in German is the name of the music composer, *Bach*.

4.1.6.5 Palatal Sounds

Palatal sounds are generated using the front of the tongue against the hard palate while the tip of the tongue is behind the lower front teeth. An example in English is the word, *hue*. Another example in German is the word, *ich*.[8]

4.1.6.6 Retroflex Sounds

These sounds are produced by curling the tip of the tongue upward and touching the back of the alveolar ridge (see Section 2.1.5) while touching the sides of the tongue against the side-upper teeth, then releasing a high-speed, steady flow of air into the enclosed cavity. This would create a low frequency flapping (compared to mean speech wave frequencies) of the tip of the tongue against the back of the alveolar ridge. An example of such sounds is one that is created by saying words with hard $/r/$ sounds in English. Of course, most British English pronunciations of words with $/r/$ do not come out as a retroflex. As mentioned, it is a hard $/r/$ sound with quick flaps of the tongue on the alveolar ridge. Irish and Scottish, and some American dialects use more pronounced retroflexes.

4.1.6.7 Alveolar Sounds

These are sounds which are made by placing the upper tip of the tongue on the alveolar ridge of the roof of the mouth (see Section 2.1.5). In English, these sounds are produced in pronouncing words that contain, $/t/$, $/d/$, $/n/$, $/s/$, $/z/$, and $/l/$.

4.1.6.8 Palato-Alveolar Sounds

These are produced by the use of the tongue blade against the back of the alveolar ridge. The production process is pretty much similar to that described for the production retroflexes, with the exception of the part of the tongue that is used and the frequency of the vibration of the tongue against the alveolar ridge. An example is the sound that the combination of characters, $/\int/$ [*sh*] makes such as in the word, *show*. Another example is the sound $/ʒ/$ [*zh*], like the sound that [s] makes in the word, *vision*.

It is interesting to note the differences and similarities between *palato-alveolar* sounds and *retroflex* sounds. In fact a good example of a sound that is in between the two which may be thought of, in the form of a transition between a *retroflex* and a *palato-alveolar* sound is the *Czech* sound of /r̝/ (*ř*) such as in the famous Czech music composer's name, *Dvořak*. It is often quite hard for English speakers to pronounce and is often pronounced as the /ʒ/ sound which is a *palato-alveolar* sound, as we mentioned earlier.

4.1.6.9 Post Alveolar Sounds

Since /ʃ/ (*sh*) and /ʒ/ (*zh*) are made more toward the front of the alveolar ridge when compared to /s/ and /z/ (*alveolar* sounds), they may also be called *post alveolar* sounds. Therefore, *palato-alveolar* sounds are also *post alveolar*. But a *retroflex* is also a *post alveolar* sound. So *post alveolar* sounds are a superset of the two types of sounds. Please note the example of *ř* given in Section 4.1.6.8 which shows another *post alveolar* sound that is in between being a *retroflex* and a *palato-alveolar* sound.

4.1.6.10 Dental Sounds

Dental sounds are produced by the touching of the tip of the tongue against the upper teeth and puffing air through so that a turbulent high frequency vibration is generated at the tip of the tongue against the upper teeth. Examples are the /ð/ (/*dh*/) sound at the beginning of the word, *though* and the /θ/ (/*th*/) sound at the beginning of the word, *thorax*.

4.1.6.11 Labiodental Sounds

Labiodental sounds, as the name suggests are produced using the lips and the teeth. It is usually the upper lip and the lower teeth with a turbulent air flow going through the small openings which are left. Examples in English are /f/ and /v/, such as in the words, *father* and *voice*.

4.1.6.12 Bilabial Sounds

Bilabial sounds are usually stops. They are produced by closing the lips, building up pressure in the mouth and then releasing the lips for an impulsive sound. Examples in English are /p/, /b/, and /m/ which are the unvoiced, voiced, and nasal versions of *bilabial sounds*, respectively.

4.1.7 Whisper

A good example of an *egressive pulmonic whisper* in English is the sound that an [h] makes in the word home. The beginning of the word starts out with a *whisper sound* /h/ and then continues to the sound of [o] which is a vowel, therefore it is voiced. Some non-native speakers of English such as the French, do not pronounce the whispered [h] and usually start with the voiced phonation of [o], so they would pronounce it as [ome].

4.1.8 Whistle

The *Spanish Silbo* (*Silbo Gomero*) language which is used by the inhabitants of La Gomera in the Canary Islands is solely based on a collection of whistles. It is based on a dialect of Spanish and only retains the tonal parts of the language. This language was developed to be able to communicate across mountainous regions. In general, spoken languages have a range of about 40 meters. If they are shouted, the range may be increased to about 200 m. However, whistles can travel about 550 m and more. It has 5 vowels and 4 consonants. The 5 vowels, /i/, /e/, /a/, /o/, and /u/ are generated by reducing the pitch from the highest (/i/) to the lowest (/u/) in the order given above. Only good whistlers can produce perceptually different sounds for /o/ and /u/ by lowering the frequency substantially. In the whistled domains, consonants are basically transient points between vowels. These are produced by modulating the prior and posterior vowels.[13]

The bird language of *Kuskoy* in Turkey is spoken by about 1000 according to a British Broadcasting Corporation (BBC) article [14] and is composed of 29 different whistle phones, one for each of the Turkish alphabet. This means that they may basically speak any Turkish word using this dialect. The name of their village, *Kuskoy*, literally means *bird village*. There are also whistled dialects of Greek, Siberian Yupik and Chepang [13]. Yupik and Chepang are non-tonal languages which by adopting whistle sounds have added tonality to these dialects. There are also whistled dialects of West African languages such as Yoruba and Ewe. In Africa some whistled dialects of French are also present.

4.1.9 Non-Pulmonic Consonants

In this section we discuss the non-pulmonic sounds including glottalic, velaric and combinations. One major group of non-pulmonic initiations are velaric airstreams which are mostly made of clicks.

4.1.9.1 Clicks

Clicks velaric sounds which are present mostly in South African languages such as Nama and Zulu. Although they are velaric phonations, they may be combined with pulmonic and glottalic phonations. Nama is a South African language. Ladefoged [8] has included a recording of 20 different variations of clicks in the accompanying CD-ROM of his book by a Nama native speaker. In Table 4.1, we have generated spectrograms for selected regions of the recordings to show the different shapes the formants take as well as the energy distribution across the different frequency bands. Also, see Figure 4.9 for an Indo-European example of a click.

Notice that there are certain clicks which display very low energy levels similar to stops and some have a voiced characteristic, last longer and activate some major formants in the speech. As far as speaker recognition is concerned, we have not had much experience dealing with languages that use clicks. However, by examining the spectrograms, it appears that they may not possess considerable speaker discriminability. Normally, long voiced phonations are much more suited for this job since they give a better reflection of the vocal tract characteristics of the speaker. Clicks would probably be as effective as most stops and flaps are.

4.1.9.2 Voiced Implosives

Voiced implosives are *ingressive glottalic* phonations which are created by the negative relative pressure developed between the larynx and any higher constriction. This is done by closing the upper articulator involved in the stricture and then lowering the larynx while the glottis is closed. The sound is created by releasing the upper stricture so that atmospheric pressure may flow into the enclosed low pressure cavity. The unvoiced version of implosives is quite rare, although it exists in some Mexican languages (see *Laver* [9]).

It may seem that the above process does not allow for any voicing, however, the voiced version is more prominent and it is achieved by combining the above process with an ingressive pulmonic process to create resonance (voicing). According to *Laver* [9], this exists in *Zulu* which is a *South African* language as well as *West*

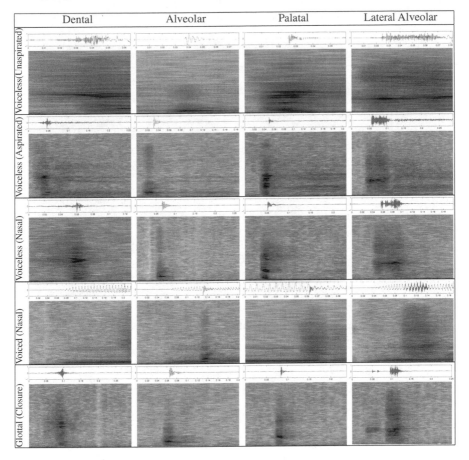

Table 4.1: Types of Clicks in Nama (South African Language)
Data was extracted from samples in the CD-ROM accompanying [8]

African languages such as *Hausa* and *Margi*.

4.1.9.3 Ejectives

Ejectives are egressive glottalic phonations which are based on the momentary closure of all the exits in the vocal tract including the glottis, the velum, and the mouth. A pressure is built up and abruptly released. If the release is followed by a stop, an ejective stop will be generated. On the other hand, having a small release at the upper stricture will generate a fricative and with combination of the two an ejective affricate. These are seen in many African languages and native languages of the

Americas.[9]

4.2 Phonology and Linguistics

In the previous section we touched upon the different aspects of phonetics. Although we had to refer to instances in different languages to clarify specific phones, in phonetics, the language dependence is only secondary. This part is left to Phonological and at the higher level to Linguistic studies and reviews.

Definition 4.5 (Phonology). *Phonology is the study of phonetics in the framework of specific languages.*

Phonology deals with the assessment of the significance of phonetics and their inter-relation in a particular language or across different languages. It comes short of studying *syntax*, *semantics* and *pragmatics* which are parts of *linguistics*. *Morphological* aspects of words are sometimes considered in Phonology as well. Here, we will take a closer look at language-specific aspects of phones and their distribution across world languages.

At the beginning of this chapter we defined phones, phonemes, and allophones, so that the proper distinction may be made among them. Since our goal is to apply our knowledge in this area to speaker recognition problem solving, we need to understand the variability of phones and phonemes across different languages so that we would understand the potential dependencies upon languages when we speak about a language-independent speaker recognition system, for instance.

4.2.1 Phonemic Utilization Across Languages

Laver [9] discusses and analyzes a vast study which was reported and discussed by *Maddieson* [12] in 1984. This study was conducted across 317 languages and found that about 70% of them possessed between 20 and 37 phonemes and averaged around 31 total of phonemes per language within the 70%. As a reference, note that Ethnologue [11] states the total number of living languages to be 6909 in May of 2009. Therefore, although the numbers presented here are probably statistically significant, they undoubtedly do not paint the whole picture, but should be just used as reference.

In this database, the least number of phonemes exists in *Rotokas* which is a *Papuan language* containing only 6 consonants and 5 vowels. *Hawaiian*, a better well known language and most likely to have been heard by the majority of the readers has the second smallest number of phonemes (8 consonants and 5 vowels). To understand the amount of restriction such a limitation would pose, it would help to look at how English words which have entered *Hawaiian* had to be modified to be pronounced with the phonemes at hand. A very popular example is the phrase, "Merry Christmas" which changes to "Mele Kalikimaka" in Hawaiian.

On the other side of the spectrum, we can see languages such as !Xū, which is a *Khoisan language* spoken in parts of *Angola* and *Namibia*, that possess the most number of phonemes in this study, 141 phonemes in total (95 consonants and 46 vowels).[9]

Of course, we have to be very careful with these comparisons since they are based on phonemes and not phones. For example, although the number of vowels is reported to be only 5, there are also 9 diphthongs in the Hawaiian language and if the long versions of the monophthongs and diphthongs are also counted, then the total number of vowels, could be as many as 25 in total. So the number of vocoids from a phonetic study is 25 versus 5 vowels in the phonemic study. In addition, the consonants /t/ and /w/ can apparently take on shapes of /k/ and /v/ which makes these allophones, increasing the number of phonetic consonants by 2 from the phonemic consonants.

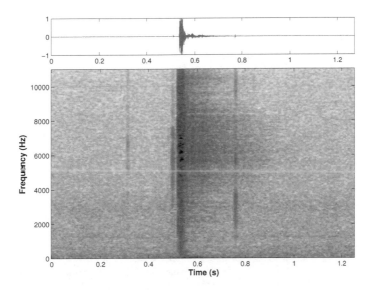

Fig. 4.9: Persian ingressive nasal velaric fricative (click), used for negation – colloquial "No"

In another example, recall that clicks mostly occur in African languages. However, clicks are also found in some unlikely languages in the form of colloquial gestures. An example (Figure 4.9) is the very informal articulation for "No" in Persian[5], an Indo-European language. This is an ingressive nasal velaric fricative which is sometimes combined with negative pulmonic pressure creating a pulmonic ingressive instance. However, this information is not included in either phonetic or phonologic descriptions of the language. Almost every language has similar inherent sounds which are ignored by phoneticians.

The evolution of languages provides us with some interesting insights regarding recognition in general (speech recognition and speaker recognition as a side-effect).

English provides a good example for clarifying the differences between phones and phonemes. In English, the number of phones is 49. Rabiner and Jung [19] list 48 and we have added the triphthong which is in the word [flower]. Surely, with more careful analysis one may come up with a few more phones and they usually vary greatly by dialect. Here is the distribution of the 49 phones as a reference,

- 19 Vocoids: 12 Vowels, 6 *Diphthongs* and one *Triphthong*
- 4 Approximants: 2 *Liquids* and 2 *Glides*
- 21 Standard Consonants
- 4 Syllabic Consonants
- 1 Glottal Stop

But, phonologically, there are only 39 phonemes in English,

- 11 Vowels
- 4 Diphthongs
- 4 Semi-Vowels – 2 Liquids (w, l) and 2 Glides (r, y)
- 3 Nasal Consonants (m, n, ng)
- 6 Voiced (b, d, g) and Unvoiced (p, t, k) Stop Consonants[6]
- 8 Fricatives: Voiced (v, th, z, zh) and Unvoiced (f, θ, s, sh)
- 2 Affricates $\hat{j}$ and $\hat{c}$
- 1 Whisper – h

Figures 4.10 through 4.24 show 15 different vowels and diphthongs in an American dialect of English.

To get an idea of other well-known languages, there are 27 (5 vowels and 16 consonants) phonemes in Japanese, 51 in Hindi (11 vowels and 40 consonants), and

---

[5] Persian is the official language of Iran, Afghanistan and Tajikistan and is spoken in many other countries such as Uzbekistan, India and China among others. The Persian word for the language is *Farsi*. Some have mistakenly referred to the language as *Farsi* in English. See the definition of *Farsi* in the Oxford English Dictionary online (http://www.oed.com) and note the 1984 reference on this incorrect usage. It states, "It may not be too late to put an end to the grotesque affectation of applying the name 'Farsi' to the language which for more than five hundred years has been known to English-speakers as Persian."

[6] In many languages, etymologically, we see transference between voiced and unvoiced phonemes.

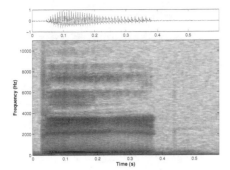

Fig. 4.10: bead /biːd/
(In an American Dialect of English)

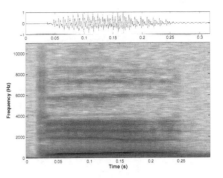

Fig. 4.11: bid /bɪd/
(In an American Dialect of English)

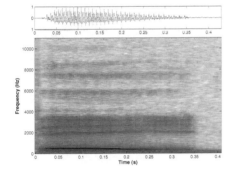

Fig. 4.12: bayed /beɪd/
(In an American Dialect of English)

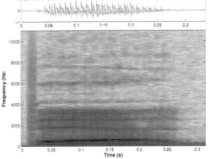

Fig. 4.13: bed /bɛd/
(In an American Dialect of English)

29 (6 vowels and 23 consonants) in Persian.

As we have seen, languages evolve using different capabilities of the human vocal system. The extreme comparison would be between Rotokas (11 phonemes) and !Xū (141 phonemes). In most cases, though, the differences are a bit more subtle. For example north Indian languages differentiate between aspirated and unaspirated stop and English does not. Let us look at one of these subtle differences in the next section which is an example that utilizes *whisper*.

4.2.2 Whisper

We discussed whisper in terms of its physiological and phonetic characteristics in Section 4.1.7. Here, we will look at its importance in a phonological and linguistic

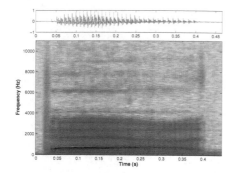

Fig. 4.14: bad /bæd/
(In an American Dialect of English)

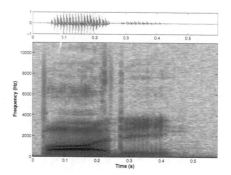

Fig. 4.15: body /bɑːdɪ/
(In an American Dialect of English)

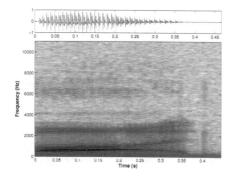

Fig. 4.16: bawd /bɔːd/
(In an American Dialect of English)

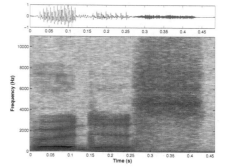

Fig. 4.17: Buddhist /bʊ dist/
(In an American Dialect of English)

sense.

Voiced Version	Meaning	Whispered Version	Meaning
ʼmayakekiʔu?	He came to play	ʼmayakekiʔu?	He played and came on
ʔuʼhanikikiʔu?	He came to fix it for him	ʔuʼhanikikiʔu?	He fixed it for him and came on
ʔuʼniʔacïkiʔu?	He came to advise him	ʔuʼniʔacïkiʔu	He advised him and came on

Table 4.2: Examples of whispered alterations in Comanche – the circle under a vocoid changes it to a whispered phonation

I most languages, whisper is only a form of phonation which is unique to a certain set of phones such as the /h/ in English and can otherwise be used for silent phonation. This form of silence is performed by relaxing the vocal folds and attempting to utter the same voiced sounds. Due to its nature, only voiced phonations are al-

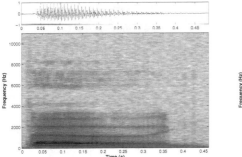

Fig. 4.18: bode /boʊd/
(In an American Dialect of English)

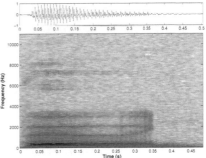

Fig. 4.19: booed /buːd/
(In an American Dialect of English)

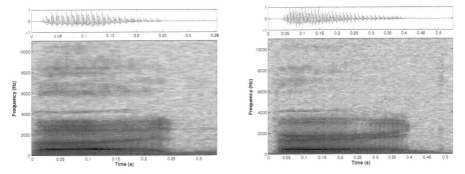

Fig. 4.20: bud /bʌd/
(In an American Dialect of English)

Fig. 4.21: bird /bɜːd/
(In an American Dialect of English)

tered. A special case exists in many Amerindian languages in which case, stating the same statement two times with a simple swap of one of the vocoids with its whispered version would produce two completely different meanings. Table 4.2 shows some examples in *Comanche*[7] [2]. Another such language is the Alaskan Tlingit [9].

4.2.3 Importance of Vowels in Speaker Recognition

Anyone who has experienced learning a new language or has spoken to non-native speakers of his/her language knows that most advanced students of the new language tend to do well with learning the proper phonation of consonants, but they

---

[7] Comanche is a North American Amerindian language.

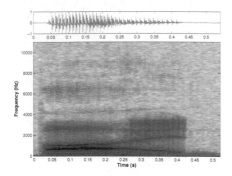

Fig. 4.22: bide /baɪd/
(In an American Dialect of English)

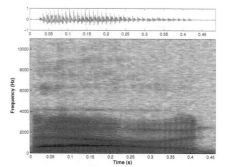

Fig. 4.23: bowed /baʊd/
(In an American Dialect of English)

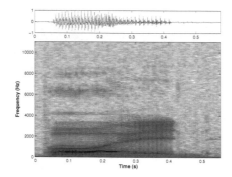

Fig. 4.24: boyd /bəːd/
(In an American Dialect of English)

would have to work much harder to master vowels.

There is an interesting saying attributed to Voltaire[8] which states that in etymology vowels count for nothing and consonants for very little."[3]

Assuming that this is a correct statement, vowels do not carry much speech content information so their variations are tolerated more by individuals speaking the language. This, in fact, makes vowels more attractive from a speaker recognition standpoint in contrast with speech recognition which will suffer from the variability.

As we mentioned in Chapter 3, in terms of spectral information, vowels provide the most discriminative features. This fact could plague speech recognition and cer-

---

[8] Apparently, this is nowhere to be found in any of Voltaire's writings and may have other origins.

tainly help speaker recognition systems.

4.2.4 Evolution of Languages toward Discriminability

The evolution of languages provides some interesting insights regarding speech recognition principally and speaker recognition as a by-product. To understand the evolution of languages and their importance to phonological features that are important for speech and speaker recognition, consider the *Persian* language. It has had three major phases in its life since over 2500 years ago, *Old Persian* [6] (*circa* 550 BCE-200 BCE), *Middle Persian* [4, 16, 17, 22] (*circa* 300 BCE-1000 CE) and *Modern Persian* (*circa* 800 CE-present). *Old Persian* was quite complex in certain linguistic terms (at least in phonology, morphology and syntax).[6] Since the *Persian Empire* (the longest running major empire), for a long time facilitated long distance communications among as many as 28 major nations from Europe and Africa to South and Central Asia, its main language became known to many nations who were not native speakers. In fact, *Flanagan* starts, on the second page of book [5], to talk about the voice communication tours of the time of Cyrus the Great in the sixth century B.C. Persian had to quickly develop into a language that would be easily learned by many foreign nationals.

This was mostly due to the intense commerce across the silk road because the Persian were running most of the major caravans, had created in excess of 1100 caravansaries, created the first major international road (the Kings' road) for commerce and communication and most of silk road passed through their territory.[1, 21] As more nations tried to learn the language, the language evolved to become simpler to learn, utter and understand. *Old Persian* [6] started out with a highly inflected grammar, having 42 possible declensions for each noun (7 cases with 3 genders and 3 numbers)[9]. An example is the gender attribute in old languages such as *French*, *Arabic*, and *German*. French and Arabic have two genders, *male* and *female* for all things including inanimate objects. German has the extra *neutral* attribute, making three genders, similar to Old Persian. In the evolution of the Persian language, this was completely dismantled in Middle Persian and never returned in the modern setting of the language, including the annihilation of all gender information. Middle Persian and Modern Persian have no notion of gender, and in most cases have even lost the number categorization (single, dual, and plural). There is not even any gender in the narrow sense that English has preserved (he and she).

---

[9] Noun Inflections In Old Persian are Nominative, Accusative, Instrumental, Ablative, Genitive, Locative, and Vocative (Avestan which is another old Iranian language has an extra Dative form in addition). The genders are Male, Female, and Neutral. The numbers are Singular, Dual, and Plural.

With respect to the phonetic aspect, this language evolution continued to the point that Persian has one of the most discriminative vowel systems among world languages[10]. For a proof of this concept, see Figure 4.25 and compare it to Figure 4.8 for English. The 6 vowels of Persian are almost at the utmost extremes of the vowel trapezoid. This means that their relative formants are at maximal separation. This maximal distance between the mean values of the relative formant locations allow for larger variances in these formants. If the trapezoid were more crowded or two of the vowels were closer to each other, then speakers would not be allowed to have too much variability in their utterance of each vowel. This lack of variability would cause a more uniform phonation of the vowels, which as we mentioned, are the most feature-rich parts of speech. Since nobody artificially designed Persian vowels to fit this profile, it is interesting to note that language evolution has done that automatically and the vowels have fallen into optimal positions for discriminability.

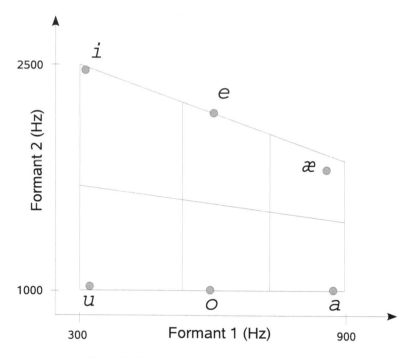

Fig. 4.25: Vowel Trapezoid for the Persian Language

In fact, today, English has the same role as Persian used to have throughout ancient history. It is becoming simplified at such a fast pace that its effects are felt within one life time. The gender in English, because of political correctness

---

[10] For example, Japanese has a similar distribution of vowels with the exception that there is no fine line between /ʌ/ and /æ/ creating a less discriminative set.

is slowly disappearing. Until the late twentieth century, to refer to a third person without knowing the gender of that person, the words [hc] or [his] would have been used. Now, we are in an awkward transition period, where a word is needed that would be neutral, but since [it] and [its] have been reserved for inanimate objects and animals[11], most people use the word [they] and [them]. So, [they] is, in an indirect way, referring to a singular third person. This transition, has come about due to new needs in the language. An older equivalent is the fact that English has lost most of the gender information from its old days when it separated from other Germanic languages such as German and Dutch. This may even be the reason for the success of English in becoming an international language. If it required the extra effort of learning gender information, maybe it would not have been as successful.

The above examples were produced in order to attain an understanding of how fluid languages may be. As long distance communication becomes more prevalent in the world, the number of languages will most likely decrease[12]. Also, as more people speak with fewer languages, those surviving languages will simplify and become easier to recognize. Easy recognition means more discriminability between different sounds. As we mentioned before, the more sounds become discriminable (especially vowels), the more optimal their positions become in the vowel trapezoid in terms of separation. Since vowels arc the most informative forms of articulation for a speaker recognition system due to their resonant content, speaker recognition of these types of languages becomes simpler. This is due to the fact that the smaller the number of vowels and the more optimally positioned the vowels, most of the variability between two speech samples would be because of the variations in speaker differences and not because of the arbitrary positioning of the formants.

4.3 Suprasegmental Features of Speech

Suprasegmental features of speech are those which surpass the segment boundaries – phone and phoneme boundaries. To be able to discuss suprasegmental features, it is important to define different segments in human speech. Phones or phonemes, as we saw earlier, are the smallest linguistic segments of audio depending on the perspective. The next segments are coarticulatory segments which are made up of two or more phones. Higher up in the chain, we have the syllabic segmentation followed by lexical (word-based) segmentation. At a higher level, the segments may go on to any larger segments all the way to a whole utterance.

---

[11] [It] is another one of those references which is, now, only being used for inanimate objects and is not seen acceptable for even animals for whom the gender is known.

[12] The number of living languages according to the 16th edition of Ethnologue [11], published in 2009, is 6909 and according to its 15th edition was 6912.

Any feature of speech that encompasses segments larger than the phonetic segments is called a suprasegmental feature. These features are categorized by type into *prosody*, metrical features and *temporal features*. In the following few subsections, we will be discussing these type of features in some detail.

4.3.1 Prosodic Features

Definition 4.6 (Prosodic Features). *Prosodic features are suprasegmental features which incorporate pitch and its variations and/or loudness and its variations.*

There are two basic categories of prosodic features. We will be discussing the pitch related features first and follow with coverage of the, so called, sonoric features (loudness and its variations).

4.3.1.1 Pitch

Definition 4.7 (Pitch). *Pitch is a perceived quantity which is related to the fundamental frequency of vibration of the vocals cords over some duration.*

Section 5.1.1 presents a quantitative description of pitch. The fundamental frequency of vibration of the vocal cords is, of course, a function of their shape and tension, the air flow characteristics and the vocal tract configuration at the moment of utterance. Therefore, pitch is an averaged concept. The duration over which pitch is averaged is fuzzy and depends on the context in which it is used. An almost instantaneous value of pitch may be used for signal processing. Since speech is a dynamic system and the frequency changes rapidly, averaging is practically unavoidable. An example is when we speak about the pitch of a syllable, we may have in mind, the mean value throughout the syllable or the moving average along the syllable. The vagueness is usually clarified in the description of the context of usage. The term, pitch, may also be used in absolute terms. For example, we may speak about a pitch range, which means all values from a lower bound in the frequency domain to the an upper bound.

Pitch levels and their variations can also be defined at different levels of segmentation. In physiological terms, the speaker has an *organic pitch range* which is from the lowest tone he can utter to the highest. However, most of us do not necessarily utilize our whole organic pitch range in conversational speech. The subrange which we habitually use in conversation is called the *linguistic pitch range*. We are all familiar with how one can change the impression of his/her voice in order to convey a paralinguistic message such as an emotion or a subtle variation in our statement.

For example, we may all recall some instance when an Electronic Mail we have sent may have conveyed a completely different message than intended since it is hard to include the paralinguistic qualities of speech in an EMail. This type of expression includes pitch variation, sonority variations and many other aspects such as body language and facial expressions. The range of pitch modification that we use to convey such paralinguistic messages is called *paralinguistic pitch range*.[9]

Figures 4.26 through 4.29 show examples of such variation in the utterance of a, grammatically classified, imperative utterance of the verb [try] expressed with 4 different paralinguistic messages. The caption explains the different messages. The pitch variation may be seen from location and variation of the formants. A careful examination of the formants shows the additional power shifted to high frequency formants and their longer sustainment. The attitudes start from an expression of dominance in Figure 4.26 and change gradually to a more submissive expression in Figure 4.29. These paralinguistic pitch variations exist in all languages with some differences in their meanings. The paralinguistic pitch variation could be of great importance in speaker recognition since a lot of the content used for producing this variation is at the subconscious level conveying emotion which is usually very speaker-dependent.

Within any one of the segments defined in Section 4.3.1.1 (phone, syllable, mora[13], lexicon, or utterance) there may be a local pitch variation, which is called the *pitch span*. It is identified by a local minimum, a local maximum and a pitch variation.

The linguistic pitch range is usually governed by linguistic constraints in relative pitch variation along different segments. It is usually modulated and modified with reference to the fundamental frequency (pitch) of the speaker's voice (the organic pitch).

When a human tries to separate speakers based on their vocal characteristics, he/she uses global pitch averages to realize the organic pitch of the individual for reference. In fact, when a person is asked to mimic another person's vocal characteristics, the first thing the impostor considers is the modification of his/her average pitch level to match the organic average pitch of the target speaker. Therefore, as a feature for speaker recognition, the mean value of the pitch is not a very reliable source of information for discriminability. In fact, the most popular features used in automatic speaker recognition systems, *cepstral features*, do not retain much of this tonal quality of the audio. This, in fact, allows the automatic speaker recognition algorithms to be more resilient to such impersonations.

---

[13] Mora is a linguistic unit that determines stress in some languages and is based on the first syllable *onset*, *nucleus*, and *coda*, final syllable or long vowels. See Figure 4.34 for the the syllable construct.

A measure of the pitch span across any segment (phone, syllable, mora, lexicon, or utterance) is called the *pitch height* which may be viewed as a moving average of the pitch level within the segment. This value is usually attributed with a fuzzy label of low, mid, high, or some intermediate levels in between. On the other hand, it is considered to vary along the segment of interest in three simple manners, *level*, *falling* and *rising*. Of course the variations may also be formed by any complex combination of these simple modes.

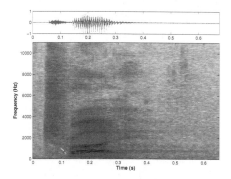

Fig. 4.26: [try] Decisive Imperative – Short and powerful

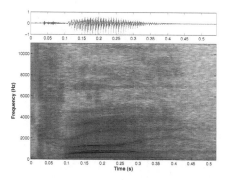

Fig. 4.27: [try] Imperative with a slight interrogative quality – short and an imperative; starts in the imperative tone and follows with an interrogative ending

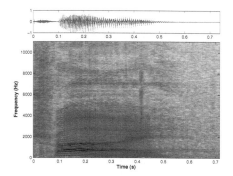

Fig. 4.28: [try] Imperative but with a stronger interrogative quality – longer and the pitch level rises, it is sustained and then it drops

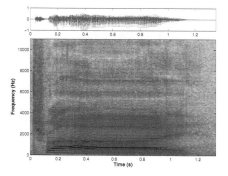

Fig. 4.29: Imperative in a grammatical sense, but certainly interrogative in tone – much longer; the emphasis is on the sustained diphthong at the end with pitch variation by rising, an alternating variation and a final drop

One type of pitch variation is that which spans a whole utterance. In general, regardless of the language, most utterances start with a higher pitch and the pitch gradually lowers. This concept is called *declination*. This is a subtlety that is usually lacking in low quality speech synthesizers giving them the expressionless quality with which we are all familiar.

There is also the local effect of pitch variation to consider. In many Indo-European languages such as English and Persian, pitch variations make little difference in the content of speech with some exceptions. In English, although intonation is present, it does not change the essential meaning of a word. One example in which pitch may change the meaning of an expression in English is the difference of the pitch variation at the end of an interrogative sentence versus an informative sentence. For example, notice the pitch height and pitch level variations for the word, "station," at the end of the following two sentences:
1. Is this your *station*?
2. This is your *station*.
In the interrogative sentence (sentence #1), the last syllable of the word *station* has higher pitch than the occurrence in sentence #2.

This constitutes high level prosodic information which is not quite paralinguistic and does change the linguistic expression. Although this is quite important in a *Natural Language Understanding (NLU)* sense, it is probably not as important in speaker recognition as paralinguistic pitch variation could be. Of course, usually it is hard to separate the two types of pitch variation in an algorithmic sense. Most likely any implementation of a pitch-based feature will include all types of pitch variations. Usually the discriminating information follows from statistical implementations.

Most linguists believe that syllabic and lexical pitch variations do not change the meaning of words in English. However, we know that syllabic and lexical pitch variations are also components of *linguistic stress* along with other concepts such as syllabic and lexical sonority variations and metrical variations. In those regards, since syllabic stress does change the meaning of words in English, then so do pitch variations – see Table 4.3.

Some languages such as different dialects of Chinese, use pitch variation to completely change the meaning of a word. Take Mandarin Chinese for example. Figures 4.30 through 4.33 are spectrograms of 4 different words in Mandarin which have completely disconnected meanings, yet they would all be transliterated as [ma] in the English alphabet.[14] The only major differences between these utterances are temporal variations (length of the vowel representing [a] and the pitch variations of those vowels). The local pitch variation and durations of the vowel may easily be seen in the spectrograms.

---

[14] I would like to thank Mr. John Mu for his utterances of the 4 variations of [ma].

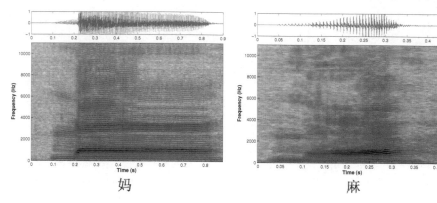

妈

Fig. 4.30: Mandarin word, Ma (Mother)

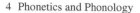

麻

Fig. 4.31: Mandarin word, Ma (Hemp)

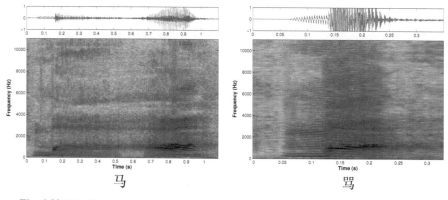

马

Fig. 4.32: Mandarin Word, Ma (Horse)

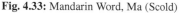

骂

Fig. 4.33: Mandarin Word, Ma (Scold)

Therefore, in *Mandarin Chinese*, intonation becomes important for speech recognition applications [7]. It is unclear if it is the case with speaker recognition. It may only be important in a text-dependent or text-prompted approach. However, since it requires longer sustainment of the vowels, and since vowels are most informative in speaker discrimination, it may present some positive effect. Of course, based on the same argument made for the generic linguistic pitch variations toward the beginning of this section, these pitch variations will blend in with paralinguistic variations in most algorithmic implementations.

Because of the importance of pitch variations in Mandarin, some mandarin speech recognizers use syllables as elementary segments of speech versus the usual phonetic segmentation. This segmentation is called a *syllabic lattice* segmentation. Apparently, there are about 1345 syllables in Mandarin [10].

Ladefoged [8] produces examples of the pitch variation effects for a word in *Cantonese Chinese*, /sɪ/, which can have 6 different meanings according to different pitch levels and their location in the word.

ínsult (noun) versus insúlt(verb)
ínsert (noun) versus insért(verb)

Table 4.3: Examples, in English, where stress changes the meaning of a word

At the most extreme case of the use of pitch in languages, there are whistle-based languages such as the *Spanish Silbo* or *Silbo Gomero* language of the *La Gomera* region of the *Canary islands*. This language is based solely on whistles. Different intonations, lengths, and power variations develop elements of the language which can be used to basically convey anything other languages convey – see Section 4.1.8.

4.3.1.2 Loudness

Definition 4.8 (Loudness (Sonority)). *Loudness (sonority) is a perceived quantity which is a function of the intensity of vibration of the vocals cords over some duration and pitch.*

Since the perceived loudness of audio is dependent on the frequency of sound, it is dependent on pitch as well. See section 5.1.2 for a more quantitative treatment of *loudness*. As in the definition of pitch, this duration is also vague and is dependent on the context.

Consider a wave. It has a frequency and an amplitude. One can think of prosody as related to the variations in mean shapes of the speech wave computed over certain durations. Therefore, most of the concepts described about pitch in Section 4.3.1.2 have a parallel counterpart for loudness. The relationships in Table 4.4 demonstrate these parallel concepts.

Organic Pitch Range	$\Longleftrightarrow$ Organic Loudness Range
Linguistic Pitch Range	$\Longleftrightarrow$ Linguistic Loudness Range
Pitch Height	$\Longleftrightarrow$ *Sonority*
Pitch Span	$\Longleftrightarrow$ *Loudness Span*

Table 4.4: Parallels between Pitch and Loudness Concepts

Similar to the declination of the pitch level discussed in Section 4.3.1.1, the loudness of an utterance also degrades from the beginning to the end of the utterance.

4.3.2 Metrical features of Speech

Other similar suprasegmental features of speech are Rate (overall tempo of the speech) and Continuity (length of stream of speech between pauses). These are metrical properties of speech which are high level features partly dependent on the language of interest and in part on individuals' habits. In fact human perception of speaker identity seems to make use of these high level features. However, implementation of these features in automatic speaker recognition systems is quite challenging.

4.3.2.1 Stress (Relative Loudness)

The difference between *stress* and *intonation* is that stress is the relative loudness of parts of speech where intonation is the variation in the pitch of different parts of speech. Linguists generally believe that there are about 3 to 4 levels of stress in the English language. In most cases, stress does not really change the meaning of words and is more or less associated with the dialect or accent being used. There are some cases where this assumption is not valid – see Table 4.3.

Although Loudness has an inherent pitch component (see Definition 4.8), stress (relative loudness) sometimes has an added pitch variation. This extra pitch variation is called a *pitch accent*. An example of a language which contains a pronounced level of pitch accent is *Turkish*. *Pitch variations* are used to change the stress level of a word mostly due to rhythmic constraints imposed by the language. See Section 4.3.2.2 for more detail.

4.3.2.2 Rhythm

Laver [9] defines *Rhythm* as "the complex perceptual pattern produced by the interaction in time of the relative prominence of stressed and unstressed syllables." Figure 4.34 shows the construct of a typical syllable. In a syllable, only the existence of nucleus is mandatory. All other parts may or may not be present. For example in English, the onset may be nonexistent or contain up to three consonants. The coda may also be nonexistent, but it can go up to four consonants for English. However, there are quite stringent constraints for the allowable consonant sequences which boil down to the language-dependent constraints on co-articulation – see Section 4.3.4.

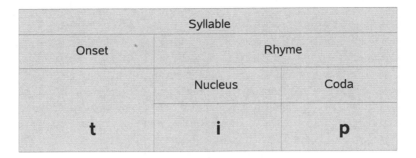

Fig. 4.34: construct of a typical syllable, [tip]

As in the case of pitch and loudness, the levels in lexical word stress are divided into three categories: unstressed, primary stress, secondary stress.[9] Languages such as English and Dutch have a stress-based rhythm. Some languages such as *Japanese*, *Turkish*, and *Telugu*[15] possess special rhythmic patterns. *Japanese* and *Telugu* seem to be very similar in their perceptual basic rhythmic construct although they are from different linguistic families. Turkish also has a syllabic rhythm which is somewhat different, perceptually, from Japanese and Telugu, but it is based on the same concept. These languages are said to have a *mora-based* rhythm.[15]

This inherent requirement for a certain rhythm forces imported words from other languages to take on a rhythmic pattern as well. All Japanese consonants with the exception of the nasal consonant, *n*, have to always be followed by at least one of the 5 basic vowels. Therefore, English words such as [program], [stress], and [software] change to [puroguramu], [sutoresu] and [sofutouea]. Turkish also requires a certain rhythm to be sustained. For example, Persian words such as [hoz] and [bazaar] have changed to [hovoz] and [bazara], when they were imported into Turkish.

Also, the rhythmic constraints of languages sometimes force them to take on special stress levels to keep the rhythm going. In a language such as Turkish, the stress, along with the usual relative sonority, also contains a high level of pitch accent variation (see Section 4.3.1.1) giving the language the musical tonality which no one could miss.

---

[15] A *Dravidian* language spoken in the South of India.

4.3.3 Temporal features of Speech

Some temporal features could also be very essential for a suprasegmental analysis of speech for speaker recognition purposes. Take the *continuity* of speech for example. Different individuals usually portray different levels of continuity in their speech. Although relative continuity is affected by fluency in a language, most native speakers also express different levels of continuity. In certain circumstances, the speaker simply pauses silently. On other occasions and much correlated with the language, the speaker may fill the pauses with *nonsense phonations* such as *mmmm, aaa, err,* etc. These features are quite speaker-dependent and could prove quite valuable for increased speaker discriminability. Of course, there are languages, such as Japanese, in which speakers learn that for social reasons they should not have silent pauses in a conversation and should also not use *mmm*s and *aaa*s. They use specific words in the Language such as [ano], [aso], etc. In this case, the continuity treatment may become similar across different speakers of the language, but still the frequency of such pauses will be a useful feature.

Another concept which may also be useful in speaker recognition is the *tempo* or the rate of speech. Different people have different speaking rates and this may be utilized to tell speakers apart. *Laver* [9] states that the average rate of speech for English is about 5.3 syllables per second for a speaker with a medium articulation rate.

4.3.4 Co-Articulation

Definition 4.9 (Co-Articulation). *Co-Articulation is the phenomenon that binds neighboring phone together and, in the process, modifies the phonation of each individual phone so that a smoother transition may occur in the utterance.*

Co-Articulation effects may happen due to different reasons. The most prominent reason is the set of limitations of the vocal tract. These limitations are both speaker- and language-dependent. The speaker-dependent limitations could be quite helpful in developing features for increased speaker discriminability. An example of language-dependent co-articulation is the fact that in some languages such as Persian and Spanish, certain consonant pairs like /st/ cannot start an utterance. For this reason, when native Persian and Spanish speakers learn languages that do allow for this type of co-articulation, they have to modify the utterance to be able to pronounce it. An example is [student] which is pronounced [estudent] by Persian and Spanish speakers until they learn to mimic the proper delivery of the word.

Another example is the mora-rhythm languages. The underlying reason for the presence of mora-rhythm in some languages such as Japanese seem to be tightly related to co-articulation – see Section 4.34.

References

1. Briant, P.: History of the Persian Empire (From Cyrus to Alexander). Eisenbrauns, Incorporated, Winona Lake, Indiana (2006). URL http://www.sil.org. Translation into English by: Peter T. Daniels, ISBN: 1-57-506120-1
2. Canonge, E.D.: Voiceless Vowels in Comanche. International Journal of American Linguistics **23**(2), 63–67 (1957). URL http://www.jstor.org/stable/1264055. Published by: The University of Chicago Press
3. Cavalli-Sforza, L.L., Cavalli-Sforza, F., Thorne, S.: The Great Human Diasporas: the History of Diversity and Evolution. Basic Books (1995). Translated into English by: Sarah Thorne, ISBN: 0-20-144231-0
4. Christopher J, B.: A Syntax of Western Middle Iran. No. 3 in Persian Studies Series. Caravan Books, New York (1977). ISBN: 0-88-206005-8
5. Flanagan, J.L.: Speech Analysis, Synthesis and Perception, 2nd edn. Springer-Verlag, New York (1972). ISBN: 0-387-05561-4
6. Kent, R.G.: Old Persian Grammar Texts Lexicon. 2nd edn. American Oriental Society, New Haven, Connecticut (1953)
7. Kratochvil, P.: Tone in Chinese. In: E.C. Fudge (ed.) Phonolgy, Selected Readings. Penguin Books, Middlesex (1973)
8. Ladefoged, P.: A Course in Phonetics, 5th edn. Wadsworth, Boston (2006). ISBN: 1-413-00688-4
9. Laver, J.: Principles of Phonetics. Cambridge Press, New York (1994). ISBN: 0-521-45031-4
10. Lee, L.S.: Voice Dictation of Mandarin Chinese. IEEE Signal Processing Magazine **14**(4), 63–101 (1997)
11. Lewis, M.P. (ed.): Ethnologue, 16th edn. SIL International (2009). URL http://www.sil.org
12. Maddieson, I.: Patterns of Sounds. Cambridge University Press, Cambridge (1984)
13. Meyer, J.: Acoustic Strategy and Topology of Whistled Langues; Phonetic COmparison and Perceptual Cases of Whistled Vowels. Journal of the International Phonetic Association **38**(1), 64–90 (2008)
14. Morris,   C.:   Whistling   Turks.   Website   (1999).   URL http://news.bbc.co.uk/2/hi/programmes/from_our_own_correspondent/506284.stm
15. Murty, L., Otake, T., Cutler, A.: Perceptual Tests of Rhythmic Similarity: I. Mora Rhythm. Language and Speech **50**(1), 77–99 (2007)
16. Nyberg, H.S.: A Manual of Pahlavi, vol. I. Otto Harrassowitz, Wiesbaden (1964). Texts, Alphabets, Index, Paradigms, Notes
17. Nyberg, H.S.: A Manual of Pahlavi, vol. II. Otto Harrassowitz, Wiesbaden (1974). Idiograms, Glossary, Abbreviations, Index, Grammatical Survey, Corrigenda
18. Peterson, G., Barney, H.L.: Control Methods Used in a Study of the Vowels. The Journal of the Acoustical Society of America (JASA) **24**(2), 175–185 (1952)
19. Rabiner, L., Juang, B.H.: Fundamentals of Speech Recognition. Prentice Hall Signal Processing Series. PTR Prentice Hall, New Jersey (1990). ISBN: 0-130-15157-2
20. SIL International. World Wide Web. URL http://www.sil.org
21. Sykes, P.: A History of Persia, 3rd edn. Macmillan and Co. Ltd., London (1958)
22. Tafazzoli, A.: Tarikhe Adabiate Iran Pish az Eslam (History of Iranian Literature Before Islam), 3rd edn. National Library of Iran, Tehran (1378 Anno Persico). In Persian, ISBN: 9-64-598814-2

Chapter 5
Signal Processing of Speech and Feature Extraction

Hark! to the hurried question of Despair: "Where is my child?"
An echo answers "Where?"

Lord Byron (George Gordon Byron)
The Bride of Abydos [10] (Canto ii, Stanza 27), 1813

In this chapter we will be reviewing signal processing techniques for speech. It is very important for the reader to have been familiarized with the contents of Chapter 3 and Chapter 24 before continuing to read this chapter. Even if the reader is a seasoned researcher in the field, it is important to at least refer to those chapters for details. To keep the main flow of the book fluid, most of the major technical details have been moved to Chapter 24.

A great majority of the information in this chapter may be used for any speech processing algorithm including, but not limited to, speaker recognition, speech recognition, speech restoration, etc. Later in the chapter, we will introduce and discuss some of the most popular features used for speaker recognition in the literature. It is important to note that there is an interesting duality to most of the features being discussed here. They are mostly used in speaker recognition and speech recognition alike in spite of the theoretical dichotomy that exists between these two branches of speech processing.

As we mentioned in the introduction (see Figure 1.6), the speech signal has an enormous capacity for carrying information, yet it only tends to be used for delivering about 50-bps worth of information in a normal conversation. Therefore, there is a hugely redundant portion of the signal which is not essential for understanding the message that comes across in a conversation. A large part of this information is related to the individual vocal characteristics of the speaker. Therefore, although the job of speech recognition is to try to filter out this irrelevant information from the signal and to convey the actual message being transmitted, speaker recognition systems must do the exact opposite. A speaker recognition system banks on the information-rich portion of the signal (physiological information) to be able to determine the custom characteristics of an individual. If you will, you may view this as filtering out what is being said to be able to determine the identity of the individual. This notion puts the two fields at odds with one-another. Despite all that has been said, still they both use the same features which are apparently designed to on one hand throw away all the information which is related to the speaker characteristics, so that one system can understand all individuals (speech recognition) and on the other hand only concentrate on what makes speakers different (speaker recognition).

In the midst of all the information that is present in the speech signal, for speaker recognition, we would like to be able to filter out some unique features with such information capacity that would be in the order of amount of information necessary to tell apart the whole total population of Earth. Of course even if we did so, since speaker characteristics are really a continuum, by catering to the whole population of Earth, we will still only be able to do well with some individuals who are significantly different from the rest of the population and not with those who are close to the norm. If we assume that speaker characteristics have a Gaussian distribution among all the speakers in the world, it is an immense challenge to separate people who fall close to the mean.

Of course the real world is not as clear-cut and well defined as we have desired in the last paragraph. The features only carry vague statistical information of both kinds and just happen to be the best we know up to now. Also, in actual implementation, to reduce the amount of audio required to perform a speaker recognition task, many systems (text-dependent and text-prompted) require that the speaker says a specific text. This will utilize both aspects of the features, namely, speech content and speaker content.

One main point to keep in mind is the statistical nature of the speech signal. The features being discussed here, are also only significant in a statistical sense. For this reason, text-independent speaker recognition systems require more data to be able to determine the speaker since they only try to dwell on the speaker characteristics and not the content. In reality, even text-independent speaker recognition systems do better if the person says the same or similar text at the enrollment and recognition tasks.

Having discussed the short-comings of the features we will be discussing in this chapter, we can now go ahead and try to do our best with what nature has provided for us to be able to perform both speech and speaker recognition using the same features.

5.1 Auditory Perception

The human auditory perception mechanism distinguishes different audio signals based on three main properties, *pitch*, *loudness* and *timbre*. In this section, we will define these properties and try to assign a quantitative description to them. This is a very difficult task due to the reliance on perception. As it will be made more clear, these are nonlinear relations and rely, heavily, on our subjective observations. In fact, Stevens [71] showed that the pitch of high tones increases with intensity, the pitch of low tones decreases with intensity, and the point where the reversal happens is dependent on the level of intensity. Despite the difficulty in separating the frequency-related and intensity-related parts of perception, since we are concerned about an automatic speaker recognition system, a good quantitative model for these measurements is crucial.

Let us try to recapitulate the process of hearing. Please refer to Section 2.2 for a clear understanding of the role and position of each component of the hearing system. The sound waves are focused into the ear canal by the auricula. Once they enter the canal, some very low frequency sounds are filtered out by the narrow shape of the canal. Then they excite the tympanic membrane, causing it to vibrate, passing its vibrations to the malleus, the incus, the stapes and from there to the inner ear through the *cochlear fenestra ovalis* (oval window of the cochlea). In the *cochlea*, the vibrations are transmitted to an incompressible fluid which fills the *scala* (the spiral part of the cochlea).

This motion excites a row of hairs at the inner lining of the scala. The motion of these hairs is transmitted to a bundle of neurons and the signal is carried to the brain for cognition. The way this part of the process works is described by the basic *place theory* in physics which says that depending on each frequency component, the hairs in a different location along the scala are excited. This means that the cochlea is actually working like a spectral analyzer somewhat in the same manner as we have been plotting spectrograms in this book. To achieve a higher resolution for distinguishing finer differences in the frequencies of the perceived signals, the brain provides feedback to the three rows of hair at the outer extreme of the curvature of the scala. This feedback sets the hairs related to specific frequencies which move the fluid inside the cochlea at a similar frequency as being processed as incoming audio. This will create some resonance, amplifying sounds in a small window apart from the central frequency being fed back. This amplification allows for a better perception of frequencies in the vicinity of that central frequency. This works as a fine-tuning mechanism which allows us to be so remarkably accurate in our perception of audio signals.

The high frequency sounds are picked up by the cilia at the beginning of the scala and the endpoint, the helicotrema, captures the lowest frequency sounds [70] Loud noise can impair our ability to hear the high frequency sounds by damaging the cilia at the beginning of the scala.

As we are younger, our ability to hear high pitched sounds is far greater. In fact, the youth can often hear the high frequency pitch coming from devices which are designed to keep pests such as mice and insects out. I have had a first hand experience with this when I was around 20 years of age. I could hear the high frequency pulse emitted by such a device in the warehouse of a company for which I was writing computer programs to do inventory control. Everyone else in the company was older by at least 10 years and could not hear the high pitched noise. Later, I heard of experiments in which someone had created a similar device to keep teenagers away from the front of a store where they were crowding at times. Of course, this device was experimental and did not meet constitutional guidelines and was later removed, but it did demonstrate our sensitivity to higher frequencies as youngsters. Therefore, quantifying pitch, loudness and timbre is quite hard and the analysis is very subjective even across different people depending on their age group and ac-cumulative experiences (such as subjection to loud noises). However, since we do need to make this quantification, we must proceed.

5.1.1 Pitch

We defined pitch in Section 4.3.1.1, Definition 4.7, as a perceived quantity related to frequency of vibration. Now, we will analyze this relation in more detail.

In terms of perception, most people do not have the ability to recognize specific pitch values. They can only recognize pitch variations. In fact, according to Ross-ing [62], only 1 in 10,000 people can do a decent job of recognizing absolute pitch values. This is in contrast with about 98% of the population who can recognize spe-cific colors without having a reference to which they can compare. This is why pitch is usually quantified by the amount it differs from a more absolute reference.

Steinberg [70] studied the sensitivity of the cilia at the different parts of the cochlea to different frequency sounds. He conducted his experiments using tones with a 60 *dB* loudness level (see Section 5.1.2) in the range of 125 Hz to 23,000 Hz. He concluded that the 125 Hz tone affected the cilia at the helicotrema. The length of the spiral in his experiment was 30mm. The results showed the tone of 1,000 Hz was detected at 10mm from the helicotrema ($\frac{1}{3}$ of the way) and the tone of 4,000 Hz was detected at 20*mm* ($\frac{2}{3}$ of the way) from the helicotrema. He plotted the relation between the distance away from the helicotrema and the frequency of the perceived tone. The plot which was done on log paper seems almost linear which is in tune with results reported by *Stevens* [75] in the same issue of the Journal of the Acous-tical Society of America in 1937. *Stevens* [75] introduced a scale called the *Melody* (*Mel*) scale which is related to the distance away from the helicotrema. In fact visual comparisons of the plot by these two papers show great likeness. Steinberg plotted the tone frequency versus the distance in mm from the helicotrema and Stevens plot-ted melody versus the frequency.

5.1.1.1 Melody (Mel) Scale

Definition 5.1 (Melody (Mel)). *Mel, an abbreviation of the word melody, is a unit of pitch. It is defined to be equal to one thousandth of the pitch ($\wp$) of a simple tone with frequency of 1000 Hz with an amplitude of 40 dB above the auditory threshold.*

The above definition is based on the experiments done by Stevens, Volkman and Newman in late 1930s. The results were published in 1937 [75] and 1940 [74]. Equation 5.1 shows the relation between frequency (f) in Hz and pitch ($\wp$) in Mels. This equation is due to O'Shaughnessy [57] who fit an equation to the data points reported by *Stevens* and *Volkman* [74]. Since the work of [74] was based on experiments, the results were presented as a graph relating the pitch to frequency. Fant, introduces another equation to estimate the relationship reported by [74]. [20] This relationship is given by equation 5.2. [83] has compared these two relations to a few others and has come to the conclusion that the best fit for the frequency range over 1000 Hz is not necessarily logarithmic, nor is it necessarily linear for frequencies below 1000 Hz, but may be estimated as such. Table 5.1 shows the estimated values of some points extracted by [83] from the graph of pitch versus frequency of [74].

$$\wp = \frac{1000}{\ln(1 + \frac{1000}{700})} \ln(1 + \frac{f}{700}) \tag{5.1}$$

$$\wp = \frac{1000}{\log(2)} \log(1 + \frac{f}{1000}) \tag{5.2}$$

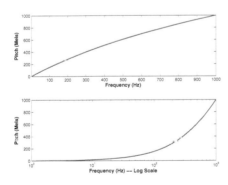 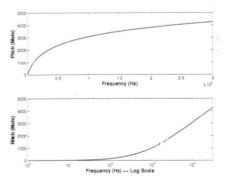

Fig. 5.1: Pitch versus Frequency for frequencies of up to 1000 Hz

Fig. 5.2: Pitch versus Frequency for the entire audible range

f (Hz)	$\wp$ (Mel)	f (Hz)	$\wp$ (Mel)	f (Hz)	$\wp$ (Mel)
40	43	867	928	4109	2314
161	257	1000	1000	5526	2600
200	300	2022	1542	6500	2771
404	514	3000	2000	7743	2914
693	771	3393	2142	12000	3228

Table 5.1: Frequency versus Pitch

5.1.1.2 Bark Scale

Zwicker [102] approached the problem of pitch perception as one that may be subdivided into several regions along the frequency domain. This is basically a filterbank modeling of the hearing system (cochlea). In his studies, he subdivided the space of frequencies into 24 basic critical bands [103], where each band has a center frequency and a bandwidth associated with it. These 24 bands start with a center frequency of 50 Hz and go up to the center frequency of 13,500 Hz for the 24th band which gives a total bandwidth of 20 Hz to 15,500 Hz based on the individual bandwidths presented in the paper. As we have seen before, the frequencies of 1000 and 4000 are important in the location of the cochlea that they excite [70]. Therefore, he assigns bands number 3, 9 and 18 center frequencies of 250, 1000 and 4000 respectively. Zwicker proposed a unit of "Bark" in memory of Bark Hausen who is attributed with the creation of the loudness level. Here is his definition for a Bark.

Definition 5.2 (Bark). *One Bark corresponds to the width of one critical band over the whole frequency range and corresponds to nearly a 100 Mel pitch interval.*

As there have been equations fit to the Mel scale results, researchers have also come up with equations that fit the table provided by Zwicker. One such relation is due to Traunmüller [79] where the center frequency is given by Equation 5.3 in Barks where f_c is the center frequency in Hz.

$$\mathscr{B}_c = \frac{26.81 f_c}{1960 + f_c} - 0.53 \quad \text{(in Barks)} \tag{5.3}$$

Then the inverse is given by,

$$f_c = \frac{1960(\mathscr{B}_c + 0.53)}{26.28 - \mathscr{B}_c} \quad \text{(in Hz)} \tag{5.4}$$

The Bandwidth for each center frequency $\mathscr{B}_c$ is given by,

$$\mathscr{B}_w = \frac{52548}{\mathscr{B}_c^2 - 52.56 \mathscr{B}_c + 690.39} \tag{5.5}$$

Figure 5.21 shows the relationship among the log(Critical Frequency in Hz), Bark, and Mel. The horizontal axis depicts the 24 critical band center frequencies in Barks. The stem plot shows the corresponding Mels for each critical center frequency with the corresponding labels at the left vertical axis. The right vertical axis corresponds to log(Critical Frequency in Hz).

5.1.2 Loudness

In Section 4.3.1.2 we defined the concept of *loudness* (Definition 4.8) to be a function of both intensity and pitch. We have already covered the quantitative analysis of pitch. Here, we will define the *Intensity* of speech followed by the quantitative analysis of *loudness*, in terms of *intensity* and *pitch*.

As its name suggests, *intensity* is a measure of power of the wave. It is actually the power per unit area which may be presented in the units of $\frac{W}{m^2}$ ($\frac{J/s}{m^2}$ or $\frac{N}{ms}$). We mentioned that our ears have different sensitivity at different frequencies. One important frequency value which is perceived at $\frac{1}{3}$ of the length of the cochlea away from the helicotrema is 1000 Hz.[70] This value is also important because it is the frequency threshold between a linear perception and a close-to logarithmic perception of the frequencies. To measure intensity, a standard threshold of hearing has been established based on the order of magnitude of the average intensity threshold of hearing.

Intensity, I, is proportional to the square of the pressure differential, P, and inversely proportional to the *specific acoustic impedance* of the sound medium, ζ. ζ is a product of the density of the medium and the speed of sound in that medium. Therefore,

$$\zeta = \rho c \tag{5.6}$$

Then the relation between Intensity, I, and the pressure differential, P, is given by the following,

$$
\begin{aligned}
I &= \frac{P^2}{\zeta} \\
&= \frac{P^2}{\rho c}
\end{aligned}
\tag{5.7}
$$

The units of ζ may be computed as follows,

$$
\begin{aligned}
\text{Unit of } \zeta &= \left(\frac{kg}{m^3}\right)\left(\frac{m}{s}\right)\left(\frac{s}{s}\right) \\
&= \frac{kg m}{s^2}\frac{s}{m^3} \\
&= \frac{N s}{m^3}
\end{aligned}
\tag{5.8}
$$

At the pressure of 1 atm ($1.01325 \times 10^5 Pa$), and room temperature ($20°C$), the density of dry air is,

$$\rho = 1.204 \ \frac{kg}{m^3} \tag{5.9}$$

and the speed of sound is,

$$c = 343.2 \ \frac{m}{s} \tag{5.10}$$

so the *specific acoustic impedance* may be computed as follows,

$$\begin{aligned} \zeta &= \rho c \\ &= 1.204 \times 343.2 \\ &= 413.21 \ \frac{Ns}{m^3} \end{aligned} \tag{5.11}$$

According to Fant [20], the root mean squared value of the pressure threshold (value at the frequency of 1000 Hz) is,

$$P_0 = 2 \times 10^{-5} \ \frac{N}{m^2} \tag{5.12}$$

Therefore, the *standard intensity threshold* may be computed by Equation 5.7 to be,

$$\begin{aligned} I_0 &= \frac{P_0^2}{\zeta} \\ &= \frac{(2 \times 10^{-5})^2}{413.21} \\ &= 9.68 \times 10^{-13} \ \frac{W}{m^2} \\ &\approx 10^{-12} \ \frac{W}{m^2} \end{aligned} \tag{5.13}$$

Therefore, Equations 5.12 and 5.13 are respectively the amount of pressure and Intensity at 1000 Hz that would be enough for us to hear that tone.

The Pressure, P, is the closely related to the amplitude returned from an analog microphone and Intensity, I, is related to the concept of *Loudness*. Given a tone with an Intensity of $I\frac{Watts}{m^2}$, we can compute the *relative intensity*[1] of this tone in deci Bells as compared to the *standard intensity threshold*. For convenience we refer to this *relative intensity*(I_r) as simply the *intensity* of the sound wave which is in dB, not to be confused with the absolute Intensity of the wave, I, which is in $\frac{Watts}{m^2}$. I_r is given by the equation for *deci Bells* (dB) in the following form,

$$I_r = 10 \log \left(\frac{I}{I_0} \right) \tag{5.14}$$

---

[1] The Relative Intensity is a dimension-less number.

Note that,

$$I_r = 10 \log \left(\frac{P^2}{P_0^2} \right)$$

$$= 20 \log \left(\frac{P}{P_0} \right) \tag{5.15}$$

The maximum intensity of hearing is at frequencies between 3000 to 4000 Hz which is coincident with the resonant frequency of the ear canal that is used to amplify sounds. It is important to note that the ear is less sensitive to intensities in the lower frequencies. The human ear can comfortably hear sounds with intensities of about $10 - 80$ dB. To get an idea about the meaning of a dB scale of loudness, a quiet library contains intensity levels of about $40 - 60$ dB and a loud rock concert can easily produce sounds of over 110 dB.

Now that we have established a quantitative definition for sound intensity, we can attempt a definition for loudness. As we saw in the previous section, pitch was related to loudness as well. Psychological tests have been conducted to assess the loudness concept in human hearing. This has been done by graphing the so called, *Equal Loudness Curves*. These curves plot loudness in terms of a unit called a *phon* on a plot of intensity versus frequency. These curves have a similar shape and relate intensity and pitch to loudness for different loudness levels.

The following two units of loudness were proposed by Stevens [72] based on his experiments.

Definition 5.3 (Phon). *A Phon is the unit of loudness which is equal to the loudness associated with the intensity of 1 dB at 1000 Hz. i.e, It is the dB level of intensity for 1000 Hz tones.*

Definition 5.4 (Sone). *A Sone is equivalent to 40 phons.*

The reason 40 phons are used is that 40 dB is what gives the general perception of doubling the loudness of a sound wave. Therefore, each sone would amount to double the loudness. *Phon* and *sone* are not SI standards. They are only used as convenient measures of the perceptual level of loudness.

5.1.3 Timbre

Timbre is a musical term which is associated with the harmonic content of the audio plus the dynamic characteristics of the audio such as complex modulations and rise and fall of the signal. In human speech, the harmonic content is related to the speaker-related locations of formants and their characteristics. The complex modulations could for example be frequency or amplitude modulations which in music

terms would be concepts like *vibrato* and *tremolo* respectively. If we wanted to categorize pitch, loudness and timbre in relation to speech, I would say that pitch and
loudness (both relative) provide the information that reflects the content of speech
and timbre would be the speaker characteristics. Therefore, in general, we are trying
to distinguish the different timbres when we do speaker recognition. Unfortunately,
this definition is quite broad and it encompasses the whole speaker recognition discipline, so aside from completing the total information we have about sound as a
complement of pitch and loudness, it does not help us in our quantitative analysis of
speaker recognition.

5.2 The Sampling Process

Let us begin where we left off in Chapter 3 when we spoke about the sampling of
the speech signal. Although, there is some freedom in choosing the order in which
the samples are generated from the analog signal, there is evidence that a certain
order would give better results. Figures 5.3 though 5.5 show three of these possible
combinations. When we design our sampling process, we should keep in mind the
type of errors which were discussed in detail in Chapter 3.

Fig. 5.3: Block Diagram of a typical Sampling Process for Speech – Best Alternative

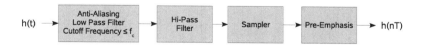

Fig. 5.4: Block Diagram of a typical Sampling Process for Speech – Alternative Option

Fig. 5.5: Block Diagram of a typical Sampling Process for Speech – Alternative Option

5.2.1 Anti-Aliasing

The first matter to anticipate is the aliasing issues discussed in Section 3.6.2. To avoid aliasing, the analog signal (output of the microphone) should pass through an anti-aliasing filter which is really a low-pass filter with a cutoff frequency which is less than the folding frequency, f_c which is defined as $\frac{f_s}{2}$. Notice that we are not calling this the Nyquist Critical Frequency since the Nyquist Critical Frequency is definitely higher than the rates which are customary for doing speaker recognition. See Section 3.4 for more information regarding this difference. Once the anti-aliasing filter is done, the new analog signal has lost most of its high frequency components. Of course, as we mentioned in Section 3.6.5, we have to be careful to use our resources wisely in order to be able to retain as much of the spectral information as possible. Figures 5.3 though 5.5, all, include the anti-aliasing filter at the same location. This is one of those blocks which has a mandatory location. Of course, note that most of the sampling process is usually done by an independent software which is contained in the sound system driver of the microphone system being used. This is true for both computer and telephony based interfaces.

5.2.2 Hi-Pass Filtering

In the best scenario, which is represented in Figure 5.3, there is a hi-pass filter immediately after the low-pass anti-aliasing filter. The reason for doing this hi-pass filtering is that all microphones are not created equal! They will have different Direct Current (DC) components which are not necessarily tied to the speech content being provided by their signals. A hi-pass filter with a low cut-off frequency will remove this DC offset and allow less variability and microphone dependence across different platforms and configurations.

If this hi-pass filter is not provided by the sampling software being used, it may be placed right after the sampler. However, it is best to keep it toward the beginning of the process and acting on the analog signal before entering the sampler. In Figures 5.3 and 5.4 the hi-pass filtering is done before the sampling and in Figure 5.5, afterwards. Of course, it is also possible to combine the low-pass anti-aliasing filter and the hi-pass filter into a single band-pass filter.

5.2.3 Pre-Emphasis

The next step is pre-emphasis. Figure 5.6 shows the power spectral density of the speech waveform that was previously represented in Figure 3.4, however, here it is

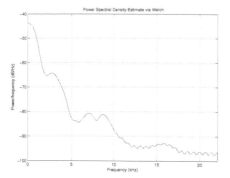

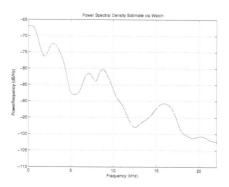

Fig. 5.6: The power spectral density of the original speech signal sampled at 44100 Hz using the Welch PSD estimation method

Fig. 5.7: The power spectral density of the pre-emphasized speech signal sampled at 44100 Hz using the Welch PSD estimation method

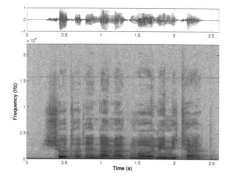

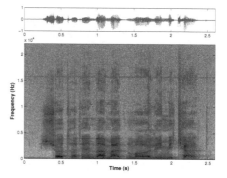

Fig. 5.8: The spectrogram of the original speech signal sampled at 44100 Hz

Fig. 5.9: The spectrogram of the pre-emphasized speech signal sampled at 44100 Hz

sampled at 44100 Hz. Notice the sharp drop in the power of the signal at higher frequencies. [17] estimates that about 80% of the power is contained within frequency components below 1000 Hz. From 1 kHz to 8 kHz, the power drops at a rate of about $-12 \, dB$/Octave[2] and it is almost negligible in frequencies higher than 8-kHz.

However, as stated in Chapters 2 and 4, the cochlea utilizes a fine-tuning mechanism based on feedback from the brain that amplifies special frequencies. It is noted that the human ear can easily recognize these low energy regions. Since we are designing an automatic speaker recognition system, we need to do something similar, to be able to utilize the important features embedded in higher frequencies such as

---

[2] An Octave is the amount of increase in the frequency that would make the new frequency double the original.

fricatives, etc. This can be achieved through pre-emphasis. One method which has been used quite often is a differentiator (which is a single zero filter – see Definition 24.44) with the following transfer function,

$$H_p(z) = 1 - \alpha z^{-1} \tag{5.16}$$

The most popular range of values for α is between 0.95 and 0.97, although values in the range of 0.9 and just less than 1.0 have also been used in different systems. Figure 5.7 shows the power spectral density of the pre-emphasized signal using $\alpha = 0.95$. Notice that the absolute power for each frequency range has been reduced, but the relative power is now better distributed along the different frequencies. To get a better feel of this process, compare Figures 5.8 (original signal) and 5.9 (pre-emphasized signal). You will notice that the high frequency part of the spectrum is significantly more prevalent. Also note the amplitude distribution in the waveform of the signal above each figure.

It is important to note that if the speech of interest (or any other audio) happens to lie below around 200-Hz, the pre-emphasis creates unwanted attenuation that may hurt the performance of speaker recognition. For normal speech, this should not be a problem. However, if the techniques which are discussed in this book are used for audio recognition from a non-speech source, such as music, audio signatures of machinery and nature or similar sources, this type of pre-emphasis may not be a good idea. More sophisticated pre-emphasis filters may be designed for those special cases, which would affect the higher frequencies and spare the low frequencies.

It would be very beneficial if we could do the pre-emphasis before the sampling block. This goes back to the quantization error which we discussed in Section 3.2. If we wait and do the pre-emphasis after the sampling block as Figures 5.4 and 5.4 suggest, the energy of the high frequency components of the audio signal may be so low that an increased relative amount of quantization error may be added to some of the high frequency information or at best the resolution may be reduced. On the other hand, if the pre-emphasis is done on the analog signal before sampling takes place (Figure 5.3), then better resolution will be attained on the high frequency components.

5.2.4 *Quantization*

We started talking about quantization in Chapter 3. Since we are dealing with digital representation of the samples, the magnitude of the samples will have a *dynamic range* as well as a *resolution* associated with it. These are usually functions of the application as well as the hardware being used. We mentioned the logarithmic nature of the human perception associated with loudness as well as pitch and spoke about the intensity component of loudness earlier in this chapter. According to [17], the average intensity of speech is about 58 dB (male speakers are about 4.5 dB louder than female speakers). However, this is an average and there is great variability across speakers, within any speaker and across different tasks. One important aspect to remember is that if we map the sample values to a linear digital scale, then we have to make sure that the resolution for that scale is enough to be able to represent the extreme ends. At one end, we have to worry about quiet speakers and at the other, we will have to worry about saturation issues. Normally, most linear quantization is done at a 16 bit resolution meaning that the values could vary from -32768 to 32767. However, these digital numbers require some normalization to happen in order to make sure that we do not saturate the scale. Anyone who has ever recorded his/her audio on tape is familiar with the saturation issue.

One method of handling the saturation issue is to utilize a dynamic sound-level or volume estimation technique. The mapping may then be adjusted quickly to make sure it is optimal for the specific instance. Section 18.2 talks about this in more detail. Also, in Section 5.2.3, we mentioned that it would better to have the pre-emphasis done on the analog signal so that the high resolution information containing low energy does not get lost due to quantization issues.

One may ask, "Why shouldn't we use much higher quantization levels?" The answer lies in the complexity that brings about. Aside for the need for more sophisticated sound digitization systems, there are other difficulties associated with using higher resolutions. One main problem would be the exponentially increasing amount of data that has to be processed for higher resolutions. As we increase the quantization resolution, the increased amount gets multiplied by the total number of samples being processed, generating higher demands on memory and processing capabilities.

One compromise has been the use of the close to logarithmic perception of volume in humans. Many techniques such as the ITU-T recommendations in ITU-T G.711 [36] use logarithmic mappings (utilizing μ-law and a-law algorithms) which increase the resolution of the volume at lower values in terms of bits and decrease the number of bits needed to represent larger changes in the volume. This is a standard in the coding of telephony audio which allows resolutions of only 8 bits to be used for the same dynamic range as a 16-bit linear scale. Other similar methods include many flavors of the, so called, Adaptive Differential Pulse Code Modulation (ADPCM) which code the difference between adjacent samples of a μ-law or a-law PCM with some extra steps in making estimates of a varying quantization step to be able to use only 4 bits per channel of audio.

5.3 Spectral Analysis and Direct Method Features

In this section, we will set up the problem of feature extraction of the speech signal for speaker recognition. These features, as we shall see, are essentially the same features that are used for speech recognition. At this point of the research into the speaker recognition problem, the most popular features are the Mel-Frequency Cepstral Coefficients (MFCC). There are many variants of these features with different normalization techniques, but the essence remains the same. There are also several approaches to computing the MFCCs. The first approach is the one shown in this section which is based on the usage of spectral estimation, Mel-scale warping and cepstral computation. This method utilizes the Discrete Fourier Transform which was shown in Section 24.14 to be a special case of the z-transform. z-transform is based on an infinite series involving an infinite number of zeros – see Definition 24.44. The special case of Fourier Transform is an infinite set of zeros which are all located on the unit circle. Moreover, the Discrete Fourier Transform which, as we shall see, is the method used for the computation of the MFCC features, is a finite series approximation with N zeros. Therefore, it is known as the *all-zero, moving average (MA)* model or the *direct model*. The next few sections show other approaches to computing cepstral features and spectral features such as *linear predictive coding* (Section 5.4) and *perceptual linear prediction* (Section 5.5).

We can view the human speech production system as a control system with the plant dynamics being related to the vocal tract characteristics and the controller being related to the constriction and air flow mechanisms discussed in detail in Chapter 4. Figure 5.10 shows the control system block diagram based on this perspective. Of course, in the real system, there are many different types of disturbances and feedback loops. We have simplified the number of disturbances with the understanding that the transfer functions may include the effect of disturbances intrinsically. Also, any feedback control system may be unraveled and written in the form of an open loop system where the inputs get modified to include the feedback dynamics. This is what we have done to simplify the control flow represented in Figure 5.10.[22]

In the figure, $U(s)$ is the Laplace transform of $u(t)$ which is the input coming from our brain to drive the speech production system into producing a specific segment of speech. The transfer functions for our nervous system (G_b) and motor control (G_m, the neuro-muscular dynamics) are combined and labeled as G_c. This block is analogous to a classical controller. The transfer function of the plant which is being controlled is G_v. This represents our vocal tract characteristics and dynamics.

The output of the vocal tract is represented by $H(s)$ in the Laplace domain, which is the Laplace transform of $h(t)$, the sound waves that are uttered. For an automated system (speaker or speech recognition system) we should capture the audio, usually using a microphone, and either store it for future processing or transport it to it final

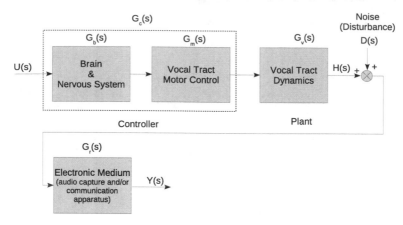

Fig. 5.10: Block diagram of the human speech production system viewed as a control system

destination where the actual recognition occurs. As the audio, $h(t)$, is propagated through its medium (usually air), other noises (disturbances) may be added to it such as environmental noises. In the Laplace domain, we are using $D(s)$ to represent these additive disturbances ($d(t)$ in the time domain). Finally, the combination of the audio and noise are modified by the audio capture and communication transfer function and the output, $Y(s)$ in the Laplace domain ($y(t)$ in the time domain) is produced.

Let us derive the relationship between the input, $U(s)$ and the output, $Y(s)$,

$$H(s) = G_v(s)G_c(s)U(s) \tag{5.17}$$

and

$$Y(s) = G_r(s)[H(s) + D(s)] \tag{5.18}$$

Therefore, combining Equations 5.17 and 5.18,

$$Y(s) = G_r(s)[G_v(s)G_c(s)U(s) + D(s)] \tag{5.19}$$

Our goal is to observe the output $Y(s)$ and try to identify the system. Speech recognition engines try to identify the G_c portion of the system. To them, G_v is considered to be a disturbance. In the perspective of text-independent speaker recognition engines, G_c is a disturbance and they are only interested in identifying G_v from the $Y(s)$.

The system identification problems of speech and speaker recognition are normally reduced to parameter estimation problems, by making certain assumptions about the transfer functions of interest (G_c and G_v respectively). The assumptions are somewhat different for the two disciplines. In speech recognition, also, some assumptions are made to be able to model $U(s)$ based on neighboring observations. That is precisely what language modeling does for speech recognition. Text-independent speaker recognition on the other hand tries to look at the problem much in the same way as a classical system identification is done on a control system with an impulse as the input and looking at the impulse response of the system. Therefore, in general text-independent speaker recognition engines, look for a more overall estimation of the system parameters.

Of course, there are also text-dependent and text-prompted speaker recognition engines that look at the system identification problem by limiting the type of input, $U(s)$ such that now they would be estimating the parameters for the combination of the control and plant blocks, namely $G = G_v G_c$. That makes the job of text-dependent speaker recognition much easier, since one of the hardest problems is the separation of G_v from G_c. Also, since the input is limited to a small set, the task of parameter estimation becomes much simpler as well. Since G_c and G_v are really quite nonlinear in nature, reducing the set of valid control inputs is tantamount to the process of linearization of a nonlinear system along a nominal path – see [22] for such linearization examples in classical control.

At this point, let us simplify the problem in Figure 5.10 by observing the signal before G_r is applied to it and even before $D(s)$ is added to it. Therefore, we will be observing $h(t)$.

In Section 24.9 we showed that the Complex Fourier Transform is a special case of the Laplace Transform where $s = i\omega$. If we assume that the amount of damping in the speech production system is negligible, then the Laplace transforms of Figure 5.10 may be converted to Fourier transforms.

In the Fourier transform domain, the expression for $H(\omega)$ will be as follows,

$$H(\omega) = G_v(\omega)G_c(\omega)U(\omega) \tag{5.20}$$

The transfer function for this system will then be,

$$G(\omega) = \frac{H(\omega)}{U(\omega)}$$
$$= G_v(\omega)G_c(\omega) \tag{5.21}$$

Here we are assuming that we can observe $H(\omega)$. Keep this control analogy in mind since we shall return to it in order to clarify certain choices that are made such as the choice of features.

Now that we have identified the speaker recognition problem as the identification of $G_v(\omega)$, we can proceed with the design of our features. Since we are assuming that we have the Fourier transform of our signal, we should convert the signal from the time domain to the frequency domain of the Fourier transform. At this point, it is quite important for the reader to be completely familiar with the contents of Chapter 24. As we mentioned earlier, to make the material in this chapter flow more smoothly without having to digress into proving theorems and stating definitions of related transforms, the core of the mathematical content has been moved to Part IV, at the end of the book.

As we mentioned in Section 24.12, the speech signal is quite dynamic. That means $G_v(\omega)$ is changing quite dramatically since it is nonlinear and its nominal state depends on the state of $G_c(\omega)$. $G_c(\omega)$ is also quite nonlinear and its nominal state is dependent on $U(\omega)$. Therefore, either we have to model a tremendously nonlinear system, or as it is done in most treatments of nonlinearities, we can linearize the system around a nominal trajectory. This is the control systems approach to the problem. The statistician's approach is worded in a slightly different way, but it really amounts to the same meaning.

Statistically speaking, as we mentioned in Chapter 3, the speech signal is a non-stationary signal. Therefore, we would like to consider portions of the signal around a specific instance of time to have a more stationary snapshot of the total signal. This was the motivation for the development of the STFT and its discrete variants, DTSTFT and DSTFT – see Section 24.12.

5.3.1 Framing the Signal

The STFT has been used throughout the previous chapters to provide us with the spectrograms which have been so helpful in elaborating our discussions. However, doing a complete STFT is not feasible for the process of feature extraction. Alternatively, a limited version has been developed, which does not compute the STFT for all the points in the time domain. Still, a windowing is done to isolate a portion of the signal and a DFT is performed on that windowed signal much in the same way as in the STFT. But only one evaluation is done per portion of the signal at the window location pertaining to the first time instance.

In an STFT, when it comes to shifting the time instance for which the STFT is performed, a single sample is shifted such that there are STFT values available for all time instances. In the speech spectral analysis of the speech signal used for recognition, it is assumed that although the speech signal is non-stationary, but it is somewhat stationary for the duration of the shift. Therefore, the window of interest moves at a much larger step size, assuming that if the window were moved by a single sample, not much change would be seen in the results of the DFT.

The window of interest is, therefore, moved by an amount in the same order of magnitude as the window width, but slightly smaller, so that there is some overlap among consequent portions of the signal. As we have done in our treatment of the DFT, we pick the window size to be N samples. Let us define the shift period to be Ξ samples. Each N-point portion of the data which is isolated to be processed by the DFT is called a *frame* and the act of picking these semi-stationary portions of the data to be spectrally processed is called *framing*. Let us assume that the index of each frame is denoted by l. The index, l is used on the left of any variable that it modifies. Of course, since the length of the signal could theoretically be infinite (e.g., a streaming audio source), there is no upper limit for l. The absolute index of a frame will then be the index of its first sample point which would be Ξl. The frame is then shifted every time by an amount of Ξ samples.

It is quite important to choose a good window size, N, and a corresponding shift period for the processing, Ξ. At first glance, one may think about requiring the window size to be in the order of about 80 ms which is the average length of a phone. However, the average has been inflated because of the relatively lengthy aspect of vowels. On the other hand, stops are quite short, in the order of 5 ms. Therefore, if the window size were to be 80 ms, the effect of stops would be almost missed, in the presence of adjacent long vowels. To account for the great variation, most systems use values of N to be such that the window width would be about $20-30$ ms. By the same argument, to make sure that the *onset* and *offset* of the phone are captured, the frames are usually shifted by about 10 ms.

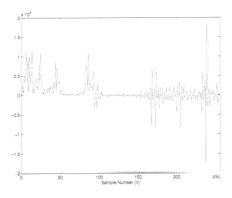
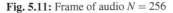

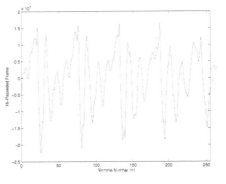

Fig. 5.11: Frame of audio $N = 256$

Fig. 5.12: Hi-Pass filtered Frame $N = 256$

5.3.2 Windowing

Once a frame of speech at time instance, $l\Xi$, and for the duration, N, has been identified, it is sent to the spectral analyzer and feature extractor. Therefore, our final
goal is to have a feature vector, $_l\mathbf{x}$ associated with every frame of audio at time instance $l\Xi$, where $l \in \{0,1,\cdots\}$ and is used as a left index to avoid the confusion
with sample indices.

The windowing process is the act of multiplying the N samples of the signal by a
window as defined in Section 24.12.2 – see Equation 5.22. Some popular windows
are Hamming, Hann (Hanning), Welch, Triangular, Gauss, Blackman and Bartlett.
There are many more windows, each with a slightly different feature. Some windows taper off (e.g., Hann) to zero and some do not (e.g., Hamming). According
to our convention of denoting the DSTFT (Section 24.12.2), we will denote the nth
sample of the lth frame by $_lh_n$ and $_lH_k$ depending on if we speak about the sample
or its DFT.

$$_l\tilde{h}_n = {}_lh_nw(n) \quad for \ n \in \{0,1,\cdots,N-1\} \tag{5.22}$$

One practical point to consider is that just like the STFT, at the start and end of
the sample sequence, when there is not enough data to cover the whole N-sample
period of a frame, the data is padded with zeros to make the total number of points
N. Naturally, when processing the beginning of the data stream, the beginning of
the sequence is padded with zeros, and when at the end, the end of the sequence is
padded.

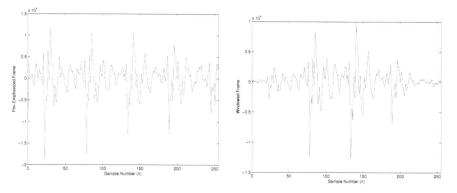

Fig. 5.13: Pre-Emphasized Frame of audio $N = 256$

Fig. 5.14: Windowed Frame $N = 256$

5.3.2.1 Hamming Window

The Hamming window is by far the most popular window used in speech processing. Equation 5.23 presents the N-point Hamming window and Figure 5.15 shows the Hamming window with $N = 64$ plus its spectrum. One reason for the popularity of the Hamming window is the fact that its spectrum falls off rather quickly, so it allows for better isolation. However, its, so called, side-lobes (higher harmonics) stay quite flat and it covers most of the spectrum. Although, it is still popular mostly due to legacy.

$$w(n) = 0.54 - 0.46 \cos\left(\frac{2\pi n}{N-1}\right) \tag{5.23}$$

5.3.2.2 Hann (Hanning) Window

The Hann window is a variation of the Hamming window and is known to belong to the Hamming window family. The main difference between the two is that Hann becomes 0 at $n = 0$ and $n = N - 1$. The zero values at the tails may or may not be desirable depending on the circumstances of signal processing. One argument is that when the tails go to zero, the full extent of the data has not been used. In speech recognition, however, this may not be a problem since there is usually sufficient frame overlap in computing the features.

$$w(n) = 0.5\left(1 - \cos\left(\frac{2\pi n}{N-1}\right)\right) \tag{5.24}$$

Figure 5.16 shows the Hann window for $N = 64$ and its spectrum. It is apparent from comparing the spectra of the two windows that the frequency response of the Hamming window drops very quickly for low frequencies and then becomes almost flat for higher frequencies. On the other hand, the Hann window drops a bit more slowly for low frequencies, but quickly drops for higher frequencies. Therefore, each of them has its own advantages and disadvantages. This was the motivation behind new families of windows that would combine the two types of responses.

5.3.2.3 Welch Window

The Welch window is quite similar to the Bartlett window. The following is the N-point Welch window. If the square in Equation 5.25 is replaced with an absolute value, the window will be identical with the Bartlett window. Both Welch and Bartlett windows become 0 at $n = 0$ and $n = N - 1$. Welch used this window for the method of PSD estimation that he proposed in 1967 [87] – see Section 24.10.4 for a description of this technique.

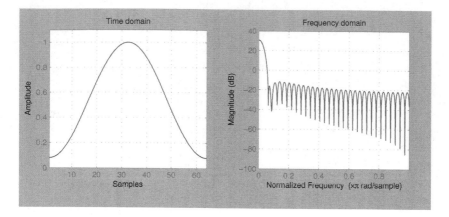

Fig. 5.15: Hamming Window

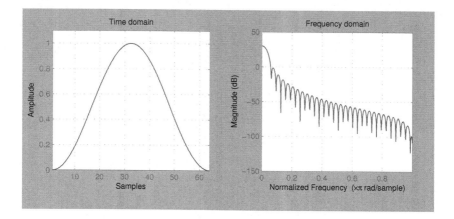

Fig. 5.16: Hann Window

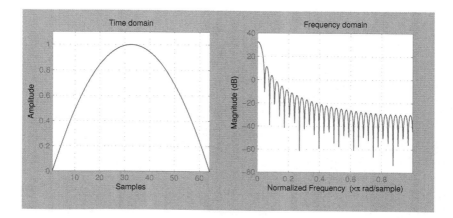

Fig. 5.17: Welch Window

$$w(n) = 1 - \left(\frac{n - \frac{N-1}{2}}{\frac{N-1}{2}} \right)^2 \tag{5.25}$$

Figure 5.17 shows the Welch window for $N = 64$ and its spectrum. Comparing the frequency response of the Welch window to Hamming and Hann, we see something in between the two responses in that Welch drops somewhat faster than Hann and has a lower power for higher frequencies compared to Hamming, however, the higher frequencies do not fall off as fast as Hann.

5.3.2.4 Triangular Window

The triangular window, as the name suggests, is just a triangularly weighted window which peaks at the center of the window ($n = \frac{N}{2}$). It is important since it is usually used in the computation of the Mel-Frequency Cepstra Coefficients – see Section 5.3.4. The following is the equation for an N-point triangular window. It is quite similar to the *Bartlett* window (see Section 5.3.2.3) with the exception that the Bartlett window is 0 at $n = 0$ and $n = N - 1$ and the triangular window is not.

$$w(n) = 1 - \left| \frac{2n - N + 1}{N - 1} \right| \tag{5.26}$$

Figure 5.18 shows the Triangular window for $N = 64$ and its spectrum. As it is apparent from the spectrum, it drops off quite abruptly. The side-lobes are much wider than the ones for the smoother-transition windows we have discussed until now. However, it still portrays an acceptable drop at higher frequencies.

5.3.2.5 Blackman Window Family

The Blackman window family is part of a larger Blackman-Harris family which is basically a multi-term cosine series. The Blackman window family uses three terms in the series. The following is the equation for the Blackman family.

$$w(n) = a_0 - a_1 \cos\left(\frac{2\pi n}{N - 1} \right) + a_2 \cos\left(\frac{4\pi n}{N - 1} \right) \tag{5.27}$$

where

$$a_0 = \frac{1 - \alpha}{2}$$

$$a_1 = \frac{1}{2}$$

$$a_2 = \frac{\alpha}{2}$$

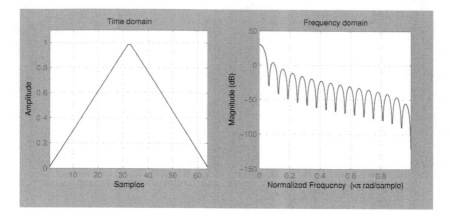

Fig. 5.18: Triangular Window and its spectrum

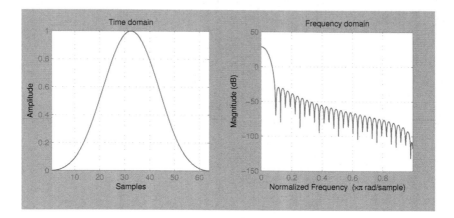

Fig. 5.19: Blackman Window ($\alpha = 0.16$) and its spectrum

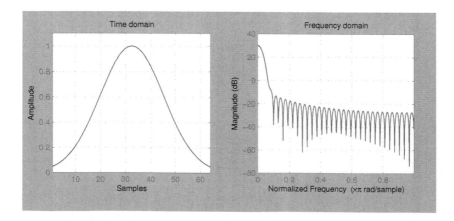

Fig. 5.20: Gauss Window ($\sigma = 0.4$) and its spectrum

When $\alpha = 0.16$, the window is known as simply the Blackman window or the classic Blackman window. Figure 5.19 shows the Classic Blackman window for $N = 64$ and its spectrum. Judging from the spectrum, the Blackman window portrays the best of Hamming and Hanning; it has a sharp drop, with a large main lobe and then its sidelobes which are quite narrow, drop off quite rapidly as well. This is achieved by the use of the higher order in a trigonometric series.

5.3.2.6 Gauss Window

There are several definitions for the Gauss or Gaussian window. All of them are based on the Gaussian function. Of course there are some that use a power of 2 instead of e. In Section 24.12, we saw a continuous form of the window, Equation 24.483, which was used for the definition of the *Gabor* transform. Here is one possible definition for an N-point discrete Gaussian window.

$$w(n) = e^{-\frac{1}{2}\left(\frac{n-\frac{N-1}{2}}{\sigma\frac{N-1}{2}}\right)^2} \tag{5.28}$$

where,

$$\sigma \leq \frac{1}{2} \tag{5.29}$$

Figure 5.20 shows the Gauss window for $N = 64$ and its spectrum with $\sigma = 0.4$. The performance of the Gauss window is quite similar to that of the Hamming window.

5.3.3 Discrete Fourier Transform (DFT) and Spectral Estimation

The next step in the processing of the speech data to be able to compute its spectral features is to take a Discrete Fourier Transform of the windowed data. This is done using the FFT algorithm – see Sections 24.10 and 24.10.5. The following is the result of the FFT,

$$_lH_k = \sum_{n=0}^{N-1} {}_l\tilde{h}_n e^{-i\frac{2\pi kn}{N}}$$

$$= \sum_{n=0}^{N-1} {}_lh_n w(n) e^{-i\frac{2\pi kn}{N}} \tag{5.30}$$

where, $k = \{0, 1, \cdots, N-1\}$ is the index of the frequency domain with $k = 0$ corresponding to the DC component and $k = \frac{N}{2}$ corresponding to the folding frequency (half of the sampling frequency).

Now that we have computed the DFT of the lth N-sample overlapping frame, we would like to estimate the spectrum of the sequence. Aside from the two points at the DC level ($\omega = 0$) and $\omega = \frac{1}{2}\omega_s$ where ω_s is the angular frequency of sampling, the $_lH_k$ are complex numbers which include a magnitude and a phase if viewed in their polar coordinates,

$$_lH_k = |_lH_k| + i\angle(_lH_k) \tag{5.31}$$

Traditionally, in most speech-related systems, the phase of the Fourier Transform is ignored and only the magnitude is used for deriving recognition features. Although, there have been some studies suggesting that this may be inappropriate [58]. Nevertheless, next, we will compute the magnitude of the FFT bins,

$$|_lH_k| = \sqrt{\mathscr{R}e(_lH_k)^2 + \mathscr{I}m(_lH_k)^2} \tag{5.32}$$

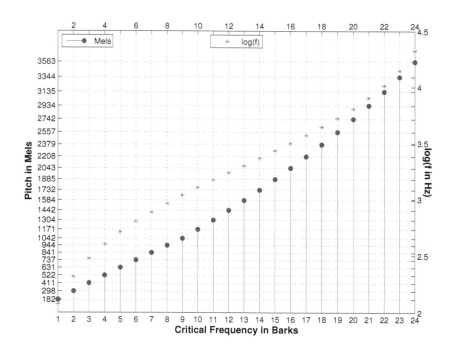

Fig. 5.21: The first 24 critical frequencies given by the Scale on the horizontal axis, On the left vertical axis, the corresponding Mels and on the right, the corresponding log(frequency in Hz)

Once we have the N FFT parameters, $|_lH_k|$ of a signal, we may define a *spectral analysis* measure which allows us to examine its variations in the frequency content of the signal, as follows.

Definition 5.5 (Spectral Flatness). *The spectral flatness of a signal is the ratio of the geometric mean to the arithmetic mean of its power spectrum.*

$$F_s(_lH_0^{N-1}) = \frac{\sqrt[N]{\prod_{k=0}^{N-1} |_lH_k|}}{\frac{1}{N}\sum_{k=0}^{N-1} |_lH_k|} \tag{5.33}$$

A large $F_s(_lH_0^{N-1})$ denotes a more uniform presence of spectral components in the signal and a small one signifies concentration of power in certain frequencies. Although Equation 5.33 uses the spectrum for the lth frame, in general, sometimes it makes more sense to use the spectrum of a larger portion of the signal, H_0^{N-1}, to evaluate its *spectral flatness*, $F_s(H_0^{N-1})$.

5.3.4 Frequency Warping

As we discussed in Section 5.1.1, the human auditory perception is based on a scale which is somewhat linear up to the frequency of 1000 Hz and then becomes close to logarithmic for the higher frequencies. This was the motivation for the definition of Pitch, $\wp$ in the Mel-scale. Also, the experiments of Zwicker [103, 102] used a 24-band filter-bank to model the auditory system. The center frequencies of these nonlinearly located filters were dubbed the Bark scale. Figure 5.21 shows the relationship among the three different measurement scales, *Bark*, *Mel* and $\log(f)$.

Since the human perception works with these critical bands, the speech features have been adapted to pertain to these specific bands. However, up to now, our signal processing has not taken this into consideration. Magnitudes of the spectra, $|_lH_k|$ which were computed in the last section pertain to the linear frequency scale. Therefore, they have to be converted to have a smaller number of values corresponding to the critical band center frequencies defined by [103, 102] using the Mel scale.

The way this can be done is to build a filter bank of, say, 24 filters as presented in Figure 5.21 and convert the total number of N DFT bin frequency centers which were equally spaced between $-f_c$ and f_c to these 24 representations. Note that since we are dealing with a real signal, based on the argument of Section 24.10.1, the values of $_lH_k$ for $k \in \{\frac{N}{2}, \cdots, N-1\}$ are complex conjugates of $_lH_k$ for $k \in \{1, \cdots, \frac{N}{2}-1\}$. Therefore, since we are not considering the phase of the transform, their amplitudes are redundant. This means that we have $\frac{N}{2}+1$ values of $|_lH_k|$ which have to be mapped to 24 critical bands.

One prevalent way of doing this mapping is to apply a critical band filter-bank with center frequencies and bandwidths given by the bark scale. Another popular method is to set these filters in equidistant centers based on the Mel scale. Although these two are similar, as noted by the graph of *Mels* versus *Barks* in Figure 5.21, the relationship is not quite linear. In practice, most systems have been reported to be using the latter. In doing so, the distance between consecutive center frequencies has to be set.

To find that distance, note that the distances between two adjacent critical frequencies in the bark scale vary from 110 to 440 Mels, averaging about 160 Mels (due to the slight nonlinearity demonstrated in Figure 5.21). The most prevalent sampling frequency is 8000 Hz which is the frequency used most frequently in the telephony industry and in the definitions of the different PCM flavors of ITU-T G.711 [36]. Considering this sampling frequency which corresponds roughly to 2840 Mels and 20 barks, we can divide the Mel frequency space into a set of filters. Some have used 20 filters [16] for this set, based on the critical locations on the bark scale, but then used a uniform spacing between center frequencies.

A more popular approach is to use the minimum difference between the first and second critical bands to model the rest of the bands. That would be almost 110 Mels. If we divide 2840 Mels by using 110 Mels in between two adjacent filters, we would end up with 24 filters. For a 16000 Hz sampling rate, an average of 145 Mels would probably be better, leaving us with 32 filters and so on.

Now that we have the filter spacing figured out, all we have to do is to pass the magnitudes of the spectra, $|_lH_k|$, through these filters and obtained the Mel-frequency versions of the spectral magnitudes. Before doing this, there is one more detail to which we have to tend and that is the shape of the filters. We spoke about window functions in Section 5.3.2. The most popular and one of the simplest approaches is to use a triangular shape for the filters (see Section 5.3.2.4).[16]

Figure 5.22 shows the filter locations based on the 24 division filter-bank with the center frequencies chosen to be 110 Mels apart. These weighting windows are applied on the linear spectral magnitudes to provide us with the Mel-scale spectral magnitudes denoted by $|_l\breve{H}_m|$,

$$|_l\breve{H}_m| = \mathscr{M}_{mk} |_lH_k| \tag{5.34}$$

where $m = \{0, 1, \cdots, M-1\}$ is the critical filter index and M is the total number of filters in the critical band filter-bank (24 in the case we are examining). The matrix $\mathscr{M}_{mk}$ is the $(m,k)^{th}$ element of matrix $\mathscr{M}$, $\{\mathscr{M} : \mathscr{R}^N \mapsto \mathscr{R}^M\}$, which is the mapping given by the triangular filters from the linear frequency scale to the Mel scale.

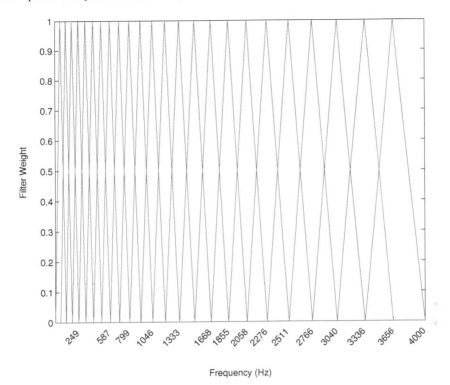

Frequency (Hz)

Fig. 5.22: Shape of the Mel filter bank weights for a 24-filter system with an 8000 Hz sampling frequency. Not all the frequency values for the lower frequency centers have been written since the numbers would have merged and would not be legible.

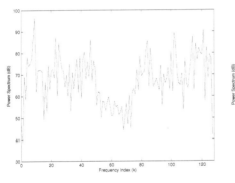

Fig. 5.23: Power Spectrum of the Frame $N = 256$

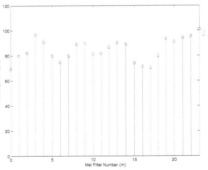

Fig. 5.24: Power Spectrum in the Mel Frequency Domain $N = 256, M = 24$

5.3.5 Magnitude Warping

Now that we have warped the magnitudes of the spectra into a Mel-frequency scale, recall the discussions we had about *loudness* in Section 5.1.2. $|{}_lH_k|^2$ is related to the *intensity*, I. Our goal is to convert it to a value that would represent *loudness* so that we may mimic human perception. As you recall, loudness is both a function of pitch, $\wp$, and intensity, I. In warping the magnitudes to the pitch scale, Mel, we built in the pitch dependency. Now we should warp $|{}_lH_k|^2$ such that the amplitude is logarithmic and similar to the DB scale. Of course to make sure it is in the DB scale, we should divide it by the *standard intensity threshold*, I_0, before taking the log. Therefore, a measure related to loudness would be,

$$
{}_l\tilde{C}_m = 10\log\left(\frac{|{}_l\breve{H}_m|^2}{I_0}\right)
\tag{5.35}
$$

Let us also define another variable, ${}_lC_m$ which lacks the normalization of ${}_l\tilde{C}_m$. ${}_lC_m$ is the log spectrum of the signal in the lth frame. We will discuss its properties in the next section.

$$
{}_lC_m = \log\left(|{}_l\breve{H}_m|^2\right)
\tag{5.36}
$$

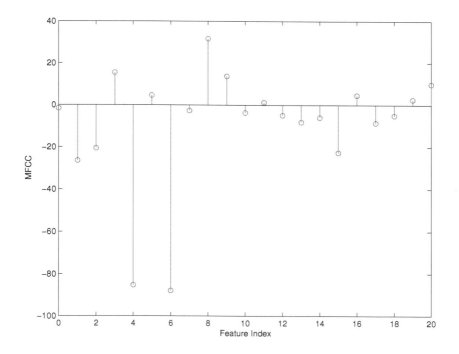

Fig. 5.25: First 21 Mel Frequency Cepstral Coefficients

5.3.6 Mel Frequency Cepstral Coefficients (MFCC)

In the previous section, we derived the log of the PSD of the signal in the Mel scale. We will refer to this as the log spectrum from here on for convenience. Also, in Section 24.15 we discussed the cepstrum of a signal in detail. In general, one of the most attractive features of the cepstrum which makes it a good candidate for usage in speaker recognition is its inherent invariance toward linear spectral distortions. Recall the control system analogy used at the beginning of Section 5.3. Equation 5.20 showed the relationship between the input and the output of the speech production system where the transfer function, $G(\omega)$ was composed of the controller and the plant transfer functions, G_c and G_v. According to Section 24.15, the convolution of the signals in the time domain which shows up as the product in the block diagram (in the spectral domain), ends up being a summation of components in the cepstral domain. This makes cepstral analysis ideal for speech and speaker recognition purposes since the vocal tract dynamics and the pulse train associated with the glottal motor control become additive components in the cepstral domain.

In Section 24.15, no assumption was made regarding the frequency scale used in the cepstral processing. However, we did discuss the short-time analysis in the cepstral domain for pitch detection as initially proposed by Noll in [54, 55]. To evaluate the cepstral coefficients, based on the analysis of the short-time Mel-Frequency log spectrum, we may compute the inverse Fourier transform of $_lC_m$ of Equation 5.36. Note that the resulting inverse is in general complex. One possibility is to take the square of the inverse as suggested by Equation 24.601. Another possibility is to take the real part of the inverse Fourier transform. Yet another possibility is to take the inverse Discrete Cosine Transform (DCT) of $_lC_m$ which provides us with real numbers. This approach was introduced by [16]. See Section 24.13, for more on the DCT.

In the inverse DCT approach, the Mel-Frequency Cepstral Coefficients will then be defined by the following using Equation 24.518,

$$_lc_d = \sum_{m=0}^{M-1} a_m \; _lC_m \; \cos\left(\frac{\pi(2d+1)m}{2M}\right) \tag{5.37}$$

where using Equation 24.519, the coefficients, a_m, are given by,

$$a_m = \begin{cases} \frac{1}{M} \; for \; m=0 \\ \\ \frac{2}{M} \; \forall \; m>0 \end{cases} \tag{5.38}$$

The Mel-Frequency Cepstral Coefficients (MFCCs) are always real and convey information about the physical aspects of the speech signal. Consider the spectrogram of the signal in Figure 5.28. Figure 5.29 is a plot of the trajectories of the short-time MFCCs c_1 and c_2 across 260 frames of overlapping speech in the signal.

Right at the location of the largest trough in c_1 (about $0.3s$ into the utterance and again at around $2.1s$), the spectrogram shows a fricative (energy in the high frequency region). A negative c_1 relates to the local minimum of the cosine in the DCT noting that the higher Mel-frequency indices in the summation of the DCT are contributing more. On the other hand, a positive peak means that there must be more power in the lower frequency range, signifying sounds such as stops which have most of their energy in the lower frequencies. The time instance of about $0.6s$ happens to be part of phoneme /t/ which demonstrates a high peak in the c_1 value and a concentration of power in the lower frequencies as depicted by the spectrogram. As the MFCC dimension d becomes larger, the number of alternating partitions in the frequency range increases. For example, c_2 is a combination of the first and third partitions of the frequency scale in contrast with the second and fourth partitions. This goes on as d increases. When $d = 0$, the cosine term of the DCT becomes 1. c_0 is then the average power of the signal. Although the computation of c_0 has involved many transformations, so it is a better idea to compute the power from the autocorrelation of the signal based on the Wiener-Khintchine theorem (see Section 24.9.11). Figure 5.30 shows the shortpass liftered version of the trajectories of c_1 and c_2. Section 18.4 describes the benefits of this type of liftering in more detail – see Section 24.15.

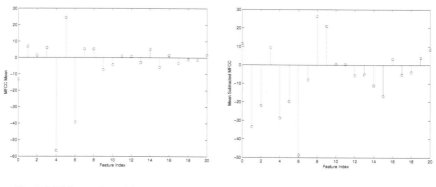

Fig. 5.26: Mean value of the MFCC vectors over 260 frames

Fig. 5.27: Mean-subtracted MFCC vector

As we have seen, Mel Frequency Cepstral Coefficients carry significant information about the structure of the signal and are ideal candidates for being used as features for speaker and speech recognition. In the next section, we will speak about extracting the dynamics of these features to complement the basic features. It is worth noting that even without any added dynamics, the windowing has already built in some local dynamics into the MFCC features.

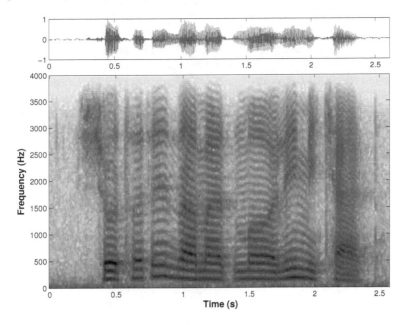

Fig. 5.28: Spectrogram of the audio being analyzed for MFCC computation

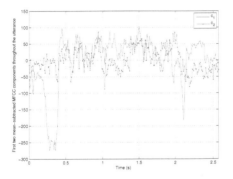 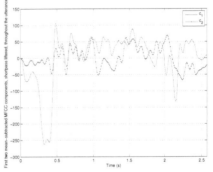

Fig. 5.29: Trajectory of the first two MFCC components over the whole utterance

Fig. 5.30: Shortpass liftered Trajectory of the first two MFCC components over the whole utterance

5.3.7 Mel Cepstral Dynamics

To capture the dynamics of the Mel Cepstral coefficients, the first and second order differences are used. In general, the first and second order differences are somewhat independent of the actual MFCC coefficients and are used to model the local dynamics of speech. These differences are known by *Delta* and *Delta-Delta Cepstral Coefficients*. In treating Mel Frequency Cepstral Coefficients and their first and second differences, some researchers have used the same number of dynamic features as actual MFCCs themselves. However, there is practical evidence that this need not be the case. Since as the number of features in a feature vector are increased, more data is needed to properly estimate the statistical parameters of the models, it is advisable to use a smaller number of delta cepstra and yet a smaller number of the delta-delta cepstra in the feature vector. The actual number is a function of the amount of data available for training and testing as well as the speed requirements. As size of the feature vector increases more data is required for both training and testing and there may be significant slow-downs associated with larger feature vectors. This is especially true for the training process.

5.4 Linear Predictive Cepstral Coefficients (LPCC)

In Section 5.3, we covered the, so called, *all-zero, moving average (MA)*, or *direct method* of computing the spectra. Then, after frequency and magnitude warping, we computed the MFCC features. In this section, we will review a counterpart which differs mostly in the way it approximates the PSD of the signal, $|H(\omega)|^2$. This method is known by the names, *all-pole* model, *maximum entropy model (MEM)*, or the *autoregressive (AR) model* and estimates the PSD by fitting with an *all-pole* model.

Several types of features are generated by the AR model. These are *Linear Predictive Coding (LPC)* coefficients, *reflection* or *Partial Correlation (PARCOR)* coefficients and *log area ratio* coefficients. As we will see, one efficient method for computing these features relies on the short-time signal autocorrelation. There is also a covariance method for solving the corresponding equations [60].

In practice, most recognizers that rely on the AR estimates convert the LPC coefficients to cepstral coefficients. First, we are going to derive the equations for finding the LPC coefficients. Then we will be exploring other related coefficients and finally convert the LPC coefficients to cepstral coefficients.

The first few steps leading to the computation of the PSD for extracting the LPC coefficients are identical to the method used in pursuing MFCC features in Section 5.3. Therefore, the *sampling* is done in the prescribed fashion, following by a *hi-pass filter* and *pre-emphasis*. Then the signal is divided into *overlapping frames* exactly in the fashion described in Section 5.3.1 followed by the application of a *low-pass window* as prescribed in Section 5.3.2. After all this processing, we will have L overlapping frames, each of which is N samples long. Therefore, each sample is denoted by $_lh_n$ where $l \in \{0, 1, \cdots, L-1\}$ is the frame number and $n \in \{0, 1, \cdots, N-1\}$ is the sample index within the frame.

5.4.1 Autoregressive (AR) Estimate of the PSD

In the Section 5.3, we reviewed a method which started with a spectral estimation (estimation of the PSD) for the sampled signal, h_n. Recall the derivation of the DFT in Section 24.10 where we decided to estimate the spectrum of the signal by truncating the approximation to the Complex Fourier Transform. This gave us the following expression for the DFT,

$$H_k = \sum_{n=0}^{N-1} h_n e^{-i\frac{2\pi k n}{N}} \tag{5.39}$$

and the power spectral density (PSD) associated with it,

$$\mathscr{P}_d^\circ(k) = \frac{1}{N^2} |H_k|^2$$

$$= \frac{1}{N^2} \left| \sum_{n=0}^{N-1} h_n e^{-i\frac{2\pi k n}{N}} \right|^2 \tag{5.40}$$

This is only an approximation, where not only time has been discretized, but the frequency has also been discretized using Equations 24.434 and 24.435. Therefore, the PSD estimation based on the DFT (*Periodogram*) is truly an estimate and not the actual PSD itself. To obtain a better estimate of the PSD, recall that we may use the DTFT of the signal which does not discretize the frequency,

$$H(\omega) = \sum_{n=-\infty}^{\infty} h_n e^{-i\omega n} \tag{5.41}$$

This, as we showed in Section 24.14, is a special case of the two-sided z-transform of the signal with $z = e^{i\omega}$,

$$H(z) = \sum_{n=-\infty}^{\infty} h_n z^{-n} \tag{5.42}$$

The PSD associated with the DTFT is given By Equation 24.465 and repeated here for convenience,

$$\mathscr{P}_d^\circ(\omega) = \frac{1}{2\pi} |H(\omega)|^2$$

$$= \frac{1}{2\pi} \left| \sum_{n=-\infty}^{\infty} h_n e^{-i\omega n} \right|^2 \tag{5.43}$$

If we write Equation 5.43 with the change of variable, $z = e^{i\omega}$, we obtain the following,

$$\mathscr{P}_d^\circ(\omega) = \frac{1}{2\pi} \left| \sum_{n=-\infty}^{\infty} h_n z^{-n} \right|^2 \tag{5.44}$$

Recall the relation between the discrete frequency, ω_k and k given by Equation 24.461 where,

$$\omega_k = \frac{2\pi k}{N} \quad 0 \le k < N \tag{5.45}$$

substituting this in Equations 5.39 and 5.40, we have,

$$H(\omega_k) = \sum_{n=0}^{N-1} h_n e^{-i\omega_k n} \tag{5.46}$$

and

$$\mathscr{P}_d^\circ(\omega_k) = \frac{1}{N^2} \left| \sum_{n=0}^{N-1} h_n e^{-i\omega_k n} \right|^2$$

$$= \frac{1}{N^2} \left| \sum_{n=0}^{N-1} h_n z^{-k} \right|^2 \tag{5.47}$$

Comparing Equations 5.44 and 5.47, we see that Equation5.47 is a polynomial estimate of the infinite series in Equation 5.44 (not bothering with the normalization). Refer to the definition of the Laurent series expansion of an analytic function in an Annular region, Definition 24.43. Note that the z-transform may be thought of as a special case of the Laurent series of the function it transforms. Therefore, Equation 5.44 is just a Laurent series expansion. In general, the Laurent series may be approximated by a rational function with a finite number of poles and zeros. Therefore, Equation 5.42 may be approximated as,

$$H(z) = \sum_{n=-\infty}^{\infty} h_n z^{-n}$$

$$\approx \frac{G_1(z)}{G_2(z)}$$

$$= \frac{\displaystyle\sum_{\hat{n}=1}^{N} \beta_{\hat{n}} z^{-\hat{n}}}{\alpha_0 + \displaystyle\sum_{q=1}^{Q} \alpha_q z^{-q}} \tag{5.48}$$

This Equation describes the, so called, *AutoregRessive Moving Average (ARMA)* model which contains both poles and zeros.

Notice that Equation 5.46 is just the special case of the approximation given by Equation 5.48 where $G_2(z) = 0$, namely, the number of poles is 0 leaving us with an *all-zero* approximation of the Laurent series expansion of the DFT of the signal, h_n. It is conceivable that we may choose to use only poles to approximate the Laurent series and to set $G_1(z) = 0$. This leads to the *all-pole* method which will be discussed here.

Assuming that $H(z)$ is approximated only by poles, which is called an *Autoregressive (AR)* model, we will have,

$$
H(z) \approx \frac{1}{\alpha_0 + \sum_{q=1}^{Q} \alpha_q z^{-q}}
$$

$$
= \frac{\frac{1}{\alpha_0}}{1 + \sum_{q=1}^{Q} \frac{\alpha_q}{\alpha_0} z^{-q}} \tag{5.49}
$$

Now let us define the following variables,

$$
\hat{G} \triangleq \frac{1}{\alpha_0}
$$

$$
a_q \triangleq -\frac{\alpha_q}{\alpha_0} \quad where \quad q = \{1, 2, \cdots, Q\} \tag{5.50}
$$

Using Definitions 5.50 in Equation 5.49, we have,

$$
\hat{H}(z) = \frac{\hat{G}}{1 - \sum_{q=1}^{Q} a_q z^{-q}} \tag{5.51}
$$

which is equivalent to

$$
\hat{H}(\omega) = \frac{\hat{G}}{1 - \sum_{q=1}^{Q} a_q e^{-i\omega q}} \tag{5.52}
$$

where $\hat{H}(\omega)$ is the *all-pole* approximation of the DTFT. Now, if we write the PSD given by Equation 5.43 in terms of the new *all-pole* estimate of DTFT ($\hat{H}(\omega)$), we have,

$$
\hat{\mathscr{P}}_d^{\circ}(\omega) = \frac{1}{2\pi} |H(\omega)|^2
$$

$$
= \frac{1}{2\pi} \frac{\hat{G}^2}{\left| 1 - \sum_{q=1}^{Q} a_q z^{-q} \right|^2} \tag{5.53}
$$

where $\hat{\mathscr{P}}_d^{\circ}(\omega)$ is the *all poles* approximation of $\mathscr{P}_d^{\circ}(\omega)$.

Let us further simplify Equation 5.53 by defining

$$
G \triangleq \frac{\hat{G}}{\sqrt{2\pi}} \tag{5.54}
$$

Plugging G into Equation 5.55 we have the final form of the *all-pole* approximation of the PSD,

$$\mathscr{P}_d^\circ(\omega) = \frac{G^2}{\left| 1 - \sum\limits_{q=1}^{Q} a_q z^{-q} \right|^2} \tag{5.55}$$

Let us examine the *all-pole* estimate of the DTFT in terms of z, given by Equation 5.51. The homogeneous part of the difference equation related to Equation 5.51 gives us the relationship between the nth sample and its prior Q samples in the following form,

$$^{(h)}\hat{h}_n = \sum\limits_{q=1}^{Q} a_q h_{n-q} \tag{5.56}$$

where $^{(h)}\hat{h}_n$ is the estimated output of the homogeneous portion of the difference equation defined by Equation 5.51. In reality, since this is an *all-pole* method, it includes no dynamics related to the input of the system. So, the particular solution of the difference equation will be dependent on the input signal which is unknown. If we assume that the input signal remains constant within Q or more samples, then we can simply add a constant DC level to the signal for each computation, namely[3],

$$\hat{h}_n = {}^{(h)}\hat{h}_n + {}^{(p)}\hat{h}_n$$
$$= \sum\limits_{q=1}^{Q} a_q h_{n-q} + {}^{(p)}\hat{h}_n \tag{5.57}$$

where $^{(p)}\hat{h}_n$ is the particular solution of the difference equation related to Equation 5.51.

The real output is the actual value (observed value) of the signal at sample n, h_n. Therefore, the modeling error would become,

$$\hat{e}_n = h_n - \hat{h}_n$$
$$= h_n - \sum\limits_{q=1}^{Q} a_q h_{n-q} + {}^{(p)}\hat{h}_n \tag{5.58}$$

Since we assumed that we are going to have a constant input signal for a period greater than or equal to Q samples, resulting in a constant particular solution, $^{(p)}\hat{h}_n$, we may define a new error function, e_n, which does not include the constant particular solution in it. Namely,

$$e_n \overset{\Delta}{=} \hat{e}_n - {}^{(p)}\hat{h}_n \tag{5.59}$$

Therefore,

---

[3] If you need to refresh your knowledge of difference equations, please see [41].

$$e_n = h_n - \sum_{q=1}^{Q} a_q h_{n-q} \tag{5.60}$$

This is the error for each sample in the frame. Now, the problem is to find the parameters, a_q, such that we have minimum error in some sense. We have to decide on the metric that we would like to choose for the minimization. A popular choice is to minimize the sum of squares of errors over all the samples in the frame. This means that our minimization objective function will be,

$$E = \sum_{n=0}^{N-1} E_n$$

$$= \sum_{n=0}^{N-1} e_n^2$$

$$= \sum_{n=0}^{N-1} \left[h_n - \sum_{q=1}^{Q} a_q h_{n-q} \right]^2 \tag{5.61}$$

where E_n is the square of the error for sample n and E is the sum of squares of errors over all N samples in the frame.

To minimize E, we can take its partial derivative with respect to the AR parameters, a_q, and set it equal to zero and then solve for the a_q,

$$\frac{\partial E}{\partial a_q} = 0 \tag{5.62}$$

The solution to problem 5.62 has been attacked from two different perspectives. Rabiner and Juang [60] use the signal samples in a similar fashion as we have discussed here, with a slight difference in handling the particular part of the solution to the difference equation. However, after taking the derivative, the same results are obtained in their approach and the one we have presented here.

Makhoul [46] has approached the problem from its *dual* perspective. Namely, he has defined the error in the spectral domain as,

$$E_s = \frac{G^2}{2\pi} \int_{-\pi}^{\pi} \frac{\mathscr{P}_d^\circ(\omega)}{\hat{\mathscr{P}}_d^\circ(\omega)} d\omega$$

$$= \frac{1}{2\pi} \int_{-\pi}^{\pi} \mathscr{P}_d^\circ(\omega) \left| 1 - \sum_{q=1}^{Q} a_q e^{-i\omega q} \right|^2 \tag{5.63}$$

where E_s is the error defined by [46] in the spectral domain, $\hat{\mathscr{P}}_d^\circ(\omega)$ is the *all-pole* estimate of the PSD given by Equation 5.55, and $\mathscr{P}_d^\circ(\omega)$ is the true PSD. Therefore, the minimization problem according to [46] becomes,

$$\frac{\partial E_s}{\partial a_q} = 0 \tag{5.64}$$

Due to Parseval's theorem, the results of both minimizations become identical and may be written in terms of the autocorrelation function of the signal, h_n, in the form of Q simultaneous linear equations in a_q [60, 46],

$$\sum_{q=1}^{Q} a_q \, _l r(|j-q|) = \, _l r(j) \quad \text{where} \quad j = \{1, 2, \cdots, Q\} \tag{5.65}$$

where $_l r(j), j \in \{0, 1, \cdots, Q\}$ is the autocorrelation of the signal in frame l for up to $Q+1$ values. The actual autocorrelation function is defined over all time, namely,

$$_l r(j) = \sum_{n=-\infty}^{\infty} h_n h_{(n+j)} \tag{5.66}$$

where $-\infty < j < \infty$. However, in the AR model we have assumed that the signal is stationary. As we have discussed before, the speech signal is a non-stationary signal. This is the reason we have chosen to use a short-time formulation, which means that we have multiplied an N-sample window with the signal with vanishing tails so that at most N samples have non-zero values. Therefore, we can rewrite the short-term autocorrelation function as,

$$_l r(j) = \sum_{n=0}^{N-1-Q} \, _l h_n \, _l h_{n+j} \tag{5.67}$$

where, $j \in \{0, 1, 2, \cdots, Q\}$ and l is the frame number (see Section 5.3.1). Note that in the short-time version, we are allowing for $Q+1$ numbers ranging from $j = 0$ to $j = Q$. Although the autocorrelation is defined past $Q+1$ points, we are only interested in $Q+1$ points to be able to solve Equation 5.65. The relations in Equation 5.65 are known as the Yule-Walker Equations [39].

Let us define the, so called, *autocorrelation matrix*, $_l \mathbf{R}$,

$$_l \mathbf{R} \triangleq \begin{bmatrix} _l r(0) & _l r(1) & _l r(2) & \cdots & _l r(Q-1) \\ _l r(1) & _l r(0) & _l r(1) & \cdots & _l r(Q-2) \\ _l r(2) & _l r(1) & _l r(0) & \cdots & _l r(Q-3) \\ \vdots & \vdots & \vdots & \ddots & \vdots \\ _l r(Q-1) & _l r(Q-2) & _l r(Q-3) & \cdots & _l r(0) \end{bmatrix} \tag{5.68}$$

and the autocorrelation vector,

$$_l \mathbf{r} \triangleq \begin{bmatrix} _l r(1) \\ _l r(2) \\ \vdots \\ _l r(Q) \end{bmatrix} \tag{5.69}$$

Then we can rewrite the *Yule-Walker* equations, Equation 5.65, in matrix form as follows,

$$_l\mathbf{R}\,_l\mathbf{a} = {}_l\mathbf{r} \tag{5.70}$$

where $_l\mathbf{a} : \mathscr{R}^1 \mapsto \mathscr{R}^Q$ is the vector of the *all-pole* parameters for frame l.

Therefore, the least squares solution to the *all-pole* estimate is given by,

$$_l\mathbf{a} = {}_l\mathbf{R}^{-1}\,_l\mathbf{r} \tag{5.71}$$

Note the structure of $_l\mathbf{R}$ in the definition of Equation 5.68. $_l\mathbf{R}$ is a *Toeplitz* matrix which shows up in many solutions to difference equations, especially in control systems and signal processing. Its structure makes it quite simple to solve for $_l\mathbf{a}$ using the *Levinson-Durbin* algorithm which was introduced by *Levinson* [42] and then modified by *Durbin* [18, 19, 8, 60]. There is also another algorithm called *Schür recursion* [66], which is more efficient for parallel implementations.

From the spectral perspective, due to Parseval's theorem (see Section 24.9.7), element q of $_l\mathbf{r}$, denoted by $_l\mathbf{r}|_q$ is as follows [46],

$$
\begin{aligned}
_lr(j) &= \frac{1}{2\pi} \int_{-\pi}^{\pi} {}_l\mathscr{P}_d^\circ(\omega)\,\cos(j\omega)d\omega \\
&= \sum_{n=-\infty}^{\infty} h_n h_{(n+j)}
\end{aligned} \tag{5.72}
$$

Note that in the spectral form of Equation 5.72, the angular frequency, ω, is treated as a continuous variable. It is possible to discretize the frequency as we did in the DFT method by Equation 24.461 so that we obtain the discretized version of Equation 5.72, namely,

$$
\begin{aligned}
_lr(j) &= \frac{0}{N-1}{}_l\mathscr{P}_d^\circ(\omega)\,\cos(2\pi kj) \\
&= \sum_{n=0}^{N-1-Q} {}_lh_n\,{}_lh_{n+j}
\end{aligned} \tag{5.73}
$$

In the discretized version,

$$E = \frac{G^2}{N} \sum_{n=0}^{N-1} \frac{\mathscr{P}(\omega_k)}{\hat{\mathscr{P}}(\omega_k)} \tag{5.74}$$

Equation 5.74 means that only discrete values of the frequency are contributing to the error being computed. This means that the minimum error is only valid for the discrete frequencies within the range of ω_k. This re-iterates the fact that if there are higher frequencies present, they will not be modeled. Note that one other possible method for computation of the *all-pole* estimate is to use a discrete cosine transform through an FFT as apparent by Equation 5.74 See [46] for a complete treatment of the spectral path of the solution of the Linear Predictive Coefficients.

Let us return to the Yule-walker equations. As we mentioned, up to the windowing step, the LPC method is identical to the *direct method* of Section 5.3. In the next step, we will use the short-time autocorrelation (Equation 5.67) for the N-sample frame and solve Equation 5.71 to compute the LPC coefficients and then follow on to compute the LPCC features.

5.4.2 LPC Computation

At this point we have arrived at Equation 5.71 which should be solved for every frame of the signal. As we noted, the Toeplitz structure of the autocorrelation matrix, $_l\mathbf{R}$, allows us to use the efficient Levinson-Durbin method to solve the Yule-walker Equations (Equation 5.70) directly, without having to compute $_l\mathbf{R}^{-1}$, for the, so called, Linear Predictive Coding (LPC) coefficients, $_la_q, q \in \{1, 2, \cdots, Q\}$ and $l \in \{0, 1, \cdots, L-1\}$.

The following pseudo-code represents the steps of the Levinson-Durbin method as stated by *Rabiner and Juang* [60] (some typographical errors which existed in [60] have been corrected here),

Initialize E:

$$E^{(0)} = {}_lr(0) \tag{5.75}$$

for (q = 1 to Q),

1.

$$_l\kappa_q = \frac{_lr(q) - \sum_{j=1}^{q-1} \alpha_j^{(q-1)} {}_lr(q-j)}{E^{(q-1)}} \tag{5.76}$$

2.

$$\alpha_q^{(q)} = {}_l\kappa_q \tag{5.77}$$

3. for (j = 1 to Q),

$$\alpha_j^{(q)} = \alpha_j^{(q-1)} - {}_l\kappa_q \alpha_{q-j}^{(q-1)} \tag{5.78}$$

endfor

$$E^{(q)} = (1 - {}_l\kappa_i^2)E^{(i-1)} \tag{5.79}$$

endfor

Once the above recursion is completed, the LPC parameters are extracted from the results as follows,

$$_l a_q = \alpha_q^{(Q)} \tag{5.80}$$

By plugging in the LPC parameters given by Equation 5.80 into Equation 5.55 we are able to compute values which are proportional to the estimated short-time PSD of the signal. It is proportional to it since we still have not computed the value of G. Well, since G^2 is actually the error of the estimate, it may be written as

$$_l G^2 = {}_l r(0) - \sum_{q=1}^{Q} {}_l a_q \; {}_l r(q) \tag{5.81}$$

where $_l a_q$ are the newly computed LPC parameters.

Now we have everything needed to be able to write the estimate of the PSD using the *all-pole* method, Equation 5.55.

Aside from recognition, LPC has been used to compress audio. One application of the LPC coefficients is in the underlying coding of *GSM* signal compression used in most cellular (mobile) telephones. GSM is sampled at 8 kHz and the frames are 20 ms long. It actually uses the Log Area Ratios (LAR) which are coefficients derived from the LPC computation process. We will speak about them shortly.

In the next few sections, we will study a series of different features which have been derived from the LPC coefficients or have been computed in the process of computing the LPC coefficients. Some of them have been given physical interpretations.

5.4.3 Partial Correlation (PARCOR) Features

There are several other features with physical interpretation which come out of the computation of the LP coefficients. One such feature is the κ_q which was generated as a part of the computation of the LP coefficients. The κ_q are known as the *Partial Correlation (PARCOR)* coefficients (see Definition 6.60) and an interpretation is made that they only contain the non-redundant part of the autocorrelation which is used in the computation of the LP parameters, hence the name. Another expression for κ_k is[4],

---

[4] The Source of this equation is a lecture by Stéphane Maes at IBM T.J. Watson Research Center in 1995.

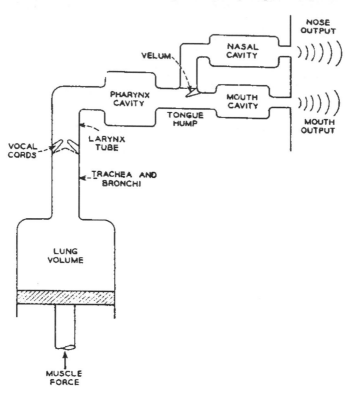

Fig. 5.31: Cylinder Model of the Vocal Tract; *Source: Flanagan [21]*

$$\kappa_q = \frac{\mathscr{E}\{E_f(q)E_b(q)\}}{\sqrt{\mathscr{E}\{E_f(q)^2\}\mathscr{E}\{E_b(q)^2\}}} \tag{5.82}$$

where $E_f(q)$ and $E_b(q)$ are the errors associated with the LPC computation of the signal. $E_f(q)$ is the error for a forward linear prediction and $E_b(q)$ is the error for a backward prediction, which means that the causality of the signal has been reversed. In the backward case, the last sample of the signal is treated as its first and the first sample as its last.

The above is the mathematical description of the PARCOR features. There is also a version based on physical attributes. Figure 5.31 has been reproduced here from *Flanagan's* book [21] by the author's permission. It is a simple model of the major cavities and constrictions along the vocal tract by a series of cylinders with different cross-sectional areas. The κ_q are also known as *reflection* coefficients. They are related to the reflection between consecutive cylinders in a resonance chamber model which flattens the cylinders of Figure 5.31 into a long series of concentric cylinders with different areas. This enables the use of a one dimensional form of the wave equation, derived for modeling wave propagation through a nonuniform pipe (with a varying cross-section)– see Figures 5.32 and 5.33. Equation 5.83 is Webster's equation in terms of the pressure in the pipe which is a one-dimensional form of the wave equation [21].

$$\frac{\partial^2 p(x,t)}{\partial x^2} + \frac{1}{A(x)} \frac{\partial p(x,t)}{\partial x} \frac{\partial A(x)}{\partial x} = \frac{1}{c^2} \frac{\partial^2 p(x,t)}{\partial t^2} \tag{5.83}$$

where $p(x,t)$ is the pressure and it is a function of time, t and the longitudinal dimension of the pipe, x. $A(x)$ is the cross-sectional area, and $c = \sqrt{\frac{k}{\rho}}$ is the speed of sound in the conditions of the compressible fluid, where ρ is the density of air and $k = -V \frac{\partial p}{\partial V}$ is its bulk which gives a notion of the compressibility of the compressible fluid (air).[23] Equation 5.83 assumes a constant temperature in the fluid in which case the $B \approx 10^5 Pa$ and ρ for air at $37°C$ is about $1.145 \frac{kg}{m^3}$. This makes $c^2 = 87336 \frac{m^2}{s^2}$ ($c = 296 \frac{m}{s}$).

As we mentioned, Equation 5.83 is a one dimensional wave equation and may be solved with the following boundary conditions,

$$p(0,t) = P_0$$
$$p(L,t) = 0 \tag{5.84}$$

where $x = 0$ signifies the glottis and $x = L$ is the opening of the mouth. Since the opening of the mouth terminates into ambient pressure and since the pressure differentials are computed against ambient pressure, $P(L,t) = 0$. Also, we may assume that at the onset of a phone, the pressure at the glottis has some constant value, P_0.

The wave equation in terms of the volume velocity, $U(x,t)$, is as follows,

$$A(x) \frac{\partial}{\partial x} \left(\frac{1}{A(x)} \frac{\partial U(x,t)}{\partial x} \right) - \frac{1}{c^2} \frac{\partial^2 U(x,t)}{\partial t^2} \tag{5.85}$$

where the volume velocity,

$$U(x,t) = u(x,t)A(x) \tag{5.86}$$

where u(x,t) is the velocity of an infinitesimal control volume of the fluid at position x and time t.

The steady state solutions to Equations 5.83 and 5.85 may be computed. If we assume the model given by Figure 5.33, then the area is constant within each cylinder. If the equations are solved for each cylinder along the way and the boundary conditions are matched between solutions obtained by the left cylinder and the right cylinder, then the impedance at any transition point may be given in terms of the pressure differentials.

Flanagan [21] shows the reflection coefficient, κ_q based on this solution in terms of the pressure differential and eventually in terms of the cross sectional areas of two adjacent cylinders, by Equation 5.87.

$$\kappa_q = \frac{A_{q-1} - A_q}{A_{q-1} + A_q} \tag{5.87}$$

where $q \in \{1, 2, \cdots, Q\}$, A_1 signifies the area at the glottis and A_{Q+1} stands for the area of the pipe at the lips (see Chapter 2). The boundary condition at the ambient, $x = x_{Q+1}$ may be computed by setting $A_{Q+1} = \infty$ and by applying l'Hôpital's rule to Equation 5.87, we will have $\kappa_{Q+1} = -1$. If we set $A_0 = A_1$ at the glottis, we get $\kappa_0 = 0$ which is the other boundary condition.

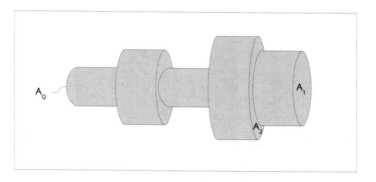

Fig. 5.32: Concentric cylinder model of the vocal tract

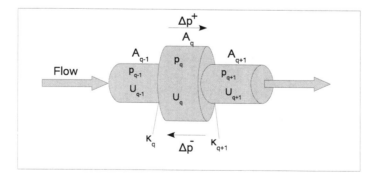

Fig. 5.33: Physical interpretation of the reflection coefficients

5.4.4 *Log Area Ratio (LAR) Features*

From Equation 5.87, consider the following ratio,

$$G_q = \frac{1 - \kappa_q}{1 + \kappa_q}$$

$$= \frac{\frac{A_{q-1} + A_q - A_{q-1} + A_q}{A_{q-1} + A_q}}{\frac{A_{q-1} + A_q + A_{q-1} - A_q}{A_{q-1} + A_q}}$$

$$= \frac{A_q}{A_{q-1}} \qquad (5.88)$$

G_q is known as the *Area Ratio*, but it is usually used in its log form,

$$g_q = \log(G_q)$$

$$= \log(\frac{A_q}{A_{q-1}}) \qquad (5.89)$$

$g_q, q \in \{1, 2, \cdots, Q\}$ are known as the *Log Area Ratio (LAR)* coefficients.

5.4.5 *Linear Predictive Cepstral Coefficient (LPCC) Features*

In Section 5.3.6 we made the case for the robustness of cepstra as features and the fact that they are more reliable than spectral features. Rabiner and Juang [60] present a recursive algorithm for computing the cepstral coefficients from the LPC coefficients. As we know, the c_0 is the energy term, so it will be given in terms of G which was defined in Equation 5.81 as follows,

$$c_0 = \log(G^2) \qquad (5.90)$$

The rest of the coefficients are given by the following recursive algorithm,

1. For $1 \leq d \leq Q$, the following recursion is used,

$$c_d = a_d + \sum_{j=1}^{d-1} \left(\frac{j}{d}\right) c_j a_{d-j} \qquad (5.91)$$

2. For $d > Q$,

$$c_d = \sum_{j=1}^{d-1} \left(\frac{j}{d}\right) c_j a_{d-j} \qquad (5.92)$$

N.B., In Equation 5.92, it is possible for $d - j$ to become larger than Q, in which case, $a_{d-j} = 0 \ \forall \ (d - j) > Q$.

These are called Linear Predictive Cepstral Coefficients (LPCC). As you recall, the order, Q, of the AR model is physically related to the number of concentric cylinders that are used to model the vocal tract. [60] recommends between $8 \leq Q \leq 16$. If we set the upper limit to the number of critical filters, determined by [103] (see Section 5.1.1.2), then the ratio of D which is the dimension of the cepstral coefficient vector becomes $\frac{3}{2}Q$.

Note that the LPCC are not based on any perceptual frequency scale (Mel or Bark), although they may be warped to do so. However, historically, the pure computation of LPCC does not warp the frequency domain. This led to the development of the Perceptual Linear Predictive (PLP) analysis which borrows ideas from MFCC and LPCC computations, but uses the Bark scale instead of the Mel scale. It also includes some other subtleties. PLP will be discussed in detail in the next section.

Also, we will see in Chapter 18 that the cepstral features are seldom used as they are. To attain a higher level of robustness to noise and condition variabilities, several liftering (filtering in the cepstra domain) techniques have been used. These techniques plus other signal enhancement methods form the topic of Chapter 18.

5.5 Perceptual Linear Predictive (PLP) Analysis

Perceptual Linear Prediction (PLP) was introduced by Hermansky [30] to work in the warping of the Frequency and spectral magnitude, based on auditory perception tests (Section 5.1), into pitch and loudness to be used mainly as a preprocessor for the Linear Prediction method (Section 5.4). The preprocessing that leads to the LP stage is very similar to the preprocessing which was discussed in the process leading to the MFCCs. There are some minor differences which will be made apparent once we go through the steps. Figure 5.34 shows the block diagram of the PLP method which was first introduced in [30].

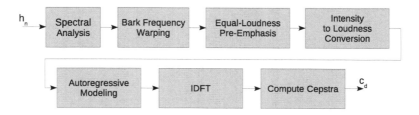

Fig. 5.34: Perceptual Linear Predictive Analysis due to Hermansky [30]

5.5.1 Spectral Analysis

The first step in the PLP is identical to the spectral analysis which was described in the computation of MFCC features in Section 5.3.3, Equation 5.32, repeated here for convenience,

$$|_lH_k| = \mathscr{R}e(_lH_k)^2 + \mathscr{I}m(_lH_k)^2 \tag{5.93}$$

5.5.2 Bark Frequency Warping

Next, analogous to the Mel Frequency Warping that was done in Section 5.3.4, [30] chooses to use the Bark scale as we had noted to be a possible alternative. Therefore, the new Spectral Magnitude in the bark scale will be given by the following conversion formula due to Schroeder [65] which converts the linear version of the angular frequency, ω, to the *Bark frequency*, $\mathscr{B}$,

$$\mathscr{B} = 6 \ \ln\left(\frac{\omega}{1200\pi} + \left[\left(\frac{\omega}{1200\pi}\right)^2 + 1\right]^{\frac{1}{2}}\right) \tag{5.94}$$

and does a warping somewhat similar to what we did in Section 5.3.4.

Of course, one should be able to just apply the Equations of Section 5.1.1.2 due to [79] which were published in the same year that *Hermansky's* paper on PLP was published. Then, the triangular filters used in Section 5.3.4 may be redesigned for the center frequencies given by Equation 5.3 and the critical bandwidths given by Equation 5.5 in terms of Barks yielding the warped Spectral Magnitudes,

$$|_l\breve{H}_m| = \tilde{\mathscr{M}}_{mk}|_lH_k| \tag{5.95}$$

where $\tilde{\mathscr{M}}_{mk}$ is the critical filter weights based on barks, $m : 0 \leq m < M$ is the new index in the Bark domain and l is the frame number. Of course, although this is the concept, [30] has not exactly taken this approach. See [30] for detailed implementation. Also note that we have been using the spectral magnitude and not the Power Spectral Density (PSD), which is the square of the magnitude. Hermansky uses the PSD instead. However, from here on, we shall use the PSD as well by squaring the term given by Equation 5.95 to produce the warped PSD, $_l\mathscr{P}_m$ defined in Equation 5.96,

$$_l\mathscr{P}_m = |_l\breve{H}_m|^2 \tag{5.96}$$

for frame index, l, and Bark scale index, m.

5.5.3 Equal-Loudness Pre-emphasis

This step is designed to do some pre-emphasis in the spirit of combining the concept of *Equal Loudness Curves*, discussed in Section 5.1.2, and the concept of pre-emphasis which changes the weights of the spectral magnitudes. Reference [30] uses an approximation to the *Equal Loudness Curves*, due to Makhoul and Cosell [47] to compute the pre-emphasis factor, $F_{p.e.}$, as a function of the angular frequency, ω, for band-limited signals with an upper cut-off frequency of $5kHz$, as,

$$F_{p.e.}(\omega) = \frac{(\omega^2 + 5.68 \times 10^7)\omega^4}{(\omega^2 + 6.3 \times 10^6)^2(\omega^2 + 3.8 \times 10^8)} \tag{5.97}$$

This pre-emphasis filter causes a $12dB/Octave$ drop in the signal strength for frequencies of up to $400Hz$. For this frequency range, the drop is similar to that used in the pre-emphasis of the signal in Section 5.2.3 for frequencies higher than $1kHz$. However, here, the drop is only $6dB/octave$ for frequencies between $1200Hz$ and $3100Hz$ and 0 for all other frequencies up to the Nyquist Critical Frequency of $5kHz$ for bandlimited signals.

For signals with a higher frequency content, an additional term is utilized which adds a sharp drop of $18dB/octave$ in the power for frequencies higher than $5kHz$.

$$F_{p.e.}(\omega) = \frac{(\omega^2 + 5.68 \times 10^7)\omega^4}{(\omega^2 + 6.3 \times 10^6)^2(\omega^2 + 3.8 \times 10^8)(\omega^6 + 9.58 \times 10^{26})} \tag{5.98}$$

The pre-emphasized and frequency-warped power spectral density, $_l\Phi_m$, for frame l and bark scale index m is given by,

$$_l\Phi_m = F_{p.e.}(\mathscr{B}) \; _l\breve{\mathscr{P}}_m \tag{5.99}$$

where the pre-emphasis filter transfer function, $F_{p.e.}(\mathscr{B})$, is the transfer function, $F_{p.e.}(\omega)$ written in terms of the Bark scale using the identity given in equation 5.94.

One practical note is the fact that the value of $_l\Phi_m$ is not well-defined for $m = 0$ (corresponding to the D.C. component) and $m = M - 1$ (corresponding to the Nyquist Critical Frequency). Reference [30] sets $_l\Phi_0 = {_l\Phi_1}$ and $_l\Phi_{M-1} = {_l\Phi_{M-2}}$ to alleviate this problem.

5.5.4 Magnitude Warping

Hermansky [30] calls this step *Intensity to Loudness Conversion*. It is in the same spirit as the Magnitude warping, discussed in Section 5.3.5, for computing the MFCC features. The difference is that instead of using the *dB* scale and normalizing against the Standard Intensity Threshold, I_0, [30] uses a cube root approximation to the *Psychophysical Power Law of Hearing* due to Stevens [73], namely,

$$_l\check{\Phi}_m = \sqrt[3]{_l\Phi_m} \tag{5.100}$$

$_l\check{\Phi}_m$ is the Bark-warped, pre-emphasized, magnitude warped power spectral density for frame l and Bark index m.

5.5.5 Inverse DFT

At this point, the transformed PSD values are run through an inverse Discrete Fourier Transform to obtain the autocorrelation function to be used in an autoregressive analysis of an all-pole model (Linear Predictive model) much in the same way as we discussed in the section on LPC (Section 5.4). [30] uses a 34-point IDFT (Inverse Discrete Fourier Transform) which produces the autocorrelation values needed for solving the Yule-Walker equations (Equation 5.70). These equations are solved to compute the autoregressive coefficients using a method such as the *Levinson-Durbin* method [60].

As we saw in Section 5.4, these coefficients may then be used to compute the cepstra. Cepstra derived by this process are called *PLP Cepstra*. In Reference [32], a *RelAtive SpecTrAl* (RASTA) filter is added to the PLP process. The new process is called *RASTA-PLP*. In Section 18.4.4, RASTA and its variations are discussed and their effects are studied on clean and noisy speech data. Also, their effectiveness levels for speech and speaker recognition applications are reviewed.

5.6 Other Features

Many other features have been proposed and tested by researchers in the past few decades. Some are new audio features [86, 64, 27, 97, 33, 29, 48, 81, 82, 38, 37, 88, 4, 51, 99, 68, 3], others are based on visual cues [11, 43, 63] and some others are just variations and transformations of established features [14, 61, 5, 101, 94, 95, 91, 40, 31, 53]. In this section we will review a few of these features. There are also many other features which have not been covered here. For example, there are many different features based on filterbank concepts spaced in uniformly, non-uniformly and uniformly in nonlinear scales. An example of one such filterbank is the Ensemble Interval Histogram (EIH) model [26] which uses a non-uniform filterbank designed with having a special ear model in mind.

5.6.1 Wavelet Filterbanks

For the past two decades, many pattern recognition approaches have been utilizing wavelet transforms for feature extraction. Some examples are ear [98, 56, 89, 84, 96], multimodal ear and face [85, 90, 93], speaker recognition in general [1, 15, 25, 24, 69], speaker identification [7, 13, 34, 52, 52, 59, 67, 78, 92], speaker verification [6, 80], and speaker segmentation [12].

In Section 24.7, we reviewed a brief introduction to wavelet series expansion. As we saw, there are an infinite number of possible wavelets that may be used. Therefore, given the broader scope of this textbook, we will not be able to cover all these different wavelets. Also, in Section 24.12 we saw that for example the Gabor wavelet is nothing but a Short Time Fourier Transform which chooses a Gaussian window, given by Equation 24.483.

Therefore, given the STFT interpretation of the wavelet and the fact that they may be used in the capacity of a filter bank to process the audio and create features somewhat akin to the STFT-based spectra, their implementation would not deviate much from the other methods which we have covered in this chapter.

[44] did a comparison study in speaker recognition using SVM among 17 different types of wavelets and found that there is not much variation in the results based on the choice of wavelet. The families of wavelets, they compared, were *Coiflets*, *Daubechies*, *Symlets*, *biorthogonal*, *reverse biorthogonal*, *Haar*, and *DMeyer*. However, they found biorthogonal-3.5 to perform the best across the board and Haar and DMeyer performed the worst, although the difference between and the best and the worst cases was nominal.

5.6.1.1 Mel-Frequency Discrete Wavelet Coefficients

Mel-Frequency Discrete Wavelet Coefficients (MFDWCs) are computed in much the same way as the MFCC features were computed in Section 5.3.6 with the exception of the final step where the DCT of the log of the spectrum is computed. The only difference in this case is that instead of using the DCT, a Discrete Wavelet

Transform (DWT) is used – see Section 24.7.

[28] proposed the usage of these features in speech recognition. [80] used them in speaker verification and reported some improvement using these features compared to standard MFCC features in aggressively noisy environments. The justification for this improvement is based on two arguments,

1. Since the basis vector of DCT encompasses all frequencies, the corruption of one frequency band affects all the MFCCs. DWT works by having the spill over from one frequency band only affect a few coefficients.
2. In the development of MFCCs the assumption was made that each frame only carries information from one phoneme. If a frame contains two consecutive phonemes, then effectively, the information of the dominant phoneme will prevail. This is especially true when a voiced phoneme is adjacent to an unvoiced phoneme and the window includes the boundary between these two phonemes. It is suggested that this may be alleviated by processing subbands separately.

5.6.1.2 Wavelet Octave Coefficients Of Residues (WOCOR)

Also known as *vocal source features*, *Wavelet Octave Coefficients of Residues* (*WOCOR*) is a feature set which is extracted by doing a pitch-synchronous wavelet transform of the *linear predictive* (*LP*) residual signals. Since Wavelet analysis may be seen as a variable window version of an analysis similar to *Fourier spectral analysis* [35], the idea of using *WOCOR* features started by using a window which is in line with the *pitch cycle* of the speech signal instead of being a fixed length window. According to [12], *WOCOR features* are less sensitive to the content of speech than *MFCC features*. *WOCOR features* are called *vocal source features* because they are dependent on the source of the speech, namely the *pitch* being generated by the vocal folds. On the other hand, *MFCC features* are thought to model the *vocal tract characteristics* since they model *fundamental frequencies* based on the *shape* of the vocal tract.

These features are well suited for *pitch-based languages* such as different dialects of Chinese (e.g., *Mandarin* and *Cantonese*). See Section 4.3.1.1 for more discussion on such languages and the role of pitch.

The following is the process of extracting *WOCOR features* used in [12] which has used these features for performing speaker segmentation.

1. Hipass filter
2. Pre-Emphasis
3. Pitch Extraction: For pitch extraction be able to create the pitch-synchronous wavelet transform, the Talkin's Robust Algorithm for Pitch Tracking (RAPT) [76]

was used to extract only voiced segments. RAPT segments the speech signal into variable length frames with common pitch values (Larynx frequencies).[5]

4. Only the voiced segments of the speech signal are processed similar to the standard overlapping windowing used in MFCC and LPCC computations. Then, the residual of the Auto Regressive method given by Equation 5.60 is computed for each frame and its amplitude is normalized to the range $[-1,1]$ for each frame.

5. In the residual signal, the pitch pulses are detected by finding the maximum amplitude in each pitch period. Then a pitch-synchronous wavelet transform is computed for each pitch period using a Hamming window which spans from the pitch pulse to the left of the current pitch pulse to the one on its right, covering two pitch periods.

6. [12] uses a fourth order Daubechies wavelet basis function [9] to expand the residual signal in the window discussed in the last step. Equation 5.101 represents this expansion.

$$w(a,b) = \frac{1}{\sqrt{a}} \sum_{n=0}^{N-1} e_n \psi \left(\frac{n-b}{a} \right) \tag{5.101}$$

where N is the window size, $a = \{2^k | k = 1, 2, \cdots, K\}$ is the scale parameter and $b \in \{0, 1, \cdots, N-1\}$ is the shift parameter for the discrete wavelet transform as defined in the original definition by Morlet [50] in order to generate the child wavelets. Note that ψ is the mother wavelet for the fourth order Daubechies wavelet basis function, ψ.

7. Assuming that there are K octave groups, the wavelet coefficients are,

$$W_k = \{w(2^k, b) | b = 0, 2, \cdots, N-1\} \ \text{where} \ k \in \{1, 2, \cdots, K\} \tag{5.102}$$

Since the W_k do not contain the temporal information about the signal, they are subdivided into M subgroups,

$$W_k^M(m) = \left\{ w(2^k, b) | b \in \left(\frac{(m-1)N}{M}, \frac{mN}{M} \right] \right\} \tag{5.103}$$

where, $m \in \{1, 2, \cdots, M\}$.

8. In this final step, the KM dimensional WOCOR features are computed as follows,

$$WOCOR = \{\|W_k^M(m)\|_2\} \tag{5.104}$$

where, $m \in \{1, 2, \cdots, M\}$ and $k \in \{1, 2, \cdots, K\}$.

[12] relies on the content independence of these features to be able to use shorter segments for determining segments. It uses $K = 4$ and $M = 4$ for the speaker seg-

---

[5] Only voiced segments of speech are retained and the unvoiced signal is considered to be approximated by random noise. As we have noted in many occasions in this book, there is great evidence that most of the speaker-dependent characteristics are within the voiced sections of the signal.

mentation implementation. In another paper [100], the authors of [12] showed that increasing K to larger than 4 does not produce any better results in speaker recognition. In [12], the domain of interest was telephone speech with a band-limitation between 300 Hz and 3400 Hz. The sub-band groups were defined to be $f > 2000$ Hz for $k = 1$, $1000 < f \leq 2000$ Hz for $k = 2$, $500 < f \leq 1000$ Hz for $k = 3$, and $250 < f \leq 500$ for $k = 4$. M is used to retain some temporal information. If $M = 1$, then all the WOCOR features are bundled into one temporal group losing all temporal information.

Since WOCOR features are derived in trying to model the vocal source characteristics and the MFCC features are related to the shape of the vocal tract, it is conceivable that combining these two sets of features would produce better results.

5.6.2 Instantaneous Frequencies

Recently, some researchers have been working with the idea that aside from extracting magnitude information for different spectral or cepstral bands, it may be beneficial to explore other information such as phase [2], the amplitude and frequencies of instantaneous spectral components [4]. One method is to use the so called, modulated features which have been shown to improve speech recognition results in both amplitude modulated (AM) and frequency modulated (FM) forms. There is also another set of features which are related to the extraction of the instantaneous frequencies. These are the, so called, *Empirical Mode Decomposition* (EMD) features proposed by Huang, et al. in [35].

5.6.2.1 Modulation Features

As we mentioned earlier, power cepstra lose all phase information. One possible improvement may come from using the phase information by using the complex cepstrum. However, another method would be the avoidance of the usage of Fourier transforms directly. Instead, one may represent the non-stationary speech signal as a sum of amplitude modulated (AM) and frequency modulation (FM) signals [45, 4]. In these techniques it is essential to be able to estimate the instantaneous frequencies of the signal to be able to devise a filter bank with these frequencies as the center frequencies of the filters. This idea resonates in many of the techniques discussed in Section 5.6 including EMD and WOCOR features. The use of modulation features in the speaker recognition field is quite new and exploratory [4, 77, 81, 82]. There have been some special names assigned to these features as well, such as *FEPSTRUM* which was coined by [82] to refer to the feature vector made up of the lower modulation frequency spectrum of the downsampled AM signal used in the

AM-FM decomposition. Another name is the *Mel Cepstrum Modulation Spectrum (MCMS)* coined by [81] which is another variation using a filter bank in the Mel domain.

5.6.3 Empirical Mode Decomposition (EMD)

Empirical Mode Decomposition (EMD) features were introduced by *Huang, et al.* [35] in 1998 and since then, they have been used in speech recognition with considerable benefits. Their merit has not been totally proven in speaker recognition yet, since their use is quite new to the field. The main idea behind EMD is the decomposition of a non-stationary signal such as speech, h_n, into a sum of band-limited functions called *Intrinsic Mode Functions (IMF)* denoted by $C_m(t)$ where $m \in \{0, 1, \cdots, M-1\}$. The only condition is that these functions must satisfy the following basic conditions,

1. The number of extrema and zero-crossings must differ by at most one throughout the whole signal.
2. At any point, the mean value between the values of the envelopes defined by the local maxima and the local minima should be equal to 0.

The following procedure is used to evaluate these functions from the signal, $h(t)$ [49].

1. Extract the extrema of the signal, $h(t)$
2. Create two envelopes, $u(t)$ and $l(t)$ which would go through the maxima and minima respectively, using a cubic spline interpolation.
3. Compute the mean value of the two envelope values for each instant of time, t,

$$\mu(t) = \frac{u(t) + l(t)}{2} \tag{5.105}$$

4. Since based on the second condition listed above, the mean value computed by the last step should be zero, subtract it from the signal,

$$g(t) = h(t) - \mu(t) \tag{5.106}$$

5. Check to see if $g(t)$ abides by the two conditions set previously for IMFs. If it does not, repeat the previous steps until it does.
6. Once the conditions are met by $g(t)$, make it the first IMF, $C_1(t) = g(t)$.
7. Now, compute the residue of the signal, by subtracting $C_1(t)$ from the original signal, $h(t)$, and perform the above steps on the residue, $r(t) = h(t) - C_1(t)$ to find the next IMF.
8. Continue the above step until the final residue is a monotonic function.

Therefore,

$$h(t) = \sum_{m=1}^{M} C_m(t) + r_M(t) \tag{5.107}$$

These, so called, empirical modes of the signal act in such a way that as we obtain a new IMF for the signal, we are covering the higher frequency parts of the signal. Therefore, the lower the indices of the IMFs, the higher their frequency contents would be. The idea is then to apply regular feature extraction techniques such as MFCC or LPCC on the subbands created by the EMD. This is essentially a filterbank technique, using the bands generated by the EMD as the bands with which the filtering is done. [49, 48] has presented a neural network-based EMD system for performing speaker identification.

5.7 Signal Enhancement and Pre-Processing

It is really important to discuss signal enhancement and other pre-processing techniques at this point to make sure that the features are not tainted by noise and other disturbances. However, since we have not yet covered some essential topics on speaker recognition which are pre-requisite to this topic, we will have to defer this discussion to Chapter 18. You may want to skim through Chapter 18, just to see what will be discussed. However, it is recommended to go through the rest of this book and once you have read Chapter 18, to return to this chapter and review it with a new perspective.

References

1. Alkhaldi, W., Fakhr, W., Hamdy, N.a.: Automatic speech/speaker recognition in noisy environments using wavelet transform. In: Circuits and Systems, 2002. MWSCAS-2002. The 2002 45th Midwest Symposium on, vol. 1, pp. I–463–6 (2002)
2. Alsteris, L.D., Paliwal, K.K.: ASR on Speech Reconstructed from Short-Time Fourier Phase Spectra. In: Proceedings of the International Conference on Spoken Language Processing (ICSLP) (2004)
3. Alsteris, L.D., Paliwal, K.K.: Evaluation of the Modified Group Delay Feature for Isolated Word Recognition. In: Proceedings of the International Symposium on Signal Processing and its Applications, vol. 2, pp. 715–718 (2005)
4. Ambikairajah, E.a.: Emerging features for speaker recognition. In: Information, Communications and Signal Processing, 2007 6th International Conference on, pp. 1–7 (2007)
5. Assaleh, K., Mammone, R.: New LP-derived features for speaker identification. IEEE Transactions on Speech and Audio Processing 2(4), 630–638 (1994)
6. Badri, N., Benlahouar, A., Tadj, C., Gargour, C., Ramachandran, V.a.: On the use of wavelet and Fourier transforms for speaker verification. In: Circuits and Systems, 2002. MWSCAS-2002. The 2002 45th Midwest Symposium on, vol. 3, pp. III–344–7 (2002)

7. Bovbel, E., Kheidorov, I., Chaikou, Y.a.: Wavelet-based speaker identification. In: Digital Signal Processing, 2002. DSP 2002. 2002 14th International Conference on, vol. 2, pp. 1005–1008 (2002)
8. Broersen, P.: Accurate ARMA models with Durbin's second method. In: IEEE International Conference on Acoustics, Speech and Signal Processing (ICASSP 1999), vol. 3, pp. 15–19 (1999)
9. Burrus, C.S., Gopinath, R.A., Guo, H.: Introduction to Wavelets and Wavelet Transforms: A Primer. Prentice Hall, New york (1997). ISBN: 0-134-89600-9
10. Byron, G.G.: The Bride of Abydos. A Turkish tale. The British Library, London (1813). Reprint: 2010
11. Cetingul, H., Yemez, Y., Erzin, E., Tekalp, A.: Discriminative Analysis of Lip Motion Features for Speaker Identification and Speech-Reading. IEEE Transactions on Image Processing **15**(10), 2879–2891 (2006)
12. Chan, W.N., Zheng, N., and, T.L.: Discrimination Power of Vocal Source and Vocal Tract Related Features for Speaker Segmentation. Audio, Speech, and Language Processing, IEEE Transactions on **15**(6), 1884–1892 (2007)
13. Chen, W.C., Hsieh, C.T., and, C.H.H.: Two-Stage Vector Quantization Based Multi-band Models for Speaker Identification. In: Convergence Information Technology, 2007. International Conference on, pp. 2336–2341 (2007)
14. Cheung, R., Eisenstein, B.: Feature selection via dynamic programming for text-independent speaker identification. IEEE Transactions on Audio, Speech and Signal Processing **26**(5), 397–403 (1978)
15. Chunrong, X., Jianhuan, Z., and, L.F.: A Dynamic Feature Extraction Based on Wavelet Transforms for Speaker Recognition. In: Electronic Measurement and Instruments, 2007. ICEMI '07. 8th International Conference on, pp. 1–595–1–598 (2007)
16. Davis, S., Mermelstein, P.: Comparison of Parametric Representations for Monosyllabic Word Recognition in Continuously Spoken Sentences. IEEE Transactions on Acoustics, Speech and Signal Processing **28**(4), 357–366 (1980)
17. Deng, L., O'Shaughnessy, D.: Speech Processing, A Dynamic and Optimization-Oriented Approach. Marcel Dekker, Inc., New york (2003). ISBN: 0-824-74040-8
18. Durbin, J.: Efficient Estimation of Parameters in Moving Average Models. Biometrika **46**, 306–316 (1959)
19. Durbin, J.: The Fitting of Time Series Models. Revue Institute International de Statistic **28**, 233–243 (1960)
20. Fant, G.: Acoustic Theory of Speech Production – with Calculations based on X-Ray Studies of Russian Articulations, revised edn. Mouton De Gruyter, The Hague (1970). ISBN: 978-9-027-91600-6
21. Flanagan, J.L.: Speech Analysis, Synthesis and Perception, 2nd edn. Springer-Verlag, New York (1972). ISBN: 0-387-05561-4
22. Fortmann, T.E., Hitz, K.L.: An Introduction to Linear Control Systems. Marcel Dekker, Inc., New York (1977). ISBN: 0-824-76512-5
23. Fox, R.W.: Introduction to Fluid Mechanics, 2nd edn. Addison-Wesley Publishing Company, New york (1978). ISBN: 0-417-01909-7
24. George, N., Evangelos, D.a.: Hands-free continuous speech recognition in noise using a speaker beam-former based on spectrum-entropy. In: Acoustics, Speech, and Signal Processing, 2002. Proceedings. (ICASSP '02). IEEE International Conference on, vol. 1, pp. I–889–I–892 (2002)
25. George, S., Dibazar, A., Liaw, J.S., Berger, T.a.: Speaker recognition using dynamic synapse based neural networks with wavelet preprocessing. In: Neural Networks, 2001. Proceedings. IJCNN '01. International Joint Conference on, vol. 2, pp. 1122–1125 (2001)
26. Ghitza, O.: Auditory Models and Human Performance in Tasks Related to Speech Coding and Speech Recognition. IEEE Transactions on Speech and Audio Processing **2**(1), 115–132 (1994)

27. Gopalan, K., Anderson, T., Cupples, E.: A comparison of speaker identification results using features based on cepstrum and Fourier-Bessel expansion. IEEE Transactions on Speech and Audio Processing **7**(3), 289–294 (1999)

28. Gowdy, J.N., Tufekci, Z.: Mel-scaled discrete wavelet coefficients for speech recognition. pp. 1351–1354 (2000)

29. Haydar, A., Demirekler, M., Yurtseven, M.: Speaker identification through use of features selected using genetic algorithm. IEE Electronic Letters **34**(1), 39–40 (1998)

30. Hermansky, H.: Perceptual linear predictive (PLP) analysis of speech. The Journal of the Acoustical Society of America (JASA) **87**(4), 1738–1752 (1990)

31. Hermansky, H., Morgan, N.: RASTA Processing of Speech. IEEE Transactions on Speech and Audio Processing **2**(4), 578–589 (1994)

32. Hermansky, H., Morgan, N., Bayya, A., Kohn, P.: RASTA-PLP speech analysis technique. In: IEEE International Conference on Acoustic, Speech, and Signal Processing, vol. 1, pp. 121–124 (1992)

33. Hsieh, C.T., Lai, E., Wang, Y.C.: Robust speech features based on wavelet transform with application to speaker identification. IEE Proceedings - Vision, Image and Signal Processing **149**(2), 108–114 (2002)

34. Hsieh, C.T., Lai, E., Wang, Y.C.a.: Robust speech features based on wavelet transform with application to speaker identification. Vision, Image and Signal Processing, IEE Proceedings - **149**(2), 108–114 (2002)

35. Huang, N.E., Shen, Z., Long, S.R., Wu, M.C., Shih, H.H., Zheng, Q., Yen, N.C., Tung, C.C., Lui, H.H.: The Empirical Mode Decomposition and the Hilbert Spectrum for Nonlinear and Non-Stationary Time Series Analysis. Proceedings of the Royal Society of London **454**(1971), 903–995 (1998)

36. ITU-T: G.711: Pulse Code Modulation (PCM) of Voice Frequencies. ITU-T Recommendation (1988). URL http://www.itu.int/rec/T-REC-G.711-198811-I/en

37. Jankowski C.R., J., Quatieri, T., Reynolds, D.a.: Formant AM-FM for speaker identification. In: Time-Frequency and Time-Scale Analysis, 1994., Proceedings of the IEEE-SP International Symposium on, pp. 608–611 (1994)

38. Kajarekar, S.S., Ferrer, L., Stolcke, A., Shriberg, E.: Voice-Based Speaker Recognition Combining Acoustic and Stylistic Features. In: N.K. Ratha, V. Govindaraju (eds.) Advances in Biometrics: Sensors, Algorithms and Systems, pp. 183–201. Springer, New York (2008)

39. Kendall, M.G.: The Estimation of Parameters in Linear Autoregressive Time Series. Econometrica **17**, 44–57 (1949). Supplement: Report of the Washington Meeting

40. Kim, M.S., Yu, H.J.: A New Feature Transformation Method Based on Rotation for Speaker Identification. pp. 68–73 (2007)

41. Kuo, B.C.: Digital Control Systems, 2nd edn. Oxford University Press, New York (1992). ISBN: 0-195-12064-7

42. Levinson, N.: The Wiener RMS (Root-Mean-Square) Error Criterion in Filter Design and Prediction. Journal of Mathematics and Physics **25**, 261–278 (1947)

43. Lim, Y.S., Choi, J.S., and, M.K.: Particle Filter Algorithm for Single Speaker Tracking with Audio-Video Data Fusion. In: Robot and Human interactive Communication, 2007. RO-MAN 2007. The 16th IEEE International Symposium on, pp. 363–367 (2007)

44. Lin, C.C., Chen, S.H., Lin, T.C., Truong, T.a.: Feature Comparison among Various Wavelets in Speaker Recognition Using Support Vector Machine. In: Multimedia, 2006. ISM'06. Eighth IEEE International Symposium on, pp. 811–816 (2006)

45. Loughlin, P.J., Tacer, B.: On the Amplitude- and Frequency-Modulation Decomposition of Signals. The Journal of the Acoustical Society of America (JASA) **100**(3), 1594–1601 (1996)

46. Makhoul, J.: Spectral Linear Prediction: Properties and Applications. IEEE Transactions on Audio, Speech and Signal Processing **23**(3), 283–296 (1975)

47. Makhoul, J., Cosell, L.: LPCW: An LPC Vocoder with Linear Predictive Spectral Warping. International Conference on Acoustics, Speech, and Signal Processing (ICASSP) **1**, 466–469 (1976)

48. Molla, M., Hirose, K., Minematsu, N.: Robust speaker identification system using multi-band dominant features with empirical mode decomposition. In: 10th International Conference on Computer and Information technology (ICCIT 2008), pp. 1–5 (2008)
49. Molla, M., Hirose, K., Minematsu, N.a.: Robust speaker identification system using multi-band dominant features with empirical mode decomposition. In: Computer and Information Technology, 2008. ICCIT 2008. 10th International Conference on, pp. 1–5 (2007)
50. Morlet, J., Arens, G., Fourgeau, I., Giard, D.: Wave Propagation and Sampling Theory. Geophysics **47**, 203–236 (1982)
51. Muroi, T., Takiguchi, T., Ariki, Y.a.: Speaker Independent Phoneme Recognition Based on Fisher Weight Map. In: Multimedia and Ubiquitous Engineering, 2008. MUE 2008. International Conference on, pp. 253–257 (2008)
52. Nghia, P.T., Binh, P.V., Thai, N.H., Ha, N.T., Kumsawat, P.a.: A Robust Wavelet-Based Text-Independent Speaker Identification. In: Conference on Computational Intelligence and Multimedia Applications, 2007. International Conference on, vol. 2, pp. 219–223 (2007)
53. Nguyen, P.C., Akagi, M., and, T.B.H.: Temporal decomposition: a promising approach to VQ-based speaker identification. In: Multimedia and Expo, 2003. ICME '03. Proceedings. 2003 International Conference on, vol. 3, pp. III–617–20 (2003)
54. Noll, A.M.: Short-Time Spectrum and 'Cepstrum' Techniques for Vocal-Pitch Detection. The Journal of the Acoustical Society of America (JASA) **36**(2), 296–302 (1964)
55. Noll, A.M.: Cepstrum Pitch Determination. The Journal of the Acoustical Society of America (JASA) **41**(2), 293–309 (1967)
56. Nosrati, M.S., Faez, K., Faradji, F.a.: Using 2D wavelet and principal component analysis for personal identification based On 2D ear structure. In: Intelligent and Advanced Systems, 2007. ICIAS 2007. International Conference on, pp. 616–620 (2007)
57. O'Shaughnessy, D.: Speech communications : human and machine, 2nd edn. IEEE Press, New York (2000). ISBN: 978-0-780-33449-6
58. Paliwal, K., Atal, B.: Frequency-related representation of speech. In: Proceedings of the European Conference on Speech Communication and Technology (EUROSPEECH-03), pp. 65–68 (2003)
59. Phan, F., Micheli-Tzanakou, E., Sideman, S.a.: Speaker identification using neural networks and wavelets. Engineering in Medicine and Biology Magazine, IEEE **19**(1), 92–101 (2000)
60. Rabiner, L., Juang, B.H.: Fundamentals of Speech Recognition. Prentice Hall Signal Processing Series. PTR Prentice Hall, New Jersey (1990). ISBN: 0-130-15157-2
61. Reynolds, D.A.: Experimental evaluation of features for robust speaker identification. IEEE Transactions on Speech and Audio Processing **2**(4), 639–643 (1994)
62. Rossing, T.D.: The Science of Sound, 3rd edn. Addison Wesley (2001). ISBN: 0-80-538565-7
63. Roy, A., Magimai-Doss, M., Marcel, S.: Boosted Binary Features for Noise-Robust Speaker Verification. In: International Conference on Acoustics, Speech, and Signal Processing (ICASSP), vol. 6, pp. 4442–4445 (2010)
64. Sambur, M.: Selection of acoustic features for speaker identification. IEEE Transactions on Audio, Speech and Signal Processing **23**(2), 390–392 (1975)
65. Schroeder, M.R.: Recognition of Complex Acoustic Signals. In: T.H. Bullock (ed.) Life Sciences Research Report; 5, p. 324. Abacon Verbag, Berlin (1977)
66. Schür, I.: On power series which are bounded in the interior of the unit circle. In: I. Gohberg (ed.) Methods in Operator Theory and Signal Processing, Operator Theory: Advances and Applications, vol. 18, pp. 31–59. Abacon Verbag (1986). Original in German in J. Reine Angew. Math., 147 (1917), pp. 205–232
67. Senapati, S., Chakroborty, S., Saha, G.a.: Log Gabor Wavelet and Maximum a Posteriori Estimator in Speaker Identification. In: Annual India Conference, 2006, pp. 1–6 (2006)
68. Seo, J., Hong, S., Gu, J., Kim, M., Baek, I., Kwon, Y., Lee, K., and, S.I.Y.: New speaker recognition feature using correlation dimension. In: Industrial Electronics, 2001. Proceedings. ISIE 2001. IEEE International Symposium on, vol. 1, pp. 505–507 (2001)

69. Siafarikas, M., Ganchev, T., Fakotakis, N.a.: Wavelet Packet Bases for Speaker Recognition. In: Tools with Artificial Intelligence, 2007. ICTAI 2007. 19th IEEE International Conference on, vol. 2, pp. 514–517 (2007)

70. Steinberg, J.C.: Position of Stimulatioin in Cochlea by Pure Tones. Journal of the Acoustical Society of America **8**(3), 176–180 (1937)

71. Stevens, S.S.: The Relation of Pitch to Intensity. Journal of the Acoustical Society of America **6**(3), 150–154 (1935)

72. Stevens, S.S.: A Scale for the Measurement of the Psychological Magnitude: Loudness. Psychological Review of the American Psychological Association **43**(5), 405–416 (1936)

73. Stevens, S.S.: On the Psychophysical Law. Psychological Review **64**(3), 153–181 (1957)

74. Stevens, S.S., Volkmann, J.E.: The Relation of Pitch to Frequency. Journal of Psychology **53**(3), 329–353 (1940)

75. Stevens, S.S., Volkmann, J.E., Newman, E.B.: A Scale for the Measurement of the Psychological Magnitude Pitch. Journal of the Acoustical Society of America **8**(3), 185–190 (1937)

76. Talkin, D.: A Robust Algorithm for Pitch Tracking (RAPT). In: W.B. Kleijn, K.K. Paliwal (eds.) Speech Coding and Synthesis. Elsevier Publishing Company, New York (1995). ISBN: 0-44-482169-4

77. Thiruvaran, T., Ambikairajah, E., Epps, J.a.: Normalization of Modulation Features for Speaker Recognition. In: Digital Signal Processing, 2007 15th International Conference on, pp. 599–602 (2007)

78. Torres, H., Rufiner, H.a.: Automatic speaker identification by means of Mel cepstrum, wavelets and wavelet packets. In: Engineering in Medicine and Biology Society, 2000. Proceedings of the 22nd Annual International Conference of the IEEE, vol. 2, pp. 978–981 (2000)

79. Traunmüller, H.: Analytical Expressions for the Tonotopic Sensory Scale. Journal of the Acoustical Society of America **88**(1), 97–100 (1990)

80. Tufekci, Z., Gurbuz, S.a.: Noise Robust Speaker Verification Using Mel-Frequency Discrete Wavelet Coefficients and Parallel Model Compensation. In: Acoustics, Speech, and Signal Processing, 2005. Proceedings. (ICASSP '05). IEEE International Conference on, vol. 1, pp. 657–660 (2005)

81. Tyagi, V., Mccowan, L., Misra, H., Bourlard, H.: Mel-Cepstrum Modulation Spectrum (MCMS) Features for Robust ASR. In: IEEE Workshop on Automatic Speech Recognition and Understanding (2003)

82. Tyagi, V., Wellekens, C.: Fepstrum Representation of Speech Signal. In: IEEE Workshop on Automatic Speech Recognition and Understanding, pp. 11–16 (2005)

83. Umesh, S., Cohen, L., Nelson, D.: Fitting the Mel scale. In: IEEE International Conference on Acoustics, Speech, and Signal Processing (ICASSP99), vol. 1, pp. 217–220 (1999)

84. Wang, Y., chun Mu, Z., Hui Zeng, a.: Block-based and multi-resolution methods for ear recognition using wavelet transform and uniform local binary patterns. In: Pattern Recognition, 2008. ICPR 2008. 19th International Conference on, pp. 1–4 (2008)

85. Wang, Y., Mu, Z.C., Liu, K., and, J.F.: Multimodal recognition based on pose transformation of ear and face images. In: Wavelet Analysis and Pattern Recognition, 2007. ICWAPR '07. International Conference on, vol. 3, pp. 1350–1355 (2007)

86. Wasson, D., Donaldson, R.: Speech amplitude and zero crossings for automated identification of human speakers. IEEE Transactions on Audio, Speech and Signal Processing **23**(4), 390–392 (1975)

87. Welch, P.: The use of fast Fourier transform for the estimation of power spectra: A method based on time averaging over short, modified periodograms. IEEE Transactions on Audio and Electroacoustics **15**(2), 70–73 (1967)

88. Wenndt, S., Shamsunder, S.a.: Bispectrum features for robust speaker identification. In: Acoustics, Speech, and Signal Processing, 1997. ICASSP-97., 1997 IEEE International Conference on, vol. 2, pp. 1095–1098 (1997)

89. Xie, Z.X., and, Z.C.M.: Improved locally linear embedding and its application on multi-pose ear recognition. In: Wavelet Analysis and Pattern Recognition, 2007. ICWAPR '07. International Conference on, vol. 3, pp. 1367–1371 (2007)

90. Xu, X., and, Z.M.: Feature Fusion Method Based on KCCA for Ear and Profile Face Based Multimodal Recognition. In: Automation and Logistics, 2007 IEEE International Conference on, pp. 620–623 (2007)

91. Yingle, F., Li, Y., Qinye, T.: Speaker gender identification based on combining linear and nonlinear features. In: 7th World Congress on Intelligent Control and Automation. (WCICA 2008), pp. 6745–6749 (2008)

92. Youssif, A., Sarhan, E., El Behaidy, W.a.: Development of automatic speaker identification system. In: Radio Science Conference, 2004. NRSC 2004. Proceedings of the Twenty-First National, pp. C7–1–8 (2004)

93. Yuan, L., and, Z.C.M.: Ear Detection Based on Skin-Color and Contour Information. In: Machine Learning and Cybernetics, 2007 International Conference on, vol. 4, pp. 2213–2217 (2007)

94. Yuan, Z.X., Xu, B.L., Yu, C.Z.: Binary quantization of feature vectors for robust text-independent speaker identification. IEEE Transactions on Speech and Audio Processing 7(1), 70–78 (1999)

95. Yuo, K.H., Hwang, T.H., Wang, H.C.: Combination of autocorrelation-based features and projection measure technique for speaker identification. IEEE Transactions on Speech and Audio Processing 13(4), 565–574 (2005)

96. Zhang, H., and, Z.M.: Compound Structure Classifier System for Ear Recognition. In: Automation and Logistics, 2008. ICAL 2008. IEEE International Conference on, pp. 2306–2309 (2008)

97. Zhang, L., Zheng, B., Yang, Z.: Codebook design using genetic algorithm and its application to speaker identification. IEE Electronic Letters 41(10), 619–620 (2005)

98. long Zhao, H., chun Mu, Z., Zhang, X., jie Dun and, W.: Ear recognition based on wavelet transform and Discriminative Common Vectors. In: Intelligent System and Knowledge Engineering, 2008. ISKE 2008. 3rd International Conference on, vol. 1, pp. 713–716 (2008)

99. Zhen, Y., and, L.C.: A new feature extraction based the reliability of speech in speaker recognition. In: Signal Processing, 2002 6th International Conference on, vol. 1, pp. 536–539 (2002)

100. Zheng, N., Ching, P., Lee, T.: Time Frequency Analysis of Vocal Source Signal for Speaker Recognition. In: Proceedings of the International Conference on Spoken Language Processing, pp. 2333–2336 (2004)

101. Zilovic, M., Ramachandran, R., Mammone, R.: Speaker identification based on the use of robust cepstral features obtained from pole-zero transfer functions. IEEE Transactions on Speech and Audio Processing 6(3), 260–267 (1998)

102. Zwicker, E.: Subdivision of the Audible Frequency Range into Critical Bands (Frequenzgruppen). Journal of the Acoustical Society of America 33(2), 248–249 (1961)

103. Zwicker, E., Flottorp, G., Stevens, S.S.: Critical Band Width in Loudness Summation. Journal of the Acoustical Society of America 29(5), 548–557 (1957)

Chapter 6
Probability Theory and Statistics

*One might perhaps say that this very thing is probable, that
many things happen to men that are not probable.*
Agathon, 445 B.C.
*(Quoted by Aristotle in the Art of
Rhetoric Chapter 24: 1402a)*

In this chapter we will review some basic Probability Theory and Statistics to
the level that applies to speaker recognition. This coverage is by no means complete. For a more complete treatment of these subjects, the avid reader is referred
to [27, 37, 39, 42, 31, 22].

To attain a good understanding of *Probability Theory*, some basic *Measure Theory* is presented with coverage of elementary set theory. This framework is then used
to define a probability measure and its related concepts as a special case.

6.1 Set Theory

To define the basics of *measure theory*, some pre-requisite definitions in set theory
are given. Let us begin with the definition of a *Sample Space*.

Definition 6.1 (Sample Space). *The Sample Space is the set $\mathscr{X}$ of all possible
values that a random variable, X, may assume. It is the set of all possible outcomes
of an experiment. In other words, it is the domain of the probability function.*

Definition 6.2 (Event). *An Event ε is any subset of outputs of an experiment involving a random variable. Namely, $\varepsilon \subset \mathscr{X}$.*

Before defining the concept of a *measure*, consider Figures 6.1, 6.2, 6.3, and 6.4.
These figures are *Venn diagram* representations of different scenarios involving one
or two subsets ($\mathscr{A}, \mathscr{B}$). In the figures, $\mathscr{X}$ is the *superset*, which in measure theory, is
known as the *sample space*. Each of the sets $\mathscr{A}$ and $\mathscr{B}$ may represents an event. $\mathscr{A}$
and $\mathscr{B}$ are called *measurable subsets* of $\mathscr{X}$ – see Definition 6.20 and its follow-up
notes on the *completion of a σ-field*. Figure 6.1 corresponds to mutually exclusive
or disjoint events. In figure 6.2, the two events $\mathscr{A}$ and $\mathscr{B}$ have a common space. This
intersection ($\mathscr{A} \cap \mathscr{B}$) relates to a *Logical And*, namely, ($\mathscr{A} \wedge \mathscr{B}$) in the sense that an
outcome is a member of both sets, $\mathscr{A}$ and $\mathscr{B}$. In some circles, this relationship may

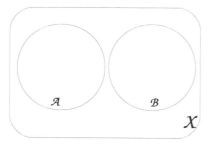

Fig. 6.1: $\mathscr{A} \cap \mathscr{B} = \{\varnothing\}$
(Disjoint) (Mutually Exclusive)

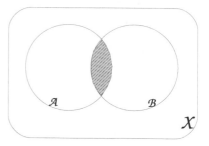

Fig. 6.2: Intersection, $\mathscr{A} \cap \mathscr{B}$
$(\mathscr{A} \wedge \mathscr{B})(\mathscr{A}, \mathscr{B})$

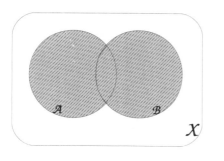

Fig. 6.3: Union, $\mathscr{A} \cup \mathscr{B}$
$(\mathscr{A} \vee \mathscr{B})(\mathscr{A} \mid \mathscr{B})$

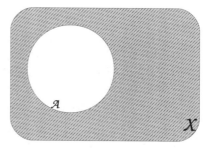

Fig. 6.4: Complement of $\mathscr{A}$, $\mathscr{A}^{\complement} =$
$\mathscr{X} \setminus \mathscr{A}$
$(!\mathscr{A})(\mathscr{A}')$

be denoted as $(\mathscr{A}, \mathscr{B})$ [31] and some others use, $(\mathscr{A}\mathscr{B})$ [42]. In this book, we will be using $(\mathscr{A} \cap \mathscr{B})$ to refer to sets and $(\mathscr{A}, \mathscr{B})$ or $(\mathscr{A} \cap \mathscr{B})$, interchangeably, when referring to events. Figure 6.3 indicates the union of two events, $(\mathscr{A} \cup \mathscr{B})$. It is related to the *Logical Or*, namely, $(\mathscr{A} \vee \mathscr{B})$, and is sometimes written as $(\mathscr{A} + \mathscr{B})$ [42]. The case described by figure 6.4 refers to any outcome which is not a part of event $\mathscr{A}$. It is denoted as $\mathscr{A}^{\complement}$ in this text. Also, whenever there is ambiguity about the universe, then the complement is denoted in terms of the difference between the universe, $\mathscr{X}$ and the set $\mathscr{A}$ written by the following notation, $\mathscr{X} \setminus \mathscr{A}$, which may be viewed as the complement of $\mathscr{A}$ with respect to the universe, $\mathscr{X}$. It is related to the *Logical Not* and in the notation of logic it may be written as $(!\mathscr{A})$. It is sometimes written as $\mathscr{A}'$ or $\overline{\mathscr{A}}$.

Now let us examine the very intuitive but powerful, *De Morgan's law*.

Law 6.1 (De Morgan's Law). *De Morgan's law states the following two identities,*

$$\{\bigcup_{i=1}^{N} \mathscr{A}_i\}^{\complement} = \bigcap_{i=1}^{N} \mathscr{A}_i^{\complement} \tag{6.1}$$

and

$$\{\bigcap_{i=1}^{N} \mathscr{A}_i\}^\complement = \bigcup_{i=1}^{N} \mathscr{A}_i^\complement \tag{6.2}$$

Equations 6.1 and 6.2 may be easily verified by observing Figure 6.3 and extending the results to more than two subset.

Another useful definition in *set theory* which will come in handy in working toward the definition of a *measure* is the concept of a *class*.

Definition 6.3 (Class). *A class, $\mathfrak{C}$, is a collection of sets which have something in common allowing them to be specifically identified by this specific trait.*

For example, one may speak of the class of Indo-European Languages. In this case, each language belonging to this class may be a set of different dialects. In application to probability theory, any subset which is a member of a class may also be thought of as a representation of one or more events.

As we shall see later, the following concepts will be needed for a rigorous definition of a *measure* and for that of *random variables*.

Definition 6.4 (Countable Base). *see [39]*
A countable base at x is a countable class, $\mathfrak{C}$, of neighborhoods of x such that every neighborhood of x contains a member of $\mathfrak{C}$.

Definition 6.5 (Countable Space). *A countable space, $\mathscr{X}$, is a space which is separable and has a countable base at every point $\{x : x \in \mathscr{X}\}$.*

Note that space $\mathscr{X}$ may be *separable* into an *infinite* number of points, x_i. However, at every point, x_i, X has a countable base. If the number of points becomes *infinite*, then the superset representing that space is known to be a *countably infinite set*. Here is a simple definition of a *countably infinite space* which also presents an alternative perspective of a *countably infinite set*.

Definition 6.6 (Countably Infinite Space). *A countably infinite space is a space which has a finite number of points in a finite interval. Note that a set representing a countably infinite space may have an infinite number of members overall, such as the set of natural numbers, $\mathbb{N}$. Such a sample set with an infinite number of overall members is said to be a countably infinite set and the corresponding space is said to be a countably infinite space.*

One of the properties, that certain classes of sets may possess, is the notion of *closure* (being closed) under certain set operations.

Definition 6.7 (Closure under a set operation). *A class,* $\mathfrak{C}$*, is closed under a set operation such as a union, an intersection, a complement, et cetera, if the operation as applied to its members would generate partitions which are also contained in that class.*

6.1.1 Equivalence and Partitions

Often we speak of members of a set being *equivalent*. The concept of equivalence has to be defined with respect to an *equivalence relation*. Basically, this means that every time we speak of equivalence of objects, we would have to qualify this equivalence by defining an *equivalence relation* which gives us the logic for the equivalence at hand. Consider an *equivalence relation* given by the symbol, R. Then $x \overset{R}{=} y$ means that x and y are equivalent as far as the logic in the equivalence relation R dictates.

Definition 6.8 (Equivalence Relation). *An equivalence relation,* R*, is a relation which generally allows for a binary response to the question of equivalence between objects. All equivalence relations must obey the following three properties,*

  1. *Reflexivity:* $x \overset{R}{=} y$
  2. *Symmetry:* $x \overset{R}{=} y \Longleftrightarrow y \overset{R}{=} x$
  3. *Transitivity:* $x \overset{R}{=} y \wedge y \overset{R}{=} x \Longrightarrow x \overset{R}{=} z$

Therefore, any relation that maintains the above three properties is called an equivalence relation.

   Another way of looking at equivalence is the amount of indiscernibility[1] between objects. Therefore, R *is also called an indiscernibility relation [43].*

Definition 6.9 (Equivalence Class). *If we pool all the objects that are equivalent into distinct classes of objects in the universal set* $\mathcal{X}$*, each distinct class containing only equivalent objects is called an equivalence class and is denoted by* $[\xi]_R$ *for equivalence relation* R*. A formal mathematical definition of* $[\xi]_R$ *is as follows,*

$$\mathcal{X}_\xi = [\xi]_R \tag{6.3}$$

$$\overset{\Delta}{=} \{x \in \mathcal{X} : x \overset{R}{=} \xi\} \tag{6.4}$$

---

[1] *Indiscernible* means "not distinct", hence *indiscernible* objects are objects that are *similar*.

Definition 6.10 (Quotient Set). *A quotient set of universe $\mathscr{X}$, with respect to equivalence relation* R, *is the set of all equivalence classes of $\mathscr{X}$ due to* R *and it is denoted by $\mathscr{Q} = \mathscr{X}/R$.*

Definition 6.11 (Parition). *A partition, $\mathscr{P}$, is a quotient set of $\mathscr{X}$ according to a partition equivalence relation,* P. *Therefore,*

$$\mathscr{P} = \mathscr{X}/P \tag{6.5}$$

where P *is generally designed to split the universal set into equivalence classes having some desired features.*

As we shall see later, an *equivalence relation* is quite similar to the concept of a *measure* which will be defined in Section 6.2. It will become more clear as we cover more equivalence concepts in this section and when we continue with the treatment of *measure theory* and the concept of a *measurable space*. In fact, we shall see that an *equivalence relation* may be viewed as a discrete measure in space $\mathscr{X}$, creating a *measure space*, $(\mathscr{X}, \mathfrak{X}, R)$. $\mathfrak{X}$ would then be a *Borel field* of $\mathscr{X}$ and R is the measure.

All the set theoretic definitions up to this point have made the assumption that objects either belong to a set or they do not. This, so called, *crisp logic*, responds to the question of equivalence in a binary fashion. Other types of sets have been developed in the past few decades which handle the concept of *uncertainty* in membership. This *uncertainty* may be viewed as the existence of objects in the boundary that a set shares with its complement, such that the membership of these objects into the set $\mathscr{A}$, and its complement $\mathscr{A}^{\complement} = \mathscr{X} \setminus \mathscr{A}$, is defined by *non-crisp logic* or in other words through an *uncertain membership*.

A generic two-class *crisp partition* may be denoted as follows, $\mathscr{P} = \{\mathscr{A}, \mathscr{X} \setminus \mathscr{A}\}$. In a *crisp* partitioning logic, we may define a binary function associated with set $\mathscr{A}$, called the *characteristic function* and denoted by $\Upsilon_{\mathscr{A}}(x)$. This function defines the membership of object x to set $\mathscr{A}$ and is defined as follows,

$$\Upsilon_{\mathscr{A}}(x) \triangleq \begin{cases} 1 \ \forall \ \{x : x \in \mathscr{A}\} \\ 0 \ \forall \ \{x : x \notin \mathscr{A}\} \end{cases} \tag{6.6}$$

As we shall see later, the *characteristic function* of a general set need not be binary for sets that allow *soft membership* such as *rough sets* and *fuzzy sets*. However, the definition does require that in a *universe* consisting of Γ disjoint sets denoted by $\mathscr{A}_\gamma, \gamma \in \{0, 1, \cdots, \Gamma\}$,

$$0 \leq \Upsilon_{\mathscr{A}_\gamma}(x) \leq 1 \ \forall \ \gamma \in \{1, 2, \cdots, \Gamma\} \tag{6.7}$$

and

$$\sum_{\gamma=1}^{\Gamma} \Upsilon_{\mathscr{A}_\gamma}(x) = 1 \tag{6.8}$$

In reality, it is possible that there exist objects which are *indiscernible* against x and yet some of them belong to $\mathscr{A}$ while some others belong to $\mathscr{A}^{\complement} = \mathscr{X} \setminus \mathscr{A}$. There are two different approaches for handling these cases, the *rough set* approach and the *fuzzy set* approach. In the next two sections, we will provide some of the fundamental definitions for these two approaches. As we will see soon, the two approaches are quite similar and it is possible to map the two approaches back and forth.

6.1.2 R-Rough Sets (Rough Sets)

R-rough sets (*rough sets* in short) were introduced by *Pawlak* [43], in 1982, to handle the non-exact memberships of objects to sets based on the *indiscernibility* of those objects. In the rough set approach to handling uncertainty in set boundaries, one defines an *approximation* [44, 46] to *R-exact* set $\mathscr{A}$, induced by *equivalence relation* R, called the *lower approximation*, $\underline{R}(\mathscr{A})$, which includes those objects, x, whose *equivalence class*, based on *equivalence relation* R, and denoted by $[x]_R$, is completely contained in $\mathscr{A}$. Therefore, $x \in \mathscr{A}$ in the traditional (*crisp*) sense, or in other words, $x \in \mathscr{A}$ with *certainty*.

On the other hand, there are those objects which are members of the equivalence class $[x]_R : [x]_R \cap \mathscr{A} \neq \{\varnothing\} \wedge [x]_R \not\subset \mathscr{A}$. These objects make up the, so called, *upper approximation* of $\mathscr{A}$, $\overline{R}(\mathscr{A})$. Therefore, we may define the *boundary* of $\mathscr{A}$ with its complement, based on equivalence relation R, as follows,

$$\mathscr{B}_R(\mathscr{A}) \overset{\Delta}{=} \overline{R}(\mathscr{A}) \setminus \underline{R}(\mathscr{A}) \tag{6.9}$$

In other words, $\mathscr{B}_R(\mathscr{A})$ is the set of all objects in $[x]_R$ which are *possibly* a part of $\mathscr{A}$, but not those which are members of $\mathscr{A}$ with *certainty*.

A *rough set* associated with the *crisp set*, $\mathscr{A}$ is represented by its lower approximation (lower bound) and upper approximation (upper bound) sets, $\underline{R}(\mathscr{A})$ and $\overline{R}(\mathscr{A})$ [44, 46], respectively.

6.1.3 Fuzzy Sets

In the *fuzzy set* approach to handling uncertainty, instead of defining lower and upper approximations to sets, the *characteristic function* of a set, $\mathscr{A}$, given by $\Upsilon_{\mathscr{A}}(x)$, is redefined from the binary function given by Equation 6.6 to a function that may take on any value in $[0,1]$.

We may split the closed interval, $[0,1]$, into two separate intervals, $[0,1)$ and 1. Then all objects $\{x : \Upsilon_{\mathscr{A}}(x) = 1\}$ belong to the *crisp set $\mathscr{A}$ with certainty*, and all objects $\{x : \Upsilon_{\mathscr{A}}(x) \in (0,1)\}$ belong to set $\mathscr{A}$ *probably*. Of course there is another set of objects, $\{x : \Upsilon_{\mathscr{A}}(x) \in 0\}$, which do not belong to set $\mathscr{A}$ in any form.

We noted that there is a mapping between *R-rough* and *fuzzy sets*. Let us examine these mappings. Assume that we have a partition logic, P, that would generate the lower and upper approximations for set $\mathscr{A}$ ($\underline{R}(\mathscr{A})$ and $\overline{R}(\mathscr{A})$) in a rough set approach. Then the *fuzzy set characteristic function* may be calculated based on these approximations as follows [46],

$$\Upsilon_{\mathscr{A}}(x) = \frac{|\mathscr{A} \cap [x]_{\mathsf{P}}|}{|[x]_{\mathsf{P}}|} \tag{6.10}$$

where $|\mathscr{A}|$ denotes the number of objects in $\mathscr{A}$ or any other defined magnitude function. The fuzzy membership due to the characteristic function in Equation 6.10 will then produce results which are equivalent to the *R-rough set partition logic* P.

Similarly, if we are presented with the fuzzy set characteristic function, $\Upsilon_{\mathscr{A}}(x)$, we may deduce the R-rough partition logic by the following dual relation [46],

$$[x]_{\mathsf{P}} = [\xi]_{\mathsf{P}} \Longleftrightarrow \Upsilon_{\mathscr{A}}(x) = \Upsilon_{\mathscr{A}}(\xi) \tag{6.11}$$

Polkowski [46] presents *hybrid uncertain sets* in the form of *rough-fuzzy sets* and *fuzzy-rough sets*. An elaborate treatment of these *hybrid sets*, along with a full treatment of the two topics is given in [46].

6.2 Measure Theory

At this point, we develop the definition of a *measure*, en route to a further refinement in order to be able to define a specific type of measure, namely, a *probability measure*.

6.2.1 Measure

Let us start by defining a fundamental *class* of subsets called a *Field* which in some circles is known as an *Algebra*.

Definition 6.12 (Field, Algebra, or Boolean Algebra). *See [39, 28, 42]. Consider the superset, $\mathcal{X}$, over which a Field, $\mathfrak{F}$, is defined as a non-empty class of subsets of $\mathcal{X}$ such that it is closed under finite complementation and union operations. Namely, a Field, $\mathfrak{F}$, possesses the following properties:*

1. *Field $\mathfrak{F}$ is non-empty:*

$$\mathfrak{F} \neq \{\varnothing\} \tag{6.12}$$

2. *Field $\mathfrak{F}$ is closed under complementation:*

$$\mathcal{A} \in \mathfrak{F} \implies \mathcal{A}^{\complement} \in \mathfrak{F} \tag{6.13}$$

3. *A Field is closed under union. In other words,*

$$\{\mathcal{A}_i : i \in \{1, 2, \cdots, N\}\} \in \mathfrak{F} \implies \bigcup_{i=1}^{N} \mathcal{A}_i \in \mathfrak{F} \tag{6.14}$$

Note that from the fundamental properties of closure under complementation and finite union, other properties of the Field follow including closure under all finite set operations.

4. *A Field is closed under finite intersection. Namely,*

$$\{\mathcal{A}_i : i \in \{1, 2, \cdots, N\}\} \in \mathfrak{F} \implies \bigcap_{i=1}^{N} \mathcal{A}_i \in \mathfrak{F} \tag{6.15}$$

Rationale:
This is easy to show by noting that since $\mathcal{A}_i \in \mathfrak{F}$, then based on property 2,

$$\mathcal{A}_i^{\complement} \in \mathfrak{F} \tag{6.16}$$

Also, by applying property 3, on the $\mathcal{A}_i^{\complement}$,

$$\mathcal{B} = \bigcup_{i=1}^{N} \mathcal{A}_i^{\complement} \in \mathfrak{F} \tag{6.17}$$

Again, applying property 2 to $\mathcal{B}$,

$$\mathscr{B}^{\complement} = \{\bigcup_{i=1}^{N} \mathscr{A}_i^{\complement}\}^{\complement} \in \mathfrak{F} \tag{6.18}$$

According to De Morgan's law (see Law 6.1), $\mathscr{B}^{\complement}$ is just the intersection of $\mathscr{A}_i$, namely,

$$\mathscr{B}^{\complement} = \bigcap_{i=1}^{N} \mathscr{A}_i \in \mathfrak{F} \tag{6.19}$$

5. *A Field contains the complete superset, $\mathscr{X}$, also known as the certain event.*

Rationale:
Based on property 1, $\mathfrak{F}$ is non-empty, so it must contain at least one event. Let us call this event (member), $\mathscr{A}$. Then,

$$\mathscr{A} \in \mathfrak{F} \tag{6.20}$$

Based on property 2,

$$\mathscr{A}^{\complement} \in \mathfrak{F} \tag{6.21}$$

Due to property 3, then we can write,

$$\mathscr{A} \cup \mathscr{A}^{\complement} \in \mathfrak{F} \tag{6.22}$$

But $\mathscr{A} \cup \mathscr{A}^{\complement} = \mathscr{X}$. Therefore,

$$\mathscr{X} \in \mathfrak{F} \tag{6.23}$$

6. *$\mathfrak{F}$ contains the empty set, $\{\varnothing\}$ also known as the impossible event.*

Rationale:
Let us apply property 4 on $\mathscr{A}$ and $\mathscr{A}^{\complement}$,

$$\mathscr{A} \cap \mathscr{A}^{\complement} \in \mathfrak{F} \implies \{\varnothing\} \in \mathfrak{F} \tag{6.24}$$

Based on the extended properties which followed the three main properties listed in Definition 6.12, namely being a non-empty class and being closed under complementation and finite union operations, we may make a more convenient statement of the definition of a *Field* as follows:

Definition 6.13 (Field). *A Field, $\mathfrak{F}$, defined in the space of a superset, $\mathscr{X}$, is a non-empty class of subsets of $\mathscr{X}$ such that it is closed under all finite set operations.*

Definition 6.13 specifies closure under *all finite* operations. Let us increase N which is the number of subclasses of $\mathscr{X}$ to become *infinite*, but apply the restriction that $\mathscr{X}$ must remain *countable*. The new category of Field which emerges from this extension is a class which remains closed under *all countable set operations*, known as a *σ-Field* which is, of course, still a *Field*. Therefore, we formally define a *σ-Field* as follows,

Definition 6.14 (σ-Field (σ-algebra)). *A σ-Field (also known as σ-algebra),* $\mathfrak{X}$, *defined in the space of a superset,* $\mathscr{X}$, *consisting of a countably infinite number of subsets,* $\{\mathscr{A}_i : i \in \mathbb{N}\}$ *is a non-empty class of subsets of* $\mathscr{X}$ *such that it is closed under all countable set operations. Namely,*

1. $\mathfrak{X} \neq \{\varnothing\}$
2. $\{\mathscr{A}_i : i \in \mathbb{N}\} \in \mathfrak{X} \implies \bigcup\limits_{i=1}^{\infty} \mathscr{A}_i \in \mathfrak{X}$
3. $\mathscr{A} \in \mathfrak{X} \implies \mathscr{A}^{\mathsf{C}} \in \mathfrak{X}$, *namely,* $\mathscr{X} - \mathscr{A} \in \mathfrak{X}$
4. $\{\mathscr{A}_i : i \in \mathbb{N}\} \in \mathfrak{X} \implies \bigcap\limits_{i=1}^{\infty} \mathscr{A}_i \in \mathfrak{X}$

Note that based on the definition of $\mathfrak{X}$, $\mathscr{X}$ and $\{\varnothing\}$ are also members of $\mathfrak{X}$, namely, $\mathscr{X} \in \mathfrak{X}$ and $\{\varnothing\} \in \mathfrak{X}$.

Now consider the open set of all *finite real numbers*, namely, $\{\mathscr{X} : -\infty < x < \infty\}$, known as the *Real Line* [39]. Then these definitions may follow:

Definition 6.15 (Borel Field). *See [39, 27]. In a space,* $\{\mathscr{X} : \mathscr{X} = \mathbb{R}\}$, *the smallest σ-Field,* $\mathfrak{X}$, *of all the closed intervals in* X, $\{\{x : a \leq x \leq b\} \,\forall\, a, b \in \mathbb{R}\}$, *is known as a Borel field or a Borel σ-field in* $\mathscr{X}$.

Note that the definition of a *Borel field* may very well have been made using *open intervals* in X, $\{\{x : a < x < b\} \,\forall\, a, b \in \mathbb{R}\}$, since the difference between the two definitions is only the inclusion of the end-points. Because these end-points are arbitrary, the resulting Field would be identical. By the same argument, the definition is true even for *half-closed intervals*, $\{\{x : a \leq x < b\} \,\forall\, a, b \in \mathbb{R}\}$ and $\{\{x : a < x \leq b\} \,\forall\, a, b \in \mathbb{R}\}$. In fact, the *Borel field* contains all closed intervals, open intervals, half-closed intervals and their countably infinite unions and intersections [39].

Definition 6.16 (Borel Sets). *The subsets making up a Borel field in* $\{\mathscr{X} : \mathscr{X} = \mathbb{R}\}$ *are known as Borel Sets of space* $\mathscr{X}$.

Definition 6.17 (Measurable Space). *The combination of the superset,* $\mathscr{X}$, *and its σ-Field,* $\mathfrak{X}$, *is known as a measurable space and is written as,* $(\mathscr{X}, \mathfrak{X})$ *[29, 28]. If* $\mathfrak{X}$ *is a Borel field, this space is known as a Borel Line [39].*

Definition 6.18 (Measure). [2] *Let us assume that $\{\mathcal{A}_i : i \in \mathbb{N}\}$ are disjoint members of a σ-Field, $\mathfrak{X}$, of a superset, $\mathscr{X}$. Then, a measure, $\rho(\mathcal{A})$ for $\{\mathcal{A} : \mathcal{A} \in \mathfrak{X}\}$ is defined over the measurable space $(\mathscr{X}, \mathfrak{X})$ if it meets the following criteria:*[3]

1. *$\{\rho(\mathcal{A}) \in \mathbb{R} \lor \rho(\mathcal{A}) = \infty\} \ \forall \ \{\mathcal{A} : \mathcal{A} \in \mathfrak{X}\}$*
2. *$\rho(\mathcal{A}) \geq 0 \ \forall \ \{\mathcal{A} : \mathcal{A} \in \mathfrak{X}\}$*
3. *If $\{\mathcal{A}_i : i \in \mathbb{N}\}$ are disjoint subsets of $\mathscr{X}$, namely,*
   $\mathcal{A}_i \subset \mathscr{X} \ \forall \ \{\mathcal{A}_i : \mathcal{A}_i \in \mathfrak{X}, i \in \mathbb{N}\}$, then,

$$\bigcap_{i=1}^{\infty} \mathcal{A}_i = \{\varnothing\} \implies \rho(\bigcup_{i=1}^{\infty} \mathcal{A}_i) = \sum_{i=1}^{\infty} \rho(\mathcal{A}_i) \tag{6.25}$$

If Definition 6.18 includes the possibility of $\rho(\mathcal{A}) = \infty$, then measure $\rho(\mathcal{A})$ is known as an *unbounded measure*. However, $\rho(\mathcal{A})$ may also be defined with the restriction that $\rho(\mathcal{A}) \in \mathbb{R}$ and that it may not become infinite. In that case, it is referred to as a *bounded measure*.

Definition 6.19 (Lebesgue Measure). *See [38, 20, 14, 27]. If $\mathfrak{X}$ is a Borel field, as described by Definition 6.15, of all half-closed intervals in X signified by the subset, $\mathcal{A}$ such that $\{\{x : a \leq x < b\} \ \forall \ a, b \in \mathbb{R}\}$, then the length of each interval, $\lambda(\mathcal{A}) = |b - a|$, adhering to the criteria set by the definition of a measure (Definition 6.18), is known as the Lebesgue measure for that interval.*

Lebesgue measure is defined and is equivalent for half-closed, open and closed subsets as discussed following Definition 6.15 of a *Borel field*.

Definition 6.20 (Lebesgue Measurable Subsets). *The class of Lebesgue measurable subsets of $\mathscr{X}$ is the smallest σ-Field of Borel Subsets of $\mathscr{X}$ plus an additional subset made up of a set contained in any of the Borel Sets with a zero Lebesgue measure.*

The methodology for the generation of the σ-field with respect to the Lebesgue measure given in Definition 6.20 may be applied to any measure. It is called the *completion* of the σ-field with respect to a measure [27].

Definition 6.21 (Lebesgue Field). *The completed Borel field with respect to the Lebesgue measure, λ, of all Lebesgue measurable subsets of $\mathscr{X}$ is known as the*

---

[2] In most measure theory literature, a measure is signified by $\mu(\mathcal{A})$, but in this book, since we use μ to signify the mean, we have chosen to use $\rho(\mathcal{A})$ instead. This should not cause much trouble since this notation is only used in passing in order to create the foundation needed for defining a probability measure and then it will no longer be needed.

[3] A number $\{\rho : \rho \in \mathbb{R} \lor \rho = \pm\infty\}$ is called an *extended Real number*

Lebesgue Field, $\mathfrak{L}$, in $\mathscr{X}$.

Definition 6.22 (Lebesgue Measure Space). *The combination of the superset, $\mathscr{X}$, and the Lebesgue field of Lebesgue measurable subsets of $\mathscr{X}$, $\mathfrak{L}$, and the Lebesgue measure given by Definition 6.19 denoted by $\lambda(\mathscr{A})$ for any subset $\mathscr{A}$ representing an interval $\{x : a \leq x < b\}$ is known as a Lebesgue measure space and is written as $(\mathscr{X}, \mathfrak{L}, \lambda)$.*

In Chapter 15, as well as later in this chapter, when we speak about the *Hilbert space* (Section 6.2.6.2), we will use the concept of a *complete space*. Definition 24.53 was the formal mathematical definition of a *complete space*. The following is a useful verbal definition.

Definition 6.23 (Complete Space). *A complete space is a vector space in which every fundamental sequence[4] (Definition 6.71) converges – see Definition 24.53.*

6.2.2 Multiple Dimensional Spaces

To be able to define a *topological product space*, let us first define a *Cartesian product* ($\times$).

Definition 6.24 (Cartesian Product). *If we have two sets, $\mathscr{X}$ and $\mathscr{Y}$ which may generally be defined in different spaces, then the Cartesian product of $\mathscr{X}$ and $\mathscr{Y}$, written as $\mathscr{X} \times \mathscr{Y}$ is defined as the set of all ordered pairs, $\{(x,y) : x \in \mathscr{X} \wedge y \in \mathscr{Y}\}$.[28]*

The *Euclidean space* is a good example for a *Cartesian Product Space*. For example the two dimensional Euclidean space may be written as the space created by the *Cartesian product* of *real lines* $\mathscr{X}$ and $\mathscr{Y}$, written as $\{\mathscr{X} \times \mathscr{Y} : \mathscr{X} = \mathbb{R}, \mathscr{Y} = \mathbb{R}\}$. This two-dimensional case of the Euclidean space is known as the *Euclidean plane*. Other descriptions of *Euclidean geometry* follow. For example, if we have two subsets of $\mathscr{X}$ and $\mathscr{Y}$ such that, $\{\mathscr{A} : \mathscr{A} \subset \mathscr{X}\}$ and $\{\mathscr{B} : \mathscr{B} \subset \mathscr{Y}\}$, then the *Cartesian product*, $\mathscr{A} \times \mathscr{B} \subset \mathscr{X} \times \mathscr{Y}$ is the set of ordered pairs which make up a *rectangle* in the *Euclidean plane*, $\mathscr{X} \times \mathscr{Y}$. With the same analogy, $\mathscr{A}$ and $\mathscr{B}$ are known as the *sides* of this *rectangle* in the *Euclidean plane*.

In general, we may consider a *Cartesian product space* of *M Real Lines* such that,

---

[4] A *fundamental sequence* is also known as a *Cauchy sequence*.

$$\mathscr{X} = \underset{m=1}{\overset{M}{\times}} \mathscr{X}_m \tag{6.26}$$

where $\{\mathscr{X}_m : \mathscr{X}_m = \mathbb{R} \ \forall \ m \in \{1, 2, \cdots, M\}\}$. $\mathscr{X}$ is known as a *Real Space*. As a short-hand, we state that, $\{\mathscr{X} : \mathscr{X} = \mathscr{R}^m\}$ If there is a *Borel field* associated with each $\mathscr{X}_m$, denoted as $\mathfrak{X}_m$, then the Borel Field corresponding to the subsets of space $\mathscr{X}$, denoted as $\mathfrak{X}$, may be written in terms of the product of Borel fields, $\mathfrak{X}_m$, as follows,

$$\mathfrak{X} = \underset{m=1}{\overset{M}{\times}} \mathfrak{X}_m \tag{6.27}$$

The resulting product space, $(\mathscr{X}, \mathfrak{X})$ is known as a *Borel Space*.

In general, the *Lebesgue measure space*, $(\mathscr{X}, \mathfrak{L}, \lambda)$, given by Definition 6.22, may be defined in the product space such that $\{\mathscr{X} : \mathscr{X} = \mathscr{R}^m, m \geq 1\}$, with the *Borel field* of all *Lebesgue measurable subsets* of $\mathscr{X}$ written as $\mathfrak{L}$, and the *Lebesgue measure*, λ. However, generally speaking, not all subsets of $\mathscr{R}^m$ are measurable. For example, if we consider $m > 2$, we come across some subsets which are not measurable, leading to anomalies such as the *Banach-Tarski paradox* [6]. Therefore, we are only concerned with measurable subsets of the space. This is also the rationale for specifying the need to use the *Borel field* which is the smallest σ-field in the space to avoid such non-measurable subsets.

6.2.3 Metric Space

Definition 6.25 (Metric Space). *A Metric Space, $\mathscr{M}(\mathscr{X}, d)$, is a space, $\mathscr{X}$, for which a function, $\{d : \{\mathscr{X} \times \mathscr{X}\} \mapsto \mathbb{R}\}$, is defined and associated with some prescribed properties. Since the real-valued metric function d (also known as a distance) has, for its domain, the Cartesian product space of $\{\mathscr{X} \times \mathscr{X}\}$, it will be written in terms of a pair of values of X, namely, $d(x_1, x_2)$. The prescribed properties of the distance, $d(x_1, x_2)$, are as follows,*

1. *$d(x_1, x_2) \in \mathbb{R}$*
2. *$d(x_1, x_2) \geq 0 \ \forall \ x_1, x_2 \in \mathbb{R}$*
3. *$d(x_1, x_2) = 0 \iff x_1 = x_2$*
4. *$d(x_1, x_2) = d(x_2, x_1)$ (Symmetry)*
5. *$d(x_1, x_2) \leq d(x_1, x_3) + d(x_2, x_3)$ triangular inequality*

Definition 6.26 (Distance Between Subsets). *The distance, $d(\mathscr{A}_1, \mathscr{A}_2)$, between two subsets of $\mathscr{X}$, $\{\mathscr{A}_1 : \mathscr{A}_1 \subset \mathscr{X}\}$ and $\{\mathscr{A}_2 : \mathscr{A}_2 \subset \mathscr{X}\}$, is given by,*

$$d(\mathscr{A}_1, \mathscr{A}_2) = \inf\{d(x_1, x_2)\} \tag{6.28}$$

where $\{x_1 : x_1 \in \mathscr{A}_1\}$ and $\{x_2 : x_2 \in \mathscr{A}_2\}$.

Definition 24.23 introduces the concept of convexity, used for defining a convex function. A similar definition may be made for a *metric space* and more specifically for special subsets of the *Euclidean space* as follows,

Definition 6.27 (Convex Metric Space). *A metric space, $\mathscr{M}(\mathscr{X}, d)$, is said to be a convex metric space if for every two of its distinct members, $x_0, x_2 \in \mathscr{X} : x_0 \neq x_2$, there exists a member, $x_1 \in \mathscr{X}$ such that,*

$$d(x_0, x_2) = d(x_0, x_1) + d(x_1, x_2) \qquad (6.29)$$

Definition 6.28 (A Convex subset of the Euclidean Space). *A Convex subset of the n-dimensional Euclidean Space, denoted by $\text{conv}\mathscr{A} \subset \mathscr{R}^n$, is one in which all the points on any straight line between any two members of the subset are also members of that subset.*

Definitions 24.53 and 6.23 presented the concept of a *complete space*. The following definition is an extension which applies to *metric spaces*.

Definition 6.29 (Complete Metric Space). *A complete metric space is a metric space in which every fundamental sequence (Cauchy sequence) converges.*

The difference between this definition and the general statement of Definition 6.23 is the fact that here, a distance measure, d, is defined in conjunction with the *space*, making it a *metric space*.

Chapter 8, presents a number of well-known *metrics* (*distances*) as well as *divergences*. A detailed definition of a *divergence* will be given in Chapter 7.

6.2.4 Banach Space (Normed Vector Space)

A *Banach space*, $(\mathscr{B}, \mathfrak{B}, \|.\|)$, is a general *measure space* which encompasses many other multidimensional and infinite dimensional spaces such as the *Hilbert space* (Definition 6.33) and the *Lebesgue space* (Definition 6.49). The most general *Banach space* is defined as follows.

Definition 6.30 (Banach Space). *A Banach space, $(\mathscr{B}, \mathfrak{B}, \|.\|)$, is a measurable vector space which has a norm ($\|.\|$) associated with it.*

See Section 23.2 and Equation 6.107 for norms defined on finite and infinite dimensional spaces. A *Banach space* may be defined over the *Borel field* of Real or Complex numbers. In Section 6.2.6, we will discuss infinite dimensional spaces. The *Banach space* may be defined over a *finite* or *infinite* dimensional space. It was named after *Stéfan Banach* who studied the work of *Hilbert* and others, and provided thorough generalizations on the theory of *integral equations* [5]. *Banach* [5], also defines *norms* in an axiomatic form. He studied *weak* and *strong completeness* of the space, in the sense of *weak* and *strong convergence* (see Definitions 6.69 and 6.70) of the *Cauchy sequences* in the space.

It is important to note that all finite dimensional spaces defined over the *Borel field* of Real or Complex numbers are *complete*. Therefore, all finite dimensional *Banach spaces* are also *complete spaces* (see Definition 6.23). However, this is not true for *infinite dimensional* spaces. As we will see in Section 6.2.6, the *completeness* of a space becomes material in infinite dimensions.

6.2.5 Inner Product Space (Dot Product Space)

As a special case of the *Banach space*, one may define an *inner product* function between the elements of the space. This new vector space is known as an *inner product space* or a *dot product space*. It may be a finite dimensional space as well as infinite dimensional. The infinite dimensional inner product space is generally known as the *pre-Hilbert space* and its completed form is known as the *Hilbert space*. Both of these will be defined and discussed in more detail in Section 6.2.6.1.

Therefore, the formal definition of an inner product space is as follows,

Definition 6.31 (Inner Product Space (Dot Product Space)). *An inner product space (also known as a dot product space) is a vector space which provides an inner product defined between all its elements,* $(\mathscr{X} \times \mathscr{X}, \mathfrak{X} \times \mathfrak{X})$, *namely,* $\langle \mathbf{x}, \mathbf{y} \rangle : (\mathscr{X} \times \mathscr{X}, \mathfrak{X} \times \mathfrak{X}) \mapsto \mathfrak{X}$, *where* $\mathbf{x}, \mathbf{y} \in (\mathscr{X}, \mathfrak{X})$ *and* $\mathfrak{X}$ *is a Borel field which may either be* $\mathfrak{X} = \mathbb{R}$ *or* $\mathfrak{X} = \mathbb{C}$.

In Chapter 15, while studying the, so called, *kernel trick*, we will see that the space associated with the kernel, known as the *feature space*, is actually an inner product space which may theoretically become a *Hilbert space*, in the limit, when the feature space takes on infinite dimensions.

6.2.6 Infinite Dimensional Spaces (Pre-Hilbert and Hilbert)

In Section 6.2.2, we visited multiple dimensional vector spaces. In this section, we will review an extension of this concept to infinite dimensions. Infinite dimensional vector spaces are used in a variety of problems including *integral equations* and *integral transforms*, as well as applications of these techniques to many problems in mathematics and physics. For the topic of this book, the most important infinite dimensional vector space is the *Hilbert space* which shows up, implicitly in *support vector machines* (Chapter 15) and explicitly in *integral transforms* (Chapter 24).

6.2.6.1 Pre-Hilbert Space

Let us begin with a general extension of an N-dimensional *vector space* such as a *Cartesian product space* into infinite dimensions. A general extension to infinite dimensions is called a *pre-Hilbert space*[5]. A formal definition is as follows:

Definition 6.32 (Pre-Hilbert Space). *An extension of a multi-dimensional vector space to infinite dimensions with an inner product function defined between elements of the space is known as a pre-Hilbert space* $(\mathscr{H}_p, \mathfrak{H}_p)$.

See Definition 24.49 of a an *inner product*; also note the text immediately after the definition. In general, it is possible to have a *pre-Hilbert space* which is not *complete* (see Definition 6.23). If the *inner product* is defined over a complete *Borel field*, such as $\mathbb{R}$ or $\mathbb{C}$, then the space is known as a *Hilbert space*. However, not all *pre-Hilbert spaces* are *complete*. In the next section, we will discuss the *Hilbert space* in more detail.

6.2.6.2 Hilbert Space

Hilbert space [30] is extremely useful in allowing for the extension of vector concepts to an *infinite dimensional space*. *Hilbert space* was named after *David Hilbert* [18] who was instrumental in the development of *integral equations* and *integral transforms*. The following is a formal definition of the *Hilbert space*.

Definition 6.33 (Hilbert Space). *A Hilbert space is an infinite dimensional complete space which provides an inner product defined between all its elements, in the vector product space,* $(\mathscr{H} \times \mathscr{H}, \mathfrak{H} \times \mathfrak{H})$, *namely,* $\langle \mathbf{x}, \mathbf{y} \rangle : (\mathscr{H} \times \mathscr{H}, \mathfrak{H} \times \mathfrak{H}) \mapsto \mathfrak{H}$, *where* $\mathbf{x}, \mathbf{y} \in (\mathscr{H}, \mathfrak{H})$.

See Definitions 6.23 and 24.49 for definitions of a *complete space* and an *inner product*, respectively.

---

[5] Naturally, *pre-Hilbert space* is sometimes loosely called an *inner product space* or a *dot product space*.

In Definition 6.33, the *inner product* was defined on the Borel field $\mathfrak{H}$ which may either be the *Real line*, $\mathbb{R}$ or the *Complex plane*, $\mathbb{C}$. A special case would be one where all the elements of the *Hilbert space*, and consequentially the *inner product*, are only defined on the Real line, $\mathbb{R}$. If the measurable space is one where the *inner product* is defined in $\mathbb{R}$, then we will have a *Hilbert metric space*. The following is a formal definition.

Definition 6.34 (Hilbert Metric Space). *A Hilbert metric space* $(\mathcal{H}, \mathfrak{H}, \langle .,. \rangle)$ *is an infinite dimensional complete space which provides an inner product defined between all its elements, mapping the vector product space,* $(\mathcal{H} \times \mathcal{H}, \mathfrak{H} \times \mathfrak{H})$, *to the real line,* $\mathbb{R}$. *Namely,* $\langle \mathbf{x}, \mathbf{y} \rangle : (\mathcal{H} \times \mathcal{H}, \mathfrak{H} \times \mathfrak{H}) \mapsto \mathbb{R}$, *where* $\mathbf{x}, \mathbf{y} \in (\mathcal{H}, \mathfrak{H})$.

The difference between Definitions 6.34 and 6.33 is in the fact that a general *Hilbert space* may have a complex or real inner product depending on if the infinite dimensional *Hilbert space* has been constructed from elements which are on an infinite dimensional complex planes or real lines respectively. In the special case where the elements are made up of real numbers (producing an inner product which is real), the space becomes a *metric space* (see Definition 6.25).

Recall Definition 6.30 of a *Banach space*. *Hilbert space* may be viewed as a special case of an infinite dimensional *Banach space* where it is associated with the *Euclidean norm* (2-norm).

A good example for revealing the practical usage of the *Hilbert space* is the extension of the *generalized secant method* [8, 13] to the infinite dimensional space. *Barnes* [8] introduced the *generalized secant method* for the purpose of solving *linear equations* with *unknown coefficients*. This technique, in the finite dimensional Euclidean space, was used by *Beigi*, et al [40, 12, 9, 10, 11] to formulate new *iterative learning control* and *adaptive control* techniques. The *generalized secant method* was then extended to the *Hilbert space* in order to be able to handle continuous time *adaptive control* and *iterative learning control* [3, 4, 2]. Comparing the *Euclidean space* versus *Hilbert space* versions of the *generalized secant method* will provide a practical perspective of the *Hilbert space*.

6.3 Probability Measure

Consider a bounded measure, as a special case, in Definition 6.18 of a *measure*. Furthermore, add one more criterion requiring that,

$$\rho(\mathcal{X}) = 1 \tag{6.30}$$

The new *bounded measure* is called a *probability measure* and is denoted as $P(\mathscr{A})$ for any event $\{\mathscr{A} : \mathscr{A} \subset \mathscr{X} \ \wedge \ \mathscr{A} \in \mathfrak{X}\}$.

At this time, we can proceed with the definition of probability from an axiomatic point of view. The following axioms of probability are apparent from observations of nature and cannot be simplified any further. Any discussion of probability theory has to deduce its propositions either directly or indirectly from these axioms.

These axioms were first formalized by *Kolmogorov*, in a German publication, in 1933 [34]. An English version of this formalization was published in 1950. Later, a second English edition was published [35] which includes many fundamental measure-theoretic treatments of probability theory. This work is still the basis for most modern probability textbooks.

Definition 6.35 (Probability Measure). *Let us assume that $\{\mathscr{A}_i : i \in \mathbb{N}\}$ represent disjoint subsets of the sample space $\mathscr{X}$, each embodying a mutually exclusive event in $\mathscr{X}$. Then a probability measure, $P(\mathscr{A})$, for any event represented by $\{\mathscr{A} : \mathscr{A} \in \mathfrak{X}\}$ where $\mathfrak{X}$ is a Borel field of all the measurable subsets of $\mathscr{X}$ is defined over the measurable space $(\mathscr{X}, \mathfrak{X})$ if it meets the following axioms:*

1. *$P(\mathscr{A}) \in \mathbb{R} \ \forall \ \{\mathscr{A} : \mathscr{A} \in \mathfrak{X}\}$*
2. *$P(\mathscr{A}) \geq 0 \ \forall \ \{\mathscr{A} : \mathscr{A} \in \mathfrak{X}\}$*
3. *If $\{\mathscr{A}_i : i \in \mathbb{N} \ \wedge \ i \leq N\}$ are disjoint subsets of $\mathscr{X}$, where $N \in \mathbb{N}$ and may be finite or infinite. Namely,*
   $\mathscr{A}_i \subset \mathscr{X} \ \forall \ \{\mathscr{A}_i \in \mathfrak{X}, i \in \mathbb{N} \ \wedge \ i \leq N\}$, then,

$$\bigcap_{i=1}^{N} \mathscr{A}_i = \{\varnothing\} \implies P(\bigcup_{i=1}^{N} \mathscr{A}_i) = \sum_{i=1}^{N} P(\mathscr{A}_i) \tag{6.31}$$

4. *$P(\mathscr{X}) = 1$*

The following properties follow directly from the above axioms.

Property 6.1 (Probability of the Impossible Event). *The probability of the impossible event, $\{\varnothing\}$, is zero. In other words,*

$$P(\{\varnothing\}) = 0 \tag{6.32}$$

Proof.

$$P(\mathscr{A}) = P(\mathscr{A} \cup \{\varnothing\}) \tag{6.33}$$
$$= P(\mathscr{A}) + P(\{\varnothing\}) \tag{6.34}$$

since $\mathscr{A} \cap \{\varnothing\} = \{\varnothing\}$ and $\mathscr{A} \cup \{\varnothing\} = \mathscr{A}$.

$$\therefore \ P(\{\varnothing\}) = 0$$

$\square$

Property 6.2 (Probability of the Complement).

$$P(\mathscr{A}) = 1 - P(\mathscr{A}^{\mathsf{C}}) \tag{6.35}$$
$$\leq 1 \tag{6.36}$$

Proof.
Setting $N = 2$ in *axiom 3* of Definition 6.35, where $\mathscr{A}_1 = \mathscr{A}$ and $\mathscr{A}_2 = \mathscr{A}^{\mathsf{C}}$, since $\mathscr{A} \cup \mathscr{A}^{\mathsf{C}} = \mathscr{X}$,

$$P(\mathscr{X}) = P(\mathscr{A}) + P(\mathscr{A}^{\mathsf{C}}) \tag{6.37}$$
$$= 1 \tag{6.38}$$

Therefore using *axiom 2*, $P(\mathscr{A}) \leq 1$ and $1 - P(\mathscr{A}^{\mathsf{C}}) \leq 1$.

$\square$

Property 6.3 (Probability of a Union).

$$P(\mathscr{A} \cup \mathscr{B}) = P(\mathscr{A}) + P(\mathscr{B}) - P(\mathscr{A} \cap \mathscr{B}) \tag{6.39}$$
$$= P(\mathscr{A}) + P(\mathscr{B}) - P(\mathscr{A}, \mathscr{B}) \tag{6.40}$$

Proof.
Note that,

$$\mathscr{A} \cup \mathscr{B} = \mathscr{A} \cup (\mathscr{A}^{\mathsf{C}} \cap \mathscr{B}) \tag{6.41}$$

however,

$$\mathscr{A} \cap (\mathscr{A}^{\mathsf{C}} \cap \mathscr{B}) = \{\varnothing\} \tag{6.42}$$

therefore,

$$P(\mathscr{A} \cup \mathscr{B}) = P(\mathscr{A}) + P(\mathscr{A}^{\mathsf{C}} \cap \mathscr{B}) \tag{6.43}$$

In the same manner,

$$\mathscr{B} = (\mathscr{A} \cap \mathscr{B}) \cup (\mathscr{A}^{\mathsf{C}} \cap \mathscr{B}) \tag{6.44}$$

where

$$(\mathscr{A} \cap \mathscr{B}) \cap (\mathscr{A}^{\mathsf{C}} \cap \mathscr{B}) = \{\varnothing\} \tag{6.45}$$

Therefore,

$$P(\mathscr{B}) = P(\mathscr{A} \cap \mathscr{B}) + P(\mathscr{A}^{\mathsf{C}} \cap \mathscr{B}) \tag{6.46}$$

and

$$P(\mathscr{A}^{\complement} \cap \mathscr{B}) = P(\mathscr{B}) - P(\mathscr{A} \cap \mathscr{B}) \tag{6.47}$$

Plugging Equation 6.47 into Equation 6.43 we have,

$$\begin{aligned} P(\mathscr{A} \cup \mathscr{B}) &= P(\mathscr{A}) + P(\mathscr{B}) - P(\mathscr{A} \cap \mathscr{B}) \\ &= P(\mathscr{A}) + P(\mathscr{B}) - P(\mathscr{A}, \mathscr{B}) \end{aligned} \tag{6.48}$$

□

Property 6.3 may be extended to three events, $\mathscr{A}_1$, $\mathscr{A}_2$, and $\mathscr{A}_3$ as follows,

$$\begin{aligned} P(\mathscr{A}_1 \cup \mathscr{A}_2 \cup \mathscr{A}_3) &= P(\mathscr{A}_1) + P(\mathscr{A}_2) + P(\mathscr{A}_3) - \\ &\quad P(\mathscr{A}_1, \mathscr{A}_2) - P(\mathscr{A}_2, \mathscr{A}_3) - P(\mathscr{A}_1, \mathscr{A}_3) - \\ &\quad P(\mathscr{A}_1, \mathscr{A}_2, \mathscr{A}_3) \end{aligned} \tag{6.49}$$

This can be easily extended to any number of events, but the generalization notation would be somewhat complicated, so it is not shown here. Keep in mind that for a larger number of events, as in the 3-event case, the probabilities of all possible combinations of intersections of events must be subtracted from the sum of the probabilities of all individual events.

Definition 6.36 (Conditional Probability). *If $\mathscr{A} \subset \mathscr{X}, \mathscr{B} \subset \mathscr{X}$, and $P(\mathscr{B}) > 0$, then the probability of event $\mathscr{A}$ given that event $\mathscr{B}$ has occurred is called the conditional probability of $\mathscr{A}$ given $\mathscr{B}$ and is written as,*

$$P(\mathscr{A}|\mathscr{B}) = \frac{P(\mathscr{A}, \mathscr{B})}{P(\mathscr{B})} \tag{6.50}$$

or equivalently,

$$P(\mathscr{A}, \mathscr{B}) = P(\mathscr{A}|\mathscr{B})P(\mathscr{B}) = P(\mathscr{B}|\mathscr{A})P(\mathscr{A}) \tag{6.51}$$

where $P(\mathscr{A}, \mathscr{B})$ (or $P(\mathscr{A} \cap \mathscr{B})$) is called the joint probability of events $\mathscr{A}$ and $\mathscr{B}$. Note that Equation 6.51 does not need the requirements that $P(\mathscr{B}) > 0$ or $P(\mathscr{A}) > 0$.

Consider the two cases where $\mathscr{A} \subset \mathscr{B}$ and $\mathscr{B} \subset \mathscr{A}$. The condition probability, $P(\mathscr{A}|\mathscr{B})$, will have the following properties for each case,

1. $\mathscr{A} \subset \mathscr{B}$,

$$\begin{aligned} P(\mathscr{A}|\mathscr{B}) &= \frac{P(\mathscr{A})}{P(\mathscr{B})} \\ &\geq P(\mathscr{A}) \end{aligned} \tag{6.52}$$

2. $\mathscr{B} \subset \mathscr{A}$,

$$P(\mathscr{A}, \mathscr{B}) = P(\mathscr{B}) \implies P(\mathscr{A}|\mathscr{B}) = 1 \tag{6.53}$$

Theorem 6.1 (Total Probability). *Let* $\mathcal{X} = \{\mathcal{A}_1, \mathcal{A}_2, \cdots, \mathcal{A}_N\}$ *where* $\mathcal{A}_i$, $i = \{1, 2, \cdots, N\}$ *are disjoint events completely partitioning sample space* $\mathcal{X}$ *into* N *sections. Also, let* $\mathcal{B}$ *be any event in* $\mathcal{X}$. *Then, the total probability* $P(\mathcal{B})$ *is given as follows,*

$$P(\mathcal{B}) = \sum_{i=1}^{N} P(\mathcal{B}|\mathcal{A}_i)P(\mathcal{A}_i) \tag{6.54}$$

Proof.
Since $\mathcal{A}_i, i = \{1, 2, \cdots, N\}$ are disjoint,

$$\bigcap_{i=1}^{N} \mathcal{A}_i = \{\varnothing\} \tag{6.55}$$

then,

$$\mathcal{B} = \bigcup_{i=1}^{N} [\mathcal{B} \cap \mathcal{A}_i] \tag{6.56}$$

Using Equations 6.51, 6.56, and *axiom 3* of Definition 6.35,

$$P(\mathcal{B}) = \sum_{i=1}^{N} P(\mathcal{B}|\mathcal{A}_i)P(\mathcal{A}_i) \tag{6.57}$$

$\square$

Next, we will examine one of the most fundamental theorems in probability theory. Together with the expression for *Total Probability* (Equation 6.54), they constitute the basic tools for pattern recognition.

Theorem 6.2 (Bayes Theorem). [6] *Let* $\mathcal{X} = \{\mathcal{A}_1, \mathcal{A}_2, \cdots, \mathcal{A}_N\}$ *where* $\mathcal{A}_i, i = \{1, 2, \cdots, N\}$ *are disjoint events completely partitioning sample space* $\mathcal{X}$ *into* N *sections. Also, let* $\mathcal{B}$ *be any event in* $\mathcal{X}$ *where* $P(\mathcal{B}) > 0$. *Then,*

$$P(\mathcal{A}_j|\mathcal{B}) = \frac{P(\mathcal{B}|\mathcal{A}_j)P(\mathcal{A}_j)}{\sum_{i=1}^{N} P(\mathcal{B}|\mathcal{A}_i)P(\mathcal{A}_i)} \tag{6.58}$$

Proof.
The proof is somewhat simple. Let us write Equation 6.51 for any event $\mathcal{A}_j$ with respect to another event $\mathcal{B}$.

$$P(\mathcal{A}_j, \mathcal{B}) = P(\mathcal{A}_j|\mathcal{B})P(\mathcal{B}) = P(\mathcal{B}|\mathcal{A}_j)P(\mathcal{A}_j) \tag{6.59}$$

If we substitute the *total probability* expression for $P(\mathcal{B})$ from Equation 6.54 into Equation 6.59, we will have the following form of *Bayes' Theorem,*

---

[6] Named after the English mathematician, *Thomas Bayes* (1702-1761), who initiated the thought process of a special case of this theorem.

$$P(\mathscr{A}_j|\mathscr{B}) \sum_{i=1}^{N} P(\mathscr{B}|\mathscr{A}_i)P(\mathscr{A}_i) = P(\mathscr{B}|\mathscr{A}_j)P(\mathscr{A}_j) \tag{6.60}$$

Equation 6.60 may be written in terms of $P(\mathscr{A}_j|\mathscr{B})$ to form the more popular statement of *Bayes' Theorem* given by Equation 6.58.

□

Definition 6.37 (Statistical Independence). *Any two events $\mathscr{A} \subset \mathscr{X}$ and $\mathscr{B} \subset \mathscr{X}$ are said to be statistically independent if and only if,*

$$P(\mathscr{A}, \mathscr{B}) = P(\mathscr{A})P(\mathscr{B}) \tag{6.61}$$

Explanation:
For events $\mathscr{A}$ and $\mathscr{B}$ to be independent, the following must hold,

$$P(\mathscr{A}|\mathscr{B}) = P(\mathscr{A}) \tag{6.62}$$

Plugging in Equation 6.62 in Equation 6.51, the statement of Equation 6.61 follows.

Definition 6.38 (Mutual Statistical Independence). $\{\mathscr{A}_i : i \in \{1, 2, \cdots, N\}\}$ *are mutually independent events if and only if,*

$$P\left(\bigcap_{j \in \mathscr{M}_k} \mathscr{A}_j\right) = \prod_{j \in \mathscr{M}_k} P(\mathscr{A}_j) \tag{6.63}$$

where $\mathscr{M}_k$ is a set of exhaustive k combinations of numbers $\in \{1, 2, \cdots, N\}$, where $k \leq N$. This is a generalization of the two-event version given by Definition 6.37.

It is interesting to note that if we consider an experiment on a *real line*, defined by the *measure space*, $(\mathscr{X}_1, \mathfrak{X}_1, P)$ with $\{\mathscr{X}_1 : \mathscr{X}_1 = \mathbb{R}\}$, then, the repetition of the experiment by n times is equivalent to the outcome of the experiment in the *measure space* of the n-dimensional *Cartesian product space* of the initial single dimensional space, namely, $(\mathscr{X}, \mathfrak{X}, P)$ such that,

$$\mathscr{X} = \underset{i=1}{\overset{n}{\times}} \mathscr{X}_i \tag{6.64}$$

and

$$\mathfrak{X} = \underset{i=1}{\overset{n}{\times}} \mathfrak{X}_i \tag{6.65}$$

For example, throwing a dye ten times is tantamount to an experiment in the 10-dimensional space.

Now let us examine a *continuous random variable* which is the underlying characteristic of most natural processes, although quantization effects for easier manipulation of the variables may sometimes render them discrete in practice. This is not to say that there are no discrete random processes in nature, but we certainly deal with more of them due to quantization for convenience, such as sampling of a continuous signal into a discrete representation.

Definition 6.39 (Continuous Random Variable). *A continuous random variable, X, is a variable which may take on a continuous set of values in a finite interval* $[a,b]$ *with the exception of a finite number of discontinuities. See the definition of a continuous function in Section 24.17.*

6.4 Integration

As it will become clear in the next few sections, it is important to be able to compute the *integral* of a *function* of a *random variable*, $g(x)$, with respect to a *measure*, $\rho(x)$. Here, we will use ideas from Definition 24.39 of a *Riemann integral* with reference to [27] to develop a methodology for computing the *integral* of a function, $g(x)$, with respect to a measure, over a *measure space*, $(\mathcal{X}, \mathfrak{X}, \rho)$. More detailed information on the topic may be found in [39, 27, 28].

Let us consider a function $\{g(x) : \mathbb{R} \mapsto \mathbb{R}\}$ defined over the *measure space*, $(\mathcal{X}, \mathfrak{X}, \rho)$. Initially, we assume $g(x)$ to be a *statistic* (see Definition 6.72), bounded from both sides,

$$G_L < g(x) < G_H \tag{6.66}$$

Also, $\rho(x)$ is a bounded measure, $\rho(\mathscr{A}) \in \mathbb{R}$. Note that $\mathbb{R}$ does not include $\pm\infty$, making $\rho(x)$ bounded. Therefore, we may define an ordered sequence [20, 27],

$$G_L = G_0 < G_1 < \cdots < G_n = G_H \tag{6.67}$$

where,

$$G_i - G_{i-1} \le \varepsilon \ \forall \ i \in \{1, 2, \cdots, n\} \tag{6.68}$$

Then, the integral over the whole *measure space*, $\mathcal{X}$, is defined as,

$$\int_{\mathcal{X}} g(x)d\rho(x) = \lim_{\substack{n \to \infty \\ \varepsilon \to 0}} \sum_{i=1}^{n} G_i \, \rho(\{x : G_{i-1} < g(x) \le G_i\}) \tag{6.69}$$

Notice the resemblance of this result to that of a *Riemann integral* given by Equation 24.119. As it stands, this is indeed a *Riemann integral* due to the definite domain which has been defined and the nature of the function being integrated.

At this stage, we may relax the boundedness requirement for $g(x)$. Let us define a new function, $f(x)$ where,

$$f(x) \triangleq \begin{cases} G_L & \forall \{x : g(x) \leq G_L\} \\ f(x) & \forall \{x : G_L < g(x) < G_H\} \\ G_H & \forall \{x : g(x) \geq G_H\} \end{cases} \tag{6.70}$$

then, for an unbounded $g(x)$, we can write the integral,

$$v = \int_{\mathscr{X}} g(x) d\rho(x) \tag{6.71}$$

$$= \lim_{\substack{G_L \to -\infty \\ G_H \to \infty}} \int_{\mathscr{X}} f(x) d\rho(x) \tag{6.72}$$

Therefore, if the limit on the right hand side of Equation 6.72 is finite as $G_L \to -\infty$ and $G_H \to \infty$, we can deduce that the integral of interest, v, exists.

At this point, we need to change the domain of integration from the whole sample space, $\mathscr{X}$ to an arbitrary subset of the *Borel field* of our *measurable space*, $\mathscr{A} \in \mathfrak{X}$. This may be simply achieved through the introduction of the *characteristic function* of $\mathscr{A}$ (Equation 6.6). Since we have defined $\Upsilon_{\mathscr{A}}(x) \forall x$, the *characteristic function* is *measurable*. Therefore, we may write the *integral* of $g(x)$ with respect to the *measure*, $d\rho(x)$, over the region defined by subset $\mathscr{A}$ of the *measurable space* as follows,

$$\int_{\mathscr{A}} g(x) d\rho(x) = \int_{\mathscr{X}} \Upsilon_{\mathscr{A}}(x) g(x) d\rho(x) \tag{6.73}$$

6.5 Functions

In this section, we will review functions, metrics and function spaces related to *measures* in general, but also as special forms relating to a *probability measure*. The related functions include *density functions, cumulative distribution functions, variances*, etc. Let us start with the *probability density function*.

We generally speak of *real valued functions* which are functions with the following property, $\{g : \mathscr{X} \mapsto \mathbb{R}\}$. However, as we shall see later, sometimes we need to consider the, so called, *extended real valued functions*.

Definition 6.40 (Extended Real Valued Function). *An extended real valued function is a function such that, $\{g : \mathscr{X} \mapsto \{\mathbb{R}, -\infty, \infty\}\}$. Namely, its domain is extended to include the points at $\pm\infty$.*

6.5.1 Probability Density Function

Before defining a *probability density function*, we need to introduce a few more concepts in *measure theory*, the first of which is the *absolute continuity of measures*.

Definition 6.41 (Absolute Continuity of Two Measures). *Given two measures of a set $\mathscr{A}$, defined in the measurable space $(\mathscr{X}, \mathfrak{X})$, and represented by $\rho(\mathscr{A})$ and $v(\mathscr{A})$, $v(\mathscr{A})$ is said to be absolutely continuous with respect to $\rho(\mathscr{A})$ if for all events where $\rho(\mathscr{A}) = 0$, also $v(\mathscr{A}) = 0$. This is written in the following mathematical notation,*

$$v(\mathscr{A}) \ll \rho(\mathscr{A}) \tag{6.74}$$

Definition 6.42 (Equivalence). *If two measure are absolutely continuous with respect to one another, namely, if $v(\mathscr{A}) \ll \rho(\mathscr{A})$ and $\rho(\mathscr{A}) \ll v(\mathscr{A})$, then they are said to be equivalent and the following notation is used[7],*

$$v(\mathscr{A}) \equiv \rho(\mathscr{A}) \tag{6.75}$$

Another notation which is useful in the context of measure theory is that of the *modulo* of a measure, $[\rho]$.

Definition 6.43 (Almost Everywhere (Modulo)). *Let us assume that a proposition in x is defined as $\pi(x)$ over the measurable space $(\mathscr{X}, \mathfrak{X})$. Then, stating the proposition with respect to a measure, $\rho(x)$, may be written as, $\pi(x)[\rho]$ (read as $\pi(x)$ modulo ρ), meaning that $\pi(x)$ is true for all x where $\rho(x) \neq 0$. For example, if proposition $\pi(x)$ is defined as $\pi(x) : f(x) = g(x)$, then $\pi(x)[\rho]$ or $f(x) = g(x)[\rho]$ means that $f(x) = g(x) \ \forall \ \{x : \rho(x) \neq 0\}$. Another way is to say that the proposition $\pi(x)$ is true almost everywhere with respect to the measure $\rho(x)$.*

At this point, we can state the very powerful *Radon-Nikodým Theorem* [29, 28] which defines a *generalized density function* in connection with *absolute continuity* of measures. This statement of the *Radon-Nikodým Theorem* is an alternative statement which is useful in defining a *generalized density function* and its uniqueness. For the original statement of the theorem see [28].

Theorem 6.3 (Radon-Nikodým). *Given two measures on any measurable subset $\mathscr{A} \in \mathfrak{X}$, defined in the measurable space $(\mathscr{X}, \mathfrak{X})$, and represented by $\rho(\mathscr{A})$ and $v(\mathscr{A})$, a necessary and sufficient condition for $v(\mathscr{A}) \ll \rho(\mathscr{A})$ is that there exists a function, $\{g(x) : g(x) \geq 0 \ \forall \ x \in \mathscr{A} \in \mathfrak{X}\}$ such that,*

---

[7] Although notations, $\ll$, $\gg$, and $\equiv$ are used for denoting *much less than*, *much greater than*, and *equivalent* under *normal circumstances*, it is understood that within the context of comparing measures, they refer to *absolute continuity of two measures*. *Equivalence* in the context of *measure theory* is also only related to continuity relations and not numerical equivalence.

$$v(\mathscr{A}) = \int_{\mathscr{A}} g(x)d\rho(x) \tag{6.76}$$

Furthermore, $g(x)$ is unique almost everywhere with respect to ρ – see Definition 6.43.

Proof.
The proof is somewhat lengthy and is available in [28].

□

The uniqueness statement of the *Radon-Nikodým Theorem* means that any two functions satisfying the theorem may be different only at points where $\rho(x) = 0$. This unique function, $g(x)$, is called a *generalized density function*. $g(x)$ may be obtained by taking the derivative of $v(\mathscr{A})$ with respect to $\rho(x)$,

$$g(x) = \frac{dv}{d\rho}[\rho] \tag{6.77}$$

Based on Equation 6.77, $g(x)$ is also called the *Radon-Nikodým derivative*.

Consider any three measures defined in the *measurable space* $(\mathscr{X},\mathfrak{X})$ such that $v \ll \rho$ and $\rho \ll \lambda$. Then, it is apparent that $v \ll \lambda$. It follows from this property that the *Radon-Nikodým derivatives* with respect to different measures satisfy,

$$\frac{dv}{d\lambda} = \frac{dv}{d\rho}\frac{d\rho}{d\lambda}[\lambda] \tag{6.78}$$

See [28] for a proof of Equation 6.78.

As a consequence of Equation 6.78, we may write Equation 6.76 as follows,

$$v(\mathscr{A}) = \int_{\mathscr{A}} g(x)d\rho(x)$$
$$= \int_{\mathscr{A}} g(x)\frac{d\rho(x)}{d\lambda(x)}d\lambda(x) \tag{6.79}$$

Equation 6.79 allows us to express a measure in terms of a third measure given its relation to a second measure. This powerful result enables us to change the variable of integration for example from the measure associated with a probability distribution to say the Lebesgue measure.

Now let us define a *probability density function* as a special case of the *Radon-Nikodým derivative*.

Definition 6.44 (Probability Density Function). *Consider the special case where $P(\mathscr{A})$ is a probability measure defined in the measurable space, $(\mathscr{X},\mathfrak{X})$, of a real line $(\{\mathscr{X} : \mathscr{X} = \mathbb{R}\})$ with respect to a probability distribution measure, $\lambda(x)$. Then $g(x)$ is defined as the generalized probability density function with respect to measure $\lambda(x)$ and is defined such that,*

$$P(\mathscr{A}) = \int_{\mathscr{A}} g(x)d\lambda(x) \tag{6.80}$$

where $\lambda(x)$ is the measure associated with the probability distribution function of random variable X.

Note a very important consequence of Equation 6.79. A specific measure is associated with every *probability distribution*. Equation 6.79 allows us to take the description of a *probability measure* given by Equation 6.80 in terms of one probability distribution (base on its probability measure) and to rewrite it in terms of a different probability distribution (with a different probability measure) as long as the probability measure associated with the first distribution is *absolutely continuous* with respect to the probability measure associated with the second distribution.

For example, the *binomial distribution* has a probability measure which is *absolutely continuous* with respect to the probability measure of a *Poisson distribution*. However, it is not *absolutely continuous* with respect to the measure of the *Normal distribution*.[27] This means that we will be able to take a probability measure defined on a Binomial distribution and express it in terms of Poisson distributions.

A very important outcome is due to the fact that every probability distribution, defined in the *measurable space* of a *real line*, $(\mathscr{X}, \mathfrak{X})$ where $\{\mathscr{X} : \mathscr{X} = \mathbb{R}\}$, is *absolutely continuous* with respect to the *Lebesgue measure*. This is quite powerful, since it allows us to define a probability measure in terms of a new distribution, $p(x)$ which is really the product of the generalized probability density, $g(x)$ in terms of the original probability distribution measure λ, and the *Radon-Nikodým derivative* of that distribution measure with respect to the*Lebesgue measure*. This gives rise to the definition of the *probability density function* $p(x)$ which satisfies the following Equation,

$$P(\mathscr{A}) = \int_{\mathscr{A}} p(x)dx \tag{6.81}$$

where $p(x) \stackrel{\Delta}{=} g(x)\frac{d\lambda(x)}{dx}$.

Here is an alternative statement of Definition 6.44 for the *probability density function,*

Definition 6.45 (Probability Density Function). *The probability density function of a random variable X, $\{p(x) \in \mathbb{R} : \mathbb{R} \mapsto \mathbb{R} \wedge p(x) \geq 0 \forall x \in \mathbb{R}\}$ is a function such that the probability of the closed interval $\mathscr{A} = [a, b] \subset \mathbb{R}$ is given by,*

$$P(\mathscr{A}) = \int_{a}^{b} p(x)dx \tag{6.82}$$

Equation 6.82 is a *Riemann integral* as given by Definition 24.39 which also provides insight into the meaning behind this integral and its computation. As such, it also possesses all the properties listed for the *Riemann integral* after the statement of Definition 24.39.

Even if the *probability density function, p(x)*, does not exist, the integral in Equation 6.80 may still be computed using *standard Lebesgue integration* [35].

If $p(x)$ does exist, then using the *Radon-Nikodým theorem*, it may be written in terms of *probability measure P* and the *Lebesgue measure*, in the following form,

$$p(x) = \frac{dP}{dx}[x]$$ (6.83)

Equation 6.83 holds for all probability measures since all *probability measures* are *absolutely continuous* with respect to the *Lebesgue measure* [27].

Sufficient conditions for any function of a continuous variable, $p(x)$, to be a *probability density function* is that $p(x) \geq 0$ and that the probability of the domain of $p(x)$ be 1, or formally,

$$P(\mathscr{X}) = \int_{-\infty}^{\infty} p(x)dx$$

$$= 1$$ (6.84)

Note that these conditions do not require $p(x)$ to have an upper limit of 1, only the integral of $p(x)$ in any interval has to be less than or equal to 1.

6.5.2 Densities in the Cartesian Product Space

If we consider the *n-dimensional Euclidean space* as a *Cartesian product* of *real lines*, as discussed earlier, the *probability density function* will then be $\{p : \mathscr{R}^n \mapsto \mathbb{R}\}$. In that case, one way of writing the random variable X is to express it in terms of a vector, $\{\mathbf{x} : \mathscr{R}^1 \mapsto \mathscr{R}^n\}$ in which case,

$$P(\mathscr{A}) = \int_{\mathscr{A}} g(\mathbf{x})d\lambda(\mathbf{x})$$ (6.85)

Based on the statement of Equation 6.79, Equation 6.85 may be written in terms of a generalized density function for λ as follows,

$$P(\mathscr{A}) = \int_{\mathscr{A}} g(\mathbf{x})\frac{d\lambda(\mathbf{x})}{d\mathbf{x}}d\mathbf{x}$$ (6.86)

$$= \int_{\mathscr{A}} p(\mathbf{x})d\mathbf{x}$$ (6.87)

An important observation is that the *probability density function* is only defined for *continuous random variables*. However, as we stated earlier, although the *probability density function* may not exist, the integral in Equation 6.85 may still be computed using standard *Lebesgue integration*. Therefore, a *probability measure* is still defined for *discrete random variables*. We shall discuss *discrete random variables* later in the chapter.

Assuming that the *probability density function* is defined in the *Cartesian Product space* for the random variable of interest, we can define the *joint probability density function* as follows,

Definition 6.46 (Joint Probability Density Function). *Given a differential probability measure, dP, defined in the Cartesian product probability space* $\{\mathscr{X} : \mathscr{X} = \mathscr{R}^n\}$ *(see Equation 6.26), the joint probability density function,* $p_X(x_1, x_2, \cdots, x_n)$ *is defined as an extended form of the Radon-Nikodým derivative such that*

$$p_X(x_1, x_2, \cdots, x_n) \overset{\Delta}{=} \frac{\partial^n P}{\partial x_1 \partial x_2 \cdots \partial x_n} \tag{6.88}$$

According to this definition, the Probability measure, P, evaluated over an event $\mathscr{A}$ may be written as follows,

$$P(\mathscr{A}) = \underset{\mathscr{A}}{\iiint} \cdots \int p_X(x_1, x_2, \cdots, x_n) dx_1 dx_2 \cdots dx_n \tag{6.89}$$

Another useful probability density function in the *Cartesian product space* is known as the *marginal probability density function* which is defined as follows,

Definition 6.47 (Marginal Probability Density Function). *Given a differential probability measure, dP, defined in the Cartesian product probability space* $\{\mathscr{X} : \mathscr{X} = \mathscr{R}^n\}$*, there are n marginal probability density functions associated with the n variables,* $\{X_i : x_i \in \mathscr{X}_i\}$ *respectively (see Equation 6.26). The marginal probability density function is defined in terms of the joint probability density function,* $p_X(x_1, x_2, \cdots, x_n)$*, as follows,*

$$p_{X_i}(x_i) \overset{\Delta}{=} \frac{\partial P}{\partial x_i} \ \forall \ i \in \{1, 2, \cdots, n\} \tag{6.90}$$

$$= \iiint \cdots \int p_X(x_1, x_2, \cdots, x_n) dx_1 dx_2 \cdots dx_{i-1} dx_{i+1} \cdots dx_n \tag{6.91}$$

It is important to note that if the $\{X_i, \{i \in \{1, 2, \cdots, n\}\}$ are *mutually independent*, then the *joint probability density function* is equal to the product of all *marginal probability density functions*. Therefore,

$$p_X(x_1, x_2, \cdots, x_n) = \prod_{i=1}^{n} p_{X_i}(x_i) \tag{6.92}$$

6.5.2.1 Combining two Random Variables

Let us consider the *probability density function* of a *function* of two random variables. In other words, let us assume that,

$$\xi \overset{\Delta}{=} g(x_1, x_2) \tag{6.93}$$

is a function of the two random variables, $\{X_1 : x_1 \in \mathscr{X}_1\}$ and $\{X_2 : x_2 \in \mathscr{X}_2\}$, itself a random variable, $\{X : \xi \in \mathscr{X}\}$. Then, by Theorem 6.3, the differential probability measure, dP, may be written as a product of the *probability density function (Radon-Nikodým derivative)* in terms of the random variable X and the differential *Lebesgue measure* associated with ξ,

$$dP = p(\xi)d\xi \tag{6.94}$$

dP is the probability measure evaluated over some $\{\mathscr{A} \in \mathscr{X}\}$ associated with the differential increment $d\xi$. If we write the *joint probability density function* defined for X_1 and X_2, as $p_X(x_1, x_2)$, then, according to Equation 6.89, dP may be written in terms of x_1 and x_2 as follows,

$$\begin{aligned} dP &= p(\xi)d\xi \\ &= \int_{\mathscr{A}} p_X(x_1, x_2) dx_1 dx_2 \end{aligned} \tag{6.95}$$

If we consider the case where

$$\begin{aligned} \xi &= g(x_1, x_2) \\ &= x_1 + x_2 \end{aligned} \tag{6.96}$$

then Equation 6.95 may be written in terms of x_2 and ξ only, by using the relation given by Equation 6.96,

$$p(\xi)d\xi = \int_{-\infty}^{\infty} p_X(\xi - x_2, x_2) dx_2 d\xi \tag{6.97}$$

Now, let us make the assumption that x_1 and x_2 are *mutually independent*. Therefore, according to Equation 6.92, *joint probability density function*, $p_X(x_1, x_2)$, is equal to the product of the *marginal probability density functions*, $p_{X_1}(x_1)$ and $p_{X_2}(x_2)$,

$$p_X(x_1, x_2) = p_{X_1}(x_1) p_{X_2}(x_2) \tag{6.98}$$

If we use the relationship given by Equation 6.98 to rewrite Equation 6.97, dividing both sides of Equation 6.97 by $d\xi$,

$$p(\xi) = \int_{-\infty}^{\infty} p_{X_1}(\xi - x_2) p_{X_2}(x_2) dx_2 \tag{6.99}$$

$$= (p_{X_1} * p_{X_2})(\xi) \tag{6.100}$$

$$\tag{6.101}$$

Notice that Equation 6.99 is just the expression for the *convolution* of the two *marginal probability density functions* , $(p_{X_1} * p_{X_2})(\xi)$. See Section 24.2.1 for more information about *convolution*.

Similarly, it is easy to show that if

$$\xi = g(x_1, x_2)$$
$$= x_1 - x_2 \tag{6.102}$$

then

$$p(\xi) = \int_{-\infty}^{\infty} p_{X_1}(\xi + x_2) p_{X_2}(x_2) dx_2 \tag{6.103}$$

$$= (p_{X_1} \circ p_{X_2})(\xi) \tag{6.104}$$

Equation 6.97 is the expression for the *correlation* between the two *marginal probability density functions*, $(p_{X_1} \circ p_{X_2})(\xi)$ – see Section 24.2.2.

6.5.3 Cumulative Distribution Function

Definition 6.48 (Cumulative Distribution Function). *The Cumulative Distribution Function of a continuous random variable, X on the measure space defined on a real line, $(\mathscr{X}, \mathfrak{X}, \lambda)$ where $\{\mathscr{X} : \mathscr{X} = \mathbb{R}\}$, is defined as follows,*

$$F(x) = \int_{-\infty}^{x} g(\xi) d\lambda(\xi) \tag{6.105}$$

$$= \int_{-\infty}^{x} p(\xi) d\xi \tag{6.106}$$

$$= P(X \le x)$$

where $g(\xi)$ and $p(\xi)$ are the density functions presented in Equations 6.80 and 6.81 respectively.

An important note in defining the *Cumulative Distribution Function* is that $F(-\infty) = 0$ and $F(\infty) = P(\mathscr{X}) = 1$. Definition 6.48 may easily be generalized to apply to any bounded measure $\rho(x)$ in *measure space* $(\mathscr{X}, \mathfrak{X}, v)$ which does not necessarily need to be a Probability measure.

6.5.4 Function Spaces

Definition 6.49 ($\mathfrak{L}_p$ Class of p-integrable Functions (Lebesgue space)). *Let us consider a measure space $(\mathscr{X}, \mathfrak{X}, \lambda)$. $\mathfrak{L}_p(\lambda)$ is defined as the class of extended real valued p-integrable functions associated with measure λ for $\{p : p \in \mathbb{R}, p \geq 1\}$ – definition 6.40 describes an extended real valued function. For $g(x)$ to be p-integrable,*

$$\|g\|_p \stackrel{\Delta}{=} \left| \left(\int |g(x)|^p d\lambda(x) \right)^{\frac{1}{p}} \right| \tag{6.107}$$

should exist and be finite, namely, $\|g\|_p \in \mathbb{R}$.

Equation 6.107 describes the, so called, *p-norm* of a function. The special case, where $p = 2$ is equivalent to the positive square root of the inner product of the function and itself,

$$\|g\| \stackrel{\Delta}{=} \|g\|_2$$
$$= \sqrt{\langle g, g \rangle} \tag{6.108}$$

Definition 24.38, presented in Chapter 24 and used for the development of integral transforms, is a special case of definition 6.49, where $p = 1$.

Note that the definition given in Equation 6.107 is somewhat more precise than that presented in the literature (e.g., see Halmos [28]). Here, we have taken the *absolute value* of the pth root of the integral. This has been overlooked in the literature. It is important to include the *absolute value* since for example for $p = 2$, the square root may be either *positive* or *negative*, but we are only interested in the positive roots, since as we will see, one of the properties we will seek is for this function to be always *positive semi-definite* (see property number one, below). With this said and done, we may at times omit the absolute value sign, for simplicity of the notation, knowing that the *positive semi-definiteness* of $\|g\|_p$ is well understood.

We define the functional,

$$\rho_p(g_1, g_2) \stackrel{\Delta}{=} \|g_1 - g_2\|_p \tag{6.109}$$

Let us examine the properties of $\rho_p(g_1, g_2)$. It is apparent from Equations 6.107 and 6.109 that,

1. $\rho_p(g_1, g_2) \geq 0 \ \forall \ g_1, g_2$
2. $\rho_p(g_1, g_2) = 0 \iff g_1 = g_2 [\lambda]$

3. $\rho_p(g_1,g_2) = \rho_p(g_2,g_1)$

The above three items match the properties stated for a *metric space* in Definition 6.25. Halmos [28] shows that $\rho_p(g_1,g_2)$, in addition to having the above three properties, also possesses the following properties,

4. $\rho_p(g_1,g_2) \in \mathbb{R}$
5. $\rho_p(g_1,g_2)$ satisfies *Hölder's inequality* (Theorem 6.4)
6. $\rho_p(g_1,g_2)$ satisfies *Minkowski's inequality* (Theorem 6.5)

We will not attempt to prove the last three properties, but we will define *Hölder's inequality* and *Minkowski's inequality* to understand the consequences and to show that these inequalities are general cases of other inequalities used elsewhere in this book.

Theorem 6.4 (Hölder's Inequality). *If $g_1 \in \mathcal{L}_p$ and $g_2 \in \mathcal{L}_q$ where $\{p,q \in \mathbb{R}\}$, $\{p,q \geq 1\}$, and if*

$$\frac{1}{p} + \frac{1}{q} = 1 \tag{6.110}$$

then, $g_1 g_2 \in \mathcal{L}_1$ and

$$\|g_1 g_2\|_1 \leq \|g_1\|_p \ \|g_2\|_q \tag{6.111}$$

Equation 6.111 is known as Hölder's inequality.

Proof. See Halmos [28].

□

Theorem 6.5 (Minkowski's Inequality or Triangular Inequality). *If $g_1, g_2 \in \mathcal{L}_p$ where $\{p \in \mathbb{R}, p \geq 1\}$, then,*

$$\|g_1 + g_2\|_p \leq \|g_1\|_p + \|g_2\|_p \tag{6.112}$$

Equation 6.112 is known as Minkowski's inequality and it is a generalization of the Triangular inequality as applied to the function space.

Proof. See Halmos [28].

□

Definition 6.50 (Schwarz's Inequality). *The special case of Hölder's inequality where $p = 2$ and $q = 2$ is known as Schwarz's inequality.*

Schwarz's inequality comes in handy in many occasions, since for the special case of matrix norms it applies to the popular *Euclidean norm* (see Section 23.2) and for functions, it helps in the definition of the variance (see Section 6.6.2).

Given Definition 6.49 of the class, $\mathfrak{L}_p(\lambda)$, of *p-integrable extended real valued functions* and the resulting properties, all the requirements of a *metric space* per Definition 6.25 have been met. Therefore, $\rho_p(g_1, g_2)$ may be considered to be a *distance*, and it is known as the L_p *distance* [17]. See Definition 8.2 for more on the L_p distance.

6.5.5 Transformations

In this section, we will make a few definitions and work our way toward the definition of a *measurable transformation*. As we will see, it will become necessary for a few definitions such as that of the *conditional expectation*. For more detailed definitions and treatment of transformations, refer to Halmos [28].

Definition 6.51 (Transformation). *A transformation,* $T : (\mathscr{X}, \mathfrak{X}) \mapsto (\mathscr{Y}, \mathfrak{Y})$, *is a function, defined for all the elements in* $\mathscr{X}$ *such that its values would be in* $\mathscr{Y}$, *if* $\mathscr{Y}$ *includes only elements which are the results of transforming elements in* $\mathscr{X}$.

Definition 6.52 (One-to-One Transformation). *If every element in* $\mathscr{X}$ *is transformed into a distinct element in* $\mathscr{Y}$ *or to a unique member of disjoint subsets of* $\mathscr{Y}$, *then there exists an inverse transformation such that,*

$$x \in \mathscr{X} = T^{-1}(y \in \mathscr{Y}) \forall y \in \mathscr{Y} \tag{6.113}$$

Such a transformation (T) is known as a one-to-one transformation.

Definition 6.53 (Product of Transformation). *If* $y = T(x) \forall x \in \mathscr{X}$ *and* $z = U(y) \forall y \in \mathscr{Y}$, *are one-to-one transformations, then the product of the two transformations is defined as follows,*

$$(UT)(x) \triangleq U(T(x)) : (\mathscr{X}, \mathfrak{X}) \mapsto (\mathscr{Z}, \mathfrak{Z}) \tag{6.114}$$

Definition 6.54 (Inverse Image of a Transformation). *If we define a one-to-one transformation,* $T : (\mathscr{X}, \mathfrak{X}) \mapsto (\mathscr{Y}, \mathfrak{Y})$, *and any Borel subset,* $\mathscr{B} \in \mathfrak{Y}$, *then the inverse image of Borel subset* $\mathscr{B}$ *is the set of elements in* $\mathscr{X}$, $\mathscr{A} \in \mathfrak{X}$ *such that their transformation lies in* $\mathscr{B}$. *Namely,*

$$T^{-1}(\mathscr{B}) = \mathscr{A} = \{x : T(x) \in \mathscr{B}\} \tag{6.115}$$

Definition 6.55 (Measurable Transformation). *A measurable transformation, defined as,* $T : (\mathscr{X}, \mathfrak{X}, \rho) \mapsto (\mathscr{Y}, \mathfrak{Y}, \nu)$, *maps every value of measure* ρ *in measure space* $(\mathscr{X}, \mathfrak{X}, \rho)$ *to a value of measure* ν *in measure space,* $(\mathscr{Y}, \mathfrak{Y}, \nu)$ *where* ν *is defined for every Borel subset,* $\mathscr{B} \in \mathfrak{Y}$ *as follows,*

$$v(\mathscr{B}) = \rho(T^{-1}(\mathscr{B}))$$ (6.116)

Following Halmos's notation [28], we may, at times, use the shorthand,

$$\rho T^{-1}(\mathscr{B}) \overset{\Delta}{=} \rho(T^{-1}(\mathscr{B}))$$ (6.117)

6.6 Statistical Moments

Statistical moments are quantities which have been inspired by concepts in mechanics. They quantify the nature of random variables, their behaviors and their functions. In the following few sections, we will define some of the moments which are used most often. Theoretically, one may define an infinity of statistical moments, but the first four are most widely used.

6.6.1 Mean

Definition 6.56 (Expected Value (Expectation)). *The expected value of a function of a random variable, $\{g(x) : \mathscr{X} \mapsto \mathbb{R}\}$, denoted by $\mathscr{E}\{g(x)\}$, is defined by the following integral,*

$$\mathscr{E}\{g(x)\} \overset{\Delta}{=} \int_{\mathscr{X}} g(x) dP(x)$$ (6.118)

It is important to note two important points. The expected value may or may not exist. For the expected value of $g(x)$ to exist, the integral in Equation 6.118 must exist. Also, in Equation 6.118 the *expected value* of $g(x)$ has been defined over the entire *sample space*, $\mathscr{X}$. It is possible to compute $\mathscr{E}\{g(x)\}$ over any subset of the *Borel field* of the probability space $(\mathscr{X}, \mathfrak{X}, P)$, such as $\{\mathscr{A} \in \mathfrak{X}\}$. In general, when the domain of the *expected value* is not specified, the entire *sample space* is used.

If the *random variable* of interest is defined on the *real line*, $\{\mathscr{X} = \mathbb{R}\}$, then based on the *Radon-Nikodým theorem* (Theorem 6.3), we may rewrite Equation 6.118 in terms of the *Lebesgue measure* and the *Radon-Nikodým derivative* of the *probability measure* with respect to the *Lebesgue measure*. Namely,

$$\mathscr{E}\{g(x)\} \overset{\Delta}{=} \int_{-\infty}^{\infty} g(x) p(x) dx$$ (6.119)

Now that we know the definition for the *expected value* of a function of a random variable, let us consider the expected value of the *random variable* itself. This would be $\mathscr{E}\{g(x)\}$ where $g(x) = x$, or $\mathscr{E}\{X\}$.

Definition 6.57 (Expected Value or Mean). *Consider a random variable X defined in the probability space $(\mathscr{X}, \mathfrak{X}, P)$. The expected value $(\mathscr{E}\{X\})$ or the mean (μ) of $\{X\}$ is defined by the following integral,*

$$\mu = \mathscr{E}\{X\}$$
$$\stackrel{\Delta}{=} \int_{\mathscr{X}} x dP(x) \tag{6.120}$$

Note that X may be continuous, discrete or may take on any other general form as long as it is defined in the probability space, $(\mathscr{X}, \mathfrak{X}, P)$.

Let us consider the more specific class of continuous random variables defined on the real line, $\{\mathscr{X} : \mathscr{X} = \mathbb{R}\}$, also known as the one-dimensional Euclidean space. Then, using the Radon-Nikodým theorem (Theorem 6.3), we may write the integral in Equation 6.120 in terms of the Lebesgue measure. The following is the expression for the expected value (mean) of random variable $\{X : x \in \mathbb{R}\}$.

$$\mathscr{E}\{X\} = \int_{-\infty}^{\infty} x p(x) dx \tag{6.121}$$

where $p(x)$ is given by Equation 6.83 in terms of the probability measure P and the Lebesgue measure.

As it may be deduced from an analogy with a property in *mechanics*, the *mean* is sometime referred to as the *first statistical moment* of the random variable, X. More will be said about the mean and methods for its *estimation* after we have covered the concept of a *discrete random variable*.

Definition 6.58 (Conditional Expectation). *Recall Equation 6.71 for an unbounded real valued function $g(x)$. Let us rewrite this equation for the measure $v(\mathscr{B})$ over Borel subset $\mathscr{B} \in \mathfrak{Y}$, defined in measure space $(\mathscr{Y}, \mathfrak{Y}, v)$ in terms of the measurable transformation (Definition 6.55), $T : (\mathscr{X}, \mathfrak{X}, \rho) \mapsto (\mathscr{Y}, \mathfrak{Y}, v)$, as follows,*

$$v(\mathscr{B}) = \int_{T^{-1}(\mathscr{B})} g(x) d\rho(x) \tag{6.122}$$

Based on the definition of measurable transformation T (Definition 6.55), $v(\mathscr{B}) \ll \rho T^{-1}(\mathscr{B})$ (see Definition 6.41 and Equation 6.117). Therefore, from the Radon-Nikodým theorem (Theorem 6.3),

$$v(\mathcal{B}) = \int_{T^{-1}(\mathcal{B})} g(x)d\rho(x) \tag{6.123}$$

$$= \int_{\mathcal{B}} f(y)d\rho T^{-1}(y) \tag{6.124}$$

Note that $f(y)$ in Equation 6.124 is a measurable function which is dependent on the function g and is conditioned upon the variable y. This special function is called the conditional expectation or conditional expected value and is written as,

$$\mathscr{E}\{g|y\} \overset{\Delta}{=} f(y) \tag{6.125}$$

Using the notation of Equation 6.125 in Equation 6.124, we have,

$$\int_{T^{-1}(\mathcal{B})} g(x)d\rho(x) = \int_{\mathcal{B}} \mathscr{E}\{g|y\}d\rho T^{-1}(y) \tag{6.126}$$

where $\mathscr{E}\{g|y\}$ is known as the conditional expectation [27, 28] of g given y.

As we will see, the *conditional expectation* plays an important role in the derivation of *expectation maximization* which is a method for solving the *maximum likelihood estimation* problem – see Sections 11.3.1 and 10.1.

At this point, let us examine a very important inequality associated with *convex functions* and their expectations.

Theorem 6.6 (Jensen's Inequality). *Consider a probability density function, $p(x)$, defined as a Radon-Nikodým derivative of a probability measure P with respect to the Lebesgue measure as prescribed by Definition 6.45, in the probability measure space, $(\mathcal{X}, \mathfrak{X}, P)$. Furthermore, consider a function, $f : \mathcal{Y} \mapsto \mathbb{R}$ which is convex[8] in $\mathcal{X}$ – see Definition 24.23. Then for any real-valued function, $y(x)$, where $y : \mathcal{X} \mapsto \mathcal{Y}$, Jensen's inequality may be written as follows,*

$$\int_{\mathcal{X}} f(y(x))p(x)dx \geq f\left(\int_{\mathcal{X}} y(x)p(x)dx\right) \tag{6.127}$$

Furthermore, for concave functions, the direction of the inequality in Equation 6.127 is reversed.

Proof. See [1].

□

Note that *Jensen's inequality* basically states that

$$\mathscr{E}\{f(y)\} \geq f(\mathscr{E}\{y\}) \tag{6.128}$$

---

[8] Since all convex functions are also continuous (see Definitions 24.12 and 24.23, the continuity of f is implied.

for any *convex function*, f.

A special case of *Jensen's inequality* is one where $y(x) = x$, in which case,

$$\int_{-\infty}^{\infty} f(x)p(x)dx \geq f\left(\int_{-\infty}^{\infty} xp(x)dx\right) \qquad (6.129)$$

for any *convex function*, f.

6.6.2 Variance

Consider a *general random variable* X defined in the *probability space* $(\mathscr{X}, \mathfrak{X}, P)$. Let us examine the $\mathfrak{L}_2$ function space with respect to *probability measure P*. Based on Equation 6.107, the metric $\rho_2(g_1, g_2)$ is,

$$\rho_2(g_1, g_2) \stackrel{\Delta}{=} \|g_1 - g_2\|_2 \qquad (6.130)$$

where,

$$\|g\|_2 = \left|\left(\int |g(x)|^2 dP(x)\right)^{\frac{1}{2}}\right| \qquad (6.131)$$

Note that $\|g\|_2$ must exist and be finite for the arguments in this section to hold.

Recall that *Hölder's inequality* will have to hold for this metric. Since we are considering the $\mathfrak{L}_2$ class, then the special case of *Hölder's inequality* with $p = 2$ in Equation 6.111, namely *Schwarz's inequality* must hold.

Let us write the expression for distance $\rho_2(x, \mu)$ between x and μ,

$$\rho_2(x, \mu) = \left|\left(\int |x - \mu|^2 dP(x)\right)^{\frac{1}{2}}\right| \qquad (6.132)$$

Schwarz's inequality may then be written for $\rho_2(x, \mu)$ as follows,

$$\|x\mu\|_1 \leq \|x\|_2 \|\mu\|_2 \qquad (6.133)$$

The expression on the left hand side of the inequality in Equation 6.133 may be expanded as such,

$$\|x\mu\|_1 = \int_{\mathscr{X}} |x\mu| dP(x)$$

$$= \int_{\mathscr{X}} |x\mu| dP(x)$$

$$= |\mu| \int_{\mathscr{X}} |x| dP(x)$$

$$= \mu^2 \tag{6.134}$$

By the same token, here is the expansion of the expression on the right hand side of Equation 6.133,

$$\|x\|_2 \|\mu\|_2 = \left| \left(\int_{\mathscr{X}} x^2 dP(x) \right)^{\frac{1}{2}} \right| \left| \left(\int_{\mathscr{X}} \mu^2 dP(x) \right)^{\frac{1}{2}} \right|$$

$$= \left| \left(\int_{\mathscr{X}} x^2 dP(x) \right)^{\frac{1}{2}} \right| \left| \left(\mu^2 \int_{\mathscr{X}} dP(x) \right)^{\frac{1}{2}} \right|$$

$$= \left| \left(\int_{\mathscr{X}} x^2 dP(x) \right)^{\frac{1}{2}} \right| |\mu| \, (1)^{\frac{1}{2}}$$

$$- \left| \left(\int_{\mathscr{X}} x^2 dP(x) \right)^{\frac{1}{2}} \right| |\mu| \tag{6.135}$$

Plugging in the expanded forms of the left and right sides of the inequality of Equation 6.133 using Equations 6.134 and 6.135 respectively, and dividing both sides of the inequality by μ, *Schwarz's inequality* for the metric of Equation 6.132 may be expressed as,

$$|\mu| \leq \left| \left(\int_{\mathscr{X}} x^2 dP(x) \right)^{\frac{1}{2}} \right| \tag{6.136}$$

We can square both sides of the inequality in Equation 6.136 without affecting the inequality,

$$\mu^2 \leq \int_{\mathscr{X}} x^2 dP(x) \tag{6.137}$$

We call the distance function in function space $\mathscr{L}_2$, as expressed in Equation 6.132, the *standard deviation* of random variable X, and denote it by $\sigma(X)$,

$$\sigma(X) \overset{\Delta}{=} \rho_2(x, \mu)$$

$$= \left| \left(\int_{\mathscr{X}} |x - \mu|^2 dP(x) \right)^{\frac{1}{2}} \right|$$

$$= \left| \left(\int_{\mathscr{X}} (x - \mu)^2 dP(x) \right)^{\frac{1}{2}} \right| \tag{6.138}$$

Note that up to now, we have not restricted random variable X to have any specific form. However, if we consider the *continuous random variable*, X, defined on the *real line*, $\{\mathscr{X} = \mathbb{R}\}$, then by the *Radon-Nikodým theorem* (Theorem 6.3), we may write the *standard deviation*, $\sigma(X)$ using the *Lebesgue measure* as follows,

$$\sigma(X) \overset{\Delta}{=} \rho_2(x, \mu)$$

$$= \left| \left(\int_{-\infty}^{\infty} (x - \mu)^2 p(x) dx \right)^{\frac{1}{2}} \right| \tag{6.139}$$

The *variance* of the generic random variable X is defined as the square of its *standard deviation*. From Equation 6.138, the expression for the *variance* becomes,

$$\sigma^2\{X\} \overset{\Delta}{=} \int_{\mathscr{X}} (x - \mu)^2 dP(x) \tag{6.140}$$

$$= \mathscr{E}\left\{ (x - \mathscr{E}\{x\})^2 \right\} \tag{6.141}$$

where, μ (the mean) is given by Equation 6.120.

The integral expression for variance given by Equation 6.140 may be expanded as follows,

$$\sigma^2\{X\} = \int_{\mathscr{X}} (x - \mu)^2 dP(x)$$

$$= \int_{\mathscr{X}} x^2 dP(x) + \int_{\mathscr{X}} \mu^2 dP(x) - 2\int_{\mathscr{X}} x\mu dP(x)$$

$$= \int_{\mathscr{X}} x^2 dP(x) + \mu^2 \int_{\mathscr{X}} dP(x) - 2\mu \int_{\mathscr{X}} x dP(x)$$

$$= \int_{\mathscr{X}} x^2 dP(x) + \mu^2 (1) - 2\mu(\mu)$$

$$= \int_{\mathscr{X}} x^2 dP(x) - \mu^2 \tag{6.142}$$

Comparing Equation 6.142 with *Schwarz's inequality* given by Equation 6.137, it is interesting to note that the variance, $\sigma^2\{X\}$, of random variable X turns out to be a measure of the residue between the two sides of *Schwarz's inequality* for the distance defined in the metric space expressed in the $\mathfrak{L}_2$ class.

Once again, if we restrict our attention to the *continuous random variable* defined on the *real line*, $\{\mathscr{X} = \mathbb{R}\}$, we may use the *Radon-Nikodým theorem* (Theorem 6.3) to express the *variance* expression of Equation 6.142, in terms of the *probability density function* and the *Lebesgue measure*,

$$\sigma^2\{X\} = \int_{-\infty}^{\infty} x^2 p(x) dx - \mu^2 \tag{6.143}$$

where $\{X : x \in \mathbb{R}\}$.

Using an analogy with a property in *mechanics*, the *variance* is sometimes referred to as the *second statistical moment* of the random variable, X.

Let us define a couple of statistics which are used for comparing two random variables, namely, the *covariance* and the *Correlation Coefficient*. These *statistics* are *joint statistical moments* which are designed to signify the level of *dependence* of two random variables on one another.

Definition 6.59 (Covariance). *Consider two random variables, X_1 and X_2, with expected values, μ_1 and μ_2, respectively. Then the covariance, $Cov(X_1, X_2)$, is*

$$Cov(X_1, X_2) \triangleq \mathcal{E}\{(X_1 - \mu_1)(X_2 - \mu_2)\} \tag{6.144}$$
$$= \mathcal{E}\{X_1 X_2\} - \mu_1 \mu_2 \tag{6.145}$$

If X_1 and X_2 are *statistically independent*, then $Cov(X_1, X_2) = 0$ since $\mathcal{E}\{X_1 X_2\} = \mathcal{E}\{X_1\}\mathcal{E}\{X_2\}$ for *independent* random variables.

Definition 6.60 (Correlation Coefficient). *The correlation coefficient of two random variables, X_1 and X_2, is defined as,*

$$\kappa(X_1, X_2) \triangleq \frac{Cov(X_1, X_2)}{\sigma(X_1)\sigma(X_2)} \tag{6.146}$$

where $Cov(X_1, X_2)$ is given by Equation 6.145 and $\sigma(X)$ (the standard deviation of a random variable, X) is given by Equation 6.138.

The *correlation coefficient* is a dimensionless measure of the dependence of two random variables on one another. $|\kappa(X_1, X_2)| \leq 1$ (see [42] for the proof) with equality when a variable is completely dependent on another variable, namely when $X_1 = X_2$. $\kappa(X_1, X_2) = 0$ means that the two random variables are statistically independent. This can be easily seen since their *covariance* will be zero.

6.6.3 Skewness (skew)

The *skew* of a random variable $\{X : x \in \mathbb{R}\}$ is defined by the following integral.

$$s\{X\} \triangleq \frac{1}{\sigma^3} \int_{-\infty}^{\infty} (x - \mu)^3 p(x) dx \tag{6.147}$$

where, μ is given by Equation 6.121 and σ is given by Equation 6.138.

The definition in Equation 6.147 may have easily defined *skew* in the *general probability space* using the *Probability measure P* instead of the *Lebesgue measure* as we saw in the definition of Mean and Variance in the previous sections.

Using the mechanics analogy, the *skew* is also referred to as the *third standardized statistical moment* of the random variable X. The term *standardized* refers to the *normalization* done about the variance term in Equation 6.147. Its physical meaning in the case of a *Normal density* is a measure of the *lopsidedness* of the distribution (the *asymmetry* of it).

6.6.4 Kurtosis

Kurtosis of a random variable $\{X : x \in \mathbb{R}\}$ is generally defined by the following integral,

$$k\{X\} \triangleq \frac{1}{\sigma^4} \int_{-\infty}^{\infty} (x-\mu)^4 p(x)dx \tag{6.148}$$

where, μ is given by Equation 6.121 and σ is given by Equation 6.138.

As with definitions of the *mean* and the *variance* we could have easily defined *kurtosis* in the *general probability space* using the *Probability measure P* instead of the *Lebesgue measure*.

Using the mechanics analogy, *kurtosis* is also referred to as the *fourth standardized statistical moment* of the random variable X. As in the case of the definition of *skewness*, the term, *standardized* refers to the *normalization* about the variance-related term in Equation 6.148.

Definition 6.61 (Excess Kurtosis). *The most popular usage of kurtosis is one which is defined relative to a Normal density. This definition is called excess kurtosis because it computes the extra kurtosis compared to a Normal density. It has been defined such that the kurtosis of a Normal density becomes 0. In this definition, a value of 3 (kurtosis of a Normal density) is subtracted from the expression in Equation 6.148 producing the following definition,*

$$k\{X\} \triangleq \frac{1}{\sigma^4} \int_{-\infty}^{\infty} (x-\mu)^4 p(x)dx \; - \; 3 \tag{6.149}$$

Kurtosis is a measure of the *sharpness of the peak* of the distribution and the *flatness of its tails*. The higher the *kurtosis*, the sharper the distribution peak and the flatter its tails. The *Bernoulli random variable* (see Definition 6.63) with a *maximum entropy distribution*, (see Section 7.3.1), where $p_1 = p_2 = 0.5$, has the lowest *excess kurtosis* (-2). There is no upper bound for *kurtosis*.

6.7 Discrete Random Variables

Definition 6.62 (Discrete Random Variable). *A discrete random variable is a random variable which has a finite number of values in a finite interval. Namely, it has at most a countably infinite sample space, $\mathscr{X}$ – see Definition 6.6. This means that the random variable may still take on an infinite number of values, but there are any finite number of such values in a finite interval.*

Definition 6.63 (Bernoulli Random Variable). *A very special discrete random variable is one that may only take on binary values, $\{X : x \in \{0, 1\}\}$. Such a random variable is called a Bernoulli random variable with the sample space $\mathscr{X} = \{0, 1\}$.*

The *Bernoulli random variable* is especially interesting because of its relevance to the *Von Neumann* digital computer used in everyday life, specially in *signal processing*.

The following definition has been a cause for confusion in many circles. The loose utilization of the term, "distribution" as applied to different situations has been the culprit. Note that the *probability distribution* which is about to be defined for the *discrete random variable* is quite different from the *probability distribution* as defined for *continuous random variables* later in this chapter. We shall address these differences in more detail at a later stage.

Definition 6.64 (Probability Distribution (Probability Mass Function)). *The probability distribution of a discrete random variable is $p(x) = P(X = x)$ where X is a discrete random variable and x is any value it may assume (across sample space $\mathscr{X}$). Note that,*

$$\sum_{x \in \mathscr{X}} p(x) = 1 \tag{6.150}$$

Definition 6.65 (Cumulative Probability Distribution). *The cumulative probability distribution function,*

$$F(x) \triangleq P(X \le x) \ \forall x \in \mathscr{X}$$
$$= \sum_{\xi \le x} p(\xi) \tag{6.151}$$

where $\mathscr{X}$ is the sample space of the discrete random variable X and $F(x)$ is the probability of the ordered set of all X with less than or equal to x.

Definition 6.66 (Expected Value (Mean) of a Discrete Random Variable). *The expected value or mean of a discrete random variable X is defined as,*

$$\mu = \mathscr{E}\{X\}$$
$$\overset{\Delta}{=} \sum_{x\in\mathscr{X}} xp(x) \tag{6.152}$$

The expected value or mean is also known as the first statistical moment of random variable X.

Definition 6.67 (Expected Value of a Function of a Discrete Random Variable). *Assume that we are interested in the expected value of a function $f(X)$ of the discrete random variable X. X is the observed variable. Then we may write the expected value of $f(X)$ as follows,*

$$\mathscr{E}\{f(X)\} = \sum_{x\in\mathscr{X}} f(x)p(x) \tag{6.153}$$

As an example, take the linear function, $f(x) = \alpha X + \beta$, then,

$$\mathscr{E}\{\alpha X + \beta\} = \sum_{x\in\mathscr{X}} (\alpha x + \beta)p(x)$$
$$= \alpha \sum_{x\in\mathscr{X}} xp(x) + \beta \sum_{x\in\mathscr{X}} p(x)$$
$$= \alpha\mathscr{E}\{X\} + \beta \tag{6.154}$$

since $\sum_{x\in\mathscr{X}} p(x) = 1$ and $\sum_{x\in\mathscr{X}} xp(x) = \mathscr{E}\{X\}$.

This reveals two main properties of the *mean*, namely,

Property 6.4 (Scaling).

$$\mathscr{E}\{\alpha X\} = \alpha\mu \tag{6.155}$$

and

Property 6.5 (Translation).

$$\mathscr{E}\{X + \beta\} = \mu + \beta \tag{6.156}$$

If the *sample space* for the *discrete random variable*, X, has the following set of finite possible outcomes, $\mathscr{X} = \{X_1, X_2, \cdots, X_N\}$, with the probability mass function, $p(x)$ defined in that set, the *Jensen's inequality*, stated by Theorem 6.6, may be written for a *convex function*, $f(x)$,

$$\sum_{i=1}^{N} p(X_i)f(X_i) \geq f\left(\sum_{i=1}^{N} p(X_i)X_i\right) \tag{6.157}$$

The proof for the discrete case is fairly straight forward and is given in [19].

Definition 6.68 (Variance of a Discrete Random Variable). *The variance of X is defined as the expected value of the square of deviation of x from its mean μ. In mathematical terms,*

$$Var(X) = \sigma^2(X)$$
$$\triangleq \mathscr{E}\left\{(X-\mu)^2\right\}$$
$$= \mathscr{E}\left\{X^2 - 2\mu X + \mu^2\right\}$$
$$= \sum_{x \in \mathscr{X}} x^2 p(x) - 2\mu \sum_{x \in \mathscr{X}} x p(x) + \mu^2 \tag{6.158}$$

$$\tag{6.159}$$

Equation 6.158 may be rewritten in terms of expectations in the following form,

$$\sigma^2(X) = \mathscr{E}\left\{X^2\right\} - 2\mu^2 + \mu^2$$
$$= \mathscr{E}\left\{X^2\right\} - \mu^2$$
$$= \mathscr{E}\left\{X^2\right\} - [\mathscr{E}\left\{X\right\}]^2 \tag{6.160}$$

$\sigma(X)$ is known as the *standard deviation* of random variable X.

Let us examine the variance of a linear function of X:

$$\sigma^2(\alpha X + \beta) = \mathscr{E}\left\{(\alpha X + \beta)^2\right\} - [\mathscr{E}\left\{(\alpha X + \beta)\right\}]^2$$
$$= \mathscr{E}\left\{(\alpha^2 X^2 + 2\alpha\beta X + \beta^2)\right\} - (\alpha\mu + \beta)^2$$
$$= \alpha^2 \mathscr{E}\left\{X^2\right\} + 2\alpha\beta\mu + \beta^2 - (\alpha^2\mu^2 + 2\alpha\beta\mu + \beta^2)$$
$$= \alpha^2 \mathscr{E}\left\{X^2\right\} - \alpha^2\mu^2$$
$$= \alpha^2 \sigma^2(X) \tag{6.161}$$

The results from Equation 6.161 convey the following properties of the variance, $\sigma^2(X)$,

Property 6.6 (Scaling).

$$\sigma^2(\alpha X) = \alpha^2 \sigma^2(X) \tag{6.162}$$

and

Property 6.7 (Translation). *Translation:*

$$\sigma^2(X + \beta) = \sigma^2(X) \tag{6.163}$$

Note the similarities between the *variance* of a *discrete variable* and that of a *continuous variable* defined in Section 6.6. Higher order statistical moments such as *skewness (skew)* and *kurtosis* may be defined much in the same manner as for the *mean* and *variance*, from their more general forms described in Section 6.6.

6.7.1 Combinations of Random Variables

The *convolution* results of Section 6.5.2.1 regarding the combination of two random variables apply to *discrete random variables*. If the *discrete random variable* $\{X : \xi \in \mathscr{X}\}$ is produced from the summation of two *independent random variables*, X_1 and X_2 such that,

$$
\begin{aligned}
\xi &= g(x_1, x_2) \\
&= x_1 + x_2
\end{aligned}
\tag{6.164}
$$

then the *probability mass function* $p(k)$ is given in terms of the *discrete convolution* of the *marginal probability mass functions*, $p_{X_1}(i)$ and $p_{X_2}(j)$ as follows,

$$
\begin{aligned}
p(i) &= \sum_{k=-\infty}^{\infty} p_{X_1}(i-k) p_{X_2}(k) \\
&= p_{X_1} * p_{X_2}
\end{aligned}
\tag{6.165}
$$
$$
\tag{6.166}
$$

As in most cases, results related to the *probability density function* of a *continuous random variable* have parallels with the *probability mass function* of a *discrete random variable*.

6.7.2 Convergence of a Sequence

Let us assume that there is an *ordered set* (*sequence* [20]) of random variables, $\{X\}_1^n = \{X_i : 1 \le i \le n\}$.[9] In this section, we examine different definitions of the *convergence* of this sequence to a specific *random variable*, X, as $n \to \infty$. The criteria of convergence will be presented here in the order of their strictness. First let us start with the definition of a *weak convergence*.

Definition 6.69 (Weak Convergence (Convergence in Probability)). *Weak convergence of a sequence of random variables, $\{X\}_1^n$, to X, also known as convergence in probability is defined as follows [33],*

---

[9] Note that, here, we are using X_i to denote different random variables, which is different from the notation used for samples of the same random variable (x_i). Also, here, X_i does not signify disjoint members of the sample space, as the notation does in the rest of the book. As much care as has been afforded, unfortunately, because of the limitations on the number of symbols available, maneuvers like this become necessary in parts of the book.

$$\lim_{n \to \infty} P\left(|X_n - X| < \varepsilon\right) = 1 \ \forall \ \varepsilon > 0 \tag{6.167}$$

The statement of Equation 6.167 may alternatively be presented as follows [35],

$$\lim_{n \to \infty} P\left(|X_n - X| \geq \varepsilon\right) = 0 \ \forall \ \varepsilon > 0 \tag{6.168}$$

The following notation is sometimes used to denote a weak convergence,

$$X_n \overset{prob.}{\longrightarrow} X \tag{6.169}$$

Definition 6.70 (Strong Convergence (Almost Sure Convergence or Convergence with Probability 1)). *Strong convergence of a sequence of random variables, $\{X\}_1^n$, to X, also known as almost surely or with probability 1 is defined as follows,*

$$P\left(\lim_{n \to \infty} X_n = X\right) = 1 \tag{6.170}$$

The following notation is sometimes used to denote a strong convergence,

$$X_n \overset{a.s.}{\longrightarrow} X \tag{6.171}$$

In Definition 6.23, we spoke about a *complete space* which used the *fundamental sequence* of a metric space. The following is a formal definition of such a sequence.

Definition 6.71 (Fundamental Sequence or a Cauchy Sequence). *Consider an ordered set (sequence), $\{X\}_1^n = \{X_i : 1 \leq i \leq n\}$, defined in a metric space, $(\mathscr{X}, \mathfrak{X}, \rho)$, where ρ is a bounded metric, $\rho \in \mathbb{R}$, such that $\exists \varepsilon(r) > 0 : \rho(X_l, X_m) < \varepsilon(r) \ \forall \ r < l, m$. Such a sequence is known as a fundamental sequence or a Cauchy sequence.*

6.8 Sufficient Statistics

In this section, we will discuss the concept of *sufficient statistics* which was introduced by *Fisher* [26], discussed in detail by *Darmois* [21], *Koopman* [36] and *Pitman* [45], and later generalized by many including Barankin and Mitra [7].

The definition of *sufficient statistics* is necessary for the undertaking of *statistical estimation* and model selection when we deal with a finite sample size. To be able to define *sufficient statistics*, we should first define the notion of a *statistic*. Later, we will introduce *efficient statistics* through the definition of *efficiency of a statistic*. These concepts were first introduced by Fisher [26] for the general case and then elaborated for the very special exponential family of densities [21, 36, 45, 7].

Definition 6.72 (Statistic). *A statistic is any function of a sample which is indepen-dent of the sample's distribution.*

A statistic may have different purposes. Any statistic that is used to estimate a statistical parameter is called an *estimator*. For example, the *sample mean* defined by Equation 6.173 is a statistic and an estimator of the statistical parameter, the true mean, of a distribution.

An important statistic is the *sufficient statistic* associated with a random variable, X.

Definition 6.73 (Sufficient Statistic). *A sufficient statistic, $v(x)$, is a statistic de-fined for a random variable, $X : x \in \mathscr{X}$ which contains in itself, all the relevant information in the X.*

Definition 6.74 (Efficiency of a Statistic). *The efficiency of a statistic is the ra-tio of the intrinsic accuracy of its random sampling distribution to the amount of information in the data from which it has been derived.*

This definition is valid for small samples of data with any distribution and is not limited to *Normal distributions*. For *large samples*, this is the *relevant information* utilized by the statistic of interest. For large samples with an underlying *Normal (Gaussian) distribution*, if we know the *variance* of any *sufficient statistic*, then we may compute the *efficiency* of any other statistic by the following ratio,

$$E_i \stackrel{\Delta}{=} \frac{\sigma_s^2}{\sigma_i^2} \tag{6.172}$$

where E_i denotes the *efficiency* of statistic s_i, σ_i^2 is the variance computed based on statistic s_i and σ_s^2 is the variance computed from any *efficient statistic*, s_s. Note that *efficiency*, E_i, is the fraction of *relevant information* utilized by the statistic of interest, s_i (for large samples).

Definition 6.75 (Statistical Efficiency Criterion). *The efficiency criterion requires that the variance of a statistic weighed by the number of samples used for computing the statistic approaches the smallest possible value for the underlying distribution.*

This is apparent from the special case related to *large number of samples*, given in Equation 6.172.

Definition 6.76 (Efficient Statistic). *Any statistic that meets the statistical effi-ciency criterion is known as an efficient statistic and any such parameter estimate is an efficient estimate.*

When we deal with the problem of estimating the statistics from a finite sample size where the data may not even be distributed normally and estimation errors may be large, then different *efficient statistics* may exist and they will have different efficiencies. It will be important to distinguish between these different *efficient statistics*.

Fisher [26] proved that when *sufficient statistics* exist, they are the solutions of the equations of *maximum likelihood* – see Section 6.9. Therefore, for a random variable for which there is a *normal* underlying distribution, the *sufficient statistic* would be the *maximum likelihood estimate* of the mean of the distribution which is given by the *sample mean* for large samples (Equation 6.173). Based on *Fisher*'s terminology [26] and the *law of large numbers* (Section 6.9.2), we see that the *sample mean* is also considered to be a *consistent statistic*. *Fisher* [26] defines a *consistent statistic* as follows.

Definition 6.77 (Consistent Statistic). *A consistent statistic is an estimate of any parameter such that when computed from an indefinitely large sample, it converges to that parameter with high accuracy.*

The estimated parameter which is a consistent statistic is known as a *consistent estimate*.

6.9 Moment Estimation

The expressions for the *mean* and the *variance* are both based on knowing the *probability density function*, $p(x)$, for the continuous case (Equations 6.121 and 6.143) and the *probability mass function*, $p(x)$, for the discrete case (Equations 6.152 and 6.158). In this section, we will examine methods for estimating these *statistical moments* when $p(x)$ is unknown.

6.9.1 Estimating the Mean

Take the *expected value* (or *mean*) of X given by Definition 6.66. In most cases, $p(x)$ is not known in advance. So, how do we compute μ, without having the *probability mass function* of X? One method is to use the *sample mean*, $\overline{X}$, to *estimate* the *true mean*, μ. The *sample mean* for a *discrete random variable*, X, is defined as,

$$\overline{X}|_n \overset{\Delta}{=} \frac{1}{n}\sum_{i=1}^{n} x_i \tag{6.173}$$

where n is the number of samples taken in the experiment for estimating the *mean*, μ. Equation 6.173 provides the *maximum likelihood estimate* of the mean of X. The idea is that if n becomes large, the number of times, each unique value of X shows up, is proportional to its *probability of appearance*. Therefore, by dividing the sum by n, the *normalization* allows for estimating the *true mean* (μ) of X. To clarify this argument, take a *Bernoulli random variable* which, based on Definition 6.63, can only take on one of two possible values, namely, $\mathcal{X} = \{0, 1\}$.

For the moment, let us assume that the underlying *probability mass function* for the system of interest is,

$$p(x) = \begin{cases} 0.25 \; for \; x = 0 \\ 0.75 \; for \; x = 1 \end{cases} \tag{6.174}$$

Then, the *true mean* computed based on Definition 6.66 will be

$$\begin{aligned}
\mu &= \mathcal{E}\{X\} \\
&= \sum_{x \in \mathcal{X}} xp(x) \\
&= (0.25)\,(0) + (0.75)\,(1) \\
&= 0.75
\end{aligned} \tag{6.175}$$

Now, let us assume that $p(x)$ is unknown, but we may do an *experiment* to estimate μ. Assume that we do 1000 trials and based on the underlying nature of the system at hand, for 725 times, $X = 1$, and for 275 times, $X = 0$, in which case the *sample mean* will be,

$$\begin{aligned}
\overline{X}|_{1000} &= \frac{1}{1000}(725 \times 1 + 275 \times 0) \\
&= 0.725
\end{aligned} \tag{6.176}$$

As n increases, the nature of X drives the *relative frequency* of each possible value of X toward the underlying *probability distribution*, so

$$\overline{X}|_n \to \mu \quad as \quad n \to \infty \tag{6.177}$$

Equation 6.177, known as the *law of large numbers*, is valid under certain assumptions made about the random variable, X. These assumptions will be discussed in some detail momentarily.

6.9.2 Law of Large Numbers (LLN)

The Mean Value Theorem (the Law of Mean), for analytic functions is presented in Theorem 24.10. Similar incarnation of this theorem exist for *random variables*. Depending on the nature of the conditions and limitations imposed by the specific case, two basic laws are described and proven in the form of theorems.

The first law is known as the *weak law of large numbers* and in one of its most complete forms, it was expressed in the form of a theorem by the Russian mathematician, *Khintchine* [33, 24, 20]. The second law with more strict conditions on the moments of the random variables of interest is known as the *strong law of large numbers* which was stated by many people with one of the best forms being the theorem which was stated and proven by *Khintchine* [20] and another Russian mathematician, *Kolmogorov* [35, 25].

6.9.2.1 Weak Law of Large Numbers (WLLN)

Theorem 6.7 (Khintchine's Theorem (Weak Law of Large Numbers – WLLN)).
See [37].   If $X_1, X_2, \cdots$ are independent and identically distributed (i.i.d.) random variables with a finite mean, μ,

$$\mu = \int_{\mathscr{X}} x \, dP(x) \tag{6.178}$$

Furthermore, if we define a number, $\overline{X}|_n$, evaluated over n samples as

$$\overline{X}|_n \triangleq \frac{1}{n} \sum_{i=1}^{n} X_i \tag{6.179}$$

then,

$$\lim_{n \to \infty} P\left(|\overline{X}|_n - \mu| < \varepsilon\right) = 1 \ \forall \ \varepsilon > 0 \tag{6.180}$$

Equation 6.180 states that the expression in Equation Equation 6.179 converges, in probability, to μ. The convergence stated in Equation 6.180 is known as weak convergence. In short,

$$X_n \xrightarrow{prob.} X \tag{6.181}$$

See Definition 6.69.

Proof. See [20, 24].

$\square$

6.9.2.2 Strong Law of Large Numbers (SLLN)

Under the additional constraint that the set of all n random variables have finite variances, the results of Theorem 6.7 may enjoy a much *stronger* statement called the *strong law of large numbers.* Many different versions of the *strong law of large numbers* exist, according to *Borel* [15], *Cantelli* [16], *Khintchine* [32], *Kolmogorov* [32], etc. [41]

Theorem 6.8 (Strong Law of Large Numbers (SLLN)). *See [24].   In general most of the statements of the law of large numbers state that the sample mean ($\overline{X}|_n$) approaches the true mean, μ, of the random variable X almost surely (a.s.), if the variance (σ_n^2) of X evaluated for a sequence of n trials is finite for all n – see Definition 6.70. Namely,*

$$X_n \xrightarrow{a.s.} X \tag{6.182}$$

or in other words,

$$\sigma_n^2 < \sigma_{max}^2 \ \forall \, n \in \mathbb{N} \implies P\left(\lim_{n \to \infty} \overline{X}|_n = \mu\right) = 1 \tag{6.183}$$

where $\sigma_{max}^2 < \infty$ and

$$\mu = \int_{\mathscr{X}} x \, dP(x) \tag{6.184}$$

Proof.
For the different convergence criteria, there are slightly different limitations imposed in the statement of SLLN. The following are some of the references with slightly different treatments: *Borel* [15], *Cantelli* [16], *Khintchine* [32], *Kolmogorov* [32], *Feller* [23].

□

The *strong law of large numbers* is quite complex and detailed. The different versions, with slight variations in the conditions met by the random variables of interest, make this topic out of reach. In the scope of our discussion, the *weak law of large numbers*, stated in *Khintchine's* theorem is sufficient.

In essence, for the purposes of speaker recognition, we are concerned with the following loosely defined limit,

$$\lim_{n \to \infty} \overline{X}|_n = \lim_{n \to \infty} \frac{1}{n} \sum_{i=1}^{n} X_i$$
$$= \mu \tag{6.185}$$

$\overline{X}|_n$, defined by Equation 6.179, is also known as the *sample mean*.

6.9.3 Different Types of Mean

Depending on the problem of interest, there have been many different concepts representing the *mean* of a set of samples. Here, we will define four different such *measures* which have different usages, depending on our objectives. For these definitions, we assume that we have observed a set of N samples of $X : x \in \mathscr{X} = \mathbb{R}$, denoted by, $\{X\}_1^N = \{x_n\}, n \in \{1, 2, \cdots, N\}$.

Definition 6.78 (Arithmetic Mean). *The arithmetic mean, $\mu_A(X_1^N)$, of a set of N samples of random variable X is given by the following equation,*

$$\mu_A(X_1^N) = \frac{1}{N} \sum_{n=1}^{N} x_n \tag{6.186}$$

The *sample mean*, estimating the true mean of X, given by Equation 6.179 is *arithmetic mean* of the set of N observed samples. In most cases, when one speaks of the mean of a sequence, the *arithmetic mean* is intended.

Definition 6.79 (Geometric Mean). *The geometric mean, $\mu_G(\{x\}_1^N)$, of a set of N samples of random variable X is given by the following equation,*

$$\mu_G(\{x\}_1^N) = \sqrt[N]{\prod_{n=1}^{N} x_n} \tag{6.187}$$

The *geometric mean* has the tendency of accentuating the effects of non-conforming members of the set. As we saw in Definition 5.5, this contrast with the *arithmetic mean* is used to define spectral flatness of a spectrum.

Definition 6.80 (Harmonic Mean). *The harmonic mean, $\mu_H(\{x\}_1^N)$, of a set of N samples of random variable X is only defined for a set of positive real numbers, $x_n > 0 \; \forall \, n \in \{1, 2, \cdots, N\}$, and it is given by the following equation,*

$$\mu_A(\{x\}_1^N) = \frac{N}{\sum\limits_{n=1}^{N} \frac{1}{x_n}} \tag{6.188}$$

The *harmonic mean* has the property that it takes on a small value if x_n for any n is small. In this sense, it is more related to the *minimum* of the set than its *arithmetic mean*. As we shall see, the *harmonic mean* is used in the *k-harmonic means algorithm*, which is used for *unsupervised clustering* and described in Section 11.2.9.

Definition 6.81 (Quadratic Mean (Root Mean Square – RMS)). *The quadratic mean or RMS, $\mu_Q\{x\}_1^N$, of a set of N samples of random variable X is defined as,*

$$\mu_Q(\{x\}_1^N) = \frac{\sqrt{\sum_{n=1}^{N} x_n^2}}{N} \tag{6.189}$$

The *quadratic mean* has the property that it is concerned with the magnitude of the samples rather than just their values. It is actually a second order statistic. It provides a good indication of the deviation of the samples from the origin, treating positive and negative samples in the same manner.

6.9.4 Estimating the Variance

There are several techniques for estimating the variance of a random variable. The first *estimator* (see Definition 6.72 for the definition of an *estimator*) is known as the *biased sample variance* or the *maximum likelihood estimator* and is defined as,

Definition 6.82 (Sample Variance (Biased Estimator)). *The biased sample variance* $\hat{\sigma}^2|_n$ *(see Section 6.9.4 for a definition of variance) is defined as,*

$$\hat{\sigma}^2|_n \triangleq \frac{1}{n} \sum_{i=1}^{n} \left(X_i - \overline{X}|_n \right)^2 \tag{6.190}$$

The *sample variance* given by Equation 6.190 is biased since it uses the same *sample* for computing the *sample mean* as it uses for computing the *sample variance*. In fact the *expected value* of the *biased sample variance* is

$$\mathscr{E}\left\{ \hat{\sigma}^2|_n(X) \right\} = \frac{n-1}{n} \sigma^2(X) \tag{6.191}$$

Definition 6.83 (Sample Variance (Unbiased Estimator)). *The unbiased sample variance* $\tilde{\sigma}^2|_n$ *(see Section 6.9.4 for a definition of variance) is defined as,*

$$\tilde{\sigma}_n^2 \triangleq \frac{1}{n-1} \sum_{i=1}^{n} \left(X_i - \overline{X}|_n \right)^2 \tag{6.192}$$

$\tilde{\sigma}_n^2(X)$ is known as the *unbiased sample variance* of X and has the property that its expected value is,

$$\mathscr{E}\left\{ \tilde{\sigma}_n^2 \right\} = \sigma^2(X) \tag{6.193}$$

6.10 Multi-Variate Normal Distribution

In most *pattern recognition* problems, and generally in most practical statistical problems, we are interested in *probability density functions* which belong to a special family of functions known as the *Darmois-Koopman-Pitman exponential family* or sometimes as the *regular exponential family* [7]. The following is a representation of this parametric family of density functions.

$$p(\mathbf{x}|\boldsymbol{\varphi}) = a(\mathbf{x}) \exp\left(\boldsymbol{\varphi}^T \mathbf{v}(\mathbf{x}) - b(\boldsymbol{\varphi})\right) \tag{6.194}$$

where $\mathbf{x}$ is a possible value of the random variable, $X : \mathbf{x} \in \mathcal{X} = \mathcal{R}^D$ and the parameter vector, $\{\Phi : \boldsymbol{\varphi} \in \phi \subset \mathcal{R}^M\}$. $\boldsymbol{\varphi}$ is sometimes called the vector of *natural parameters* of the density function and it is a function of another parameter which is called the *conventional parameter vector*. The *natural parameter vector* is used to present a canonical form of the *exponential family*. In our discussion, we will avoid complicating matters by steering away from the *conventional parameters* and only use the *natural parameters*.

In Equation 6.194, $\{\mathbf{v}(\mathbf{x}) : \mathcal{X} \mapsto \mathcal{V} = \mathcal{R}^M\}$ is the vector of *sufficient statistics* of X. $a(\mathbf{x})$ is called the *base measure*. It is an arbitrary function of $\mathbf{x}$ to be able to formalize different members of the family in a common normalized form. $b(\boldsymbol{\varphi})$ is a function of the parameter vector, which is specific to members of the family. It is another normalization term which may be written in terms of the rest of the parameters in the following way and is designed to offset the weight of $a(\mathbf{x})$.

$$b(\boldsymbol{\varphi}) = \ln\left(\int_{x \in \mathcal{X}} \exp\left(\boldsymbol{\varphi}^T \mathbf{v}(\mathbf{x}) a(\mathbf{x})\right) d\mathbf{x}\right) \tag{6.195}$$

In speaker recognition we mostly deal with *multi-variate random variables*. Equation 6.196 is the expression for the *multi-dimensional Normal density* function which is a member of the *exponential family* of Equation 6.194.

$$p(\mathbf{x}|\boldsymbol{\mu}, \boldsymbol{\Sigma}) = \frac{1}{(2\pi)^{\frac{d}{2}} |\boldsymbol{\Sigma}|^{\frac{1}{2}}} \exp\left(-\frac{1}{2}(\mathbf{x}-\boldsymbol{\mu})^T \boldsymbol{\Sigma}^{-1}(\mathbf{x}-\boldsymbol{\mu})\right) \tag{6.196}$$

$$where \begin{cases} \mathbf{x}, \boldsymbol{\mu} \in \mathcal{R}^d \\ \boldsymbol{\Sigma} : \mathcal{R}^d \mapsto \mathcal{R}^d \end{cases}$$

In 6.196, $\boldsymbol{\mu}$ is the *mean vector* where,

$$\boldsymbol{\mu} = \mathcal{E}\{\mathbf{x}\}$$

$$\stackrel{\Delta}{=} \int_{-\infty}^{\infty} \mathbf{x}\, p(\mathbf{x}) d\mathbf{x} \tag{6.197}$$

The integral in Equation 6.197 is really equivalent to the vector of expected values of individual components $\mathbf{x}$.

The, so called, *Sample Mean*[10] approximation for 6.197 is,

$$\bar{\mathbf{x}}|_N = \frac{1}{N} \sum_{i=1}^{N} \mathbf{x}_i \tag{6.198}$$
$$\approx \boldsymbol{\mu}$$

where N is the number of samples.

The *Variance-Covariance matrix* of a multi-dimensional random variable is defined as,

$$\begin{aligned}
\boldsymbol{\Sigma} &\triangleq \mathscr{E}\left\{ (\mathbf{x} - \mathscr{E}\{\mathbf{x}\})(\mathbf{x} - \mathscr{E}\{\mathbf{x}\})^T \right\} \tag{6.199} \\
&= \mathscr{E}\left\{ (\mathbf{xx}^T - \mathbf{x}(\mathscr{E}\{\mathbf{x}\})^T - \mathscr{E}\{\mathbf{x}\}\mathbf{x}^T + \mathscr{E}\{\mathbf{x}\}(\mathscr{E}\{\mathbf{x}\})^T) \right\} \\
&= \mathscr{E}\left\{ \mathbf{xx}^T - \mathbf{x}\boldsymbol{\mu}^T - \boldsymbol{\mu}\mathbf{x}^T + \boldsymbol{\mu}\boldsymbol{\mu}^T \right\} \\
&= \mathscr{E}\left\{ \mathbf{xx}^T \right\} - \mathscr{E}\left\{ \mathbf{x}\boldsymbol{\mu}^T \right\} - \boldsymbol{\mu}(\mathscr{E}\{\mathbf{x}\})^T + \boldsymbol{\mu}\boldsymbol{\mu}^T \\
&= \mathscr{E}\left\{ \mathbf{xx}^T \right\} - \boldsymbol{\mu}\boldsymbol{\mu}^T - \boldsymbol{\mu}\boldsymbol{\mu}^T + \boldsymbol{\mu}\boldsymbol{\mu}^T \\
&= \mathscr{E}\left\{ \mathbf{xx}^T \right\} - \boldsymbol{\mu}\boldsymbol{\mu}^T \tag{6.200}
\end{aligned}$$

This matrix is called the *Variance-Covariance* since the diagonal elements are the *variances* of the individual dimensions of the multi-dimensional vector, $\mathbf{x}$. The *off-diagonal elements* are the *covariances* across the different dimensions. Some have called this matrix the *Variance matrix*. Mostly in the field of *Pattern Recognition* it has been referred to simply as the *Covariance matrix* which is the name we will adopt from this point onward due to its popularity and brevity.

The *unbiased estimate* of $\boldsymbol{\Sigma}$, $\tilde{\boldsymbol{\Sigma}}$ is given by the following expression (see Definition 6.83),

---

[10] Here, we are using $\bar{\mathbf{x}}|_N$ to denote the N-sample, *sample mean* of X where X is a random variable in the d-dimensional *Euclidean space*. If we had used $\overline{X}|_N$, we would not be able to portray the multi-dimensional nature of X. If we made X bold, it would be confused with a matrix. So, although it is technically sloppy, we would have to stick with the notation of $\bar{\mathbf{x}}|_N$ for the sample mean of a multidimensional random variable.

$$\tilde{\boldsymbol{\Sigma}}|_N \triangleq \frac{1}{N-1} \sum_{i=1}^{N} (\mathbf{x}_i - \overline{\mathbf{x}}|_N)(\mathbf{x}_i - \overline{\mathbf{x}}|_N)^T \tag{6.201}$$

$$= \frac{1}{N-1} \sum_{i=1}^{N} \left(\mathbf{x}_i \mathbf{x}_i^T - \mathbf{x}_i \overline{\mathbf{x}}^T|_N - \overline{\mathbf{x}}|_N \mathbf{x}_i^T + \overline{\mathbf{x}}|_N \overline{\mathbf{x}}^T|_N \right)$$

$$= \frac{1}{N-1} \mathbf{S}|_N + \frac{N}{N-1} \overline{\mathbf{x}}|_N \overline{\mathbf{x}}^T|_N - \frac{N}{N-1} \overline{\mathbf{x}}|_N \overline{\mathbf{x}}^T|_N - \frac{N}{N-1} \overline{\mathbf{x}}|_N \overline{\mathbf{x}}^T|_N$$

$$= \frac{1}{N-1} \mathbf{S}|_N - \frac{N}{N-1} \left(\frac{\mathbf{s}|_N}{N} \frac{\mathbf{s}^T|_N}{N} \right)$$

$$= \frac{1}{N-1} \left[\mathbf{S}|_N - \frac{1}{N} \mathbf{s}|_N \mathbf{s}^T|_N \right] \tag{6.202}$$

$$= \frac{1}{N-1} \left[\mathbf{S}|_N - N(\overline{\mathbf{x}}|_N \overline{\mathbf{x}}^T|_N) \right] \tag{6.203}$$

where the *sample mean*, $\overline{\mathbf{x}}|_N$, is given by equation 6.198, the first order sum is

$$\mathbf{s}|_N \triangleq \sum_{i=1}^{N} \mathbf{x}_i \tag{6.204}$$

$$= N \overline{\mathbf{x}}|_N \tag{6.205}$$

and the *second order sum matrix*, $\mathbf{S}|_N$, is given by,

$$\mathbf{S}|_N \triangleq \sum_{i=1}^{N} \mathbf{x}_i \mathbf{x}_i^T \tag{6.206}$$

If the components of random variable $\mathbf{x}$ are *statistically independent*, namely if the covariance matrix, $\boldsymbol{\Sigma}$, is diagonal, then,

$$\left(\tilde{\boldsymbol{\Sigma}}|_N \right)_{[i][j]} = \begin{cases} (\mathbf{S}|_N)_{[i][j]} = 0 & \forall\, i \neq j \\[2mm] \frac{1}{N-1} \left[(\mathbf{S}|_N)_{[i][i]} - \frac{1}{N} \mathbf{s}_i|_N \mathbf{s}_i^T|_N \right] & \forall\, i = j \end{cases} \tag{6.207}$$

In the next chapter, we will examine the different, but related concepts of *Information* as defined by *Fisher, Shannon, Wiener* and and many others. This leads to the whole concept of *Information Theory*.

References

1. Antezana, J., Massey, P., Stojanoff, D.: Jensen's Inequality and Majorization. Web (2004). URL arXiv.org:math.FA/0411442
2. Avrachenkov, K.E., Beigi, H.S., Longman, R.W.: Operator-Updating Procedures for Quasi-Newton Iterative Learning Control in Hilbert Space. In: IEEE Conference on Decision and Control (CDC99), vol. 1, pp. 276–280 (1999). Invited Paper

3. Avrachenkov, K.E., Beigi, H.S., Longman, R.W.: Updating Procedures for Iterative Learning Control in Hilbert Space. Intelligent Automation and Soft Computing Journal **8**(2) (2002). Special Issue on Learning and Repetitive Control

4. Avrachenkov, K.E., Beigi, H.S.M., Longman, R.W.: Updating Procedures for Iterative Learning Control in Hilbert Space. Technical Report (1999)

5. Banach, S.: Sur les Opérations dans les Ensembles Abstraits et leur Application aux Équations Intégrales. Fundamenta Mathematicae **3**, 133–181 (1922)

6. Banach, S., Tarski, A.: Sur la Décomposition des Ensembles de Points en Parties Respective-ment Congruentes. Fundamenta Mathematicae **6**, 244–277 (1924)

7. Barankin, E.W., Mitra, A.P.: Generalization of the Fisher-Darmois-Koopman-Pitman Theorem on Sufficient Statistics. The Indian Journal of Statistics **25**(3), 217–244 (1963)

8. Barnes, J.: An Algorithm for Solving Nonlinear Equations Based on the Secant Method. Computer Journal **8**, 66–72 (1965)

9. Beigi, H.S.: An Adaptive Control Scheme Using the Generalized Secant Method. In: Proceedings of the Canadian Conference on Electrical and Computer Engineering, vol. II, pp. TA7.21.1–TA7.21.4 (1992)

10. Beigi, H.S.: A Parallel Network Implementation of The Generalized Secant Learning-Adaptive Controller. In: Proceedings of the Canadian Conference on Electrical and Computer Engineering, vol. II, pp. MM10.1.1–MM10.1.4 (1992)

11. Beigi, H.S.: Adaptive and Learning-Adaptive Control Techniques based on an Extension of the Generalized Secant Method. Intelligent Automation and Soft Computing Journal **3**(2), 171–184 (1997)

12. Beigi, H.S., Li, C.J., Longman, R.: Learning Control Based on Generalized Secant Methods and Other Numerical Optimization Methods. In: Sensors, Controls, and Quality Issues in Manufacturing, the ASME Winter Annnual Meeting, vol. 55, pp. 163–175 (1991)

13. Beigi, H.S.M.: Neural Network Learning and Learning Control Through Optimization Techniques. Columbia University, New York (1991). Doctoral Thesis: School of Engineering and Applied Science

14. Bingham, N.H.: Studies in the History of Probability and Statistics XLVI. Measure into Probability: From Lebesgue to Kolmogorov. Biometrika **87**(1), 145–156 (2000)

15. Borel, E.: Les Probabilités Dénombrables et Leurs Applications Arithmétique. Rend. Circ. Mat. Palermo **2**(27), 247–271 (1909)

16. Cantelli, F.: Sulla Probabilità Come Limite della Frequenza. Atti Accad. Naz. Lincei **26**(1), 39–45 (1917)

17. Cortes, C., Mohri, M., Rastogi, A.: Lp Distance and Equivalence of Probabilistic Automata. International Journal of Foundations of Computer Science (IJFCS) **18**(4), 761–779 (2007)

18. Courant, R., Hibert, D.: Methods of Mathematical Physics, 1st english edn. Volume I. WILEY-VCH Verlag GmbH & Co. KGaA, Weinheim (2009). Second reprint of English version Translated and Revised from German Original, 1937, ISBN: 0-471-50447-4

19. Cover, T.M., Thomas, J.A.: Elements of Information Theory, 2nd edn. John Wiley & Sons, New Jersey (2006). ISBN-13: 978-0-471-24195-9

20. Cramér, H.: Mathematical Methods of Statistics. Princeton University Press (1999). ISBN: 0-691-00547-8

21. Darmois, G.: Sur les lois de probabilites a estimation exhaustive. Comptes Rendus de l'Académie des Sciences **200**, 1265–1266 (1935)

22. Devore, J.L.: Probability and Statistics for Enineering and the Sciences, 3rd edn. Cole Publishing Company, Pacific Grove, CA, USA (1990). ISBN: 0-534-14352-0

23. Feller, W.: The fundamental limit theorems in probability. Bulletin of the American Mathematical Society **51**(11), 800–832 (1945)

24. Feller, W.: An Introduction to Probability Theory and Its Applications, 3rd edn. John Wiley & Sons (1968). Volume I

25. Feller, W.: An Introduction to Probability Theory and Its Applications. John Wiley & Sons (1968). Two Volumes

26. Fisher, R.A.: Theory of Statistical Estimation. Proceedings of Cambridge Philosophical Society **22**, 700–725 (1925)

27. Fraser, D.A.S.: Nonparametric Methods in Statistics. John Wiley & Sons, New York (1957)
28. Halmos, P.R.: Elements of Information Theory, 2nd printing edn. Springer-Verlag, New York (1974). ISBN: 0-387-90088-8
29. Halmos, P.R., Savage, L.J.: Application of the Radon-Nikodym Theorem to the Theory of Sufficient Statistics. The Annals of Mathematical Statistics **20**(2), 225–241 (1949)
30. Hilbert, D.: Grundzüge Einer Allgemeinen Theorie der Linearen Integralgleichungen (Outlines of a General Theory of Linear Integral Equations). Fortschritte der Mathematischen Wissenschaften, heft 3 (Progress in Mathematical Sciences, issue 3). B.G. Teubner, Leipzig and Berlin (1912). In German. Originally published in 1904.
31. Hogg, R.V., Craig, A.T.: Introduction to Mathematical Statistics, 4th edn. Macmillan Publishing Co., Inc., New York (1979). ISBN: 0-023-55710-9
32. Khintchine, A.Y.: Fundamental Laws of Probability Theory, russian edn. The University Series in Higher Mathematics. Moscow (1927)
33. Khintchine, A.Y.: Sur la Loi des Grands Nombres. Comptes rendus de l'Acadmie des Sciences **189**, 477–479 (1929)
34. Kolmogorov, A.N.: Grundbegriffe der Wahrscheinlichkeitsrechnung (1933)
35. Kolmogorov, A.N.: Foundations of the Theory of Probability, 2nd english edn. The University Series in Higher Mathematics. Chelsea Publishing Company, New York (1956)
36. Koopman, B.: On Distribution Admitting a Sufficient Statistic. Transactions of the American Mathematical Society **39**(3), 399–409 (1936)
37. Kullback, S.: Information Theory and Statistics. Dover Publications, Inc., New York (1997). Unabridged publication of the original text published by John Wiley, New York (1959), ISBN: 0-486-69684-7
38. Lebesgue, H.: Intégrale, Longueue, Airc. Université de Paris, Paris, France (1902). PhD Thesis
39. Loève, M.: Probability Theory, 3rd edn. The University Series in Higher Mathematics. C. Scribnet's Sons, New York (1902)
40. Longman, R.W., Beigi, H.S., Li, C.J.: Learning Control by Numerical Optimization Methods. In: Proceedings of the Modeling and Simulation Conference, Control, Robotics, Systems and Neural Networks, vol. 20, pp. 1877–1882 (1989)
41. Michiel Hazewinkel, E.: Encyclopaedia of Mathematics. Kluwer Academic Publishers, Amsterdam (2002). ISBN: 1-4020-0609-8
42. Papoulis, A.: Probability, Random Variables and Stochastic Processes, 3rd edn. McGraw-Hill, Inc., New York (1991). ISBN: 0-070-48477-5
43. Pawlak, Z.: Rough Sets. International Journal of Computer and Information Sciences **11**, 341–356 (1982)
44. Pawlak, Z.: Rough Sets: Theoretical Aspects of Reasoning About Data. Kluwer Academic Publishers, Norwell, Massachusetts (1991)
45. Pitman, E.: Sufficient Statistics and Intrinsic Accuracy. Mathematical Proceedings of the Cambridge Philosophical Societ **32**, 567–579 (1936)
46. Polkowski, L.: Rough Sets, 1st edn. Advances in Soft Computing. Physica-Verlag, A Springer-Verlag Company, Heidelberg (2002). ISBN: 3-7908-1510-1

Chapter 7
Information Theory

Where is the wisdom we have lost in knowledge?
Where is the knowledge we have lost in information?

T.S. Eliot
The Rock, 1934

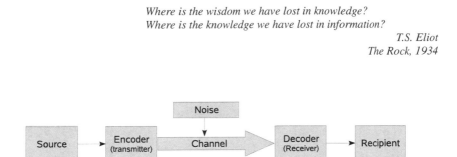

Fig. 7.1: Information Flow in a Single Direction of a Generic Communication System

Figure 7.1 presents a generic communication system. In this figure, the message is composed at the *source*. Then, it undergoes some *coding* to become suitable for transmission through the *channel*. Most channels are noisy and this is signified by the *noise* that affects the channel from the block on top. At the receiving end of the channel, a decoder must *decode* the encoded and noisy message into one that the *recipient* will understand. This is the basis for the development of the topic of *Information Theory* which started with the advent of the *telegraph* and *telephone* systems. Fisher [7], Nyquist [14, 15], Hartley [9], Shannon [18], Wiener [22], and Kullback [12] were among some of the early developers of *Information Theory*. A lot of this work was developed in response to the *encryption* and *decryption* needs for sensitive communications during the *second world war*.

As we have seen in Section 6.8, Fisher [7] used the concept of *information* in his development of *sufficient statistics*. In the definition of *efficiency of statistics* (Definition 6.74), he referred to the amount of *information* in the sampled data and used it to compute *efficiency of a statistic*. His definition had to do with the convergence of variance of *sampled data* to the variance of the *underlying distribution*. This concept is somewhat different from later definitions of *information* (although very related [12]). Shannon [18] and Wiener [22] defined *information* in a manner quite similar to each other, although they had different motivations. Nyquist [14] specified two factors for determining the speed of transmission of *intelligence* and did occasionally call what was being transmitted, *information*; but he mostly referred to it as *intelligence*. Hartley [9] developed a concept of *information* but not in much of a rigorous fashion. Nyquist had started talking about the concept of *symbols with equal probability*, which as we will see later, have to do with *entropy* and *information* concepts. He also used a *logarithmic form* for their representation. The most complete work was done by Wiener [22] and Shannon [18]. Shannon [18] expanded on the works of Nyquist [14] and Hartley [9] and this work provided the basis for future developments in the field of *Information Theory*.

7.1 Sources

Before attempting to produce a formal definition for *information*, let us examine the components of Figure 7.1. The whole essence of communication is to reproduce the output of the *source* block at the receiving end, once it undergoes some encoding to become compatible with the *transmission channel*. Generally, the signal is affected by the *noise* in the channel before a decoding process takes place to attempt to reproduce the original signal.

Shannon [18] qualifies the *source* as an *information source* and *states* that it may take on the following different forms,

1. A sequence of letters from an alphabet – an example is the *telegraph*, *Telex*, *electronic mail*, or even any digital transmission, since it is essentially a source of data from a *binary alphabet*. From a *language modeling* standpoint, many regard *speech communication* as such a source where the *alphabet* is the list of *phonemes* of the language.

2. A function of time, $f(t)$ – such as the analog signal from a *telephone* or a *radio*. From a *signal processing* standpoint, speech may be regarded as having such a source in its analog form.

3. A function of time and other variables – an example is *black and white television* which produces a function of not only time, but the two-dimensional pixel value at coordinates x and y.

4. Multiple functions of time, $\{f_1(t), f_2(t), \cdots, f_n(t)\}$. An example would be the transmission from a *microphone array* recording a conversation in a meeting, or even *multiplexed audio* from two different sources of a conversation on one side of a transmission such as a conference call where there are multiple speakers in one office, communicating with a few speakers in another office, using analog telephony.

5. Multiple functions of time and other variables. This is an extension of the function of time and other variables listed above – as the previous (multi-function

of time) source was an extension of the single function of time source. An example of this kind of source is the source of a *color video* transmission where each color is transmitted as a separate function of time in addition to the x and y coordinate information.

6. Any combination of the above sources such as the multiplexing of video and audio.

Depending on the definition of the boundary between the *source* and the encoder/transmitter, the *source* may be interpreted differently. For example, in Figure 1.6, any number of blocks at the transmission end may be combined to form the *source* as defined by Figure 7.1 and the remaining blocks then become contained in the encoder of Figure 1.6. The same is true at the receiving end. Based on what components one bundles into the *source*, the *encoder*, the *decoder* and the *receiver*, the *information source* for the speech production process would be one of the many possible types of sources listed above.

From a statistical standpoint, all sources may be viewed as *mechanisms* based on some underlying statistics which generate *random numbers* associated with the *information* being conveyed. Let us examine the different concepts of *measurement of information* and the related terminology. Once we have defined *information* and its related concepts, we will be able to analyze it at different stages of the process depicted by Figure 7.1.

In another perspective, *information sources* may be categorized into the following types. These categories, as we shall see later, are quite pertinent to the topic of our discussion.

Definition 7.1 (Discrete Source). *A discrete source is one whose output is constrained to take on values from a finite set (atoms or alphabet),*
i.e., $\mathcal{X} = \{X_1, X_2, \cdots, X_n\}$.

Definition 7.2 (Discrete Memoryless Source). *Let us assume that the sample space associated with the discrete random variable X, defined in probability space* $(\mathcal{X}, \mathfrak{X}, P)$, *is given by* $\mathcal{X} - \{X_1, X_2, \cdots, X_n\}$, $n \in \mathbb{N}$, *known as the atoms (alphabet) of X. Also, consider the time sequence* $\{T\}_0^t = \{0, 1, 2, \cdots, t\}$ *such that the value of X at time t is denoted by* x_t. *If we denote the output history of a discrete memoryless source by the following output sequence,* $\{x\}_0^t = \{x_0, x_1, \cdots x_t\}$, *then a memoryless source is a discrete source such that its output at any time t is independent of all its previous outputs. Namely,*

$$P(\{x\}_0^t) = P(x_t, x_{t-1}, \cdots, x_0)$$

$$= \prod_{\tau=0}^{t} P(x_\tau) \tag{7.1}$$

Definition 7.3 (Discrete Markov (Markoff) Source). *Let us assume that the sample space associated with discrete random variable X, defined in probability space* $(\mathscr{X}, \mathfrak{X}, P)$, *is given by* $\mathscr{X} = \{X_1, X_2, \cdots, X_n\}, n \in \mathbb{N}$. *Also consider the time sequence* $\{T\}_0^t = \{0, 1, 2, \cdots, t\}$ *such that the value of X at time t is denoted by* x_t, *then we may denote the output history of a discrete source by the following output sequence,* $\{x\}_0^t = \{x_0, x_2, \cdots x_t\}$. *Such a source is called a Discrete Markov source if*

$$P(x_t | \{x\}_0^{t-1}) = P(x_t | x_{t-1}) \tag{7.2}$$

Equation 7.2 is known as Markov property. Furthermore, the output sequence, $\{x\}_0^t$, *is known as a Discrete Markov Process.*

Therefore, a *Markov process* is one that possesses some memory and that memory is summarized in the latest state of the system. As we will see later, either this memory is present in its entirety in the last output of the system (*unifilar Markov source*) or in a hidden state (*non-unifilar Markov source*). There are other modeling mechanisms which are based on time series, portraying similar memory modeling such as *auto regression* (e.g., *LPC* – Section 5.4), *moving average* (Section 5.3), the combination of the two, *auto regressive moving average ARMA* and its extension, known as the *Kalman filter* (Section 18.4.7.5).

Definition 7.4 (Unifilar Markov Source). [1] *A unifilar Markov source [1, 20] is one that has no hidden states, meaning that the state sequence is completely determined by knowing the initial state and observing the output sequence.*

The details of a *unifilar Markov source* will become more clear in Section 13.3. The *unifilar Markov source* is in contrast with a *non-unifilar Markov source*. The following is a simple definition of the *non-unifilar Markov source*.

Definition 7.5 (Non-Unifilar Markov Source). *A non-unifilar Markov source [1, 20] is one whose states are not uniquely determined based on the observation of the output sequence and the knowledge of the initial state. An example of a non-unifilar Markov source is the hidden Markov model (Section 13.4).*

In Section 13.4 we will see alternative definitions for the unifilar (Definition 13.2) and non-unifilar (Definition 13.3) Markov sources.

Another important categorization of *Markov sources* is in terms of their *ergodictiy*.

---

[1] Unifilar, literally, means single-threaded. It is composed of the Latin expressions, uni: one and filum: thread.

Definition 7.6 (Ergodic Sources). [2] *An ergodic source is a type of discrete Markov source such that every sequence it produces has the same statistical properties.*[18]

All *natural languages* including *English* may be modeled by *ergodic Markov sources.*[21] Also, every *memoryless source* is *ergodic*.

Definition 7.7 (Continuous Source). *A continuous source is one whose output is a continuous random variable, X, given by Definition 6.39.*

7.2 The Relation between Uncertainty and Choice

To be able to define *information*, we should first try to understand some related concepts. One such concept is *uncertainty*. Consider the *uncertainty* attached to the outcome of a fair coin (one which has an equal probability, $\frac{1}{2}$, of producing heads or tails, $\mathscr{X} = \{h, t\}$). Now, without quantifying this *uncertainty*, consider the *uncertainty* of the outcome of a fair die with six sides (having the probability of $\frac{1}{6}$ in producing each of its faces, $\mathscr{X} = \{1, 2, 3, 4, 5, 6\}$). In each of these examples, $\mathscr{X}$ (*sample space*) is also called the set of *possible symbols* or the *alphabet* of the process. Although in both examples, each outcome is equally probable among the possible outcomes for that *mechanism (source)*, the *uncertainty* associated with the outcome of the *coin* is less than that associated with the outcome of the *die*.

Therefore, *uncertainty* is related to the structure of *possible choices* in a process. Shannon [18] presented a formal definition of this entity (*uncertainty*) by requiring it to possess a set of specific properties. In his treatment of *uncertainty* and *information*, *Shannon* only considered *discrete sources*. Therefore, at this point we will focus our attention toward *discrete sources*. After making certain observations, we will discuss the properties of *uncertainty* per Shannon's remarks. Later, we shall extend the definition of *uncertainty* as applied to *continuous sources*.

7.3 Discrete Sources

If our *information source* is considered to be a *discrete Markov process* capable of producing a finite set of n distinct outcomes, then the *sample space*, $\mathscr{X}$, of random variable X may be written as $\mathscr{X} = \{X_1, X_2, \cdots, X_n\}$, with a probability of $p(X_i)$ for

---

[2] *Ergodic* is a term which was coined by Josiah S. Gibbs from the two Greek words, $\varepsilon\rho\gamma o\nu$ (work) and $o\delta o'\varsigma$ (path).[22]

the ith possible outcome, $\{X_i : i \in \{1, 2, \cdots, n\}\}$. $\{p(x) : x \in \mathscr{X}\}$ is known as the *probability mass function* of the discrete random variable X – see Definition 6.64. The X_i are known as the *atoms* of the *sample space* $\mathscr{X}$. Then, a measure of *uncertainty*, $\mathscr{H}$, would be a function of the only *knowledge* we have about this *Markov process*, namely, its *output probabilities* for each of the n outcomes. As a shorthand notation, we will denote the *probability* of the *distinct outcome*, X_i $(p(X_i))$ as just p_i. Therefore,

$$\mathscr{H}(X) = \mathscr{H}(p_1, p_2, \cdots, p_n) \tag{7.3}$$

7.3.1 Entropy or Uncertainty

For clarification, let us review some notation being used in this book, related to this topic.

- X: A Random Variable
- $\mathscr{X} = \{X_i\}$ *where* $i \in \{1, 2, \cdots, n\}$ is the set of *possible outcomes* or *sample space* of X
- x is a single sample $(X = x)$ which may take on any of the values X_i with probability $p(X = X_i) = p(X_i) = p_i$

Shannon determined that a set of reasonable properties to demand for *uncertainty* is the following,

1. $\mathscr{H}$ should be a *continuous function* of the probabilities of the outcomes, $\{p_i : i \in \{1, 2, \cdots, n\}\}$.

2. If $p_i = \frac{1}{n}$ $\forall$ $i \in \{1, 2, \cdots, n\}$, then $\mathscr{H}$ should increase monotonically as n increases. This is to achieve the property stated earlier when we compared the *uncertainty* of the outcome of a *fair coin* with that of a *fair die* with 6 sides. This means that as we have different sources, each emitting equally likely outcomes, the *source* with *more choices* contains more *uncertainty*. In other words, the *amount of choice* in a process is a measure of its *uncertainty*.

3. If a choice is split into two successive choices, the original $\mathscr{H}$ should be equivalent to the *weighted sum* of the subsequent values of $\mathscr{H}$ for the new choices. Shannon [18] illustrates this property with the example depicted in Figure 7.2. In the figure, there are 3 original choices with the following probabilities, $p_1 = \frac{1}{2}$, $p_2 = \frac{1}{3}$, and $p_3 = \frac{1}{6}$. The process on the right is equivalent in the sense that the final choices still have the same cumulative probabilities. However, in the system on the right, the second choice has been split into two subsequent choices. Since the final *path probabilities* are the same in the two systems, the *uncertainty* must be the same. Namely,

$$\mathscr{H}\left(\frac{1}{2},\frac{1}{3},\frac{1}{6}\right) = \mathscr{H}\left(\frac{1}{2},\frac{1}{2}\right) + \frac{1}{2}\mathscr{H}\left(\frac{2}{3},\frac{1}{3}\right) \tag{7.4}$$

The $\frac{1}{2}$ in front of $\mathscr{H}(\frac{2}{3},\frac{1}{3})$ means that the whole set, containing the second choice and the third choice, only happens half the time.

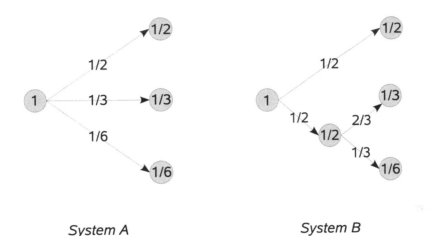

Fig. 7.2: Decomposition of three choices into a binary set of choices – after Shannon [18]

Theorem 7.1 (Entropy). *The only entity, $\mathscr{H}$, that satisfies all the properties listed above, would have the following form,*

$$\mathscr{H} = -k\sum_{i=1}^{n} p_i \log_b p_i \tag{7.5}$$

where b is the choice of the base being used for the logarithm and $k > 0$ is a constant which depends on the unit of $\mathscr{H}$.

If the *logarithm* of base e is used ($b = e$), the expression in Equation 7.5 will be identical with *Gibbs' formula* for *entropy* [2] defined in the *second law of Thermodynamics*, where k would be the *Boltzmann constant*.

Due to its similarity to *Gibbs' formula*, Shannon [18] called $\mathscr{H}$, *entropy*, which is a measure of the amount of *choice*, or *uncertainty*, and is eventually related to *information*. Shannon's definition uses $k = 1$ which concentrates on the *logarithmic* units of *entropy* and removes the proportionality constant of *Boltzmann*, which was designed to give *entropy* a physical interpretation (as *energy*) to fit in the statement of the *second law of Thermodynamics*.

Definition 7.8 (Entropy).

$$\mathcal{H} \stackrel{\Delta}{=} -\sum_{i=1}^{n} p_i \log_b p_i \tag{7.6}$$

$$= -\mathcal{E}\{\log_b p(X)\} \tag{7.7}$$

Depending on the base, b, of the *logarithm*, *entropy* has different *logarithmic units*. In Equation 7.6, if log ($\log_{10}$) is used, the *unit of entropy* is one *ban*, *dit* (a contraction of *decimal digit*), or *Hartley* due to the *logarithmic* definition of *information* which he introduced in his 1928 paper [9].

If $\log_2$ is used, then the *unit of entropy* is a *bit* which is a contraction for *binary digit*, first introduced by J.W. Tukey [18] to represent the *information* needed to qualify the *choice* between two equally likely events (as a binary digit does).

Finally, if the *Napierian* (*natural*) *logarithm* (ln) is used in Equation 7.6, then the *unit of entropy* is a *nat* (a contraction of *natural*) which is the unit used in *Thermodynamics* when quantifying *entropy*.[3]

Sometimes the *entropy* of a *random number* is written as $\mathcal{H}(X)$. This does not mean that *entropy* is a function of X. It only specifies that the *entropy* of interest is the *entropy* related to the outcome of random variable X in contrast with any other random variable, say Y or Z.

For a *Bernoulli process* (*binary source* – see Definition 6.63), if $p_1 = p$, then $p_2 = q = 1 - p$. Therefore,

$$\mathcal{H} = -(p \log_2 p + q \log_2 q) \tag{7.8}$$

$$= -(p \log_2 p + (1 - p) \log_2 (1 - p)) \tag{7.9}$$

Figure 7.3 shows the *entropy* of such a process as a function of p. Notice the fact that this process has *maximum entropy* at $p = 0.5$, which is the point where outcomes 0 and 1 are as likely (such as in a fair coin toss).

The *entropy*, $\mathcal{H}$, has some very interesting properties. Let us examine some of these properties as well as try to define some important concepts based on *entropy*.

Property 7.1 (Zero Entropy). $\mathcal{H} = 0$ *if and only if all but one of the probabilities on which it depends are zero. This means that there is really no uncertainty, since the output of the system is always the same with probability 1 (this is the probability which is nonzero and which has to be 1 for the sum to be 1).*

---

[3] Kullback [11] also cites the use of a *nit* from MacDonald who suggested it in his 1952 paper [13]. MacDonald made this suggestion to be, in essence, a contraction for *natural digit*. He even states in his paper that this has an unattractive etymological significance. Later, *nats* became a much more popular choice.

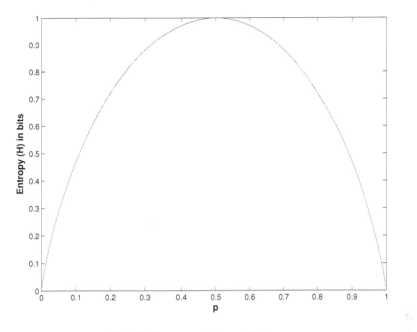

Fig. 7.3: Entropy of a Bernoulli Process

Property 7.2 (Maximum Entropy). *For a process with n outcomes,*

$$\mathcal{H}_{max}(X) = \max \mathcal{H}$$
$$= \log_2 n \tag{7.10}$$

where $p_i = \frac{1}{n} \ \forall \ i \in \{1, 2, \cdots, n\}$. This is the most uncertain situation since there are no outcomes that would have a higher probability than the rest.

Definition 7.9 (Relative Entropy – as defined by Shannon). [4] *See [18].*
   The relative entropy of a process is defined as its entropy divided by its maximum entropy while it is still confined to the same alphabet,

$$\mathcal{H}_R(X) \triangleq \frac{\mathcal{H}(X)}{\mathcal{H}_{max}(X)}$$
$$= \frac{-\sum\limits_{i=1}^{n} p(X_i) \log_2 p(X_i)}{\log_2 n} \tag{7.11}$$

---

[4] This entity, called relative entropy by Shannon [18], is rarely used today. There is another *relative entropy* which is the same as the *Kullback-Leibler divergence* and which will be discussed in quite a detail later.

The relative entropy, as defined by Shannon [18], may be interpreted as the compression rate achieved by the said process using the same alphabet. This is different from what, today, most people call relative entropy. We will present a definition for the more popular relative entropy, or the Kullback-Leibler Directed Divergence, later. For now, we have stated the Definition 7.9 only for the purpose of presenting the following definition of redundancy.

Definition 7.10 (Redundancy). *The redundancy of a process is defined as the fraction of unnecessary bits used to code a sequence relative to the maximum entropy which the process may have, with the same alphabet. Therefore,*

$$R(X) \triangleq 1 - \mathscr{H}_R(X) \tag{7.12}$$

Shannon [18] reports the redundancy of the written English language to be around 50%. This means that if we throw out every other letter in an English sentence, we will still be able to reconstruct the sentence based on the remaining half of the letters.

In *Khintchine's theorem*, Theorem 6.7, also known as the *weak law of large numbers*, there is a corresponding statement of the convergence of an estimate of the *entropy* of a source stemming from that law. It is called the *asymptotic equipartition property*. The following is a statement of the corresponding theorem.

Theorem 7.2 (Asymptotic Equipartition Property (AEP)). *If $X_1, X_2, \cdots$ are independent and identically distributed (i.i.d.) random variables with a finite mean, μ,*

$$\mu = \int_{\mathscr{X}} x \, dP(x) \tag{7.13}$$

Furthermore, if we define a number, $\tilde{\mathscr{H}}|_n$, in terms of the joint probability mass function,

$$\tilde{\mathscr{H}}|_n \triangleq -\frac{1}{n} \log_b p_X(X_1, X_2, \cdots, X_n) \tag{7.14}$$

then,

$$\lim_{n \to \infty} P\left(|\tilde{\mathscr{H}}|_n - \mathscr{H}(X)| < \varepsilon\right) = 1 \ \forall \ \varepsilon > 0 \tag{7.15}$$

In other words, Equation 7.14 converges, in probability, to the entropy of X, $\mathscr{H}(X)$. This, similar to Khintchine's theorem, is a weak convergence statement.

Proof.
Since $\{X_i : 1 \leq i \leq n\}$ are *statistically independent*, we may write,

$$\tilde{\mathcal{H}}|_n = -\frac{1}{n}\sum_{i=1}^{n}\log_b p_{X_i}(X_i) \tag{7.16}$$

where $p_{X_i}(X_i)$ are *marginal probability mass functions*. Also, by Theorem 6.7,

$$\lim_{n\to\infty} -\frac{1}{n}\sum_{i=1}^{n}\log_b p_{X_i}(X_i) = -\mathcal{E}\{\log_b p(X)\} \tag{7.17}$$

The right hand side of Equation 7.17 is just $\mathcal{H}(X)$.

$\square$

AEP, as in the case of WLLN and SLLN for the computation of the mean, is used on many occasions when dealing with practical aspects of computing the *entropy* of a source of a process. For more on the *asymptotic equipartition property*, see [3].

Definition 7.11 (Joint Entropy). *The joint entropy, $\mathcal{H}(X,Y)$, of two random variables, X and Y is*

$$\mathcal{H}(X,Y) \triangleq -\sum_{i=1}^{n}\sum_{j=1}^{m} p(X_i,Y_j)\log_2 p(X_i,Y_j) \tag{7.18}$$

$$= -\mathcal{E}\{log_2 p(X,Y)\} \tag{7.19}$$

$$\tag{7.20}$$

Note that,

$$p(X_i) = \sum_{j=1}^{m} p(X_i,Y_j) \tag{7.21}$$

$$p(Y_j) = \sum_{i=1}^{n} p(X_i,Y_j) \tag{7.22}$$

Therefore,

$$\mathcal{H}(X) = -\sum_{i=1}^{n} p(X_i)\log_2 p(X_i)$$

$$= -\sum_{i=1}^{n}\sum_{j=1}^{m} p(X_i,Y_j)\log_2 \sum_{j=1}^{m} p(X_i,Y_j) \tag{7.23}$$

$$\mathcal{H}(Y) = -\sum_{j=1}^{m} p(Y_j)\log_2 p(Y_j)$$

$$= -\sum_{j=1}^{m}\sum_{i=1}^{n} p(X_i,Y_j)\log_2 \sum_{i=1}^{n} p(X_i,Y_j) \tag{7.24}$$

Based on Equations 7.18, 7.23, and 7.24, we may write the following *triangular inequality* for *joint entropy*,

$$\mathscr{H}(X,Y) \leq \mathscr{H}(X) + \mathscr{H}(Y) \tag{7.25}$$

In Equation 7.25, equality holds when X and Y are *statistically independent* (see Definition 6.37) in which case, the *joint probability* may be written in terms of the product of the individual probabilities, namely, $p(X_i, Y_j) = p(X_i)p(Y_j)$.

The inequality of Equation 7.25 states that the *entropy* of the joint event is less than or equal to the sum of the *entropies* of the individual events. In other words, considering the *joint event* removes *uncertainty* if the events are in any way *dependent* on one another.

The amount of *uncertainty* which is removed, based on *partial dependence* of X and Y, may be viewed as the *information* shared by these two *random processes*. We call this information, *mutual information*, which will be defined in more detail in Section 7.6.1.

Property 7.3 (Averaging increases uncertainty). *If there exists a set of constants such that,*

$$\sum_{j=1}^{n} a_j = 1 \tag{7.26}$$

and

$$a_j \geq 0 \ \forall \ j \in \{1, 2, \cdots, n\} \tag{7.27}$$

Then, if p_i goes through a linear transformation such that,

$$p'(X_i) = \sum_{j=1}^{n} a_j p(X_j) \tag{7.28}$$

then the new entropy associated with the transformed probabilities will be greater than or equal to the original entropy, namely,

$$\mathscr{H}' = -\sum_{i=1}^{n} p'(X_i) \log_2 p'(X_i)$$
$$\geq \mathscr{H} \tag{7.29}$$

where

$$\mathscr{H} = -\sum_{i=1}^{n} p(X_i) \log_2 p(X_i) \tag{7.30}$$

The transformation of Equation 7.28 describes any general averaging technique, an example of which is the case of joint entropy mentioned previously. Therefore, Equation 7.29 means that averaging increases uncertainty. This is quite intuitive since any averaging tends to cause some loss of information which is akin to increasing uncertainty (entropy).

Definition 7.12 (Conditional Entropy). *For two random variables, X and Y, with output probabilities $p(X_i)$ and $p(Y_j)$ respectively, based on Equation 6.50,*

$$p(Y_j|X_i) = \frac{p(X_i, Y_j)}{p(X_i)}$$

$$= \frac{p(X_i, Y_j)}{\sum\limits_{j=1}^{m} p(X_i, Y_j)} \qquad (7.31)$$

Therefore, the conditional entropy of Y given X is defined as,

$$\mathcal{H}(Y|X) \triangleq -\sum_{i=1}^{n}\sum_{j=1}^{m} p(X_i, Y_j)\log_2 p(Y_j|X_i) \qquad (7.32)$$

which tells us about the uncertainty of the random variable Y on the average, when X is known.

If we plug in for the *conditional probability*, $p(Y_j|X_i)$, from Equation 7.31 into Equation 7.32, we will have,

$$\mathcal{H}(Y|X) = -\sum_{i=1}^{n}\sum_{j=1}^{m} p(X_i, Y_j)\log_2 p(X_i, Y_j) +$$

$$\sum_{i=1}^{n}\sum_{j=1}^{m} p(X_i, Y_j)\log_2 \sum_{j=1}^{m} p(X_i, Y_j) \qquad (7.33)$$

$$= -\sum_{i=1}^{n}\sum_{j=1}^{m} p(X_i, Y_j)\log_2 \frac{p(X_i, Y_j)}{p(X_i)} \qquad (7.34)$$

Substituting Equations 7.18 and 7.23 into Equation 7.33, we have the following Equation, known as the *chain rule for conditional entropy*,

$$\mathcal{H}(Y|X) = \mathcal{H}(X, Y) - \mathcal{H}(X) \qquad (7.35)$$

Equation 7.35 may also be written as

$$\mathcal{H}(X, Y) = \mathcal{H}(X) + \mathcal{H}(Y|X) \qquad (7.36)$$

Therefore, the uncertainty (entropy) of the joint events is equal to the uncertainty (entropy) of the first event plus the uncertainty (entropy) of the second event when the first event is known.

Property 7.4 (Information reduces uncertainty). *Based on Equation 7.25, $\mathcal{H}(X) + \mathcal{H}(Y) \geq \mathcal{H}(X,Y)$. Also, based on the chain rule (Equation 7.36), $\mathcal{H}(X,Y) = \mathcal{H}(X) + \mathcal{H}(Y|X)$. Therefore,*

$$\mathcal{H}(Y) \geq \mathcal{H}(Y|X) \tag{7.37}$$

which is quite intuitive since it says that the uncertainty of an event decreases once more information is provided, in this case, in the form of the outcome of event X. The equality in Equation 7.37 happens when the two events are statistically independent, in which case, since they are independent, there is no uncertainty removed from Y by knowing anything about X.

7.3.2 Generalized Entropy

As with many other entities in mathematics, *Shannon's entropy* has also been generalized based on certain criteria. Réyni [16] started with stating Fadeev's [6] four proven postulates (axioms) on *Shannon's entropy*. Let us consider a *probability mass function*, $\{p(X) : p(X) = \{p_1, p_2, \cdots, p_n\}\}$. Then the *entropy*, $\mathcal{H}(p(X))$ may alternatively be written as, $\mathcal{H}(p_1, p_2, \cdots, p_n)$. Then the following are the *four postulates* (*axioms*) of Fadeev for *entropy*:

1. $\mathcal{H}(p_1, p_2, \cdots, p_n)$ is a symmetric function of its variables for $n \geq 2$.
2. $\mathcal{H}(p, 1-p)$ is *continuous* $\forall\ 0 \leq p \leq 1$
3. $\mathcal{H}(\frac{1}{2}, \frac{1}{2}) = 1$
4. For any distribution, $p(X)$ and any $\{q : 0 \leq q \leq 1\}$,

$$\mathcal{H}(\ qp_1, (1-q)p_1, p_2, \cdots, p_n) = $$
$$\mathcal{H}(p_1, p_2, \cdots, p_n) + p_1 \mathcal{H}(q, 1-q) \tag{7.38}$$

Next, [16] relaxes the last *postulate* (*axiom* 4) to a weaker condition which is quite significant in practice, but is not as strict as *postulate* 4. Let us define a second *probability mass function*, $\{q(X) : q(X) = \{q_1, q_2, \cdots, q_m\}\}$. *Postulate* 4′ is the *additivity* of *entropy*, namely,

$$\mathcal{H}(p(X) \times q(X)) = \mathcal{H}(p(X)) + \mathcal{H}(q(X)) \tag{7.39}$$

where $(p(X) \times q(X))$ is the *direct product* of $p(X)$ and $q(X)$. The *direct product* is a *Cartesian product* of the two sets (Definition 6.24), creating a new distribution whose elements are $p_i q_j$ where $i = \{1, 2, \cdots, n\}$ and $j = \{1, 2, \cdots, m\}$.

Given postulates 1 through 3 and replacing postulate 4 with the weaker postulate 4′, given by Equation 7.39, Reýni defined a new *generalized entropy* which meets all the new postulates,

$$\mathcal{H}_\alpha(p(X)) = \frac{1}{1-\alpha} \log_2 \left(\frac{\sum\limits_{i=1}^{n} p_i^\alpha}{\sum\limits_{i=1}^{n} p_i} \right) \tag{7.40}$$

where $\{p(X) : p(X) = \{p_1, p_2, \cdots, p_n\}\}$ and $\{\alpha : \alpha > 0 \wedge \alpha \neq 1\}$. Réyni [16] calls $\mathcal{H}_\alpha(p(X))$ *entropy of order* α. Note that

$$\lim_{\alpha \to 1} \mathcal{H}_\alpha = \mathcal{H}_1 \tag{7.41}$$

$$= \frac{-\sum\limits_{i=1}^{n} p_i \log_2 p_i}{\sum\limits_{i=1}^{n} p_i} \tag{7.42}$$

Also, for the case when

$$\sum_{i=1}^{n} p_i = 1 \tag{7.43}$$

we have,

$$\mathcal{H}_1 = -\sum_{i=1}^{n} p_i \log_2 p_i \tag{7.44}$$

$$= \mathcal{H}(p(X)) \quad \text{(Shannon's entropy)} \tag{7.45}$$

7.3.3 Information

Fisher, in his 1925 paper [7] gave a definition of *information*. He introduced the concept of *sufficient statistics* (see Section 6.8) in which he stated that *sufficient statistics* "contain in themselves the whole of the relevant *information* available in the data." Although this definition of *information* is somewhat different from that defined by others (starting with Hartley [9] in 1928), as stated by [8] and shown by [12], it is still related in a limiting sense. In other words, the *Wiener-Shannon information*, which will be defined momentarily, becomes invariant if and only if *sufficient statistics* are used – see Section 7.7 for more details. One important feature which is necessary for the definition of *information* is its *invariance* to the used statistic.

Information, the way it is used today in *Information Theory*, was probably first introduced by *Hartley* in his 1928 paper [9] which influenced *Claude Shannon* [18] in his definition of *uncertainty* and *information* from the telecommunication standpoint. In parallel, *Norbert Wiener* [22] reached the same definition from the control and statistical system prediction point of view. *Hartley* gave a rough definition of *information*. He defined it for n equally likely *hypotheses* as,

$$\mathscr{I}_{Hartley}(X) = \log n \tag{7.46}$$

where $x \in X : x \in \mathscr{X} = \{X_i\}, i \in \{1, 2, \cdots, n\}$.[9]

In his book, *Wiener* [22] stated, correctly, that *information* is the opposite of *entropy* and therefore is the negative of entropy. At this point, we will clarify what seems to be a discrepancy in terminology and which is really not so.

7.3.4 The Relation between Information and Entropy

There has been great confusion from the day *Shannon* defined the, so called, *entropy of a discrete source* in 1948 [18]. Although he referred to *Gibbs' Equation* for *entropy*, as used in *Thermodynamics* and *statistical mechanics*, he did not qualify the fact that in the statement of the *second law of Thermodynamics*, one speaks of the change in *entropy*, ΔS, and not the *absolute entropy*. When we speak of *information*, we refer to the change of information due to an observation. With the same analogy, we may only quantify the change in *entropy* and not the *absolute entropy*. We may, however, speak of *information* or *Entropy* relative to some reference point. For example, this reference may be a state of nature.

As *MacDonald* stated in his 1952 paper [13], many scientists who were present at the 1950 *Symposium on Information Theory* in London were confused about why *entropy* and *Information* should have the same sign when they measure opposing qualities (disorder and order, respectively). Wiener [22] also stated that *information* is the negative of *entropy*.

I recall a similar confusion and discussion in my *Thermodynamics* class many years ago, as a student of *mechanical engineering*. Most of this confusion stems from the relative measure of *entropy* and the choice of the reference point used as origin. For the sake of simplicity and only for this discussion, let us confine our discussion to *discrete sources*.

In *Shannon*'s definition of *entropy* (Equation 7.6), the *probability distribution*, $p(X_i)$ is a set of *a*-priori probabilities associated with possible outcomes. If we consider the *a*-posteriori state of a statistical system when an outcome is revealed, then the *a*-posteriori probability, $p(X_j)$, becomes 1 and all the complementary outcomes will have $p(X_i) = 0 \ \forall \ i \neq j$. We may call the *entropy* of the *a*-posteriori state of the process, $\mathscr{H}(1;X)$ and its *a*-priori *entropy*, $\mathscr{H}(0;X)$. Then, the change in the *entropy* based on the observed outcome becomes,

$$\Delta \mathscr{H}(0:1;X) \triangleq \mathscr{H}(1;X) - \mathscr{H}(0;X) \tag{7.47}$$

Equation 7.47 defines the amount of change in the *entropy* based on the order (*information*) made available from observing the outcome of the system. This average gain in *information*, may be written as,

$$\Delta \mathscr{I}(0:1;X) = -\sum_{i=1}^{n} p(X_i) \log_2 p(X_i) \tag{7.48}$$

Also, as it is stated in [22],

$$\Delta \mathscr{H}(0:1;X) = -\Delta \mathscr{I}(0:1;X) \tag{7.49}$$

Combining Equations 7.47 and 7.48 using the relation in Equation 7.49, we will have,

$$\Delta \mathscr{H}(0:1;X) = \mathscr{H}(1;X) - \mathscr{H}(0;X)$$

$$= -\left(-\sum_{i=1}^{n} p(X_i) \log_2 p(X_i) \right)$$

$$= \sum_{i=1}^{n} p(X_i) \log_2 p(X_i) \tag{7.50}$$

If we consider the *a*-posteriori state, one of complete *information* and no *uncertainty*[5], then,

$$\Delta \mathscr{H}(0:1;X) = 0 - \mathscr{H}(0;X)$$

$$= \sum_{i=1}^{n} p(X_i) \log_2 p(X_i) \tag{7.51}$$

which leads to the following expression for $\mathscr{H}(0;X)$,

$$\mathscr{H}(0;X) = -\sum_{i=1}^{n} p(X_i) \log_2 p(X_i)$$

$$= \mathscr{H}(X) \tag{7.52}$$

Equation 7.52 coincides with *Shannon*'s definition, but it should be noted that the direction of the change of *entropy* and change of *information* were indeed opposite as prescribed by Equation 7.49 and *information* is indeed the negative of *entropy* as *Wiener* correctly stated in multiple occasions in his book [22]. Therefore, *Wiener*'s definition of *information* coincides with that of *Shannon*, namely, *Gibbs' formula* with $k = 1$.

---

[5] Of course this is only used as a reference point and no more. In fact, Kullback [11], as we will see later, assumes *absolute continuity* between two hypothetical distributions related to the *a-priori* and *a-posteriori* hypotheses. He also quotes from Savage [17] who states, "... definitive observations do not play an important part in statistical theory, precisely because statistics is mainly concerned with uncertainty, and there is no uncertainty once an observation definitive for the context at hand has been made." Of course, this statement only re-iterates the relevance of a limiting case where there is no uncertainty, even if it is not used directly.

As we mentioned, the definition of *Information* makes sense for a specific outcome, revealing the amount of *information* based on the observation of an event.
If the source is a *discrete source*, then we will have a *discrete random variable*, X,
as defined in the previous section. Therefore, for a discrete source, the amount of
information gained for every event, X_i is,

$$\mathscr{I}_{Wiener-Shannon}(X_i) = -\log_2 p(X_i) \tag{7.53}$$

where, $X_i \in \mathscr{X}, i \in \{1, 2, \cdots, n\}$ are distinct subsets (*atoms*) of the sample space
$\mathscr{X}$.

In comparison, *entropy* (Equation 7.6) may be thought of, as the expected value
of *information*. For the discrete case, this will be,

$$\mathscr{H}(X) = -\sum_{i=1}^{n} p(X_i) \log_2 p(X_i)$$

$$= \sum_{i=1}^{n} p(X_i) \mathscr{I}(X_i) \tag{7.54}$$

$$= \mathscr{E}\{\mathscr{I}(X)\} \tag{7.55}$$

The unit of *information* as defined by Equation 7.54 is a bit. If we changed the
base of the *logarithm*, we would get different units – as discussed earlier in the definition of *entropy*.

If X_i is one of n equally probable outcomes, then the probability of any of these
outcomes will just be $\frac{1}{n}$ which produces the following expression for *information*,

$$\mathscr{I}(X_i) = -\log_2 p(X_i)$$

$$= -\log_2 \frac{1}{n}$$

$$= \log_2 n \tag{7.56}$$

in bits, coinciding with *Hartley*'s definition of *information*, if the *logarithm* were to
be taken in base 10 (Equation 7.46).

7.4 Discrete Channels

We started this chapter by speaking about the onset of *information theory* because
of the development of communication infrastructures. In general, *Information Theory* is concerned with transmission of information through *channels*. In the past few
decades, with the advancement of digital communication systems, *discrete channels*
have become the most prevalent types of channels discussed in this field. Here, we
will define a *discrete channel* and continue with the definition *memoryless discrete
channels*. In future chapters we shall see the use of discrete channel analogies for

handling concepts such as *hidden Markov models* and *speech transmission*.

Definition 7.13 (Discrete Channel). *A discrete channel is a mechanism which outputs a discrete random variable, $Y : y \in \mathscr{Y}$, for every discrete random variable, $X : x \in \mathscr{X}$, that is presented to it with the conditional probability mass function $p(Y|X) : p(y|x) \geq 0 \ \forall x \wedge \sum_Y p(Y|X) = 1$. Such a discrete channel is represented by $(\mathscr{X}, p(Y|X), \mathscr{Y})$.*

Definition 7.14 (Discrete Memoryless Channel). *Consider two discrete random variables, $\{X : x \in \{X_1, X_2, \cdots, X_M\}\}$ and $\{Y : y \in \{Y_1, Y_2, \cdots, Y_N\}\}$. Furthermore, Y is dependent on X according to the conditional probability mass function, $p(Y|X)$, given by the stochastic matrix (see Definition 23.18) $\mathbf{A} : \mathscr{R}^N \mapsto \mathscr{R}^M$, such that the elements of $\mathbf{A}$ are defined as follows,*

$$\mathbf{A}_{ij} \overset{\Delta}{=} p(y = Y_j | x = X_i) \tag{7.57}$$

Also, consider the time sequence, $\{T\}_0^t = \{0, 1, 2, \cdots, t\}$ such that the value of X and Y at time t are denoted by x_t and y_t respectively. Then a discrete memoryless channel is a channel such that, given an input random variable X at time t, its output is random variable Y with the following output probability,

$$P(y_t = Y_j) = \sum_{i=1}^{M} p(y_t = Y_j | x_t = X_i) P(x_t = X_i) \tag{7.58}$$

$$= \sum_{i=1}^{M} \mathbf{A}_{ij} P(x_t = X_i) \tag{7.59}$$

Therefore, the output of a discrete memoryless channel at time t is independent of its input at previous times, $\{x_\tau : \tau < t\}$.

Note that $\mathbf{A}_{ij}$ is a *stochastic matrix* since the following should always be true,

$$\sum_{j=1}^{N} P(y_t - Y_j) - 1 \tag{7.60}$$

$\mathbf{A}$ is also known as the *channel matrix*. Figure 7.4 shows a diagram of a *memoryless channel*.

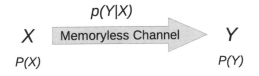

$$p(Y|X)$$

$$X \quad \text{Memoryless Channel} \quad Y$$

$$P(X) \qquad\qquad\qquad P(Y)$$

Fig. 7.4: A Memoryless (Zero-Memory) Channel

Definition 7.15 (Binary Symmetric Channel). *Take the discrete memoryless channel described in Definition 7.14. A binary symmetric channel [21] is a popular configuration for a discrete channel which is a discrete memoryless channel with the following channel matrix,*

$$\mathbf{A}_{ij} = p(y = Y_j | x = X_i)$$

$$= \begin{bmatrix} 1-p & p \\ p & 1-p \end{bmatrix} \tag{7.61}$$

where p is the probability of a mistake in transmission. For example, if the channel is a perfect channel, then $p = 0$ in which case, there will be no mistakes made in transmission. Therefore, the channel matrix would become the identity matrix, $\mathbf{I}$, outputting a 0 for every input 0 and a 1 for every input 1 with probability 1. A less perfect channel would transmit output 1 for an input of 0 and a 0 for an input of 1 with probability p.

Generally, the performance of a channel may be assessed using a distance measure such as the *Hamming distance* allowing the comparison of the output *sequence*, $\{y\}_{t_0}^{t_1}$ with the corresponding input *sequence*, $\{x\}_{t_0}^{t_1}$ – see Definition 8.1.1.

7.5 Continuous Sources

For a *continuous source*, the output is a *continuous random variable, X*. For this source, every possible value of X may be given by x with a probability density, $p(x)$ – see Definition 6.45.

7.5.1 Differential Entropy (Continuous Entropy)

Shannon [18] only defined *entropy* for a discrete source, in terms of the probability mass function of the source. It is, however, conceivable to define a similar entity in the more general probability space, $(\mathscr{X}, \mathfrak{X}, P)$, and then to consider a *continuous random variable* with a relation to the newly defined entity. This new entity is called *differential entropy* or *continuous entropy*. Generally, in the literature, $h(X)$ is used to denote *differential entropy*. However, since we have reserved $h(t)$ for signals, to reduce confusion, we utilize the symbol for *Planck's constant, $\hbar$* to denote *entropy differential*, knowing that we will not be using *Planck's constant* anywhere in this text.

To simplify matters, let us define the *differential entropy* in terms of the *measurable Euclidean probability space*, for a *continuous random variable, X*, using the *Lebesgue measure,*

$$\hbar(X) \overset{\Delta}{=} - \int\limits_{-\infty}^{\infty} p(x) \log_2 p(x) dx \qquad (7.62)$$

where $p(x)$ is the *probability density function* associated with X.

In the same spirit, the *joint entropy*, $\hbar(X,Y)$ is,

$$\hbar(X,Y) = - \int\limits_{-\infty}^{\infty} \int\limits_{-\infty}^{\infty} p(x,y) \log_2 p(x,y) dx dy \qquad (7.63)$$

where $p(x,y)$ is the *joint probability density* of X and Y such that,

$$\int\limits_{-\infty}^{\infty} \int\limits_{-\infty}^{\infty} p(x,y) dx dy = 1 \qquad (7.64)$$

The properties and definitions stated in the previous section extend to the case of the *continuous source* except for a couple of exceptions. The first thing to note is that *differential entropy* is only defined when the integral in Equation 7.62 exists and when X has a *probability density*. Another important distinction is due to the fact that we are using the *probability density function* which may be larger than 1. Unlike *discrete entropy* which is *positive semi-definite*, due to the fact that the *probability mass function* may never be more than 1, the *differential entropy* may become negative. Take the example of a *uniform distribution*, with the *probability density function*,

$$p(x) = \begin{cases} 0 & \forall \quad x < a \\ \frac{1}{b-a} & \forall \ a \leq x \leq b \\ 0 & \forall \quad x > b \end{cases} \qquad (7.65)$$

Then, the *differential entropy* is

$$\hbar(X) = - \int\limits_{-\infty}^{\infty} \frac{1}{b-a} \log_2 \frac{1}{b-a} dx$$

$$= - \int\limits_{a}^{b} \frac{1}{b-a} \log_2 \frac{1}{b-a} dx$$

$$= \log_2(b-a) \qquad (7.66)$$

which is negative for all $(b-a) < 1$.

In *speaker recognition* we are usually interested in *random variables* which have a *normal density function*. Since *differential entropy* is really a function of the *probability density*, $p(x)$, we can compute it for a generic *normal density* in terms of its parameters, μ and σ^2,

$$p(x) = \mathcal{N}(\mu, \sigma^2)$$

$$= \frac{1}{\sqrt{2\pi}\sigma} e^{-\frac{1}{2}\left(\frac{(x-\mu)^2}{\sigma^2}\right)} \tag{7.67}$$

Let us consider the *expectation* interpretation of *differential entropy*,

$$\hbar(X) = -\int_{-\infty}^{\infty} p(x) \log_2 p(x) dx$$

$$= -\mathcal{E}\{\log_2 p(x)\} \tag{7.68}$$

Therefore, the problem reduces to the computation of the expected value of $\log_2 p(x)$. Let us evaluate $\log_2 p(x)$,

$$\log_2 p(x) = -\frac{1}{2}\log_2(2\pi\sigma^2) - \frac{(x-\mu)^2}{2\sigma^2}\log_2 e \tag{7.69}$$

Now we have to compute the expected value. Since the two parts of $\log_2 p(x)$ have negative signs and since we are interested in the negative expected value of this expression, we can write,

$$\hbar(X) = -\mathcal{E}\{\log_2 p(x)\}$$

$$= \mathcal{E}\{-\log_2 p(x)\}$$

$$= \mathcal{E}\left\{\frac{1}{2}\log_2(2\pi\sigma^2) + \frac{(x-\mu)^2}{2\sigma^2}\log_2 e\right\}$$

$$= \frac{1}{2}\log_2(2\pi\sigma^2) + \frac{1}{2}\log_2 e \, \mathcal{E}\left\{\frac{(x-\mu)^2}{\sigma^2}\right\}$$

$$= \frac{1}{2}\log_2(2\pi\sigma^2) + \frac{1}{2}\frac{\log_2 e}{\sigma^2}\mathcal{E}\left\{(x-\mu)^2\right\}$$

$$= \frac{1}{2}\log_2(2\pi\sigma^2) + \frac{1}{2}\frac{\log_2 e}{\sigma^2}\sigma^2$$

$$= \frac{1}{2}\log_2(2\pi e\sigma^2) \tag{7.70}$$

7.6 Relative Entropy

We will now cover some important concepts leading to the definition of *relative entropy* (also known as *Kullback-Leibler directed divergence*) and *mutual information*. Following the derivations in Kullback and Leibler [12, 11], we will examine different concepts of *Information Theory* as well as provide insight into the meaning of *relative entropy*, which of course they do not call by that name and simply present it as *directed divergence*.

Let us assume for the moment that we have two hypotheses H_0 and H_1 – see Section 9.1. Furthermore, assume that the *probability spaces* associated with H_0 and H_1 are given by $(\mathscr{X}, \mathfrak{X}, P_0)$ and $(\mathscr{X}, \mathfrak{X}, P_1)$ respectively where *probability measures* P_0 and P_1 are *equivalent* $(P_0 \equiv P_1)$ – see Definition 6.42. The *equivalence* assumption is such that it would avoid any *a-priori* hypothesis for which the *a-posteriori* hypothesis would not be allowed and vice-versa.

Recall the *Radon-Nykodým theorem* (Theorem 6.3) and consider a *measure*, $\lambda(x)$ in the *measurable space*, $(\mathscr{X}, \mathfrak{X})$, where

$$\{P_i, i \in \{0,1\}\} \ll \lambda \tag{7.71}$$

Then,

$$P_i(\mathscr{A}) = \int_{\mathscr{A}} g_i(x) d\lambda(x) \quad \forall i \in \{0,1\} \tag{7.72}$$

In Equation 7.72, $g_i(x)$ is the *generalized probability density* (Radon-Nikodým derivative, Equation 6.77) for hypothesis $\{H_i : i \in \{0,1\}\}$. An important application of the use of *information* is its assessment in favor of picking, say, a *hypothesis H_0* (see Section 9.1) when compared to *hypothesis H_1* after observing $X = x$. In this case, the index i in Equation 7.72 may be $i \in \{0,1\}$.

According to the definition of *information* (Equation 7.53), the increment of *information* gained by observing $X = x$ in favor of *hypothesis H_0* against H_1 may be written as,

$$\mathscr{I}(0:1,x) = \log_2 g_0(x) - \log_2 g_1(x)$$
$$= \log_2 \left(\frac{g_0(x)}{g_1(x)} \right) \tag{7.73}$$

The mean value of the *information* in the observation of an event which is made up of a subspace of X, $\mathscr{A} \in \mathfrak{X}$, in favor of *hypothesis H_0* against H_1 for $P_0(\mathscr{A})$ is

$$\mathscr{I}(0:1,\mathscr{A}) = \begin{cases} \frac{1}{P_0(\mathscr{A})} \int_{\mathscr{A}} \log_2 \frac{g_0(x)}{g_1(x)} dP_0(x) \ \forall \ P_0(\mathscr{A}) > 0 \\ \\ 0 \quad\quad\quad\quad\quad\quad \forall \ P_0(\mathscr{A}) = 0 \end{cases} \tag{7.74}$$

Let us write Equation 7.74 for the whole sample space, $\mathscr{X}$. In this case, we can drop the $\mathscr{X}$ and denote the mean value of *information* gained by observing $\{X = x\}$ in favor of *hypothesis H_0* against H_1 by $\mathscr{I}(0:1)$. Since the probability of the entire sample space, $P_0(\mathscr{X}) = 1$, then $\mathscr{I}(0:1)$ may be written as follows,

$$\mathscr{I}(0:1) = \int_{\mathscr{X}} \log_2 \frac{g_0(x)}{g_1(x)} dP_0(x) \tag{7.75}$$

But we know from the definition of the *Radon-Nikodým derivative*,

$$dP_0(x) = g_0(x)d\lambda(x) \tag{7.76}$$

Using Bayes' theorem (Theorem 6.2), we can write the following expression for the *conditional probability* of hypothesis H_i where $i \in \{0,1\}$ [11],

$$P(H_i|x) = \frac{g_i(x)P(H_i)}{\sum\limits_{j=0}^{1} g_j(x)P(H_j)} [\lambda] \quad \forall\, i \in \{0,1\} \tag{7.77}$$

Therefore, the integrand of Equation 7.75 may be written, *almost everywhere*, in terms of the gain in the *relative information* from the *prior state* of the two hypotheses (before knowing x) to the *posterior state* (after x is known). Namely,

$$\begin{aligned}
\mathscr{I}(0:1,x) &= \log_2 \frac{g_0(x)}{g_1(x)} \\
&= \log_2 \frac{P(H_0|x)}{P(H_1|x)} - \log_2 \frac{P(H_0)}{P(H_1)} [\lambda]
\end{aligned} \tag{7.78}$$

We can define the entity given by Equation 7.78 as the *information gained* by knowing that $X = x$ to decide in favor of the *null hypothesis* (Definition 9.1), H_0, versus the *alternative hypothesis* (Definition 9.2), H_1. Note that $g_0(x)$ and $g_1(x)$ are actually *conditional probability densities*, conditioned upon hypotheses H_0 and H_1 respectively. They are also known as the *likelihoods* of H_0 and H_1 given the observation, $X = x$. Therefore, $\mathscr{I}(0:1,x)$ is the *log-likelihood ratio* of hypothesis H_0 against H_1. As we shall see, the *log-likelihood ratio* is used to make a *Bayesian decision* in favor of hypothesis H_0 against H_1 – see Section 9.2.

Let us use Equation 7.76 to write $\mathscr{I}(0:1)$ in terms of the measure, λ,

$$\mathscr{I}(0:1) = \int_{\mathscr{X}} g_0(x) \log_2 \frac{g_0(x)}{g_1(x)} d\lambda(x) \tag{7.79}$$

$\mathscr{I}(0:1)$, given by Equation 7.79 is a *directed divergence* derived by Kullback and Leibler in [12]. It is a *divergence* of the null hypothesis against the alternative based on the information gain in observing $X = x$, but it does not obey the *triangular property*, so it is not a true metric. In addition, it is a *directed divergence*, since it is not *symmetric*. Namely, $\mathscr{I}(0:1) \neq \mathscr{I}(1:0)$.

Although the notation, $\mathscr{I}(0:1)$ was used briefly in the above derivation based on Kullback and Leibler's [12] convention, the entity in Equation 7.79 is really the *expected value* of the *relative information* presented by Equation 7.78, computed over the entire sample space, $\mathscr{X}$. Therefore, being the expected value of *relative information*, it should really be called *relative entropy*[6] – see the interpretation of Equation 7.7 for the definition of *entropy* (Definition 7.8).

---

[6] Not to be confused with *Shannon's relative entropy* (Definition 7.9).

Since we will be using the *Kullback-Leibler directed divergence* in the rest of the book, we will refer to it by the following notation,

$$\mathscr{D}_{KL}(0 \rightarrow 1) \overset{\Delta}{=} \mathscr{I}(0:1) \tag{7.80}$$

$$= \int_{\mathscr{X}} g_0(x) \log_2 \frac{g_0(x)}{g_1(x)} d\lambda(x) \tag{7.81}$$

and

$$\mathscr{D}_{KL}(1 \rightarrow 0) \overset{\Delta}{=} \mathscr{I}(1:0) \tag{7.82}$$

$$= \int_{\mathscr{X}} g_1(x) \log_2 \frac{g_1(x)}{g_0(x)} d\lambda(x) \tag{7.83}$$

$\mathscr{D}_{KL}(0 \rightarrow 1)$ and $\mathscr{D}_{KL}(1 \rightarrow 0)$ do, however, obey all other properties of a metric (see Definition 6.25).

Note that if X is defined on the *real line*, $\{\mathscr{X} = \mathbb{R}\}, \lambda(x)$, the $\lambda(x)$ is the *Lebesgue measure*, so that $g_0(x)$ and $g_1(x)$ will become the same as $p_0(x)$ and $p_1(x)$ which are the *probability density functions* of X with respect to probability measures P_0 and P_1. In that case, the variable of integration becomes x. So we can write the following,

$$\mathscr{D}_{KL}(0 \rightarrow 1) = \int_{\mathscr{X}} p_0(x) \log_2 \frac{p_0(x)}{p_1(x)} dx \tag{7.84}$$

and

$$\mathscr{D}_{KL}(1 \rightarrow 0) = \int_{\mathscr{X}} p_1(x) \log_2 \frac{p_1(x)}{p_0(x)} dx \tag{7.85}$$

Recall the statement of *Jensen's inequality* given by Equation 6.128. If we define $y(x)$ as follows,

$$y(x) \overset{\Delta}{=} \frac{p_1(x)}{p_0(x)} \tag{7.86}$$

since $y(x)$ is always *positive*, because $p_1(x)$ and $p_0(x)$ are *probability density functions*, $\log_2(y) = \log_2(\frac{p_1(x)}{p_0(x)})$ is *concave*, which means that we may say $f(y)$ is *convex* if,

$$f(y) = -\log_2(y) \tag{7.87}$$

Then, Equation 6.128 may be written out in integral form in terms of x and $p_0(x)$ as follows,

$$-\int_{\mathscr{X}} p_0(x) \left(\log_2 \frac{p_1(x)}{p_0(x)} \right) dx \geq -\log_2 \left(\int_{\mathscr{X}} p_0(x) \frac{p_1(x)}{p_0(x)} dx \right) \tag{7.88}$$

If we invert the argument of the $\log 2$ at the left side of Equation 7.88, we will have

$$\int_{\mathscr{X}} p_0(x) \left(\log_2 \frac{p_0(x)}{p_1(x)} \right) dx \geq -\log_2(1) \tag{7.89}$$

The left hand side of Equation 7.89, based on Equation 7.84, is just $\mathscr{D}_{KL}(0 \to 1)$. Therefore, we can say,

$$\mathscr{D}_{KL}(0 \to 1) \geq 0 \tag{7.90}$$

which is a very important result, proving an important property of a *divergence*.

Note that Equation 7.90 may be written in terms of the *expected values* of the $f(q(x))$ and $f(p(x))$, where $f(x)$ is given by Equation 7.87,

$$-\int_{\mathscr{X}} p_0(x) \log_2 p_1(x) dx \geq -\int_{\mathscr{X}} p_0(x) \log_2 p_0(x) dx \tag{7.91}$$

where the right hand side of Equation 7.94 is known as the *cross entropy* of the true density of X with any other density, $p_1(x)$, and is denoted by $\hbar(p_0 \to p_1)$ for the continuous case and $\mathscr{H}(p_0 \to p_1)$ for the discrete case. Note the following formal definitions of *cross entropy*:

Definition 7.16 (Differential Cross Entropy). *The differential cross entropy,* $\hbar(p_0 \to p_1)$, *of two probability density functions,* $p_0(x)$ *and* $p_1(x)$ *is given by the following expression, when the Lebesgue measure is used,*

$$h(p_0 \to p_1) \stackrel{\Delta}{=} -\int_{-\infty}^{\infty} p_0(x) \log_2 p_1(x) dx \tag{7.92}$$

Definition 7.17 (Cross Entropy). *Consider the discrete source of Section 7.3. The cross entropy,* $\mathscr{H}(p_0 \to p_1)$, *of two different probability mass functions,* $p_0(X)$ *and* $p_1(X)$, *for the discrete random variable* X *is given by,*

$$\mathscr{H}(p_0 \to p_1) \stackrel{\Delta}{=} -\sum_{i=1}^{n} p_0(X_i) \log_2 p_1(X_i) \tag{7.93}$$

Therefore,

$$\hbar(p_0) \leq \hbar(p_0 \to p_1) \tag{7.94}$$

for the continuous case and

$$\mathscr{H}(p_0) \leq \mathscr{H}(p_0 \to p_1) \tag{7.95}$$

for the discrete case.

Equation 7.94 is known as *Gibb's inequality* and it states that the Entropy is always less than or equal to the *cross entropy*, where $p_0(x)$ is the true probability density function of X and $p_1(x)$ is any other density function.

Before *Kullback and Leibler* [12], *Jeffreys* [10] defined a measure, now known as *Jeffreys' divergence*, which is related to the *Kullback-Leibler directed divergence* as follows,

$$\mathscr{D}_J(0 \leftrightarrow 1) = \int_{\mathscr{X}} \log_2 \frac{dP_0}{dP_1} d(P_0 - dP_1) \tag{7.96}$$

Jeffreys called it an invariant for expressing the difference between two distributions and denoted it as I_2. It is easy to see that this integral is really the sum of the two Kullback and Leibler directed divergences, one in favor of H_0 and the other in favor of H_1. Therefore,

$$\mathscr{D}_J(0 \leftrightarrow 1) = \mathscr{D}_{KL}(0 \rightarrow 1) + \mathscr{D}_{KL}(1 \rightarrow 0) \tag{7.97}$$

$$\int_{\mathscr{X}} (p_0(x) - p_1(x)) \log_2 \frac{p_0(x)}{p_1(x)} dx \tag{7.98}$$

It is apparent that $\mathscr{D}_J(0 \leftrightarrow 1)$ is symmetric with respect to hypotheses H_0 and H_1, so it is a measure of the *divergence* between these hypotheses. Although $\mathscr{D}_J(0 \leftrightarrow 1)$ is *symmetric*, it still does not obey the *triangular inequality* property, so it cannot be considered to be a *metric*.

Throughout this book, we use $\mathscr{D}(0 \rightarrow 1)$ to denote a *directed divergence*, $\mathscr{D}(0 \leftrightarrow 0)$ to denote a (symmetric) *divergence* and $d(0,1)$ for a *distance*. The subscripts, such as the *KL* in $\mathscr{D}_{KL}(0 \rightarrow 1)$, specify the type of *directed divergence*, *divergence* or *distance*.

It was mentioned that the nature of the measure is such that it may specify any type of random variable including a *discrete random variable*. In that case, the *KL-divergence* may be written as,

$$\mathscr{D}_{KL}(0 \rightarrow 1) = \sum_{x_i \in X} P_0(x_i) \log_2 \frac{P_0(x_i)}{P_1(x_i)} \tag{7.99}$$

See Section 8.2.1 for the expression for the *KL-divergence* between two normal density probability density functions.

7.6.1 Mutual Information

Consider a special case of *relative entropy* for a random variable defined in the *two-dimensional Cartesian product space* $(\mathscr{X}, \mathfrak{X})$, where $\{\mathscr{X} = \mathscr{R}^2\}$ – see Section 6.2.2. Then the *relative entropy* (*KL-divergence*) in favor of hypothesis H_0 versus H_1 is given by

$$\mathscr{D}_{KL}(0 \to 1) = \int\limits_{-\infty}^{\infty} \int\limits_{-\infty}^{\infty} p_0(x_1, x_2) \log_2 \frac{p_0(x_1, x_2)}{p_1(x_1, x_2)} dx_1 dx_2 \qquad (7.100)$$

Now, let us assume that the *null hypothesis*, H_0, states that x_1 and X_2 are dependent on one another and the *alternative hypothesis* states that they are *mutually indepen-dent*. Based on the statement of H_1, the joint *probability density function*, $p_1(x_1, y_2)$, is given by the product of the *marginal probability density functions*, $p_{X_1}(x_1)$ and $p_{X_2}(x_2)$ (see Equation 6.92),

$$p_1(x_1, y_2) = p_{X_1}(x_1) p_{X_2}(x_2) \qquad (7.101)$$

Therefore, the *expected value* of the *relative information* in favor of the *depen-dence* of X_1 and X_2 versus their independence is given by,

$$\mathscr{D}_{KL}(0 \to 1) = \int\limits_{-\infty}^{\infty} \int\limits_{-\infty}^{\infty} p_0(x_1, x_2) \log_2 \frac{p_0(x_1, x_2)}{p_{X_1}(x_1) p_{X_2}(x_2)} dx_1 dx_2 \qquad (7.102)$$

We can omit the subscript 0 since for the true X_1 and X_2, there is only one joint probability density function with respect to the *Lebesgue measure*. Therefore,

$$\mathscr{D}_{KL}(0 \to 1) = \int\limits_{-\infty}^{\infty} \int\limits_{-\infty}^{\infty} p(x_1, x_2) \log_2 \frac{p(x_1, x_2)}{p_{X_1}(x_1) p_{X_2}(x_2)} dx_1 dx_2 \qquad (7.103)$$

Note that Equation 7.103 is symmetric about X_1 and X_2. Also, it may be viewed as a measure of the *expected value* of the information in X_1 about X_2, or by symmetry in X_2 about X_1. Therefore, it is called the *mutual information* between X_1 and X_2, and it is denoted as,

$$\mathscr{I}(X_1; X_2) = \mathscr{I}(X_2; X_1)$$

$$= \int\limits_{-\infty}^{\infty} \int\limits_{-\infty}^{\infty} p(x_1, x_2) \log_2 \frac{p(x_1, x_2)}{p_{X_1}(x_1) p_{X_2}(x_2)} dx_1 dx_2 \qquad (7.104)$$

Examining Equation 7.104, we can also say that,

$$\mathscr{I}(X_1; X_2) = \mathscr{I}(X_2; X_1) \qquad (7.105)$$
$$= \mathscr{D}_{KL}(p(x_1, y_2) \to p_{X_1}(x_1) p_{X_2}(x_2)) \qquad (7.106)$$

Based on Equation 7.106, the discrete version of *mutual information* would be given by the following,

$$\mathscr{I}(X; Y) = \mathscr{I}(Y; X) \qquad (7.107)$$

$$\overset{\Delta}{=} \sum_{i=1}^{n} \sum_{j=1}^{m} p(X_i, Y_j) \ln \frac{p(X_i, Y_j)}{p_X(X_i) p_Y(Y_j)} \qquad (7.108)$$

As we shall see in Problem 7.2, mutual information may be viewed as the amount of information which is present when comparing the entropy of a random variable with its entropy conditioned upon the observation of another random variable. Namely,

$$\mathscr{I}(X;Y) = \mathscr{H}(X) - \mathscr{H}(X|Y) \tag{7.109}$$

$$= \mathscr{H}(Y) - \mathscr{H}(Y|X) \tag{7.110}$$

According to Equation 7.98, Jeffreys' divergence between the hypothesis of dependence versus independence of X_1 and X_2 would be given by the following,

$$\mathscr{I}_J(X_1;X_2) = \mathscr{D}_J(p(x_1,y_2) \leftrightarrow p_{X_1}(x_1)p_{X_2}(x_2))$$

$$\int\limits_{-\infty}^{\infty}\int\limits_{-\infty}^{\infty} \quad (p(x_1,x_2) - p_{X_1}(x_1)p_{X_2}(x_2))$$

$$\log_2 \frac{p(x_1,x_2)}{p_{X_1}(x_1)p_{X_2}(x_2)}dx_1 dx_2 \tag{7.111}$$

Example 7.1 (Mutual Information of two Normally Distributed Random Variables). *Recall the definition of a multi-dimensional Normal density function (Equation 6.196). Let us consider the special case of a two-dimensional Euclidean space. We shall treat each dimension as a random variable and denote them as X_1 and X_2, with correlation coefficient $\kappa(X_1,X_2)$ between them (see Definition 6.60). Then we can write the mean vector, $\boldsymbol{\mu}$, and the Covariance matrix, $\boldsymbol{\Sigma}$, of the two-dimensional random variable, $\{X : \mathbf{x} \in \mathscr{R}^2\}$, in terms of the mean, the variance and the correlation coefficient of its two components, as follows,*

$$\boldsymbol{\mu} = \begin{bmatrix} \mu_{X_1} \\ \mu_{X_2} \end{bmatrix} \tag{7.112}$$

$$\boldsymbol{\Sigma} = \begin{bmatrix} \sigma_{X_1}^2 & \kappa\sigma_{X_1}\sigma_{X_2} \\ \kappa\sigma_{X_1}\sigma_{X_2} & \sigma_{X_2}^2 \end{bmatrix} \tag{7.113}$$

Therefore, we may write the joint probability density function for X_1 and X_2, which are normally distributed random variables, in terms of their individual means (μ_{X_1}, μ_{X_2}), variances $(\sigma_{X_1}^2, \sigma_{X_2}^2)$, and their correlation coefficient, $\kappa(X_1,X_2)$,

$$p(x_1,x_2) = \frac{1}{2\pi\sqrt{(1-\kappa^2)}\sigma_{Y_1}\sigma_{Y_2}}$$

$$\exp\left\{ -\frac{1}{2(1-\kappa^2)} \left(\frac{(x_1-\mu_{X_1})^2}{\sigma_{X_1}^2} - 2\kappa\frac{(x_1-\mu_{X_1})(x_2-\mu_{X_2})}{\sigma_{X_1}\sigma_{X_2}} + \frac{(x_2-\mu_{X_2})^2}{\sigma_{X_2}^2} \right) \right\} \tag{7.114}$$

Also, the marginal probability density functions may be written in terms of one-dimensional Normal density functions as follows,

$$p_{X_1}(x_1) = \frac{1}{\sqrt{2\pi}\sigma_{X_1}} \exp\left(-\frac{1}{2}\frac{(x_1 - \mu_{X_1})^2}{\sigma_{X_1}^2} \right) \tag{7.115}$$

and

$$p_{X_2}(x_2) = \frac{1}{\sqrt{2\pi}\sigma_{X_2}} \exp\left(-\frac{1}{2}\frac{(x_2 - \mu_{X_2})^2}{\sigma_{X_2}^2}\right) \tag{7.116}$$

Substituting Equations 7.114, 7.115 and 7.116 into Equation 7.104, we have the expression for the mutual information of normally distributed random variables, X_1 and X_2, as a function of their correlation coefficient only,

$$\begin{aligned}\mathscr{I}(X_1;X_2) &= \mathscr{I}(X_2;X_1) \\ &= -\frac{1}{2}\log\left(1 - [\kappa(X_1,X_2)]^2\right)\end{aligned} \tag{7.117}$$

7.7 Fisher Information

In Section 10.1, we will discuss the idea behind *maximum likelihood estimation* of parameters which would model a statistical system. At the beginning of this chapter, we discussed the fact that in the limit, the *Fisher information* is related to the information defined by *Wiener* and *Shannon* (Section 7.3.3). In this section, we will start with the *Kullback-Leibler directed divergence* and *Jeffreys' divergence* and show that under certain limiting conditions and in relation with statistical parameter estimation, they are related to the *Fisher information matrix*, hence the relation between this information measure and the *Wiener-Shannon* concept of information.

The topic of this section is quite related to the concept of *maximum likelihood parameter estimation* which will be discussed in Section 10.1. Therefore, it is recommended that the two sections be studied side-by-side.

Consider the random variable X, defined in *measure space* $(\mathscr{X}, \mathfrak{X}, \lambda)$, where $\{X : \mathbf{x} \in \mathscr{X} = \mathscr{R}^D\}$. Also, for simplicity, let us assume that the measure, $\lambda(\mathbf{x})$ is just the *Lebesgue measure*. In addition, let us say that we have a parametric model which estimates the *probability density function* associated with X, $p(\mathbf{x})$, by $\hat{p}(\mathbf{x}|\boldsymbol{\varphi})$, where $\boldsymbol{\varphi} \in \phi = \mathscr{R}^M$ is the *parameter vector*. Furthermore, let us assume that another model exists, with a *probability density function* of the same form, with the exception that it has a slightly different *parameter vector*, $\tilde{\boldsymbol{\varphi}}$, such that the difference between the two parameter vectors may be written as a deviation vector, $\Delta\boldsymbol{\varphi}$, as follows,

$$\tilde{\boldsymbol{\varphi}} = \boldsymbol{\varphi} + \Delta\boldsymbol{\varphi} \tag{7.118}$$

Using the *natural logarithm*, the *Kullback-Leibler directed divergence* from the first model to the second, may be written as follows,

$$\mathscr{D}_{KL}(\boldsymbol{\varphi} \to \tilde{\boldsymbol{\varphi}}) = \int_{\mathscr{X}} \hat{p}(\mathbf{x}|\boldsymbol{\varphi}) \ln\left(\frac{\hat{p}(\mathbf{x}|\boldsymbol{\varphi})}{\hat{p}(\mathbf{x}|\tilde{\boldsymbol{\varphi}})}\right) d\mathbf{x} \tag{7.119}$$

If we defined the following two notations,

$$\Delta \hat{p}(\mathbf{x}|\boldsymbol{\varphi}) \stackrel{\Delta}{=} \hat{p}(\mathbf{x}|\tilde{\boldsymbol{\varphi}}) - \hat{p}(\mathbf{x}|\boldsymbol{\varphi}) \tag{7.120}$$

and

$$\Delta \ln(\hat{p}(\mathbf{x}|\boldsymbol{\varphi})) \stackrel{\Delta}{=} \ln(\hat{p}(\mathbf{x}|\tilde{\boldsymbol{\varphi}})) - \ln(\hat{p}(\mathbf{x}|\boldsymbol{\varphi})) \tag{7.121}$$

Then, using the definition in Equation 7.121, we may write Equation 7.119 as follows,

$$\mathscr{D}_{KL}(\boldsymbol{\varphi} \to \tilde{\boldsymbol{\varphi}}) = -\int_{\mathscr{X}} \hat{p}(\mathbf{x}|\boldsymbol{\varphi})\Delta \ln(\hat{p}(\mathbf{x}|\boldsymbol{\varphi})) \, d\mathbf{x} \tag{7.122}$$

Since we have made the fundamental assumption that $\Delta\boldsymbol{\varphi}$ is a small deviation vector, we may try to approximate $\mathscr{D}_{KL}(\boldsymbol{\varphi} \to \tilde{\boldsymbol{\varphi}})$ by writing the *Taylor series expansion* (Definition 24.42) of $\Delta \ln(\hat{p}(\mathbf{x}|\boldsymbol{\varphi}))$, about the parameter vector, $\boldsymbol{\varphi}$[11]. If we write the terms of the series, up to the second order term, the remainder [19] may be added in terms of the third order, evaluated at $\boldsymbol{\varphi} + \boldsymbol{\alpha}^T \Delta \boldsymbol{\varphi}$ such that $0 < (\boldsymbol{\alpha})_{[m]} < 1$, $\forall m \in \{1, 2, \cdots, M\}$.

$$\Delta \ln(\hat{p}(\mathbf{x}|\boldsymbol{\varphi})) = \sum_{i=1}^{M} (\Delta\boldsymbol{\varphi})_{[i]} \frac{\partial \ln \hat{p}(\mathbf{x}|\boldsymbol{\varphi})}{\partial (\boldsymbol{\varphi})_{[i]}}$$

$$+ \frac{1}{2!} \sum_{i=1}^{M} \sum_{j=1}^{M} (\Delta\boldsymbol{\varphi})_{[i]} (\Delta\boldsymbol{\varphi})_{[j]} \frac{\partial^2 \ln \hat{p}(\mathbf{x}|\boldsymbol{\varphi})}{\partial (\boldsymbol{\varphi})_{[i]} \partial (\boldsymbol{\varphi})_{[j]}}$$

$$+ \frac{1}{3!} \sum_{i=1}^{M} \sum_{j=1}^{M} \sum_{k=1}^{M} (\Delta\boldsymbol{\varphi})_{[i]} (\Delta\boldsymbol{\varphi})_{[j]} (\Delta\boldsymbol{\varphi})_{[k]}$$

$$\frac{\partial^3 \ln \hat{p}(\mathbf{x}|\boldsymbol{\varphi})}{\partial (\boldsymbol{\varphi})_{[i]} \partial (\boldsymbol{\varphi})_{[j]} \partial (\boldsymbol{\varphi})_{[k]}} \Big|_{(\boldsymbol{\varphi} + \boldsymbol{\alpha}^T \Delta \boldsymbol{\varphi})} \tag{7.123}$$

Therefore, Equation 7.122 may be written in terms of the *Taylor expansion* of Equation 7.123. Also, we may use the chain rule to write the first and second order partial derivatives of the log of the density in terms of the density function itself as follows,

$$\frac{\partial \ln \hat{p}(\mathbf{x}|\boldsymbol{\varphi})}{\partial (\boldsymbol{\varphi})_{[i]}} - \frac{1}{\hat{p}(\mathbf{x}|\boldsymbol{\varphi})} \frac{\partial \hat{p}(\mathbf{x}|\boldsymbol{\varphi})}{\partial (\boldsymbol{\varphi})_{[i]}} \tag{7.124}$$

$$\frac{\partial^2 \ln \hat{p}(\mathbf{x}|\boldsymbol{\varphi})}{\partial (\boldsymbol{\varphi})_{[i]} \partial (\boldsymbol{\varphi})_{[j]}} = \frac{1}{\hat{p}(\mathbf{x}|\boldsymbol{\varphi})} \frac{\partial^2 \hat{p}(\mathbf{x}|\boldsymbol{\varphi})}{\partial (\boldsymbol{\varphi})_{[i]} \partial (\boldsymbol{\varphi})_{[j]}} - \frac{1}{(\hat{p}(\mathbf{x}|\boldsymbol{\varphi}))^2} \frac{\partial \hat{p}(\mathbf{x}|\boldsymbol{\varphi})}{\partial (\boldsymbol{\varphi})_{[i]}} \frac{\partial \hat{p}(\mathbf{x}|\boldsymbol{\varphi})}{\partial (\boldsymbol{\varphi})_{[j]}} \tag{7.125}$$

Using Equations 7.124 and 7.125 in conjunction with the Taylor series expansion of Equation 7.123, we may write the expression for $\mathscr{D}_{KL}(\boldsymbol{\varphi} \to \tilde{\boldsymbol{\varphi}})$ from Equation

7.122 as follows,

$$
\mathscr{D}_{KL}(\boldsymbol{\varphi} \to \tilde{\boldsymbol{\varphi}}) = -\sum_{i=1}^{M} (\Delta\boldsymbol{\varphi})_{[i]} \int_{\mathscr{X}} \frac{\partial \hat{p}(\mathbf{x}|\boldsymbol{\varphi})}{\partial (\boldsymbol{\varphi})_{[i]}} d\mathbf{x}
$$
$$
-\frac{1}{2}\sum_{i=1}^{M}\sum_{j=1}^{M} (\Delta\boldsymbol{\varphi})_{[i]} (\Delta\boldsymbol{\varphi})_{[j]}
$$
$$
\int_{\mathscr{X}} \left(\frac{\partial^2 \hat{p}(\mathbf{x}|\boldsymbol{\varphi})}{\partial (\boldsymbol{\varphi})_{[i]} \partial (\boldsymbol{\varphi})_{[j]}} - \frac{1}{\hat{p}(\mathbf{x}|\boldsymbol{\varphi})} \frac{\partial \hat{p}(\mathbf{x}|\boldsymbol{\varphi})}{\partial (\boldsymbol{\varphi})_{[i]}} \frac{\partial \hat{p}(\mathbf{x}|\boldsymbol{\varphi})}{\partial (\boldsymbol{\varphi})_{[j]}} \right) d\mathbf{x}
$$
$$
-\frac{1}{6}\sum_{i=1}^{M}\sum_{j=1}^{M}\sum_{k=1}^{M} (\Delta\boldsymbol{\varphi})_{[i]} (\Delta\boldsymbol{\varphi})_{[j]} (\Delta\boldsymbol{\varphi})_{[k]}
$$
$$
\int_{\mathscr{X}} \hat{p}(\mathbf{x}|\boldsymbol{\varphi}) \left. \frac{\partial^3 \ln \hat{p}(\mathbf{x}|\boldsymbol{\varphi})}{\partial (\boldsymbol{\varphi})_{[i]} \partial (\boldsymbol{\varphi})_{[j]} \partial (\boldsymbol{\varphi})_{[k]}} \right|_{(\boldsymbol{\varphi}+\boldsymbol{\alpha}^T \Delta\boldsymbol{\varphi})} d\mathbf{x} \qquad (7.126)
$$

As we saw in Section 7.6 and we will see in more detail in Section 10.1, the *probability density function*, $\hat{p}(\mathbf{x}|\boldsymbol{\varphi})$, is also known as the *likelihood* of $\boldsymbol{\varphi}$ given $\mathbf{x}$. *Cramér* [4] specifies a set of *regularity conditions* after *Dugué* [5] which are used for showing the asymptotic properties of *maximum likelihood estimation* 10.1. These conditions are required by *Kullback* [11], in order to show the relation between the *Kullback-Leibler directed divergence* and the *Fisher information matrix* which is being discussed here. Therefore, we require the *likelihood*, $\hat{p}(\mathbf{x}|\boldsymbol{\varphi})$, to meet these *three regularity conditions* in order to be able to simplify Equation 7.126.

Regularity conditions on $\hat{p}(\mathbf{x}|\boldsymbol{\varphi})$:

1. $\hat{p}(\mathbf{x}|\boldsymbol{\varphi})$ is $\mathfrak{C}^3$ *continuous* (see Definition 24.19) with respect to $\boldsymbol{\varphi}$ in the interval $[\boldsymbol{\varphi}, \boldsymbol{\varphi} + \Delta\boldsymbol{\varphi}]$ and for *almost all* $\mathbf{x}$ ($\mathbf{x}[\mathbf{x}]$).

2. $\frac{\partial \hat{p}(\mathbf{x}|\boldsymbol{\varphi})}{\partial (\boldsymbol{\varphi})_{[i]}}$ and $\frac{\partial^2 \hat{p}(\mathbf{x}|\boldsymbol{\varphi})}{\partial (\boldsymbol{\varphi})_{[i]} \partial (\boldsymbol{\varphi})_{[j]}}$ are *absolutely integrable* (see Definition 24.38) in the interval $[\boldsymbol{\varphi}, \boldsymbol{\varphi} + \Delta\boldsymbol{\varphi}]$ and $\forall \mathbf{x}[\mathbf{x}]$. Also, if $G(\mathbf{x})$ is the upper-bound, on the third partial derivative of $\hat{p}(\mathbf{x}|\boldsymbol{\varphi})$, i.e.,

$$
\left| \frac{\partial^3 \ln \hat{p}(\mathbf{x}|\boldsymbol{\varphi})}{\partial (\boldsymbol{\varphi})_{[i]} \partial (\boldsymbol{\varphi})_{[j]} \partial (\boldsymbol{\varphi})_{[k]}} \right| < G(\mathbf{x}) \qquad (7.127)
$$

then

$$
\int_{\mathscr{X}} \hat{p}(\mathbf{x}|\boldsymbol{\varphi}) G(\mathbf{x}) d\mathbf{x} < \beta < \infty \qquad (7.128)
$$

3. Finally,

$$
\int_{\mathscr{X}} \frac{\partial \hat{p}(\mathbf{x}|\boldsymbol{\varphi})}{\partial (\boldsymbol{\varphi})_{[i]}} d\mathbf{x} = 0 \ \ \forall\, i \in \{1, 2, \cdots, M\} \qquad (7.129)
$$

and

$$\int_{\mathscr{X}} \frac{\partial^2 \hat{p}(\mathbf{x}|\boldsymbol{\varphi})}{\partial (\boldsymbol{\varphi})_{[i]} \partial (\boldsymbol{\varphi})_{[j]}} d\mathbf{x} = 0 \ \ \forall \ i,j \in \{1,2,\cdots,M\} \tag{7.130}$$

It is important to note that since due to the *regularity assumption* 1, $\hat{p}(\mathbf{x}|\boldsymbol{\varphi})$ is a $\mathfrak{C}^3$ continuous function in the interval $[\boldsymbol{\varphi}, \boldsymbol{\varphi} + \Delta\boldsymbol{\varphi}]$, then based on Definition 24.19 and Property 24.6 all its derivatives up to the third derivative are bounded. Therefore, this property is implied and need not be listed.[7]

Given the above *regularity conditions*, Equation 7.126 may be simplified as follows,

$$\mathscr{D}_{KL}(\boldsymbol{\varphi} \to \hat{\boldsymbol{\varphi}}) = \frac{1}{2} \sum_{i=1}^{M} \sum_{j=1}^{M} (\mathscr{I}_F)_{[i][j]} (\Delta\boldsymbol{\varphi})_{[i]} (\Delta\boldsymbol{\varphi})_{[j]} \tag{7.131}$$

where $\mathscr{I}_F$ is the *Fisher information matrix* and its elements are given by the following definition,

$$(\mathscr{I}_F)_{[i][j]} \overset{\Delta}{=} \int_X \hat{p}(\mathbf{x}|\boldsymbol{\varphi}) \left(\frac{1}{\hat{p}(\mathbf{x}|\boldsymbol{\varphi})} \frac{\partial \hat{p}(\mathbf{x}|\boldsymbol{\varphi})}{\partial (\boldsymbol{\varphi})_{[i]}} \right) \left(\frac{1}{\hat{p}(\mathbf{x}|\boldsymbol{\varphi})} \frac{\partial \hat{p}(\mathbf{x}|\boldsymbol{\varphi})}{\partial (\boldsymbol{\varphi})_{[j]}} \right) d\mathbf{x} \tag{7.132}$$

Using Equation 7.124, we may write the *Fisher information matrix* in terms of the *log-likelihood* as follows,

$$(\mathscr{I}_F)_{[i][j]} = \int_X \hat{p}(\mathbf{x}|\boldsymbol{\varphi}) \left(\frac{\partial \ln(\hat{p}(\mathbf{x}|\boldsymbol{\varphi}))}{\partial (\boldsymbol{\varphi})_{[i]}} \right) \left(\frac{\partial (\hat{p}(\mathbf{x}|\boldsymbol{\varphi}))}{\partial (\boldsymbol{\varphi})_{[j]}} \right) d\mathbf{x} \tag{7.133}$$

Equation 7.133 may be seen as the *expected value* of the product of partial derivatives of the *log-likelihood*, namely,

$$(\mathscr{I}_F)_{[i][j]} = \mathscr{E} \left\{ \left(\frac{\partial \ln(\hat{p}(\mathbf{x}|\boldsymbol{\varphi}))}{\partial (\boldsymbol{\varphi})_{[i]}} \right) \left(\frac{\partial (\hat{p}(\mathbf{x}|\boldsymbol{\varphi}))}{\partial (\boldsymbol{\varphi})_{[j]}} \right) \right\} \tag{7.134}$$

Then the *Fisher information matrix* may be written in matrix form as follows,

$$\mathscr{I}_F = \mathscr{E} \left\{ (\nabla_{\boldsymbol{\varphi}} \hat{p}(\mathbf{x}|\boldsymbol{\varphi})) (\nabla_{\boldsymbol{\varphi}} \hat{p}(\mathbf{x}|\boldsymbol{\varphi}))^T \right\} \tag{7.135}$$

where $\nabla_{\boldsymbol{\varphi}} \hat{p}(\mathbf{x}|\boldsymbol{\varphi})$ is known as the *Fisher score* or *score statistic* – see 10.1.

Also, we may write the *Kullback-Leibler divergence* of Equation 7.131 in matrix form as,

---

[7] *Cramér* [4] and *Kullback* [11] include these conditions as a part of the *second regularity condition*, but aside from having a role in clarity and completeness, they do not technically need to be specified as conditions, since they are implied.

$$\mathscr{D}_{KL}(\boldsymbol{\varphi} \to \hat{\boldsymbol{\varphi}}) = \frac{1}{2}(\Delta\boldsymbol{\varphi})^T \mathscr{I}_F(\Delta\boldsymbol{\varphi}) \tag{7.136}$$

In a similar fashion, *Kullback* [11] shows the intuitively apparent result that *Jeffreys' divergence* may be approximated, using the above assumptions as follows,

$$\mathscr{D}_J(\boldsymbol{\varphi} \leftrightarrow \boldsymbol{\varphi}+\Delta\boldsymbol{\varphi}) = \int_{\mathscr{X}} (\hat{p}(\mathbf{x}|\boldsymbol{\varphi}) - \hat{p}(\mathbf{x}|\boldsymbol{\varphi}+\Delta\boldsymbol{\varphi})) \ln \frac{\hat{p}(\mathbf{x}|\boldsymbol{\varphi})}{\hat{p}(\mathbf{x}|\boldsymbol{\varphi}+\Delta\boldsymbol{\varphi})} d\mathbf{x} \tag{7.137}$$

$$\approx \sum_{i=1}^{M}\sum_{j=1}^{M} (\mathscr{I}_F)_{[i][j]}(\Delta\boldsymbol{\varphi})_{[i]}(\Delta\boldsymbol{\varphi})_{[j]} \tag{7.138}$$

$$= 2\mathscr{D}_{KL}(\boldsymbol{\varphi} \to \boldsymbol{\varphi}+\Delta\boldsymbol{\varphi}) \tag{7.139}$$

$$= (\Delta\boldsymbol{\varphi})^T \mathscr{I}_F(\Delta\boldsymbol{\varphi}) \tag{7.140}$$

It is important to remember that the above derivations made the basic assumption that the divergences of interest are from a point in the parameter space, $\boldsymbol{\varphi}$, to one which is a small distance away, at $\boldsymbol{\varphi}+\Delta\boldsymbol{\varphi}$. Therefore, we have shown that in a limiting sense, *Fisher information* is related to *Wiener-Shannon information*.

Earlier we showed (Equation 7.90) that the *Kullback-Leibler directed divergence* is positive semi-definite. Also, due to Equation 7.135 we may arrive at the conclusion that the *Fisher information matrix* should also be positive semi-definite.

Problems

For solutions to the following problems, see the Solutions section at the end of the book.

Problem 7.1 (Conditional Entropy).
Show that

$$\mathscr{H}(X,Y|Z) = \mathscr{H}(X|Z) + \mathscr{H}(Y|X,Z) \tag{7.141}$$

Problem 7.2 (Mutual Information).
If $\mathscr{I}(X;Y)$ is the mutual information between X and Y, show that,

$$\mathscr{I}(X;Y) = \mathscr{H}(X) - \mathscr{H}(X|Y) \tag{7.142}$$
$$= \mathscr{H}(Y) - \mathscr{H}(Y|X) \tag{7.143}$$

References

1. Ash, R.B.: Information Theory. Dover Publications, New York (1990). ISBN: 0-486-6652-16
2. Boltzmann, L.: Vorlesungen über Gastheorie (Lectures on gas theory). Dover Publications, New York (1896–1898). Translated into English by: Stephen G. Brush, ISBN: 0-486-68455-5
3. Cover, T.M., Thomas, J.A.: Elements of Information Theory, 2nd edn. John Wiley & Sons, New Jersey (2006). ISBN-13: 978-0-471-24195-9
4. Cramér, H.: Mathematical Methods of Statistics. Princeton University Press (1999). ISBN: 0-691-00547-8
5. Dugué, D.: Application des propriétés de la Limite au sens du Calcul des Probabilités à lÉtude de Diverses Questions d'estimation. Journal de lÉcole Polytechnique p. 305 (1937)
6. Fadeev, D.K.: Zum Begriff der Entropie einer endlichen Wahrscheinlichkeitss. Deutscher Verlag der Wissenschaften pp. 85–90 (1957)
7. Fisher, R.A.: Theory of Statistical Estimation. Proceedings of Cambridge Philosophical Society **22**, 700–725 (1925)
8. Halmos, P.R., Savage, L.J.: Application of the Radon-Nikodym Theorem to the Theory of Sufficient Statistics. The Annals of Mathematical Statistics **20**(2), 225–241 (1949)
9. Hartley, R.V.L.: Transmission of Information. Bell System Technical Journal **7**, 535–563 (1928)
10. Jeffreys, H.: An Invariant Form for the Prior Probability in Estimation Problems. Proceedings of the Royal Society of London **186**(1007), 453–461 (1946)
11. Kullback, S.: Information Theory and Statistics. Dover Publications, Inc., New York (1997). Unabridged publication of the original text published by John Wiley, New York (1959), ISBN: 0-486-69684-7
12. Kullback, S., Leibler, R.A.: On Information and Sufficiency. The Annals of Mathematical Statistics **22**(1), 79–86 (1951)
13. MacDonald, D.K.C.: Information Theory and Its Application to Taxonomy. Journal of Applied Physics **23**(5), 529–531 (1952)
14. Nyquist, H.: Certain Factors Affecting Telegraph Speed. Bell System Technical Journal **3**, 324–346 (1924)
15. Nyquist, H.: Certain Topics in Telegraph Transmission Theory. Transactions of the American Institute of Electrical Engineers (AIEE) **47**, 617–644 (1928). Reprint in Proceedings of the IEEE (2002), Vol. 90, No. 2, pp. 280–305

16. Réyni, A.: On Measure of Entropy and Information. In: Proceedings of the 4th Berkeley Symposium on Probability Theory and Mathematical Statistics, pp. 547–561 (1961)
17. Savage, L.J.: The Foundations of Statistics. John Wiley & Sons, New York (1954)
18. Shannon, C.E.: A Mathematical Theory of Computation. The Bell System Technical Journal **27**, 379–423,623–656 (1948). Reprint with corrections
19. Stewart, J.: Calculus, 6th edn. Brooks Cole, New York (2007). ISBN: 0-495-01160-6
20. Tjalkens, T.: State Dependent Coding: How to find the State? In: 43rd annual Allerton Conference on Communication, Control, and Computing (2005). URL http://cslgreenhouse.csl.illinois.edu/allertonarchives/allerton05/PDFs/Papers/V_A_2.pdf
21. Welsh, D.: Codes and Cryptography. Oxford University Press, New York (1990). ISBN: 0-198-53287-3
22. Wiener, N.: Cybernetics: or Control and Communication in the Animal and the Machine, 2nd edn. The M.I.T. Press, Cambridge (1976). First Edition was published by John Wiley and Sons, New York (1948), ISBN: 0-262-73009-X

Chapter 8
Metrics and Divergences

All places are distant from the heaven alike, ...

Robert Burton
The Anatomy of Melancholy, 1621

In this chapter, we continue the treatment of *distances* (*metrics*) and *divergences* by a introducing the terminology and the notation which will be used throughout this book for these two concepts. Comparing feature vectors and model parameters are at the heart of speaker recognition algorithms. As it will become more clear with the *definition* and *coverage* of *distance*, *divergence*, and *directed divergence*, we will use the notations, $d(.,.)$, $\mathscr{D}(. \leftrightarrow .)$, and $\mathscr{D}(. \rightarrow .)$ respectively.

8.1 Distance (Metric)

In the definition of a *metric space* (Definition 6.25), the concept of a *metric* or a *distance* was formally introduced. In general, different *metrics* may be defined for different entities. For example, the distance between complex numbers was defined in Chapter 24 (Definition 24.4).

First, the *distance* between *sequences* is introduced, which is of some importance in comparisons related to binary sequence corruption, encryption, search, decision, etc. Then the general concept of the distance between measurable subsets is revisited with the introduction of many different types of *metrics* being used, or having the potential of being used, in *speaker recognition*.

Consider two random variables X_1 and X_2 in the *measurable spaces*, $(\mathscr{X}_1, \mathfrak{X}_1)$ and $(\mathscr{X}_2, \mathfrak{X}_2)$ respectively where we may define a *measurable space*, $(\mathscr{X}, \mathfrak{X})$ such that $\{\mathscr{X}_1, \mathscr{X}_2 \subset \mathscr{X}\}$ and $\{\mathfrak{X}_1, \mathfrak{X}_2 \in \mathfrak{X}\}$. The *metric space*, $\mathscr{M}(\mathscr{X}, d)$, may be then defined (Definition 6.25) according to *measurable space* $(\mathscr{X}, \mathfrak{X})$. The *distance* between these X_1 and X_2 according to this *metric space* would be denoted by $d(X_1, X2)$. More specifically, the *distance* between two instances of X_1 and X_2 is written as, $d(x_1, x_2)$ and the *distance* between two subsets of the *Borel fields*, $\mathscr{A}_1 \in \mathfrak{X}_1$ and $\mathscr{A}_2 \in \mathfrak{X}_2$ may be written in the space of the supersets including these two *Borel fields*, $\mathfrak{X}$, as $d\left(\mathscr{A}_1, \mathscr{A}_2\right)$ – see Definition 6.26.

8.1.1 Distance Between Sequences

It sometimes makes sense to talk about *distance between sequences*. One such example is when we consider a *binary sequence* of length n. For example, consider a *pattern* which is stored in *binary code* in a *sequence* of n bits and call this sequence, in the form of a binary vector, $\mathbf{x}$. Now let us consider a second pattern, $\mathbf{y}$ of the same length. The *Hamming* distance [20], $d_H(\mathbf{x}, \mathbf{y})$ is defined as the *number of bits* where $\mathbf{x}$ and $\mathbf{y}$ differ. The *Hamming distance* becomes important, for example, in the analysis of the *corruption* of a *binary message* when it passes through a *noisy channel*.

8.1.2 Distance Between Vectors and Sets of Vectors

In this section, we will be defining several distance measures which meet all the properties stated in the definition of a *metric space* (Section 6.2.3). First, we define two random variables in the same sample space, $\{X : \mathbf{x} \in \mathscr{R}^d\}$ and $\{Y : \mathbf{y} \in \mathscr{R}^d\}$ Now, let us consider a set of outcomes for each of these random variables, $\mathscr{A} = \{\mathbf{x}_i, i = \{1, 2, \cdots, m\}$ and $\mathscr{B} = \{\mathbf{y}_j = \{1, 2, \cdots, n\}$. Furthermore, the *sample means* for these two sets are denoted by $\bar{\mathbf{x}}$ and $\bar{\mathbf{y}}$.

We are generally interested in computing distances between two outcomes, $d(\mathbf{x}, \mathbf{y})$, between an outcome and a set, $d(\mathbf{x}, \mathscr{A})$, or between two sets, $d(\mathscr{A}, \mathscr{B})$. There are different ways of interpreting the meaning of distances between two sets of random variables. We introduce one method in Definition 6.26 based on the *infimum* of the distance between their members. However, another method is to use the *sufficient statistics* (Section 6.8), as representatives of the sets and compute the distance based on these sufficient statistics such as the *sample mean* and *sample covariance*.

Definition 8.1 (Euclidean Distance). *The Euclidean distance between two vectors, $\mathbf{x}$ and $\mathbf{y}$ is defined as the Euclidean norm (Definition 23.7) of their difference vector,*

$$d_E(\mathbf{x}, \mathbf{y}) = \|\mathbf{x} - \mathbf{y}\|_{\mathscr{E}}$$
$$= \sqrt{(\mathbf{x} - \mathbf{y})^T (\mathbf{x} - \mathbf{y})} \tag{8.1}$$

Generally, when one speaks of the Euclidean distance between two sets of vectors, $\mathscr{A}$ and $\mathscr{B}$, the understanding is that this would be a distance between their sample means, $\bar{\mathbf{x}}$ and $\bar{\mathbf{y}}$.

$$d_E(\mathscr{A},\mathscr{B}) = d_E(\overline{\mathbf{x}},\overline{\mathbf{y}})$$

$$= \sqrt{(\overline{\mathbf{x}}-\overline{\mathbf{y}})^T(\overline{\mathbf{x}}-\overline{\mathbf{y}})} \tag{8.2}$$

As the *Euclidean norm* is generalized into the *p-norm*, the Euclidean distance may also be generalized into the L_p distance as follows,

Definition 8.2 (L_p Distance). *The L_p distance between two vectors, $\mathbf{x}$ and $\mathbf{y}$ is defined as the p-norm (Definition 23.7) of their difference vector,*

$$d_{L_p}(\mathbf{x},\mathbf{y}) = \|\mathbf{x}-\mathbf{y}\|_p \tag{8.3}$$

$$= \left(\sum_{i=1}^{D}((\mathbf{x})_{[i]} - (\mathbf{y})_{[i]})^p\right)^{\frac{1}{p}} \tag{8.4}$$

where $\mathbf{x},\mathbf{y}:\mathscr{R}^1 \mapsto \mathscr{R}^D$.

Therefore, the *Euclidean distance* is just the L_2 distance.

As in the case of the *Euclidean distance*, the L_p distance between two sets is normally understood to be,

$$d_{L_p}(\mathscr{A},\mathscr{B}) = d_{L_p}(\overline{\mathbf{x}},\overline{\mathbf{y}}) \tag{8.5}$$

Also, the L_p distance between two probability density functions was given by Equation 6.109. Therefore,

$$d_{L_p}(g_1,g_2) = \rho_p(g_1,g_2) \tag{8.6}$$

In Section 6.5.4 we examined some properties of the L_p distance between PDFs. See Section 23.2 for more on the L_p norm. Also, more on the L_p distance may be found in [5].

Definition 8.3 (Weighted Euclidean Distance). *At times, some dimensions of a multidimensional vector may possess more relevant information than the rest in a distance study. In such cases, we may weigh the different dimensions differently. A weighted Euclidean distance is a Euclidean distance where each dimension has a different weight in the distance computation and it may be written as follows*

$$d_{WE}(\mathbf{x},\mathbf{y}) = \sqrt{(\mathbf{x}-\mathbf{y})^t\,\mathbf{\Lambda}(\mathbf{x}-\mathbf{y})} \tag{8.7}$$

where $\mathbf{\Lambda}:\mathscr{R}^d \mapsto \mathscr{R}^d$ *is a diagonal matrix with diagonal elements,* $\{\Lambda_{ii} = \lambda_i, i \in \{1,2,\cdots,d\}\}$ *are the weights of the distance measure.*

Definition 8.4 (Mahalanobis Distance). *Mahalanobis [13] defines a general weighted distance which is based on the same idea as that of Definition 8.3, except here the waiting matrix is the covariance matrix. He also adds a scalar weight, k, for normalization. The normalization factor, k, may always be used in any distance or* *divergence. It is especially important to keep the proper units and to stay within expected range. In addition, Mahalanobis makes the assumption that the two random variables, X and Y, have the same estimated covariance. We have dropped the $\sim$ above the estimate of the covariance for convenience. Σ is actually the estimated covariance. It may either be the biased or unbiased estimate.*

$$\Sigma = \Sigma_X = \Sigma_Y \tag{8.8}$$

Then, the Mahalanobis' General Distance is defined as,

$$d_M(\mathbf{x}, \mathbf{y}) \triangleq k\boldsymbol{\delta}^T \Sigma^{-1} \boldsymbol{\delta} \tag{8.9}$$

where,

$$\boldsymbol{\delta} \triangleq \mathbf{x} - \mathbf{y} \tag{8.10}$$

When $k = 1$, the General Mahalanobis Distance is known, simply, as the Mahalanobis distance.

8.1.3 Hellinger Distance

The *Hellinger distance* is used to compute the distance between two different distributions. It is a distance since it obeys all the properties of a *metric* presented in Section 6.2.3 including *symmetry* and satisfying the *triangular inequality*. To understand the *Hellinger distance*, please read Sections 8.2.3, 8.2.4, and 8.2.5.

To make this metric a distance, it is defined based on the *Hellinger Integral* [15], the form of *Bhattacharyya measure* (Equation 8.22), $\rho_{Bc}(0,1)$, such that the *triangular inequality* is satisfied. Here is the definition of the *Hellinger distance* between two distributions [10],

$$d_{He}(0,1) = k\sqrt{1 - \rho_{Bc}(0,1)} \tag{8.11}$$

where k is a normalization constant. Reference [15] defines $d_{He}(0,1)$ with $k = 2$, whereas [10] uses $k = 1$.

8.2 Divergences and Directed Divergences

We have already introduced two types of *divergences*, *Kullback-Leibler directed divergence* and *Jeffreys' divergence*, in Section 7.6. Here, we shall review a few more and present simple comparisons among them. Let us start with a review of the Kullback-Leibler, Jeffreys and a few more divergences. Then, we will continue by looking at the f-divergence which is a class including most others.

8.2.1 Kullback-Leibler's Directed Divergence

The Kullback-Leibler directed divergence [11] has also been called a *distance* or a *divergence*, but it is neither. We saw, in Section 7.6, that it does not satisfy the *triangular inequality* nor is it generally *symmetric*. It is really a measure of *relative entropy* which has many uses including the comparison of two related distributions. There have been many attempts to symmetrize the Kullback-Leibler divergence [18]. Also, *Jeffreys' divergence* is a way of symmetrizing the Kullback Leibler divergence (see Equation 7.96). Here is the *Kullback-Leibler directed divergence* in favor of hypothesis H_0 against H_1, repeated for convenience (see Section 7.6 for more detail),

$$\mathscr{D}_{KL}(0 \rightarrow 1) = \int_{-\infty}^{\infty} p_0(x) \log_2 \frac{p_0(x)}{p_1(x)} dx \tag{8.12}$$

$$= \mathscr{E}\left\{ \log_2 \frac{p_0(x)}{p_1(x)} \right\} \tag{8.13}$$

Consider two N-dimensional *normal density* probability density functions, where the means and covariances are denoted by $\boldsymbol{\mu}_0$, $\boldsymbol{\mu}_1$, $\boldsymbol{\Sigma}_0$, and $\boldsymbol{\Sigma}_1$. Then the *Kullback-Leibler directed divergence*, from density 0 to density 1 may be written as follows,

$$\mathscr{D}_{KL}(0 \rightarrow 1) = \frac{1}{2\ln(2)} \left(tr(\boldsymbol{\Sigma}_1^{-1}\boldsymbol{\Sigma}_0) - N + (\boldsymbol{\mu}_1 - \boldsymbol{\mu}_0)^T \boldsymbol{\Sigma}_1^{-1}(\boldsymbol{\mu}_1 - \boldsymbol{\mu}_0) \right.$$
$$\left. - \ln\left(\frac{\det(\boldsymbol{\Sigma}_0)}{\det(\boldsymbol{\Sigma}_1)} \right) \right) \tag{8.14}$$

where $tr(\mathbf{A})$ stands for the trace of matrix A (see Section 23.2).

8.2.2 Jeffreys' Divergence

Jeffreys [9] defined two different measures which he dubbed *invariants for expressing the difference between two distributions*, I_1 and I_2,

$$I_1 = \int_{\mathscr{X}} \left(\sqrt{dP_0} - \sqrt{dP_1} \right)^2 \tag{8.15}$$

and

$$I_2 = \int_{\mathscr{X}} \log_2 \frac{dP_0}{dP_1} d(P_0 - P_1) \tag{8.16}$$

We discussed I_2 in some detail in Section 7.6. It is known as *"the divergence"* and is basically the sum of the two *Kullback-Leibler directed divergences*, one in favor of H_0 and the other in favor of H_1. For convenience, here, we repeat Equation 7.98 which is the form of *Jeffreys' divergence* where we use the Lebesgue measure and therefore we end up with *probability density functions* in the integrand.

$$\mathscr{D}_J (0 \leftrightarrow 1) = \mathscr{D}_{KL} (0 \rightarrow 1) + \mathscr{D}_{KL} (1 \rightarrow 0) \tag{8.17}$$

$$\int_{-\infty}^{\infty} (p_0(x) - p_1(x)) \log_2 \frac{p_0(x)}{p_1(x)} dx \tag{8.18}$$

Based on Equation 8.17 and Equation 8.14, the expression for $\mathscr{D}_J (0 \leftrightarrow 1)$, where the probability density functions are normal densities, may be written as follows,

$$\begin{aligned}
\mathscr{D}_J (0 \leftrightarrow 1) &= \mathscr{D}_{KL} (0 \rightarrow 1) + \mathscr{D}_{KL} (1 \rightarrow 0) \\
&= \frac{1}{2\ln(2)} \left(tr(\boldsymbol{\Sigma}_1^{-1} \boldsymbol{\Sigma}_0) - 2N \right. \\
&\quad \left. + (\boldsymbol{\mu}_1 - \boldsymbol{\mu}_0)^T (\boldsymbol{\Sigma}_1^{-1} + \boldsymbol{\Sigma}_0^{-1})(\boldsymbol{\mu}_1 - \boldsymbol{\mu}_0) \right)
\end{aligned} \tag{8.19}$$

Kullback [11] often called the *Jeffreys divergence*, "the divergence." In fact, because of its symmetry, it has been used in many speaker recognition applications [2, 16, 4, 7]. However, it is usually called a symmetric Kullback-Leibler divergence by mistake. In this book, we have tried to address it by its proper name where we have referred to the corresponding publications.

8.2.3 Bhattacharyya Divergence

Although the *Bhattacharyya divergence* [3] is sometimes called a *distance*, it does not satisfy the *triangular inequality*. However, it is *symmetric*, therefore, it is a *divergence*. It is usually used to compare two different distributions.

In July of 1946, Bhattacharyya [3] defined a *divergence* between two *multinomial distributions* by first defining a *statistic* which is a *measure of the dissimilarity* of two sets of samples, in discrete form. The *continuous* form of this measure is known as the *Bhattacharyya measure* [10, 19]. We have written it here in its most general form in terms of the two probability measures, P_0 and P_1.

$$\rho_{Bc}(0, 1) = \int_{\mathscr{X}} \sqrt{d(P_0 P_1)} \tag{8.20}$$

If we assume that the *real line* is the sample space, then Equation 8.20 may be written in terms of the *Lebesgue measure* and the *probability density functions* for the two distributions. It will be the *Bhattacharyya measure* [10],

$$\rho_{Bc}(0,1) = \int\limits_{-\infty}^{\infty} \sqrt{p_0(x)p_1(x)}\,dx \qquad (8.21)$$

$\rho_{Bc}(0,1)$ has also shown up in the work of *Hellinger* while studying integral equations.

It is quite interesting to note that we may rewrite the expression for $\rho_{Bc}(0,1)$ in the following form,

$$\rho_{Bc}(0,1) = \int\limits_{-\infty}^{\infty} p_1(x)\sqrt{\frac{p_0(x)}{p_1(x)}}\,dx \qquad (8.22)$$

$$= \mathcal{E}\left\{\sqrt{\frac{p_0(x)}{p_1(x)}}\right\} \qquad (8.23)$$

which is just the *expected value* of the *square root* of the ratio of the probability density functions. This may be compared to the Kullback-Leibler divergence 8.13 which is the expected value of the negative of the log of the same ratio. In fact this idea gives rise to a generalization called f-*divergence* (Definition 8.6).

Definition 8.5 (Bhattacharyya Divergence). *It is also important to note that* $0 \leq \rho_{Bc} \leq 1$. *Since it has the same range as a probability measure, one possible way of defining the Bhattacharyya Divergence is,*

$$\mathscr{D}_{Bc}(0 \leftrightarrow 1) \stackrel{\Delta}{=} -\log_2 \rho_{Bc}(0,1) \qquad (8.24)$$

Therefore, the defined *Bhattacharyya divergence* may take on any value from 0 to ∞. The *Bhattacharyya divergence* defined in Equation 8.24 does not always obey the *triangular inequality property*, so it is not considered a distance [10]. However, a variation of it called the *Hellinger* distance does obey all the properties of a *metric*. See Definition 8.1.3.

8.2.1 Matsushita Divergence

Matsushita's divergence [14] is really the same as I_1 given by *Jeffreys* [9]. It may be written on the *real line* with respect to the *Lebesgue measure* as follows,

$$\mathscr{D}_{Ms}(0 \leftrightarrow 1) \stackrel{\Delta}{=} \int\limits_{-\infty}^{\infty} \left(\sqrt{p_0(x)} - \sqrt{p_1(x)}\right)^2 dx \qquad (8.25)$$

$\mathcal{D}_{Ms}(0 \leftrightarrow 1)$ does not necessarily obey the *triangular inequality property* of a metric, but it is *symmetric*.

It is interesting to see that this distance could be viewed just as a variation on possible *Bhattacharyya divergences*. Take the statement of Equation 8.25. If we write all the terms of the square, we shall have,

$$\mathcal{D}_{Ms}(0 \leftrightarrow 1) = \int_{-\infty}^{\infty} \left(\sqrt{p_0(x)} - \sqrt{p_1(x)}\right)^2 dx$$

$$= \int_{-\infty}^{\infty} p_0(x) - 2\sqrt{p_0(x)p_1(x)} + p_1(x)dx$$

$$= \int_{-\infty}^{\infty} p_0(x)dx + \int_{-\infty}^{\infty} p_1(x)dx - 2\int_{-\infty}^{\infty} \sqrt{p_0(x)p_1(x)}dx$$

$$= 2\left(1 - \rho_{Bc}(0,1)\right) \tag{8.26}$$

We also see that

$$\mathcal{D}_{Ms}(0 \leftrightarrow 1) = 2\left(1 - \rho_{Bc}(0,1)\right)$$

$$= 2\left(d_{He}(0,1)\right)^2 \tag{8.27}$$

where the k in $d_{He}(0,1)$ (Equation 8.11) is chosen to be 1.

8.2.5 F-Divergence

As we saw in Section 7.3.2 Réyni [17] introduced a general *class* of *entropy* which includes the *Shannon entropy* as one of its limiting cases ($\alpha \to 1$). Using this *generalized entropy*, [17] defined a general class of divergences as well, of which the *Kullback-Leibler divergence* would be a special case, when ($\alpha \to 1$).

Also, Csiszár [6] and [1] introduced a more general class called f-divergence which includes many different divergences. [12] provides a thorough treatment of the f-divergence and speaks about the special cases. Of course these generalizations are good for compact notation, but most of the, so called, special cases included in these generalizations carry different properties which may or may not be desired in the particular application of interest.

Definition 8.6 (f-Divergence). *Recall the similarity between the Bhattacharyya measure, Equation 8.23, and the Kullback-Leibler directed divergence, Equation 8.13. Comparing these two, it is easy to see that we can define a general divergence measure,*

$$\mathscr{D}_f\left(0 \rightarrow 1\right) = \int_{\mathscr{X}} p_1(x) f\left(\frac{p_0(x)}{p_1(x)}\right) d\lambda(x) \tag{8.28}$$

where $\{f : [0,\infty) \mapsto \mathbb{R}\}$ *is a measurable function of choice.* $\mathscr{D}_f\left(0 \rightarrow 1\right)$ *is called the f directed divergence, or simply f-divergence [6, 8]. If the sample space is taken to be the real line, then* $\lambda(x)$ *will just be the Lebesgue measure and* $p_0(x)$ *and* $p_1(x)$ *will be the probability density functions associated with hypotheses* H_0 *and* H_1, *respectively. Equation 8.29 shows this special case which is of most interest,*

$$\mathscr{D}_f\left(0 \rightarrow 1\right) = \int_{-\infty}^{\infty} p_1(x) f\left(\frac{p_0(x)}{p_1(x)}\right) dx \tag{8.29}$$

We have already shown that $\mathscr{D}_{KL}\left(0 \rightarrow 1\right)$ and $\rho_{Bc}(0,1)$ are special cases of the f-divergence. Since we have also shown the relationships between $\mathscr{D}_{Bc}\left(0 \leftrightarrow 1\right)$, $\mathscr{D}_{Ms}\left(0 \leftrightarrow 1\right)$, $d_{He}\left(0,1\right)$ and $\rho_{Bc}(0,1)$, relations of these divergences to the f-divergence are also shown.

8.2.6 δ-Divergence

Much in the same way as *Jeffreys' divergence* is related to the *Kullback-Leibler directed divergence*, we can think of generalizing *Jeffreys' divergence* much in the same way as we did in the definition of the *f-divergence* from the *Kullback-Leibler directed divergence*.

Definition 8.7 (δ-Divergence). *For a measurable function* $\{g : [0,\infty) \mapsto \mathbb{R}\}$, *a general divergence called the δ-divergence may be defined as follows,*

$$\mathscr{D}_{\delta_g}\left(0 \leftrightarrow 1\right) = \int_{\mathscr{X}} [p_0(x) - p_1(x)] g\left(\frac{p_0(x)}{p_1(x)}\right) d\lambda(x) \tag{8.30}$$

If the sample space is the real line, then a special case of the δ-divergence may be written in terms of the Lebesgue measure and the probability density functions of H_0 *and* H_1,

$$\mathscr{D}_{\delta_g}\left(0 \leftrightarrow 1\right) = \int_{-\infty}^{\infty} [p_0(x) - p_1(x)] g\left(\frac{p_0(x)}{p_1(x)}\right) dx \tag{8.31}$$

8.2.7 χ^α Directed Divergence

A special subclass of the *f-divergence* is the χ^α *directed divergence* or simply χ^α-*divergence* [8]. It is defined such that,

$$f(u) = \left| u - 1 \right|^\alpha \tag{8.32}$$

where, $\{ f : [0, \infty) \mapsto \mathbb{R} \}$ and $1 \le \alpha < \infty$. Therefore,

$$\mathscr{D}_{\chi^\alpha} (0 \to 1) = \int_{\mathscr{X}} p_1(x) \left| \frac{p_0(x)}{p_1(x)} - 1 \right|^\alpha d\lambda(x) \tag{8.33}$$

$$= \int_{\mathscr{X}} (p_1(x))^{(1-\alpha)} |p_0(x) - p_1(x)|^\alpha d\lambda(x) \tag{8.34}$$

Setting $\alpha = 2$, $\mathscr{D}_{\chi^\alpha} (0 \to 1)$ produces a widely used divergence called the *Pearson divergence* or *Pearson's statistic* which is used widely for testing *statistical independence*.

References

1. Ali, S., Silvey, S.: A General Class of Coefficients of Divergence of One Distribution from Another. Journal of the Royal Statistical Society **28**(1), 131–142 (1966)
2. Beigi, H.S., Maes, S.S.: A Distance Measure Between Collections of Distributions and its Application to Speaker Recognition. In: International Conference on Acoustics, Speech, and Signal Processing (ICASSP98) (1998)
3. Bhattacharyya, A.: On a Measure of Divergence between Two Multinomial Populations. Sankhya: The Indian Journal of Statistics **7**(4), 401–406 (1946)
4. Campbell, W., Sturim, D., Reynolds, D.a.: Support vector machines using GMM supervectors for speaker verification. Signal Processing Letters, IEEE **13**(5), 308–311 (2006)
5. Cortes, C., Mohri, M., Rastogi, A.: Lp Distance and Equivalence of Probabilistic Automata. International Journal of Foundations of Computer Science (IJFCS) **18**(4), 761–779 (2007)
6. Csiszár, I.: Eine Informationstheoretische Ungleichung und ihre Anwendung auf den Beweis der Ergodizität on Markofschen Ketten. Publication of the Mathemtical Institute of Hungarian Academy of Science **A:8**, 84–108 (1963)
7. Dehak, N., Kenny, P., Dehak, R., Glembek, O., Dumouchel, P., Burget, L., Hubeika, V., Castaldo, F.: Support vector machines and Joint Factor Analysis for speaker verification. In: International Conference on Acoustics, Speech, and Signal Processing (ICASSP), pp. 4237–4240 (2009)
8. Dragomir, S.S.: Some General Divergence Measures for Probability Distributions. Acta Mathematica Hungarica **109**(4), 331–345 (2005)
9. Jeffreys, H.: An Invariant Form for the Prior Probability in Estimation Problems. Proceedings of the Royal Society of London **186**(1007), 453–461 (1946)
10. Kailath, T.: The Divergence and Bhattacharyya Distance Measure in Signal Selection. IEEE Transactions on Communication Technology **15**(1), 52–60 (1967)
11. Kullback, S., Leibler, R.A.: On Information and Sufficiency. The Annals of Mathematical Statistics **22**(1), 79–86 (1951)
12. Liese, F., Vajda, I.: On Divergences and Informations in Statistics and Information Theory. IEEE Transactions on Information Theory **52**(10), 4394–4412 (2006)

13. Mahalanobis, P.C.: On the Generalized Distance in Statistics. Proceedings of the National Institute of Sciences of India **12**, 49–55 (1936)
14. Matsushita, K.: Decision Rule, Based on Distance, for the Classification Problem. Annals of the Institute of Statistical Mathematics **8**, 67–77 (1956)
15. Michiel Hazewinkel, E.: Encyclopaedia of Mathematics. Kluwer Academic Publishers, Amsterdam (2002). ISBN: 1-4020-0609-8
16. Moreno, P.J., Ho, P.P., Vasconcelos, N.: A Kullback-Leibler Divergence Based Kernel for SVM Classification in Multimedia Applications. In: Advances in Neural Information Processing Systems 16. MIT Press (2004)
17. Réyni, A.: On Measure of Entropy and Information. In: Proceedings of the 4th Berkeley Symposium on Probability Theory and Mathematical Statistics, pp. 547–561 (1961)
18. Sinanović, S., Johnson, D.H.: Toward a Theory of information processing. Signal Processing **87**(6), 1326–1344 (2007)
19. Toussaint, G.T.: Comments on "The Divergence and Bhattacharyya Distance Measure in Signal Selection". IEEE Transactions on Communications **20**(3), 485 (1972)
20. Welsh, D.: Codes and Cryptography. Oxford University Press, New York (1990). ISBN: 0-198-53287-3

Chapter 9
Decision Theory

The door must either be shut or be open.

Oliver Goldsmith
The Citizens of the World, 1760

Decision theory is one of the most basic underlying theories which is crucial for the creation, understanding and implementation of a successful speaker recognition algorithm. To begin covering this topic, we need to understand the process of formalizing a hypothesis and testing it. Then, we will continue to talk about Bayesian decision theory. We also talk about hypotheses in the development of information theoretic concepts of Chapter 7.

9.1 Hypothesis Testing

Let us begin with the process of making a binary decision (one for which the outcome will be either true or false). *Speaker verification* is just that kind of decision. In *speaker verification*, the question is if the *test speaker* is the same as the *target speaker* or not. In Section 1.2.1 we briefly described the meaning of a *target speaker* and a *test speaker*. Shortly, we will present a formal definition for both.

In a *binary decision*, there are two *hypotheses*, a main hypothesis to which there may be a true or false answer and an *alternative hypothesis* which is the opposite of the main hypothesis. The main hypothesis (the first supposition) is called the *null hypothesis*. The logic of *hypothesis testing* was first formalized by Fisher in 1935.[5]

Definition 9.1 (Null Hypothesis). *A null hypothesis, H_0, is a hypothesis (supposition) which is tested for correctness. If the test result negates the null hypothesis, then it will be in favor of the alternative hypothesis. Null hypothesis is the hypothesis which generally agrees with the normal trend of nature.*

Definition 9.2 (Alternative Hypothesis). *The alternative hypothesis, H_1, is the hypothesis which is complementary to the null hypothesis. If the null hypothesis is* *rejected, the alternative hypothesis is validated and vice versa. The alternative hypothesis signifies an un-natural state and its truth generally signals an alarm.*

In order to understand the logic of hypothesis testing, we would need to use an example. The most appropriate example in the context of this book is *speaker verification*. So, before we continue in pursuit of formalizing the logic of *binary hypothesis testing*, we should define some key concepts in *speaker verification*.

Definition 9.3 (Target Speaker (Reference Model)). *When a speaker is enrolled into the speaker recognition database, the audio of that speaker is presented to the system with a unique identifier (ID) associated with that speaker. This is the true identifier of the enrolled speaker. At the time of testing of a speaker verification hypothesis, a segment of audio is presented to the system in conjunction with a claimed identity. The speaker in the database who has been previously enrolled using this identity is called the target speaker or the reference model.*

In case of an *identification* task, there are many target speakers (usually as many as there are speakers in the database). The identification task hypothesizes the likelihood of the *test speaker* being any one of these *target speakers* and it is no longer a binary hypothesis. For the *verification* task, there is only one *target speaker*, making the decision *binary*.

Definition 9.4 (Test Speaker). *In a verification hypothesis, a segment of audio is presented to the recognition engine with a claimed identity. This claimed identity may be correct or otherwise misrepresented. However, there is a true identity associated with the audio being used for the test hypothesis. That true identity is the identity of the test speaker. Namely, it is the identity of the true owner of the test audio being presented for verification.*

For the *identification* task, the *test speaker* is defined in the same manner as done for *verification*. Namely, it is the *true identity* of the speaker of the *test audio*.

Definition 9.5 (Impostor). *In a verification task, an Impostor is the test speaker when the test speaker is different from the target speaker.*

There is some degree of freedom in testing a binary hypothesis, depending on which aspect of the main question is deemed the *null hypothesis* and which the *alternative*. Fisher [5] picked the *null hypothesis* to be that hypothesis which is due to chance and not due to any systematic cause. In *speaker verification*, we normally pose the main question as, "does the presented audio come from the target speaker?" Another way to pose this question is, "whether the *test speaker* is the same as the *target speaker*?" Therefore, we pick this question to be the "*null hypothesis*." Then, the *alternative hypothesis* becomes, "is the *test speaker* different from the *target speaker?*" or "Is the *test speaker* a *non-target speaker?*" Answering yes to the *alternative hypothesis* (no to the *null hypothesis*) is equivalent to signaling an *alarm* which states that something has not gone right with the natural state of the system (the assumption that the audio being presented at the time of testing was produced by the *target speaker*).

Figure 9.1 summarizes the logic of the *hypothesis testing* used in *speaker verification*. There are two possible *ground truths* for each test; either the *test speaker* is the same as the *target speaker* or not. These two ground truths are represented by two boxes in the flow diagram of the *test logic*.

The *null hypothesis* may have two different *ground truths* or *natures*. The first possible ground truth is that the *test speaker* is the same person as the *target speaker*, (Test = Target). The second possibility is that they are different (Test $\neq$ Target). For the case where the truth is (Test = Target), if the verification system returns a true result (*Verified*), then it is in tune with the ground truth, therefore, the result is correct. If, however, it returns a false result (Not Verified), then its result is different from the ground truth, which means that it has made an *error*. This *type of error* is known by many different names, depending on the field of use. Traditionally, it is called a *Type I error* or an α *error.*

On the other hand, if the ground truth is such that the *test speaker* is not the same person as the *target speaker*, then the *alternative hypothesis* (H_1) should be true. However, if the speaker is *verified*, it means that the *null hypothesis* was deemed true, which is an *error*. This is a *Type II error* or a β *error.*

Figure 9.1 shows all the alternate names for these two types of errors. There is great confusion in this terminology and it is important to try to understand the logic behind the terminology before using it. The figure also shows the two popular names for the two types of error in the speaker verification community, *false acceptance* and *false rejection*. Let us give a formal definition for these two errors.

Definition 9.6 (False Acceptance (Miss)).
The false acceptance rate (FAR) is the number of verified identities for which the test speaker was different from the target speaker normalized against the total number of acceptances. This ratio is also known as the miss probability.

Definition 9.7 (False Rejection (False Alarm)). *The false rejection rate (FRR) is the number of identities which were not verified for which the test speaker was the same as the target speaker normalized against the total number of rejections. This ratio is also known as the false alarm probability.*

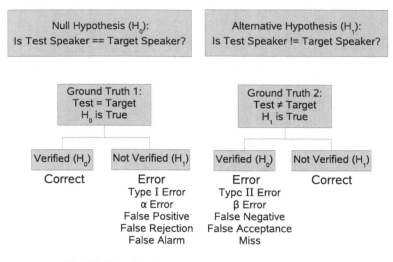

Fig. 9.1: Hypothesis Testing Logic for Speaker Verification

9.2 Bayesian Decision Theory

In the previous section, we discussed the *hypothesis testing* process for a *binary decision*. In mathematical terms, we denote the *null hypothesis* by H_0 and the *alternative hypothesis* by H_1. Since we are speaking of a binary decision and there are only two possible hypotheses, then based on the *Bayesian decision theory*, we may assign an *a-priori* probability [4] to the two possible states, for each of which one of the hypotheses would be correct. The *a-priori* probabilities of the two states of truth associated with the two hypotheses are denoted by $P(H_0)$ and $P(H_1)$. Note that since there are only two possible states (binary decision), then

$$P(H_0) + P(H_1) = 1 \tag{9.1}$$

$P(H_0)$ and $P(H_1)$ are the values of a *Bernoulli distribution* related to the nature of the two hypotheses and are not related to any specific sample. However, if we consider representing *reality* or *nature* of an experiment in the form of a random variable, X, then there will exist two *conditional probability densities*, $p(x|H_0)$ and $p(x|H_1)$. These densities represent the probability density function of sample x given that the state of nature is such that either H_0 or H_1 is true respectively.

Another way of viewing these conditional densities is as *likelihoods*. Namely, $p(x|H_i), i \in \{0,1\}$ would be viewed as the *likelihood* of hypothesis H_i being true given that random variable X has taken on the value, $X = x$. Therefore, using the *total probability theorem*, Theorem 6.1, we may write the *total probability density function* of X as,

$$p(x) = \sum_{j=0}^{1} p(x|H_j)P(H_j) \tag{9.2}$$

As we saw in the development of *Bayes theorem*, Theorem 6.2, for each hypothesis, we may write the *a-posteriori* probability of that hypothesis, *almost everywhere*, with respect to sample, $X = x$, in terms of the *likelihood* and the *a-priori* probability of that hypothesis, normalized with respect to the *total probability* of the sample, $p(x)$, where $p(x) \neq 0$,

$$P(H_i|x) = \frac{p(x|H_i)P(H_i)}{p(x)} \quad [x] \quad \forall \, i \in \{0,1\} \tag{9.3}$$

Equation 9.3 describes the change in probability from the time before sample $X = x$ is observed to after its observation. Namely, it is the relationship between the *a-posteriori* probability of hypothesis H_i and its *a-priori* probability, where the reference is the observation of sample $X = x$. As we saw in Section 7.6, observing $X = x$ results in a gain in *information* about the state of nature – See Equation 7.78.

Note that for every value, $X = x$, the *a-posteriori* probabilities of all the possible states would have to sum up to 1,

$$\sum_{j=0}^{1} P(H_j|x) = 1 \tag{9.4}$$

Although we have been talking about the binary decision process, all the results of this section are applicable to any Γ number of states related to Γ hypotheses. These hypotheses may be viewed as *functions* of the *subsets* of the *Borel field*, $\mathfrak{X}$, of the *sample space*, $\mathscr{X}$, of the random variable X. Although we are free to choose these hypotheses any way we wish, it makes sense to define them such that they do not have any overlap and such that they cover all possible states of the system. In fact, let us define a *discrete random variable*, $\{S : s \in \mathscr{S} = \{H_i, i \in \{1, 2, \cdots, \Gamma\}\}\}$, which is indicative of the true *state* of nature in the problem of interest such that,

$$\sum_{j=1}^{\Gamma} P(S = H_j) = 1 \tag{9.5}$$

For testing with Γ hypotheses,

$$P(S = H_i|x) = \frac{p(x|S = H_i)P(S = H_i)}{p(x)} \quad [x] \quad \forall \, i \in \{1, 2, \cdots, \Gamma\} \tag{9.6}$$

where $p(x) \neq 0$ and

$$p(x) = \sum_{j=1}^{\Gamma} p(x|S = H_j)P(S = H_j) \tag{9.7}$$

subject to the constraint,

$$\sum_{j=1}^{\Gamma} P(S = H_j | x) = 1 \qquad (9.8)$$

For the identification problem, we would like to associate each hypothesis with the likelihood of the target speaker being the person who has generated the test audio sample. Here, for convenience, we have started the hypothesis index from 1. It is customary to start with 1 when we speak of more than 2 hypotheses and to use 0 and 1 for binary hypotheses.

For every observation, $X = x$, using Equation 9.6, we can compute the *a-posteriori* probability for the Γ hypotheses. It is intuitively apparent that the best guess for the correct state of nature is the one which matches with the most likely hypothesis. Therefore, for a given observation, $X = x$, the state associated with the most likely hypothesis is $S = H_i$ such that $i = \{i : P(S = H_i|x) \geq P(S = H_j|x) \, \forall \, j \in \{1, 2, \cdots, \Gamma\}\}$. This is called the *maximum a-posteriori solution* and it is the *minimum error solution* given by *Bayesian decision theory*.

Let us examine this intuitive decision in more detail. Refer to Figure 9.1. The figure shows the logic behind a binary hypothesis testing scenario. In this case, there are two hypotheses which may constitute the two different possible ground truths (states). In every one of these cases, the decision of picking the hypothesis associated with the ground truth is the correct decision and picking the hypothesis would constitute an error. This logic may be extended to any Γ set of hypotheses.

We denote the discrete random variable associated with the decision to pick a state from the set of all states, $\mathscr{S}$, by $\{O : o \in \mathscr{S}\}$. Therefore, O and S are different random variables, defined in a common sample space, $\mathscr{S}$. In order to assess the quality of our choices for any decision method, we define a *measure* (see Definition 6.18) of the *penalty* (*loss*), conditioned on the state $S = s$, $\varpi(o|s)$, such that it is equal to 0 for *correct choices* and 1 for *errors* [4],

$$\varpi(O = H_i | S = H_j) = \begin{cases} 1 \; \forall \, i \neq j \\ 0 \; \forall \, i = j \end{cases} \qquad (9.9)$$

Then, using Equation 6.79, according to the *Radon-Nikodým theorem* (Theorem 6.3), we may define a *measure* of the *risk* involved in making decision $\{O = o\}$, as,

$$R(o) = \int_{\mathscr{X}} \varpi(o|x) dP(x) \qquad (9.10)$$

where R(o) is basically considered to be a measure which is written in terms of the probability measure $P(x)$ and the *Radon-Nikodým derivative*,

$$\varpi(o|x) = \frac{dR}{dP}[P] \tag{9.11}$$

Therefore, as seen in Definition 6.44, if the probability density $p(x)$ exists, then Equation 9.10 may be written as,

$$R(o) = \int_{\mathscr{X}} \varpi(o|x)p(x)dx \tag{9.12}$$

For a finite number of hypotheses, Γ, given an observation, $X = x$, we may use the *total probability theorem* (Theorem 6.1), to write the *conditional risk* [4], $\varpi(o|x)$, as follows,

$$\varpi(o|x) = \sum_{i=1}^{\Gamma} \varpi(o|S = H_i)P(S = H_i|x) \tag{9.13}$$

$R(o)$ may be viewed as the *expected value* of the *penalty* associated with the decision, $O = o$, or the *risk* associated with decision $o(x)$ – See Equation 6.118.

The objective of *Bayesian decision theory* is to find $O = o_B$ such that

$$o_B = \arg\min R(o) \tag{9.14}$$

$R(o_B)$ is known as *Bayes risk* [4].

Since $p(x)$ is independent of decision $o(x)$, $R(o)$ will be minimized if,

$$\varpi(o_B|x) = \min_i \varpi(O = H_i|x) \tag{9.15}$$

where $i \in \{1, 2, \cdots, \Gamma\}$.

Given the *penalty* (*loss*) function as defined by Equation 9.9, the conditional risk for hypothesis H_i may then be written as,

$$\varpi(O = H_i|x) = \sum_{j=1}^{\Gamma} \varpi(O = H_i|S = H_j)P(S = H_j|x) \tag{9.16}$$

$$= 1 - P(S = H_i|x) \tag{9.17}$$

This means that for maximizing $P(S = H_i|x)$, we would have to minimize $\varpi(O = H_i|x)$. This is the *maximum a-posteriori* (*minimum error*) solution that was stated earlier.

Equation 9.16 may be written in matrix form as follows,

$$\varpi(o|x) = \boldsymbol{\Pi} \; \mathbf{p}(s|x) \tag{9.18}$$

where the $\boldsymbol{\Pi} : \mathscr{R}^{\Gamma} \mapsto \mathscr{R}^{\Gamma}$ is the *penalty matrix* such that

$$(\mathbf{\Pi})_{[i][j]} \stackrel{\Delta}{=} \varpi(O = H_i | S = H_j) \tag{9.19}$$

and the vector $\mathbf{p}(s|x) : \mathscr{R}^1 \mapsto \mathscr{R}^\Gamma$ is the *a-posteriori* probability vector of the possible states of nature,

$$(\mathbf{p}(s|x))_{[j]} \stackrel{\Delta}{=} P(S = H_j | x) \tag{9.20}$$

It is conceivable for more than one hypothesis to have a *maximum a-posteriori* probability among all possible hypotheses. In that case, any one of those hypotheses with the maximum posterior probability may be chosen. This may be done either randomly or through adding other criteria to the decision process.

9.2.1 Binary Hypothesis

If we consider the *binary hypothesis* problem ($\Gamma = 2$), the conditional risk, $\varpi(o|x)$ for the two possible hypotheses, H_0 and H_1, may be written in terms of the penalty, $\varpi(o|s)$, and the *a-posteriori* probabilities, $P(s|x)$. Let us use the matrix element representation of $\varpi(o|x)$ from Equation 9.19 for brevity,

$$\varpi(O = H_0 | x) = (\mathbf{\Pi})_{[0][0]} P(S = H_0 | x) + (\mathbf{\Pi})_{[0][1]} P(S = H_1 | x) \tag{9.21}$$

$$\varpi(O = H_1 | x) = (\mathbf{\Pi})_{[1][0]} P(S = H_0 | x) + (\mathbf{\Pi})_{[1][1]} P(S = H_1 | x) \tag{9.22}$$

In order to make a decision in favor of the *null hypothesis*, H_0, we saw that based on Equation 9.17, $\varpi(o = H_0 | x)$ had to be less than $\varpi(O = H_1 | x)$. This means that the right hand side of Equation 9.21 would have to be smaller than the right hand side of Equation 9.22,

$$(\mathbf{\Pi})_{[0][0]} P(S = H_0 | x) + (\mathbf{\Pi})_{[0][1]} P(S = H_1 | x) <$$
$$(\mathbf{\Pi})_{[1][0]} P(S = H_0 | x) + (\mathbf{\Pi})_{[1][1]} P(S = H_1 | x) \tag{9.23}$$

Rearranging the terms in Equation 9.23 such that factors of $P(S = H_0 | x)$ are on one side and factors of $P(S = H_1 | x)$ are on the other side of the inequality, we have,

$$((\mathbf{\Pi})_{[1][0]} - (\mathbf{\Pi})_{[0][0]}) P(S = H_0 | x) >$$
$$((\mathbf{\Pi})_{[0][1]} - (\mathbf{\Pi})_{[1][1]}) P(S = H_1 | x) \tag{9.24}$$

Using *Bayes theorem* (Theorem 6.2), we may write $P(S = H_i | x) \ \forall \ i \in \{0, 1\}$ in terms of the likelihoods, $p(x|S = H_i)$, and the *a-priori* probabilities, $P(S = H_i)$, such that the inequality in Equation 9.24 would become,

$$((\mathbf{\Pi})_{[1][0]} - (\mathbf{\Pi})_{[0][0]}) P(x|S = H_0) P(S = H_0) >$$
$$((\mathbf{\Pi})_{[0][1]} - (\mathbf{\Pi})_{[1][1]}) P(x|S = H_1) P(S = H_1) \tag{9.25}$$

Divide both sides of Equation 9.25 by $((\boldsymbol{\Pi})_{[1][0]} - (\boldsymbol{\Pi})_{[0][0]})p(x|S = H_1)P(S = H_0)$,

$$\frac{p(x|S = H_0)}{p(x|S = H_1)} > \frac{(\boldsymbol{\Pi})_{[0][1]} - (\boldsymbol{\Pi})_{[1][1]}}{(\boldsymbol{\Pi})_{[1][0]} - (\boldsymbol{\Pi})_{[0][0]}} \frac{P(S = H_1)}{P(S = H_0)}[x] \tag{9.26}$$

Equation 9.26 should hold, *almost everywhere*, and presents a *lower bound* on the *likelihood ratio* for making the decision, $O = H_0$. The right hand side of Equation 9.26 does not depend on x. It is a constant which is only dependent on the associated penalties and the *a-priori* probability ratio of the two hypotheses H_0 and H_1.

Let us define the threshold,

$$\theta_{H_0} \overset{\Delta}{=} \log_b\left(\frac{(\boldsymbol{\Pi})_{[0][1]} - (\boldsymbol{\Pi})_{[1][1]}}{(\boldsymbol{\Pi})_{[1][0]} - (\boldsymbol{\Pi})_{[0][0]}} \frac{P(S = H_1)}{P(S = H_0)}\right) \tag{9.27}$$

as the *log-likelihood ratio threshold* for deciding in favor of the *null hypothesis*, H_0, against the *alternative hypothesis*, H_1. Deciding in favor of H_0, if

$$\log_b \frac{p(x|S = H_0)}{p(x|S = H_1)} > \theta_{H_0}[x] \tag{9.28}$$

and choosing H_1 otherwise, will give the *minimum error solution* or the, so called, *maximum a-posteriori solution* for a *binary hypothesis*. This result is also known as the *Neyman-Pearson lemma* [3]. As in before, any base may be picked for the computation of the logarithm – see Section 7.3.1.

Note that if we use the penalty matrix defined in Equation 9.9, then Equation 9.26 would be simplified as,

$$\frac{p(x|S = H_0)}{p(x|S = H_1)} > \frac{P(S = H_1)}{P(S = H_0)}[x] \tag{9.29}$$

or in other words,

$$\theta_{H_0} = \frac{P(S = H_1)}{P(S = H_0)} \tag{9.30}$$

9.2.2 Relative Information and Log Likelihood Ratio

Referring to Section 7.6, recall the expression for the *relative information* gained by observing $X = x$ in favor of hypothesis H_0 against H_1. As we saw, this *relative information* was given by Equation 7.78 as the difference between the logarithms of the *generalized probability densities*, $g_0(x)$ and $g_1(x)$ associated with H_0 and H_1 respectively. We also saw that if X is defined on the *real line* ($\mathcal{X} = \mathbb{R}$), the relative information, $\mathcal{I}(0:1,x)$, would just be the log of the ratio of the *probability density functions* of X given hypotheses H_0 and H_1 evaluated at $X = x$ ($p_0(x)$ and $p_1(x)$).

In Section 7.6, we defined H_0 and H_1 to be the *a-posteriori* and *a-priori* states of the system. Without any loss of generality, we can redefine H_0 and H_1 to stand for any *null* and *alternative hypotheses*, for example, as defined in Section 9.1.

The *probability density functions* $p_0(x)$ and $p_1(x)$ are density functions conditioned upon hypotheses H_0 and H_1 respectively. Therefore, we may think of them as *likelihoods* of being in the states associated with hypotheses H_0 and H_1, given the observation of $X = x$. In other words, we may write them as $p(x|S = H_0)$ and $p(x|S = H_1)$. This makes the *relative information* in favor of state $S = H_0$ against $S = H_1$, the logarithm of the ratio of their *likelihoods*, or the so called, *log-likelihood ratio* (LLR) of H_0 against H_1,

$$\mathcal{I}(S = H_0 : S = H_1, x) = \log_b \frac{p(x|S = H_0)}{p(x|S = H_1)} \tag{9.31}$$

Based on Equation 7.78, Equation 9.31 may be written as follows,

$$\begin{aligned} \mathcal{I}(S = H_0 : S = H_1, x) &= \log_b \frac{p(x|S = H_0)}{p(x|S = H_1)} \\ &= \log_b P(x|S = H_0) - \log_b P(x|S = H_1) \tag{9.32} \\ &= \log_b \frac{P(S = H_0|x)}{P(S = H_1|x)} - \log_b \frac{P(H_0)}{P(H_1)} \ [x] \tag{9.33} \end{aligned}$$

Equations 9.31- 9.33 give physical interpretations for the argument in Section 9.2.1.

9.3 Bayesian Classifier

As we saw in Section 9.2, the *maximum a-posteriori* solution is also the *minimum error* solution of *Bayesian decision theory* for the hypothesis testing problem. The problem of *classification* may also be viewed as a *hypothesis testing problem*, where, in the case of a Γ-class classification problem, there are Γ hypotheses associated with the Γ classes, each asking whether an observation belongs to its associated class.

To choose class i, associated with the Bayesian decision, the *a-posteriori* probability of the hypothesis associated with that class, H_i, would have to be greater than that of every other class, H_j, where $j \neq i$,

$$o_B = H_i$$
$$= \arg\max_j P(O = H_j|x) \tag{9.34}$$

where $j \in \{1, 2, \cdots, \Gamma\}$.

Based on Equation 9.17, the *minimum error* equivalent of Equation 9.34 is,

$$o_B = H_i$$
$$= \arg\min_j \varpi(O = H_j|x) \tag{9.35}$$

In order to change the minimization problem in Equation 9.35 to one of maximization, we can write Equation 9.35 as,

$$o_B = H_i$$
$$= \arg\max_j (-\varpi(O = H_j|x)) \tag{9.36}$$

The resulting classification would be the same regardless of whether Equation 9.34, 9.35, or 9.36 is used.

Therefore, $g_j(x) = P(O = H_j|x)$ or $g_j(x) = -\varpi(O = H_j|x)$ are known as *discriminant functions* since in both cases,

$$o_B = H_i$$
$$= \arg\max_j g_j(x) \tag{9.37}$$

The $\arg\max$ function, used in Equation 9.37, will give the same result with many different choices of $g_j(x), j \in \{1, 2, \cdots, \Gamma\}$. If we expand the *a-posteriori* probability according to Equation 9.6, since the denominator, $p(x)$, is the same across the board for all hypothesis, then it may be eliminated and an incarnation of the *discriminant function*, $g_j(x)$, may be defined in terms of the *likelihood* of the hypothesis associated with the class of interest and its *a-priori* probability.

$$g_j(x) \overset{\Delta}{=} p(x|O = H_j)P(O = H_j) \tag{9.38}$$

In fact, we can also define $g_j(x)$ in logarithmic form,

$$g_j(x) \overset{\Delta}{=} \log_b p(x|O = H_j) + \log_b P(O = H_j) \tag{9.39}$$

If we use any of the noted *discriminant functions*, $g_j(x)$, such as the one given by Equation 9.39, the Γ-class classification problem simply becomes a maximization problem, choosing the class which maximizes the discriminant function, $g_j(x)$.

In the special case of a *two-category* classification problem, since we would always try to decide whether $g_0(x) > g_1(x)$, we may subtract $g_1(x)$ from both sides and ask whether,

$$g_0(x) - g_1(x) > 0 \tag{9.40}$$

If Equation 9.40 is true, then we chose in favor of the *null hypothesis* and otherwise choose the *alternative*. Therefore, we may define a new *discriminant function* for the *binary* case, as a test of the binary hypothesis,

$$g(x) > 0 \tag{9.41}$$

In the *two-category* case, if we use Equation 9.39 for the discriminant function, then the *null hypothesis* combined discriminant function would be given by,

$$
\begin{aligned}
g(x) &= \log_b p(x|O = H_0) + \log_b P(O = H_0) \\
&\quad - \log_b p(x|O = H_1) - \log_b P(O = H_1) \\
&= \log_b \frac{p(x|O = H_0)}{p(x|O = H_1)} + \log_b \frac{P(O = H_0)}{P(O = H_1)}
\end{aligned}
\tag{9.42}
$$

Equation 9.42 Using Equation 9.42 in the null hypothesis test of Equation 9.41, we get the following test,

$$\log_b \frac{p(x|O = H_0)}{p(x|O = H_1)} > \log_b \frac{P(O = H_1)}{P(O = H_0)} \tag{9.43}$$

which is the same as the criterion of Equation 9.29.

Consider a Γ-class classification problem where the likelihoods of the different class hypotheses are distributed normally. Then, the likelihood for each class hypothesis would be defined by its mean and standard deviation,

$$p(x|O = H_i) = \frac{1}{\sqrt{2\pi}\sigma_i} \exp\left(-\frac{1}{2}\frac{(x - \mu_i)^2}{\sigma_i^2}\right) \tag{9.44}$$

Then, one set of discriminant functions, $g_i(x)$, may be written as,

$$g_i(x) = \ln p(x|O = H_i) + \ln P(O = H_i) \tag{9.45}$$

Plugging in the value of the likelihood from Equation 9.44 into Equation 9.45, the expression for the discriminant function with normal density likelihood would become,

$$g_i(x) = \ln P(O = H_i) - \ln\left(\sqrt{2\pi}\sigma_i\right) - \frac{1}{2}\frac{(x - \mu_i)^2}{\sigma_i^2} \tag{9.46}$$

Let us consider an example where there are 3 classes and their likelihoods are normally distributed. Furthermore, let us assume that these classes have equal *a-priori* probabilities ($P(O = H_i) = \frac{1}{3}$). We saw that the discriminant function may have different forms. To get a more intuitive graphic representation, assume the following form for the discriminant functions,

$$g_i(x) = p(x|O = H_i)P(O = H_i)$$

$$= \frac{1}{3} \frac{1}{\sqrt{2\pi}\sigma_i} \exp\left(-\frac{1}{2}\frac{(x-\mu_i)^2}{\sigma_i^2}\right) \tag{9.47}$$

Figure 9.2 shows a plot of these discriminant functions versus the samples, x, for

$$g_1(x) = \frac{1}{3}\mathcal{N}(-80, 30)$$

$$g_2(x) = \frac{1}{3}\mathcal{N}(0, 20) \tag{9.48}$$

$$g_3(x) = \frac{1}{3}\mathcal{N}(40, 25)$$

The figure shows classification results based on the discriminant functions of Equation 9.47. The classification results would of course be identical if we defined the following discriminant functions,

$$\tilde{g}_i(x) = \ln g_i(x) \tag{9.49}$$

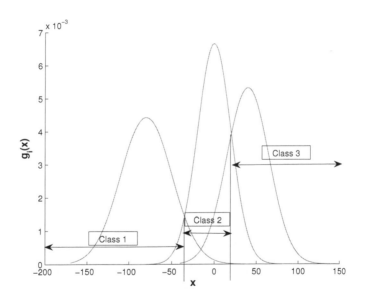

Fig. 9.2: Boundaries of three discriminant functions based on normal density likelihoods and equal a-priori probability classes

In the next chapter we will have a detailed look into methods of estimating the statistics for classes based on clustering algorithms which look for congregations of samples and try to model them into clusters with given distributions. In speaker recognition, for simplicity, mostly *normal distributions* are considered.

9.3.1 Multi-Dimensional Normal Classification

In speaker recognition, as well as many other pattern recognition fields, the *sample space* of choice is the *multi-dimensional Euclidean space*, containing representative features of the physical phenomenon of interest. In speaker recognition, we are generally interested in using feature vectors described in Chapter 5. Generally, we are interested in the random variable of features, X, defined in the *probability measure space*, $(\mathscr{X}, \mathfrak{X}, P)$, such that

$$\mathscr{X} = \bigtimes_{i=1}^{D} \mathscr{X}_i \tag{9.50}$$

and

$$\mathfrak{X} = \bigtimes_{i=1}^{D} \mathfrak{X}_i \tag{9.51}$$

Therefore, any sample of X would be $\{\mathbf{x} : \mathscr{R}^1 \mapsto \mathscr{R}^D\}$ where D is the dimension of the feature vector $\mathbf{x}$.

Referring to Section 6.10, assuming that the likelihoods of classes are *normally distributed*, for the general Γ-class classification problem with D-dimensional samples, the *discriminant functions*, $g_i(\mathbf{x})$, may be written as follows,

$$g_i(\mathbf{x}) = \ln p(\mathbf{x}|O = H_i) + \ln P(O = H_i) \tag{9.52}$$

The *natural logarithm* of the *likelihood* of the class i, associated with hypothesis H_i is given by the natural logarithm of the expression of the *multivariate normal density function* given by Equation 6.196. Namely,

$$P(\mathbf{x}|O = H_i) = \underbrace{-\ln\left((2\pi)^{\frac{D}{2}} |\boldsymbol{\Sigma}_i|^{\frac{1}{2}}\right)}_{\text{bias term}} \underbrace{-\frac{1}{2}(\mathbf{x} - \boldsymbol{\mu}_i)^T \boldsymbol{\Sigma}_i^{-1}(\mathbf{x} - \boldsymbol{\mu}_i)}_{d_M(\mathbf{x}, \boldsymbol{\mu}_i)} \tag{9.53}$$

The first term in Equation 9.53 is called the *bias term* and the second term is just the *Mahalanobis distance* between the sample, $\mathbf{x}$, and the mean of class i ($d_M(\mathbf{x}, \boldsymbol{\mu}_i)$), given by Definition 8.4, where the normalization constant of Equation 8.9 is chosen to be $k = \frac{1}{2}$.

The *discriminant functions* are then

$$g_i(\mathbf{x}) = -\frac{1}{2}(\mathbf{x} - \boldsymbol{\mu}_i)^T \boldsymbol{\Sigma}_i^{-1}(\mathbf{x} - \boldsymbol{\mu}_i) - \frac{D}{2}\ln(2\pi) - \frac{1}{2}|\boldsymbol{\Sigma}_i| + \ln P(O = H_i) \tag{9.54}$$

Since $\frac{D}{2}\ln(2\pi)$ is common to all $g_i(\mathbf{x})$, it does not contribute to changing the discrimination capabilities of $g_i(\mathbf{x})$. Therefore, it may be dropped, resulting in the following *discriminant functions*,

$$g_i(\mathbf{x}) = -\frac{1}{2}(\mathbf{x} - \boldsymbol{\mu}_i)^T \boldsymbol{\Sigma}_i^{-1} (\mathbf{x} - \boldsymbol{\mu}_i) - \frac{1}{2}\ln\left(|\boldsymbol{\Sigma}_i|\right) + \ln P(O = H_i) \qquad (9.55)$$

Equation 9.55 is quadratic in $\mathbf{x}$,

$$g_i(\mathbf{x}) = \frac{1}{2}\mathbf{x}^T \mathbf{A}_i \mathbf{x} + \mathbf{b}_i^T \mathbf{x} + c_i \qquad (9.56)$$

where,

$$
\begin{aligned}
\mathbf{A}_i &= -\boldsymbol{\Sigma}_i^{-1} && \longleftarrow \text{Quadratic Term} \\
\mathbf{b}_i &= \boldsymbol{\Sigma}_i^{-1}\boldsymbol{\mu}_i && \longleftarrow \text{Linear Term} \qquad (9.57)\\
c_i &= -\tfrac{1}{2}\left(\boldsymbol{\mu}_i^T \boldsymbol{\Sigma}_i^{-1} \boldsymbol{\mu}_i + \ln(|\boldsymbol{\Sigma}_i|)\right) + \ln P(O = H_i) && \longleftarrow \text{Constant Term}
\end{aligned}
$$

In practice, since the *mean* and the *covariance matrix* have to be estimated from training data (see Section 6.10), there is generally not enough data to be able to estimate a full covariance matrix. Most implementations resort to assuming that the sample spaces, $\mathscr{X}_j, j \in \{1,2,\cdots,D\}$, defined in Equation 9.50 are statistically independent. This means that the variance matrices, $\boldsymbol{\Sigma}_i, i \in \{1,2,\cdots,\Gamma\}$ are diagonal matrices,

$$(\boldsymbol{\Sigma}_i)_{[j][k]} = \begin{cases} \sigma_{ij} & \forall\ j \neq k \\ 0 & \forall\ j = k \end{cases} \qquad (9.58)$$

where $j,k \in \{1,2,\cdots,D\}$.

Recall the *unbiased estimator* of Section 6.10. Equation 6.202 shows the relation between the *biased estimate* of the *covariance matrix* and the first and second order sums, $\mathbf{s}|_{N_i}$ and $\mathbf{S}|_{N_i}$, defined by Equations 6.204 and 6.206. N_i is the number of samples used for estimating $\boldsymbol{\Sigma}_i$. Making the assumption of diagonal covariances, we may estimate the diagonal elements of the unbiased estimate of the covariance matrices as follows,

$$(\tilde{\boldsymbol{\Sigma}}_i|_{N_i})_{[j][j]} = \frac{1}{N_i - 1}\left[(\mathbf{S}_i|_{N_i})_{[j][j]} - \frac{1}{N_i}\left((\mathbf{s}_i|_{N_i})_j\right)^2\right] \qquad (9.59)$$

where $j \in \{1,2,\cdots,D\}$.

Using the independence assumption, the discriminant functions of Equation 9.55 may be written in terms of the sample mean and sample variance,

$$g_i(\mathbf{x}) = -\frac{1}{2}(\mathbf{x} - \bar{\mathbf{x}}_i|_{N_i})^T \tilde{\boldsymbol{\Sigma}}_i^{-1}|_{N_i}(\mathbf{x} - \bar{\mathbf{x}}_i|_{N_i}) - \frac{1}{2}\ln\left(|\tilde{\boldsymbol{\Sigma}}_i|_{N_i}|\right) + \ln P(O = H_i) \quad (9.60)$$

and the following relations simplify the computations of the terms in Equation 9.60,

$$\ln\left(|\tilde{\boldsymbol{\Sigma}}_i|_{N_i}|\right) = \sum_{j=1}^{D} \ln\left(\tilde{\boldsymbol{\Sigma}}_i|_{N_i}\right)_{[j][j]} \tag{9.61}$$

$$\left(\tilde{\boldsymbol{\Sigma}}_i^{-1}|_{N_i}\right)_{[j][j]} = \frac{1}{\left(\tilde{\boldsymbol{\Sigma}}_i|_{N_i}\right)_{[j][j]}} \tag{9.62}$$

9.3.2 Classification of a Sequence

Up to now, we have been concerned with the classification of single *samples* (*feature vectors*). However, it is quite important to be able to extent the results to a sequence of feature vectors. This is specifically important in speaker recognition, since usually we are concerned with obtaining a single identity from a sequence of feature vectors stemming from a person's speech.

Consider a T-sample sequence (see Section 6.7.2) of random variables, $\{X\}_0^{T-1}$, where each $X_t, t \in \{0, 1, \cdots, T-1\}$ is considered to be an independent random variable, itself consisting of a D-dimensional set of independent scalar random variables making up each D-dimensional sample, $\mathbf{x}_t$. In essence, we will have a T-dimensional Cartesian product space (Section 6.5.2) consisting of T independent D-dimensional Cartesian product spaces, making the total number of random scalar random variables, TD.

If we assume independence, we may use Equation 6.92 to compute the *joint probability density function* conditioned upon each hypothesis (likelihood of hypothesis H_i given a sequence of samples, $\{\mathbf{x}\}_0^{T-1}$). Equation 6.92 gives us an estimate for the likelihood, assuming independence of the samples in the sequence.

$$p(\{\mathbf{x}\}_0^{T-1}|O = H_i) = \prod_{t=0}^{T-1} p(\mathbf{x}_t|O = H_i) \tag{9.63}$$

Therefore, the *a-posteriori* probability of the sequence, $\{\mathbf{x}\}_0^{T-1}$, given each hypothesis associated with class i, will be,

$$P(O = H_i|\{\mathbf{x}\}_0^{T-1}) = \frac{\prod_{t=0}^{T-1} p(\mathbf{x}_t|O = H_i)P(O = H_i)}{p(\{\mathbf{x}\}_0^{T-1})} \tag{9.64}$$

Furthermore, we may use Equation 6.92 once more on the denominator of Equation 9.64,

$$P(O - H_i|\{\mathbf{x}\}_0^{T-1}) = \frac{\prod\limits_{t=0}^{T-1} p(\mathbf{x}_t|O = H_i)P(O = H_i)}{\prod\limits_{t=0}^{T-1} p(\mathbf{x}_t)} \tag{9.65}$$

At this point, we may attempt to construct the expression for the *discriminant functions* for classifying the sequence, $\{\mathbf{x}\}_0^{T-1}$. Therefore a possible discriminant function associated with the class hypothesis, H_j, would be,

$$
\begin{aligned}
g_j(\{\mathbf{x}\}_0^{T-1}) &= P(O = H_i|\{\mathbf{x}\}_0^{T-1}) \\
&= \frac{\prod\limits_{t=0}^{T-1} p(\mathbf{x}_t|O = H_i)P(O = H_i)}{\prod\limits_{t=0}^{T-1} p(\mathbf{x}_t)}
\end{aligned} \tag{9.66}
$$

Recall the mean-value theorem (Theorem 24.10). Based on this theorem, there will exist an *expected value* of the *marginal probability density* for each random variable, in the sequence, $\{X\}_0^{T-1}$. If we denote this value by $\tilde{p}(\mathbf{x})$, then we may write Equation 9.66 in terms of this value, as follows,

$$g_j(\{\mathbf{x}\}_0^{T-1}) = \frac{\prod\limits_{t=0}^{T-1} p(\mathbf{x}_t|O = H_i)P(O = H_i)}{(\tilde{p}(\mathbf{x}))^T} \tag{9.67}$$

We mentioned that the discriminant functions are not unique. Also, in deciding the class based on the highest *a-posteriori* probability of different class hypotheses, we may ignore the *joint probability density* of the sequence since it is the same for all hypotheses and same length sequences, giving us the following alternative set of discriminant functions,

$$g_j(\{\mathbf{x}\}_0^{T-1}) = \prod\limits_{t=0}^{T-1} p(\mathbf{x}_t|O = H_i)P(O = H_i) \tag{9.68}$$

If we look closely at the expression for the discriminant function in Equation 9.68, we see that it would generate numbers of different orders depending on the length of the sequence, T. This causes two major problems. First, as the length of the sequence, T, gets larger, the value of $g_i(\{\mathbf{x}\}_0^{T-1})$ has a smaller magnitude. Although, we will be using logarithms to handle this reduction in the magnitude, still, with sequences which are long enough, we do risk loss of significance. The second problem is that if we have any other source of information such as *context*, we will not be able to compare numbers associated with sequences of different lengths. Of course, as long as we are only using the discriminant function to compute the likelihood of a hypothesis among several hypotheses associated with the same sequence, this will not be a problem.

To help reduce the effects of the loss of numerical significance, we may compute use *log-likelihood function* for the sequence, allowing us to construct the logarithmic version of the discriminant function of Equation 9.68, where the product will change to a summation.

$$g_j(\{\mathbf{x}\}_0^{T-1}) = \sum_{t=0}^{T-1} \ln(p(\mathbf{x}_t|O = H_i)) + \ln(P(O = H_i)) \tag{9.69}$$

In Equation 9.69, the second term is independent of the length of the sequence. However, the first term depends on T. To remove this dependence, we may again assume that based on the mean-value theorem, there exists an expected value of the marginal likelihoods of the class hypotheses. Then, based on the argument of Section 6.7.2 on the convergence of a sequence, we may write a new discriminant function which is based on the expected value of the marginal likelihoods, $p(\mathbf{x}|O = H_i)$, and is independent of the length of the sampled sequence,

$$g_j(\mathbf{x}) = \ln p(\mathbf{x}|O = H_i) + \ln P(O = H_i) \tag{9.70}$$

However, since we do not know $p(\mathbf{x}|O = H_i)$, we will need to approximate it based on the marginal likelihoods of the samples in a T-long sequence, using the sample-mean approximation to $p(\mathbf{x}|O = H_i)$, given the T-long sequence of samples, $\{\mathbf{x}\}_0^{T-1}$, as follows,

$$\ln \tilde{p}(\mathbf{x}|O = H_i) = \frac{1}{T} \sum_{t=0}^{T-1} \ln(p(\mathbf{x}_t|O = H_i)) \tag{9.71}$$

such that based on *Khintchine's theorem* (Theorem 6.7), as $T \to \infty$, $\ln \tilde{p}(\mathbf{x}|O = H_i)$ would converge to $\ln p(\mathbf{x}|O = H_i)$ in probability, giving us the following approximation to the discriminant function of Equation 9.70 which is independent of the length of the sequence, T,

$$g_j(\mathbf{x}) = \frac{1}{T} \sum_{t=0}^{T-1} \ln p(\mathbf{x}|O = H_i) + \ln P(O = H_i) \tag{9.72}$$

For the two-category classification case (see Equation 9.42), we would have the following approximation for the log-likelihood, which is one possible discriminant function for the *null hypothesis*,

$$g(\mathbf{x}) = \frac{1}{T} \sum_{t=0}^{T-1} [\ln(p(\mathbf{x}_t|O = H_0)) - \ln(p(\mathbf{x}_t|O = H_1))]$$
$$+ \ln(P(O = H_0)) - \ln(P(O = H_1)) \tag{9.73}$$

where $g(\mathbf{x}) > 0$ would result in choosing class hypothesis H_0 and otherwise, H_1 would be chosen. We may interpret Equation 9.73 in terms of the comparison of the log-likelihood ratio to a threshold,

$$\frac{1}{T}\sum_{t=0}^{T-1}[\ln(p(\mathbf{x}_t|O=H_0))-\ln(p(\mathbf{x}_t|O=H_1))] > \theta_{H_0} \qquad (9.74)$$

where

$$\theta_{H_0} = \ln(P(O=H_1))-\ln(P(O=H_0)) \qquad (9.75)$$

9.4 Decision Trees[1]

In speech and speaker recognition, many conditional models and conditional statistics are used. This is the basis for using decision trees which ask questions at different levels of a decision process and based on the condition, take a specific path or a few possible paths with probabilities attached to them.

For example, if the speech is modeled using a hidden Markov model, then usually, different models and configurations are used for different phone contexts. Another example is the whole concept of a statistical language model which provides a probability distribution for the next linguistic unit based on the current context. A suboptimal search through different paths is another great example of decision trees. A most important example is the hierarchical solution to the large-scale speaker identification problem [1]. Basically, there is a need for a decision tree whenever the number of hypotheses becomes large.

Definition 9.8 (Decision Tree). *A decision tree is a function which has the objective of establishing effective equivalence classes. In other words, it is a function that maps a large number of conditions to a smaller number of equivalence classes.*

An equivalence class is defined as the class (subset) of all the elements in a set such that they are equivalent using some predefined equivalence relation. One way to look at this is the act of clustering which has been used throughout this book. The limit on the distance which is used to cluster the data into specific number of classes is an equivalence relation and the clusters are equivalence classes. Another example is that of an N-gram in which all the Nth words that follow a specific $N-1$ sequence of words are an equivalence class with the N-gram being the equivalence relation.

Therefore, the job of a decision tree is to start with an array of conditions, group them into a smaller number of equivalence classes and then take action only based on conditions derived from those equivalence classes. Thus, the goal of constructing a decision tree is to find the best possible equivalence classes, so it is really a search

---

[1] Some of the information in this section is based on my notes on several lectures by Peter V. DeSouza at IBM Research in the mid 1990s. No IBM-confidential information has been included. Permission was requested from Dr. Peter V. DeSouza and IBM research and was granted on July 9, 2010. I am thankful for this generous offer.

problem. For a decision tree to be effective in practical terms, it will be constructed based on a *suboptimal search technique* [7].

Usually, we will have to limit our number of choices in the equivalence classes to some number, N. This has mostly related to practical issues such as the amount of available data and the time limitations. If the we try to create too many equivalence classes with little data, the created decision tree will do very well with the given training data, but it will not generalize well. On the other hand, if we use too few, then we are effectively not reducing the complexity of the problem which is our goal based on Definition 9.8.

One practical approach is to find those N classes that maximize the likelihood of the training data (hence *maximum likelihood estimation*). When we are trying to use conditional models, we will not only have to look for N classes, but also N corresponding models that would maximize the class-dependent likelihood of the training data. On the other hand, when we are working with conditional probabilities, we should look for N classes and N corresponding probability distributions that maximize the class-conditional probability of the training data.

An example of the conditional model case is the hierarchical clustering of speakers in a speaker tree and the corresponding models at each node of the tree to maximize the likelihood of arriving at the leaves with the training data.[1] An example for the conditional probability case is the construction of a language model [6] where we are maximizing the conditional probability of the words in a sequence. In both cases, our goal is to maximize the likelihood of some criterion.

9.4.1 Tree Construction

In general, while trying to construct a tree, we will have to take the following steps,

1. Establish the best question (the one that maximizes the likelihood of the criterion of interest) for partitioning the data into two equivalence classes and keep doing this on each of the emerging classes until the stopping criterion in the next step is met.
2. Stop when there is either insufficient data with which to proceed or when the best question is not sufficiently informative.

This kind of tree is called a binary decision tree since at any point in the process it asks a single question with a yes or no answer (hence the division of the data into *two* equivalence classes at every stage of the process). Therefore, we can define the underlying fundamental operation of constructing a tree as the establishment of the best question that would partition the data into a subset of 2 smaller equivalent classes.

Thus, constructing a tree is a *greedy* process. By this we mean that although at any step of the process a split into smaller equivalence classes may be locally optimal, the final product is not necessarily globally optimal. In other words, finding a globally optimal tree is an *NP*-complete [8] problem.[2]

9.4.2 Types of Questions

There is always a question that must be asked for any non-terminal node of a decision tree. The terminal nodes of the tree are called the *leaves* of the tree. There are several possible types of questions that may be asked. The classes of questions which will be discussed here are not necessarily disjoint classes. For example a question may be discrete and be a fixed question. We will discuss different types of questions below.

9.4.2.1 Discrete

Definition 9.9 (Discrete Question). *A discrete question is a question that is asked about a discrete random variable, X.*

For example, if $\mathcal{U}$ is the set of possible outcomes of X and if $\mathcal{X} \subset \mathcal{U}$, then a question of the form, "Is $x \in \mathcal{X}$?" is a discrete question.

9.4.2.2 Continuous

Definition 9.10 (Continuous Question). *A continuous question is one that is asked about a continuous random variable, X.*

For example, if $X \in \mathbb{R}$, then a question of the form, "Is $x < \theta$ where $\theta \in \mathbb{R}$?" is a continuous question.

---

[2] *NP*-complete means a problem that requires a nondeterministic polynomial time or in other words it is combinatorially intractable (or it is not practical).

9.4.2.3 Fixed

Definition 9.11 (Fixed Question). *A fixed question refers to the selection of a question from a collection of predefined questions that would optimize the objective function of choice for building the tree.*

For example, sometimes it is conceivable that a complete list of possible classes are constructed and their membership is used as valid questions. Then, in the building of a tree, all of the questions (possibly in a specific order) may be tried and the one that maximizes the likelihood of the training data may be picked as the question for the node of interest. In a phonetic context, an example of such classes for handling *stops* in English are,

- Is the phone a stop ($x \in \{/p/, /t/, /k/, /b/, /d/, /g/\}$)?
- Is the phone an unvoiced stop ($x \in \{/p/, /t/, /k/\}$)?
- Is the phone a voiced stop ($x \in \{/b/, /d/, /g/\}$)?
- Is the phone /p/?
- Is the phone /t/?
- Is the phone /k/?
- Is the phone /b/?
- Is the phone /d/?
- Is the phone /g/?

9.4.2.4 Dynamic

Definition 9.12 (Dynamic Question). *A dynamic question refers to the generation of a question dynamically as the tree nodes are traversed. A search problem is solved to create the question that optimizes the equivalence classes given the training data.*

Building decision trees with *dynamic* questions has two important downfalls. First, the search problem could become quite complex and take up tremendous amount of energy (for example in terms of CPU utilization). The second (more important) problem is that there are usually too many possible classifications at any level of the tree and the amount of freedom available to us could easily create an over-trained tree[3].

Therefore, generally, dynamic questions are not recommended. In practice, fixed and discrete questions perform better. Of course, the choice is completely case-dependent.

---

[3] An overtrained decision tree is one that fits the training data very well, but does not generalize well for unseen data.

9.4.2.5 Binary

Definition 9.13 (Binary Question). *A binary question is one that may be answered in a Boolean (true or false) response.*

Generally, any question may have two or more answers. For example, in building a phone level N-gram model [6], at any level of the decision tree (state of the N-gram), we may ask, "What is the next phone?" The response to this question may be any one of 49 English phones and *none* totaling 50 responses.

Any n-ary question (a question with n responses), may be converted to series of binary questions. For instance, take the example of the phone-level N-gram model. We may first ask, "is there a phone next?" The response to this will eliminate the *none* equivalence class and will leave us with *none* on one side and all the 49 phones on the other. Then we may ask "Is it a /t/?" and then split into two classes again, each time eliminating one possibility. Therefore, any n-ary tree may be converted into a binary tree.

One problem with n-ary trees of $n > 2$ is that they typically fragment the data too quickly in the onset or earlier stage of the decision process and will have less of a chance to achieve close to global optimality. In the example that we gave for the conversion to a binary tree, we did not really choose the best questions. The questions, as posed, would generate an unbalanced tree. It is better to have a balanced tree, by asking questions that will split the data in a more uniform fashion. This will, in general, reduce the number of layers of questions in the conversion. The first, unbalanced technique, would basically use brute force to go through all the possibilities. A more intelligent set of questions would be able to achieve results much quicker.

Binary questions come in two forms, *simple* binary questions and *complex* binary questions. A *simple* binary question is one that is made up of a single Boolean question with no Boolean operators. For example, "$x \subset \mathcal{X}_1$?" is a simple binary question, where "$(x \in \mathcal{X}_1 \wedge (x \in \mathcal{X}_2)$?" is a complex binary question. Figure 9.3 shows the possible outcomes of this example. Note that any complex binary question may only have two outcomes. However, as it can seen in the figure, some of the outcomes are tied. If the states with the same outcome are tied to each other to present a single node and then continue from that node, the resulting system is called a *decision network* or a *decision graph*.

There is no evidence that complex questions make the decision tree any more effective than simple questions. An example of complex questions is a language model, which asks questions regarding the different combinations of states.

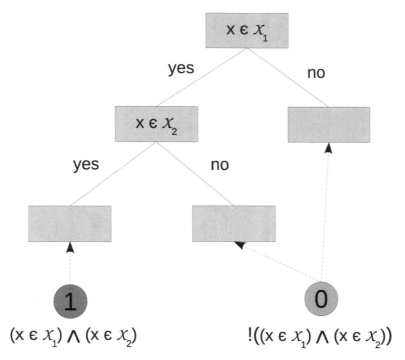

Fig. 9.3: Path of a sample complex binary question – it always has only two final outcomes.

9.4.3 *Maximum Likelihood Estimation (MLE)*

At the beginning of our discussion on decision trees, we said that we may maximize the likelihood of the training data to be able to construct the best question to ask at each node of the tree. Let us do a more formal treatment of the maximum likelihood technique.

Let $x_1 x_2 \cdots x_n$ be a sequence of a discrete random variable X. Then, we can define the following variables,

$a_i \overset{\Delta}{=} i^{th}$Unique outcome, $i \in \{1, 2, \cdots, N\}$

$c_i \overset{\Delta}{=} i^{th}$Frequency of occurrence of a_i

$P_i \overset{\Delta}{=} i^{th} p(X|a_i)$

$Q \overset{\Delta}{=}$ A question that partitions the sample into two classes

$n_l \overset{\Delta}{=}$ Number of samples that fall into the class on the left

$n_r \overset{\Delta}{=}$ Number of samples that fall into the class on the right (9.76)

$_lc_i \overset{\Delta}{=}$ Frequency of occurrence of the outcome a_i in the left partition

$_rc_i \overset{\Delta}{=}$ Frequency of occurrence of the outcome a_i in the right partition

$_lP_i \overset{\Delta}{=} i^{th}$Probability of the outcome a_i in the left partition

$_rP_i \overset{\Delta}{=} i^{th}$Probability of the outcome a_i in the right partition

$x_j^K \overset{\Delta}{=}$ The sample sequence, $x_j x_{j+1} \cdots x_{k-1} x_k$

The most effective question, Q, which may be asked is that question which has maximum likelihood conditioned upon the sample. If we assume that the members of the n-long sequence of instances of X, $\{x\}_1^n$, are *independent and identically distributed (i.i.d.)*, then the likelihood of the question Q given the observations of the sequence is as follows,

$$\mathscr{L}(Q|\{x\}_1^n) = p(\{x\}_1^n|Q)$$
$$= \prod_{j=1}^{n} p(X = x_j|Q)$$
$$= \prod_{j=1}^{n} \mathscr{L}(Q|X = x_j) \qquad (9.77)$$

Note that maximizing the log of $\mathscr{L}(Q|\{x\}_1^n)$ produces the same results.

The likelihood of the Question Q, condition on $\{x\}_1^n$, may then be written in terms of the defined variables of Equation 9.76, in the following manner,

$$\ell(Q|\{x\}_1^n) = \log_2 (\mathscr{L}(Q|\{x\}_1^n))$$
$$= \sum_{i=1}^{N} {_lc_i} \log_2({_lP_i}) + \sum_{i=1}^{N} {_rc_i} \log_2({_rP_i}) \qquad (9.78)$$

Note that we are using $\log_2$ for convenience since we are dealing with binary decision trees.

If we use the *maximum likelihood estimate* (Equation 6.173) for $_lP_i$ and $_rP_i$, we will have the maximum likelihood solution,

$$\ell(Q|\{x\}_1^n) = \sum_{i=1}^{N} {}_lc_i \, \log_2(\frac{{}_lc_i}{n_l}) + \sum_{i=1}^{N} {}_rc_i \, \log_2(\frac{{}_rc_i}{n_r}) \tag{9.79}$$

$$= \sum_{i=1}^{N} {}_lc_i \, \log_2({}_lc_i) - \log_2(n_l) \sum_{i=1}^{N} {}_lc_i +$$

$$\sum_{i=1}^{N} {}_rc_i \, \log_2({}_rc_i) - \log_2(n_r) \sum_{i=1}^{N} {}_rc_i \tag{9.80}$$

$$= \sum_{i=1}^{N} \left({}_lc_i \, \log_2({}_lc_i) + {}_rc_i \, \log_2({}_rc_i) \right) - n_l \log_2(n_l) - n_r \log_2(n_r) \tag{9.81}$$

In Equation 9.81, since ${}_lc_i$, ${}_rc_i$, n_l and n_r are all non-negative integers, the components may be precomputed and stored in a table so that they may be retrieved using a table lookup, for the computation of the log-likelihood.

The joint log-likelihood of the whole sample is

$$\mathcal{L}(Q|\{x\}_1^n) = \prod_{i=1}^{N} P_i^{c_i} \tag{9.82}$$

Using the maximum likelihood estimate for P_i, just like before, we may write the joint log-likelihood for the whole sample as follows,

$$\ell(Q|\{x\}_1^n) = \sum_{i=1}^{N} c_i \log_2 \hat{P}_i$$

$$= \sum_{i=1}^{N} c_i \log_2 \frac{c_i}{n} \tag{9.83}$$

If we multiply $\ell(Q|\{x\}_1^n)$ by $-\frac{1}{n}$, we will have entropy,

$$-\frac{1}{n}\ell(Q|\S 1n) = -\sum_{i=1}^{N} \frac{c_i}{n} \log_2 \frac{c_i}{n} \tag{9.84}$$

$$= -\sum_{i=1}^{N} \hat{P}_i \log_2 \hat{P}_i \tag{9.85}$$

$$= \mathcal{H}(\hat{P}) \tag{9.86}$$

Since n, the number of samples, is a constant, maximizing the log-likelihood, would be equivalent to minimizing the entropy. For more on the topic of decision trees, see [2].

References

1. Beigi, H.S., Maes, S.H., Chaudhari, U.V., Sorensen, J.S.: A Hierarchical Approach to Large-Scale Speaker Recognition. In: EuroSpeech 1999, vol. 5, pp. 2203–2206 (1999)

2. Breiman, L., Fiedman, J., Stone, C.J., Olshen, R.A.: Classification and Regression Trees, 1st edn. Chapman & Hall, New york (2002). ISBN: 0-412-04841-8

3. Cover, T.M., Thomas, J.A.: Elements of Information Theory, 2nd edn. John Wiley & Sons, New Jersey (2006). ISBN-13: 978-0-471-24195-9

4. Duda, R.O., Hart, P.E.: Pattern Classification and Scene Analysis. John Wiley & Sons, New York (1973). ISBN: 0-471-22361-1

5. Fisher, R.A.: The Design of Experiments. Oliver and Boyd, Edinburgh (1935). 8th Edition, Hafner, New York, 1966

6. Manning, C.D.: Foundations of Statistical Natural Language Processing. The MIT Press, Boston (1999). ISBN: 0-262-13360-1

7. Nilsson, N.J.: Problem Solving Methods in Artificial Intelligence. McGraw-Hill Inc., New York (1971). ISBN: 0-070-46573-2

8. Sedgewick, R.: Algorithms in C. Addison-Wesley Publishing Company, New york (1990). ISBN: 0-201-51425-7

Chapter 10
Parameter Estimation

With five free parameters, a theorist could fit the profile of an elephant.

George Gamow (1904-1968)
Quoted in Nature, June 21, 1990

In Section 9.3, we discussed the definition of hypotheses which are designed to classify data into different categories. The main idea is to be able to categorize data into pieces, each of which would be represented by a statistical family of parametric models, using a parameter vector, $\boldsymbol{\varphi}$. This is the problem of *model selection* in general and *parameter estimation* in specific. We have already discussed several techniques for estimating the *mean*, *variance*, and *higher statistics* in Chapter 6. We needed to develop the concepts of *entropy* and *information* in order to be able to continue the parameter estimation techniques of this chapter. For this reason, the concepts covered here, had to be deferred until now, instead of being included with the moment estimation concepts covered in Chapter 6. This chapter looks into the parameter estimation problem in more detail, specifically in the realm of estimating model parameters to represent observed samples.

At times, such as in Section 9.2.1, we have touched upon *maximum likelihood estimation* which is geared toward estimating such parameters in order to produce the *maximum likelihood solution* to the modeling problem. There are also other similar objectives, based on the problem at hand, such as *maximum a-posteriori estimation* and *maximum entropy* estimation. We shall study these techniques in more detail in this chapter.

The estimation techniques, presented here, are general overviews which do not include much detail about the implementation issues. Also, here, we are not considering the constraints that may be present in the optimization problems. One such set of constraints is the fact that all probabilities have to be positive and should sum up to 1. Other constraints may exist which are generally related to the type of model being used or the input parameters to the model. Most of the time, the constraints are introduced into the optimization objective function, using *Lagrange multipliers* (see Chapter 25).

10.1 Maximum Likelihood Estimation

Consider the random variable $\{X : \mathbf{x} \in \mathscr{X} = \mathscr{R}^D\}$. Furthermore, assume that we have a model which estimates the *probability density function* associated with X, $p(\mathbf{x})$, by $\hat{p}(\mathbf{x}|\boldsymbol{\varphi})$, where $\{\Phi : \boldsymbol{\varphi} \in \phi = \mathscr{R}^M\}$ is the parameter vector. The likelihood of the model, given $\mathbf{x}$, is then defined as,

$$\mathscr{L}(\boldsymbol{\varphi}|\mathbf{x}) = \hat{p}(\mathbf{x}|\boldsymbol{\varphi}) \tag{10.1}$$

The maximum likelihood estimate is defined as the set of parameters, $\hat{\boldsymbol{\varphi}}$, such that,

$$\hat{\boldsymbol{\varphi}} \overset{\Delta}{=} \underset{\boldsymbol{\varphi}}{\arg\max}\,\mathscr{L}(\boldsymbol{\varphi}|\mathbf{x}) \tag{10.2}$$

$$= \underset{\boldsymbol{\varphi}}{\arg\max}\,\hat{p}(\mathbf{x}|\boldsymbol{\varphi}) \tag{10.3}$$

In most cases of interest, we are concerned with the *exponential family* of density functions. Therefore, it is natural, due to the *concave* and *monotone* nature of the *logarithm* (see Definition 24.25), to speak in terms of the *log* of the *likelihood* which becomes maximum when the likelihood itself is at its maximum. Hence, we may define the *log-likelihood*, $\ell_b(\boldsymbol{\varphi}|\mathbf{x})$, of the model parameters as follows,

$$\ell_b(\boldsymbol{\varphi}|\mathbf{x}) \overset{\Delta}{=} \log_b \mathscr{L}(\boldsymbol{\varphi}|\mathbf{x}) \tag{10.4}$$

$$= \log_b \hat{p}(\mathbf{x}|\boldsymbol{\varphi}) \tag{10.5}$$

and the maximum likelihood estimate may be written in terms of the log-likelihood of the parameters,

$$\hat{\boldsymbol{\varphi}} = \underset{\boldsymbol{\varphi}}{\arg\max}\,\mathscr{L}(\boldsymbol{\varphi}|\mathbf{x}) \tag{10.6}$$

$$= \underset{\boldsymbol{\varphi}}{\arg\max}\,\ell(\boldsymbol{\varphi}|\mathbf{x}) \tag{10.7}$$

$$\tag{10.8}$$

Here, we are using $b = e$ as the base of the logarithm for which we omit the writing of the base all together, for simplicity.

Due to the *concavity* and *monotonicity* of the *log-likelihood* function (Definition 24.25), to maximize the log-likelihood, we may compute the gradient of the log-likelihood, $\nabla_{\boldsymbol{\varphi}}\ell(\boldsymbol{\varphi}|\mathbf{x})$ and set it equal to *zero*, solving for $\hat{\boldsymbol{\varphi}}$,

$$\varsigma(\hat{\boldsymbol{\varphi}}|\mathbf{x}) \overset{\Delta}{=} \nabla_{\boldsymbol{\varphi}}\ell(\boldsymbol{\varphi}|\mathbf{x})\Big|_{\boldsymbol{\varphi}=\hat{\boldsymbol{\varphi}}} \tag{10.9}$$

$$= \frac{1}{\mathscr{L}(\boldsymbol{\varphi}|\mathbf{x})}\nabla_{\boldsymbol{\varphi}}\mathscr{L}(\boldsymbol{\varphi}|\mathbf{x}) \tag{10.10}$$

$$= 0 \tag{10.11}$$

where $\mathscr{L}(\boldsymbol{\varphi}|\mathbf{x})$ may not be *zero*. Also, we agree to disregard any roots of Equation 10.11 of the form, $\boldsymbol{\varphi} = constant$, and only keep those roots which are dependent on the sample, $\mathbf{x}$.[7] $\varsigma(\boldsymbol{\varphi}|\mathbf{x})$ is known as the *score statistic* of the parameter vector (also known as the *Fisher score* [13]), $\boldsymbol{\varphi}$, given the observation, $\mathbf{x}$.

Recall the *log-likelihood* of a set of independent observations, $\{\mathbf{x}\}_1^N$, of X, given by Equation 9.71. In the notation of this section, we may write the *log-likelihood* of the set of observed vectors as follows,

$$\ell(\boldsymbol{\varphi}|\{\mathbf{x}\}_1^N) = \sum_{i=1}^N \ell(\boldsymbol{\varphi}|\mathbf{x}_i) \tag{10.12}$$

The *score statistic* for the set of observations, $\{\mathbf{x}\}_1^N$, is then,

$$\varsigma(\hat{\boldsymbol{\varphi}}|\{\mathbf{x}\}_1^N) = \left.\nabla_{\boldsymbol{\varphi}}\ell(\boldsymbol{\varphi}|\{\mathbf{x}\}_1^N)\right|_{\boldsymbol{\varphi}=\hat{\boldsymbol{\varphi}}} \tag{10.13}$$

$$= \left.\sum_{i=1}^N \nabla_{\boldsymbol{\varphi}}\ell(\boldsymbol{\varphi}|\mathbf{x}_i)\right|_{\boldsymbol{\varphi}=\hat{\boldsymbol{\varphi}}} \tag{10.14}$$

$$\tag{10.15}$$

The *variance-covariance matrix* of the *score statistic* is a measure of the *Fisher information* in the model parameters. When the parameter vector has dimension one, it is a simple *variance* and is known as the *Fisher information*. When the parameters are of dimension $M > 1$, then the matrix is known as the *Fisher information matrix* (see Section 7.7),

$$\mathscr{I}_F = Cov\left(\varsigma(\boldsymbol{\varphi}|\mathbf{x})\right) \tag{10.16}$$

$$= \mathscr{E}\left\{\varsigma(\boldsymbol{\varphi}|\mathbf{x})\varsigma^T(\boldsymbol{\varphi}|\mathbf{x})\right\} \tag{10.17}$$

Under *Dugué's regularity conditions*, discussed in Section 7.7, the *Fisher information matrix* may be written as the expected value of the *Hessian matrix* (see Chapter 25) of the *log-likelihood* with respect to $\boldsymbol{\varphi}$ [20]. Namely,

$$\mathscr{I}_F = \mathscr{E}\left\{\nabla_{\boldsymbol{\varphi}}^2 \ell(\boldsymbol{\varphi}|\mathbf{x})\right\} \tag{10.18}$$

To simplify the notation, we may write the *likelihood*, *log-likelihood*, and *score statistic* of $\{\mathbf{x}\}_1^N$, by omitting the conditionality on $\{\mathbf{x}\}_1^N$, as $\mathscr{L}(\boldsymbol{\varphi})$, $\ell(\boldsymbol{\varphi})$, and $\varsigma(\boldsymbol{\varphi})$, respectively. The conditionality notation is included when we speak about a single sample vector, $\mathbf{x}$.

Therefore, the *Fisher information matrix* for a set of independent observations, $\{\mathbf{x}\}_1^N$, may be written as,

$$\mathscr{I}_F = Cov\left(\varsigma\right) \tag{10.19}$$

$$= \mathscr{E}\left\{\varsigma\varsigma^T\right\} \tag{10.20}$$

$$= \mathscr{E}\left\{\nabla_{\boldsymbol{\varphi}}^2 \ell\right\} \tag{10.21}$$

See *Cramér* [7] for a review of the asymptotic properties of *maximum likelihood estimation*.

Given the above set up of the problem of *maximum likelihood estimation (MLE)*, we may use any of the techniques, discussed in detail, in Chapter 25. The *Quasi-Newton* techniques discussed in that chapter will be well suited, since the estimation of the *Hessian matrix* provides a method for computing the *Fisher information* as well (Equation 10.18). However, as we shall see in Section 11.3, the *expectation maximization (EM)* algorithm is very well suited for solving the *MLE* problem.

10.2 Maximum A-Posteriori (MAP) Estimation

Let us recall the maximum likelihood objective function of Equation 10.3. This objective function was used under the basic premise that the model parameters, $\boldsymbol{\varphi}$, are distributed uniformly, in the parameter space, ϕ. Namely, any $\boldsymbol{\varphi}$ would have the same prior probability denoted by $P(\boldsymbol{\varphi})$. However, we can imagine cases where an approximation of the a-priori probability of parameter vector, $\boldsymbol{\varphi}$, is available to us through some information source. This prior distribution may also be made available by a mechanism such as an unsupervised clustering algorithm.

For example, as we shall see in Chapter 13, the sample data may be modeled by an *HMM* or simply by a *mixture model* (mostly a *Gaussian mixture model*). In this case, the actual model is a linear combination of several models, where the a-priori probability of every model in the combination may be estimated based on the model memberships of individual samples. These memberships may either be prescribed by a hard membership function or a soft one. The important point is that we may estimate the prior probabilities of individual models, $P(\boldsymbol{\varphi}_\gamma)$.

Another source of estimating such priors may be an external model such as a language model. For instance, a language model may dictate the prior probability of a parameter vector based on context. Whatever the source of the estimate for the *a-priori* probability of parameters, $\boldsymbol{\varphi}$, it is apparent that using this information will help the ultimate goal of learning. The true goal of learning is to maximize the *a-posteriori* probability of the parameters given an observation, $\mathbf{x}$,

$$\hat{\boldsymbol{\varphi}} = \arg\max_{\boldsymbol{\varphi}} P(\boldsymbol{\varphi}|\{\mathbf{x}\}_1^N) \tag{10.22}$$

Using Bayes theorem (Equation 6.58),

$$P(\boldsymbol{\varphi}|\{\mathbf{x}\}_1^N) = \frac{p(\{\mathbf{x}\}_1^N|\boldsymbol{\varphi})P(\boldsymbol{\varphi})}{P(\{\mathbf{x}\}_1^N)} \tag{10.23}$$

Since the denominator of Equation 10.23 is the same for all the different parameter vectors, we may drop it from the maximization objective functions. Namely,

$$\hat{\boldsymbol{\varphi}} = \underset{\boldsymbol{\varphi}}{\arg\max} \frac{p(\{\mathbf{x}\}_1^N|\boldsymbol{\varphi})P(\boldsymbol{\varphi})}{P(\{\mathbf{x}\}_1^N)}$$

$$= \underset{\boldsymbol{\varphi}}{\arg\max}\, p(\{\mathbf{x}\}_1^N|\boldsymbol{\varphi})P(\boldsymbol{\varphi}) \qquad (10.24)$$

It is easy to see that if the prior probabilities, $P(\boldsymbol{\varphi})$, are unknown, then the *principle of maximum entropy* (see Section 7.3.1) suggests that the best solution would be one where the priors are set to have a uniform distribution. In that case, $P(\boldsymbol{\varphi})$ may be dropped and we will be left with the *maximum likelihood estimate*.

However, if the prior probabilities are known, or may be estimated from the training data, then it makes sense to use them. The solution to the problem of Equation 10.24 is the *maximum a-posteriori estimate* or the *MAP estimate* for short. At different occasions in the book, we will cover the formulation of the MAP solution to estimation and adaptation problems. Depending on the details of the problem, the solution may be slightly different, but the main difference between MAP estimation and MLE is in the estimation of the prior probabilities. This will be different for *Gaussian mixture models*, *Hidden Markov models*, etc.

10.3 Maximum Entropy Estimation

In Section 7.3.1 we saw that in the absence of any information, the *maximum entropy solution* is the proper solution. Given the available information, the least biased estimate is the *maximum entropy estimate*. According to *Jaynes* [15], the *maximum entropy estimate* is maximally noncommittal with regard to missing information.

Basically the *maximum entropy principle* states that if we have a choice to pick a parameter vector among all the possible vectors, $\{\Phi : \boldsymbol{\varphi} \in \phi\}$, we should choose the parameter vector which is the least biased, given the possible choices. This would be,

$$\hat{\boldsymbol{\varphi}} = \underset{\boldsymbol{\varphi}}{\arg\max}\, \mathscr{H}\left(P(\boldsymbol{\varphi}|\mathbf{x})\right) \qquad (10.25)$$

Maximum entropy techniques have been used in statistical mechanics [15, 16] for long and in fact even *Laplace* stated the *principle of insufficient reason*[1], discovering intuitively about the validity of using uniform priors in the absence of any information. *Maximum entropy estimation* was introduced to the field of speech related research through the paper of *Berger, et al.* [5], in 1996 through an application to statistical language translation of English to French.

---

[1] *Laplace's principle of insufficient reason* states that two events must be assigned equal probabilities if there is no reason to think otherwise.[15]

Of course we already knew the fact that entropy is maximized when there is uniform probability. However, in practice, what does that mean? Of course we are not going to be happy with cases when no information is available. The attractive part of the principle of maximum entropy in practice appears when there are *constraints* present. Therefore, in the presence of constraints, we would have to find a solution to Equation 10.25 subject to the constraints imposed on the parameter vector, $\boldsymbol{\varphi}$ [10]. As with the case of *MLE* and *MAP* in the last two sections, we will have different techniques associated with the problems of interest, such as the *HMM*, *GMM*, language modeling, etc.

Good [9] restated the *principal of maximum entropy* which was originally stated by *Jaynes* [15] (applied to statistical mechanics) in different words, relating it to the selection of *null hypotheses* which would be used in testing. Here is his statement,

Definition 10.1 (Maximum Entropy (Restatement According to Good)). *See [9]. Let X be a random variable whose distribution is subject to some set of restraints. Then entertain the null hypothesis that the distribution is one of maximum entropy, subject to these conditions.*

Noting that entropy is the expected value of information, *Good* [9] clarifies the statement in Definition 10.1 as the following alternate statement, "entertain the *null hypothesis*, H_0 that maximizes the expected amount of selective information per observation."

10.4 Minimum Relative Entropy Estimation[2]

The principle of *minimum relative entropy* is a generalization of the principle of *maximum entropy* [9] and is quite intuitive in the sense that in the presence of constraints, it would be best to choose a distribution which is least discriminable from the best distribution. This would be the distribution which has the smallest *Kullback-Leibler directed divergence*, $\mathscr{D}_{KL}(p_0 \rightarrow p)$, with respect to the best possible hypothesis, when there is no constraint present. Starting with *Good* [9], many publications [25, 17, 26, 8] have been referring to this technique as *minimum cross entropy*, assuming the name, *cross entropy* for what we are calling *relative entropy*. As we shall see later in this section, our concept of *minimum cross entropy* is indeed

----

[2] *Minimum relative entropy* is also known as *minimum cross-entropy* by some researchers[9, 25, 8], following the legacy of terminology which was introduced for the first time by Good in 1963 [9], referring to the *Kullback-Leibler directed divergence* as *cross entropy*. However, recently, cross entropy has been used to denote a different information theoretic entity, $\mathscr{H}(p \rightarrow q)$ in the discrete case and $\hbar(p \rightarrow q)$ in the continuous case – see Section 7.6. *Relative entropy* is a more appropriate term for the entity which is described by the *Kullback-Leibler directed divergence* and therefore, it is what we use in this book. We will see, toward the end of this section, that although the two concepts are different, the *principle of minimum relative entropy* will coincide with the *principle of minimum cross entropy* for a different reason.

the same as *minimum relative entropy*, although the actual quantities of the minima differ, but they both give the same minimizing distributions.

Let us present a formal definition of the *principle of minimum relative entropy*, also known as the *principle of minimum discriminability* [9].

Definition 10.2 (Principle of minimum relative entropy (minimum discriminability)).

Consider a random variable, X, defined in the probability space, $(\mathscr{X}, \mathfrak{X}, P_0)$ where the Radon-Nykodým derivative of the probability measure, P_0, is given by $p_0(x)$. Furthermore, assume that this measure may be related to an excellent hypothesis[3]. Now, let us assume that we are interested in defining a null hypothesis restrained by certain constraints on the Radon-Nykodým derivative where the new probability measure space would be defined by a class of constrained density functions, amounting to the probability space, $(\mathscr{X}, \mathfrak{X}, P)$ with corresponding Radon-Nykodým derivatives, $p(x)$. Then the principle of minimum relative entropy states that the best null hypothesis would be one such that, if true, it may be discriminated from the excellent hypothesis defined earlier, the least. This works out to the null hypothesis that minimizes the relative entropy (Kullback-Leibler directed divergence),

$$\hat{p}(x) = \underset{p}{\arg\min}\, \mathscr{D}_{KL}(p_0 \to p)$$

$$= \underset{p}{\arg\min} \int_{\mathscr{X}} \ln \frac{p_0(x)}{p(x)} dP_0 \qquad (10.26)$$

For the case where the sample space is the Euclidean space, $X = \mathscr{R}^n$, we may write Equation 10.2 in terms of the *Lebesgue measure*, using the *Radon-Nykodým derivative* of P_0, given by $p_0(\mathbf{x})$, such that,

$$\hat{p}(\mathbf{x}) = \underset{p}{\arg\min}\, \mathscr{D}_{KL}(p_0 \to p)$$

$$= \underset{p}{\arg\min} \int_{\mathscr{X}} p_0(\mathbf{x}) \ln \frac{p_0(\mathbf{x})}{p(\mathbf{x})} d\mathbf{x} \qquad (10.27)$$

$$= \underset{p}{\arg\min} (\hbar(p_0 \to p) - \hbar(p)) \qquad (10.28)$$

where $p(x)$ is subject to the constraints stated in Definition 10.2.

Note that based on the fact that we are seeking a density function, $p(x)$, which would be least discriminable from a fixed best hypothesis density function, $p_0(x)$, the same $\hat{p}(x)$ will minimize the *relative entropy* and the *cross entropy*, since across the different $p(x)$, $p_0(x)$ is kept unchanged. Therefore, Equation 10.28 may be written as follows,

---

[3] Good [9] uses the German expression, *ausgezeichnet hypothesis*, which has been translated here to mean *excellent hypothesis*.

$$\hat{p}(\mathbf{x}) = \arg\min_{p} \mathscr{D}_{KL}(p_0 \to p)$$

$$= \arg\min_{p}(\hbar(p_0 \to p) - \hbar(p)) \tag{10.29}$$

$$= \arg\min_{p} \hbar(p_0 \to p) \tag{10.30}$$

Of course $\mathscr{D}_{KL}(p_0 \to \hat{p}) \neq \hbar(p_0 \to \hat{p})$, but the same value which minimizes, $p(x) = \hat{p}(\mathbf{x})$, minimizes both $\mathscr{D}_{KL}(p_0 \to p)$ and $\mathscr{D}_{KL}(p_0 \to p)$. Good [9] and other researchers referring to his paper, such as references [25, 17, 26, 8], are referring to *relative entropy* when they say *cross entropy*.

Jaakkola, in Section 2.1.1 of [14], shows that *support vector machines* (Chapter 15), for a two-class separable case, are a special case of the *minimum relative entropy* technique. In addition, [14] shows that in the general case, the results are also quite similar between the two methods. See the introduction of Chapter 15 for more on this comparison.

10.5 Maximum Mutual Information Estimation (MMIE)

As we saw in Section 7.6.1, *mutual information* (Equation 7.104) is defined as a special case of *relative entropy* where it is a measure of the amount of information between two random variables such that it would become zero for two random variables, X and Y which are *statistically independent*. In other words, the *mutual information* $\mathscr{I}(X_1; X_2)$ may be viewed as a measure of the *statistical dependence* between the two random variables.

Recall the discussion of the optimal choice of the model parameters, $\boldsymbol{\varphi} = \hat{\boldsymbol{\varphi}}$ for constructing a parametric model that would estimate the true distribution of the output of a system, $p(\mathbf{x})$. We have examined different optimization criteria in the last few sections, *MLE*, *MAP*, and *maximum entropy*. Another possible objective would be the maximization of the *mutual information* between the observed output of the system, $X = \mathbf{x}$, and the output of the parametric estimate of the system based on a class of parametric models, $Y(\boldsymbol{\varphi}) = \mathbf{y}(\boldsymbol{\varphi})$. The output of this class of parametric models would be dependent on the model parameters, $\{\Phi : \boldsymbol{\varphi} \in \phi = \mathscr{R}^M\}$. Therefore, the dependent variable of the output of the parametric model would be the choice of $\boldsymbol{\varphi}$.

The *mutual information* would then be written as,

$$\mathscr{I}(X; Y(\boldsymbol{\varphi})) = \int_{\mathscr{X}} \int_{\mathscr{Y}(\phi)} p(\mathbf{x}, \mathbf{y}(\boldsymbol{\varphi})) \ln \frac{p(\mathbf{x}, \mathbf{y}(\boldsymbol{\varphi}))}{p(\mathbf{x})p(\mathbf{y}(\boldsymbol{\varphi}))} d\mathbf{x} d\mathbf{y}(\boldsymbol{\varphi}) \tag{10.31}$$

Note that since the output of the parametric model is dependent on the parameter vector, $\boldsymbol{\varphi}$, Equation 10.31 may be rewritten in terms of the parameter vector, $\boldsymbol{\varphi}$. So the objective would be to find that parameter vector, $\boldsymbol{\varphi}$, which would maximize the mutual information between the output of the model and the true output of the system. Although, generally in parameter estimation problems, the integration may not be done for the entire sample space, $\mathcal{X}$, it may be approximated using the law of large numbers (Section 6.9.2). The following is the formal expression for the MMIE in the continuous domain,

$$\hat{\boldsymbol{\varphi}} = \arg\max_{\varphi} \mathcal{I}\left(X; Y(\boldsymbol{\varphi})\right)$$

$$= \arg\max_{\varphi} \int_{\mathcal{X}} \int_{\mathcal{Y}(\varphi)} p(\mathbf{x}, \mathbf{y}(\boldsymbol{\varphi})) \ln \frac{p(\mathbf{x}, \mathbf{y}(\boldsymbol{\varphi}))}{p(\mathbf{x}) p(\mathbf{y}(\boldsymbol{\varphi}))} d\mathbf{x} d\mathbf{y}(\boldsymbol{\varphi}) \qquad (10.32)$$

$\mathbf{y}(\boldsymbol{\varphi})$ is generally dependent on parameter vector, $\boldsymbol{\varphi}$, and it may be computed knowing the equations of the model. It may be dependent on other parameters as well, such as the history of the output of the system, a language model, etc. Depending on the form of the model, the maximization problem given by Equation 10.32 would be solved differently and would yield different results. In addition, as we mentioned in the beginning of the estimation section, in most cases, there will be constraints which would have to be worked out into the main objective function. These may be constraints on the probabilities to make sure they sum up to 1, or they may be related to the form of the model or other inputs to the model.

Bahl, et al. [4, 3], followed by others [22], have used the *maximum mutual information estimation* technique in optimizing the parameters of a *hidden Markov model (HMM)* in training speech recognition models. We will discuss this technique in more detail in the chapter on *hidden Markov models*.

10.6 Model Selection

One of the important quests in statistical analysis is to decide on the complexity of the model which is used to represent the data. An important indicator for the complexity is the model dimension. Starting with the work of *Akaike* in 1972 [1], a number of information theoretic techniques were developed to aid in the statistical model selection process, by presenting minimum information criteria which are dependent on the model dimension.

Akaike [2] defines the AIC^4, which is an *efficient statistic* (see Definition 6.76), as a criterion for model selection. Four years later, *Schwarz* [24] extended the work

---

[4] In his paper [2], *Akaike* states that AIC stands for *An Information Criterion* and he forecasts the future development of *BIC, DIC*, et cetera, as future versions of Information Criteria. His forecast was at least correct in one case, since BIC *Bayesian Information Criterion* was presented by Schwarz [24], only 4 years later.

of *Akaike* by employing *Bayesian* concepts and introduced a *consistent statistic* (see Definition 6.77) which he called the *Bayesian information criterion* (*BIC*). There have been other information criteria dealing with specific classes of problems, such as the *residual information criterion* (*RIC*) [21], which was designed for *single-index models* [12], widely used in the *financial industry*.

Another method is the *structural risk minimization* technique [11], which is premised on the tuning of the complexity of the model by adjusting a measure of the *capacity* of the model. This measure is known by the *Vapnik-Chervonenkis* (*VC*) *dimension* [27]. The *VC dimension* is a non-zero integer, generally related to the amount of training data used for training the model of interest. In most practical cases, its analytical derivation is not an easy task [11]. Therefore, it is generally approximated to be the dimension of the free parameter vector being used in an optimal parametric model associated with the given training data (Example 2 of Section 4.11 in [27]). Of course, there are examples which are contrary to this assumption. Examples 3 and 4 in Section 4.11 of [27] present models for which the *VC dimension* is less and more than the number of free parameters of the model, respectively. We defer the coverage of this topic to Chapter 15, while covering the background material for the introduction of *support vector machines* (*SVM*).

In the next two sections we will cover *AIC* and *BIC* criteria in some detail. Both *AIC* and *BIC* are criteria which verify the *principle of parsimony*[5] in model selection. As we will see, both criteria will give preference to models which have similar likelihood, but use a smaller number of parameters. *BIC* has specifically been used as the basis of many audio segmentation algorithms – see Sections 16.5 and 17.4.

10.6.1 Akaike Information Criterion (AIC)

Akaike [2] made the argument that the *null hypothesis*, as described and used in the *Neyman-Pearson lemma* (see Section 9.2.1), for a binary hypothesis testing scenario leading to a *log-likelihood ratio* test, is only an approximate hypothesis when it comes to practice. Furthermore, he noted that these hypotheses are almost always different from reality, hence inadequate for usage in approximate identification procedures.

Consider the *Kullback-Leibler directed divergence*, $\mathscr{D}_{KL}(p \rightarrow \hat{p})$, which is a measure of the expected value of information which is lost by modeling the true *probability density function*, $p(\mathbf{x})$, of a system using a model with parameter vector, $\boldsymbol{\varphi}$, represented by the probability density function of the model, $\hat{p}(\mathbf{x}|\boldsymbol{\varphi})$.

---

[5] The *principle of parsimony* is the principle of preferring the simplest model representation of a system.

$$\mathscr{D}_{KL}(p \to \hat{p}) = \hbar(p \to \hat{p}) \; - \; \hbar(p) \tag{10.33}$$

In the discrete case, this would be written as,

$$\mathscr{D}_{KL}(p \to \hat{p}) = \mathscr{H}(p \to \hat{p}) \; - \; \mathscr{H}(p) \tag{10.34}$$

It is apparent that our objective, in model selection, would have to be one that would minimize $\mathscr{D}_{KL}(p \to \hat{p})$. If we would have to choose among models with different parameter vectors, $\boldsymbol{\varphi}$, then the best model would be the one that would have *minimum differential cross entropy*, since $\hbar(p)$ ($\mathscr{H}(p)$) is fixed and would not change based on the choice of the model. Namely,

$$\min \; \mathscr{D}_{KL}(p \to \hat{p}) = \min \; \hbar(p \to \hat{p}) \tag{10.35}$$

Therefore, the parameter vector selection may be viewed as the following minimization problem in the Euclidean space $\mathscr{X} = \mathscr{R}^D$, using a Lebesgue measure in the integration,

$$\min \; \hbar(p \to \hat{p}) = -\int_{\mathscr{X}} p(\mathbf{x}) \ln(\hat{p}(\mathbf{x}|\boldsymbol{\varphi})) d\mathbf{x} \tag{10.36}$$

In solving the *model identification* problem, one is faced with deciding the best model from a set of different models, $\hat{p}_i(\mathbf{x}|\boldsymbol{\varphi}_i), i = \{1, 2, \cdots, M\}$. Akaike maximizes the negative of $\hbar(p \to \hat{p}_i)$,

$$\hat{\boldsymbol{\varphi}} = \arg\max_{\boldsymbol{\varphi}_i} \; -\hbar(p \to \hat{p}_i)$$

$$= \arg\max_{\boldsymbol{\varphi}_i} \; \int_{\mathscr{X}} p(\mathbf{x}) \ln(\hat{p}_i(\mathbf{x}|\boldsymbol{\varphi}_i)) d\mathbf{x} \tag{10.37}$$

across different models with probability density functions, $\hat{p}_i(\mathbf{x}|\boldsymbol{\varphi}_i)$, to find the model with the most optimal parameters, $\hat{\boldsymbol{\varphi}}$.

Note that $p(x)$ (the true density of the system) in Equation 10.37 is fixed, regardless of the model which is used. Also, the integral in Equation 10.37 is really the *expected value* of the *likelihood* of the model parameters and may be estimated without knowing $p(x)$, when a large number of observations are available. The optimization objective function of Equation 10.37 would also coincide with the *maximum mean likelihood estimate* where the likelihood of the model parameters is given by $\hat{p}(\mathbf{x}|\boldsymbol{\varphi})$.

In Section 7.7, we made the assumption that we can estimate the true probability density, $p(\mathbf{x})$, by a parametric version, $\hat{p}(\mathbf{x}|\boldsymbol{\varphi})$. Let us assume the same thing here. Therefore, we are assuming that the best we can do in estimating $p(\mathbf{x})$ is to find the closest parametric version of the density function, $\hat{p}(\mathbf{x}|\boldsymbol{\varphi})$. This assumption in itself introduces certain error which would not be correctable with the upcoming formulation. Furthermore, we make the assumptions that we made in Section 7.7, regarding the *proximity* to the optimal parameter vector and *regularity conditions* listed in that section.

Therefore, we may write the expression for the approximation to the *Kullback-Leibler directed divergence* from a point in the parameter space to another point near by, using Equation 7.136. As we stated, we are interested in minimizing $\mathscr{D}_{KL}(\boldsymbol{\varphi} \to \hat{\boldsymbol{\varphi}})$, which is the *directed divergence* from a point, $\boldsymbol{\varphi}$, in the vicinity of the parameter vector associated with the model which produces the *maximum mean likelihood estimate*, $\hat{\boldsymbol{\varphi}}$, to that point such that,

$$\hat{\boldsymbol{\varphi}} = \boldsymbol{\varphi} + \Delta\boldsymbol{\varphi} \tag{10.38}$$

$\hat{\boldsymbol{\varphi}}$ is the parameter vector associated with the *maximum mean likelihood estimate*. Let us rewrite Equations 7.136 and 7.135 here for simplicity.

$$\mathscr{D}_{KL}(\boldsymbol{\varphi} \to \hat{\boldsymbol{\varphi}}) = \frac{1}{2}(\Delta\boldsymbol{\varphi})^T \mathscr{I}_F(\Delta\boldsymbol{\varphi}) \tag{10.39}$$

where

$$\mathscr{I}_F = \mathscr{E}\left\{ (\nabla_{\boldsymbol{\varphi}} \hat{p}(\mathbf{x}|\boldsymbol{\varphi})) (\nabla_{\boldsymbol{\varphi}} \hat{p}(\mathbf{x}|\boldsymbol{\varphi}))^T \right\} \tag{10.40}$$

Akaike [2] goes through an exercise of restricting the parameter space, $\phi = \mathscr{R}^M$, to a lower-dimensional space, where every model with parameters $\boldsymbol{\varphi}_\gamma$ of the total Γ competing models, is defined in the parameter space, $\phi_\gamma = \mathscr{R}^{M_\gamma} : M_\gamma < M$. Then he computes an approximation to the minimum $\mathscr{D}_{KL}(\boldsymbol{\varphi}_\gamma \to \hat{\boldsymbol{\varphi}})$ for different models (different values of $\boldsymbol{\varphi}_\gamma \in \phi_\gamma$). As we saw earlier, this amounts to minimizing the *cross entropy*. Equation 10.40 requires the computation of an *expected value*. This may be done using a large sample size (say with N samples) without knowing the actual probability density function, only by using the *law of large numbers* for computing an approximation to the expected value, namely, the *sample mean*. Consider the discussion in Section 6.5.2, for probability densities defined in a *Cartesian product space*. With the effect of dimensionality in the definition of the probability density function in mind, for a large sample size, N, this leads to minimizing the following criterion [2],

$$AIC(\boldsymbol{\varphi}_\gamma) = -2\ell(\hat{\boldsymbol{\varphi}}_\gamma) + 2M_\gamma \tag{10.41}$$

where

$$\hat{\boldsymbol{\varphi}}_\gamma = \arg\max_{\boldsymbol{\varphi}_\gamma} \mathscr{E}\left\{ \ln \hat{p}(\mathbf{x}|\boldsymbol{\varphi}_\gamma) \right\} \tag{10.42}$$

and $\ell(\hat{\boldsymbol{\varphi}}_\gamma)$ is the *log-likelihood* of $\hat{\boldsymbol{\varphi}}_\gamma$ given the set of observations, $\{\mathbf{x}\}_1^N$. Equation 10.41 is the *AIC* criterion per [2].

The model with parameters, $\hat{\boldsymbol{\varphi}}$, which minimizes $AIC(\hat{\boldsymbol{\varphi}}_\gamma)$, is the model that approximates the *maximum mean likelihood estimate* the best, given the above restrictions on the parameter space. This model is called the *MAICE* which stands for the *minimum Akaike information criterion estimate* model and is given by the following equation,

$$MAICE = \min_{\boldsymbol{\varphi}_\gamma} AIC(\hat{\boldsymbol{\varphi}}_\gamma) \tag{10.43}$$

Note that since the problem of Equation 10.43 is one of *minimization*, we may factor out the constant, 2, from the definition of *AIC* and define a new *AIC* ↓ associated with this minimization problem as follows,

$$AIC \downarrow (\boldsymbol{\varphi}_\gamma) \overset{\Delta}{=} -\ell(\hat{\boldsymbol{\varphi}}_\gamma) + M_\gamma \tag{10.44}$$

In another interpretation, we may also rewrite the *AIC*, to form a new criterion, say *AIC* ↑, such that it may be *maximized*, as follows,

$$AIC \uparrow (\boldsymbol{\varphi}_\gamma) \overset{\Delta}{=} \ell(\hat{\boldsymbol{\varphi}}_\gamma) - M_\gamma \tag{10.45}$$

Therefore, the MAICE may be written in terms of any of the above AIC criteria as follows,

$$MAICE - \min_{\boldsymbol{\varphi}_\gamma} AIC(\hat{\boldsymbol{\varphi}}_\gamma) \tag{10.46}$$

$$= \min_{\boldsymbol{\varphi}_\gamma} AIC \downarrow (\hat{\boldsymbol{\varphi}}_\gamma) \tag{10.47}$$

$$= \max_{\boldsymbol{\varphi}_\gamma} AIC \uparrow (\hat{\boldsymbol{\varphi}}_\gamma) \tag{10.48}$$

The *AIC* may be seen as an *efficient statistic* (see Definition 6.76). This criterion may also be viewed as a *minimum cross-entropy*. If the restriction on the dimensionality of the parameters of model γ are removed, then all the models will have the same number of parameters, $M_{gamma} = M$. In that case, the problem formulation would just become one of *maximum likelihood estimation*.

10.6.2 Bayesian Information Criterion (BIC)

Schwarz [24] takes the Bayesian approach toward the representation of the models in lower dimensional space to present a new criterion called the *Bayesian information criterion (BIC)*. Reference [24] makes several assumptions which are generally satisfied in *speaker recognition* approaches using this criterion, such as many *speaker segmentation* algorithms [6, 18, 19, 23]. The following is a list of these assumptions.

Assumptions made in the development of BIC:

1. The observed samples are assumed to be generated by a parametric model belonging to the *Darmois-Koopman-Pitman exponential family* (Equation 6.194).
2. There are Γ competing models, each having a parameter vector, $\boldsymbol{\varphi}_\gamma \in \phi_\gamma$, such that $\phi_\gamma \subset \mathcal{R}^{M_\gamma}, M_\gamma \leq M$.
3. The *a-priori* probability of $\boldsymbol{\varphi}$ has the following Bayesian form,

$$P(\boldsymbol{\varphi}) = \sum_{\gamma=1}^{\Gamma} p(\boldsymbol{\varphi}|\boldsymbol{\varphi}_\gamma)P(\boldsymbol{\varphi}_\gamma)[\boldsymbol{\varphi}] \tag{10.49}$$

4. We may use the Bayesian approach, picking a model with the parameter vector $\boldsymbol{\varphi}_\gamma$ such that the *a-posteriori* probability of the model is maximal.

Using the above assumptions, *Schwarz* [24] derives the following *Bayesian information criterion (BIC)* which is comparable to the definition of *AIC* ↑ and must be at its *maximum* for the best model,

$$BIC(\boldsymbol{\varphi}_\gamma) \triangleq \ell(\hat{\boldsymbol{\varphi}}_\gamma) - M_\gamma(\frac{1}{2}\ln N) \tag{10.50}$$

Just as in the definition of *AIC*, $\ell(\hat{\boldsymbol{\varphi}}_\gamma)$ is the *log-likelihood* of $\hat{\boldsymbol{\varphi}}_\gamma$ (Equation 10.42) given the set of observations, $\{\mathbf{x}\}_1^N$. Therefore, we may define the *maximum Bayesian information criterion estimate (MBICE)* model in the spirit of *Akaike*, as follows,

$$MBICE \triangleq \max_{\boldsymbol{\varphi}_\gamma} BIC(\hat{\boldsymbol{\varphi}}_\gamma) \tag{10.51}$$

Note that

$$BIC(\boldsymbol{\varphi}_\gamma) > AIC \uparrow (\boldsymbol{\varphi}_\gamma) \ \forall \ N > 7 \tag{10.52}$$

In practice, N is chosen to be quite large for any analysis to be statistically significant. Therefore, practically, *BIC* and *AIC* produce very different results.

References

1. Akaike, H.: Information Theory and an Extension of the Maximum Likelihood Principle. In: Proceedings of the 2nd International Symposium on Information Theory, Supp. to Problems of Control and Informatioin Theory, pp. 267–281 (1972)
2. Akaike, H.: A new look at the statistical model identification. IEEE Transactions on Automatic Control **19**(6), 716–723 (1974)
3. Bahl, L., Brown, P., deSouza, P., Mercer, R.: A new algorithm for the estimation of hidden Markov model parameters. In: International Conference on Acoustics, Speech, and Signal Processing (ICASSP), pp. 493–497 (1988)
4. Bahl, L.R., Brown, P.F., deSouza, P.V., Mercer, R.L.: Maximum Mutual Information Estimation of Hidden Markov Model Parameters for Speech Recognition. In: International Conference on Acoustics, Speech, and Signal Processing (ICASSP), vol. 11, pp. 49–52 (1986)
5. Berger, A., Della-Pietra, S., Della-Pietra, V.: A Maximum Entropy Approach to Natural Language Processing. Computational Linguistics **22**(1) (1996)

6. Chen, S.S., Gopalakrishnan, P.S.: Speaker, Environment and Channel Change Detection and Clustering via the Bayesian Inromation Criterion. In: IBM Techical Report, T.J. Watson Research Center (1998)

7. Cramér, H.: Mathematical Methods of Statistics. Princeton University Press (1999). ISBN: 0-691-00547-8

8. Dumitrescu, M.E.B.: The Application of the Principle of Minimum Cross-Entropy to the Characterization of the Exponential-Type probability Distributions. Annals of the Institute of Statistical Mathematics **38**(1), 451–457 (1986)

9. Good, I.J.: Maximum Entropy for Hypothesis Formulation, Especially for Multidimensional Contingency Tables. The Annals od Mathematical Statistics **34**(3), 911–934 (1963)

10. Guiasu, S., Shenitzer, A.: The Principle of Maximum Entropy. The Mathematical Intelligencer **7**(1), 42–48 (1985)

11. Guyon, I.M., Vapnik, V.N., Boser, B.E., Solla, S.A.: Structural Risk Minimization for Character Recognition. In: T. David S (ed.) Neural Information Processing Systems, vol. 4. Morgan Kaufmann Publishers, San Mateo, CA (1992)

12. Horowitz, J.L.: Semiparametric Models in Econometrics. Springer, New York (1998). ISBN: 0-387-98477-1

13. Jaakkola, T., Haussler, D.: Exploiting Generative Models in Discriminative Classifiers. In: Advances in Neural Information Processing Systems, vol. 11, pp. 487–493. MIT Press (1998)

14. Jaakkola, T., Meila, M., Jebara, T.: Maximum Entropy Discrimination. In: Advances in Neural Information Processing Systems, vol. 12, pp. 470–476. MIT Press (1999)

15. Jaynes, E.T.: Information Theory and Statistical Mechanics. The Physical Review **106**(4), 620–630 (1957)

16. Jaynes, E.T.: Information Theory and Statistical Mechanics II. The Physical Review **108**(2), 171–190 (1957)

17. Johnson, R.W., Shore, J.E.: Comments on and Corrections to 'Axiomatic Derivation of the Principle of Maximum Entropy and the Principle of Minimum Cross-Entropy'. IEEE Transactions on Information theory **IT-29**(6), 942–943 (1983)

18. Kotti, M., Benetos, E., Kotropoulos, C.a.: Computationally Efficient and Robust BIC-Based Speaker Segmentation. Audio, Speech, and Language Processing, IEEE Transactions on **16**(5), 920–933 (2008)

19. Lagrange, M., Martins, L., Teixeira, L., Tzanetakis, G.a.: Speaker Segmentation of Interviews Using Integrated Video and Audio Change Detectors. In: Content-Based Multimedia Indexing, 2007. CBMI '07. International Workshop on, pp. 219–226 (2007)

20. McLachlan, G.J., Krishnan, T.: The EM Algorithm and Extensions, 2nd edn. Wiley Series in Probability and Statistics. John Wiley & Sons, New York (2008). ISBN: 0-471-20170-7

21. Naik, P., Tsai, C.L.: Residual Information Criterion for Single-Index Model Selections. Journal of Nonparametric Statistics **16**(1–2), 187–195 (2004)

22. Normandin, Y., Cardin, R., DeMori, R.: High-Performance Connected Digit Recognition using Maximum Mutual Information Estimation. IEEE Transactions on Speech and Audio Processing **2**(2), 299–311 (1994)

23. Reynalds, D.A., Torres-Carrasquillo, P.: Approaches and Applications of Audio Diarization. In: International Conference on Acoustics, Speech, and Signal Processing (ICASSP), vol. 5, pp. 953–956 (2005)

24. Schwarz, G.: Estimating the Dimension of a Model. Annals of Statistics **6**(2), 461–464 (1978)

25. Shore, J.E., Johnson, R.W.: Axiomatic Derivation of the Principle of Maximum Entropy and the Principle of Minimum Cross-Entropy. IEEE Transactions on Information theory **IT-26**(1), 26–37 (1980)

26. Shore, J.E., Johnson, R.W.: Properties of Cross-Entropy Minimzation. IEEE Transactions on Information theory **IT-27**(4), 472–482 (1981)

27. Vapnik, V.N.: Statistical learning theory. John Wiley, New York (1998). ISBN: 0-471-03003-1

Chapter 11
Unsupervised Clustering and Learning

All generalizations are dangerous, even this one!
Alexandre Dumas (1824–1895)

In Chapter 10 we discussed parameter estimation and model selection. In this chapter, we will review different techniques for partitioning the total sample space, $\mathcal{X}$, of a random variable, X, into different subsets as classes, $\mathcal{X}_\gamma$, which are Γ sample spaces associated with Γ random variables, X_γ. Therefore, based on the notation of Definition 6.11, the partition, $\mathcal{P}$ may be written in several representations,

$$\mathcal{P} = \mathcal{X}/\mathsf{P} \tag{11.1}$$
$$= \{\mathcal{X}_\gamma\} \quad \gamma \in \{1, 2, \cdots, \Gamma\} \tag{11.2}$$
$$= \{[\xi_\gamma]_\mathsf{P}\} \quad \gamma \in \{1, 2, \cdots, \Gamma\} \tag{11.3}$$

In Equation 11.3, ξ_γ is the representative element for partition class γ which based on equivalence relation P of the partitioning logic of choice will be equivalent to a set of elements in the sample space, $\mathcal{X}$.

This partitioning is called *unsupervised clustering* since it does not use any labels associated with the samples in $\mathcal{X}$ and it is only conditioned upon the *hypotheses* which are designed by the *classification techniques*, similar to those of Section 9.3. If the clustering is done based on *a-priori* labels, the clustering is known as supervised clustering.

Generally, in most unsupervised clustering techniques, the number of classes, Γ, is assumed to be known. Of course, as we will see, there are techniques for starting with an approximate Γ and then to use certain criteria to modify that number.

We start with *unsupervised clustering* techniques which concentrate on partitioning a collection of samples into bins and estimating parameters that would best represent the *sufficient statistics* for each of these bins. The clustering problem is an *NP*-complete problem, similar to the decision tree generation problem discussed in Section 9.4.1. Generally, the first set of techniques use the *Euclidean distance* as a metric to evaluate their estimation performance and they rely on estimating first-order statistics (means) in the process of clustering. We call these techniques, *basic clustering techniques.*

Once the clusters are completed, higher order statistics are generally computed. They also use some kind of averaging technique to define the overall objective function associated with their performance. Depending on the metric, type of averaging and the process of computing statistics and cluster assignments for each sample vector, different techniques have been developed.

The second set of techniques, covered in this chapter, work on optimizing more advanced objective functions such as the *likelihood function* and *entropy* in order to be able to estimate the cluster parameters and to assign samples to these clusters. In general, these techniques work on estimating the parameters, that would represent the *sufficient statistics*, by using information from a set of observed random samples. These observed samples are in a space which may not completely map the original space of the *sufficient statistics* of the problem of interest. In other words, the system may not be *fully observable* [16].[1] Therefore, we refer to this approach as *estimation using incomplete data*.

Algorithms of this set of techniques generally work by utilizing higher order statistics about the clusters. Some modified versions of these algorithms even provide estimates for such higher order statistics, such as the *covariance matrix* in the process of clustering.

The *basic clustering techniques*, which will be discussed in Section 11.2, generally possess a dual usage. They have been used to simply create *codewords* for conducting *vector quantization*, as well as clustering data for further, more sophisticated modeling, using higher order statistics. We will examine the idea behind vector quantization. Later, we will discuss and develop the basic clustering techniques which provide us with clusters and which will use the first order statistics of the data in the corresponding clusters to optimize a global objective function. The final results may be further refined to estimate higher order statistics about the determined clusters or to be used as initial conditions for more sophisticated techniques which will follow in the rest of this chapter.

11.1 Vector Quantization (VQ)

Generally speaking, *vector quantization* stems from a clustering mind-set wherein some *label* for each cluster called a *codeword* is used to signify the presence of each of the feature vectors that fall in the neighborhood of (are classified as) that

---

[1] The observability concept is one that has been used in many parameter estimation communities including control and signal processing. An example is the observability relations in system theory [36] and control system theory [16].

specific cluster. This *codeword* associated with each cluster, $\mathcal{X}_\gamma$, is the ξ_γ which was discussed in the beginning of this chapter, defining the equivalence classes of the clustering (partitioning) scheme. The collection of these clusters (*codewords*), which has been determined by a common distance or distortion measure or any feature dependent criterion, is called a *codebook*. The *codebook* may be viewed as the *partition* defined by Definition 6.11. There may be multiple *codebooks* used in a specific task and each *codebook* has a set of *codewords* associated with it.

Vector quantization is mostly used in cases where the small deviations from a mean value are not very important in the decision process, but they will cause confusion and error instead. It also tends to have an inherent noise retardation. In Section 11.2, we will see an array of different unsupervised clustering techniques. Some of these techniques such as *k-means* clustering may be used to come up with the *quanta*, as shown by an example in the original paper of *Lloyd* [32] on finding *optimal quantization levels* for *PCM* representation of an audio signal.

Once the overlap between clusters becomes pronounced, more sophisticated categorization is needed. In fact in a variety of cases, instead of discrete categorization, as in the case of general vector quantization, a more distributed and smooth clustering may be desired. In the latter case, instead of quantizing a vector which would associate it with a specific cluster (*codeword*), a membership level is assigned to the vector based on how likely or how close it is to individual clusters (codewords). An example of this type of assignment is the *Gaussian Mixture Modeling* (*GMM*), discussed in detail in Section 13.7.

Usually, regardless of whether pure vector quantization is done or a smooth multi-class membership, such as a *GMM*, the initial stage, computing the *centroids*, is quite similar. Once the centroids are found, a *distortion measure* is used to compute the relative memberships and to do the fine-tuning on the statistical estimates (*means* and in some case *variances* and higher order statistics).

In the next section, we will discuss several different vector quantization algorithms which are used in speech-related clustering. Also, in Chapter 8, we reviewed an array of different metrics and distortion measures that would allow us to estimate the *centroids* and eventually the *variances* associated with the data in addition to *membership functions*.

11.2 Basic Clustering Techniques

The most popular basic technique which has been used successfully in many applications for decades, including speech-related ones, is the *k-means* unsupervised clustering technique. Hundreds of refinements to the *k-means* clustering technique have been done in the more than 5 decades in which it has been used. It is still one of the most simple-to-implement techniques with good practical results.

In the next few subsections, we will study this and related techniques. Of course, it is impossible to go into the details of all the techniques of this sort. However, some important references will be provided for the avid reader to continue examining the different possibilities.

It is also important to have *outliers* in mind, while reading this chapter. *Outliers* are samples which are not well suited to belong to any specific cluster in the technique of choice. They are usually the cause of many unwanted distortions in the definition of the clusters. In other words, they tend to bias the value of the representative of the equivalence class in such a way that the correct underlying portioning may be disrupted. It is generally important to have some mechanism in mind to detect and handle outliers. In most cases, we would like to keep the outliers separate from the well-behaved set of samples, which are those samples that conform to the underlying model of interest. There are many methods available in the literature for the detection and handling of outliers [25]. Later in the chapter (see Section 11.2.3) we will speak about starting with a larger number of clusters (*overpartitioning*) to ensure that outliers will not affect the main clusters of interest with many members. In Section 11.2.4 we will discuss merging of clusters to some degree. A decision may be made to throw away outliers or to merge them with the populated clusters.

11.2.1 Standard k-Means (Lloyd) Algorithm

Lloyd [32] presented a classification technique geared for *vector quantization* used in determining the quantized signal levels so that the *quantization error* in the *PCM coding* of *audio signals* for *telephony applications* would be optimized. The basic premise behind this algorithm was to minimize the sum of squares of errors, computed using the *Euclidean distance* between the signal and the *quanta* represented by the means of the discovered clusters. According to the footnote in [32], the contents of the paper were first presented in 1957 at a meeting organized by the *Institute of Mathematical Statistics*. This work was, however, not published until 1982 [32].

In 1967, *MacQueen* [33] described the *k-means* algorithm and called it by that name for the first time. [33] went further in presenting several applications of the algorithm as well as extensions to general *metric spaces* (see Section 6.2.3).

Generally, the standard *k-means algorithm*, used today, is very similar to that described by *Lloyd* [32] and for that reason it is sometimes called the *Lloyd algorithm*. There have been a plethora of methods derived from the basic *k-means* algorithm, some of which will be touched upon, in passing. However, the existence of an astonishing number of techniques, with only minor changes, prohibits the possibility of their complete coverage in this book. Let us examine the standard *k-means clustering* algorithm, by first defining the associated *minimization objective function*.

Consider a random variable, $X : \mathbf{x} :\in \mathscr{X} = \mathscr{R}^D$. Therefore, $\mathbf{x} : \mathscr{R}^1 \mapsto \mathscr{R}^D$. Let us assume that we are interested in partitioning the sample space, $\mathscr{X}$ into Γ partitions. The *k-means* clustering algorithm assumes prior knowledge of the number of clusters, Γ, and it is concerned with finding the *parameter matrix, $\Phi : \mathscr{R}^\Gamma \mapsto \mathscr{R}^D$*, that would minimize the objective function given by Equation 11.5. The γth column of Φ is the *parameter vector, φ_γ*, associated with cluster γ.

$$E(X,\Phi) = \sum_{x \in \mathscr{X}} \min_{\gamma} \|\mathbf{x} - \varphi_\gamma\|_{\mathscr{E}}^2 \tag{11.4}$$

In practice, we do not have complete knowledge of the *sufficient statistics* of X. However, we are able to approximate the partitioning by considering an observed set of samples of X. Let us assume that we have a set of observed samples, $\{\mathbf{x}\}_1^N = \{\mathbf{x}_n, n \in \{1,2,\cdots,N\}\}$. Then the objective function of the minimization problem quantified by the objective function of Equation 11.5 may be approximated by a new objective function described in terms of the observed samples, $\mathbf{x}_n$, as follows,

$$E(X,\Phi) \approx E_{KM}(\{\mathbf{x}\}_1^N, \Phi)$$
$$= \sum_{n=1}^N \min_{\gamma} \|\mathbf{x}_n - \varphi_\gamma\|_{\mathscr{E}}^2 \tag{11.5}$$

In *k-means clustering*, the parameter vector, φ_γ, is just the sample mean of the members of cluster γ. We may define the matrix of means of the clusters, as $\mathbf{M} : \mathscr{R}^\Gamma \mapsto \mathscr{R}^D$ whose γ^{th} column is φ_γ. Therefore,

$$\varphi_\gamma = \hat{\boldsymbol{\mu}}_\gamma \tag{11.6}$$
$$\Phi = \mathbf{M}_\gamma \tag{11.7}$$

The following steps describe the *k-means algorithm*.
The k-means algorithm:

1. Choose the number of clusters, Γ. Initialize the sample mean vectors, $\hat{\boldsymbol{\mu}}_\gamma^{(k)}, k = 0$, for the initial clusters, $\mathscr{X}_\gamma^{(k)}, k = 0$, using some randomization logic or based on any *a priori* information about the clusters. The parameters are then signified by $\varphi_\gamma^{(k)}, k = 0$, making up the parameter matrix, $\Phi^{(k)}, k = 0$.
2. Given the current cluster definitions, $\mathscr{X}_\gamma^{(k)}$, go through all the observed samples, $\mathbf{x}_n, n \in \{1,2,\cdots,N\}$, and assign each sample vector to one (and only one) of the clusters, $\mathscr{X}_\gamma^{(k)}$, by using the following *membership index function*,

$$\hat{\gamma}_{nk} \stackrel{\Delta}{=} \eth(\mathbf{x}_n, \boldsymbol{\Phi}^{(k)})$$

$$= \eth(\mathbf{x}_n, \mathbf{M}^{(k)})$$

$$= \underset{\gamma}{\arg\min}\ d_E\left(\mathbf{x}_n, \hat{\boldsymbol{\mu}}_\gamma^{(k)}\right)$$

$$= \underset{\gamma}{\arg\min}\ \|\mathbf{x}_n - \hat{\boldsymbol{\mu}}_\gamma^{(k)}\|_{\mathscr{E}} \tag{11.8}$$

3. Compute the objective function of *k-means* based on the current clusters and memberships,

$$E_{KM}{}^{(k)} = E_{KM}(\{X\}_1^N, \mathbf{M}^{(k)})$$

$$= \sum_{n=1}^{N} \|\mathbf{x}_n - \hat{\boldsymbol{\mu}}_{\hat{\gamma}_{nk}}\|_{\mathscr{E}}^2 \tag{11.9}$$

4.

$$\delta^{(k)} \stackrel{\Delta}{=} E_{KM}{}^{(k)} - E_{KM}{}^{(k-1)} \tag{11.10}$$

If $\left((k > 0 \wedge \delta^{(k)} \le \delta_{min}) \vee (k > 1 \wedge (\delta^{(k)} - \delta^{(k-1)}) \le \varepsilon)\right)$ then *terminate*.
   In the above logic, δ_{min} and ε are two small numbers.

5. Based on the new memberships, compute the means, $\mathbf{M}^{(k+1)}$ of the new clusters, $\mathscr{X}_\gamma^{(k)}$, using the *maximum likelihood estimate* (sample mean), given by Equation 6.173.

6. Increment k and repeat the procedure, starting from step 2.

Bottou and Bengio [10] show that the *k*-means algorithm is really a gradient descent algorithm (see Section 25.1.1) for the *quantization error*, Equation 11.5. Therefore, every iteration will decrease this error, although when a local minimum is reached, this convergence could be arbitrarily slow.

Once the *k-means* procedure has completed, generally, it is interesting to compute higher order statistics on the clusters which have been computed by the algorithm. In most cases, especially in speaker recognition and other pattern recognition problems, we assume that the clusters will have *Gaussian (normal)* distributions. For such problems, once the clustering algorithm has converged to a set of *means*, $\hat{\boldsymbol{\mu}}_\gamma$, it associates each sample vector, $\mathbf{x}_n$ with a corresponding cluster, $\hat{\gamma}_n$, through the defined *membership index function*, $\hat{\gamma}_n = \eth(\mathbf{x}_n, \mathbf{M})$, and provides the number of vectors, N_γ, associated with each cluster, $\mathscr{X}_\gamma$. Therefore, all the information necessary for estimating the covariance matrix of the samples in cluster $\mathscr{X}_\gamma$ is available. Using this information, Equation 6.203 may be used to compute the *unbiased* estimate of the *sample covariance* matrix, $\tilde{\boldsymbol{\Sigma}}_\gamma$,

$$\tilde{\boldsymbol{\Sigma}}_\gamma = \frac{1}{N_\gamma - 1} \sum_{\{n:\hat{\gamma}_n = \gamma\}} (\mathbf{x}_n - \hat{\boldsymbol{\mu}}_\gamma)(\mathbf{x}_n - \hat{\boldsymbol{\mu}}_\gamma)^T \tag{11.11}$$

In the same manner, the *maximum likelihood estimate* of the covariance matrix, $\hat{\Sigma}_\gamma$ may be computed for each of the clusters, using the following equation,

$$\hat{\Sigma}_\gamma = \frac{1}{N_\gamma} \sum_{\{n:\hat{\gamma}_n=\gamma\}} (\mathbf{x}_n - \hat{\boldsymbol{\mu}}_\gamma)(\mathbf{x}_n - \hat{\boldsymbol{\mu}}_\gamma)^T \tag{11.12}$$

In most practical applications, including speaker recognition, since the amount of data is limited, it is assumed that the values of the D dimensions of the sample data $\mathbf{x}_n$ are uncorrelated to be able to assume a diagonal *sample covariance* matrix. This is of course, certainly not true, however, the increased estimation accuracy, in most cases, outweighs the side-effects of this independence assumption. If we assume *statistical independence* of the coordinates of $\mathbf{x}_n$, then we may use Equation 6.207 to computed the diagonal sample covariance estimate, $\tilde{\Sigma}_\gamma$.

11.2.2 Generalized Clustering

As we shall see later, while reviewing the *k-harmonic means algorithm*, [43] introduces a generalized form for all *centroid-based* clustering algorithms and generalizes the parameter vector computation for these clustering algorithms in one compact form. The following is the generalized expression for the γth cluster,

$$\boldsymbol{\varphi}_\gamma^{(k)} = \frac{\sum\limits_{n=1}^{N} \Upsilon_{\mathscr{X}_\gamma^{(k)}}(\mathbf{x}_n) w^{(k)}(\mathbf{x}_n) \mathbf{x}_n}{\sum\limits_{n=1}^{N} \Upsilon_{\mathscr{X}_\gamma^{(k)}}(\mathbf{x}_n) w^{(k)}(\mathbf{x}_n)} \tag{11.13}$$

where $\Upsilon_{\mathscr{X}_\gamma^{(k)}}(\mathbf{x}_n)$ is the *characteristic function (membership function)* of sample vector against set (cluster) $\mathscr{X}_\gamma^{(k)}$. See Equations 6.6 and 6.10 for examples of *hard* and *soft characteristic (membership) functions*, respectively. $w^{(k)}(\mathbf{x}_n)$ is a weighting function which may weight different samples differently, or alternatively treat them all the same if it is set to the constant, $w(\mathbf{x}_n) = 1$.

For the *standard Lloyd algorithm*,

$$\Upsilon_{\mathscr{X}_\gamma^{(k)}}(\mathbf{x}_n) \stackrel{\Delta}{=} \begin{cases} 1 \ \forall \ \gamma = \hat{\gamma}_{nk} \\ 0 \ \forall \ \gamma \neq \hat{\gamma}_{nk} \end{cases} \tag{11.14}$$

where $\hat{\gamma}_{nk}$ is prescribed by Equation 11.8, and

$$w^{(k)}(\mathbf{x}_n) = 1 \ \forall \ n \in \{1, 2, \cdots, N\} \tag{11.15}$$

Hamerly and *Elkan* [20] have used this generalization to compare several different clustering algorithms.

11.2.3 Overpartitioning

One practical method for limiting the effect of outliers on the convergence of the *k-means algorithm* is to start with a slightly larger number of clusters, $\Gamma_o > \Gamma$ and then to reduce the number of clusters to the final goal of Γ by merging clusters, two at a time, based on the procedure described in Section 11.2.4. This will allow outliers to initially fall into dedicated clusters to stop them from presenting unwanted influence on the convergence of the rest of the clusters. Once the algorithm has somewhat converged, the extra clusters may be combined with their neighboring clusters, reducing their effect on the rest of the clusters. Practically, this allows for better global convergence.

As we mentioned earlier, a decision may have to be made to either throw away extraneous partitions formed from the outliers or to merge them with neighboring partitions. It is important to note that keeping partitions with few samples will create an ill-conditioned representation of the cluster by defying the *law of large numbers* (Section 6.9.2), which is necessary for many of the assumptions made in this book regarding statistics.

11.2.4 Merging

Sometimes it becomes important to reduce the number of clusters from say Γ_1 to Γ_2 where $\Gamma_2 < \Gamma_1$. To do this, usually, a *metric* is used to compare the distance between neighboring clusters. Then the two clusters with the smallest distance get merged into one. One reason for doing such merging may be because the amount of data which has been associated with each individual cluster is too small to produce meaningful statistics. Of course this type of reduction should be done sparingly. Such outliers usually change the statistics of a cluster, making it somewhat less appropriate for modeling the majority of data points which are associated with it. Therefore, it is important to be able to reject outliers as well.

In general, the merging may be done on more than two clusters at once, although it is seldom done so in practice. Usually, we try to merge two clusters and then re-evaluate cluster *memberships* and *statistics*. Let us assume that we have picked out Γ_m clusters to be merged. Each of these clusters is identified by three statistics which are the number of vectors in each cluster, N_γ, their *sample means*, $\hat{\boldsymbol{\mu}}_\gamma$, and their *sample covariance* matrices, $\hat{\boldsymbol{\Sigma}}_\gamma$, where $\gamma_m \in \{1, 2, \cdots, \Gamma_m\}$.

The new cluster will have the following statistics after merging its component clusters,

$$N_\gamma = \sum_{\gamma_m=1}^{\Gamma_m} N_{\gamma_m} \tag{11.16}$$

$$\hat{\boldsymbol{\mu}}_\gamma = \frac{\displaystyle\sum_{\gamma_m=1}^{\Gamma_m} N_{\gamma_m}\hat{\boldsymbol{\mu}}_{\gamma_m}}{N_\gamma} \tag{11.17}$$

$$\hat{\boldsymbol{\Sigma}}_\gamma = \frac{\displaystyle\sum_{\gamma_m=1}^{\Gamma_m} N_{\gamma_m}\hat{\boldsymbol{\Sigma}}_{\gamma_m}}{N_\gamma} \tag{11.18}$$

$$\tilde{\boldsymbol{\Sigma}}_\gamma = \frac{\displaystyle\sum_{\gamma_m=1}^{\Gamma_m} (N_{\gamma_m}-1)\tilde{\boldsymbol{\Sigma}}_{\gamma_m}}{N_\gamma - 1} \tag{11.19}$$

11.2.5 Modifications to the k-Means Algorithm

Although the *k-means* algorithm is quite useful in many applications including *vector quantization* and cluster initialization for *mixture models*, it presents a few stumbling blocks which have been the focus of many research projects since the inception of the algorithm. The two most important problems with the *k-means* algorithm are its extremely slow convergence at times and more importantly its major dependence on the initial guesses for the cluster parameters, $\boldsymbol{\Phi}^{(0)}$.

Both of these problems have been tackled by many researchers in different capacities. In this section, we will examine some of the solutions for reducing the effects of the initial parameters and convergence acceleration techniques. Some of these techniques are discussed in the following subsections and some, which are significantly different, are presented in their own sections later on.

11.2.5.1 Gaussian k-Means

As we noted, one of the problems with the *k-means* algorithm is the *hard decision* (*binary membership*) it makes regarding its membership function since it uses the *Euclidean distance*. One way to somewhat address this problem is to change the distance measure to the *Mahalanobis distance*, assuming a normal distribution within each cluster. Then $\boldsymbol{\Phi}^{(k)}$ will actually include the estimate of the *mean vector*,

$\hat{\boldsymbol{\mu}}_\gamma^{(k)}$, and the *sample covariance matrix*, $\hat{\boldsymbol{\Sigma}}_\gamma^{(k)}$. Therefore, the objective function of Equation 11.5 will be modified to include the *covariance* information, as follows,

$$E_{GKM}(\{\mathbf{x}\}_1^N, \boldsymbol{\Phi}) = \sum_{n=1}^{N} \min_{\gamma} \left[\left(\mathbf{x}_n - \hat{\boldsymbol{\mu}}_\gamma\right)^T \left(\hat{\boldsymbol{\Sigma}}_\gamma\right)^{-1} \left(\mathbf{x}_n - \hat{\boldsymbol{\mu}}_\gamma\right) \right] \quad (11.20)$$

Similarly, the membership for the kth iteration is then given by the following,

$$\hat{\gamma}_{nk} \overset{\Delta}{=} \eth(\mathbf{x}_n, \boldsymbol{\Phi}^{(k)})$$

$$= \arg\min_{\gamma} \; d_M\left(\mathbf{x}_n, \hat{\boldsymbol{\mu}}_\gamma^{(k)}\right)$$

$$= \arg\min_{\gamma} \; \left[\left(\mathbf{x}_n - \hat{\boldsymbol{\mu}}_\gamma^{(k)}\right)^T \left(\hat{\boldsymbol{\Sigma}}_\gamma^{(k)}\right)^{-1} \left(\mathbf{x}_n - \hat{\boldsymbol{\mu}}_\gamma^{(k)}\right) \right] \quad (11.21)$$

One of the major problems with this technique is that it increases the computational load of the algorithm, slowing down even more. Since the *k-means* algorithm is already a very slow procedure, the increased complexity is usually not tolerable. For this reason, in most practical cases, the regular *Lloyd algorithm* is used until convergence is achieved. Then, a few steps of *Gaussian k-means* are performed.

In the *Gaussian k-means* algorithm, the *characteristic function* and *weight*, defined in Section 11.2.2, may be written as follows,

$$\Upsilon_{\mathscr{X}_\gamma^{(k)}}(\mathbf{x}_n) \overset{\Delta}{=} \begin{cases} 1 \; \forall \; \gamma = \hat{\gamma}_{nk} \\ 0 \; \forall \; \gamma \neq \hat{\gamma}_{nk} \end{cases} \quad (11.22)$$

where $\hat{\gamma}_{nk}$ is given by Equation 11.21, and

$$w^{(k)}(\mathbf{x}_n) = 1 \; \forall \; n \in \{1, 2, \cdots, N\} \quad (11.23)$$

Note that the only difference between the characteristic functions of the *Gaussian* and *standard k-means* algorithms is the difference in the membership index functions in Equation 11.21 and 11.8. Also, the weights are identical between the two algorithms.

11.2.5.2 Modified k-Means

Many modifications have been made to the *k-means algorithm*, basically making nominal changes, to be able to either increase the convergence speed of the algorithm or to make it converge to a better local minimum. In 1990, *Green, et. al* [19] reviewed the literature, up to that date, for several *optimal weighting k-means* procedures and showed that the performance of these techniques is also very much tied to the initial partitioning of the space.

Among these modifications to the *k-means algorithm*, there is one known, simply, as *modified k-means* which has set out to achieve both of these improvements. This modification was introduced by *Lee* [28] in 1997. One problem with the *modified k-means* approach is that it lacks sound theoretical backing and it deviates from a basic premise, the, so called, *centroid condition*, requiring that the *centroids* at each iteration be computed according to step 5 given in the statement of the *k-means algorithm* (Section 11.2.1).

This idea follows the opinion that perturbing the location of the centroids would act toward the stabilization of the global convergence of statistical systems. This idea has been inherited from *material science* in the, so called, *annealing technique* used for growing stable crystals, especially in producing strong *steel alloys*. In traditional crystal optimization, instead of quickly cooling the material, while building the crystal, gradual cooling is sometimes followed by a marginal amount of heating to allow the crystals to be formed in a globally optimal structure rather than one which is locally optimized. The same idea was introduced into *global optimization techniques* used for training *neural networks* by *Kirkpatrick* in 1983 [26] (see Sections 25.4.1 and 25.6). These types of techniques are known as *stochastic relaxation methods* [42].

Therefore, the modification introduced by [28] is a perturbation technique for the *means* of the clusters. As we stated at the beginning of this section, it amounts to modifying step 5 of the *k-means algorithm* as follows,
Modified step 5 for the k-means algorithm:
Compute the Γ mean vectors (columns of $\mathbf{M}^{(k+1)}$) as prescribed by step 5 of the basic *k-means algorithm*. Since in the modified algorithm these will not be the actual centers, let us call these vectors, $\tilde{\boldsymbol{\mu}}_\gamma^{(k+1)}$ instead of $\hat{\boldsymbol{\mu}}_\gamma^{(k+1)}$. Then the new centroids will be picked as

$$\hat{\boldsymbol{\mu}}_\gamma^{(k+1)} = \hat{\boldsymbol{\mu}}_\gamma^{(k)} + \alpha \left(\tilde{\boldsymbol{\mu}}_\gamma^{(k+1)} - \hat{\boldsymbol{\mu}}_\gamma^{(k)} \right) \tag{11.24}$$

where α is proposed by [28] to be $1 \leq \alpha < 2$, based on empirical results. $\alpha = 1$ coincides with the basic *Lloyd algorithm*. When $\alpha = 2$, the new center will be the reflection of the centroid at iteration k about the centroid at iteration $k+1$ given by the *Lloyd algorithm*. When $\alpha = 2$, the reflected centroid coincides with the proposal of *Jancey* in 1966 while performing botanical clustering [23, 24].

Lee [28] empirically shows that when α approaches 2 (the *Jancey algorithm*), the convergence begins to slow down and the objective function values become less optimal according to tests with pixel clustering applications on standard images. [28] suggests that $\alpha = 1.8$ produces optimal results for these pixel clustering applications and that up to this value, by increasing α, the rate of convergence increases and better local minima are attained.

11.2.6 k-Means Wrappers

k-Means wrappers [20] are techniques that try to improve *unsupervised clustering* by using some version of the *k-means algorithm* as a component of the overall clustering scheme. Some, like the *global k-means* [29] approach, operate by proliferating different initial conditions to find a global optimum. Here we will discuss a few such techniques.

11.2.6.1 Efficient k-Means

The algorithm introduced by [45], is a *wrapper* which splits the sample data, $\{\mathbf{x}\}_1^N$ into M random sets, $\mathscr{A}_m, m \in \{1, 2, \cdots, M\}$ and conducts *overpartitioned k-means* clustering (see Section 11.2.3) on each of these sets. Then it uses the M sets of Γ_o clusters to evaluate the *k-means* objective function (Equation 11.5) against all the data. The set of clusters that produces the smallest objective function is chosen as the initial set of clusters for the *overpartitioned k-means* procedure applied on the total sample set, $\{\mathbf{x}\}_1^N$. The merging techniques of Section 11.2.4 are then used to reduce the number of clusters to the target Γ clusters as described in Section 11.2.3.

Zhang, et al. [45] show that this technique helps improve convergence into a more global optimum, in some occasions. Reference [27], also, shows some improvement over the standard *k-means* algorithm in its study.

11.2.6.2 Global k-Means

Likas, et al. [29] introduced a *k-means wrapper* which they call *global k-means* to combat convergence into a local minimum, when using the *k-means* algorithm. The basic technique, introduced in [29], starts with one cluster and builds up the number of clusters through a global search, starting with a prescribed initial location.

The following is the procedure for the *global k-means* clustering algorithm.

1. Set the first cluster center, $\hat{\boldsymbol{\mu}}_{\gamma=1}^{(0)}$ equal to the sample mean of the whole sample data, $\{\mathbf{x}\}_1^N$.

2. Choose the next $\Gamma - 1$ clusters, one-by-one, such that $\hat{\boldsymbol{\mu}}_{\gamma+1}^{(0)}$ is chosen from the remaining pool of samples, $\mathscr{A}_\gamma = \{\mathbf{x} \in \{\{\mathbf{x}\}_1^N \setminus \bigcup_{i=1}^{\gamma} \hat{\boldsymbol{\mu}}_i^{(0)}\}\}$ by running one full execution of the *k-means* algorithm and picking the sample point which produces the smallest value of the objective function.

Unfortunately, at every cluster evaluation, this technique ends up doing N executions of the *k-means* algorithm for a set of samples, $\{\mathbf{x}\}_1^N$. This is quite impractical for problems of interest in *speaker recognition*. The number of feature vectors, N, which are used as data samples being clustered in D-dimensional space is usually quite large and this prohibits the use of this technique. For this reason, [29] provides an acceleration method which avoids running the *k-means algorithm* so many times, by making an assessment of the upper-bound of the error based on different iterations. Also, a *k-d tree*[2] is proposed in the paper to facilitate the initialization of the clusters. The avid reader is referred to [29] for the details, as well as to [40] for the description of a variant of this algorithm called the *kernel-based global k-means* algorithm which uses kernel ideas discussed in the next section.

11.2.6.3 Kernel k-Means

Several *kernel k-means* techniques have been described in the literature. These techniques use the, so called, *kernel trick*, discussed in detail in Chapter 15. In summary the, so called, *kernel trick* was first described by *Aizerman* [1] in 194. *Burges* [11] called it by that name. This *kernel mapping technique* (see Section 15.3) uses a function, $\boldsymbol{\psi}$, to map the observation space $(\mathscr{X}, \mathfrak{X})$, to a complete inner product space of higher dimensions, $(\mathscr{H}, \mathfrak{H})$,

$$\boldsymbol{\psi} : (\mathscr{X}, \mathfrak{X}) \mapsto (\mathscr{H}, \mathfrak{H}) \tag{11.25}$$

Most of the techniques in this chapter may be converted to their kernel alternatives by replacing $\mathbf{x}$ in the objective function associated with the clustering technique with $\boldsymbol{\psi}(\mathbf{x})$. For the *standard k-means (Lloyd) algorithm* (Section 11.2.1), this would mean the modification of Equation 11.5 in a manner that would lead to the *kernel k-means* algorithm described by *Dhillon, et al.* [14]. As in the case of *support vector machines* (Chapter 15), the computation of the *objective function* and the *mean* will result in terms which only contain the *inner products* (Definition 24.49) of the mapping functions, $\boldsymbol{\psi}(\mathbf{x})$. For example, by defining a linear kernel,

$$(\mathscr{K})_{[i][j]} \stackrel{\Delta}{=} \mathscr{K}(\boldsymbol{\psi}_i, \boldsymbol{\psi}_j) \tag{11.26}$$
$$= \mathscr{K}(i, j) \tag{11.27}$$
$$= \langle \boldsymbol{\psi}_i, \boldsymbol{\psi}_j \rangle \tag{11.28}$$

the objective function may be represented in terms of the *kernel function* (Definition 24.56).

Let us formulate the above statement. The objective function for the *standard k-means* clustering algorithm is given by plugging Equation 11.6 into Equation 11.5

---

[2] A *k-d tree*, also known as a *k-dimensional tree*, is a multidimensional binary tree structure which is used for conducting search. It was introduced in 1975 by *Bentley* [6].

and replacing $\mathbf{x}_n$ with $\boldsymbol{\psi}(\mathbf{x}_n)$, to produce the following,

$$E(X,\boldsymbol{\Phi}) = \sum_{n=1}^{N} \min_{\gamma} \|\boldsymbol{\psi}(\mathbf{x}_n) - \hat{\boldsymbol{\mu}}_{\gamma}\|_{\mathscr{E}}^{2} \tag{11.29}$$

where the mean vector is now computed in the *feature space* (see Chapter 15), $\boldsymbol{\psi} \in \mathscr{H}$.

The sample mean vector, $\hat{\boldsymbol{\mu}}_{\gamma}$, for every cluster, γ, may be written in terms of the new mapping, $\boldsymbol{\psi}$ as follows,

$$\hat{\boldsymbol{\mu}}_{\gamma} = \frac{1}{N_{\gamma}} \sum_{\{n:\hat{\gamma}_n = \gamma\}} \boldsymbol{\psi}(\mathbf{x}_n) \tag{11.30}$$

For simplicity, let us define the following shorthand notation,

$$\boldsymbol{\psi}_n \triangleq \boldsymbol{\psi}(\mathbf{x}_n) \tag{11.31}$$

Using this shorthand notation and the expression for the sample mean vector given by Equation 11.30, we may write the square of the norm in Equation 11.29, only in terms of dot products, as follows,

$$\|\boldsymbol{\psi}(\mathbf{x}_n) - \hat{\boldsymbol{\mu}}_{\gamma}\|_{\mathscr{E}}^{2} = \langle \boldsymbol{\psi}_n, \boldsymbol{\psi}_n \rangle$$
$$- \frac{2}{N_{\gamma}} \sum_{\{m:\hat{\gamma}_m = \gamma\}} \langle \boldsymbol{\psi}_n, \boldsymbol{\psi}_m \rangle$$
$$+ \frac{1}{N_{\gamma}^2} \sum_{\{l:\hat{\gamma}_l = \gamma\}} \sum_{\{m:\hat{\gamma}_m = \gamma\}} \langle \boldsymbol{\psi}_l, \boldsymbol{\psi}_m \rangle \tag{11.32}$$

Using the definition of Equation 11.27, we may rewrite Equation 11.29 in terms of the Kernel function as follows,

$$E(X,\boldsymbol{\Phi}) = \sum_{n=1}^{N} \min_{\gamma} \mathscr{K}(n,n)$$
$$- \frac{2}{N_{\gamma}} \sum_{\{m:\hat{\gamma}_m = \gamma\}} \mathscr{K}(n,m)$$
$$+ \frac{1}{N_{\gamma}^2} \sum_{\{l:\hat{\gamma}_l = \gamma\}} \sum_{\{m:\hat{\gamma}_m = \gamma\}} \mathscr{K}(l,m) \tag{11.33}$$

Equation 11.33 may be used to perform the *k-means* steps as we did before. By choosing a higher dimensional space, we may be able to achieve linear separability in the *feature space*, $\mathscr{H}$, where the *sample space*, $\mathscr{X}$, may be nonlinearly separable. A lot more will be said about this concept in Chapter 15.

As we mentioned before, *kernel* extensions may be applied to most other clustering techniques. An example is the *kernel global k-means* technique [40].

11.2.6.4 k-Means++

In 2006, [2] introduced a new *initialization technique* for the clusters of the *k-means* algorithm. This algorithm basically replaces step 1 of the standard *Lloyd algorithm*. Reference [2] presents an elaborate review of the technique and provides details of the theory behind this selection. Here, we have reproduced the step-by-step procedure for the *k-means++* algorithm using the nomenclature of this textbook.

1. Set the first cluster center, $\hat{\boldsymbol{\mu}}_{\gamma=1}^{(0)}$ equal to a sample chosen in random from the sample data, $\{\mathbf{x}\}_1^N$.

2. Choose the next $\Gamma - 1$ clusters, one-by-one, such that $\hat{\boldsymbol{\mu}}_{\gamma+1}^{(0)}$ is chosen from the remaining pool of samples, $\mathscr{A}_\gamma = \{\mathbf{x} \in \{\{X\}_1^N \setminus \bigcup_{i=1}^{\gamma} \hat{\boldsymbol{\mu}}_i^{(0)}\}\}$ using the following probability distribution,

$$p(\mathbf{x}_n) = \frac{\left(d_{KM}\left(\mathbf{x}_n, \mathscr{X}_1^{\gamma(0)}\right)\right)^2}{\sum_{i=1}^N \left(d_{KM}\left(\mathbf{x}_i, \mathscr{X}_1^{\gamma(0)}\right)\right)^2} \quad \forall \mathbf{x} \tag{11.34}$$

where $\mathscr{X}_1^{\gamma(0)}$ is the set of clusters, $\{\mathscr{X}_j^{(0)} : j \in \{1, 2, \cdots, \gamma\}\}$ and,

$$d_{KM}\left(\mathbf{x}_n, \mathscr{X}_1^{\gamma(0)}\right) \triangleq \min_{1 \le j \le \gamma} \|\mathbf{x}_n - \hat{\boldsymbol{\mu}}_j^{(0)}\|_{\mathscr{E}} \tag{11.35}$$

Note that based on Equation 11.34,

$$p(\mathbf{x}) = 0 \; \forall \mathbf{x} = \hat{\boldsymbol{\mu}}_i^{(0)}, i \in \{1, 2, \cdots, \gamma\}. \tag{11.36}$$

Therefore, any data point will never be picked as a center more than once.

3. Perform the standard *Lloyd algorithm* of Section 11.2.1, starting from Step 2 onward.

Note that this technique is somewhat similar to the *global k-means* algorithm discussed in Section 11.2.6.2, with the exception that instead of running the *k-means* procedure N times, a probability distribution is used. This probability distribution which is based on the Euclidean distance, reduces the computation load. The difference in the computational load is significant.

Kumar and Wasan [27] have shown that the *k-means++ algorithm* has superior accuracy on their selected data, when compared to *Lloyd* (Section 11.2.1), *global k-means* (Section 11.2.6.2), *efficient k-means* (Section 11.2.6.1), and *x-means* (Section

11.2.6.6) algorithms. *k-means++* is basically an initialization modification wrapper. As we will see, there are techniques which make fundamental changes to the *k-means algorithm*, by changing the objective function and adding adaptive weighting techniques, such as the *general k-harmonic means* algorithm which will be discussed in Section 11.2.9.

11.2.6.5 Linde-Buzo-Gray (LBG) Algorithm

The *Linde-Buzo-Gray (LBG)* algorithm was introduced in 1980 [30]. It was intended as a wrapper around the *Lloyd* algorithm. However, it may be used to wrap any unsupervised clustering algorithm. It is a hierarchical partitioning technique which due to its binary partitioning, will always create clusters which are based on powers of two, in number. Therefore, the input to the algorithm would have to be a sample data set, $\{\mathbf{x}\}_1^N$, as in any other clustering algorithm, but the number of target clusters, Γ, would have to be perfect powers of two.

Here is how the *LBG* algorithm works:

1. Starting with a sample data set, $\{\mathbf{x}\}_1^N$, create one main cluster with its mean being the sample mean of the whole set and set the iteration $k = 0$.
2. Given the current number of clusters, $\Gamma_k = 2^k$, split each cluster into two new clusters, the mean of each new cluster being given by,

$$\boldsymbol{\mu}_{\gamma_k}^{(k+1)} = \boldsymbol{\mu}_{\gamma_k}^{(k)} + \boldsymbol{\varepsilon} \tag{11.37}$$

$$\boldsymbol{\mu}_{(\gamma_k+\Gamma_k)}^{(k+1)} = \boldsymbol{\mu}_{\gamma_k}^{(k)} - \boldsymbol{\varepsilon} \tag{11.38}$$

   where $\boldsymbol{\varepsilon}$ is any small perturbation vector.
3. Increment k.
4. Use the new Γ_k cluster centers as initial conditions for the *k-means* algorithm and run *k-means* until convergence.
5. If $\Gamma_k \geq \Gamma$, terminate.

The *LBG* algorithm has similar performance to that of the standard *Lloyd algorithm*, both in terms of accuracy and speed, as it relates to clustering features in the speaker recognition problem.

11.2.6.6 x-Means

In 2000, *Pelleg and Moore* [38] introduced a clustering technique in order to address some of the problems associated with *k-means* clustering. The claim was that the new method, called the *x-means* algorithm, would possess better computational scalability, would not require the prescription of the number of clusters, Γ, and would have better global convergence.

Since one of the claims of *x-means* clustering is to estimate the number of clusters, Γ, from the data set, a lower bound, Γ_{min} and an upper bound, Γ_{max} are provided to the algorithm. It starts with Γ_{min} clusters and keeps increasing the number of clusters, keeping track of its performance for each $\Gamma_k : \Gamma_{min} \leq \Gamma_k \leq \Gamma_{max}$. Once it reaches Γ_{max} clusters, it chooses that Γ_k which had produced the best results.

Therefore, beginning with $\Gamma_{(k=0)} = \Gamma_{min}$, the basic steps are as follows,

1. *Improve parameters* – Run the *k-means* clustering algorithm on the data, using Γ_k until convergence.
2. *Improve structure* – Split some of the clusters to obtain a new set of $\Gamma_{(k+1)}$ clusters.
3. If $\Gamma_k < \Gamma_{max}$, increment k and repeat from step 1. Otherwise, terminate.

Steps 1 and 3 are quite well-defined. Reference [38] gives the following proposal for the step which is used to improve the structure.

1. Split every cluster into two.
2. Move out the centers of the two children to the means of the two split spaces.
3. Run a local k-means clustering procedure for each two-cluster set until converged.
4. For each new two cluster group, compute a *Bayesian information criterion* (*BIC*) score for the two child clusters and the original parent cluster and drop the proposition with the worse score. Therefore, for each cluster, either the parent or the pair of children is kept.

Pelleg [38] makes several assumptions regarding the *x-means* formulation. The first assumption is that all clusters will have an identical variance which is given by the following *maximum likelihood estimate*,

$$\tilde{\sigma}^2 = \frac{1}{N - \Gamma_k} \sum_{n=1}^{N} \left(\mathbf{x}_n - \mu_{\hat{\gamma}_{nk}} \right) \tag{11.39}$$

where $\hat{\gamma}_{nk}$ is defined by Equation 11.21. This is a simple-minded view of the variance of the points, where it is assumed that every dimension of the data has an identical spherical distribution and that all the dimensions are statistically independent from one-another. Namely,

$$\left(\tilde{\mathbf{\Sigma}}_\gamma \right)_{[i][j]} = \begin{cases} 0 & \forall \, i \neq j \\ \tilde{\sigma}^2 & \forall \, i = j \end{cases} \tag{11.40}$$

The *BIC* score (see Section 10.6.2) is given by Equation 10.50, which is repeated here for convenience,

$$BIC(\mathbf{\Phi}^{(k)}) \triangleq \ell(\hat{\mathbf{\phi}}_{\Gamma_k}) - M^{(k)}\left(\frac{1}{2} \ln N \right) \tag{11.41}$$

Using the assumptions made by [38], the number of free model parameters would be given as follows,

$$M^{(k)} = (\Gamma_k - 1) + D\Gamma_k + 1 \tag{11.42}$$

The first term of Equation 11.42 is related to the class probabilities (priors). Since there is a constraint such that the sum of these probabilities would be 1, only $\Gamma_k - 1$ of them are considered to be free parameters. The second term corresponds to the Γ_k D-dimensional mean vectors which would have to be estimated. The last term is a single parameter related to the estimation of the shared variance given by Equation 11.39. If we allow for a more realistic variance computation in contrast to the one proposed by [38], we should change this number from 1 to whatever number of variance parameters that would be estimated. Alternatives could be

1. estimation of D shared variance parameters associated with the different dimensions of the data, but shared across all clusters
2. estimation of $\Gamma_k D$ variance parameters associated with the diagonal elements of the γ_k clusters, each being D-dimensional
3. estimation of $\Gamma_k D^2$ variance parameters associated with full covariance matrices for every one of the Γ_k clusters

At this point, to compute the *BIC* for parametric model, we need to compute the *maximum likelihood estimate* of the data associated with each model. Note that we are working with discrete random variables. Therefore, assuming a normal distribution, we may compute the, so called, *point probability* for each of the data points given the parametric model of choice. This is done by dividing the continuous normal probability density function of Equation 6.196 by the total number of points in the cluster,

$$\hat{P}(\mathbf{x}_n | \boldsymbol{\varphi}_{\hat{\gamma}_{nk}}) = \frac{1}{N_{\hat{\gamma}_{nk}}} \hat{p}(\mathbf{x}_n | \boldsymbol{\varphi}_{\hat{\gamma}_{nk}}) \tag{11.43}$$

$$= \frac{1}{N_{\hat{\gamma}_{nk}}} \frac{1}{(2\pi)^{\frac{D}{2}} |\boldsymbol{\Sigma}_{\hat{\gamma}_{nk}}|^{\frac{1}{2}}} \exp\left(-\frac{1}{2}(\mathbf{x}_n - \boldsymbol{\mu}_{\hat{\gamma}_{nk}})^T \boldsymbol{\Sigma}_{\hat{\gamma}_{nk}}^{-1}(\mathbf{x}_n - \boldsymbol{\mu}_{\hat{\gamma}_{nk}})\right) \tag{11.44}$$

Therefore, the total likelihood of the data would be given by the following

$$\mathcal{L}(\boldsymbol{\Phi}^{(k)} | \{\mathbf{x}\}_1^N) = \prod_{n=1}^{N} \hat{P}(\mathbf{x}_n | \boldsymbol{\varphi}_{\hat{\gamma}_{nk}}) \tag{11.45}$$

The log-likelihood of the data with respect to the parametric model k may then be written as follows,

$$\ell(\mathbf{\Phi}^{(k)}|\{\mathbf{x}\}_1^N) = \sum_{n=1}^N \ell(\boldsymbol{\varphi}_{\hat{\gamma}_{nk}}|\mathbf{x}_n) \tag{11.46}$$

$$= -\left(\ln(N_{\hat{\gamma}_{nk}}) + \frac{d}{2}\ln(2\pi) + \frac{1}{2}\ln|\mathbf{\Sigma}_{\hat{\gamma}_{nk}}| + \frac{1}{2}(\mathbf{x}_n - \boldsymbol{\mu}_{\hat{\gamma}_{nk}})^T \mathbf{\Sigma}_{\hat{\gamma}_{nk}}^{-1}(\mathbf{x}_n - \boldsymbol{\mu}_{\hat{\gamma}_{nk}})\right) \tag{11.47}$$

which makes the *BIC* associated with each model selection given by the following,

$$BIC(\mathbf{\Phi}^{(k)}) = \ell(\mathbf{\Phi}^{(k)}|\{\mathbf{x}\}_1^N) - \frac{1}{2}M^{(k)}\ln(N) \tag{11.48}$$

where $\ell(\mathbf{\Phi}^{(k)}|\{\mathbf{x}\}_1^N)$ is given by Equation 11.47, $M^{(k)}$ is given by Equation 11.42 in the simplest case and is modified by the pertaining case following right after Equation 11.42.

Equation 11.48 was derived for the whole data depending on which parametric model is selected. It is, however, easily modified to relate to any one or more chosen clusters when it comes to the splitting step of the algorithm.

Note that [38] has suggested the simple symmetrically, independently and identically distributed case. Here, we have extended the method to include the more general cases including any normally distributed clusters.

Similar to *global k-means*, [38] also proposes the use of *k-d trees* (see Section 11.2.6.2) to accelerate the *x-means algorithm*.

11.2.7 Rough k-Means

One other problem which indirectly helps the slow convergence and sensitivity to initial parameters in *k-means* clustering is the inherent tendency to create a strong bond between the cluster centers and their member samples, not allowing samples to change their cluster membership that easily. Solving this issue will, indirectly, help alleviate the major problems listed in Section 11.2.5. [31] introduced a *k-means* algorithm where the clusters were modified to be *fuzzy and rough sets* instead of traditional sets with *crisp boundaries*. In Section 6.1.2, we introduced the concept of a *rough set*.

According to [31], since the subsets of $\mathscr{X}$ may, theoretically, not be completely represented by the sampled data, not all of *Pawlak's rough set properties* [37, 39] may be verifiable. However, [31] proposes that the following three basic properties must be met. Let us define the *equivalence relation*, P, associated with the partitioning of the sample space, $\mathscr{X}$, due to this approach. Then,

1. $\mathbf{x}_n$ may be a member of no more than one *lower approximation*,

$$\mathbf{x}_n \in \underline{P}(\mathscr{X}_{\hat{\gamma}_n}) \tag{11.49}$$

   with the following lower approximation membership,

$$\hat{\gamma}_n \overset{\Delta}{=} \eth(\mathbf{x}_n, \boldsymbol{\Phi}) \tag{11.50}$$

   where $\eth(\mathbf{x}_n, \underline{\mathbf{M}})$ will be defined later.

2. $\mathbf{x}_n \in \underline{P}(\mathscr{X}_{\hat{\gamma}_n}) \implies \mathbf{x}_n \in \overline{P}(\mathscr{X}_{\hat{\gamma}_n})$
3. if $\nexists \hat{\gamma}_n : \mathbf{x}_n \in \underline{P}(\mathscr{X}_{\hat{\gamma}_n})$ then $\mathbf{x}_n \in \{\overline{P}(\mathscr{X}_{\gamma_\zeta})\} \forall \zeta = \{1, \cdots, Z\} : Z \geq 2$

Assuming Γ clusters, the objective of *rough k-means* clustering is to determine the lower and upper approximations of the Γ clusters given the partitioning scheme at hand, P. These *rough set* approximations, representing the Γ clusters are denoted by, $(\underline{P}(\mathscr{X}_\gamma), \underline{P}(\mathscr{X}_\gamma))$ where $\gamma = \{1, 2, \cdots, \Gamma\}$.

Then *Lingras* [31] proposes the following expression for computing the parameter vector, $\boldsymbol{\varphi}_\gamma^{(k)}$, for each of the Γ clusters,

$$\boldsymbol{\varphi}_\gamma^{(k)} = \begin{cases} \underline{\omega} \sum\limits_{\mathbf{x} \in \underline{P}(\mathscr{X}_\gamma^{(k)})} \dfrac{\mathbf{x}}{|\underline{P}(\mathscr{X}_\gamma^{(k)})|} + \\ \qquad\qquad\qquad\qquad\qquad \forall \gamma : \mathscr{B}_P(\mathscr{X}_\gamma^{(k)}) \neq \{\varnothing\} \\ \overline{\omega} \sum\limits_{\mathbf{x} \in \mathscr{B}_P(\mathscr{X}_\gamma^{(k)})} \dfrac{\mathbf{x}}{|\mathscr{B}_P(\mathscr{X}_\gamma^{(k)})|} \\ \\ \underline{\omega} \sum\limits_{\mathbf{x} \in \underline{P}(\mathscr{X}_\gamma^{(k)})} \dfrac{\mathbf{x}}{|\underline{P}(\mathscr{X}_\gamma^{(k)})|} \qquad \forall \gamma : \mathscr{B}_P(\mathscr{X}_\gamma^{(k)}) = \{\varnothing\} \end{cases} \tag{11.51}$$

where $\mathscr{B}_P(\mathscr{X}_\gamma^{(k)})$ is the boundary of $\underline{P}(\mathscr{X}_\gamma^{(k)})$ with $\overline{P}(\mathscr{X}_\gamma^{(k)})$, as defined by Equation 6.9. In Equation 11.51, $\underline{\omega}$ and $\overline{\omega}$ are weighting parameters which specify the relative importance of the lower and upper approximations of the sets. In this equation, if the boundary, $\mathscr{B}_P(\mathscr{X}_\gamma^{(k)})$, is empty, it means that the lower and upper approximations of every cluster are identical, reducing the clustering technique to the conventional *k-means clustering* algorithm.

Since the *rough k-means algorithm* has devised a method for computing the cluster parameters at every iteration, $\boldsymbol{\Phi}^{(k)}$, it also needs to devise a membership function for the lower and upper approximations. Let us define the following two index sets for each sample, $\mathbf{x}_n$. First we should define the following set for each sample, $\mathbf{x}_n$ at iteration k,

$$\mathscr{G}_{nk} \overset{\Delta}{=} \{\gamma : \|\mathbf{x} - \boldsymbol{\varphi}_\gamma\|_{\mathscr{E}}^2 - \|\mathbf{x} - \boldsymbol{\varphi}_{\hat{\gamma}}\|_{\mathscr{E}}^2 \leq \overline{\theta} \; \forall \; \gamma, \hat{\gamma} = \{1, 2, \cdots, \Gamma\} : \gamma \neq \hat{\gamma}\} \quad (11.52)$$

where $\overline{\theta}$ is a similarity threshold which is provided to the *rough k-means algorithm* as an input parameter. Based on Equation 11.52, the membership of $\mathbf{x}_n$ into lower and upper approximations of the clusters may be computed. Note that the number of members in the set of indices for lower approximations may never be more than one. Also, to meet the third basic property listed at the beginning of this section, when $\mathscr{G}_{nk} \neq \{\varnothing\}$, then $\mathbf{x}_n$ may not belong to any lower approximations, and it will belong to two or more upper approximations. Namely,

$$\underline{\mathscr{G}_{nk}} \overset{\Delta}{=} \begin{cases} \{\hat{\gamma}_{nk}\} \; \forall \; \mathscr{G}_{nk} = \{\varnothing\} \\ \{\varnothing\} \; \forall \; \mathscr{G}_{nk} \neq \{\varnothing\} \end{cases} \quad (11.53)$$

Also,

$$\overline{\mathscr{G}_{nk}} \overset{\Delta}{=} \begin{cases} \{\hat{\gamma}_{nk}\} \; \forall \; \mathscr{G}_{nk} = \{\varnothing\} \\ \mathscr{G}_{nk} \; \forall \; \mathscr{G}_{nk} \neq \{\varnothing\} \end{cases} \quad (11.54)$$

where $\hat{\gamma}_{nk}$ is given by the *k-means membership index function*, Equation 11.8.

Therefore, the three parameters, $\overline{m}$, $\underline{m}$, and $\overline{\theta}$ are design parameters which must be chosen for the application and the data of interest. Using these parameters, for each iteration, the parameter vectors (centroids) of the clusters may be computed using Equation 11.51. Then, the membership of each sample vector, $\mathbf{x}_n$, is determined by using Equations 11.52, 11.53, and 11.54.

Lingras [31] uses *rough k-means clustering* to cluster the users of websites into different overlapping categories. It is easy to see that this non-empty boundary of the lower approximation with its complement could easily take place in many incarnations of clustering used for speaker recognition as well as any other pattern recognition technique.

11.2.8 Fuzzy k-Means

Much in the same spirit as the *rough k-means algorithm*, several *fuzzy k-means algorithms* have been proposed. [20] cites one by *Bezdek* [7] which provides expressions for the *fuzzy membership function* and the *parameter vector evaluation* for the clusters. The objective function of the *fuzzy k-means algorithm* proposed by [7], in contrast with Equation 11.5 for the *k-means algorithm*, may be written as follows,

$$E_{FKM}(X, \boldsymbol{\Phi}) = \sum_{n=1}^{N} u_{n\gamma}^r \|\mathbf{x}_n - \boldsymbol{\varphi}_\gamma\|_{\mathscr{E}}^2 \quad (11.55)$$

where $u_{n\gamma} : \sum_{\gamma=1}^{\Gamma} u_{n\gamma} = 1, u_{n\gamma} \geq 0 \ \forall \ n, \gamma$ and $r : r \geq 1$ is the fuzzification parameter such that a larger r would make the membership more fuzzy.

In Section 6.1.3, we saw how to transform a problem from the *rough set* formulation to a *fuzzy set* formulation. The formulation of the *rough k-means* which was discussed in Section 11.2.7 may also be reformulated using the methodology of Section 6.1.3 to produce a *fuzzy k-means* algorithm.

Based on the generalization of Section 11.2.2, we may write the *characteristic function* for the *fuzzy k-means algorithm* as follows [20],

$$
\Upsilon_{\mathscr{X}_\gamma^{(k)}}(\mathbf{x}_n) \triangleq \frac{\left\| \mathbf{x}_n - \boldsymbol{\varphi}_\gamma^{(k)} \right\|_{\mathscr{E}}^{\frac{-2}{(r-1)}}}{\sum\limits_{\gamma=1}^{\Gamma} \left\| \mathbf{x}_n - \boldsymbol{\varphi}_\gamma^{(k)} \right\|_{\mathscr{E}}^{\frac{-2}{(r-1)}}}
\tag{11.56}
$$

Although the *fuzzy k-means algorithm* uses soft memberships, the weight, $w^{(k)}(\mathbf{x}_n)$ is still 1. In the next section, we will discuss a different technique which seeks to use different weighting for the sample vectors as well as using a soft membership.

11.2.9 k-Harmonic Means Algorithm

The *k-harmonic means* algorithm was introduced by *Zhang et al.* [44] in 1999 and later, in [43] a generalized version was introduced, replacing the *Euclidean distance* with a more general $\mathfrak{L}_p$ distance. This algorithm is influenced by the *k-means* algorithm, based upon two main distinguishing characteristics. First, as the name suggests, it replaces the min function in the objective function of the *k-means* algorithm (Equation 11.5), with the *harmonic mean* of the *Euclidean distances* between the samples and all the cluster centers.

As we noted in Definition 6.80, the *harmonic mean* does have a somewhat minimum-like characteristic in the sense that it tends to become small when there is an exceptionally small member within its samples. This is designed to handle one of the inherent problems in *k-means* clustering, which is tendency to create a strong bond between the cluster centers and their member samples, not allowing samples to change their cluster membership that easily. *Zhang* [43] refers to this feature as a *winner-takes-all* strategy. There have been other approaches toward alleviating this problem, such as the use of *fuzzy sets theory* to soften the cluster membership of samples – see Section 11.2.8.

The objective function for the basic *k-harmonic means algorithm* may be written as follows,

$$E_{KHM}(\{\mathbf{x}\}_1^N, \boldsymbol{\Phi}) = \sum_{n=1}^N \frac{\Gamma}{\displaystyle\sum_{\gamma=1}^{\Gamma} \frac{1}{\|\mathbf{x}_n - \boldsymbol{\varphi}_\gamma\|_{\mathscr{E}}^2}} \tag{11.57}$$

In the generalized version of the algorithm, introduced in [43], the power of the norm[3] in Equation 11.58 was changed from 2 to a prescribed parameter, p. To avoid confusion of notation, since we generally use p for probabilities, in this book, we shall use ρ to denote this parameter. Therefore, the objective function for the generalized *k-harmonic means algorithm* may be written as,

$$E_{KHM\rho}(\{\mathbf{x}\}_1^N, \boldsymbol{\Phi}) = \sum_{n=1}^N \frac{\Gamma}{\displaystyle\sum_{\gamma=1}^{\Gamma} \frac{1}{\|\mathbf{x}_n - \boldsymbol{\varphi}_\gamma\|_{\mathscr{E}}^\rho}} \tag{11.58}$$

To obtain the optimal parameter vectors for the clusters of the *k-harmonic means algorithm*, [43] takes the gradient of Equation 11.58 with respect to the parameter vectors and sets it to *zero*,

$$\nabla_{\boldsymbol{\varphi}_\gamma} E_{KHM\rho}(\{\mathbf{x}\}_1^N, \boldsymbol{\Phi}) = -\Gamma \sum_{n=1}^N \frac{\rho(\mathbf{x}_n - \boldsymbol{\varphi}_\gamma)}{d_{\gamma n}^{(\rho+2)} \left(\displaystyle\sum_{\tilde{\gamma}=1}^{\Gamma} \frac{1}{d_{\tilde{\gamma}n}^\rho}\right)^2} \tag{11.59}$$

$$= \mathbf{0} \ \forall \ \gamma \in \{1, 2, \cdots, \Gamma\} \tag{11.60}$$

where,

$$d_{\gamma n} \overset{\Delta}{=} \|\mathbf{x}_n - \boldsymbol{\varphi}_\gamma\|_{\mathscr{E}} \tag{11.61}$$

Solving the set of equations stemming from Equation 11.60 for the parameter vectors produces results which fit into the form of Equation 11.13. To make sure that the properties of the characteristic equations given by Equations 6.7 and 6.8 hold, the characteristic Equation and weight are given by the following two relations.

$$\Upsilon_{\mathscr{X}_\gamma^{(k)}}(\mathbf{x}_n) = \frac{\dfrac{1}{d_{\gamma n}^{(\rho+2)}}}{\displaystyle\sum_{\tilde{\gamma}=1}^{\Gamma} \frac{1}{d_{\tilde{\gamma}n}^{(\rho+2)}}} \tag{11.62}$$

and

---

[3] [43] states that there is a possibility of using an L_p-norm (see Definition 23.8), however, doing so will not provide us with centroids. Use of the Euclidean norm will preserve the centroid properties of the parameter vectors for the clusters.

$$
w^{(k)}(\mathbf{x}_n) = \frac{\sum\limits_{\tilde{\gamma}=1}^{\Gamma} \frac{1}{d_{\tilde{\gamma}n}^{(\rho+2)}}}{\left(\sum\limits_{\tilde{\gamma}=1}^{\Gamma} \frac{1}{d_{\tilde{\gamma}n}^{(\rho)}} \right)^2}
\tag{11.63}
$$

Zhang [43] compares the performance of the general *k-harmonic means algorithm* with that of *k-means* and the *expectation maximization* (see Section 11.3.1) algorithms, using different values for ρ. Based on these results, $3.0 \leq \rho \leq 3.5$ generally outperforms other algorithms and its convergence shows much less dependence on the choice of the initial conditions. [43] attributes some of the success of the results to the fact that the weighting function (Equation 11.63) becomes much smaller for samples which are close to one of the centroids, when $\rho > 2$. This behavior acts as a *boosting function* used in supervised learning techniques [17]. It amplifies the effects of points that are not close to centers in determining the clusters in the next iterations. This will reduce the chance of multiple clusters becoming stagnant in the presence of a local data cluster.

11.2.10 Hybrid Clustering Algorithms

[20] uses a combination of features from the *k-means* and *k-harmonic means* algorithms to create *hard membership* and *soft membership* hybrid clustering algorithms. This is done by creating the two other possible combinations of mixing the characteristic equation and weight from the *k-means algorithm* given by Equations 11.14 and 11.15 with those of the *k-harmonic means algorithm* in Equations 11.62 and 11.63.

Therefore, the *hard membership* hybrid algorithm is the one using Equation 11.14 for its *characteristic equation* and Equation 11.63 for its weight. Similarly, the *soft membership* hybrid algorithm uses Equation 11.62 for its *characteristic equation* and Equation 11.15 for its weight.

It is easy to imagine that a prolific array of different clustering algorithms may be generated by mixing the characteristic equations and weights of the different algorithms discussed in this chapter. However, it is important to try and understand the physical interpretations of these different combinations and not to be carried away with all the different possibilities at hand.

11.3 Estimation using Incomplete Data

The techniques in this section are designed to estimate higher order statistics as well as the means of the clusters being identified. They generally use similar objectives as in the previous section, with somewhat more refined requirements and assumptions about the distribution of the data in the clusters they identify.

For example, the techniques in Section 11.2 estimated the means of the clusters using some *maximum likelihood criteria* (for example by computing the sample mean which was seen to be the *maximum likelihood solution*). However, they generally did not make any assumption about the distributions of the data within each cluster. Of course, after the centroids are computed, most practical implementations compute higher order statistics for the found clusters.

In the methods of this section, more assumptions are made about the distribution of the data within each cluster as well as across the different clusters. This is generally in the form of assuming a distribution of the *exponential family* such as a *Gaussian (normal)* distribution. The most popular technique is the *expectation maximization (EM)*, which will be discussed here in detail.

11.3.1 Expectation Maximization (EM)

Expectation maximization (EM)[4] is an iterative methodology for the *maximum likelihood estimation* of the parameters that best represent the *sufficient statistics* of a random variable based on the observation of *incomplete data*. By *incomplete data*, we mean that the true data samples leading to the computation of the *sufficient statistics* of the data are not completely observed. The unseen data is also known as *hidden data* or *missing data*. This concept is used in formulating *Gaussian mixture models* and *hidden Markov models* and will be examined, in detail, later.

The *EM algorithm* was first properly introduced by *Hartley* in 1958 [21] and later formalized by *Dempster, et al.* in 1977 [13]. Although there was earlier work (1956) [22] presenting special cases of the *EM* algorithm and many intermediate works between 1958 and 1977 [34], still [13] presented the most complete treatment of *EM*. In the interim, in 1966 [3] and then later in 1970 [4] Baum, et al. presented the same algorithm for the special case of estimating the parameters of Markov chains in an efficient manner. The name, *expectation maximization*, stems from the two main steps of *expectation* computation and *maximization* (maximum

---

[4] By some accounts, *EM* is not an algorithm. It is simply a methodology which may be used to produce many different algorithms [34]. However, in accordance with popular references, we occasionally refer to it as an algorithm.

likelihood estimation) in the structure of each of its iterations.

Expectation maximization (EM) is an iterative process which is highly dependent on the initial selection of the cluster statistics. In addition, per iteration, it is more costly than any of the other unsupervised clustering techniques presented here. However, it is quite attractive since it uses the *Mahalanobis distance* in its objective function and since it may use *incomplete data*. Given these facts, the *EM* algorithm is generally used as a fine tuning step, after the means of the clusters have somewhat converged, using any one of *k-means*, *LBG*, *k-harmonic means*, or other standard algorithms. Since it uses the *Mahalanobis* distance, a few steps at the end of another clustering regimen, using the *EM algorithm*, will result in a better estimate of the *means* and *variances* of the clusters.

The original *EM algorithm*, as described by *Dempster, et al.* [13], did not prescribe any *estimate* for the *covariance matrix* of the *maximum likelihood estimate* (*MLE*), but later modifications added such an estimate to the algorithm.[34] Here, we will also examine the use of this estimate. Since *EM*, in its raw form, is concerned with *maximizing the likelihood*, it applies to both the *frequentist* [35] and the *Bayesian* [15] communities. However, in the approach taken by most of the *speaker recognition* community, the *Bayesian* concept of *a-posteriori* probability needs to be maximized in order to be able to make a recognition decision. This leads to a slight addition to the original *EM algorithm* to maximize the *a-posteriori* probability instead of just the *likelihood*. We have also seen that in some cases, such as the *binary hypothesis* theory, we are simply interested in the *likelihood* estimates.

Although the *EM algorithm* may be used for any random variable X, for simplicity, [13] considers the *regular exponential family* (Section 6.10). Later, *Dempster, et al.* [13] generalize the algorithm for the more general *curved exponential family* and, eventually, with further generalization the handling of any *likelihood function* becomes possible. In speaker recognition, we are mostly interested in a special case of the *regular exponential family*, namely the *normal density family*. We will formulate the problem for the whole *Darmois-Koopman-Pitman exponential family* discussed in Section 6.10, and then look at the generalized version of the algorithm.

Later, we will consider the Bayesian *maximum a-posteriori (MAP)* formulation by introducing the prior probability density of the parameter vector. In Section 13.7, we shall study the *normal density* case in more detail.

Let us examine the *EM algorithm* in a light which would be in tune with our concerns within the field of speaker recognition.

11.3.1.1 Formulation

As we noted in Section 10.1, we are interested in learning the parameters of a parametric model which models the behavior of a random variable, $\{X : \mathbf{x} \in \mathcal{X} = \mathcal{R}^D\}$, in a *maximum likelihood* sense. Namely, we would like to solve the following problem,

$$\hat{\boldsymbol{\varphi}} = \arg\max_{\boldsymbol{\varphi}} \ell(\boldsymbol{\varphi}|\{\mathbf{x}\}_1^N) \tag{11.64}$$

where $\{\mathbf{x}\}_1^N$ is a set of underlying (*complete*) data for the problem of interest. However, often we are not able to observe the true sample vectors, $\{\mathbf{x}\}_1^N$. Instead, we may have to base our decision on the observed vectors which may be defined by a *transformation* (Section 6.5.5) of the complete data, also known as the *incomplete data* [13]. Let us consider the following *transformation*, T, from the *complete data*, $\mathbf{x}$ to the *observed (incomplete data)*, $\mathbf{y}$, where $\{Y : y \in \mathcal{Y} = \mathcal{R}^D\}$. Here, for the sake of simplicity, we are assuming that sample spaces $\mathcal{X}$ and $\mathcal{Y}$ are defined within the same *Cartesian product space* (see Section 6.2.2). However, this assumption is by no means necessary. We can describe transformation T as follows,

$$\mathbf{y} = T(\mathbf{x}) \ \forall \ \mathbf{x} \in \mathcal{X} \tag{11.65}$$

where $T : (\mathcal{X}, \mathfrak{X}) \mapsto (\mathcal{Y}, \mathfrak{Y})$ is a *measurable transformation* which is not necessarily *one-to-one* and *invertible* – see Definition 6.52. That is to say that having observed the samples of Y may not provide us with all the information necessary to be able to deduce X. In this case, the system is said not to be *fully observable* – see [16].

Note that in Section 10.1, the observation was coincident with the *complete data* and there was no distinction made between the two. One of the strengths of the *EM* algorithm is its use of an observation which may not be the same as X or the true *state of the system*.

Since the *observed data*, $\mathbf{y}$, and the *complete data*, $\mathbf{x}$, generally occupy different parts of their corresponding spaces, we may write the parametric approximation to the probability density of $\mathbf{y}$, denoted by $p_Y(\mathbf{y}|\boldsymbol{\varphi})$, in terms of the parametric approximation to the probability density of $\mathbf{x}$, denoted by $p_X(\mathbf{x}|\boldsymbol{\varphi})$ as follows,

$$p_Y(\mathbf{y}|\boldsymbol{\varphi}) = \int_{T(\mathbf{y})} p_X(\mathbf{x}|\boldsymbol{\varphi})d\mathbf{x} \tag{11.66}$$

Using the expression for the *regular exponential family* given by Equation 6.194, we may write the parametric form of $p_X(\mathbf{x}|\boldsymbol{\varphi})$ as follows,

$$p_X(\mathbf{x}|\boldsymbol{\varphi}) = a(\mathbf{x})\exp\left(\boldsymbol{\varphi}^T \mathbf{v}(\mathbf{x}) - b(\boldsymbol{\varphi})\right) \tag{11.67}$$

Our objective is to maximize the likelihood of the set of underlying vectors, $\{\mathbf{x}\}_1^N$, based on the observation of the corresponding incomplete data, $\{\mathbf{y}\}_1^N$. We may write the objective of Equation 11.64 as follows,

$$\hat{\boldsymbol{\varphi}} = \arg\max_{\boldsymbol{\varphi}} \ell_X(\boldsymbol{\varphi}|\{\mathbf{x}\}_1^N) \tag{11.68}$$

$$= \sum_{n=1}^N \ln(p_X(\mathbf{x}_n|\boldsymbol{\varphi})) \tag{11.69}$$

For the moment, let us consider the maximum log likelihood for an individual vector, $\mathbf{x}_n$ and the regular exponential family of Equation 11.67,

$$\arg\max_{\boldsymbol{\varphi}} \ell_X(\boldsymbol{\varphi}|\mathbf{x}_n) = \arg\max_{\boldsymbol{\varphi}} \ln(p_X(\mathbf{x}_n|\boldsymbol{\varphi})) \tag{11.70}$$

$$= \arg\max_{\boldsymbol{\varphi}} \ln(a(\mathbf{x}_n)) + \boldsymbol{\varphi}^T \mathbf{v}(\mathbf{x}) - b(\boldsymbol{\varphi}) \tag{11.71}$$

In Equation 11.71, we are seeking a parameter vector which maximizes $\ell_X(\boldsymbol{\varphi}|\mathbf{x}_n)$, for a specific complete data vector, $\mathbf{x}_n$. Therefore, since the first term is only dependent on $\mathbf{x}_n$ and has no $\boldsymbol{\varphi}$ dependency, it may be eliminated. Therefore, Equation 11.71 may be written as

$$\arg\max_{\boldsymbol{\varphi}} \ell_X(\boldsymbol{\varphi}|\mathbf{x}_n) = \arg\max_{\boldsymbol{\varphi}} \boldsymbol{\varphi}^T \mathbf{v}(\mathbf{x}) - b(\boldsymbol{\varphi}) \tag{11.72}$$

We see that the terms in Equation 11.72 only depend on $\mathbf{x}_n$, independently through the *sufficient statistics* term, $\mathbf{v}(\mathbf{x}_n)$. Therefore, if we have a decent estimate of the *sufficient statistics*, $\mathbf{v}(\mathbf{x})$, whether through the explicit knowledge of $\mathbf{x}$ or any other way, we may be able to solve the maximization problem of Equation 11.72.

Now let us go back to the true problem which is the *maximum mean likelihood estimation* discussed in Section 10.6.1 and estimated by maximizing the likelihood of a set of vectors, $\{\mathbf{x}\}_1^N$, formed by Equation 11.69. In this case, due to the fact that the logarithm function is *strictly concave* and the use of *Jensen's inequality* (similar to the arguments made in Section 7.6), the maximization problem reduces to maximizing the terms in Equation 11.72 using the expected value of the *sufficient statistics*. This provides the grounds for the *expectation* step in *EM*.

The *expectation* step is the computation of an estimate of the expected value of the *sufficient statistics* of the *complete data* from the *conditional expectation* of its *sufficient statistics* given the observed data and the current estimates of the parameter vector for the parametric representation of the underlying density function. Therefore, the *expectations* step of *EM* may be written as follows,

Expectation Step:

$$\mathbf{v}^{(k)} = \mathscr{E}\left\{\mathbf{v}(\mathbf{x})|\mathbf{y}, \boldsymbol{\varphi}^{(k)}\right\} \tag{11.73}$$

Note that in Equation 11.73, the expectation is conditioned upon the observation vector, $\mathbf{y}$, since we do not know the values of the complete data, $\mathbf{x}$ – see Definition 6.58.

Furthermore, assume that we pick the new parameters at the next iteration, $\boldsymbol{\varphi}^{(k+1)}$, to be such that the *expected value* of the *complete data sufficient statistics*, $\mathcal{E}\{\mathbf{v}(\mathbf{x})|\boldsymbol{\varphi}\}$ (not conditioned upon the observation vector, $\mathbf{y}$) is equal to that given by the *expectation step* (Equation 11.73). Therefore, the *maximization step* would be

Maximization Step:
Pick $\boldsymbol{\varphi}^{(k+1)}$ such that,

$$\mathcal{E}\{\mathbf{v}(\mathbf{x})|\boldsymbol{\varphi}\} = \mathbf{v}^{(k)} \tag{11.74}$$

Let us hold on to this thought and examine the log of likelihood of the density function due to the observation, $\ln(p_Y(\mathbf{x}|\boldsymbol{\varphi}))$.

Note that we may write the *conditional log likelihood* of the complete data, $\mathbf{x}$ given the observed data, $\mathbf{y}$, and the parameter vector of the parametric model as follows,

$$\ln\left(p_{(X|Y)}(\mathbf{x}|\mathbf{y},\boldsymbol{\varphi})\right) = \ln(p_X(\mathbf{x}|\boldsymbol{\varphi})) - \ln(p_Y(\mathbf{y}|\boldsymbol{\varphi})) \tag{11.75}$$

For the *regular exponential family*,

$$\ln\left(p_{(X|Y)}(\mathbf{x}|\mathbf{y},\boldsymbol{\varphi})\right) = \ln(a(\mathbf{x})) + \boldsymbol{\varphi}^T\mathbf{v}(\mathbf{x}) - b(\boldsymbol{\varphi}|\mathbf{y}) \tag{11.76}$$

where $b(\boldsymbol{\varphi}|\mathbf{y})$ is given by Equation 6.195 as follows,

$$b(\boldsymbol{\varphi}|\mathbf{y}) = \ln\left(\int_{T(\mathbf{y})} \exp\left(\boldsymbol{\varphi}^T\mathbf{v}(\mathbf{x})a(\mathbf{x})\right)\right) \tag{11.77}$$

If we compare Equations 11.71 and 11.76, we see that the density function for the *complete data* as well as that of the complete data conditioned upon the observed data, are described by the same parameter vector, $\boldsymbol{\varphi}$, and *sufficient statistics*, $\mathbf{v}(\mathbf{x})$. However, the first is defined over the sample space, $\mathcal{X}$, and the second one, over the sample space formed by the transformation, $T(\mathcal{Y})$. Therefore, we may write the expression for the log likelihood with respect to the observed data in terms of the log likelihoods given the complete data and the conditional log likelihood using Equation 11.75 as follows,

$$\ln(p_Y(\mathbf{y}|\boldsymbol{\varphi})) = \ln(p_X(\mathbf{x}|\boldsymbol{\varphi})) - \ln\left(p_{(X|Y)}(\mathbf{x}|\mathbf{y},\boldsymbol{\varphi})\right)$$
$$= b(\boldsymbol{\varphi}|\mathbf{y}) - b(\boldsymbol{\varphi}) \tag{11.78}$$

where,

$$b(\boldsymbol{\varphi}) = \ln \left(\int_{\mathbf{x} \in \mathcal{X}} \exp \left(\boldsymbol{\varphi}^T \mathbf{v}(\mathbf{x}) a(\mathbf{x}) \right) \right) \tag{11.79}$$

Since we would like to maximize the expression in Equation 11.78, we would have to compute the stationary point, $\nabla_{\boldsymbol{\varphi}} \ln \left(p_Y(\mathbf{y}|\boldsymbol{\varphi}) \right) = 0$. Note that,

$$\nabla_{\boldsymbol{\varphi}} b(\boldsymbol{\varphi}) = \mathscr{E} \left\{ \mathbf{v}(\mathbf{x}) | \boldsymbol{\varphi} \right\} \tag{11.80}$$

and

$$\nabla_{\boldsymbol{\varphi}} b(\boldsymbol{\varphi}|\mathbf{y}) = \mathscr{E} \left\{ \mathbf{v}(\mathbf{x}) | \mathbf{y}, \boldsymbol{\varphi} \right\} \tag{11.81}$$

Taking the gradient of Equation 11.78 by using Equations 11.80 and 11.81, we have the following expression,

$$\begin{aligned} \nabla_{\boldsymbol{\varphi}} \ln \left(p_Y(\mathbf{y}|\boldsymbol{\varphi}) \right) &= \nabla_{\boldsymbol{\varphi}} b(\boldsymbol{\varphi}|\mathbf{y}) - \nabla_{\boldsymbol{\varphi}} b(\boldsymbol{\varphi}) \\ &= \mathscr{E} \left\{ \mathbf{v}(\mathbf{x}) | \mathbf{y}, \boldsymbol{\varphi} \right\} - \mathscr{E} \left\{ \mathbf{v}(\mathbf{x}) | \boldsymbol{\varphi} \right\} \end{aligned} \tag{11.82}$$

Therefore, if we take the two iterative steps prescribed by Equations 11.73 and 11.74 while assuming that the algorithm converges to $\hat{\boldsymbol{\varphi}} = \boldsymbol{\varphi}^{(k+1)} = \boldsymbol{\varphi}^{(k)}$, then the combination of these two steps provides equal estimates of the *expected value* of the *sufficient statistics* of $\mathbf{x}$ and its *conditional expected value* given the observation $\mathbf{y}$. Therefore, the derivative in Equation 11.82 approaches 0 in the limit, maximizing the log likelihood in Equation 11.78. As we have noted before, due to the strictly concave nature of the log function, this means that the *likelihood* itself is maximized.

We will see during the application of the EM algorithm to estimating the parameters of *Gaussian mixture models*, the EM algorithm may actually be viewed as a *variable metric* (Chapter 25) optimization technique with no line-search, akin to those techniques discussed in Section 25.1.3.8. In fact Xu and Jordan [41] present the expressions for the positive definite matrix $\mathbf{H}$ which is used for the weighting of the gradients in each ascent iteration toward the maximum likelihood.

We should note that it is possible for the *EM* algorithm not to converge to a point where the expectation and the conditional expectation are equal. This would be the case, for example, if the observation, $\mathbf{y}$, and the complete data, $\mathbf{x}$, do not span the same space. In that case, the maximum likelihood solution is still found.

At a later stage, [13] expands the scope of the *EM* by removing the reference to any *exponential family*. In this case, a new function is defined based on the conditional expected value of the log likelihood of the parameter vector given the complete data conditioned upon the observed data. Namely,

$$Q(\boldsymbol{\varphi}_1|\boldsymbol{\varphi}_2) \overset{\Delta}{=} \mathscr{E} \left\{ \ln \left(p(\mathbf{x}|\boldsymbol{\varphi}_1) \right) | \mathbf{y}, \boldsymbol{\varphi}_2 \right\} \tag{11.83}$$

which should exist for all pairs of vectors, $(\boldsymbol{\varphi}_1, \boldsymbol{\varphi}_2) : \boldsymbol{\varphi}_1, \boldsymbol{\varphi}_2 \in \phi$. Also, it is assumed that the likelihood $p(\mathbf{x}|\boldsymbol{\varphi}_1)$ is positive *almost everywhere* (Definition 6.43). Namely,

$$p(\mathbf{x}|\boldsymbol{\varphi}_1) > 0 \; [\mathbf{x}] \tag{11.84}$$

Therefore, the two steps of *EM* become, *Expectation Step*:

$$Q(\boldsymbol{\varphi}|\boldsymbol{\varphi}^{(k)}) = \mathscr{E}\left\{\ln\left(p(\mathbf{x}|\boldsymbol{\varphi})\right)|\mathbf{y},\boldsymbol{\varphi}^{(k)}\right\} \tag{11.85}$$

and *Maximization Step*:

$$\boldsymbol{\varphi}^{(k+1)} = \arg\max_{\boldsymbol{\varphi}\in\phi} Q(\boldsymbol{\varphi}|\boldsymbol{\varphi}^{(k)})) \tag{11.86}$$

Similar to the idea of the computation of the expected value of the sufficient statistics discussed previously, Equations 11.85 and 11.86 attempt to maximize $\ell(\boldsymbol{\varphi}|\mathbf{x})$ with the limited knowledge available from the observation of $\mathbf{y}$.

11.3.1.2 Generalized Expectation Maximization (GEM)

In Section 25.3 we spoke about the fact that in modern optimization techniques, the requirements for the optimization are often relaxed to a level that would only guarantee an improvement in the value of the objective function instead of setting out to optimize it right away. A similar idea has been used in special formulations of the *EM* algorithm which only ask for an improvement to the likelihood and not necessarily the greatest increase in the likelihood to achieve maximum likelihood in one shot. These techniques go by the name of *generalized expectation maximization* (*GEM*) and are preferred for complex problems for which a quick convergence is not practical. *Dempster, et al.* [13] define this guaranteed improvement as the only requirement of the *GEM*.

11.3.1.3 Maximum A-Posteriori (MAP) Estimation

In Section 11.3.1.1 we formulated the *EM* algorithm for solving the *maximum log likelihood problem* using the observed data. While introducing the *EM algorithm*, we stated that it is not too hard to modify it to handle the *maximum a-posteriori* (*MAP*) estimation problem. In fact [13] goes through this exercise briefly. Here, we will formulate the *MAP* problem which is for example used in the *Gaussian mixture modeling* of *speaker recognition*.

Let us denote the *prior probability density* of the parameter vector in the *maximum a-posteriori* formulation by $p(\boldsymbol{\varphi})$. Therefore, the *posterior probability densities* of the parameter vector with respect to the observed (incomplete) data, Y, and the complete data, X, may be denoted by $p(\boldsymbol{\varphi}|\mathbf{y})$ and $p(\boldsymbol{\varphi}|\mathbf{x})$ respectively.

Therefore, the two steps of *EM* would be modified for the *MAP* problem in the following manner: *Expectation Step*:

$$Q_{MAP}(\boldsymbol{\varphi}|\boldsymbol{\varphi}^{(k)}) = Q(\boldsymbol{\varphi}|\boldsymbol{\varphi}^{(k)}) + \ln(p(\boldsymbol{\varphi})) \tag{11.87}$$

and *Maximization Step*:

$$\boldsymbol{\varphi}^{(k+1)} = \arg\max_{\boldsymbol{\varphi} \in \phi} Q_{MAP}(\boldsymbol{\varphi}|\boldsymbol{\varphi}^{(k)})) \tag{11.88}$$

McLachlan and Krishnan [34] argue that the introduction of the log of the prior density will make the objective function of the maximization problem of the *MAP* *almost always* more *concave* (Definition 24.25), when compared to the objective function of the *MLE*. In Section 13.7, we will formulate the *EM* algorithm as applied to the *MAP* problem in the *Gaussian mixture modeling* (*GMM*) of speaker models. That formulation will clearly expose the relevance of the *EM* algorithm to *unsupervised clustering*.

In addition to the original papers [21, 13], references [34, 8, 46, 12, 9] provide good overviews and tutorials of the *EM algorithm*. [34] is a recent book entirely devoted to the *EM algorithm*. References [8] and [46] specifically address its application to *Gaussian mixture models* as well as *hidden Markov models*.

11.4 Hierarchical Clustering

In this chapter, we have examined many different unsupervised clustering techniques. Most of the techniques were based on the *k-means clustering* algorithm, which by itself provides the tools for one level of bottom-up clustering. Although, some of the clustering algorithms based on *k-means*, actually use the results of the k-means clustering to create a *top-down* (*divisive*) set of clusters. An example of a divisive algorithm is *LBG* – see Section 11.2.6.5.

In *hierarchical clustering*, one retains the intermediate results of the algorithm such that there is a hierarchical structure developed in the course of clustering. This structure may be used for many different purposes including the definition of the logic of a *decision tree* (Section 9.4). Hierarchical structures need not be based on a tree structure. They may very well have more complicated structures and may go as far as a fully connected graph in which case, the direction of the hierarchy may not be so well defined. However, in most cases, the structures resemble trees.

One important use of hierarchical clustering algorithms is the organization of speakers in a *speaker tree* [5] for the purpose of reducing the computation load of *large-scale speaker identification* systems. Another useful application, which is often used in *speech recognition*, is the hierarchical clustering of the elements of *hidden Markov models* which resemble phonemes at a lower level of the modeling, namely at the HMM state levels (Chapter 13). These states, transitions, or mixture models may be clustered using a hierarchical technique to allow for better *tying* or *combination* of these parameters in a hierarchy. Normally, such hierarchical clustering techniques are used to create decision trees for such combinations and tied parameters.

11.4.1 Agglomerative (Bottom-Up) Clustering (AHC)

Agglomerative clustering, also known as *bottom-up clustering*, deals with the merging of alike clusters into common clusters which make up the clusters for a level higher in a tree structure. This is done level-by-level, merging close clusters into one and moving up the tree until a single node is reached. The merging may be done at a binary pace or based on some threshold or other criteria. An example of such merging was discussed in Section 11.2.4.

To be able to achieve an agglomerative cluster, a distance measure is required for comparing different clusters at each level to be able to decide whether to merge them or not. Also, a merging method is needed such as that of Section 11.2.4.

Beigi, et al. [5] have used such agglomeerative clustering techniques for clustering speaker models in a hierarchical structure. The resulting tree provides the basis for a decision tree to be able to reduce the amount of computation for a large-scale speaker identification system from linear to logarithmic times with respect to the number of speaker models in the database.

11.4.2 Divisive (Top-Down) Clustering (DHC)

Divisive or *top-down* clustering techniques work much in the same way as *agglomerative clustering*, but in the opposite direction. They generally start with one or a handful of clusters and divide them either in a binary fashion or based on a certain threshold or other metric criteria into a variable number of clusters of other fixed numbers at each level of the hierarchy. *Divisive* techniques require a splitting algorithm or method in contrast to the *merging* methods needed for *agglomerative techniques*.

An example of a divisive clustering algorithm is the *LBG* algorithm (Section 11.2.6.5) which used a binary splitting approach in conjunction with the *k*-

means clustering algorithm to create an even larger number of smaller clusters. If the history of these divisions is retained, the *LBG* will create a hierarchical cluster, much in the same shape as those created by *agglomerative* techniques of the previous sections.

11.5 Semi-Supervised Learning

In this chapter we discussed methods of unsupervised clustering. We will treat the supervised learning problem throughout the rest of this book, in conjunction with unsupervised clustering used for the underlying structure. However, in a supervised learning system, the data points are often tied to labels which have been assigned to data through a laborious process. In supervised learning, data is clearly labeled and the learning process happens with the usage of these labels. There is, however, some middle ground in which some of the data is labeled, but, generally, the majority of the data is unlabeled. This case is known as the semi-supervised learning process [47].

In most cases, the semi-supervised learning problem is approached by using an iterative process in which the unlabeled data is first labeled through the use of the outcome of a supervised learning scheme. Once labels are assigned to the unlabeled data, the learning process is repeated until convergence. More detail about semi-supervised learning may be found in [47] and [18].

References

1. Aizerman, A., Braverman, E.M., Rozoner, L.I.: Theoretical foundations of the potential function method in pattern recognition learning. Automation and Remote Control **25**, 821–837 (1964). Translated into English from Russian journal: Avtomatika i Telemekhanika, Vol. 25, No. 6, pp. 917–936, Jun 1964
2. Arthur, D., Vassilvitskii, S.: k-means++: The Advantages of Careful Seeding. Technical Report 2006-13, Stanford InfoLab (2006). URL http://ilpubs.stanford.edu:8090/778/
3. Baum, L.E., Petrie, T.: Statistical Inference for Probabilistic Functions of Finite State Markov chains. The Annals of Mathematical Statistics **37**(6), 1554–1563 (1966)
4. Baum, L.E., Petrie, T., Soules, G., Weiss, N.: A Maximization Technique Occurring on the Statistical Analysis of Probabilistic Functions of Markov Chains. The Annals of Mathematical Statistics **41**(1), 164–171 (1970)
5. Beigi, H.S., Maes, S.H., Chaudhari, U.V., Sorensen, J.S.: A Hierarchical Approach to Large-Scale Speaker Recognition. In: EuroSpeech 1999, vol. 5, pp. 2203–2206 (1999)
6. Bentley, J.L.: Multidimensional Binary Search Trees Used for Associative Searching. Communications of the ACM **18**(9), 509–517 (1975)
7. Bezdek, J.: Pattern Recognition with Fuzzy Objective Function Algorithms. Plenum Press, New York (1981)

8. Bilmes, J.A.: A Gentle Tutorial of the EM Algorithm and its Applications to Parameter Estimation for Gaussian Mixture and Hidden Markov Models (1998). Technical Report Number TR-97-021

9. Borman, S.: The Expectation Maximization Algorithm – A Short Tutorial (2004)

10. Bottou, L., Bengio, Y.: Convergence Properties of the k-Means Algorithm. In: G. Tesauro, D. Touretzky (eds.) Advances in Neural Information Processing Systems, vol. 7. MIT Press, Denver (1995). URL http://leon.bottou.org/papers/bottou-bengio-95

11. Burges, C.J.: A Tutorial on Support Vector Machines for Pattern Recognition. Data Mining and Knowledge Discovery **2**, 121–167 (1998)

12. Dellaert, F.: The Expectation Maximization Algorithm (2002). Technical Report Number GIT-GVU-02-20

13. Dempster, A.P., Laird, N.M., Rubin, D.B.: Maximum Likelihood from Incomplete Data via the EM Algorithm. Journal of the Royal Statistical Society. Series B (Methodological) **39**(1), 1–38 (1977)

14. Dhillon, I.S., Kulis, Y.G.B.: Weighted Graph Cuts without Eigenvectors: A Multilevel Approach. IEEE Transactions on Pattern Analysis and Machine Intelligence (PAMI) **29**(11), 1944–1957 (2007)

15. Fienberg, S.E.: When Did Bayesian Inference Become Bayesian? Bayesian Analysis **1**(1), 1–40 (2006)

16. Fortmann, T.E., Hitz, K.L.: An Introduction to Linear Control Systems. Marcel Dekker, Inc., New York (1977). ISBN: 0-824-76512-5

17. Freund, Y., Schapire, R.E.: Experiments with a new boosting algorithm. In: Proceedings of the Thirteenth International Conference on Machine Learning (ICML), pp. 148–156 (1996)

18. Grandvalet, Y., Bengio, Y.: Entropy Regularization. In: Semi-Supervised Learning. The MIT Press, Boston (2006). Chapter 9

19. Green, P.E., Kim, J., Carmone, F.J.: A preliminary study of optimal variable weighting in k-means clustering. Journal of Classification **7**(2), 271–285 (1990)

20. Hamerly, G., Elkan, C.: Alternatives to the k-Means Algorithm that Find Better Clusterings. In: Proceedings of the Eleventh International Conference on Information and Knowledge Management, pp. 600–607 (2002)

21. Hartley, H.O.: Maximum Likelihood Estimation from Incomplete Data. Biometrics **14**(2), 174–194 (1958)

22. Healy, M.J., Westmacott, M.: Missing Values in Experiments Analysed on Automatic Computers. Journal of the Royal Statistical Society, Series C (Applied Statistics) **5**(3), 203–206 (1956)

23. Jancey, R.C.: Multidimensional Group Analysis. Australian Journal of Botany **14**, 127–130 (1966)

24. Jancey, R.C.: Algorithm for Detection of Discontinuities in Data Sets. Vegetatio **29**(2), 131–133 (1974)

25. Jolliffe, I.: Principal Component Analysis, 2nd edn. Springer, New york (2002)

26. Kirkpatrick, S., Gelatt, C.D., Vecchi, M.P.: Optimization by Simulated Annealing. Science **220**(4598), 671–680 (1983)

27. Kumar, P., Wasan, S.K.: Comparative Analysis of k-means Based Algorithms. IJCSNS International Journal of Computer Science and Network Security **10**(4), 1–13 (2010)

28. Lee, D., Baek, S., Sung, K.: Modified k-Means Algorithm for Vector Quantizer Design. IEEE Signal Processing Letters **4**(1), 2–4 (1997)

29. Likas, A.C., Vlassis, N., Verbeek, J.: The Global k-Means Clustering Algorithm. Pattern Recognition **36**(2), 451–461 (2003)

30. Linde, Y., Buzo, A., Gray, R.M.: An Algorithm for Vector Quantizer Design. IEEE Transactions on Communications **28**(1), 84–95 (1980)

31. Lingras, P., West, C.: Interval Set Clustering of Web Users with Rough K-Means. Journal of Intelligent Information Systems **23**(1), 5–16 (2004)

32. Lloyd, S.: Least squares quantization in PCM. IEEE Transactions on Information Theory **28**(2), 129–137 (1982)

33. MacQueen, J.: Some methods for classification and analysis of multivariate observations. In: Proceedings of the Fifth Berkeley Symposium on Mathematical Statistics and Probability, vol. 1, pp. 281–297 (1967)
34. McLachlan, G.J., Krishnan, T.: The EM Algorithm and Extensions, 2nd edn. Wiley Series in Probability and Statistics. John Wiley & Sons, New York (2008). ISBN: 0-471-20170-7
35. Neyman, J.: Frequentist probability and frequentist statistics. The Synthese Journal **360**(1), 97–131 (2004)
36. Oppenheim, A.V.: Superposition in a Class of Nonlinear Systems. Massachussetts Institute of Technology, Cambridge, Massachussetts (1964). Ph.D. Dissertation
37. Pawlak, Z.: Rough Sets: Theoretical Aspects of Reasoning About Data. Kluwer Academic Publishers, Norwell, Massachussetts (1991)
38. Pelleg, D., Moore, A.: x-Means: Extending k-Means with Efficient Estimation of the Number of Clusters. In: Proceedings of the Seventeenth International Conference on Machine Learning, pp. 261–265 (2000)
39. Polkowski, L.: Rough Sets, 1st edn. Advances in Soft Computing. Physica-Verlag, A Springer-Verlag Company, Heidelberg (2002). ISBN: 3-7908-1510-1
40. Tzortzis, G., Likas, A.C.: The Global Kernel k-Means Algorithm for Clustering in Feature Space. IEEE Transactions on Neural Networks **20**(7), 1181–1194 (2009)
41. Xu, L., Jordan, M.I.: On Convergence Properties of the EM Algorithm for Gaussian Mixtures. Neural Computation **8**(1), 129–151 (1996)
42. Zeger, K., Vaisey, K., Gersho, A.: Globally optimal vector quantizer design by stochastic relaxation. IEEE Transactions on Signal Processing **40**(2), 310–322 (1992)
43. Zhang, B.: Generalized k-Harmonic Means – Boosting in Unsupervised Learning (2000). Technical Report Number HPL-2000-137
44. Zhang, B., Hsu, M., U.Dayal: k-Harmonic Means – A Data Clustering Algorithm (1999). Technical Report Number HPL-1999-124
45. Zhang, Y.F., Mao, J.L., Xiong, Z.Y.: An efficient clustering algorithm. In: International Conference on Machine Learning and Cybernetics, vol. 1, pp. 261–265 (2003)
46. Zhang, Z., Dai, B.T., Tung, A.K.: Estimating Local Optimums in EM Algorithm over Gaussian Mixture Model. In: The 25th International Conference on Machine Learning (ICML), pp. 1240–1247 (2008)
47. Zhu, X.: Semi-Supervised Learning Literature Survey. Technical Report 1530, Computer Sciences, University of Wisconsin-Madison (2008). URL http://pages.cs.wisc.edu/~jerryzhu/pub/ssl_survey.pdf. Originally written in 2005, but modified in 2008.

Chapter 12
Transformation

What is hell? Hell is oneself, Hell is alone, the other figures in it
Merely projections. There is nothing to escape from and nothing
to escape to. One is always alone.

T.S. Eliot
The Cocktail Party (Act I, Scene 3), 1950

In Chapter 10, we discussed techniques for clustering and the estimation of model parameters. We have also discussed the concept of *sufficient statistics*. In this chapter, we examine techniques for finding *transformations* (see Section 6.5.5) which would operate on the parameter space, ϕ, or the sample space, $\mathcal{X}$. Both are related to the concept of *model selection* and *sufficient statistics* which were discussed in earlier chapters. In both cases, we have noted that the *principle of parsimony* (see the footnote in Section 10.6) applies. Therefore, the objective is to find the smallest space which would convey the most information. This may be in the form of rotation and scaling of the parameters, but it is often accompanied by some method of projection or other transformation which reduces the dimensionality of the model parameters to achieve the desired increased concentration of information.

The objective is to be able to extract as much information out of the parameters which are found by the methods of Chapter 10 while reducing the number of model parameters. The desired reduction in the number of parameters is associated with two different practical issues.

First, in *statistical parameter estimation* and *modeling*, we are always faced with shortage of data. In fact, statistically, we can have complete information about a process, only if we have seen all possible data samples. Seeing all possible samples is of course impossible, so reducing the number of parameters being estimated ensures that, per estimated parameter, an ample amount of data has been observed.

The second practical issue is a matter of computation. As the number of parameters are increased, generally more memory, computation time, and computational effort is required. This is especially important for a problem such as *speaker recognition* which deals with speech (a *high capacity* signal containing much *redundancy* – see Section 1.5.14).

12.1 Principal Component Analysis (PCA)

Principal component analysis (PCA) is quite a simple idea and has been used in many different fields. It is a *linear orthogonal transformation* that transforms the space of interest into one that has its basis along the *principal components* of the space. Here, we are interested in a special transformation called the linear *Karhunen-Loève Transformation (KLT)* which operates on the covariance of a statistical distribution instead of any general matrix – consult [13] for the general *PCA* problem.

At this point we will call this special case by its general name, simply, *PCA*. The basic idea is that if we start with a set of features which are somehow correlated to each other, then we should be able to reduce the dimensionality of the *feature vector* without losing much information. It is an immediate consequence of an *Eigensystem decomposition*. *PCA* generates a *linear orthogonal transformation matrix* which will transform the original feature vector to a lower-dimensional space through a built in *rotation* and *projection*. It works by transforming the features to a new space such that in the new space, the coordinates are ordered in terms of the variance of the data in their corresponding dimensions. Namely, the first coordinate will have the largest variance, then the second coordinate and so on.

The last few coordinates will have the least variance, which means that they carry the least amount of information (much in the way that *Fisher* [9] defined information – see Section 7.7).

12.1.1 Formulation

Let us assume that we have a set of N feature vectors, $\{\breve{\mathbf{x}}\}_1^N$. For the *PCA* procedure, the first requirement would be the *translation* of the *sample space* to one where the mean of the sample distribution of $\breve{\mathbf{x}}_n$ would be 0. This can be easily achieved by computing the *sample mean*, $\breve{\boldsymbol{\mu}}$, of the data, $\{\breve{\mathbf{x}}\}_1^N$, and subtracting it from all the samples,

$$\mathbf{x}_n = \breve{\mathbf{x}}_n - \breve{\boldsymbol{\mu}} \ \forall \ n \in \{1, 2, \cdots, N\} \tag{12.1}$$

From here on, we will assume that the data samples are the translated samples, $\{\mathbf{x}\}_1^N$, which have a mean that is *approximately zero*[1].

The coordinates in the transformed *Cartesian product space* (Section 6.2.2) to be incident with the *Eigenvectors* of the *covariance matrix* and the magnitudes of their

---

[1] The *sample mean* of the translated data is identically zero, but since the *sample mean* is an approximation of the *true mean*, we cannot say that the data is from a *zero-mean sample space*.

corresponding variances are related to the *Eigenvalues* of the *covariance matrix*. Therefore, the problem of *PCA* reduces to the following *Eigenvalue problem*,

$$\Sigma \mathbf{v} = \lambda \mathbf{v} \tag{12.2}$$

where $\mathbf{v} : \mathscr{R}^1 \mapsto \mathscr{R}^D$ is an *Eigenvector* associated with the feature vectors of interest, $\mathbf{x} : \mathscr{R}^1 \mapsto \mathscr{R}^D$, and $\Sigma : \mathscr{R}^D \mapsto \mathscr{R}^D$ is the associated variance-covariance (*covariance*) matrix (see Section 6.10). λ is known as an *Eigenvalue* associated with *Eigenvector*, $\mathbf{v}$ and the matrix, Σ.

We may rearrange Equation 12.2 in the following form,

$$(\lambda \mathbf{I} - \Sigma)\mathbf{v} = \mathbf{0} \; \forall \; \mathbf{v} \tag{12.3}$$

For any general matrix, Σ, Equation 12.3 may only be true if the determinant of $\lambda \mathbf{I} - \Sigma$ is zero. There will generally be D values of λ (the *Eigenvalues*) for which the *determinant* can become zero, given any general matrix. $\lambda_i, i \in \{1, 2, \cdots, D\}$ are said to be the solutions to the characteristic equation,

$$|\lambda \mathbf{I} - \Sigma| = \prod_{i=1}^{D} (\lambda - \lambda_i)$$
$$= \mathbf{0} \tag{12.4}$$

where $|\lambda \mathbf{I} - \Sigma|$ denotes the determinant of $\lambda \mathbf{I} - \Sigma$. Note that in general $\lambda_i \in \mathbb{C}$.

for every *Eigenvalue*, λ_i, Equation 12.2 must be true. There is a single *Eigenvector*, $\mathbf{v}_i$, associated with every λ_i which makes Equation 12.2 valid, namely,

$$\Sigma \mathbf{v}_i = \lambda_i \mathbf{v}_i \; \forall \; i \in \{1, 2, \cdots, D\} \tag{12.5}$$

Solving Equation 12.2 for all i will produce a set of D *Eigenvectors* associated with the D Eigenvalues, λ_i. Let us construct a matrix, $\mathbf{V} : \mathscr{R}^D \mapsto \mathscr{R}^D$ whose columns are the *Eigenvectors*, $\mathbf{v}_i$, such that the first *Eigenvector* is associated with the largest *Eigenvalue* and the last one is associated with the smallest *Eigenvalue*. Also, note that Equation 12.5 may be multiplied by any constant from both sides. Therefore, the magnitude of the *Eigenvectors*, $\mathbf{v}_i$ is arbitrary. However, in here, we assume that all $\mathbf{v}_i$ have been normalized to have unit magnitude,

$$\|\mathbf{v}_i\|_{\mathscr{E}} = 1 \; \forall \; i \in \{1, 2, \cdots, D\} \tag{12.6}$$

This may be simply achieved by dividing the computed *Eigenvectors* by their corresponding Euclidean norms. After applying the normalization, we will have the following relation, based on Equation 12.2,

$$\Sigma \mathbf{V} = \Lambda \mathbf{V} \tag{12.7}$$

where $\boldsymbol{\Lambda} : \mathcal{R}^D \mapsto \mathcal{R}^D$ is a *diagonal matrix* whose diagonal elements are the λ_i, arranged in the manner such that λ_1 is the largest *Eigenvalue* and λ_D is the smallest,

$$\lambda_1 \geq \lambda_2 \geq \cdots \geq \lambda_D \tag{12.8}$$

Due to the fact that the Eigenvectors produce a *linearly independent*[2] basis [13], **V** has an inverse and therefore, Equation 12.7 may be written in the following form,

$$\mathbf{V}^{-1}\boldsymbol{\Sigma}\mathbf{V} = \boldsymbol{\Lambda} \tag{12.9}$$

According to Equation 12.9, the matrix of Eigenvectors, **V**, is said to diagonalize $\boldsymbol{\Sigma}$. Also, because of the normalization that was done to the Eigenvectors earlier, matrix **V** is unitary (see Definition 23.11), namely,

$$\mathbf{V}^T\mathbf{V} = \mathbf{I} \tag{12.10}$$

Therefore,

$$\begin{aligned}\mathbf{V}^{-1}\boldsymbol{\Sigma}\mathbf{V} &= \mathbf{V}^T\boldsymbol{\Sigma}\mathbf{V} \\ &= \boldsymbol{\Lambda}\end{aligned} \tag{12.11}$$

The *Karhunen-Loève Transformation (KLT)* of any feature vector, $\mathbf{x}_n, n \in \{0, 1, \cdots, N\}$ is then given by,

$$\begin{aligned}\mathbf{y}_n &= T_{KL}(\mathbf{x}_n) \\ &= \mathbf{V}^T\mathbf{x}_n\end{aligned} \tag{12.12}$$

The new vectors, $\mathbf{y}_n$, are represented in the space which is defined by the *orthonormal basis* of the *Eigenvectors*. Considering the fact that the *transformed space* is such that all its constituents have the highest variance in the first dimension (associated with the first *Eigenvalue* and the first *Eigenvector*), and the lowest variance in the Dth dimension (associated with the Dth *Eigenvalue* and the Dth *Eigenvector*), we may be selective about which dimensions we consider as informative. This allows us to reduce the dimensionality of the space.

To do this reduction, first we would have to have a criterion by which we can prune the space. One important assumption is that the importance of the feature dimensions has a one-to-one relationship with their corresponding variances. We saw evidence of the truth behind this assumption in Sections 10.1 and 7.7, in the way *Fisher information matrix*, $\mathcal{I}_F$, was defined and computed. Using this assumption, we may define a criterion based on the *minimum tolerated ratio* between any one Eigenvalue and some norm of $\boldsymbol{\Lambda}$. This norm may be, for instance, any of $\{\|\boldsymbol{\Lambda}\|_1, \|\boldsymbol{\Lambda}\|_{\mathcal{F}}, \cdots, \|\boldsymbol{\Lambda}\|_\infty\}$. Also, since the Eigenvalues are ordered by magnitude, we will basically be removing the last $(D - M)$ dimensions associated with

---

[2] We are assuming that we have distinct Eigenvalues. If λ_i repeats, then there will be a *generalized Eigenvector* v_i^r associated with the rth *repeated Eigenvalue*, where $\boldsymbol{\Lambda}$ will have a *Jordan block form* [10]. A repeated Eigenvalue is also known as a *multiple Eigenvalue* or a *degenerate Eigenvalue* [6].

the smallest Eigenvalues. Then, we will modify $\mathbf{V}$ so that only the first M columns are kept, leaving us with a new Eigenvector matrix, $\mathbf{U} : \mathscr{R}^M \mapsto \mathscr{R}^D$ and the corresponding reduced Eigenvalue Matrix,

$$(\mathbf{L})_{[i][j]} = \begin{cases} \lambda_i \ \forall \ i = j \\ 0 \ \forall \ i \neq j \end{cases} \tag{12.13}$$

where $i, j \in \{1, 2, \cdots, M\}$.

The projection, causing this space reduction, is then given by the following relation,

$$\mathbf{y} = \mathbf{U}^T \mathbf{x} \tag{12.14}$$

where $\mathbf{y} : \mathscr{R}^1 \mapsto \mathscr{R}^M$ is the new, reduced length, *feature vector*. In practice, the computation of Λ and $\mathbf{V}$ may be done using a *singular value decomposition* [2, 14] – see Definition 23.15.

The *Karhunen Loève transformation* (*KLT*) has been successfully used in a variety of fields such as *online handwriting recognition* [22], *image recognition* (so called *EigenFaces* [17], and of course *speaker recognition* [3, 5, 21, 4]. However, extreme care should be taken to make sure the *nonlinear effects* of the features are not taken too lightly. If there is an unbalanced importance level in a *perceptual sense* between the higher variance and lower variance feature dimensions, the *PCA* can hurt. It is important to make sure that proper *inter-dimension normalization* is performed, based on *physiological data* before performing *KLT*.

12.2 Generalized Eigenvalue Problem

The *Eigenvalue problem*, described by Equation 12.2 may be generalized as follows,

$$\lambda_\alpha \Sigma_\alpha \mathbf{v} = \lambda_\beta \Sigma_\beta \mathbf{v} \tag{12.15}$$

Equation 12.15 describes a new problem called the *generalized Eigenvalue problem*. The *simple Eigenvalue problem* of Equation 12.2 may be viewed as the special case of the *generalized Eigenvalue problem*, where $\Sigma_\alpha = \Sigma$ and $\Sigma_\beta = \mathbf{I}$. Therefore, in the same way that we wrote Equation 12.3, Equation 12.15 may be rearranged as follows,

$$(\lambda_\beta \Sigma_\beta - \lambda_\alpha \Sigma_\alpha) \mathbf{v} = \mathbf{0} \tag{12.16}$$

Much in the same way as in the simple Eigenvalue problem, we may write the following relations,

$$\Lambda_\alpha \Sigma_\alpha V = \Lambda_\beta \Sigma_\beta V \tag{12.17}$$

$$\lambda_{\alpha_n} \Sigma_\alpha v_n = \lambda_{\beta_n} \Sigma_\beta v_n \tag{12.18}$$

with similar definitions as in Section 12.1.

If Λ_α has *full rank*, then we may define a new Eigenvalue matrix,

$$\Lambda \overset{\Delta}{=} \Lambda_\alpha^{-1} \Lambda_\beta \tag{12.19}$$

translating to the following definition for each of the general Eigenvalues,

$$\lambda_n \overset{\Delta}{=} \frac{\lambda_{\beta_n}}{\lambda_{\alpha_n}} \tag{12.20}$$

In this special case, we may rewrite equations 12.17 and 12.18 in terms of Λ and λ_n as follows,

$$\Sigma_\alpha V = \Lambda \Sigma_\beta V \tag{12.21}$$

$$\Sigma_\alpha v_n = \lambda_n \Sigma_\beta v_n \tag{12.22}$$

λ_n and v_n are known as the *generalized Eigenvalues* and the *right generalized Eigenvectors* of Equation 12.22.[3]

Furthermore, in the special case when Σ_β has full rank, the generalized Eigenvalue problem reduces to the simple Eigenvalue problem where,

$$\Sigma_\beta^{-1} \Sigma_\alpha V = \Lambda V \tag{12.23}$$

$$\Sigma_\beta^{-1} \Sigma_\alpha v_n = \lambda_n v_n \tag{12.24}$$

On the other hand, if $|\Sigma_\beta| = 0$, then there are p Eigenvectors associated with p Eigenvalues such that p is the rank of Σ_β.[4] The rest of the Eigenvalues will be *infinite*. In that case, if $|\Sigma_\beta| \neq 0$, then we may rewrite the Eigenvalue problem by defining a different Eigenvalue such that,

$$\hat{\Lambda} \overset{\Delta}{=} \Lambda_\beta^{-1} \Lambda_\alpha \tag{12.25}$$

translating to the following definition for each of the generalized Eigenvalues,

$$\hat{\lambda}_n \overset{\Delta}{=} \frac{\lambda_{\alpha_n}}{\lambda_{\beta_n}} \tag{12.26}$$

---

[3] The *Left generalized Eigenvalue* problem is also possible which gives rise to *left generalized Eigenvectors*. The problem statement for the left Eigenvector problem is written in the following general form, $\lambda_\alpha v^H \Sigma_\alpha = \lambda_\beta v^H \Sigma_\beta$.

[4] $|\Sigma_\beta| = 0$ means that $\Sigma_\beta v = 0 \; \forall v$.

in which case, the new Eigenvalues corresponding to the infinite Eigenvalues in the previous variation are 0.

If there are repeated Eigenvalues, the directions are the generalized Eigenvectors and the Eigenvalue matrix, Λ, will have a *Jordan block* form [10, 14]. As an example, a *Jordan block* associated with a repeated eigenvalue, λ_i, which is repeated r times would have r rows and r columns and is written as,

$$
\begin{bmatrix}
\lambda_i & 1 & 0 & \cdots & & 0 \\
0 & \lambda_i & 1 & & \cdots & 0 \\
\vdots & \vdots & \vdots & & \cdots & \vdots \\
\cdots & \cdots & \cdots & & \ddots & \vdots \\
0 & 0 & 0 & \lambda_i & 1 \\
0 & 0 & 0 & 0 & \lambda_i
\end{bmatrix}
\tag{12.27}
$$

Let us assume that under normal circumstances we would have N non-repeating Eigenvalues. In the special case when r of them would be repetitions of λ_i, then the block of Equation 12.27 would take the place of the $r \times r$ block that would have had 0s on the off diagonal and r different Eigenvalues on the diagonal for the non-repeating case.

Due to the limited space, we will not go through the details of the case with repeated Eigenvalues. *Fortmann* [10] presents a complete treatment of this in the appendix of his book, as well as the body, while treating the theory of linear control systems.

12.3 Nonlinear Component Analysis

In the formulation of the *PCA* (Section 12.1.1), we briefly touch upon the danger of treating the different dimensions of the feature vector ($\mathbf{x}$) the same. A suggestion was made to make sure that some normalization is done to account for possible perceptual variations across dimensions of the feature vector. Also, it is possible for the feature vector to be created from a combination of different sources which may not have much in common in terms of their behavior. Therefore, there is no single recipe which may be applied to treat the different dimensions in a feature vector in order to make them behave similarly and to have similar effects in our pattern recognition analyses.

The normalization which was advised in Section 12.1.1, would normally be a linear normalization attempt. However, as we know, for example in the case of perceptual mapping discussed in Section 5.1, there are logarithmic effects present in human perception, both in the frequency domain and the amplitude domain. Since speech includes such inherent nonlinearities, a nonlinear normalization of the feature space may be needed, before or while we reduce the space through a projection. This may be viewed in two different lights. The first option is to do a *nonlinear transformation* to the feature vectors and then do a linear PCA for achieving a reduction in the dimensionality of the problem. A second option is to perform a *nonlinear transformation* which includes a reduction in the dimensionality of the original feature vector.

Jolliffe [13] discusses several different applications, presented by different researchers, which utilize the nonlinear transformation followed by a linear PCA. There have also been other alternatives, such as that proposed by Gnanadesikan [11]. [11] augments the original feature vector with elements which are nonlinear functions of the original elements in the vector. Then, a linear PCA is done on the augmented feature vector and the dimensionality is reduced. Although any nonlinear function of the elements may be augmented, [11] uses quadratic functions as examples. This, in essence is similar to the previous proposal of using a nonlinear transformation, followed by a linear PCA. The difference lies in the fact that the method of [11] combines a linear PCA on the original feature vector, with one that operates on a nonlinear transformation of the original vector, through augmentation. It is easy to see that the two are basically identical and one may devise a nonlinear transformation that would include the original nonlinear mapping followed by the space reducing linear projection.

12.3.1 Kernel Principal Component Analysis (Kernel PCA)

Schölkopf [23, 24] proposes a computationally intensive *kernel*-based PCA which operates in a dot-product space. The dot product is computed over all the N observations, producing a dot product matrix which is N by N. A *nonlinear kernel* of this operation may be chosen, possibly producing more than D non-zero Eigenvalues and Eigenvectors, even up to N such values! *Kwok* [19] has used this *kernel PCA* technique for *speaker adaptation* and named it *Eigenvoice speaker adaptation*, which is based on the *standard Eigenvoice speaker adaptation* technique [18]. *Kernel PCA* is also used by [25] in helping visualize the data separation in an *SVM* formulation (see Chapter 15).

The concept of kernel PCA is basically similar to the *support vector machines* discussed in detail in Chapter 15, where a *nonlinear Mercer Kernel* (Definition 24.61) is used to transform the, so called, *input space* to the *feature space*, using a nonlinear function. In the new *feature space*, a regular PCA is carried out. This process will become more clear as we cover this concept while discussing *support vector machines* in Chapter 15. Since we need to review some basic topics related to *support vector machines*, in order to be able to describe the kernel PCA in more detail, this topic has been deferred to Section 15.7.

12.4 Linear Discriminant Analysis (LDA)

Linear Discriminant Analysis (LDA) is a statistical pattern classification technique which makes the following basic assumptions [8, 12],

1. Samples vectors which are considered to belong to the same class, are considered to be identically distributed about their own class mean according to a *covariance matrix*, Σ_W, known as the *within class covariance matrix*.
2. Mean values of all classes are themselves distributed according to a probability density function of the same family as the within class distributions, but with different sufficient statistics. Means of these classes, therefore, have a *central mean* and their own *covariance matrix*, known as the *between class covariance matrix*, Σ_B.

As we shall see, LDA is somewhat related to PCA, in the sense that it generally includes similar techniques for reducing the dimensionality of the samples. Most applications of LDA [1, 26] are designed to initially take on many different sources of information at the initial cost of increasing the dimensionality of the features (samples) and then using LDA to reduce the number of dimensions down to a more manageable set, but aligned in the direction of higher information. This is the essential similarity between PCA and LDA.

Given the assumptions at the beginning of this section, LDA amounts to estimating Σ_W for the different classes based on the high-dimensional samples, doing linear transformation in the form of a rotation followed by scaling of the spaces in these classes, so that they point along principal directions. Then, a global linear transformation based on the matrix of Eigenvectors of Σ_B or Σ_T (total covariance matrix) may be done. In practice, the final transformation will be a projection which is designed to reduce the dimensionality of the space, as was done in the *PCA* implementation of Section 12.1.

In order to estimate the *means* and *covariance* matrices, we may use different techniques which were discussed in Section 6.9. Here, we choose the *maximum likelihood estimate* for estimating these statistics. Other methods such as the *unbiased estimate* of the variance may also be chosen.

Let us assume that we have observed N samples, $\mathbf{x}_n : \mathcal{R}^1 \mapsto \mathcal{R}^D$, $n \in \{1, 2, \cdots, N\}$. Furthermore, suppose that we have used a classification technique as discussed in Section 9.3 to determine Γ classes and to associate each observed sample with one of these Γ classes. N_γ denotes the number of samples associated with each class,

$\mathscr{X}_\gamma$. Then, the *maximum likelihood estimate (MLE)* of the *mean* of each class, $\mathscr{X}_\gamma$, would be given by,

$$\boldsymbol{\mu}_\gamma = \frac{1}{N_\gamma} \sum_{\mathbf{x} \in \mathscr{X}_\gamma} \mathbf{x} \tag{12.28}$$

Also, the MLE of the covariance of class $\mathscr{X}_\gamma$ may be written as follows,

$$\boldsymbol{\Sigma}_\gamma = \frac{1}{N_\gamma} \sum_{\mathbf{x} \in \mathscr{X}_\gamma} \left(\mathbf{x} - \boldsymbol{\mu}_\gamma\right) \left(\mathbf{x} - \boldsymbol{\mu}_\gamma\right)^T \tag{12.29}$$

Equations 12.28 and 12.29 are estimates of the *class* statistics for class $\mathscr{X}_\gamma$.

Let us define a *scatter matrix* which is related to the maximum likelihood estimate of the covariance matrix in the following way,

$$\mathbf{S} \triangleq N\boldsymbol{\Sigma} \tag{12.30}$$

It is somewhat simpler to maintain the statistics using the *scatter matrix* in lieu of the covariance matrix and we can always convert from one to the other using the relation in Equation 12.31. The *scatter matrix* for each class, $\mathscr{X}_\gamma$, may then be written as follows,

$$\mathbf{S}_\gamma = \sum_{\mathbf{x} \in \mathscr{X}_\gamma} \left(\mathbf{x} - \boldsymbol{\mu}_\gamma\right) \left(\mathbf{x} - \boldsymbol{\mu}_\gamma\right)^T \tag{12.31}$$

Based on the first assumption of this section, the within class scatter matrix would be,

$$\mathbf{S}_W = \sum_{\gamma=1}^{\Gamma} \mathbf{S}_\gamma \tag{12.32}$$

$$= \sum_{\gamma=1}^{\Gamma} \sum_{\mathbf{x} \in \mathscr{X}_\gamma} \left(\mathbf{x} - \boldsymbol{\mu}_\gamma\right) \left(\mathbf{x} - \boldsymbol{\mu}_\gamma\right)^T \tag{12.33}$$

The *total mean* can be computed either from the individual observations, or from the class means of Equation 12.28 as follows,

$$\boldsymbol{\mu} = \frac{1}{N} \sum_{n=1}^{N} \mathbf{x}_n \tag{12.34}$$

$$= \frac{1}{N} \sum_{\gamma=1}^{\Gamma} N_\gamma \boldsymbol{\mu}_\gamma \tag{12.35}$$

and in the same manner[8], the total scatter matrix would be,

$$\mathbf{S}_T = \sum_{n=1}^{N} (\mathbf{x}_n - \boldsymbol{\mu}) (\mathbf{x}_n - \boldsymbol{\mu})^T \tag{12.36}$$

$$= \sum_{\gamma}^{\Gamma} \sum_{\mathbf{x} \in \mathcal{X}_\gamma} (\mathbf{x} - \boldsymbol{\mu}_\gamma + \boldsymbol{\mu}_\gamma - \boldsymbol{\mu}) (\mathbf{x} - \boldsymbol{\mu}_\gamma + \boldsymbol{\mu}_\gamma - \boldsymbol{\mu})^T \tag{12.37}$$

$$= \sum_{\gamma}^{\Gamma} \sum_{\mathbf{x} \in \mathcal{X}_\gamma} (\mathbf{x} - \boldsymbol{\mu}_\gamma) (\mathbf{x} - \boldsymbol{\mu}_\gamma)^T +$$

$$\sum_{\gamma}^{\Gamma} N_\gamma (\boldsymbol{\mu}_\gamma - \boldsymbol{\mu}) (\boldsymbol{\mu}_\gamma - \boldsymbol{\mu})^T \tag{12.38}$$

The first term in Equation 12.38 is identical to $\mathbf{S}_W$ of Equations 12.33. The second term seems to have the properties stated at the beginning of this section. In fact, we can call that the *between class* scatter matrix, giving us the following definition of S_B [8],

$$\mathbf{S}_B \overset{\Delta}{=} \sum_{\gamma}^{\Gamma} N_\gamma (\boldsymbol{\mu}_\gamma - \boldsymbol{\mu}) (\boldsymbol{\mu}_\gamma - \boldsymbol{\mu})^T \tag{12.39}$$

Therefore, based on Equations 12.38, 12.33, and 12.39, the *total scatter* matrix may be written in terms of the *within class* and *between class* scatter matrices as follows,

$$\mathbf{S}_T = \mathbf{S}_W + \mathbf{S}_B \tag{12.40}$$

At this point, we have the between and within scatter matrices (hence covariances) of random variable, X, estimated based on the observation of N samples of X. The objective of linear discriminant analysis [5] is to maximize discriminability by performing a transformation $(T : \mathcal{R}^{(\Gamma-1)} \mapsto \mathcal{R}^D)$ on X such that the transformed random variable, $Y : \mathbf{y} \in \mathcal{R}^{(\Gamma-1)}$ would have maximal ratio of the determinant of the *between class scatter matrix* to the determinant of the *within class scatter matrix*. The determinant is used since it is equal to the product of the Eigenvalues of the matrix and an indication of the *geometric mean* (Definition 6.79), of the Eigenvalues.

Let us define the following *linear transformation* for discriminant analysis of X,

$$\mathbf{y}_n = T(\mathbf{x}_n) \tag{12.41}$$

$$\overset{\Delta}{=} \mathbf{U}^T \mathbf{x}_n \tag{12.42}$$

where $\mathbf{U} : \mathcal{R}^{(\Gamma-1)} \mapsto \mathcal{R}^D$ is a rectangular matrix and $(\Gamma - 1) < D$. Therefore, the between class and within class scatter matrices of Y may be written in terms of those of X and the transformation matrix $\mathbf{U}$, as follows,

$$\mathbf{S}_B(Y) = \mathbf{U}^T \mathbf{S}_B(X) \mathbf{U} \tag{12.43}$$

$$\mathbf{S}_W(Y) = \mathbf{U}^T \mathbf{S}_W(X) \mathbf{U} \tag{12.44}$$

---

[5] Called *multiple discriminant analysis*[8] when there are more than two classes involved – $\Gamma > 2$.

Therefore, the maximization problem of interest would be,

$$\hat{\mathbf{U}} = \arg\max_{\mathbf{U}} E(\mathbf{U}) \tag{12.45}$$

$$= \arg\max_{\mathbf{U}} \frac{|\mathbf{S}_B(Y)|}{|\mathbf{S}_W(Y)|} \tag{12.46}$$

$$= \arg\max_{\mathbf{U}} \frac{|\mathbf{U}^T \mathbf{S}_B(X)\mathbf{U}|}{|\mathbf{U}^T |\mathbf{S}_W(X)| \mathbf{U}|} \tag{12.47}$$

The solution to Equation 12.47 is given by a special case of the *generalized Eigenvalue problem* (see the assumptions made for Equation 12.21 in Section 12.2),

$$\mathbf{S}_B\mathbf{U} = \Lambda \mathbf{S}_W\mathbf{U} \tag{12.48}$$

12.4.1 Integrated Mel Linear Discriminant Analysis (IMELDA)

When LDA is used on the output of a *Mel-scale filterbank*, some have called the combined system IMELDA [12] which stands for Integrated Mel Linear Discriminant Analysis. *Hunt* [12] describes two systems based on static spectral analysis (IMELDA-1) and dynamic spectral analysis (IMELDA-2).

As an example in speech recognition, [26] has used IMELDA in reducing a 44 dimensional feature vector to 24 dimensions. The original 44 features were a combination of 8 *spectral features*, the first 12 *cepstral* coefficients, the first 12 *delta cepstral* features, and the first 12 *delta-delta cepstral* features. The between and within covariance matrices are then computed [12] and the top transformed 24 features are used as the features for each frame of speech.

12.5 Factor Analysis

So far, we have discussed *principal component analysis (PCA)* and *linear discriminant analysis (LDA)* which provided transformations to rotate and project the random variable of interest to one that would generally have a smaller number of components and would yet retain most of the information in the original variable. *Factor analysis (FA)* is yet another such technique with basically the same objectives. As *Jolliffe* [13] puts it, "for most practical purposes PCA differs from factor analysis in having *no explicit model*." The *explicit model*, of which [13] speaks, will be discussed in more detail, shortly.

As with any other transformation in this chapter, the objective of *FA* is to model the observed samples, $\{\mathbf{x}\}_1^N$, as a transformation of a simpler underlying model with a lower dimensionality compared to the observed vectors.

$$\mathbf{y}_n = T(\mathbf{x}_n) \ \forall \, n \in \{1,2,\cdots,N\} \tag{12.49}$$

where $\mathbf{y}_n : \mathscr{R}^1 \mapsto \mathscr{R}^D$ is an observed sample and $\mathbf{x}_n$ is the underlying random sample which is not necessarily observable. The *explicit model* of *FA* is a *linear transformation model* which makes an assumption by which it differentiates itself from *PCA* and *LDA* techniques. In fact in some perspective, as we will see, it may be seen as a more general version of *PCA*. *FA* assumes that the underlying random variable is composed of two different components.

The first component is a random variable, containing the, so called, *common factors*. It has a lower dimensionality compared to the combined random state, X, and the observation, Y. It is called the vector of *common factors* since the same vector, $\Theta : \boldsymbol{\theta} : \mathscr{R}^1 \mapsto \mathscr{R}^M, M <= D$, is a component of all the samples of $\mathbf{y}_n$.

The second component is known as the vector of *specific factors*, or sometimes called the *error* or the *residual* vector. It is denoted by $E : \mathbf{e} : \mathscr{R}^1 \mapsto \mathscr{R}^D$. Therefore, this linear *FA* model for a specific random variable, $\tilde{Y} : \tilde{\mathbf{y}} : \mathscr{R}^1 \mapsto \mathscr{R}^D$, related to the observed random variable Y may be written as follows,

$$\tilde{\mathbf{y}}_n = \mathbf{V}\boldsymbol{\theta}_n + \mathbf{e}_n \tag{12.50}$$

where $\mathbf{V} : \mathscr{R}^M \mapsto \mathscr{R}^D$ is known as the *factor loading* matrix and its elements, $(\mathbf{V})_{[d][m]}$, are known as the *factor loadings*. Samples of random variable $\Theta : (\boldsymbol{\theta}_n)_{[m]}, n \in \{1,2,\cdots,N\}$ are known as the vectors of *common factors*, since due to the linear combination nature of the factor loading matrix, each element, $(\boldsymbol{\theta})_{[m]}$, has a hand in shaping the value of (generally) all $(\tilde{\mathbf{y}}_n)_{[d]}, d \in \{1,2,\cdots,D\}$. Samples of random variable $E : \mathbf{e}_n, n \in \{1,2,\cdots,N\}$ are known as vectors of *specific factors*, since each element, $(\mathbf{e}_n)_{[d]}$ is specifically related to a corresponding, $(\tilde{\mathbf{y}}_n)_{[d]}$.

In Equation 12.50, the *sufficient statistics* of the elements of the Equation are assumed to have the following forms [13],

$$\mathscr{E}\{\tilde{Y}\} = \mathbf{0} \tag{12.51}$$

$$\mathscr{E}\{\Theta\} = \mathbf{0} \tag{12.52}$$

$$\mathscr{E}\{E\} = \mathbf{0} \tag{12.53}$$

$$\mathscr{E}\{\Theta E\} = \mathbf{0} \tag{12.54}$$

$$\mathscr{E}\{\Theta\Theta\} = \mathbf{I}_M \tag{12.55}$$

$$\mathscr{E}\{EE\} = \boldsymbol{\Lambda}^2 \tag{12.56}$$

In Equation 12.50, the *FA* model was written for $\tilde{Y}$ and not Y. The reason behind this notation is the assumption of Equation 12.51, which assumes that the observation random variable has standard mean (*zero mean*). Also, the random variables of the *common* and *specific factors* are also assumed to have a standard (zero) mean.

The assumption of Equation 12.54, basically states that the *common factors* and the *specific factors* are *uncorrelated*. This is one of the basic assumptions of the model, which is the ground for separating the two terms in the first place. Equation 12.55 states that the *common factors* have *standard covariance*. Finally, In Equation 12.56, Λ is defined as the following diagonal matrix,

$$\left(\Lambda^2\right)_{[i][j]} \triangleq \begin{cases} \lambda_i^2 \ \forall \ i = j \\ 0 \ \ \forall \ i \neq j \end{cases} \tag{12.57}$$

and it states what we said in words earlier. Namely, the fact that each element of $(\mathbf{e}_n)_{[m]}$ is only responsible for affecting the element, $(\tilde{\mathbf{y}}_n)_{[m]}$, making the covariance matrix of E diagonal. λ_i^2 are known as the *uniquenesses* of the *factor analysis* formulation [7].

Of course our observations usually do not have *standard (zero)* mean. Also, it may make sense to redefine the residual term so that the actual residual vectors would have a *standard variance* of 1. These two requirements will modify Equation 12.50 to the following form,

$$\mathbf{y}_n = \boldsymbol{\mu} + \mathbf{V}\boldsymbol{\theta}_n + \Lambda\mathbf{r}_n \tag{12.58}$$

In Equation 12.58, we have introduced the mean vector, $\boldsymbol{\mu}$, such that the observation samples, $\mathbf{y}_n$, would have the mean vector $\boldsymbol{\mu}$ as their *sample mean*. Also, we have rewritten the *specific factor* term with the following identity,

$$\mathbf{e}_n \equiv \Lambda\mathbf{r}_n \tag{12.59}$$

where,

$$\left(\Lambda\right)_{[i][j]} \triangleq \begin{cases} \lambda_i \ \forall \ i = j \\ 0 \ \ \forall \ i \neq j \end{cases} \tag{12.60}$$

and

$$\mathcal{E}\{RR\} = \mathbf{I}_D \tag{12.61}$$

$R : \mathbf{r} : \mathcal{R}^1 \mapsto \mathcal{R}^D$ is the new residual random variable with standard variance. Equation 12.58 is the general form of the *Factor Analysis* model which will be used here.

We stated earlier that *FA* may be viewed as a generalized *PCA*. To see this point, consider a case where $\Lambda = 0$ and the equations will reduce to that of *principal component analysis*. Of course, this is only in the form and the underlying assumptions are not the same.

Note that based on all the assumptions that have been made in formulating the *FA* model, the following important result follows,

$$\Sigma_Y = \mathbf{V}\mathbf{V}^T + \Lambda^2 \tag{12.62}$$

where Σ_Y is the covariance of the observation random variable, Y. Since we do not know the true *covariance* of Y, we may substitute it with the value of one of the estimates discussed in Section 6.9.4, namely the biased sample covariance, $\hat{\Sigma}$, or the *biased sample covariance*, $\tilde{\Sigma}$. Assuming that we use the *biased estimator* (Definition 6.82), we will have the following approximation,

$$\hat{\Sigma}_Y \approx \mathbf{V}\mathbf{V}^T + \Lambda^2 \tag{12.63}$$

For the sake of simplicity of notation, we will replace the approximation sign with an equal sign and understand that we are using an approximation,

$$\hat{\Sigma}_Y = \mathbf{V}\mathbf{V}^T + \Lambda^2 \tag{12.64}$$

At this point, the objective of *factor analysis* becomes the estimation of the parameters of the *FA* model, namely,

$$\Phi = [\mathbf{V} \; \Lambda] \tag{12.65}$$

where $\Phi : \mathcal{R}^{M+D} \mapsto \mathcal{R}^D$. It is also possible to arrange the elements of Φ in a vector, $\varphi : \mathcal{R}^1 \mapsto \mathcal{R}^{D(M+1)}$, which is a column vector composed of the M columns of $\mathbf{V}$ followed by the D diagonal elements of Λ.

As a first step in estimating φ, we may make the assumption that we pick φ such that Equation 12.64 is satisfied. The left hand side of Equation 12.64 may easily be computed, using a training data set, $\{\mathbf{y}\}_1^N$.

Almost all the estimation techniques discussed in Chapter 10 may be used in estimating the parameter vector, φ, including *maximum likelihood estimation* and *maximum a-posteriori estimation* which are the two most popular techniques used in the literature. In fact, *Dempster, et al.*, in their original paper on *expectation maximization* [7], present the estimation problem of *factor analysis* as one of the examples of this algorithm. In Section 11.3.1 we presented a detailed treatment of the use of the *EM* for *maximum likelihood* and *maximum a-posteriori estimation*. The *maximum likelihood estimate* of the parameter vector sometimes takes a long time to converge [20].

Most of the above approaches employ iterative techniques which require an initial estimate of the parameter vector. As we saw in Section 11, the initial estimate is quite important in ensuring a speedy convergence. An example is the *EM* approach.

We noted that if $\boldsymbol{\Lambda} = \mathbf{0}$, then the problem of *FA* will be similar to the *PCA* problem. Therefore, a good initial estimate of the *factor loading matrix*, $\mathbf{V}$, would be the reduced Eigenvector matrix (Equation 12.14), used in the formulation of the *PCA* problem.

A detailed maximum likelihood treatment of $\boldsymbol{\varphi}$ for *factor analysis* may be found in [20]. *Kenny* [15] works out the maximum likelihood and *MAP* estimation of the parameters of the *FA* problem as applied to *speaker recognition*. As we will see in Section 16.4.2, the treatment in [15] is known by the name of *joint factor analysis*, since it splits the models into speaker and channel models, each of which is treated as a factor analysis problem, hence the qualifier, "*joint*". Other techniques have also been used for the estimation of the parameters, $\boldsymbol{\varphi}$ [15]. [16] uses *Baum-Welch* which is a special case of the *generalized expectation maximization* (*GEM*) algorithm (Section 11.3.1.2), as we will see in Section 13.6.4.

It is important to note that, given Equation 12.64, the choice of $\boldsymbol{\varphi}$ is not unique.[13] To see this, let us assume that we make a choice, $\mathbf{V}$, for the estimate of the *factor loadings*. Furthermore, let us assume that we may choose any other matrix, $\tilde{\mathbf{V}}$, which has the following relation with our original choice,

$$\tilde{\mathbf{V}} = \mathbf{VU} \tag{12.66}$$

where $\mathbf{U} : \mathscr{R}^M \mapsto \mathscr{R}^M$ is any arbitrary *unitary matrix* (see Definition 23.16). Then it follows that,

$$\tilde{\mathbf{V}}\tilde{\mathbf{V}}^T = \mathbf{VUU}^T\mathbf{V}^T \tag{12.67}$$

Since $\mathbf{U}$ is a *unitary matrix*, $\mathbf{UU}^T = \mathbf{I}_M$, resulting in the following identity,

$$\tilde{\mathbf{V}}\tilde{\mathbf{V}}^T = \mathbf{VV}^T \tag{12.68}$$

Therefore $\mathbf{V}$ is not unique. For this reason, in estimating the parameter vector, certain criteria and restrictions may be placed on the transformation of $\mathbf{V}$, through the different possible *unitary transformations*, $\mathbf{U}$, to be able to narrow down the number of choices. Many such approaches are referenced in [13].

One such choice is the *variance maximization criterion* (*varimax*) [13] given by the following,

$$\hat{\mathbf{V}} = \underset{\tilde{\mathbf{V}}}{\arg\max} \sum_{m=1}^{M} \left(\sum_{d=1}^{D} v_{dm}^4 - \frac{1}{D} \left(\sum_{d=1}^{D} v_{dm}^2 \right)^2 \right) \tag{12.69}$$

where each v_{dm} is $\left(\tilde{\mathbf{V}}\right)_{[d][m]}$, normalized about the Euclidean norm of the dth column of $\tilde{\mathbf{V}}$, namely,

$$v_{dm} \triangleq \frac{\left(\tilde{\mathbf{V}}\right)_{[d][m]}}{\sqrt{\sum_{m=1}^{M} \left(\left(\tilde{\mathbf{V}}\right)_{[d][m]}\right)^2}} \tag{12.70}$$

The *varimax criterion* (the maximization objective function of Equation 12.70) is proportional to the variance of $\tilde{V}\tilde{V}^T$.

As with *PCA* and *LDA*, a sufficient number of observed samples need to be collected in order to be able to estimate the parameter vector. This may be a problem, when small amounts of training data are collected and models with a large number of free parameters have been chosen.

In experiments, it has been shown that for small amounts of data, the *specific factors* do not provide much advantage. This is mostly due to the fact that they are modelled in a diagonal variance matrix, whereas the number of parameters to be estimated in **V** are far more than those in **Λ**. In fact, [16] states that the *specific factor* term only starts becoming useful for speaker data larger than 15 minutes, which is far more than the average of 1 minute required for a practical enrollment session. Therefore, it is arguably alright to remove that term and only estimate the term related to the *common factors*.

References

1. Bahl, L.R., Brown, P.F., deSouza, P.V., Mercer, R.L.: Speech Recognition with Continuous-Parameter Hidden Markov Models. Computer Speech and Language **2**(3-4), 219–234 (1987)
2. Beltrami, E.: Sulle Funzioni Bilineari. Giornale di Mathematiche di Battaglini **11**, 98–106 (1873)
3. Chen, C., Chen, C., Cheng, P.a.: Hybrid KLT/GMM approach for robust speaker identification. Electronics Letters **39**(21), 1552–1554 (2003)
4. Chen, C.C., Chen, C.T., and, C.M.T.: Hard-limited Karhunen-Loeve transform for text independent speaker recognition. Electronics Letters **33**(24), 2014–2016 (1997)
5. Chen, C.T., Chiang, C.T., and, Y.H.C.: Efficient KLT based on overlapped subframes for speaker identification. In: Wireless Communications, 2001. (SPAWC '01). 2001 IEEE Third Workshop on Signal Processing Advances in, pp. 376–379 (2001)
6. Courant, R., Hibert, D.: Methods of Mathematical Physics, 1st english edn. Volume I, WILEY-VCH Verlag GmbH & Co. KGaA, Weinheim (2009). Second reprint of English version Translated and Revised from German Original, 1937, ISBN: 0-471-50447-4
7. Dempster, A.P., Laird, N.M., Rubin, D.B.: Maximum Likelihood from Incomplete Data via the EM Algorithm. Journal of the Royal Statistical Society. Series B (Methodological) **39**(1), 1–38 (1977)
8. Duda, R.O., Hart, P.E.: Pattern Classification and Scene Analysis. John Wiley & Sons, New York (1973). ISBN: 0-471-22361-1
9. Fisher, R.A.: Theory of Statistical Estimation. Proceedings of Cambridge Philosophical Society **22**, 700–725 (1925)
10. Fortmann, T.E., Hitz, K.L.: An Introduction to Linear Control Systems. Marcel Dekker, Inc., New York (1977). ISBN: 0-824-76512-5

11. Gnanadesikan, R.: Methods for Statistical Data Analysis of Multivariate Observations, 2nd edn. Wiley-Interscience, New York (1997). ISBN: 0-471-1611-95
12. Hunt, M., Richardson, S., Bateman, D., Piau, A.: An Investigation of PLP and IMELDA Acoustic Representations and of Their Potential for Combination. In: International Conference on Acoustics, Speech, and Signal Processing (ICASSP), vol. 2, pp. 881–884 (1991)
13. Jolliffe, I.: Principal Component Analysis, 2nd edn. Springer, New york (2002)
14. Jordan, M.C.: Mémoire sur les Formes Bilinéaires. Journal de Math'ematiques Pures et Appliquées **19**, 35–54 (1874)
15. Kenny, P.: Joint Factor Analysis of Speaker and Session Varaiability: Theory and Algorithms. Technical report, CRIM (2006). URL http://www.crim.ca/perso/patrick.kenny/FAtheory.pdf
16. Kenny, P., Ouellet, P., Dehak, N., Gupta, V., Dumouchel, P.a.: A Study of Interspeaker Variability in Speaker Verification. Audio, Speech, and Language Processing, IEEE Transactions on **16**(5), 980–988 (2008)
17. Kirby, M., Sirovich, I..: Application of the Karhunen-Loeve Procedure for the Characterization of Human Faces. IEEE Transactions on Pattern Analysis and Machine Intelligence **12**(1), 103–108 (1990)
18. Kuhn, R., Junqua, J.C.: Rapid Speaker Adaptation in Eigenvoice Space. IEEE Transaction of Speech and Audio Processing **8**(6), 695–707 (2000)
19. Kwok, J.T., Mak, B., Ho, S.: Eigenvoice Speaker Adaptation via Composite Kernel PCA. In: Advances in Neural Information Processing Systems 16. MIT Press (2003)
20. Lawley, D., Maxwell, A.: Factor Analysis as a Statistical Method. Butterworths Mathematical Texts, London (1971)
21. Lung, S.Y., Chen, C.C.a.: Further reduced form of Karhunen-Loeve transform for text independent speaker recognition. Electronics Letters **34**(14), 1380–1382 (1998)
22. Nathan, K.S., Beigi, H.S., Clary, G.J., Subrahmonia, J., Maruyama, H.: Real-Time On-Line Unconstrained Handwriting Recognition using Statistical Methods. In: International Conference on Acoustics, Speech, and Signal Processing (ICASSP95), vol. 4, pp. 2619–2622 (1995)
23. Schölkopf, B., Smola, A., Müller, K.R.: Nonlinear Component Analysis as a Kernel Eigenvalue Problem (1996). Technical Report No. 44
24. Schölkopf, B., Smola, A.J., Müller, K.R.: Kernel Principal Component Analysis. In: B. Schölkopf, C. Burges, A.J. Smola (eds.) Advances in Kernel Methods. MIT Press, Cambridge, MA (2000)
25. Solomonoff, A., Campbell, W., Quillen, C.: Channel Compensation for SVM Speaker Recognition. In: The Speaker and Language Recognition Workshop Odyssey 2004, vol. 1, pp. 57–62 (2004)
26. Weber, F., Peskin, B., Newman, M., Corrada-Emmanuel, A., Gillick, L.: Speaker Recognition on Single and Multispeaker Data. Digital Signal Processing **10**, 75–92 (2000)

Chapter 13
Hidden Markov Modeling (HMM)

It's often better to be in chains than to be free.

Franz Kafka
The Trial (Chapter 2), 1925

In *speaker recognition*, as well as other related topics such as *speech recognition*, *speech synthesis*, *language modeling*, *language recognition*, *language understanding*, *language translation* and so on, we often look for sequences of objects. Some examples are *sequences* of *feature vectors*, *words*, *phonemes*, *phones*, *acoustic labels*, et cetera. In most of these cases, there is also a need for modeling the *duration* of the output sequence as well as the content.

Hidden Markov models (*HMM*) are basically first-order *discrete time series* with some *hidden information*. Namely, the states of the *time series* are not the observed information, but they are related through an abstraction to the observation. This abstraction was described in somewhat detail in the formulation of the *expectation maximization* (*EM*) in Section 11.3.1. The existence of this indirect abstraction is what gives *HMM* the *hidden* qualifier. Take away the hidden aspect and we are left with basic *Markov chains* (Section 13.2). In modeling, the abstraction is designed to be able to reduce the number of parameters needed for describing that information – see Equation 11.65. In fact, any *hidden Markov model* may be expanded to generally create a larger *transparent Markov model*, which would be the equivalent of an inverse transformation in the sense of Equation 6.52.

Usage of *Hidden Markov models* peaked with the tremendous attention given to them from researchers in the field of speech recognition. Two major players in the development of the practical theories and architecture of *HMM* were *AT&T Bell Laboratories* [23] and *IBM T.J. Watson Research Center* [4]. In fact, I have used some of the examples from my notes of a variety of lectures I attended while working at *IBM Research*, in the 1990s.[1]

---

[1] My notes, while attending the lectures by *Lalit Bahl* at IBM Research in the mid 1990s were used for some of the examples in this chapter. No IBM-confidential information has been included. Permission was requested from Dr. Lalit Bahl and IBM research and granted on July 9, 2010. I am thankful for this generous offer.

Although the process is a *discrete time series*, the observations, which are the interesting part of the information, may be in the form of *discrete* as well as *continuous* random variables. In other words, an *HMM* is a *Markov chain* with the capacity to contain *extra information*, either associated with its *states* or its *transitions*.

In *speaker recognition*, we have limited use for *HMM* in the conventional sense. They are generally used for *text-dependent* and *text-prompted* modalities of *speaker verification*, where a whole phrase is matched. For *speaker identification* and *text-independent speaker verification*, we are not interested in modeling the phone sequence as much, so we tend to use a special case of *HMM*, which is really a *finite mixture model*. In fact, in most cases, since we assume a normally distributed underlying probability density, the finite mixture model of choice would be a *Gaussian mixture model* (*GMM*). As we shall see in Section 13.7, the *GMM* may be considered as a *single state degenerate HMM*. The treatment of the *GMM* is much simpler than the general *HMM*, since the *duration* aspect of the model has been suppressed.

In this chapter, we start by defining *zero-memory* (*memoryless*) *models*. Then, we shall continue with the introduction of *single memory* models, namely *Markov chains* (*Markov models*). We will follow with the introduction of the hiding mechanism used in *HMM* to be able to simplify the model. We will also see that *mixture models* may be represented as *single state degenerate HMM*. It was mentioned earlier that *HMM* are cases of *first-order time-series expansions*. In fact, there is a direct analogy between *HMM* and *linear systems*, where the *hidden* concept is related to the *observability* criteria [11]. Finally, we shall talk about different types models, their training, and their evaluation scenarios.

There are generally two different evolutionary paths to the conception of an *HMM*. Both approaches start on a clean slate with a *memoryless model*. Figure 13.1 shows the block diagram of these two evolution schemes.

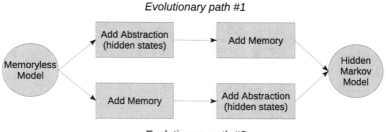

Fig. 13.1: Alternative Evolutionary Paths of the Hidden Markov Model

The first path [22, 23] adds an abstraction in the form of states which are really different from the actual observations. The observations are the points of interest, not the hidden states. At this point, the model at hand may be used to convey information in a probability mixture model such as a *Gaussian Mixture Model*. *Poritz* [22] gives an example of taking a *Bernoulli random process* (see Definition 6.63) and associating its output to such a model. Next, *memory* is added to create a *hidden Markov model*.

A second approach [4, 2], similarly, starts with a *memoryless model*. But it continues, by first adding a single *memory* to the model, creating a *transparent Markov model* or a *Markov chain*. Next, it adds the abstraction of hiding information and ends up with an *HMM*.

As we define the concepts used in these two evolutionary paths, we will see that it will be possible to associate observable outputs with the *states* or the *transitions* alike. Advocates of the two different methods happen to use one of these two techniques in practice. Most advocates of the *first path* [22, 23] associate the *outputs* with *states* whereas advocates of the *second path* [4, 2] associate them with *transitions*. This is just a small design variation and creates slightly different looking models. Most of the approaches in this chapter follow the second path.[2]

13.1 Memoryless Models

A very useful parallel is the *information theoretic* aspect of *HMM*. The *states* or *transitions* of an *HMM* may be thought of, as *information channels* with inputs and outputs – see Figure 7.1.

Consider this *information theoretic* analogy. Then, a *memoryless model* may be represented by a *memoryless source* with inputs, X, and outputs, Y. It was noted that the distributions may be *discrete* or *continuous* in which case we would be left with a *discrete* or *continuous memory channel* respectively. Figure 7.4 shows a diagram of a generic *memoryless channel* which may be associated with discrete or continuous *random variables*.

A *memoryless model* is basically a *memoryless source* (see Definition 7.2) whose parameters are learned to mimic a certain output distribution. Let us take the following example where the process being modeled is a *Bernoulli random process*.

---

[2] There is no specific reason for this other than the author's background going back to the IBM T.J. Watson Research Center which was the main advocate of this path [2, 23]. In fact it seems from the survey of the literature that although there is a definite balance, more of the references may even go back to the first evolutionary path.

Example 13.1 (Memoryless model of a fair coin toss).

Let us assume that we have a fair coin, capable of generating heads (1) or tails (0). This is a binary alphabet, making the process a Bernoulli random process – see Definition 6.63. Let us call the output random variable, $Y : y \in \mathcal{Y} = \{0,1\}$. Furthermore, let us assume that the probability that this model produces heads (1) is p. Then the probability of outputting a head (0) would be $1 - p$. Recall the definition of a discrete memoryless source (Definition 7.2). Then, using Equation 7.1, the probability of the output sequence $\{y\}_1^N$ is given by

$$P(\{y\}_1^N) = \prod_{n=1}^{N} P(y_n) \tag{13.1}$$

However, the probability mass function, $p(Y)$ is only dependent on the the the length of the sequence, N, and the number of heads, N_1 or tails, N_0. Knowing either N_1 or N_0, the other number may be found by subtracting it from N, the length of the sequence since $N_1 + N_0 = N$. Therefore, order of the output does not matter at all.

We stated that a discrete memoryless model is actually a discrete memoryless source whose parameters would have to be learned. In this case, there is really just one parameter to learn, p. We would like to estimate p from possible experiments. Let us write the probability of a N-long sequence.

$$\begin{aligned} P(\{y\}_1^N) &= p^{N_1}(1-p)^{N_0} \\ &= p^{N_1}(1-p)^{(N-N_1)} \end{aligned} \tag{13.2}$$

Since $p \leq 1$, for long sequences, $P(\{y\}_1^N)$ could become very small, with possible underflows, in practice. We have seen that maximizing the log of the likelihood is equivalent to maximizing the likelihood – see Section 10.1. Let us write the log likelihood of the sequence $\{y\}_1^N$ as a function of p and maximize it through setting its partial derivative with respect to p equal to 0. Following the notation of Section 10.1, the log likelihood will be $\ell(p|\{y\}_1^N)$, then,

$$\begin{aligned} \ell(p|\{y\}_1^N) &= \ln(p^{N_1}(1-p)^{(N-N_1)}) \\ &= N_1 \ln(p) + (N - N_1)\ln(1-p) \end{aligned} \tag{13.3}$$

Now, we take the partial derivative of ℓ, evaluated at the optimal parameter, $p = p^*$, and set it to zero, solving for p^*,

$$\begin{aligned} \left.\frac{\partial \ell}{\partial p}\right|_{p^*} &= \frac{N_1}{p^*} - \frac{N - N_1}{1 - p^*} \\ &= 0 \end{aligned} \tag{13.4}$$

Based on Equation 13.4,

$$\frac{N_1}{p^*} = \frac{N - N_1}{1 - p^*} \tag{13.5}$$

Therefore,

$$N_1(1 - p^*) = p^*(N - N_1) \tag{13.6}$$

Solving for p^*,

$$p^* = \frac{N_1}{N} \tag{13.7}$$

As we see, the maximum likelihood estimate is the relative frequency estimate for p.

Example 13.1 shows that for any *frequency* of occurrence, N_1 for outcome 1 and $N_0 = N - N_1$ for outcome 0, the *maximum likelihood* would be achieved with the *relative frequency estimate* of p given by Equation 13.7. If we use this estimate, we can write the *likelihood* of an N-long sequence in terms of N_1 and N only,

$$P(\{y\}_1^N) = \left(\frac{N_1}{N}\right)^{N_1} \left(1 - \frac{N_1}{N}\right)^{(N-N_1)} \tag{13.8}$$

and the *log likelihood* of the sequence would be written as,

$$
\begin{aligned}
\ell(p|\{y\}_1^N) &\triangleq \ln(P(\{y\}_1^N|p)) \\
&= N_1 \ln(N_1) + (N - N_1) \ln(N - N_1) - N \ln(N) \tag{13.9}
\end{aligned}
$$

13.2 Discrete Markov Chains

In Section 13.1, we discussed *memoryless models* as *memoryless sources*. In this Section, we will examine the effect of introducing *memory* into a random process. The *discrete Markov source* and the *Markov process* were defined in Section 7.1. Here, we will examine an example which helps motivate the study of models based on the *Markov process*, namely the *discrete Markov chain*.

In 1913, in a lecture at the *physical-mathematical faculty* of the *Royal Academy of Sciences* in *Saint Petersburg, Russia, Markov* presented a study of the *character sequence* in a selected part of the text of *Eugene Onegin* by Russian novelist, *Alexander Pushkin*. He picked the first 20,000 characters[3] of this novel, consisting of the first chapter and sixteen stanzas of the second chapter. Then he studied the sequences of letters in the larger sequence, referring to them as chains, in accordance to definitions from his two papers in 1907 [17] and 1911 [18] – *Markov chains*. The modeling of the character sequence may be referred to, as a *Markov model*.

Let us define the underlying structure of a *Markov model*, namely the *Markov chain* [19, 14, 21].

---

[3] In [19], Markov excluded the two letters in Russian which make no specific sound and which are associated with the stress on the previous letter.

Definition 13.1 (Discrete Markov Chain). *A discrete Markov chain or a discrete-time Markov chain is a discrete Markov process (see Definition 7.3) with a countable number of states.*

The major difference between the output of a *Markov chain* and that of a *memoryless source* (Section 13.1) is the introduction of *memory* or the dependence of the current output of the model on the previous output. This is known as a *single memory Markov chain*. In general, it would be possible to have an m-memory Markov chain, in which case Equation 7.2 would be changed to rely on the m previous observations and not just the last observation,

$$P(x_n|\{x\}_1^{n-1}) = P(x_n|\{x\}_{n-m}^{n-1}) \ \forall \ n > m \tag{13.10}$$

where n is the index of the sample in time and m is amount of memory in the Markov chain.

In most acoustic models of *speaker recognition* (as with *speech recognition*), we generally assume a single memory Markov process. However, in language modeling, an multiple memory Markov model may be used. An example of such a model is the N-gram [16] which uses an $(N-1)$-memory Markov model to predict the probability of the Nth word given the last $N-1$ words.

13.3 Markov Models

At the beginning of this chapter we spoke about two possible paths of arriving at the structure of a *hidden Markov model* (Figure 13.1), mentioning that we will be mostly entertaining the second path. The fist evolutionary step in the second path is the introduction of memory, resulting in a, so called, *unifilar Markov source* [1, 25]. As we noted in its definition (Definition 7.4), the state sequence of a *unifilar Markov source* is completely determined by knowing the *initial state* and the *observation sequence*.

The following example is a related to Example 13.1, where we no longer have fair coins and that the usage of each coin is completely determined on the previous outcome. It provides some insight into the *single-memory unifilar Markov source*.

Example 13.2 (Single-memory model for two unfair coins).
Let us consider two unfair coins with with probability characteristics listed in Figure 13.2. In this problem, we assume that the state is $\{X : x \in \mathcal{X} = \{X_1, X_2\}\}$, where $x = X_1$ corresponds to coin 1 and $x = X_2$ to coin 2. Furthermore, the output of the system is the result of a coin toss, so it is $Y : y \in \mathcal{Y} = \{Y_1 = H, Y_2 = T\}\}$, where $y = Y_1 = H$ stands for the output being head and $y = Y_2 = T$ would mean a tail.

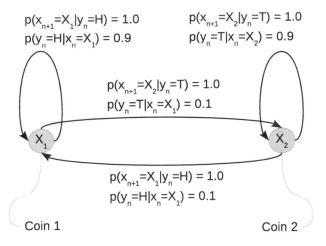

$$p(x_{n+1}=X_1|y_n=H) = 1.0 \qquad\qquad p(x_{n+1}=X_2|y_n=T) = 1.0$$
$$p(y_n=H|x_n=X_1) = 0.9 \qquad\qquad p(y_n=T|x_n=X_2) = 0.9$$

$$p(x_{n+1}=X_2|y_n=T) = 1.0$$
$$p(y_n=T|x_n=X_1) = 0.1$$

X₁   X₂

$$p(x_{n+1}=X_1|y_n=H) = 1.0$$
$$p(y_n=H|x_n=X_1) = 0.1$$

Coin 1   Coin 2

Fig. 13.2: Two Coins, Unifilar Single Memory Markov Model

Figure 13.2 is a finite state representation of the Markov model being used. As we saw with the definition of a unifilar source (Definition 7.4), we need to know the initial state of the source in order to be able to determine its state sequence (in addition to the observed output sequence). Let us choose the following initial state, $x_1 = X|_{n=1} = X_1$, where x_1 denotes the state at time sample, $n = 1$ (initial state), which in this case is coin 1 ($x = X_1$). Up to this point, we know that we shall be flipping coin 1 at the first trial, due to the dictated initial state of $x_1 = X_1$.

As we already know, for a single memory Markov chain, the Markov property (Definition 7.3) must be upheld. Therefore, any output is only dependent on the last output. Based on the conditional probability densities of the figure, we may assess the probability of a specific output (observation) sequence. Before trying to compute the probability of any specific output sequence, let us write the expressions for the state transition and output transition probabilities.
The state transition may be written in terms of the conditional state/output and output/state probabilities as follows,

$$p(x_{n+1}|x_n) = \sum_{y \subset \mathcal{Y}} p(x_{n+1}|y)p(y|x_n) \tag{13.11}$$

In the same spirit, the output transition probabilities may be written as follows,

$$p(y_{n+1}|y_n) = \sum_{x \in \mathcal{X}} p(y_{n+1}|x)p(x|y_n) \tag{13.12}$$

Let us compute the probability of the sequence, $\{y\}_1^4 = \{HHTT\} = \{Y_1 Y_1 Y_2 Y_2\}$, as an example, having in mind that the initial state is $x_1 = X_1$.

$$p(\{y\}_1^4 = \{Y_1Y_1Y_2Y_2\}) = p(y = Y_1|x = X_1)p(y_2 = Y_1|y_1 = Y_1)$$
$$p(y_3 = Y_2|y_2 = Y_1)p(y_4 = Y_2|y_3 = Y_2)$$
$$= 0.9 \times 0.9 \times 0.1 \times 0.9$$
$$= 0.0729 \tag{13.13}$$

For comparison, let us compute the probability of a slightly different sequence where
y_2 and y_3 are transposed, namely,

$$p(\{y\}_1^4 = \{Y_1Y_2Y_1Y_2\}) = p(y = Y_1|x = X_1)p(y_2 = Y_2|y_1 = Y_1)$$
$$p(y_3 = Y_1|y_2 = Y_2)p(y_4 = Y_2|y_3 = Y_1)$$
$$= 0.9 \times 0.1 \times 0.1 \times 0.1$$
$$= 0.0009 \tag{13.14}$$

Notice that although there are the same number of heads and tails in the two
tested output sequences, the first sequence (Equation 13.13) is 81 times more probable than the second sequence (Equation 13.14). Therefore, due to the built-in memory, the order of the sequence becomes quite important. It shows the tendency of the
system to produce more like outputs. Therefore, the initial conditions will also be
quite important in the progression of the sequence.

Since the conditional probabilities of the state at sample $n+1$ given the output
at sample n, $p(x_{n+1}|y_n)$, are deterministic (i.e., $p(x_{n+1}|y_n) \in \{0, 1\}$), then the model
in Figure 13.2 is considered to be a *unifilar Markov source*. If this condition did
not hold, the model would have become *nonunifilar* and the state sequence would
no longer be uniquely determinable from the observed output sequence – see Section 13.4.

13.4 Hidden Markov Models

Following the second evolutionary path in Figure 13.1, we started with a memoryless model, based on a memoryless source. A fair coin toss example was given for
this type of model. Then, in Section 13.3, a single memory was added to create a
Markov model based on a *unifilar Markov source*. The example for that case was a
pair of coins which would be tossed one at a time and whose choice was related to
the last output sample.

In this section, we go further and introduce the *hidden state* aspect. Therefore,
the source is no longer *unifilar*. We gave a brief description of a *non-unifilar Markov*
source in Definition 7.5. In essence, it is a *discrete Markov source* whose states are
not uniquely determined from the observation of the output sequence. Because of
this feature, the model which is created, using this type of source is called a *hidden*

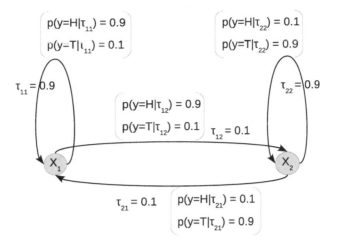

Fig. 13.3: Two Coins, Non-Unifilar Single Memory Markov Model (HMM)

Markov model (HMM).

There are two different ways we can tie the observed output of an *HMM* to its components. The first case would be if an output were to be generated at an instance when a *state* is reached. This would mean that the model would emit an observable output at each state, hence associating that output with the current state.

A second method (Figure 13.3), which is the method of choice in this book is to emit an output when a *transition* is taken from one state to the next. In this case, while going from one state to the next state, the model emits an output. Therefore, the observed output is associated with the transition from the current state to the next, and not with the current state itself.

At the first glance, the two options are similar in their capacities. However, the introduction of another concept (*null transitions*) will make the second choice a bit more compact. *Null transitions* are transitions which result in a time-lapse without any output being generated. Namely, they will change the state without emitting an output – see Figure 13.4. Soon we will discuss the different types of transitions in more detail.

Based on the assumption that the output is associated with the transition out of a state, we may produce the following restatements of Definitions 7.4 and 7.5.

Definition 13.2 (Unifilar Markov Source). *A necessary and sufficient condition for a Markov source to be unifilar is that every transition leaving a state would produce a distinct output which is not produced by any other transition leaving the*

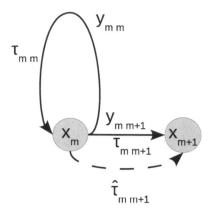

Fig. 13.4: Basic HMM Element

same state.

The above definition may be verified by examining Figure 13.2 and noting the fact that each transition may only produce a distinct output.

Definition 13.3 (Non-Unifilar Markov Source). *A necessary and sufficient condition for a Markov source to be non-unifilar is that two or more transition leaving a state would be capable of producing the same output.*

So through evolutionary path 2 of Figure 13.1, we have arrived at the *hidden Markov model*. *HMM* have two main components [22],

1. A Markov chain for synthesizing the state sequence or path.
2. A finite set of output probability distributions which may be discrete or continuous. These output distributions translate the state sequence in the first component to an output sequence, acting as a time series.

For clarification, let us revisit the problem of Example 13.2. This time, we will use an *HMM* to model the same problem. Therefore, based on our choice, the outputs will be associated with the transitions and not the states.

Example 13.3 (Hidden Markov model for two unfair coins).
Figure 13.3 shows the architecture for the HMM representing the same process as in Example 13.2. Note that in this implementation, X_1 and X_2 are no longer directly associated with coin 1 and coin 2. In fact, although they may end up having some intrinsic physical meaning, the values are no longer uniquely tied to the output sequence. Therefore, to compute the probability of any specific output sequence, $\{y\}_1^N$, we would need to consider many different state sequences which may give rise to the same output sequence and then combine the results of these probabilities.

In this HMM representation, each transition is capable of producing both outputs, $y \in \{H,T\}$. Therefore, based on Definition 13.3, the model is a non-unifilar Markov source, producing a hidden Markov model. In close examination, it is apparent that there are four transitions and that two of the transitions may be viewed as having coin 1 as the output production mechanism and the other two transitions are associated with coin 2.

As we have seen, the multiple association of the output probability distributions to the different parts of the state sequence (transition sequence) create a hiding mechanism which makes the states unobservable. Since the different output sequences may be generated from a combination of state sequences, in general, hidden Markov models tend to be much smaller than their transparent Markov model counterparts. This means that a smaller number of parameters will need to be estimated. These hidden states sometimes have underlying physical characteristics. For example, they may be related to the positions of the articulators (see Section 4.1.3) which in turn produce different sounds through the vocal tract. Therefore, the output sequence is still very much dependent upon the state sequence, but not in a direct, observable way.

In the next section, we will study the different aspects of hidden Markov model design and the arrangement of the states for modeling speech. In the following sections the training and decoding of *HMM* in conjunction with different popular algorithms are examined in detail with some simple examples.

13.5 Model Design and States

In Section 7.1, we defined the basic concept of *ergodicity*. Definition 7.6 described an *ergodic source* as a special case of a *discrete Markov sources*. Most of the models in the literature of speaker and speech recognition [4, 23], including those for models of languages [10] assume the *HMM* of choice to be *ergodic* (see Definition 7.6). This simplifies the design and implementation of these models.

Let us start by making the following assumptions:

1. States do not produce any output; instead, some transitions produce outputs based on output probability distributions associated with them. The output probability distribution may be either discrete or continuous.
2. There are a finite number of states. This assumption follows a bound on the definition of Markov chain (Definition 13.1), which originally calls for a countable number of states. Here, for practicality, we have restricted the number to be finite, since countable may theoretically refer to countably infinite as well.
3. Final states are states which have no outgoing transitions.

4. Null transitions may exist between two states. These transitions do not produce any output. They only allow the progression of their origination state to their destination state with no output. The introduction of *null transitions* allows for much more simplified model architectures.
5. No *null transition* is allowed from any state to itself.
6. By the simple serial concatenation of the basic architectural unit in Figure 13.4, most output sequences are made possible. Note that for simplification, we only allow a null transition from any state to the next state, as seen in Figure 13.4. There is no need to handle the more complex case where a transition would be possible from any state to any other state. The reason is that by reassigning transition probabilities, such transitions are also possible using the basic architecture made from the concatenation of the simple units of Figure 13.4. This will, however, simplify the practical implementation of the model.

The fourth assumption for the basic structure of the HMM suggested that the HMM be made up of a concatenation of elements of the form listed in Figure 13.4. Any forward HMM model with the restriction of having no self null transitions may be created from a series of basic elements as shown in Figure 13.4. With this in mind, the model design is simplified. Figure 13.5 shows how a model may be simplified by using a null transition to reduce the number of transitions. The complex design problem, therefore, reduces to one of simply choosing the number of states. In that regard also, it is simple to make certain assumption on the average number of states which are necessary to model basic elements of the acoustic sample sequence.

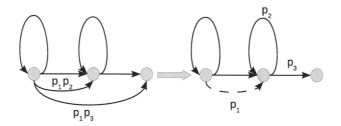

Fig. 13.5: Simplification using a Null Transition

As we see from Figure 13.4, the basic element includes three possible transitions: a *self transition*, a *forward transition* and a *null transition*. If we assume that the basic sound component may be modeled by a *GMM*, then a single state would be adequate for the average length in the simplest unit of sound.

By the simplest unit we do not necessarily mean a phone (Definition 4.2). Although the phone is the simplest unit of sound that occurs in a language, but phones themselves may be broken into different segments, by being modeled using a few states. Most speech recognition implementations use *three states* to model an average phone – see Figure 13.6. Other elongated phones, for example *diphthongs* and

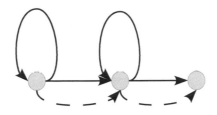

Fig. 13.6: HMM of an Average Phone

triphthongs (see Section 4.1.3) may be composed of more states, say 5 and 7 respectively.

The fact that there is a *self transition*, allows for accounting for elongated phonetic symbols and the null transition allows for the having shorter occurrences of the same symbol. The basic idea behind this variation is that there are natural variations in the length of a phone being uttered by different individuals and even within different enunciations used by the same individual.

Therefore, the number of states would generally be in the same order of magnitude as the length of the phonetic sequence the model is supposed to represent. Of course, this is not an exact prescription, however, it does allow for an approximate design criterion on the number of states to use.

A perturbation of the number of states around the above nominal length would allow for the optimization of the lengths of the sequences. In fact, this may be done by using held-out data at the training stage. Namely, a portion of the training data may be held out to test the different length models. The rest of the training data may be used to train the models based on some assumption of the length. Then the held out data may be used to optimize the length of the models. This is quite similar to the other optimization processes used throughout this chapter.

13.6 Training and Decoding

We have arrived at the basic structure of an *HMM*. At this point, we need to examine the parameter estimation and decoding techniques for the devised models. In general, there are three different problems which interest us in the event of observing a random output sequence, w, as a sample of possible output sequences modeled by the random variable, $W : w \in \mathscr{W} = \{W_1, W_2, \cdots, W_\Omega\}$. Notice, that every sample, w, of W is itself a sequence of observations emitted by the *HMM*. Therefore, we will write the generic sequence, $w = \{y\}_1^N$ and for a every distinct output sequence we qualify the sequence as follows, $w = W_\omega = \{y_\omega\}_1^{N_\omega}$ where, $\omega \in \{1, 2, \cdots, \Omega\}$. Any instance of a sequence may be written as w_k, where k is the kth sample of W.

The three said problems may then be stated as follows,

1. *Training*: Estimate parameters of the Markov source with highest likelihood of producing output sequences, $\{w\}_1^K$.
2. *State sequence*: Find the most probable state sequence that would produce the output sequence, $W_\omega = \{y_\omega\}_1^{N_\omega}$ as its output for Markov source with parameter vector, $\boldsymbol{\varphi}$. This problem may be solved using a special case of *dynamic programming* [6] called the *Viterbi algorithm* [26].
3. *Decoding*: Compute the probability of an output sequence for the Markov source with parameter vector, $\boldsymbol{\varphi}$,

$$P(W_\omega|\boldsymbol{\varphi}) = P(\{y_\omega\}_1^{N_\omega}|\boldsymbol{\varphi}) \tag{13.15}$$

In the second and third problems, the model parameters $\boldsymbol{\varphi}$ (*transition* and *output* probabilities) are known, however in the training stage (*problem 1*), the objective is to compute the parameters that would best fit a number of output sequences, $\{w\}_1^K$.

Let us consider the third problem, which is the decoding of the probability of the output sequence, $P(W_\omega = \{y_\omega\}_1^{N_\omega}|\boldsymbol{\varphi})$. Since we are interested in a *non-unifilar Markov source*, the output sequence may be produced through many different state sequences (transition sequences). Therefore, the probability of the output sequence would be given by the sum of the joint probabilities of the output sequence and the different possible state sequences. In notation,

$$P(W_\omega|\boldsymbol{\varphi}) = \sum_{\{x_\omega\}_1^{N_\omega}:\{y_\omega\}_1^{N_\omega}} P(\{y_\omega\}_1^{N_\omega}, \{x_\omega\}_1^{N_\omega}|\boldsymbol{\varphi}) \tag{13.16}$$

where the notation, $\{x_\omega\}_1^{N_\omega} : \{y_\omega\}_1^{N_\omega}$, refers to all the state sequences that can produce $\{y_\omega\}_1^{N_\omega}$.

The simplest way to compute $P(\{y_\omega\}_1^{N_\omega}|\boldsymbol{\varphi})$ is to enumerate all the possible state sequences that can produce $\{y_\omega\}_1^{N_\omega}$, computing each corresponding joint probability and then adding them all, according to Equation 13.16. The problem is that this computation will grow exponentially with the length of each output sequence!

To illustrate this exponential growth, let us take the following example. In the example, we have dropped the index, ω, since we are only concerned with demonstrating the decoding of a single output sequence, $w = \{y\}_1^N$ to find the related state sequence, $\{x\}_1^N$. The same procedure may be applied to any other output sequence, $W_\omega = \{y_\omega\}_1^{N_\omega}$, with its corresponding state sequence, $\{x_\omega\}_1^{N_\omega}$.

Example 13.4 (Decoding the Output Sequence).
Let us consider the HMM described in the finite state diagram of Figure 13.7 and

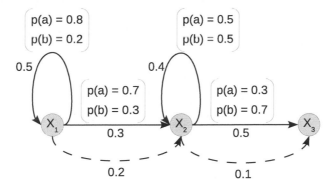

Fig. 13.7: HMM of Example 13.4

compute the output probability of the sequence, $\{y\}_1^4 = \{aabb\}$ where $Y : y \in \{a, b\}$. First, consider all the different possible transitions which may output the first output in the sequence, $y_1 = a$. Figure 13.8 shows the different paths for generating $y_1 = a$, by the model in Figure 13.7.

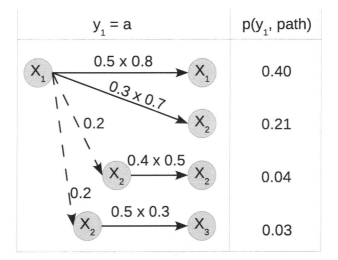

Fig. 13.8: Possible Paths for generating $y_1 = a$ in Example 13.4

The probability of $y_1 = a$ is therefore the sum of all the joint probabilities with the corresponding paths,

$$P(y_1 = a|\boldsymbol{\varphi}) = 0.40 + 0.21 + 0.04 + 0.03$$
$$= 0.68 \tag{13.17}$$

If we continue for one more time instance, namely if we enumerate the paths that output $\{y\}_1^2 = \{aa\}$, we see that the possible paths grow exponentially. Figure 13.9 shows all the paths that would lead to outputting this sequence. Note that the probability of the output sequence will then be computed by summing the joint probabilities of all the possible paths which output this sequence, as follows,

$$P(\{y\}_1^2 = \{aa\}|\boldsymbol{\varphi}) = 0.16 + 0.084 + 0.016 + 0.012 +$$
$$0.042 + 0.0315 + 0.008 + 0.006$$
$$= 0.3595 \tag{13.18}$$

In Figure 13.9, note the second and third main paths from the initial state. They

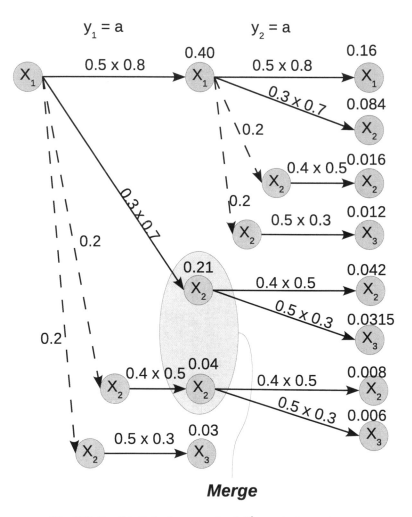

Fig. 13.9: Possible Paths for generating $\{y\}_1^2 = aa$ in Example 13.4

have been marked on the figure with an oval marker, suggesting a merger. Since their continuation into the output, $y_2 = a$, take an identical set of paths, according to Markov property (Equation 7.2), they may be merged, since once we are in that state, it is not important how we have arrived at that junction. Figure 13.10 shows this merger. Note that merging these two paths does not affect the total output probability for the sequence, nor will it alter the possible extensions of the live paths for further output. The same procedure may be continued to arrive at the total proba-

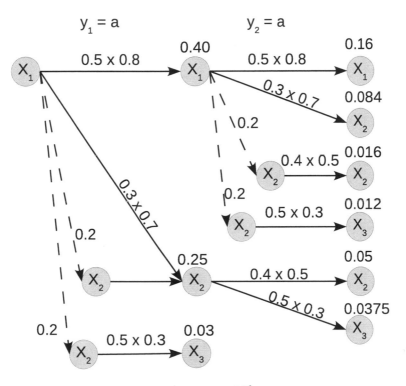

Fig. 13.10: Merged paths for generating $\{y\}_1^2 = aa$ in Example 13.4

bility of the whole sequence, $\{y\}_1^4 = \{aabb\}$.

As we saw from Example 13.4, it is not very efficient to compute all the possible paths, only to find out that certain paths may be merged. In the next section, we will introduce a *trellis diagram* which helps manage the different paths and outputs. It will also give insight into methods for reducing the computation load from exponential to linear.

13.6.1 Trellis Diagram Representation

The *trellis diagram* is a powerful representation of a *finite state diagram*. It is used to relate the state and output sequences together. The number of samples is drawn along the horizontal axis from left to right and the states are marked on the vertical axis from top to bottom. There is a one-to-one relationship between the type of transition and the direction of motion along the trellis. The following three notes relate the different kinds of transition to the direction of motion.

1. A pure vertical descent in the diagram is equivalent to a *null transition*, since the state is changed, but no output is generated, hence no time has elapsed.
2. A pure horizontal motion corresponds to a self-transition since the state does not change, but an output is generated, elapsing one sample.
3. A diagonal motion is akin to moving along a forward transition from any state to the next state, while generating an output sample.

Figure 13.11 shows the trellis diagram and the corresponding products of the transition probabilities and conditional output probabilities for the corresponding outputs. An important feature of the *trellis diagram* is that every transition sequence is only

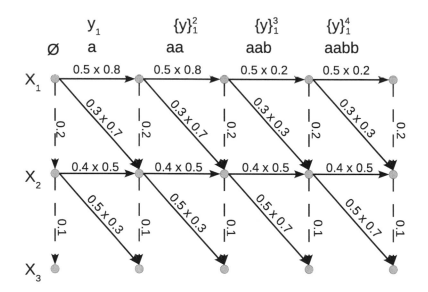

Fig. 13.11: Trellis diagram for the output sequence, $\{y\}_1^4 = \{aabb\}$ being generated by the HMM of Figure 13.7

present exactly one time, hence the repeated transition sequences such as in Figure 13.9 are automatically merged. Although in the diagram of Figure 13.11, we only have the three types of transitions listed in the previous section (*self, forward* and *null*), the *trellis diagram* is also capable of representing backward transition, say from X^2 to X^1. These would just show up as diagonal connections from lower left to upper right. However, it is not suitable for handling backward and self null transitions. As we mentioned earlier, we are only concerned with *left-right forward architectures* [22], in this book. Specifically, we will be concentrating on architectures which are made up of the basic elements of the kind shown in Figure 13.4.

In the next three sections, we will review efficient algorithms for solving the three problems of interest, listed in Section 13.6. Although these problems were listed in the order of being used in the natural implementation of any system, the order of covering them in the next few sections has been reversed. The reason for this reversed order of discussion is that the concepts for the *decoding* problem (third problem) will be needed for describing the *Viterbi algorithm* [26], used for finding the most likely path through the model for a given sequence (second problem) and the *Baum-Welch algorithm* [5] used for estimating the parameters (training) of the *HMM* (first problem).

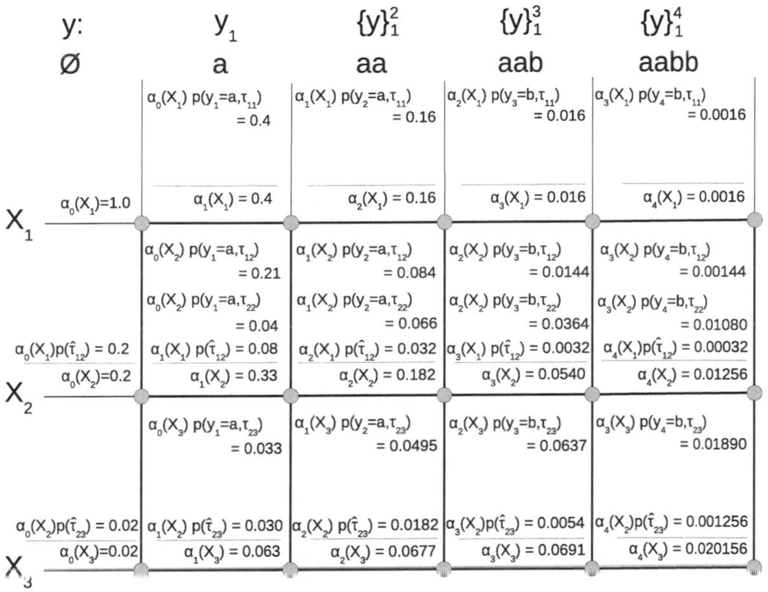

y:	y_1	$\{y\}_1^2$	$\{y\}_1^3$	$\{y\}_1^4$
∅	a	aa	aab	aabb
	$\alpha_0(X_1)\,p(y_1{=}a,\tau_{11})$ $=0.4$	$\alpha_1(X_1)\,p(y_2{=}a,\tau_{11})$ $=0.16$	$\alpha_2(X_1)\,p(y_3{=}b,\tau_{11})$ $=0.016$	$\alpha_3(X_1)\,p(y_4{=}b,\tau_{11})$ $=0.0016$
X_1 $\alpha_0(X_1){=}1.0$	$\alpha_1(X_1)=0.4$	$\alpha_2(X_1)=0.16$	$\alpha_3(X_1)=0.016$	$\alpha_4(X_1)=0.0016$
	$\alpha_0(X_2)\,p(y_1{=}a,\tau_{12})$ $=0.21$	$\alpha_1(X_2)\,p(y_2{=}a,\tau_{12})$ $=0.084$	$\alpha_2(X_2)\,p(y_3{=}b,\tau_{12})$ $=0.0144$	$\alpha_3(X_2)\,p(y_4{=}b,\tau_{12})$ $=0.00144$
	$\alpha_0(X_2)\,p(y_1{=}a,\tau_{22})$ $=0.04$	$\alpha_1(X_2)\,p(y_2{=}a,\tau_{22})$ $=0.066$	$\alpha_2(X_2)\,p(y_3{=}b,\tau_{22})$ $=0.0364$	$\alpha_3(X_2)\,p(y_4{=}b,\tau_{22})$ $=0.01080$
$\alpha_0(X_1)p(\hat{\tau}_{12})=0.2$ $\alpha_0(X_2){=}0.2$ **X_2**	$\alpha_1(X_1)\,p(\hat{\tau}_{12})=0.08$ $\alpha_1(X_2)=0.33$	$\alpha_2(X_1)\,p(\hat{\tau}_{12})=0.032$ $\alpha_2(X_2)=0.182$	$\alpha_3(X_1)\,p(\hat{\tau}_{12})=0.0032$ $\alpha_3(X_2)=0.0540$	$\alpha_4(X_1)p(\hat{\tau}_{12})=0.00032$ $\alpha_4(X_2)=0.01256$
	$\alpha_0(X_3)\,p(y_1{=}a,\tau_{23})$ $=0.033$	$\alpha_1(X_3)\,p(y_2{=}a,\tau_{23})$ $=0.0495$	$\alpha_2(X_3)\,p(y_3{=}b,\tau_{23})$ $=0.0637$	$\alpha_3(X_3)\,p(y_4{=}b,\tau_{23})$ $=0.01890$
$\alpha_0(X_2)p(\hat{\tau}_{23})=0.02$ $\alpha_0(X_3){=}0.02$ **X_3**	$\alpha_1(X_2)\,p(\hat{\tau}_{23})=0.030$ $\alpha_1(X_3)=0.063$	$\alpha_2(X_2)\,p(\hat{\tau}_{23})=0.0182$ $\alpha_2(X_3)=0.0677$	$\alpha_3(X_2)p(\hat{\tau}_{23})=0.0054$ $\alpha_3(X_3)=0.0691$	$\alpha_4(X_2)p(\hat{\tau}_{23})=0.001256$ $\alpha_4(X_3)=0.020156$

Fig. 13.12: α computation for the HMM of Example 13.4

13.6.2 Forward Pass Algorithm

The *forward pass algorithm*, *match algorithm*, or commonly referred to as the *forward algorithm* is an efficient technique for computing the probability of an output sequence, $w = \{y\}_1^N$, given an *HMM*. The term *forward pass* stems from the fact that it computes the probability of the observed sequence by doing forward (and top-down) sweep through the *trellis diagram* associated with the *HMM* and the output sequence of interest.

Let us examine the sample trellis diagram of Figure 13.11, associated with the HMM of Figure 13.7 and the desired output, $\{y\}_1^N = \{aabb\}$. The idea behind the *forward algorithm* is that if we compute all the probabilities at the different lattice points of the trellis, moving from the top left to the bottom right, by the time we arrive at the most extreme point of the lower right of the trellis, we have the probability of the sequence, $\{y\}_1^N$, as generated by the corresponding *HMM*, $P(\{y\}_1^N | \boldsymbol{\varphi})$, where $\boldsymbol{\varphi}$ is the parameter vector associated with the *HMM*.

The probabilities which are computed at each lattice point (node) are usually denoted by $\alpha_n(X_m)$ and are defined as follows,

$$\alpha_n(X_m | \boldsymbol{\varphi}) \overset{\Delta}{=} P(X_m, \{y\}_1^n | \boldsymbol{\varphi}) \ \forall X_m \in \mathscr{X}, \ 1 \leq n \leq N \tag{13.19}$$

Since it is well understood that all the probabilities, computed for the forward pass, are based on the knowledge of the model parameter vector, $\boldsymbol{\varphi}$, the conditionality has been suppressed from the notation in this and next section. For example, it should be understood that by $P(X_m, \{y\}_1^n | \boldsymbol{\varphi})$ we really mean $P(X_m, \{y\}_1^n)$ and that $\alpha_n(X_m | \boldsymbol{\varphi})$ is written as $\alpha_n(X_m)$.

Before writing the expression for $\alpha_n(X_m)$, let us define a notation for designating the transitions. Refer to Figure 13.4 which shows the simplest unit used in our design structure. The output producing transitions are denoted by $\tau_{m(m+1)}$, which signifies the transition from state X_m to X_{m+1}. The null (non-output-producing) transitions are denoted by $\hat{\tau}_{m(m+1)}$. Since the null transitions, in our restricted left-right architecture, may only go from a state to the next state, sometimes we may use the shorthand which only includes the index associated with the originating state, namely, $\hat{\tau}_m \equiv \hat{\tau}_{m(m+1)}$.

Using this notation, the expression for $\alpha_n(X_m)$, as defined by Equation 13.19, may be written as follows,

$$\alpha_n(X_m) = P(X_m, \{y\}_1^n)$$

$$= \sum_{m':m' \xrightarrow{\tau_{m'm}} m} P(X_{m'}, \{y\}_1^{n-1}) P(y_n, \tau_{m'm}) +$$

$$\sum_{m':m' \xrightarrow{\hat{\tau}_{m'm}} m} P(X_{m'}, \{y\}_1^n) P(\hat{\tau}_{m'm}) \qquad (13.20)$$

$$= \sum_{m':m' \xrightarrow{\tau_{m'm}} m} \alpha_{n-1}(X_{m'}) P(y_n | \tau_{m'm}) P(\tau_{m'm}) +$$

$$\sum_{m':m' \xrightarrow{\hat{\tau}_{m'm}} m} \alpha_n(X_{m'}) P(\hat{\tau}_{m'm}) \qquad (13.21)$$

If we denote the node on the trellis, associated with the state X_m and the output sequence, $\{y\}_1^n$, by the pair, $(X_m, \{y\}_1^n)$, then the summation over $m' : m' \xrightarrow{\tau_{m'm}} m$ denotes the summation over all the output producing transitions which arrive at state X_m, while being involved in the production of the sequence, $\{y\}_1^n$. The $X_{m'}$ are the states that originate these transitions. Similarly, the summation over $m' : m' \xrightarrow{\hat{\tau}_{m'm}} m$ is the summation over all null transitions arriving at the state X_m, while being involved in the path to generating the output sequence, $\{y\}_1^n$, although not directly generating any output themselves.

Equation 13.21 is computed iteratively, with the initial probability equal to 1, namely,

$$\alpha_0(X_1) = 1.0 \qquad (13.22)$$

For a better grasp of the forward algorithm based on the recursion of Equation 13.21 subject to the initial condition of Equation 13.22, we have computed the path probabilities and their summation at every node of the trellis, the $\alpha_n(X_m)$. Figure 13.12 shows this table for the problem of Example 13.4. For notational convenience, the conditionality on φ as been suppressed in the figure, namely, $P(y, \tau | \varphi)$ has been written as $P(y, \tau)$.

The forward algorithm is used for the decoding of a speech sequence using an *HMM*. The amount of computation for the *forward algorithm* is linearly proportional to the number of outputs in the observation sequence. This is in contrast with the exponential relation for the case where all the possible paths are computed independently. As we mentioned in the Section 13.6, sometimes we are interested in knowing the maximum likelihood path for the sequence and not the total path. In the next section we will examine an algorithm which is quite similar to the forward algorithm, with the difference that it uses the *maximum* function instead of *summation*, to find the most likely path.

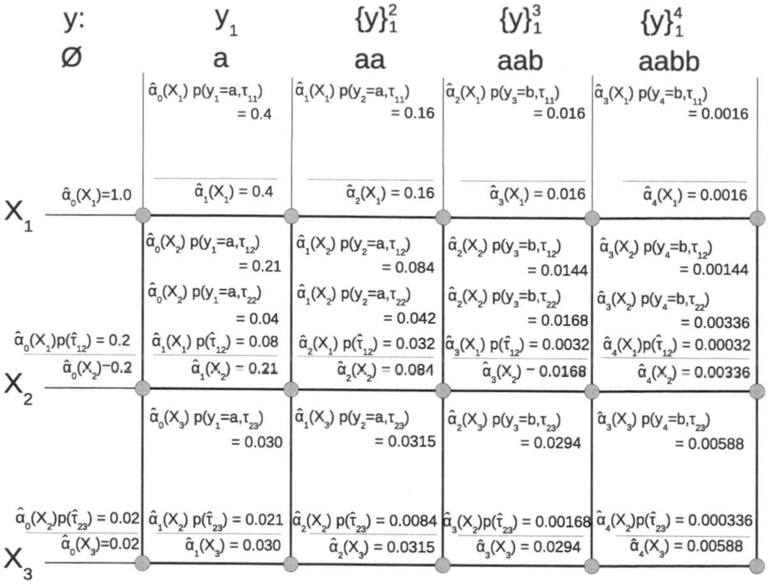

Figure 13.13 (Viterbi trellis):

y:	y_1	$\{y\}_1^2$	$\{y\}_1^3$	$\{y\}_1^4$
∅	a	aa	aab	aabb

Row X_1: $\hat{a}_0(X_1)=1.0$

- $\hat{a}_0(X_1) p(y_1=a,\tau_{11}) = 0.4$; $\hat{a}_1(X_1) = 0.4$
- $\hat{a}_1(X_1) p(y_2=a,\tau_{11}) = 0.16$; $\hat{a}_2(X_1) = 0.16$
- $\hat{a}_2(X_1) p(y_3=b,\tau_{11}) = 0.016$; $\hat{a}_3(X_1) = 0.016$
- $\hat{a}_3(X_1) p(y_4=b,\tau_{11}) = 0.0016$; $\hat{a}_4(X_1) = 0.0016$

Row X_2:

- $\hat{a}_0(X_2) p(y_1=a,\tau_{12}) = 0.21$
- $\hat{a}_1(X_2) p(y_2=a,\tau_{12}) = 0.084$
- $\hat{a}_2(X_2) p(y_3=b,\tau_{12}) = 0.0144$
- $\hat{a}_3(X_2) p(y_4=b,\tau_{12}) = 0.00144$

- $\hat{a}_0(X_2) p(y_1=a,\tau_{22}) = 0.04$
- $\hat{a}_1(X_2) p(y_2=a,\tau_{22}) = 0.042$
- $\hat{a}_2(X_2) p(y_3=b,\tau_{22}) = 0.0168$
- $\hat{a}_3(X_2) p(y_4=b,\tau_{22}) = 0.00336$

- $\hat{a}_0(X_1)p(\hat{\tau}_{12}) = 0.2$; $\hat{a}_0(X_2)-0.2$
- $\hat{a}_1(X_1) p(\hat{\tau}_{12}) = 0.08$; $\hat{a}_1(X_2) - 0.21$
- $\hat{a}_2(X_1) p(\hat{\tau}_{12}) = 0.032$; $\hat{a}_2(X_2) = 0.084$
- $\hat{a}_3(X_1) p(\hat{\tau}_{12}) = 0.0032$; $\hat{a}_3(X_2) - 0.0168$
- $\hat{a}_4(X_1)p(\hat{\tau}_{12}) = 0.00032$; $\hat{a}_4(X_2) = 0.00336$

Row X_3:

- $\hat{a}_0(X_3) p(y_1=a,\tau_{23}) = 0.030$
- $\hat{a}_1(X_3) p(y_2=a,\tau_{23}) = 0.0315$
- $\hat{a}_2(X_3) p(y_3=b,\tau_{23}) = 0.0294$
- $\hat{a}_3(X_3) p(y_4=b,\tau_{23}) = 0.00588$

- $\hat{a}_0(X_2)p(\hat{\tau}_{23}) = 0.02$; $\hat{a}_0(X_3)=0.02$
- $\hat{a}_1(X_2) p(\hat{\tau}_{23}) = 0.021$; $\hat{a}_1(X_3) = 0.030$
- $\hat{a}_2(X_2) p(\hat{\tau}_{23}) = 0.0084$; $\hat{a}_2(X_3) = 0.0315$
- $\hat{a}_3(X_2)p(\hat{\tau}_{23}) = 0.00168$; $\hat{a}_3(X_3) = 0.0294$
- $\hat{a}_4(X_2)p(\hat{\tau}_{23}) = 0.000336$; $\hat{a}_4(X_3) = 0.00588$

Fig. 13.13: Viterbi maximum probability path computation for the HMM of Example 13.4

13.6.3 Viterbi Algorithm

In Section 13.6.2 we reviewed the *forward pass algorithm* which computes the total probability of a sequence given an HMM, using the trellis diagram. Sometimes it is necessary to compute the most likely path through an HMM, for generating a sequence. In 1967 Viterbi [26] proposed a nonsequential decoding algorithm applied to the transmission of optimal *convolutional codes* over a *memoryless channel* (Definition 7.14). This turns out to be a dynamic programming style algorithm, related to Bellman's work [6] of a decade earlier. It is quite similar to the forward algorithm, described in Section 13.6.2, with the difference that at every lattice point, instead of using the total path probability up to that point, $\alpha_n(X_m)$, only the maximum probability path is kept alive and every other path is discarded. The maximum probability path at any node is denoted by $\hat{\alpha}_n(X_m)$. As with the argument of Section 13.6.2, $\hat{\alpha}_n(X_m)$ is really a short-hand notation for $\hat{\alpha}_n(X_m|\boldsymbol{\varphi})$. The same contraction applies to other probabilities in this section. Therefore, $\hat{\alpha}_n(X_m)$ may be written as follows,

$$\hat{\alpha}_n(X_m) \overset{\Delta}{=} P_{max}(X_m, \{y\}_1^n) \tag{13.23}$$

$$= \max\left(\max_{\substack{m':m' \overset{\hat{\tau}_{m'm}}{\longmapsto} m}} \left(\hat{\alpha}_{n-1}(X_{m'})P(y_n|\tau_{m'm})P(\tau_{m'm})\right), \right.$$

$$\left. \max_{\substack{m':m' \overset{\hat{\tau}_{m'm}}{\longmapsto} m}} \left(\hat{\alpha}_n(X_{m'})P(\hat{\tau}_{m'm})\right)\right) \tag{13.24}$$

Figure 13.13 shows the table of $\hat{\alpha}_n(X_m)$ computations for the problem of Example 13.4. The maxima listed in the last row of the table are the maximum path probabilities for the different partial output strings and the entry in the last row and last column is the maximum probability path for the full output, $\{aabb\}$. In the implementation of this algorithm, one needs to only save the maximum path information and the rest of the paths may be discarded, reducing the computational load from exponential to linear in relation to the length of the output sequence.

13.6.4 Baum-Welch (Forward-Backward) Algorithm

The *forward-backward algorithm* is essentially the *maximum likelihood estimation* algorithm for a hidden Markov model, maximizing the likelihood of observing the training data through a given HMM structure. Extensions [12] have also been made to solve the *Bayesian maximum a-posteriori (MAP)* problem. As we noted in Section 13.6.2, the *trellis diagram* and the α computations of that section will be used as a part of the training as well. In this section, we will review the, generically called, *forward-backward algorithm*, also known as the *Baum* [5] or *Baum-Welch*. It is an application of *expectation maximization (EM)* (Section 11.3.1) to the computation of the *HMM* parameters and in that regard it is sometimes referred to as the *EM algorithm*.

The basic premise for this and other such algorithms (Section 7.1) is to take advantage of the fact that all the memory of the process, to the point of interest, is summarized in the recent history. In the *HMM*, we use the *Markov property*, which makes that summary contained only in the last time step. In other similar time series algorithms, such as *ARMA* [15] and *Kalman filtering* [13], the length of the memory is decided based on the dynamic system being modeled and a *forgetting factor* [15] is used to reduce the effect of the prior history in the formation of new states.

As we saw in the introduction of the trellis diagram (Section 13.6.1), the enumerative method of evaluating different paths of the *HMM* would lead to exponentially large number of calculations, with respect to the length of the output sequence. The same problem plagues the training of HMM parameters to an even larger extent, since for training, a large amount of data will be required and for each training sequence the same exponential behavior will be present. The most apparent way of reducing this computation is to revisit the *trellis diagram* representation of the HMM. The *forward-backward algorithm* is an iterative algorithm which utilizes the trellis diagram to reduce the parameter estimation problem to a linear level relative to the output sequence.

To understand this algorithm, let us revisit the *trellis diagram* of Figure 13.11. For any desired output sequence, $W_\omega = \{y_\omega\}_1^{N_\omega}$, if we take any node of the trellis, denoted by its history, arriving at a state X_m while generating the output sequence, $\{y_\omega\}_1^n$, by the pair, $(X_m, \{y_\omega\}_1^n)$, then there are two different possibilities to continue to the end of the sequence, $(X_M, \{y_\omega\}_1^{N_\omega})$. The first choice is to take any of the possible output generating transitions and the second is to take a null transition, when available.

Based on the *Markov property* (Definition 7.3), we saw in the α computations (Equation 13.21) that the path to the end of any transition may be broken into the two multiplicative components of the α up to that transition and the effect of the transition itself. Based on the same *Markov property*, what happens after any transition, only depends on the input of the transition which would be the α leading up to the transition and the transition itself. Therefore, the effect of transition, in the total path of generating the sequence of choice, may be broken up into three parts, what happens up to the transition, the transition itself and what happens after the transition. To account for what happens after the transition is taken, we may start at the end of the sequence and work our way back toward the destination of the transition and in the process compute all the path probabilities, denoted by $\beta_n(X_m)$, associated with the path, (X_m, W_ω) in the trellis. Note that for the $\beta_n(X_m)$ designation, we are using the path from that output node, $y_{\omega n}$, to the end of the sequence, $y_{\omega N}$.

The following Equation gives the expression for $\beta_n(X_m)$ based upon the above argument while dropping the ω index for notational convenience,

$$
\begin{aligned}
\beta_n(X_m) &= P(\{y\}_{n+1}^N | X_m) \\
&= \sum_{m':m \xrightarrow{\tau_{mm'}} m'} \beta_{n+1}(X_{m'}) P(y_{n+1} | \tau_{mm'}) P(\tau_{mm'}) + \\
&\quad \sum_{m':m \xrightarrow{\hat{\tau}_{mm'}} m'} \beta_n(X_{m'}) P(\hat{\tau}_{mm'})
\end{aligned}
\tag{13.25}
$$

with the understanding that when it not specifically stated, $w = \{y\}_1^N$ is assumed to be a sample of W where the total number of samples used for the purpose of estimating the parameters is $\{w\}_1^K$, where $w_k \in \{W_\omega\} = \{y_\omega\}_1^{N_\omega}$.

Then for any output generating transition-output combination in the trellis,

$$
P(y_n, \tau_{mm'} \mid w = \{y\}_1^N) = \underbrace{\alpha_{n-1}(X_{m'})}_{\text{up to the transition}} \quad \underbrace{P(y_n | \tau_{mm'}) P(\tau_{mm'})}_{\text{the transition}} \quad \underbrace{\beta_n(X_m)}_{\text{to the end of output}}
\tag{13.26}
$$

and for any null transition,

$$
P(\hat{\tau}_{mm'}, t = n | w = \{y\}_1^N) = \underbrace{\alpha_n(X_{m'})}_{\text{up to the transition}} \quad \underbrace{P(\tau_{mm'})}_{\text{the transition}} \quad \underbrace{\beta_n(X_m)}_{\text{to the end of output}}
\tag{13.27}
$$

In Equation 13.26, the joint conditional probability, $P(y_n, \tau_{mm'} | w = \{y\}_1^N)$, signifies the probability of transiting from a node in the trellis to a neighboring node while outputting the nth output of the *HMM*. However, since *null transitions* do not output anything, in Equation 13.27, a variable, t, has been introduced which signifies the time step associated with the output. Therefore, $t = n$ means that a transition from a node in the trellis prior to outputting the nth output is of interest. Since it is important to note which instance of the traversal of the null transition is being considered in the computation of probability, $P(\hat{\tau}_{mm'}, t = n | w = \{y\}_1^N)$, this variable (t) complements the the present knowledge that the transition is emanating from state X_m, based on the transition notation.

From Equations 13.26 and 13.27, one may re-estimated *a-posteriori* transition probabilities for output generating transitions and null transition using the following two expressions, respectively.

$$P(\tau_{mm'} \mid w) = \frac{\displaystyle\sum_{n=1}^{N} P(y_n, \tau_{mm'} | w)}{\displaystyle\sum_{m':m \xrightarrow{\tau_{mm'}} m'} \sum_{n=1}^{N} P(y_n, \tau_{mm'} | w) + \sum_{m':m \xrightarrow{\hat{\tau}_{mm'}} m'} P(\hat{\tau}_{mm'}, t = n | w)} \tag{13.28}$$

$$P(\hat{\tau}_{m'm} \mid w) = \frac{\displaystyle\sum_{n=1}^{N} P(\hat{\tau}_{mm'}, t = n | w)}{\displaystyle\sum_{m':m \xrightarrow{\tau_{mm'}} m'} \sum_{n=1}^{N} P(y_n, \tau_{mm'} | w) + \sum_{m':m \xrightarrow{\hat{\tau}_{mm'}} m'} P(\hat{\tau}_{mm'}, t = n | w)} \tag{13.29}$$

In practice, each of the summations in the numerators and denominators of Equations 13.28 and 13.29 are accumulated into variables, as the trellis is swept from left to right, to compute the conditional probabilities of all the transitions at the end of the sweep.

So far, we have developed the expressions for recomputing the transition probabilities given an output sequence, which is part of the parameter vector for the *HMM*. The missing part is now the computation of the output distributions at the output generating transitions. Again, let us consider the case of a discrete output distribution. If the possible output values for Y are given as, $y \in \{Y_1, Y_2, \cdots, Y_Q\}$, then the joint probability in Equation 13.28 may be written for a specific output, Y_q, as follows,

$$P(y_n = Y_q, \tau_{mm'} \mid w) = \underbrace{\alpha_{n-1}(X_{m'})}_{\text{up to the transition}} \quad \underbrace{P(y_n = Y_q | \tau_{mm'}) P(\tau_{mm'})}_{\text{the transition}} \quad \underbrace{\beta_n(X_m)}_{\text{to the end of output}} \tag{13.30}$$

Therefore, the conditional probability distribution for the outputting generating transition $\tau_{mm'}$, producing the output $y = Y_q$ would be given by the following expression,

which is a summation of all the instances of this output having been produced by
this specific transition for the given sequence, w.

$$P(y = Y_q | \tau_{mm'}, w) = \frac{\sum\limits_{n=1}^{N} P(y_n = Y_q, \tau_{mm'} | w)}{\sum\limits_{n=1}^{N} P(y_n, \tau_{mm'} | w)} \tag{13.31}$$

Equations 13.28, 13.29, and 13.31 provide us with the re-estimation of the elements of the parameter vector $\boldsymbol{\varphi}$ based on a single output sequence, w. Since we
need to train the *HMM* to be able to model a large set of training data, say Q sequences, the total probabilities need to be computed across all observed sequences
(training data). The following two equations express these relations for the transition
probabilities.

$$\begin{aligned}
P(\tau_{mm'}) &= \sum\limits_{\omega=1}^{\Omega} P(\tau_{mm'}, w = W_{\omega}) \\
&= \sum\limits_{\omega=1}^{\Omega} P(\tau_{mm'} | w = W_{\omega}) P(w = W_{\omega}) \\
&\approx \frac{1}{K} \sum\limits_{k=1}^{K} P(\tau_{mm'} | w_k)
\end{aligned} \tag{13.32}$$

$$\begin{aligned}
P(\hat{\tau}_{mm'}) &= \sum\limits_{\omega=1}^{\Omega} P(\hat{\tau}_{mm'}, w = W_{\omega}) \\
&= \sum\limits_{\omega=1}^{\Omega} P(\hat{\tau}_{mm'} | w = W_{\omega}) P(w = W_{\omega}) \\
&\approx \frac{1}{K} \sum\limits_{k=1}^{K} P(\hat{\tau}_{mm'} | w_k)
\end{aligned} \tag{13.33}$$

In Equations 13.32 and 13.33, we have used the *law of large numbers* (Section 6.9.2)
to estimate the total probabilities of the transitions based on the assumptions that K
is large and that the training sample distribution is representative of the real-world
distribution of W.

Much in the same way as with the approximation of the transition probabilities,
we may approximate the output distributions conditioned on their corresponding
transitions as follows, by making use of Equation 13.31.

$$\begin{aligned}
P(y = Y_q | \tau_{mm'}) &= \sum\limits_{\omega=1}^{\Omega} P(y = Y_q | \tau_{mm'}, w = W_{\omega}) P(w = W_{\omega}) \\
&\approx \frac{1}{K} \sum\limits_{k=1}^{K} P(y = Y_q | \tau_{mm'}, w_k)
\end{aligned} \tag{13.34}$$

For a better understanding of this algorithm, let us take the following simple example with discrete output distributions.

Example 13.5 (Simple Training).

To make the computations more manageable, let us start with a simplified version of the model in Example 13.4 which was used for the illustration of the decoding process. This time, we will remove the null transition from the second state to the third. Since we are still left with one null transition from the first state to the second state, we will not be impairing our coverage; we will only be simplifying the number of calculations that have to be done for the illustration of the Baum-Welch algorithm.

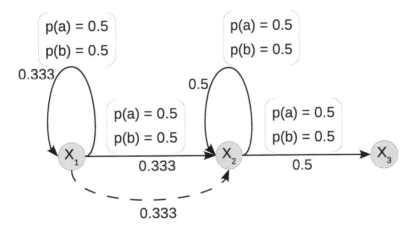

Fig. 13.14: HMM of Example 13.5 with maximum entropy initial distributions

Figure 13.14 shows the model used for this example. In this case, we are seeking the maximum likelihood estimation of the model parameters, given a single sequence, $\{y\}_1^4 = \{abaa\}$. Notice that the model and the sequence are both slightly different from those of Example 13.4. In the figure, for a lack of any information, we choose the maximum entropy solution of equally probable initial choices for the output distributions and transition probabilities of the model – see Section 10.3.

Figure 13.15 shows the trellis diagram for the initial parameters of Figure 13.14. For each node on the trellis, the α and β values have been computed using Equations 13.21 and 13.25 respectively. Using these values in Equations 13.26, 13.27, and 13.30 we may compute the so called, counts, related to the computation of the transition probabilities and associated output distributions. These values are used in Equations 13.28, 13.29, and 13.31 to compute the output generating and null transition probabilities as well as the output probability distributions conditioned on the output producing transitions, respectively. These recomputed values for the

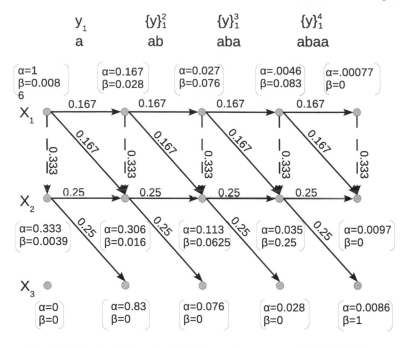

$$y_1 \qquad \{y\}_1^2 \qquad \{y\}_1^3 \qquad \{y\}_1^4$$
$$a \qquad ab \qquad aba \qquad abaa$$

Fig. 13.15: Trellis of Example 13.5 with maximum entropy initial distributions

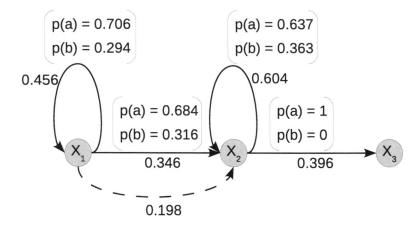

Fig. 13.16: HMM of Example 13.5 with recomputed distributions and transition probabilities after one iteration of Baum-Welch

first iteration are shown in Figure 13.16.

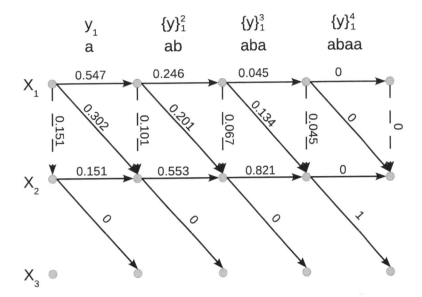

Fig. 13.17: Trellis of Example 13.5 with recomputed *a*-posteriori output-transition probabilities

Figure 13.17 shows the trellis diagram with its values computed based on the new parameters. The likelihood of the new model parameters given the output string, {abaa} is increased as expected by the expectation maximization, of which the Baum-Welch algorithm is a special case Equation 13.35 shows this increase.

$$P^{(1)}(\{y\}_1^4 = \{abaa\}) = 0.02438$$
$$> P^{(0)}(\{y\}_1^4 = \{abaa\})$$
$$= 0.008632 \qquad (13.35)$$

Figure 13.18 shows a plot of the likelihood as a function of the iteration number (in log scale). Given the initial conditions from which we started, a local maximum is reached where only one path has a non-zero probability, giving rise to the configuration of Figure 13.19 with a likelihood of $P^{(*)}(\{abaa\}|\boldsymbol{\varphi}) = 0.037\overline{037}$.

To examine the dependence of the solution to the initial conditions, let us perturb the initial distribution of the first output generating transition just slightly from the maximum entropy distribution of Figure 13.14. Figure 13.20 shows this slightly modified initial set of parameters. Let us repeat the execution of the Baum-Welch algorithm with this new model. Figure 13.21 shows the model parameters for the

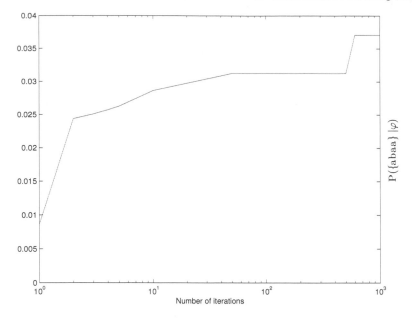

Fig. 13.18: Convergence of the likelihood of the HMM given the sequence, {*abaa*}, to a local maximum, as related to Example 13.5

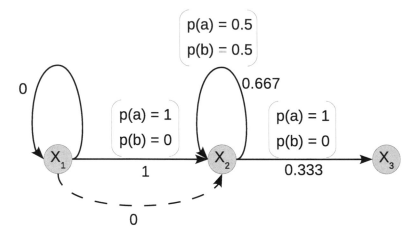

Fig. 13.19: Configuration of the locally converged HMM model for sequence {*abaa*} in Example 13.5

converged model.

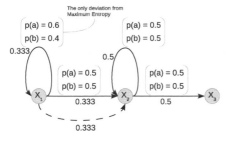
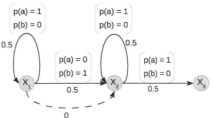

Fig. 13.20: A slight deviation from maximum entropy in the initial distribution of the HMM of Example 13.5

Fig. 13.21: Configuration of the globally converged HMM model for sequence {*abaa*} in Example 13.5

This time, the algorithm has converged to a globally maximal likelihood. Figure 13.22 shows the convergence of the algorithm with the new initial conditions against plotted relative to the number of iterations of the Baum-Welch. The plot also shows the likelihood values of the previous attempt for comparison. The new maximum for the likelihood is $P(\{abaa\}|\boldsymbol{\varphi}) = 0.0625$ which is considerably larger than the previously found local maximum of $0.03\overline{7037}$, although it only took 16 iterations to converge to this global maximum versus the 600+ iterations it took for the previous attempt.

Therefore, we see that the convergence of the *forward-backward* (*Baum-Welch*) algorithm is quite dependent on the initial conditions and that there are many local maxima which can trap the algorithm. This was expected based on the discussion of the general *EM* approach discussed in Section 11.3.1.

For an extension to handle the *MAP* (Section 10.2) estimation of the parameters of an *HMM* with *continuous output densities* see [12]. The algorithms shown here, may be easily extended to tied and shared mixture *HMM* where the transitions and output densities may be tied and shared across different units. In Section 9.4 we discussed decision trees in some detail. Also, in Section 11.4, while discussing hierarchical clustering, we noted the usage of such clustering techniques in typing and combination of parameters in a hierarchy. In reducing the computational load and handling the *tractability problem* (see Section 13.7.2), most practical architectures using *HMM*, create hierarchical structures which utilize decision trees where the questions are in the form of likelihood comparisons of state sequences. For example, due to small amounts of data available for training *HMM*, one solution is to share substructural units (partial *HMM*) and to train these substructures using the global data. Decision trees are used to create the sharing logic, where the leaves are the substructures and the hierarchy of the tree shows the combination of these

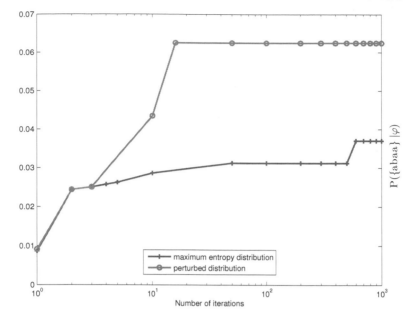

Fig. 13.22: Convergence of the likelihood of the HMM given the sequence, {*abaa*}, for two different initial conditions: 1. maximum entropy and 2. slight perturbation from maximum entropy

substructures into larger units.

13.7 Gaussian Mixture Models (GMM)

The *Gaussian mixture model* (*GMM*) is a model that expresses the probability density function of a random variable in terms of a weighted sum of its components, each of which is described by a *Gaussian* (*normal*) density function. In other words,

$$p(\mathbf{x}|\boldsymbol{\varphi}) = \sum_{\gamma=1}^{\Gamma} p(\mathbf{x}|\boldsymbol{\theta}_\gamma)P(\boldsymbol{\theta}_\gamma) \tag{13.36}$$

where the supervector of parameters, $\boldsymbol{\varphi}$, is defined as an augmented set of Γ vectors constituting the free parameters associated with the Γ mixture components, $\boldsymbol{\theta}_\gamma, \gamma \in \{1, 2, \cdots, \Gamma\}$ and the $\Gamma - 1$, and the mixture weights, $P(\boldsymbol{\theta} = \boldsymbol{\theta}_\gamma), \gamma = \{1, 2, \cdots, \Gamma - 1\}$, which are the prior probabilities of each of these mixture models known as the *mixing distribution* [20].

The parameter vectors associated with each mixture component, in the case of the Gaussian mixture model, is the parameters of the normal density function,

$$\boldsymbol{\theta}_\gamma = \begin{bmatrix} \boldsymbol{\mu}_\gamma^T & \mathbf{u}^T(\boldsymbol{\Sigma}_\gamma) \end{bmatrix}^T \tag{13.37}$$

where the *unique parameters* vector is an invertible transformation that stacks all the free parameters of a matrix into vector form. For example, if $\boldsymbol{\Sigma}_\gamma$ is a full covariance matrix, then $\mathbf{u}(\boldsymbol{\Sigma}_\gamma)$ is the vector of the elements in the upper triangle of $\boldsymbol{\Sigma}_\gamma$ including the diagonal elements. On the other hand, if $\boldsymbol{\Sigma}_\gamma$ is a diagonal matrix, then,

$$\left(\mathbf{u}(\boldsymbol{\Sigma}_\gamma)\right)_{[d]} \stackrel{\Delta}{=} \left(\boldsymbol{\Sigma}_\gamma\right)_{[d][d]} \ \forall \, d \in \{1, 2, \cdots, D\} \tag{13.38}$$

Therefore, we may always reconstruct $\boldsymbol{\Sigma}_\gamma$ from $\mathbf{u}_\gamma$ using the inverse transformation,

$$\boldsymbol{\Sigma}_\gamma = \mathbf{u}_\gamma^{-1} \tag{13.39}$$

The parameter vector for the mixture model may be constructed as follows,

$$\boldsymbol{\varphi} \stackrel{\Delta}{=} \begin{bmatrix} \boldsymbol{\mu}_1^T & \cdots & \boldsymbol{\mu}_\Gamma^T & \mathbf{u}_1^T & \cdots & \mathbf{u}_\Gamma^T & p(\boldsymbol{\theta}_1) & \cdots & p(\boldsymbol{\theta}_{\Gamma-1}) \end{bmatrix}^T \tag{13.40}$$

where only $(\Gamma - 1)$ mixture coefficients (prior probabilities), $p(\boldsymbol{\theta}_\gamma)$, are included in $\boldsymbol{\varphi}$, due to the constraint that

$$\sum_{\gamma=1}^{\Gamma} p(\boldsymbol{\varphi}_\gamma) = 1 \tag{13.41}$$

Thus the number of free parameters in the prior probabilities is only $\Gamma - 1$.

For a sequence of *independent and identically distributed (i.i.d.)* observations, $\{\mathbf{x}\}_1^N$, the log of likelihood of the sequence may be written as follows,

$$\begin{aligned} \ell(\boldsymbol{\varphi}|\{\mathbf{x}\}_1^N) &= \ln\left(\prod_{n=1}^{N} p(\mathbf{x}_n|\boldsymbol{\varphi})\right) \\ &= \sum_{n=1}^{N} \ln p(\mathbf{x}_n|\boldsymbol{\varphi}) \end{aligned} \tag{13.42}$$

Assuming the mixture model, defined by Equation 13.36, the likelihood of may be written in terms of the mixture components,

$$\ell(\boldsymbol{\varphi}|\{\mathbf{x}\}_1^N) = \sum_{n=1}^{N} \ln\left(\sum_{\gamma=1}^{\Gamma} p(\mathbf{x}_n|\boldsymbol{\theta}_\gamma) P(\boldsymbol{\theta}_\gamma)\right) \tag{13.43}$$

Since maximizing Equation 13.43 requires the maximization of the logarithm of a sum, we can utilize the incomplete data approach that was used in the development of the *EM algorithm* in Section 11.3.1.1 to simplify the solution. In the next section, we will derive the incomplete data equivalent of the maximization of Equation 13.43 using the *EM algorithm*.

13.7.1 *Training*

To be able to maximize the likelihood of a sequence, $\{x\}_1^N$, given by Equation 13.43, we will define the expectation and maximization steps associated with the incomplete data approach to this problem. The unobserved data, in this case, is the membership of each data point to the corresponding Gaussian, that data point helps define, in the training step. For more detailed coverage, see [8, 27].

The expectation step may be formulated based on the *a-posteriori* probability of each component of the mixture, given the observed data,

$$p(\boldsymbol{\theta}_\gamma|\mathbf{x}_n) = \frac{p(\boldsymbol{\theta}_\gamma, \mathbf{x}_n)}{p(\mathbf{x}_n)} \tag{13.44}$$

$$= \frac{p(\mathbf{x}_n|\boldsymbol{\theta}_\gamma)p(\boldsymbol{\theta}_\gamma)}{\displaystyle\sum_{\gamma'=1}^{\Gamma} p(\mathbf{x}_n|\boldsymbol{\theta}_{\gamma'})p(\boldsymbol{\theta}_{\gamma'})} \tag{13.45}$$

For the *maximization step*, we would like to maximize the expected log-likelihood of the joint event of the observations, $\{x\}_1^N$, and the parameter vector, $\boldsymbol{\varphi}$. Therefore, the maximization function becomes,

$$\begin{aligned}
Q(\boldsymbol{\varphi}) &= \mathscr{E}\left\{\ln\left(\prod_{n=1}^N p(\mathbf{x}_n, \boldsymbol{\varphi})\right)|\mathbf{x}\right\} \\
&= \mathscr{E}\left\{\sum_{n=1}^N \ln\left(p(\mathbf{x}_n, \boldsymbol{\varphi})\right)|\mathbf{x}\right\} \\
&= \sum_{n=1}^N \mathscr{E}\left\{\ln\left(p(\mathbf{x}_n, \boldsymbol{\varphi})\right)|\mathbf{x}\right\} \\
&= \sum_{n=1}^N \sum_{\gamma=1}^\Gamma p(\boldsymbol{\theta}_\gamma^{(k)}|\mathbf{x}_n)\ln\left(p(\boldsymbol{\theta}_\gamma, \mathbf{x}_n)\right) \\
&= \sum_{n=1}^N \sum_{\gamma=1}^\Gamma p(\boldsymbol{\theta}_\gamma^{(k)}|\mathbf{x}_n)\ln\left(p(\mathbf{x}_n|\boldsymbol{\theta}_\gamma)p(\boldsymbol{\theta}_\gamma)\right)
\end{aligned} \tag{13.46}$$

In order to solve the constrained optimization problem with the constraint on the summation of the prior probabilities, we may redefine $\boldsymbol{\varphi}$ to include all Γ prior probabilities instead of the compact form which was defined in Equation 13.40. Let us call the new parameter vector, $\tilde{\boldsymbol{\varphi}}$, to avoid any confusion. Then

$$\tilde{\boldsymbol{\varphi}} \stackrel{\Delta}{=} \left[\boldsymbol{\mu}_1^T \ \cdots \ \boldsymbol{\mu}_\Gamma^T \ \mathbf{u}_1^T \ \cdots \ \mathbf{u}_\Gamma^T \ p(\boldsymbol{\theta}_1) \ \cdots \ p(\boldsymbol{\theta}_\Gamma)\right]^T \tag{13.47}$$

Therefore, we would like to solve the following iterative maximization problem,

$$\tilde{\boldsymbol{\varphi}}^{(k+1)} = \underset{\tilde{\boldsymbol{\varphi}}}{\arg\max} \, Q(\tilde{\boldsymbol{\varphi}}|\tilde{\boldsymbol{\varphi}}^{(k)}) \tag{13.48}$$

subject to the following constraints,

$$\sum_{\gamma=1}^{\Gamma} p(\gamma_k) = 1 \tag{13.49}$$

In solving the constrained maximization problem of Equations 13.48 and 13.49, we assume a multi-dimensional normal density for $p(\mathbf{x}|\boldsymbol{\theta})$ of the form of Equation 6.196. We may use the method of *Lagrange multipliers* discussed in detail in Section 25.5. Let us define the *Lagrangian*, $\mathscr{L}(\tilde{\boldsymbol{\varphi}}, \lambda)$,

$$\mathscr{L}(\tilde{\boldsymbol{\varphi}}, \lambda) = \sum_{n=1}^{N} \sum_{\gamma=1}^{\Gamma} p(\boldsymbol{\theta}_{\gamma}^{(k)}|\mathbf{x}_n)$$
$$\left(-\tfrac{D}{2} \ln(2\pi) - \tfrac{1}{2} \ln|\boldsymbol{\Sigma}_{\gamma}| \right.$$
$$\left. -\tfrac{1}{2}(\mathbf{x}_n - \boldsymbol{\mu}_{\gamma})^T \boldsymbol{\Sigma}_{\gamma}^{-1}(\mathbf{x}_n - \boldsymbol{\mu}_{\gamma}) + \ln\left(p(\boldsymbol{\theta}_{\gamma})\right) \right) \tag{13.50}$$
$$-\lambda \left(\sum_{\gamma} p(\boldsymbol{\theta}_{\gamma}) - 1 \right)$$

To solve for $\tilde{\boldsymbol{\varphi}}^{(k+1)}$, we should set the gradient of the *Lagrangian* equal to zero and solve for $\tilde{\boldsymbol{\varphi}}$,

$$\nabla_{\tilde{\boldsymbol{\varphi}}} \mathscr{L}(\tilde{\boldsymbol{\varphi}}, \lambda) = \mathbf{0} \tag{13.51}$$

However, since $\tilde{\boldsymbol{\varphi}}$ is made up of different partitions, with common characteristics, we may break up the problem in Equation 13.51 into three different problems, each having Γ subproblems as follows,

$$\nabla_{\boldsymbol{\mu}_{\gamma}} \mathscr{L}(\tilde{\boldsymbol{\varphi}}, \lambda) = \mathbf{0} \tag{13.52}$$

$$\frac{\partial \mathscr{L}(\tilde{\boldsymbol{\varphi}}, \lambda)}{\partial \boldsymbol{\Sigma}_{\gamma}} = \mathbf{0} \tag{13.53}$$

$$\frac{\partial \mathscr{L}(\tilde{\boldsymbol{\varphi}}, \lambda)}{p(\boldsymbol{\theta}_{\gamma})} = \mathbf{0} \tag{13.54}$$

Equations 13.52, 13.53, and 13.54 should be met for all $\gamma \in \{1, 2, \cdots, \Gamma\}$. Also, the partial derivative of $\mathscr{L}(\tilde{\boldsymbol{\varphi}}, \lambda)$ with respect to the covariance matrix is defined to be a matrix of the same size as the covariance matrix, where each element is given by,

$$\left(\frac{\partial \mathscr{L}(\tilde{\boldsymbol{\varphi}}, \lambda)}{\partial \boldsymbol{\Sigma}_{\gamma}} \right)_{[i][j]} \overset{\Delta}{=} \frac{\partial \mathscr{L}(\tilde{\boldsymbol{\varphi}}, \lambda)}{\partial (\boldsymbol{\Sigma}_{\gamma})_{[i][j]}} \forall i, j \in \{1, 2, \cdots, D\} \tag{13.55}$$

It would have been possible to solve Equation 13.51 as a whole in one attempt, since there are formulations of the problem which have consolidated the components of $\tilde{\boldsymbol{\varphi}}$ to be handled as one variable – see [27]. However, solving these three sets of subproblems separately gives us another advantage. The estimates from solving Equation 13.52 for $\boldsymbol{\mu}_{\gamma}^{(k+1)}$, may be used in Equation 13.53 of iteration k to speed up the convergence process.

Let us start with the gradient of Equation 13.52,

$$\nabla_{\boldsymbol{\mu}_\gamma} \mathscr{L}(\tilde{\boldsymbol{\varphi}}, \lambda) = \sum_{n=1}^{N} p(\boldsymbol{\theta}_\gamma^{(k)} | \mathbf{x}_n) \left(-\frac{\partial}{\partial \boldsymbol{\mu}_\gamma} \left(\frac{1}{2} (\mathbf{x}_n - \boldsymbol{\mu}_\gamma)^T \boldsymbol{\Sigma}_\gamma^{-1} (\mathbf{x}_n - \boldsymbol{\mu}_\gamma) \right) \right) \quad (13.56)$$

$$= \sum_{n=1}^{N} p(\boldsymbol{\theta}_\gamma^{(k)} | \mathbf{x}_n) \left(\boldsymbol{\Sigma}_\gamma^{-1} (\mathbf{x}_n - \boldsymbol{\mu}_\gamma) \right) \quad (13.57)$$

$$= \mathbf{0} \quad (13.58)$$

Assuming that the covariance matrix has full rank, then it may be factored out of Equation 13.58. Also, since $\boldsymbol{\mu}_\gamma$ is not directly dependent on the summation index in the equation, it may be factored out of one part of the summation, producing the following result,

$$\boldsymbol{\mu}_\gamma \sum_{n=1}^{N} p(\boldsymbol{\theta}_\gamma^{(k)} | \mathbf{x}_n) = \sum_{n=1}^{N} p(\boldsymbol{\theta}_\gamma^{(k)} | \mathbf{x}_n) \mathbf{x}_n \quad (13.59)$$

Therefore, we may solve for $\boldsymbol{\mu}_\gamma^{(k+1)}$ from Equation 13.59, giving the following intuitive solution to the value of the mean vector at iteration $k+1$,

$$\boldsymbol{\mu}_\gamma^{(k+1)} = \frac{\sum\limits_{n=1}^{N} p(\boldsymbol{\theta}_\gamma^{(k)} | \mathbf{x}_n) \mathbf{x}_n}{\sum\limits_{n=1}^{N} p(\boldsymbol{\theta}_\gamma^{(k)} | \mathbf{x}_n)} \quad (13.60)$$

Likewise, let us evaluate the update for the covariance matrix by solving 13.53,

$$\frac{\partial}{\partial \boldsymbol{\Sigma}_\gamma} \mathscr{L}(\tilde{\boldsymbol{\varphi}}, \lambda) = \sum_{n=1}^{N} p(\boldsymbol{\theta}_\gamma^{(k)} | \mathbf{x}_n)$$

$$\left(-\frac{\partial}{\partial \boldsymbol{\Sigma}_\gamma} \left(\frac{1}{2} \ln(|\boldsymbol{\Sigma}_\gamma|) \right) \right.$$

$$\left. -\frac{\partial}{\partial \boldsymbol{\Sigma}_\gamma} \left(\frac{1}{2} (\mathbf{x}_n - \boldsymbol{\mu}_\gamma)^T \boldsymbol{\Sigma}_\gamma^{-1} (\mathbf{x}_n - \boldsymbol{\mu}_\gamma) \right) \right) \quad (13.61)$$

$$= \sum_{n=1}^{N} p(\boldsymbol{\theta}_\gamma^{(k)} | \mathbf{x}_n)$$

$$\left(-\frac{1}{2} \boldsymbol{\Sigma}_\gamma^{-1} + \frac{1}{2} \boldsymbol{\Sigma}_\gamma^{-1} (\mathbf{x}_n - \boldsymbol{\mu}_\gamma)(\mathbf{x}_n - \boldsymbol{\mu}_\gamma)^T \boldsymbol{\Sigma}_\gamma^{-1} \right) \quad (13.62)$$

$$= \mathbf{0} \quad (13.63)$$

Again, assuming that Σ_γ has full rank, we may factor one instance of it out of Equation 13.63, leaving us with the following equality,

$$\sum_{n=1}^{N} p(\boldsymbol{\theta}_\gamma^{(k)} | \mathbf{x}_n) = \sum_{n=1}^{N} p(\boldsymbol{\theta}_\gamma^{(k)} | \mathbf{x}_n) \boldsymbol{\Sigma}_\gamma^{-1} (\mathbf{x}_n - \boldsymbol{\mu}_\gamma)(\mathbf{x}_n - \boldsymbol{\mu}_\gamma)^T \quad (13.64)$$

From Equation 13.64 we may solve for the covariance matrix in iteration $k+1$,

$$\Sigma_\gamma^{(k+1)} = \frac{\sum_{n=1}^{N} p(\theta_\gamma^{(k)}|\mathbf{x}_n)(\mathbf{x}_n - \mu_\gamma)(\mathbf{x}_n - \mu_\gamma)^T}{\sum_{n=1}^{N} p(\theta_\gamma^{(k)}|\mathbf{x}_n)} \tag{13.65}$$

which again seems pretty intuitive, especially when compared to the updated expression for $\mu_\gamma^{(k+1)}$, given by Equation 13.60.

As we mentioned earlier, since we already have the expression for $\mu_\gamma^{(k+1)}$, instead of using the value of $\mu_\gamma^{(k)}$, we can actually use the value of $\mu_\gamma^{(k+1)}$ in Equation 13.65 to increase the speed of convergence. Therefore,

$$\Sigma_\gamma^{(k+1)} = \frac{\sum_{n=1}^{N} p(\theta_\gamma^{(k)}|\mathbf{x}_n)(\mathbf{x}_n - \mu_\gamma^{(k+1)})(\mathbf{x}_n - \mu_\gamma^{(k+1)})^T}{\sum_{n=1}^{N} p(\theta_\gamma^{(k)}|\mathbf{x}_n)} \tag{13.66}$$

Let us move on to the computation of the new estimates for the a-priori probabilities of the mixture components, $p(\theta_\gamma^{(k+1)})$.

$$\frac{\partial \mathcal{L}(\tilde{\varphi}, \lambda)}{\partial p(\theta_\gamma)} = \left(\sum_{n=1}^{N} p(\theta_\gamma^{(k)}|\mathbf{x}_n) \frac{\partial \ln(p(\theta_\gamma))}{\partial p(\theta_\gamma)} \right) -$$
$$\lambda \left(\frac{\partial p(\theta_\gamma)}{\partial p(\theta_\gamma)} \right) \tag{13.67}$$

$$= \left(\sum_{n=1}^{N} p(\theta_\gamma^{(k)}|\mathbf{x}_n) \frac{1}{p(\theta_\gamma)} \right) - \lambda \tag{13.68}$$

$$= 0 \tag{13.69}$$

Using Equation 13.69, we may solve for $p(\theta_\gamma)$ in terms of the *Lagrange multiplier*, λ and the sum of the posterior probabilities,

$$p(\theta_\gamma) = \frac{1}{\lambda} \sum_{n=1}^{N} p(\theta_\gamma^{(k)}|\mathbf{x}_n) \tag{13.70}$$

To solve for the *Lagrange multiplier*, we may use the constraint of Equation 13.49 in conjunction with Equation 13.70,

$$\sum_{\gamma=1}^{\Gamma} p(\gamma_k) = 1$$

$$= \sum_{\gamma=1}^{\Gamma} \frac{1}{\lambda} \sum_{n=1}^{N} p(\theta_\gamma^{(k)}|\mathbf{x}_n) \tag{13.71}$$

Since $\hat{\lambda}$ is not a function of the summation index, γ, it may be factored out and solved for, in terms of the sum of the posterior probabilities over the total output sequence and all mixture components,

$$\hat{\lambda} = \sum_{\gamma=1}^{\Gamma} \sum_{n=1}^{N} p(\boldsymbol{\theta}_{\gamma}^{(k)}|\mathbf{x}_n) \tag{13.72}$$

We may plug for $\hat{\lambda}$ from Equation 13.72 into Equation 13.70, to solve for the *prior probabilities* (*mixture coefficients*) at iteration $k+1$,

$$p(\boldsymbol{\theta}_{\gamma}^{(k+1)}) = \frac{\sum_{n=1}^{N} p(\boldsymbol{\theta}_{\gamma}^{(k)}|\mathbf{x}_n)}{\sum_{\gamma'=1}^{\Gamma} \sum_{n=1}^{N} p(\boldsymbol{\theta}_{\gamma'}^{(k)}|\mathbf{x}_n)}$$

$$= \frac{1}{N} \sum_{n=1}^{N} p(\boldsymbol{\theta}_{\gamma}^{(k)}|\mathbf{x}_n) \tag{13.73}$$

Here, we will repeat the expectation step of Equation 13.45 and Equations 13.60, 13.66, and 13.73, which are the steps needed for forming an iterative *expectation maximization* process of computing the elements of the *parameter vector*, $\boldsymbol{\varphi}$, as defined by Equation 13.47.

Expectation step:

$$p(\boldsymbol{\theta}_{\gamma}|\mathbf{x}_n) = \frac{p(\mathbf{x}_n|\boldsymbol{\theta}_{\gamma})p(\boldsymbol{\theta}_{\gamma})}{\sum_{\gamma'=1}^{\Gamma} p(\mathbf{x}_n|\boldsymbol{\theta}_{\gamma'})p(\boldsymbol{\theta}_{\gamma'})} \tag{13.74}$$

Maximization step:

$$\boldsymbol{\mu}_{\gamma}^{(k+1)} = \frac{\sum_{n=1}^{N} p(\boldsymbol{\theta}_{\gamma}^{(k)}|\mathbf{x}_n)\mathbf{x}_n}{\sum_{n=1}^{N} p(\boldsymbol{\theta}_{\gamma}^{(k)}|\mathbf{x}_n)} \tag{13.75}$$

$$\boldsymbol{\Sigma}_{\gamma}^{(k+1)} = \frac{\sum_{n=1}^{N} p(\boldsymbol{\theta}_{\gamma}^{(k)}|\mathbf{x}_n)(\mathbf{x}_n - \boldsymbol{\mu}_{\gamma}^{(k+1)})(\mathbf{x}_n - \boldsymbol{\mu}_{\gamma}^{(k+1)})^T}{\sum_{n=1}^{N} p(\boldsymbol{\theta}_{\gamma}^{(k)}|\mathbf{x}_n)} \tag{13.76}$$

$$p(\boldsymbol{\theta}_{\gamma}^{(k+1)}) = \frac{1}{N} \sum_{n=1}^{N} p(\boldsymbol{\theta}_{\gamma}^{(k)}|\mathbf{x}_n) \tag{13.77}$$

See Section 13.8.3 for an alternative perspective of the computation of the *mixture coefficients* of a GMM.

Xu and Jordan [27] among others have studied the convergence of the *EM algorithm* for estimating the parameters of the *GMM* in quite detail, arriving at the conclusions stated in Section 11.3.1 that the algorithm increases the likelihood with every iteration. However, they also show the slow convergence of the *EM algorithm* which was stated earlier. In addition, we showed in a numerical example (Example 13.5) that the EM algorithm can sometimes converge very slowly and end up in a local maximum and in times it can be quite fast. Generally, though, it is known to be a slowly converging algorithm. For this reason, mostly, it is used in combination with other algorithms which would first provide it with good initial guesses of the parameter vector such as the *k-means* and similar algorithms discussed in Chapter 11.

13.7.2 Tractability of Models

One of the main problems with computing $\boldsymbol{\varphi}$ is the problem of intractability. Namely, the number of parameters is

$$\dim(\boldsymbol{\varphi}) = \Gamma \left(\underbrace{D}_{\mu_\gamma} + \underbrace{\frac{D(D+1)}{2}}_{\Sigma_\gamma} + \underbrace{1}_{p(\boldsymbol{\theta}_\gamma)} \right) \underbrace{-1}_{\Sigma_\gamma p(\boldsymbol{\theta}_\gamma)=1}$$

$$= \frac{\Gamma}{2} \left(D^2 + 3D + 2 \right) - 1 \tag{13.78}$$

which for large D would require large amounts of data in order to be able to get a reliable estimate of the parameters. Let us assume the *speaker recognition* problem where D is generally computed by taking somewhere in the order of about 15 *Mel frequency Cepstral coefficients* (Section 5.3.6), *delta Cepstra* and *delta-delta Cepstra* (Section 5.3.7). If the same number of delta and delta-delta Cepstra are chosen to be included in the feature vector, then $D = 45$. If we decide to use $\Gamma = 256$ *Gaussians* in the *mixture model*, then the number of parameter which need to be estimated would be,

$$\dim(\boldsymbol{\varphi}) = \frac{\Gamma}{2} \left(D^2 + 3D + 2 \right) - 1$$

$$= \frac{256}{2} \left(45^2 + 3 \times 45 + 2 \right) - 1$$

$$= 276,735 \tag{13.79}$$

For tractability, if we take the rule of thumb which says that we should have at least 10 points per parameter being estimated, then we would need at least $2,767,350$

frames of speech. Considering that each frame spans about $10ms$, then, this would be equivalent to having $27,673.5s$ or 561.2 minutes!

Of course it is quite difficult to come up with 561 minutes of speech for training. There are different remedies to this problem. One is to do away with the diagonal elements of the covariance matrix and only consider the variances. This will reduce the number of required parameters to

$$
\begin{aligned}
\dim(\boldsymbol{\varphi}) &= \frac{\Gamma}{2}(2D+1) - 1 \\
&= \frac{256}{2}(2 \times 45 + 1) - 1 \\
&= 11,647
\end{aligned}
\tag{13.80}
$$

which is 95.8% less than the number of parameters needed in a full covariance model. The number of minutes of audio for estimating these parameters would then only be 1.94 minutes which is much more practical.

However, we usually do not use the data from only one speaker to estimate the parameters of the whole set of parameters in a model. Generally, a large training set is used, which consists of hundreds and maybe even thousands of speakers to compute the basic model parameters. This is sometimes called the *speaker independent model* and in some circumstances it is referred to as the *universal model, universal background model*, or simply *background model*. As we will see in chapter 21, the individual target speaker's enrollment audio is generally used to adapt new model parameters from this *speaker independent model*.

Another approach is to reduce the dimensionality of the data. This would mean the reduction of D which greatly affects the total number of free parameters. In Chapter 12 we discussed a few techniques designed to optimize the amount of information which is preserved by reducing the dimensionality of the free parameters in the system. Some such techniques are *PCA*, *LDA*, and *FA*. We shall see more about such dimensionality reduction as applied specifically to the speaker recognition problems, such as *joint factor analysis* (Section 16.4.2) and *nuisance attribute projection* (Section 15.8).

In Chapter 16, we will speak more about the *speaker model* and how it is devised. In the chapters following Chapter 16, more detail will be given regarding different aspects of model and parameter optimization.

13.8 Practical Issues

In this section, we will examine a few practical issues involved in the successful training of *HMM* and *GMM*. The first and one of the foremost problems with training a statistical system, regardless of whether it is an HMM or a GMM is the shortage of training data. Most of the theory that has been developed in this book revolves around different definitions of the *law of large numbers* (Section 6.9.2). We have repeatedly spoken about *sufficient statistics*. However, in practice, it is quite impossible to have accounted for all the different possible observations which may be encountered. In this section, we will examine possible preemptive methods for avoiding gross failures (events leading to 0 output probabilities). Also, we will devise ways of making our limited training data go as far as possible.

An inherent problem with using HMM is related to the discrepancies in the components of the terms for computing the likelihood of an output sequence given an HMM. There are two major components, one of which is related on logs of transition probabilities and the other is based on the output distribution or density given a state. Unfortunately, these two components have different inherent statistics. The transition probabilities have a large mean value among them and a small variance compared to the output distributions or densities which have a much smaller mean, but relatively larger variance. The addition of the logs of these two components is highly biased toward the output distributions/densities for their larger variances, hence downplaying the role of the transitions (dynamics) of the sequence – see p. 215 of [9], p. 48 of [28], and [7].

13.8.1 Smoothing

Let us assume that we have two sets of data, one of which we call the *training data* and is used for training ($\{\mathbf{y}\}_1^{N_1}$) and a second set which we call *test data* that is intended to be used for validation ($\{\boldsymbol{\eta}\}_1^{N_2}$). As we saw in both the *HMM* and *GMM* training procedures with the EM algorithm, we estimate $\boldsymbol{\varphi}$ from $\{\mathbf{y}\}_1^{N_1}$ to try to maximize the likelihood, $p(\{\mathbf{y}\}_1^{N_1}|\boldsymbol{\varphi})$.

As we mentioned in the beginning of this discussion, in practice we are faced with data shortage. This problem was also discussed in reviewing the tractability the *GMM*, in Section 13.7.2. When the training data is small, the parameters will be overtrained such that it will be quite possible for the likelihood of the model given the *test data*, $p(\{\mathbf{y}\}_1^{N_1}|\boldsymbol{\varphi})$ to be almost or even identically 0. One way to preemptively adjust the model parameters in order to avoid such situations is to *smooth* them with parameters which would allow for any output to be produced, such as the *maximum entropy solution* we used as the initial conditions of the parameters of the

HMM in Section 13.6.4 – see Figure 13.14.

Example 13.6 (Smoothing).

For the sake of simplicity, let us take the example of a memoryless model (as discussed in Section 13.1). Let us take the following die toss problem. As we know, the possible outputs are $\mathscr{Y} = \{1,2,3,4,5,6\}$. If we assume the training sequence, $\{y\}_1^8 = \{5,6,5,3,6,5,6,1\}$, and the test sequence, $\{\eta\}_1^4 = \{1,5,6,2\}$, then the maximum likelihood solution to the parameter estimation based on the observed training data is given by the relative frequency provided from the training data (see Example 13.1),

$$\hat{\boldsymbol{\phi}} = \{P(1) = \frac{1}{8}, P(2) = 0, P(3) = \frac{1}{8}, P(4) = 0, P(5) = \frac{3}{8}, P(6) = \frac{3}{8}\} \qquad (13.81)$$

Therefore, the likelihood of the training data would be given by,

$$p(\{\mathbf{y}\}_1^{N_1} | \hat{\boldsymbol{\phi}}) = (\frac{3}{8})^6 \times (\frac{1}{8})^2$$

$$= 4.35 \times 10^{-5} \qquad (13.82)$$

But the likelihood of the test data would be 0, since the test data contains a 2, which has not been seen in the training data and has a 0 probability of occurrence as far as the model parameters are concerned. The maximum entropy solution is the uniform distribution,

$$\boldsymbol{\phi}_{me} = \{P(1) = \frac{1}{6}, P(2) = \frac{1}{6}, P(3) - \frac{1}{6}, P(4) - \frac{1}{6}, P(5) = \frac{1}{6}, P(6) = \frac{1}{6}\} \quad (13.83)$$

Therefore, we may smooth $\hat{\boldsymbol{\phi}}$ with $\boldsymbol{\phi}_{me}$ using a linear combination with weighting parameter, α,

$$\boldsymbol{\phi}_s = \alpha \hat{\boldsymbol{\phi}} + (1 - \alpha) \boldsymbol{\phi}_{me} \qquad (13.84)$$

Therefore, the problem of smoothing reduces to choosing the smoothing parameter, α. An intuitive choice would be one that would be dependent on the size of the training data, N_1, such that for very large N_1, the effect of the smoothing distribution would fade away and the maximum likelihood estimate would dominate. Therefore, one solution would be if we pick α as follows,

$$\alpha = 1 - \frac{\varepsilon}{N_1} \qquad (13.85)$$

where ε is a small number in comparison with N_1. For example, if we pick ε to be $\varepsilon = 1$, then for the above example,

$$\alpha = 1 - \frac{\varepsilon}{N_1}$$

$$= \frac{7}{8} \qquad (13.86)$$

The new smoothed parameters would be,

$$\boldsymbol{\varphi}_s = \frac{7}{8}\boldsymbol{\hat{\phi}} + (1 - \frac{1}{8})\boldsymbol{\varphi}_{me}$$
$$= \{0.13, 0.02, 0.13, 0.02, 0.35, 0.35\} \quad\quad (13.87)$$

Using $\boldsymbol{\varphi}_s$ to compute the likelihoods for the model given the training and test data, we have,

$$p(\{\mathbf{y}\}_1^6|\boldsymbol{\hat{\phi}}) = 0.35^6 \times 0.13^2$$
$$= 3.1 \times 10^{-5} \quad\quad (13.88)$$

and

$$p(\{\boldsymbol{\eta}\}_1^4|\boldsymbol{\hat{\phi}}) = 0.13 \times 0.35^2 \times 0.02$$
$$= 3.2 \times 10^{-4} \qu\quad (13.89)$$

Therefore, although the likelihood of the model given the training data has been reduced with this smoothing, the model has become more versatile and produces a viable likelihood for an unseen sequence. Of course, just picking the value of α or ε may not be so optimal. In Section 13.8.3, we will talk about the method of held-out estimation, which may be used to optimize the value of α according to some held-out data.

13.8.2 *Model Comparison*

Notice from the results at the end of Section 13.8.2, that we cannot really compare the two likelihoods in order to assess the performance of the model, since the likelihoods are dependent on the length of the observed sequence. We will encounter this in performing speaker recognition as well. Sometimes, we would like to compare the likelihood of two sequences. For example, as we will see, for the verification problem, we would like to compare the performance (likelihoods) of two different models, one of which is trained on a speaker-independent set, sometimes called a background model, and the other on a model based on the speaker-dependent observations. In this case, it would make sense to speak of the *unit likelihood* or the *normalized likelihood* which is the likelihood of the model given a sequence, but normalized over the number of observations.

In the case of speaker recognition, each observation is generally a frame of speech (see Chapter 5), for which a *feature vector* has been computed. To find the *normalized likelihood*, we would have to take the Nth root of the likelihood where N is the length of the observed sequence. However, as we have discussed in the past and will do so in future references, for many reasons, we will be using the log of the likelihood, including *significance (underflow)* problems as well as handling normalizations like this. If we use the log likelihood instead, then the Nth root will translate into dividing the log by N which is a much simpler operation. This computation will act as the computation of the *expected value* of the log of likelihood, sometimes called the *expected log likelihood*. Therefore, let us take the two likelihoods, presented in Equations 13.88 and 13.89 and compute them in log form, while normalizing them against the length of the corresponding observed sequence that produced those likelihoods (finding the their expected values).

For the training data,

$$
\begin{aligned}
\ell(\hat{\phi}|\{\mathbf{y}\}_1^8) &= \ln\left(p(\{\mathbf{y}\}_1^8|\hat{\phi})\right) \\
&= 6\ln(0.35) + 2\ln(0.13) \\
&= -10.38
\end{aligned}
\tag{13.90}
$$

with the *expected log likelihood* of,

$$
\begin{aligned}
\tilde{\ell}(\hat{\phi}|\{\mathbf{y}\}_1^8) &= \frac{1}{8}\ell(\hat{\phi}|\{\mathbf{y}\}_1^8) \\
&= \frac{1}{8} - 10.38 \\
&= -1.30
\end{aligned}
\tag{13.91}
$$

For the test data,

$$
\begin{aligned}
\ell(\hat{\phi}|\{\boldsymbol{\eta}\}_1^4) &= \ln\left(p(\{\boldsymbol{\eta}\}_1^4|\hat{\phi})\right) \\
&= \ln(0.13) + 2\ln(0.35) + \ln(0.02) \\
&= -8.05
\end{aligned}
\tag{13.92}
$$

with the *expected log likelihood* of,

$$
\begin{aligned}
\tilde{\ell}(\hat{\phi}|\{\boldsymbol{\eta}\}_1^4) &= \frac{1}{4}\ell(\hat{\phi}|\{\boldsymbol{\eta}\}_1^4) \\
&= \frac{1}{4} - 8.05 \\
&= -2.01
\end{aligned}
\tag{13.93}
$$

As expected, the *normalized likelihood* of the model given training data is larger than that given the test data.

In fact, if we are interested in comparing the performance of a model against two different observed data sets or similarly comparing the performance of two different models given the same observed data, we may use a *likelihood ratio*. We discussed the *likelihood ratio* in some detail in the context of *decision theory* in Chapter 9. Here, we will define it one more time, in the context of the aspects encountered by the above example.

Definition 13.4 (Likelihood Ratio). *Likelihood ratio is a measure of the comparison of the performances of two models which may have produced a sequence, or the comparison of the performances two sequences of the same model which may have been used to produce the same sequence.*

$$r \stackrel{\Delta}{=} \frac{\sqrt[N_1]{\mathscr{L}(\boldsymbol{\varphi} | \{y\}_1^{N_1})}}{\sqrt[N_2]{\mathscr{L}(\boldsymbol{\varphi} | \{\eta\}_1^{N_2})}} \tag{13.94}$$

or

$$r \stackrel{\Delta}{=} \frac{\sqrt[N_1]{\mathscr{L}(\boldsymbol{\varphi}_1 | \{y\}_1^{N_1})}}{\sqrt[N_1]{\mathscr{L}(\boldsymbol{\varphi}_2 | \{y\}_1^{N_1})}} \tag{13.95}$$

In the case where the comparison is being made between two different models for producing the same observation sequence, normalization against the length of the sequence may not matter as much, since the same length sequence is being used to do the comparison. However, if the same model is being used for compare its performances of producing different sequences, then the likelihood needs to be normalized against the length of the sequences. As a rule, it is better to do the normalization at all times, to reduce confusion.

As we mentioned, we generally use the log of the likelihood for many reasons which have been discussed. Therefore, another measure, the *log likelihood ratio* (*LLR*) is used for this comparison. It was rigorously defined in the context of the gain in *relative information* between the prior states of two hypotheses to one after observing the data (Equation 7.78 and Sections 9.2.1 and 9.2.2). Here, we reiterate its definition, in relation to the problem described here, to assess the performance of a model.

$$
\begin{aligned}
r_l &= \ln(r) \\
&\stackrel{\Delta}{=} \frac{\ell(\boldsymbol{\varphi} | \{y\}_1^{N_1})}{N_1} - \frac{\ell(\boldsymbol{\varphi} | \{\eta\}_1^{N_2})}{N_2}
\end{aligned} \tag{13.96}
$$

or

$$
\begin{aligned}
r_l &= \ln(r) \\
&\stackrel{\Delta}{=} \frac{\ell(\boldsymbol{\varphi}_1 | \{y\}_1^{N_1})}{N_1} - \frac{\ell(\boldsymbol{\varphi}_2 | \{y\}_1^{N_1})}{N_1}
\end{aligned} \tag{13.97}
$$

The *LLR* will be used in speaker verification, when we compare the speaker dependent model against a competing model to see which one is more likely to have generated the test data. Generally, a more positive *LLR* would be in favor of the first model and one in the negative direction would advocate the second model. For example, the *LLR* of the test data against the training data in the above example would be,

$$
\begin{aligned}
r_l &= -2.01 - (-1.30) \\
&= -0.71
\end{aligned} \tag{13.98}
$$

which means that the model fits the training data better than the test data. Comparing the performance of our model which was generated using the training data against the same training data would produce an *LLR* of 0, which may be used as baseline. In Section 9.2.1, however, we saw the need for defining a threshold to be used in making this decision.

13.8.3 Held-Out Estimation

Toward the end of Section 13.8.2, we manually picked a smoothing factor, α, by choosing ε which was used to compute α in relation with the training sample size. The value of α may, of course, not be optimal, since it was chosen based, mostly, on intuition and not any data-driven information. In this section, we will use some *held-out data* to compute the optimal value of α based on a *maximum likelihood* technique.

Let us take the *training data* of Example 13.6, $\{y\}_1^8 = \{5,6,5,3,6,5,6,1\}$, and split it into two parts, $\{y\}_1^4 = \{5,6,5,3\}$ and $\{y\}_5^8 = \{6,5,6,1\}$. In a more general form, the *training data*, $\{y\}_1^N$ is split into $\{y\}_1^{N_t}$ and $\{y\}_{N_t+1}^N$. We may then go ahead with estimating the parameter vector, $\hat{\boldsymbol{\phi}}$, from the first part of the training data, $\{y\}_1^{N_t}$, in the same way as we did in Example 13.6. The only difference, here, is that only part of the training data is used for this computation. Then, we can look for a smooth model given by Equation 13.84, repeated here for convenience.

$$\boldsymbol{\varphi}_s = \alpha\hat{\boldsymbol{\phi}} + (1-\alpha)\boldsymbol{\varphi}_{me} \tag{13.99}$$

This time, instead of guessing the value of α, we compute it such that the new smooth parameters, $\boldsymbol{\varphi}_s$, would maximize the likelihood of the second part of the training data, $\mathscr{L}(\boldsymbol{\varphi}_s|\{y\}_{N_t+1}^N)$.

The parameter vector (probability) estimate using the first half of the data would be,

$$\hat{\boldsymbol{\phi}} = \{0,0,\frac{1}{4},0,\frac{1}{2},\frac{1}{4}\} \tag{13.100}$$

Using $\hat{\boldsymbol{\phi}}$, we may compute the likelihood of the smooth parameters, $\boldsymbol{\varphi}_s$, given the second sequence, $\{y\}_{N_t+1}^N$, in terms of α. Equation 13.99 is used to compute the likelihood of the smoothed parameter vector given each output observation in the $\{y\}_{N_t+1}^N$.

$$P(\{6,5,6,1\}|\boldsymbol{\varphi}_s) = \prod_{n=5}^{8} P(y_n|\boldsymbol{\varphi}_s)$$

$$= \underbrace{\left(\frac{\alpha}{4} + \frac{1-\alpha}{6}\right)}_{P(y=6|\boldsymbol{\varphi}_s)} \times \underbrace{\left(\frac{\alpha}{2} + \frac{1-\alpha}{6}\right)}_{P(y=5|\boldsymbol{\varphi}_s)} \times$$

$$\underbrace{\left(\frac{\alpha}{4} + \frac{1-\alpha}{6}\right)}_{P(y=6|\boldsymbol{\varphi}_s)} \times \underbrace{\left(0 + \frac{1-\alpha}{6}\right)}_{P(y=1|\boldsymbol{\varphi}_s)} \tag{13.101}$$

$$= -\frac{1}{5184}\left(2\alpha^4 + 7\alpha^3 + 3\alpha^2 - 8\alpha - 4\right) \tag{13.102}$$

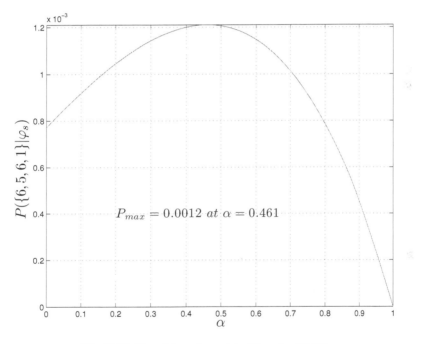

Fig. 13.23: Plot of the polynomial of Equation 13.102

Figure 13.23 shows a plot of the likelihood as a polynomial in α, given by Equation 13.102. Since α is restricted to exist between 0 and 1, it has only been plotted in that domain. Of course the polynomial may exist everywhere else. We can see that this likelihood has only one maximum in the closed interval, $[0,1]$, and it lies at $P_{max}(\{6,5,6,1\}|\boldsymbol{\varphi}_s) = 1.2 \times 10^{-3}$, for $\alpha = 0.461$.

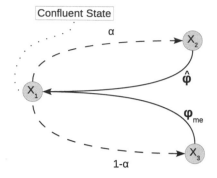

Fig. 13.24: State diagram of the smoothing model of Equation 13.99

Equation 13.99 may be expressed in a Markov chain, represented in a finite state diagram, shown in Figure 13.24. Therefore, one approach to finding the α which provides the maximum likelihood solution for the smooth model is to use the *forward-backward algorithm* discussed in Section 13.6.4. Figure 13.25 shows the trellis diagram associated with the model in Figure 13.24 and the output sequence, $\{y\}_5^8 = \{6, 5, 6, 1\}$.

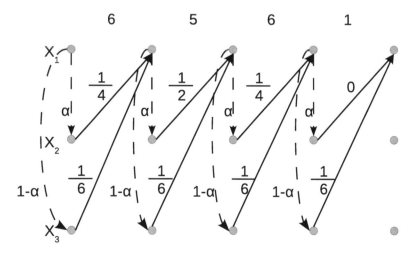

Fig. 13.25: Trellis diagram of the smoothing model of Equation 13.99

This trellis has a very interesting structure. After every output, y_n is produced, all the possible paths converge on a single lattice associated with state X_1 of the model in Figure 13.24. Such a state is called a *confluent state* in the finite state machinery terminology. It has the property that allows us to break up the trellis diagram into repeating individual sections associated with each output, y_n.

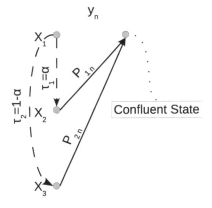

Fig. 13.26: Repetitive trellis component ending in a confluent state

Figure 13.26 shows the diagram for each of these trellis components. In the figure, the two null transitions associated with $alpha$ and $(1-\alpha)$ are named transitions, τ_1 and τ_2, respectively. Also, the simplified notation of $p_{1n} \triangleq P(y_n|\hat{\boldsymbol{\phi}})$ and $p_{2n} \triangleq P(y_n|\boldsymbol{\phi}_{me})$ has been used in the figure, for convenience.

Due to the property of the confluent state, we may write the following simple closed form expression for $P(\tau_1, t = n|\{y\}_{N_t+1}^N)$, in terms of α, $P(y_n|\hat{\boldsymbol{\phi}})$ and $P(y_n|\boldsymbol{\phi}_{me})$,

$$P(\tau_1, t = n|\{y\}_{N_t+1}^N) = \frac{\alpha P(y_n|\hat{\boldsymbol{\phi}})}{\alpha P(y_n|\hat{\boldsymbol{\phi}}) + (1-\alpha)P(y_n|\boldsymbol{\phi}_{me})} \tag{13.103}$$

and at any time $t = n$, $P(\tau_2, t = n|\{y\}_{N_t+1}^N)$ may be computed as follows,

$$P(\tau_2, t = n|\{y\}_{N_t}^N) = 1 - P(\tau_1, t = n|\{y\}_{N_t}^N) \tag{13.104}$$

Therefore, the estimate of α at iteration $k+1$ may be computed using Equation 13.29, which may be written by the following expression,

$$\alpha^{(k+1)} = \frac{\displaystyle\sum_{n=N_t+1}^{N} P(\tau_1^{(k)}, t = n|\{y\}_{N_t+1}^N)}{\displaystyle\sum_{n=N_t+1}^{N} P(\tau_1^{(k)}, t = n|\{y\}_{N_t+1}^N) + \sum_{n=N_t+1}^{N} P(\tau_2^{(k)}, t = n|\{y\}_{N_t+1}^N)} \tag{13.105}$$

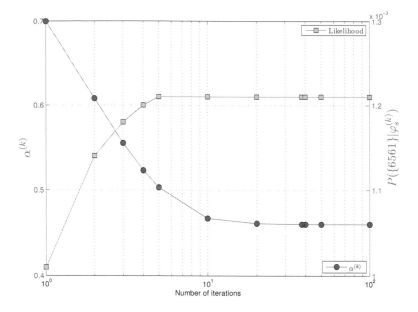

Fig. 13.27: Convergence of the forward-backward estimate of α to the value found earlier in Figure 13.23

If we start with an initial guess for α, say $\alpha^{(0)} = 0.7$, by applying Equation 13.105, the new estimate, $\alpha^{(1)}$, would be $\alpha^{(1)} = 0.608$. The likelihood for the models based on $\alpha^{(1)}$ increases from that based on $\alpha^{(0)}$, given the training sequence, as follows,

$$
\begin{aligned}
P(\{6,5,6,1\}|\boldsymbol{\varphi}^{(1)}) &= 1.14 \times 10^{-3} \\
&> 1.01 \times 10^{-3} \\
&= P(\{6,5,6,1\}|\boldsymbol{\varphi}^{(0)})
\end{aligned}
\tag{13.106}
$$

Figure 13.27 shows the convergence of the *forward-backward algorithm*, to $\alpha = 0.46$ with the likelihood of the smooth model being 1.21×10^{-3}, which was found earlier from the graphical inspection of the polynomial form of the likelihood in terms of α derived in Equation 13.102.

The approach of optimizing the mixture of two distributions using the forward-backward algorithm may be generalized to linear combinations of more than two probability distributions. In fact, the results may be used to optimize the *mixture coefficients* of a *GMM*, much in the same way. Figure 13.28 shows such a model. The smooth parameters for Γ distributions would then be given by,

$$
\boldsymbol{\varphi}_s = \alpha_1 \boldsymbol{\varphi}_1 + \alpha_2 \boldsymbol{\varphi}_2 + \cdots + \alpha_\Gamma \boldsymbol{\varphi}_\Gamma
\tag{13.107}
$$

where

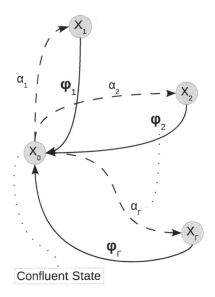

Confluent State

Fig. 13.28: State diagram of the mixture coefficients for more than two distributions

$$\sum_{\gamma=1}^{\Gamma} = 1 \qquad\qquad (13.108)$$

and $0 \le \alpha_\gamma \le 1$.

Bahl, et al. [3] present a fast algorithm based on binary search which may be used for mixtures of two distributions, as presented in the above example. It will produce the optimal solution much faster than the convergence of the forward-backward algorithm. The basis of this approach was demonstrated earlier in the polynomial approach.

13.8.4 Deleted Estimation

In the previous section, we arrived at a technique for performing *held-out estimation*. If we use *deleted estimation*, we may reuse the same data for both purposes of computing the parameter vector for the model and estimating the smoothing parameter, α. *Delete estimation* is sometimes referred to as *deleted interpolation* or *k-fold*[4] *cross-validation* [24]. The idea is to partition the training data into many equal sets. Then we may follow these steps for *deleted estimation*,

---

[4] Although it is generally known as k-fold cross-validation, here, we use the variable L instead of k in the identifier, k-fold.

1. Remove set l from the total L partitions and compute the parameter vector, $\boldsymbol{\varphi}_l$ using the data in the L partitions, not containing the data in the lth partition.
2. Once all the $\boldsymbol{\varphi}_l, l \in \{1, 2, \cdots, L\}$ have been computed, for each data segment, l, write the following linear combination, $\alpha \boldsymbol{\varphi}_l + (1 - \alpha) \boldsymbol{\varphi}_{me}$.
3. Use the forward-backward or binary search technique to find α which optimizes all L smoothing mixtures.
4. Once the optimal α is computed, use all the data to compute the parameter vector for the model, $\hat{\boldsymbol{\varphi}}$.
5. Use the optimal value of α computed over all partitions as prescribed earlier to find the new parameter vector using Equation 13.99.

Using the above algorithm, all the data has been used for both purposes of parameter vector estimation and the estimation of the smoothing parameter, α.

The limiting case of the above estimation is the case where $L = N$. This case is called the *leave-one-out cross-validation* or *deleted estimation*, sometimes written as *LOO*. Of course in most cases, since N (number of training samples) is quite large, the computation would be too much to be practical, unless in certain cases where short-cuts may be available.

References

1. Ash, R.B.: Information Theory. Dover Publications, New York (1990). ISBN: 0-486-6652-16
2. Bahl, L.R., Brown, P.F., deSouza, P.V., Mercer, R.L.: Speech Recognition with Continuous-Parameter Hidden Markov Models. Computer Speech and Language **2**(3-4), 219–234 (1987)
3. Bahl, L.R., Brown, P.F., deSouza, P.V., Mercer, R.L., Nahamoo, D.: A fast algorithm for deleted interpolation. In: European Conference on Speech Communication and Technology (Eurospeech), pp. 1209–1212 (1991)
4. Bahl, L.R., Jelinek, F., Mercer, R.L.: A Maximum Likelihood Approach to Continuous Speech Recognition. IEEE Transactions on Pattern Analysis and Machine Intelligence (PAMI) **5**(2), 179–190 (1983)
5. Baum, L.E., Petrie, T., Soules, G., Weiss, N.: A Maximization Technique Occurring on the Statistical Analysis of Probabilistic Functions of Markov Chains. The Annals of Mathematical Statistics **41**(1), 164–171 (1970)
6. Bellman, R.: Dynamic Programming, dover paperback edition edn. Princeton University Press, Princeton, New Jersey (2010). ISBN: 0-691-1466-83
7. Bengio, S., Keshet, J.: Introduction. Automatic Speech and Speaker Recognition. John Wiley and Sons, Ltd., West Sussex, UK (2009). ISBN: 978-0-470-69683-5
8. Bilmes, J.A.: A Gentle Tutorial of the EM Algorithm and its Applications to Parameter Estimation for Gaussian Mixture and Hidden Markov Models (1998). Technical Report Number TR-97-021
9. Boulard, H., Hermansky, H., Morgan, N.: Towards Increasing Speech Recognition Error Rates. Speech Communications **18**, 205–231 (1996)
10. Brown, P.F., Pietra, S.A.D., Pietra, V.J.D., Lai, J.C., Mercer, R.L.: An Estimate of an Upper Bound for the Entropy of English. Computational Linguistics **18**(1), 32–40 (1992)
11. Fortmann, T.E., Hitz, K.L.: An Introduction to Linear Control Systems. Marcel Dekker, Inc., New York (1977). ISBN: 0-824-76512-5

12. Gauvain, J.L., Lee, C.H.: Maximum a Posteriori Estimation for Multivariate Gaussian Mixture Observation of Markov Chains. IEEE Transactions on Speech and Audio Processing **2**(2), 291–298 (1994)

13. Karman, R.E.: A New Approach to Linear Filtering and Prediction Problems. Transactions of the ASME–Journal of Basic Engineering **82**(Series D), 35–45 (1960)

14. Kolmogorov, A.N.: Foundations of the Theory of Probability, 2nd english edn. The University Series in Higher Mathematics. Chelsea Publishing Company, New York (1956)

15. Li, C.J., Beigi, H.S., Li, S., Liang, J.: Nonlinear Piezo-Actuator Control by Learning Self-Tuning Regulator. ASME Transactions, Journal of Dynamic Systems, Measurement, and Control **115**(4), 720–723 (1993)

16. Manning, C.D.: Foundations of Statistical Natural Language Processing. The MIT Press, Boston (1999). ISBN: 0-262-13360-1

17. Markov, A.A.: Investigation of a Remarkable Case of Dependent Samples. pp. 61–80 (1907)

18. Markov, A.A.: On a Case of Samples Connected in Comples Chains. pp. 171–186 (1911)

19. Markov, A.A.: An Example of Statistical Investigation of the Text *Eugene Onegin* Concerning the Connection of Samples in Chains. In: Izd. Red. Upr. Svyazi RKKA (The First All-Union Conference on Questions of Communications), vol. 7, pp. 153–162 (1913). Lecture – English Translation: Classical Text in Translation, Science in Context, **19**(4), pp. 591–600 (2006)

20. McLachlan, G.J., Peel, D.: Finite Mixture Models, 2nd edn. Wiley Series in Probability and Statistics. John Wiley & Sons, New York (2000). ISBN: 0-471-00626-2

21. Papoulis, A.: Probability, Random Variables and Stochastic Processes, 3rd edn. McGraw-Hill, Inc., New York (1991). ISBN: 0-070-48477-5

22. Poritz, A.B.: Hidden Markov models: a guided tour. In: International Conference on Acoustics, Speech, and Signal Processing (ICASSP-1988), vol. 1, pp. 7–13 (1988)

23. Rabiner, L.R.: A tutorial on hidden Markov models and selected applications in speech recognition. Proceedings of the IEEE **77**(2), 257–286 (1989)

24. Ramussen, C.E., Williams, C.K.I.: Gaussian Processes for Machine Learning (Adaptive Computation and Machine Learning). The MIT Press, Boston (2006). ISBN: 978-026218253-9

25. Tjalkens, T.: State Dependent Coding: How to find the State? In: 43rd annual Allerton Conference on Communication, Control, and Computing (2005). URL http://cslgreenhouse.csl.illinois.edu/allertonarchives/allerton05/PDFs/Papers/V_A_2.pdf

26. Viterbi, A.J.: Error bounds for convolutional codes and an asymptotically optimum decoding algorithm. IEEE Transactions on Information Theory **13**(2), 260–269 (1967)

27. Xu, L., Jordan, M.I.: On Convergence Properties of the EM Algorithm for Gaussian Mixtures. Neural Computation **8**(1), 129–151 (1996)

28. Young, S.: A Review of Large-Vocabular Continuous Speech Recognition. IEEE Signal Processing Magazine **13**(5), 45–57 (1996)

Chapter 14
Neural Networks

> *The Brain is wider than the Sky*
> *For put them side by side*
> *The one the other will contain*
> *With ease and You beside*

Emilly Dickinson
The Brain [14] (No. 632), 1862

Neural network (*NN*) models have been studied for many years with the hope that the superior learning and recognition capability of the human brain could be emulated by man-made machines. Similar massive networks, in the human brain, make the complex pattern and speech recognition of humans possible. In contrast to the *Von Neumann computers* which compute sequentially, neural nets employ huge parallel networks of many densely interconnected computational elements called neurons. *Neural networks* have been used in many different applications such as *adaptive* and *learning control*, *pattern recognition*, *image processing*, *signature recognition*, *signal processing* and *speech recognition, finance*, etc.

A *neuron* is the most elementary computational unit in a neural network which sums a number of weighted inputs and passes the result through (generally) a *nonlinear activation function*. *Multi-layer neural networks* (Figure 14.2) consist of a large number of neurons. Before a neural network can be used for any purpose, the weights connecting inputs to neurons and the parameters of the activation functions of neurons should be adjusted so that outputs of the network will match desired values for specific sets of inputs. The methods used for adjusting these weights and parameters are usually referred to as learning algorithms.

The architectures of different neural networks being used for solving problems greatly vary. The design of the architecture is quite important and it depends on the nature of the problem. Several different types of neural networks have been tested for conducting Speaker Recognition [8, 40, 9, 11]. Different architectures and neuron types have been used such as the *time-delay neural network* (*TDNN*) architecture which captures the inherent dynamics of speech as a time-varying signal. *TDNNs* have successfully been combined with *hidden Markov models* (*HMM*) in speech recognition [35, 15]. *TDNNs* have also been applied to *speaker identification* with some success.

Most of the *speaker identification* trials have been done on extremely small sets with clean data and small populations of speakers [8]. Theoretically, the temporal features used in the *TDNNs* should possess some useful information compared to *Gaussian mixture model* (*GMM*) based systems, however, they have shown to be quite practical. Aside from architectural variations such as *TDNNs* [8] and *hierarchical mixtures of experts* (*HMEs*) [9], different neuron types such as the *probabilistic random access memory* (*pRAM*) have been used, again with small populations, so not much conclusive results are seen [12, 11].

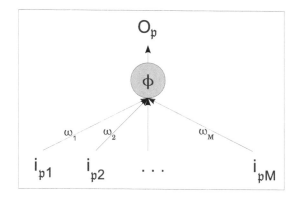

Fig. 14.1: A Perceptron

14.1 Perceptron

The concept of a *perceptron* was first introduced by Rosenblatt [30]. A perceptron is a linear neuron with an *activation function*, ϕ which depends on the input weights, ω_m, the input samples, i_{pm} and a threshold, θ, where p is the index of the presentation and m is the input number.

The activation function for a perceptron is given by the following relation,

$$\phi(\boldsymbol{\omega}, \mathbf{i}_p, \theta) = \begin{cases} 1 & \forall \quad \boldsymbol{\omega}^T \mathbf{i}_p + \theta > 0 \\ 0 & \forall \quad \boldsymbol{\omega}^T \mathbf{i}_p + \theta \leq 0 \end{cases} \tag{14.1}$$

14.2 Feedforward Networks

In 1986, Rumelhart et al. [31] introduced a generalized learning theory for multi-layer neural nets and to some extent demonstrated that a general learning algorithm

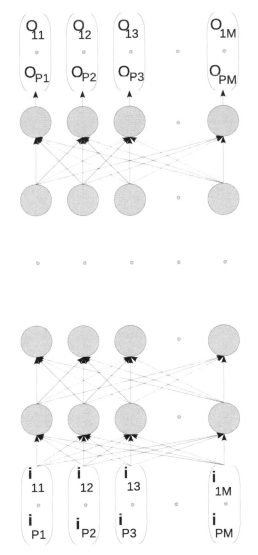

Fig. 14.2: Generic L-Layer Feedforward Neural Network

for multilayer Neural Networks with the task of learning arbitrary patterns is pos-sible. This was contradictory to earlier remarks in 1969 by *Minsky* and *Papert* [25] that such generalization was not possible. This started a flood of publications by the end of the 1980*s*, in re-addressing the learning problem.

Rumelhart et al [31], in their learning algorithm called the *back-propagation technique*, used the so called generalized delta rule to calculate an approximate gra-dient vector which is then used by a steepest descent search to minimize the difference between the *NN* output and the desired output. However, as they demonstrated by simulations, low rates of convergence were seen in practically every problem. *Lippman* [24] states, "One difficulty noted by the Back-Propagation algorithm is that in many cases the number of presentations of training data required for convergence has been large (more than 100 passes through all the training data.)"

In Speaker, Speech and Handwriting recognition, the one percent error margin of the Back-Propagation learning [31] is not always acceptable; a better convergence is desirable. In the literature, several methods [24, 27] have been proposed to increase the rate of convergence of learning by making strong assumptions such as linearity for multilayer networks. In addition, other more practical methods have recently been proposed for speeding the convergence of the Back-Propagation technique.[16, 38, 34, 26]

The Back-Propagation technique is a special case of the steepest descent technique with some additional assumptions. In general, steepest descent techniques perform well while away from local minima and require many iterations to converge when close to the minima. On the other hand, Newton's method usually converges fast in the vicinity of the minima. In addition, Newton's minimization technique handles functions with ill-conditioned Hessian matrices elegantly.[18] It would be desirable to take advantage of the properties of steepest descent when the state is far from minimal and then to use Newton's method in the vicinity of the minimum.

To use Newton's method, the first gradient and the matrix of second partial derivatives (Hessian) matrix should be evaluated. The moment one talks about evaluating the Hessian matrix, it becomes clear that a layer by layer adjustment of the weights is not possible because there are elements of the Hessian which are related to neurons in different layers. In general, two difficulties have prohibited the use of Newton's method for neural network learning: 1) the complexity of the evaluation of the Hessian, and 2) the inversion of the Hessian. One way to alleviate these difficulties is to use a momentum method which would approximate the diagonal elements of the Hessian matrix and would stay ignorant of the off-diagonal elements.[1]

On the other hand, Quasi-Newton methods provide another solution to the problem [6, 3, 5], by providing an iterative estimate for the inverse of the Hessian matrix. If one selects the initial estimate to be an identity matrix, initially, the method coincides with the steepest descent technique and gradually changes into Newton's method as the estimate approaches the inverse of the Hessian.

In addition, it is also possible to achieve quadratic convergence without any direct use of gradients[4]. These techniques are ideal for cases when the evaluation of forward pass is much less expensive than computing gradients. This would be the case when a hardware version of the feedforward network is available to compute the forward passes, but extra computation would have to be carried out to compute or estimate gradients.

In another approach, to simplify the training procedure and to be able to analyze it more rigorously, certain restrictions were placed on the architecture such as limiting the number of hidden layers to one. This analysis is what led to the development of a kernel method called *support vector machines* [37] which will be discussed in detail in Chapter 15.

14.2.1 Auto Associative Neural Networks (AANN)

Auto associative neural networks (AANN) are a special branch of *feedforward neural networks* which try to learn the *nonlinear principal components* of a *feature vector*. the way this is accomplished is that the network consists of three layers, an input layer, an output layer of the same size and a hidden layer with a smaller number of neurons. The input and output neurons generally have linear activation functions and the hidden (middle) layer has nonlinear functions.

In the training phase, the input and target output vectors are identical. This is done to allow for the system to learn the principal components that built the patterns which most likely have built-in redundancies. One such a network is trained, a feature vector undergoes a dimensional reduction and is then mapped back to the same dimensional space as the input space. If the training procedure is able to achieve a good reduction in the output error over the training samples and if the training samples are representative of the reality and span the operating conditions of the true system, the network can achieve learning the essential information in the input signal. Autoassociative networks have successfully been used in speaker verification.[23]

14.2.2 Radial Basis Function Neural Networks (RBFNN)

Radial basis function neural networks[1] are feedforward networks that utilize any class of *radial basis functions (RBF)* as their activation functions. In general, the output of a neuron would then be given by the following,

$$\phi(\boldsymbol{\omega}, \mathbf{i}_{\mathrm{p}}, \theta) = \boldsymbol{\omega}^T \boldsymbol{\varphi} + \theta \tag{14.2}$$

where $\boldsymbol{\varphi} : \varphi : \mathcal{R}^1 \mapsto \mathcal{R}^M$ is a vector of radial function values such that

---

[1] Sometimes these networks are simply called *radial basis function networks (RBFN)*, leaving out the *neural* identifier.

$$(\boldsymbol{\varphi})_{[m]} = \Phi_m(\xi_{\text{p}m}) \; \forall \, m \in \{1,2,\cdots,M\} \tag{14.3}$$

and

$$\xi_{\text{p}m} \overset{\Delta}{=} \|\mathbf{i}_{\text{p}} - \boldsymbol{\mu}_m\| \tag{14.4}$$

where the $\boldsymbol{\mu}_m$ turn out to be means of the domains of radial basis functions, $\Phi_m(\xi_{\text{p}m}) \; \forall \, m \in \{1,2,\cdots,M\}, \text{p} = \{1,2,\cdots,P\}$. Although any norm may be used in Equation 14.4, most applications of the theory use the *Euclidean norm*. Radial basis functions are functions which have been used for performing multivariate interpolation [28] and are designed so that they would fit all the points in the training data. In most pattern recognition applications, the Gaussian radial basis function, give by the following equation is used.

$$\Phi(\xi_m) = \exp\left\{-\frac{\xi_m^2}{2*\sigma_m^2}\right\} \tag{14.5}$$

where, σ_m is the standard deviation of the data, associated with the training data. Although, there are many other types of RBFs which may be utilized[33]. *Kiernan et al.* [21] show that if a *Gaussian radial basis function (GRBF)* is used, then the initial means and variances of the data may be estimated using an unsupervised clustering technique such as *k*-means (see Section 11.2.1). Using these initial parameters will also help reduce the number of neurons necessary to achieve the same results [32].

14.2.3 Training (Learning) Formulation[2]

The objective of learning, in the problem of learning in neural networks, is to minimize the output error of the top (output) layer (layer L in Figure 14.2) of a neural network over a set of P input-output patterns.

Let us define the following variables,

$l \in [1,L]$ (Layer number in the network)
$n_l \in [1,N_l]$ (Neuron number in layer l)
$\text{p} \in [1,P]$ (Pattern number)
ω_{nm}^l (Weighting factor between the mth input and neuron n in layer l)
$o_{\text{p}n}^l$ (Output of neuron n of layer l for input pattern p)
$t_{\text{p}n}$ (Desired output of neuron n in layer L)
$i_{\text{p}m}$ (Input m of pattern p to the network)

Then, the objective of learning becomes,

---

[2] The learning formulation discussed in this section is drawn from an earlier publication by the author [7].

$$minimizeE = \sum_{p-1}^{P} \sum_{n_L-1}^{N_L} \left(o_{pn_L}^L - t_{pn_L} \right)^2 \tag{14.6}$$

Define,

$$E_p \overset{\Delta}{=} \sum_{n_L=1}^{N_L} \left(o_{pn_L}^L - t_{pn_L} \right)^2 \tag{14.7}$$

so that,

$$E = \sum_{p=1}^{P} E_p \tag{14.8}$$

Next, we will define a few variables and eventually a state vector which would include all the variables to be optimized for minimum E:

$$\boldsymbol{\phi}^l = \left[\phi_1^l, \phi_2^l, \cdots, \phi_{N_l}^l \right]^T \qquad \text{(Activation function parameter vector for level } l\text{)}$$

$$\boldsymbol{\omega}_{n_l}^l = \left[\omega_{n_l 1}^l, \omega_{n_l 2}^l, \cdots, \omega_{n_l N_{l-1}}^l \right]^T \qquad \text{(Vector of intercellular weights to neuron } n_l \text{ at layer } l\text{)}$$

$$\boldsymbol{\omega}^l = \left[\boldsymbol{\omega}_1^{l\,T}, \boldsymbol{\omega}_2^{l\,T}, \cdots, \boldsymbol{\omega}_{N_l}^{l\,T} \right]^T \qquad \text{(Supervector of intercellular weights of level } l\text{)}$$

$$\mathbf{x}^l = \left[\boldsymbol{\phi}^{l\,T}, \boldsymbol{\omega}^{l\,T} \right]^T \qquad \text{(State vector for level } l\text{)}$$

$$\mathbf{x} = \left[\mathbf{x}^{1\,T}, \mathbf{x}^{2\,T}, \cdots, \mathbf{x}^{L\,T} \right]^T \qquad \text{(Super state vector)}$$

Let $(\mathbf{x})_{[j]}$ denote any element of the state vector, $\mathbf{x}$. Then,

$$\frac{\partial E}{\partial (\mathbf{x})_{[j]}} = \sum_{p=1}^{P} \frac{\partial E_p}{\partial (\mathbf{x})_{[j]}} \tag{14.9}$$

and by applying the *chain rule*,

$$\frac{\partial E_p}{\partial (\mathbf{x})_{[j]}} = 2 \sum_{n_L=1}^{N_L} \left(o_{pn_l}^L - t_{pn_L} \right) \frac{\partial o_{pn_L}^L}{\partial (\mathbf{x})_{[j]}} \tag{14.10}$$

where,

$$\frac{\partial o_{pn_L}^L}{\partial \phi_{n_l}^l} = \frac{\partial o_{pn_L}^L}{o_{pn_l}^l} \frac{o_{pn_l}^l}{\partial \phi_{n_l}^l} \tag{14.11}$$

$$\frac{\partial o_{pn_L}^L}{\partial \omega_{n_l n_{l-1}}^l} = \frac{\partial o_{pn_L}^L}{o_{pn_l}^l} \frac{o_{pn_l}^l}{\partial \omega_{n_l n_{l-1}}^l} \tag{14.12}$$

and

$$\frac{\partial o_{\mathrm{p}n_L}^L}{o_{\mathrm{p}n_l}^l} = \prod_{k=l+1}^{L} \frac{\partial o_{\mathrm{p}n(L+l-k+1)}^{(L+l-k+1)}}{\partial o_{\mathrm{p}n(L+l-k)}^{(L+l-k)}} \quad \text{(Using \textit{index notation})} \tag{14.13}$$

In Equation 14.13, the *index notation* [36], which is customary in *tensor algebra*, has been employed. The following Equation shows the expansion of an statement in *index notation* once written in the *traditional notation*,

$$\underbrace{\frac{\partial o_{\mathrm{p}n(l+1)}^{l+1}}{\partial o_{\mathrm{p}n_l}^l} \frac{\partial o_{\mathrm{p}n_l}^l}{\partial o_{\mathrm{p}n(l-1)}^{l-1}}}_{\text{Index Notation}} = \underbrace{\sum_{n_l=1}^{N_l} \frac{\partial o_{\mathrm{p}n(l+1)}^{l+1}}{\partial o_{\mathrm{p}n_l}^l} \frac{\partial o_{\mathrm{p}n_l}^l}{\partial o_{\mathrm{p}n(l-1)}^{l-1}}}_{\text{Traditional Notation}} \tag{14.14}$$

Take, for example, the *logistic function*,

$$o_{\mathrm{p}n_l}^l = \frac{1}{1 + e^{-s_{\mathrm{p}n_l}^l}} \tag{14.15}$$

where,

$$s_{\mathrm{p}n_l}^l \overset{\Delta}{=} \sum_{n_{(l-1)}=1}^{N_{(l-1)}} \left(o_{\mathrm{p}n(l-1)}^{(l-1)} \omega_{n_l n(l-1)}^l \right) + \phi_{n_l}^l \tag{14.16}$$

We may compute the elements of the gradient of the output at layer l about the state vector and the output of the previous layer by the following three equations,

$$\frac{\partial o_{\mathrm{p}n_l}^l}{\partial \phi_{n_l}^l} = d_{\mathrm{p}n_l}^l \tag{14.17}$$

$$\frac{\partial o_{\mathrm{p}n_l}^l}{\partial \omega_{n_l n(l-1)}^l} = d_{\mathrm{p}n_l}^l o_{\mathrm{p}n(l-1)}^{(l-1)} \tag{14.18}$$

$$\frac{\partial o_{\mathrm{p}n_l}^l}{\partial o_{\mathrm{p}n(l-1)}^{(l-1)}} = d_{\mathrm{p}n_l}^l \omega_{n_l n(l-1)}^l \tag{14.19}$$

where,

$$d_{\mathrm{p}n_l}^l \overset{\Delta}{=} \frac{e^{-s_{\mathrm{p}n_l}^l}}{\left(1 + e^{-s_{\mathrm{p}n_l}^l}\right)^2} \tag{14.20}$$

However,

$$o_{\mathrm{p}n_l}^l = \frac{1}{1 + e^{-s_{\mathrm{p}n_l}^l}} \tag{14.21}$$

Therefore, we may rearrange Equation 14.21 as follows,

$$e^{-s_{pn_l}^l} = \frac{1}{o_{pn_l}^l} - 1 \tag{14.22}$$

Using the identity in Equation 14.22, we may rewrite Equation 14.20 in terms of the output of the corresponding unit as follows,

$$d_{pn_l}^l = o_{pn_l}^l (1 - o_{pn_l}^l) \tag{14.23}$$

Using Equation 14.23, we may rewrite Equations 14.17, 14.18, and 14.19 in terms of o_{pn_l},

$$\frac{\partial o_{pn_l}^l}{\partial \phi_{n_l}^l} = o_{pn_l}^l (1 - o_{pn_l}^l) \tag{14.24}$$

$$\frac{\partial o_{pn_l}^l}{\partial \omega_{n_l n_{(l-1)}}^l} = o_{pn_{(l-1)}}^{(l-1)} o_{pn_l}^l (1 - o_{pn_l}^l) \tag{14.25}$$

$$\frac{\partial o_{pn_l}^l}{\partial o_{pn_{(l-1)}}^{(l-1)}} = \omega_{n_l n_{(l-1)}}^l o_{pn_l}^l (1 - o_{pn_l}^l) \tag{14.26}$$

Equations 14.24- 14.26 and 14.9-14.13 may be used to evaluate the elements of the gradient of E with respect to supervector $\mathbf{x}$, denoted by $\nabla_{\mathbf{x}}E$. If we defined the *gradient* and the *Hessian* of E with respect to $\mathbf{x}$ as follows,

$$\mathbf{g} \stackrel{\Delta}{=} \nabla_{\mathbf{x}}E = \sum_{p=1}^{P} \nabla_{\mathbf{x}}E_p \quad \text{(Gradient Vector)} \tag{14.27}$$

$$\mathbf{G} \stackrel{\Delta}{=} \nabla_{\mathbf{x}}^2 E = \sum_{p=1}^{P} \nabla_{\mathbf{x}}^2 E_p \quad \text{(Hessian Matrix)} \tag{14.28}$$

then we may use any of the optimization techniques discusses in Chapter 25 to learn the elements in $\mathbf{x}$ that would minimize Equation 14.8. See Problem 14.1 for an example regarding the computation of the gradient of the minimization objective function of a feedforward neural network.

14.2.4 Optimization Problem

In Section 14.2.3, we formulated the learning problem of a generic feedforward network as a minimization problem. The expressions for the *gradient* vector and the *Hessian* matrix of the minimization objective function were derived for any feedforward network. Quadratically convergent optimization techniques have been applied to the problem of learning in neural networks to create learning techniques with superior speeds and precision compared to previously used first order methods such as the backpropagation technique. These methods have shown many orders of magnitude faster convergences and higher final accuracies [6, 3, 5, 4].

At their core, most of these methods basically use some kind of *conjugate gradient* or *quasi Newton* technique (see Chapter 25). To ensure convergence, most quadratic learning techniques discussed in the literature require some sort of *exact* or *inexact line search* to be conducted. For each step in a line search, usually a set of patterns should be presented to the network. One measure of the speed of learning is the number of pattern presentations to the network to attain proper learning state.

Learning algorithms based on *conjugate gradient* techniques and *quasi Newton* Techniques such as *Broyden, Davidon-Fletcher-Powell* (*DFP*) and *projected Newton-Raphson* (Chapter 25) require *exact line searches* to be done in order to satisfy theoretical convergence criteria and also to practically converge in many cases. [6]

Broyden-Fletcher-Goldfarb-Shanno (*BFGS*) update, *Pearson's* two   Exact line searches are very costly and slow down the learning process. Some quasi Newton techniques get away with inexact line searches. Among these methods we could name quasi Newton techniques based on the updates, *Greenstadt's* two updates and *self-scaling variable metric* (*SSVM*) updates [6, 3, 5] – also see Chapter 25. Some of these techniques such as *SSVM* and *BFGS* meet their promise of *inexact line searches* better than others. However, none of the above methods produce learning algorithms which do not require line searches.

Hoshino presented a *variable metric* (*quasi Newton*) method which theoretically requires only *inexact line searches* to be done [20]. *Davidon* on the other hand introduced a quasi Newton method which does not require any line search to be done most of the time [13]. *Davidon's* algorithm only needs a very crude line search to be done in some special cases. The theory behind these two methods is discussed, including some practical information, in Chapter 25.

Neural network related objective functions normally possess many local minima by nature and getting trapped in these local minima usually makes the job of learning much more difficult. In Section 14.2.5, we will describe a technique for providing a better chance of arriving at a global minimum rather than being trapped in local minima, through restructuring the network. In addition to this technique, there are other methods such as *simulated annealing* which are discussed in Chapter 25.

14.2.5 Global Solution

Local convergence has always been a major problem in *neural network learning algorithms*. In most cases, to avoid this, in practical applications, different initial conditions are used until one case would converge to the desirable *global minimum*. This is, however, not very practical and some supervision is necessary. Also, due to the high degrees of nonlinearity of neural network objective functions, picking *optimal initial conditions* is next to impossible.

Classically, when training a neural network, the architecture of the network is fixed to some intuitively sound architecture with certain number of hidden units. Then, a *learning algorithm* is used with different initial conditions until convergence to a near global minimum is achieved. Some researchers have also used global optimization techniques such as the statistical method of *simulated annealing* [22] (see Section 25.6). *Simulated annealing*, however, is known, due to its statistical nature, to be very costly in terms of number of presentations needed for convergence.

In *line search free learning (LSFL)* algorithm [2], a method of restructuring the network in an adaptive fashion is used to achieve better global convergence. In this method the network starts out with one hidden neuron. Then, the *LSFL* algorithm is applied to the network until a *local minimum* is obtained. Once a *local minimum* is reached, another neuron is added to the hidden layer and the weights of corresponding to this new neuron are initialized to some random values. However, weights and activation function parameters corresponding to the original hidden neuron are not changed. Therefore, for the original hidden unit, the final state of the previous run is the new initial condition.

The idea behind this theory is that a new sets of dimensions are introduced by the introduction of this new neuron and the algorithm can once again start moving. Here, the assumption is made that the weights of the original neuron do not need to be changed much. Philosophically, in this way, the problem is being broken up into subproblems and each hidden neuron handles a part of the problem.

This methodology is then applied each time a new local minimum is reached until there is no longer any improvement with the addition of new hidden neurons. At this point, it is assumed that either the algorithm has reached a global minimum, or with the given set of initial conditions, there is no further improvement possible. In their examples, [6, 3, 5] have noted many of cases when the *global minimum* was not reached given a fixed set of hidden units. Most of the cases discussed in those references have reached a *global minimum* by applying this new neuron addition technique [2]. In addition, this method of finding the minimum reduces the amount of overall computation, since at the beginning, a smaller network is used to do part of the learning.

14.3 Recurrent Neural Networks (RNN)

In trying to model a dynamic system, recurrent neural networks (RNN) use a feedback look which uses the output of a neuron or a collection of neurons as input to the same neuron(s). The feedback allows for better modeling of dynamic systems with feedback inherently present in their nature. Take control system for example. Recurrent neural networks do a much better modeling of these systems. The neurons try to model the system while the feedback is used to model the inherent feedback in such systems. One can imagine that as stability is an issue a control system, introducing feedback could cause stability issues in the networks. In general, a recurrent network is much smaller than a feedforward network since the complexity is modeled by the feedback.

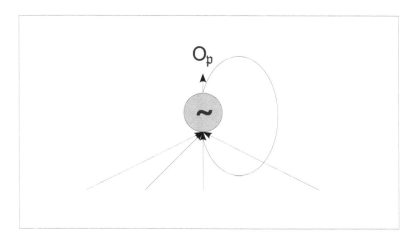

Fig. 14.3: Generic Recurrent Neuron

If the recurrent network is built using a physical system such as a circuit, there will be inherent delay involved when the output is fed back to act as the input to the system. Stability requirements sometimes dictate the introduction of an additional delay between the output of the network and the input signal.

RNNs bring something to the table which is missing in most other learning techniques introduced in this book. They include a continuous internal state. *hidden Markov models* (Chapter 13) have internal states, and they may be modeled by continuous parameters, however, the states themselves are not continuous. Feedforward neural networks (Section 14.2) and *support vector machines* (Chapter 15) do not even possess any internal states. Early versions of RNNs only looked into the past for a limited time and therefore were not able to handle patterns which extended for a long time, especially in the presence of noise; also, they had displayed some shortcomings in modeling complex periodic trajectories, for example, required in speech processing [29]. This is due to the fact that traditional RNNs are cursed with the problem of *vanishing gradient* which means that they are not able to keep long-term information. In the next section, we will briefly look at a more recent design that has provisions that would allow it to handle short-term as well as long-term information.

The *long short-term memory* (*LSTM*) [19] *RNN* changed that and allowed for handling these shortcomings. Other research followed which used the *LSTM* in problems such as *language modeling* [17].

14.4 Time-Delay Neural Networks (TDNNs)

Time-delay neural networks (*TDNN*) have been used as the predominant form of neural network architecture for the purpose of speech [35] and speaker recognition [41, 9, 10]. They were introduced by *Waibel et al.* [39] in 1989 and were designed to handle the frame-based analysis of speech, which was discussed in detail in Chapter 5. Figure 14.4 shows a *TDNN* architecture.

The *TDNN* borrows some ideas from recurrent neural networks. In fact, it is similar to the unfolding snapshots of a recurrent network through its state transitions in time. It is usually similar to a feedforward network in each instance. The basic difference between a unit of *TDNN* and a unit of a *feedforward neural network* (*FFNN*)is as follows. In the *FFNN*, each input is multiplied by a weight, then the weighted inputs are summed up before they enter the nonlinearity of the neuron whose output is the output of the unit. In a *TDNN* unit, each input goes through $N+1$ delays, $\{d_0, d_1, \cdots, d_N\}$. $d_0 = 0$, which means that the 0 index uses the input as it is. For a constant delay, this may be written as,

$$d_n = nd \tag{14.29}$$

where d is the increment in delay and n is the index. d is usually translates into the delay of one frame of audio, which is 10ms in most implementations. Each delayed input is treated just like a separate input in a feedforward unit. Namely, each delayed input is multiplied by a weight, the results from the $N+1$ delayed versions of an input are added, together with other delayed inputs coming from other inputs, totaling $M(N+1)$, where M is the number of actual inputs to the unit (for the first layer, it is equal to the dimension of the cepstral feature vector, D – see Section 5.3.6). The $M(N+1)$ weighted inputs are then summed and passed through the nonlinearity to produce the output of the *TDNN* unit. In out analogy with the *recurrent NN*, we can see that it is similar in design to the old versions of *recurrent network* when certain number of delays were used to unfold the *RNN*.

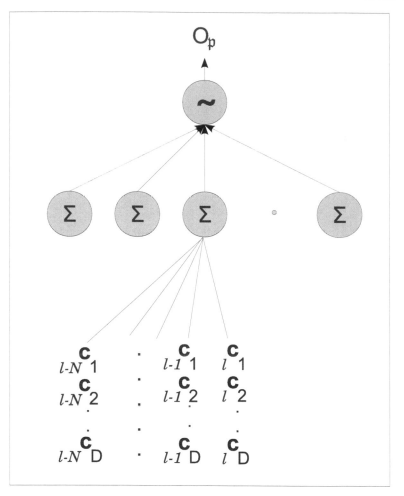

Fig. 14.4: Generic Time-Delay Neural Network

The implementation described by [39] tries to discriminate among the three phones, /b/, /d/ and /g/. It uses a 16 dimensional MFCC vector as the raw features for each frame. These 16 features are passed through the delay structure with $N = 2$. This means that there will be $16 \times 3 = 48$ inputs to the neuron associated with each frame of the first layer. 16 of them are the MFCC features of the frame itself and the rest ($2 \times 16 = 32$ features) are MFCC features associated with the two previous frames. Essentially, this process is replicating the overlapping frame process discussed in Chapter 5. This means that each unit processes a frame of 30ms worth of data and that there is one unite for each 10ms interval.

The outputs of the units at the first layer are fed to the middle layer for which $M = 8$ and $N = 5$. This means that in the first hidden layer, there is a wider perspective looking at more temporal data. There is a second hidden layer which feeds on the output of the this layer and uses $M = 3$ and $N = 2$ which then outputs its results into a final layer which has three possible outputs.

One may use this *TDNN* classifier to conduct *speaker* and *speech recognition* somewhat in the same manner as other standard techniques. In practice, most systems for speech recognition, combine the *TDNN* with a basic *HMM* model [15].

14.5 Hierarchical Mixtures of Experts (HME)

[9] (also see errata [10]) uses an *HME* architecture and trains the parameters of the network by using the *expectation maximization* (*EM*) (Section 11.3.1) algorithm for doing *text-independent speaker identification*. This suggests that the statistical model is treated as a *maximum likelihood* (*ml*) problem. The application of this technique is only presented for a small problem dealing with 10 male speakers and only isolated digits. This is in tune with the fact that the complexity of neural network systems increases exponentially and the training becomes impractical, at times. Section 14.4 discusses a technique which uses the dynamics of speech hence increasing the complexity even more.

14.6 Practical Issues

Neural networks usually work pretty well in learning most complex and nonlinear training data. However, they have a great tendency to *overtrain*[3]. The *overtraining* issue is also seen in the discussion of the construction of decision trees in Section 9.4 and the derivation and formulation of *support vector machines* (*SVM*) in Chapter 15. In general, with most learning mechanisms that have a flexible mapping, capable of learning great detail, the problem of overtraining exists and is a serious problem. It is up to the designer to make sure the great ability of these systems to learn complex functions is not misused. This problem is especially true when the training sample is small. Normally, the training method is quite instrumental in the generalization capability of such systems. Neural networks, in general, do not determine crisp class boundaries.

---

[3] Sometimes it is referred to as *overfitting*.

In contrast, there are other techniques that do, such as *support vector machines* (*SVM*) which are the topic of discussion of chapter 15. Neural networks provide smooth functions that change along the *state-space*. This is basically an attributing factor related to the *overtraining* issue.

Of courses these are general statements and there are certainly countless methods that have been used to increase the generalization capabilities of neural networks and to reduce their overtraining effects. Examples of these methods are systems which combination neural networks with other classifiers such as *hidden Markov models* (Chapter 13) as they have been done in many *time delay neural network* (*TDNN*) implementations for speech-related problems [39].

Problems

For solutions to the following problems, see the Solutions section at the end of the book.

Problem 14.1 (Exclusive OR).

Figure 14.5 shows the architecture of a feedforward neural network used to produce an exclusive OR logic. Table 14.1 shows the input/output relationship for a two-input exclusive OR unit for the four possible combinations of patterns. Write the expression for the objective function in terms of true output and the expected output of the system. Also, write the expressions for the state vector and the gradient of the objective function with respect to the state vector.

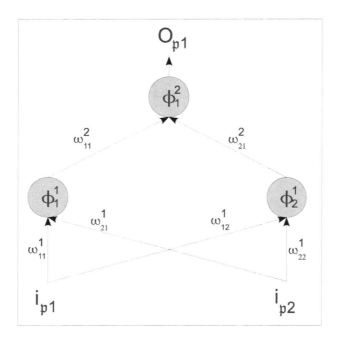

Fig. 14.5: Feedforward neural network architecture used for the exclusive OR (XOR) logic

References

1. Becker, S., Cunn, Y.L.: Improving the Convergence of Back-Propagation Learning with Second Order Methods. In: 1988 Connectionist Models Summer School, pp. 29–37 (1988)

p	i_{p1}	i_{p2}	t_{p1}
1	0	0	0
2	1	0	1
3	0	1	1
4	1	1	0

Table 14.1: XOR input/output for Problem 14.1

2. Beigi, H.: Neural Network Learning Through Optimally Conditioned Quadratically Convergent Methods Requiring NO LINE SEARCH. In: IEEE-36th Midwest Symposium on Circuits and Systems, vol. 1, pp. 109–112 (1993)
3. Beigi, H.S., Li, C.J.: Neural Network Learning Based on Quasi-Newton Methods with Initial Scaling of Inverse Hessian Approximate. In: The IEEE 1990 Long Island Student Conference on Neural Networks, pp. 49–52 (1990). Recipient of Best Paper Award
4. Beigi, H.S., Li, C.J.: New Neural Network Learning Based on Gradient-Free Optimization Methods. In: The IEEE 1990 Long Island Student Conference on Neural Networks, pp. 9–12 (1990). Recipient of Best Paper Award
5. Beigi, H.S., Li, C.J.: Learning Algorithms for Neural Networks Based on Quasi-Newton Methods with Self-Scaling. ASME Transactions, Journal of Dynamic Systems, Measurement, and Control **115**(1), 38–43 (1993)
6. Beigi, H.S., Li, C.J.: Learning Algorithms for Feedforward Neural Networks Based on Classical and Initial-Scaling Quasi-Newton Methods. ISMM Journal of Microcomputer Applications **14**(2), 41–52 (1995)
7. Beigi, H.S.M.: Neural Network Learning and Learning Control Through Optimization Techniques. Columbia University, New York (1991). Doctoral Thesis: School of Engineering and Applied Science
8. Berger, G., Gowdy, J.a.: TDNN based speaker identification. In: System Theory, 1993. Proceedings SSST '93., Twenty-Fifth Southeastern Symposium on, pp. 396–399 (1993)
9. Chen, K., Xie, D., Chi, H.: A modified HME architecture for text-dependent speaker identification. IEEE Transactions on Neural Networks **7**(5), 1309–1313 (1996)
10. Chen, K., Xie, D., Chi, H.: Correction to A modified HME architecture for text-dependent speaker identification. IEEE Transactions on Neural Networks **8**(2), 455–455 (1997)
11. Clarkson, T., Christodoulou, C., Guan, Y., Gorse, D., Romano-Critchley, D.A., Taylor, J.: Speaker identification for security systems using reinforcement-trained pRAM neural network architectures. IEEE Transactions on Systems, Man and Cybernetics **31**(1), 65–76 (2001)
12. Clarkson, T., Gorse, D., Taylor, J.: Hardware realisable models of neural processing. In: First IEE International Conference on Neural Networks, pp. 242–246 (1989)
13. Davidon, W.C.: Optimally Conditioned Optimization Algorithms Without Line Searches. Mathematical Programming **9**, 1–30 (1975)
14. Dickinson, E.: The Complete Poems. Back Bay Books, Boston, MA (1976). The Brain: Poem number 632
15. Dugast, C., Devillers, L., Aubert, X.: Combining TDNN and HMM in a hybrid system for improved continuous-speech recognition. IEEE Transactions on Speech and Audio Processing **2**(1), 217–223 (1994)
16. Fahlman, S.E.: Faster-Learning Variations on Back-Propagation: An Empirical Study. In: 1988 Connectionist Models Summer School, pp. 38–51 (1988)
17. Gers, F.A., Schmidhuber, J.: LSTN Recurrent Networks Learn Simple Context-Free and Context-Sensitive Languages. IEEE Transactions on Neural Networks **12**(6), 1333–1340 (2001)
18. Himmelblau, D.M.: Applied Nonlinear Programming. McGraw-Hill Book Company, New York (1972)
19. Hochreiter, S., Schmidhuber, J.: Long Short-Term Memory. Neural Computing **8**, 1735–1780 (1997)

20. Hoshino, S.: A Formulation of Variable Metric Methods. Journal of the Institue of Mathemtics and its Applicatiions **10**, 394–403 (1972)
21. Kiernan, L., Mason, J., Warwick, K.: Robust Initialisation of Gaussian Radial Basis Function Networks using Partitioned k-Means Clustering. IET Electronics Letters **32**(7), 671–673 (1996)
22. Kirkpatrick, S., Gelatt, C.D., Vecchi, M.P.: Optimization by Simulated Annealing. Science **220**(4598), 671–680 (1983)
23. Kishore, S., Yegnanarayana, B.a.: Speaker verification: minimizing the channel effects using autoassociative neural network models. In: Acoustics, Speech, and Signal Processing, 2000. ICASSP '00. Proceedings. 2000 IEEE International Conference on, vol. 2, pp. II1101–II1104 (2000)
24. Lippmann, R.P.: An Introduction to Computing with Neural Nets. The IEEE Acoustic, Speech, and Signal Processing Magazine **4**(2), 4–22 (1987). Part 1
25. Minsky, M., Papert, S.: Perceptrons: An Introduction to Computation Geometry. Massachusetts Institute of Technology, Cambridge, MA (1969)
26. Owens, A.J., Filkin, D.L.: Efficient Training of the Back Propagation Network by Solving a System of Stiff Ordinary Differential Equations. In: IEEE/INNS International Conference on Neural Networks (1989)
27. Parker, D.B.: A Comparison of Algorithms for Neuron-Like Cells. In: Neural Networks for Computing, AIP conference Proceeding 151, pp. 327–332 (1986)
28. Powell, M.: Radial Basis Functions for Multivariablc Intcrpolation: a Review. In: Algorithms for Approximation. Clarendon Press, New York (1987)
29. Robinson, A., Fallside, F.: A Recurrent Error Propation Speech Recognition System. Computer Speech and Language **5**, 259–274 (1991)
30. Rosenblatt, F.: Principles of Neurodynamics: Perceptrons and the Theory of Brain Mechanisms. Spartan Books, Washington, D.C. (1962)
31. Rumelhart, D.E., Hinton, G.E., Williams, R.J.: Learning Internal Representations by Error Propagation. In: D.E. Rumelhart, J.L. McClelland (eds.) Parallel Distributed Processing: Explorations in the Microstructure of Cognition, vol. 1, pp. 675–695. MIT Press (1986). ISBN: 0-262-63112-1
32. Sahin, F.: A Radial Basis Function Approach to a Color Image Classificatioin Problem in a Real Time Industrial Application. Virginia Tech University, Blacksburg, Virginia (1997). URL http://scholar.lib.vt.edu/theses/available/etd-6197-223641. Masters Thesis
33. Sheela, K.A., Prasad, K.S.: Linear Discriminant Analysis F-Ratio for Optimization of TESPAR & MFCC Features for Speaker Recognition. Journal of Multimedia **2**(6), 34–43 (2007)
34. Solla, S.A., Levin, E., Fleisher, M.: Accelerated Learning in Layered Neural Networks. Complex Systems **2**, 625–640 (1988)
35. Sugiyama, M., Sawai, H., Waibel, A.: Review of TDNN (time delay neural network) architectures for speech recognition. In: IEEE International Symposium on Circuits and Systems, vol. 1, pp. 582–585 (1991)
36. Timoshenko, S., Goodier, J.: Theory of Elasticity. McGraw-Hill Book Company, Inc., New York (1951)
37. Vapnik, V.N.: Estimation of Dependences Based on Empirical Data, russian edn. Nauka, Moscow (1979). English Translation: Springer-Verlag, New York, 1982
38. Vogl, T.P., Mangis, J.K., Rigler, A.K., Zink, W.T., Alkon, D.L.: Accelerating the Convergence of the Back-Propagation Method. Biological Cybernetics **59**, 257–263 (1988)
39. Waibel, A., Hanazawa, T., Hinton, G., Shikano, K., Lang, K.: Phoneme Recognition Using Time-Delay Neural Networks. IEEE Transactions on Acoustics, Speech and Signal Processing **37**(3), 328–339 (1989)
40. Wang, L., Chen, K., Chi, H.: Capture interspeaker information with a neural network for speaker identification. IEEE Transactions on Neural Networks **13**(2), 436–445 (2002)
41. Wang, X.: Text-Dependent speaker verification using recurrent time delay neural networks for feature extraction. In: IEEE Signal Processing Workshop – Neural Netrowks for Signal Processing – III, pp. 353–361 (1993)

Chapter 15
Support Vector Machines

Nature has neither kernel, Nor shell
She is everything, All at once!

Johann Wolfgang von Goethe
Morphologie, 1820

In Section 24.4 we defined a *kernel function*, $\mathscr{K}(s,t)$ – see Definition 24.56. Recently, quite a lot of attention has been given to *kernel methods* for their inherent discriminative abilities and the capability of handling nonlinear decision boundaries with good discrimination scalability, in relation with increasing dimensions of observations vectors. Of course using kernel techniques is nothing new. Integral transforms, for example, are some of the oldest techniques which use kernels to be able to transform a problem from one space to another space which would be more suitable for a solution. Eventually, the solution is transformed back to the original space. Chapter 24 is devoted to the details of such techniques and we have already used different transforms in other parts of the book, especially in doing feature extraction. It is highly recommended to the reader to review Chapter 24 entirely. However, if that is not possible, at least Sections 24.4 and 24.5 should be read prior to continuing with this chapter.

Among kernel methods used for pattern recognition, some of the best scalability has been demonstrated by *support vector machines (SVM)*. Although, other kernel techniques [1, 29] have also been used along the same mind-set. In this chapter, we will discuss *SVM* because of their usage, in the past decade, by the speaker recognition community.

The claim-to-fame of *support vector machines* is that they determine the boundaries of classes based on the training data, and they have the capability of *maximizing the margin* of class separability in the, so called, *feature space* (see Section 15.1). *Boser et al.* [7] state that the number of parameters used in an SVM is automatically computed (see *VC-Dimension* below) to present a solution in terms of a linear combination of a subset of observed (training) vectors, which are located closest to the decision boundary. These vectors are called *support vectors*, hence the model is known as a *support vector machine*.

This idea follows the basic premise that if a learning machine includes a large number of parameters, then it would, most likely, be able to learn a small number of patterns without error. However, it would suffer from *overtraining* (Section 14.6), which will reduce its capability to generalize to new and unseen patterns, seen after the training has been completed. On the other hand, reducing the number of parameters in the model will provide a smoother model, but it will possess a higher error rate for the training data. Balancing between these two goals is an important objective of learning algorithms. There have been different techniques used to reach this balance. For example, Beigi [5] starts with a small number of hidden neurons in a neural network classifier and increases this number to achieve such a goal.

Vapnik [45] pioneered the statistical learning theory of *SVM* to minimize the classification error of both the training data and some unknown (held-out) data. Such a classifier would, while being capable of learning new arbitrary training data, still retain good performance on the main training data based on which it was constructed [8]. The capability of learning new unseen training data may be thought of as a form of *learning capacity* for the classifier of choice. As we will see later, this learning capacity has been quantified by a *non-negative integer* and is dubbed the *Vapnik-Chervonenkis (VC) dimension* [46, 8]. More will be said about this dimension, as we delve into the formulation of *SVM*.

Of course, as we pointed out earlier, the core of support vector machines and other *kernel* techniques stems from much earlier work on setting up and solving *integral equations*. Hilbert [26] was one of the main developers of the formulation of *integral equations*. In Chapter 24, we categorize *integral equations* into three different kinds. Support vector machines are concerned with the *linear integral equation* of the *second kind*. In his 1904 book [26], *David Hilbert* used the German word *kern* for $\mathscr{K}(s,t)$ (see Definition 24.54), when defining a *linear integral equation* of the *second kind*.[1] This word has been translated into the English word, *kernel*, which is where *kernel methods* get their name.

SVM techniques have been used for the past two decades on many different pattern recognition applications including *handwriting recognition* [25, 2], *signature verification* [24], *fault detection* in analog electronics [31], *anomaly detection* [49], (Penicillin) *fermentation process modeling* [23], *digital libraries* [47], *image recognition* (e.g. *cloud detection* in satellite images [30]), *speech recognition* [22, 21], *speaker recognition* [19, 9], and so on.

The group at Bell Laboratories under the management of *Larry Jackel*, in the early 1990*s*, including *Vladimir Vapnik*, *Isabelle Guyon*, and others did most of the initial practical development and implementation of SVM, in the context of *handwriting recognition*. Their main focus was the use of *neural networks* (NN) in this arena. However, due to the complexity of NN structures and to be able to analyze such networks mathematically, they looked at simple structures possessing single

---

[1] See the bottom of page IX in the original 1904 print version of the book.

hidden layers.

In general, *SVM* are formulated as *two-class* classifiers. Γ-class classification problems are usually reduced to Γ two-class problems [46], where the γth two-class problem compares the γth class with the rest of the classes combined. There are also other generalizations of the SVM formulation which are geared toward handling Γ-class problems directly. *Vapnik* has proposed such formulations in Section 10.10 of his book [46]. He also credits *M. Jaakkola* and *C. Watkins, et al.* for having proposed similar generalizations independently. For such generalizations, the constrained optimization problem becomes much more complex. For this reason, the approximation using a set of Γ two-class problems has been preferred in the literature. It has the characteristic that if a data point is accepted by the decision function of more than one class, then it is deemed as *not classified*. Furthermore, it is not classified if no decision function claims that data point to be in its class. This characteristic has both positive and negative connotations. It allows for better rejection of outliers, but then it may also be viewed as giving up on handling outliers.

One of the major problems with SVM is their intensive need for memory and computation power at the training stage. Training of SVM for speaker recognition also suffers from these limitations. This is due to the fact that often the training problem is a *quadratic programming* problem, with the number of unknown variables equaling the number of training samples. Chapter 25 provides insight into high speed methods for solving this optimization problem. To address this issue, new techniques have been developed to split the problem into smaller subproblems which would then be solved in parallel, as a network of problems. One such technique is known as *cascade SVM* [44] for which certain improvements have also been proposed in the literature [51].

Similar to the case of *neural networks*, some of the shortcomings of *SVM* have been addressed by combining them with other learning techniques such as *fuzzy logic* and *decision trees*. Also, to speed up the training process, several techniques based on the decomposition of the problem and selective use of the training data have been proposed. More on these techniques will be discussed, later in the chapter. There have also been developments of semi-supervised learning (Section 11.5) to create semi-supervised support vector machines [35].

In application to speaker recognition, experimental results have shown that *SVM* implementations of speaker recognition are slightly inferior to *GMM* approaches. However, it has also been noted that systems which combine *GMM* and *SVM* approaches often enjoy a higher accuracy, suggesting that part of the information revealed by the two approaches may be complementary [40]. We will study this in more detail in Chapter 22.

We noted toward the end of Section 10.4 of this book that Section 2.1.1 of *Jaakkola* [28] shows the relation between SVM and the principal of *minimum rela-*

tive entropy. [28] shows that for the separable case, the two methods coincide where the parameter for the *potential term* in [28] approaches *infinity*. Also, it is shown in that paper that even for the general case, the difference between the two techniques is summarized in a *potential term* related to the different penalties for misclassification. This term does not contribute much to the objective function being optimized, making the two techniques very similar in essence.

15.1 Risk Minimization

In Section 9.2, we discussed the concept of *risk* as defined by Equation 9.14. *Bayes risk* was denoted by $R(o_B)$, signifying the risk associated with decision o_B, which minimizes the *risk* or the *expected value* of a *penalty* function, $\varpi(o|\mathbf{x})$, given by Equation 9.13.

Consider the risk given by Equation 9.10. Let us define *differential probability measure dP* in the *Cartesian product space* of the observations $\{X : x \in \mathscr{X}\}$ and the true state of nature $\{S : s \in \mathscr{S}\}$, represented by the following *probability measure space*, $(\mathscr{X} \times \mathscr{S}, \mathfrak{X} \times \mathfrak{S}, P)$ – see Definitions 6.24, 6.22 and 6.46. Furthermore, let us define a *transformation* Ω (see Definition 6.51) from the space of observations to the space of states, $\Omega : (\mathscr{X}, \mathfrak{X}) \mapsto (\mathscr{S}, \mathfrak{S})$. Ω is chosen to coincide with the transformation which maps the observations to a decision in the space associated with the state of the system. Therefore, it may also be written as, $\Omega : \mathbf{x} \in \mathscr{X} \mapsto o(\mathbf{x}) \in \mathscr{S}$.

If we rewrite Equation 9.10 to include the dependence on the true state of nature, as well as the observations which are represented by random variable $\{X : x \in \mathscr{X}\}$, we will have the following *functional* [15] for the *risk* associated with decision function $o(\mathbf{x})$,

$$R(o) = \int_{\mathscr{X} \times \mathscr{S}} \varpi(o|\mathbf{x})dP \tag{15.1}$$

$$= \int_{\mathscr{S}} \int_{\mathscr{X}} \varpi(o|\mathbf{x})p(\mathbf{x},s)d\mathbf{x}ds \tag{15.2}$$

In the transition from Equation 15.1 to Equation 15.2, we have used the concepts discussed in Section 6.5.2, with the common existence assumption for the *joint probability density*, $p(\mathbf{x},s)$. Of course, as mentioned in Section 6.5.1, Equation 15.1 may still be evaluated, even if $p(\mathbf{x},s)$ does not exist.

Let us consider a *parametric deterministic machine* which makes a decision, o, based on the *observation vector*, $\mathbf{x}$, as well a *parameter vector*, $\boldsymbol{\varphi}$. Then, the transformation from the *observation space* to the *state space* would be based on the parametric function $o(\mathbf{x}, \boldsymbol{\varphi})$. Namely, this transformation is defined as $\Omega : \mathbf{x} \in \mathscr{X} \mapsto$

$o(\mathbf{x}, \boldsymbol{\varphi}) \in \mathcal{S}$. Therefore, the evaluation of the *risk functional* becomes much easier and allows us to write the risk as a function of the parameter vector $\boldsymbol{\varphi}$, for this class of parametric functions as follows,

$$R(\boldsymbol{\varphi}) = \int_{\mathcal{S}} \int_{\mathcal{X}} \varpi(o|\mathbf{x}, \boldsymbol{\varphi}) p(\mathbf{x}, s) d\mathbf{x} ds \tag{15.3}$$

In Section 14.6, we discussed the important *overtraining* (*overfitting*) problem, associated with neural networks. Indeed, this problem plagues many learning techniques, and it has been one of the driving factors for the development of support vector machines. In the process of developing the concept of *capacity* and eventually SVM, *Vapnik* considered the generalization capacity of learning machines, especially *neural networks*. The main goal of support vector machines is to maximize the generalization capability of the learning algorithm, while keeping good performance on the training patterns. As we will see later, this is the basis for the *Vapnik-Chervonenkis theory* (*CV theory*) [46], which computes bounds on the risk, $R(o)$, according to the definition of the *VC dimension* (Section 15.1.2) and the *empirical risk* (Section 15.1.1).

We may use any number of decision functions, $o(\mathbf{x}, \boldsymbol{\varphi})$, in SVM. Regardless of the decision function, let us consider a penalty function which is based on the error between the *true state of the system*, $s(\mathbf{x})$, and the *decision*, $o(\mathbf{x}, \boldsymbol{\varphi})$. There are many ways such a penalty may be devised.

As we saw in Section 14.1, a *perceptron* is designed to fit a hyperplane which would separate *linearly separable* observations. Recall Equation 14.1, in which the decision function for a perceptron was given by a binary function – taking on the value of 0 or 1. The values of this decision function were dependent on the value of $\boldsymbol{\omega}^T \mathbf{i}_p + \theta$, which is a linear relationship based on the weight vector, $\boldsymbol{\omega}$, the threshold (bias), θ, and the input vector, $\mathbf{i}_p$ for pattern (sample) index p. Let us consider a similar decision function with some minor modifications. First, we change the variable indicating the pattern (sample) index, from p to n_c, where $n_c \in \mathcal{N}_c = \{1, 2, \cdots, N_c\}$. The choice of index n_c becomes clear later in the chapter and is related to the constraint index described in Section 25.5. Also, we change the definition of the weight vector ($\boldsymbol{\omega}$) and the threshold (θ) such that the following relation would hold,

$$o(\mathbf{i}_{n_c}) = \begin{cases} 1 & \forall \ \boldsymbol{\omega}^T \mathbf{i}_{n_c} + \theta > 0 \\ 0 & \forall \ \boldsymbol{\omega}^T \mathbf{i}_{n_c} + \theta = 0 \\ -1 & \forall \ \boldsymbol{\omega}^T \mathbf{i}_{n_c} + \theta < 0 \end{cases} \tag{15.4}$$

$$= \text{sgn}(\boldsymbol{\omega}^T \mathbf{i}_{n_c} + \theta) \tag{15.5}$$

Furthermore, let us assume that sample $\mathbf{i}_{n_c}$ is a function of some *observation sample* $\mathbf{x}_{n_c}$, through the following function,

$$\mathbf{i}_{n_c} = \boldsymbol{\psi}(\mathbf{x}_{n_c}) \tag{15.6}$$

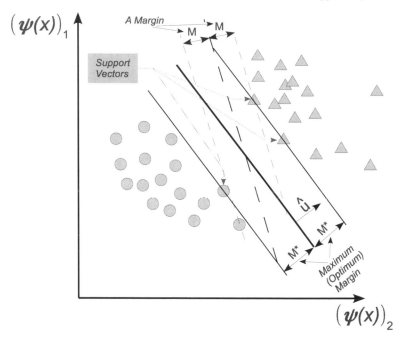

Fig. 15.1: A two-class problem which is linearly separable in the space of $\boldsymbol{\psi}(\mathbf{x})$. $(\boldsymbol{\psi}(\mathbf{x}))_{[1]}$ and $(\boldsymbol{\psi}(\mathbf{x}))_{[2]}$ are the two components (dimensions) of $\boldsymbol{\psi}(\mathbf{x})$. A simple case could be the linear case where $\boldsymbol{\psi}(\mathbf{x}) = \mathbf{x}$.

A special case of $\boldsymbol{\psi}(\mathbf{x}_{n_c})$ would be $\boldsymbol{\psi}(\mathbf{x}_{n_c}) = \mathbf{x}_{n_c}$, which results in a linear decision function, with respect to the observation pattern (sample), $\mathbf{x}_{n_c}$. The general case would provide a decision function which, although linear in $\boldsymbol{\psi}(\mathbf{x}_{n_c})$, may be nonlinear with respect to the observation, $\mathbf{x}_{n_c}$.

In addition, for the general case, as we shall see, the mapping, $\boldsymbol{\psi}(\mathbf{x})$ is defined in the *complete inner product space*, $\mathscr{H}$ (see Definition 6.31) which may have infinite dimension. The infinite dimensional case would be a *Hilbert space* (Definition 6.33), however, in general $\mathscr{H}$ may be of finite dimensionality. Therefore, $\boldsymbol{\psi} : (\mathscr{X}, \mathfrak{X}) \mapsto (\mathscr{H}, \mathfrak{H})$. Based on the discussions of Chapter 6 and Chapter 24, $\mathscr{H}$ is a *complete* space, with an *inner product* defined over the entire space. The *complete inner product space*, $\mathscr{H}$, is known as the *feature space*.

An example of the nonlinearity, discussed earlier, is the case where $\mathbf{i}_{n_c}$ is the output of a neuron in a previous layer. $\mathbf{x}_{n_c}$ is the input to the neuron and $\mathbf{i}_{n_c} = \boldsymbol{\psi}(\mathbf{x}_{n_c})$ is the output of such neuron, in the previous layer, feeding into the perceptron of interest, as input pattern $\mathbf{i}_{n_c}$.

Therefore, Equation 15.5 may be written in terms of the, so called, *observation vectors*, $\mathbf{x}_{n_c}$, using the *feature space* representation, as follows,

$$o(\mathbf{x}_{n_c}, \boldsymbol{\varphi}) = o(\mathbf{x}_{n_c}, \boldsymbol{\omega}, \theta)$$
$$= \text{sgn}(\boldsymbol{\omega}^T \boldsymbol{\psi}(\mathbf{x}_{n_c}) + \theta) \tag{15.7}$$
$$= \text{sgn}(\zeta(\mathbf{x}_{n_c}, \boldsymbol{\omega}, \theta)) \tag{15.8}$$

where,

$$\zeta(\mathbf{x}_{n_c}, \boldsymbol{\omega}, \theta) \overset{\Delta}{=} \boldsymbol{\omega}^T \boldsymbol{\psi}(\mathbf{x}_{n_c}) + \theta \tag{15.9}$$

For simplicity, we may define the following shorthand notation,

$$\boldsymbol{\psi}_{n_c} \overset{\Delta}{=} \boldsymbol{\psi}(\mathbf{x}_{n_c}) \tag{15.10}$$

Equation 15.7 is a valid decision function for a linearly separable two-class problem, where an infinite number of hyperplanes may separate the data points into the two corresponding classes – see Figure 15.1.

Note that it is possible to eliminate the threshold (bias) from Equation 15.9, by merging the information into the weights. This is done by increasing the dimensionality of $\boldsymbol{\omega}$, $\boldsymbol{\psi}$, and $\mathbf{x}_{n_c}$ by one, defining the first element of these vectors such that $(\boldsymbol{\psi}_{n_c})_{[1]} = 1$. Therefore, $(\boldsymbol{\omega})_{[1]} = \theta$ and the θ term may be eliminated [25]. This is just a convenience for making inner products somewhat simpler, but it increases the complexity of the observation pattern and the nonlinear function, $\boldsymbol{\psi}(\mathbf{x})$.

In general, the state of nature is given by the random variable, $\{S : s \in \mathscr{S} = \{H_i, i \in \{1, 2, \cdots, \Gamma\}\}\}$. Therefore, S encompasses Γ possible hypotheses. For a two-class problem, $\Gamma = 2$. In the following section, we discuss a technique for arriving at an estimate of the *minimum risk*. This minimization problem may be stated by the following equation,

$$\boldsymbol{\varphi}^* = \underset{\boldsymbol{\varphi}}{\arg\min} R(\boldsymbol{\varphi})$$
$$= \arg\min \int_{\mathscr{S}} \int_{\mathscr{X}} \varpi(o|\mathbf{x}, \boldsymbol{\varphi}) p(\mathbf{x}, s) d\mathbf{x} ds \tag{15.11}$$

where,

$$\boldsymbol{\varphi} \overset{\Delta}{=} \left[\boldsymbol{\omega}^T, \theta \right]^T \tag{15.12}$$

15.1.1 Empirical Risk Minimization

If the decision portrayed in Equation 9.17 is learned from a limited training sam-
ple, $\{X : \mathbf{x} \in \mathscr{X}_{train} \subset \mathscr{X}\}$, then the minimization discussed earlier is called an
empirical risk minimization [25, 46], since it is decided based on some empirical
data given by the training sample $\mathscr{X}_{train}$. This may be viewed as the *sample mean*
(Equation 6.173) of the penalty function instead of its *true mean* (Definition 6.56).
The *risk* or *expected penalty* which is estimated in this fashion may be denoted by
$R_{emp}(o)$.

Consider the separable two-class problem where the possible hypotheses for the
two classes are $H_i = \{1, -1\}, i \in \{1, 2\}$. Therefore, s_{n_c} and $o(\mathbf{x}_{n_c}, \boldsymbol{\varphi})$ may take on the
values of 1 or -1 for each observed sample, $\mathbf{x}_{n_c}$. In the computation of the *empir-
ical risk*, $R_{emp}(\boldsymbol{\varphi})$, we are presented with N_c such training samples, each of which
corresponds to a pair of *observations*, $\mathbf{x}_{n_c}$, and labels, s_{n_c}.

The parametric decision function, $o(\mathbf{x}_{n_c}, \boldsymbol{\varphi})$, provides the label of 1 or -1 which
should be compared to the true label, s_{n_c}, for a correctness assessment. One possible
penalty function may then be defined as,

$$\varpi(o|\mathbf{x}_{n_c}) \stackrel{\Delta}{=} \frac{1}{2} |s_{n_c} - o(\mathbf{x}_{n_c}, \boldsymbol{\varphi})| \tag{15.13}$$

The weight of $\frac{1}{2}$ in Equation 15.13 ensures that a penalty of 1 is assigned to an
incorrect classification and that a 0 penalty is assigned to a correct classification.
This stems from the fact that we have chosen class labels, $H_i = \{1, -1\}$ instead of
$H_i = \{0, 1\}$.

The integral in Equation 15.3 may then be written as follows, using the penalty
function of Equation 15.13 and assuming that the *joint probability distribution*,
$p(\mathbf{x}, s)$, exists.

$$R(\boldsymbol{\varphi}) = \int_{\mathscr{S}} \int_{\mathscr{X}} \frac{1}{2} |s - o(\mathbf{x}, \boldsymbol{\varphi})| \, p(\mathbf{x}, s) d\mathbf{x} ds \tag{15.14}$$

Furthermore, Equation 15.14 may be estimated by its *sample mean* approximation, for a set of N_c observed samples, leading to the *empirical risk*,

$$R_{emp}(\boldsymbol{\varphi}) = \frac{1}{N_c} \sum_{n_c=1}^{N_c} \frac{1}{2} |s_{n_c} - o(\mathbf{x}_{n_c}, \boldsymbol{\varphi})| \tag{15.15}$$

$$= \frac{1}{2N_c} \sum_{n_c=1}^{N_c} |s_{n_c} - o(\mathbf{x}_{n_c}, \boldsymbol{\varphi})| \tag{15.16}$$

The training sample is a subset of all possible observations ($\mathscr{X}_{train} \subset \mathscr{X}$), so the
true risk cannot be readily computed, yet it may be cross-validated by using some

held-out data (a *test set*). In Sections 13.8.3 and 13.8.4, we discussed *held-out* and *deleted estimation* (k-fold cross-validation) techniques which allow for using all the data to get a better estimate of the risk. *Vapnik* [45] uses the limiting case, *leave-one-out estimation*, for this task – see end of Section 13.8.4.

15.1.2 *Capacity and Bounds on Risk*

In Section 10.6, we briefly touched upon the learning capacity of a model and the *Vapnik-Chervonenkis dimension*, also known as the *VC dimension* [46, 25]. The following is a formal definition which has been adapted from one of Vapnik's definitions [46].

Definition 15.1 (Vapnik-Chervonenkis Dimension – VC Dimension). *The VC dimension, h, of a set of parametric decision functions, $o(\mathbf{x}, \boldsymbol{\varphi})$, $\boldsymbol{\varphi} \in \phi$, is the largest number of vectors, $\{\mathbf{x}_n\}, \{n = 1, 2, \cdots, h\}$, which these functions may classify into two different classes in all 2^h possible ways.*

In other words, h is the maximum number of vectors which may be *shattered* by the set of decision functions, $o(\mathbf{x}, \boldsymbol{\varphi})$. A simple example of the *VC dimension*, presented by [37], is the number of points which may be shattered by any hyperplane in two-dimensional Euclidean space ($\mathscr{R}^2$). In this case, $h = 3$, since a hyperplane in two dimensions can *shatter* all possible combinations of three points. However, once the number of points is increased to 4, then a hyperplane in $\mathscr{R}^2$ is no longer capable of classifying all the possible combinations without error.

15.1.3 *Structural Risk Minimization*

In Section 10.6, *structural risk minimization* was viewed in the context of model selection, in the way it adjusts the capacity of a model in relation with the the *VC dimension*. Here, we take a closer look at *structural risk minimization* which uses the bounds set on the capacity of a model in order to create a hierarchy in the risk assessment of the model.

15.2 The Two-Class Problem

In Section 15.1 (Equation 15.9), we introduced the function, $\zeta(\mathbf{x}, \boldsymbol{\omega}, \theta)$, which is linear with respect to $\boldsymbol{\psi}(\mathbf{x})$, but not necessarily with respect to $\mathbf{x}$ itself. Sometimes, for brevity, we will write $\zeta(\mathbf{x}, \boldsymbol{\omega}, \theta)$ as $\zeta(\mathbf{x}, \boldsymbol{\varphi})$, based on Equation 15.12. Also, note that $\zeta(\mathbf{x}, \boldsymbol{\varphi})$ may be thought of as a functional in terms of $\boldsymbol{\psi}$, namely $\zeta(\boldsymbol{\psi}, \boldsymbol{\varphi})$, since it really depends on the mapping, $\boldsymbol{\psi}$. Therefore, depending on the intention, it may also be written in its functional form, $\zeta(\boldsymbol{\psi}, \boldsymbol{\varphi})$. Figure 15.1 shows a case where the transformation of the patterns, $\mathbf{x}_{n_c}$, is linearly separable in the *feature space*, or the space of the transformation, $\boldsymbol{\psi}(\mathbf{x}_{n_c})$. If the patterns are still linearly separable for the special case where $\boldsymbol{\psi}(\mathbf{x}_{n_c}) = \mathbf{x}_{n_c}$, then they are known as *linearly separable*. However, we only consider the general case, where the patterns are linearly separable in the transformed space of the mapping, $\boldsymbol{\psi}(\mathbf{x})$.

Figure 15.1 shows that for any given mapping, $\boldsymbol{\psi}$, there are an infinite number of hyperplanes given by Equation 15.17 such that they separate the input patterns (transformed observations) in the feature ($\boldsymbol{\psi}$) space.

$$\Psi_h(\boldsymbol{\varphi}) \overset{\Delta}{=} \{\boldsymbol{\psi} : \zeta(\mathbf{x}, \boldsymbol{\varphi}) = \zeta(\boldsymbol{\psi}, \boldsymbol{\varphi}) = 0\} \tag{15.17}$$

where the parameter vector, $\boldsymbol{\varphi}$, is defined by Equation 15.12.

The shortest vector from *hyperplane*, $\Psi_h(\boldsymbol{\varphi})$, to any transformed pattern, $\boldsymbol{\psi}(\mathbf{x})$, in the transformed space of $\boldsymbol{\psi}(\mathbf{x})$ may be written as follows,

$$\Psi_h(\boldsymbol{\varphi}) \rightarrow \boldsymbol{\psi}(\mathbf{x}) = \frac{\zeta(\mathbf{x}, \boldsymbol{\varphi})}{\|\boldsymbol{\omega}\|_{\mathscr{E}}} \hat{\mathbf{u}} \tag{15.18}$$

where $\hat{\mathbf{u}}$ is the unit vector pointing from the separating hyperplane toward the side with a $\{+1\}$ label – see Figure 15.1.

Therefore, the Euclidean distance (magnitude of the vector) between a transformed pattern, $\boldsymbol{\psi}(\mathbf{x})$, and hyperplane $\Psi_h(\boldsymbol{\varphi})$ would be the magnitude of Equation 15.18, given by the following relation,

$$d_E(\boldsymbol{\psi}(\mathbf{x}), \Psi_h(\boldsymbol{\varphi})) = \frac{|\zeta(\mathbf{x}, \boldsymbol{\varphi})|}{\|\boldsymbol{\omega}\|_{\mathscr{E}}} \tag{15.19}$$

By the same token, the distance between the hyperplane and the origin of the transformed space is given by

$$d_E(\mathbf{0}, \Psi_h(\boldsymbol{\varphi})) = \frac{|\theta|}{\|\boldsymbol{\omega}\|_{\mathscr{E}}} \tag{15.20}$$

If the data is linearly separable in the *feature space*, $\mathscr{H}$, for all $\mathbf{x}_{n_c}$, then

$$\frac{s_{n_n} \zeta(\mathbf{x}_{n_n}, \boldsymbol{\varphi})}{\|\boldsymbol{\omega}\|_{\mathscr{E}}} \geq M \tag{15.21}$$

where M is the margin, as illustrated in Figure 15.1, and s_{n_c} is the true state of nature associated with observation pattern $\mathbf{x}_{n_c}$, $s_{n_c} \in \{+1, -1\}$.

We would like to find the hyperplane $\Psi_h(\boldsymbol{\varphi})$ that maximizes the margin, M. As stated earlier, there will generally be an infinite number of solutions to the following problem,

$$M^* = \max_{\varphi} M(\boldsymbol{\varphi}) \tag{15.22}$$

Initially, let us confine ourselves to those direct parameters, $\boldsymbol{\varphi}$, such that the weights, $\boldsymbol{\omega}$, have unit length. Then, the optimization problem will reduce to

$$M^* = \max_{\varphi} M(\boldsymbol{\varphi}) \tag{15.23}$$

subject to

$$\|\boldsymbol{\omega}\|_{\mathscr{E}} = 1 \tag{15.24}$$

and

$$s_{n_c} \zeta(\mathbf{x}_{n_c}, \boldsymbol{\varphi}) \geq M \; \forall \; n_c \in \mathscr{N}_c \tag{15.25}$$

where the maximization is taking place for some labeled training set with N_c observation patterns, $\mathbf{x}_{n_c}, n \in \mathscr{N}_c = \{1, 2, \cdots, N_c\}$, associated with N_c true states of nature, $s_{n_c}, n_c \in \mathscr{N}_c$.

Those patterns which lie exactly at a distance of M^* away from the optimal separating hyperplane $\Psi_h(\boldsymbol{\varphi})$ are called the *support vectors* with indices given by,

$$\mathscr{S}_{s.v.} \stackrel{\Delta}{=} \mathscr{N}_a(\boldsymbol{\varphi}^*) \tag{15.26}$$
$$= \mathscr{N}_a^* \tag{15.27}$$
$$= \arg\min_{n_c} \; s_{n_c} \zeta(\mathbf{x}_{n_c}, \boldsymbol{\varphi}^*) \tag{15.28}$$

where $\boldsymbol{\varphi}^*$ is the solution to Equation 15.23, subject to Equations 15.24 and 15.25, and is known as the *optimal direct parameter vector* $(\{\boldsymbol{\omega}^{*I}, \theta^*\}^I)$. Note that $\mathscr{S}_{s.v.}$ is the set of *active constraints* $(\mathscr{N}_a^*)$ associated with the constrained optimization problem of *SVM*, defined in Section 25.5.1.2. See Equations 25.216 and 25.217.

As it may be seen from Equations 15.23 and 15.28, the optimization problem is only dependent on the *support vectors*, and not all the other vectors in the sample space. Theoretically, this reduces the problem to the following,

$$\boldsymbol{\varphi} = \arg\max_{\varphi} \; \min \; s_{n_a} \zeta(\mathbf{x}_{n_a}, \boldsymbol{\varphi}) \tag{15.29}$$

where $n_a \in \mathscr{S}_{s.v.} = \mathscr{N}_a^*$. Equation 15.29 describes a *saddle point* – see the beginning of Chapter 25.

In Equation 15.29, since only the *support vectors* are needed, practical techniques may be employed to reduce the sample space to those that have a higher probability of meeting the minimization condition. Of course, in practice, finding the *support vectors* may not be so straight forward.

To further simplify the optimization problem, *Boser, et al.* [7] considered a different normalization for the direct parameters, $\boldsymbol{\varphi}$, by requiring that,

$$M\|\boldsymbol{\omega}\|_{\mathscr{E}} = 1 \tag{15.30}$$

instead of $\|\boldsymbol{\omega}\|_{\mathscr{E}} = 1$ which was considered earlier. With this new normalization, since

$$M = \frac{1}{\|\boldsymbol{\omega}\|_{\mathscr{E}}} \tag{15.31}$$

then the maximization of M would be equivalent to minimizing $\|\boldsymbol{\omega}\|_{\mathscr{E}}$, which allows us to rewrite the optimization problem as follows,

$$\boldsymbol{\omega}^* = \arg\min_{\boldsymbol{\omega}} \|\boldsymbol{\omega}\|_{\mathscr{E}} \tag{15.32}$$

$$= \arg\min_{\boldsymbol{\omega}} \frac{1}{2}\|\boldsymbol{\omega}\|_{\mathscr{E}}^2 \tag{15.33}$$

$$= \arg\min_{\boldsymbol{\omega}} \frac{1}{2}\boldsymbol{\omega}^T\boldsymbol{\omega} \tag{15.34}$$

subject to

$$s_{n_a}\zeta(\mathbf{x}_{n_a}, \boldsymbol{\varphi}) \geq 1 \ \forall \ n_a \in \mathscr{S}_{s.v.} = \mathscr{N}_a^* \tag{15.35}$$

after which, M^* may be computed by,

$$M^* = \frac{1}{\|\boldsymbol{\omega}^*\|_{\mathscr{E}}} \tag{15.36}$$

Note that in the above optimization problem, the bias term, θ, only shows up in the constraints and does not appear in the objective function. Therefore, as we will see, the bias will affect the *Lagrangian* which will be constructed, to represent the optimization problem.

Due to legacy and for aesthetics of the mathematics while computing gradients, in the optimization procedure, the minimization problem was transformed from its basic form of Equation 15.32 to an alternative *quadratic form* of Equation 15.34. From here onward, the *quadratic form* will be used in most discussions.

The optimization problem described by Equations 15.34, 15.35, and 15.36 is known as the *optimal margin classification* problem and is discussed in detail in the seminal paper by Boser, et al. [7]. It is one of the important formulations in the topic of *support vector machines*.

Of course the above formulation assumes that the training data is linearly separable. If that is not the case, then the margin with a maximal magnitude may be negative in which case Equation 15.30 would be rewritten as follows,

$$M\|\boldsymbol{\omega}\|_{\mathscr{E}} = -1 \tag{15.37}$$

from which the optimization problem of Equation 15.34 becomes,

$$\boldsymbol{\omega}^* = \arg\max_{\boldsymbol{\omega}} \frac{1}{2}\|\boldsymbol{\omega}\|_{\mathscr{E}}^2 \tag{15.38}$$

$$= \arg\max_{\boldsymbol{\omega}} \frac{1}{2}\boldsymbol{\omega}^T\boldsymbol{\omega} \tag{15.39}$$

In the next section, we will set up the problem of Equation 15.34 in its *Wolfe dual form* – see Section 25.5.

15.2.1 Dual Representation

Chapter 25 presents a detailed discussion of optimization theory, with Section 25.5 concentrating on *constrained optimization*. The concept of *duality* has also been discussed in detail in Section 25.5.2. It is recommended that the reader would start by reading Section 25.5 and then follow the rest of the discussion in this section.

As we state in Section 25.5.2, the dual representation often allows for a simpler computational effort. There are many different types of duality. Here, we are interested in the *Wolfe dual* [48] representation which is especially formulated to handle *convex* objective functions. The *primal problem* is recapped here from the last section,

$$\boldsymbol{\varphi}^* = \arg\min_{\boldsymbol{\omega}} \frac{1}{2}\boldsymbol{\omega}^T\boldsymbol{\omega} \tag{15.40}$$

subject to

$$s_{n_c}\zeta(\mathbf{x}_{n_c}, \boldsymbol{\varphi}) \geq 1 \ \forall \ n_c \in \mathcal{N}_c \tag{15.41}$$

where the optimal margin, M^* is computed by,

$$M^* = \frac{1}{\|\boldsymbol{\omega}^*\|_{\mathscr{E}}} \tag{15.42}$$

Of course, the constraints may have also been written such that they would only include the *support vectors (active constraints)*, $n_a \in \mathcal{N}_a^* = \mathscr{S}$ s.v..

The *Lagrangian function* of the above problem may be written as follows,

$$\mathscr{L}(\boldsymbol{\varphi}, \boldsymbol{\lambda}) = \frac{1}{2}\boldsymbol{\omega}^T \boldsymbol{\omega} - \sum_{n_c=1}^{N_c} (\boldsymbol{\lambda})_{[n_c]} \left(s_{n_c} \zeta(\mathbf{x}_{n_c}, \boldsymbol{\varphi}) - 1 \right) \tag{15.43}$$

$$= \frac{1}{2}\boldsymbol{\omega}^T \boldsymbol{\omega} - \sum_{n_c=1}^{N_c} (\boldsymbol{\lambda})_{[n_c]} \left(s_{n_c} (\boldsymbol{\omega}^T \boldsymbol{\psi}_{n_c} + \theta) - 1 \right) \tag{15.44}$$

As we discussed earlier in this chapter and in Section 25.5.1.2, the only constraints that are important in determining the solution of the optimization problem are the *active constraints* or those constraints associated with the *support vectors* with indices, $n_a = \{n_c : n_c \in \mathscr{S}_{s.v.} = \mathcal{N}_a^*\}$. This is because of the fact that

$$(\boldsymbol{\lambda})_{[n_c]} = 0 \;\; \forall \;\; n_c \in \mathscr{S}_{s.v.}^{\complement} \tag{15.45}$$

Therefore, Equation 15.44 may be written with the sole inclusion of the active constraints,

$$\mathscr{L}(\boldsymbol{\varphi}, \boldsymbol{\lambda}^{(a)}) = \frac{1}{2}\boldsymbol{\omega}^T \boldsymbol{\omega} - \sum_{n_a \in \mathscr{S}_{s.v.}} (\boldsymbol{\lambda})_{[n_a]} \left(s_{n_a} (\boldsymbol{\omega}^T \boldsymbol{\psi}_{n_a} + \theta) - 1 \right) \tag{15.46}$$

The formulation of the *dual problem* begins with the following maximization,

$$\boldsymbol{\varphi}^*, \boldsymbol{\lambda}^* = \arg\max_{\boldsymbol{\varphi}, \boldsymbol{\lambda} \in \bar{\Omega}} \mathscr{L}(\boldsymbol{\varphi}, \boldsymbol{\lambda}) \tag{15.47}$$

where

$$\bar{\Omega} = \{\boldsymbol{\varphi}, \boldsymbol{\lambda} : (\nabla_{\boldsymbol{\varphi}} \mathscr{L}(\boldsymbol{\varphi}, \boldsymbol{\lambda}) = \mathbf{0} \;\; \wedge \;\; \boldsymbol{\lambda} \succeq \mathbf{0})\} \tag{15.48}$$

In Section 25.5.2.1 we go through the derivation of the *Wolfe dual* problem in terms of the *Lagrange multipliers*, $\boldsymbol{\lambda}$, for a quadratic primal objective function. Let us write the expression for the *Lagrangian* in vector form,

$$\mathscr{L}(\boldsymbol{\varphi}, \boldsymbol{\lambda}) = \mathscr{L}(\boldsymbol{\omega}, \theta, \boldsymbol{\lambda})$$
$$= \frac{1}{2}\boldsymbol{\omega}^T \boldsymbol{\omega} - (\boldsymbol{\omega}^T \mathbf{J}_{\boldsymbol{\omega}} + \theta \; \mathbf{j}_\theta^T - \bar{\mathbf{e}}_{N_a}^T) \boldsymbol{\lambda} \tag{15.49}$$

The *Jacobian matrix* associated with the active constraints relative to $\boldsymbol{\omega}$, $\mathbf{J}_{\boldsymbol{\omega}} : \mathscr{R}^{N_a} \mapsto \mathscr{R}^N$, is such that,

$$(\mathbf{J}_{\boldsymbol{\omega}})_{[n_a]} = s_{n_a} \boldsymbol{\psi}_{n_a} \tag{15.50}$$

In Equation 15.50, the notation $(\mathbf{J}_{\boldsymbol{\omega}})_{[n_a]}$ denotes column n_a of $\mathbf{J}_{\boldsymbol{\omega}}$. Furthermore, the *Jacobian vector* associate with the bias term, $\mathbf{j}_\theta : \mathscr{R}^1 \mapsto \mathscr{R}^{N_a}$, is such that,

$$(\mathbf{j}_\theta)_{[n_a]} = s_{n_a} \tag{15.51}$$

Also, the all-ones vector, $\bar{\mathbf{e}}_{N_a} : \mathscr{R}^1 \mapsto \mathscr{R}^{N_a}$, is defined as follows,

$$(\bar{\mathbf{e}}_{N_a})_{[n_a]} = 1 \ \forall \ n_a \in \mathscr{N}_a \tag{15.52}$$

The *dual feasibility* condition, $\nabla_\varphi \mathscr{L} = 0$, may then be split into its two components relating to $\boldsymbol{\omega}$ and θ,

$$\nabla_\omega \mathscr{L} = \boldsymbol{\omega} - \mathbf{J}_\omega \boldsymbol{\lambda} \tag{15.53}$$
$$= 0 \tag{15.54}$$

and

$$\nabla_\theta \mathscr{L} = \mathbf{j}_\theta^T \boldsymbol{\lambda} \tag{15.55}$$
$$= 0 \tag{15.56}$$

We may use the *dual feasibility* relation of Equation 15.54 to solve for the primal variable $\boldsymbol{\omega}$ in terms of the *dual variable* $\boldsymbol{\lambda}$,

$$\boldsymbol{\omega} = \mathbf{J}_\omega \boldsymbol{\lambda} \tag{15.57}$$

If we substitute for $\boldsymbol{\omega}$ from Equation 15.57 and for $\mathbf{j}_\theta^T \boldsymbol{\lambda}$ from Equation 15.56 into Equation 15.49, we will have the following expression for the *Lagrangian*, solely in terms of the *dual variables*, $\boldsymbol{\lambda}$ by eliminating the *primal variables*, $\boldsymbol{\varphi}$,

$$\mathscr{L}(\boldsymbol{\lambda}) = \frac{1}{2}\boldsymbol{\lambda}^T \mathbf{J}_\omega^T \mathbf{J}_\omega \boldsymbol{\lambda} - \boldsymbol{\lambda}^T \mathbf{J}_\omega^T \mathbf{J}_\omega \boldsymbol{\lambda} - 0 + \bar{\mathbf{e}}_{N_a}^T \boldsymbol{\lambda} \tag{15.58}$$

$$= -\frac{1}{2}\boldsymbol{\lambda}^T \mathbf{J}_\omega^T \mathbf{J}_\omega \boldsymbol{\lambda} + \bar{\mathbf{e}}_{N_a}^T \boldsymbol{\lambda} \tag{15.59}$$

Therefore, we may write the simplified *Wolfe dual* problem as the following quadratic maximization problem in terms of $\boldsymbol{\lambda}$,

$$\boldsymbol{\lambda}^* = \underset{\substack{\boldsymbol{\lambda} \succeq 0 \\ \mathbf{j}_\theta^T \boldsymbol{\lambda} = 0}}{\arg\max} -\frac{1}{2}\boldsymbol{\lambda}^T \mathbf{J}_\omega^T \mathbf{J}_\omega \boldsymbol{\lambda} + \bar{\mathbf{e}}_{N_a}^T \boldsymbol{\lambda} \tag{15.60}$$

Once $\boldsymbol{\lambda}^*$ has been computed by performing the constrained maximization problem of Equation 15.60, it may be used to compute the corresponding primal variables. Using Equation 15.57, the optimal weight vector, $\boldsymbol{\omega}^*$ is computed as follows,

$$\boldsymbol{\omega}^* = \mathbf{J}_\omega \boldsymbol{\lambda}^* \tag{15.61}$$

Also, by using the *complementarity condition* of Equation 25.232 which is recapped here,

$$\boldsymbol{\lambda}^T \mathbf{c}(\boldsymbol{\varphi}) = 0 \tag{15.62}$$

we will have

$$\left(\boldsymbol{\omega}^{*T}\mathbf{J}_\omega + \theta^* \ \mathbf{j}_\theta^T - \bar{\mathbf{e}}_{N_a}^T\right)\boldsymbol{\lambda}^* = 0 \tag{15.63}$$

We may solve for θ^* in Equation 15.63,

$$\theta^* = \frac{\boldsymbol{\omega}^{*T}\mathbf{J}_\omega\boldsymbol{\lambda}^* - \bar{\mathbf{e}}_{N_a}^T\boldsymbol{\lambda}^*}{\mathbf{j}_\theta^T\boldsymbol{\lambda}^*} \tag{15.64}$$

$$= \frac{\boldsymbol{\lambda}^{*T}\mathbf{J}_\omega^T\mathbf{J}_\omega\boldsymbol{\lambda}^* - \bar{\mathbf{e}}_{N_a}^T\boldsymbol{\lambda}^*}{\mathbf{j}_\theta^T\boldsymbol{\lambda}^*} \tag{15.65}$$

If we plug in for $\boldsymbol{\omega}$ and θ from Equations 15.61 and 15.65 into Equation 15.7, the decision function, $o(\mathbf{x},\boldsymbol{\varphi},\theta)$ would be given by the following expression,

$$o(\mathbf{x},\boldsymbol{\varphi}^*) = o(\mathbf{x},\boldsymbol{\omega}^*,\theta^*)$$

$$= \mathrm{sgn}(\boldsymbol{\omega}^{*T}\boldsymbol{\psi}(\mathbf{x}) + \theta^*) \tag{15.66}$$

$$= \mathrm{sgn}(\boldsymbol{\lambda}^{*T}\mathbf{J}_\omega^T\boldsymbol{\psi}(\mathbf{x}) + \frac{\boldsymbol{\lambda}^{*T}\mathbf{J}_\omega^T\mathbf{J}_\omega\boldsymbol{\lambda}^* - \bar{\mathbf{e}}_{N_a}^T\boldsymbol{\lambda}^*}{\mathbf{j}_\theta^T\boldsymbol{\lambda}^*}) \tag{15.67}$$

15.2.2 Soft Margin Classification

The formulation discussed up to now assumed that the data is completely separable, at least in the *feature space*, if not in the *observation space*. This type of classification is hence called hard margin classification. But what happens if the data is not completely separable by a linear hyperplane, even in the *feature space*? The problem with the hard margin formulation is that this would be result in no solution. In fact, hard margin classification is not at all forgiving when it comes to outliers. The algorithm actually calls for dropping such points.

To make a more practical two-class classifier, we need to allow some flexibility. This may be done by using, so called, *slack variables*, which have been used extensively in the optimization literature [34, 20, 33], in order to solve simple constraints. These variables generally increase the dimensionality of the optimization problem, much in the same way as Lagrange multipliers do. This added dimensionality allows for more flexibility on the optimal solution such that certain constraints are met.

If we call the *slack variable* vector $\boldsymbol{\xi}$, defines such that, $\boldsymbol{\xi} : \mathcal{R}^1 \mapsto \mathcal{R}^{N_a}$, we may write the new primal optimization problem akin to Equation 15.40 and its constraints of Equation 15.41, as follows,

$$\{\boldsymbol{\varphi}^*,\boldsymbol{\xi}^*\} = \underset{\boldsymbol{\omega},\boldsymbol{\xi}}{\arg\min}\,\frac{1}{2}\boldsymbol{\omega}^T\boldsymbol{\omega} + C\|\boldsymbol{\xi}\|_p \tag{15.68}$$

subject to

$$s_{n_c} \zeta(\mathbf{x}_{n_c}, \boldsymbol{\varphi}) = s_{n_c}(\boldsymbol{\omega}^T \boldsymbol{\psi}(\mathbf{x}_{n_c}) + \theta)$$

$$\geq 1 - (\boldsymbol{\xi})_{[n_c]} \ \forall \, n_c \in \mathcal{N}_c \tag{15.69}$$

and

$$\boldsymbol{\xi} \succeq \mathbf{0} \tag{15.70}$$

where p specifies that the L_p norm (see Definition 23.8) of the slack variable vector should be used in the optimization functions. C is the *margin parameter* and it provides a trade-off between maximizing the margin, M and minimizing the classification error [1]. The two most popular norms used in the literature are the L_1 and the L_2 norms. In the next two sections, we will examine special features of each of these classifiers.

15.2.2.1 L_1 classifier

Let us set $p = 1$ in Equation 15.68. To be able to write the Lagrangian function, we would have to define two after defining two sets of Lagrange multipliers. The first is $\boldsymbol{\lambda}$ which is associated with the original parameters, $\boldsymbol{\varphi}$, and $\boldsymbol{\lambda}_\xi$ which is associated with the slack variables, $\boldsymbol{\xi}$. Therefore, the soft margin version of the dual optimization problem involving the Lagrangian function, given by Equation 15.47 may be written as follows,

$$\{\boldsymbol{\varphi}^*, \boldsymbol{\lambda}^*, \boldsymbol{\xi}^*, \boldsymbol{\lambda}_\xi^*\} = \arg\max_{\substack{\boldsymbol{\varphi}, \boldsymbol{\lambda}, \\ \boldsymbol{\xi}, \boldsymbol{\lambda}_\xi \in \bar{\Omega}}} \mathcal{L}(\boldsymbol{\varphi}, \boldsymbol{\lambda}, \boldsymbol{\xi}, \boldsymbol{\lambda}_\xi) \tag{15.71}$$

where

$$\bar{\Omega} = \{\boldsymbol{\varphi}, \boldsymbol{\lambda}, \boldsymbol{\xi}, \boldsymbol{\lambda}_\xi : \begin{cases} \nabla_{\boldsymbol{\varphi}} \mathcal{L} = \mathbf{0} \\ \nabla_{\boldsymbol{\xi}} \mathcal{L} = \mathbf{0} \\ \boldsymbol{\lambda} \succeq \mathbf{0} \\ \boldsymbol{\lambda}_\xi \succeq \mathbf{0} \\ \boldsymbol{\xi} \succeq \mathbf{0} \\ \left(\boldsymbol{\omega}^T \mathbf{J}_\omega + \theta \ \mathbf{j}_\theta^T - (\bar{\mathbf{e}}_{N_a} - \boldsymbol{\xi})^T\right)\boldsymbol{\lambda} = 0 \\ \boldsymbol{\xi}^T \boldsymbol{\lambda}_\xi = 0 \end{cases} \tag{15.72}$$

where the last two statements in Equation 15.72 are due to the *complementarity condition* which was discussed in Section 15.2.1. The Lagrangian function for the above maximization problem is given by,

$$\mathcal{L}(\boldsymbol{\varphi}, \boldsymbol{\lambda}, \boldsymbol{\xi}, \boldsymbol{\lambda}_\xi) = \frac{1}{2}\boldsymbol{\omega}^T \boldsymbol{\omega} - \left(\boldsymbol{\omega}^T \mathbf{J}_\omega + \theta \ \mathbf{j}_\theta^T - (\bar{\mathbf{e}}_{N_a} - \boldsymbol{\xi})^T\right)\boldsymbol{\lambda}$$

$$+ C\bar{\mathbf{e}}_{N_a}^T \boldsymbol{\xi} - \boldsymbol{\xi}^T \boldsymbol{\lambda}_\xi \tag{15.73}$$

Just like the case of the hard margin classification,

$$\nabla_{\boldsymbol{\omega}} \mathscr{L} = 0 \implies \boldsymbol{\omega} = \mathbf{J}_{\boldsymbol{\omega}} \boldsymbol{\lambda} \tag{15.74}$$

and

$$\nabla_{\theta} \mathscr{L} = \mathbf{j}_{\theta}^{T} \boldsymbol{\lambda} \tag{15.75}$$
$$= 0 \tag{15.76}$$

Also, the dual feasibility condition for the slack variables gives us,

$$\nabla_{\xi} \mathscr{L} = 0 \implies \boldsymbol{\lambda}^{*} + \boldsymbol{\lambda}_{\xi}^{*} = C \bar{\mathbf{e}}_{N_a} \tag{15.77}$$

If we use these relations to reduce the Lagrangian of Equation 15.73 to the *Wolfe dual form*, we have,

$$\boldsymbol{\lambda}^{*} = \underset{\substack{0 \succeq \boldsymbol{\lambda} \succeq C \bar{\mathbf{e}}_{N_a} \\ \mathbf{j}_{\theta}^{T} \boldsymbol{\lambda} = 0}}{\arg \max} \; \mathscr{L}(\boldsymbol{\lambda}) - \frac{1}{2} \boldsymbol{\lambda}^{T} \mathbf{J}_{\boldsymbol{\omega}}^{T} \mathbf{J}_{\boldsymbol{\omega}} \boldsymbol{\lambda} + \bar{\mathbf{e}}_{N_a}^{T} \boldsymbol{\lambda} \tag{15.78}$$

Note that the only difference between Equation 15.78 for the L_1 soft margin classifier and Equation 15.60 for the hard margin classifier is the first constraint on $\boldsymbol{\lambda}$, requiring that each Lagrange multiplier associated with the primal variable, $\boldsymbol{\varphi}$, must be positive and less than C, where as the upper limit of C does not exist for the hard margin classifier. The decision function, $o(\mathbf{x}, \boldsymbol{\varphi}, \theta)$ also stays the same as that of the hard margin case.

15.2.2.2 L_2 classifier

As we did in the beginning of Section 15.2, for legacy and aesthetics of the mathematics, we add a factor of $\frac{1}{2}$ to the term involving the slack variables, for the L_2 norm version. Also, we do not take the square root as prescribed by Definition 23.8. Instead, we will use the inner product of the slack variable vectors. Also, we do not need to have any Lagrange multipliers associated with the slack variables [1]. Therefore, soft margin version of the Lagrangian function, given by Equation 15.73 may be written as follows,

$$\mathscr{L}(\boldsymbol{\varphi}, \boldsymbol{\lambda}, \boldsymbol{\xi}, \boldsymbol{\lambda}_{\xi}) = \frac{1}{2} \boldsymbol{\omega}^{T} \boldsymbol{\omega} - \left(\boldsymbol{\omega}^{T} \mathbf{J}_{\boldsymbol{\omega}} + \theta \; \mathbf{j}_{\theta}^{T} - (\bar{\mathbf{e}}_{N_a} - \boldsymbol{\xi})^{T} \right) \boldsymbol{\lambda}$$
$$+ \frac{C}{2} \boldsymbol{\xi}^{T} \boldsymbol{\xi} \tag{15.79}$$

In fact, it is easy to show that with a simple change of variables, the soft margin L_2 classifier reduces to the same problem as the hard margin classifier. *Abe* [1] derives the following change of variables,

$$\hat{\boldsymbol{\omega}} = \begin{bmatrix} \boldsymbol{\omega} \\ \sqrt{C} \boldsymbol{\xi} \end{bmatrix} \tag{15.80}$$

$$\hat{\theta} \overset{\Delta}{=} \theta \tag{15.81}$$

$$\hat{\psi}\mathbf{x}_{n_a} \overset{\Delta}{=} \begin{bmatrix} \boldsymbol{\psi}(\mathbf{x}_{n_a}) \\ \sqrt{C}\hat{\mathbf{e}}_{n_a} \end{bmatrix} \tag{15.82}$$

Note that in Equation 15.82, $\hat{\mathbf{e}}_m : \mathscr{R}^{N_a} \mapsto \mathbb{R}$ is the unit vector defined in the *nomenclature*, namely,

$$(\hat{\mathbf{e}}_{n_a})_{[m]} = \begin{cases} 1 \ \forall \ m = n_a \\ 0 \ \forall \ m \neq n_a \end{cases} \tag{15.83}$$

Using the above variables, the problem reduces to that of the hard margin case where the variables, $\hat{\boldsymbol{\omega}}$, $\hat{\theta}$, $\hat{\psi}$ replace $\boldsymbol{\omega}$, θ, ψ respectively, in the hard margin formulation. Once the problem is solved, the original L_2 variables may be recomputed.

15.3 Kernel Mapping

Definition 24.56 defines a *kernel function*, $\mathscr{K}(s,t)$. Also, per *Hilbert's expansion theorem* 24.21 and its extensions by Schmidt (Theorem 24.22) and Mercer (Theorem 24.23), a positive definite kernel, $\mathscr{K}(s,t)$, may be expressed in term of an infinite series based on its Eigenfunctions and Eigenvalues. The kernel provides the means for mapping the infinite set of possible values of t and another set of infinite values of s, in different spaces. It is elaborated in that definition that this mapping is a projection from a point by an infinite set of orthogonal basis functions (producing the kernel).

First, let us recall that the forms of the optimization problems and the decision functions for both hard margin and soft margin classification were shown to be the same, in the last section. Therefore, we will concentrate on the derivations from Section 15.2.1 which apply to both cases.

Let us re-examine the results of Section 15.2.1. Take the Jacobian matrix, $\mathbf{J}_{\boldsymbol{\omega}}$, defined by Equation 15.50. If we plug in from this equation into the results obtained in Section 15.2.1, we will get some interesting results. For example, take the statement of the Lagrangian given by Equation 15.59 and used in setting up the *dual problem* of optimization. We may write it in terms of summation on the number of active constraints (support vectors) as follows,

$$\begin{aligned} \mathscr{L}(\boldsymbol{\lambda}) &= -\frac{1}{2}\boldsymbol{\lambda}^T \mathbf{J}_{\boldsymbol{\omega}}^T \mathbf{J}_{\boldsymbol{\omega}}\boldsymbol{\lambda} + \bar{\mathbf{e}}_{N_a}^T \boldsymbol{\lambda} \\ &= -\frac{1}{2}\sum_{l=1}^{N_a}\sum_{m=1}^{N_a} (\boldsymbol{\lambda})_{[l]}(\boldsymbol{\lambda})_{[m]}\, s_l s_m \langle \boldsymbol{\psi}_l, \boldsymbol{\psi}_m \rangle + \bar{\mathbf{e}}_{N_a}^T \boldsymbol{\lambda} \end{aligned} \tag{15.84}$$

Notice that Equation 15.84 only includes *inner products* of $\boldsymbol{\psi}$ and does not have any term in terms of any standalone $\boldsymbol{\psi}$.

Let us continue by writing the expanded form of the optimum bias term of Equation 15.65,

$$
\theta^* = \frac{\boldsymbol{\lambda}^{*T} \mathbf{J}_\omega^T \mathbf{J}_\omega \boldsymbol{\lambda}^* - \bar{\mathbf{e}}_{N_a}^T \boldsymbol{\lambda}^*}{\mathbf{j}_\theta^T \boldsymbol{\lambda}^*}
$$

$$
= \frac{\left(\sum_{l=1}^{N_a} (\boldsymbol{\lambda}^*)_{[l]} \, s_l \boldsymbol{\psi}_l \right)^T \left(\sum_{m=1}^{N_a} (\boldsymbol{\lambda}^*)_{[m]} \, s_m \boldsymbol{\psi}_m \right) - \bar{\mathbf{e}}_{N_a}^T \boldsymbol{\lambda}^*}{\mathbf{j}_\theta^T \boldsymbol{\lambda}^*} \tag{15.85}
$$

$$
= \frac{\left(\sum_{l=1}^{N_a} \sum_{m=1}^{N_a} (\boldsymbol{\lambda}^*)_{[l]} (\boldsymbol{\lambda}^*)_{[m]} \, s_l s_m \langle \boldsymbol{\psi}_l, \boldsymbol{\psi}_m \rangle \right) - \bar{\mathbf{e}}_{N_a}^T \boldsymbol{\lambda}^*}{\mathbf{j}_\theta^T \boldsymbol{\lambda}^*} \tag{15.86}
$$

Again, we see that Equation 15.87 only involves *inner products* of $\boldsymbol{\psi}$.

Lastly, let us write the expanded expression for the *optimal decision function*, $o(\mathbf{x}, \boldsymbol{\varphi}^*)$,

$$
o(\mathbf{x}, \boldsymbol{\varphi}^*) = \operatorname{sgn} \left(\boldsymbol{\lambda}^{*T} \mathbf{J}_\omega^T \boldsymbol{\psi}(\mathbf{x}) + \frac{\boldsymbol{\lambda}^{*T} \mathbf{J}_\omega^T \mathbf{J}_\omega \boldsymbol{\lambda}^* - \bar{\mathbf{e}}_{N_a}^T \boldsymbol{\lambda}^*}{\mathbf{j}_\theta^T \boldsymbol{\lambda}^*} \right)
$$

$$
= \operatorname{sgn} \left(\begin{array}{c} \sum_{n_a}^{N_a} (\boldsymbol{\lambda}^*)_{[n_a]} \, s_{n_a} \langle \boldsymbol{\psi}_{n_a}, \boldsymbol{\psi}(\mathbf{x}) \rangle \\ + \frac{\left(\sum_{l=1}^{N_a} \sum_{m=1}^{N_a} (\boldsymbol{\lambda}^*)_{[l]} (\boldsymbol{\lambda}^*)_{[m]} s_l s_m \langle \boldsymbol{\psi}_l, \boldsymbol{\psi}_m \rangle \right) - \bar{\mathbf{e}}_{N_a}^T \boldsymbol{\lambda}^*}{\mathbf{j}_\theta^T \boldsymbol{\lambda}^*} \end{array} \right) \tag{15.87}
$$

Once again, only *inner products* of $\boldsymbol{\psi}$ show up in the expression of the *optimal decision function*.

15.3.1 The Kernel Trick

There is a very useful direct consequence of *Hilbert's expansion theorem* (Theorem 24.21) and its extensions by Schmidt (Theorem 24.22) and Mercer (Theorem 24.23). We saw from Mercer's version of the theorem (Theorem 24.23) that if we have a positive semi-definite kernel, or even one that has a finite number of negative Eigenvalues with an infinite set of positive ones, we may expand it in terms of the inner product of a set of functions which happen to be the Eigenfunctions of that kernel. In fact, we also saw from the definition of a degenerate kernel (Definition 24.62) that we may indeed write a degenerate kernel in terms of a finite sum of inner products of functions. Therefore, by choosing the set of orthogonal functions that are used to describe a kernel of choice, we may write any kernel as an inner product in some other space. Namely,

$$\mathscr{K}(s,t) = \langle \psi(s), \psi(t) \rangle \tag{15.88}$$

This is a very powerful result, called the *kernel trick*, which was first exploited by Aizerman [3] in 1964.[2] Let us go back to the case where the training data is linearly separable. In fact, we can increase the dimensionality of $\psi(\mathbf{x})$ to the point that this is indeed the case. As we stated before, the observation vectors, $\mathbf{x}$, need not be linearly separable for this to take place.

This, so called, *kernel trick* can help in two ways. The fist is that we may now include any, preferably, positive definite kernel in the place of the inner products seen in Equations 15.84, 15.87, 15.87, and similar occurrences. Secondly, we may use different kernels in different situations, using the nonlinearity in the kernel to help with special scenarios. As a bonus, the amount of computation is also reduced. For example, the kernel values for the different training data may be tabulated and used without having to perform the expensive inner products.

In foreseeing the capability of mapping to a different space, we have been treating the problem in its general form in the feature space ($\mathscr{H}$). A simplification, by setting $\psi(\mathbf{x}) = \mathbf{x}$, would suffice for problems which are linearly separable in the *observation space*. However, as we saw in Section 15.2.2, we may also use a *soft margin formulation* with the introduction of some *slack variables* [34], where although a *feature space* representation of the observation vectors is chosen, still slight overlaps are allowed through the introduction of these *slack variables*.

Here are Equations 15.84, 15.87, 15.87, where we have replaced the inner products with kernels,

$$\mathscr{L}(\boldsymbol{\lambda}) = -\frac{1}{2} \sum_{l=1}^{N_a} \sum_{m=1}^{N_a} (\boldsymbol{\lambda})_{[l]} (\boldsymbol{\lambda})_{[m]} s_l s_m \mathscr{K}(\mathbf{x}_l, \mathbf{x}_m) + \bar{\mathbf{e}}_{N_a}^T \boldsymbol{\lambda} \tag{15.89}$$

$$\theta^* = \frac{\left(\sum_{l=1}^{N_a} \sum_{m=1}^{N_a} (\boldsymbol{\lambda}^*)_{[l]} (\boldsymbol{\lambda}^*)_{[m]} s_l s_m \mathscr{K}(\mathbf{x}_l, \mathbf{x}_m) \right) - \bar{\mathbf{e}}_{N_a}^T \boldsymbol{\lambda}^*}{\mathbf{j}_\theta^T \boldsymbol{\lambda}^*} \tag{15.90}$$

$$o(\mathbf{x}, \boldsymbol{\varphi}^*) = \text{sgn} \left(\sum_{n_a}^{N_a} (\boldsymbol{\lambda}^*)_{[n_a]} s_{n_a} \mathscr{K}(\mathbf{x}_{n_a}, \mathbf{x}) + \frac{\left(\sum_{l=1}^{N_a} \sum_{m=1}^{N_a} (\boldsymbol{\lambda}^*)_{[l]} (\boldsymbol{\lambda}^*)_{[m]} s_l s_m \mathscr{K}(\mathbf{x}_l, \mathbf{x}_m) \right) - \bar{\mathbf{e}}_{N_a}^T \boldsymbol{\lambda}^*}{\mathbf{j}_\theta^T \boldsymbol{\lambda}^*} \right) \tag{15.91}$$

---

[2] *Burges* [8] has called this the *kernel trick*.

15.4 Positive Semi-Definite Kernels

We spoke about positive semi-definite kernels in detail in Section 24.5. These kernels are also known in the literature as *Mercer kernels* – see Section 24.5.1. Of course, as *Mercer* shows in his version of the expansion theorem (Theorem 24.23, it is not necessary for the kernel to be positive definite in order for *Hilbert's expansion* (Equation 24.243) to converge uniformly and absolutely. However, the positive definiteness becomes essential when the number of Eigenfunctions used for the expansion are finite, which is the case with most of the applications discussed here, including support vector machines. In the same spirit, some have called the positive definiteness of the kernel, the *Mercer condition* (Equation 24.240).

In general, most kernels of interest are *Mercer kernels* (i.e., they are positive semi-definite). However, there are cases when non-positive semi-definite kernels are used. In Section 15.5 we will discuss some such kernels. For now, let us look at some popular *Mercer kernels*.

15.4.1 Linear Kernel

As we have discussed earlier, the simplest kernel is the linear kernel, which is simply the inner product of the vectors in the observation space. The linear kernel may be used for cases where the data is linearly separable.

$$\mathscr{K}_{Lin}(\mathbf{x}_l, \mathbf{x}_m) \overset{\Delta}{=} \langle \mathbf{x}_l, \mathbf{x}_m \rangle \tag{15.92}$$

$$= \mathbf{x}_l^T \mathbf{x}_m \tag{15.93}$$

This amounts to setting $\psi(\mathbf{x}) = \mathbf{x}$.

Dehak, et al. [17] have used the *linear kernel* for *speaker verification*, using *speaker factor coefficients* (Section 16.4.3) as the observed vector, $\mathbf{x}$. They have compared its performance to that of the *GRBF kernel* (Section 15.4.3) and the *Cosine kernel* (Section 15.4.4). As we shall see, the *Cosine kernel* is just the linear kernel which has been normalized by the product of the norms of the two independent variables, $\mathbf{x}_l$ and $\mathbf{x}_m$.

15.4.2 Polynomial Kernel

There are two different definitions of the polynomial kernel. The simplest definition is as follows,

$$\mathcal{K}_{Poly}(\mathbf{x}_l, \mathbf{x}_m) \overset{\Delta}{=} \left(\mathbf{x}_l^T \mathbf{x}_m\right)^d \tag{15.94}$$

Since the kernel given by Equation 15.94 will have a many to one mapping for even degrees ($d = 2n, n \in \{1, 2, \cdots\}$), the following formulation is generally preferred [1],

$$\mathcal{K}_{Poly}(\mathbf{x}_l, \mathbf{x}_m) \overset{\Delta}{=} \left(\mathbf{x}_l^T \mathbf{x}_m + 1\right)^d \tag{15.95}$$

When $d = 1$, the form in Equation 15.95 will result in the linear kernel (Equation 15.93) plus a deviation of 1. This deviation may be corrected for, by adjusting the bias term, θ [1].

As an example, if we take $N = 2$ as the dimension of $\mathbf{x}$ and use a second degree ($d = 2$) polynomial kernel according to the definition of Equation 15.95, then we see the following,

$$\mathcal{K}(\mathbf{x}_l, \mathbf{x}_m) = \langle \boldsymbol{\psi}(\mathbf{x}_l), \boldsymbol{\psi}(\mathbf{x}_m) \rangle \tag{15.96}$$

$$= \left(\mathbf{x}_l^T \mathbf{x}_m + 1\right)^2 \tag{15.97}$$

Therefore, the higher dimensional transformation function may be computed by the above equation [1],

$$\boldsymbol{\psi}(\mathbf{x}) = \begin{bmatrix} 1 \\ \sqrt{2}\,(\mathbf{x})_{[1]} \\ \sqrt{2}\,(\mathbf{x})_{[2]} \\ \sqrt{2}\,(\mathbf{x})_{[1]}\,(\mathbf{x})_{[2]} \\ \left((\mathbf{x})_{[1]}\,(\mathbf{x})_{[2]}\right)^2 \end{bmatrix} \tag{15.98}$$

Bocklet, et al. [6] show their best results using *MAP estimation* (Section 11.3.1.3) in conjunction with *polynomial kernels*. They compare it to other kernels such as the pure *Jeffreys kernel* (Section 15.5.1), and the *GRBF kernel* (Section 15.4.3).

15.4.3 Gaussian Radial Basis Function (GRBF) Kernel

The *Gaussian radial basis function* (GRBF) is given by the following definition,

$$\mathcal{K}_{GRBF}(\mathbf{x}_l, \mathbf{x}_m) \overset{\Delta}{=} \exp\left(-\frac{1}{2\sigma^2}\|\mathbf{x}_l - \mathbf{x}_m\|^2\right) \tag{15.99}$$

Because of its relevance to probability modeling, it has been successfully used in many applications such as *age and gender speaker classification* [6], *online hand-writing recognition* [2], and *speaker verification* [17].

15.4.4 Cosine Kernel

The *cosine kernel* of two vectors is a basically the *linear kernel* of the two which is normalized by the product of the norms of both vectors. Namely,

$$\mathscr{K}_{Cos}(\mathbf{x}_l, \mathbf{x}_m) \triangleq \frac{\langle \mathbf{x}_l, \mathbf{x}_m \rangle}{\|\mathbf{x}_l\| \|\mathbf{x}_m\|} \tag{15.100}$$

Dehak, et al. [16, 17] have used the *cosing kernel* for *speaker verification.* [16] uses what it calls the *total variability vectors* as the the observation vectors, $\mathbf{x}$. As we will see in Section 16.4.3, these are really factors which are related to Principal component analysis and not factor analysis since they combine the common factor and the residual space, reverting to PCA – also see the discussion in Section 12.5 on this issue.

On the other hand, [17] uses the *speaker factor coefficients* (Section 16.4.3) for observation vectors. This kernel's performance has been compared by [17] to that of the *linear kernel* (Section 15.4.1) and the *GRBF kernel.* The results show that it slightly outperforms the linear kernel for that task, but it is slightly inferior to the *GRBF kernel.*

15.4.5 Fisher Kernel

The *Fisher kernel* [27], named after *Fisher information* which was discussed and derived in detail in Section 7.7. Equation 7.135 gives the expression for the *Fisher information matrix,* $\mathscr{I}_F$. We also showed the relation between the *Kullback-Leibler directed divergence* and the Fisher information matrix in Equation 7.136. Also, in Equation 7.140, we showed that the *Jeffreys divergence* may also be approximated with respect to the *Fisher information matrix.* Also, at the end of Section 7.7 we showed that the *Fisher information matrix* will always be positive definite. [27] uses the above information to arrive at a kernel based on the *Fisher information matrix* which is give as follows,

$$\mathscr{K}(\mathbf{x}_l, \mathbf{x}_m) \triangleq \nabla_{\boldsymbol{\varphi}} \hat{p}(\mathbf{x}_l|\boldsymbol{\varphi})^T \mathscr{I}_F^{-1} \nabla_{\boldsymbol{\varphi}} \hat{p}(\mathbf{x}_l|\boldsymbol{\varphi}) \tag{15.101}$$

where

$$\varsigma(\boldsymbol{\varphi}|\mathbf{x}) \triangleq \nabla_{\boldsymbol{\varphi}} \hat{p}(\mathbf{x}_l|\boldsymbol{\varphi}) \tag{15.102}$$

is known as the *Fisher score* or the *score statistic* of the parameter vector. The Fisher score has an important role in maximum likelihood estimation, as shown in Section 10.1.

Later in their original paper [27] on this kernel, *Jaakkola and Haussler* proposed that the $\mathscr{I}_F$, in the context of logistic regression models of their interest, becomes less significant and that a simpler kernel may be used as follows,

$$\mathscr{K}(\mathbf{x}_l, \mathbf{x}_m) \triangleq \nabla_{\boldsymbol{\varphi}} \hat{p}(\mathbf{x}_l|\boldsymbol{\varphi})^T \nabla_{\boldsymbol{\varphi}} \hat{p}(\mathbf{x}_l|\boldsymbol{\varphi}) \tag{15.103}$$

The simple form shows that using the Fisher kernel is similar to finding a separating linear hyperplane in the *Fisher score* space [27]. This attests to the use of kernel mapping in order to linearize an otherwise nonlinear classification boundary, discussed at the beginning of this chapter. In Section 7.7 and Chapter 10 we discussed the invariability of the Fisher information matrix to invertible transformations. This gives the Fisher kernel a very important invariance property [27].

The Fisher kernel has been successfully used by many researchers. Reference [19] uses it in building a hybrid GMM/SVM system for conducting speaker identification (see Section 15.9). Also, [36] has used the *Fisher kernel* to do audio classification on audio files on the *world wide web*. The application spiders through audio files on the *Internet* and classifies them into three categories of *speech*, *music*, and *other*. These two papers seem to be among some of the earliest papers that have used support vector machines for speaker recognition.

15.4.6 GLDS Kernel

The *generalized linear discriminant sequence* (GLDS) Kernel was proposed by *Campbell* [12]. As we will see momentarily, it seems to be very similar to the *Fisher kernel* discussed in Section 15.4.5. In fact it almost seems like an implementation of the *Fisher kernel* to the speaker recognition domain. It uses generalized linear discriminant function of the form,

$$o(\mathbf{x}) = \boldsymbol{\omega}^T \boldsymbol{\psi}(\mathbf{x}) \tag{15.104}$$

where $\boldsymbol{\omega}$ is the vector of classifier parameters, much in the same way as we had defined it in Equation 15.7, with the bias term set to 0. [12] then defines the *GLDS kernel* as follows,

$$\mathscr{K}(\mathbf{x}_l, \mathbf{x}_m) = \overline{\boldsymbol{\psi}(\mathbf{x}_l)}^T \left(\frac{1}{N_{imp}} \boldsymbol{\Psi}^T \boldsymbol{\Psi} \right)^{-1} \overline{\boldsymbol{\psi}(\mathbf{x}_m)} \tag{15.105}$$

where N_{imp} is the number of impostors, $\overline{\boldsymbol{\psi}(\mathbf{x}}$ specifies the mean value of $\boldsymbol{\psi}$ over all the speakers and $\boldsymbol{\Psi}$ is a matrix whose rows are the transpose of $\boldsymbol{\psi}$ for all speakers including target and impostor speakers.

As we mentioned at the beginning of this section, the *GLDS kernel* seems to be quite related to the *Fisher kernel*. In fact it seems to be an estimate of that kernel, using the speaker data.

The *GLDS kernel* has also been used by [42] in conjunction with *NAP* (Section 15.8) in speaker verification across different channels and compared to a *GMM technique* (Section 13.7) using *FA* (Section 16.4.2). The results of the *GMM* set up were slightly more favorable.

15.4.7 GMM-UBM Mean Interval (GUMI) Kernel

Recall the expression for the *Jeffreys divergence* between two normal density functions. The last term in that expression carries most of the information. It also shows up in the Bhattacharyya divergence (Section 8.2.3). Reference [50] takes that portion and uses it as the basis for building the, so called, *GUMI kernel*. This partial divergence is given by

$$\mathscr{D}_G\left(0 \leftrightarrow 1\right) \frac{1}{2\ln(2)}(\boldsymbol{\mu}_1 - \boldsymbol{\mu}_0)^T (\boldsymbol{\Sigma}_1^{-1} + \boldsymbol{\Sigma}_0^{-1})(\boldsymbol{\mu}_1 - \boldsymbol{\mu}_0) \tag{15.106}$$

where we have included the $\ln(2)$ term to have the result in *bits* versus *nats* (see Section 7.3.1).

Therefore, dropping the constant $\ln(2)$ term, the *GUMI kernel* for two GMMs would be given by the following expression for the two random variables, X_l and X_m, where the number of mixtures is Γ,

$$\mathscr{K}\left(X_l, X_m\right) = \sum_{\gamma=1}^{\Gamma} \left(\left[\left(\frac{\boldsymbol{\Sigma}_\gamma^{(l)} + \boldsymbol{\Sigma}_\gamma^{(u)}}{2} \right)^{-\frac{1}{2}} (\boldsymbol{\mu}^l - \boldsymbol{\mu}^{(u)}) \right]^T \right.$$
$$\left. \left[\left(\frac{\boldsymbol{\Sigma}_\gamma^{(m)} + \boldsymbol{\Sigma}_\gamma^{(u)}}{2} \right)^{-\frac{1}{2}} (\boldsymbol{\mu}^m - \boldsymbol{\mu}^{(u)}) \right] \right) \tag{15.107}$$

where $\boldsymbol{\mu}^{(u)}$ and $\boldsymbol{\Sigma}^{(u)}$ are the mean and covariance of the *universal background model* (Section 16.2).

15.5 Non Positive Semi-Definite Kernels

In the treatment of *support vector machines*, we are mostly concerned with Mercer kernels, although other kernels are also sometimes used. Examples of kernels which do not satisfy the *Mercer condition* are those which are based on the *Kullback-Leibler divergence* [32, 10] and the three-layer *neural network kernel* [1]. We will introduce these Kernels in the next two subsections.

15.5.1 Jeffreys Divergence Kernel

In Chapters 7 and 8, we discussed the *Jeffreys divergence* and *Kullback-Leibler directed divergence* in quite a bit of detail. *Moreno, et al.* [32] refer to the kernel discussed in this section as the *symmetric Kullback-Leibler divergence*, which is actually the *Jeffreys divergence*. They start by using a *GMM* model to transform the observation vectors from the observation space to the probability densities given the parameters, $\boldsymbol{\theta}$, where $\boldsymbol{\theta}$ includes the statistics of the GMM, such as the means, variances, and priors. Therefore, using the *kernel trick* (Section 15.3.1), the kernel would be computed in the new space of probability densities as follows

$$\mathcal{K}\left(p(\mathbf{x}|\boldsymbol{\theta}_l), p(\mathbf{x}|\boldsymbol{\theta}_m)\right) = e^{\{-a\mathcal{D}_J(l \leftrightarrow m) + b\}} \tag{15.108}$$

where $\mathcal{D}_J(l \leftrightarrow m)$ (Equation 8.18) is the *Jeffreys divergence* and a and b are arbitrary constants. In Section 8.2.2 we presented the expression for the Jeffreys divergence between two normal density functions (Equation 8.19).

Moreno, et al. [32] exponentiate the divergence to force the kernel to be positive semi-definite, since they state that *Jeffreys divergence* does not meet the *Mercer condition* (Equation 24.240). They state that this kernel is appropriate for variable length sequential data and that it works well in conjunction with established probabilistic models such as GMM. Experiments are presented on speaker and image recognition data. The speaker data used in [32] is the *KING* database (Section 22.6.2.3).

However, [17] and [6] use the Jeffreys divergence directly, without doing the exponentiation. [17] uses it in doing speaker verification. [6] applies this kernel as well as other kernels such as the *polynomial kernel* (Section 15.4.2) and the *GRBF kernel* (Section 15.4.3) on the problem of *gender* and *age speaker classification* (Section 17.5.1).

Reference [50] proposes a kernel for SVM which is based on the Bhattacharyya divergence (Section 8.2.3) and works out the expression using a GMM, much in the same manner as the Jeffreys divergence kernel was produced, above. The reader is referred to [50] for more.

15.5.2 Fuzzy Hyperbolic Tangent (tanh) Kernel

Camps-Valls, et al. [13] uses a fuzzy activation (sigmoid) function, adopted from [41], as the kernel for support vector machine classification. The following Equation summarizes this kernel,

$$\mathscr{K}_{FT}(\mathbf{x}_l, \mathbf{x}_m) \triangleq \begin{cases} -1 & \text{when } \mathbf{x}_l^T \mathbf{x}_m \text{ is low} \\ 1 & \text{when } \mathbf{x}_l^T \mathbf{x}_m \text{ is high} \\ m\mathbf{x}_l^T \mathbf{x}_m & \text{when } \mathbf{x}_l^T \mathbf{x}_m \text{ is medium} \end{cases} \tag{15.109}$$

where m is a constant, representing the smoothness of the sigmoid function. To produce a continuous kernel function, [13] expresses the kernel in terms of two numbers, r and a, where the location of the threshold is given by,

$$\gamma = -\frac{r}{a} \tag{15.110}$$

and hence the membership limits are given by $\gamma \pm \frac{1}{a}$.

The new smooth kernel function may then be written in terms of r and a as follows,

$$\mathscr{K}_{FT}(\mathbf{x}_l, \mathbf{x}_m) = \begin{cases} -1 & \forall \mathbf{x}_l^T \mathbf{x}_m \leq \gamma - \frac{1}{a} \\ 1 & \forall \mathbf{x}_l^T \mathbf{x}_m \geq \gamma + \frac{1}{a} \\ 2(\mathbf{x}_l^T \mathbf{x}_m - \gamma) - a^2(\mathbf{x}_l^T \mathbf{x}_m - \gamma)|\mathbf{x}_l^T \mathbf{x}_m - \gamma| & \text{Otherwise} \end{cases}$$

$$\tag{15.111}$$

The *fuzzy tanh kernel*, given by Equation 15.111, is not positive semi-definite. The paper claims that this kernel provides more positive eigenvalues in the *Gram matrix* (Definition 24.59) and a lower computation cost compared to a regular *tanh* kernel. Also it demonstrates the relation between *support vector machines* and *neural networks* to a certain extent. We discussed this relation at the beginning of this chapter, in some detail.

15.5.3 Neural Network Kernel

At the beginning of this chapter, we touch upon the relation between *neural network theory* and *support vector machines*. A *neural network kernel* is basically designed after the *logistic activation function* of Equation 14.15. Namely,

$$\mathcal{K}_{NN}(\mathbf{x}_l, \mathbf{x}_m) = \frac{1}{1 + \exp\left(a\mathbf{x}_l^T \mathbf{x}_m - b\right)} \tag{15.112}$$

where a and b are constant parameters, defining the kernel. In general, $\mathcal{K}_{NN}$ is not a *Mercer kernel*. However, by choosing the proper combination of a and b, it may be made positive semi-definite [1]. This kernel simulates a neural network with one hidden layer. [43] presents another neural network kernel based on the hyperbolic tangent (tanh).

15.6 Kernel Normalization

As a practical note, depending on the size of the problem and the number of variables involved, the value of the *Kernel* may become either too small or too large for the precision at hand. With kernels as with any other function, proper normalization may make a significant difference in the practical results, although it may not make much of a difference in the theoretical sense. We have already seen an example of such normalization. As we saw in Section 15.4.4, the *cosine kernel* is basically a normalized *linear kernel*. We saw that for example, [17] has obtained better verification results with the cosine kernel in contrast with the linear kernel.

The main idea with most normalization techniques is to divide the variable entity either by its maximum value (if it is always positive) or some scale which is related to the difference between its maximum and minimum operating values. We will see many examples of this in Chapter 18. In this spirit, for example, Abe [1] suggests using the maximum values for the polynomial and GRBF kernel to normalize them. These work out to the following two normalized equations respectively,

$$\mathcal{K}_{Poly}(\mathbf{x}_l, \mathbf{x}_m) = \frac{\left(\mathbf{x}_l^T \mathbf{x}_m + 1\right)^d}{(N+1)^d} \tag{15.113}$$

and

$$\mathcal{K}_{Gauss}(\mathbf{x}_l, \mathbf{x}_m) \triangleq \exp\left(-\frac{1}{2\sigma^2 N}\|\mathbf{x}_l - \mathbf{x}_m\|^2\right) \tag{15.114}$$

where N is the dimension of the observation vector, $\mathbf{x}$. Similar ideas may be applied to other kernels in order to ensure that they would have values within the precision of our computations.

15.7 Kernel Principal Component Analysis (Kernel PCA)

In Section 12.3.1 we briefly touch upon the idea behind *kernel PCA* which was introduced by [38]. It uses ideas from support vector machines to transform the observation into a feature space which may possibly become infinite and form a *Hilbert space*. Recall the Eigenvalue problem associated with the original PCA formulation, given by Equation 12.2, which we are rewriting here for convenience.

$$\Sigma \mathbf{v} = \lambda \mathbf{v} \tag{15.115}$$

Let us assume that we will be using a maximum likelihood estimate for the covariance matrix given by Equation 11.12. In fact, we have simplified that Equation to one with standard (**0**) mean and having a single density ($\Gamma = 1$). Then Equation 11.12 may be written as,

$$\hat{\Sigma} = \frac{1}{N} \sum_{n=1}^{N} \mathbf{x}_n \mathbf{x}_n^T \tag{15.116}$$

Now let us refer to Section 15.1 and recall the definition of $\boldsymbol{\psi}(\mathbf{x})$ such that $\boldsymbol{\psi}$: $(\mathscr{X}, \mathfrak{X}) \mapsto (\mathscr{H}, \mathfrak{H})$. In addition, let us assume that the data in the $\mathscr{H}$ space also has the standard mean of 0. Therefore,

$$\sum_{n=1}^{N} \boldsymbol{\psi}(\mathbf{x}_n) = \mathbf{0} \tag{15.117}$$

Therefore, the covariance of the transformed data in the feature space is given by,

$$\bar{\Sigma} = \frac{1}{N} \sum_{n=1}^{N} \boldsymbol{\psi}(\mathbf{x}_n) \boldsymbol{\psi}(\mathbf{x}_n)^T \tag{15.118}$$

Therefore, the Eigenvalue problem of for the transformed covariance matrix becomes,

$$\bar{\Sigma} \mathbf{v} = \lambda \mathbf{v} \tag{15.119}$$

Since all the Eigenvectors of $\boldsymbol{\psi}(\mathbf{x})$ are spanned in the same space as $\boldsymbol{\psi}$, we may write the above equation by applying an inner product of $\boldsymbol{\psi}$ to both sides of Equation 15.119,

$$\langle \boldsymbol{\psi}(\mathbf{x}_n), \bar{\Sigma} \mathbf{v} \rangle = \lambda \langle \boldsymbol{\psi}(\mathbf{x}_n), \mathbf{v} \rangle \quad \forall n \in \{1, 2, \cdots, N\} \tag{15.120}$$

Therefore, we may expand the Eigenvectors with respect to the transformations, $\boldsymbol{\psi}(\mathbf{x}_n)$ as follows,

$$\mathbf{v} = \sum_{n=1}^{N} c_n \boldsymbol{\psi}(\mathbf{x}_n) \tag{15.121}$$

Therefore, we may define a kernel as

$$\mathcal{K}(\mathbf{x}_l, \mathbf{m}) = \langle \boldsymbol{\psi}(\mathbf{x}_l), \boldsymbol{\psi}(\mathbf{x}_m) \rangle \tag{15.122}$$

If we call the *Gram matrix* (Definition 24.59) associated with the above kernel, $\mathcal{K}$, then it produces the following new Eigenvector problem,

$$N \lambda \mathcal{K} \mathbf{c} = \mathcal{K}^2 \mathbf{c} \tag{15.123}$$

where $\mathbf{c} : \mathcal{R}^1 \mapsto \mathcal{R}^N$ is such that

$$(\mathbf{c})_{[n]} = c_n \tag{15.124}$$

associated with Equation 15.122.

[38] argues that we may only concern ourselves with the subset of solutions that would satisfy the following Equation,

$$N \lambda \mathbf{c} = \mathcal{K} \mathbf{c} \tag{15.125}$$

Note that Equation 15.125 is an Eigenvalue problem which may be solved by the methods discussed in Chapter 12. Also note that for PCA, we only need the projection of the feature vectors ($\boldsymbol{\psi}(\mathbf{x}_n)$ onto the principal axes, so we only need the *Karhunen-Loève Transformation* (KLT) given by Equation 12.12,

$$\mathbf{y}_n = \mathbf{V}^T \boldsymbol{\psi}(\mathbf{x}_n) \tag{15.126}$$

Therefore, we may write this transformation as

$$\langle \mathbf{v}, \boldsymbol{\psi}(\mathbf{x}) \rangle = \sum_{n=1}^{N} c_n \langle \boldsymbol{\psi}(\mathbf{x}_n), \boldsymbol{\psi}(\mathbf{x}) \rangle \tag{15.127}$$

$$= \sum_{n=1}^{N} c_n \mathcal{K}(\mathbf{x}_n, \mathbf{x}) \tag{15.128}$$

As we see, Equation 15.128 only involves the kernel and as we have seen before, we may replace this kernel with any other kernel. This is the basis for *kernel PCA* which allows similar properties as we have seen with support vector machines to be applied to the projection task.

15.8 Nuisance Attribute Projection (NAP)

The name *Nuisance attribute projection* has been attributed to a method which is related to the kernel PCA technique discuss in Section 15.7, with some supervision involved, relating to moving certain data points away from each other and moving some closer, depending on some predefined attribute. It was first introduced by [40] and since then it has become somewhat of a standard technique to use on the speaker verification problem, when modeling the system with support vector machines.

As the reader would have probably guessed by now, the data attributes which need to be made closer in the feature space are the microphone and channel attributes. This would basically take away the variation of the data based on microphone and channel types. On the other hand, the aspects of the data relating to the identity of speakers should be separated by as much as possible. This will reduce channel variability effects while providing better speaker separability. It is somewhat in the same train of thought as *discriminant analysis*.

The way [40] approaches this problem is to first assume that there will a *rank-one* removal of some attributes from the identity matrix. This is somewhat similar to the *rank one* update techniques discussed in detail in Chapter 25, with the exception that a positive definite matrix is subtracted from the identity matrix to create the new projection matrix. This is possibly the basis for the name suggesting the removal of nuisance information. This amounts to the following, projection matrix,

$$\mathbf{P} \triangleq \mathbf{I} - \mathbf{w}\mathbf{w}^T \tag{15.129}$$

where the norm of $\mathbf{w}$ should be bounded by 1,

$$\|\mathbf{w}\|_{\mathscr{E}} = 1 \tag{15.130}$$

Using this projection matrix, the *feature vectors*, $\boldsymbol{\psi}(\mathbf{x})$, may be transformed as follows,

$$\hat{\boldsymbol{\psi}}(\mathbf{x}) = \mathbf{P}\boldsymbol{\psi}(\mathbf{x}) \tag{15.131}$$

where $\hat{\boldsymbol{\psi}}(\mathbf{x}$ is the projected feature vector.

Now let us consider the matrix, $\boldsymbol{\Psi}$, such that,

$$(\boldsymbol{\Psi})_{[n_a]} = \boldsymbol{\psi}(\mathbf{x}_{n_a}) \ \forall \ n_a \in \mathcal{N}_a \tag{15.132}$$

Then, we may write the *Gram matrix* (Definition 24.59) for the kernel associated with the *feature vectors*, $\boldsymbol{\psi}(\mathbf{x}_{n_a})$ as

$$\mathcal{K} = \boldsymbol{\Psi}^T \boldsymbol{\Psi} \tag{15.133}$$

Therefore, the *transformed Gram matrix* (associated with $\boldsymbol{\psi}(\mathbf{x}_{n_a})$) may be written as follows,

$$\dot{\mathscr{K}} = (\mathbf{P}\boldsymbol{\Psi})^T (\mathbf{P}\boldsymbol{\Psi}) \tag{15.134}$$

$$= \mathscr{K} - (\mathscr{K}\mathbf{v})(\mathscr{K}\mathbf{v})^T \tag{15.135}$$

where $\mathbf{v}$ satisfies the following relation,

$$\mathbf{w} = \boldsymbol{\Psi}\mathbf{v} \tag{15.136}$$

The relation of Equation 15.136 will ensure that the vectors $\mathbf{v}$ are normalized with respect to Gram matrix $\mathscr{K}$, since if we start with Equation 15.130, we will have the following relation,

$$\mathbf{w}^T \mathbf{w} = 1$$

$$= (\boldsymbol{\Psi}\mathbf{v})^T (\boldsymbol{\Psi}\mathbf{v}) \tag{15.137}$$

$$= \mathbf{v}^T (\boldsymbol{\Psi}^T \boldsymbol{\Psi})\mathbf{v} \tag{15.138}$$

$$= \mathbf{v}^T \mathscr{K} \mathbf{v} \tag{15.139}$$

As a first objective, [40] assumes that there are only two types of channels, one which is associated with *electret microphones* and one associated with *carbon button microphones* (Section 22.3). Therefore, if we call these two sets, $\mathscr{M}_e$ and $\mathscr{M}_c$ respectively, the objective is to find $\mathbf{w}$ in the following optimization problem,

$$\mathbf{w}^* = \underset{\substack{l \in \mathscr{M}_e \\ m \in \mathscr{M}_c}}{\arg\min} \| \mathbf{P}\left(\boldsymbol{\psi}(\mathbf{x}_l) - \boldsymbol{\psi}(\mathbf{x}_l)\right) \|_{\mathscr{E}}^2 \tag{15.140}$$

[40] shows that the solution of the above optimization problem is coincident with the solution to the following Eigenvalue problem,

$$\mathscr{K}\mathbf{Z}\mathscr{K}\mathbf{v} = \hat{\lambda}\mathscr{K}\mathbf{v} \tag{15.141}$$

where $\mathbf{Z}$ is a contrast matrix which is defined as follows,

$$\mathbf{Z} \overset{\Delta}{=} diag(\mathbf{W}\bar{\mathbf{e}}^T) - \mathbf{W} \tag{15.142}$$

such that

$$(\mathbf{W})_{[l][m]} = \begin{cases} 1 & if\, \mathbf{x}_l \ and \ \mathbf{x}_m \ \text{had the different channels} \\ 0 & if\, \mathbf{x}_l \ and \ \mathbf{x}_m \ \text{had the same channel} \end{cases} \tag{15.143}$$

Much in the same way as it was done in Section 15.7, a somewhat similar Eigenvalue problem may be solved in lieu of Equation 15.144, as long as we ensure that Equation 15.140 holds.

Once the Eigenvalue problem is solved for the Eigenvectors, v, the w may be computed using Equation 15.136. This will then provide the projection matrix by Equation 15.129. At this point, provisions have been made to reduce the effect of channel variability. However, this may make matters worse by reducing the speaker separation information since there was no provision made to avoid such change. For this reason, [40] proposes a new Eigenvalue problem where there is a *contrast matrix*, $\mathbf{Z}$, associated with each action. Namely, $\mathbf{Z}_s$ is the contrast matrix which ensures the separation of speakers and $\mathbf{Z}_c$ is one that ensures the closeness of the channel identities. The new Eigenvalue problems becomes,

$$\mathscr{K}\left(\alpha\mathbf{Z}_c - \beta\mathbf{Z}_s\right)\mathscr{K}\mathbf{v} = \lambda\mathscr{K}\mathbf{v} \tag{15.144}$$

where α and β are two positive weighting parameters which should be empirically determined. $\mathbf{Z}_c$ is defined using $\mathbf{W}_c$ which is defined by Equation 15.145. In the same manner, $\mathbf{Z}_s$ is defined using the same relation as in Equation 15.142 with $\mathbf{W}_s$ given by the following,

$$(\mathbf{W}_s)_{[l][m]} = \begin{cases} 1 & if \mathbf{x}_l \ and \ \mathbf{x}_m \text{ are from different speakers} \\ 0 & if \mathbf{x}_l \ and \ \mathbf{x}_m \text{ are from the same speaker} \end{cases} \tag{15.145}$$

As a modification, [40] suggests increasing the rank of the update. This is a natural extension, as we saw in Chapter 25. Therefore, Equation 15.129 may be modified as follows,

$$\mathbf{P} \overset{\Delta}{=} \mathbf{I} - \sum_{q=1}^{Q} \mathbf{w}\mathbf{w}^T \tag{15.146}$$

where Q denotes the rank of the update.

This technique has been used in many systems. [11] uses NAP with the *GUMI kernel*. [14] covers a variety of kernels on the speaker verification problem in conjunction with gender-dependent NAP. [4] and [39] use NAP on a two-wire communication[3].

15.9 The multiclass (Γ-Class) Problem

The multiclass classification problem is quite important, since it is the basis for the speaker identification problem. In Section 10.10 of his book, *Vapnik* [46] proposes a simple approach where one class is compared to all other classes and then this is done for each class. This will convert a Γ-class problem to Γ two-class problems. Abe [1] refers to this as the one-against-all multiclass SVM. This method may be

---

[3] Two-wire communication means that the conversation from the two ends of a telephone call have been multiplexed into one common channel, in contrast with a 4-wire communication in which the two speakers at the different ends may be separated – see Section 16.5.

efficiently implemented using a decision tree.

Of course there are other ways of keeping track of such comparisons. One such method is due to [18], which provides a technique based on error correcting codes for creating a multi-class classifier out of binary classifiers. Since the basic SVM classifier is a binary classifier, this method may be used to keep track of the different comparisons. Fine [19] uses the Fisher kernel and builds a hybrid GMM/SVM system for speaker identification and discusses the multiclass classification problem which is needed for speaker identification, using an N-bit *error correcting code* after the work of [18].

One of the important problems that arises from the, so called, one-against-all method is the existence of regions in the space which may not be classifiable. As we have seen, this is a problem with two-class linearly inseparable problems as well. However, the extent of the problem grows as the number of classes increase. This is a direct problem being addressed by many applications of speaker identification which use support vector machines.

References

1. Abe, S.: Support Vector Machines for Pattern Classification. Advances in Pattern Recognition. Springer-Verlag, London (2005). ISBN: 1-85233-929-2
2. Ahmad, A.R., Khalia, M., Viard-Gaudin, C., Poisson, E.: Online handwriting recognition using support vector machine. In: IEEE Region 10 Conference, vol. 1, pp. 311–314 (2004)
3. Aizerman, A., Braverman, E.M., Rozoner, L.I.: Theoretical foundations of the potential function method in pattern recognition learning. Automation and Remote Control 25, 821–837 (1964). Translated into English from Russian journal: Avtomatika i Telemekhanika, Vol. 25, No. 6, pp. 917–936, Jun 1964
4. Aronowitz, H., Solewicz, Y.A.: Speaker Recognition in Two-Wire Test Sessions. In: Inter-Speech, pp. 865–868 (2008)
5. Beigi, H.: Neural Network Learning Through Optimally Conditioned Quadratically Convergent Methods Requiring NO LINE SEARCH. In: IEEE-36th Midwest Symposium on Circuits and Systems, vol. 1, pp. 109–112 (1993)
6. Bocklet, T., Maier, A., Bauer, J.G., Burkhardt, F., Nöth, E.: Age and gender recognition for telephone applications based on GMM supervectors and support vector machines. In: International Conference on Acoustics, Speech, and Signal Processing (ICASSP), pp. 1605 1608 (2008)
7. Boser, B.E., Guyon, I.M., Vapnik, V.N.: A Training Algorithm for Optimal Margin Classifiers. In: Proceedings of the fifth annual workshop on Computational learning theory, pp. 144–152 (1992)
8. Burges, C.J.: A Tutorial on Support Vector Machines for Pattern Recognition. Data Mining and Knowledge Discovery 2, 121–167 (1998)
9. Campbell, W., Assaleh, K., Broun, C.a.: Speaker recognition with polynomial classifiers. Speech and Audio Processing, IEEE Transactions on 10(4), 205–212 (2002)
10. Campbell, W., Gleason, T., Navratil, J., Reynolds, D., Shen, W., Singer, E., Torres-Carrasquillo, P.: Advanced Language Recognition using Cepstra and Phonotactics: MITLL

System Performance on the NIST 2005 Language Recognition Evaluation. In: The Speaker and Language Recognition Workshop, 2006. IEEE Odyssey 2006, pp. 1–8 (2006)

11. Campbell, W., Sturim, D., Reynolds, D.a.: Support vector machines using GMM supervectors for speaker verification. Signal Processing Letters, IEEE **13**(5), 308–311 (2006)

12. Campbell, W.a.: Generalized linear discriminant sequence kernels for speaker recognition. In: Acoustics, Speech, and Signal Processing, 2002. Proceedings. (ICASSP '02). IEEE International Conference on, vol. 1, pp. I–161–I–164 (2002)

13. Camps-Valls, G., Martin-Guerrero, J., Rojo-Alvarez, J., Soria-Olivas, E.: Fuzzy Sigmoid Kernel for Support Vector Classifiers. Neurocomputing **62**, 501–506 (2004)

14. Chao, J., Huang, W., Zhang, Y.: SVM Based Speaker Verification and Gender Dependent NAP Variability Compensation. In: Bioinformatics and Biomedical Engineering, 2008. ICBBE 2008. The 2nd International Conference on, pp. 710–713 (2008)

15. Courant, R., Hibert, D.: Methods of Mathematical Physics, 1st english edn. Volume I. WILEY-VCH Verlag GmbH & Co. KGaA, Weinheim (2009). Second reprint of English version Translated and Revised from German Original, 1937, ISBN: 0-471-50447-4

16. Dehak, N., Dehak, R., Kenny, P., Brummer, N., Ouellet, P., Dumouchel, P.: Support Vector Machines versus Fast Scoring in the Low-Dimensional Total Variability Space for Speaker Verication. In: InterSpeech, pp. 1559–1562 (2009)

17. Dehak, N., Kenny, P., Dehak, R., Glembek, O., Dumouchel, P., Burget, L., Hubeika, V., Castaldo, F.: Support vector machines and Joint Factor Analysis for speaker verification. In: International Conference on Acoustics, Speech, and Signal Processing (ICASSP), pp. 4237–4240 (2009)

18. Dietterich, T., Bakiri, G.: Solving Multiclass Learning Problems via Error-Correcting Output Codes. Journal of Artificial Intelligence Research **2**, 263–286 (1995)

19. Fine, S., Navratil, J., Gopinath, R.a.: A hybrid GMM/SVM approach to speaker identification. In: Acoustics, Speech, and Signal Processing, 2001. Proceedings. (ICASSP '01). 2001 IEEE International Conference on, vol. 1, pp. 417–420 (2001)

20. Fletcher, R.: Practical Methods of Optimization, 2nd edn. J. Wiley and Sons, New york (2000). ISBN: 0-471-49463-1

21. Ganapathiraju, A.: Support Vector Machines for Speech Recognition. Mississippi State University, Mississipi (2002). PhD Thesis

22. Ganapathiraju, A., Hamaker, J., Picone, J.: Support Vector Machines for Speech Recognition. In: International Conference on Spoken Language Processing, pp. 2348–2355 (1998)

23. Gao, X., Wang, P., Qi, Y., Yan, A., Zhang, H., Gong, Y.: Comparison Studies of LS-SVM and SVM on Modeling for Fermentation Processes. In: International Conference on Natural Computation (ICNC), pp. 478–484 (2009)

24. Gruber, C., Gruber, T., Krinninger, S., Sick, B.: Online Signature Verification With Support Vector Machines Based on LCSS Kernel Functions. IEEE Transactions on Systems, Man, and Cybernetics, Part B **40**(4), 1088–1100 (2010)

25. Guyon, I.M., Vapnik, V.N., Boser, B.E., Solla, S.A.: Structural Risk Minimization for Character Recognition. In: T. David S (ed.) Neural Information Processing Systems, vol. 4. Morgan Kaufmann Publishers, San Mateo, CA (1992)

26. Hilbert, D.: Grundzüge Einer Allgemeinen Theorie der Linearen Integralgleichungen (Outlines of a General Theory of Linear Integral Equations). Fortschritte der Mathematischen Wissenschaften, heft 3 (Progress in Mathematical Sciences, issue 3). B.G. Teubner, Leipzig and Berlin (1912). In German. Originally published in 1904.

27. Jaakkola, T., Haussler, D.: Exploiting Generative Models in Discriminative Classifiers. In: Advances in Neural Information Processing Systems, vol. 11, pp. 487–493. MIT Press (1998)

28. Jaakkola, T., Meila, M., Jebara, T.: Maximum Entropy Discrimination. In: Advances in Neural Information Processing Systems, vol. 12, pp. 470–476. MIT Press (1999)

29. Kim, J., Scott, C.: L2 Kernel Classification. IEEE Transactions on Pattern Analysis and Machine Intelligence **32**(10), 1822–1831 (2010)

30. Latry, C., Panem, C., Dejean, P.: Cloud detection with SVM technique. In: IEEE International Geoscience and Remote Sensing Symposium (IGARSS, pp. 448–451 (2007)

31. Li, H., Zhang, Y.X.: An algorithm of soft fault diagnosis for analog circuit based on the optimized SVM by GA. In: International Conference on Electronic Measurement and Instruments (ICMEI), vol. 4, pp. 1023–1027 (2009)

32. Moreno, P.J., Ho, P.P., Vasconcelos, N.: A Kullback-Leibler Divergence Based Kernel for SVM Classification in Multimedia Applications. In: Advances in Neural Information Processing Systems 16. MIT Press (2004)

33. Nocedal, J., Wright, S.J.: Numerical Optimization, 2nd edn. Springer, New York (2000)

34. Pierre, D.A.: Optimization Theory with Applications. Dover Publications, Inc., New york (1986). ISBN: 0-486-65205-X

35. Reddy, S., Shevadea, S., Murty, M.: A Fast Quasi-Newton Method for Semi-Supervised Classification. Pattern Recognition **In Press, Corrected Proof** (2010). DOI DOI: 10.1016/j.patcog.2010.09.002. URL http://www.sciencedirect.com/science/article/B6V14-5100HJG-3/2/ddad1fa09bf51c3e3b7754415566b061

36. arid Rjjan Rafkin, P.I.M.: Using the Fisher Kernel Method for Web Audio Classification. In: International Conference on Acoustics, Speech, and Signal Processing (ICASSP), vol. 4, pp. 2417 – 2420 (2000)

37. Schölkopf, B., Smola, A.J.: Learning with Kernels: Support Vector Machines, Regularization, Optimization, and Beyond (Adaptive Computation and Machine Learning). MIT Press, Cambridge, MA (2002). ISBN: 978-0-262-19475-4

38. Schölkopf, B., Smola, A.J., Müller, K.R.: Kernel Principal Component Analysis. In: B. Schölkopf, C. Burges, A.J. Smola (eds.) Advances in Kernel Methods. MIT Press, Cambridge, MA (2000)

39. Solewicz, Y.A., Aronowitz, H.: Two-Wire Nuisance Attribute Projection. In: InterSpeech, pp. 928–931 (2009)

40. Solomonoff, A., Campbell, W., Quillen, C.: Channel Compensation for SVM Speaker Recognition. In: The Speaker and Language Recognition Workshop Odyssey 2004, vol. 1, pp. 57–62 (2004)

41. Soria-Olivas, E., Martin-Guerrero, J.D., Camps-Valls, G., Serrano-López, A.J., Calpe-Maravilla, J., Gómez-Chova, L.: A Low-Complexity Fuzzy Activation Function for Artificial Neural Networks. IEEE Transactions on Neural Networks **14**(6), 1576–1579 (2003)

42. Sturim, D.E., Campbell, W.M., Reynolds, D.A., Dunn, R.B., Quatieri, T.: Robust Speaker Recognition with Cross-Channel Data: MIT-LL Results on the 2006 NIST SRE Auxiliary Microphone Task. In: International Conference on Acoustics, Speech, and Signal Processing (ICASSP), vol. IV, pp. 49–52 (2007)

43. Suykens, J., Vandewalle, J.: Training Multilayer Perceptron Classification Based on a Modified Support Vector Method. IEEE Transactions on Neural Networks **10**(4), 907–9011 (1999)

44. Tveit, A., Engum, H.: Parallelization of the Incremental Proximal Support Vector Machine Classifier using a Heap-Based Tree Topology. In: Workshop on Parallel Distributed Computing for Machine Learning (2003)

45. Vapnik, V.N.: Estimation of Dependences Based on Empirical Data, russian edn. Nauka, Moscow (1979). English Translation: Springer-Verlag, New York, 1982

46. Vapnik, V.N.: Statistical learning theory. John Wiley, New York (1998). ISBN: 0-471-03003-1

47. Wang, Y.: A Tree-based Multi-class SVM Classifier for Digital Library Document. In: International Conference on MultiMedia and Information Technology (MMIT), pp. 15–18 (2008)

48. Wolfe, P.: A Duality Theorem for Nonlinear Programming. Quarterly Journal of Applied Mathematics **19**(3), 239–244 (1961)

49. Wu, R., Su, C., Xia, K., Wu, Y.: An approach to WLS-SVM based on QPSO algorithm in anomaly detection. In: World Congress on Intelligent Control and Automation (WCICA), pp. 4468–4472 (2008)

50. You, C.H., Lee, K.A., Li, H.: A GMM Supervector Kernel with the Bhattacharyya Distance for SVM Based Speaker Recognition. In: International Conference on Acoustics, Speech, and Signal Processing (ICASSP), vol. 1, pp. 4221–4224 (2009)

51. Zhang, L., Zheng, B., Yang, Z.: Codebook design using genetic algorithm and its application to speaker identification. IEE Electronic Letters **41**(10), 619–620 (2005)

Part II
Advanced Theory

Chapter 16
Speaker Modeling

The basic objective of speaker modeling is to be able to be able to associate an identifier to the speech of an individual speaker which is different from all other unique speakers, if not in the world, at least in the database of interest. Once this is achieved, all the different branches of speaker recognition, discussed in Chapter 1, may come to fruition. In other words, speaker modeling lies at the heart of the speaker recognition task. This may not necessarily be true with many other seemingly similar fields. For example, speech recognition which is very closely related to speaker recognition requires many different stages, many of which are of similar importance. For instance, in speech recognition, the phonetic modeling, language modeling, and search are almost of similar importance. In speaker recognition, on the other hand, if a good model of the speaker is built, the rest of the work becomes extremely easy.

Of course, with different branches of speaker recognition, we may use different modeling techniques, just because we may want to take advantage of special shortcuts in the said branch. An example is the contrast between speaker verification speaker identification. Of course if we had a good model of each speaker, then speaker verification would just be solvable exactly the same way as we do speaker identification. Consider that you know the identity of an individual as soon as you hear that person, meaning that you can immediately assign a unique ID to the individual. By definition, this is the description of a speaker identification problem. However, if we are sure of the ID, then we may use the information, as it stands, for doing speaker verification, by just seeing if the ID we ascertained is the same as the ID being produced by the speaker or not. Of course since we do not have such great models and since it is quite more expensive to absolutely identify a person, we resort to a simpler problem of deciding if the audio we are hearing is closer to the audio we have heard from the target speaker or if it is closer to some complementary model of the speaker space.

Much in the same way, consider the task of speaker segmentation, say in a 2-wire telephone communication (see Section 16.5), where the speech of the parties at both end of the telephone channel have been multiplexed. Again, if we had a way of identifying each piece of audio specifically, then the identification task would be sufficient to allow us to separate the two speakers in the conversation. Again, since we do not have this ability, we resort to special modeling techniques that would only allow us to tell two voices apart in a small window, giving rise to the specific branch of audio segmentation.

In summary, we may think of speaker identification as the most difficult task and one that if it were to be solved, it would cover all the different branches of speaker recognition. However, as we mentioned, we are no where close to that stage in this technology. So, we would have to look at different modeling techniques. I have personally believed in the past, and still do, that we should work toward the goal of attaining the best speaker identification. [5] shows an approach for establishing such models and to be able to cluster speakers in a hierarchical fashion, so that we may simply conduct other branches of speaker recognition through identification. And the search goes on!

Up to now, we have been dealing with a lot of mathematical modeling and we have tried to set the mathematical foundation needed for being able to deal with all aspects of speaker recognition. From this point on, we will not see much in the way of mathematical models. Most of these models have already been covered in the first 15 chapters of the book or in the extensive background material in Part IV, at the end of the book. Most of what will come, is in the way of reference to content from previous chapters and the chapters of Part IV.

16.1 Individual Speaker Modeling

The speaker recognition community is not very big. Although there are many models which have been proposed for modeling an individual speaker, most of them are somewhat similar and are somehow related to features that stem from speech recognition. At the beginning of Chapter 5, we presented an extensive discussion about the theoretical dichotomy between speech and speaker recognition features. Assuming that we are content with why we should use speech recognition features, we continue to use these features for building models which would carry information about a specific individual.

When we speak of speaker models, it is important, first, to see if we are interested in a text-dependent model, a text-independent model, or a hybrid. In Chapter 1 we spoke about the difference between these different modalities, which are most applicable to speaker verification.

Initially assuming that we are interested in a purely text-independent model, the most common and successful model, and possibly one of the oldest, is the Gaussian mixture model which was discussed in detail in Section 13.7. A text-independent model may of course be used for all branches of speaker recognition. In Chapter 13, we discussed how the GMM is actually a degenerate HMM model. There is definitely no shortage of papers which use a GMM model for modeling speakers [41, 5, 38, 12, 46, 46, 6, 2, 51, 32, 34, 52, 25, 22, 58, 8, 36, 59, 11, 56, 48, 35, 31, 20, 3, 37, 43].

It is interesting to note that the above references are just a fraction of all the papers which report on using GMM for speaker modeling. Besides, I have tried very hard to make sure I do not include more than one paper from each group of researchers. Of course, each and every one of these papers is talking about some different combination of techniques with GMM as its base. Therefore, it is not even possible to address each and every flavor or the usage of GMM for modeling speakers.

One way to quantify a speaker's model is to establish the parameters associated with the GMM which best models the speakers speech. In Section 13.7, we called this supervector, $\boldsymbol{\varphi}$, given by Equation 13.40. Therefore, assuming that we have N speakers in general, we may specify the model associated with speaker n as $\boldsymbol{\varphi}_n$.

Usually, we do not have much data from a single speaker. In most practical cases related to text-independent scenarios, the enrollment data is at best in the order of a minute. Assuming that the number of Gaussians used in the mixture mode, $\Gamma \sim 100$, then it is easy to understand why there is no where close to enough data to be able to estimate the mixture parameters for the speaker. This is one of the reasons why we have to begin with an existing speaker-independent model and then adjust the parameters slightly, based on the information in the limited enrollment data for the said speaker.

This and other reasons that will be made more clear in the next few sections is why we have to build a speaker-independent model. Assuming we have one, using an adaptation technique such as MAP (Sections 10.2 and 11.3.1.3), we are able to adjust the parameter vector, $\boldsymbol{\varphi}_n$, to give us the best performance for the speaker of choice. In the next section we will speak about speaker-independent models.

16.2 Background Models and Cohorts

In the last section we said that we need to have a speaker-independent model which may be used as a starting point in order to be able to make a speaker model. We discussed this in relation to GMM, however, it is true with basically any statistical

model. We are only using GMM as an example. This is one of the reasons for need-
ing to have a speaker-independent model, but there are more reasons which are as
important and which will become more clear as we continue this discussion.

The speaker-independent model has been called by many names. A popular name
for it is a *universal background model* [40, 1] (UBM). Generally, the UBM model
is similar to having a speaker for which the speech of thousands of individuals have
been pooled to create a model of the average features of everyone in the database. Of
course, in the next section we will speak about how the number of Gaussians used in
a UBM may be different and usually much larger than the speaker-dependent model.

16.2.1 Background Models

As we started to say, the background model may be used as a reference model which
has been trained on many thousands of speakers to be able to have a robust and rich
model of speech, and to be able to use it for adapting the individual speaker model.
Usually this initial estimation of the GMM parameters for the background model is
known as training.

In that capacity, for the speaker-independent model, since an abundance of data
is usually available, we may use a lot more Gaussians for this model. Then, assuming that an adaptation technique such as MAP (Sections 10.2 and 11.3.1.3) is used
to adjust the parameters, initially a survey may be done to deactivate certain mixtures which do not seem to be receiving an counts from the enrollment data of the
speaker of choice. Of course this technique usually creates more problems than it
alleviates. For instance, it is easy to see how by having a lot of rich data for a background model, we may end up with a disproportionately large number of Gaussians
compared to the amount of data available from the individual speaker. Then theoretically, in the limit it is possible to have almost each frame associated with a different
Gaussian. Therefore, it is an art to be able to come up with the number of Gaussians
to use with the amount of data at hand.

A second role for a background model is for it to be used as a complementary
model. In Section 1.2.1 we talk about the need for such a complementary model
which would be the anti-model if you will. By having the UBM model trained on a
large population, we have a model of the so called average speaker. However, this
smooth model may become too smooth at times and not be able to provide enough
discrimination. Having such a model allows us to compare a new instance of speech
to the target speaker's model, as well as the background model to decide whether to
accept the test speaker or not, in a verification scenario.

16.2.2 Cohorts

There is a trade-off to using a universal background model or a set of cohorts. Cohort models are models of a population of speakers whose parameters resemble a target speaker's model. At the time of training, different cohort models may be associated with each model. When a test segment is received, it may be compared to the target speaker's model as well as the cohort models. The idea behind having cohort models is that if we have have models for individuals who sound similar to the target speaker and yet are not the target speaker, we have a good understanding of the amount deviation that would tip the authentication from favorable to not. Of course, one we have cohort models, we do not need other models anymore, since it is very likely that if any test segment is rejected by a cohort model, then it would have been rejected by models which are farther from the target speaker's mode.

The trade-off comes in the amount of computation that is need to create and maintain the cohort speaker models as well as the housekeeping effort involved in associating speakers with their cohorts [5]. Also having a single UBM means that for every verification, only two matches need to be done, one to the target model and another one to the UBM. However, as the number of cohorts grows, the computation load increases. However, if we have a set of good cohort models and if we know how to associate cohort models with target speakers, then in the limit we would approach the golden speaker identification scenario of which we spoke at the beginning of this chapter.

16.3 Pooling of Data and Speaker Independent Models

As we mentioned in the previous two sections, there is a trade-off between having a UBM or cohort models. In case of cohort models, the trade-off continues to get larger once we increase the number of these models. The is a compromise which reduces the smoothness of the UBM to the point that at least some distinction is made between competing background models. This is the selective pooling of the training data. A natural choice is to pool all the male speakers into one UBM and all the females into another. In this specific case, we will have to do three comparisons for each verification. Of course this may continue by making further classifications where in the limit we would go from a UBM-based system to a cohort-base verification system.

Of course it is important to note that as we discussed in Chapter 1, background models are not only used for verification. They may be used with any branch of speaker recognition in order to have a rejection scheme. For example, in the speaker identification case, they provide the, so called, open-set speaker identification platform (Section 1.2.2). In the case of segmentation, also, they provide a rejection scheme in order to be able to tag a segment of audio as an unknown segment or one that has come from an unknown or unenrolled speaker.

16.4 Speaker Adaptation

The methodology for adapting the model of a speaker has a lot in common with the actual identification and verification task. The first step before adaptation is done is to decide what model to use as the basis for the adaptation. This is where identification may come into place. An initial identification, although crude and inaccurate, may be able to help us choose the basic model to use for the adaptation. For example, let us say that we happen to have chosen a gender-based UBM technique. In that case, we will have a general UBM which may be trained on all types of speakers. We will then have two individual UBMs, each of which may have been trained on data from males or females. Once we have a new enrollment, we may decide to ask the gender of the individual and to use that information to assign a basic model for the starter model. However, this is not at all practical.

First, obtaining the gender information will produce a lot of extra work and it may create practical blocks as well. In most cases, it would not be permitted to ask the gender of the individual. Also, even if it is legally feasible to do so, the venue for obtaining this information may be costly. In addition, not all female speakers would pass for a female and not all male speakers would pass for a male. It is far better to conduct an initial gender classification and once the gender of the individual is obtained from the models (even if incorrect), the closest model may be used as the basis for the adaptation.

In Chapter 21 we will show a case study of speaker adaptation along the time line. This is most useful for alleviating time-lapse effects which are the topic of Chapter 20.

16.4.1 Factor Analysis (FA)

In Section 12.5 we went through a detailed description and formulation of factor analysis. [28] started using factor analysis ideas in order to separate channel dependent and speaker dependent features through projection. Kenny [28] calls this usage for factor analysis, *joint factor analysis* (JFA). In the next section, we briefly describe it in relation with the content of Section 12.5. Also, we will see in Section 16.4.3 that although JFA is expected to identify channel and speaker variability, the information still leaks from one information source to the other. In fact this was enough to have some researchers revert back to a single projection solution, namely PCA.

16.4.2 Joint Factor Analysis (JFA)

Factor analysis has been used in pattern recognition for a long time [42]. Recently, it has been used to handle the channel variation problem in the form of an adaptation technique that would model the target speaker's vocal characteristics in conjunction with the channel effects in the form, called *joint factor analysis*, as it is apparent in a large number of recent publications on the subject [27, 9, 51, 32, 23, 57]. There have been treatments of different features, including prosodic ones [17] in recent literature.

Most of the attempts on modeling speakers have been along the line of modeling the mean vectors associated with Gaussian mixture models. As we mentioned in Section 12.5, *joint factor analysis* models the data by splitting it into two components: a *speaker-dependent component* and a *channel-dependent component*.

$$\boldsymbol{\mu} = \boldsymbol{\mu}_s + \boldsymbol{\mu}_c \qquad (16.1)$$

where $\boldsymbol{\mu}$ is the overall mean vector, $\boldsymbol{\mu}_s$ is the *speaker-dependent mean vector* and $\boldsymbol{\mu}_c$ is the *channel-dependent mean vector*. In Equation 16.2, we assume that the speaker model consists of a Gaussian mixture with Γ Gaussians of D-dimension. Of course in JFA, the means are not simply added, as will be seem shortly.

The reason this technique is called joint factor analysis is that it produces a factor analysis based on the speaker and one based on the channel. Please refer to our thorough treatment of factor analysis in Section 12.5. Especially, refer to Equation 12.58 which basically corresponds to one of the Equations in JFA, namely the one for the speaker side of things. In addition to this Equation, the result, which is known to be the speaker dependent mean, is used to do a separate analysis with the channels, where in this case it is really a PCA with the mean vector being the speaker mean computed from Equation 12.58. Here, we repeat equation 12.58 with a slight modification to the notation. We will call the output of that equation $\mu^{(s)}$ to show its relation to the speaker side of things,

$$\boldsymbol{\mu}_n^{(s)} = \boldsymbol{\mu} + \mathbf{V}\boldsymbol{\theta}_n + \boldsymbol{\Lambda}\mathbf{r}_n \qquad (16.2)$$

$$\boldsymbol{\mu}_n^{(c)} = \boldsymbol{\mu}_n^{(s)} + \mathbf{U}\mathbf{x}_{n,c} \qquad (16.3)$$

The elements of Equation 16.2 have been defined in detail in Section 12.5. In Equation 16.3, c is the index into the number of channel index. $\mathbf{U} : \mathscr{R}^C \mapsto \mathscr{R}^D$ is similar to the role of $\mathbf{V}$, only it operates on the channel data. Also, $\mathbf{x}_{n,c} : \mathscr{R}^1 \mapsto \mathscr{R}^C$ is similar to $\boldsymbol{\theta}_n$, only it is with respect to the channel, where $\boldsymbol{\theta}_n$ pertains to the speaker. $\mathbf{x}_{n,c}$ has two indices since it is dependent on the recording from speaker n with channel c. JFA basically solves the combination of Equations 16.2 and 16.3 simultaneously, trying to separate speaker and channel effects.

16.4.3 Total Factors (Total Variability)

In Section 12.5 we presented a full treatment of *factor analysis*. In that section, Equation 12.54 basically states that the common factors and the specific factors are uncorrelated. However, [16] postulates that the *mean vector* associated with the channel variability, $\boldsymbol{\mu}_c$, still contains some speaker-related information, by showing, empirically, that speakers may still be somewhat identified using this information.

For this reason, *Dehak, et al.* [16] propose recombining the *speaker variability* and the *channel variability spaces* back into one, essentially ending with,

$$\mathbf{y}_n = \boldsymbol{\mu} + \mathbf{V}\boldsymbol{\theta}_n \tag{16.4}$$

They call $\boldsymbol{\theta}_n$, *total variability*, and the space associated with it, the *total variability space* [14]. However, as we discussed in detail, in Section 12.5, they are basically reverting back to *PCA*, since they are removing the residual term which is one of the main factors that differentiates factor analysis from PCA. Nevertheless, aside from the problem with the terminology, it is perfectly fine to use PCA techniques for performing speaker recognition. In later incarnations of their work [15, 45, 18], *Dehak et al.* have called these vectors *i-vectors*. They have used these vectors in conjunction with support vector machines, employing the *Cosine kernel* (Section 15.4.4) as well as others.

speaker factor coefficients are related to the speaker coordinates in the where each speaker is represented as a point. This space is defined by the *Eigenvoice matrix*. These speaker factor vectors are relatively short, having in the order of about 300 elements [19], which makes them desirable for use with *support vector machines*, as the observed vector in the observation space ($\mathbf{x}$).

16.5 Audio Segmentation

Audio segmentation is one of the challenges faced in processing telephone or recorded speech. Consider telephone speech for the moment. Majority of telephone conversations take place between two individuals at the different ends of a channel. It is conceivable to record a two-party conversation into two separate channels using digital telephony channel such as ISDN-PRI channels (also known as *four-wire telephone channel*). However, this kind of recording or interception is not available in most scenarios. Most conversations take place on a multiplexed audio channel (also known as *two-wire telephone channel*) with no access to the separate audio channels that make up the conversation. In many cases, even when there is a T-1, E-1 or any other digital telephony service available, the speaker recognition process may not have access to the individual channels.

One solution is to devise techniques for separating the multiplexed audio into individual streams. In the two-conversation example, there would be an assumption that two distinct speakers are on the line. In that case, the goal would be the generation of two separate streams, each only containing the voice of one of the parties on the telephone call.

If we are definitely sure that there are two speakers in a conversation, we can address the problem of segmentation by first deploying a *turn detection* algorithm which would split the original audio into small segments where interesting changes happen at the border lines. Examples of these lower level segmentation techniques are the *Bayesian Information Criterion* (BIC) [13] and *model-based segmentation* [26, 4].

Once we have this over-segmented audio stream, we may utilize any of an array of different techniques which mostly utilize speaker recognition methods to tag and merge these small segments. At the end these techniques arrive at a stream with the individual segments alternatively tagged as speakers A and B. Some methods use prior knowledge of the speakers' identities to recognize the segments and to tag them with proper labels. Others only postulate segments with alternate tags (A and B).

Just as *open-set speaker identification* does, it is important for a *two-speaker segmentation* system to have the capability of tagging parts of the audio as *unknown*. It is possible to use an two-speaker *open-set speaker identification* system to talk the different parts of speech. In most cases, speaker models are made on-the-fly and once identification models are created, they are used to tag the different segments. Another possibility is to run a *classification* system through the over-segmented audio and to through out parts of the audio which are tagged as non-speech. After this, the *close-set limited identification* system can tag the rest of the audio as speakers A and B only, ignoring the rejection task.

Of course, in the couple of decades, there has been an increase in the number of multi-party conference calls that take place. These conversations make the job of audio segmentation quite difficult. With multi-party conversations, in most cases the number of speakers on the conference call is unknown. Even if one can think up scenarios where the number of speakers is predetermined (say through a registration process to access the conference call), still it is unclear how many speakers would actually participate in the conversation. In most such multi-part conversations, the distribution is far from uniform. In the majority of cases, one or two people conduct most of the conversation while others just listen in.

Not knowing the exact number of speakers in a conference call and specifically lack of knowledge about the distribution of length of speech from different individuals in the call, makes it an extremely challenging problem. For this reason, most of the effort in the literature has been focused on solving the two-speaker segmentation problem. Multi-speaker analysis has been greatly ignored due to its difficulty.

Factor analysis has been used for speaker diarization as well. [10] uses simple factor analysis only containing Eigenvoices. [10] presents a comparison among different techniques deploying *joint factor analysis* and *variational Bayesian* techniques.

16.6 Model Quality Assessment

It is desirable to classify models by their quality. This allows for taking action when necessary. There are different perspectives in what causes quality issues in models. Some assumptions point toward the existence of outlier utterances in the enrollment data and followers of this belief generally try to identify these outliers and to remove them in order to provide a better quality model. In general, this approach, if more than one outlier exists, they rate the whole model as a *low quality* model.

In a different light some researchers, in biometrics as a whole, entertain the idea that different people perform differently and that they may be clustered into groups with varying verification qualities. In this mindset, an analogy has been made to animals, initially to farm animals and later to more exotic animals, in their behavior. Therefore the, so-called, *biometric menagerie*[1] was born. In Section 16.6.2 we will discuss some of these ideas.

16.6.1 Enrollment Utterance Quality Control

Gu, et al. [24] address the *utterance quality control* problem by using the *leave-one-out* (Section 13.8.4) technique in order to identify a trouble or outlier utterance among N utterances using in an enrollment session. In this approach, each time a

---

[1] Before the advent of zoos, a *menagerie* was a place where exotic animals were trained and kept for exhibition.

distinct list of $N-1$ utterances out of the N available utterances is used for building a model and the Nth utterance is used to do a cross validation. This will leave us with N models and N corresponding scores. A *threshold* is used to throw away the utterance associated with the worst scoring model. If more than one utterance is identified as an *outlier*, then the model is simply tagged as being a low quality model. At this time, different decisions may be made about the use of the model, which depend on the application.

Saeta and Hernando [44] present a different quality measure to throw away utterance in the enrollment data which are not deemed of high quality, hence replacing them with newly solicited sentences from the target speaker at the time of enrollment. They show a 40% increase in their baseline accuracy by doing this for a test in Spanish over a population of 184 speakers. The approach uses at least 5 utterances for each speaker ($N == 5$), but practically many more (in the order of 16 to 48). The score that [44] uses is related to the discrimination capability of the model between target and impostor utterances, but instead of impostors, they use the different utterances from the target speaker's enrollment utterance pool. To build the theory, [44] builds upon concepts discussed in [30], using a discrimination measure derived from the *zero normalization (z-norm)* (Section 18.5.1) of the mean of the target *log-likelihood ratios (LLR)* (Section 7.6) which may be given by the following formula [29, 30],

$$Z_m = \frac{\max\{0, \mu_t|_m - \mu_i|_m\}}{\sigma_i|_m} \tag{16.5}$$

where $\mu_t|_m$ is the mean value of the *LLR* with the model of interest (model m) using data from the target speaker, $\mu_i|_m$ is the analogous mean for model m when using data from impostors and $\sigma_i|_m$ is the standard deviation of the *LLR* for the impostor data. Therefore, the larger the measure Z_m is for a model m, the more discriminative that model would be. For models with very little discrimination capability, Z_m becomes nearly 0.

The main difference between Equation 16.5 and the definition of the *z-norm* (Equation 18.9) is that here, the mean value of the *LLR* is used rather than a single *LLR*. The max operation in Equation 16.5 only ensures that no positive numbers are produced.

However, since [44] assumes that data from impostors may not be readily available, the score is defined as a similar score, only using different utterances from the enrollment data of the target speaker as follows,

$$S_n \geq \mu_t - \alpha \sigma_t \tag{16.6}$$

where $\mu_t|_n$ and $\sigma_t|_n$ are the mean and standard deviation of the log-likelihood scores of all the utterances used to train model $n \in \{1, 2, \cdots, N\}$ and S_n is the *LLR* of each utterance against a model, trained from this utterance and other utterances from the same data. In other words,

$$\mu_t|_n = \mathscr{E}\left\{\{S_i\}_{i=1}^n\right\} \tag{16.7}$$

Once S_n meets the criteria imposed by Equation 16.6, its model is deemed of acceptable quality. [44] continues with the introduction of a threshold and an algorithm for starting with a small number of utterances and going up to $N/5$ iterations by increasing the number of utterances used in producing a model until the expression in Equation 16.6 is satisfied.

This perspective assumes that different parts of data coming from the same speaker may have positive or negative contribution to the enrollment, hence verification results. Of course both of the above examples assume that the system is provided with more utterances that it needs to be able to build a model. In most practical cases this is indeed not the case. In practice, the users are always resistant to giving large amounts of enrollment data. Both papers listed above were dealing with short utterances in the form of digit sequences. Since text-dependent recognition systems can deal with much less training data, this may be somewhat practical. However, for the text-independent case, it would be very hard to get any more data that is needed for a bare-bone creation of a speaker model.

In fact, one of the problems with most of the NIST trials on speaker recognition is the fact that the amount of enrollment data used in the tests is unrealistically large (2+ minutes in most scenarios). Realistic systems seldom have access to any more than 30 seconds of usable data for enrollment. In the next section, we will see a different approach which assumes that different speakers have inherently different qualities, related to the verification task.

16.6.2 Speaker Menagerie

In biometrics, often, candidates are categorized into the two main categories of *sheep* and *goats* [49]. Those candidates who are categorized as *sheep* are the ones that show great performance and consistency across different samples of their biometric of choice. On the other hand, there are certain candidates who have a high variance in their scores from one sample to another.

According to [29], there are other sets of animals associated with speaker recognition, such as the following set of four animals, $\{sheep, goats, lambs, rams\}$, associated with *speaker identification* or even a larger set, $\{sheep, goats, lambs, rams, wolves, badgers\}$, associated with *speaker verification. Doddington, et al.* [21] suggested the four member set of $\{sheep, goats, lambs, wolves\}$ for categorizing speakers in a verification task. Table 16.1 shows a description of these different categories of speakers according to [21].

Speaker Category	Description
Sheep	*Good target speaker acceptance*
Goats	*Not accepted so easily*
Lambs	*Vulnerable to impostors*
Wolves	*Good impostors*

Table 16.1: *The Hypothetical Menagerie* – an animal analogy of the behavior of speakers, defined by [21]

Doddington, et al. [21] used the results obtained from the NIST trials on speaker verification in 1998 to study these populations of speakers. These trials used 30 second test segments obtained from land-line telephone speech using different *electret handsets* (see Section 22.3) used for training and testing stages. The gender distribution included 200 men and 220 women. They found that the majority of speakers may be categorized as *sheep*. However, most of the errors are generated by the small complement space containing the rest of the speakers.

Of course these definitions only tell us the shortcomings of our speaker recognition algorithms and how they carry over to part of the population. The category of speakers dubbed *sheep* represent the class of speakers who match our algorithms well and the other three categories signify those speakers for which the generalization capabilities of our algorithms fail or in the least degrade. It is important to be able to identify these different categories of speakers and to be able to provide different treatments in order to increase the generalization capability of the algorithm of interest. For example, *Thompson and Mason* [49] proposed a method for finding *goats* at the enrollment stage, for a speaker recognition system. As [49] puts it, it allows for taking *appropriate action* before getting to the recognition stage. [21] proposes two non-parametric statistical tests, the *Kuskal-Wallis test* which is a variant of the χ^2 test and the *Durbin test*, which both analyze the variance of ranks.

Also, *Stoll and Doddington* [47] have used different features including *jitter*, *shimmer* [33] and other pitch related statistics in addition to spectral analysis features such as *spectral slope statistics* and *formant statistics* to be able to identify different categories of speakers. Sometimes, if this is not possible, fusion techniques may be used to supplement the source of information about each candidate, using other biometrics which possess a generally independent source of information. As an example, *Poh and Kittler* [39] provide such treatment for many biometrics, which may easily include *speaker biometrics*. In Section 1.5.13 we spoke of such multimodal approaches and presented the examples in References[50, 7, 53].

Of course, the fun of animal analogies does not end with Table 16.1. Table 16.2 shows another categorization due to *Yager and Dunstone* [54, 55]. This is a different

Speaker Category	Description
Doves	*Good recognition accuracy – have unique features*
Chameleons	*High match scores for all models – match everyone*
Phantoms	*Low match scores for all models – even against their own models*
Worms	*Inconsistent – cause of most errors*

Table 16.2: *The Biometric Menagerie* – an animal analogy of the behavior of speakers, defined by [54, 55]

perspective of the types of speakers. This categorization uses a relation between the true and impostor match scores in lieu of the scores themselves. According to this categorization, the hardest category of speakers to verify is what they call *worms*. [55] also provides tests for identifying these groups and provides results for many biometrics including *speaker biometrics*.

References

1. Ahn, S., Kang, S., and, H.K.: Effective speaker adaptations for speaker verification. In: Acoustics, Speech, and Signal Processing, 2000. ICASSP '00. Proceedings. 2000 IEEE International Conference on, vol. 2, pp. II1081–II1084 (2000)
2. Al Marashli, A., Al Dakkak, O.: Automatic, Text-Independent, Speaker Identification and Verification System Using Mel Cepstrum and GMM. In: 3rd International Conference on Information and Communication Technologies: From Theory to Applications (ICTTA 2008), pp. 1–6 (2008)
3. Aronowitz, H., Solewicz, Y.A.: Speaker Recognition in Two-Wire Test Sessions. In: Inter-Speech, pp. 865–868 (2008)
4. Beigi, H.S., Maes, S.S.: Speaker, Channel and Environment Change Detection. In: Proceedings of the World Congress on Automation (WAC1998) (1998)
5. Beigi, H.S.M., Maes, S.H., Chaudhari, U.V., Sorensen, J.S.: IBM Model-Based and Frame-By-Frame Speaker Recognition. In: Speaker Recognition and its Commercial and Forensic Appications (1998)
6. Bengherabi, M., Tounsi, B., Bessalah, H., Harizi, F.: Forensic Identification Reporting Using A GMM Based Speaker Recognition System Dedicated to Algerian Arabic Dialect Speakers. In: 3rd International Conference on Information and Communication Technologies: From Theory to Applications (ICTTA 2008), pp. 1–5 (2008)
7. Besson, P., Popovici, V., Vesin, J.M., Thiran, J.P., Kunt, M.a.: Extraction of Audio Features Specific to Speech Production for Multimodal Speaker Detection. Multimedia, IEEE Transactions on **10**(1), 63–73 (2008)
8. Bocklet, T., Maier, A., Bauer, J.G., Burkhardt, F., Nöth, E.: Age and gender recognition for telephone applications based on GMM supervectors and support vector machines. In: International Conference on Acoustics, Speech, and Signal Processing (ICASSP), pp. 1605–1608 (2008)
9. Campbell, W.M., Sturim, D.E., Reynolds, D.A., Solomonoff, A.: SVM Based Speaker Verification using a GMM Supervector Kernel and NAP Variability Compensation. In: IEEE International Conference on Acoustics, Speech and Signal Processing (ICASSP 2006), vol. 1, pp. 14–19 (2006)

10. Castaldo, F., Colibro, D., Dalmasso, E., Laface, P., Vair, C.a.: Stream-based speaker segmentation using speaker factors and eigenvoices. In: Acoustics, Speech and Signal Processing, 2008. ICASSP 2008. IEEE International Conference on, pp. 4133–4136 (2008)

11. Chan, A., Sherwani, J., Mosur, R., Rudnicky, A.: Four-Layer Categorization Scheme of Fast GMM Computation Techniques in Large Vocabulary Continuous Speech Recognition Systems. In: Proceedings of the International Conference on Spoken Language Processing (ICSLP), pp. 689–692 (2004)

12. Chen, C., Chen, C., Cheng, P.: Hybrid KLT/GMM Approach for Robust Speaker Identification. IEE Electronic Letters 39(21), 1552–1554 (2003)

13. Chen, S.S., Gopalakrishnan, P.S.: Speaker, Environment and Channel Change Detection and Clustering via the Bayesian Inromation Criterion. In: IBM Techical Report, T.J. Watson Research Center (1998)

14. Dehak, N.: Discriminative and Generative Approaches for Long- and Short-Term Speaker Characteristics Modeling: Application to Speaker Verification. École de Technologie Supériure, Montreal (2009). PhD Thesis

15. Dehak, N., Dehak, R., Glass, J., Reynolds, D., Kenny, P.: Cosine Similarity Scoring without Score Normalization Techniques. In: The Speaker and Language Recognition Workshop (Odyssey 2010), pp. 15–19 (2010)

16. Dehak, N., Dehak, R., Kenny, P., Brummer, N., Ouellet, P., Dumouchel, P.: Support Vector Machines versus Fast Scoring in the Low-Dimensional Total Variability Space for Speaker Verication. In: InterSpeech, pp. 1559–1562 (2009)

17. Dehak, N., Dumouchel, P., Kenny, P.a.: Modeling Prosodic Features With Joint Factor Analysis for Speaker Verification. Audio, Speech, and Language Processing, IEEE Transactions on 15(7), 2095–2103 (2007)

18. Dehak, N., Kenny, P., Dehak, R., Dumouchel, P., Ouellet, P.: Front-End Factor Analysis for Speaker Verification. IEEE Transactions on Audio, Speech and Language Processing 19(4), 788–798 (2011)

19. Dehak, N., Kenny, P., Dehak, R., Glembek, O., Dumouchel, P., Burget, L., Hubeika, V., Castaldo, F.: Support vector machines and Joint Factor Analysis for speaker verification. In: International Conference on Acoustics, Speech, and Signal Processing (ICASSP), pp. 4237–4240 (2009)

20. Dhanalakshmi, P., Palanivel, S., Ramalingam, V.: Classification of Audio Signals using AANN and GMM. Applied Soft Computing 11(1), 716 – 723 (2011). DOI 10.1016/j.asoc.2009.12.033. URL http://www.sciencedirect.com/science/article/B6W86-4Y3JY8D-1/2/722d39fe60e735af8ddda0be27d48057

21. Doddington, G., Liggett, W., Martin, A., Przybocki, M., Reynolds, D.: Sheep, Goats, Lambs and Wolves: A Statistical Analysis of Speaker Performance in the NIST 1998 Speaker Recognition Evaluation. In: Proceedings of the International Conference on Spoken Language Processing (ICSLP), pp. 1–5 (1998)

22. Garcia, V., Nielsen, F., Nock, R.: Hierarchical Gaussian Mixture Model (2010)

23. Glembek, O., Burget, L., Dehak, N., Brümmer, N., Kenny, P.: Comparison of Scoring Methods used in Speaker Recognition with Joint Factor Analysis. In: International Conference on Acoustics, Speech, and Signal Processing (ICASSP), pp. 4057–4060 (2009)

24. Gu, Y., Jongebloed, H., Iskra, D., den Os, E., Boves, L.: Speaker Verification in Operational Environments Monitoring for Improved Service Operation. In: Proceedings of the International Conference on Spoken Language Processing (ICSLP), pp. 450–453 (2000)

25. Hershey, J.R., Olsen, P.A.: Approximating the Kullback Leibler Divergence Between Gaussian Mixture Models. In: International Conference on Acoustics, Speech, and Signal Processing (ICASSP), vol. 4, pp. 317–320 (2007)

26. Homayoon S. M. Beigi, S.H.M.: Speaker, Channel and Environment Change Detection. Technical Report (1997)

27. Kenny, P.: Joint Factor Analysis of Speaker and Session Varaiability: Theory and Algorithms. Technical report, CRIM (2006). URL http://www.crim.ca/perso/patrick.kenny/FAtheory.pdf

28. Kenny, P., Boulianne, G., Ouellet, P., Dumouchel, P.: Factor Analysis Simplified. In: International Conference on Acoustics, Speech, and Signal Processing (ICASSP), vol. 1, pp. 637–640 (2005)
29. Koolwaaij, J., Boves, L.: A new procedure for classifying speakers in speaker verification systems. In: EUROSPEECH (1997)
30. Koolwaaij, J., Boves, L., Jongebloed, H., den Os, E.: On Model Quality and Evaluation in Speaker Verification. In: International Conference on Acoustics, Speech, and Signal Processing (ICASSP), vol. 6, pp. 3759–3762 (2000)
31. Li, Z., Jiang, W., Meng, H.: Fishervioce: A discriminant subspace framework for speaker recognition. In: International Conference on Acoustics, Speech, and Signal Processing (ICASSP), pp. 4522–4525 (2010)
32. Matrouf, D., Scheffer, N., Fauve, B., Bonastre, J.F.: A Strainghforward and Efficient Implementation of the Factor Analysis Model for Speaker Verification. In: International Conference on Speech Communication and Technology (2007)
33. Naini, A.S., Homayounpour, M.M.: Speaker age interval and sex identification based on Jitters, Shimmers and Mean MFCC using supervised and unsupervised discriminative classification methods. In: The 8th International Conference on Signal Processing, vol. 1 (2006)
34. Nelwamondo, F.V., Marwala., T.: Faults Detection Using Gaussian Mixture Models, Mel-Frequency Cepstral Coefficients and Kurtosis. In: IEEE International Conference on Systems, Man and Cybernetics, (SMC'06), vol. 1, pp. 290–295 (2006)
35. Nishida, M., Kawahara, T.: Speaker indexing and adaptation using speaker clustering based on statistical model selection. In: International Conference on Acoustics, Speech, and Signal Processing (ICASSP), vol. 1, pp. 172–175 (2004)
36. Nwe, T.L., Sun, H., Li, H., Rahardja, S.: Speaker Diarization in Meeting Audio. In: International Conference on Acoustics, Speech, and Signal Processing (ICASSP), pp. 4073–4076 (2009)
37. Omar, M.K., Pelecanos, J.: Training Universal Background Models for Speaker Recognition. In: The Speaker and Language Recognition Workshop (Odyssey 2010), pp. 52–57 (2010)
38. Pellom, B., Hansen, J.: An efficient scoring algorithm for Gaussian mixture model based speaker identification. IEEE Signal Processing Letters 5(11), 281–284 (1998)
39. Poh, N., Kittler, J.: A Methodology for Separating Sheep from Goats for Controlled Enrollment and Multimodal Fusion. In: Biometric Symposium, pp. 17–22 (2008)
40. Reynolds, D.A., Quatieri, T.F., , Dunn, R.B.: Speaker Verification Using Adapted Gaussian Mixture Models. Digital Signal Processing 10, 19–41 (2000)
41. Reynolds, D.A., Rose, R.: Robust text-independent speaker identification using Gaussian mixture speaker models. IEEE Transactions on Speech and Audio Processing 3(1), 72–83 (1995)
42. Roweis, S., Ghahramani, Z.: A Unifying Review of Linear Gaussian Models. Neural Computation 11(2), 305–345 (1999)
43. Roy, A., Magimai-Doss, M., Marcel, S.: Boosted Binary Features for Noise-Robust Speaker Verification. In: International Conference on Acoustics, Speech, and Signal Processing (ICASSP), vol. 6, pp. 4442–4445 (2010)
44. Saeta, J.R., Hernando, J.: Model quality evaluation during enrolment for speaker verification. In: Proceedings of the International Conference on Spoken Language Processing (ICSLP), pp. 1801–1804 (2004)
45. Senoussaoui, M., Kenny, P., Dehak, N., Dumouchel, P.: An i-Vector Extractor Suitable for Speaker Recognition with Both Microphone and Telephone Speech. In: The Speaker and Language Recognition Workshop (Odyssey 2010), pp. 28–33 (2010)
46. Seo, C., Lee, K.Y., Lee, J.: GMM based on local PCA for speaker identification. IEE Electronic Letters 37(24), 1486–1488 (2001)
47. Stoll, L., Doddington, G.: Hunting for Wolves in Speaker Recognition. In: The Speaker and Language Recognition Workshop (Odyssey 2010), pp. 159–164 (2010)
48. Sun, H., Ma, B., Huang, C.L., Nguyen, T.H., Li, H.: The IIR NIST SRE 2008 and 2010 Summed Channel Speaker Recognition Systems. In: Interspeech (2010)

49. Thompson, J., Mason, J.S.: The Pre-detection of Error-prone Class Members at the Enrollment Stage of Speaker Recognition Systems. In: ESCA Workshop on Automatic Speaker Recognition, Identification, and Verification, pp. 127–130 (1994)

50. Viswanathan, M., Beigi, H.S., Maali, F.: Information Access Using Speech, Speaker and Face Recognition. In: IEEE International Conference on Multimedia and Expo (ICME2000) (2000)

51. Vogt, R., Sridharan, S.: Explicit modelling of session variability for speaker verification. Computer Speech and Language **22**(1), 17–38 (2008)

52. van Vuuren, S.: Comparison of Text-Independent Speaker Recognition Methods on Telephone Speech with Acoustic Mismatch. In: Proceedings of the International Conference on Spoken Language Processing (ICSLP), pp. 784–787 (1996)

53. Xiong, Z., Chen, Y., Wang, R., Huang, T.a.: A real time automatic access control system based on face and eye corners detection, face recognition and speaker identification. In: Multimedia and Expo, 2003. ICME '03. Proceedings. 2003 International Conference on, vol. 3, pp. III–233–6 (2003)

54. Yager, N., Dunstone, T.: Worms, Chameleons, Phantoms and Doves: New Additions to the Biometric Menagerie. In: IEEE Workshop on Automatic Identification Advanced Technologies, pp. 1–6 (2007)

55. Yager, N., Dunstone, T.: The Biometric Menagerie. IEEE Transactions on Pattern Analysis and Machine Intelligence (PAMI) **32**(2), 220–230 (2010)

56. Yantorno, R.E., Iyer, A.N., Shah, J.K., Smolenski, B.Y.: Usable speech detection using a context dependent Gaussian mixture model classifier. In: International Symposium on Circuits and Systems (ISCAS), vol. 5, pp. 619–623 (2004)

57. Yin, S.C., Rose, R., Kenny, P.a.: A Joint Factor Analysis Approach to Progressive Model Adaptation in Text-Independent Speaker Verification. Audio, Speech, and Language Processing, IEEE Transactions on **15**(7), 1999–2010 (2007)

58. You, C.H., Lee, K.A., Li, H.: A GMM Supervector Kernel with the Bhattacharyya Distance for SVM Based Speaker Recognition. In: International Conference on Acoustics, Speech, and Signal Processing (ICASSP), vol. 1, pp. 4221–4224 (2009)

59. Zhang, Z., Dai, B.T., Tung, A.K.: Estimating Local Optimums in EM Algorithm over Gaussian Mixture Model. In: The 25th International Conference on Machine Learning (ICML), pp. 1240–1247 (2008)

Chapter 17
Speaker Recognition

17.1 The Enrollment Task

The objective of the enrollment process is to modify (adapt) a speaker-independent model into one that best characterizes the target speaker's vocal tract characteristics. Depending on whether the task at hand is text-dependent or text-independent, different objectives should be observed while designing the enrollment process.

As we discussed in Chapter 1, the text-dependent and text-prompted modalities are only practical as far speaker verification is concerned. For other branches of speaker recognition, we only limit our discussion to text-independent recognition. In a pure text-dependent modality, the same the same phrase will be spoken at the enrollment time as it will at the recognition time. Regardless of its practicality, the enrollment process is quite simple for text-dependent enrollment. The phrase is produced to the user and the result is used to modify the models. The coverage in this case is identical in the enrollment and verification.

In the case of text-prompted speaker verification, the enrollment phrase will possibly be different from the verification phrase. There are several approaches to designing the enrollment process for this case. The simplest approach is to anticipate all the different strings that the user may speak at the verification time and to cover all of them at enrollment time. To be a bit more sophisticated, the enrollment and verification text may be designed for each specific user. For example, the user may be asked to say phrases which also carry information specific to her/him, such as personal information.

However, one of the main attractions of speaker verification is the avoidance of asking personal information such as social security numbers, birth dates, place of birth, favorite color, etc. To alleviate this problem, most vendors of text-prompted technologies limit their systems to a subset of all possible phones. An example would be the use of digit-only strings. Some even go further into reducing the space of digits by using only a subset of digits. This way, the enrollment sequence will include the limited space of coverage which will be seen at the verification time, but at the verification time, a random process will prompt the user for a different string every time to ensure liveness. Such as system is by no means impostor proof and may theoretically be spoofed quite easily – see Section 22.11.

In a text-independent speaker recognition scenario, the text which is spoken by the user may be completely unconstrained. However, in practice, consider the exaggerated example where the user repeats the same three word phrase over and over again for the length of the enrollment. Also recall from Chapter 5 that the speech signal is non-stationary. This means that depending on the nominal configuration of the vocal tract, different resonance is produced.

Since in this book we are concerned with statistical speaker recognition, the In a text-independent speaker recognition task, the enrollment text should be designed to cover most phonemes and the most frequent phonetic transitions. The more data is captured for enrollment, the richer will the statistics become which is used for adapting models to the specific vocal characteristics of any specific user. An enrollment text with a good coverage of the phonemes of interest would certainly increase the accuracy of the recognition system. Since the scope of this book is to cover statistical recognition systems, it is important that the events which are expected at the recognition time have been observed at the enrollment time.

17.2 The Verification Task

At this point it is important to understand the general method of computing the scores involved in the decision making of verification. As we saw, in general, for doing speaker verification we need one or more competing models. We also need a target model which would be built on enrollment audio for the target speaker. The competing models are generally created in a similar fashion to the target model, with the exception that they are usually created from a select population of speakers. These could either be cohorts of the target speaker or a large general population sometimes called background population. The most popular variation of the competing model is hence called the background model.

Verification then boils down to a comparison between the likelihood of the test audio having been generated by the target model versus the competing model. Let's call the test sequence X. This is the output sequence for our front-end which takes the audio and converts it to an abstract sequence based on our front-end modeling. This sequence is conceptually the output of some model (which we can call the test model at this point). The test model is never really created. It is an abstraction to show that the sequence X must have been emitted by some model. The actual verification question is whether this sequence is more likely to have been emitted by a specific target model or the competing model or models. Let us take the simpler case where there is only one competing model and let us call it S_I. Also, since at this point we are concerned with speaker verification, the test speaker must have provided an identity along with a speech sequence. The verification task has to validate this identity. Therefore, the model for that identity which must have been created at the time of enrollment is referenced. Let's call that model S_D.

If the likelihood of the test sequence being generated by the target model, S_D, is higher than if it were generated by the competing model, S_I, then the test speaker is verified, otherwise, the test speaker is rejected. One way of setting up this logic is to divide the likelihood of the target model relative to the test audio by the likelihood of the competing model relative to the test audio. This generates a likelihood ratio such that a value greater than one would translate to verification and anything less than one would translate to rejection.

To have a better resolution and to reduce numerical errors, we generally use the log of the likelihood. Therefore, the likelihood ratio when performed in the log domain, would end up being a subtraction. Namely, the decision score, λ, would be the difference between the Log Likelihood of the Target model relative to the test audio and that of the competing model. It is customary to call this value the log likelihood ratio. It is really the log of the likelihood ratio and if it is positive, it reduces to a verification decision. A negative LLR would mean rejection.

In practice, the decision is made by comparing the LLR to a threshold which would be set by the system designer. Let's call this threshold, λ_o. The value of this threshold is dependent on the performance of the system and the domain in which it is utilized. It's picked by looking at the performance of the system on some test data which would best mimic the operating conditions of the system and it is chosen based on how flexible we would want the verification system to be. The higher the threshold, the higher the false rejection and the lower the false acceptance numbers.

Researchers typically plot the False Rejection rates of the recognizer against its False Acceptance rates. This curve is call a Receiver Operating Characteristic (ROC) curve. More recently, a different graph has been proposed by the National Institute for Standards (NIST) called the Detection Error Trade-off (DET) curve. Most recent papers use the DET curve.

In order to be able to describe the performance of a speaker verification system with one number, sometimes a single somewhat controversial error rate is given. It is called the equal error rate or EER for short. It is the operating point where the rate of false acceptance of the system is equal to its false rejection. We will talk about this later.

17.2.1 Text-Dependent

Up to now, we have mostly concentrated on text-independent verification. As we mentioned, this modality only matters with verification and generally does not make much sense within other branches such as identification, segmentation, and classification. However, in Chapter 13, we gave a detailed treatment of *hidden Markov models*. This treatment was partly done to fulfill the needs for the background. However, the most direct usage of HMM is in text-dependent and text-prompted verification. We also gave a rather detailed account of phonetics and phonology in Chapter 4 and a considerable introduction on decision trees in Section 9.4. These are some of the fundamentals needed for performing text-dependent speaker verification.

The problem, although cumbersome, is not a specifically hard problem compared to the text-independent case. Generally, an HMM model is chosen, which is most likely phoneme-based. A dictionary of phonemes is needed to look up different phonetic makeups for words in a dictionary. This part generally requires a first attempt by someone who is well-versed with the topics covered in Chapter 4. Then, by concatenating phoneme-based HMM models to build a word or phrase model, one is built and trained based on speaker independent data, initially. Upon the receipt of enrollment data from the target speaker, the HMM is trained, using the techniques described in Chapter 13, such as the Baum-Welch algorithm of Section 13.6.4 to produce the best posterior probability given the enrollment data. At the time of verification, the test data is passed through the HMM using a forward pass (Section 13.6.2) and a score is computed to decide whether to authenticate the speaker or not.

Some implementation of text-dependent verification may be found at the following references [16, 24, 81, 49, 52, 57, 61, 69, 72, 77, 79, 82, 84, 86, 11, 28, 29, 33, 48]. Of course there are many more research efforts on the subject. It is interesting that many of the research papers refer to the work they do as text-dependent speaker identification. We mentioned that this does not make sense. Of course it can be done, but it does not have many practical applications. In most cases, when one is trying to identify a person, the effort has to be done in a passive manner. This means that the speaker should be able to freely speak and be identified. It is almost comedic to ask someone to say a specific phrase to be able to identify him/her.

17.2.2 Text-Prompted

In Chapter 1 we discussed the impracticalities of using a text-dependent speaker recognition system. It reduces the chance of being able to check for liveness of the test speaker. For this reason, text-prompted systems produce a (usually) random prompt and ask the test speaker to repeat it. However, to lower the complexity of the model generation, most such systems simply concatenate digits and ask the speaker to repeat the string. With digital recording capabilities of today's devices, it would be quite simple to record a person saying the 10 digits and to produce the string or digits by simply typing on a keypad. This can easily be done, even using a smartphone. Although it is possible to created complex text strings for the prompt, the lack of adequate amount enrollment makes this impractical. For example, usually, a speech recognition system requests about 20 minutes of speech in order to be able to build a speaker-dependent model. However, there is almost no application where the speaker would be willing to speak that long in order to enroll for speaker recognition. As we mentioned, 1 minute is usually the upper limit.

One of the techniques to reduce the amount of training is to use tied mixtures [55], but this is common practice in speech recognition. Some implementations of text-prompted speaker recognition may be found in the following references [55, 15, 20, 51].

One of the reasons for moving to a text-prompted system rather than a text-independent system is usually the shorter enrollment requirement. Therefore, even text-prompter verification is not very practical!

It is conceivable that the same way phone models for the analysis of the audio are concatenated to build a model for the prompt being spoken, the synthesis counterpart may also be used to spoof the text-prompted system. These Text-to-Speech systems using the voice of an individual to build their models have been around for a while now.[38, 23] As handheld computers become more widely available, it will become easier to build a synthesized model of the speech of the individual form these baseforms and to play it back. Using a knowledge base in conjunction with the speaker verification results will be one way to handle these types of attempts. Of course this is nothing new and has been known by the industry from the first day that extra information was asked by an agent trying to assess the authenticity of a caller's claimed identity.

Another way of handling trainable text-to-speech-based impostors is to look for regularity in the phonemes being spoken. Methods using synthesis will have a regular set of synthesis model being concatenated together to form the synthesized speech. Normal speech does not possess such regularity. The phones in an utterance, once torn apart, do not look identical in natural speech. Of course, this is a non-ending war since, as the speaker recognition systems add such sophistication, the text-to-speech impostors will try to utilize random effects to modify the output of their models and then it is the speaker recognition designers to cope with the new threats; and it goes back and forth.

Still addition of a knowledge base and fusion with other biometrics will be the correct way of handling these and other issues in addition to adding sophisticated anti-spoofing techniques such as the one described in the previous paragraph.

17.2.3 Knowledge-Based

Base on what we have discuss in the past few sections, knowledge-based verification
seems to be the answer. Of course the type of knowledge needs not (and should not)
be personal information. Getting away from providing personal information is one
of the major attractions of speaker recognition. The only reason for any knowledge-
based question is to assess the *liveness* of the individual. Simple questions may be
asked to ensure that the speaker is responding in an intelligent manner and that the
speech is not just a recording. The downfall of this is that such a technique is much
more costly. It requires access to a good speaker independent speech recognizer,
possibly, as well as a limited *natural language understanding* (NLU) engine.

17.3 The Identification Task

The identification decision is made quite in a similar fashion as compared to verification. However, the identification problem involves N models, N being the number
of target speakers enrolled in the system. Of course, this would be the closed-set
identification process. The open-set identification process is an $N + M$ match problem where M is the number of competing models. In the case where we have one
background model, this would require $N + 1$ likelihood computations. The result
is usually presented in two forms, summary and detail. The summary result would
simply be the ID of the target speaker whose model has the highest likelihood of
having generated the output test sequence, X. If the highest likelihood comes from
the competing model, then the result of identification would state that the speaker is
unknown.

The detailed results are usually in the form of a sorted list of possible identities
which are sorted based on decreasing likelihood. Again, for better numerical stability and resolution, the log likelihood is used.

17.3.1 Closed-Set Identification

Closed set speaker recognition, as mentioned in Chapter 1, is a simple mode of identification where the test speaker is definitely in the database. In fact, if that is not the
case, there may be dire consequences, since such systems will generally return the
ID of the closest person in the database. Now this closest person may be very far
removed from the test speaker. The more practical, but also much harder problem is
open-set identification. The following references present some implementations of

closed-set identification [21, 31, 34, 59, 39].

17.3.2 Open-Set Identification

The open set identification problem, as we stated, is the hardest problem. If we had a perfect speaker identification system, then we would be able to implement every other branch perfectly. It is specifically difficult as the population size grows. Three problems plague open-set identification, and the first two are shared with closed-set identification.

1. As the number of speakers grows, the number of comparisons increases (theoretically in a linear fashion). We say theoretically since it is possible to use hierarchical techniques to reduce the number of comparisons to a logarithmic level from linear [7]. Still this is a curse which does not plague speaker verification. Speaker verification usually deals with the same order of number of comparisons.
2. As the number of speakers grows, the speakers get closer in the speaker space. This increases the chance that one speaker is mistaken for another. Theoretically this would not have been a problem if each speaker only occupied a single point in the space. However, generally, there are large variations within the speech of the same individual. Chapters 20 and 21 attest to this.
3. A complementary model is needed, specifically for an open-set identification problem. This is similar to the speaker verification problem, however it involves a lot more models than the two or so models in speaker verification.

Therefore, open-set identification seems to suffer from the problems of the closed-set identification and speaker verification combined. The following are some references on open-set identification [7, 3, 13, 27, 83, 88, 2].

17.4 Speaker Segmentation

Different strategies may be considered for detecting speaker changes. These techniques fall into two general categories:

1. Detection of speaker identity changes, by performing sequential speaker identifications and detecting a change of decision. Such approaches are described in [5].
2. Detection of significant changes in the acoustic behavior followed by interpretation of the nature of these changes. Such approaches have been previously proposed in [73], [8] and [17].

The algorithm described in [73] belongs to the second category. A draw-back of the method described in [73] is its extremely poor detection of the exact location of speaker changes. Furthermore, the resulting segmentation end-times rarely correspond to pauses, silences or breath noises. Unfortunately, this is unacceptable when the segmentation is followed by a speech recognition phase. Indeed, every time that a boundary is introduced in the middle of a word, it introduces one or two word errors. It is also a characteristic of methods which arbitrarily cut the input speech into small segment and then re-cluster them[42, 35].

[1] presents a two-speaker segmentation technique by tackling the change detection according to energy and statistical properties of the energy function. [30] presents a comparison between the BIC technique [17] and an adapted GMM technique which basically builds GMM models for the segments. [37] compares several techniques for segmenting the *broadcast news* corpus. [46] compares three different segmentation approaches, all of which include BIC as an integral part. [58] compares several different BIC based systems. [78] gives an overview of diarization systems. [14] uses WOCOR features (Section 5.6.1.2) for doing the segmentation. [19] uses a correlation matrix using anchor models. [12] uses the Eigenvoices model, using a simple factor analysis model, for doing segmentation. [32] looks at feature combinations for doing better diarization. [44] uses robustness analysis, based on BIC, for doing segmentation. [47] combines video and audio cues. [64] uses a *dynamic time warping (DTW)* algorithm which is basically a dynamic programming technique. [70] uses genetic algorithms and mutual information (Section 7.6.1). [67] does a review of the state of the art in 2009. [4] uses *kernel PCA* (Section 15.7) to do diarization.

17.5 Speaker and Event Classification

Sometimes either no enrollment is used for classification, or a special enrollment techniques are utilized. For example, the enrollment data for several events or speakers may be pooled together, as representative data for the class of interest. Another approach would be the use of extra features in supplemental codebooks, related to specific natural or logical aspects of the classes of interest.

In many practical audio processing systems, it is important to determine the type of audio. For instance, consider a telephone-based system which includes a speech recognizer. Such recognition engines would produce spurious results if they were presented with non-speech, say music. These results may be detrimental to the operation of an automated process. This is also true for speaker identification and verification systems which expect to receive human speech. They may be confused if they are presented with music or other types of audio such as noise. For *text-independent speaker identification* systems, this may result in mis-identifying the audio as a viable choice in the database and resulting in dire consequences!

Similarly, some systems are only interested in processing music. An example is a music search system which would look for a specific music or one resembling the presented segment. These systems may be confused, if presented with human speech, uttered inadvertently, while only music is expected.

The goal of audio classification is the development of a classification filter which would tag a segment of audio as speech, music, noise, or silence. This problem contains two separate parts. The first part is the segmentation of the audio stream into segments of similar content. This work has been under development for the past few decades with some good results [36, 8, 17].

The second part is the classification of each segment into speech, music, or the rejection of the segment as silence or noise. Furthermore, when the audio type is *human speech*, it is desirable to do a further classification to determine the gender of the individual speaker. Gender classification is helpful in choosing appropriate models for conducting better speech recognition, more accurate speaker verification, and reducing the computation load in large-scale speaker identification.

On the other hand, if the signal of interest is music, it is interesting to be able to determine the specific type of music, for instance in the form of identifying the instrument. Of course, this problem is not quite so simple due to overlap of instruments in orchestral pieces and the sheer number of possible instruments. However, a close approximation to the target instrument and categorization as orchestral or specific types of bands is also useful. We are also interested in an approach which would not require tremendous modeling efforts for every new circumstance which may arise.

To address the instrument identification problem and to be able to cover most types of music, [6] uses a set of 14 representative instruments or collections of instruments to create models. Different approaches with varying perspectives to audio source classification have been reported. One group has tried to identify individual musical instruments [54, 50, 56, 26, 25, 9, 45]; whereas another group has concentrated on classifying speakers based on gender [18, 40, 60, 71, 85, 10]. [22] reports developments in classifying the genre of audio, as stemming from different video sources, containing movies, cartoons, news, etc.

17.5.1 Gender and Age Classification (Identification)

In 1952, Peterson [65] conducted a series of experiments on the 10 common vowels in English. 33 men, 28 women, and 15 children (a total of 76 speakers) were asked to say 10 words (two times each) and their utterances were recorded. The words were designed to examine the 10 vowels in context of an "*h*" to the left and a "*d*" to the right: hid, hɪd, hɛd, hæd, hɑd, hɔd, hʊd, hud, hʌd, and hɝd.

In Figure 4.1 we saw that the mean value of the fundamental frequency for the vowel in each of the above words, displayed for men, women, and children separately. Note that the fundamental frequency (*formant 0*) does not change much among different vowels. This is the fundamental frequency of the vocal tract based on a normal opening of the vocal folds when one is producing a vowel, but it varies significantly across gender and age. Formants 1 and 2 do vary considerably depending on which vowel is being uttered, however Formant 3 does not (Section 4.1.5).

Recently, several different techniques [18, 40, 60, 71, 85, 10], based on the above premise, have been reported for identifying gender. Some effort has also been focused on determining the age groups of individuals based on the above and the concept of *jitter* (Section 3.6.4). [60] proposes using the mean MFCC as an indicator of jitter and states that it is a good indicator of the gender and age of the individual. [6] uses *Cepstral mean subtraction (CMS)*, and shows great results for gender classification, indicating that gender does not seem to be so correlated to the Cepstral mean. Of course, [6] has not been considered it in its study, it is possible that the Cepstral mean may still be related to jitter and age.

Martin [54, 53] has used pattern recognition techniques for the problem of musical instrument identification. He uses the *log-lag correlogram* which is adopted from *cochlear models*. This technique is related to the *pitch* which is usually ignored in standard speaker recognition techniques that do not use prosodic features. Since we are not using pitch here, it is fundamentally different from our approach. For a robust and universal resolution, the objective is to determine *timbre* and not be dependent on values related to *pitch* and *sonority* (Section 4.3.1).

[26] also uses cepstral coefficients for conducting musical instrument recognition. However, it uses very complex features which are connected to the dynamics of musical pieces and maps the frequencies to the Bark frequecy scale [89] which is similar to the Mel-Frequency mapping in that it is also based on the *psychophysical power law of hearing* (Section 5.5.4). However, the complex set of features as well as heuristics make this approach too impractical for the purpose of a simple and universal pre-filter. The approach of [26] is more suitable for accurate recognition of instruments and is inherently much more costly.

17.5.2 Audio Classification

[68] classifies audio collected from the *world wide web* into three different categories, 1. speech, 2. music, 3. other. It uses *support vector machines* with the *Fisher kernel* [41]. Reference [68] shows an error rate of about 19% for the classification of about 173 hours of audio collected from the *web*.

[22] talks about using autoassociative neural networks in conjunction with GMM, using LPCC and MFCC features. They try to categorize the genre of videos using the audio track. For example, they use the following categories in their experiments: advertisement, cartoon, movie, news, song and sports.

Beigi [6] uses 1400 excerpts of music in different styles from over 70 composers together with the speech of 700 male and 700 female speakers. The audio signal is telephone quality sampled at $8kHz$ with μ-law amplitude encoding for all cases including the music, which is converted to reduce its quality and bring it to par with the speech data. [6] reports 1% error rate of speech versus music classification and a 1.9% gender classification error rate at speeds of more than three times real-time on a single core of a multi-core Xeon processor.

17.5.3 Multiple Codebooks

In most cases, *classification*, more than any other branch of speaker recognition, requires the combination of different information sources for obtaining the final decision. This branch requires a greater *artistic* touch in its design, mostly dependent on the specific problem at hand.

In Section 11.1, we discussed the idea of a *codebook*. To be able to combine the different information sources, often, it makes sense to treat each source as a different *codebook* and then to combine the decisions stemming from these different codebooks. This combination may be viewed as a *fusion technique*. In Sections 1.3.4 and 1.5.13, we discussed some applications of fusion of different information sources. This combination of results from different codebooks in classification is quite similar to those cases.

17.5.4 Farfield Speaker Recognition

There has been some work done in farfield speaker recognition. This is a very challenging problem due to the added noise and the drop in the quality of audio. [76] presents a comparison among different speaker identification models such as the

GMM-UBM model and kernel based methods such as SVM and *relevance vector machines (RVM)* under adverse farfield recording with short utterances. It concludes that the GMM-UBM system yields the best results.

[43] does a study of farfield speaker recognition by playing with the features, using MFCC (Section 5.3.6) with CMS (Section 18.4.1) and RASTA (Section 18.4.4) for normalization. It compares 8 channels for variability and applies a *reverberation compensation* and warping to enhance the results.

17.5.5 Whispering Speaker Recognition

We discussed whisper from speech generation and phonological points of view in Chapter 4. We noted how in a whisper mode, the vocal folds become more relaxed such that the harmonics are changed by moving toward a more turbulent flow compared to voiced resonance. As we have seen, voice portions of the speech hold the most information about the speaker's vocal characteristics due to the presence of rich harmonics. In the absence or alteration of such harmonics, the recognition could seriously degrade.

17.6 Speaker Diarization

[87] A lot of what we have discussed in this chapter make up what is generally called speaker diarization. It is the act of organizing and tagging a stream of audio, which of course may be extracted from a video track. For diarization, many speaker recognition branches are employed. A good example of a full diarization is [80] which takes a video stream; segments the audio and video; identifies the speakers; identifies the faces; and transcribes the audio in each segment associated with a speaker. The results then become searchable by the name of the individual who was recognized, the transcribed text is made searchable, and the content is filed for future retrieval.

[67] and [66] provide two survey papers for the years 2009 and 2005 respectively. [62] looks at most aspects of diarization including nonspeech removal. [74] makes modifications to the basic NAP (Section 15.8), in order to obtain better performance for speaker verification when using a two-wire telephone channel (speaker segmentation, followed by verification). The method removes some dominant components from the inter-speaker as well as inter-session spaces and shows improvement in a two-wire verification, hence dubbed a two-wire NAP.

17.6.1 Speaker Position and Orientation

Several research projects have been looking at finding the speaker position and orientation. For example, [63] uses a linear microphone array to acoustically locate a speaker. It uses *crosspower-spectrum phase* techniques to find the position of a sound source and to assess whether it is moving or not. [75] uses only a single microphone. It uses an HMM model of clean speech presented to the microphone. Then using the results of the model, localization is deduced.

References

1. Adami, A., Kajarekar, S., Hermansky, H.a.: A new speaker change detection method for two-speaker segmentation. In: Acoustics, Speech, and Signal Processing, 2002. Proceedings. (ICASSP '02). IEEE International Conference on, vol. 4, pp. IV–3908–IV–3911 (2002)
2. and, Y.G.: Noise-robust open-set speaker recognition using noise-dependent Gaussian mixture classifier. In: Acoustics, Speech, and Signal Processing, 2002. Proceedings. (ICASSP '02). IEEE International Conference on, vol. 1, pp. I–133–I–136 (2002)
3. Ariyaeeinia, A., Fortuna, J., Sivakumaran, P., Malegaonkar, A.a.: Verification effectiveness in open-set speaker identification. Vision, Image and Signal Processing, IEE Proceedings - **153**(5), 618–624 (2006)
4. Aronowitz, H.: Trainable Speaker Diarization. In: InterSpeech, pp. 1861–1864 (2007)
5. Bakis, R., Chen, S., Gopalakrishnan, P., Gopinath, R., Maes, S., Polymenakos, L.: Transcription of Broadcast News Shows with the IBM Large Vocabulary Speech Recognition System. In: Proceedings of the Speech Recognition Workshop (1997)
6. Beigi, H.: Audio Source Classification using Speaker Recognition Techniques. World Wide Web (2011). URL http://www.recognitiontechnologies.com/~beigi/ps/RTI20110201-01.pdf. Report No. RTI-20110201-01
7. Beigi, H.S., Maes, S.H., Chaudhari, U.V., Sorensen, J.S.: A Hierarchical Approach to Large-Scale Speaker Recognition. In: EuroSpeech 1999, vol. 5, pp. 2203–2206 (1999)
8. Beigi, H.S., Maes, S.S.: Speaker, Channel and Environment Change Detection. In: Proceedings of the World Congress on Automation (WAC1998) (1998)
9. Benetos, E., Kotti, M., Kotropoulos, C.: Large Scale Musical Instrument Identification. In: Proceedings of the 4th Sound and Music Computing Conference, pp. 283–286 (2007)
10. Bocklet, T., Maier, A., Bauer, J.G., Burkhardt, F., Nöth, E.: Age and gender recognition for telephone applications based on GMM supervectors and support vector machines. In: International Conference on Acoustics, Speech, and Signal Processing (ICASSP), pp. 1605–1608 (2008)
11. Buck, J., Burton, D., Shore, J.a.: Text-dependent speaker recognition using vector quantization. In: Acoustics, Speech, and Signal Processing, IEEE International Conference on ICASSP '85., vol. 10, pp. 391–394 (1985)
12. Castaldo, F., Colibro, D., Dalmasso, E., Laface, P., Vair, C.a.: Stream-based speaker segmentation using speaker factors and eigenvoices. In: Acoustics, Speech and Signal Processing, 2008. ICASSP 2008. IEEE International Conference on, pp. 4133–4136 (2008)
13. Cetingul, H., Erzin, E., Yemez, Y., Tekalp, A.a.: On optimal selection of lip-motion features for speaker identification. In: Multimedia Signal Processing, 2004 IEEE 6th Workshop on, pp. 7–10 (2004)
14. Chan, W., Lee, T., Zheng, N., and, H.O.: Use of Vocal Source Features in Speaker Segmentation. In: Acoustics, Speech and Signal Processing, 2006. ICASSP 2006 Proceedings. 2006 IEEE International Conference on, vol. 1, pp. I–I (2006)

15. Che, C., Lin, Q., and, D.S.Y.: An HMM approach to text-prompted speaker verification. In: Acoustics, Speech, and Signal Processing, 1996. ICASSP-96. Conference Proceedings., 1996 IEEE International Conference on, vol. 2, pp. 673–676 (1996)

16. Chen, K., Xie, D., Chi, H.: A modified HME architecture for text-dependent speaker identification. IEEE Transactions on Neural Networks **7**(5), 1309–1313 (1996)

17. Chen, S.S., Gopalakrishnan, P.S.: Speaker, Environment and Channel Change Detection and Clustering via the Bayesian Inromation Criterion. In: IBM Techical Report, T.J. Watson Research Center (1998)

18. Childers, D., Wu, K., Bae, K., Hicks, D.: Automatic recognition of gender by voice. In: International Conference on Acoustics, Speech, and Signal Processing (ICASSP-1988), vol. 1, pp. 603–606 (1988)

19. Collet, M., Charlet, D., Bimbot, F.a.: A Correlation Metric for Speaker Tracking Using Anchor Models. In: Acoustics, Speech, and Signal Processing, 2005. Proceedings. (ICASSP '05). IEEE International Conference on, vol. 1, pp. 713–716 (2005)

20. Delacretaz, D., Hennebert, J.a.: Text-prompted speaker verification experiments with phoneme specific MLPs. In: Acoustics, Speech and Signal Processing, 1998. Proceedings of the 1998 IEEE International Conference on, vol. 2, pp. 777–780 (1998)

21. Deshpande, M.S., Holambe, R.S.a.: Text-Independent Speaker Identification Using Hidden Markov Models. In: Emerging Trends in Engineering and Technology, 2008. ICETET '08. First International Conference on, pp. 641–644 (2008)

22. Dhanalakshmi, P., Palanivel, S., Ramalingam, V.: Classification of Audio Signals using AANN and GMM. Applied Soft Computing **11**(1), 716 – 723 (2011). DOI 10.1016/j.asoc.2009.12.033. URL http://www.sciencedirect.com/science/article/B6W86-4Y3JY8D-1/2/722d39fe60e735af8ddda0be27d48057

23. Donovan, R.E., Woodland, P.C.: A Hidden Markov-Model-Based Trainable Speech Synthesizer. Computer Speech and Language pp. 1–19 (1999)

24. Dutta, T.: Dynamic Time Warping Based Approach to Text-Dependent Speaker Identification Using Spectrograms. In: Congress on Image and Signal Processing (CISP '08), vol. 2, pp. 354–360 (2008)

25. Eronen, A.: Comparison of Features for Musical Instrument Recognition. In: IEEE Workshop on the Applications of Signal Processing to Audio and Acoustics, pp. 19–22 (2001)

26. Eronen, A., Klapuri, A.: Musical Instrument Recognition Using Cepstral Coefficients and Temporal Features. In: International Conference on Acoustics, Speech, and Signal Processing (ICASSP), vol. 2, pp. 753–756 (2000)

27. Erzin, E., Yemez, Y., Tekalp, A.a.: Multimodal speaker identification using an adaptive classifier cascade based on modality reliability. Multimedia, IEEE Transactions on **7**(5), 840–852 (2005)

28. Finan, R., Sapeluk, A., Damper, R.a.: Comparison of multilayer and radial basis function neural networks for text-dependent speaker recognition. In: Neural Networks, 1996., IEEE International Conference on, vol. 4, pp. 1992–1997 (1996)

29. Foo, S.W., and, E.G.L.: Speaker recognition using adaptively boosted decision tree classifier. In: Acoustics, Speech, and Signal Processing, 2002. Proceedings. (ICASSP '02). IEEE International Conference on, vol. 1, pp. I–157–I–160 (2002)

30. Grasic, M., Kos, M., Zgank, A., Kacic, Z.a.: Comparison of speaker segmentation methods based on the Bayesian Information Criterion and adapted Gaussian mixture models. In: Systems, Signals and Image Processing, 2008. IWSSIP 2008. 15th International Conference on, pp. 161–164 (2008)

31. Grimaldi, M., Cummins, F.a.: Speaker Identification Using Instantaneous Frequencies. Audio, Speech, and Language Processing, IEEE Transactions on **16**(6), 1097–1111 (2008)

32. Gupta, V., Kenny, P., Ouellet, P., Boulianne, G., Dumouchel, P.a.: Multiple feature combination to improve speaker diarization of telephone conversations. In: Automatic Speech Recognition and Understanding, 2007. ASRU. IEEE Workshop on, pp. 705–710 (2007)

33. Hayakawa, S., Itakura, F.a.: Text-dependent speaker recognition using the information in the higher frequency band. In: Acoustics, Speech, and Signal Processing, 1994. ICASSP-94., 1994 IEEE International Conference on, vol. i, pp. I/137–I/140 (1994)

34. He, J., Liu, L., Palm, G.a.: A discriminative training algorithm for VQ-based speaker identification. Speech and Audio Processing, IEEE Transactions on **7**(3), 353–356 (1999)

35. Heck, L., Sankar, A.: Acoustic Clustering and Adaptation for Improved Speech Recognition. In: Proceedings of the Speech Recognition Workshop (1997)

36. Homayoon S. M. Beigi, S.H.M.: Speaker, Channel and Environment Change Detection. Technical Report (1997)

37. Huang, R., Hansen, J.a.: Advances in unsupervised audio classification and segmentation for the broadcast news and NGSW corpora. Audio, Speech, and Language Processing, IEEE Transactions on **14**(3), 907–919 (2006)

38. Huang, X., Acero, A., Adcock, J., wuen Hon, H., Goldsmith, J., Liu, J., Plumpe, M.: A Trainable Text-to-Speech System. In: Proceedings of the International Conference on Spoken Language Processing (ICSLP), pp. 2387–2390 (1996)

39. Inal, M., Fatihoglu, Y.a.: Self organizing map and associative memory model hybrid classifier for speaker recognition. In: Neural Network Applications in Electrical Engineering, 2002. NEUREL '02. 2002 6th Seminar on, pp. 71–74 (2002)

40. J., A.: Effect of age and gender on LP smoothed spectral envelope. In: The IEEE Odyssey Speaker and Language Recognition Workshop, pp. 1–4 (2006)

41. Jaakkola, T., Haussler, D.: Exploiting Generative Models in Discriminative Classifiers. In: Advances in Neural Information Processing Systems, vol. 11, pp. 487–493. MIT Press (1998)

42. Jin, H., Kubala, F., Schwartz, R.: Automatic Speaker Clustering. In: Proceedings of the Speech Recognition Workshop (1997)

43. Jin, Q., Pan, Y., Schultz, T.a.: Far-Field Speaker Recognition. In: Acoustics, Speech and Signal Processing, 2006. ICASSP 2006 Proceedings. 2006 IEEE International Conference on, vol. 1, pp. I–I (2006)

44. Kadri, H., Lachiri, Z., Ellouze, N.a.: Robustness Improvement of Speaker Segmentation techniques Based on the Bayesian Information Criterion. In: Information and Communication Technologies, 2006. ICTTA '06. 2nd, vol. 1, pp. 1300–1301 (2006)

45. Kitahara, T., Goto, M., Komatani, K., Ogata, T., Okuno, H.G.: INSTROGRAM: A New Musical Instrument Recognition Technique Without Using Onset Detection Nor F0 Estimation. In: International Conference on Acoustics, Speech, and Signal Processing (ICASSP), vol. 5, pp. 229–232 (2006)

46. Kotti, M., Martins, L., Benetos, E., Cardoso, J., Kotropoulos, C.a.: Automatic Speaker Segmentation using Multiple Features and Distance Measures: A Comparison of Three Approaches. In: Multimedia and Expo, 2006 IEEE International Conference on, pp. 1101–1104 (2006)

47. Lagrange, M., Martins, L., Teixeira, L., Tzanetakis, G.a.: Speaker Segmentation of Interviews Using Integrated Video and Audio Change Detectors. In: Content-Based Multimedia Indexing, 2007. CBMI '07. International Workshop on, pp. 219–226 (2007)

48. Laxman, S., Sastry, P.a.: Text-dependent speaker recognition using speaker specific compensation. In: TENCON 2003. Conference on Convergent Technologies for Asia-Pacific Region, vol. 1, pp. 384–387 (2003)

49. Li, B., Liu, W., and, Q.Z.: Text-dependent speaker identification using Fisher differentiation vector. In: Natural Language Processing and Knowledge Engineering, 2003. Proceedings. 2003 International Conference on, pp. 309–314 (2003)

50. Livshin, A.A., Rodet, X.: Musical Instrument Identification in Continuous Recordings. In: Proceedings of the 7th International Conference on Digital Audio Effects (DAFX-04), pp. 1–5 (2004)

51. Lodi, A., Toma, M., Guerrieri, R.a.: Very low complexity prompted speaker verification system based on HMM-modeling. In: Acoustics, Speech, and Signal Processing, 2002. Proceedings. (ICASSP '02). IEEE International Conference on, vol. 4, pp. IV–3912–IV–3915 (2002)

52. Mak, M., Allen, W., Sexton, G.a.: Speaker identification using radial basis functions. In: Artificial Neural Networks, 1993., Third International Conference on, pp. 138–142 (1993)

53. Martin, K.D.: Sound-Source Recognition: A Theory and Computational Model. Massachusetts Institute of Technology, Cambridge, MA (1999). PhD Thesis

54. Martin, K.D., Kim, Y.E.: Musical instrument identification: A pattern-recognition approach. In: 136th Meeting of the Acoustical Society of America (1998)

55. Matsui, T., Furui, S.a.: Speaker adaptation of tied-mixture-based phoneme models for text-prompted speaker recognition. In: Acoustics, Speech, and Signal Processing, 1994. ICASSP-94., 1994 IEEE International Conference on, vol. i, pp. I/125–I/128 (1994)

56. Mazarakis, G., Tzevelekos, P., Kouroupetroglou, G.: Musical Instrument Recognition and Classification Using Time Encoded Signal Processing and Fast Artificial Neural Networks. In: Advances in Artificial Intelligence, *Lecture Notes in Computer Science*, vol. 3955, pp. 246–255. Springer, Berlin/Heidelberg (2006). URL http://dx.doi.org/10.1007/11752912_26. 10.1007/11752912_26

57. Molla, K., Hirose, K.a.: On the effectiveness of MFCCs and their statistical distribution properties in speaker identification. In: Virtual Environments, Human-Computer Interfaces and Measurement Systems, 2004. (VECIMS). 2004 IEEE Symposium on, pp. 136–141 (2004)

58. Moschou, V., Kotti, M., Benetos, E., Kotropoulos, C.a.: Systematic comparison of BIC-based speaker segmentation systems. In: Multimedia Signal Processing, 2007. MMSP 2007. IEEE 9th Workshop on, pp. 66–69 (2007)

59. Murthy, H., Beaufays, F., Heck, L., Weintraub, M.a.: Robust text-independent speaker identification over telephone channels. Speech and Audio Processing, IEEE Transactions on **7**(5), 554–568 (1999)

60. Naini, A.S., Homayounpour, M.M.: Speaker age interval and sex identification based on Jitters, Shimmers and Mean MFCC using supervised and unsupervised discriminative classification methods. In: The 8th International Conference on Signal Processing, vol. 1 (2006)

61. Nelwamondo, F., Mahola, U., Marwala, T.a.: Improving Speaker Identification Rate Using Fractals. In: Neural Networks, 2006. IJCNN '06. International Joint Conference on, pp. 3231–3236

62. Nwe, T.L., Sun, H., Li, H., Rahardja, S.: Speaker Diarization in Meeting Audio. In: International Conference on Acoustics, Speech, and Signal Processing (ICASSP), pp. 4073–4076 (2009)

63. Omologo, M., Svaizer, P.: Acoustic event localization using a crosspower-spectrum phase based technique. In: International Conference on Acoustics, Speech, and Signal Processing (ICASSP), vol. II, pp. 273–276 (1994)

64. Park, A., Glass, J.a.: A NOVEL DTW-BASED DISTANCE MEASURE FOR SPEAKER SEGMENTATION. In: Spoken Language Technology Workshop, 2006. IEEE, pp. 22–25 (2006)

65. Peterson, G., Barney, H.L.: Control Methods Used in a Study of the Vowels. The Journal of the Acoustical Society of America (JASA) **24**(2), 175–185 (1952)

66. Reynalds, D.A., Torres-Carrasquillo, P.: Approaches and Applications of Audio Diarization. In: International Conference on Acoustics, Speech, and Signal Processing (ICASSP), vol. 5, pp. 953–956 (2005)

67. Reynolds, D., Kenny, P., Castaldo, F.: A Study of New Approaches to Speaker Diarization. In: InterSpeech (2009)

68. arid Rjjan Rafkin, P.I.M.: Using the Fisher Kernel Method for Web Audio Classification. In: International Conference on Acoustics, Speech, and Signal Processing (ICASSP), vol. 4, pp. 2417 – 2420 (2000)

69. Sae-Tang, S., Tanprasert, C.a.: Feature windowing-based Thai text-dependent speaker identification using MLP with backpropagation algorithm. In: Circuits and Systems, 2000. Proceedings. ISCAS 2000 Geneva. The 2000 IEEE International Symposium on, vol. 3, pp. 579–582 (2000)

70. Salcedo-Sanz, S., Gallardo-Antolin, A., Leiva-Murillo, J., Bousono-Calzon, C.a.: Offline speaker segmentation using genetic algorithms and mutual information. Evolutionary Computation, IEEE Transactions on **10**(2), 175–186 (2006)

71. Scheme, E., Castillo-Guerra, E., Englehart, K., Kizhanatham, A.: Practical Considerations for Real-Time Implementation of Speech-Based Gender Detection. In: Proceesings of the 11th Iberoamerican Congress in Pattern Recognition (CIARP 2006) (2006)

72. Shahin, I., Botros, N.a.: Speaker identification using dynamic time warping with stress compensation technique. In: Southeastcon '98. Proceedings. IEEE, pp. 65–68 (1998)
73. Siegler, M.A., Jain, U., Raj, B., Stern, R.M.: Automatic Segmentation, Classification and Clustering of Broadcast News Audio. In: Proceedings of the DARPA Speech Recognition Workshop (1997)
74. Solewicz, Y.A., Aronowitz, H.: Two-Wire Nuisance Attribute Projection. In: InterSpeech, pp. 928–931 (2009)
75. Takashima, R., Takiguchi, T., Ariki, Y.: HMM-based separation of acoustic transfer function for single-channel sound source localization. In: International Conference on Acoustics, Speech, and Signal Processing (ICASSP), pp. 2830–2833 (2010)
76. Tang, H., Chen, Z., Huang, T.a.: Comparison of Algorithms for Speaker Identification under Adverse Far-Field Recording Conditions with Extremely Short Utterances. In: Networking, Sensing and Control, 2008. ICNSC 2008. IEEE International Conference on, pp. 796–801 (2008)
77. Tanprasert, C., Wutiwiwatchai, C., and, S.S.T.: Text-dependent speaker identification using neural network on distinctive Thai tone marks. In: Neural Networks, 1999. IJCNN '99. International Joint Conference on, vol. 5, pp. 2950–2953 (1999)
78. Tranter, S., Reynolds, D.a.: An overview of automatic speaker diarization systems. Audio, Speech, and Language Processing, IEEE Transactions on **14**(5), 1557–1565 (2006)
79. Venayagamoorthy, G., Sundepersadh, N.a.: Comparison of text-dependent speaker identification methods for short distance telephone lines using artificial neural networks. In: Neural Networks, 2000. IJCNN 2000, Proceedings of the IEEE-INNS-ENNS International Joint Conference on, vol. 5, pp. 253–258 (2000)
80. Viswanathan, M., Beigi, H.S., Dharanipragada, S., Maali, F., Tritschler, A.: Multimedia document retrieval using speech and speaker recognition
81. Wang, X.: Text-Dependent speaker verification using recurrent time delay neural networks for feature extraction. In: IEEE Signal Processing Workshop – Neural Netrowks for Signal Processing – III, pp. 353–361 (1993)
82. Webb, J., Rissanen, E.a.: Speaker identification experiments using HMMs. In: Acoustics, Speech, and Signal Processing, 1993. ICASSP-93., 1993 IEEE International Conference on, vol. 2, pp. 387–390 (1993)
83. Wolf, M., Park, W., Oh, J., Blowers, M.a.: Toward Open-Set Text-Independent Speaker Identification in Tactical Communications. In: Computational Intelligence in Security and Defense Applications, 2007. CISDA 2007. IEEE Symposium on, pp. 7–14 (2007)
84. Wutiwiwatchai, C., Achariyakulporn, V., Tanprasert, C.a.: Text-dependent speaker identification using LPC and DTW for Thai language. In: TENCON 99. Proceedings of the IEEE Region 10 Conference, vol. 1, pp. 674–677 (1999)
85. Yingle, F., Li, Y., Qinye, T.: Speaker gender identification based on combining linear and nonlinear features. In: 7th World Congress on Intelligent Control and Automation. (WCICA 2008), pp. 6745–6749 (2008)
86. Zheng, Y.C., Yuan, B.Z.a.: Text-dependent speaker identification using circular hidden Markov models. In: Acoustics, Speech, and Signal Processing, 1988. ICASSP-88., 1988 International Conference on, pp. 580–582 (1988)
87. Zhu, X., Barras, C., Meignier, S., Gauvain, J.L.: Combining Speaker Identification and BIC for Speaker Diarization. In: InterSpeech (2005)
88. Zigel, Y., Wasserblat, M.a.: How to Deal with Multiple-Targets in Speaker Identification Systems? In: Speaker and Language Recognition Workshop, 2006. IEEE Odyssey 2006: The, pp. 1–7 (2006)
89. Zwicker, E., Flottorp, G., Stevens, S.S.: Critical Band Width in Loudness Summation. Journal of the Acoustical Society of America **29**(5), 548–557 (1957)

Chapter 18
Signal Enhancement and Compensation

In Chapter 5 we had to defer the topic of signal enhancement since we had not yet covered speaker modeling. As we shall see, some of the arguments for different techniques refer to the act of recognition. It is recommended that after reading this chapter, the reader would return to Chapter 5 and quickly glance at the topics which were discussed. Some aspects in that chapter may sink in better with accumulated knowledge.

In this chapter we will examine robustness to noise and conditions via two different perspectives. The first part of this chapter will treat either the raw signal, the power spectrum or the cepstrum of the signal. The methods which are discussed here, attempt to remove or at least reduce components of the signal related to noise, special conditions such as microphone characteristics, or do some kind of normalization of the signal based on the signal (or its derived attributes) alone.

In the second part, we will discuss an array of normalization techniques which basically have the same goal, but utilize speaker-specific knowledge to achieve that goal.

18.1 Silence Detection, Voice Activity Detection (VAD)

Silence Detection, sometimes referred to, as its complementary scheme, Voice Activity Detection (VAD), is a very important practical step in doing speaker or speech recognition. Normally, when there is silence, the relative signal to noise ratio could be quite small and consequently all types of unexpected results may happen affecting both disciplines.

Also, theoretically speaking, as we saw in Section 9.2.1 for the binary hypothesis testing problem (e.g., associated with speaker verification), the likelihood ratio test

(Equation 9.26) should be true *almost everywhere*, [**x**], which means that the feature vector, **x** may not be 0.

In practice, in regard to speaker recognition, even when $\mathbf{x} \neq 0$, the features of noise will occupy part of the feature space, forcing low power speech features to congregate, reducing their variances. This can mainly happen when a spectrally rich noise is on the line. VAD can basically remove these near-silence portions and un-wanted sections to avoid such problems.

Based on experience, close to 30% of the audio frames in a normal audio record-ing are silence frames. This means that through silence removal, the recognition process may become faster by the same rate. In *speech recognition*, the extrane-ous silence segments will produce spurious nonsense words by taking leaps through different arcs of Hidden Markov Models. It is much simpler to use some energy thresholding to cut out moments of silence (or only consider moments of speech) than to have to model the silence, due to the variability of its noise content. The elimination of silence will not only increase the accuracy, but it will also reduce un-necessary processing energy and some cases bandwidth utilization.

The most efficient silence detection is energy thresholding due to its simplicity and effectiveness. The actual threshold may be variable. Although some even use *phone models* developed for *speech recognition* to *detect silence*. Sophisticated al-gorithms may be used to estimate it along the time line. However, in principal, it remains quite simple. Once a threshold is known, the signal power is computed and if it falls below a certain threshold, it is considered to be silence. The signal power may be simply computed using Parseval's theorem which means that the total power in a window is the sum of the squares of its sampled values. Another alternative is to use the c_0 value (see Section 5.3.6), but the value based on the actual samples is much more reliable since it has not undergone the many transformations that were used to attain the MFCCs.

Another piece of information which is somewhat complementary to the energy level and which may be used to detect voice activity is the zero crossing rate of the signal. The zero crossing rate is defined as,

$$\zeta \triangleq \frac{1}{N} \sum_{n=0}^{N-1} \chi(h_n h_{n-1}) \tag{18.1}$$

where $\chi(x)$ is an *indicator function* and is defined as follows,

$$\chi(x) \triangleq \begin{cases} 1 \ if \ x < 0 \\ 0 \ if \ x >= 0 \end{cases} \tag{18.2}$$

ζ is used in conjunction with the energy since usually, the energy for low frequency periodic signals is high compared to high frequency aperiodic cases. The zero cross-ing rate on the other hand works in the opposite manner. See [12] for a sample algorithm which uses both energy and zero crossing rate.

Many different algorithms have been devised to use this information in order to detect the voice activity in a speech signal. Examples are maximum likelihood techniques [9], neural network techniques and discrete wavelet transform methods [33]. In general, all the classification techniques which are used for speech and speaker recognition based on speech features may be used for detecting voice activity.

In some cases, silence detection does not have to be perfect. An example is the two-speaker channel normalization scheme that [39] uses. In this case, only frames with energy levels above the 50% mark are used to do a CMN for a two-speaker detection and tracking task. Of course since the task provides 60s of speech for each case, this luxury may be afforded. Under normal and most practical circumstances, the utterance sections could be much shorter in which case one needs be so choosy. However, as [39] reports later, a fixed energy threshold still gives much better performance even for this type of task. This is partly due to the fact that in this case there are two speakers in a conversation and if it happens (as it usually does) that one speaker's channel energy level is higher than the other (depending on the telephone channel and the speaker's style), then the speaker with the higher energy level will dominate the whole computation.

Another class of competing methods simply train one or more models for the silence and then match against that model as if speaker detection is being done. Sometimes, the feature codebooks used for the silence detection models may be quite different and may use information specific to detecting silence which is not used in regular speaker recognition models. Some of these features are things like pitch detection and the zero crossing rate [3, 12]. Also there are models which are based on non-Gaussian distributions which govern the presence of silence (such as the exponential distribution). This method works well in cases where we have a rich database of different silence situations. Otherwise, the threshold method should suffice and sometimes produce better results.

Sometimes VAD is conducted at the communication channel level, such as in a VoIP channel, to reduce bandwidth utilization. Since telephony applications are designed mainly to be used in two-way conversations, the moments of silence, created as a side-effect of VAD in the communication channel, are sometimes filled with a certain noise called *comfort noise*. This noise is designed to assure the listener, on the call, that the remote party has not been disconnected. This is not a natural noise and is usually generated by a pre-designed model called a *comfort noise generation (CNG)* model [3]. These models may cause problems in the speaker recognizer since the recognizer will try to learn the characteristics of the *CNG model* as part of the speaker's model. Also, sometimes the switch-over between this noise and actual speech may create high-frequency components. In Section 18.4 we will discuss several techniques for removing such anomalies.

18.2 Audio Volume Estimation

In Section 5.2.4 we spoke about the importance of estimating the audio level in real-time to be able to minimize quantization errors and to be able to achieve a better spread of the audio in the dynamic range of the task at hand. When we are dealing with a stream of audio being processed in real time, it becomes important to be able to do audio level normalization. To do this, we need to be able to estimate the average audio level throughout the process of doing our task. Techniques that do this estimation are called *automatic gain control* techniques. The International Telecommunications Union has produced a recommendation, ITU-T-P.56 [17], which provides two methods for the measurement of the active speech level. An ANSI-c code is also available from the ITU.

18.3 Echo Cancellation

Echo cancellation was not included as any of the standard blocks in the sampling process because it is not always necessary. Echo cancellation becomes especially important in telephony applications. As we mentioned in the Introduction, one important advance of speaker recognition to other biometrics is the existence of the telephony infrastructure. Therefore, telephony applications are of utmost importance to the subject.

Anyone who has ever use an international telephone line, should have heard this anomaly, sometimes to an annoying level and at instances to the point where the conversation becomes completely unintelligible. Aside from the annoying part, it also changes the spectral characteristics of the audio which can jeopardize speaker recognition efforts in two different ways. The spectral distortion affects the vocal characteristic analysis and the audible echo can also degrade recognition rates for text-dependent systems that rely on the utterance of a specific text.

In general, echo may be produced by two different means, acoustic-based and network-based. Acoustic-based echoes are those which are quite familiar to anyone who has been in a large ballroom or in a valley surrounded by mountains. It can also happen from reflections off any objects around us when we speak. It may also be present due to the reflections within the handset and feedback.

On the other hand, network delays are caused by the feedback of the speaker's own voice which travels through network and gets reflected from the remote end, returning to the speaker's handset. These delays are characterized by the level difference between the original signal and the echoed signal. This level difference is measure by, "Talker Echo Loudness Rating" (TELR).[16] ITU-T G.122 [15] describes how TELR may be determined in a 4wire/2wire hybrid telephone system and the Terminal Coupling Loss of a telephone set. Most local loops are made up of two wires (wires going from the telephone company's Central Office to homes and businesses) and most transmission lines between different Central Offices are four-wire lines. The hybrid echo happens whenever the two meet. In these scenarios, there are multiple reflection points which add to the complexity of the echo. In addition, Voice Over IP (VoIP) systems have added echo due to the use of the Internet for transporting the audio signal.

There are an array of echo cancellation programs available. Most high-end telephony cards, such as voice T1 and E1 cards have sophisticated echo cancellation software built into them. VoIP suffers more than others in this arena. This is partly due to the fact that most client platforms run on non-realtime operating systems such as Microsoft® Windows® which with built-in delays which add to the network delays already present in these systems.[3]

Echo Cancellation software in essence tries to compare the original (sampled) signal with the returning signal after it has been through the loop going to the far-end and back. Then, they search, with an estimated window or delay, for two signals that resemble each other. The echoed signal is then shifted in phase by 180 degrees and removed from the returning signal prior to sending the returned signal to the loudspeaker. Most of the challenge lies nonlinear signal changes along the way. This includes the addition of colored noise and non-uniform changes to the spectrum of the signal.

Another challenge in the newer telephony environments is the full-bandwidth effect of the echoes which are harder to remove than the original band-limited versions of the signals based on Public Switched Telephone Network (PSTN) lines which operate at bandwidths lower than 4kHz and generally higher than 300Hz.

18.4 Spectral Filtering and Cepstral Liftering

In this section we will discuss filtering techniques used on the PSD and liftering of the cepstra to remove components associated with noise. The main idea behind these techniques is that the range of frequencies which may be produced by the human vocal system is dependent on the motor control of the vocal tract. These frequencies are usually in a central band. As we mentioned in Chapter 5, different microphones introduce different DC components into the signal. By doing a hi-pass filter as the first step of our signal processing, we removed some of this DC component from the spectrum of the signal. There is evidence that slow varying components in the cepstral domain may still exist which are not always associated with the human speech production system. Removal of such slow-moving dynamics have produced good performance results.

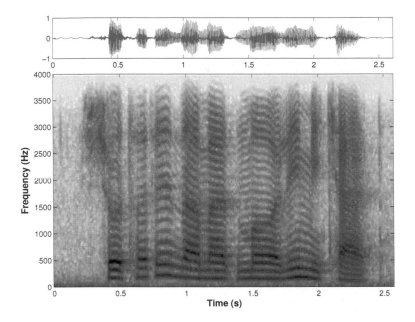

Fig. 18.1: Spectrogram of audio used in the filtering and liftering exercises

In addition, we discussed the introduction of high frequency components due to discontinuities in the signal related to CNG in Section 18.1 and other discontinuities which may not be producible by the limited dynamics of the human vocal system. These discontinuities show up as a high-frequency component in the spectral domain and sharp edges in the cepstral domain. In other words, considering the interpretation of the cepstral coefficients which was given in Section 5.3, high frequency cepstral changes much have to do with a system which much faster dynamics than the human vocal tract is capable of achieving. Therefore, by removing very high frequency components in the cepstral domain, the effects of noise sources capable of such aggressive dynamics are reduced. These two ideas have led to the development of several spectral filtering and cepstral liftering techniques which will be discussed here.

Throughout this chapter, to see the effects of these different techniques, we will be using a short representative segment of speech which was used in Chapter 5. The spectrogram of this audio clip is given in Figure 18.1.

18.4.1 Cepstral Mean Normalization (Subtraction) – CMN (CMS)

Cepstral mean normalization (CMN) is usually conducted by the act of cepstral mean subtraction, so the two terminologies are used interchangeably. It is essentially a longpass lifter[1]. However, the importance of it in speaker recognition is that it removes the DC component only. This is equivalent to having a longpass lifter with a cutoff of 0+ s. As we know, dynamics of filters (lifters) do not allow such sharp contrast. However, since we work in a digital domain, it can readily be done by estimating the mean of the cepstral features in the different dimensions and simply subtracting it form the features in all the frames. Chances are that using a normal longpass lifter subject to dynamic limitations would remove some significant components of G_v which was discussed in Section 5.3.

Although CMN may still remove some speaker characteristics, it has been shown that under noisy conditions, it may improve results for both text-dependent and text-independent speaker recognition implementations [11, 22, 18, 14]. This improvement seems to be due to the fact that the dynamics of the human vocal system is quite rich and that the gain in removing the slow moving components of the cepstra outweigh the loss of part of the vocal tract dynamics. This is why it is crucial not to use a conventional longpass lifter and to try to take care of the DC component through CMN alone. Speech recognition is not as sensitive to this as speaker recognition, since in speech recognition, the goal is to also remove the slow varying dynamics due to the slow vocal tract variations which are more characteristic of the vocal tract itself that they are of the speech content which tends to have faster dynamics.

One important hurdle in conducting CMN is that for short segments, the cepstral mean estimation, for example using a method such as sample mean computation, can be quite far from the actual mean. It is usually recommended that the mean be computed over an entire utterance so that a robust estimate is available. Incorrect estimates could severely degrade the features. As a practical matter, since CMN is one of the first manipulations, if we deal with audio streams, we cannot afford to wait until the end of the utterance until we can start processing it. A moving average method with a forgetting factor somewhat like the moving average technique in *ARMA* modeling [21] may be used to create a running estimate of the cepstral mean. Equation 18.3 is an exponential relation which works reasonably well [3].

$$l\bar{c}_d = (1 - \bar{\lambda}) \; (l-1)\bar{c}_d + \bar{\lambda} \; {}_l c_d \tag{18.3}$$

where $l\bar{c}_d$ signifies the estimated cepstral mean for dimension d and frame l. $(l-1)\bar{c}_d$ is the estimated mean at frame $l - 1$. Also, $_l c_d$ is the value of d dimension of the MFCC vector at the lth frame. $\bar{\lambda}$ is called the forgetting factor or the exponential factor. [3] suggests a forgetting factor of $\bar{\lambda} = 0.999$ to be used to give us a time

---

[1] It is a lifter since it operates in cepstral domain which essentially has unit time – see Chapter 5 for more on this.

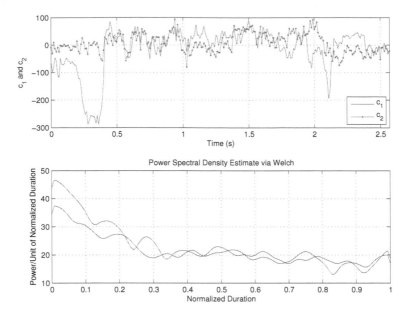

Fig. 18.2: Top: MFCC components (c_1 and c_2) throughout the utterance; Bottom: Power Spectral Density of c_1 and c_2

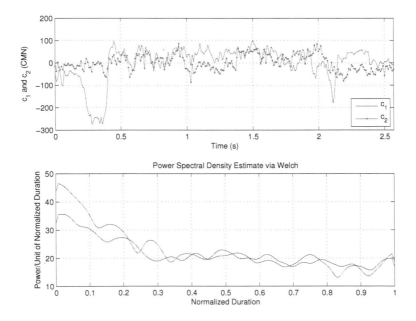

Fig. 18.3: Top: c_1 and c_2 after CMN; Bottom: Power Spectral Density of c_1 and c_2 after CMN

constant of about 7 ms for a 10 ms frame advancement period.

In general, it is quite important to use a good Silence Detection or Voice Activity Detection (VAD) scheme before using CMN. VAD can help in a better estimation of the cepstral mean by reducing the effects of noise, low energy echoes and similar artifacts.

18.4.2 Cepstral Mean and Variance Normalization (CMVN)

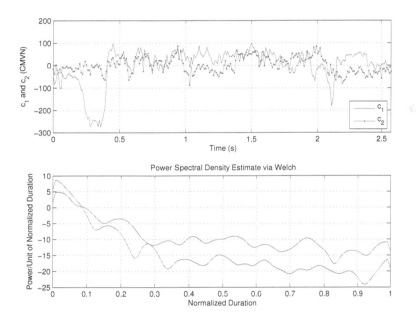

Fig. 18.4: Top: c_1 and c_2 after CMVN; Bottom: Power Spectral Density of c_1 and c_2 after CMVN

The cepstral mean subtraction of the previous section, makes the mean of the cepstral features, 0. A natural extension is to normalize their variance as well. If we assume that the dimensions of the MFCC are statistically independent, we may normalize the variance of each dimension without affecting the other dimensions. However, we may be making the dimensions more compatible such that a multi-dimensional Gaussian cluster would have a more well rounded set of clusters. The *maximum likelihood estimate* of the *variance* of each dimension of the *MFCC* vector is given by the following equation – see Section 6.9.4,

$$Var(c_d) = \frac{1}{N} \sum_{n=0}^{N-1} {}_nc_d^2 - \overline{c}_d^2 \tag{18.4}$$

where N is a large number of frames used to compute the sample mean and the maximum likelihood estimate of the variance. Therefore, assuming a Gaussian distribution, to change the distribution of c_d to a Gaussian with zero mean and a unit variance, we will have to subtract the mean of each dimension of the MFCC vector from the element in that dimension and divide the result by the standard deviation. Namely,

$$_l\hat{c}_d = \frac{_l\hat{c}_d - \overline{c}_d}{\sqrt{\frac{1}{N}\sum_{n=0}^{N-1} {}_nc_d^2 - \overline{c}_d^2}} \tag{18.5}$$

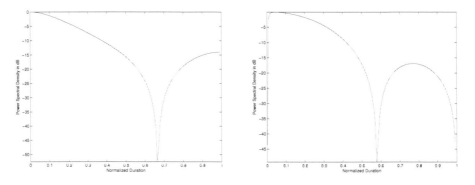

Fig. 18.5: Impulse response of the ARMA lifter

Fig. 18.6: Impulse response of the RASTA lifter

18.4.3 Cepstral Histogram Normalization (Histogram Equalization)

[36] introduced Histogram Equalization (HEQ) for usage in speech processing. The concept was derived from the histogram equalization technique used in image processing. This technique is applied to the features, independently in every feature dimension, to normalize the speech. When MFCCs are used as features, this method is called Cepstral Histogram Normalization (CHN) [3].

In essence, CHN also produces a zero-mean and unit-variance set of features when applied to the MFCCs. It also produces very similar results to those obtained by CMVN. Other similar techniques such as *spectral subtraction* [4] work in a similar manner, but in the spectral domain.

18.4.4 RelAtive SpecTrAl (RASTA) Filtering

[38] shows that for text-independent RASTA filtering reduces the accuracy of the system in the absence of noise. However, it increases its accuracy significantly in the presence of severe noise. This supports the intuitive notion that since RASTA performs band-pass filtering of the spectrum, it removes some speaker characteristics as well. This is quite good for speech recognition, but not so acceptable for speaker recognition, especially the text-independent kind. This is also confirmed by the tests reported by [11] on text-dependent cases. In that case, we see great performance by RASTA and it may be attributed to the fact that text-dependent speaker recognition is really a combination of text-independent speaker recognition and speech recognition. Therefore, the gain due to noise removal greatly outweighs the loss due to speaker characteristic removal.

18.4.4.1 J-RASTA Filtering

[11] has shown great performance for text-dependent speaker recognition using J-RASTA. As we mentioned in the previous section, it may very well be the speech content related part of the system that has benefited greatly from the J-RASTA filtering as it is expected based on speech recognition results. Since both RASTA and J-RASTA basically filter out slow moving spectral characteristics, theoretically, they should remove substantial speaker characteristics as well. Under extremely noisy conditions, this may be alright since the improvements due to noise reduction could be more than the damages incurred due to lost speaker characteristics. This is probably why J-RASTA did substantially better than RASTA in [11]. In J-RASTA, the J factor is learned and it can therefore improve performance. It was especially true for the adaptive J-factor estimation case.

18.4.5 Other Lifters

[28] makes similar arguments for the justification of the band-pass lifter and suggests a sinusoidal window function to be applied to the cepstral coefficients derived from the LPC coefficients. This window is given by Equation 18.6,

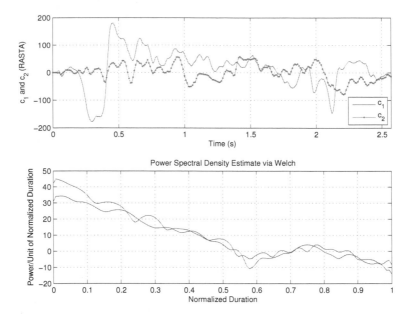

Fig. 18.7: Top: c_1 and c_2 after RASTA Liftering; Bottom: Power Spectral Density of c_1 after RASTA Liftering

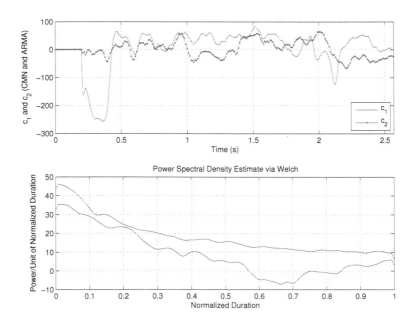

Fig. 18.8: Top: c_1 and c_2 after CMN and ARMA Liftering; Bottom: Power Spectral Density of c_1 after CMN and ARMA Liftering

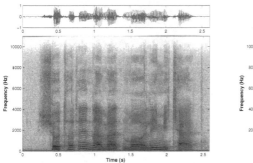

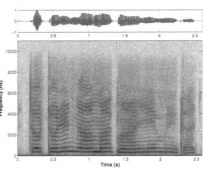

Fig. 18.9: Adult male (44 years old)   **Fig. 18.10:** Male child (2 years old)

$$w_d = 1 + \frac{D}{2} \sin\left(\frac{\pi d}{D}\right) \tag{18.6}$$

Note that classic filter design techniques may be used to design bandpass lifter which would perform as well or better than any of the techniques discussed here. It is important to design the lifter in accordance with the conditions at hand. Figure 18.12 shows an example of such lifteration on the first two MFCCs. Figure 18.11 represents the impulse response of this shortpass lifter.

18.4.6 Vocal Tract Length Normalization (VTLN)

In Chapter 4 we noted that their fundamental frequencies are on the average at 130 Hz, 220 Hz, and 265 Hz respectively. The idea behind Vocal Tract Length Normalization (VTLN) is the fact that the relative position of the formants compared to the fundamental frequency is different for different people. An extreme example is the comparison between an adult male and a child. Consider the two Figures 18.9 and 18.10 which were shown once in Chapter 3 and have been shown here again for convenience. This is the same utterance that was spoken by the adult male and a 2-year old male child. Compare the locations of the formants. If we assume that the vocal tract is modeled by a long tube with varying cross-sections (5.31), then the length of the tube has an inverse relationship with the fundamental frequency and the location of the formants [5].

Vocal Tract Length Normalization (VTLN) has stemmed out of this variation and originally due to the fact that usually there is much more training data for speech recognition obtained from adults, specifically adult males, and that it would be a good idea to be able to transform the features such that they would map one class of individuals to another. In this case, they would map the vocal tract characteristics of a child to an adult male or vice versa to be able to train on one and test on the

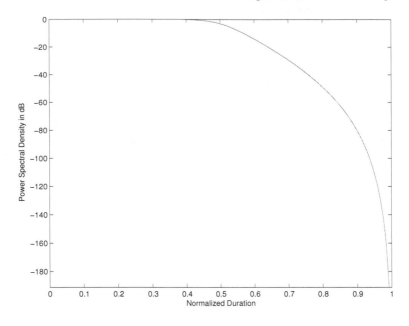

Fig. 18.11: Impulse response of the shortpass lifter

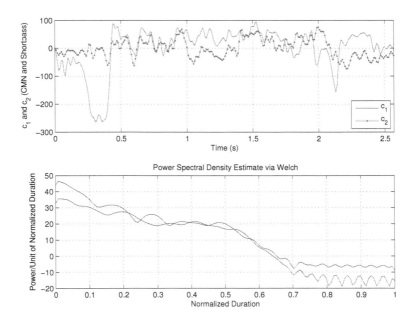

Fig. 18.12: c_1 and c_2 after applying CMN followed by a shortpass lifter

other. This operation may be done in two different ways. The first is the warping of the feature to match a nominal vocal tract length. The second method is the modification of the model to match the speech being recognized. In the case of speaker recognition, though, things are quit the opposite.

As you may recall, at the beginning of Section 5.3, we presented a control system overview of the speech production system. Speaker recognition concentrates on finding dissimilarities among speakers and speech recognition banks on their similarities. It does not mean that VTLN cannot be used for speaker recognition; in fact it has [10]. It just has to be done with a slightly different objective.

In VTLN, it is assumed that the simple model of Figure 5.31 is simplified even more to the extent of the whole vocal tract being modeled as a single hollow tube of length L. This assumption is based on the fact that most of the resonant information about the vocal tract comes from the vowels (see Chapter 4) which are represented by a combination of formants (harmonics). The range of the voice, may then be measure as starting from the fundamental frequency, f_0 (formant 0) and ending at the Nyquist frequency, $\hat{f}_c$. The reason we have called the Nyquist frequency $\hat{f}_c$ and not f_c is that we usually have no indication of the Nyquist frequency. All we have is the folding frequency, f_f which we had marked as f_c in the previous treatments of the features. Most research papers and reports have been quite sloppy in using the term Nyquist frequency for this value. We would like to be very strict about this definition since it causes many misunderstandings and incorrect results for that reason. See Chapter 3 for a detailed description of the difference. Since speech is usually sampled at far lower rates than the Nyquist frequency and is treated, as mentioned at the beginning of this chapter, by a low-pass filter to avoid aliasing, we have no idea what the range of an individual's speech may be. We really just have the lower limit, being the fundamental frequency, f_0. f_0 is then inversely proportional to the length of the tube L.

As we mentioned, [10] assumes misses the difference between f_f and f_c and assumes that the f_f is actually the Nyquist frequency for the speaker. With that (incorrect) assumption, a linear mapping is made from the source frequency range (starting from f_0 and going to f_f, to a target frequency range. This linear mapping is signified by multiplication of the frequency by a parameter, α.

[10] uses a hybrid approach to combine the results of a speaker verification system with that of a speaker classification system. The speaker classifier provides the information for the class of the individual speaker which in turn establishes the warping factor that has to be used to do a VTLN on the features of the individual's speech. Then a normal verification system goes to work to do the actual verification on the warped features. In this scenario, as opposed to a speech recognition scenario, no warping is actually done. The warping factor, α is computed for each individual in the training. Then an average is established. At the verification stage, the warping factor is estimated again and compared to the mean, weighted and then added to the

results of the classical Log-Likelihood Ratio classifier (see Chapter 17) to come up with the results.

A glance at the Figures 18.9 and 18.10 shows the fact that for the adult male, much more of the audio range falls within the sampling range and for the child it is cut-off, signifying again that the f_f is not a valid estimation of the Nyquist frequency. Therefore, a more accurate VTLN approach would be to estimate the first few formants and then estimate the VTLN parameter based on the relative distances between pairs of formats and their distances from the fundamental frequency. A linear, or possibly nonlinear mapping based on perceptual bases, may be used to extract a dissimilarity parameter which may then be used in a similar fashion as in [10]. It is also possible to use this parameter as a separate feature codebook and use it in training and testing the system just like any other feature. The advantage of this approach would be the fact that the formants will still be within the range of the sampling and the problem we noted regarding the difference between f_f and f_c does not exist anymore. This is especially true for low sampling rates such as telephony applications.

Of course, we should also note that this information is already present in the MFCC, PLP or LPC features that have been used to model the system, although not explicitly. As we have seen, the VTLN is used in quite the opposite way. In speech recognition it is used to normalize features. In speaker recognition it is used as a feature to signify the variation.

18.4.7 Other Normalization Techniques

There have been many different normalization techniques which have been used in the literature. Of course it will be impossible to list all of them. Here are some that have shown marginal improvements.

18.4.7.1 Feature Warping

[22] introduced a method for feature warping to be able to handle channel mis-match conditions and additive noise. As we mentioned in Section 18.4.2, one can modify the distribution of the features if it is believed that the distribution has been modified by a noisy process. However, the *CMVN* method uses an arbitrary destination for the variance, namely, 1. In the feature warping approach of [22], it is assumed that the features have been corrupted by noise and that their distribution has been modified from an original distribution to a new distribution, $g_d(\hat{c}_d)$ where d is the index into each dimension of the MFCC vector for a set of frames to be defined later. This

technique may be combined with other techniques that we have discussed earlier.

The method describe in [22] seeks to warp the cepstral features such that over a specific interval, the cumulative distribution of the new features matches a target cumulative distribution. The ideal target distribution would be a multi-modal distribution that would model the true underlying distribution of the speaker. However, to simplify matters, a single mode distribution, namely a Gaussian distribution is chosen.

As we mentioned, the cumulative distribution is only matched in a set time interval. This is defined by a window of N frames. Assuming there are d feature in each *MFCC feature vector* for each frame of audio, there will be $d \times N$-frame windows, each operating on a specific dimension. The features in each N-frame window are ranked according to their value, with the largest (most positive) raking as 1 and the most negative ranking as N. Also, the rank of the feature that falls in the middle of the ranked list (median of the features) is recorded and the value is saved in R. In practice, N is usually set to about 300.

Given a target probability density function, $g_d(\hat{c}_d)$, then the cumulative distribution of the selected N-frame window is computed to fall at the center of the non-parametric distribution seen by the ranking, namely,

$$\int_{-\infty}^{\hat{c}_d} g_d(\xi)d\xi = \frac{N + \frac{1}{2} - R}{N} \qquad (18.7)$$

This really means that the *measured feature*, c_d has the same *cumulative probability distribution* as the *warped feature*, $\hat{c}_d$, but it is based on a different *probability density function*. Namely,

$$\int_{-\infty}^{c_d} h_d(\zeta)d\zeta = \int_{-\infty}^{\hat{c}_d} g_d(\xi)d\xi \qquad (18.8)$$

[35] also uses this technique for warping *Frequency Modulation* features – see Section 5.6.2.1.

Pelecanos assumes that the original distribution of the features should have been a *Gaussian distribution* with 0 mean and *unit* variance. However, the trouble is that it really comes down to the selection of the density function and hence the distribution which is quite arbitrary. In fact the warping concept, here, is not very different form the types of warping that we have done based on the frequency and magnitude of the features according to perceptual distribution models. However, in the perception approaches, the distributions were based on an experiment results designed to mimic human perception.

18.4.7.2 Short-Time Gaussianization

Short-time Gaussianization was introduced [6] to be aid in density estimation for high number of dimensions. It works by transforming the high dimensional data into the *least dependent* coordinates and marginally Gaussuanizng each coordinate separately. [6] claims that the results of the density estimation are sharper than traditional kernel methods.

The process may be done in two steps of a linear transformation, followed by the nonlinear transofrmation of short-time windowed feature warping. [40] has used short-time Gaussianization for formulating a robust speaker verification algorithm.

[41] has also used it by basically adding a *kurtosis normalization* (Section 6.6.4) step right after what was done by [40]. The improvements seem to be quite nominal in the *EER* (Section 19.1.1)!

18.4.7.3 Feature Mapping

[31] introduced a feature mapping technique which works on a similar idea as the SMS (Section 18.4.7.4). However, in feature mapping, the features are not tied to any specific model. In fact all features are mapped to a common space so that they would work best with the same background model. This is done through a transformation which is derived based on the premise that by mapping the features, effectively the Gaussian Mixture Model representing the new features would have been transformed from a Model representing the original features to a universal model representing the new features and being more compatible with all other transformed features.

18.4.7.4 Speaker Model Synthesis (SMS)

Speaker Model Synthesis (SMS) was introduced by [27]. It uses different background models for different channel conditions such that there are essentially as many background models as there are seen channels. Once these models are available, it uses the parameters of these channel models such as the mixture weights and the Gaussian parameters to map any model from one channel to other channels.

Let's assume that we only have a speaker's voice on a cellular network. The transformation will allow the synthesis of different enrollment models for this speakers as if recordings were available from other types of channels like a land-line carbon microphone and a land-line *electret microphone channel*. One major problem with SMS is that it relies on having a rich background model for every channel which may be encountered.

18.4.7.5 Kalman Filtering

Kalman filtering [19] has been used in many different scenarios from control to signal processing. *Sequential Interacting Multiple Models* (*SIMM*) [20] is a method of signal enhancement which uses an adaptive Kalman filter to compensate for nonstationary noise, which is the type noise associated with audio signals. This technique was used for performing isolated digit recognition. It is interesting to see how it would perform for speaker recognition applications.

18.4.8 Steady Tone Removal (Narrowband Noise Reduction)

Steady tones or *narrowband noise* generally exist in many different scenarios, but they are prevalent in any audio that has at some point gone through an analog incarnation. These are steady bands which appear due to a variety of reasons. Sometimes they appear in multiple locations in the spectrum. Of course they can seriously affect the performance of a speaker recognition system, since they will show up as features.

18.4.8.1 Conventional Notch Filters

Traditionally, these types of anomalies have been removed using a *notch filter* which is a filter designed to remove part of the spectrum, associated with the narrow band of interest. Of course, there is a need for recognizing the existence of this band and its bandwidth. Also, such filters will basically remove all the information in that narrow spectral band, including the speaker information. In addition, there will always be modifications made to the adjacent frequencies, mostly in the form of amplitude reduction.

18.4.8.2 Single Value Decomposition (SVD)

Rice, et al. [32] use singular value decomposition for the detection of steady multitones. Then they subtract the resulting components in the spectral domain and reconstruct the signal. This is basically a *notch filter* with a method for detecting the narrowband intrusion.

18.4.8.3 Other Notch-Based Tone Suppression

As another application of a notch filter, Strurim, et al. [34] describe a tone removal scheme which was used in the *MIT-LL*[2] *system* performing the *NIST 2006 speaker recognition* task for *auxiliary microphones*. They use a long duration *Hamming window*[3] (8s) to compute the *Fourier transform* of the signal. Then, they use a low pass filter on the original signal to obtain a smooth version of the signal and subsequently subtract it from the original signal to get a whitened spectrum. Then, by applying a threshold to this whitened spectrum, culprit tones are detected and removed by subtracting a $2 - Hz$ bandwidth *Gaussian* from the amplitude (effectively a *notch filer*). Then the 8 second segments are reconstructed and using an overlap, the 8 second windows are joined to reconstruct the complete signal. [34] shows some positive improvements by using the above technique on the said data.

18.4.8.4 Constant Modulus Algorithm (CMA)

Treichler and Agee [37] present the *constant modulus algorithm* (*CMA*) and an adaptive version of the same algorithm, *constant modulus adaptive algorithm*, which are based on the premise that utilize the existence of an incidental amplitude modulation which is generated by multipath reception of the signal and tonal interference. As we discussed earlier, the *cepstrum*, itself tends to be robust to most multipath reception issues. However, they do still show up in the features. Also, features that are not *cepstrum-based* may even be more affected by the multipath effects. The tonal addition generally affects all features to some degree.

18.4.9 Adaptive Wiener Filtering

Wiener filters have been used in enhancing the quality of speech with some success. *Quatieri and Baxter* [25] proposed one such technique for specific usage with speech signals. Their technique is an adaptive *Wiener filter* which is designed to remove non-stationary noise from the background of speech recordings. The filter was designed with the sensitivity of biological receptors in mind, when dealing with spectral change. Biological receptors mask quick dynamics. This helps eliminate the

---

[2] MIT-LL: *Massachusetts Institute of Technology's Lincoln Laboratories*
[3] Hamming window: see Section 5.3.2.1

removal of valid signal components and aids in the removal of noise. An enhancement to this filter was made in [26]. This filter was used by the *MIT-LL system* [34], in conjunction with the tone removal filter discussed in Section 18.4.8.3.

18.5 Speaker Model Normalization

There are also methods that are based on normalization techniques applied to the scores being returned by the recognition engine. Since decisions are based on these scores and since the scores are channel dependent, certain normalization techniques have been developed to use the channel information in order to transform the score into a normalized score which would be more easily compared across the board.

Another group of methods for handling channel mismatch conditions concentrates on normalizing the scores that are returned by the recognition engine. These techniques have been mainly applied to verification and are called score normalization techniques. In these techniques, *zero normalization* or *Z-Norm* and *test normalization* or *T-Norm* are the bases.

18.5.1 Z-Norm

The *zero normalization*, more widely known as the *z-norm* [30], uses a set of impostor utterances, at the time of enrollment, to get the impostor distribution for each target model and this distribution is saved with the target models. Then the score is modified against the relevant impostor distribution. The following equation is the expression for the *zero normalization* of the log likelihood score, which tries to increase the discriminability of the score.

$$Z = \frac{S - \mu_i}{\sigma_i} \tag{18.9}$$

where S is the score, which is usualy the log likelihood ratio, μ_i is the mean value of S with the model of interest using data from many impostor speakers, and σ_i is the standard deviation of the score, S, for the impostor data. On the other hand, S is the *log-likelihood ratio* of the data being evaluated against the target model.

18.5.2 T-Norm (Test Norm)

Test norm (T-Norm) was first introduced by [1]. It estimates the impostor score distribution for each test utterance against a set of impostor models. Therefore the impostor distribution is specific to each test utterance instead of the target model in the case of Z-Norm.

18.5.3 H-Norm

Handset normalization (H-Norm) was introduced by [8]. It is basically a Z-Norm where there is a different distribution for each handset type.

18.5.4 HT-Norm

Handset Test normalization or HT-Norm is in the same spirit as H-Norm, but is a T-Norm algorithm which is dependent on different handsets.

18.5.5 AT-Norm

AT-Norm is a T-Norm with an adaptive cohort model selection for the impostor models.

18.5.6 C-Norm

C-Norm is basically an H-Norm which tries to cluster lots of enrollment data to come up with clusters associated with unknown, but learned handset types.

18.5.7 D-Norm

D-Norm, introduced by [2] is a technique that normalizes the score by an approximation to the Jeffreys divergence (Section 8.2.2) between the target model and the background model.

18.5.8 F-Norm (F-Ratio Normalization)

[23] describes a group specific normalization called F-Ratio Normalization. The paper shows a relation between the so called F-Norm and the equal error rate. It is described as interpolating between target dependent and target independent information by introducing a mixture parameter which may be optimized by maximizing the class description which is a measure of the separability between target and impostor distributions.

18.5.9 Group-Specific Normalization

This normalization technique was proposed by Poh, et al. [24]. It uses the groups of speakers in the, so called, *biometric menagerie*, defined by *Yager and Dunstone* [42, 43] – see table 16.2 in Section 16.6.2.

18.5.10 Within Class Covariance Normalization (WCCN)

Hatch, et al. [13] proposed *within class covariance normalization (WCCN)* to be used in conjunction with support vector machine based recognizers. *WCCN* uses the *one-against-all* multiclass approach (Section 15.9) where *one* refers to the target speaker and *all* refers to a collection of impostors. It minimizes the expected error rate in both false acceptance and false rejection senses in the training step of the SVM [7].

18.5.11 Other Normalization Techniques

There are also many other techniques such as Joint Factor Analysis (JFA) (Section 16.4.2), Nuisance Attribute Projection (NAP) (Section 15.8), World Maximum

A-Posteriori (WMAP), Eigenchannel, etc. JFA is usually used in conjuction with GMM models and NAP with SVM. However, as we have mentioned in the respective sections, some hybridization has also been done. Eigenchannel estimation is similar to PCA and WMAP is an extension of MAP (Sections 10.2 and 11.3.1.3).

Also, [29] plays around with some new and some combination compensations such as *emotional stress compensation*, *Speaker and stress information based compensation* (*SSIC*), *Compensation by removal of stress vectors* (*CRSV*), and *Selection of compensation by stress recognition* (*scsr*). These are all related to speakers operting under stress.

References

1. Auckenthaler, R., Carey, M., Lloyd-Thomas, H.: Score Normalization for Text-Independent Speaker Verification Systems. Digital Signal Processing **10**(1–3), 42–54 (2000)
2. Ben, M., Blouet, R., Bimbot, F.: A Monte-Carlo method for score normalization in Automatic Speaker Verification using Kullback-Leibler distances. In: International Conference on Acoustics, Speech, and Signal Processing (ICASSP), vol. 1, pp. 689–692 (2002)
3. Benesty, J., Sondhi, M.M., Huang, Y.: Handbook of Speech Processing. Springer, New york (2008). ISBN: 978-3-540-49125-5
4. Boll, S.F.: Suppression of Acoustic Noise in Speech using Spectral Subtraction. IEEE Transactions on Acoustics, Speech, and Signal Processing (ASSP) **27**(2), 113–120 (1979)
5. Chau, C.K., Lai, C.S., Shi, B.E.: Feature vs. Model Based Vocal Tract Length Normalization for a Speech Recognition Based Interactive Toy. In: Active Media Technology, Lecture Notes in Computer Science, pp. 134–143. Springer, Berlin/Heidelberg (2001). ISBN: 978-3-540-43035-3
6. Chen, S., Gopinath, R.A.: Gaussianization. In: Neural Information Processing Systems (NIPS) (2000)
7. Dehak, N., Kenny, P., Dehak, R., Dumouchel, P., Ouellet, P.: Front-End Factor Analysis for Speaker Verification. IEEE Transactions on Audio, Speech and Language Processing **19**(4), 788–798 (2011)
8. Dunn, R.B., Reynolds, D.A., Quatieri, T.F.: Approaches to Speaker Detection and Tracking in Conversational Speech. Digital Signal Processing **10**, 92–112 (2000)
9. Gauci, O., Debono, C., Micallef, P.: A maximum log-likelihood approach to voice activity detection. In: International Symposium on Communications, Control and Signal Processing (ISCCSP-2008), pp. 383–387 (2008)
10. Grashey, S., Geibler, C.a.: Using a Vocal Tract Length Related Parameter for Speaker Recognition. In: Speaker and Language Recognition Workshop, 2006. IEEE Odyssey 2006: The, pp. 1–5 (2006)
11. Hardt, D., Fellbaum, K.a.: Spectral subtraction and RASTA-filtering in text-dependent HMM-based speaker verification. In: Acoustics, Speech, and Signal Processing, 1997. ICASSP-97., 1997 IEEE International Conference on, vol. 2, pp. 867–870 (1997)
12. Harsha, B.: A noise robust speech activity detection algorithm. In: Intelligent Multimedia, Video and Speech Processing, pp. 322–325 (2004)
13. Hatch, A.O., Kajarekar, S., Stolcke, A.: Within-Class Covariance Normalization for SVM-based Speaker Recognition. In: Interspeech (2006)
14. Hermansky, H., Morgan, N.: RASTA Processing of Speech. IEEE Transactions on Speech and Audio Processing **2**(4), 578–589 (1994)

15. ITU-T: G.122: Influence of national systems on stability and talker echo in international connections. ITU-T Recommendation (2003). URL http://www.itu.int/rec/T-REC-G.122/e
16. ITU-T: G.131: Talker echo and its control. ITU-T Recommendation (2003). URL http://www.itu.int/rec/T-REC-G.131/e
17. ITU-T: P.56: Objective measurement of active speech level. ITU-T Recommendation (2003). URL http://www.itu.int/rec/T-REC-P.56/e
18. Johnsen, M.H., Svendsen, T., Harborg, E.:
19. Karman, R.E.: A New Approach to Linear Filtering and Prediction Problems. Transactions of the ASME–Journal of Basic Engineering 82(Series D), 35–45 (1960)
20. Kim, N.S.: Feature domain compensation of nonstationary noise for robust speech recognition. Speech Communication 37(3–4), 59–73 (2002)
21. Li, C.J., Beigi, H.S., Li, S., Liang, J.: Nonlinear Piezo-Actuator Control by Learning Self-Tuning Regulator. ASME Transactions, Journal of Dynamic Systems, Measurement, and Control 115(4), 720–723 (1993)
22. Pelecanos, J., Sridharan, S.: Feature Warping for Robust Speaker Verification. In: A Speaker Odyssey - The Speaker Recognition Workshop, pp. 213–218 (2001)
23. Poh, N., Bengio, S.: F-Ratio and Client-Dependent Normalisation for Biometric Authentication tasks. In: International Conference on Acoustics, Speech, and Signal Processing (ICASSP), vol. 1, pp. 721–724 (2005)
24. Poh, N., Kittler, J., Rattani, A., Tistarelli, M.: Group-Specific Score Normalization for Biometric Systems. In: IEEE Computer Society Conference on Computer Vision and Pattern Recognition Workshops (CVPRW), pp. 38–45 (2010)
25. Quatieri, T., Baxter, R.A.: Noise reduction based on spectral change. In: IEEE ASSP Workshop on Applications of Signal Processing to Audio and Acoustics (1997)
26. Quatieri, T.F., Dunn, R.B.: Speech enhancement based on auditory spectral change. In: International Conference on Acoustics, Speech, and Signal Processing (ICASSP), vol. 1, pp. 257–260 (2002)
27. R., T., B., S., L., H.: A Model-Based Transformational Approach to Robust Speaker Recognition. In: International Conference on Spoken Language Processing, vol. 2, pp. 495–498 (2000)
28. Rabiner, L., Juang, B.H.: Fundamentals of Speech Recognition. Prentice Hall Signal Processing Series. PTR Prentice Hall, New Jersey (1990). ISBN: 0-130-15157-2
29. Raja, G.S., Dandapat, S.: Speaker Recognition Under Stress Condition. International Journal of Speech Technology 13 (2010). Springerlink Online: DOI 10.1007/s10772-010-9075-z
30. Reynolds, D.A.: Comparison of Background Normalization Methods for Text-Independent Speaker Verification. In: Eurospeech 1997, pp. 963–966 (1997)
31. Reynolds, D.a.: Channel robust speaker verification via feature mapping. In: Acoustics, Speech, and Signal Processing, 2003. Proceedings. (ICASSP '03). 2003 IEEE International Conference on, vol. 2, pp. II–53–6 (2003)
32. Rice, B.F., Fechner, R.M., Wilhoyte, M.E.: A new approach to multipath correction of constant modulus signals. In: IEEE, vol. 1, pp. 325–329 (1983)
33. Stadtschnitzer, M., Van Pham, T., Chien, T.T.: Reliable voice activity detection algorithms under adverse environments. In: International Conference on Communications and Electronics, pp. 218–223 (2008)
34. Sturim, D.E., Campbell, W.M., Reynolds, D.A., Dunn, R.B., Quatieri, T.: Robust Speaker Recognition with Cross-Channel Data: MIT-LL Results on the 2006 NIST SRE Auxiliary Microphone Task. In: International Conference on Acoustics, Speech, and Signal Processing (ICASSP), vol. IV, pp. 49–52 (2007)
35. Thiruvaran, T., Ambikairajah, E., Epps, J.a.: Normalization of Modulation Features for Speaker Recognition. In: Digital Signal Processing, 2007 15th International Conference on, pp. 599–602 (2007)
36. de la Torre, A., Peinado, A.M., Segura, J.C., Perez-Cordoba, J.L., Benitez, M.C., Rubio, A.J.: Histogram Equalization of Speech Representation for Robust Speech Recognition. IEEE Transaction of Speech and Audio Processing 13(3), 355–366 (2005)

37. Treichler, J.R., Agee, B.G.: A new approach to multipath correction of constant modulus signals. IEEE Transactions on Audio, Speech and Signal Processing **31**(2), 459–472 (1983)
38. van Vuuren, S.: Comparison of Text-Independent Speaker Recognition Methods on Telephone Speech with Acoustic Mismatch. In: Proceedings of the International Conference on Spoken Language Processing (ICSLP), pp. 784–787 (1996)
39. Weber, F., Peskin, B., Newman, M., Corrada-Emmanuel, A., Gillick, L.: Speaker Recognition on Single and Multispeaker Data. Digital Signal Processing **10**, 75–92 (2000)
40. Xiang, B., Chaudhari, U., Navratil, J., Ramaswamy, G., Gopinath, R.a.: Short-time Gaussianization for robust speaker verification. In: Acoustics, Speech, and Signal Processing, 2002. Proceedings. (ICASSP '02). IEEE International Conference on, vol. 1, pp. I–681–I–684 (2002)
41. Xie, Y., Dai, B., and, J.S.: Kurtosis Normalization after Short-Time Gaussianization for Robust Speaker Verification. In: Intelligent Control and Automation, 2006. WCICA 2006. The Sixth World Congress on, vol. 2, pp. 9463–9467 (2006)
42. Yager, N., Dunstone, T.: Worms, Chameleons, Phantoms and Doves: New Additions to the Biometric Menagerie. In: IEEE Workshop on Automatic Identification Advanced Technologies, pp. 1–6 (2007)
43. Yager, N., Dunstone, T.: The Biometric Menagerie. IEEE Transactions on Pattern Analysis and Machine Intelligence (PAMI) **32**(2), 220–230 (2010)

Part III
Practice

Chapter 19
Evaluation and Representation of Results

Given all the discussions in this book, when it is time to present recognition results for the sake of performance evaluation and comparison among different techniques, some quantitative evaluation standards are deemed necessary. This chapter discusses the different evaluation metrics and jargon used in the speaker recognition discipline.

19.1 Verification Results

Speaker verification and speaker identification are inherently different in the way the results are presented. In this section we will describe different concepts which help us in presenting verification results. As we will see, the results may be and have been presented in many occasions by common platforms which are used by other verification communities. However, recently, subtle changes have been adopted by the speaker verification community which create a more customized methodology for presenting the results in this field. Please refer to Definitions 9.6 and 9.7 before continuing.

19.1.1 Equal-Error Rate

Equal Error Rate (EER) is a popular and yet a much despised method of reporting the performance of biometric techniques. In biometric applications, there is a trade-off between the tolerated *false rejection rate (FRR)* and *false acceptance rate (FAR)* of the system. In most implementations, there is a threshold which may be chosen to change this trade-off toward either side. Most biometric engine providers try to solve the general biometric problem and are not necessarily designing a specific application with pre-dictated requirements. For this reason, they will have to report their performance results at some operating condition. In order to be able to quote percentages for the error rate, a popular operating point is the Equal Error Rate which is the point where there is an equal chance of false rejections (FRR) and false acceptances (FAR).

Therefore, the EER provides the means for producing a convenient percent error. However, since normal operating conditions seldom lie at a point where there will be as many false rejections made as there are false acceptances, the user community does not like results produced for this operating point. To remedy this problem, the *National Institute of Standards and Technology* (NIST) has proposed more informative means of reporting accuracies (error rates) which will be described later in this chapter.

19.1.2 Half Total Error Rate

Sometimes it is desirable to compare performance by considering both *false acceptance* and *false rejection* numbers. The *total error rate* [3] is one such measure. The *total error rate (TER)* is defined as the sum of the *false rejection (FRR)* and the *false acceptance (FAR)* rates. Also, it makes sense to talk about a number which is of the same order as FRR and FAR. Therefore, sometimes researchers talk about the *half total error rate (HTER)* which is basically computed as a half of the *total error rate*. In other words,

$$HTER = \frac{FRR + FAR}{2} \tag{19.1}$$

In systems where the *TER* (or HTER) stays constant, it is a good measure for the performance of the system. However, it is possible that there is an imbalance between the FAR and FRR such that, for example, the FAR changes with very little change in FRR and vice versa. At those operating points, the TER (or HTER) is not a good performance measure. The HTER is more popular in non-speaker biometrics.

19.1.3 Receiver Operating Characteristic (ROC) Curve

Receiver Operating Characteristic (ROC) curves were developed in the mid-20th century by radio engineers to show noise levels of signals. These curves were adopted by biometrics researchers up to the late 1990s, in order to show the different operating conditions of their engines. The ROC curves plot the *false positive rate* on the *abscissa* and the *true positive rate* on the *ordinate*. For such curves, the objective is to approach a performance such that the area under the curve would be nearly 1. Note that if a random guess is taken for the results, the mean area under the curve would be 0.5.

In the *speaker recognition* circles, ROC curves are normally plotted with the *false acceptance rate* on the *abscissa* and the *false rejection* on the *ordinate*. Figure 19.1 shows an example ROC curve comparing two different verification methods and their combination. In this kind of ROC, the objective is to have the area under the curve approach 0. Generally, a random guess should produce an area of about 0.5.

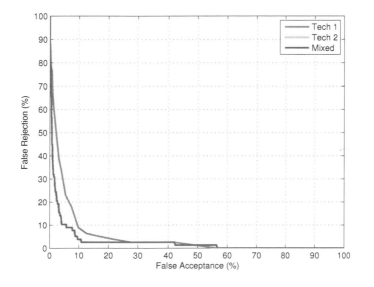

Fig. 19.1: Sample Receiver Operating Characteristic (ROC) curve representing the same data which was used to plot the DET Curve of Figure 19.2

Here are two sample ROC curves showing false rejection numbers at the ordinate and false acceptance at the abscissa. They are plotted for the same system performing two different tasks each having different quality of audio. Note that the ROC curve is usually plotted in a linear scale. It may also be plotted in logarithmic scale to avoid the resolution problems seen in the figure at the right. The performance of the system is so good in that figure that it is hard to see the graph. The location of the equal error rate is marked in both of these curves. Notice that the three different colors signify difference techniques.

19.1.4 Detection Error Trade-Off (DET) Curve

The DET curve has been favored over traditional ROC curves since its introduction by NIST in 1997.[1] The DET curve plots the "Miss Probability" in percentage form verses the "False Alarm Probability," also in percentages. Figure 19.2 shows the an example DET curve comparing two different verification methods and their combination.

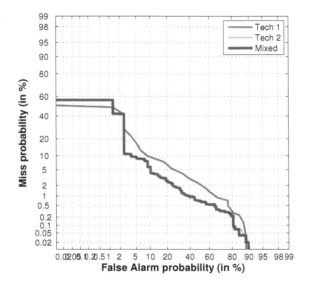

Fig. 19.2: Sample Detection Error Tradeoff (DET) curve representing the same data which was used to generate the ROC Curve of Figure 19.1

The DET curve always uses a logarithmic scale and it plots two different entities against each other. In this case the abscissa is the false alarm probability which happens to be the same number as the false rejection. The ordinate is the miss probability which is the same as the false acceptance. The logarithmic scale usually makes this curve more linear and the ROC more concave. The operating point for the equal error rate is marked in these figures.

Note that simply knowing the EER does not really provide much insight into other possible operating points. It is almost never desirable to operate at the EER point. Depending on how much the security of the system being protected by the verification engine is important to us versus the ease of use, we would use different operating points. If we would definitely not want anyone to get into the system without proper authority, then we would operate at a point where we may have to disturb the clients by falsely rejecting some valid users in order to maintain high security. This is usually alleviated by requesting more audio data from the users or using some supplemental information to make our decision.

19.1.5 Detection Cost Function (DCF)

NIST [2] defines a *Detection Cost Function (DCF)*, C_{Det}, as

$$\mathscr{C}_{Det} \triangleq (\mathscr{C}_{Miss}\ p(Miss|Target)\ P(Target)) +$$
$$(\mathscr{C}_{FalseAlarm}\ p(FalseAlarm|NonTarget)\ P(NonTarget)) \qquad (19.2)$$

where, $\mathscr{C}_{Miss}$ is the cost of missed detection, $\mathscr{C}_{FalseAlarm}$ is the cost of a false alarm, p(Target) is the *a-priori* probability of a *target speaker*, and p(NonTarget) is the *a-priori* probability of a *non-target speaker*. [2] uses the following values for the 1999 NIST evaluations,

$$\mathscr{C}_{Miss} = 10$$
$$\mathscr{C}_{FalseAlarm} = 1$$
$$P(Target) = 0.01$$
$$P(NonTarget) = 1 - P(Target)$$
$$= 0.99$$

19.2 Identification Results

In this section, we show sample results of identification. It is important to show the *rank* of an *identification* result as well as the *top result*. The figure shown here provides insight into how close the identification results were to the truth.

Figure 19.3 shows the result of an identification test performed on a telephony task using call-center data. In this example, there are 78 speakers who have been enrolled. A total of 78 tests have been conducted. 1 test speakers have not been identified. This is due to the fact that the *open-set* option was chosen for the *identification engine*. 72 tests have rank of 1. This means that the *test speaker* was the *top result*. In one test, the *test speaker's* true identity showed up as the *second* in the list (hence having rank 2). Similarly, two tests ranked 3 and one test ranked 7. 7 was the largest rank, meaning that there was no test in which the rank was higher than 7 compared to the theoretic maximum of 78. Another number which is important for understanding the performance of the identification process is the average rank. In this case, the average rank was 1.15. The smaller this number, the better has the engine performed. Of course, the minimum rank is 1, so the objective is to have the *average rank* approach 1.

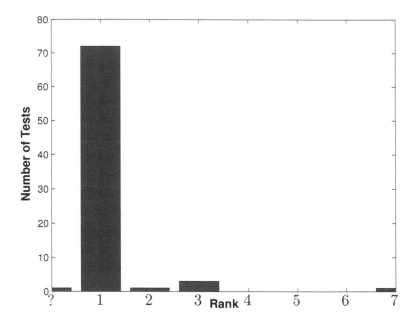

Fig. 19.3: Sample Histogram of the rank of the target speaker from a 78-speaker database with 78 tests using the same data as was used for the verification task in Figures 19.1 and 19.2

The use of the average rank and the error rate are somewhat similar to giving an EER for the verification case in contrast with the presentation of the DET or ROC curve. Eventually, the DET curve in verification and the *rank distribution* in *identification* are more important than single numbers, although single number results are still used for convenience.

References

1. Martin, A., Doddington, G., Kamm, T., Ordowski, M., Przybocki, M.: The DET Curve in Assessment of Detection Task Performance. In: Eurospeech 1997, pp. 1–8 (1997)
2. Martin, A., Przybocki, M.: The NIST 1999 Speaker Recognition Evaluation – An Overview. Digital Signal Processing **10**, 1–18 (2000)
3. Toh, K.A., Kim, J., Lee, S.: Biometric scores fusion based on total error rate minimization. Pattern Recognition **41**(3), 1066 – 1082 (2008). DOI DOI: 10.1016/j.patcog.2007.07.020. Part Special issue: Feature Generation and Machine Learning for Robust Multimodal Biometrics

Chapter 20
Time Lapse Effects (Case Study)

The effect of time-lapse has not been studied well in biometrics. Although the literature is full of brief discussions about time-lapse effects in speaker recognition, no proper quantitative study has been done on the subject. [9, 8] There are two main types of time-lapse effects: *short-term* and *long-term* (aging). Here, short-term effects are studied for speaker recognition (speaker identification and speaker verification). The RecoMadeEasy[1] speaker recognition engine has been used to obtain baseline results for 22 speakers who have been involved in a persistent (ongoing) study.

Speakers were involved in language proficiency testing where they had to repeat their tests due to undesirable scores. The speakers used here retook their tests two more times after the original testing was accomplished. The time lapse between consecutive tests was on the average between 1 to 2 months. Figures 20.1 and 20.2 show the distribution of the time lapse in days between tests 1 and 2 and tests 2 and 3 respectively. The words test and trial will be used interchangeably from this point on. Each test consists of multiple audio segments which are each about 1 minute long. These segments are free-form responses to questions to assess the candidates' proficiency in the English language. Unfortunately, due to the fact that this type of study is quite rare, no standard corpus is available.

The RecoMadeEasy® Speaker Recognition engine uses a Gaussian Mixture Model (**GMM**) approach to conduct identification and verification of the speakers. Under normal circumstances, the first response of the first test (first trial) is used to enroll the speaker in the database. Consequent segments are identified or verified against the enrollment data captured from the first response. This scenario is used to conduct the rest of the test without any need for a proctor, therefore reducing the cost of testing. The data is obtained from a real-world application and has not been manipulated or specifically collected for this purpose. These are real candidates tak-

---

[1] RecoMadeEasy® is the Commercial Speaker Recognition Engine of Recognition Technologies, Inc.

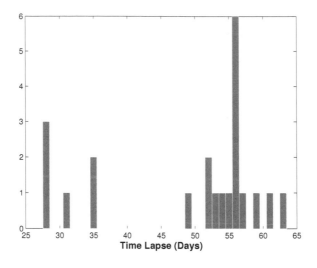

Fig. 20.1: Number of Days Between Test 1 and 2

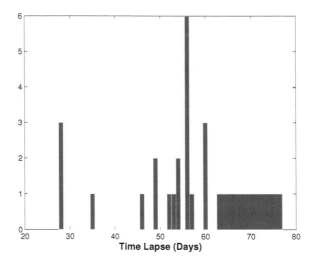

Fig. 20.2: Number of Days Between Test 2 and 3

ing tests to be evaluated for English proficiency.

Tests have revealed that there is substantial degradation in the results of both identification and verification from one seating to the next. There are two main reasons for this degradation. The first well-known reason is known as channel mismatch between the enrollment session and the recognition of consequent tests. Many different approaches have been taken to reduce the effects of this mismatch. Channel mismatch has in the past been mostly connected to handset mismatch, however, it is quite more complicated than that. In addition to handset mismatch, changes in the ambient noise (sometimes called source noise [14]), acoustic properties of the ambiance (such as echo and reverberation), microphone distance and positioning (angle), strain on the vocal tracts (holding the handset on one's shoulder) and many more are also responsible for these types of mismatches.[15].

Different techniques have been proposed for handling this type of mismatch by considering specific sources of mismatch. These include Handset Score Normalization (H-Norm) [13], feature mapping [12], and speaker model synthesis (SMS) [11]. Others have approached the problem by suppressing the effects through using the Test Normalization (T-Norm), the Z-Norm [1, 2] and Feature Warping techniques [10].

The second reason for degradation is a combination of other factors such as physiological changes, environmental changes, emotional changes, etc.[4, 3] These effects are not very well understood and are bundled here in one category called time-lapse effects. Among these changes, there are some which get worse with time. We are interested in these effects which we lump together into time-lapse effects. Note that we are not dealing with what the literature calls aging, since aging deals with much longer effects, outside the range of these shorter-term studies. Aging effects deal with more of the physiological changes that affect speakers as substantial time progresses.[4, 3]

To see these effects of interest we have considered three consecutive tests per candidate. The changes between the first test and the second test include both channel-mismatch and time-lapse effects. However, by doing a third test and seeing further degradation of the recognition results, we can conclude that time-lapse effects have caused most of the extra degradation seen from the first trial to the third trial as compared to the changes from the first trial to the second trial.

First, a description of the data is given in the following section. Then, we discusses these degradations in more detail by doing a quantitative analysis of the Identification and Verification results. Following this discussion, we try to reduce the effects of time-lapse using several adaptation techniques and the results are reported for identification and verification tasks followed by concluding remarks.

20.1 The Audio Data

The audio data was collected using the μ-Law amplitude coding technique [5] at a sampling rate of 8 kilo Hertz (kHz). The audio was then immediately converted to the High-Efficiency Advanced Audio Coding Format (**HE-AAC**) [7] which is a very aggressive, lossy and low-bit-rate audio compression technique. **HE-AAC** was used to stream the audio to a server through flash. The audio, in turn, was converted back to μ-Law 8-kHz audio and subsequently converted to a 16-bit linear Pulse Code Modulation (**PCM**) which was used in the recognizer for enrollment, identification and verification purposes.

The RecoMadeEasy® speaker recognition engine was used for obtaining results. This engine is a **GMM**-based text-independent and language-independent engine. It uses models for the speaker and the competing models to conduct the identification and verification tasks. The population in the identification task is the 22 speakers described in the next section plus competing models. The models are parameters for collections of multi-variate normal density functions which describe the distribution of the Mel-Cepstral features for speakers' enrollment data.

20.2 Baseline Speaker Recognition

As previously mentioned, each candidate goes through a testing procedure in which questions are asked and responses from the candidate are recorded. Under usual circumstances, the first audio response is used to enroll the speaker into the system. All the responses average to about 1 minute of audio. Figure 20.3 shows the results of identification of individuals among the 22 candidates in our database. All subsequent audio responses are identified at a rate of 100% or an error rate of 0%. In this case, although the enrollment and recognition data differ, there is no channel mismatch. These results are expected from a good commercial recognition system. However, as conditions change and the candidates return to be tested for a second or third time, a substantial degradation is noted, see figure 20.3. For the second trial (test) and the third trial, there is channel mis-match [13] present as well as time-lapse effects. Since the channels are chosen at a completely random manner in both second and third trials, the extra degradation seen between trial 2 and trial 3 is most likely due to time-lapse effects.

Figure 20.4 shows similar results for the verification process. In this figure, three Detection Error Tradeoff (**DET**) curves [6] are presented using the first response in the first test for enrollment, consequent data in the first test for verification of trial 1 and the second response in the second and third trials for verification of those trials. The plot shows results which are similar to those seen in the identification case. Namely, the Equal Error Rate (**EER**) increases from about 2.5% to nearly an order

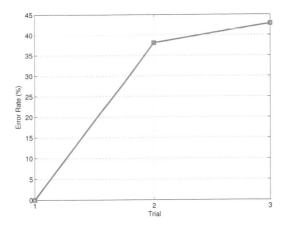

Fig. 20.3: Identification Time Lapse – Usual Enrollment

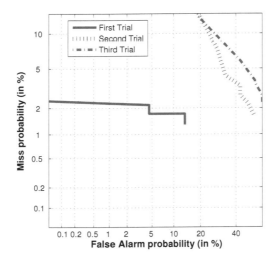

Fig. 20.4: Verification Time Lapse using Usual Enrollment

of magnitude higher for the second and third trials. In consistence with the results for the identification tests, the performance of the verification system also degrades in time as we move from the second to the third trial, whereas, the two channel conditions in these trials are statistically as distant from the channel conditions of the first trial. Therefore, the extra degradation from the second to the third trial may be attributed to the time-lapse aspects.

In the next chapter we will explore some adaptation techniques to allow for the modification of the speaker models. Adaptation, as it will be seen later, allows for significant reduction in the time lapse effects, although it does degrade the baseline results slightly.

References

1. Auckenthaler, R., Carey, M., Lloyd-Thomas, H.: Score Normalization for Text-Independent Speaker Verification Systems. Digital Signal Processing **10**(1–3), 42–54 (2000)
2. Barras, C., Gauvain, J.L.a.: Feature and score normalization for speaker verification of cellular data. In: Acoustics, Speech, and Signal Processing, 2003. Proceedings. (ICASSP '03). 2003 IEEE International Conference on, vol. 2, pp. II–49–52 (2003)
3. Endres, E., Bambach, W., Flösser, G.: Voice Spectrograms as a Function of Age, Voice Disguise and Voice Imitation. Journal of the Acoustical Society of America (JASA) **49**, 1842–1848 (1971)
4. Gurbuz, S., Gowdy, J., Tufekci, Z.: Speech spectrogram based model adaptation for speaker identification. In: Southeastcon 2000. Proceedings of the IEEE, pp. 110–115 (2000)
5. ITU-T: G.711: Pulse Code Modulation (PCM) of Voice Frequencies. ITU-T Recommendation (1988). URL http://www.itu.int/rec/T-REC-G.711-198811-I/en
6. Martin, A., Doddington, G., Kamm, T., Ordowski, M., Przybocki, M.: The DET Curve in Assessment of Detection Task Performance. In: Eurospeech 1997, pp. 1–8 (1997)
7. Meltzer, S., Moser, G.: MPEG-4 HE-AAC v2 – audio coding for today's digital media world. World Wide Web (2005). URL http://www.ebu.ch/fr/technical/trev/trev_305-moser.pdf
8. Naini, A.S., Homayounpour, M.M.: Speaker age interval and sex identification based on Jitters, Shimmers and Mean MFCC using supervised and unsupervised discriminative classification methods. In: The 8th International Conference on Signal Processing, vol. 1 (2006)
9. Nolan, F.: The Phonetic Bases of Speaker Recognition. Cambridge University Press, New York (1983). ISBN: 0-521-24486-2
10. Pelecanos, J., Sridharan, S.: Feature Warping for Robust Speaker Verification. In: A Speaker Odyssey - The Speaker Recognition Workshop, pp. 213–218 (2001)
11. R., T., B., S., L., H.: A Model-Based Transformational Approach to Robust Speaker Recognition. In: International Conference on Spoken Language Processing, vol. 2, pp. 495–498 (2000)
12. Reynolds, D.a.: Channel robust speaker verification via feature mapping. In: Acoustics, Speech, and Signal Processing, 2003. Proceedings. (ICASSP '03). 2003 IEEE International Conference on, vol. 2, pp. II–53–6 (2003)
13. Reynolds, D.A., Quatieri, T.F., , Dunn, R.B.: Speaker Verification Using Adapted Gaussian Mixture Models. Digital Signal Processing **10**, 19–41 (2000)
14. Shannon, C.E.: Communication in the Presence of Noise. Proceedings of the Institute of Radio Engineers **37**(1), 10–21 (1949). Reprint available at: Proceedings of the IEEE, Vol. 86, No. 2, Feb. 1998
15. Vogt, R., Sridharan, S.: Explicit modelling of session variability for speaker verification. Computer Speech and Language **22**(1), 17–38 (2008)

Chapter 21
Adaptation over Time (Case Study)

In the previous chapter we noted the degrading effects of time lapse. The text- and language-independent RecoMadeEasy® speaker recognition engine was used to obtain baseline results for 22 speakers who have been involved in a long-term study. The speakers generated data in three seatings with 1 to 2 months delay between consecutive collections. The speakers were actual proficiency test candidates who were asked to speak in response to prompts. At each seating, several recordings were made in response to different prompts. In this chapter, we will explore different adaptation techniques which allow us to correct for some of these effects.

Adaptation techniques have been discussed in the literature and they mostly try to adapt a speaker's model to a *universal background model* (*UBM*), see [12, 1]. Here, we will further use adaptation to change the model for a candidate from the originally adapted model based on the first enrollment data to a new model which will be more resilient to changes in the channel and the time-lapse effects. The first technique is data augmentation.

21.1 Data Augmentation

To modify the model for a speaker using data augmentation, the original enrollment data is retained for the candidate. At a point when a positive ID of the candidate is made, extra data is appended to the original enrollment data to provide a more universal enrollment model for the candidate matching different channel conditions and time-lapse changes. Figures 21.1 and 21.2 show the identification and verification results respectively, using the new models enrolled by utilizing the augmented data. The results are in-tune with expectations. The identification performance degrades from a perfect performance to a 5% error rate. This is due to the contamination of the enrollment data which now contains channel information from trial 2 as well as trial 1. It also results, quite expectedly, in a large improvement for the second

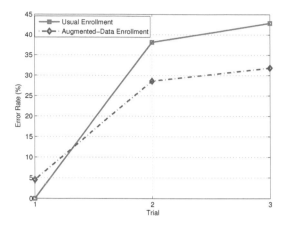

Fig. 21.1: Identification Time Lapse – Augmented-Data Enrollment

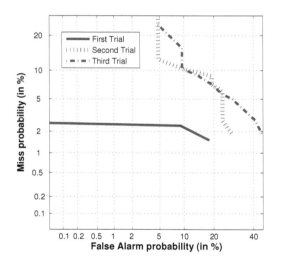

Fig. 21.2: Verification using Augmented-Data Enrollment

trial. This is the case, since now the model contains channel information from the second trial. However, there is also a very big improvement in the third trial which is partly attributed to the smoother information content about channels contained in the enrollment data. This smoothness is apparently attributing to better time-lapse performance since no specific information is contained in the this enrollment about the channel dynamics of the third trial.

Similar results are seen in figure 21.2 for verification of the second and third trials. In fact, comparing figures 20.4 and 21.2 shows that there is no degradation in the verification of the first trial with the EER still being around 2.5%. However, the EER of the second and the third trial have been reduced to only 10%, though the overall performance in the second trial is still better than the third trial as expected.

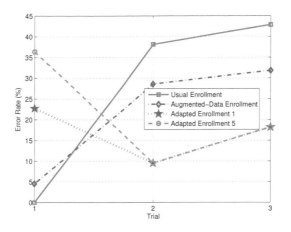

Fig. 21.3: Identification Time Lapse

21.2 Maximum A Posteriori (MAP) Adaptation

One of the problems with the augmented-data approach of Section 21.1 is that the original audio data has to be maintained to be able to do a re-enrollment by adapting from the speaker-independent model to the speaker model for the augmented data. Also, conceptually, the same weight is given to the old data as is given to the new data. One remedy is to use the adaptation techniques which were used to adapt from the speaker-independent model to the speaker model, to adapt from the speaker model to a new speaker model considering the new data at hand. The adaptation technique which was used here is the Maximum A-Posteriori adaptation method. Other techniques such as *maximum likelihood linear regression (MLLR)* may have very well been used for this purpose. [1]

In doing the MAP adaptation, the number of iterations dictate the forgetting factor of the technique. The higher the number of iterations, the more the new data is considered in contrast to the old data. Normally, about 5 iterations are used to go from the speaker-independent model to the speaker model. Initially, this number was used to further transform the prototypes from the old model to the new model for the speaker.

Figure 21.3 shows the results for identification using the new models compared to the usual enrollment and the augmented-data enrollment. The MAP adapted enrollment using 5 iterations shows much better overall performance than both of the usual enrollment and augmented-data enrollment models. However, because it overtrains on the data of the second trial, the results of the first trial are highly degraded. To remedy this problem, the number of iterations for this MAP adaptation was reduced to 1. The results are shown in the same figure (21.3). The results show that no degradation is reported for the second and third trials, however, the identification performance of the first trial is greatly improved.

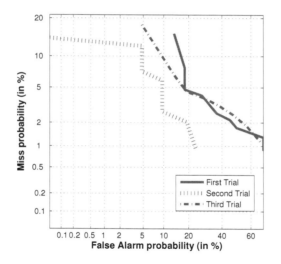

Fig. 21.4: Verification using 5 iteration Adaptation

Figures 21.4 and 21.5 show similar performance for the verification case. Again, using 5 iterations degrades the first trial by over-training. Figure 21.5 portrays much better performance across the different trials. In addition, the third trial has quite an acceptable performance although no channel information has been included in the speaker model from this trial. The EER for all three trials using a MAP adaptation with one iteration varies from about 10% for the best case which is trial two (the trial for which we have adapted) to a respectable maximum of about 14% for the worst case which is trial one from which we deviated. Trial three has an EER of about 12% which seems to coincide with the average EER across the different trials.

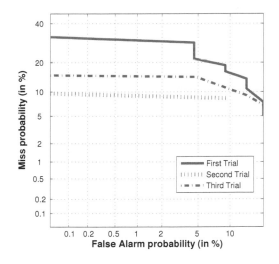

Fig. 21.5: Verification using 1 iteration Adaptation

21.3 Eigenvoice Adaptation

The *Eigenvoice* technique is an incarnation of Eigenfaces which has been quite successful in the image recognition world. It is a PCA (Section 12.1)technique and just line PCA, it has a linear and many nonlinear versions. [9] uses the linear version to speed up adaptation by projection. [10] has implemented the nonlinear version of Eignevoice adaptation based on *kernel PCA* (Section 15.7).

21.4 Minimum Classification Error (MCE)

Another adaptation technique which has been reported for adapting the continuous density HMM parameters is the *minimum classification error (MCE)* technique [8]. The authors claim that this adaptation technique is more suitable for model adaptations with small amount of data.

21.5 Linear Regression Techniques

Starting with Leggetter and Woodland [11] in 1995, many techniques were introduced which follow the simple idea of using linear regression to transform the parameters of the mixture models of speaker-independent models, using some small adaptation data, into speaker-dependent parameters. the following sections are basically a proliferation of different combinations of ML, MAP, Eigenspace, and MCE techniques of the previous sections with linear regression for parameter transformation. in essence most of these techniques start with transforming the means and then are extended to include transformations for the covariance matrices as well.

21.5.1 Maximum Likelihood Linear Regression (MLLR)

In 1995, Leggetter and Woodland [11] proposed to use *linear regression* to adapt the *mean vectors* of the *probability density functions* associated with the *HMM* models used in modeling speakers. This was specifically related to speech recognition and the adaptation of the speaker-independent model to generate speaker-dependent versions of the model with limited data. The method, they proposed, does not modify the covariance matrices. It assumes that a small amount of data may be used to estimate the transformation matrix which would then transform the speaker-independent *mean vectors* into the speaker-dependent ones. To deal with small amounts of adaptation data, some tying may be done. Tying may be done either at the state level of the *HMM*, or the probability density functions may be tied. The latter would means that the same transformation matrix may be used for several densities so that if any one density has not had enough representation in the limited quantity adaptation data, it would still be transformable. All the tied densities which share a transformation matrix are said to be within the same *regression class*. In fact [11] proposes the usage of single transformation matrix when the adaptation data is extremely small.

The transformation matrix is designed such that it would maximize the likelihood of the adaptation data. For this reason, the techniques is called *maximum likelihood linear regression* (*MLLR*. For the normal probability density family, the likelihood of a model given a sample, $\mathbf{x}$ and parameters, $\boldsymbol{\varphi}$, would be given by Equation 6.196, which is repeated here for convenience.

$$\mathscr{L}(\boldsymbol{\varphi}|\mathbf{x}) = p(\mathbf{x}|\boldsymbol{\varphi})$$
$$= \frac{1}{(2\pi)^{\frac{D}{2}}|\boldsymbol{\Sigma}|^{\frac{1}{2}}} \exp\left(-\frac{1}{2}(\mathbf{x}-\boldsymbol{\mu})^T\boldsymbol{\Sigma}^{-1}(\mathbf{x}-\boldsymbol{\mu})\right) \qquad (21.1)$$
$$where \begin{cases} \mathbf{x}, \boldsymbol{\mu} \in \mathscr{R}^D \\ \boldsymbol{\Sigma} : \mathscr{R}^D \mapsto \mathscr{R}^D \end{cases}$$

Since in the original *MLLR* approach, the covariance matrices are untouched, the shapes of the density functions are not changed. Only their locations are translated in the feature space.

First, the *MLLR* approach augments the *mean vector* with a parameter called the *offset term*, $\omega_\gamma : 0 \leq \omega \leq 1$. ω_γ This parameter is designed to allow for added input. *Christensen* [6] has a whole thesis dedicated to the implementation of the *MLLR* in speaker adaptation using *HMMs*.

21.6 Maximum a-Posteriori Linear Regression (MAPLR)

Using the Bayesian framework, variance estimation in *MAPLR* [5] is similar to the method used in *MLLR*, with the difference that the *a-priori* distribution of the transformation matrix is also multiplied by the likelihood and the *argmax function* is computed. In other words, the maximization is done over the a-posteriori probability rather than the likelihood.

21.6.1 Other Adaptation Techniques

There are many other adaptation techniques some of which are designed for special tasks and situations. WMLLR [7] is a modification of the MLLR technique. It is designed to speed up MLLR. *Discounted likelihood linear regression (DLLR)* is yet another similar technique due to [2]. It uses the EM algorithm (Section 11.3.1). Then there are two variations on MLLR and MAP by the names of *Eigenspace maximum likelihood linear regression (EMLLR)* [3] and [4], respectively.

21.7 Practical Perspectives

We have seen that there seem to be other effects in addition to channel mismatch which further degrade the identification and verification performance of a statistical speaker recognition system. We have lumped all these effects into a category called time-lapse effects. These effects along the side of channel mismatch effects were somewhat suppressed using an augmented data approach where the enrollment audio data is always kept around and augmented with new data whenever a positive ID is made and this way the overall performance increases. Although this degrades the best case performance of the engine.

One of the problems with keeping the enrollment audio data is the memory-intensive nature of the solution. In addition security breaches may occur including legal and constitutional issues with keeping audio data around on a server. Some constitutions including that of the United States of America attach an ownership to the raw audio of a person. In addition, compromised access to the server holding the audio data will cause security breaches such as spoofing capabilities, etc. In order to remedy these problems and the performance degradation issues, we used a MAP adaptation technique to adapt an existing speaker model to a new model using new enrollment data. It was shown that using non-aggressive adaptation works a lot better since over-training causes an overall degradation in the performance of both identification and verification engines. From the results we may further deduce that there is indeed a time-dependent degradation which may be remedied by using smoother models with more information across the time-line as well as different channels.

We have only scratched the surface of the time-lapse issue and plan to do much further research in this area to do better speaker model smoothing using other compensation techniques such as MLLR [1] and Latent Factor Analysis (LFA) [13]. At the present, the study is being expanded to include over 100 speakers and to experiment with more re-takes to see how the time-lapse effects and the adaptation results follow the trends seen here.

References

1. Ahn, S., Kang, S., and, H.K.: Effective speaker adaptations for speaker verification. In: Acoustics, Speech, and Signal Processing, 2000. ICASSP '00. Proceedings. 2000 IEEE International Conference on, vol. 2, pp. II1081–II1084 (2000)
2. Byrne, W., Gunawardana, A.: Discounted Likelihood Lineaer Regression for Rapid Adaptation. In: Proceedings of the European Conference on Speech Communication and Technology (EUROSPEECH), pp. 203–206 (1999)
3. Chen, K.T., Liau, W.W., , Lee, H.M.W.L.S.: Fast Speaker Adaptation Using Eigenspace-based Maximum Likelihood Linear Regression. In: Proceedings of the International Conference on Spoken Language Processing (ICSLP), pp. 742–745 (2000)
4. Chen, K.T., Wang, H.M.: Eigenspace-based Maximum A Posteriori Linear Regression for Rapid Speaker Adaptation. In: International Conference on Acoustics, Speech, and Signal Processing (ICASSP), pp. 317–320 (2001)
5. Chou, W.: Maximum A Posteriori Linear Regression (MAPLR) Variance Adaptation for Continuous Density HMMS. In: Proceedings of the European Conference on Speech Communication and Technology (EUROSPEECH-2003), pp. 1513–1516 (2003)
6. Christensen, H., Andersen, O.: Speaker Adaptation of Hidden Markov Models using Maximum Likelihood Linear Regression. (1996)
7. Ding, I.J.: Improvement of MLLR Speaker Adaptation Using a Novel Method. International Journal of Information Technology 5(1), 12–17 (2009)

8. He, X., Chou, W.: Minimum Classification Error (MCE) Model Adaptation of Continuous Density HMMS. In: Proceedings of the European Conference on Speech Communication and Technology (EUROSPEECH-2003), pp. 1629–1632 (2003)

9. Kuhn, R., Junqua, J.C.: Rapid Speaker Adaptation in Eigenvoice Space. IEEE Transaction of Speech and Audio Processing 8(6), 695–707 (2000)

10. Kwok, J.T., Mak, B., Ho, S.: Eigenvoice Speaker Adaptation via Composite Kernel PCA. In: Advances in Neural Information Processing Systems 16. MIT Press (2003)

11. Leggetter, C.J., Woodland, P.C.: Maximum Likelihood Linear Regression for Speaker Adaptation of Continuous Density Hidden Markov Models. Computer Speech and Language 9(2), 171–185 (1995)

12. Reynolds, D.A., Quatieri, T.F., , Dunn, R.B.: Speaker Verification Using Adapted Gaussian Mixture Models. Digital Signal Processing 10, 19–41 (2000)

13. Vogt, R., Sridharan, S.: Explicit modelling of session variability for speaker verification. Computer Speech and Language 22(1), 17–38 (2008)

Chapter 22
Overall Design

22.1 Choosing the Model

As we have seen, there are quite a number of modeling tools which are available for utilization in different branches speaker recognition. There are many different variables which help us decide on the type of algorithms or techniques. These are mostly application dependent. In fact one of the main purposes of writing this textbook was to bring the many methods together and provide enough information to the reader so that the art of choosing different components for a specific problem would be supported by some a-priori knowledge about these available techniques.

The core of a speaker model has basically been designed based on statistical analysis and modeling of the data. A GMM system, intrinsically, requires a smaller number of parameters to be learned. This may be understood by remembering that most HMM-based techniques use GMM as their underlying distributions. Also, as we mentioned, a GMM is really a single-state degenerate HMM. Which means that it must reside at the lower limit of the number of parameters.

Of course there are also some fundamental problems associated with the choice of GMM and HMM models. One such problem is associated with the problem of local optima which are returned by the training algorithms used in training the parameters of the GMM or HMM, such as the k-means algorithm and its variations and the EM algorithm. We have talked in detail about these issues in Chapter 11.

Another problem which is more related to the *text-dependent speaker verification* and *speech recognition* is the problem associated with the proper capturing of the dynamics associated with the transitions within the said phrase. We discussed this issue in detail at the end of Section 13.8.

Another model that has been discussed, is based on SVM. We also saw that most SVM implementations require a large number of parameters to be solved across an extended time-line. Features are usually concatenated to create large supervectors which act as the input to large optimization problems requiring more computation than, say GMM. There have also been combinations of these models to be able to use the best features of each of technique.

Generally speaking, most of the decision comes to the fundamental question of the amount of data that is available for doing the recognition. In forensic and broadcast-type situations where the recognition task usually runs in a passive, background mode, large amounts of data could be available based on long conversations. Then, more complex techniques such as Large Vocabulary Continuous Speech Recognition (LVCSR) techniques [48] may even be used, which require long segments (at least 30s) of speech by the speaker. However, in most everyday speaker recognition problems such as authentication applications and quick identifications, even the 10s or so which is recommended for GMM-based text-independent recognizers is hard to attain. For any lengths in between, we can consider HMM-based and most SVM-based techniques.

When we need to seriously lower the length of speech input down to the order of 4s or less, then we really need to resort to text-dependent or text-prompted models which in themselves may have quite an array of different modeling techniques.

According to a comparison done across different models, Campbell, et al. [9] report that the best performance is achieved on the NIST SRE 8c task by fusing many different models. However, within individual models, *GMM* with factor analysis seems to have the best overall performance. It is hard to make any conclusions from these tests though. There are so many different considerations to be made while choosing a model. Also, all the results have a statistical margin of error that could be quite data dependent. However, it may be safely concluded that the state of the art at the present lies within some combination of GMM and factor analysis or SVM and NAP or MLLR or other transformations. Therefore, regardless of which classifier is used (GMM, HMM, SVM, etc.), it seems like there is a need to handle the session variability through some feature transformation to separate the effects of the channel from the speaker characteristics.

22.1.1 Phonetic Speaker Recognition

Phonetic models have been used in a text-independent setting to utilize the information associated with expected phones in a given language. Using such models limits the text-independent recognizer to being somewhat language-dependent. Due to phonetic similarities, some languages or dialects may be lumped together to reduce the language-dependency, but certain languages have inherently different phones. Examples of phonetic speaker recognition models are [23, 28, 33, 41, 1, 16, 15, 38,

31, 10].

22.2 Choosing an Adaptation Technique

We discussed several different speaker adaptation techniques, *MAP* (Section 21.2), *MLLR* (Section 21.5.1) and Eigenvoice (Section 21.3) adaptation. The main claim of Eigenvoice is that it needs less data for the adaptation than *MAP* or *MLLR*.[34] However, based on the comparison reported by [40], *MAP* seems to do better with the *GMM*-based approach, when tested against the 1999 *NIST speaker recognition evaluation* database. Of course, as we all know, there is no general rule and there is usually no best approach, only the most appropriate approach given the data, the conditions and the task.

22.3 Microphones

There are several aspects that make microphones fundamentally different from one another. One distinguishing characteristic is the directionality and sensitivity of the microphone. In terms of directionality they are often categorized into directional (unidirectional) or omnidirectional types. As for sensitivity, there are close-talk and sensitive microphones.

There is no single microphone which would be suitable for all conditions, however, as a general rule, unidirectional close-talk microphones are better suited for most speaker recognition applications. Most microphones are sensitive to the audio signal present in a space in the from a cardioid (heart shape) region around the microphone. Unidirectional microphones can be more efficiently used in controlled recognition systems.

Very sensitive, omnidirectional microphones can increase the possibility of intercepting unwanted spurious signals such as ambient noise and other speakers. Specifically, when other voices are superimposed on the voice of interest, the recognition task may quite easily be compromised. However, there are conditions in which a unidirectional close-talk microphone is simply not suitable for the application at hand. An example of such conditions is the case of teleconferencing in which multiple individuals share a microphone in a conference room and the minutes of the conference are automatically taken using speaker and speech recognition software – see Section 1.4.6. Therefore, the choice of a proper microphone may be essential in the results of recognition.

There are two major technologies used in microphone manufacturing, *electret* and *carbon-button* types. The name, electret is a composed of parts of the words in the phrase *electrostatic magnet*. An electret microphone is a condenser type microphone.[45] Its design is based on the use of stable dielectric materials which stay charged for a long time (in order of centuries). Because of this type of charged material, it does not require any polarizing power. Electret microphones generally come in three different forms, diaphragm, back-electret and front-electret. The latter is the most recent technology. The charged material is normally made of special plastic.

Carbon button or carbon microphones are becoming harder to find these days. They were used in telephones in the early days until the mid 1980s. These microphones are generally made of carbon granules which are sandwiched between two metal plates. A direct current is passed through the sandwich. The compression of the carbon granules caused by the audio wave excitation is translated into a varying resistance in the carbon sandwich. This varying resistance changes the flow across the two metal plates causing in the creation of an electric signal related to the audio excitation.

All microphones have their own characteristics which become interleaved with the vocal characteristics of the speaker being monitored. A good microphone is one that has a large dynamic range and that will add the least amount of distortion to the audio being captured. Those with experience with photography will note the analogy of a good microphone being similar to a good lens in a camera, one that produces the least distortion to the photograph being taken.

Unfortunately, once the audio signal has been distorted by the characteristics of the microphone, it is almost impossible to extract the true signal even if the exact characteristics of the microphone are known. This is due to the highly nonlinear effects possessed by audio capture devices and the fact that some types of distortion destroy the original dynamics of the signal in an irreversible fashion. A simple example is band limitation which cuts off higher and lower frequency contents of the audio signal getting rid of some essential parts of speech such as fricatives.

However, the shear number of different types of microphones makes the problem even harder. It is next to impossible to be able to recognize all the different makes and models of microphones and to be able to account for their distortions on the signal.

Pop screens filter the pops (impulses) that are associated with stops such the b^h and p^h sounds. To some degree these impulses should be preserved to allow distinguishing ability between regular and aspirated stops. However, since the energy level of these aspirations is quite large they can present saturation conditions in the microphone output making screens useful.

One thing which is known for sure is that one of the most attractive aspects of speech as a biometric is the fact that it can use existing infrastructure. Therefore, for speaker recognition to be successful, it needs to cope with all kinds of microphones and acoustic conditions. Therefore, researchers will always strive for more robust algorithms in the hope of reducing the effects of microphone effects and mismatches – see Section 22.4.

22.4 Channel Mismatch

[32] uses microphone-specific background models to reduce channel handset mismatch effects. It creates two models which are the equivalent of the Universal Background Model for the two major sets of microphones, carbon-button and electret. Similar number of speakers (46 in this case) are used to create each of these background models which are used to provide the competing model in the speaker verification test. This achieved some tangible improvement in the performance of matched channel tests, but did not really improve the mismatched channel results. Using an Auto Associative Neural Network for a nonlinear principal component analysis (PCA) in conjunction with these specialized background models a tangible reduction in the error rate of the mismatched microphone and channel was observed. However, from the results it is hard to hypothesize the amount of improvement due to the PCA and that of the specialized models since no results for the PCA with the Universal Background Model has been reported.

22.5 Voice Over Internet Protocol (VoIP)

The dominant codec using in VoIP systems is the ITU-T G.711. This codec has been the standard fall-back codec which means that all Voice over IP (VoIP) systems are required to support it. Because of its low bandwidth requirements, G.729 has been increasingly utilized in VoIP systems where bandwidth is at a premium. A typical G.711 system (PCMU or PCMA) uses 64-kbps transmission and G.729 only uses 8-kbps. See Chapter 26 for more information on these standards.

One of the problems that plagues VoIP systems is echo. Usually, because of large network delays and hardware delays, the echo in VoIP systems becomes one of the most important problems. ITU-T G.131 [26] defines acceptable round-trip delays based on the strength of the signal difference between the main signal and the echoed signal, on average, round-trip delays of about 30ms are tolerated, but most VoIP systems have much longer delays, amplifying this issue. The recommended maximum delay on the total round-trip delay is 150ms where a maximum 400ms, one-way delay for all circumstances has been recommended by ITU-T G.114 [25]. See Section 18.3 for more information.

22.6 Public Databases

There have been quite a few public databases collected in the past decade or so. It is not only impossible to list them all here, but it is also outside the realm of this book. We will briefly describe a few of the databases which have been used in the literature.

22.6.1 NIST

The *National Institute of Standards and Technology* (NIST) has been putting together many corpora which have been used in their regular evaluations. Most of these are available from the LDC [36]. In the recent years, these corpora have included many different languages. NIST has also been generating corpora for other speech-related evaluations which have sometime been used for the *training* and the *evaluation* of speaker recognition systems by independent authors.

With the exception of the years 2007 and 2009, some kind of a *speaker recognition* evaluation was conducted by NIST in every year from 1996 to 2010 (when this book is being written). The details of these evaluations and are available from the NIST website [42] for all these years with the exception of 1996. The website includes information about the evaluation plan for each year. It also includes some useful software for processing the data and the results.

22.6.2 Linguistic Data Consortium (LDC)

Linguistic Data Consortium (LDC) [36] is a nonprofit organization which works on preparing data in most speech and language dependent tasks. It works by membership with a different price structure for education, nonprofit, government, and commercial institutions. Of course some of the data may be purchased without being a member. They have put out many corpora related to speech and speaker recognition in the recent past. Here are some of these data sets.

22.6.2.1 The Fisher Corpus

The *Fisher corpus* is a collection of 16,000 telephone conversations in English, about 2,000 hours of which were transcribed by 2004.[12] Although the *Fisher corpus* was collected for text-to-speech efforts, it has been used in speaker recognition benchmarks [9] as well. It is available from the *Linguistic Data Consortium* (LDC) [36].

22.6.2.2 TIMIT and NTIMIT

The TIMIT database is one of most widely used speaker recognition corpora [2, 3, 4, 5, 11, 18, 19, 46, 49]. It was developed for DARPA and is distributed by the Linguistic Data Consortium (LDC). In TIMIT, speech samples have been obtained from 630 speakers using 8 different dialects of English, across the United States. The gender distribution in TIMIT is not balanced. There are 438 male speakers and only 192 female speakers. Each speaker has been asked to read out 10 sentences based on a phonetically balanced text. [22] The recordings were done in a controlled clean environment and there is only one speaker per session. So it is ideal for performing speaker identification and verification tests. However, it cannot be used for segmentation evaluations.

The TIMIT recordings were done on a clean wideband microphone at a sampling rate of rate of 16 kHz was used for recording the audio, producing an 8 kHz speech bandwidth in a controlled clean environment.

NTIMIT is another database which is basically very similar to the TIMIT database, but it was recorded using telephone handsets over a PSTN with half of the calls going through long distance carriers.

22.6.2.3 KING

The KING database is distributed by the LDC and has been used for many years as the basic comparison of speaker recognition algorithms [13, 43]. It is not being used very often these days due to many limitations. It only contains the speech of 51 male speakers and no female presence. It was recorded using a combination of wideband microphones as well as some electret telephone handsets through a PSTN. The donors were asked to either read out or repeat prompted words, digit strings, and sentences. They were also shown photos which they had to describe in free form phrases and sentences.

The KING speech data was recorded in a clean environment and contains only clean speech. One attractive feature of the database is that the recordings were done in 10 sessions which were spread out over several weeks. This allows for studying limited variations due to time lapse. In general, the KING corpus is not very useful for any serious analysis and testing.

22.6.2.4 YOHO

The YOHO database is distributed by the LDC. It is a collection of low bandwidth recordings produced in a clean office environment containing clean speech. It only includes prompted digit strings and has been organized into 4 enrollment sessions and 10 verification sessions per speaker. There are a total of 138 speakers in the database which includes an unbalanced mix of 106 male speakers and 32 female speakers.

Although YOHO was used in many results produced in the early days of the development of speaker verification [14, 50], it is not very appropriate for evaluations in real conditions. It is especially not very useful for text-independent speaker verification evaluations. The one attractive feature of the database is its low sampling rate which is only 3.8 kHz.

22.6.2.5 Switchboard I & II

There are two *switchboard* databases which are distributed by the LDC. They have been recorded over various types of telephone handsets through PSTN. The main feature of these databases is the fact that they have a balanced distribution of male and female speakers (about 50% in each gender). The recordings depict conversational speech, although the channels are separated so that there is only a single speaker in each channel. There are from 1 to 25 sessions of speech per speaker with each session containing about 5 minutes of speech in either a home or an office environment.

The *switchboard I* database contains a total 543 speakers and it has been used extensively in evaluating speaker recognition tasks.[8, 35, 20, 21, 44] The *switchboard II* database has also been used extensively for such evaluations.[9, 47] In addition, a subset if the *switchboard I* database is known as the *SPIDRE corpus* and it is often used for the evaluation of *speaker identification* algorithms.

22.6.2.6 Cellular Switchboard

In the last section, we discussed two switchboard corpora distributed by the LDC. However, those databases were recorded over PSTN. A cellular version is also available from the LDC, called simply, the *cellular switchboard* database. It is also made up of recordings in the form of conversation speech, but through various cellular handsets using the *GSM 1900* cellular standard. The recordings are done over normal settings in which cellular telephony is used. There are at least 10 sessions per speaker, recorded in the course of several days. Each session contains about 5 minutes of speech and there are a total of 190 speakers in the database with an equal distribution of males and females.

22.6.2.7 Tactical Speaker Identification

The tactical speaker identification (TSID) database is distributed by the LDC. It has been collected in order to test the performance of speaker identification algorithms over military radio handsets. The database is quite unbalanced with regard to gender. There are a total of 40 speakers in the database, one of whom is female, the rest of the 39 speakers coming from male speakers. A single session of speech data per person was recorded outdoors, using 4 military radio handsets and an electret microphone. The military handsets were transmitted over HF, UHF and VHF bands and wideband. The speakers were asked to read out sentences and digits and also to utter free-style speech.

22.6.3 *European Language Resources Association (ELRA)*

The European Language Resources Association (ELRA) [17] is the European counterpart of LDC which has been quite active in capturing speech in most official European Union languages. The data is also made available for sale to interested parties in the same manner as it is with the LDC. The ELRA databases are quite extensive. It presents quite a wide variety of speakers in different languages. Most of the ELRA databases contain large numbers of speakers under different conditions. Over 140 corpora with large number of speakers (up to a maximum of about 4000 speakers) are available from ELRA, recorded through PSTN and cellular networks with various handsets. These recordings have been done with sessions spanning over many months in British English, German, Spanish, Italian, French, Danish, and Finish. In the following few sections, we will describe a few of these databases.

22.6.3.1 SIVA

SIVA is a gender-balanced Italian corpus which contains two different sets of telephony data recorded over PSTN. The first set is a collection of 18 sessions per
individual, recorded in the span of 3 days for 40 speakers. The second set is a series
of single session recordings from 800 different speakers. The recordings were done
in office and home environments and contain short sentences of prompted words
and digits.

22.6.3.2 POLYVAR

POLYVAR is a French telephony corpus from ELRA which was recorded over PSTN
and ISDN, in clean home and office environments. It includes

22.7 High Level Information

High level information such as phonology, suprasegmental information (see Section 4.3), syntax and dialog may be used to reduce the speaker recognition error
rates [30]. However, as with any other improvement, there is a price to pay. In
these systems, usually, significantly larger training data is need to be able to account
for the extra information. Also, these types of clues are usually quite language-
dependent and come at a high cost in terms of processing and analysis which in
most cases does not justify the amounts of improvement seen in the results. Also,
the reported results in the literature could be quite biased based on the database used
for doing these evaluations.

It is also important to try to learn from the experience of other related fields.
According to Lalit Bahl, when we were working together in the speech group of
IBM research in 1995, there has been a paradox surrounding the role of linguistic
knowledge in the advancement of speech recognition performance. Around 1972
when serious research efforts began in the field of speech recognition by some major players such as IBM and AT&T, a natural assumption was made that the more
linguistic knowledge one uses, the better results will be obtained for speech recognition. However, Figure 22.1 shows a qualitative relationship according to Lalit Bahl
over the course of about 30 years. This may be counter-intuitive, but a major reason
behind this behavior is the fact that linguistic knowledge is a rigid set of rules created to give order to an otherwise free-form natural phenomenon which is speech.
Although there is order in speech so that it may be understood, the order is not so
clear cut and it is very hard to model in a global sense so that all possible variations
are considered.

Also, looking at Figure 22.1, one may then mistakenly conclude that we can do away with all linguistic knowledge. That is certainly not the case. If were able to account for every possibility, then having more linguistic knowledge would have produces better performance. In fact it has been shown that N-Gram language models [39] gain quite considerable performance for speech recognizers. In addition, it has been shown that a class-based N-Gram model [7] performs even better. The reason for the better performance of the class-based N-Gram versus normal N-Gram and N-Gram versus more strict language models is the smoothing effect that occurs. This is a consequence of maximum entropy in Information Theory 7 which states that in the absence of any knowledge about a system, the maximum entropy solution is the most optimal. Here, we are operating between the two extremes of knowing nothing, which would then require equal probability (maximum entropy) solutions and *thinking* that we know everything there is to know about the linguistic content in which case, everything would be modeled to the 't'. In this mid-point operating condition, it is best to use some information, but to smooth that information with a maximum entropy solution which would allow for exceptions. This is true in any statistical system.

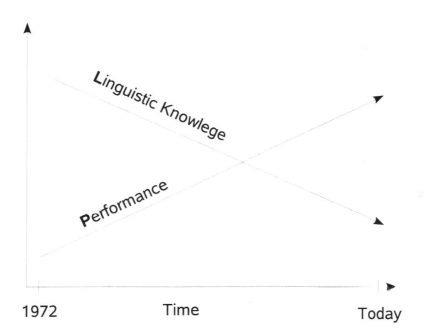

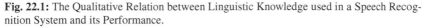

Fig. 22.1: The Qualitative Relation between Linguistic Knowledge used in a Speech Recognition System and its Performance.

Given this experience from the speech recognition field, we should also try to use some suprasegmental information together with traditional features based on local information, such as MFCCs. This combination is sure to work better than any one sided approach as evidenced by such experiments such as [30]. The design objective then will have two levels. At the first level, each of the local and suprasegmental models should be optimally generated and at the second level, they should be optimally combined to produce the best possible results.

We also have to be careful about using terms such as we just did, "the best possible results." Best results would always have to be determined relative to the deployment environment. This is one thing that has always disturbed me about tests and results that are requested and reported. Users request numbers such as error rates and accuracies and researchers, reluctantly, provide them. Based on my own experience, the best of breed systems that I have developed under certain conditions have not necessarily performed the best under natural circumstances. It is important that this thesis is considered when deployments of the speaker recognition system are made. It is quite feasible that a higher level decision making apparatus be developed whose expertise is the automatic configuration of the recognition engine with the requested environment. This should be an area of research and would most likely render speaker recognition much more practical and reliable.

22.7.1 *Choosing Basic Segments*

One important thing to remember is that segmentation may also be language dependent. Phonetic segmentation has worked well with many languages of Indo-European origin. However, we should remember that there are many syllabic and mora-based languages which could benefit from those boundaries as the basic elements of the speech or at least to be considered as a suprasegmental division which would add further information to the process.

[27] uses a *syllabic lattice* based algorithm for speaker verification – see Section 4.3.1.1. In this approach, a large vocabulary continuous speech recognizer (LVCSR) is used to approximate the likelihoods of the background model and the target-speaker. In this approach the optimal state alignment of the best speech recognition result which is the speaker independent HMM model and the alignment results from the target-speaker's HMM model are used to estimate the likelihoods of the background model and the target speaker model respectively. This is done in the place of the regular GMM models for the background and target models (as compared to the GMM approach). This approach is similar to the Dragon Systems' approach used in the NIST-1999 evaluations [48] for the two-speaker detection and tracking task, with the difference that [27] used a syllabic lattice model (Chinese) and [48] used a basic phonetic model (English).

22.8 Numerical Stability

Once it gets down to the actual implementation of a recognizer, there are many different aspects of design that should be considered. One of the most important, is numerical stability. About 15 years ago, I remember spending a great deal of my time on optimizing for integer-based processors. I guess this should not be an issue today, since unlike those days, math co-processors are quite standard even on the smallest handheld and embedded processors. As we have seen, quite a lot of logarithms have to be computed, both in the front-end and back-end of such systems. For an integer implementation, the best remedy is to use logarithm tables and some handy decompositions. The most important table is one that adds two logs (a logadd function). Of course, even with the existence of fancy math co-processors, it sometime makes sense to use tables due to their stability. There is a great renewed interest in being able to implement embedded acoustic models [37] in handheld devices such as smartphones.

Take the MFCC computation step. At the stage when we should take the log of the spectrum, it is possible, due to precision problems to end up with a very small number whose log may approach $-\infty$. There are some possible remedies for these situations. One is to use tables as mentioned. A more appropriate technique is to set a floor for the domain of the log function and to check the argument and set it to the chosen floor value if it happens to be less.

Another very important tool for avoiding numerical instabilities is proper scaling. In general, it is quite important to assess the dynamic range of different variables and in use, both in the domain and range of the functions related to these variables. Then a proper scaling may buy us quite consider stability. This is especially true for probabilities. It is important to use logs of probabilities as much as possible. Sometimes it makes sense to take the logarithm of any set of numbers is a computation involving varying ranges of variables. Then after the final computation, we may exponential and go back to the original space.

In the process of designing a recognizer. We deal with covariance-like matrices quite a bit. These may be true covariance matrices or other matrices of a similar nature, such as the Hessian matrix, its inverse and their numerical updates. As we have seen in Chapter 25, many rank-two update techniques bank on formulating the updates in terms of square-roots of the final update. A great example of a similar technique for using covariance matrices is the *Cholesky factorization (Cholesky decomposition)* [6] or *square-root filtering* which is widely used in signal processing techniques whenever covariances are used. For instance, square-root filtering is used in the estimation of the *ARMA model* parameters (or *Kalman filtering*).

Take matrix such as a covariance matrix, $\Sigma : \mathcal{R}^N \mapsto \mathcal{R}^N$. Cholesky factorization gives us the following decomposition for the matrix,

$$\Sigma = LL^T \tag{22.1}$$

The *condition number* of the Σ is defined as,

$$c_\Sigma = \frac{\lambda_1}{\lambda_N} \tag{22.2}$$

where λ_1 and λ_N are the largest and smallest Eigenvalues of Σ, respectively. When Σ is factorized, the Eigenvalues of L are equal to $\sqrt{\lambda_n}, n = \{1, 2, \cdots, N\}$. This means that the condition number of L is also $\sqrt{c}$.

This could be very significant for cases where there is relative discrepancy in the sizes of the different dimensions of a vector. For example, if $\lambda_1 = 1000$ and $\lambda_N = 0.001$, then the condition number of Σ would be $1,000,000$ and the one for L would only be 1000. This could determine the success level of the whole recognizer. Once the computations are completed in the square-root space, they may be converted back.

Another very important strategy, for attaining and maintaining numerical stability, is normalization. We spent some time discussing signal enhancement and normalization techniques in Chapter 18. Many of those techniques tried to remove the variability of the data due to conditions. However, sometimes, normalization may be used to simply bring variables into more valid and consistent dynamic ranges. Take two codebooks of features, for instance. Let us say one is based on the MFCCs and the other on a new feature with which we are experimenting. It is important to consider the dynamic ranges of these two feature sets and then normalize them so that they would be comparable without changing the intra-variability of these features. These techniques are quite successful in practice.

22.9 Privacy

Privacy is also an important problem which can manifest itself in different forms. For example, speaking in populated settings may not be suitable, making it hard to use speaker recognition in some cases. Also, some individuals may not be happy with supplying voice samples. This problem seems to be less prevalent in speaker recognition than other biometrics as we touched on it earlier in the tutorial.

22.10 Biometric Encryption

Some believe it may be important to do extra encryption of the models to ensure that there is no reversible information deduced from the models. Usually statistical models are not reversible since due to the change in entropy it is generally impossible to build any audio from statistical models. Care must be given that non-essential data is not stored in the models and only an irreversible model is kept. Even then, it is important to store and send these models back and forth using utmost encryption.

22.11 Spoofing

Spoofing is the act of presenting a verification system with the audio of the target speaker by an impostor which has been either collected using interception or generated by other means. If the possible response is limited to a small set of phones, then it may be possible to concatenate enrollment audio to create the string needed to fool the verification system. Also, many speech synthesis systems are increasingly becoming capable of using enrollment data from a speaker and then produce speech with that speaker's characteristics [29].

Spoofing is a serious concern for speaker recognition platform designers. At the early stages of the utilization of speaker recognition systems, enough attention may not have been paid to this subject. Most of the effort has been devoted to solving basic problems such as recognition results and condition mismatch problems. However, as the discipline matures, anti-spoofing is beginning to move up in the to-do lists of engine and application designers as well implementors. Here, we will discuss possible spoofing scenarios for different modalities of speaker verification. We also build the case toward the abandonment of text-dependent and text-prompted systems for more security conscious applications. Instead we show that text-independent verification systems present more freedom in being able to build in a sophisticated knowledge-base into the complete verification solution to avoid spoofing.

22.11.1 Text-Prompted Verification Systems

In Chapter 17 we discussed the text-prompted scenario in some detail. Limiting the space to be able to handle text-prompted enrollment and verification has its disadvantages. Let us say we limit the coverage to only digits. Then, an impostor may presumably have a recording of the voice of the target speaker (say from intercepting the enrollment process) and can segment that audio to retain samples for individual digits. Since the number of digits is limited to 10, this would be quite an easy task.

Then by hearing the prompt and the aid of an audio editing software, the impostor may either type in the digits which are requested at the verification prompt and have the software play back that sequence with the target speaker's voice. This will get the impostor in.

As a counter measure, one may attempt an anti-spoofing mechanism such that at the verification time, exceptionally good scores are also rejected. This would usually happen when the verification string is made up of the enrollment data. However, this anti-spoofing technique may not quite work. For starters, since the audio may have been intercepted, if it were intercepted by a different recording device from, say, a telephone line, any characteristics of the recording apparatus may modify the audio. This modification may be enough to conceal the spoofing characteristic of the signal. Also, a more sophisticated impostor may even modify the enrollment data to make it seem different, fooling the simple anti-spoofing mechanism discussed here.

In general, as in any other system, there will always be a struggle between biometric engine and application designers and professional hackers. Our conclusion is that a sophisticated knowledge-based recognition would be much more practical. As the knowledge-base becomes more sophisticated, text-dependent and text-prompted systems will no longer be practical and only text-independent recognition is recommended to be able to handle the more complete space of vocal characterization dictated by the advanced question and answer sessions.

22.11.2 Text-Independent Verification Systems

At this beginning of the spoofing discussion we noted that there are increasingly more sophisticated systems capable of synthesizing speech using the vocal characteristics of an individual [29]. To be able to do this, these synthesizers require enrollment data. It is very important to guard the audio which is used for the enrollment of speaker recognition systems. Practical systems would capture the audio over a secure line and then discard it immediately after the enrollment for the recognizer has been completed. Many implementors are hanging on the enrollment audio to be able to do further research and tuning of their systems. It is important that serious live implementations of speaker recognition demand the immediate destruction of the audio as soon as the enrollment process is completed.

Even with taking such security measures, it is conceivable that the avid impostor would find a way of intercepting the target speaker's audio from other sources and to train a synthesizer to mimic the speech of that individual. As we mentioned earlier, from the viewpoint of engine developers and implementors, it is important to build in sophisticated question and answer sessions which would utilize a speech recognizer in conjunction with the speaker verification system and to test the knowledge level of the test speaker for knowledge that may only be know to the target individual. In this manner, even if the impostor has the means for generating perfect replicas of the speech of the target speaker, the knowledge-base is not readily available to the impostor.

However, the verification engine designer should assume that the impostor will be able to ascertain the knowledge-specific responses and will also be able to generate the audio on-the-fly to spoof the system. In this regard, it will become important for verification engine designers to understand speech synthesis engines and to try to bank on their vulnerabilities to be able to avoid being defeated. One possible techniques is the recognition of the, so called, *annoying effects* of synthesizers. Customizable synthesizers usually possess discontinuities and clicks at boundaries of speech segments. Also, silence generated by these synthesizers is usually unnatural (see Section 18.1). Verification engines may be tooled with components that would recognize such anomalies. In addition, until these synthesizers become more sophisticated, the speech being generated will be repeatable. Anti-spoofing mechanisms may be utilized to look for these regular repetitions which are not possible through natural speech generation means.

As synthesizers become more sophisticated, verification engines should keep up with the synthesizer advancements and use counter-measures to avoid being spoofed using these synthesizers. Using these techniques together with destroying the raw enrollment audio and a smart and sophisticated knowledge-base which would be ever-adapting to test for liveness, speaker verification using text-independent techniques would be quite viable. The technology will, however, always have to be in alert mode to counter new techniques for spoofing.

22.12 Quality Issues

Data quality is another problem. The training and enrollment data are of utmost importance since if they are not of representative quality the models will almost be useless for doing further recognition. Also, the quality of the candidate audio segment should be good and representative to be able to assess the identity of the individual based on stored models. We will discuss data quality a bit more in the next few slides.

Another problem is the effect of aging on an individual which was somewhat described earlier. In addition to the long-term aging, there are short-term time-lapse discrepancies which are usually due to a combination of different changes in the data capture scenario, changes in the behavior of the speaker, anatomical changes, channel mismatch, variations in the distance and angle of the microphone, strains on the person's vocal tract, and many other issues. We will talk about these problems and possible solutions later.

22.13 Large-Scale Systems

A last but in no means least problem is the handling of a large-scale database. As we mentioned, the matching may take a long time. Inter-speaker variation may become smaller than the intra-speaker variations and a central location may be needed for the storage of the large database together with all sorts of associated problems such as redundancy, integrity, etc.

22.14 Useful Tools

Some useful open source tools have been develped recently. These tools are geared toward allowing for segmentation and cleaning of the speech data such as *Praat* which has been developed by people from University of Amsterdam. It runs on many different operating systems and allows for a quite and accurate segmentation of the speech data.

Another toolkit is the very useful Matlab toolbox for speech processing. It is written by people from the Imperial College of London and it is made available under the GPL public license. It includes a rich set of functions and is a good first pass for trying something new.

The *hidden Markov toolkit (HTK)* [24] is a set of *C* libraries which includes many functions for building and manipulating hidden Markov models. Its development started in 1989 as a project at *Cambridge University* as a byproduct of their large vocabulary speech recognition system [51]. It was made available as an open source package under its own license agreement which, although similar to the GNU license, it has its own quirks. Nevertheless, it has been used by many academic research organizations to develop speech and speaker recognition systems. Its HMM format is also quite standard and is used by several other engines, or at least is included as an import/export format by such engines such as *the Julius speech recognition* software. According to the HTK website, at the time of the writing of this book, the latest version of the HTK was *Version 3.4.1* which was released in March of 2009. No new releases have been available up to May of 2011. To be able to download the HTK, one must register at the product website. At this time, the HTK is owned by the *Microsoft corporation* through the purchase of its owner, *Entropic*, however, at the moment the license is delegated back to *Cambridge University*.

References

1. Auckenthaler, R., Parris, E., Carey, M.a.: Improving a GMM speaker verification system by phonetic weighting. In: Acoustics, Speech, and Signal Processing, 1999. ICASSP '99. Proceedings., 1999 IEEE International Conference on, vol. 1, pp. 313–316 (1999)

2. Bennani, Y., Gallinari, P.a.: A modular connectionist architecture for text-independent talker identification. In: Neural Networks, 1991., IJCNN-91-Seattle International Joint Conference on, vol. ii, pp. 857–860 (1991)

3. Bennani, Y., Gallinari, P.a.: On the use of TDNN-extracted features information in talker identification. In: Acoustics, Speech, and Signal Processing, 1991. ICASSP-91., 1991 International Conference on, pp. 385–388 (1991)

4. Besacier, L., Grassi, S., Dufaux, A., Ansorge, M., Pellandini, F.a.: GSM speech coding and speaker recognition. In: Acoustics, Speech, and Signal Processing, 2000. ICASSP '00. Proceedings. 2000 IEEE International Conference on, vol. 2, pp. II1085–II1088 (2000)

5. Bimbot, F., Magrin-Chagnolleau, I., Mathan, L.: Second-order statistical measures for text-independent speaker identification. Speech Communication 17(1–2), 177–192 (1995)

6. Brezinski, C.: The Life and Word of André Cholesky. Numer Algor 43, 279–288 (2006)

7. Brown, P.F., Pietra, S.A.D., Pietra, V.J.D., Lai, J.C., Mercer, R.L.: An Estimate of an Upper Bound for the Entropy of English. Computational Linguistics 18(1), 32–40 (1992)

8. Byrne, W., Gunawardana, A.: Discounted Likelihood Lineaer Regression for Rapid Adaptation. In: Proceedings of the European Conference on Speech Communication and Technology (EUROSPEECH), pp. 203–206 (1999)

9. Campbell, W., Sturim, D., Shen, W., Reynolds, D., Navratil, J.a.: The MIT-LL/IBM 2006 Speaker Recognition System: High-Performance Reduced-Complexity Recognition. In: Acoustics, Speech and Signal Processing, 2007. ICASSP 2007. IEEE International Conference on, vol. 4, pp. IV–217–IV–220 (2007)

10. Campbell, W.M., Campbell, J.P., Reynolds, D.A., Jones, D.A., Leek, T.R.: Phonetic Speaker Recognition with Support Vector Machines. In: Advances in Neural Information Processing Systems, pp. 1377–1384 (2003)

11. Chen, C., Chen, C., Cheng, P.a.: Hybrid KLT/GMM approach for robust speaker identification. Electronics Letters 39(21), 1552–1554 (2003)

12. Cieri, C., Miller, D., Walker, K.: The Fisher Corpus: a Resource for the Next Generations of Speech-to-Text. In: Fourth International Conference on Language Resources and Evaluation (2004). Available at Linguistic Data Consortium (LDC)

13. Colombi, J., Anderson, T., Rogers, S., Ruck, D., Warhola, G.a.: Auditory model representation for speaker recognition. In: Acoustics, Speech, and Signal Processing, 1993. ICASSP-93., 1993 IEEE International Conference on, vol. 2, pp. 700–703 (1993)

14. Colombi, J., Ruck, D., Anderson, T., Rogers, S., Oxley, M.a.: Cohort selection and word grammar effects for speaker recognition. In: Acoustics, Speech, and Signal Processing, 1996. ICASSP-96. Conference Proceedings., 1996 IEEE International Conference on, vol. 1, pp. 85–88 (1996)

15. El Hannani, A., Petrovska-Delacretaz, D.a.: Comparing Data-driven and Phonetic N-gram Systems for Text-Independent Speaker Verification. In: Biometrics: Theory, Applications, and Systems, 2007. BTAS 2007. First IEEE International Conference on, pp. 1–4 (2007)

16. El Hannani, A., Toledano, D., Petrovska-Delacretaz, D., Montero-Asenjo, A., Hennebert, J.a.: Using Data-driven and Phonetic Units for Speaker Verification. In: Speaker and Language Recognition Workshop, 2006. IEEE Odyssey 2006: The, pp. 1–6 (2006)

17. ELRA: European Language Resources Association. Web (2010). URL http://www.elra.info. Speech Data Resource

18. Eskidere, O., Ertas, F.a.: Impact of Pitch Frequency on Speaker Identification. In: Signal Processing and Communications Applications, 2007. SIU 2007. IEEE 15th, pp. 1–4 (2007)

19. Eskidere, O., Ertas, F.a.: Parameter Settings for Speaker Identification using Gaussian Mixture Model. In: Signal Processing and Communications Applications, 2007. SIU 2007. IEEE 15th, pp. 1–4 (2007)

20. Ganapathiraju, A., Hamaker, J., Picone, J.: Support Vector Machines for Speech Recognition. In: International Conference on Spoken Language Processing, pp. 2348–2355 (1998)
21. Ganapathiraju, A., Hamaker, J., Picone, J.: Hybrid SVM/HMM Architectures for Speech Recognition. In: SPeech Transcription Workshop, pp. 504–507 (2000)
22. Garofalo, J., Lamel, L., Fisher, W., Fiscus, J., Pallett, D., Dahlgren, N.: Darpa TIMIT: Acoustic-Phonetic Continuous Speech Corpus. CD-ROM (1993). Linguistic Data Consortium (LDC)
23. Hatch, A., Peskin, B., Stolcke, A.a.: Improved Phonetic Speaker Recognition Using Lattice Decoding. In: Acoustics, Speech, and Signal Processing, 2005. Proceedings. (ICASSP '05). IEEE International Conference on, vol. 1, pp. 169–172 (2005)
24. HTK. Web (2011). URL http://htk.eng.cam.ac.uk
25. ITU-T: G.114: One-way transmission time. ITU-T Recommendation (2003). URL http://www.itu.int/rec/T-REC-G.114/e
26. ITU-T: G.131: Talker echo and its control. ITU-T Recommendation (2003). URL http://www.itu.int/rec/T-REC-G.131/e
27. Jin, M., Soong, F., Yoo, C.a.: A Syllable Lattice Approach to Speaker Verification. Audio, Speech, and Language Processing, IEEE Transactions on **15**(8), 2476–2484 (2007)
28. Jin, Q., Navratil, J., Reynolds, D., Campbell, J., Andrews, W., Abramson, J.a.: Combining cross-stream and time dimensions in phonetic speaker recognition. In: Acoustics, Speech, and Signal Processing, 2003. Proceedings. (ICASSP '03). 2003 IEEE International Conference on, vol. 4, pp. IV–800–3 (2003)
29. Junqua, J.C., Perronnin, F., Kuhn, R.: Voice Personalization of Speech Synthesizer. U.S. Patent (2005). Patent No. U.S. 6,970,820
30. Kajarekar, S.S., Ferrer, L., Stolcke, A., Shriberg, E.: Voice-Based Speaker Recognition Combining Acoustic and Stylistic Features. In: N.K. Ratha, V. Govindaraju (eds.) Advances in Biometrics: Sensors, Algorithms and Systems, pp. 183–201. Springer, New York (2008)
31. Kao, Y.H., Rajasekaran, P., Baras, J.a.: Free-text speaker identification over long distance telephone channel using hypothesized phonetic segmentation. In: Acoustics, Speech, and Signal Processing, 1992. ICASSP-92., 1992 IEEE International Conference on, vol. 2, pp. 177–180 (1992)
32. Kishore, S., Yegnanarayana, B.a.: Speaker verification: minimizing the channel effects using autoassociative neural network models. In: Acoustics, Speech, and Signal Processing, 2000. ICASSP '00. Proceedings. 2000 IEEE International Conference on, vol. 2, pp. II1101–II1104 (2000)
33. Kohler, M., Andrews, W., Campbell, J., Herndndez-Cordero, J.a.: Phonetic speaker recognition. In: Signals, Systems and Computers, 2001. Conference Record of the Thirty-Fifth Asilomar Conference on, vol. 2, pp. 1557–1561 (2001)
34. Kuhn, R., Junqua, J.C.: Rapid Speaker Adaptation in Eigenvoice Space. IEEE Transaction of Speech and Audio Processing **8**(6), 695–707 (2000)
35. Lamel, L., Gauvain, J.L.a.: Speaker recognition with the Switchboard corpus. In: Acoustics, Speech, and Signal Processing, 1997. ICASSP-97., 1997 IEEE International Conference on, vol. 2, pp. 1067–1070 (1997)
36. LDC: Linguiostic Data Consortium. Web (2010). URL http://www.ldc.upenn.edu/. Speech Data Resource
37. Lévy, C., Linarès, G., Bnastre, J.F.: Compact Acoustic Models for Embedded Speech Recognition. EURASIP Journal on Audio, Speech, and Music Processing **2009** (2009). Article ID 806186, 12 pages, 2009. doi:10.1155/2009/806186
38. Liou, H.S., Mammone, R.a.: Speaker verification using phoneme-based neural tree networks and phonetic weighting scoring method. In: Neural Networks for Signal Processing [1995] V. Proceedings of the 1995 IEEE Workshop, pp. 213–222 (1995)
39. Manning, C.D.: Foundations of Statistical Natural Language Processing. The MIT Press, Boston (1999). ISBN: 0-262-13360-1
40. Mariéthoz, J., Bengio, S.: A Comparative Study of Adaptation Methods for Speaker Verification. In: Proceedings of the International Conference on Spoken Language Processing (ICSLP), pp. 25–28 (2002)

41. Navratil, J., Jin, Q., Andrews, W., Campbell, J.a.: Phonetic speaker recognition using maximum-likelihood binary-decision tree models. In: Acoustics, Speech, and Signal Processing, 2003. Proceedings. (ICASSP '03). 2003 IEEE International Conference on, vol. 4, pp. IV–796–9 (2003)

42. NIST: Speaker Recognition Evaluation Site. Web (2010). URL http://www.nist.gov/itl/iad/mig/sre.cfm

43. Pelecanos, J., Slomka, S., Sridharan, S.a.: Enhancing automatic speaker identification using phoneme clustering and frame based parameter and frame size selection. In: Signal Processing and Its Applications, 1999. ISSPA '99. Proceedings of the Fifth International Symposium on, vol. 2, pp. 633–636 (1999)

44. Reynolds, D.a.: The effects of handset variability on speaker recognition performance: experiments on the Switchboard corpus. In: Acoustics, Speech, and Signal Processing, 1996. ICASSP-96. Conference Proceedings., 1996 IEEE International Conference on, vol. 1, pp. 113–116 (1996)

45. Sessler, G.M., West, J.E.: Self-Biased Condenser Microphone with High Capacitance. The Journal of the Acoustical Society of America (JASA) 34(11), 293–309 (1962)

46. Stadtschnitzer, M., Van Pham, T., Chien, T.T.: Reliable voice activity detection algorithms under adverse environments. In: International Conference on Communications and Electronics, pp. 218–223 (2008)

47. Vogt, R., Baker, B., Sridharan, S.: Modelling Session Variability in Text-Independent Speaker Verification. In: Interspeech, pp. 3117 3120 (2005)

48. Weber, F., Peskin, B., Newman, M., Corrada-Emmanuel, A., Gillick, L.: Speaker Recognition on Single and Multispeaker Data. Digital Signal Processing 10, 75–92 (2000)

49. Yanguas, L., Quatieri, T.a.: Implications of glottal source for speaker and dialect identification. In: Acoustics, Speech, and Signal Processing, 1999. ICASSP '99. Proceedings., 1999 IEEE International Conference on, vol. 2, pp. 813–816 (1999)

50. kwong Yiu, K., Mak, M.W., and, S.Y.K.: Speaker verification with a priori threshold determination using kernel-based probabilistic neural networks. In: Neural Information Processing, 2002. ICONIP '02. Proceedings of the 9th International Conference on, vol. 5, pp. 2386–2390 (2002)

51. Young, S.: A Review of Large-Vocabular Continuous Speech Recognition. IEEE Signal Processing Magazine 13(5), 45–57 (1996)

Part IV
Background Material

Chapter 23
Linear Algebra

23.1 Basic Definitions

Definition 23.1 (Identity Matrix). *The N dimensional it identity matrix is denoted by $\mathbf{I}_N$ (or sometimes $\mathbf{I}$) and is defined as follows,*

$\mathbf{I}_N : \mathscr{R}^N \mapsto \mathscr{R}^N$ *is the matrix such that*

$$\mathbf{I}_{ij} = \begin{cases} 1 \ \forall \ i = j \\ 0 \ \forall \ i \neq j \end{cases} \tag{23.1}$$

where $i, j \in \{1, 2, ..., N\}$ are the row number and column number of the corresponding element of matrix $\mathbf{I}_N$.

Definition 23.2 (Transpose of a Matrix). *The transpose of a matrix $\mathbf{A} : \mathscr{R}^N \mapsto \mathscr{R}^M$ is given by $\mathbf{A}^T : \mathscr{R}^M \mapsto \mathscr{R}^N$ such that,*

$$\mathbf{A}^T_{ji} = \mathbf{A}_{ij} \tag{23.2}$$

where indices $i \in \{1, 2, ..., M\}$ and $j \in \{1, 2, ..., N\}$ denote the location of elements of the matrix such that the first index corresponds to the row and the second index corresponds to the column number.

Definition 23.3 (Hermitian Transpose). *The Hermitian transpose of a matrix $\mathbf{A} : \mathscr{C}^N \mapsto \mathscr{C}^M$ is given by $\mathbf{A}^H : \mathscr{C}^M \mapsto \mathscr{C}^N$ such that,*

$$\mathbf{A} = \mathbf{A}_R + i\mathbf{A}_I \tag{23.3}$$
$$\mathbf{A}_R, \mathbf{A}_I : \mathscr{R}^N \mapsto \mathscr{R}^M$$

and

$$\mathbf{A}^H = \mathbf{A}_R^T - i\mathbf{A}_I^T \tag{23.4}$$

Matrix $\mathbf{A}^H$ is also known as the adjoint matrix of matrix $\mathbf{A}$.

Definition 23.4 (Hermitian Matrix). *A Hermitian matrix $\mathbf{A} : \mathscr{C}^N \mapsto \mathscr{C}^N$ is the matrix for which,*

$$\mathbf{A} = \mathbf{A}^H \tag{23.5}$$

Definition 23.5 (Inverse of a Square Matrix). *The Inverse of a Square Matrix $\mathbf{A} : \mathscr{R}^N \mapsto \mathscr{R}^N$ (if it exists) is denoted by $\mathbf{A}^{-1} : \mathscr{R}^N \mapsto \mathscr{R}^N$ and is that unique matrix such that,*

$$\mathbf{A}^{-1}\mathbf{A} = \mathbf{A}\mathbf{A}^{-1} = \mathbf{I}_N \tag{23.6}$$

Definition 23.6 (Kronecker Product). *The Kronecker product of two matrices, $\mathbf{A} : \mathscr{R}^N \mapsto \mathscr{R}^M$ and $\mathbf{B}$ of arbitrary dimension is denoted by $\mathbf{A} \otimes \mathbf{B}$ and is defined as follows,*

$$\mathbf{A} \otimes \mathbf{B} \overset{\Delta}{=} \begin{bmatrix} (\mathbf{A})_{[1][1]}\,\mathbf{B} & (\mathbf{A})_{[1][2]}\,\mathbf{B} & \cdots & (\mathbf{A})_{[1][N]}\,\mathbf{B} \\ (\mathbf{A})_{[2][1]}\,\mathbf{B} & (\mathbf{A})_{[2][2]}\,\mathbf{B} & \cdots & (\mathbf{A})_{[2][N]}\,\mathbf{B} \\ \vdots & \vdots & \vdots & \vdots \\ \cdots & \cdots & \cdots & \ddots \\ (\mathbf{A})_{[M][1]}\,\mathbf{B} & (\mathbf{A})_{[M][2]}\,\mathbf{B} & \cdots & (\mathbf{A})_{[M][N]}\,\mathbf{B} \end{bmatrix} \tag{23.7}$$

23.2 Norms

To quantify the magnitude of a vector or a matrix, several norms have been used in the literature. In this book, predominantly, the *Euclidean norm* [8] is used. Let us examine the norm of a vector and following by looking at one special norm used for matrices.

Definition 23.7 (Euclidean Norm of a Vector). *The Euclidean norm of a vector $\mathbf{x} \in \mathscr{R}^N$ is denoted by $\|\mathbf{x}\|_{\mathscr{E}}$ and is defined as,*

$$\|\mathbf{x}\|_{\mathscr{E}} = \left(\sum_{i=1}^{N} (\mathbf{x})_{[i]}^2 \right)^{\frac{1}{2}} \tag{23.8}$$

where, $(\mathbf{x})_{[i]}, i \in \{1, 2, ..., N\}$ *is the* i^{th} *element of vector* $\mathbf{x}$.

The norm of a vector may be generalized in the form of an L_p-*norm* where $\{p : p \in \mathbb{R}, p \geq 1\}$. Here is the definition of an L_p-*norm*:

Definition 23.8 (L_p-**norm of a vector**). *The* L_p-*norm of a vector* $\mathbf{x} \in \mathscr{R}^N$, *where* $\{p : p \in \mathbb{R}, p \geq 1\}$, *is denoted by* $\|\mathbf{x}\|_p$ *and is defined as,*

$$\|\mathbf{x}\|_p = \left(\sum_{i=1}^{N} | (\mathbf{x})_{[i]} |^p \right)^{\frac{1}{p}} \tag{23.9}$$

where, $(\mathbf{x})_{[i]}, i \in \{1, 2, ..., N\}$ *is the* i^{th} *element of vector* $\mathbf{x}$.

The *Euclidean Norm* of a vector given by Definition 23.7 is, therefore, a special case of the L_p norm where $p = 2$. Other important special cases which are often used are the L_1 norm,

$$\|\mathbf{x}\|_1 = \sum_{i=1}^{N} | (\mathbf{x})_{[i]} | \tag{23.10}$$

and the L_∞ Norm (also known as the *Maximum Norm*),

$$\|\mathbf{x}\|_\infty = \lim_{p \to \infty}$$
$$= \max_{i=1}^{N} | (\mathbf{x})_{[i]} | \tag{23.11}$$

Definition 23.9 (**Linear Dependence / Independence**). *A set of vectors* $\mathbf{s}_i \in \mathscr{R}^N, i \in \{1, 2, ..., N\}$ *is said to be a linearly dependent set if there exist numbers* $\lambda_i, i \in \{1, 2, ..., N\}$, *not all zero, such that,*

$$\sum_{i=1}^{N} \lambda_i \mathbf{s}_i = 0 \tag{23.12}$$

Definition 23.10 (**Euclidean (Frobenius) Norm of a Matrix**). *The Euclidean (Frobenius) norm of a matrix* $\mathbf{A} : \mathscr{R}^N \mapsto \mathscr{R}^M$ *is denoted by* $\|\mathbf{A}\|_\mathscr{E}$ *or* $\|\mathbf{A}\|_\mathscr{F}$ *and is defined as,*

$$\|\mathbf{A}\|_\mathscr{E} = \|\mathbf{A}\|_\mathscr{F}$$
$$= \left(\sum_{i=1}^{M} \sum_{j=1}^{N} (\mathbf{A})_{[i][j]}^2 \right)^{\frac{1}{2}} \tag{23.13}$$

where $(\mathbf{A})_{[i][j]}, (i \in \{1, 2, ..., M\}; j \in \{1, 2, ..., N\})$ *is the* $(i, j)^{th}$ *element of matrix* $\mathbf{A}$.

The Euclidean norm of a matrix can also be written in the following forms,

$$\|\mathbf{A}\|_{\mathscr{E}} = \|\mathbf{A}\|_{\mathscr{F}} = \left(\sum_{i=1}^{M} \|\mathbf{A}\mathbf{u}_i\|_{\mathscr{E}}^2 \right)^{\frac{1}{2}} \tag{23.14}$$

where $\mathbf{u}_i, i \in \{1, 2, ..., M\}$ is any orthonormal basis

$$\|\mathbf{A}\|_{\mathscr{E}} = \|\mathbf{A}\|_{\mathscr{F}} = \sqrt{tr(\mathbf{A}^T\mathbf{A})} \tag{23.15}$$

where $tr(\mathbf{A}^T\mathbf{A})$ denotes the trace of $(\mathbf{A}^T\mathbf{A})$ which is equivalent to the sum of all its diagonal elements.

In general, all matrix norms satisfy the following four conditions:
For $\mathbf{A}, \mathbf{B} : \mathscr{R}^N \mapsto \mathscr{R}^M$ and $\mathbf{C} : \mathscr{R}^M \mapsto \mathscr{R}^N$,

1. *$\|\mathbf{A}\| \geq 0$ and $\|A\| = 0$ iff $\mathbf{A} = \mathbf{0}$*
2. *$\|k\mathbf{A}\| = |k| \|A\|$ where k is any scalar*
3. *$\|\mathbf{A} + \mathbf{B}\| \leq \|\mathbf{A}\| + \|\mathbf{B}\|$ (Triangular Inequality)*
4. *$\|\mathbf{AC}\| \leq \|\mathbf{A}\| \|\mathbf{C}\|$ (Schwarz's Inequality)*

If the set is not linearly dependent, then it is said to be linearly independent.

Definition 23.11 (Unitary / Orthogonal Matrices). *A matrix, $\mathbf{U} : \mathscr{C}^N \mapsto \mathscr{C}^N$ is said to be Unitary if,*

$$\mathbf{U}^H\mathbf{U} = \mathbf{U}\mathbf{U}^H = \mathbf{I}_N \tag{23.16}$$

A special case of unitary matrices is $\mathbf{V} : \mathscr{R}^N \mapsto \mathscr{R}^N$ in which case,

$$\mathbf{V}^T\mathbf{V} = \mathbf{V}\mathbf{V}^T = \mathbf{I}_N \tag{23.17}$$

Matrices falling under this special case are called orthogonal.

Definition 23.12 (Conjugacy, Orthogonality, and Orthonormality). *Any set of linearly independent vectors,*

$$\mathbf{v}_i : \mathbf{v}_i \in \mathscr{R}^N, i \in \{1, 2, ..., M\}, M \leq N \tag{23.18}$$

is said to be mutually conjugate about a positive definite, full rank matrix, $\mathbf{Q} : \mathscr{R}^N \mapsto \mathscr{R}^N$ such that,

$$\mathbf{v}_i^T\mathbf{Q}\mathbf{v}_i = \begin{cases} a > 0 \ \forall \ i = j \\ 0 \quad \forall \ i \neq j \end{cases} \tag{23.19}$$

If $\mathbf{Q} = \mathbf{I}_N$, *then the set is a mutually orthogonal set of vectors. If in addition* $a = 1$, *then the set is mutually orthonormal (i.e., for an orthonormal set of vectors,* $\|\mathbf{v}_i\|_{\mathscr{E}} = 1$).

Definition 23.13 (Singular Values of a Matrix). *If* $\mathbf{A} : \mathscr{C}^N \mapsto \mathscr{C}^M$, *then the strictly positive square roots* σ_i *of the non-zero eigenvalues of* $\mathbf{A}^H\mathbf{A}$ *(or* $\mathbf{A}\mathbf{A}^H$*) are called the singular values of matrix* $\mathbf{A}$.

Definition 23.14 (Rank of a Matrix). *Matrix* $\mathbf{A} : \mathscr{C}^N \mapsto \mathscr{C}^M$ *has rank k if it has k singular values.*

Next, we will define *singular value decomposition (SVD)* which was introduced, independently, by *Beltrami* [2] and *Jordan* [9] in 1873 and 1874 respectively.

Definition 23.15 (Singular Value Decomposition). *If* $\mathbf{A} : \mathscr{C}^N \mapsto \mathscr{C}^M$ *has rank k and its singular values are denoted by* $\sigma_1 \geq \sigma_2 \geq ... \geq \sigma_k ¿ 0$, *then there exist two unitary matrices,*

$$\mathbf{U} = [\mathbf{u}_1, \mathbf{u}_2, ..., \mathbf{u}_M] : \mathscr{C}^M \mapsto \mathscr{C}^M \tag{23.20}$$

and

$$\mathbf{V} = [\mathbf{v}_1, \mathbf{v}_2, ..., \mathbf{v}_M] : \mathscr{C}^N \mapsto \mathscr{C}^N \tag{23.21}$$

such that,

$$\mathbf{S} = \mathbf{U}^H \mathbf{A} \mathbf{V} \, and \, \mathbf{A} = \mathbf{U} \mathbf{S} \mathbf{V}^H \tag{23.22}$$

where,

$$\mathbf{S} = \begin{bmatrix} \mathbf{D} & \mathbf{0} \\ \mathbf{0} & \mathbf{0} \end{bmatrix} : \mathscr{C}^N \mapsto \mathscr{C}^M \tag{23.23}$$

and,

$$(\mathbf{D})_{[i][j]} = \begin{cases} \sigma_i \ \forall \ i = j \\ 0 \ \forall \ i \neq j \end{cases} \tag{23.24}$$

Then,

$$\mathbf{A} = \mathbf{U}\mathbf{S}\mathbf{V}^H \tag{23.25}$$

is the singular value decomposition of matrix $\mathbf{A}$, *where, for* $1 \le i \le k$, $\mathbf{u}_i = \frac{\mathbf{A}\mathbf{v}_i}{\sigma_i}$ *and* $\mathbf{v}_i = \frac{\mathbf{A}^H \mathbf{u}_i}{\sigma_i}$ *are Eigenvectors of* $\mathbf{A}\mathbf{A}^H$ *and* $\mathbf{A}^H\mathbf{A}$ *respectively, associated with the* k *eigenvalues* $\sigma_i^2 > 0$ *and the vectors* $\mathbf{u}_i, k+1 \le i \le M$ *and* $\mathbf{v}_i, k+1 \le i \le N$ *are Eigenvectors associated with the zero eigenvalues. If* $\mathbf{A}$ *is real, then* $\mathbf{U}$ *and* $\mathbf{V}$ *will also be real and are therefore orthogonal matrices.*

Definition 23.16 (Pseudo-Inverse (Moore-Penrose Generalized Inverse)). *If* $\mathbf{A}$: $\mathscr{C}^N \mapsto \mathscr{C}^M$ *and* $\mathbf{A}^\dagger : \mathscr{C}^M \mapsto \mathscr{C}^N$, *then* $\mathbf{A}^\dagger$ *is the pseudo-inverse (Moore-Penrose generalized inverse) of A iff,*

1. $\mathbf{A}\mathbf{A}^\dagger\mathbf{A}$
2. $\mathbf{A}^\dagger\mathbf{A}\mathbf{A}^\dagger$
3. $\mathbf{A}$ *and* $\mathbf{A}^\dagger$ *are Hermitian*

Furthermore, if the singular value decomposition of A is given by,

$$\mathbf{A} = \mathbf{U}\mathbf{S}\mathbf{V}^H \tag{23.26}$$

then the pseudo-inverse of A, $A^\dagger$, *is given by,*

$$\mathbf{A}^\dagger = \mathbf{V}\mathbf{S}^\dagger\mathbf{U}^H \tag{23.27}$$

where,

$$\mathbf{S}^\dagger = \begin{bmatrix} \mathbf{E} & \mathbf{0} \\ \mathbf{0} & \mathbf{0} \end{bmatrix} : \mathscr{C}^M \mapsto \mathscr{C}^N \tag{23.28}$$

$\mathbf{E}$ *is the kxk diagonal matrix such that,*

$$\mathbf{E}_{ij} = \begin{cases} \frac{1}{\sigma_i} & \forall\, i = j \\ 0 & \forall\, i \ne j \end{cases} \tag{23.29}$$

and k is the rank of A. For more on the Pseudo-Inverse see [10].

For a real matrix, $\mathbf{A}$, $\mathbf{A}^\dagger$ may be written in terms of the following limit [4],

$$\mathbf{A}^\dagger = \lim_{\varepsilon \to 0} \left(\mathbf{A}^T\mathbf{A} + \varepsilon\mathbf{I} \right)^{-1} \mathbf{A}^T \tag{23.30}$$

Definition 23.17 (Positive Definiteness). *Let* $\mathbf{s}$ *be any vector such that* $\mathbf{s} \in \mathscr{R}^N$. *A matrix* $\mathbf{G} : \mathscr{R}^N \mapsto \mathscr{R}^N$ *is said to be positive definite if,*

$$\mathbf{s}\mathbf{G}\mathbf{s} > 0 \forall \mathbf{s} \neq 0 \tag{23.31}$$

23.3 Gram-Schmidt Orthogonalization

Two types of Gram-Schmidt orthogonalization techniques are covered here. The first one is the ordinary procedure and then a modified version with higher numerical accuracy is described.

23.3.1 Ordinary Gram-Schmidt Orthogonalization

Suppose, $\mathbf{v}_i : \mathbf{v}_i \in \mathcal{R}^N, i \in \{1, 2, ..., M\}, M \leq N$ are a set of unit vectors. Then, the following is the Gram-Schmidt procedure [12] which generates the set of vectors $\mathbf{v}_i, i \in \{1, 2, ..., M\}$ which form an *orthonormal set* spanning the same space as vectors $\mathbf{v}_i$,

$$\mathbf{u}_1 = \mathbf{v}_1 \tag{23.32}$$

$$\mathbf{u}_i = \mathbf{v}_i - \sum_{j=1}^{i-1}(\mathbf{v}_i^T \mathbf{z}_j)\mathbf{z}_j \ \ with \ \ i \in \{1, 2, ..., M\} \tag{23.33}$$

$$\mathbf{z}_i = \frac{\mathbf{u}_i}{\|\mathbf{u}_i\|_{\mathscr{E}}} \ \ with \ \ i \in \{1, 2, ..., M\} \tag{23.34}$$

23.3.2 Modified Gram-Schmidt Orthogonalization

The following pseudo-code presents a modified Gram-Schmidt orthogonalization method which, theoretically, gives the same set of vectors as the ordinary procedure (Section 23.3.1), but it is more accurate in its numerical implementation,

1.   a. $\mathbf{u}_1 = \mathbf{v}_1$
         b. $\mathbf{z}_1 = \frac{\mathbf{u}_1}{\|\mathbf{u}_1\|_{\mathscr{E}}}$
2. $\mathbf{v}_i^{(1)} = \mathbf{v}_i - (\mathbf{v}_i^T \mathbf{z}_1)\mathbf{z}_1$ for $i = 2, 3, ..., M$
3.   a. $\mathbf{u}_j = \mathbf{v}_j^{(j-1)}$
         b. $\mathbf{z}_j = \frac{\mathbf{u}_j}{\|\mathbf{u}_j\|_{\mathscr{E}}}$ for $j = 2, 3, ..., M$
         c. $\mathbf{v}_i^{(j)} = \mathbf{v}_i^{(j-1)} - (\mathbf{v}_i^{(j-1)^T} \mathbf{z}_j)\mathbf{z}_j$ for $i = j+1, ..., M$

23.4 Sherman-Morrison Inversion Formula

If $\mathbf{G}_{k+1}, \mathbf{G}_k : \mathscr{R}^N \mapsto \mathscr{R}^N$, then the rank $M(M \leq N)$ update to $\mathbf{G}_k$ for obtaining $\mathbf{G}_{k+1}$ is,

$$\mathbf{G}_{k+1} = \mathbf{G}_k + \mathbf{R}\mathbf{S}\mathbf{T}^T \tag{23.35}$$

where $\mathbf{R}, \mathbf{T} : \mathscr{R}^M \mapsto \mathscr{R}^N$ and $\mathbf{S} : \mathscr{R}^M \mapsto \mathscr{R}^M$, then the inverse of $\mathbf{G}_{k+1}$ is given by the following,

$$\mathbf{G}_{k+1}^{-1} = \mathbf{G}_k^{-1} - \mathbf{G}_k^{-1}\mathbf{R}\mathbf{U}^{-1}\mathbf{T}^T\mathbf{G}^{-1} \tag{23.36}$$

where,

$$\mathbf{U} = \mathbf{S}^{-1} + \mathbf{T}^T\mathbf{G}_k^{-1}\mathbf{R} \tag{23.37}$$

Equation 23.36 is known as the *Sherman-Morrison* formula [6]. It is used to keep track of the change in the inverse of a matrix as the original matrix is updated through 23.35.

23.5 Vector Representation under a Set of Normal Conjugate Direction

Theorem 23.1 (Conjugate Directions). *Since conjugate directions are linearly independent, any vector* $\mathbf{v} \in \mathscr{R}^N$ *can be represented in terms of a set of directions,* $\mathbf{s}_i, i \in \{1, 2, ..., N-1\}$, *conjugate about a positive-definite full rank matrix* $\mathbf{G} : \mathscr{R}^N \mapsto \mathscr{R}^N$ *as follows,*

$$\mathbf{v} = \sum_{i=1}^{N} \lambda_i \mathbf{s}_i \tag{23.38}$$

where,

$$\lambda_i = \frac{\mathbf{s}_i^T \mathbf{G} \mathbf{v}}{\mathbf{s}_i^T \mathbf{G} \mathbf{s}_i} \tag{23.39}$$

Furthermore, there always exists a full set of N directions $\mathbf{s}_i$ *conjugate directions about* $\mathbf{G}$ *[1] since the Eigenvectors of* $\mathbf{G}$ *form such a set.*

Theorem 23.2 (Inverse of a Matrix). *Consider the matrix,*

$$H = \sum_{i=1}^{N} \frac{\mathbf{s}_i \mathbf{s}_i^T}{\mathbf{s}_i^T \mathbf{G} \mathbf{s}_i} \tag{23.40}$$

where $\mathbf{s}_i \in \mathcal{R}^N, i \in \{1, 2, ..., N\}$ *are a set of directions mutually conjugate about the positive-definite full rank matrix,* $\mathbf{G} : \mathcal{R}^N \mapsto \mathcal{R}^N$. *Post multiplication by* $\mathbf{G}\mathbf{s}_k$ *gives,*

$$\sum_{i=1}^{N} \frac{1}{\mathbf{s}_i^T \mathbf{G} \mathbf{s}_i} \mathbf{s}_i \mathbf{s}_i^T \mathbf{G} \mathbf{s}_k = \frac{\mathbf{s}_k \mathbf{s}_k^T \mathbf{G} \mathbf{s}_k}{\mathbf{s}_k \mathbf{G} \mathbf{s}_k} \tag{23.41}$$

$$= \mathbf{s}_k \tag{23.42}$$

$$\text{since } \mathbf{s}_i \mathbf{G} \mathbf{s}_k = \begin{cases} 1 \ \forall \ i = k \\ 0 \ \forall \ i \neq k \end{cases} \tag{23.43}$$

Therefore, $\mathbf{H}$ *is the representation of the inverse of* $\mathbf{G}$ *[1],*

$$\mathbf{G}^{-1} = \mathbf{H} = \sum_{i=1}^{N} \frac{\mathbf{s}_i \mathbf{s}_i^T}{\mathbf{s}_i^T \mathbf{G} \mathbf{s}_i} \tag{23.44}$$

23.6 Stochastic Matrix

In *probability theory*, we often run across a special type of matrix called a *stochastic matrix*. Here is a formal definition for such matrices.

Definition 23.18 (Stochastic Matrix). *A Stochastic matrix* $\mathbf{A} : \mathcal{R}^N \mapsto \mathcal{R}^M$ *is a matrix such that* $(\mathbf{A})_{[i][j]} \geq 0$ *and*

$$\sum_{j=1}^{N} (\mathbf{A})_{[i][j]} = 1 \ \forall \ \{i : 1 \leq i \leq M\} \tag{23.45}$$

For example, *stochastic matrices* are used to denote the *transition probabilities* of *Markov chains*.

23.7 Linear Equations

Consider the following set of linear equations,

$$\mathbf{A}\boldsymbol{\lambda} = \mathbf{g} \tag{23.46}$$

where $\mathbf{A} : \mathscr{R}^M \mapsto \mathscr{R}^N$, $\boldsymbol{\lambda} : \mathscr{R}^1 \mapsto \mathscr{R}^M$, and $\mathbf{g} : \mathscr{R}^1 \mapsto \mathscr{R}^N$.

Farkas [5] published a paper examining the vector space represented by this *simple linear equation*. By definition, $\mathbf{A}\boldsymbol{\lambda}$, where $\boldsymbol{\lambda} \succeq \mathbf{0}$, defines a *polyhedral cone*, where the columns of $\mathbf{A}$ represent the *outer boundary* of the cone, in $\mathscr{R}^N$. In other words, it is said that the columns of $\mathbf{A}$ span a *cone*. Note the following definition of a *cone*.

Definition 23.19 (Cone). *A cone is a convex set, $\mathscr{C} \in \mathscr{R}^N$ (see Definition 6.28), such that for all $\mathbf{a} : \mathscr{R}^1 \mapsto \mathscr{R}^N$, where $\mathbf{a} \in \mathscr{C}$,*

$$\lambda \mathbf{a} \in \mathscr{C} \ \forall \ \lambda \geq 0 \tag{23.47}$$

Note that it is easy to show that for a cone, $\mathscr{C}$, if

$$\mathbf{a}_1, \mathbf{a}_2 \in \mathscr{C} \iff \lambda_1 \mathbf{a}_1 + \lambda_2 \mathbf{a}_2 \in \mathscr{C} \tag{23.48}$$

Definition 23.20 (Polyhedral Cone). *A polyhedral cone is a closed convex set, $\mathscr{C} \in \mathscr{R}^N$, such that for all $\mathbf{a}_i : \mathscr{R}^1 \mapsto \mathscr{R}^N, i = \{1, 2, \cdots\}$, where $\mathbf{a}_i \in \mathscr{C}$, then*

$$\sum_{i=1}^{M} \lambda_i \mathbf{a}_i \in \mathscr{C} \ \forall \ \lambda_i \geq 0 \tag{23.49}$$

Equation 23.49 may be written in matrix form as follows,

$$\mathbf{A}\boldsymbol{\lambda} \in \mathscr{C} \ \forall \ \boldsymbol{\lambda} \succeq \mathbf{0} \tag{23.50}$$

Equation 23.50 attests to the *closedness* and *convexity* of the polyhedral cone. This becomes important in the way *Farkas* approaches the establishment of the existence of a solution for the set of linear equation given by Equation 23.46. These results are at the core of linear optimization and constrained nonlinear optimization theory, as we will see in Section 25.5. Figure 23.1 shows the geometric representation of a *polyhedral cone* which is spanned by the columns of $\mathbf{A}$.

Farkas' Lemma [5] is basically a separation statement for *polyhedral cones*. A general *separation theorem* would be stated as follows,

Theorem 23.3 (General Separation Theorem). *If $\mathscr{S}_1$ and $\mathscr{S}_2$ are both closed convex sets, then they may be separated by a hyperplane if their common space is empty, i.e.,*

$$\mathscr{S}_1 \cap \mathscr{S}_2 = \{\varnothing\} \tag{23.51}$$

In words, *Farkas Lemma*[5] notes that a vector $\mathbf{g}$ would either be in a polyhedral cone defined by the columns of a spanning matrix $\mathbf{A}$, or it would be sepa-

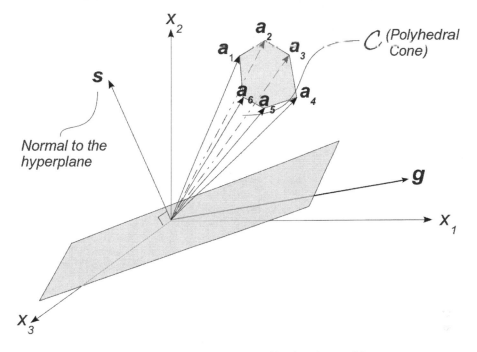

Fig. 23.1: Polyhedral Cone Spanned by the columns of **A**.

rated from the cone by a *hyperplane*. There are many different variations of this *Lemma* [5, 7, 6, 3, 11]. Here is a variation which expresses the above statement in mathematical terms. It is most useful for handling *linear inequality constraints* in an optimization problem – see Section 25.5.1.3.

Lemma 23.1 (Farkas' Lemma). *Consider matrix* $\mathbf{A} : \mathscr{R}^M \mapsto \mathscr{R}^N$*, vector* $\boldsymbol{\lambda} : \mathscr{R}^1 \mapsto \mathscr{R}^M$*, and vector* $\mathbf{g} : \mathscr{R}^1 \mapsto \mathscr{R}^N$*. Then, one and only one of the following statements can hold,*

1. $\mathbf{g} = \mathbf{A}\boldsymbol{\lambda}$
2. $\exists \; \{\mathbf{s} : \mathscr{R}^1 \mapsto \mathscr{R}^N\} : \mathbf{As} \succeq \mathbf{0} \; \wedge \; \mathbf{s}^T\mathbf{g} < 0$

Proof. see [5, 6].   □

As we mentioned, statement 1 in the *lemma* corresponds to the case when $\mathbf{g} \in \mathscr{C}$ and statement 2 corresponds to when $\mathbf{g} \notin \mathscr{C}$, where $\mathscr{C}$ is the *polyhedral cone* defined as follows,

$$\mathscr{C} = \{\mathbf{x} : \mathbf{x} = \mathbf{A}\boldsymbol{\lambda}, \boldsymbol{\lambda} \succeq \mathbf{0}\} \tag{23.52}$$

s in statement 2 is the *normal vector*, defining the *separating hyperplanes* which separates **g** from cone $\mathscr{C}$.

References

1. Beigi, H.S.M.: Neural Network Learning and Learning Control Through Optimization Techniques. Columbia University, New York (1991). Doctoral Thesis: School of Engineering and Applied Science
2. Beltrami, E.: Sulle Funzioni Bilineari. Giornale di Mathematiche di Battaglini **11**, 98–106 (1873)
3. Boyd, S., Vandenberghe, L.: Convex Optimization, 7th printing edn. Cambridge University Press (2009). First Published in 2004. ISBN: 978-0-521-83378-3
4. Duda, R.O., Hart, P.E.: Pattern Classification and Scene Analysis. John Wiley & Sons, New York (1973). ISBN: 0-471-22361-1
5. Farkas, J.: Theorie der Einfachen Ungleichungen (Theory of the Simple Inequality). Journal für die Reine und Angewandte Mathematic **124**, 1–27 (1902)
6. Fletcher, R.: Practical Methods of Optimization, 2nd edn. J. Wiley and Sons, New york (2000). ISBN: 0-471-49463-1
7. Gale, D., Kühn, H.H., Tucker, A.W.: Linear Programming and the Theory of Games. In: e.a. Tjalling C. Koopmans (ed.) Activity Analysis of Production and ALllocation, pp. 317–329 (1951). URL http://cowles.econ.yale.edu/P/cm/m13/
8. Gerald, C.F., Wheatly, P.O.: Applied Numerical Analysis, 3rd edn. Addison-Wesley Publishing Company, Reading, Massachusetts (1985). ISBN: 0-201-11577-8
9. Jordan, M.C.: Mémoire sur les Formes Bilinéaires. Journal de Math'ematiques Pures et Appliquées **19**, 35–54 (1874)
10. Kohonen, T.: Self-Organizination and Associative Memory, 3rd edn. Springer-Verlag, Berlin (1989). ISBN: 0-387-51387-6
11. Mangasarian, O.L.: Nonlinear Programming. Society for Industrial and Applied Mathematics (SIAM) (1994). Originally Published: New York: McGraw Hill, 1969; ISBN: 0-89871-341-2
12. Noble, B., Daniel, J.W.: Applied Linear Algebra, 2nd edn. Prentice-Hall, Inc., New Jersey (1977). ISBN: 0-130-41343-7

Chapter 24
Integral Transforms

This chapter is a rich section of the book which includes information usually covered in pieces in graduate courses such as Complex Variable Theory, Integral Transforms, Partial Differential Equations, Analog and Digital Signal Processing, and Control. As stated in the Preface, one of the goals of this book is to bring all the fundamental sciences and mathematics needed for doing speaker recognition into one place with a comprehensive narrative connecting all the dots in the field. The fact that speaker recognition is greatly multi-disciplinary has been the stumbling block for the development of such a textbook. Although it is impossible to be complete, but the goal is to include all the necessary information in one place. It makes this chapter ideal for students and professionals and allows for a complete understanding of the subject. It is recommended that it be treated like any other chapter of the book and not skipped. The only reason it is included in this background chapter is to keep the higher-level flow of speaker recognition smoother, but as they say, "the Devil is in the Details."

In the following section we will give some basic definitions, theorems and properties related to the Real ($\mathbb{R}$) and Complex ($\mathbb{C}$) Domains. There are different notations in the literature depending on the subject being discussed. Since in this book we are interested in time-dependent signals, we do not consider the Cartesian space. For that reason, we either speak of Time ($t : t \in \mathbb{R}$) or we speak of the complex plane in which the variable used here is $\{s = (\sigma + i\omega) : s \in \mathbb{C}, \sigma \in \mathbb{R}, \omega \in \mathbb{R}\}$. The choice of σ and ω for the real and imaginary axes of the Complex plane, as will be seen later, has to do with their coincidence with the definitions of Laplace and Fourier transforms. This may introduce some strange looking notation when we arrive at the point of handling the Laplacian and partial derivative notations used in the *Cauchy-Riemann* Conditions. There, we will use σ and ω in place of the more familiar x and y from the wave and heat equations which propagate in Cartesian-Based media and which were the historic problems of interest at the time of the development of these definitions and theorems. Strange as the notation may seem, it is rigorous and great care has gone into making sure that all the necessary definitions are present in the text.

24.1 Complex Variable Theory in Integral Transforms

In this section, we will cover many of the concepts related to the theory of complex variables, needed to formulate integral transforms. Once we have addressed these fundamental definitions, in the following sections we will move toward the definition of integral equations followed by the general form of integral transforms. Then, specific transforms are covered in consequent sections.

24.1.1 Complex Variables

Since the *theory of complex variables* is at the heart of integral equations, we will spend some time on defining basic concepts in this theory. Of course, the most interesting part of a complex variable is related to its imaginary part. Therefore, we start with the definition of an *imaginary number*.

Definition 24.1 (Imaginary Number). *i is the imaginary number and is defined as,*

$$i \overset{\Delta}{=} \sqrt{-1} \tag{24.1}$$

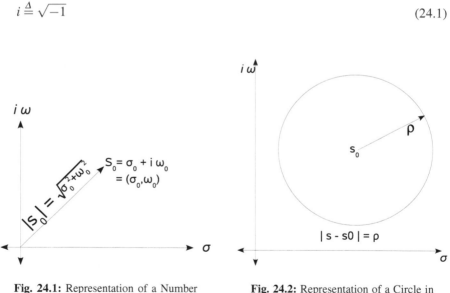

Fig. 24.1: Representation of a Number s_0 in the Complex Plane

Fig. 24.2: Representation of a Circle in $\mathbb{C}$

Definition 24.2 (Modulus or Magnitude of a Complex Number). *The Modulus or Magnitude of the complex number, $\{s : s = \sigma + i\omega \in \mathbb{C}\}$, is denoted by $|s|$ and is defined as,*

$$|s| = \sqrt{\sigma^2 + \omega^2} \tag{24.2}$$

See Figure 24.1.

N.B., In the complex plain, size is considered using the modulus, e.g., we cannot say $s_1 < s_2$, but we can say $|s_1| < |s_2|$.

Property 24.1 (Properties of Complex Variables).

$$|s| = |\bar{s}| \tag{24.3}$$

$$|s|^2 = s\bar{s} \tag{24.4}$$

$$s + \bar{s} = 2\mathscr{R}e\{s\} \tag{24.5}$$

$$s - \bar{s} = 2\mathscr{I}m\{s\} \tag{24.6}$$

$\{s_0 : s_0 \in \mathbb{C}\}$ *may be represented in polar coordinates as follows,*

$$s_0 = \rho_0 e^{i\theta_0} \tag{24.7}$$

where

$$\rho_0 = |s_0| \tag{24.8}$$

and

$$\begin{aligned} \theta &= \angle s_0 \\ &= \sin^{-1}\left(\frac{\omega_0}{\rho_0}\right) \tag{24.9} \\ &= \cos^{-1}\left(\frac{\sigma_0}{\rho_0}\right) \tag{24.10} \\ &= \tan^{-1}\left(\frac{\omega_0}{\sigma_0}\right) \tag{24.11} \end{aligned}$$

Property 24.2 (Triangular Inequality in the Complex Plane).

$$|s_1 + s_2| \le |s_1||s_2| \tag{24.12}$$

Property 24.3 (Product of Complex Variables).

$$\begin{aligned} s_1 s_2 &= \rho_1 e^{i\theta_1} \rho_2 e^{i\theta_2} \\ &= \rho_1 \rho_2 e^{i(\theta_1 + \theta_2)} \\ &= \rho_1 \rho_2 e^{i(\theta_1 + \theta_2 + 2n\pi)} \quad where \quad n = \{0, \pm 1, \pm 2, \cdots\} \tag{24.13} \end{aligned}$$

N.B., In Equation 24.13, we have added the period of the complex exponential function for generalization, see Problem 24.4.

Property 24.4 (Quotient of Complex Variables).

$$\frac{s_1}{s_2} = \frac{\rho_1 e^{i\theta_1}}{\rho_2 e^{i\theta_2}}$$

$$= \frac{\rho_1}{\rho_2} e^{i(\theta_1 - \theta_2)}$$

$$= \frac{\rho_1}{\rho_2} e^{i(\theta_1 - \theta_2 + 2n\pi)} \quad where \quad n = \{0, \pm 1, \pm 2, \cdots\} \tag{24.14}$$

N.B., In Equation 24.14, we have added the period of the complex exponential function for generalization, see Problem 24.4.

Theorem 24.1 (Modulus of the product of two Complex Numbers). *The modulus of the product of two complex numbers, s_1 and s_2, is equal to the product of the moduli of the two numbers, namely,*

$$|s_1 s_2| = |s_1| |s_2| \tag{24.15}$$

Proof.
See Problem 24.1.   □

Definition 24.3 (A Circle in the Complex Plane). *A circle is defined by its center, s_0 and its radius, ρ. In the complex place, such a circle is defined by,*

$$|s - s_0| = \rho \tag{24.16}$$

See Figure 24.2

Definition 24.4 (Distance between two Complex Variables). *The distance between two complex variables, s_1 and s_2 is defined as,*

$$d(s_1, s_2) \stackrel{\Delta}{=} |s_2 - s_1| \tag{24.17}$$

Property 24.5 (Euler identities).

$$e^{i\theta} = \cos(\theta) + i\,\sin(\theta) \tag{24.18}$$
$$e^{-i\theta} = \cos(\theta) - i\,\sin(\theta) \tag{24.19}$$

Theorem 24.2 (de Moivre's Theorem).

$$(\cos\theta + i\sin(\theta))^n = \cos(n\theta) + i\sin(n\theta) \qquad n = \{0, \pm 1, \pm 2, \cdots\} \tag{24.20}$$

Proof.
Using the Euler identities (Property 24.5),

$$s = \rho\left[\cos\theta + i\sin\theta\right]$$
$$= \rho e^{i\theta} \tag{24.21}$$

Therefore,

$$s^n = \rho^n\left[\cos\theta + i\sin\theta\right]^n$$
$$= \rho^n e^{in\theta} \tag{24.22}$$

$\square$

Definition 24.5 (A Hermitian Function). *A function* $H(s) = U(\sigma, \omega) + iV(\sigma, \omega)$, $\{s \in \mathbb{C}, \sigma \in \mathbb{R}, \omega \in \mathbb{R}, s = \sigma + i\omega\}$, *is called a Hermitian function if*

$$H(-s) = \overline{H(s)}$$
$$= U(\sigma, \omega) - iV(\sigma, \omega) \tag{24.23}$$

24.1.2 Limits

Now that we have defined the basics of an complex variable, let us review limits related to these numbers in general, but also to the special subset which is the real line.

Definition 24.6 (Limit of a Sequence of Numbers).

$$\lim_{n \to \infty} S_n = A \tag{24.24}$$

A is the limit of sequence S_n *as* $n \to \infty$. *Examples are,*

$$\lim_{n \to \infty} \frac{2n}{n+1} = 2$$

and

$$\lim_{n \to \infty} \left(1 + \frac{1}{n}\right)^n = e$$

Definition 24.7 (One Sided Limit of a Function – Right Hand Limit).

$$\lim_{t \to t_0+} h(t) = A \tag{24.25}$$

Implies that given $\varepsilon > 0$, $\exists \delta(\varepsilon) > 0 : |h(t) - A| < \varepsilon$ *if* $t_0 < t < t_0 + \delta$

Definition 24.8 (One Sided Limit of a Function – Left Hand Limit).

$$\lim_{t \to t_0-} h(t) = A \tag{24.26}$$

Implies that given $\varepsilon > 0$, $\exists \delta(\varepsilon) > 0 : |h(t) - A| < \varepsilon$ *if* $t_0 - \delta < t < t_0$

Definition 24.9 (Limit of a Function of a Continuous Variable).

$$\lim_{t \to t_0} h(t) = A \tag{24.27}$$

Implies that given $\varepsilon > 0$, $\exists\, \delta(\varepsilon) > 0 : |h(t) - A| < \varepsilon$ when $0 < |t - t_0| < \delta$
Note that all the points lying inside a circle of radius ε and center t_0 are an ε neighbor of t_0.

$$\exists \lim_{t \to t_0} h(t) = A \iff \lim_{t \to t_0+} h(t) = \lim_{t \to t_0-} h(t) = A.$$

Example 24.1 (Existence of the Limit).
Take the following question: If

$$h(t) = 1 + \frac{|t|}{t} \tag{24.28}$$

does the limit, $\lim\limits_{t \to 0} h(t)$ exist?

The right hand limit of 24.28 is,

$$\lim_{t \to 0+} 1 + \frac{|t|}{t} = 1 + 1 = 2 \quad (|t| = t) \tag{24.29}$$

The left hand limit of 24.28 is,

$$\lim_{t \to 0-} 1 + \frac{|t|}{t} = 1 - 1 = 0 \quad (|t| = -t) \tag{24.30}$$

Therefore, the limit, $\lim\limits_{t \to 0} h(t)$ does not exist since the left hand and right hand limits are not equal.

Definition 24.10 (Positive Infinite Limit). *If $\forall\, M > 0 \,\exists\, \delta > 0 : h(t) > M$ when $0 < |t - t_0| < \delta$, then,*

$$\lim_{t \to t_0} h(t) = \infty \tag{24.31}$$

Definition 24.11 (Negative Infinite Limit). *If $\forall\, M > 0 \,\exists\, \delta > 0 : h(t) < -M$ when $0 < |t - t_0| < \delta$, then,*

$$\lim_{t \to t_0} h(t) = -\infty \tag{24.32}$$

24.1.3 Continuity and Forms of Discontinuity

In Definition 24.9 we started dealing with the concept of a continuous variable. It makes sense to review some definitions of continuity of functions and later to examine different kinds of discontinuities. In fact, as we will see, it makes sense to define bounds and smoothness of functions, at the same time.

Definition 24.12 (Continuity of Functions). *If* $\forall\, \varepsilon > 0\; \exists \delta(\varepsilon) > 0$ *such that* $0 < |t - t_0| < \delta \implies |h(t) - h(t_0)| < \varepsilon$, *then, $h(t)$ is continuous at $t = t_0$, or,*

$$\lim_{t \to t_0} h(t) = h(t_0) \tag{24.33}$$

The function $h(t)$ is a continuous function if it is continuous for all t_0.

Definition 24.13 (Discontinuous Functions). *A function $h(t)$ is discontinuous if for some t_0,*

$$\lim_{t \to t_0} h(t) \neq h(t_0) \tag{24.34}$$

A function may be discontinuous at a point t_0 for two main reasons:

1. *$h(t)$ may not approach any limit as $t \to t_0$.*
2. *$h(t)$ may approach a limit different from $h(t_0)$.*

Let us classify the different points of discontinuity.

Definition 24.14 (A Point of Ordinary Discontinuity). *When $\lim_{t \to t_0} h(t)$ exists, i.e.,*

$$\lim_{t \to t_0} h(t) = \lim_{t \to t_0+} h(t) = \lim_{t \to t_0-} h(t) \tag{24.35}$$

but is not equal to $h(t_0)$ or $h(t_0)$ is not defined, then t_0 is a point of ordinary discontinuity.

Example 24.2 (Ordinary Discontinuity).
Take the following function,

$$h(t) = (t - t_0) \sin\left(\frac{1}{t - t_0}\right) \tag{24.36}$$

$h(t_{0+}) = h(t_{0-})$, but $h(t_0)$ is not defined.

Here, we may postulate that $h(t_0) = 0$ and then $h(t)$ becomes continuous. In cases where $h(t_{0+}) \neq h(t_{0-})$, regardless of the definition of $h(t_0)$, t_0 becomes a point of ordinary discontinuity – see Figure 24.3.

Example 24.3 (Ordinary Discontinuity at $t = 0$).
The following function has an ordinary discontinuity at $t - 0$, see Figure 24.4

$$h(t) = \frac{\sin(t)}{|t|} \tag{24.37}$$

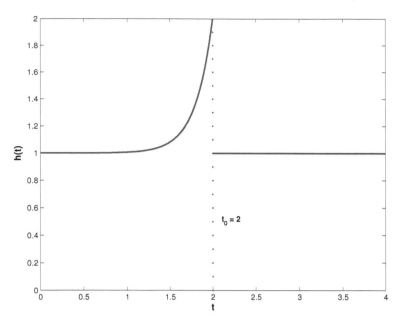

Fig. 24.3: Point of ordinary discontinuity at $t = t_0 = 2$

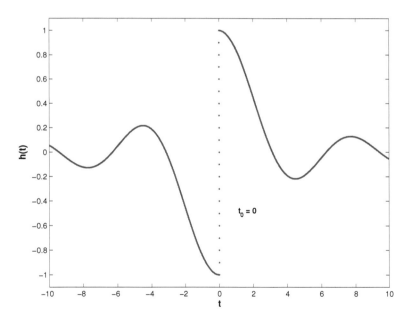

Fig. 24.4: Point of ordinary discontinuity at $t = t_0 = 0$ $(h(t) = \frac{\sin(t)}{|t|})$

$$\lim_{t \to 0^+} \frac{\sin(t)}{|t|} = \lim_{t \to 0^+} \frac{\sin(t)}{t} = 1 \tag{24.38}$$

$$\lim_{t \to 0^-} \frac{\sin(t)}{|t|} = \lim_{t \to 0^-} \frac{\sin(t)}{-t} = -1 \tag{24.39}$$

Note that in order to be able to use l'Hôpital's rule for obtaining the values of the limits, we replaced $|t|$ by its corresponding equivalent for the cases of the positive and negative limits.

Based on Equations 24.38 and 24.39,

$$\lim_{t \to 0^+} \frac{\sin(t)}{|t|} \neq \lim_{t \to 0^-} \frac{\sin(t)}{|t|} \tag{24.40}$$

Namely, the limit does not exist.

Definition 24.15 (A Point of Infinite Discontinuity). *The following are the different cases of infinite discontinuities,*

1. $h(t_{0+}) = h(t_{0-}) = \pm\infty$
2. $h(t_{0+}) = +\infty$ *and* $h(t_{0-}) = -\infty$
3. $h(t_{0+}) = \pm\infty$ *and* $h(t_{0-})$ *exists*
4. $h(t_{0+})$ *exists and* $h(t_{0-}) = \pm\infty$

Definition 24.16 (A Point of Oscillatory Discontinuity). *An oscillatory discontinuity s_0 is one where no matter how small the ε neighborhood of s_0 $\{s : |s - s_0| < \varepsilon\}$ is made, the value of s oscillates to different values with function $H(s)$ not being defined at the exact value of s_0, but it may be defined in its neighborhood (it is defined for the Finite Amplitude version – see below).*

The following are the different kinds of oscillatory discontinuities,

1. *Finite Amplitude:*

$$h(t) = \sin\left(\frac{1}{t - t_0}\right) \tag{24.41}$$

Figure 24.5 shows a plot of h(t) in Equation 24.41 where $t_0 = 0$. Figure 24.6 zooms into this plot. No matter how much one zooms in, in the vicinity of $t = 0$, there is an oscillation where at the exact point, $t = 0$, h(t) is undefined.

2. *Infinite Amplitude:*

In this case, in addition to the oscillatory nature of the discontinuity s_0, the value of the function will approach infinity in the neighborhood of the singularity. An example is,

$$h(t) = \frac{1}{(t - t_0)} \sin\left(\frac{1}{t - t_0}\right) \tag{24.42}$$

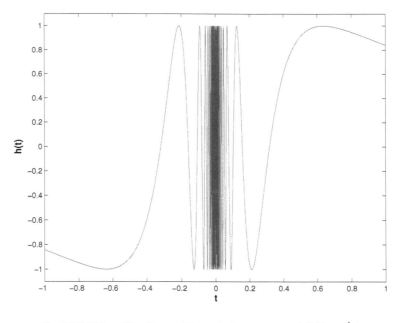

Fig. 24.5: Point of Oscillatory Discontinuity at $t = t_0 = 0$ $(h(t) = \frac{1}{t-0})$

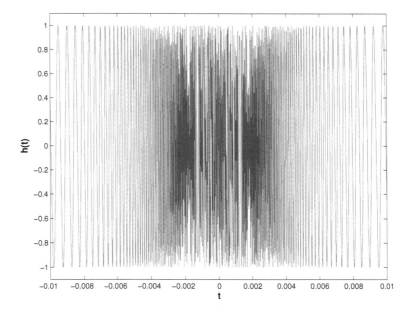

Fig. 24.6: More Detailed Viewpoint of the Oscillatory Discontinuity at $t = t_0 = 0$

Definition 24.17 (Continuity of a Function in an Interval). *A function $h(t)$ is said to be continuous in an interval $[a,b]$ if*

$$\lim_{t \to t_0} h(t) = h(t_0) \qquad a < t < b \tag{24.43}$$

$$\lim_{t \to a^+} h(t) = h(a)$$
$$\lim_{t \to b^-} h(t) = h(b)$$

Definition 24.18 (Boundedness). *A function $h(t)$ is bounded in an interval $[a,b]$, if* $\exists M : |h(t)| \leq M \, \forall \, t \in [a,b]$.

Property 24.6 (Boundedness of a Continuous Function). *A function $h(t)$ which is continuous in an interval $[a,b]$, is bounded.*

Proof.
If a function is continuous in an interval $[a,b]$, then by definition of continuity, Definition 24.12, a small change, δ, in t can only cause a small change, ε in $h(t)$, therefore, in the finite interval $[a,b]$ where

$$\max_{\substack{a \leq t \leq b \\ a \leq t_0 \leq b}} |t - t_0| = b - a \tag{24.44}$$

$|t - t_0|$ is bounded, so $\exists M : M < \infty$ so that

$$\max_{\substack{a \leq t \leq b \\ a \leq t_0 \leq b}} |h(t) - h(t_0)| < M \tag{24.45}$$

Therefore, based on Definition 24.18, $h(t)$ is bounded in interval $[a,b]$.

□

Definition 24.19 (Continuity Class (Degree of Continuity)). *A function $h(t)$ is continuous with degree 1 if it is continuous and its first derivative is continuous. First degree continuity is denoted as $\mathfrak{C}^1$.*

If function $h(t)$ is continuous and all its derivatives up to the nth derivative are continuous, then the function is a $\mathfrak{C}^n$ continuous function. N.B., If a function is up to n derivatives, then it is at least of continuity class $\mathfrak{C}^{n-1}$, namely all the derivatives up to and including degree $n - 1$ are also continuous.

A class $\mathfrak{C}^0$ function is simply continuous.

Definition 24.20 (Smoothness). *A function $h(t)$ is smooth if it is continuous and it has up to order ∞ continuous derivatives, namely it is of class $\mathfrak{C}^\infty$ continuous. N.B. All analytic functions are smooth, but since there is a requirement that analytic*

*functions be determined completely by a power series, not all smooth functions are
analytic. See Definition 24.34.*

Definition 24.21 (Piecewise Continuity). *A function $h(t)$ is piecewise continuous
if it is continuous at all points in an interval except a finite number of discontinuities
in that interval.*

Definition 24.22 (Piecewise Smoothness). *A function $h(t)$ is piecewise smooth if it
is piecewise continuous and its derivatives are piecewise continuous.*

24.1.4 Convexity and Concavity of Functions

Continuity, smoothness, and boundedness have secondary connotations, as we will
see. When we speak about these concepts, it also makes sense to further qualify the
shape of functions in the form of *convex* or *concave* functions, since these shapes
will play a role in their boundedness and convergence aspects. In this section, *convexity* and *concavity* are defined and analyzed in some detail.

Definition 24.23 (Convex Function). *A real-valued function, $h(t)$, which is continuous in the closed interval $\{t : t \in [a,b]\}$, is said to be convex if*

$$h(\alpha t_0 + (1-\alpha)t_2) \le \alpha h(t_0) + (1-\alpha)h(t_2) \;\; \forall t_0, t_2 \in [a,b] \; and \; \forall \alpha \in [0,1] \quad (24.46)$$

Note that a function which has a *non-negative second derivative* over the whole
interval, $[a,b]$, is *convex* in that interval. Figure 24.7 depicts as example of a *convex
function* $(h(t) = -\ln(t))$, with a graphic representation of the inequality in Equation 24.46.

Definition 24.24 (Strictly Convex Function). *A strictly convex function is defined
by Definition 24.23, such that the inequality in Equation 24.46 is changed to a strict
inequality, not allowing equality, except when $\{\alpha = 0 \vee \alpha = 1\}$.*

Note that a function which has a *positive second derivative* over the whole interval, $[a,b]$, is *strictly convex* in that interval.

Theorem 24.3 (Convex Function). *A real-valued function, $h(t)$, which is $\mathfrak{C}^1$ continuous in the closed interval $\{t : t \in [a,b]\}$, is said to be convex if it has a non-
negative second derivative,*

$$\frac{d^2h(t)}{dt^2} \ge 0 \qquad\qquad\qquad\qquad (24.47)$$

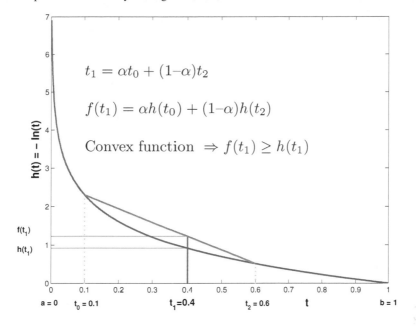

Fig. 24.7: Graphic representation of the inequality in Equation 24.46. $h(t)$ is the convex function of interest and $f(t)$ describes the line in this figure.

Proof.

Let us take two points, $t_0, t_2 \in [a, b]$ and the parameter, $\alpha \in [0, 1]$, as in Definition 24.23 such that $t_0 \leq t_2$. Furthermore, define an intermediate point, t_1, in terms of t_0 and t_1, according to Definition 24.23, based on parameter α,

$$t_1 = \alpha t_0 + (1 - \alpha)t_2 \tag{24.48}$$

Therefore, by Equation 24.46, h(t) is convex if and only if,

$$h(t_1) \leq \alpha h(t_0) + (1 - \alpha)h(t_2) \ \forall \ \alpha \in [0, 1] \tag{24.49}$$

Note that we may write $h(t_1)$ as follows,

$$h(t_1) = \alpha h(t_1) + (1 - \alpha)h(t_1) \tag{24.50}$$

Therefore, substituting from Equation 24.50 into Equation 24.49, we have

$$\alpha h(t_1) + (1 - \alpha)h(t_1) \leq \alpha h(t_0) + (1 - \alpha)h(t_2) \tag{24.51}$$

The inequality in Equation 24.51 may be rewritten as follows,

$$\alpha \left(h(t_1) - h(t_0) \right) \leq (1 - \alpha) \left(h(t_2) - h(t_1) \right) \tag{24.52}$$

If we prove the identity in Equation 24.52 from the assumption of the non-negativeness of the second derivative of $h(t)$, then we have proven the proposition in this theorem.

If we consider the two intervals, $t_l \in [t_0, t_1]$ and $t_r \in [t_1, t_2]$, on the left side and right side of Equation 24.52, then we may write the following two equations based on the mean value theorem (Theorem 24.10),

$$\left. \frac{dh(t)}{dt} \right|_{t=t_l} = \frac{h(t_1) - h(t_0)}{t_1 - t_0} \tag{24.53}$$

$$\left. \frac{dh(t)}{dt} \right|_{t=t_r} = \frac{h(t_2) - h(t_1)}{t_2 - t_1} \tag{24.54}$$

where, by definition,

$$a \le t_0 \le t_l \le t_1 \le t_r \le t_2 \le b \tag{24.55}$$

Based on the main assumption of this theorem, the second derivative of $h(t)$ is non-negative for all $t \in [a, b]$,

$$\frac{d^2 h(t)}{dt^2} \ge 0 \,\forall\, t \in \{a, b\} \tag{24.56}$$

Therefore, based on the inequalities of Equations 24.55 and 24.56,

$$\left. \frac{dh(t)}{dt} \right|_{t=t_l} \le \left. \frac{dh(t)}{dt} \right|_{t=t_r} \tag{24.57}$$

By Equation 24.53, the left hand side of Equation 24.52 may be written as follows,

$$\alpha \left(h(t_1) - h(t_0) \right) = \alpha \left. \frac{dh(t)}{dt} \right|_{t=t_l} (t_1 - t_0) \tag{24.58}$$

Due to the inequality in Equation 24.57, we may write the following inequality for the right hand side of Equation 24.58,

$$\alpha \left. \frac{dh(t)}{dt} \right|_{t=t_l} (t_1 - t_0) \le \alpha \left. \frac{dh(t)}{dt} \right|_{t=t_r} (t_1 - t_0) \tag{24.59}$$

Now let us subtract $\alpha t_0 + (1 - \alpha) t_1$ from both sides of Equation 24.48 to develop an expression for $\alpha(t_1 - t_0)$ which shows up in the right side of Equation 24.59,

$$\alpha(t_1 - t_0) = (1 - \alpha)(t_2 - t_1) \tag{24.60}$$

Substituting for $\alpha(t_1 - t_0)$ from Equation 24.60 into the right hand side of the inequality in Equation 24.59 and for the left hand side of the inequality from Equation 24.58, we may write the following inequality,

$$\alpha \left(h(t_1) - h(t_0) \right) \leq (1 - \alpha)(t_2 - t_1) \left. \frac{dh(t)}{dt} \right|_{t=t_r} \tag{24.61}$$

Using Equation 24.54, we may rewrite the above inequality, only in terms of values of $h(t)$ and α,

$$\alpha \left(h(t_1) - h(t_0) \right) \leq (1 - \alpha) \left(h(t_2) - h(t_1) \right) \tag{24.62}$$

The inequality in Equation 24.62 is equivalent to the proposition of Equation 24.52, which has been shown by using the non-negativeness of the second derivative of $h(t)$ in the closed interval, $\{t : t \in [a,b]\}$.

$\square$

It is easy to show that $h(t)$ will be *strictly convex* if and only if, all the conditions in Theorem 24.3 would hold in addition to the restriction that the inequality in Equation 24.56 would become a strict inequality, making the second derivative of $h(t)$ positive definite.

Definition 24.25 (Concave Function). *A real-valued function, $h(t)$, which is continuous in the closed interval $\{t : t \in [a,b]\}$, is said to be concave if*

$$h(\alpha t_0 + (1 - \alpha)t_2) \geq \alpha h(t_0) + (1 - \alpha)h(t_2) \quad \forall t_0, t_2 \in \lfloor a,b \rfloor \text{ and } \forall \alpha \in [0,1] \tag{24.63}$$

Note that a function, which has a *non-positive second derivative* over the whole interval, $[a,b]$, is *concave* in that interval. Figure 24.8 depicts as example of a *concave function* $(h(t) = \ln(t))$, with a graphic representation of the inequality in Equation 24.63.

Definition 24.26 (Strictly Concave Function). *A strictly concave function is defined by Definition 24.25, such that the inequality in Equation 24.63 is changed to a strict inequality, not allowing equality, except when $\{\alpha = 0 \vee \alpha = 1\}$.*

Note that a function, which has a *negative second derivative* over the whole interval, $[a,b]$, is *strictly concave* in that interval.

24.1.5 Odd, Even and Periodic Functions

There is a global perspective which categorizes functions in their shape with respect to some reference system. The, so called, odd and even nature of functions may be used to simplify their integration and related operations. In addition to these two concepts, the periodicity of a function becomes important in the same sense. In this section, we will provide the basic definitions in these regards.

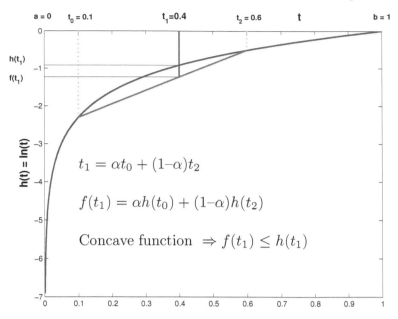

Fig. 24.8: Graphic representation of the inequality in Equation 24.63. $h(t)$ is the concave function of interest and $f(t)$ describes the line in this figure.

Definition 24.27 (Odd Functions). *A function $h(t)$ is odd if $h(-t) = -h(t) \; \forall \; t$.*

If $h(t)$ is periodic with period 2π, then oddness implies that,

$$\int_{-\pi}^{\pi} h(t)dt = 0 \tag{24.64}$$

Example 24.4 (Some Odd Functions).
Some examples of odd functions are,

$$h(t) = \sin(t) \tag{24.65}$$

and

$$h(t) = t^3 \tag{24.66}$$

Definition 24.28 (Even Functions). *A function $h(t)$ is even if $h(-t) = h(t) \; \forall \; t$.*

If $h(t)$ is periodic with period 2π, then oddness implies that,

$$\int_{-\pi}^{\pi} h(t)dt = 2 \int_{0}^{\pi} h(t)dt \tag{24.67}$$

Example 24.5 (Some Even Functions).
Some examples of even functions are,

$$h(t) = \cos(t) \tag{24.68}$$

and

$$h(t) = t^2. \tag{24.69}$$

Property 24.7 (Odd and Even Functions). *Here are some properties related to odd and even functions,*

- *odd function $\times$ odd function = odd function*
- *odd function $\times$ even function = odd function*
- *even function $\times$ even function = even function*

Definition 24.29 (Periodic Function). *Let s be a variables in the Domain $\mathscr{D} \subset \mathbb{C}$. Also, let $\tilde{\lambda}$ be a constant where $\tilde{\lambda} \neq 0$ and such that $s + \tilde{\lambda} \in \mathscr{D}$. A function $H(s)$ is said to be periodic with period $\tilde{\lambda}$ if $H(s) = H(s + \tilde{\lambda}) \forall s \in \mathscr{D}$.*

Definition 24.30 (Periodic Extension of a Function). *Let $h(t), t \in \mathbb{R}$ be defined in an interval $t_0 \leq t < t_0 + \tilde{\lambda}$, then the periodic extension of $h(t)$, $\tilde{h}(\tau)$ is defined as a function defined in $-\infty < \tau < \infty$ where $\tilde{h}(t + n\tilde{\lambda}) = h(t), t_0 \leq t < t_0 + \tilde{\lambda}; -\infty < n < \infty$.*

This is essentially the collection of $h(t)$ and its copies which have been shifted by $n\tilde{\lambda}, n = 1, 2, \cdots$ to the right and to the left. The periodic extension is a useful notion for doing manipulations on functions where the function is expected to be periodic, such as the Fourier Series expansion [18] – see Section 24.6.

24.1.6 Differentiation

We have reviewed many concepts relating to functions. In this section, differentiation will be covered, especially related to functions of complex variables.

Definition 24.31 (Differentiation of Functions of Complex Variables). *Let $H(s)$ be a single-valued function of s : $s \in \mathscr{D} \subset \mathbb{C}$. Let s_0 be any fixed point in domain $\mathscr{D}$. Then, $H(s)$ is said to have a derivative at point s_0 if the limit in Equation 24.70 exists.*

$$\left. \frac{dH(s)}{ds} \right|_{s=s_0} = \lim_{s \to s_0} \frac{H(s) - H(s_0)}{s - s_0} \tag{24.70}$$

Property 24.8 (Differentiation of Functions of Complex Variables). *The formal rules for the differentiation of functions of complex variables are similar to those* *for functions of real variables. If $s \in \mathbb{C}$, c is a constant such that $c \in \mathbb{C}$, and $G(s)$ and $H(s)$ are functions of s defined in $\mathbb{C}$, then,*

$$\frac{dc}{ds} = 0 \tag{24.71}$$

$$\frac{ds}{ds} = 1 \tag{24.72}$$

$$\frac{d\left[H(s) \pm G(s)\right]}{ds} = \frac{dH(s)}{ds} \pm \frac{dG(s)}{ds} \tag{24.73}$$

$$\frac{d\left[H(s).G(s)\right]}{ds} = H(s)\frac{dG(s)}{ds} + G(s)\frac{dH(s)}{ds} \tag{24.74}$$

and assuming $G(s) \neq 0$,

$$\frac{d\left[\frac{H(s)}{G(s)}\right]}{ds} = \frac{G(s)\frac{dH(s)}{ds} - H(s)\frac{dG(s)}{ds}}{G(s)^2} \tag{24.75}$$

Also, the chain rule still holds in the complex domain, namely,

$$w \overset{\Delta}{=} H(\eta)$$
$$= H\left(G(s)\right)$$

Therefore,

$$\frac{dw}{ds} = \frac{dw}{d\eta}\frac{d\eta}{ds} \tag{24.76}$$

Definition 24.32 (Partial Differentiation Notation). *If $u(\xi_1, \xi_2, \cdots, \xi_n)$ is a function of n variables, then the following shorthand derivative notation is used,*
Partial Derivatives,

$$u_{\xi_i} \overset{\Delta}{=} \frac{\partial u(\xi_1, \xi_2, \cdots, \xi_n)}{\partial \xi_i} \tag{24.77}$$

Partial Second Derivatives,

$$u_{\xi_i \xi_j} \overset{\Delta}{=} \frac{\partial^2 u(\xi_1, \xi_2, \cdots, \xi_n)}{\partial \xi_i \partial \xi_j} \tag{24.78}$$

Laplacian,

$$\nabla^2 u \overset{\Delta}{=} \sum_{i=1}^{n} u_{\xi_i \xi_i} \tag{24.79}$$

$$= \sum_{i=1}^{n} \frac{\partial^2 u(\xi_1, \xi_2, \cdots, \xi_n)}{\partial \xi_i^2} \tag{24.80}$$

Definition 24.33 (Laplace's Equation). *Laplace's equation states that*

$$\nabla^2 u(\xi_1, \xi_2, \cdots, \xi_n) = 0 \tag{24.81}$$

where $\nabla^2 u(\xi_1, \xi_2, \cdots, \xi_n)$ is defined by Equation 24.79. It describes many states of nature including steady-state heat conduction and potentials such as gravitation and electric potential.

24.1.7 Analyticity

The *analyticity* of functions is an important concept which shows up in the statement of the *residue theorem* and many other theorems. The residue theorem is a fundamental theorem needed for the formulation of the inverse of integral transforms. This section is devoted to analyticity and related theorems.

Definition 24.34 (Analytic Function). *A function of a complex variable, $H(s)$ where $s \in \mathcal{D} \subset \mathbb{C}$, is said to be analytic in an interval $[a, b]$ if it is single valued in that domain (only has one value for each point in the domain) and if the first derivative, $\frac{dH(s)}{ds}$, exists at every point of the domain. In addition, an analytic function may be completely described in terms of power series in a Domain $\mathcal{D} \subset \mathbb{C}$. See Definition 24.42.*

Analytic functions may alternatively be called Holomorphic or Regular. Based on a consequence of the Cauchy Integral Formula, an analytic function is class $\mathbb{C}^\infty$ continuous – see Definition 24.19 and Theorem 24.11.

Definition 24.35 (Pointwise Analyticity of Functions). *A function $H(s)$ is said to be analytic at point s_0 if $H(s)$ is analytic in neighborhood of s_0.*

Theorem 24.4 (Relation between existence of derivative and continuity). *If a function of a complex variable, $H(s)$ where $s \in \mathbb{C}$, has a derivative at $s_0 \in \mathbb{C}$, then it is continuous at s_0. All analytic functions are continuous.*

Proof.

$$\lim_{s \to s_0} [H(s) - H(s_0)] = \lim_{s \to s_0} (s - s_0) \lim_{s \to s_0} \left[\frac{H(s) - H(s_0)}{(s - s_0)} \right]$$

$$= 0 \times \left. \frac{dH(s)}{ds} \right|_{s=s_0}$$

$$= 0 \tag{24.82}$$

Hence,

$$\lim_{s \to s_0} H(s) = \lim_{s \to s_0} [H(s_0) + (H(s) - H(s_0))]$$

$$= H(s_0) \tag{24.83}$$

Equation 24.83 is just the definition of continuity, see Equation 24.33 in Definition 24.12.

□

However, not all continuous functions are analytic. i.e. Continuity does not necessarily imply differentiability. For example, take $H(s) = |s|^2$ which is continuous for all $s \in \mathbb{C}$.

$$G(s) \triangleq \frac{H(s) - H(s_0)}{s - s_0} \tag{24.84}$$

$$= \frac{|s|^2 - |s_0|^2}{s - s_0} \quad \forall \ (s \neq s_0)$$

$$= \frac{s\bar{s} - s_0\bar{s_0}}{s - s_0}$$

$$= \bar{s} + s_0 \left[\frac{\bar{s} - \bar{s_0}}{s - s_0} \right] \tag{24.85}$$

Now let us do a change of variable and use polar coordinates for convenience,

$$\rho e^{i\theta} \equiv s - s_0$$

$$= \rho (\cos(\theta) + i \sin(\theta)) \tag{24.86}$$

Then,

$$G(s) = \bar{s} + \frac{s_0 \rho e^{-i\theta}}{\rho e^{i\theta}}$$

$$= \bar{s} + s_0 e^{-i(2\theta)} \tag{24.87}$$

From Equations 24.84 and 24.87,

$$G(s) = \frac{H(s) - H(s_0)}{s - s_0}$$

$$= \bar{s} + s_0 [\cos(2\theta) - i \sin(2\theta)] \tag{24.88}$$

Consider two different ways $s \to s_0$ in the complex plane $\mathbb{C}$,

1. $s \to s_0$ along $\theta = 0 \implies G(s) = \overline{s_0} + s_0$
2. $s \to s_0$ along $\theta = \frac{\pi}{4}$ rad. $\implies G(s) = \overline{s_0} - s_0$

Therefore, in general the limit, hence the derivative, does not exist unless $s_0 = 0$ where $G(s) = \overline{s_0} = 0$. This implies that although $H(s) = |s|^2$ exists everywhere and hence is continuous, it is not analytic since its derivative does not exist except at $s = 0$.

Definition 24.36 (Cauchy-Riemann Conditions). *If $H(s)$ may be written in its real and imaginary components, namely,*

$$H(s) \equiv U(\sigma, \omega) + iV(\sigma, \omega) \tag{24.89}$$

Then, the Cauchy-Riemann conditions dictate that,

$$U_\sigma = V_\omega \tag{24.90}$$
$$U_\omega = -V_\sigma \tag{24.91}$$

Theorem 24.5 (Cauchy-Riemann Theorem). *A necessary condition for a function, $H(s) = U(\sigma, \omega) + iV(\sigma, \omega)$ to be analytic in a domain $\mathscr{D} \subset \mathbb{C}$ is that the four partial derivatives, $U_\sigma, U_\omega, V_\sigma$, and V_ω exist and satisfy the Cauchy-Riemann conditions (see Definition 24.36) at each point in $\mathscr{D}$.*

Proof.
Let $s_0 = \sigma_0 + i\omega_0$ be any fixed point in domain $\mathscr{D}$. Then,

$$\left.\frac{dH(s)}{ds}\right|_{s=s_0} = \lim_{s \to s_0} \frac{H(s) - H(s_0)}{s - s_0}$$
$$= \lim_{s \to s_0} \frac{\Delta H(s)}{\Delta s} \tag{24.92}$$

Consider two paths long which $s \to s_0$,

1. Let $s \to s_0$ along a line parallel to the $\mathbb{R}$-axis, i.e. along $\omega = \omega_0$. Therefore,

$$s - s_0 = \sigma + i\omega_0 - \sigma_0 - i\omega_0$$
$$= \sigma - \sigma_0$$
$$= \Delta\sigma$$

$$\left.\frac{dH(s)}{ds}\right|_{\substack{s \to s_0 \\ \omega = \omega_0}} = \lim_{\Delta\sigma \to 0} \frac{U(\sigma_0 + \Delta\sigma, \omega_0) - U(\sigma_0, \omega_0)}{\Delta\sigma} +$$
$$i \lim_{\Delta\sigma \to 0} \frac{V(\sigma_0 + \Delta\sigma, \omega_0) - V(\sigma_0, \omega_0)}{\Delta\sigma}$$
$$= U_\sigma(\sigma_0, \omega_0) + iV_\sigma(\sigma_0, \omega_0) \tag{24.93}$$

2. Let $s \to s_0$ along a line parallel to the $\mathbb{I}$-axis, i.e. along $\sigma = \sigma_0$. Therefore,

$$
\begin{aligned}
s - s_0 &= \sigma_0 + i\omega - \sigma_0 - i\omega_0 \\
&= i(\omega - \omega_0) \\
&= i\Delta\omega
\end{aligned}
$$

$$
\begin{aligned}
\left.\frac{dH(s)}{ds}\right|_{\substack{s \to s_0 \\ \sigma = \sigma_0}} &= \lim_{\Delta\omega \to 0} \frac{U(\sigma_0, \omega_0 + \Delta\omega) - U(\sigma_0, \omega_0)}{i\Delta\omega} + \\
&\quad i \lim_{\Delta\omega \to 0} \frac{V(\sigma_0, \omega_0 + \Delta\omega) - V(\sigma_0, \omega_0)}{i\Delta\omega} \\
&= \frac{U_\omega(\sigma_0, \omega_0)}{i} + V_\omega(\sigma_0, \omega_0) \\
&= -iU_\omega(\sigma_0, \omega_0) + V_\sigma(\sigma_0, \omega_0)
\end{aligned}
\tag{24.94}
$$

If $\left.\frac{dH(s)}{ds}\right|_{s \to s_0}$ exists, the expressions arrived at Equations 24.93 and 24.94 should be identical. Equating the real and imaginary parts of Equations 24.93 and 24.94 we get,

$$
\begin{aligned}
U_\sigma(\sigma_0, \omega_0) &= V_\omega(\sigma_0, \omega_0) \\
U_\omega(\sigma_0, \omega_0) &= -V_\sigma(\sigma_0, \omega_0)
\end{aligned}
$$

which are the Cauchy-Riemann conditions stated in Definition 24.36.

$\square$

Problem 24.2 shows that Theorem 24.5 only supplies necessary condition and not sufficient for analyticity. See Theorem 24.7 for a necessary and sufficient statement of the theorem.

Theorem 24.6 (Alternate Cauchy-Riemann Theorem). *Another way of stating Theorem 24.5 is that a necessary condition for a function, $H(s) = U(\sigma, \omega) + iV(\sigma, \omega)$ to be analytic in a domain $\mathscr{D} \subset \mathbb{C}$ is that the Laplace's Equation (see Equation 24.81) be satisfied for both Real and Imaginary parts of $H(s)$, namely,*

$$
\nabla^2 U(\sigma, \omega) = 0 \tag{24.95}
$$
$$
\nabla^2 V(\sigma, \omega) = 0 \tag{24.96}
$$

Proof.
Let us consider the Cauchy-Riemann conditions of Equations 24.90 and 24.91. If we take $\frac{\partial}{\partial\sigma}$ of Equation 24.90 and $\frac{\partial}{\partial\omega}$ of Equation 24.91 and add the two resulting Equations together we get,

$$
U_{\sigma\sigma} + U_{\omega\omega} = V_{\omega\sigma} - V_{\sigma\omega} \tag{24.97}
$$

or

$$
\nabla^2 U(\sigma, \omega) = 0 \tag{24.98}
$$

Similarly, if we take $\frac{\partial}{\partial \omega}$ of Equation 24.90 and $\frac{\partial}{\partial \sigma}$ of Equation 24.91 and add the two resulting Equations together we get,

$$U_{\sigma\omega} - U_{\omega\sigma} = V_{\sigma\sigma} + V_{\omega\omega} \tag{24.99}$$

or

$$\nabla^2 V(\sigma, \omega) = 0 \tag{24.100}$$

Equations 24.98 and 24.100 together with the proof of Theorem 24.5 prove Theorem 24.6.

$\square$

Theorem 24.7 (Necessary and Sufficient Cauchy-Riemann Theorem (General Analyticity)). *A necessary and sufficient condition for a function, $H(s) = U(\sigma, \omega) + iV(\sigma, \omega)$ to be analytic in a domain $\mathcal{D} \subset \mathbb{C}$ is that the four partial derivatives, $U_\sigma, U_\omega, V_\sigma,$ and V_ω exist, be continuous in domain $\mathcal{D}$, and satisfy the Cauchy-Riemann conditions (see Definition 24.36) at each point in $\mathcal{D}$.*

Proof.

$$\frac{H(s + \Delta s) - H(s)}{\Delta s} = \frac{U(\sigma + \Delta\sigma, \omega + \Delta\omega) + iV(\sigma + \Delta\sigma, \omega + \Delta\omega)}{\Delta\sigma + i\Delta\omega} -$$

$$\frac{U(\sigma, \omega) + iV(\sigma, \omega)}{\Delta\sigma + i\Delta\omega}$$

$$= \frac{U(\sigma + \Delta\sigma, \omega + \Delta\omega) - U(\sigma, \omega)}{\Delta\sigma + i\Delta\omega} +$$

$$\frac{i[V(\sigma + \Delta\sigma, \omega + \Delta\omega) - V(\sigma, \omega)]}{\Delta\sigma + i\Delta\omega} \tag{24.101}$$

Now,

$$U(\sigma + \Delta\sigma, \omega + \Delta\omega) - U(\sigma, \omega) = U(\sigma + \Delta\sigma, \omega + \Delta\omega) -$$

$$U(\sigma + \Delta\sigma, \omega) + U(\sigma + \Delta\sigma, \omega) - U(\sigma, \omega)$$

By the virtue of the *law of mean* (*mean value theorem*), Theorem 24.10,

$$U(\sigma + \Delta\sigma, \omega) - U(\sigma, \omega) = U_\sigma(\sigma', \omega)\Delta\sigma \tag{24.102}$$

$$\text{where } \sigma < \sigma' < \sigma + \Delta\sigma$$

and

$$U(\sigma + \Delta\sigma, \omega + \Delta\omega) - U(\sigma + \Delta\sigma, \omega) = U_\omega(\sigma + \Delta\sigma, \omega')\Delta\omega \tag{24.103}$$

$$\text{where } \omega < \omega' < \omega + \Delta\omega$$

$$U(\sigma + \Delta\sigma \;,\; \omega + \Delta\omega) - U(\sigma, \omega)$$
$$= U_\sigma(\sigma', \omega)\Delta\sigma + U_\omega(\sigma + \Delta\sigma, \omega')\Delta\omega \qquad (24.104)$$
$$\text{where } \begin{cases} \sigma < \sigma' < \sigma + \Delta\sigma \\ \omega < \omega' < \omega + \Delta\omega \end{cases}$$

Equation 24.104 is actually the extension of the *Mean Value Theorem* for two variables.

Similarly,

$$V(\sigma + \Delta\sigma \;,\; \omega + \Delta\omega) - V(\sigma, \omega)$$
$$= V_\sigma(\sigma'', \omega)\Delta\sigma + V_\omega(\sigma + \Delta\sigma, \omega'')\Delta\omega \qquad (24.105)$$
$$\text{where } \begin{cases} \sigma < \sigma'' < \sigma + \Delta\sigma \\ \omega < \omega'' < \omega + \Delta\omega \end{cases}$$

Let us define the following,

$$\varepsilon_1 \overset{\Delta}{=} U_\sigma(\sigma', \omega) - U_\sigma(\sigma, \omega)$$
$$\varepsilon_2 \overset{\Delta}{=} U_\omega(\sigma + \Delta\sigma, \omega') - U_\omega(\sigma, \omega)$$
$$\varepsilon_3 \overset{\Delta}{=} V_\sigma(\sigma'', \omega) - V_\sigma(\sigma, \omega)$$
$$\varepsilon_4 \overset{\Delta}{=} V_\omega(\sigma + \Delta\sigma, \omega'') - V_\omega(\sigma, \omega)$$

$$(24.106)$$

Using the assumption of continuity of $U_\sigma, U_\omega, V_\sigma$, and V_ω at each point in Domain $\mathscr{D}$,

$$\lim_{\substack{\Delta s \to 0 \\ (\Delta\sigma \to 0, \Delta\omega \to 0)}} \varepsilon_i = 0 \qquad i = \{1, 2, 3, 4\} \qquad (24.107)$$

Using ε_i in Equations 24.104 and 24.105,

$$U(\sigma + \Delta\sigma, \omega + \Delta\omega) - U(\sigma, \omega) = U_\sigma(\sigma, \omega)\Delta\sigma + U_\omega(\sigma, \omega)\Delta\omega +$$
$$\varepsilon_1\Delta\sigma + \varepsilon_2\Delta\omega \qquad (24.108)$$
$$V(\sigma + \Delta\sigma, \omega + \Delta\omega) - V(\sigma, \omega) = V_\sigma(\sigma, \omega)\Delta\sigma + V_\omega(\sigma, \omega)\Delta\omega +$$
$$\varepsilon_3\Delta\sigma + \varepsilon_4\Delta\omega \qquad (24.109)$$

Using Equations 24.108 and 24.108 in 24.101,

$$\frac{H(s + \Delta s)}{\Delta s} = \frac{U_\sigma\Delta\sigma + iV_\omega\Delta\omega + iV_\sigma\Delta\sigma + U_\omega\Delta\omega}{\Delta\sigma + i\Delta\omega} +$$
$$\frac{\varepsilon_1\Delta\sigma + \varepsilon_2\Delta\omega + \varepsilon_3\Delta\sigma + \varepsilon_4\Delta\omega}{\Delta\sigma + i\Delta\omega}$$

and using the Cauchy-Riemann conditions, $U_\sigma = V_\omega$ and $V_\sigma = -U_\omega$,

$$\frac{H(s+\Delta s)}{\Delta s} = \frac{U_\sigma(\Delta\sigma + i\Delta\omega)}{(\Delta\sigma + i\Delta\omega)} + i\frac{V_\sigma(\Delta\sigma + i\Delta\omega)}{(\Delta\sigma + i\Delta\omega)} +$$

$$[\varepsilon_1 + i\varepsilon_3]\frac{\Delta\sigma}{\Delta s} + [\varepsilon_2 + i\varepsilon_4]\frac{\Delta\omega}{\Delta s}$$

$$= U_\sigma(\sigma,\omega) + iV_\sigma(\sigma,\omega) + (\varepsilon_1 + i\varepsilon_3)\frac{\Delta\sigma}{\Delta s} + (\varepsilon_2 + i\varepsilon_4)\frac{\Delta\omega}{\Delta s} \quad (24.110)$$

Taking the $\lim\limits_{s\to 0}$ of Equation 24.110,

$$\frac{dH(s)}{ds} = \lim_{\Delta s\to 0}\frac{H(s+\Delta s) - H(s)}{\Delta s}$$

$$= U_\sigma(\sigma,\omega) + iV_\sigma(\sigma,\omega) +$$

$$\lim_{\Delta s\to 0}[\varepsilon_1 + i\varepsilon_3]\frac{\Delta\sigma}{\Delta s} +$$

$$\lim_{\Delta s\to 0}[\varepsilon_2 + i\varepsilon_4]\frac{\Delta\omega}{\Delta s} \quad (24.111)$$

Using the limits of ε_i from Equation 24.107 and the fact that, $\left|\frac{\Delta\sigma}{\Delta s}\right| \le 1$ and $\left|\frac{\Delta\omega}{\Delta s}\right| \le 1$, the limits on the right hand side of Equation 24.111 vanish and we are left with,

$$\frac{dH(s)}{ds} = U_\sigma(\sigma,\omega) + iV_\sigma(\sigma,\omega) \quad (24.112)$$

Therefore, since $U_\sigma(\sigma,\omega)$ and $V_\sigma(\sigma,\omega)$ exist in all points in Domain $\mathscr{D}$, then based on Equation 24.112, $\frac{dH(s)}{ds}$ must exits in all points in Domain $\mathscr{D}$ and by the Cauchy-Riemann conditions, we may further establish that,

$$\frac{dH(s)}{ds} = U_\sigma(\sigma,\omega) + iV_\sigma(\sigma,\omega) \quad (24.113)$$

$$= V_\omega(\sigma,\omega) - iU_\omega(\sigma,\omega) \quad (24.114)$$

$\square$

Definition 24.37 (Harmonic Conjugate). *If $U(\sigma,\omega)$ and $V(\sigma,\omega)$ are harmonic functions of the variables $\sigma,\omega : s = \sigma + i\omega \in \mathscr{D} \subset \mathbb{C}$, and $H(s) = U(\sigma,\omega) + iV(\sigma,\omega)$ is an analytic function in Domain $\mathscr{D}$, $V(\sigma,\omega)$ is the harmonic conjugate of $U(\sigma,\omega)$. See Problem 24.3.*

Theorem 24.8 (Analyticity of the Exponential Function). *The Exponential function, e^s is analytic.*

Proof.

$$H(s) = e^s = \underbrace{e^\sigma\cos(\omega)}_{U} + \underbrace{ie^\sigma\sin(\omega)}_{V} \quad (24.115)$$

Let us write the four partial derivatives of $H(s)$,

$$U_\sigma = e^\sigma \cos(\omega)$$
$$U_\omega = -e^\sigma \sin(\omega)$$
$$V_\sigma = e^\sigma \sin(\omega)$$
$$V_\omega = e^\sigma \cos(\omega)$$

All four partial derivatives are continuous and are defined in the $\mathbb{C}$ plane. Also, the Cauchy-Riemann conditions are satisfied. Therefore, $H(s) = e^s$ is analytic everywhere.   □

Theorem 24.9 (Analyticity of the Trigonometric Functions). *Trigonometric Functions of complex variable s are defined in terms of the Exponential Function, e^s, as follows,*

$$\sin(s) \triangleq \frac{e^{is} - e^{-is}}{2i}$$

$$\cos(s) \triangleq \frac{e^{is} + e^{-is}}{2}$$

$$\csc(s) \triangleq \frac{1}{\sin(s)}$$

$$\sec(s) \triangleq \frac{1}{\cos(s)}$$

$$\tan(s) \triangleq \frac{\sin(s)}{\cos(s)}$$

$$\cot(s) \triangleq \frac{\cos(s)}{\sin(s)}$$

All these functions are analytic everywhere in the $\mathbb{C}$ plane. $\sin(s)$ and $\cos(s)$ are periodic with period 2π.

Proof.
The proof is quite simple and since these functions are defined in terms of e^s which was proven to be analytic everywhere by Theorem 24.8, the same methodology as in the proof of Theorem 24.8 may be followed.   □

24.1.8 Integration

In this section, we will look at some ideas related to the integrals of functions. Naturally, these concepts are quite central to the idea of integral transforms. As we will see, the domain of integration and, in the case of integration on the complex plane, the contour of integration are essential to the formulation of an integral transform and its inverse. As we shall see in the statement of the residue theorem, the choice of a contour of integration highly simplifies the computation of otherwise very complex integrals.

Definition 24.38 (Absolutely Integrable). *A function $h(t)$ is absolutely integrable in a closed interval $[a,b]$ if it is piecewise continuous in that interval, if $\exists\, M : M < \infty$, and if*

$$\int_a^b |h(t)|\, dt < M \tag{24.116}$$

or in words, if the integral in Equation 24.116 is bounded.

Definition 24.39 (Riemann Integral (Definite Integral)). *Consider a function $h(t)$ defined and continuous in the closed interval, $a \le t \le b$, then the Riemann Integral,*

$$R \overset{\Delta}{=} \int_a^b h(t)\, dt \tag{24.117}$$

is a number defined by the following process:

1. *Choose an integer $n \ge 1$ and arrange the closed interval $[a,b]$ as follows,*

$$[a,b] = \{a = t_0, t_1, t_2, \cdots, t_{n-1}, t_n = b\} \tag{24.118}$$

   Such that $t_0 < t_1 < t_2 < \cdots < t_n$ and let $\Delta t_i \overset{\Delta}{=} t_i - t_{i-1}$ where $i = 1, 2, \cdots, n$
2. *In each subinterval choose an arbitrary point t_i' such that $t_{i-1} \le t_i' \le t_i$*
3. *Form the approximating sum $\sum_{i=1}^n h(t_i')\Delta t_i'$*
4. *Then,*

$$R = \lim_{\substack{n \to \infty \\ \Delta t_i \to 0}} \sum_{i=1}^n h(t_i')\Delta t_i = \int_a^b h(t)\, dt \tag{24.119}$$

Property 24.9 (Properties of the Riemann Integral (Definite Integral)).

1. *Continuous functions are always integrable.*
2. *Functions somewhere, but not everywhere, continuous*

   a. are integrable if they have finite numbers of points of discontinuity.
   b. maybe integrable if they have an ∞ number of discontinuities.

Lemma 24.1 (Riemann's Lemma). *Let $h(t)$ be piecewise continuous in the closed interval $a \le t \le b$, also assume that $h(t)$ has piecewise continuous derivatives in that interval. Then,*

$$\lim_{k \to \infty} \int_a^b h(t)\sin(kt)\, dt = \lim_{k \to \infty} \int_a^b h(t)\cos(kt)\, dt = 0 \tag{24.120}$$

Proof.
Here, we will prove the case for the sine function in Equation 24.120 and due to similarity the results may be generalized to include the case with the cos function.

1. Let us take the simplest case where $h(t)$ and $\frac{dh(t)}{dt}$ are both continuous in the closed interval $a \le t \le b$. Then, using integration by parts (i.e. $\int (f'g) = fg - \int (fg')$ where $f \equiv -\frac{\sin(kt)}{k}$ and $g \equiv h(t)$),

$$\int_a^b h(t)\sin(kt)dt = -\frac{\cos(kt)}{k}h(t)\bigg|_a^b + \int_a^b \frac{\cos(kt)}{k}\frac{dh(t)}{dt}dt \qquad (24.121)$$

Based on the assumption that $h(t)$ and $\frac{dh(t)}{dt}$ are continuous, they are bounded in that interval (see Property 24.6), namely,

$$|h(t)| \le M \qquad (24.122)$$

and

$$\left|\frac{dh(t)}{dt}\right| \le M' \qquad (24.123)$$

Combining Equations 24.121, 24.122, 24.123 and using the mean value theorem (Theorem 24.10), and taking the absolute values and the limits of both sides as $k \to \infty$,

$$\lim_{k \to \infty}\left|\int_a^b h(t)\sin(kt)dt\right| \le \lim_{k \to \infty} \frac{2M + M'(b-a)}{k} \qquad (24.124)$$

In Equation 24.124, the limit on the right hand side clearly approaches 0 for any arbitrary bounds a and b. Therefore, the left hand side of Equation 24.124 must be 0 which implies that,

$$\lim_{k \to \infty}\int_a^b h(t)\sin(kt)dt = 0 \qquad (24.125)$$

2. Now let us add a simple complication to the first case by allowing the derivative of $h(t)$ to have discontinuities at the end points of the interval. Namely, the interval of continuity of $\frac{dh(t)}{dt}$ is relaxed a bit to include only the open interval $a < t < b$, but the interval of continuity of the function $h(t)$ itself still remains as the closed interval of case 1. In this case, let us remove a neighborhood of the end points of the interval by a small amount ε. Therefore, the left side of Equation 24.121 may be written as follows,

$$\int_a^b h(t)\sin(kt)dt = \int_a^{a+\varepsilon} h(t)\sin(kt)dt$$

$$+ \int_{a+\varepsilon}^{b-\varepsilon} h(t)\sin(kt)dt$$

$$+ \int_{b-\varepsilon}^b h(t)\sin(kt)dt \tag{24.126}$$

Taking the absolute values of the two sides of Equation 24.126, $h(t)$ is still bounded in the interval $a \leq t \leq b$ due to continuity in this interval, so the absolute value of Equation 24.126 may be written as,

$$\left| \int_a^b h(t)\sin(kt)dt \right| \leq 2M\varepsilon + \left| \int_{a+\varepsilon}^{b-\varepsilon} h(t)\sin(kt)dt \right| \tag{24.127}$$

Consider the first term in the right hand side of Equation 24.127. ε may be chosen to be arbitrarily small so in the limit,

$$\lim_{\varepsilon \to 0} 2M\varepsilon = 0 \tag{24.128}$$

Let us define another closed interval, $a + \varepsilon \leq t \leq b - \varepsilon$. In that interval, the second term of Equation 24.127 is continuous in both $h(t)$ and $\frac{dh(t)}{dt}$. Thus, based on the proven hypothesis of case 1, it goes to 0 in the limit as $k \to \infty$. Therefore,

$$\lim_{k \to \infty} \left| \int_a^b h(t)\sin(kt)dt \right| \leq \lim_{\substack{k \to \infty \\ \varepsilon \to 0}} 2M\varepsilon + \lim_{\substack{k \to \infty \\ \varepsilon \to 0}} \left| \int_{a+\varepsilon}^{b-\varepsilon} h(t)\sin(kt)dt \right|$$

$$\leq 0 + 0 \tag{24.129}$$

which means that

$$\lim_{k \to \infty} \int_a^b h(t)\sin(kt)dt = 0 \tag{24.130}$$

3. Finally, we generalize to the case where there are a finite number of discontinuities of $\frac{dh(t)}{dt}$ in the interval $a \leq t \leq b$. In this case, assuming there are n such discontinuities, again, an ε neighborhood of the discontinuity may be isolated as we did in the last case. Then, we will be left with $n + 1$ closed intervals with continuity of $\frac{dh(t)}{dt}$ and n isolated integrals in a small interval, where the absolute values of each equal $M\varepsilon$. The integrals with continuity in the closed subset intervals between discontinuities go to 0 as $k \to 0$. The contribution of the discontinuities will be at most $nM\varepsilon$. Since n is finite, $\lim_{\varepsilon \to 0} nM\varepsilon = 0$. Hence,

$$\lim_{k \to \infty} \int_a^b h(t)\sin(kt)dt = 0 \tag{24.131}$$

for this case as well.

The last case is the most general case and completely proves the Lemma.

□

Theorem 24.10 (Mean Value Theorem (Law of Mean)). *Let $h(t)$ be a $\mathfrak{C}^1$ continuous function in the closed interval, $a \leq t \leq b$. Then, there is a point T in the open interval, $a < t < b$, such that,*

$$h(b) - h(a) = \frac{dh(t)}{dt}\bigg|_{t=T} (b-a) \tag{24.132}$$

Let $c = b - a$. Then $T = a + \alpha c$ where $0 < \alpha < 1$.

$$h(a+c) = h(a) + c \frac{dh(t)}{dt}\bigg|_{t=(a+\alpha c)} \qquad 0 < \alpha < 1 \tag{24.133}$$

Equation 24.104 denotes the extension of this theorem to a function of two variables. The derivation of this extension which has been shown within the body of the Proof of the general analyticity theorem (Theorem 24.7) may be extended to functions of any number of variables by following the same procedure.

Definition 24.40 (Simply Connected Domain). *A simply connected domain is one in which every closed contour, Γ in domain $\mathscr{D} \subset \mathbb{C}$, contains only points in $\mathscr{D}$. i.e. We can shrink any closed contour Γ in $\mathscr{D}$ down to a point without leaving the domain $\mathscr{D}$.*

Alternatively, a domain $\mathscr{D} \subset \mathbb{C}$ is simply connected if, given any two continuous curves, Γ_1 and Γ_2 with the same initial and terminal points, each curve is continuously deformable to the other.

Theorem 24.11 (Cauchy Integral Theorem).

- *Simply Connected Domains:*
  Let $H(s)$ be analytic in a simply connected Domain $\mathscr{D} \subset \mathbb{C}$ and let Γ be any closed contour in $\mathscr{D}$. Then,

$$\oint_{\Gamma} H(s)ds = 0 \tag{24.134}$$

- *Multiply Connected Domains:*
  Let $\Gamma, \Gamma_1, \Gamma_2, \cdots, \Gamma_n$ be simple closed contours, each described in the positive (counter clockwise) direction and such that each Γ_j is inside Γ and outside $\Gamma_k \forall j \neq k; j, k = \{1, 2, \cdots, n\}$. See Figure 24.10.

  Let $H(s)$ be analytic on each of the contours Γ and $\Gamma_j, j = \{1, 2, \cdots, n\}$ and at each point interior to Γ and exterior to all the $\Gamma_j, j = \{1, 2, \cdots, n\}$. Then,

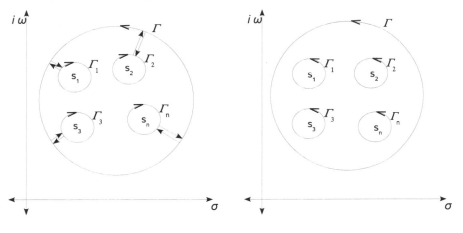

Fig. 24.9: Integration Path of Multiply Connected Contours

Fig. 24.10: Individual Contour Paths Used by the Cauchy Integral Theorem

the contour integral which contains all the said analytic points of $H(s)$ is zero, namely,

$$\oint_{\Gamma} H(s)ds + \oint_{\Gamma_1} H(s)ds + \oint_{\Gamma_2} H(s)ds +$$
$$\oint_{\Gamma_3} H(s)ds + \cdots + \oint_{\Gamma_n} H(s)ds = 0 \tag{24.135}$$

See Figure 24.9 for the path of integration and note the integration directions in each term of Equation 24.135.

Changing the direction of integration for contours $\Gamma_j, j = \{1, 2, \cdots, n\}$,

$$\oint_{\Gamma} H(s)ds = \oint_{\Gamma_1} H(s)ds + \oint_{\Gamma_2} H(s)ds + \oint_{\Gamma_3} H(s)ds + \cdots + \oint_{\Gamma_n} H(s)ds$$
$$= \sum_{j=1}^{n} \oint_{\Gamma_j} H(s)ds \tag{24.136}$$

Definition 24.41 (Length of a Contour Γ in $\mathbb{C}$). *Consider the contour Γ, defined by the parameter $\{\tau : \tau_1 \leq \tau \leq \tau_2\}$, such that,*

$$s(\tau) = \sigma(\tau) + \omega(\tau) \tag{24.137}$$

Then Length of the contour, L_{Γ} is given by the following definition,

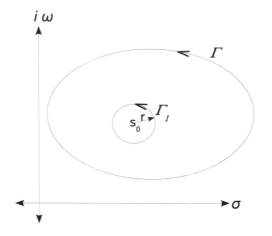

Fig. 24.11: Contour of Integration for Cauchy Integral Formula

$$L_\Gamma \triangleq \int_{\tau_1}^{\tau_2} \left[\left(\frac{d\sigma(\tau)}{d\tau} \right)^2 + \left(\frac{d\omega(\tau)}{d\tau} \right)^2 \right]^{\frac{1}{2}} d\tau \tag{24.138}$$

$$= \int_{\tau_1}^{\tau_2} \left| \frac{ds}{d\tau} \right| d\tau \tag{24.139}$$

Theorem 24.12 (Absolute Integral Bound). *Consider the complex contour Γ defined in terms of a parameter τ, i.e. Γ is defined on the curve, $g(\tau) : \tau \mapsto \mathbb{C}, \tau_1 \leq \tau \leq \tau_2$. Then,*

$$\left| \int_{\tau_1}^{\tau_2} g(\tau) d\tau \right| \leq \int_{\tau_1}^{\tau_2} |g(\tau)| d\tau \tag{24.140}$$

Proof.

$$s = \int_{\tau_1}^{\tau_2} g(\tau) d\tau$$

$$= |s| e^{i\theta} \tag{24.141}$$

where $\theta \triangleq \angle s$.

Based on the polar representation of s,

$$s = |s| e^{i\theta} \implies |s| = s e^{-i\theta} \tag{24.142}$$

Then, the left hand side of Equation 24.140 may be written as,

$$|s| = \left| \int_{\tau 1}^{\tau 2} g(\tau) d\tau \right|$$

$$= e^{-i\theta} \int_{\tau 1}^{\tau 2} g(\tau) d\tau$$

$$= \int_{\tau 1}^{\tau 2} g(\tau) e^{-i\theta} d\tau$$

$$\geq 0 \tag{24.143}$$

Note that

$$\int_{\tau 1}^{\tau 2} g(\tau) e^{-i\theta} d\tau \leq \int_{\tau 1}^{\tau 2} \left| g(\tau) e^{-i\theta} \right| d\tau$$

$$= \int_{\tau 1}^{\tau 2} |g(\tau)| \left| e^{-i\theta} \right| d\tau$$

$$= \int_{\tau 1}^{\tau 2} |g(\tau)| d\tau \tag{24.144}$$

Therefore,

$$\left| \int_{\tau 1}^{\tau 2} g(\tau) d\tau \right| \leq \int_{\tau 1}^{\tau 2} |g(\tau)| d\tau \tag{24.145}$$

□

Theorem 24.13 (Cauchy Integral Formula).

- *Part I (For Analytic Functions):*
  *Let $H(s)$ be analytic within and on a simple closed contour Γ in Domain $\mathscr{D} \subset$
  $\mathbb{C}$. If $s = s_0$ is any point inside contour Γ, then,*

$$H(s_0) = \frac{1}{2\pi i} \oint_{\Gamma} \frac{H(\zeta)}{\zeta - s_0} d\zeta \tag{24.146}$$

  N.B. The contour integration is taken in the positive (counter-clockwise) sense.

- *Part II (For Derivatives of Analytic Functions):*
  *Furthermore, based on the definition of analyticity of functions, Definition 24.34,
  these functions may be written in the form of a power series (Definition 24.42)
  and therefore, the n-th derivative of $H(s)$ evaluated at $s = s_0$ may be written as,*

$$\left. \frac{dH(s)}{ds} \right|_{s=s_0} = \frac{n!}{2\pi i} \oint_{\Gamma} \frac{H(\zeta)}{(\zeta - s_0)^{n+1}} d\zeta \tag{24.147}$$

  *for any n. A consequence of this is that any analytic function is infinitely differ-
  entiable, hence of class $\mathbb{C}^{\infty}$ continuity.*

Proof.

- Part I (*For Analytic Functions*):
  Since $H(s)$ is assumed to be analytic, then,

$$\varphi(s) = \frac{H(s)}{s - s_0} \tag{24.148}$$

is also analytic at all points inside the contour Γ except for $s = s_0$. Draw a small circle of radius r and center s_0 lying entirely inside contour Γ and call it Γ_1 – see Figure 24.11. According to the Cauchy Integral Theorem (Theorem 24.11),

$$\oint_\Gamma \varphi(s)ds = \oint_{\Gamma_1} \varphi(s)ds \tag{24.149}$$

Since $H(s)$ is analytic at $s = s_0$,

$$H(s) = H(s_0) + \left.\frac{dH(s)}{ds}\right|_{s=s_0} (s - s_0) + \varepsilon(s)(s - s_0) \tag{24.150}$$

where

$$\varepsilon(s) = \begin{cases} \left.\frac{H(s)-H(s_0)}{s-s_0} - \frac{dH(s)}{ds}\right|_{s=s_0} & \forall \quad s \neq s_0 \\ \\ 0 & \text{for} \quad s = s_0 \end{cases}$$

$$\oint_{\Gamma_1} \varphi(s)ds = \oint_{\Gamma_1} \frac{H(s)}{s - s_0}ds$$

$$= \oint_{\Gamma_1} \frac{H(s_0) + \left.\frac{dH(s)}{ds}\right|_{s=s_0}(s - s_0) + \varepsilon(s)(s - s_0)}{s - s_0}ds$$

$$= H(s_0) \underbrace{\oint_{\Gamma_1} \frac{ds}{s - s_0}}_{Integral 1} + \left.\frac{dH(s)}{ds}\right|_{s=s_0} \underbrace{\oint_{\Gamma_1} ds}_{Integral 2} + \underbrace{\oint_{\Gamma_1} \varepsilon(s)ds}_{Integral 3} \tag{24.151}$$

Now, let us evaluate the three Integrals labeled in Equation 24.151.

Integral 1:
On Γ_1, let $s - s_0 = re^{i\theta}$. Therefore, $ds = ire^{i\theta}d\theta$. Then,

$$\oint_{\Gamma_1} \frac{ds}{s - s_0} = \int_0^{2\pi} \frac{ire^{i\theta}}{re^{i\theta}}d\theta$$

$$= 2\pi i \tag{24.152}$$

Integral 2:

$$\oint_{\Gamma_1} ds = ir \int_0^{2\pi} e^{i\theta} d\theta$$
$$= ir \int_0^{2\pi} [\cos(\theta) + i\sin(\theta)] d\theta$$
$$= 0 \tag{24.153}$$

Since it is a closed contour.

Integral 3:
Let us take a look at the absolute value of Integral 3,

$$\left| \oint_{\Gamma_1} \varepsilon(s) ds \right| \leq 2\pi r\varepsilon \tag{24.154}$$

where $\varepsilon \overset{\Delta}{=} \underset{s \, on \, \Gamma_1}{\max} |\varepsilon(s)|$.

Combining Equations 24.151, 24.152, 24.153, and 24.154,

$$\left| \oint_{\Gamma_1} \varphi(s) ds - 2\pi i H(s_0) \right| \leq 2\pi r\varepsilon \tag{24.155}$$

$r = |s - s_0| \to 0 \implies \varepsilon \to 0$ (see Definition of ε). Therefore,

$$\left| \oint_{\Gamma_1} \varphi(s) ds - 2\pi i H(s_0) \right| = 0 \tag{24.156}$$

$$\therefore \oint_{\Gamma_1} \varphi(s) ds = 2\pi i H(s_0) \tag{24.157}$$

From Equations 24.157 and 24.148,

$$\boxed{H(s_0) = \frac{1}{2\pi i} \oint_{\Gamma_1} \frac{H(s)}{s - s_0} ds} \tag{24.158}$$

- Part II (*For Derivatives of Analytic Functions*):
  Let s_0 and $s_0 + \delta$ be points interior to a simple closed contour Γ. Then,

$$\frac{H(s_0 + \delta) - H(s_0)}{\delta} = \frac{1}{2\pi i \delta} \oint_{\Gamma} \left[\frac{1}{s - s_0 - \delta} - \frac{1}{s - s_0} \right] H(s) ds$$
$$= \frac{1}{2\pi i} \oint_{\Gamma} \frac{H(s)}{(s - s_0 - \delta)(s - s_0)} ds \tag{24.159}$$

$$\frac{1}{(s - s_0 - \delta)(s - s_0)} = \frac{1}{(s - s_0)^2} + \frac{\delta}{(s - s_0 - \delta)(s - s_0)^2} \tag{24.160}$$

Combining Equations 24.159 and 24.160,

$$\frac{H(s_0 + \delta) - H(s_0)}{\delta} = \frac{1}{2\pi i} \oint_\Gamma \frac{H(s)}{(s - s_0)^2} ds +$$
$$\frac{\delta}{2\pi i} \oint_\Gamma \frac{H(s)}{(s - s_0)^2(s - s_0 - \delta)} ds \qquad (24.161)$$

Let us consider,

$$\lim_{|\delta| \to 0} \frac{\delta}{2\pi i} \oint_\Gamma \frac{H(s)}{(s - s_0)^2(s - s_0 - \delta)} ds \qquad (24.162)$$

Since H(s) is continuous on Γ, it is bounded, so, $|H(s)| \leq M$ on Γ. Let L be the length of contour Γ and let r be the smallest distance from s_0 to Γ. Let us choose $|\delta| : |\delta| \leq \frac{1}{2}r$. Then, for a point s on contour Γ,

$$|s - s_0| \geq r \qquad (24.163)$$

$$|s - s_0 - \delta| \geq |s - s_0| - |\delta| \geq r - \frac{r}{2} = \frac{r}{2} \qquad (24.164)$$

Hence,

$$\frac{\delta}{2\pi i} \oint_\Gamma \frac{H(s)}{(s - s_0)^2(s - s_0 - \delta)} ds \leq \left(\frac{ML}{\pi r^3}\right) |\delta| \qquad (24.165)$$

$$\lim_{|\delta| \to 0} \left(\frac{ML}{\pi r^3}\right) |\delta| = 0 \qquad (24.166)$$

This means that,

$$\lim_{|\delta| \to 0} \frac{\delta}{2\pi i} \oint_\Gamma \frac{H(s)}{(s - s_0)^2(s - s_0 - \delta)} ds = 0 \qquad (24.167)$$

Taking the limit of Equation 24.161 as $\delta \to 0$ in conjunction with the result of Equation 24.167,

$$\frac{dH(s)}{ds}\bigg|_{s=s_0} = \lim_{\delta \to 0} \frac{H(s_0 + \delta) - H(s_0)}{\delta}$$
$$= \frac{1!}{2\pi i} \oint_\Gamma \frac{H(s)}{(s - s_0)^2} ds \qquad (24.168)$$

or

$$\oint_\Gamma \frac{H(s)}{(s - s_0)^2} ds = \frac{2\pi i}{1!} \frac{dH(s)}{ds}\bigg|_{s=s_0} \qquad (24.169)$$

The above procedure may be repeated to prove the general case,

$$\frac{d^n H(s)}{ds}\bigg|_{s=s_0} = \frac{n!}{2\pi i} \oint_\Gamma \frac{H(\zeta)}{(\zeta - s_0)^{n+1}} d\zeta \qquad (24.170)$$

□

Theorem 24.14 (Morera's Theorem (Converse of Cauchy's Integral Theorem)).
If $H(s)$ is continuous in a Domain $\mathscr{D} \subset \mathbb{C}$ and if, $\oint_\Gamma H(s)ds$ is zero for every closed contour, Γ, then $H(s)$ is analytic.

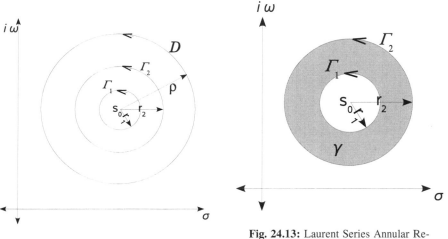

Fig. 24.12: Taylor Series Convergence for an Analytic Function

Fig. 24.13: Laurent Series Annular Region of Convergence for an Analytic Function

24.1.9 Power Series Expansion of Functions

In this section, we will review the power series expansion of analytic functions and related concepts such as convergence criteria and the definitions of poles and zeros.

Definition 24.42 (Taylor Series (Expansion of an analytic function into a Power Series)). *Let $H(s)$ be analytic within the interior of a circular Domain $\mathscr{D} \subset \mathbb{C}$ with center s_0 and radius ρ, i.e. $\mathscr{D} = \{s : |s - s_0| \le \rho\}$. Then, at each point s interior to $\mathscr{D}$, the function $H(s)$ may be written in terms of the following power series,*

$$H(s) = \sum_{n=0}^{\infty} a_n(s - s_0)^n \qquad (24.171)$$

where,

$$a_n = \frac{1}{2\pi i} \oint_C \frac{H(\zeta)}{(\zeta - s_0)^{n+1}} d\zeta \quad n = \{0,1,2,\cdots\} \tag{24.172}$$

*As a consequence of the Cauchy Integral Formula for derivatives (Equation 24.147),
Equation 24.171 may be written as the following which is the Taylor Series expansion of $H(s)$ in powers of $(s - s_0)$.*

$$H(s) = \sum_0^\infty \frac{1}{n!} \frac{d^n H(s)}{ds}\bigg|_{s=s_0} (s - s_0)^n \tag{24.173}$$

*Moreover, the series converges uniformly (Definition 24.69) for points within and
on any circle Γ with center at s_0 and radius $r < \rho$ where ρ is the radius of convergence. Given any center s_0, the radius of convergence, ρ, is the distance from s_0 to
the nearest singularity of the function. See Figure 24.12*

  N.B. When $s_0 = 0$, the Taylor Series is known as the McLauren series.

Example 24.6 (Infinite Radius of Convergence).
The radius of convergence of $H(s) = e^s$ is $\rho = \infty$ for any center s_0.

Example 24.7 (Radius of Convergence).
If the center of the domain Γ is taken to be $s_0 = 0$, then the radius of convergence of

$$H(s) = \frac{e^s}{s-1} \tag{24.174}$$

will be $\rho = 1$.

**Definition 24.43 (Laurent Series (Expansion of analytic functions in an Annular
Region)).** *Let Υ be an annular region bounded by two concentric circles, Γ_1 and Γ_2
(see Figure 24.13) with centers at s_0 and radii r_1 and r_2 where $r_1 < r_2$. Let $H(s)$ be
analytic within Υ and on Γ_1 and Γ_2. Then at each point in the interior of Υ, $H(s)$ can
be represented by a convergent power series consisting of both positive and negative
powers of $(s - s_0)$ as follows,*

$$H(s) = \sum_{n=0}^\infty a_n (s - s_0)^n + \sum_{n=1}^\infty b_n (s - s_0)^{-n} \tag{24.175}$$

where,

$$a_n = \frac{1}{2\pi i} \oint_{\Gamma_2} \frac{H(\zeta)}{(\zeta - s_0)^{n+1}} d\zeta \quad n = \{0,1,2,\cdots\} \tag{24.176}$$

$$b_n = \frac{1}{2\pi i} \oint_{\Gamma_1} \frac{H(\zeta)}{(\zeta - s_0)^{-n+1}} d\zeta \quad n = \{1,2,\cdots\} \tag{24.177}$$

N.B.

$$b_1 = \frac{1}{2\pi i} \oint_{\Gamma_1} H(\zeta)d\zeta \tag{24.178}$$

is the residue of $H(s)$ at $s = s_0$. Also, note that the Taylor Series expansion defined in Definition 24.42 is a special case of the Laurent Series where $r_1 \to 0$.

Property 24.10 (Uniqueness of Power Series). *If two power series, $\sum_{n=0}^{\infty} a_n(s - s_0)^n$ and $\sum_{n=0}^{\infty} b_n(s - s_0)^n$ both converge to the same function $H(s)$, in the same neighborhood of s_0, $|s - s_0| < \rho$, then the two series are identical. i.e. $a_n = b_n$ $\forall n = \{0, 1, 2, \cdots\}$.*

Property 24.11 (Addition and Multiplication of Power Series). *If two power series, $G(s) = \sum_{n=0}^{\infty} a_n(s - s_0)^n$ and $H(s) = \sum_{n=0}^{\infty} b_n(s - s_0)^n$ both converge with nonzero convergence radii r_1 and r_2 respectively, such that, $r_1 \leq r_2$, then,*

$$G(s) \pm H(s) - \sum_{n=0}^{\infty} (a_n \pm b_n)(s - s_0)^n \quad where\ |s - s_0| < r1 \tag{24.179}$$

and

$$G(s) \cdot H(s) = \sum_{n=0}^{\infty} (c_n)(s - s_0)^n \tag{24.180}$$

$$where\ c_n = \sum_{k=0}^{n} a_k b_{n-k}$$

$$|s - s_0| < r1$$

$$n = \{0, 1, 2, \cdots\} \tag{24.181}$$

Property 24.12 (Division of Power Series). *Consider the two power series $G(s)$ and $H(s)$ of Property 24.11. If $H(s) \neq 0$, then there exists a power series $\sum_{n=0}^{\infty} c_n(s - s_0)^n$ and a number $\zeta > 0$ such that*

$$\frac{G(s)}{H(s)} = \sum_{n=0}^{\infty} c_n(s - s_0)^n \quad \forall\ |s - s_0| < \zeta \tag{24.182}$$

where the coefficients c_n satisfy the following equations.

$$a_n = \sum_{k=0}^{n} c_k b_{n-k} \tag{24.183}$$

$$b_n = \sum_{k=0}^{n} c_k a_{n-k} \tag{24.184}$$

Definition 24.44 (Zeros of a Function). *A point s_0 is called a zero of order r of $H(s)$ if*

$$\lim_{s \to s_0} \left[(s - s_0)^{-r} H(s) \right] = M \quad \text{where } M \neq 0 \wedge M < \infty \tag{24.185}$$

Definition 24.45 (Isolated Singularities and Poles of a Function). *A point s_0 is called an isolated singularity or an isolated singular point of $H(s)$ if $H(s)$ is not analytic at s_0, but it is analytic in a deleted neighborhood of s_0. s_0 is also called a pole of order r of function $H(s)$ if,*

$$\lim_{s \to s_0} \left[(s - s_0)^{r} H(s) \right] = M \quad \text{where } M \neq 0 \wedge M < \infty \tag{24.186}$$

Example 24.8 (Isolated Singularity).
The function,

$$H(s) = \frac{s^2 + 1}{(s^2 + 4)(s^2 - 5s)} \tag{24.187}$$

is analytic over the entire $\mathbb{C}$ except at the points where it blows up. $s = \pm 2i, 0, 5$ are isolated singularities of $H(s)$.

24.1.10 Residues

In this section, we will state one of the most powerful theorems in the theory of complex variable which provides the means for inverting an integral transform, followed by another theorem which facilitates the computation of the *residues*. Let us start with the definition of a *meromorphic function* which is the type of function handled by the *Cauchy residue theorem*.

Definition 24.46 (Meromorphic Functions). *A function $H(s)$ which is analytic in a Domain $\mathscr{D} \subset \mathbb{C}$ except at some point of $\mathscr{D}$ where it has poles is said to be meromorphic in $\mathscr{D}$.*

Theorem 24.15 (The Cauchy Residue Theorem). *Let Γ be a simple closed contour and $H(s)$ be analytic on Γ and interior to Γ except at a finite number of isolated singular points, $(s_1, s_2, \cdots, s_n)$ – See Figure 24.10. Then,*

$$\oint_\Gamma H(s) ds = 2\pi i \sum_{k=1}^{n} Residue \left[H(s), s_k \right] \tag{24.188}$$

Proof.
By the Cauchy Integral Theorem for a Multiply Connected Region (Theorem 24.11),

$$\oint_\Gamma H(s) ds = \sum_{k=1}^{n} \oint_{\Gamma_k} H(s) ds \tag{24.189}$$

Recall the b_1 coefficient of the Laurent series expansion of $H(s)$ about s_k – see Equation 24.178 in Definition 24.43. Then,

$$\oint_{\Gamma_k} H(s)ds = 2\pi i b_1$$

$$= 2\pi i \, \text{Residue}[H(s), s_k] \tag{24.190}$$

Combining Equations 24.190 and 24.189, we get the statement of this theorem, namely,

$$\oint_{\Gamma} H(s)ds = 2\pi i \sum_{k=1}^{n} \text{Residue}[H(s), s_k] \tag{24.191}$$

$\square$

Theorem 24.16 (The Residue Evaluation Theorem). *Suppose $H(s)$ is analytic in a neighborhood of $s = s_0$ except at s_0 where it has a pole of order m. Then,*

$$\textit{Residue}\,[H(s), s_0] = \frac{1}{(m-1)!} \lim_{s \to s_0} \frac{d^{(m-1)}}{ds^{(m-1)}} [(s-s_0)^m H(s)] \tag{24.192}$$

Proof.
Define

$$l(s) \overset{\Delta}{=} (s - s_0)^m H(s) \tag{24.193}$$

In essence $l(s)$ is analytic everywhere including at $s = s_0$. Now let us write the power series for $l(s)$ about s_0 (see Definition 24.43)

$$l(s) = \sum_{k=0}^{m-1} b_{m-k}(s - s_0)^k + \sum_{n=0}^{\infty} a_n(s - s_0)^{n+m} \tag{24.194}$$

Differentiate $l(s)$, $(m-1)$ times and then take its limits as $s \to s_0$,

$$\lim_{s \to s_0} \frac{d^{(m-1)} l(s)}{ds^{(m-1)}} = (m-1)! b_1 \tag{24.195}$$

$$\text{Residue}\,[H(s), s_0] = b_1$$

$$= \frac{1}{(m-1)!} \lim_{s \to s_0} \frac{d^{(m-1)}}{ds^{(m-1)}} [(s-s_0)^m H(s)] \tag{24.196}$$

$\square$

N.B. For a simple pole, i.e. $m = 1$,

$$Residue[H(s), s_0] = \lim_{s \to s_0} [(s - s_0)H(s)] \qquad (24.197)$$

24.2 Relations Between Functions

There are several different definitions which convey the relationships of two or more functions in varying perspectives. These relationships manifest themselves in the form of functions over the same domain as the original functions which they relate. Two of the most important such functions used in signal processing are *convolution* and *correlation*.

24.2.1 Convolution

The convolution function $Conv(g, h)(t)$ is an indication of the amount of overlap between two functions $g(t)$ and $h(t)$ at t where one function is slid an amount t with respect to the other function. It may be defined in two different ways for Riemann integrable continuous functions (see Definition 24.39).

Definition 24.47 (Finite-Domain Convolution).

$$Conv(g, h)(t) = (g * h)(t)$$
$$\triangleq \int_0^t g(\tau)h(t - \tau)d\tau \qquad (24.198)$$

The finite domain definition is popular in control and signal processing where t stands for time.[17] It is most used in conjunction with Laplace Transforms as we shall see later. In general, however, many signal processing techniques use the Infinite-Domain Convolution when they explore the signal with Fourier Transforms. The relationship between convolution and integral transforms will become more apparent in future sections of this chapter.

Definition 24.48 (Infinite-Domain Convolution (Convolution)). *Infinite-Domain Convolution is the more popular definition of Convolution and may be expressed as follows:*

$$Conv(g, h)(t) = (g * h)(t)$$
$$\triangleq \int_{-\infty}^{\infty} g(\tau)h(t - \tau)d\tau \qquad (24.199)$$

Assume that $f(t)$, $g(t)$, and $h(t)$ are general complex-valued functions and that $\gamma = \alpha + i\beta$ is a constant in the complex domain, $\mathbb{C}$. The following properties hold and are easily verifiable by writing out the expression for convolution in terms of the definition in Equation 24.199

Property 24.13 (Commutativity of Convolution). *Convolution of functions* $g(t)$ *and* $h(t)$ *is commutative, namely,*

$$g * h = h * g \tag{24.200}$$

Property 24.14 (Associativity of Convolution). *Convolution among functions* $f(t)$, $g(t)$ *and* $h(t)$ *is associative, namely,*

$$f * (g * h) = (f * g) * h \tag{24.201}$$

Property 24.15 (Distributivity of Convolution). *Convolution is distributive, namely,*

$$(f + g) * h = (f * h) + (g * h) \tag{24.202}$$

Property 24.16 (Scaling Associativity of Convolution). *Convolution is associative with respect to scaling, namely,*

$$\gamma(g * h) = (\gamma g) * (\gamma h) \tag{24.203}$$

24.2.2 Correlation

Another relationship between functions, which is defined in the form of correlation, signifies the location at which the second function has similar features to the first function. It is usually used in cases where a long-term signal is searched for specific features. These features may be represented in a shorter signal which then is slid along the longer signal looking for locations where the similarity peaks. The correlation is defined in mathematical terms as follows,

$$
\begin{aligned}
Corr(g,h)(t) &= (g \circ h)(t) \\
&\overset{\Delta}{=} \int_{-\infty}^{\infty} \overline{g(\tau)} h(t + \tau) d\tau
\end{aligned} \tag{24.204}
$$

In Equation 24.204, the independent variable t is called *lag*.

For real-valued functions,

$$
\begin{aligned}
Corr(g,h)(t) &= (g \circ h)(t) \\
&\overset{\Delta}{=} \int_{-\infty}^{\infty} g(\tau) h(t + \tau) d\tau
\end{aligned} \tag{24.205}
$$

The correlation of a function with itself is called *autocorrelation* and it peaks at the lag $t = 0$, since at lag of zero, the two functions are identical. Correlation is also known as the *sliding inner product* (see Definition 24.49), cross-correlation and the *sliding dot product* of two functions. Like convolution it has interesting connections with the Fourier Transform of functions which will be discussed later in this chapter.

24.3 Orthogonality of Functions

As we shall see, orthogonality plays a very important role in the expansion of functions, also leading to the concept of an integral transform. In this section some of these concepts are defined and reviewed. It is recommended to review the Section on *measure theory* (Section 6.2) prior to, or in conjunction with, this section.

Definition 24.49 (Inner Product of Functions). *The inner product of two functions of a complex variable, $\varphi(s)$ and $\psi(s)$, in a closed interval $s_1 \leq s \leq s_2$, is denoted by $\langle \varphi, \psi \rangle$ and is defined by the following integral,*

$$\langle \varphi, \psi \rangle = \int_{s_1}^{s_2} \varphi(s)\overline{\psi(s)}ds \qquad (24.206)$$

Refer to the Definition 6.49 of the class of p-integrable functions. Let us assume that functions $\varphi(s)$ and $\psi(s)$ are members of the class of 2-integrable functions ($\varphi(s), \psi(s) \in \mathfrak{L}_2$). Then, according to Theorem 6.4 and Definition 6.50, they satisfy *Hölder's inequality* for $p = q = 2$, known as the *Schwarz inequality*. Namely,

$$\langle \varphi, \psi \rangle \leq [\langle \varphi, \varphi \rangle]^{\frac{1}{2}} [\langle \psi, \psi \rangle]^{\frac{1}{2}} \qquad (24.207)$$

or

$$(\langle \varphi, \psi \rangle)^2 \leq \langle \varphi, \varphi \rangle \langle \psi, \psi \rangle \qquad (24.208)$$

Note that Definition 24.49 is a general definition of the *inner product* which applies to any complex function of complex variables. Special cases of the *inner product* include the *inner product* of *complex* or *real* variables in the product space, ($\mathscr{H} \times \mathscr{H}, \mathfrak{H} \times \mathfrak{H}$). For example, consider $\varphi, \psi \in \mathscr{H}$, where the inner product is given by $\langle \varphi, \psi \rangle : \{\mathscr{H} \times \mathscr{H}\} \mapsto \mathbb{R}$, for the *Hilbert metric space* (see Definition 6.34).

Definition 24.50 (Orthogonality of Functions). *Two complex-valued functions, $\varphi(s)$ and $\psi(s)$ are orthogonal in a closed interval, $s_1 \leq s \leq s_2$ if their inner product is zero,*

$$\langle \varphi, \psi \rangle = \int_{s_1}^{s_2} \varphi(s)\overline{\psi(s)}ds = 0 \qquad (24.209)$$

Definition 24.51 (Orthogonality of a Set of Functions). *An infinite set of functions, $\varphi_n(s), n = \{1,, \cdots\}$, is an orthogonal set in the closed interval, $s_1 \leq s \leq s_2$*

if,

$$\int_{s_1}^{s_2} \varphi_n(s)\overline{\varphi_m(s)}ds = 0 \ \forall \ n \neq m \tag{24.210}$$

and

$$\int_{s_1}^{s_2} \varphi_n(s)\overline{\varphi_n(s)}ds \neq 0 \ \forall \ n = 1, 2, \cdots \tag{24.211}$$

If in addition to being orthogonal, the *norm* of each of the functions (Section 6.5.4) is equal to 1 ($\|\varphi_n\| = 1, n = \{1,, \cdots\}$), then the set is said to be an *orthonormal set of functions*.

Definition 24.52 (Orthogonality of a Set of Functions about a Weighting Function). *An infinite set of functions, $\varphi_n(s)$, is an orthogonal set about a weighting function, $w(s)$, in the closed interval, $s_1 \leq s \leq s_2$ if,*

$$\int_{s_1}^{s_2} \varphi_n(s)w(s)\overline{\varphi_m(s)}ds = 0 \ \forall \ n \neq m \tag{24.212}$$

and

$$\int_{s_1}^{s_2} \varphi_n(s)w(s)\overline{\varphi_n(s)}ds \neq 0 \ \forall \ n = 1, 2, \cdots \tag{24.213}$$

Theorem 24.17 (Bessel's Inequality).
Consider an orthonormal set of functions, denoted by $\{\varphi_n(s)\} \in \mathcal{H}_p, n = \{1, 2, \cdots\}$, where $\mathcal{H}_p$ is a pre-Hilbert space, generally defined in terms of the complex plane (see Section 6.2.6.1). Then if we denote the components of a function, $h(s)$, represented in this orthonormal basis as,

$$c_n \stackrel{\Delta}{=} \langle h, \overline{\varphi}_n \rangle \tag{24.214}$$

the following inequality known as Bessel's inequality holds for any such orthonormal system.

$$\sum_{n=1}^{N} |c_n|^2 \leq \langle h, \overline{h} \rangle \tag{24.215}$$

Proof. (due to [12])
A truncated expansion of h in terms of the basis functions of the orthonormal set, $\{\varphi_n\}$, would be one that is written for $n = \{1, 2, \cdots, N\}$. Therefore, the following expansion error seems evident,

$$\int_{\mathscr{D}} \left| h(s) - \sum_{n=1}^{N} c_n \varphi_n(s) \right|^2 ds \geq 0 \tag{24.216}$$

We may expand the terms on the left hand side of Equation 24.216 as follows,

$$\int_{\mathscr{D}} \left| h(s) - \sum_{n=1}^{N} c_n \varphi_n(s) \right|^2 ds = \int_{\mathscr{D}} |h|^2 ds - 2 \sum_{n=1}^{N} c_n \int_{\mathscr{D}} h(t) \overline{\varphi_n(s)} ds$$

$$+ \sum_{n=1}^{N} |c_n|^2 \tag{24.217}$$

Using the Equation 24.206 and the definition of c_n, given by Equation 24.214, in Equation 24.217, we may write,

$$\int_{\mathscr{D}} \left| h(s) - \sum_{n=1}^{N} c_n \varphi_n(s) \right|^2 ds = \langle h, \overline{h} \rangle - 2 \sum_{n=1}^{N} |c_n|^2 + \sum_{n=1}^{N} |c_n|^2 \tag{24.218}$$

$$= \langle h, \overline{h} \rangle - \sum_{n=1}^{N} |c_n|^2 \tag{24.219}$$

$$\geq 0 \tag{24.220}$$

The transition from Equation 24.217 to Equation 24.218 takes advantage of the orthonormality of $\{\varphi_n\}$.

Therefore, we can write,

$$\sum_{n=1}^{N} |c_n|^2 \leq \langle h, \overline{h} \rangle \tag{24.221}$$

However, $\langle h, \overline{h} \rangle$ is independent of N, so the inequality in Equation 24.221 should be valid for any N, even as $N \to \infty$. Therefore,

$$\sum_{n=1}^{\infty} |c_n|^2 \leq \langle h, \overline{h} \rangle \tag{24.222}$$

$\square$

Bessel's inequality tells us, not only that the sum of squares of the expansion coefficients of all orthonormal expansions converges, but also that it is bounded by the inner product of the function with itself, or namely, the square of its norm.

Theorem 24.18 (Least Squares Estimation (Approximation in the Mean)). *Let us consider an expansion of $h(s)$ in terms of a truncated set of orthonormal functions, $\{\varphi_n\}, n = \{1, 2, \cdots, N\}$. Then, the coefficients, $\{a\}_1^N$, of the approximation of $h(s)$ by the linear combination approximation of the basis functions, $\{\varphi\}_1^N$,*

$$h(s) \approx \sum_{n=1}^{N} a_n \varphi_n(s) \tag{24.223}$$

has the smallest sum of squares of errors,

$$S \overset{\Delta}{=} \int_{\mathscr{D}} \left| h(s) - \sum_{n=1}^{N} a_n \varphi_n(s) \right|^2 ds \tag{24.224}$$

if and only if $\{a\}_1^N = \{c\}_1^N$, where,

$$c_n = \langle h, \overline{\varphi_n} \rangle \ \forall \, n = \{1, 2, \cdots, N\} \tag{24.225}$$

Proof.
Given Equation 24.225, Equation 24.224 may be expanded as follows,

$$S = \int_{\mathscr{D}} \left| h(s) - \sum_{n=1}^{N} a_n \varphi_n(s) \right|^2 ds \tag{24.226}$$

$$= \int_{\mathscr{D}} |h(s)|^2 ds + \sum_{n=1}^{N} |a_n - c_n|^2 - \sum_{n=1}^{N} |c_n|^2 \tag{24.227}$$

$$\tag{24.228}$$

Therefore, since the c_n are functions of the basis, $\{\varphi\}_1^N$, the only term in Equation 24.227 which may be manipulated in order to make S minimal is the second term. Therefore, S would be minimum if and only if $a_n = c_n \ \forall \, n = \{1, 2, \cdots, N\}$.

$\square$

Definition 24.53 (Complete Space). *Any orthonormal function space given by the basis functions, $\varphi_n, n = \{1, 2, \cdots\}$ is said to be a complete space if for any piecewise continuous function, $h(s)$, there exists an N which may be chosen such that the sum of squares of errors, S, given by Equation 24.229 may be made arbitrarily small, i.e., $S < \varepsilon$, where $\varepsilon > 0 \in \mathbb{R}$ is an arbitrarily small positive real number.*

$$S \overset{\Delta}{=} \int_{\mathscr{D}} \left| h(s) - \sum_{n=1}^{N} c_n \varphi_n(s) \right|^2 ds < \varepsilon \tag{24.229}$$

Theorem 24.19 (Completeness Relation (Bessel's Identity in a Complete Space)).
Consider an orthonormal set of functions, denoted by $\{\varphi_n(s)\} \in \mathscr{H}, n = \{1, 2, \cdots\}$, where $\mathscr{H}$ is a Hilbert space, generally defined in terms of the complex plane (see Section 6.2.6.2). Then if we denote the components of a function, $h(s)$, represented in this orthonormal basis as, $c_n = \langle h, \overline{\varphi} \rangle$, the following completeness relation or Bessel's identity holds for any such orthonormal system.

$$\sum_{n=1}^{\infty} |c_n|^2 = \langle h, \overline{h} \rangle \tag{24.230}$$

Proof.
Since the *Hilbert space* is a *complete space* (see Definition 24.53), then ε in Definition 24.53 may be made arbitrarily small by taking N in Definition 24.53 to be large enough such that in the limit as $N \to \infty$, the sum of squares of errors, S, in Equation 24.229 vanishes, hence the inequality in Equation 24.222 turns into an equality, given the identity in Equation 24.230.

$\square$

The reason Equation 24.230 is known as the *completeness relation* is that it defines a space, given by an orthonormal set of basis functions which form a *complete space*. This is a *Hilbert space*, since it is a complete space where an inner product is also defined – see Definition 6.33.

Note that *Bessel's inequality* does not mean that any function $h(s)$ may be expanded using an orthonormal set of functions as a basis, given that infinite components are chosen. All it states is the, so called, *completeness relation* [12],

$$\lim_{N \to \infty} \int_{\mathscr{D}} \left| h(s) - \sum_{n=1}^{N} c_n \varphi_n(s) \right|^2 ds = 0 \tag{24.231}$$

which is a *necessary*, but *not sufficient*, condition for the existence of the expansion. The *sufficience condition*, however, requires that the series *converges uniformly* (Definition 24.69), namely that,

$$\int_{\mathscr{D}} \lim_{N \to \infty} \left(h(t) - \sum_{n=1}^{N} c_n \varphi_n(t) \right)^2 dt = 0 \tag{24.232}$$

The *completeness relation* (Equation 24.231) means that the series *converges in the mean* to h (see Theorem 24.18).

24.4 Integral Equations

Integral equations have been used in many different fields including problems in engineering, theoretical physics, mathematics, etc. They are at the base of integral transforms and have been used in many approaches for signal processing as well as pattern recognition such as *support vector machines*. In this section, we present the basic definition of an integral equation and an integral transform.

Definition 24.54 (Linear Integral Equations). *A general linear integral equation [24, 39, 40, 41, 30] may be written in the following form [31],*

$$A(s)\varphi(s) + \int_{\mathcal{D}} \mathcal{K}(s,t)\varphi(t)dt = f(s) \qquad (24.233)$$

where $A(s)$ is known as the coefficient, $\mathcal{K}$ is known as the kernel[1], or the kernel of the integral operator (see Definition 24.56), $f(s)$ is the free term, and $\mathcal{D} : s, t \in \mathcal{D}$ is the domain of the integral equation [31].

In definition 24.54, depending on the value of $A(s)$, different kinds of integral equations are produced as follows,

- *First Kind:* if $A(s) = 0 \ \forall \ s \in \mathcal{D}$, then Equation 24.233 is known as a linear integral equation of the *first kind*.
- *Second Kind:* if $A(s) \neq 0 \ \forall \ s \in \mathcal{D}$, then Equation 24.233 is known as a linear integral equation of the *second kind*.
- *Third Kind:* if $A(s) = 0$ only for $s \in \mathcal{D}_1 \subset \mathcal{D}$, then Equation 24.233 is known as a linear integral equation of the *third kind*.

Definition 24.55 (General Integral Transform).
The Equation pair 24.234 and 24.235 represents the general integral transform and its inverse (if it exists), respectively. In these relations, $\mathcal{K}(t,s)$ and $\mathcal{K}^{-1}(t,s)$ are the kernel and inverse kernel functions [12]. The choice of the kernel function (Definition 24.56) and the bounds of the integration define the different integral transform at hand.

$$\mathcal{T}(s) = \int_{t_1}^{t_2} \mathcal{K}(t,s)h(t)dt \qquad (24.234)$$

$$h(t) = \int_{s_1}^{s_2} \mathcal{K}^{-1}(t,s)\mathcal{T}(s)ds \qquad (24.235)$$

In the next section we will examine the kernel function in more detail. This is useful, not only for understanding integral transforms, but in general, for understanding other kernel techniques. Some important examples of kernel techniques, aside from integral transforms, are *support vector machines* (Chapter 15), *Kernel PCA* (Section 12.3.1), and *Kernel k-Means* (Sections 11.2.6.3).

---

[1] In his 1904 book [24], *David Hilbert* used the German word *kern* for $\mathcal{K}(s,t)$

24.5 Kernel Functions

The *Kernel function*, as its name implies, is at the heart of integral equations. Let us begin with its definition and then follow by examining some special such functions and their properties.

Definition 24.56 (Kernel Function). *A general kernel function provides a mapping of the form,*

$$\mathscr{K}(s,t) = \{\mathscr{K} : \mathscr{D} \times \mathscr{D} \mapsto \mathbb{C}\} \tag{24.236}$$

where $\mathscr{K}(s,t)$ satisfies Equation 24.233. As a spacial case, the range of $\mathscr{K}$ may be real ($\mathbb{R}$).

A *kernel function*, $\mathscr{K}(s,t)$, as we shall see briefly, may in some cases be expanded in terms of a set of *orthogonal*[2] *Eigenfunctions* (also known as *characteristic functions* [26]) and which provides the means for a mapping between a generally infinite sets of values, s, to another generally infinite set of values, t. The act of mapping is really a projection from the definition of a point based on one infinite set of *orthogonal basis functions* to another. This is a one-to-one mapping through the use of the kernel operator on the product space of the two infinite sets. The kernel function may or may not be *symmetric* and/or *definite*.

We may start with the general definition of a *Hermitian kernel*.

Definition 24.57 (Hermitian Kernel). *A Hermitian kernel is such that,*

$$\mathscr{K}(s,t) = \overline{\mathscr{K}(t,s)} \tag{24.237}$$

For real valued kernels, this amounts to the concept of symmetry.

Definition 24.58 (Symmetric Kernel). *A symmetric kernel [12] is such that,*

$$\mathscr{K}(s,t) = \mathscr{K}(t,s) \tag{24.238}$$

Courant and Hilbert[3] *[12]* show that every *symmetric kernel* that does not vanish identically, possesses *Eigenvalues*[4] and *Eigenfunctions*[5], also that *real symmetric kernels* have only *real Eigenvalues*. *Kellogg* [26] presents a method which begins with an arbitrary function as the first Eigenfunction and continues to determine

---

[2] see Definition 24.50.

[3] Courant was the author for this book, but as he states in his preface to volume I, being Hilbert's student, he used Hilbert's notes and papers in producing the material in the book.

[4] Mercer [30] calls them *singular values*.

[5] Courant and Hilbert [12] give a complete treatment of the evaluation of the Eigenvalues and Eigenfunctions of a kernel function in their Chapter 3. This is an elaborate procedure and is outside the scope of this textbook.

the Eigenvalues and corresponding *Eigenfunctions* (*characteristic functions*) for the kernel. This method is also described in Chapter 3 of [12]. Also, see Section 24.5.2 for the description of a method due to *Hilbert*.

Definition 24.59 (Gram Matrix (Kernel Matrix)). *Let us assume that we have a kernel function,* $\mathscr{K}(s_l, s_m)$, *defined by Definition 24.56. Furthermore, assume that there are N possible samples,* $\{s\}_1^N = \{s_n : n = \{1, 2, \cdots, N\}\}$. *Then, the Gram matrix [12] (kernel matrix) associated with the sequence,* $\{s\}_1^N$, *is* $\mathbf{K} : \mathscr{C}^N \mapsto \mathscr{C}^N$ *such that,*

$$(\mathbf{K})_{[l][m]} = \mathscr{K}(s_l, s_m) \ \forall \ l, m \in \{1, 2, \cdots, N\} \tag{24.239}$$

Definition 24.60 (Definite Kernel). $\mathscr{K}(s, t)$ *is a definite kernel if all its Eigenvalues have the same sign [12]. If they are all positive, then the kernel is known to be positive definite and if they are all negative, then it is known to be negative definite.*

On the other hand, if the kernel possesses *positive* and *negative* Eigenvalues, it is known to be an *indefinite kernel*[6]. If the Eigenvalues are either 0 or *positive*, the kernel is *positive semidefinite* and if they are either 0 or *negative*, it is *negative semidefinite*.

Definition 24.61 (Mercer Kernel). *A positive semidefinite kernel, per Definition 24.60, is sometimes called a Mercer kernel*[7] *in celebration of its relation to the extension of Hilbert's expansion theorem (Theorem 24.21) by J. Mercer [30] (Theorem 24.23).*

In essence, a positive semi-definite kernel has the property that if we set $\mathscr{K}(s, t) = \langle \boldsymbol{\varphi}(s), \boldsymbol{\varphi}(t) \rangle$, *then,*

$$\sum_{l=1}^{\infty} \sum_{m=1}^{M} c_m c_l \mathscr{K}(s, t) \geq 0 \tag{24.240}$$

Some have called Equation 24.240, the Mercer condition [42, 1].

Theorem 24.20 (Schwarz Inequality for Positive Semidefinite Kernels). *In Section 24.3, we stated the Schwarz inequality for the inner product of two functions. The following Schwarz inequality also holds [42] for positive semidefinite kernels,*

$$|\mathscr{K}(s_1, s_2)|^2 \leq \mathscr{K}(s_1, s_1)\mathscr{K}(s_2, s_2) \tag{24.241}$$

Proof.
Since the kernel is positive semidefinite, the determinant of the *Gram matrix* (Definition 24.59) associated with the two values s_1 and s_2 must be greater than or equal

---

[6] *Mercer* [30] calls such function, one of *ambiguous type.*

[7] A Mercer kernel is also known by the following names: *covariance function* [38], *reproducing kernel* [42] (see Property 24.17), and *admissible kernel* [42].

to zero,

$$
\begin{aligned}
\Gamma_{\mathscr{K}}\Big|_{\{s_1,s_2\}} &= det(\mathbf{K}) \\
&= (\mathbf{K})_{[1][1]}\,(\mathbf{K})_{[2][2]} - (\mathbf{K})_{[1][2]}\,(\mathbf{K})_{[1][2]} \\
&= (\mathbf{K})_{[1][1]}\,(\mathbf{K})_{[2][2]} - \left[(\mathbf{K})_{[1][2]}\right]^2 \\
&\geq 0
\end{aligned}
\tag{24.242}
$$

Equation 24.242 directly proves the statement of the theorem.

$\square$

24.5.1 Hilbert's Expansion Theorem

David Hilbert [12] introduced an expansion theorem for representing the kernel, $\mathscr{K}(s,t)$, in terms of a generally infinite set of orthogonal Eigenfunctions, $\varphi_n(s)$, and their corresponding Eigenvalues, λ_n, associated with $\mathscr{K}$. Equation 24.243 presents this proposed expansion.

$$
\mathscr{K}(s,t) = \sum_{n=1}^{\infty} \frac{\varphi_n(s)\varphi_n(t)}{\lambda_n}
\tag{24.243}
$$

The problem with Equation 24.243 is that it may not be proven to be true in general. Therefore, *Hilbert* presented his theorem as the first cut, covering a special case which will be expressed in Theorem 24.21. This Theorem was later strengthened by *Schmidt* and then by *Mercer* to include more general cases. These extensions will be presented in Theorems 24.22 and 24.23, respectively.

Theorem 24.21 (Hilbert's Expansion Theorem). *Any continuous function, $\mathscr{T}(s)$, which may be written as an integral transform (Equation 24.234) of a piecewise continuous function, $h(t)$, using the symmetric kernel, $\mathscr{K}(s,t)$, may be expanded using a series of the Eigenfunctions of the kernel, $\mathscr{K}(s,t)$ given by given by Equation 24.243. This series converges uniformly (Definition 24.69) and absolutely.*[12]

Theorem 24.22 (Shmidt's Extension to Hilbert's Expansion Theorem). *A symmetric kernel, $\mathscr{K}(s,t)$, may be expanded using a series of the Eigenfunctions of the kernel, given by Equation 24.243 if that series converges uniformly (Definition 24.69) – see Section 8 (page 449) of [39].*

Theorem 24.23 (Mercer's Theorem). *If $\mathscr{K}(s,t)$ is a definite continuous symmetric kernel, or if it has only a finite number of Eigenvalues of one sign, then the said kernel may be expanded by Equation 24.243 in terms of its Eigenfunctions and this* *series converges uniformly (Definition 24.69) and absolutely.*[30, 12]

Proof.
Mercer [30], on page 444, states and proves this theorem for the case where all the kernel is positive definite. Then, he handles the case where the kernel is negative definite. At the end of [30], Mercer remarks that with some slight modifications the expansion is still valid and converges uniformly if there are a finite number of Eigenvalues with an opposite sign, compared to the sign of the rest of the infinite set of Eigenvalues. Courant [12] produces the complete statement of the theorem as remarked at the end of [30] and discusses the proof.

□

An important class of kernels is known as the class of *degenerate kernels*. These kernels are used for many practical applications including *support vector machines* and intermediate developments for proving properties of more general kernels. Once such application is the development of the techniques used in Section 24.5.2 for computing the *Eigenvalues* and *Eigenfunctions* of more general kernels. Here is a definition of a *degenerate kernel*.

Definition 24.62 (Degenerate Kenrnel). *A degenerate kernel, $\mathcal{K}(s,t)$, is one which may expanded in terms of a finite series of products of two sets of functions, $\xi_n(s)$ and $\zeta_n(t)$, which are functions of s and t respectively,*

$$\mathscr{A}(s,t) = \sum_{n=1}^{N} \xi_n(s)\zeta_n(t) \tag{24.244}$$

In its compact form, we assume that $\{\xi\}_1^N$ are linearly independent among themselves and so are $\{\zeta\}_1^N$, assuming that s and t are defined in the same domain, $s,t \in \mathscr{D}$.

To be able to work with a single set of orthogonal functions, it is possible to express the two sets of functions, $\xi 1N$ and $\zeta 1N$, in terms of another orthogonal set of functions, $\{\omega\}_1^M$, leading to a double sum expression for the *degenerate kernel* as follows,

$$\mathscr{A}(s,t) = \sum_{l=1}^{M} \sum_{m=1}^{M} \sigma_{l,m}\omega_l(s)\omega_m(t) \tag{24.245}$$

Note that the *degenerate kernel* expressed by Equation 24.245 will be a *symmetric kernel* if $\sigma_{l,m} = \sigma_{m,l}$. Therefore, for a *symmetric degenerate kernel*, let us define new coefficients, c_m, in terms of $\sigma_{m,l}$ such that,

$$\sigma_{m,l} = \sigma_{l,m} = c_m c_l \ \forall \ l,m \in \{1,2,\cdots,M\} \tag{24.246}$$

Then, we may write the double sum expression for the symmetric degenerate kernel as follows,

$$\mathscr{A}(s,t) = \sum_{l=1}^{M}\sum_{m=1}^{M} c_l c_m \omega_l(s)\omega_m(t) \tag{24.247}$$

$$= \sum_{m=1}^{M}\sum_{l=1}^{M} c_m c_l \omega_m(s)\omega_l(t) \tag{24.248}$$

$$= \mathscr{A}(t,s) \tag{24.249}$$

$$\tag{24.250}$$

24.5.2 Eigenvalues and Eigenfunctions of the Kernel

In this section we only concern ourselves with real symmetric kernels (Definition 24.58). We will present a summary of the method of finding Eigenvalues and Eigenfunctions given by [12], based upon the proof given by *Eric Holmgren*. Here, we have elaborated some of the steps which were simply stated in [12].

Let us begin with the introduction of the following *quadratic integral form*,

$$\mathscr{J}(\varphi,\varphi) \triangleq \int_{\mathscr{D}}\int_{\mathscr{D}} \mathscr{K}(s,t)\varphi(s)\varphi(t)\,ds\,dt \tag{24.251}$$

where $\varphi : \mathscr{D} \mapsto \mathbb{R}$ is any piecewise continuous function. We may write Equation 24.251 in terms of *inner products*, defined by Equation 24.206,

$$\mathscr{J}(\varphi,\varphi) \triangleq = \int_{\mathscr{D}}\left(\int_{\mathscr{D}} \mathscr{K}(s,t)\varphi(s)ds\right)\varphi(t)dt \tag{24.252}$$

$$= \langle\langle K,\varphi\rangle,\varphi\rangle \tag{24.253}$$

Let us write the Schwarz inequality (Equation 24.208) for Equation 24.253,

$$[\mathscr{J}(\varphi,\varphi)]^2 \leq [\langle\mathscr{K},\mathscr{K}\rangle\langle\varphi,\varphi\rangle]\langle\varphi,\varphi\rangle \tag{24.254}$$

$$\leq \langle\mathscr{K},\mathscr{K}\rangle\langle\varphi,\varphi\rangle^2 \tag{24.255}$$

$$= \langle\varphi,\varphi\rangle^2 \int_{\mathscr{D}}\int_{\mathscr{D}} \mathscr{K}^2(s,t)\,ds\,dt \tag{24.256}$$

In summary,

$$[\mathscr{J}(\varphi,\varphi)]^2 \leq \langle\varphi,\varphi\rangle^2 \int_{\mathscr{D}}\int_{\mathscr{D}} \mathscr{K}^2(s,t)\,ds\,dt \tag{24.257}$$

If we restrict the function, φ, to be bounded and, without any loss of generality, if we normalize the function, φ, such that

$$\langle \varphi, \varphi \rangle = 1 \tag{24.258}$$

then the quadratic integral form, $\mathscr{I}(\varphi, \varphi)$, is bounded by the double integral on the right hand side of Equation 24.257. Naturally,

$$\int\limits_{\mathscr{D}} \int\limits_{\mathscr{D}} \mathscr{K}^2(s,t)\, ds\, dt \geq 0 \tag{24.259}$$

Therefore, $\mathscr{I}(\varphi, \varphi) = 0$, if and only if the kernel is zero everywhere.

The entity $\mathscr{I}(\varphi, \varphi)$ is quite important, since it is related to the Eigenvalues of the kernel function. In fact, the smallest Eigenvalue of the kernel, λ_1 is give by the following [12],

$$\lambda_1 = \frac{1}{\kappa_1} \tag{24.260}$$

where,

$$\kappa_1 = \max_{\varphi} \mathscr{I}(\varphi, \varphi) \tag{24.261}$$

for an appropriate function, φ, and where the maximizer of the functional, $\mathscr{I}$, in Equation 24.261,

$$\psi_1(s) = \arg\max_{\varphi} \mathscr{I}(\varphi, \varphi) \tag{24.262}$$

is the Eigenfunction associated with λ_1.

Since $\mathscr{I}(\varphi, \varphi)$ is a functional, a form for φ should be assumed in order to go on with the maximization effort. *Courant and Hilbert* [12] solve the maximization problem by assuming that the kernel may be approximated uniformly, by a sequence of *degenerate symmetric kernels* (see Definition 24.62 and Section 24.1.2), such that the nth sequence is given by,

$$\mathscr{A}_n(s,t) = \sum_{l=1}^{M_n} \sum_{m=1}^{M_n} \sigma_{l,m}^{(n)} \omega_l(s) \omega_m(t) \tag{24.263}$$

where

$$\sigma_{l,m}^{(n)} = \sigma_{m,l}^{(n)} \tag{24.264}$$

and where the products, $\omega_l(s)\omega_m(t)$ are linearly independent in the domain of s and t – the rectangle of limits of the double integration, $(\{s_1 \leq s \leq s_2\} \times \{t_1 \leq t \leq t_2\})$.

Note that here we have used the double sum expression for expressing the degenerate kernel sequence, as defined in Definition 24.62.

Therefore, the approximation for the nth approximation of the quadratic integral form, $\mathscr{I}_n(\varphi, \varphi)$, given by the following,

$$\mathscr{I}_n(\varphi, \varphi) = \int_{\mathscr{D}} \int_{\mathscr{D}} \mathscr{A}_n(s,t) \varphi(s) \varphi(t) \, ds \, dt \tag{24.265}$$

$$= \int_{\mathscr{D}} \int_{\mathscr{D}} \sum_{l=1}^{M_n} \sum_{m=1}^{M_n} \sigma_{l,m}^{(n)} \omega_l(s) \omega_m(t) \varphi(s) \varphi(t) \, ds \, dt \tag{24.266}$$

$$= \sum_{l=1}^{M_n} \sum_{m=1}^{M_n} \sigma_{l,m}^{(n)} \int_{\mathscr{D}} \omega_l(s) \varphi(s) ds \int_{\mathscr{D}} \omega_m(t) \varphi(t) dt \tag{24.267}$$

$$= \sum_{l=1}^{M_n} \sum_{m=1}^{M_n} \sigma_{l,m}^{(n)} \langle \omega_l, \varphi \rangle \langle \omega_m, \varphi \rangle \tag{24.268}$$

$$= \sum_{l=1}^{M_n} \sum_{m=1}^{M_n} \sigma_{l,m}^{(n)} x_l \, x_m \tag{24.269}$$

where

$$x_m \overset{\Delta}{=} \langle \omega_m, \varphi \rangle \tag{24.270}$$

Therefore, $\mathscr{I}_n(\varphi, \varphi)$ is a quadratic function of the M_n variables, $\{x\}_1^{M_n}$.

If we apply *Bessel's inequality* (Theorem 24.17) to Equation 24.270 for the expansion of the orthogonal sequence of functions, $\{\omega\}_1^{M_n}$, we will have the following inequality,

$$\sum_{m=1}^{M_n} x_m^2 \leq \langle \varphi, \varphi \rangle \tag{24.271}$$

Earlier, we have imposed the a restriction on function φ, given by Equation 24.258. This presents the following limit on the quadratic summation in Equation 24.272,

$$\sum_{m=1}^{M_n} x_m^2 \leq 1 \tag{24.272}$$

Since Equation 24.269 has a quadratic form, $\mathscr{I}_n(\varphi, \varphi)$ will be maximal in relation to the choice of x_m, when the equality holds in Equation 24.272,

$$\sum_{m=1}^{M_n} x_m^2 = 1 \tag{24.273}$$

Therefore, we will pick the values of x_m such that the Equality of Equation 24.273 holds. Notice that Equation 24.269 subject to Equation 24.258 will then be maximized by doing a *principal component analysis* (Section 12.1) where the Eigenvalue problem of Equation 12.2 is such that the elements of *covariance matrix*, $\Sigma^{(n)}$, are,

$$\left(\Sigma^{(n)}\right)_{[l][m]} = \sigma^{(n)}_{l,m} \; \forall \; l,m \in \{1,2,\cdots,M_n\} \tag{24.274}$$

and the elements of the vector in Equation 12.2 are given by

$$\left(\mathbf{x}^{(n)}\right)_{[m]} = x_m \; \forall \; m \in \{1,2,\cdots,M_n\} \tag{24.275}$$

where the superscript (n) specifies the fact that the vector, $\mathbf{x}^{(n)}$, is associated with the summation associated with the nth sequence which has M_n terms. The Eigenvalue problem of Equation 12.2 may then be written as follows,

$$\Sigma^{(n)}\mathbf{x}^{(n)} = \kappa^{(n)}_1\mathbf{x}^{(n)} \tag{24.276}$$

Then based on the discussion of Section 12.1, the maximum value for $\mathscr{I}_n(\varphi,\varphi)$ is given by the first Eigenvalue of the solution to the Eigenvalue problem of Equation 24.276 which gives the largest principal component,

$$\max_{\langle\varphi,\varphi\rangle=1} \mathscr{I}_n(\varphi,\varphi) = \kappa^{(n)}_1 \tag{24.277}$$

Since $\varphi_n(s)$, associated with the nth sequence may be any piecewise continuous function and since the set of basis functions, $\{\omega\}^{M_n}_1$, which were used to express the degenerate kernel, $\mathscr{A}(s,t)$ are linearly independent, let us use this set to describe $\varphi_n(s)$. Furthermore, since the set, $\{x\}^{M_n}_1$, that maximize $\mathscr{I}_n(\varphi,\varphi)$ obey the normalization of Equation 24.273, we know that we can use these values as coefficients for expressing $\varphi_n(s)$ such that we are left with a normalized function. Therefore, we write $\varphi_n(s)$ in terms of the basis functions, $\{\omega\}^{M_n}_1$, and the normalized coefficients, $\{x\}^{M_n}_1$, as follows,

$$\varphi_n(s) = \sum_{l=1}^{M_n} x_l \omega_l(s) \tag{24.278}$$

Let us rewrite the Eigenvalue problem of Equation 24.276 in scalar form,

$$\kappa^{(n)}_1 x_l = \sum_{m=1}^{M_n} \sigma^n_{l,m} x_m \tag{24.279}$$

Now if we multiply both sides of Equation 24.279 by $\omega_l(s)$ and sum both sides over l, we will have the following,

$$\kappa_1^{(n)} \sum_{l=1}^{M_n} x_l \, \omega_l(s) = \sum_{l=1}^{M_n} \sum_{m=1}^{M_n} \sigma_{l,m}^n \, x_m \, \omega_l(s) \tag{24.280}$$

$$= \sum_{l=1}^{M_n} \sum_{m=1}^{M_n} \sigma_{l,m}^n \int_{t_1}^{t_2} \omega_m(t) \varphi_n(t) dt \, \omega_l(s) \tag{24.281}$$

Note that based on our choice of $\varphi_n(s)$, given by Equation 24.278, the left hand side of Equation 24.281 is just equal to $\kappa_1^{(n)} \varphi_n(s)$. Therefore, by rearranging the terms on the right hand side of Equation 24.281, we may rewrite it as follows,

$$\kappa_1^{(n)} \varphi_n(s) = \int_{t_1}^{t_2} \left(\sum_{l=1}^{M_n} \sum_{m=1}^{M_n} \sigma_{l,m}^n \omega_l(s) \omega_m(t) \right) \varphi_n(t) dt \tag{24.282}$$

$$= \int_{t_1}^{t_2} \mathscr{A}_n(s,t) \varphi_n(t) \, dt \tag{24.283}$$

Therefore, we may solve for $\varphi_n(s)$ from Equation 24.283,

$$\varphi_n(s) = \frac{1}{\kappa_1^{(n)}} \int_{t_1}^{t_2} \mathscr{A}_n(s,t) \varphi_n(t) \, dt \tag{24.284}$$

$$= \lambda_1^{(n)} \int_{t_1}^{t_2} \mathscr{A}_n(s,t) \varphi_n(t) \, dt \tag{24.285}$$

Since we have use an orthogonal set of functions to express the degenerate kernel approximation, $\mathscr{A}_n(s,t)$, to the original kernel, $\mathscr{K}(s,t)$, we may write the following convergence bound,

$$|\mathscr{K}(s,t) - \mathscr{A}_n(s,t)| < \varepsilon \tag{24.286}$$

Let us revisit Equation 24.253. If we write the expression for $\mathscr{J}(\varphi, \varphi) - \mathscr{J}_n(\varphi, \varphi)$ in the same manner, we will have the following,

$$\mathscr{J}(\varphi, \varphi) - \mathscr{J}_n(\varphi, \varphi) = \langle \langle (\mathscr{K} - \mathscr{A}_n), \varphi \rangle, \varphi \rangle \tag{24.287}$$

Then we may write the Schwarz inequality (Equation 24.208) for Equation 24.287, much in the same as we did in Equation 24.254 and its subsequent equations,

$$[\mathscr{J}(\varphi, \varphi) - \mathscr{J}_n(\varphi, \varphi)]^2 \le [\langle (\mathscr{K} - \mathscr{A}_n), (\mathscr{K} - \mathscr{A}_n) \rangle \langle \varphi, \varphi \rangle] \langle \varphi, \varphi \rangle \tag{24.288}$$

$$\le \langle (\mathscr{K} - \mathscr{A}_n), (\mathscr{K} - \mathscr{A}_n) \rangle \langle \varphi, \varphi \rangle^2 \tag{24.289}$$

$$= \langle \varphi, \varphi \rangle^2 \int_{t_1}^{t_2} \int_{s_1}^{s_2} [\mathscr{K}(s,t) - \mathscr{A}_n(s,t)]^2 \, ds \, dt \tag{24.290}$$

If we Equations 24.258 and 24.286, we may write Equation 24.290 as follows,

$$[\mathscr{J}(\varphi, \varphi) - \mathscr{J}_n(\varphi, \varphi)]^2 \le \int_{t_1}^{t_2} \int_{s_1}^{s_2} \varepsilon^2 \, ds \, dt \tag{24.291}$$

$$= \varepsilon^2 \int_{t_1}^{t_2} \int_{s_1}^{s_2} ds \, dt \tag{24.292}$$

Let us assume that the limits on the integration are bounded by a and b from the bottom and the top. Namely, we may write,

$$a \leq s_1 < s_2 \leq b \tag{24.293}$$

and

$$a \leq t_1 < t_2 \leq b \tag{24.294}$$

Then, we may write Equation 24.291 as follows,

$$[\mathscr{J}(\varphi,\varphi) - \mathscr{J}_n(\varphi,\varphi)]^2 \leq \varepsilon^2(b-a)^2 \tag{24.295}$$

We see from Equation 24.295 that due to the arbitrary nature of the right hand side, the sequence, $\mathscr{J}_n(\varphi,\varphi)$ converges to $\mathscr{J}(\varphi,\varphi)$ as n increases,

$$\lim_{n\to\infty} \mathscr{J}_n(\varphi,\varphi) = \mathscr{J}(\varphi,\varphi) \tag{24.296}$$

Therefore, the maximum values of the two sides of Equation 24.296 must also abide by this convergence,

$$\lim_{n\to\infty} \max \mathscr{J}_n(\varphi,\varphi) = \max \mathscr{J}(\varphi,\varphi) \tag{24.297}$$

Equation 24.297 may be written in terms of κ,

$$\lim_{n\to\infty} \kappa_1^{(n)} = \kappa_1 \tag{24.298}$$

Similarly, if we consider the limit of convergence of the sequence, $\varphi_n(s)$, as n grows, we may call this limit, $\psi_1(s)$,

$$\lim_{n\to\infty} \varphi_n(s) = \psi_1(s) \tag{24.299}$$

Therefore, in the limit, Equations 24.284 and 24.285 may be written as follows,

$$\psi_1(s) = \frac{1}{\kappa_1} \int_{t_1}^{t_2} \mathscr{K}(s,t)\psi_1(t)\,dt \tag{24.300}$$

$$= \lambda_1 \int_{t_1}^{t_2} \mathscr{K}(s,t)\psi_1(t)\,dt \tag{24.301}$$

where

$$\langle \psi_1, \psi_1 \rangle = 1 \tag{24.302}$$

$\psi_1(s)$ is the maximizer of $\mathscr{J}(\varphi,\varphi)$,

$$\mathscr{J}(\psi_1, \psi_1) = \kappa_1 \tag{24.303}$$

Using Equation 24.302, we can write Equation 24.303 as follows,

$$\mathcal{J}(\psi_1, \psi_1) = \kappa_1 \langle \psi_1, \psi_1 \rangle \tag{24.304}$$

Therefore, $\mathcal{J}$ evaluated at any other function, ψ, would have a value which is smaller than or equal to that evaluated at the Eigenfunction associated with κ_1, ψ_1. In other words,

$$\mathcal{J}(\psi, \psi) \leq \kappa_1 \langle \psi, \psi \rangle \tag{24.305}$$

In essence, we have provided a method for computing λ_1 which is the *smallest Eigenvalue*, λ_1 of the kernel, $\mathcal{K}(s,t)$, and its associated Eigenfunction, $\psi_1(s)$. Now, let us consider the rest of the Eigenvalues and Eigenfunctions. The simplest way of computing those values is to eliminate the known Eigenvalue and Eigenfunction from the objective function of maximization and then, recursively, compute the next smallest Eigenvalue and its associated Eigenfunction. To do this, we have to add to the conditions that we use for finding the Eigenfunction. For the first Eigenfunction, we had the condition given by Equation 24.258. Now, we should add another condition, which requires that the new Eigenfunction being sought is orthogonal to the ones that have already been found. Namely, if we are seeking the mth smallest Eigenvalue and its corresponding Eigenfunction,

$$\langle \varphi, \psi_l \rangle = 0 \ \forall \ l < m \tag{24.306}$$

The set of equations given by Equation 24.306 for the mth Eigenvalue and Eigenfunction are called *orthogonality conditions*, which should be satisfied in conjunction with the normalization condition of Equation 24.258.

Essentially, according to the Hilbert, Schmidt, and Mercer expansion theorems (Theorems 24.21, 24.22, and 24.23), we may eliminate the known terms associated with the already computed Eigenvalues and Eigenfunctions as follows,

$$\mathcal{K}^{(m-1)}(s,t) = \mathcal{K}(s,t) - \sum_{l=1}^{m-1} \frac{\psi_l(s)\psi_l(t)}{\lambda_l} \tag{24.307}$$

Therefore, the new maximization objective function, $\mathcal{J}^{(m-1)}(\varphi, \varphi)$, may be written as follows,

$$\mathcal{J}^{(m-1)}(\varphi, \varphi) \overset{\Delta}{=} \int_{\mathcal{D}} \int_{\mathcal{D}} \mathcal{K}^{(m-1)}(s,t)\varphi(s)\varphi(t) \, ds \, dt \tag{24.308}$$

Therefore, the mth Eigenvalue and Eigenfunction may be computed by maximizing Equation 24.308, much in the same way as we did for the first Eigenvalue and Eigenfunction.

$$\begin{aligned} \kappa_m &= \frac{1}{\lambda_m} \\ &= \max_{\varphi} \mathcal{J}^{(m-1)}(\varphi, \varphi) \end{aligned} \tag{24.309}$$

$$\psi_m(s) = \arg\max_{\varphi} \mathscr{J}^{(m-1)}(\varphi, \varphi) \tag{24.310}$$

It is easy to see that due to the orthogonality conditions, the succeeding Eigenvalues and Eigenfunctions also satisfy the Equivalent of Equation 24.301, namely,

$$\psi_m(s) = \frac{1}{\kappa_m} \int_{t_1}^{t_2} \mathscr{K}(s,t)\psi_m(t)\, dt \tag{24.311}$$

$$= \lambda_m \int_{t_1}^{t_2} \mathscr{K}(s,t)\psi_m(t)\, dt \tag{24.312}$$

where

$$\langle \psi_l, \psi_m \rangle = \begin{cases} 1 \ \forall \ l = m \\ 0 \ \forall \ l \neq m \end{cases} \tag{24.313}$$

Eventually, we will be left with the expression for the Kernel, given by the Hilbert expansion theorem (Theorem 24.21),

$$\mathscr{K}(s,t) = \lim_{m \to \infty} \sum_{n=1}^{m} \frac{\psi_n(s)\psi_n(t)}{\lambda_n} \tag{24.314}$$

where the λ_n are the Eigenvalues of $\mathscr{K}(s,t)$, such that,

$$\lambda_1 \leq \lambda_2 \leq \cdots \leq \lambda_m \tag{24.315}$$

and the ψ_n are the corresponding Eigenfunctions.

At this point, let us present a very important kernel property which is a direct consequence of Equation 24.314 or equally from the equivalent which is Hilbert's expansion theorem. Assuming that a kernel may be given by Equation 24.314, then we may write define $\boldsymbol{\psi}(\mathbf{x}$ as follows,

$$\left(\boldsymbol{\psi}\right)_{[n]}(x) \overset{\Delta}{=} \frac{\psi_n(x)}{\sqrt{\lambda_n}} \tag{24.316}$$

Therefore, we may write any such kernel by the following inner product,

$$\mathscr{K}(\mathbf{x}_l, \mathbf{x}_m) = \boldsymbol{\psi}^T(\mathbf{x}_l)\boldsymbol{\psi}(\mathbf{x}_m) \tag{24.317}$$

Then the following property holds for such a kernel.

Property 24.17 (Reproducing Kernel). *A kernel which meets the criteria of Mercer's expansion theorem (Theorem 24.23) has the following property called the reproducing kernel property.*

$$\langle \mathscr{K}(\mathbf{x}, \mathbf{x}_l), \mathscr{K}(\mathbf{x}, \mathbf{x}_m) \rangle = \mathscr{K}(\mathbf{x}_l, \mathbf{x}_m) \tag{24.318}$$

Proof.
According to Mercer's expansion theorem and if we use the shorthand version of

$\psi(\mathbf{x})$, $\psi(\mathbf{x}_l)$, and $\psi(\mathbf{x}_m)$, written as ψ, ψ_l, and ψ_m respectively,

$$\langle \mathcal{K}(\mathbf{x},\mathbf{x}_l), \mathcal{K}(\mathbf{x},\mathbf{x}_m)\rangle = \left(\psi^T\psi_l\right)^T\left(\psi^T\psi_m\right)$$
$$= \left(\psi_l^T\psi\psi^T\psi_m\right) \tag{24.319}$$

In Equation 24.319, we know from Equation 24.313 that $\psi\psi^T = \mathbf{I}$. Therefore, Equation 24.319 reduces to

$$\langle \mathcal{K}(\mathbf{x},\mathbf{x}_l), \mathcal{K}(\mathbf{x},\mathbf{x}_m)\rangle = \left(\psi_l^T\psi\psi^T\psi_m\right)$$
$$= \left(\psi_l^T\psi_m\right)$$
$$= \mathcal{K}(\mathbf{x}_l,\mathbf{x}_m) \tag{24.320}$$

$\square$

This property states that the product of two kernels is itself a kernel.

In this chapter, we will examine several different kernel functions and limits which define the most useful transforms in speaker recognition, namely, *Laplace*, *Fourier* and *z transforms*. Before doing this, we will have to review review certain series expansions such as *Fourier* and *wavelet series*.

24.6 Fourier Series Expansion

Historically, most of the theory behind the Fourier Series expansion of functions has been developed while studying the heat equation [18], Equation 24.321, and the associated Eigensystem which came out of solving these boundary value problems. In 1822, Fourier published a book on the theory of heat transfer [18], in which he introduced the series expansion of functions. Although some of his assumptions were incorrect, the main idea was quite revolutionary and lead to the further development of the Fourier Series expansion and handling different boundary conditions including Dirichlet, Neumann and the more general mixed boundary conditions (Equations 24.322 and 24.323).

$$\frac{\partial u(x,t)}{\partial t} = \frac{1}{w(x)}\frac{\partial}{\partial x}\left[\kappa(x)\frac{\partial u(x,t)}{\partial x}\right] + q(x)u(x,t) \begin{cases} \forall & -L < x < L \\ \forall & t > 0 \end{cases} \tag{24.321}$$

Boundary conditions,

$$a_{11}u(-L,t) + a_{12}\frac{\partial u(x,t)}{\partial t}\bigg|_{(x=-L,t)} = 0 \ \forall\, t > 0 \tag{24.322}$$

$$a_{21}u(L,t) + a_{22}\frac{\partial u(x,t)}{\partial t}\bigg|_{(x=L,t)} = 0 \ \forall\, t > 0 \tag{24.323}$$

Initial Condition,

$$u(x,0) = f(x) \;\; \forall \;\; -L < x < L \tag{24.324}$$

By doing a separation of variables, the time-dependent and space-dependent segments will be separated into different equations and the general *Sturm-Liouville problem* arises. Although we are interested in time-dependent speech signals here, the processing resembles the space-dependent *Sturm-Liouville problem*. Therefore, at the first stage we will change the variable x to t for consistency with the rest of the text. The most general Eigenvalue problem which evolves from the separation of variables of the one-dimensional heat equation is the general *Sturm-Liouville problem* – see Equations 24.337 and 24.338 which are shown with a change of the independent variable from x to t, for this class of Eigenvalue problems and boundary conditions.

Before we can define the *complex Fourier series* of a function of time, $h(t)$, the boundary of the signal will have to also be defined. As we shall see later, the Fourier Scrics is only valid for a periodic function. This period in the Cartesian space is in the interval $[-L, L]$ as it was defined in Equation 24.321. In the time domain we use the interval, $[-T, T]$. You may have noticed that here were are talking about the closed interval $[-L, L]$ or $[-T, T]$ whereas Equation 24.321 was defined in the open interval $(-L, L)$. To clarify this distinction, note that the Boundary conditions handle the points at $-L$ and L and eventually, the differential equation results have to become continuous between point pairs $-L^+$ and $-L$ and L^- and L.

A function $h(t)$, which is at least piecewise $\mathcal{C}^1$ continuous in the interval $[-T, T]$, may be represented in terms of a *Complex Fourier Series* expansion if it meets the Dirichlet convergence conditions. In dealing with the heat equation, Dirichlet studied the convergence of the *Fourier Series* and in the process defined some necessary and sufficient conditions for the convergence of the series.

Definition 24.63 (Dirichlet Conditions). *Consider the function, $h(t)$ in the interval $-T \le t \le T$. The following are the Dirichlet Conditions for the convergence of the Fourier Series of that function,*

1. *$h(t)$ is absolutely integrable in the interval $[-T, T]$ – see Definition 24.38*
2. *$h(t)$ is periodic with period $2T$*
3. *$h(t)$ is at least $\mathcal{C}^1$ continuous in the interval $[-T, T]$*

   N.B., See Definition 24.38. It happens that for $h(t)$ to be absolutely integrable in $[-T, T]$, it has to be piecewise continuous in that interval with no infinite discontinuities in that interval – see Definition 24.15.

Definition 24.64 (Complex Fourier Series Expansion). *If $h(t)$ meets all the Dirichlet conditions, it may be represented by an infinite sum of orthogonal func-*

tions (see Definition 24.50), namely, the family of exponential functions in the following form,

$$h(t) \approx \sum_{n=-\infty}^{\infty} c_n e^{i\left(\frac{n\pi t}{T}\right)} \tag{24.325}$$

where,

$$c_n = \frac{1}{2T} \int_{-T}^{T} h(t) e^{-i\left(\frac{n\pi t}{T}\right)} dt \tag{24.326}$$

At some occasions in the text, in order to simplify the Fourier Series expansion, we will attempt to use an equivalent form with a normalized period. Take the following new variable,

$$\hat{t} \stackrel{\Delta}{=} \frac{\pi}{T} t \tag{24.327}$$

such that

$$t = \frac{T}{\pi} \hat{t} \tag{24.328}$$

$\hat{t}$ is defined such that the function $h(t)$ in the interval $-T \leq t \leq T$ is linearly redistributed as $h(\hat{t})$ such that the points for t are mapped to $-\pi \leq \hat{t} \leq \pi$. Using this new mapping, the Dirichlet conditions may be rewritten as,

Definition 24.65 (Dirichlet Conditions with Period Normalization).

1. *$h(\hat{t})$ is absolutely integrable in the interval $[-\pi, \pi]$*
2. *$h(\hat{t})$ is periodic with period 2π*
3. *$h(\hat{t})$ is at least $\mathfrak{C}^1$ continuous in the interval $[-\pi, \pi]$*

The statement of the Fourier Series expansion in the new variable, $\hat{t}$ is,

Definition 24.66 (Complex Fourier Series Expansion with Period Normalization).

$$h(\hat{t}) \approx \sum_{n=-\infty}^{\infty} c_n e^{i(n\hat{t})} \tag{24.329}$$

where,

$$c_n = \frac{1}{2\pi} \int_{-\pi}^{\pi} h(\hat{t}) e^{-i(n\hat{t})} d\hat{t} \tag{24.330}$$

Note that Equations 24.327 and 24.328 may be used to go back and forth between the periods of $2T$ and 2π. Without any loss of generality, let us drop the ˆ with the understanding that the new variable t used for the rest of this section is really the normalized t for which the corresponding function has a period of 2π instead of $2T$, so that,

$$h(t) \approx \sum_{n=-\infty}^{\infty} c_n e^{i(nt)} \qquad (24.331)$$

where,

$$c_n = \frac{1}{2\pi} \int_{-\pi}^{\pi} h(t) e^{-i(nt)} dt \qquad (24.332)$$

We will use the un-normalized and normalized versions of the Fourier series expansion interchangeably depending on the needs of the subject being discussed.

Equation 24.331 is known as the *Complex Fourier Series* expansion of $h(t)$ and is based on the premise that the complex exponential trigonometric function, $e^{i(nt)}$ forms a set of orthogonal functions (see Definition 24.51), namely,

$$
\begin{aligned}
\int_{-\pi}^{\pi} e^{i(kt)} \overline{e^{i(nt)}} dt &= \int_{-\pi}^{\pi} e^{i(kt)} e^{-i(nt)} dt \\
&= \int_{-\pi}^{\pi} e^{i(k-n)t} dt \\
&= \frac{-i e^{i(k-n)t}}{k-n} \Bigg|_{-\pi}^{\pi} \\
&= \begin{cases} 0 & \forall \ k \neq n \\ 2\pi & \text{for } k = n \end{cases}
\end{aligned}
\qquad (24.333)
$$

Dirichlet also showed that for the limited number of discontinuities of $h(t)$, in the interval $[-\pi, \pi]$, the series in 24.331 converges to $\frac{h(t_{0+}) + h(t_{0-})}{2}$ where t_0 are the points of finite discontinuities of $h(t)$.

Using Property 24.5 of the complex trigonometric function, e^{int}, the Complex Fourier Series expansion of Equation 24.331 may be written in terms of series in sines and cosines in the following form,

$$h(t) \approx \frac{u_0}{2} + \sum_{n=1}^{\infty} (a_n \cos(nt) + b_n \sin(nt)) \qquad (24.334)$$

$$a_n = \frac{1}{\pi} \int_{-\pi}^{\pi} h(t) \cos(nt) dt \quad n = 0, 1, 2, \cdots$$

$$b_n = \frac{1}{\pi} \int_{-\pi}^{\pi} h(t) \sin(nt) dt \quad n = 1, 2, \cdots$$

Note that if the function $h(t)$ is odd, then the expansion will reduce to one including sines only, since the $\{a_n\}$ vanish. Similarly, if $h(t)$ is even, then the $\{b_n\}$

vanish and the function may be expanded in terms of $\frac{a_0}{2}$ and the cosine terms in Equation 24.334 – see Definitions 24.27 and 24.28.

Property 24.18 (Fourier Coefficients of a Real function are Real). *Note that in 24.334, $h(t) \in \mathbb{R} \iff a_n \in \mathbb{R} \wedge b_n \in \mathbb{R}$. The equivalent of this statement for Equation 24.331 is that $h(t) \in \mathbb{R} \iff c_{-n} = \overline{c_n}$.*

In fact, the *Fourier Series* is a general series which may be written in terms of any infinite system of functions $\{\varphi_0, \varphi_1, \varphi_2, \cdots\}$ orthogonal relative to a weight function, $w(t)$ in the closed interval $[t_1, t_2]$.

Definition 24.67 (General Fourier Series). *See [2]. Let $\{\varphi_0, \varphi_1, \varphi_2, \cdots\}$ be an infinite system of functions orthogonal relative to the weight function $w(t)$ in the closed interval $[t_1, t_2]$. If $h(t)$ is piecewise continuous, then,*

$$h(t) \approx \sum_{k=0}^{\infty} a_k \varphi_k(t) \tag{24.335}$$

where

$$a_n = \frac{\int_a^b h(t)\varphi(t)w(t)dt}{\int_a^b \varphi_n^2(t)w(t)dt} \tag{24.336}$$

is the General Fourier Series expansion of $h(t)$ relative to the system of orthogonal functions, $\{\varphi_n(t)\}$. The $\{a_n\}$ are called the General Fourier Coefficients of $h(t)$ relative to the orthogonal basis, $\{\varphi_n(t)\}$.

The Complex Fourier Series is a special case of the General Fourier Series. In this special case, the *orthogonal set of functions* (Definition 24.51), $\varphi_n(t)$, are the family of *exponential functions*, $e^{i(nt)}$ and the weighting function, $w(t) = 1$. A more general set of orthogonal functions happen to be the *Eigenfunctions* of the *Sturm-Liouville Eigenvalue problem* which stems from the problem of cooling of a uniform rod with mixed boundary conditions (Equations 24.321, 24.322, 24.323 and 24.324).

The General Sturm-Liouville Eigenvalue problem is as follows,

$$\frac{1}{w(t)} \frac{d}{dt}\left[\kappa(t)\frac{d\varphi(t)}{dt}\right] + q(t)\varphi(t) + \lambda\varphi(t) = 0 \quad \forall t_1 < t < t_2 \tag{24.337}$$

$$\mathbf{C_1}\begin{bmatrix} \varphi(t_1) \\ \frac{\varphi(t)}{dt}\Big|_{t=t_1} \end{bmatrix} + \mathbf{C_2}\begin{bmatrix} \varphi(t_2) \\ \frac{\varphi(t)}{dt}\Big|_{t=t_2} \end{bmatrix} = 0 \tag{24.338}$$

Equation 24.338 which is a set of general boundary conditions (a linear combination governed by the constant mixture matrices $\mathbf{C_1}, \mathbf{C_2} : \mathcal{R}^2 \mapsto \mathcal{R}^2$ of the function and its derivative at the boundaries) should be satisfied by the function, $h(t)$. Also, $h(t)$ should be $\mathfrak{C}^2$ continuous in the interval $[t_1, t_2]$ for uniform convergence of the series of Equation 24.335. The set of orthogonal functions in Equation 24.335 are the *Eigenfunctions* of the *Eigenvalue problem* 24.337 which are *orthogonal* about the weighting function $w(t)$ of the same Eigenvalue problem. The λs in Equation 24.337 are the corresponding Eigenvalues. The use of the independent variable t in Equation 24.337 is to make it relevant to the discussion in this section. In order to better understand the physics behind the original problem, note that the independent variable is x which is the position in the length of a uniform rod and the weighting function w is the specific heat per unit volume, κ is the conductivity of the beam being considered and q is the rate of heat generation (heat input into the rod).

24.6.1 Convergence of the Fourier Series

There are different notions of convergence. To examine the convergence of the Complex Fourier Series, let us define two different convergence criteria:

First, we define a partial sum approximation of an infinite series expansion of a function, $h(t)$, in terms of a set of orthogonal functions, $\varphi_k(t)$ as,

$$s_n(t) \overset{\Delta}{=} \sum_{k=0}^{n} a_k \varphi_k(t) \tag{24.339}$$

Definition 24.68 (Pointwise Convergence in an Interval). *A pointwise convergence of an infinite series expansion of a function, $h(t)$, in an interval $[t_1, t_2]$ means that the partial sum $s_n(t)$ (Equation 24.339) of the infinite series (Equations 24.335 and 24.336) converges to $h(t)$ for every point $t \in [t_1, t_2]$ independently as $n \to \infty$.*

For *pointwise convergence* of the Complex Fourier Series of a function, the function has to be $\mathfrak{C}^1$ continuous in the interval $[t_1, t_2]$ except for a finite number of discontinuities – see Definition 24.19.

Definition 24.69 (Uniform Convergence in an Interval). *For uniform convergence of the infinite series expansion described in Equations 24.335 and 24.336, the partial sum, $s_n(t)$ (Equation 24.339) should converge to $h(t)$ in the interval $[t_1, t_2]$ in the following manner. Let us define,*

$$\{\varepsilon : \varepsilon \in \mathbb{R} \ \wedge \ \varepsilon > 0\} \tag{24.340}$$

such that,

$$\forall \varepsilon \ \exists \{N(\varepsilon) : N(\varepsilon) \in \{1, 2, \cdots\}\} \tag{24.341}$$

with the property that,

$$|h(t) - s_n(t)| \leq \varepsilon \ \forall \, n \geq N(\varepsilon) \tag{24.342}$$

Definition 24.69 implies *uniform convergence* in the interval $[t_1, t_2]$ since no matter how small we make ε, the partial sum still converges to $h(t)$ in the interval. This means that two adjacent points in the interval cannot have very different convergence rates. This neighborhood of continuity of convergence is described by ε which may be taken to be as small as needed. To achieve this type of convergence, of course, it is intuitively apparent that as ε is made smaller, $N(\varepsilon)$ should grow. For this relatively stringent convergence, the function $h(t)$ must possess an extra feature and that is class $\mathcal{C}^2$ continuity instead of the $\mathcal{C}^1$ continuity which was needed for pointwise convergence. See [2] for a *uniform convergence* proof for the General Fourier Series expansion of a function, $h(t)$.

24.6.2 Parseval's Theorem

Parseval's theorem was developed before Fourier series were defined, but was later applied to Fourier Series and Fourier and other transforms. It was originally defined for real-value functions. The original motivation of the theorem roughly translate to the fact that the total spectral energy of a signal (over all frequencies) is equal to the total energy of the signal over all time. Parseval's Theorem for the general case of complex functions is stated as follows,

Theorem 24.24 (Parseval's Theorem – Fourier Series). *Assume that there are two functions, $g(t)$ and $h(t)$ which may be expanded using Fourier Series Expansion as follows,*

$$g(t) = \sum_{m=-\infty}^{\infty} G_m e^{i(mt)} \ \text{where} \ G_m = \frac{1}{2\pi} \int_{-\pi}^{\pi} g(t) e^{-i(mt)} dt \tag{24.343}$$

$$h(t) = \sum_{n=-\infty}^{\infty} H_n e^{i(nt)} \ \text{where} \ H_n = \frac{1}{2\pi} \int_{-\pi}^{\pi} h(t) e^{-i(nt)} dt \tag{24.344}$$

Then,

$$\sum_{n=-\infty}^{\infty} G_n \overline{H_n} = \frac{1}{2\pi} \int_{-\pi}^{\pi} g(t) \overline{h(t)} dt \tag{24.345}$$

Proof.
Before we start the proof, note that if $s = \sum_i s_i$ where $s \in \mathbb{C} \forall i$, then

$$\bar{s} = \sum_i \overline{s_i} \tag{24.346}$$

Also, examine the product, $se^{i\theta}$, where $s = \sigma + i\omega$ and $s \in \mathbb{C}, \sigma \in \mathbb{R}, \omega \in \mathbb{R}$, and $\theta \in \mathbb{R}$. Then, the following is the conjugate of this product,

$$
\begin{aligned}
\overline{se^{i\theta}} &= \overline{(\sigma + i\omega)e^{i\theta}} \\
&= \overline{(\sigma + i\omega)(\cos(\theta) + i\sin(\theta))} \\
&= \overline{(\sigma\cos(\theta) - \omega\sin(\theta)) + i(\sigma\sin(\theta) + \omega\cos(\theta))} \\
&= (\sigma\cos(\theta) - \omega\sin(\theta)) - i(\sigma\sin(\theta) + \omega\cos(\theta)) \\
&= \cos(\theta)(\sigma - i\omega) - \sin(\theta)(\omega + i\sigma) \\
&= \cos(\theta)(\sigma - i\omega) - i\sin(\theta)(\sigma - i\omega) \\
&= (\sigma - i\omega)(\cos(\theta) - i\sin(\theta)) \\
&= \bar{s}e^{-i\theta}
\end{aligned}
\tag{24.347}
$$

Let us start the proof from the right hand side of Equation 24.345 and having in mind Equations 24.346 and 24.347,

$$
\begin{aligned}
\frac{1}{2\pi}\int_{-\pi}^{\pi} g(t)\overline{h(t)}dt &= \frac{1}{2\pi}\int_{-\pi}^{\pi}\sum_{m=-\infty}^{\infty}G_m e^{i(mt)}\sum_{n=-\infty}^{\infty}\overline{H_n}e^{-i(nt)}dt \\
&= \frac{1}{2\pi}\int_{-\pi}^{\pi}\sum_{m=-\infty}^{\infty}\sum_{n=-\infty}^{\infty}G_m e^{i(mt)}\overline{H_n}e^{-i(nt)}dt \\
&= \frac{1}{2\pi}\sum_{m=-\infty}^{\infty}\sum_{n=-\infty}^{\infty}\int_{-\pi}^{\pi}G_m\overline{H_n}e^{i(mt)}e^{-i(nt)}dt \\
&= \frac{1}{2\pi}\sum_{m=-\infty}^{\infty}\sum_{n=-\infty}^{\infty}G_m\overline{H_n}\int_{-\pi}^{\pi}e^{i(mt)}e^{-i(nt)}dt
\end{aligned}
\tag{24.348}
$$

From the orthogonality of the set of complex exponential functions (see Definition 24.51 and Equation 24.333),

$$
\int_{-\pi}^{\pi}e^{i(mt)}e^{-i(nt)}dt = \begin{cases} 0 & \forall\ m \neq n \\ 2\pi & \text{for}\ m = n \end{cases}
\tag{24.349}
$$

Equation 24.349 means that one of the infinite sums in Equation 24.348 reduces to a single value, at the point where $m = n$ and it becomes zero for all other m. Therefore, Equation 24.348 may be written as follows,

$$
\frac{1}{2\pi}\int_{-\pi}^{\pi}g(t)\overline{h(t)}dt = \sum_{n=-\infty}^{\infty}G_n\overline{H_n}
\tag{24.350}
$$

proving the original statement of the Theorem.

$\square$

Note that if $g(t) = h(t)$,

$$
\sum_{n=-\infty}^{\infty}H_n\overline{H_n} = \frac{1}{2\pi}\int_{-\pi}^{\pi}h(t)\overline{h(t)}dt
\tag{24.351}
$$

or

$$\sum_{n=-\infty}^{\infty} |H_n|^2 = \frac{1}{2\pi} \int_{-\pi}^{\pi} |h(t)|^2 \, dt \tag{24.352}$$

24.7 Wavelet Series Expansion

Recall the Generalized Fourier Series expansion described by Equations 24.335 and 24.336. This series is a linear combination of a set of orthogonal functions which map each point in time $h(t)$ (either *pointwise* (Definition 24.68) or *uniformly* (Definition 24.69) in a neighborhood) to points in the frequency domain defined by the combination coefficients.

The following is a basic wavelet expansion using the basis functions, $\{\psi_{jk}(t)\}$ which are derived from the stretching and translation of a basic function for the set called the *Mother Wavelet*, $\psi(t)$. The coefficients of this expansion form a two-dimensional infinite matrix of numbers $\{a_{ij}\}$ which are called the *Discrete Wavelet Transform (DWT)* of $h(t)$ and are analogous to the *Discrete Fourier Transform* discussed in section 24.10. Although theoretically this matrix has infinite elements, based on a property of wavelets only a few of these elements are enough to provide a good approximation of the function being expanded.

$$h(t) = \sum_{k=-\infty}^{\infty} a_{jk} \psi_{jk}(t) \tag{24.353}$$

Usually, the transformation of the mother wavelet to the individual members of the basis function looks something like Equation 24.354.

$$\psi_{jk}(t) = 2^{\frac{j}{2}} \psi(2^j t - k) \tag{24.354}$$

Note that the wavelet expansion is not unique. In fact there are infinitely many different wavelet bases which may be used to expand any function $h(t)$. Although wavelets had been studied in as early as 1910 ([21] and [6]), the 1980s brought about a revolution in their expansion and usage in different engineering problems. [32] [48] and [5] present useful tutorials on the history of wavelets and treat the subject in detail.

In his early work, Haar showed that taking certain functions and modifying them through scaling and shifting to form orthogonal bases can represent many different types of signals using a two-dimensional infinite series given by Equation 24.355.

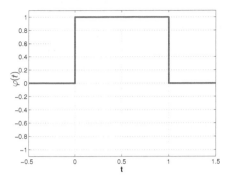

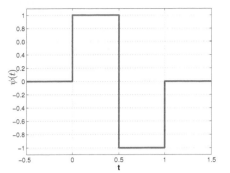

Fig. 24.14: The Haar Scale Function $\varphi(t)$

Fig. 24.15: The Haar Mother Wavelet $\psi(t)$

$$h(t) = \sum_{k=-\infty}^{\infty} \left(c_k \underbrace{\varphi(t-k)}_{\text{Haar Scale Function}} + \sum_{j=0}^{\infty} d_{jk} \underbrace{\psi(2^j t - k)}_{\text{Haar Wavelet}} \right) \tag{24.355}$$

where,

$$\varphi(t) = \begin{cases} 1 \ \forall \ 0 \leq t < 1 \\ 0 \ \forall \ t < 0 \vee t \geq 1 \end{cases} \tag{24.356}$$

and

$$\psi(t) = \begin{cases} 0 \ \forall \ t < 0 \vee t \geq 1 \\ 1 \ \forall \ 0 \leq t < \frac{1}{2} \\ -1 \ \forall \ \frac{1}{2} \leq t < 1 \end{cases} \tag{24.357}$$

In Section 24.12 we will see a very interesting connection between wavelets transforms and different forms of Fourier Transforms.

For a list of popular wavelets and their properties see [5].

24.8 The Laplace Transform

To be able to understand the nature of integral transforms used in most signal processing applications such as Speaker Recognition, we start with a general class of *Integral Transforms* , namely, the *Laplace Transform* of which the *Fourier Transform* and the *z-Transform* are special cases.

Consider the General Integral Transform and its inverse given by Equation pair 24.234 and 24.235. The General Laplace Transform is defined by setting

$\mathscr{K}(t,s) = e^{-st}$, $t_1 = -\infty$, and $t_2 = \infty$ in Equation 24.234. This is the most general integral transform using the exponential family of orthogonal functions. As we will see later, by making certain restrictions on the definition of s, other transforms will be generated.

Definition 24.70 (Laplace Transform). *The general Laplace Transform is defined by the following,*

$$H(s) = \mathscr{L}\{h\} \overset{\Delta}{=} \int_{-\infty}^{\infty} h(t)e^{-st}dt \quad s \in \mathbb{C} \tag{24.358}$$

Although it is defined for $-\infty < t < \infty$; for speaker recognition, or signal processing in general, we are only interested in functions of time where we may ignore the history of the signal. Assume that a signal started τ seconds prior to the point where our origin of time is defined, then without any loss of generality, we may shift the origin with which we define t to the past by τ seconds so that the function of interest $h(t) = 0 \; \forall \; t < 0$. With this new definition of t, the integral limit t_1 of Equation 24.234 may be set to 0. In fact, to be able to allow for the definition of the *Laplace Transform* of an Impulse function, we will set $t_1 = 0^-$. This issue will be visited later, when we explore the Laplace Transform of the more important functions. Based on this new definition of the origin, we define the, so called, *Unilateral (one-sided) Laplace Transform*.

Definition 24.71 (Unilateral (One-Sided) Laplace Transform). *The Unilateral or One-Sided Laplace Transform is defined by the following,*

$$H(s) = \mathscr{L}(h) \overset{\Delta}{=} \int_{0^-}^{\infty} h(t)e^{-st}dt \quad s \in \mathbb{C} \tag{24.359}$$

In general, the Laplace Transform of Equation 24.359 does not exist for the whole Complex plane. However, it does exist for functions which grow at most as fast as exponential functions.

Theorem 24.25 (Existence and Boundedness of the Unilateral Laplace Transform). *For the Unilateral Laplace Transform to exist, the integral of Equation 24.359 should be bounded, or,*

$$\left| \int_{0}^{\infty} h(t)e^{-st}dt \right| < M \quad where \; M : M < \infty \tag{24.360}$$

A sufficient condition for this to happen is that the function $h(t)$ would be bounded by an exponential family of functions, namely, that $\exists \; \{k, s_c\}$ such that,

$$|h(t)| \begin{cases} \leq k e^{s_c t} & \forall \; t \geq 0 \\ = 0 & \forall \; t < 0 \end{cases} \tag{24.361}$$

Proof.
For the Laplace transform to exist, we should show that,

$$\left| \int_0^\infty h(t) e^{-st} dt \right| < M \tag{24.362}$$

We know from Theorem 24.12 that,

$$\left| \int_0^\infty h(t) e^{-st} dt \right| \leq \int_0^\infty \left| h(t) e^{-st} \right| dt \tag{24.363}$$

$$= \int_0^\infty |h(t)| \left| e^{-st} \right| dt \quad \text{(From Theorem 24.1)} \tag{24.364}$$

$$\leq \int_0^\infty k e^{\sigma_c t} \left| e^{-st} \right| dt \quad \text{(From Equation 24.361)} \tag{24.365}$$

$$= \int_0^\infty k e^{\sigma_c t} \left| e^{-\sigma t} \right| \left| e^{i\omega_c t} \right| \left| e^{-i\omega t} \right| dt \tag{24.366}$$

$$= \int_0^\infty k e^{\sigma_c t} e^{-\sigma t} dt \quad \left(\text{Since } \left| e^{i\omega_c t} \right| = \left| e^{-i\omega t} \right| = 1 \right) \tag{24.367}$$

$$= \int_0^\infty k e^{(\sigma_c - \sigma)t} dt \tag{24.368}$$

Therefore, if $\exists \, \{k, s_c = \sigma_c + i\omega_c\}$ such that

$$k \int_0^\infty e^{(\sigma_c - \sigma)t} dt < M \tag{24.369}$$

Consequently, $\exists \, \{s_c = \sigma_c + i\omega_c\}$ such that

$$\int_0^\infty e^{(\sigma_c - \sigma)t} dt < \tilde{M} \tag{24.370}$$

$$\tag{24.371}$$

where $\tilde{M}$ is another number such that $\tilde{M} < \infty$, or

$$\left| \int_0^\infty h(t) e^{-st} dt \right| < \tilde{M} \tag{24.372}$$

which means that the Laplace Transform exists.

□

Figure 24.16 shows the region of convergence of the Laplace Transform. The transform converges everywhere to the right of the, so called, abscissa of convergence of the transform, the line parallel to the imaginary axis and going through σ_c as defined in Theorem 24.25.

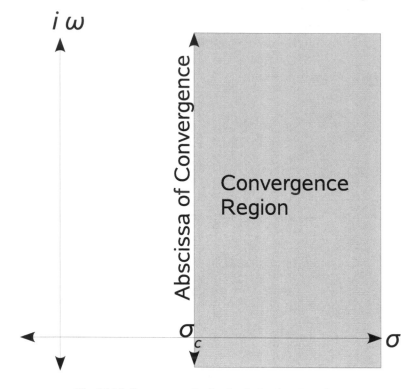

Fig. 24.16: Convergence Region for the Laplace Transform

24.8.1 Inversion

$$h(t) = \frac{1}{2\pi i} \left[\int_{\sigma_c - i\infty}^{\sigma_c + i\infty} H(s) e^{st} ds \right] \tag{24.373}$$

σ_c is chosen so that all the singularities are to the left of the line going parallel to the Imaginary access. Based on the Residue Theorem (Theorem 24.15) , integrating over a semicircle in the left of the s-plane,

$$h(t) = \frac{1}{2\pi i} 2\pi i \sum (\text{Residue inside Contour}) \tag{24.374}$$

e.g.,

$$H(s) = \frac{1}{(s+a)(s+b)} \tag{24.375}$$

The Residue at $s = -a$ is,

$$\left[\frac{(s+a)e^{st}}{(s+a)(s+b)} \right]_{s \to -a} \tag{24.376}$$

and the Residue at $s = -b$ is,

$$\left[\frac{(s+b)e^{st}}{(s+a)(s+b)} \right]_{s \to -b} \tag{24.377}$$

Therefore,

$$h(t) = \frac{1}{b-a} \left[e^{-at} - e^{-bt} \right] \tag{24.378}$$

24.8.2 Some Useful Transforms

Unit Step Function,

$$h(t) = \begin{cases} 1 \ \forall \ t \geq 0 \\ 0 \ \forall \ t < 0 \end{cases} \tag{24.379}$$

$$H(s) = \int_0^\infty e^{-st} dt = \left[\frac{e^{-st}}{-s} \right]_0^\infty = \frac{1}{s} \tag{24.380}$$

hence a simple pole at the origin.

Unit Ramp Function,

$$h(t) = \begin{cases} t \ \forall \ t \geq 0 \\ 0 \ \forall \ t < 0 \end{cases} \tag{24.381}$$

$$H(s) = \int_0^\infty t e^{-st} dt \tag{24.382}$$

Using Integration by Parts,

$$H(s) = \left[\frac{t e^{-st}}{-s} \right]_0^\infty + \frac{1}{s} \int_0^\infty e^{-st} dt = \frac{1}{s^2} \tag{24.383}$$

24.9 Complex Fourier Transform (Fourier Integral Transform)

Fourier Transform may be viewed from two main perspectives. One view would be as the extension of the Fourier series expansion of Section 24.6. This track of thinking is geared toward the removal of the periodicity dependence of the expansion. Recall that one of the Dirichlet conditions necessary for the convergence of the Fourier Series states that the function $h(t)$ must be periodic with period, $2T$, and that it should be absolutely integrable in the interval, $[-T, T]$. In fact, to be able to use the Fourier Series expansion, of a function, we had to resort to the definition of the periodic extension of the function, Definition 24.30 and performed the series expansion for that extension of the function. Then we stated that would only be interested in the interval $[-T, T]$ of the constructed series. If we take the limit of $T \to \infty$, we arrive at the Fourier Integral in place of the Fourier Sum. Rewriting Equations 24.329 and 24.330 which define the Period Normalized Complex Fourier Series Expansion of a periodic function $h(t)$, and taking the period from 2π to ∞, using ω for the continuous version of the integration variable, and using $H(\omega)$ as the continuous version of c_n, we have,[8]

$$h(t) = \int_{-\infty}^{\infty} H(\omega)e^{i(\omega t)}d\omega \tag{24.384}$$

where,

$$H(\omega) = \frac{1}{2\pi}\int_{-\infty}^{\infty} h(t)e^{-i(\omega t)}dt \tag{24.385}$$

The pair $h(t)$ and $H(\omega)$ are the Complex Fourier Integral Transform pair and their relation is denoted as follows,

$$h(t) \leftrightarrow H(\omega) \tag{24.386}$$

Note that there is some freedom in the definition of the pair of functions in 24.386. The following is a more general statement of these functions,

$$H(\omega) = \alpha \int_{-\infty}^{\infty} h(t)e^{-i(\omega t)}dt \tag{24.387}$$

and,

$$h(t) = \beta \int_{-\infty}^{\infty} H(\omega)e^{i(\omega t)}d\omega \tag{24.388}$$

where,

---

[8] Here we have replaced the more proper $\equiv$ sign with $=$ for the sake of convenience.

$$\alpha \cdot \beta = \frac{1}{2\pi} \tag{24.389}$$

To end up with the version of the Fourier Transform used in the historic paper of Shannon [43], we chose $\alpha = 1$ and $\beta = \frac{1}{2\pi}$ which will give us the following definition,

$$H(\omega) = \int_{-\infty}^{\infty} h(t)e^{-i(\omega t)}dt \tag{24.390}$$

and,

$$h(t) = \frac{1}{2\pi}\int_{-\infty}^{\infty} H(\omega)e^{i(\omega t)}d\omega \tag{24.391}$$

Some have also defined the Fourier Transform pair in terms of the Frequency $f = \frac{\omega}{2\pi}$ which removes the need for carrying over the $\frac{1}{2\pi}$ factor, but complicates the exponential function,

$$H(f) = \int_{-\infty}^{\infty} h(t)e^{-i(2\pi ft)}dt \tag{24.392}$$

and,

$$h(t) = \int_{-\infty}^{\infty} H(f)e^{i(2\pi ft)}df \tag{24.393}$$

In Equations 24.392 and 24.393,

$$\underbrace{\omega}_{rad/s} \equiv 2\pi \underbrace{f}_{Hz} \tag{24.394}$$

Hence,

$$[H(f)]_{f=\frac{\omega}{2\pi}} \equiv H(\omega) \tag{24.395}$$

Note that in Equations 24.392 and 24.393, if $h = h(x)$ where x is measured in meters, then $H = H(f)$ where f is in cycles per meter or in the units of $\frac{1}{\lambda}$ where λ is the *wavelength*, measured in meters.

There is yet more freedom where one may choose to use $e^{i(\omega t)}$ in 24.390 which then forces the choice of $e^{-i(\omega t)}$ in 24.391. All these different forms are used and each may be preferred at certain circumstances for the sake of convenience. As stated earlier, we have chosen the form which is used in Shannon's paper [43] (Equations 24.390 and 24.391) so that the statement of the Sampling Theorem in

Section 3.1.1 will follow Shannon's definition more closely.

Another way of arriving at the Complex Fourier Integral Transforms is as a special case of the Laplace Transform (Equations 24.358 and 24.373) in which the Laplace variable $s = i\omega$, namely, the transform happens along the imaginary axis. This will identically produce Equations 24.390 and 24.391.

24.9.1 Translation

Fourier transform is a linear operation

    i.e.,

$$\mathscr{F}\{h \pm g\} = \mathscr{F}\{h\} \pm \mathscr{F}\{g\} \tag{24.396}$$

24.9.2 Scaling

$$\mathscr{F}\{ah\} = a\mathscr{F}\{h\} \tag{24.397}$$

24.9.3 Symmetry Table

Let "$h(t) \leftrightarrow H(\omega)$" denote a Fourier Transform Pair. The following table provides a set of statements where given the statement on the left, the statement on the right is true

$h(t) \in \mathbb{R}$	$\overline{H(\omega)} = H(-\omega) (H(\omega) \text{ is Hermitian})$
$h(t) \in \mathbb{I}$	$\overline{H(\omega)} = -H(-\omega)$
$h(t) = h(-t) (h(t) \text{ is even})$	$H(\omega) = H(-\omega) (H(\omega) \text{ is even})$
$h(t) = -h(-t) (h(t) \text{ is odd})$	$H(\omega) = -H(-\omega) (H(\omega) \text{ is odd})$
$h(t) \in \mathbb{R} \wedge h(t) = h(-t)$	$H(\omega) \in \mathbb{R} \wedge H(\omega) = H(-\omega)$
$h(t) \in \mathbb{R} \wedge h(t) = -h(-t)$	$H(\omega) \in \mathbb{I} \wedge H(\omega) = -H(-\omega)$
$h(t) \in \mathbb{I} \wedge h(t) = h(-t)$	$H(\omega) \in \mathbb{I} \wedge H(\omega) = H(-\omega)$
$h(t) \in \mathbb{I} \wedge h(t) = -h(-t)$	$H(\omega) \in \mathbb{I} \wedge H(\omega) = -H(-\omega)$

24.9.4 Time and Complex Scaling and Shifting

The following are Fourier transform pairs:

- $h(\alpha t) \quad \leftrightarrow \quad \frac{1}{|\alpha|} H(\frac{\omega}{\alpha})$ (Time Scaling)
- $\frac{1}{|\gamma|} h(\frac{t}{\gamma}) \quad \leftrightarrow \quad H(\gamma \omega)$ (Frequency Scaling)
- $h(t - t_0) \quad \leftrightarrow \quad H(\omega) e^{i\omega t_0}$ (Time Shifting)
- $h(t) e^{i\omega_0 t} \quad \leftrightarrow \quad H(\omega - \omega_0)$ (Frequency Shifting)

24.9.5 Convolution

Let $g(t)$ and $h(t)$ be two functions defined in the time domain with their Fourier transforms as $G(\omega)$ and $H(\omega)$ respectively,

$$g(t) \quad \leftrightarrow \quad G(\omega) \tag{24.398}$$
$$h(t) \quad \leftrightarrow \quad H(\omega) \tag{24.399}$$

The convolution is defined in the time domain as follows,

$$g * h \ \overset{\Delta}{=} \ \int_{-\infty}^{\infty} g(\tau)h(t-\tau)d\tau \tag{24.400}$$

property of convolution,

$$g * h \ = \ h * g \tag{24.401}$$

Theorem 24.26 (Convolution Theorem). *The Fourier Transform of the convolution of two functions is the product of their individual Fourier Transforms. Namely, the following is a Fourier Transform pair,*

$$g(t) * h(t) \ \leftrightarrow \ G(\omega)H(\omega) \tag{24.402}$$

$$\mathscr{F}\{g(t) * h(t)\} = \mathscr{F}\{g(t)\}\mathscr{F}\{h(t)\} \tag{24.403}$$

24.9.6 Correlation

The correlation of two functions, $f(t)$ and $g(t)$ is defined as follows (see Section 24.2.2),

$$\underbrace{g(t) \circ h(t)}_{\text{function of lag t}} \ \overset{\Delta}{=} \ \int_{-\infty}^{\infty} g(\tau + \underbrace{t}_{lag})h(\tau)d\tau \tag{24.404}$$

Theorem 24.27 (Correlation Theorem). *The Fourier Transform of the correlation of two functions is the product of the Fourier Transform of the first one and the Fourier Transform of the negative frequency for the second one. Namely, the following is a Fourier Transform pair,*

$$g(t) \circ h(t) \ \leftrightarrow \ G(\omega)H(-\omega) \tag{24.405}$$

Usually, $g \in \mathbb{R}$ *and* $h \in \mathbb{R} \implies H(-\omega) = \overline{H(\omega)}$

or

$$\mathscr{F}\{g \circ h\} = \mathscr{F}\{g\}\mathscr{F}\{h\} \tag{24.406}$$

24.9.7 Parseval's Theorem

Theorem 24.28 (Parseval's Theorem – Fourier Transform). *The Total power in a signal is the same when computed in the time or Frequency domain. In other words, the Total Power P is given by,*

$$P \triangleq \int_{-\infty}^{\infty} |h(t)|^2 dt$$

$$= \frac{1}{2\pi} \int_{-\infty}^{\infty} |H(\omega)|^2 d\omega \tag{24.407}$$

Proof.
Consider the definition of the Complex Fourier Transform and its inverse given by
Equation pair 24.390 and 24.391. Also, note the proof of Parseval's theorem for the
Complex Fourier Series in Section 24.6.2. Take any two complex, time dependent
functions, $g(t)$ and $h(t)$, and let us define the integral,

$$\hat{P} \triangleq \int_{-\infty}^{\infty} g(t)\overline{h(t)}dt \tag{24.408}$$

Note that based on the argument of Section 24.6.2, if

$$h(t) = \frac{1}{2\pi} \int_{-\infty}^{\infty} H(\omega)e^{i\omega t} d\omega \tag{24.409}$$

then,

$$\overline{h(t)} = \frac{1}{2\pi} \int_{-\infty}^{\infty} \overline{H(\omega)}e^{-i\omega t} d\omega \tag{24.410}$$

Also, the inverse Complex Fourier transform expression for $G(\omega)$ is,

$$g(t) = \frac{1}{2\pi} \int_{-\infty}^{\infty} G(\omega)e^{i\omega t} d\omega \tag{24.411}$$

Plugging in $g(t)$ and $\overline{h(t)}$ into Equation 24.408 from Equations 24.411 and 24.410
respectively,

$$\hat{P} = \frac{1}{(2\pi)^2} \int_{-\infty}^{\infty} \int_{-\infty}^{\infty} G(\omega_1)e^{i\omega_1 t} d\omega_1 \int_{-\infty}^{\infty} \overline{H(\omega_2)}e^{-i\omega_2 t} d\omega_2 dt$$

$$= \frac{1}{(2\pi)^2} \int_{-\infty}^{\infty} \int_{-\infty}^{\infty} \int_{-\infty}^{\infty} G(\omega_1)e^{i\omega_1 t}\overline{H(\omega_2)}e^{-i\omega_2 t} d\omega_2 d\omega_1 dt$$

$$= \frac{1}{(2\pi)^2} \int_{-\infty}^{\infty} \int_{-\infty}^{\infty} \int_{-\infty}^{\infty} G(\omega_1)\overline{H(\omega_2)}e^{i\omega_1 t}e^{-i\omega_2 t} d\omega_2 d\omega_1 dt$$

$$= \frac{1}{(2\pi)^2} \int_{-\infty}^{\infty} \int_{-\infty}^{\infty} G(\omega_1)\overline{H(\omega_2)} \int_{-\infty}^{\infty} e^{i\omega_1 t}e^{-i\omega_2 t} dt d\omega_2 d\omega_1 \tag{24.412}$$

Recall the orthogonality of the exponential function,

$$\int_{-\infty}^{\infty} e^{i\omega_1 t}e^{-i\omega_2 t} dt = \begin{cases} 0 & \forall \ \omega_1 \neq \omega_2 \\ 2\pi & \text{for } \omega_1 = \omega_2 \end{cases} \tag{24.413}$$

Using the orthogonality relation of Equation 24.413 in Equation 24.412,

$$\hat{P} = \int_{-\infty}^{\infty} g(t)\overline{h(t)}dt$$

$$= \frac{1}{2\pi} \int_{-\infty}^{\infty} G(\omega)\overline{H(\omega)}d\omega \tag{24.414}$$

Now, if we set $g(t) = h(t)$, we have,

$$P = \int_{-\infty}^{\infty} |h(t)|^2 dt$$

$$= \frac{1}{2\pi} \int_{-\infty}^{\infty} |H(\omega)|^2 d\omega \tag{24.415}$$

$\square$

24.9.8 Power Spectral Density

Take the integrand of the right hand side of Equation 24.415. $|H(\omega)|^2$ is the *power spectral density* which means that it is the power for the infinitesimal portion of the spectrum of the signal, $h(t)$, from frequency f to $f + df$ or in angular form from $\frac{\omega}{2\pi}$ to $\frac{\omega + d\omega}{2\pi}$. Of course it may be defined in terms of the spectral space from ω to $\omega + d\omega$ as well. The former would be in the units of $\frac{\text{unit of Power}}{Hz}$ and the latter would be in $\frac{\text{unit of Power}}{rad.}$. To recapitulate, the power spectral density in the linear frequency is given by,

$$\mathscr{P}_d(f) = |H(f)|^2 \tag{24.416}$$

where $f = \frac{\omega}{2\pi}$ and the power spectral density in angular frequency is given by,

$$\mathscr{P}_d^{\circ}(\omega) \triangleq \frac{|H(\omega)|^2}{2\pi} \tag{24.417}$$

24.9.9 One-Sided Power Spectral Density

Most of the time, we do not distinguish between negative and positive frequencies. Therefore, the one-sided PSD is defined to contain both ω and $-\omega$.

The one-sided Power Spectral Density (PSD) for the linear frequency of signal $h(t)$ is defined as,

$$\mathscr{P}_{dh}(f) \triangleq |H(f)|^2 + |H(-f)|^2 \quad 0 \le f < \infty \tag{24.418}$$

and the one-sided PSD for the angular frequency is,

$$\mathscr{P}_{dh}^{\circ}(\omega) \triangleq \frac{|H(\omega)|^2 + |H(-\omega)|^2}{2\pi} \quad 0 \leq \omega < \infty \tag{24.419}$$

Note that for $h(t) \in \mathbb{R}$,

$$H(-f) = \overline{H(f)} \leftrightarrow \mathscr{P}_d(f) = 2|H(f)|^2$$
$$H(-\omega) = \overline{H(\omega)} \leftrightarrow \mathscr{P}_d^{\circ}(\omega) = \frac{1}{\pi}|H(\omega)|^2 \tag{24.420}$$

Therefore the total power $\mathscr{P}$ may be given as,

$$\mathscr{P} = \int_0^{\infty} \mathscr{P}_d(f) df$$
$$= \int_0^{\infty} \mathscr{P}_d^{\circ}(\omega) d\omega \tag{24.421}$$

24.9.10 PSD-per-unit-time

If $h(t)$ is in general non-trivial (non-zero) for $-\infty < t < \infty$ then the total power of the signal will be infinite.

Thus, we are interested in PSD-per-unit-time which is computed by taking a long, but finite portion of $h(t)$, computing the PSD for $h(t)$ for a finite time and assuming that $h(t) = 0$ everywhere else. Then, we divide the resulting PSD by total time T or $\frac{PSD}{T}$.

By *Parseval's Theorem*,

$$\mathscr{P}_{dt} \triangleq \underbrace{mean(|h(t)|^2)}_{\text{mean of the square of amplitude of h(t)}} \tag{24.422}$$

24.9.11 Wiener-Khintchine Theorem

Theorem 24.29 (Wiener-Khintchine Theorem). *The autocorrelation of a signal is the Inverse Fourier transform of its Power Spectral Density. Alternatively, the Power Spectral Density of a signal is the Fourier transform of its autocorrelation. In mathematical terms,*

$$(h \circ h)(t) = \mathscr{F}^{-1}\{\mathscr{P}_d^{\circ}\} \tag{24.423}$$

or

$$\mathscr{P}_d^{\circ}(\omega) = \mathscr{F}\{(h \circ h)\} \tag{24.424}$$

Proof.
Let us start with the autocorrelation of $h(t)$ (see Section 24.2.2) and write it in terms of the inverse Fourier transform of its spectra, somewhat in the same manner as we did for the proof of Parseval's theorem, Theorem 24.28.

$$(h \circ h)(\tau) = \int_{-\infty}^{\infty} \overline{h(t)} h(t + \tau) dt \qquad (24.425)$$

Recall that,

$$h(t) = \frac{1}{2\pi} \int_{-\infty}^{\infty} H(\omega) e^{i\omega t} d\omega \qquad (24.426)$$

and

$$\overline{h(t)} = \frac{1}{2\pi} \int_{-\infty}^{\infty} \overline{H(\omega)} e^{-i\omega t} d\omega \qquad (24.427)$$

Therefore, Equation 24.425 may be written as follows,

$$
\begin{aligned}
(h \circ h)(\tau) &= \frac{1}{(2\pi)^2} \int_{-\infty}^{\infty} \int_{-\infty}^{\infty} \overline{H(\omega_1)} e^{-i\omega_1 t} d\omega_1 \int_{-\infty}^{\infty} H(\omega_2) e^{i\omega_2(t+\tau)} d\omega_2 dt \\
&= \frac{1}{(2\pi)^2} \int_{-\infty}^{\infty} \int_{-\infty}^{\infty} \int_{-\infty}^{\infty} \overline{H(\omega_1)} H(\omega_2) e^{-i\omega_1 t} d\omega_1 e^{i\omega_2(t+\tau)} d\omega_2 dt \\
&= \frac{1}{(2\pi)^2} \int_{-\infty}^{\infty} \int_{-\infty}^{\infty} \overline{H(\omega_1)} H(\omega_2) \\
&\qquad \left[\int_{-\infty}^{\infty} e^{-i\omega_1 - \omega_2 t} dt \right] e^{i\omega_2 \tau} d\omega_1 d\omega_2 \qquad (24.428)
\end{aligned}
$$

Note that by definition,

$$\int_{-\infty}^{\infty} e^{-i(\omega_1 - \omega_2)t} dt = \delta(\omega_1 - \omega_2) \qquad (24.429)$$

Plugging the identity of Equation 24.429 for the bracketed expression in Equation 24.428,

$$(h \circ h)(\tau) = \frac{1}{(2\pi)^2} \int_{-\infty}^{\infty} \int_{-\infty}^{\infty} \overline{H(\omega_1)} H(\omega_2) e^{i\omega_2 \tau} \delta(\omega_1 - \omega_2) d\omega_1 d\omega_2 \qquad (24.430)$$

An important property of the Delta function, $\delta(x)$, is that for any function $\varphi(x)$,

$$\int_{-\infty}^{\infty} \varphi(x) \delta(x - x_0) dx = \varphi(x_0) \qquad (24.431)$$

Using this property,

$$\int_{-\infty}^{\infty} \overline{H(\omega_1)} \delta(\omega_1 - \omega_2) d\omega_1 = \overline{H(\omega_2)} \qquad (24.432)$$

Therefore, Equation 24.430 becomes,

$$(h \circ h)(\tau) = \frac{1}{(2\pi)^2} \int_{-\infty}^{\infty} \overline{H(\omega_2)} H(\omega_2) e^{i\omega_2 \tau} d\omega_2$$

$$= \frac{1}{2\pi} \int_{-\infty}^{\infty} \frac{|H(\omega_2)|^2}{2\pi} e^{i\omega_2 \tau} d\omega_2$$

$$= \mathscr{F}^{-1}\{\frac{|H(\omega)|^2}{2\pi}\}$$

$$= \mathscr{F}^{-1}\{\mathscr{P}_d^\circ\} \tag{24.433}$$

□

24.10 Discrete Fourier Transform (DFT)

To work with a finite set of sampled data points, we need a mechanism that would allow us to obtain a transform of the sampled data into a finite set of finite frequency components. Let us consider discretizing the Complex Fourier transform given by Equation pair 24.390 and 24.391. Before attempting the discretization, let us assume that we have a finite set of data points and we would like to map them to a finite set of frequencies. Let us assume that we have a finite set of N samples from time $t = 0$ to $t = N - 1$. Then, we can write the discrete time instances as,

$$t_n = nT \quad n = \{0, 1, \cdots, N - 1\} \tag{24.434}$$

Furthermore, let us assume that we would like to have a frequency resolution of N as well. This means that we should be able to take the whole frequency space and discretize it into N equi-distant values. We know that our signal has been sampled using the concepts discussed in Chapter 3 based on the sampling theorem. Therefore, the signal is band-limited and sampled such that there are only spectral components present that have frequencies less than f_c. Therefore, referring to the inverse Complex Fourier transform given by Equation 24.391, instead of ranging from $-\infty$ to ∞, the frequency will range from $-f_c$ to f_c. This means that the resolution of the discrete frequency is such that there is a step of $\frac{2f_c}{N}$ frequency levels. There fore, the discrete frequency would be,

$$f_k = \frac{k}{NT} \quad k = \{0, 1, \cdots, N - 1\}$$

$$\omega_k = \frac{2\pi k}{NT} \quad k = \{0, 1, \cdots, N - 1\} \tag{24.435}$$

Now let us start from Equation 24.390 which is the Complex Fourier Transform of signal $h(t)$ and discretize the integral with the N number of sample points using the following definitions,

$$h_n = h(nT)$$

$$\hat{H}_k = H\left(\omega = \frac{2\pi k}{NT}\right) \tag{24.436}$$

Then, in discrete form, the integral of Equation 24.390 changes to a finite sum of the N values of the signal with $dt \to T$. Therefore,

$$\hat{H}_k = \sum_{n=0}^{N-1} h_n e^{-i\frac{2\pi k}{NT}nT} T$$

$$= \sum_{n=0}^{N-1} h_n e^{-i\frac{2\pi kn}{N}} T \tag{24.437}$$

To make the Discrete Fourier Transform independent of the sampling frequency, we define the Discrete Fourier Transform H_k such that

$$\hat{H}_k = H_k T \tag{24.438}$$

Therefore, the Discrete Fourier Transform (DFT) is defined as,

$$H_k = \sum_{n=0}^{N-1} h_n e^{-i\frac{2\pi kn}{N}} \tag{24.439}$$

Note that there is also another type discretized Fourier transform called *Discrete-Time Fourier Transform* and it should not be confused with the subject of this section which is *Discrete Fourier Transform*.

24.10.1 Inverse Discrete Fourier Transform (IDFT)

Now, to compute the inverse Discrete Fourier Transform, consider the inverse Complex Fourier Transform, Equation 24.391, and discretize it by having $d\omega \to \frac{2\omega_c}{N} = \frac{2\pi}{NT}$, then, the discretized version of Equation 24.391 will become,

$$h_n = \frac{1}{2\pi} \hat{H}_k e^{i\frac{2\pi k}{NT}nT} \frac{2\pi}{NT}$$

$$= \frac{1}{2\pi} H_k T e^{i\frac{2\pi k}{N}n} \frac{2\pi}{NT}$$

$$= \frac{1}{N} H_k e^{i\frac{2\pi kn}{N}} \tag{24.440}$$

Therefore, Equations 24.441 and 24.442 are the DFT and IDFT respectively,

$$H_k = \sum_{n=0}^{N-1} h_n e^{-i\frac{2\pi kn}{N}} \tag{24.441}$$

$$h_n = \frac{1}{N} H_k e^{i\frac{2\pi kn}{N}} \tag{24.442}$$

It is customary to define a factor called the *twiddle factor* in the following way,

$$W_N \overset{\Delta}{=} e^{i\frac{2\pi}{N}} \tag{24.443}$$

Therefore, Equations 24.441 and 24.442 may be expressed in terms of W_N as follows,

$$H_k = \sum_{n=0}^{N-1} h_n W_N^{-kn} \tag{24.444}$$

$$h_n = \frac{1}{N} H_k W_N^{kn} \tag{24.445}$$

The DFT is generally a set of complex numbers. If we have a real signal (which is the case for speech samples), then, since the h_n are real, when $k = 0$, the exponent in Equation 24.441 becomes 0, making the exponential term 1 for all the elements of the summation. Therefore, H_0 becomes real. This term is called the DC term. Also, when N is even (which is usually the case with DFT implementations), then for $k = \frac{N}{2}$, the exponential term of the summation may be written as,

$$e^{-i\frac{2\pi kn}{N}} = e^{-i\frac{2\pi(\frac{N}{2})n}{N}}$$
$$= e^{-i\pi n}$$
$$= \cos(n\pi) + i\sin(n\pi)$$
$$= \pm 1$$

This means that the value of the DFT for the folding frequency, f_c is also real.

Also, the real-ness of the signal means that,

$$H_k = \overline{H_{N-k}} \quad \forall \ 0 < k < \frac{N}{2} \tag{24.446}$$

Since $H_0, H_{\frac{N}{2}} \in \mathbb{R}$ as shown earlier, we may add these two cases to the list in Equation 24.446 so that,

$$H_k = \overline{H_{N-k}} \quad \forall \ 0 \leq k \leq \frac{N}{2} \tag{24.447}$$

where N is even.

Note the similarity of the IDFT to DFT. In a practical sense, with slight modifications, the DFT may be used to compute the IDFT. This is done in practice. Therefore, there is only need for the implementation of one side of the algorithm, the DFT. Later, we will see that there are efficient techniques for computing the DFT. Fast Fourier Transform is one such algorithm which will be discussed later.

24.10.2 Periodicity

Now, consider H_{k+N},

$$
\begin{aligned}
H_{k+N} &= \sum_{n=0}^{N-1} h_n e^{-i\frac{2\pi(k+N)n}{N}} \\
&= \sum_{n=0}^{N-1} h_n e^{-i\frac{2\pi kn}{N}} e^{-i\frac{2\pi \not{N} n}{\not{N}}} \\
&= \sum_{n=0}^{N-1} h_n e^{-i\frac{2\pi kn}{N}} \\
&= H_k
\end{aligned}
\tag{24.448}
$$

Equation 24.448 suggests that the set of H_k is periodic with period N.

In the case where we have a real signal (such as speech), then by only knowing the first $\frac{N}{2}+1$ elements, we will know the information for any index since the elements from $\frac{N}{2}+1$ to $N-1$ are complex conjugates and easily determined by the first $\frac{N}{2}+1$ elements and the indices for N and higher are just periodically related to the first N numbers.

To recapitulate, H_0 corresponds to the DC level, $H_{\frac{N}{2}}$ corresponds to f_c. Indices $0 < k < \frac{N}{2}-1$ corresponds to $0 < f < f_c$ and $\frac{N}{2}+1 < f < N$ correspond to $f_c < f < 0$.

24.10.3 Plancherel and Parseval's Theorem

Following the example of Parseval's Theorem for the Complex Fourier Series (Section 24.6.2) and the Complex Fourier Transform (Section 24.9.7), it may easily be shown that for two sampled signals, g_n and h_n,

$$
\sum_{n=0}^{N-1} g_n \overline{h_n} = \frac{1}{N} \sum_{k=0}^{N-1} G_k \overline{H_k}
\tag{24.449}
$$

Some call this general theorem, Plancherel's Theorem and the case when $g_n = h_n$, Parseval's Theorem.

$$
\sum_{n=0}^{N-1} |h_n|^2 = \frac{1}{N} \sum_{k=0}^{N-1} |H_k|^2
\tag{24.450}
$$

Equation 24.450 is the statement of Parseval's theorem for the Discrete Fourier Transform (DFT).

24.10.4 Power Spectral Density (PSD) Estimation

The definition of the Power Spectral density for the angular frequency may be extended from the continuous case, Equation 24.417, using Parseval's Theorem for the Discrete Fourier Transform, Equation 24.450 and the methods of discretization discussed in Section 24.10, we can write the discrete Power Spectral Density as,

$$\mathscr{P}_d^\circ(k) = \frac{1}{N^2} |H_k|^2 \tag{24.451}$$

Since,

$$H_k = \sum_{n=0}^{N-1} h_n e^{-i\frac{2\pi kn}{N}} \tag{24.452}$$

then the power spectral density may be written in terms of the original signal as,

$$\mathscr{P}_d^\circ(k) = \frac{1}{N^2} \left| \sum_{n=0}^{N-1} h_n e^{-i\frac{2\pi kn}{N}} \right|^2 \tag{24.453}$$

Equation 24.453 is known as the *Periodogram* estimation of the power spectral density.

Note that the PSD for DTFT (considering an infinite length for the signal) is slightly different. See Section 24.11.1 for more on the PSD of a DTFT.

One of the problems with the Periodogram estimate is that we will not be able to increase the accuracy (decrease the variance) of the estimate by simply taking a larger number of points in our DFT. The variance is independent of N and is always 1. To address this problem, several different methods have superseded the Periodogram estimation technique. One very effective method is due to Welch [49].

Welch proposed blocking the N-point sampled data into K, $2M$-point data segments. Then we may do a *Periodogram estimate* of the windowed version of each of the K segments and average all of them to come up with the spectrogram for M+1-frequency values. See the Welch window definition in Section 5.3.2.3. He reports that the variance of this estimate is $\frac{1}{K}$ of the variance of the larger spectrogram. Also, $K \times 2M$ *FFT* evaluations are less computationally intensive than one larger, $2MK$ *FFT* evaluation. Note that N and M should be powers of 2 as usual.

In this case, we would have to sacrifice some frequency resolution, or if we have access to more data, we can increase our N so that when split into K bins, it would still provide a decent number of frequency values. Of course, this will not always be the case. For example in speech, due to the non-stationary nature of the signal, we will not be able to increase N arbitrarily and the value of N is dependent on the dynamics of the speech segment – see Section 5.3. In this case, Welch proposes that we may break up the data into overlapping segments of half their original lengths. The variance reduction factor will then be $11/9$.

24.10.5 Fast Fourier Transform (FFT)

The DFT, when computed directly from Equation 24.441, requires an $O(N^2)$ operations to be done. In most applications such as speaker recognition, the DFT has to be computed many many times and any improvement in the efficiency of its computation will be hugely amplified in practice. An idea which is based on splitting the data into subsections over and over has been used to derive an array of algorithms for accelerating the computation of the DFT, called Fast Fourier Transform (FFT) algorithms.

The first known FFT algorithm was developed by Carl Gauss.[19] According to [23] it is estimated to have been written in 1805. Although, the manuscript at hand is from a collection of his works published in 1866 [19]. Right after Gauss, others such as Carlini, Smith, Everett, and Runge published variations, all within the remainder of the 19th century.[23] These extensions (and sometimes re-inventions) continued until a famous paper by Cooley and Tukey [11] which published the most widely used version of the algorithm to that date. However, according to [23], Gauss' algorithm and that of Cooley and Tukey are equivalent, once the notation is changed and some conversions are done. All these algorithms reduce the number of operations from $O(N^2)$ down to $O(N\log_2(N))$ arithmetic operations.

One widely used algorithm which was first developed by Yavne [50] in 1968 and then by Duhamel and Hollmann [16], Martens [29], Vetterli and Nussbaumer [47] independently in 1984, is known as the Split-Radix algorithm. This algorithm requires $N\log_2(N) - 3N + 4$ multiplications and $3N\log_2(N) - 3N + 4$ additions [3]. Apparently, the name was first coined by [16]. An efficient version of the split-radix algorithm was presented by Sorensen and Heideman in 1986 [44] which was used for a long time in practical applications.

It is interesting that these improvements come in waves. At this point, there is another wave of improvements in three other papers which came out in 2007 and they have reduced the number of operations to $3.78N\log_2(N)$ operations. This improvement has brought the number of operations down from that of the split-radix which was in the order of $4N\log_2(N)$ to $3.78N\log_2(N)$. There three algorithms are the algorithm of Lundy and Van Buskirk [28] the FFT of the West (FFTW) algorithm [25], and Bernstein's Tangent FFT algorithm [3].

The main idea behind most of these algorithms is the same. They take the N point FFT and split it into $2 \times \frac{N}{2}$ problems. Then they keep repeating that until they are left with single leaves of tree, hence the $O(N \log_2(N))$ operations. Let us examine one such algorithm due to Danielson and Lanczos [13, 14] as described by [37]. Consider the DFT expression given by Equation 24.443 written again here for convenience,

$$H_k = \sum_{n=0}^{N-1} h_n W_N^{-kn} \tag{24.454}$$

H_k may be written as the following two smaller sums,

$$H_k = \underbrace{\sum_{n=0}^{\frac{N}{2}-1} h_{2n} W_N^{-k(2n)}}_{even} + \underbrace{\sum_{n=0}^{\frac{N}{2}-1} h_{2n+1} W_N^{-k(2n+1)}}_{odd}$$

$$= \sum_{n=0}^{\frac{N}{2}-1} h_{2n} W_{\frac{N}{2}}^{-kn} + W_N^{-k} \sum_{n=0}^{\frac{N}{2}-1} h_{2n+1} W_{\frac{N}{2}}^{-kn}$$

$$= H_k^{(e)} + W_N^{-k} H_k^{(o)} \tag{24.455}$$

where,

$$H_k^{(e)} \triangleq \sum_{n=0}^{\frac{N}{2}-1} h_{2n} W_{\frac{N}{2}}^{-kn} \tag{24.456}$$

$$H_k^{(o)} \triangleq \sum_{n=0}^{\frac{N}{2}-1} h_{2n+1} W_{\frac{N}{2}}^{-kn} \tag{24.457}$$

As you can see, this process may be repeated for $\log_2(N)$ times until leaves are reached. The leaves will just be the data points themselves. Therefore, at the leaves, we will have the data points shuffled in a certain way and they will be tagged by a sequence of odd and even tags such as for instance, $H_k^{(oeooe\cdots eo)}$. The algorithm states that we would replace every o with a 1 and every e with a 0, then do a bit reversal (i.e. making the most significant bit the least significant and so on), and the binary number that gets generated will be the index of the original sampled data. Thus, we are left with a binary decision tree where the leaves are the sampled data points and the indices of these points related to the leaves are known by the described binary substitution and bit reversal.

In addition, the bit reversal may be done on the indices of the original sampled points in memory, so that all we have to do is to combine adjacent points to get 2-point transforms and then combine another layer and get 4-point transforms, and continue until we obtain the N-point transform. Each combination of this kind requires $O(N)$ operations. Since there are are $\log_2(N)$ layers in the tree, there will be $N \log_2(N)$ operations required all together.

As we mentioned in Section 24.10, for real signals,

$$H_k = \overline{H_{N-k}} \ \ \forall \ 0 \le k \le \frac{N}{2} \tag{24.458}$$

therefore, the DFT problem may be split into two other problems, a DCT (Discrete Cosine Transform) and a DST (Discrete Sine Transform). This is usually done in practice to reduce the complexity of the computation. A good implementation for real samples is given by [45, 46].

24.11 Discrete-Time Fourier Transform (DTFT)

In the definition of the *DiscreteFourierTransform(DFT)*, we started with the continuous time and frequency of the Complex Fourier Transform and discretized both of them. Now, let us take another approach toward the development of the *Discrete-Time Fourier Transform* (DTFT).

Recall the Complex Fourier Series expansion discussed in detail in Section 24.6. In that expansion, the continuous domain of time was mapped to an infinite number of discrete frequencies. Now, imagine doing the exact opposite. let us keep the frequency domain, ω continuous and discretize the time axis. In that case, all that will happen is that the Fourier Integral, given by Equation 24.390, is sampled in time. To do this, we can just pass the signal, $h(t)$, through an ideal sampler providing us with samples, $h_n = h(nT)$, for the nth time instance with the sampling period T (see Chapter 3). The Discrete-Time Fourier Transform may then be written as follows,

$$H(\omega) \stackrel{\Delta}{=} \sum_{-\infty}^{\infty} h_n e^{-i\omega n} \tag{24.459}$$

In the definition of Equation 24.459, we have defined $H(\omega)$ without the T multiplier which would have come out of the approximation of the dt in the integral. This makes the DTFT dimensionless and somewhat independent of the sampling frequency, as was in the case of the definition for DFT in Section 24.10.

Notice that the angular frequency, ω is still continuous. Now, realizing the duality between the Complex Fourier Series and the DTFT, consider the normalized version of the series and its coefficients given by the Equation pair 24.329 and 24.330. We will immediately see that Equation 24.459 is in the form of that normalized Complex Fourier Series given by Equation 24.329 with the following change of variables, $\omega \longleftrightarrow \hat{t}$ and $h_n \longleftrightarrow c_n$. Therefore, we may use the expression for the coefficients, c_n, given by Equation 24.330 to solve for h_n. Therefore, the inverse Discrete-Time Fourier Transform may be written as,

$$h_n = \frac{1}{2\pi} \int_{-\pi}^{\pi} H(\omega)e^{-i(\omega n)} d\omega \tag{24.460}$$

What is quite interesting is that if we truncated the series in Equation 24.459 and quantize the angular frequency such that,

$$\omega_k = \frac{2\pi k}{N} \quad 0 \leq k < N \tag{24.461}$$

and plug the new quantized frequency into Equation 24.459, then,

$$H(\omega_k) = H_k$$
$$= \sum_{n=0}^{N-1} h_n e^{-i(\frac{2\pi k}{N})n} \tag{24.462}$$

which is exactly the expression for the DFT given in Equation 24.441.

Another interesting relation is that much in the same way that Fourier transform is a special case of Laplace transform for which $s = i\omega$, if we set $z = e^{i\omega}$ in the definition of the z-transform we get the expression for the DTFT. See Section 24.14 for more information on the z-transform.

24.11.1 Power Spectral Density (PSD) Estimation

We presented the PSD for the DFT case in Section 24.11.1. Now, let us consider the DTFT case. For a DTFT, The power spectral density will be a function of a continuous frequency. Therefore, following the same trend in the differences among continuous, Discrete and Discrete-Time versions, the PSD for the angular frequency and DTFT is given by the following equation,

$$\mathscr{P}_d^{\circ}(\omega) = \frac{1}{2\pi} |H(\omega)|^2 \tag{24.463}$$

Since,

$$H(\omega) = \sum_{n=-\infty}^{\infty} h_n e^{-i\omega n} \tag{24.464}$$

then the power spectral density may be written in terms of the original signal as,

$$\mathscr{P}_d^{\circ}(\omega) = \frac{1}{2\pi} \left| \sum_{n=-\infty}^{\infty} h_n e^{-i\omega n} \right|^2 \tag{24.465}$$

Equation 24.465 is known as the *Periodogram* of the power spectral density and it is usually reported in dB per radian. For this, it would have to be offset by the power at 1000 Hz which is the hearing threshold – see Section 5.1.2.

24.12 Complex Short-Time Fourier Transform (STFT)

Complex Fourier Transform is usually concerned with stationary signals – see Definition 3.4. However, as have discussed at the beginning of Chapter 3, the speech signal is an non-stationary signal, which means that its parameters are constantly changing with time. According to [15], an average phone lasts for about 80 ms. Even within that time, the vocal tract varies in length and configuration, changing its spectral properties. the main idea behind the *short-time Fourier transform* is to only consider part of a signal at any moment. Therefore, the STFT is not only a function of the frequency, ω, but it is also a function of time, t, around which the STFT has been computed.

Imagine a signal $h(t)$ which is non-stationary. Take the moment in time, t. We may choose to only analyze part of the signal in the vicinity of t and ignore the signal far to the right or left of this time instance. Then, we can analyze the spectrum of the selected portion of the signal. One way to do this is to just clip some Δt to the right and left of t. However, this is equivalent to multiplying the signal by a square window function. Let us assume that we have a window function that would look to the left of $t = 0$ by an amount δt and to the right by the same amount, δt. The window function may be written as a linear combination of two step functions, $u(t)$, shifted to the left and right of the origin by an amount δt. Namely,

$$w(t) = u(t + \delta t) - u(t - \delta t) \tag{24.466}$$

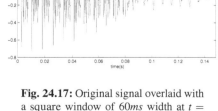
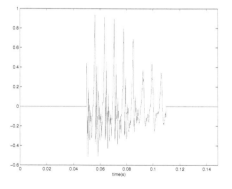

Fig. 24.17: Original signal overlaid with a square window of 60*ms* width at $t = 80ms$

Fig. 24.18: Windowed Signal

Therefore, if we are interested in the signal, but only in the vicinity of $t = \tau$, all we have to do is to shift $w(t)$ given by Equation 24.466 so that its center is at time $t = \tau$ and then multiply it by the original signal creating a new signal, $\hat{h}(t)$, which only has a portion of the original signal,

$$\hat{h}(t) = h(t)w(t - \tau) \tag{24.467}$$

Now, let us write the Complex Fourier Transform and its inverse for the new signal,

$$\hat{H}(\omega) = \int_{-\infty}^{\infty} \hat{h}(t)e^{-i(\omega t)}dt \tag{24.468}$$

$$\hat{h}(t) = \frac{1}{2\pi} \int_{-\infty}^{\infty} \hat{H}(\omega)e^{i(\omega t)}d\omega \tag{24.469}$$

By doing this, we have added one more dimension to the Fourier Analysis. Equations 24.468 and 24.469 are only for the version of the function windowed at $t = \tau$. So in general, we may write a transform that represents all possible locations of the window function. This is called the Short-Time Fourier Transform of the function, $h(t)$ and may be written as follows,

$$H(\omega, \tau) = \int_{-\infty}^{\infty} h(t)w(t - \tau)e^{-i(\omega t)}dt \tag{24.470}$$

Also, its inverse is,

$$h(t)w(t - \tau) = \frac{1}{2\pi} \int_{-\infty}^{\infty} H(\omega, \tau)e^{i(\omega t)}d\omega \tag{24.471}$$

Note that at this point, we have not made any restrictions on the size and shape of the window function. Therefore, we have many choices for possible windows and their corresponding sizes. Let us try to use this advantage to find a class of windows such that the Fourier transform of the signal $h(t)$ would encompass all the short-time Fourier transforms along the time domain, namely, let us require that,

$$H(\omega) = \int_{-\infty}^{\infty} H(\omega, \tau)d\tau \tag{24.472}$$

Let us examine Equation 24.472, by plugging in for $H(\omega, \tau)$ from Equation 24.470.

$$H(\omega) = \int_{-\infty}^{\infty} \left[\int_{-\infty}^{\infty} h(t)w(t - \tau)e^{-i\omega t}dt \right] d\tau$$

$$= \int_{-\infty}^{\infty} h(t) \int_{-\infty}^{\infty} w(t - \tau)d\tau e^{-i\omega t}dt \tag{24.473}$$

Equation 24.473 means that for Equation 24.472 to be true, the area under the window function must be 1, namely,

$$\int_{-\infty}^{\infty} w(t - \tau)d\tau = 1 \quad \forall \, t \tag{24.474}$$

We will use this restriction in choosing the window function. It still leaves us with infinite number of possibilities. Now, to recapitulate, the Short-Time Fourier Transform of $h(t)$ is as follows,

$$H(\omega, \tau) = \int_{-\infty}^{\infty} h(t)w(t - \tau)e^{-i\omega t}dt \tag{24.475}$$

where,

$$\int_{-\infty}^{\infty} w(t - \tau)d\tau = 1 \quad \forall \, t \tag{24.476}$$

and as a property,

$$H(\omega) = \int_{-\infty}^{\infty} H(\omega, \tau)d\tau \tag{24.477}$$

Figures 24.19 and 24.20 show the effect of the modified square window such that the area under the window is 1. Note that we are only using a square window here for simplicity of demonstrating the idea behind STFT. Using a square window, because of the sharp drop at its edges, creates strong, high frequency components which will show up in the transform. For this reason, practical windows are chosen to have a very smooth and gradual transition between the center of the window and the sides which tend to zero. This will act like a low-pass filter which reduces the leakage effect.

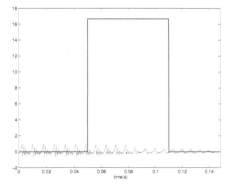

Fig. 24.19: Original signal overlaid with a square window of 60*ms* width at $t = $ 80*ms* with a normalized window using Equation 24.474

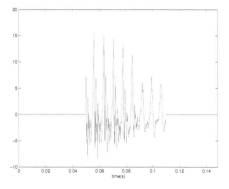

Fig. 24.20: Windowed signal with a normalized window using Equation 24.474

Now, let us consider the inverse Fourier transform for Equation 24.473,

$$h(t) = \frac{1}{2\pi} \int_{-\infty}^{\infty} H(\omega) e^{i\omega t} d\omega$$

$$= \frac{1}{2\pi} \int_{-\infty}^{\infty} \int_{-\infty}^{\infty} H(\omega, \tau) e^{i\omega t} d\omega$$

$$= \int_{-\infty}^{\infty} \left[\frac{1}{2\pi} \int_{-\infty}^{\infty} H(\omega, \tau) e^{i\omega t} d\omega \right] d\tau \tag{24.478}$$

$$= \int_{-\infty}^{\infty} h(t) w(t - \tau) d\tau \tag{24.479}$$

The function,

$$h(t, \tau) \overset{\Delta}{=} h(t) w(t - \tau) \tag{24.480}$$

is known as a *wavelet* of $h(t)$ and is given by the inverse Fourier transform of the short-time Fourier transform of $h(t)$ at time τ,

$$h(t, \tau) = \frac{1}{2\pi} \int_{-\infty}^{\infty} H(\omega, \tau) e^{i\omega t} d\omega \tag{24.481}$$

Therefore, since $h(t, \tau)$ is a wavelet of $h(t)$, then $H(\omega, \tau)$, the short-term Fourier transform at τ is also a *wavelet transform* of $h(t)$. We, briefly, touched upon the *wavelet series expansion* of functions in Section 24.7. Now, we have see the connection between the *Fourier transform* and the *Wavelet transform*.

In fact, one special wavelet transform, known as the *Gabor transform* is defined as a special case of short-time Fourier transform where the window function has been chosen to be a *Gaussian function*. Since

$$\int_{-\infty}^{\infty} e^{-t^2} = \sqrt{\pi} \tag{24.482}$$

the window was chosen to meet the requirement of Equation 24.476 which means,

$$w_g(t) = e^{-\pi t^2} \tag{24.483}$$

Therefore, the *Gabor transform* becomes,

$$H_G(\omega, \tau) = \int_{-\infty}^{\infty} h(t) e^{-\pi(t-\tau)^2} e^{-i\omega t} dt \tag{24.484}$$

and its inverse (known as the *Gabor wavelet*) is,

$$h_g(t, \tau) = \frac{1}{2\pi} \int_{-\infty}^{\infty} H_g(\omega, \tau) e^{i\omega t} d\omega \tag{24.485}$$

and $h(t)$ is then given by Equation 24.479 or,

$$h(t) = \int_{-\infty}^{\infty} h_g(t, \tau) d\tau$$

$$= \int_{-\infty}^{\infty} \frac{1}{2\pi} \int_{-\infty}^{\infty} H_g(\omega, \tau e^{i\omega t} d\omega d\tau \tag{24.486}$$

As a practical note,

$$w_G(t) \approx 0 \ \forall \ |t| > 2 \tag{24.487}$$

Therefore, one may write an approximation to the Gabor transform as,

$$\tilde{H}_G(\omega, \tau) = \int_{-2}^{2} h(t) e^{-\pi(t-\tau)^2} e^{-i\omega t} dt \tag{24.488}$$

where, $\tilde{H}_G(\omega, \tau) \approx H_G(\omega, \tau)$.

24.12.1 Discrete-Time Short-Time Fourier Transform DTSTFT

In this section, we will derive the *Discrete-Time Short-Time Fourier Transform (DT-STFT)* whose relationship to the continuous version is analogous to the relationship of the DTFT and the Complex Fourier Transform.

Let us start with the same objective as we did for the continuous STFT, which is the isolation of our analysis to a portion of a non-stationary signal to be able to capture local stationary effects. Consider, as did with the continuous case, a windowed version of the sampled signal, h_n at time instance m,

$$\hat{h}_n = h_n w(n - m) \tag{24.489}$$

The DTFT of the new signal is,

$$\hat{H}(\omega) = \sum_{n=-\infty}^{\infty} \hat{h}_n e^{-i\omega n} \tag{24.490}$$

and using Equation 24.462, the inverse transform is

$$\hat{h}_n = \frac{1}{2\pi} \int_{-\pi}^{\pi} \hat{H}(\omega) e^{i\omega n} d\omega \tag{24.491}$$

Equations 24.490 and 24.491 are only for the original function windowed at the discrete time instance, m. Therefore, we may write the transform that represents all the possible locations of the windowed signal as,

$$H(\omega, m) = \sum_{n=-\infty}^{\infty} h_n w(n-m) e^{-i\omega n} \tag{24.492}$$

with its inverse,

$$h_n w(n-m) = \frac{1}{2\pi} \int_{-\infty}^{\infty} H(\omega, m) e^{i\omega n} d\omega \tag{24.493}$$

Now, let us exert the following restriction,

$$H(\omega) = \sum_{m=-\infty}^{\infty} H(\omega, m) \tag{24.494}$$

which may be rewritten out as,

$$H(\omega) = \sum_{m=-\infty}^{\infty} \sum_{n=-\infty}^{\infty} h_n w(n-m) e^{-i\omega n}$$
$$= \sum_{n=-\infty}^{\infty} h_n e^{i\omega n} \sum_{m=-\infty}^{\infty} w(n-m) \tag{24.495}$$

One way For Equation 24.495 to hold, is if the window function $w(n)$ has the following property,

$$\sum_{m=-\infty}^{\infty} w(n-m) = 1 \quad \forall \; n \tag{24.496}$$

Then,

$$H(\omega, m) = \sum_{n=-\infty}^{\infty} h_n w(n-m) e^{-i\omega n} \tag{24.497}$$

where,

$$\sum_{m=-\infty}^{\infty} w(n-m) = 1 \quad \forall \; n \tag{24.498}$$

and $H(\omega, m)$ has the following property,

$$H(\omega) = \sum_{m=-\infty}^{\infty} H(\omega, m) \tag{24.499}$$

Now, let us consider the case of the inverse transform,

$$h_n = \frac{1}{2\pi} \int_{-\infty}^{\infty} H(\omega)e^{i\omega n} d\omega$$

$$= \frac{1}{2\pi} \int_{-\infty}^{\infty} \sum_{m=-\infty}^{\infty} H(\omega,m)e^{i\omega n} d\omega$$

$$= \sum_{m=-\infty}^{\infty} \frac{1}{2\pi} \int_{-\infty}^{\infty} H(\omega,m)e^{i\omega n} d\omega$$

$$= \sum_{m=-\infty}^{\infty} h_n w(n-m) \tag{24.500}$$

Then we can define,

$$h_{nm} \overset{\Delta}{=} h_n w(n-m) \tag{24.501}$$

which is a discrete wavelet of h_n and is given by,

$$h_{nm} = \frac{1}{2\pi} \int_{-\infty}^{\infty} H(\omega,m)e^{i\omega n} d\omega \tag{24.502}$$

Therefore, the DTSTFT of h_n, $H(\omega,m)$, evaluated at time instant m is a discrete-time wavelet transform of h_n.

24.12.2 Discrete Short-Time Fourier Transform DSTFT

In this section, we are going to derive the Discrete Short-Time Fourier Transform and its inverse much in the same manner as we did for the Discrete-Time Short-Time Fourier Transform. As you may recall, the difference between these two is analogous to the difference between DTFT and DFT where in DTSTFT and DTFT, the frequency is continuous and the time domain is sampled, but in a *countably infinite* fashion (see Section 6.7). Since we have gone through similar exercises for the STFT and DTSTFT, here, we will simply derive the equations for the DSTFT case this section, without much discussion.

$$\hat{h}_n \overset{\Delta}{=} h_n w(n-m) \tag{24.503}$$

$$\hat{H}_k = \sum_{n=0}^{N-1} h_n e^{-i\frac{2\pi nk}{N}} \tag{24.504}$$

$$\hat{h}_n = \frac{1}{N} \sum_{k=0}^{N-1} \hat{H}_k e^{i\frac{2\pi nk}{N}} \tag{24.505}$$

$$H_{km} = \sum_{n=0}^{N-1} h_n w(n-m) e^{-i\frac{2\pi nk}{N}} \tag{24.506}$$

$$h_n w(n-m) = \frac{1}{N} \sum_{k=0}^{N-1} H_{km} e^{i\frac{2\pi nk}{N}} \tag{24.507}$$

Similar to our previous restriction, let us require that,

$$H_k = \sum_{m=0}^{N-1} H_{km} \tag{24.508}$$

Therefore,

$$H_k = \sum_{m=0}^{N-1}\sum_{n=0}^{N-1} h_n w(n-m) e^{-i\frac{2\pi nk}{N}}$$

$$= \sum_{n=0}^{N-1} h_n e^{-i\frac{2\pi nk}{N}}$$

$$\sum_{m=0}^{N-1} w(n-m) \tag{24.509}$$

Equation 24.509 will hold in general if we choose our window function, $w(n)$ such that,

$$\sum_{m=0}^{N-1} w(n-m) = 1 \quad \forall\ n \tag{24.510}$$

This gives us the following Discrete Short-Time Fourier Transform,

$$H_{km} = \sum_{n=0}^{N-1} h_n w(n-m) e^{-i\frac{2\pi nk}{N}} \tag{24.511}$$

where,

$$\sum_{m=0}^{N-1} w(n-m) = 1 \quad \forall\ n \tag{24.512}$$

and H_k has the property that,

$$H_k = \sum_{m=0}^{N-1} H_{km} \tag{24.513}$$

Now, let us derive the inverse transform,

$$h_n = \frac{1}{N} \sum_{k=0}^{N-1} H_k e^{i\frac{2\pi nk}{N}}$$

$$= \frac{1}{N} \sum_{k=0}^{N-1} \sum_{m=0}^{N-1} H_{km} e^{i\frac{2\pi nk}{N}}$$

$$= \sum_{m=0}^{N-1} \frac{1}{N} \sum_{k=0}^{N-1} H_{km} e^{i\frac{2\pi nk}{N}}$$

$$= \sum_{m=0}^{N-1} h_n w(n-m) \tag{24.514}$$

Then we can define,

$$h_{nm} \stackrel{\Delta}{=} h_n w(n-m) \tag{24.515}$$

which is given by,

$$h_{nm} = \frac{1}{N} \sum_{k=0}^{N-1} H_{km} e^{i\frac{2\pi nk}{N}} \tag{24.516}$$

24.13 Discrete Cosine Transform (DCT)

At the end of Section 24.10, we touched upon the fact that the DFT may be split up into its real and imaginary parts (DCT and DST). The Discrete Cosine Transform, because of it relation to the real axis, becomes important in speech processing. One place it is used is as the last stage of the computation of Mel Frequency Cepstral Features (MFCC) – see Section 5.3.6. Besides speech processing, some familiar image compression techniques such as Joint Photographic Expert Group (JPEG) compression use a two dimensional version of the DCT. In this book, we will only be concerned with the one dimensional case pertaining to the time axis.

Since the cosine transform is a component of the Fourier transform, it can also take on a discrete form in time and frequency (DCT) or a just in time which would be called the Discrete-Time Cosine Transform (DTCT). Since we will not really use the DTCT in this book and since the relation between DTCT and DCT is quite similar to that of the DTFT and DFT, we will only discuss the DCT.

There are several different ways the DCT may be defined. This is due to the fact that for indices which are zero, the sin component of the Fourier transform vanishes, but the cos becomes 1 which produces a DC value for the transform. There are different ways this DC component may be define. We will not go into the details. Here, we have picked a popular form of the Discrete Cosine Transform (DCT) which is as follows,

$$H_k = \sum_{n=0}^{N-1} h_n \cos\left(\frac{\pi(2n+1)k}{2N}\right) \tag{24.517}$$

where $k = \{0, 1, \cdots, N-1\}$ is the frequency index.

The inverse DCT (IDCT) for the form we have chosen (Equation 24.517), is the following,

$$h_n = \sum_{k=0}^{N-1} a_k H_k \cos\left(\frac{\pi(2n+1)k}{2N}\right) \tag{24.518}$$

where

$$a_k = \begin{cases} \frac{1}{N} & for\ k = 0 \\ \frac{2}{N} & \forall\ k > 0 \end{cases} \tag{24.519}$$

Normally, the DCT is computed using an FFT algorithm and the real part is extracted. However, there are special cases where an FFT will not be the most efficient way of approaching the problem. In the following section, we will examine some methods that take advantage of the problem at hand and increase the efficiency of computation of the DCT.

24.13.1 Efficient DCT Computation

In relation to the topic of this book, we are interested in processing of a long sequence of speech. As it is made apparent in Chapter 5, the speech is processed in equal size frames and for each frame, the cos values to be determined, and which are the highest costing part of the computation, may be computed only once and stored in a vector. Then the cosine transform part of the process can recall these values without having to compute them for every frame. In this case, we can do the processing in much more efficient manner compared to computing the FFT values for each frame.

Another example where more efficient algorithms may be used to compute the DCT is the case of Dual Tone Multi-Frequency (DTMF) detection, also known as touch tone detection. In this problem and other related cases, we are interested in assessing the power associated with specific frequencies in a signal (the DTMF part) and not necessary the whole signal (the whole telephone conversation). The Goertzel Algorithm [20] provides one such mechanism. It is a recursive algorithm used for computing finite trigonometric series, such as the Discrete Cosine Transform (DCT). This algorithm is well-known for its application to DTMF detection. DTMF signals are the touch tone signal with which we are mostly familiar through our daily lives. The goal of a DTMF detector is to assess the power associated with specific DTMF frequencies. This job could always be done using FFT (see Section 24.10.5), however, the FFT algorithms are still too slow for this specific task which has to be almost instantaneous. The Goertzel Algorithm provides a much faster approach, since only a limited number of spectra are needed. It is designed to evaluate a finite trigonometric series, in this case a Discrete Cosine Transform (DCT), by conducting only N multiplications and about $2N$ additions [8]. Goertzel's algorithm is a recursive algorithm which is structured such that it may be parallelized quite easily. This allows efficient VLSI implementations [7].

In the DTMF application, only eight frequencies and their second harmonics are needed for the detection. For each frequency, there is a second harmonic frequency (hence the dual tone descriptor in the name). The second harmonic frequency is used to make sure that there is no confusion between the DTMF and voice or other audio components on the call. The sampling frequencies for these frequency components are assumed to be 8 kHz.

[8] also present another algorithm which is designed to be implemented in a parallel setting. It also presents a circuit for doing this computation.

24.14 The z-Transform

Assume that we have a signal, $h(t)$, which has been sampled using an ideal sampler (see Section 3.6.1). Therefore, the sampled signal, if represented in the continuous domain, is given by Equation 3.48. Let us take the one-sided Laplace transform of $h^*(t)$, which is the output of an ideal sampler with a sampling frequency T,

$$\mathscr{L}\{h^*\} = H^*(s) = \sum_{n=0}^{\infty} h(nT)e^{-nTs} \tag{24.520}$$

Now, since the general form of a Laplace transform applied to the output of an ideal sampler will have an exponential component (based on Equation 24.520, and since the exponential function is not a rational function, taking the inverse Laplace transform would be quite difficult. This is the motivation behind the definition of the z-transform. Let us define,

$$z \triangleq e^{Ts} \tag{24.521}$$

Then, the s may be written in terms of z as follows,

$$s = \frac{1}{T} \ln z \tag{24.522}$$

If we write the real and imaginary components of the complex variable s,

$$s = \sigma + i\omega \tag{24.523}$$

then the real and imaginary components of z may be given by,

$$\mathscr{R}e(z) = e^{Ts}\cos(\omega T) \tag{24.524}$$
$$\mathscr{I}m(z) = e^{Ts}\sin(\omega T) \tag{24.525}$$
$$\tag{24.526}$$

Also, note that the z-transform of h^* is as follows,

$$H(z) = H^*\left(s = \frac{1}{T}\ln z\right)$$

$$= \sum_{n=0}^{\infty} h(nT)z^{-n} \tag{24.527}$$

Since we started with one-sided Laplace transform, the definition of z-transform which is derived from it is also the one-sided z-transform.

Notice the relation between Laplace and z transforms:

$$H(s) = \int_0^{\infty} h(t)e^{-st}dt \longleftrightarrow H(z) = \sum_{n=0}^{\infty} h(nT)z^{-n} \tag{24.528}$$

But if we let $T \to 0$,

$$\lim_{T \to 0} H(z) \neq H(s) \tag{24.529}$$

since the output of an ideal sample does not approach the signal as T approaches 0, namely,

$$\lim_{T \to 0} h^*(t) \neq h(t) \tag{24.530}$$

In fact, as $T \to 0$, we will end up with an infinite number of impulses at $t = 0$. Let us examine the z-transform of some prevalent functions.

Also, much in the same way that Fourier transform is a special case of Laplace transform for which $s = i\omega$, if we set $z = e^{i\omega}$ and use the definition of the two-sided z-transform, we get,

$$H(\omega) = \sum_{-\infty}^{\infty} h(nT)e^{-i\omega n} \tag{24.531}$$

which is called the *Discrete Time Fourier Transform* (DTFT) – not to be confused with Discrete Fourier Transform (DFT). Just like the relation between Fourier Transform and Laplace, since the DTFT is evaluation at $z = e^{i\omega}$, and since $\left|e^{i\omega}\right| = 1$, the points of this mapping lie on the perimeter of the unit circle which means that there is no damping. See Section 24.11 for more on the DTFT.

Now take the unit step function given by Equation 24.532,

$$u(t) \triangleq \begin{cases} 1 \ \forall \ t \geq 0 \\ 0 \ \forall \ t < 0 \end{cases} \tag{24.532}$$

Then the output of the ideal sampler would be,

$$u^*(t) = \sum_{n=0}^{\infty} \delta(t - nT) \tag{24.533}$$

and

$$\mathscr{L}\{u^*\} = \sum_{n=0}^{\infty} e^{-nTs}$$

$$= \frac{1}{1 - e^{-Ts}} \quad for \ \left| e^{-Ts} \right| < 1 \tag{24.534}$$

$$\tag{24.535}$$

Writing Equation 24.535 using the substitution in Equation 24.521,

$$\mathscr{Z}\{u\} = \frac{1}{1 - z^{-1}}$$

$$= \frac{z}{z - 1} \quad for \ |z^{-1}| < 1(|z| > 1) \tag{24.536}$$

Now, let us consider the decaying exponential function,

$$h(t) = e^{-at} \tag{24.537}$$

Then,

$$H(z) = \sum_{n=0}^{\infty} e^{-anT} z^{-n} \quad \text{converges for all} \ \left| e^{-aT} z^{-1} \right|$$

$$= \frac{1}{1 - e^{-aT} z^{-1}}$$

$$= \frac{z}{z - e^{-aT}} \quad \forall \ \left| e^{-aT} z^{-1} \right| < 1 \quad \text{or} \quad |z^{-1}| < e^{aT} \tag{24.538}$$

Now consider the periodic function,

$$h(t) = \sin(\omega t) \tag{24.539}$$

Again, by using the ideal sampler output,

$$H(z) = \sum_{n=0}^{\infty} \sin(\omega nT) z^{-n}$$

$$= \sum_{n=0}^{\infty} \frac{e^{i\omega nT} - e^{-i\omega nT}}{2i} z^{-n}$$

$$= \frac{1}{2i} \left[\sum_{n=0}^{\infty} e^{i\omega nT} z^{-n} - \sum_{n=0}^{\infty} e^{-i\omega nT} z^{-n} \right]$$

$$= \frac{1}{2i} \left[\frac{z}{z - e^{i\omega T}} - \frac{z}{z - e^{-i\omega T}} \right]$$

$$= \frac{z \sin(\omega T)}{z^2 - 2z \cos(\omega T) + 1} \tag{24.540}$$

The *Ramp* function is defined as,

$$h(t) = t \tag{24.541}$$

We can also write the Ramp function in the following form:

$$h(t) = t u(t) \tag{24.542}$$

where $u(t)$ is the unit step function given by Equation 24.532. Therefore,

$$H(z) = \sum_{n=0}^{\infty} nT z^{-n}$$

$$= Tz^{-1} + 2Tz^{-2} + \cdots \tag{24.543}$$

If we multiply both sides of Equation 24.543 by z^{-1}, we will have,

$$H(z) = Tz^{-2} + 2Tz^{-3} + \cdots \tag{24.544}$$

Now subtract Equation 24.544 from Equation 24.543,

$$(1 - z^{-1})H(z) = Tz^{-1} + Tz^{-2} + \cdots \tag{24.545}$$

If we multiply Equation 24.545 by z_{-1} and subtract the result from Equation 24.545, we will get,

$$(1 - z^{-1}) \left[1 - z^{-1} \right] H(z) = Tz^{-1} \tag{24.546}$$

Therefore,

$$H(z) = \frac{Tz^{-1}}{(1 - z^{-1})^2}$$

$$= \frac{Tz}{(z - 1)^2} \tag{24.547}$$

Note that,

$$H(s) = \frac{1}{s^2} \hspace{7cm} (24.548)$$

Figure 3.24 shows the effect of folding in the Laplace domain when applied to the output of an ideal sampler. In the z-plane, the portion that is highlighted in the figure is mapped to the unit circle with its center at the origin. Since the circle is cyclic, the higher harmonics (reflections) show up as the addition of $\pm 2n\pi i$ in the the polar coordinates. See Section 3.6.1 for more information about folding.

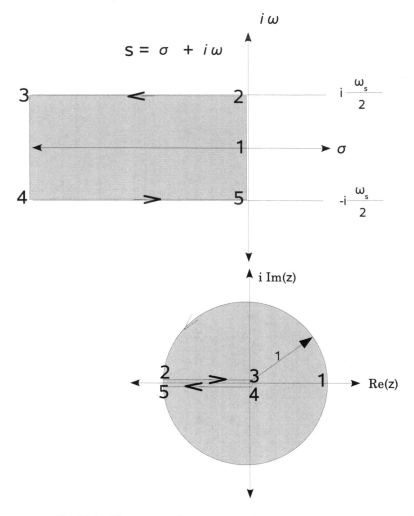

Fig. 24.21: The corresponding extrema in the Laplace and z planes

Consider the extreme cases of the values of s and z:

1.

$$s = 0 \longleftrightarrow z = 1 \qquad\qquad (24.549)$$

2.

$$s = i\frac{\omega_s}{2} \longleftrightarrow z = e^{T\left(i\frac{\omega_s}{2} \pm 2\pi i m\right)} \qquad\qquad (24.550)$$

since $\omega_s = \frac{2\pi}{T}$,

$$z = e^{i\pi \pm 2\pi i m}$$
$$= \cos(\pi) \pm i\sin(\pi)$$
$$= -1 + i0$$

$$(24.551)$$

Therefore,

$$s = i\frac{\omega_s}{2} \longleftrightarrow z - -1 \qquad\qquad (24.552)$$

3.

$$s = -\infty + i\frac{\omega_s}{2} \longleftrightarrow z = e^{T\left(-\infty + i\frac{\omega_s}{2}\right) \pm 2\pi i m} \qquad\qquad (24.553)$$

Let us write z in polar coordinates, then,

$$z = \rho e^{i\theta \pm 2\pi i m}$$
$$= e^{-\infty} e^{i\frac{T\omega_s}{2} \pm 2\pi i m}$$
$$\therefore \rho = e^{-\infty} = 0 \qquad\qquad (24.554)$$

Therefore,

$$s = -\infty + i\frac{\omega_s}{2} \longleftrightarrow z \text{ is at the origin} \qquad\qquad (24.555)$$

4. By the same token as the previous point,

$$s = -\infty - i\frac{\omega_s}{2} \longleftrightarrow z \text{ is at the origin} \qquad\qquad (24.556)$$

5.

$$For s = 0 - i\frac{i\omega_s}{2} \qquad\qquad (24.557)$$

since

$$z = e^{Ts \pm 2\pi im}$$
$$= e^{0T} e^{-i\frac{i\omega_s T}{2} \pm 2\pi im}$$
$$= 1e^{-\pi \pm 2\pi im}$$

$$(24.558)$$

Therefore, in a polar representation of z, $\rho = 1$ and $theta = -\pi$.

Figure 24.21 shows the locations of the corresponding 5 extrema on the Laplace and z planes.

24.14.1 Translation

The z-transform is a linear operation

i.e.,

$$\mathscr{L}\{h \pm g\} = \sum_{n=0}^{\infty} ah(nT)z^{-n} \pm \sum_{n=0}^{\infty} ag(nT)z^{-n}$$
$$= \mathscr{L}\{h\} \pm \mathscr{L}\{g\} \tag{24.559}$$

This may be shown as the follows,

$$h(t) \pm g(t) \longleftrightarrow H(z) \pm G(z) \tag{24.560}$$

24.14.2 Scaling

$$\mathscr{L}\{ah\} = a\mathscr{L}\{h\} \tag{24.561}$$

$$\mathscr{L}\{ah\} = \sum_{n=0}^{\infty} ah(nT)z^{-n}$$
$$= a\sum_{n=0}^{\infty} h(nT)z^{-n}$$
$$= aH(z) \tag{24.562}$$

Therefore,

$$h(t) = ag(t) \Longleftrightarrow H(z) = aG(z) \tag{24.563}$$

24.14.3 Shifting – Time Lag

Consider $h(t)$ which is shifted to the right by an amount kT (a time lag of kT seconds). Then its z-transform may be written as follows,

$$\mathscr{Z}\{h(t-kT)u(t-kT)\} = \sum_{n=0}^{\infty} h(nT-kT)z^{-(n-k)}$$

$$= z^{-k}\sum_{n=0}^{\infty} h(nT-kT)z^{-(n-k)}$$

$$= z^{-k}\sum_{\bar{n}=0}^{\infty} h(\bar{n}T)z^{-(\bar{n})}$$

$$= z^{-k}H(z) \tag{24.564}$$

Therefore, the *time lag* theorem becomes,

$$\mathscr{Z}\{h(t-kT)\} = z^{-k}H(z) \tag{24.565}$$

24.14.4 Shifting – Time Lead

Now, consider $h(t)$ which is shifted to the left by an amount kT (a time lead of kT seconds). Then its z-transform is as follows,

$$\mathscr{Z}\{h(t+kT)u(t+kT)\} = \sum_{n=0}^{\infty} h(nT+kT)z^{-(n+k)}$$

$$= z^{k}\sum_{n=0}^{\infty} h(nT+kT)z^{-(n+k)}$$

$$= z^{k}\sum_{\bar{n}=k}^{\infty} h(nT)z^{-n}$$

$$= z^{k}\left[H(z)\sum_{\bar{n}=0}^{k-1} h(nT)z^{-n}\right] \tag{24.566}$$

Therefore, the *time lead* theorem becomes,

$$\mathscr{Z}\{h(t+kT)\} = z^{k}\left[H(z)\sum_{\bar{n}=0}^{k-1} h(nT)z^{-n}\right] \tag{24.567}$$

24.14.5 Complex Translation

We know that a complex translation in the Laplace domain relates to scaling with the exponential,

$$H(s \pm a) \longleftrightarrow e^{\mp at} h(t) \tag{24.568}$$

Now, let us take the z-transform of this scaled signal,

$$\mathscr{Z}\{e^{\mp at} h(t)\} = \sum_{n=0}^{\infty} h(nT) e^{\mp anT} z^{-n}$$

Now, if we define,

$$\bar{z} \overset{\Delta}{=} z e^{\pm aT} \tag{24.569}$$

then,

$$\begin{aligned} \mathscr{Z}\{e^{\mp at} h(t)\} &= \sum_{n=0}^{\infty} h(nT) \bar{z}^{-n} \\ &= H(\bar{z}) \\ &= H\left(z e^{\pm aT}\right) \end{aligned} \tag{24.570}$$

24.14.6 Initial Value Theorem

Theorem 24.30 (Initial Value Theorem).

$$\lim_{n \to 0} h(nT) = \lim_{z \to \infty} H(z) \tag{24.571}$$

Proof.

$$\begin{aligned} H(z) &= \sum_{n=0}^{\infty} h(nT) z^{-n} \\ &= h(0) + h(T) z^{-1} + \cdots \end{aligned}$$

$$\tag{24.572}$$

If we take the limit of $H(z)$ as $z \to \infty$, all the terms but $h(0)$ in Equation 24.572 vanish. Therefore,

$$\begin{aligned} \lim_{z \to \infty} H(z) &= h(0) \\ &= \lim_{n \to 0} h(nT) \end{aligned} \tag{24.573}$$

$\square$

24.14.7 Final Value Theorem

Theorem 24.31 (Final Value Theorem). *The final value theorem states that,*

$$\lim_{n \to \infty} h(nT) = \lim_{z \to 1} (1 - z^{-1})H(z) \qquad (24.574)$$

if $(1 - z^{-1})H(z)$ has no poles outside the unit circle.

Proof.
Take the truncated z-transform of $h(t)$ up to the kth term,

$$\sum_{n=0}^{k} h(nT)z^{-n} \qquad (24.575)$$

Now, consider the truncated z-transform of $h(t)$ with a lag of one period $(h(t - T))$ and again truncated to the kth term,

$$\sum_{n=0}^{k} h((n-1)T)z^{-n} \qquad (24.576)$$

Here, if $n = 0$, the first term of the series becomes $h(-T)$ which is zero since, $h(t) = 0 \; \forall \; t < 0$.

Therefore, writing out all the terms of the summation Expression 24.576, we see that,

$$\sum_{n=0}^{k} h((n-1)T)z^{-n} = z^{-1} \sum_{n=0}^{k-1} h(nT)z^{-n} \qquad (24.577)$$

Notice the similarity between the right hand side of Equation 24.577 and the summation in Expression 24.575. Subtract the right side of Equation 24.577 from Expression 24.575, and take the limit of $z \to 1$ in the result,

$$\lim_{z \to 1} \left[z^{-1} \sum_{n=0}^{k-1} h(nT)z^{-n} - \sum_{n=0}^{k} h(nT)z^{-n} \right] = \sum_{n=0}^{k} h(nT) - \sum_{n=0}^{k-1} h(nT)$$

$$= h(kT) \qquad (24.578)$$

Since we had truncated the z-transforms, now, we would like to take the limit of $k \to \infty$ for both sides of Equation 24.578 and replacing the infinite series with its z-transform equivalent,

$$\lim_{k \to \infty} h(kT) = \lim_{z \to 1} (1 - z^{-1})H(z) \qquad (24.579)$$

□

24.14.8 Real Convolution Theorem

Theorem 24.32 (Real Convolution). *If $h(t) = 0 \; \forall \; t < 0$ and $g(t) = 0 \; \forall \; t < 0$,*

$$\mathscr{L}\{(h*g)\} = \mathscr{L}\{\sum_{k=0}^{n} h(kT)g(nT-kT)\} \qquad (24.580)$$

$$= H(z)G(z) \qquad (24.581)$$

Proof.
Take the right hand side of Equation 24.580 which is the definition of convolution and write our the infinite sum z-transform,

$$\mathscr{L}\{\sum_{k=0}^{n} h(kT)g(nT-kT)\} = \sum_{n=0}^{\infty}\sum_{k=0}^{n} h(kT)g(nT-kT)z^{-n} \qquad (24.582)$$

For the right hand side of Equation 24.582, because the terms $g(t-\tau)=0 \ \forall \ \tau > t$, we can change the upper limit of the inner summation from n to ∞. Therefore,

$$\mathscr{L}\{\sum_{k=0}^{n} h(kT)g(nT-kT)\} = \sum_{n=0}^{\infty}\sum_{k=0}^{\infty} h(kT)g(nT-kT)z^{-n} \qquad (24.583)$$

Let us define,

$$\bar{n} \overset{\Delta}{=} n-k \qquad (24.584)$$

and rewrite Equation 24.583 using the Definition 24.584,

$$\mathscr{L}\{\sum_{k=0}^{n} h(kT)g(nT-kT)\} = \sum_{k=0}^{\infty} h(kT)z^{-k} \sum_{\bar{n}=-k}^{\infty} g(\bar{n}T)z^{-\bar{n}} \qquad (24.585)$$

However, as we stated earlier, $g(t-\tau)=0 \ \forall \ \tau > t$, so the lower limit of the second summation (involving $g(t)$) may be changed to $\bar{n}=0$ which gives,

$$\mathscr{L}\{\sum_{k=0}^{n} h(kT)g(nT-kT)\} = H(z)G(z) \qquad (24.586)$$

$\square$

24.14.9 Inversion

To derive the inversion formula for the z-transform, we will take the same approach as we took in deriving the expression for the transform itself. Namely, we will start with the inversion expression for the Laplace transform and substitute for $t=nT$ using the ideal sampler. The inverse Laplace transform is given by Equation 24.373 which is repeated here, in Equation 24.587, for convenience,

$$h(t) = \frac{1}{2\pi i}\left[\int_{\sigma_c-i\infty}^{\sigma_c+i\infty} H(s)e^{st}ds\right] \qquad (24.587)$$

where σ_c denotes the abscissa of convergence (see Section 24.8.1). The integral in Equation 24.587 is evaluated along the abscissa of convergence from $-i\infty$ to $i\infty$ (see Figure 24.16).

Note the folding phenomenon which was described in Section 3.6.1 when we deal with an ideal sampler. The effects of the extra harmonic reflections due to folding were shown in Figure 3.24. The integration in Equation 24.587, when applied to the output of an ideal sampler may be written as a sum of the integrals for each of the harmonic components of the integration, namely,

$$h(nt) = \frac{1}{2\pi i} \sum_{k=-\infty}^{\infty} \left[\int_{\sigma_c + i\frac{2k-1}{2}\omega_s}^{\sigma_c + i\frac{2k+1}{2}\omega_s} H(s)e^{nTs}ds \right] \tag{24.588}$$

Now, we can do a change of variables to isolate each section of the integration using

$$s \equiv \hat{s} + ik\omega_s \tag{24.589}$$

Therefore, using Equation 24.589 in 24.588, and moving the summation inside the integral, we have,

$$h(nt) = \frac{1}{2\pi i} \left[\int_{\sigma_c - i\frac{\omega_s}{2}}^{\sigma_c + i\frac{\omega_s}{2}} \sum_{k=-\infty}^{\infty} H(\hat{s} + ik\omega_s)e^{nT\hat{s}}d\hat{s} \right] \tag{24.590}$$

Using the results for the Fourier transform of the output of an ideal sampler, Equation 3.49, and using the fact that a Fourier transform is only a special case of the Laplace transform with $s = i\omega$, we have the following expression for the Laplace transform of an ideal sampler,

$$H^*(s) = \frac{1}{T} \sum_{n=-\infty}^{\infty} H(s + in\omega_s) \tag{24.591}$$

Using Equation 24.591 in Equation 24.590, we have,

$$h(nt) = \frac{T}{2\pi i} \left[\int_{\sigma_c - i\frac{\omega_s}{2}}^{\sigma_c + i\frac{\omega_s}{2}} H^*(\hat{s})e^{nT\hat{s}}d\hat{s} \right] \tag{24.592}$$

Plug in from Equation 24.521 into Equation 24.592 and change $H^*(\hat{s})$ with $H(z)$ and use the fact that

$$d\hat{s} = \frac{1}{T}d(\ln(z))$$

$$= z^{-1}\frac{dz}{T} \tag{24.593}$$

Then we have,

$$h(nT) = \frac{1}{2\pi i} \oint_{\Gamma_c} H(z)z^{n-1}dz \tag{24.594}$$

which is the inversion formula for the z-transform. Note that the abscissa of convergence, $s = \sigma_c$ translates to $\rho = e^{\sigma_c T}$ where $\rho = |z|$ is the radius of z in a polar representation. This means that the convergence space of the Laplace domain maps into $|z| < e^{\sigma_c T}$. This circle is the Γ_c in Equation 24.594. Therefore, all the singularities (poles) of $H(z)z^{n-1}$ should lie within this contour. The *Cauchy residue theorem* (Theorem 24.15) may be used for computing the inverse z-transform. In most cases, though, the functions of interest are rational, in which case the method of partial fractions may be utilized.

For a more complete coverage of the z-transform, see [27].

24.15 Cepstrum

In studying echoes, Bogert, et al. [4] considered analyzing the log of the power spectral density of the echo-contaminated signal to estimate the echo delay. Equation 24.595 represents a signal with an additive echo of itself after a delay of τ seconds. The reflection factor is denoted by a.

$$x(t) = h(t) + a\,h(t - \tau) \tag{24.595}$$

If we compute the power spectral density of the new signal, $x(t)$ which contains the original signal, $h(t)$ and a reflected echo with a reflection factor, a, arriving τ seconds later, we will have,

$$\begin{aligned}
|X(\omega)|^2 &= |H(\omega)|^2 + a^2|H(\omega)|^2 + 2a\cos(\omega\tau)|H(\omega)|^2 \\
&= |H(\omega)|^2\left[1 + a^2 + 2a\cos(\omega\tau)\right]
\end{aligned} \tag{24.596}$$

Let us take the log of both sides of Equation 24.596,

$$\begin{aligned}
\log(|X(\omega)|^2) &= \log(|H(\omega)|^2) + \log\left(1 + a^2 + 2a\cos(\omega\tau)\right) \\
&= \log(|H(\omega)|^2) + \log\left(\left[1 + a^2\right]\left[1 + \frac{2a\cos(\omega\tau)}{1 + a^2}\right]\right) \\
&= \log(|H(\omega)|^2) + \log(1 + a^2) + \\
&\quad \log\left(1 + \frac{2a}{1 + a^2}\cos(\omega\tau)\right)
\end{aligned} \tag{24.597}$$

The last term in Equation 24.597 may be expanded using an infinite power series, except at the points, $a = \pm 1$ and $\cos(\omega\tau) = \pm 1$.[9, 22]

$$\log(1 + \alpha\cos(\omega\tau)) = \sum_{j=1}^{\infty} \frac{(-1)^{j+1}}{j}(\alpha\cos(\omega\tau))^j \tag{24.598}$$

where,

$$\alpha \triangleq \frac{2a}{1+a^2} \tag{24.599}$$

Equation 24.598 suggests a spectral modulation in the log of the power spectral density of the signal in the form of a cosinusoidal ripple. As we can see, the amplitude and frequency of this ripple are related to the reflection factor, a, and the echo delay, τ, respectively.

To analyze this spectral modulation, [4] computed the power spectral density of the log of power spectral density of the signal. This computation is related to the time domain and has units of time, however, it also possesses all the aspects of a spectral analysis. Therefore, in their paper [4], Bogert, et al. decided to utilize some play on words to derive new terminology which would be analogous to spectral analysis, but would apply to the time domain instead of the frequency domain. For instance, the new domain was called the "cepstral" domain, by flipping the first four letters of the word, "spectral." Table 24.1 shows these analogs.

Spectral Domain	Cepstral Domain
Frequency	Quefrency
Spectrum	Cepstrum
Phase	Saphe
Amplitude	Gamnitude
Filter	Lifter
Harmonic	Rahmonic
Period	Repiod
Lowpass	Shortpass
Hipass	Longpass

Table 24.1: Terminology analogs of the spectral domain, used in the cepstral domain

[9] defines the general expression for the *power Cepstrum* in terms of the z-transform of the signal $H(z)$ as follows,

$$\tilde{h}_{pc} \triangleq \left[\mathscr{Z}^{-1}\{\log\left(|H(z)|^2\right)\} \right]^2$$
$$= \left[\frac{1}{2\pi i} \oint_{\Gamma_c} \log\left(|H(z)|^2\right) z^{n-1} dz \right]^2 \tag{24.600}$$

[9] calls $\tilde{h}_{pc}$ the power Cepstrum to differentiate it from two other Cepstrum definitions, namely, the *complex Cepstrum* and the *phase Cepstrum*. The original definition of *Cepstrum* presented in [4] is a special case of this entity, evaluated on the

unit circle, to coincide with its Fourier transform version of the definition.[9]

The definition given by Equation 24.600, for $\tilde{h}_{pc}$, uses the square of the inverse z-transform. The square was used by [9] in order to be consistent with the original definition of Cepstrum by [4]. There are many different variations of the Cepstrum in the literature. most of them analyze the log of the power spectral density, but they differ in the final definition related to the spectral analysis of the log of the PSD. Equation 24.600 squares the inverse z-transform. At this point let us limit ourselves to the values on the unit circle in the z-plane, leading to a Fourier transform version of the power Cepstrum.

In the continuous-time, the power Cepstrum may be defined in terms of the inverse of the Fourier transform of the log of the PSD as follows,

$$\tilde{h}_{pc} = \left[\frac{1}{2\pi} \int_{-\pi}^{\pi} \log\left(|H(\omega)|^2 \right) e^{i\omega t} \right]^2 \tag{24.601}$$

Example 24.9 (Cepstrum of an Echo).
Consider the periodic signal of Equation 24.602 which is a combination of 4 sinusoidal components,

$$h(t) = 0.2 \left(\sin(\omega t) + \sin(2\omega t) + \sin(3\omega t) + \sin(4\omega t) \right) \tag{24.602}$$

where $\omega = 2\pi f$ and $f = 80 \, Hz$.

Furthermore, consider a simple echo of this signal which arrives at a delay of $\tau = 0.3s$ with a reflection factor of $a = 0.4$ governed by Equation 24.595. The new signal will include the original signal plus the echo. Equation 24.603 describes this new signal.

$$x(t) = h(t) + 0.4h(t - 0.3) \tag{24.603}$$

Figure 24.22 shows the waveform and spectrogram representations of $x(t)$ of Equation 24.603. Figure 24.23 is a zoomed portion of the signal around the point in time when the echo arrives ($t = 0.3s$). In figure 24.24, we see the 4 peaks associated with the 4 different frequencies in the basic periodic signal. Finally, Figure 24.25 shows a plot of the power Cepstrum of the echo-contaminated signal of Equation 24.603. Note the peak at the arrival of the echo, $t = 0.3s$. The 1977 paper by Childers, et al. and its correction [9, 10] present a thorough treatment of the power Cepstrum, the complex Cepstrum and the phase Cepstrum. [35] also presents a good historical account of the Cepstrum for further reference.

---

[9] [4] simply used *Cepstrum* since it only defined the power Cepstrum. The *complex Cepstrum* was defined later by Oppenheim [36] in the process of developing the subject of *homomorphic system theory*.

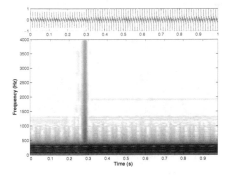

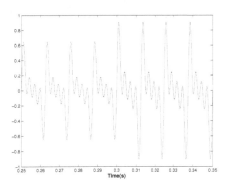

Fig. 24.22: Waveform and spectrogram of the signal and its echo at a delay of 0.3 s and a reflection factor of 0.4, given by Equations 24.602 and 24.603

Fig. 24.23: A small portion of the signal in the vicinity of the arrival of the echo

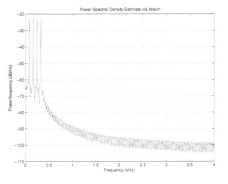

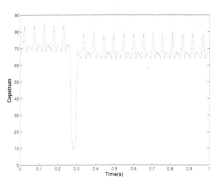

Fig. 24.24: Spectrum of the signal, showing the 4 periodic components

Fig. 24.25: Cepstrum of the signal, showing a peak at the moment of arrival of the echo

The problem of echo arrival-time detection is similar to the problem of pitch detection in speech, since speech may be viewed as the convolution of the impulse response of the vocal tract with the set of glottal pulses which manifest themselves in a quasi-periodic manner.[35] Noll [33, 34] applied a short-time version of the power Cepstrum to the speech signal to deconvolve the speech signal and to determine the pitch which is the basis for the cepstral analysis of speech signals used in Chapter 5. [35] demonstrates the pitch detection problem with an example. Consider a signal, $h(t)$ of this type which is formed from the convolution of two other signals, $h_1(t)$ and $h_2(t)$, see Equation 24.604.

$$h(t) = h_1(t) * h_2(t) \tag{24.604}$$

The Fourier transform of $h(t)$, $H(\omega)$ may be written as the product of the Fourier transforms of the convolved signals, namely,

$$H(\omega) = H_1(\omega)H_2(\omega) \tag{24.605}$$

Since the Cepstrum of $h(t)$ is computed from the $\log(H(\omega))$, if we use $\hat{h}(t)$ to denote the Cepstrum of $h(t)$, then,

$$\hat{h}(t) = \hat{h}_1(t) + \hat{h}_2(t) \tag{24.606}$$

Therefore, convolution in the time domain is represented by the summation in the cepstral domain.

Liftering (filtering in the cepstral domain) has been used by [4] to enhance the determination of the echo arrival. Using the complex Cepstrum, Oppenheim [36] introduced the concept of homomorphic deconvolution to separate different components of a signal which is made up of the convolution of several underlying signals. Examples of such signals is the speech signal viewed as a convolution of the vocal tract impulse response and the glottal pulses as described earlier. See Section 5.3, specifically Figure 5.10 for more information on the convolution of the speech signal.

Equation 24.607 presents the definition of the *complex Cepstrum*.

$$\hat{h} \triangleq \mathscr{Z}^{-1}\{\log(H(z))\}$$
$$= \frac{1}{2\pi i} \oint_{\Gamma_c} \log(H(z)) z^{n-1} dz \tag{24.607}$$

Note that the complex Cepstrum does not use the power spectrum of the log PSD. Instead, it is the inverse z-transform of the complex logarithm (hence complex Cepstrum) of the z-transform of the signal. This preserves the phase information which allows us to reconstruct the signal from its complex Cepstrum ([9] describes the procedure for this reconstruction in detail). The same is not true for the power Cepstrum, since the power Cepstrum loses the phase information of the signal. Upon modification of the complex Cepstrum, by removing certain additive components in the cepstral domain, the resulting complex Cepstrum may be used to reconstruct the version of the signal which has deconvolved the unwanted component of the time-domain signal. This process is called *homomorphic deconvolution*. This deconvolution may be done in the form of *liftering* the Cepstrum using similar concepts to the filtering of the spectral domain. Therefore, one may use a long-pass, short-pass or notch lifter in the cepstral domain. See Table 24.1 for the analogs of these filters in the spectral domain.

Since the complex logarithm is a periodic function with a period of $2\pi i$, the imaginary part of log of the z-transform of the signal, $\log H(z)$ has the period $\frac{2\pi}{T}$ where T is the sampling period. Furthermore, by definition, the complex Cepstrum of a real function is itself a real function which requires that the imaginary part of the complex Cepstrum be an odd periodic function of the ω. See *phase unwrapping* in [9].

[9] shows that the power Cepstrum at time, nT is related to the complex Cepstrum at times nT and $-nT$ as follows,

$$\tilde{h}_{pc}(nT) = \left(\hat{h}(nT) + \hat{h}(-nT)\right)^2 \tag{24.608}$$

In speaker recognition we are usually not concerned with the reconstruction of the signal, so we use the power Cepstrum instead of the complex Cepstrum. In most cases, the power Cepstrum is defined without squaring the inverse transform. Using this definition, the power Cepstrum would become,

$$\hat{h}_{pc} \triangleq \mathscr{Z}^{-1}\{\log\left(|H(z)|^2\right)\}$$
$$= \frac{1}{2\pi i} \oint_{\Gamma_c} \log\left(|H(z)|^2\right) z^{n-1} dz \tag{24.609}$$

or in the Fourier domain,

$$\hat{h}_{pc} = \frac{1}{2\pi} \int_{-\pi}^{\pi} \log\left(|H(\omega)|^2\right) e^{i\omega t} \tag{24.610}$$

Another possible definition of Cepstrum is that of the *phase Cepstrum* which is analogous to the power Cepstrum with the difference that in the phase Cepstrum, the inverse z-transform of the phase of the complex logarithm is computed instead of the inverse z-transform of its magnitude. Ripples are generated in the phase of the complex logarithm much in the same way as the appear in its magnitude. See [9] for more on this subject.

Problems

For solutions to the following problems, see the Solutions section at the end of the book.

Problem 24.1 (Modulus of the Product).
Prove Theorem 24.1, namely, show that,

$$|s_1 s_2| = |s_1| |s_2| \tag{24.611}$$

Problem 24.2 (Cauchy-Riemann Conditions not Sufficient).
Consider,

$$H(s) - U(\sigma, \omega) + iV(\sigma, \omega) \tag{24.612}$$

where

$$U(\sigma, \omega) = \frac{\sigma^3 - \omega^3}{\sigma^2 + \omega^2}$$

$$V(\sigma, \omega) = \frac{\sigma^3 + \omega^3}{\sigma^2 + \omega^2}$$

and show that meeting the Cauchy-Riemann Conditions is not sufficient for a function, $H(s)$, to be analytic. This problem relates to Theorem 24.5.

Problem 24.3 (Harmonic Conjugate).
Given $U(\sigma, \omega) = e^{\sigma} \cos(\omega)$, find the harmonic conjugate $V(\sigma, \omega)$. This problem is related to Definition 24.37

Problem 24.4 (Perid of Exponential).
What is the period of the exponential function,

$$H(s) = e^s \tag{24.613}$$

Problem 24.5 (Zeros of Complex Functions).
Show that the only zeros of complex functions $\sin(s)$ and $\cos(s)$ are the zeros of the real sine and cosine functions.

Problem 24.6 (z-Transform).
Find the z-transform of,

$$h(t) = \cos(\omega t) \tag{24.614}$$

References

1. Abe, S.: Support Vector Machines for Pattern Classification. Advances in Pattern Recognition. Springer-Verlag, London (2005). ISBN: 1-85233-929-2
2. Berg, P.W., McGregor, J.L.: Elementary Partial Differential Equations. Holden-Day Series in Mathematics. Holden-Day, San Francisco (1966)
3. Bernstein, D.J.: The Tangent FFT. In: Applied Algebra, Algebraic Algorithms and Error-Correcting Codes, pp. 291–300. Springer, Berlin (2007)
4. Bogert, B.P., Healy, M.J.R., Tukey, J.W.: The Quefrency Alanysis of Time Series for Echoes: Cepstrum, Pseudo-Autocovariance, Cross-Cepstrum, and Saphe Cracking. In: M. Rosenblatt (ed.) Time Series Analysis, pp. 209–243 (1963). Ch. 15
5. Burrus, C.S., Gopinath, R.A., Guo, H.: Introduction to Wavelets and Wavelet Transforms: A Primer. Prentice Hall, New york (1997). ISBN: 0-134-89600-9
6. Calderon, A.P., Zygmund, A.: On the Existence of Certain Singular Integrals. Acta Math. **88**, 85–129 (1952)
7. Canaris, J.: A VLSI architecture for the real time computation of discrete trigonometric transforms. The Journal of VLSI Signal Processing **5**(1), 95–104 (1993)
8. Chau, L.P., Siu, W.C.: Recursive Algorithm for the Discrete Cosine Transform with General Lengths. Electronics Letters **30**(3), 197–198 (1994)
9. Childers, D.G., Skinner, D.P., Kemerait, R.C.: The cepstrum: A guide to processing. Proceedings of the IEEE **65**(10), 1428–1443 (1977)
10. Childers, D.G., Skinner, D.P., Kemerait, R.C.: Corrections to "The cepstrum: A guide to processing". Proceedings of the IEEE **66**(10), 1290–1290 (1978)
11. Cooley, J.W., Tukey, J.W.: An Algorithm for the Machine Calculation of Complex Fourier Series. Mathematics of Computation **19**(2), 297–301 (1965)
12. Courant, R., Hibert, D.: Methods of Mathematical Physics, 1st english edn. Volume I. WILEY-VCH Verlag GmbH & Co. KGaA, Weinheim (2009). Second reprint of English version Translated and Revised from German Original, 1937, ISBN: 0-471-50447-4
13. Danielson, G.C., Lanczos, C.: Some Improvements in Practical Fourier Analysis and Their Application to X-ray Scattering From Liquids – Part I. Journal of the Franklin Institute **233**(4), 365–380 (1942)
14. Danielson, G.C., Lanczos, C.: Some Improvements in Practical Fourier Analysis and Their Application to X-ray Scattering From Liquids – Part II. Journal of the Franklin Institute **233**(5), 435–452 (1942)
15. Deng, L., O'Shaughnessy, D.: Speech Processing, A Dynamic and Optimization-Oriented Approach. Marcel Dekker, Inc., New york (2003). ISBN: 0-824-74040-8
16. Duhamel, P., Hollmann, H.: Aplit-Radix FFT Algorithm. Electronics Letters **20**, 14–16 (1984)
17. Fortmann, T.E., Hitz, K.L.: An Introduction to Linear Control Systems. Marcel Dekker, Inc., New York (1977). ISBN: 0-824-76512-5
18. Fourier, J.B.J.: Théorie Analytique de la Chaleur. Chez Firmin Didot, Père et Fils, Paris, France (1822). Digitized by Google from an Astor Library, New York, Copy: http://books.google.com/books?id=TDQJAAAAIAAJ&printsec=frontcover&dq=Th%C3%A9orie+analytique+de+la+chaleur&q=&hl=en#v=onepage&q&f=false
19. Gauss, C.F.: Nachlass, Theoria Interpolationis Methodo Nova Tractata. In: Carl Friedrich Gauss Werke, Band 3, Königlichen Gesellschaft der Wissenschaften: Göttingen, pp. 265–330 (1866). Note: Volume 3 of the collective works of Gauss.
20. Goertzel, G.: An Algorithm for the Evaluation of Finite Trigonometric Series. American Mathematical Monthly **65**, 34–35 (1958)
21. Haar, A.: Theorie der Orthogonalen Funktionensysteme (Theory of the Orthohonal System of Functions. Mathematische Annalen (Annals of Mathematics) **69**(3), 331–371 (1910). Original in German
22. Hassab, J.C.: On the Convergence Interval of the Power Cepstrum. IEEE Transactions on Information Theory **1**, 111–112 (1974)

23. Heideman, M.T., Johnson, D.H., Burrus, C.S.: Gauss and the History of Fast Fourier Transform. Archive for History of Exact Sciences **34**(3), 265–277 (1985)
24. Hilbert, D.: Grundzüge Einer Allgemeinen Theorie der Linearen Integralgleichungen (Outlines of a General Theory of Linear Integral Equations). Fortschritte der Mathematischen Wissenschaften, heft 3 (Progress in Mathematical Sciences, issue 3). B.G. Teubner, Leipzig and Berlin (1912). In German. Originally published in 1904.
25. Johnson, S.G., Frigo, M.: A Modified Split-Radix FFT with Fewer Arithmetic Operations. IEEE Transactions on Signal Processing **55**(1), 111–119 (2007)
26. Kellogg, O.: On the Existence and Closure of Sets of Characteristic Functions. Mathematische Annalen **86**(1002), 14–17 (1922)
27. Kuo, B.C.: Digital Control Systems, 2nd edn. Oxford University Press, New York (1992). ISBN: 0-195-12064-7
28. Lundy, T., Buskirk, J.V.: A new matrix approach to real FFTs and convolutions of length 2k. Computing **80**(1), 23–45 (2007)
29. Martens, J.B.: Recursive Cyclotomic Factorization – A New Algorithm for Calculating the Discrete Fourier Transform. IEEE Transactions on Acoustic, Speech and Signal Processing **32**(4), 750–761 (1984)
30. Mercer, J.: Functions of Positive and Negative Type, and their Connection with the Theory of Integral Equations. Philosophical Transactions of the Royal Society of London. Series A, Containing Papers of a Mathematical or Physical Character **209**, 415–446 (1909)
31. Michiel Hazewinkel, E.: Encyclopaedia of Mathematics. Kluwer Academic Publishers, Amsterdam (2002). ISBN: 1-4020-0609-8
32. Morlet, J., Arens, G., Fourgeau, I., Giard, D.: Wave Propagation and Sampling Theory. Geophysics **47**, 203–236 (1982)
33. Noll, A.M.: Short-Time Spectrum and 'Cepstrum' Techniques for Vocal-Pitch Detection. The Journal of the Acoustical Society of America (JASA) **36**(2), 296–302 (1964)
34. Noll, A.M.: Cepstrum Pitch Determination. The Journal of the Acoustical Society of America (JASA) **41**(2), 293–309 (1967)
35. Oppenheim, A., Schafer, R.: From frequency to quefrency: a history of the cepstrum. IEEE Signal Processing Magazine **21**(5), 95–106 (2004)
36. Oppenheim, A.V.: Superposition in a Class of Nonlinear Systems. Massachussetts Institute of Technology, Cambridge, Massachussetts (1964). Ph.D. Dissertation
37. Press, W.H., Teukolsky, S.A., Vetterling, W.T., Flannery, B.P.: Numerical Recipes in C++. Cambridge University Press, New York (2002). ISBN: 0-521-75033-4
38. Ramussen, C.E., Williams, C.K.I.: Gaussian Processes for Machine Learning (Adaptive Computation and Machine Learning). The MIT Press, Boston (2006). ISBN: 978-026218253-9
39. Schmidt, E.: Zur Theorie der Linearen und Nichtlinearen Integralgleichungen. I. Teil: Entwicklung Willkürlicher Funktionen nach Systemen Vorgeschriebener. Mathematische Annalen **63**, 433–476 (1906)
40. Schmidt, E.: Zur Theorie der linearen und nichtlinearen Integralgleichungen. II. Auflösung der Allgemeinen Linearen Integralgleichung. Mathematische Annalen **64**, 161–174 (1907)
41. Schmidt, E.: Zur Theorie der Linearen und Nichtlinearen Integralgleichungen. III. Teil: Über die Auflösung der nichtlinearen Integralgleichung und die Verzweigung ihrer Lösungen. Mathematische Annalen **65**, 370–399 (1907)
42. Schölkopf, B., Smola, A.J.: Learning with Kernels: Support Vector Machines, Regularization, Optimization, and Beyond (Adaptive Computation and Machine Learning). MIT Press, Cambridge, MA (2002). ISBN: 978-0-262-19475-4
43. Shannon, C.E.: Communication in the Presence of Noise. Proceedings of the Institute of Radio Engineers **37**(1), 10–21 (1949). Reprint available at: Proceedings of the IEEE, Vol. 86, No. 2, Feb. 1998
44. Sorensen, H., Heideman, M., Burrus, C.: On computing the split-radix FFT. IEEE Transactions on Acoustics, Speech and Signal Processing **34**(1), 152–156 (1986)
45. Sorensen, H., Jones, D., Heideman, M., Burrus, C.: Real-valued Fast Fourier Transform Algorithms. IEEE Transactions on Acoustics, Speech and Signal Processing **35**(6), 849–863 (1987). See Correction

46. Sorensen, H., Jones, D., Jones, D., Burrus, C.: Correction to Real-valued Fast Fourier Transform Algorithms. IEEE Transactions on Acoustics, Speech and Signal Processing **35**(9), 849–863 (1987). Correction to earlier publication in Jun 1987
47. Vetterli, M., Nussbaumer, H.J.: Simple FFT and DCT Algorithms with Reduced Number of Operations. Signal Processing **6**, 262–278 (1984)
48. Weiss, L.G.: Wavelets and wideband correlation processing. IEEE Signal Processing Magazine **11**(1), 13–32 (1994)
49. Welch, P.: The use of fast Fourier transform for the estimation of power spectra: A method based on time averaging over short, modified periodograms. IEEE Transactions on Audio and Electroacoustics **15**(2), 70–73 (1967)
50. Yavne, R.: An Economical Method for Calculating the Discrete Fourier Transform. In: Proceedings of the AFIPS Fall Joint Computer Conference, vol. 33, pp. 115–125 (1968)

Chapter 25
Nonlinear Optimization

Throughout this chapter, we only treat the minimization problem for *convex functions* (see Definition 24.23). Furthermore, in most cases, we assume that the objective function being minimized is a *quadratic function*. These minimization assumptions may easily cover the cases where a function needs to be maximized. In the case of *concave functions* (Definition 24.25), where we are interested in the maxima, the function may be multiplied by -1 which inverts it into a convex function such that the location of the maximum now points to the minimum of the new function. So, the maximization function is changed to a minimization function.

The treatment of a quadratic function is undertaken since it is a much simpler optimization problem. Since we are concerned with a quadratic or higher order function, the optimization is known as *nonlinear optimization*. In general, if the objective function or any of the constraints is nonlinear, then the problem is known as a *nonlinear optimization problem*. On the other hand, to have a *linear optimization problem*, the objective function as well as all the constraints would have to be linear with respect to the dependent variable.

Most *smooth functions* (see Definition 24.20) may be approximated by quadratic functions in a small interval. Therefore, a general function, depending on whether it is locally *concave* or *convex*, may be approximated by a *quadratic function* in small intervals around the optimal points and then the total solution may be reached based on the solution to the smaller quadratic convex minimization problems.

A very important problem is one of finding the *global optimum* versus the *local optima* and *saddle points*. A general multivariate function, $E(\mathbf{x})$, may have several *minima*, *maxima* and *saddle points* (see Figure 25.1). These, so called, *stationary points*, are points at which the gradient of the function approaches zero,

$$\nabla_{\mathbf{x}} E(\mathbf{x}) = 0 \tag{25.1}$$

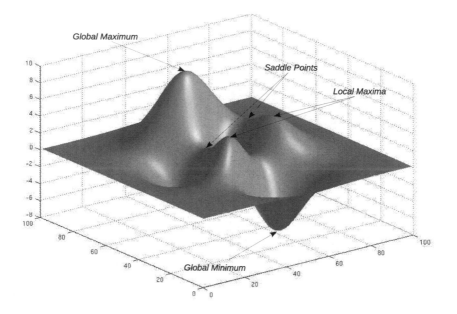

Fig. 25.1: Stationary Points of a Function of Two Variables

Generally, at the location of a maximum, the matrix of its *second partial derivatives (Hessian)*, G, is negative definite (has all-negative *Eigenvalues* – see Section 12.1). It is *positive* definite (has all positive Eigenvalues) for a *minimum* and it is *indefinite* (has zero or both positive and negative eigenvalues) for a *saddle point*.

The following definitions identify the *states* (values of the dependent variable) for which we will be searching, in this chapter.

Definition 25.1 (Global Minimizer). *A state* $\mathbf{x}^*$ *is the global minimizer of the objective function* $E(\mathbf{x})$ *if,*

$$E(\mathbf{x}^*) \leq E(\mathbf{x}) \ \forall \ \mathbf{x} \tag{25.2}$$

Definition 25.2 (Strict Local Minimizer (Strong Local Minimizer). *A state* $\mathbf{x}_i^*$ *is a strict local minimizer of* $E(\mathbf{x})$ *if,*

$$E(\mathbf{x}_i^*) \leq E(\mathbf{x}) \ \forall \ \mathbf{x} \in \mathscr{X}_i, i = \{1, \cdots, N\} \tag{25.3}$$

where $\mathscr{X}_i \subset \mathscr{D}_E, i = \{1, \cdots, N\}$, $\mathscr{D}_E$ *is the Domain of E and N is the number of strict local minimizers in that domain.*

25.1 Gradient-Based Optimization

This section describes a set of gradient based minimization techniques which require the evaluation of the objective function, its gradient and sometimes the Hessian for certain states. Another group of minimization techniques are described in Section 25.2 which require evaluations of the objective function only. In the latter methods, no direct gradient evaluations are needed.

Let us define the following short-hand notation,

$$\mathbf{g}_k \overset{\Delta}{=} \nabla_{\mathbf{x}} E \Big|_{\mathbf{x}_k} \tag{25.4}$$

$$\mathbf{G}_k \overset{\Delta}{=} \nabla_{\mathbf{x}}^2 E \Big|_{\mathbf{x}_k} \tag{25.5}$$

As it will be made more apparent later, this notation will be useful in writing out the equations in the following section in a more compact form.

25.1.1 The Steepest Descent Technique

The steepest descent technique is a gradient-minimization method which uses a first order approximation to the objective function (E) to generate directions of descent with the knowledge of the gradient at each iteration k. The steepest descent method is very reliable since it always provides a descent direction. This technique is ideal for points which are far away from the local minima. However, close to the local minima, the steepest descent technique will generally require lots of iterations to converge due to its nature of approximating the objective function with a linear function.

Let $\mathbf{x}^*$ denote the state vector which results in a minimum objective function E. Further, assume that the current state is $\mathbf{x}_k$ and define $\Delta \mathbf{x}_k$ to be the difference between the states at k and $k+1$ iterations, namely,

$$\Delta \mathbf{x}_k = \mathbf{x}_{k+1} - \mathbf{x}_k \tag{25.6}$$

Decompose the step $\Delta \mathbf{x}_k$ into a direction $\mathbf{s}_k$ and a magnitude η_k,

$$\Delta \mathbf{x}_k = \eta_k \mathbf{s}_k \tag{25.7}$$

Since the gradient of E points in the ascent direction, the steepest descent is given by the direction opposite to the gradient direction at every step k. Therefore for steepest descent minimization,

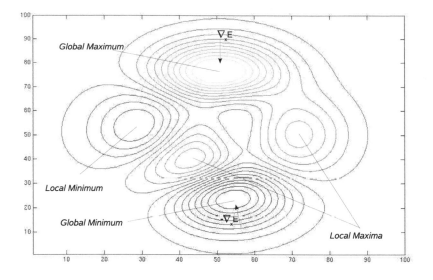

Fig. 25.2: Contour plot of the function of Figure 25.1

$$s_k = -\frac{\mathbf{g}_k}{\|\mathbf{g}_k\|} \tag{25.8}$$

Figure 25.2 shows the contour plot of the function displayed earlier in Figure 25.1. Note the direction of the gradient relative to the contour lines. The gradient points to the increasing value of the function (the local maximum). Therefore, the negative of the gradient points to the local minimum in the vicinity of the location of the gradient evaluation.

However, a *line search* should be used to provide an optimum step size η_k^* to minimize E in the s_k direction. This will produce the following recursive steepest descent technique for minimization of E,

$$\mathbf{x}_{k+1} = \mathbf{x}_k - \eta_k^* \frac{\mathbf{g}_k}{\|\mathbf{g}_k\|} \tag{25.9}$$

The steepest descent method only provides good minimizing directions if the condition number of the Hessian of the objective function is close to 1. As this condition number gets larger, the steepest descent becomes slower and more advanced minimization techniques (preferably with quadratic convergence) will be preferred. To further enhance the speed of learning of the network, let us consider the following more advanced minimization techniques.

25.1.2 Newton's Minimization Technique

Write the *Taylor series expansion* (Definition 24.42) of E at $\mathbf{x}^*$ and about $\mathbf{x}_k$ assuming that every update of the state vector should drive the state vector to optimal value $\mathbf{x}^*$, namely, $\mathbf{x}^* = \mathbf{x}_k + \Delta\mathbf{x}_k$,

$$E(\mathbf{x}^*) = E(\mathbf{x}_k) + \nabla_\mathbf{x}^T E \Big|_{\mathbf{x}_k} \Delta\mathbf{x}_k + \frac{1}{2}\Delta\mathbf{x}_k^T \nabla_\mathbf{x}^2 E \Big|_{\mathbf{x}_k} \Delta\mathbf{x}_k + \mathcal{O}(\Delta\mathbf{x}_k^3) \tag{25.10}$$

where $\mathcal{O}(\Delta\mathbf{x}_k^3)$ is the *Bachmann-Landau asymptotic notation*[1], showing that the rest of terms are of $\mathcal{O}(\Delta\mathbf{x}_k^3)$.

Using the definitions of Equations 25.4 and 25.5, Equation 25.10 may be rewritten as follows,

$$E(\mathbf{x}^*) = E(\mathbf{x}_k) + \mathbf{g}_k^T \Delta\mathbf{x}_k + \frac{1}{2}\Delta\mathbf{x}_k^T \mathbf{G}_k \Delta\mathbf{x}_k + \mathcal{O}(\Delta\mathbf{x}_k^3) \tag{25.11}$$

If we disregard the higher than second order terms in 25.11 and thus approximate E with a quadratic function in the vicinity of $\mathbf{x}_k$ and $\mathbf{x}^*$, then the quadratic approximation of Equation 25.11 may be written as follows,

$$E(\mathbf{x}^*) \approx E(\mathbf{x}_k) + \mathbf{g}_k^T \Delta\mathbf{x}_k + \frac{1}{2}\Delta\mathbf{x}_k^T \mathbf{G}_k \Delta\mathbf{x}_k \tag{25.12}$$

Note that for a minimum of $E(\mathbf{x}^*)$, a necessary condition is that $\nabla_\mathbf{x} E$ be zero. However, keeping the current state, $\mathbf{x}_k$, constant and then taking the gradient of both sides of Equation 25.12, since $\mathbf{x}^* = \mathbf{x}_k + \Delta\mathbf{x}_k$,

$$\nabla_\mathbf{x} E \Big|_{\mathbf{x}^*} \approx \mathbf{g}_k + \mathbf{G}_k \Delta\mathbf{x}_k \tag{25.13}$$

Setting the gradient of E at $\mathbf{x}^*$ equal to zero, gives,

$$\mathbf{g}_k + \mathbf{G}_k \Delta\mathbf{x}_k \approx 0 \tag{25.14}$$

or,

$$\Delta\mathbf{x}_k \approx -\mathbf{G}_k^{-1}\mathbf{g}_k \tag{25.15}$$

Since E is generally not a quadratic function in $\mathbf{x}_k$, the following recursive update may be used for the computing the state vector,

---

[1] The Bachmann-Landau notation for $\mathcal{O}$ was first introduced by *Bachmann* in 1894 [2], and later in 1909, it was adopted by *Landau* [42] and included in a more elaborate *asymptotic notation* which included $o(.)$, etc.

$$\mathbf{x}_{k+1} = \mathbf{x}_k + \eta_k^* \mathbf{s}_k \tag{25.16}$$

where η_k^* is the optimum step size in the direction, $\mathbf{s}_k$, and $\mathbf{s}_k$ is given by the following,

$$\mathbf{s}_k = -\frac{G_k^{-1}\mathbf{g}_k}{\|G_k^{-1}\mathbf{g}_k\|} \tag{25.17}$$

A *line search* method could be used to determine η_k^* along direction $\mathbf{s}_k$.

This method provides quadratic convergence and is very efficient in the vicinity of the minima. However, there are three problems that are faced when trying to use this algorithm. The first problem is that in order for E to always descend in value, the matrix $\mathbf{G}^{-1}$ should be positive definite. Since E is generally not quadratic, $\mathbf{G}^{-1}$ could become indefinite or even negative definite. There are many techniques developed to keep a positive definite approximation of the inverse Hessian matrix, $\mathbf{G}^{-1}$, such that the quadratic information in the Hessian matrix will be used. Using the quadratic information generally provides a better direction of descent than the steepest descent direction, especially in the vicinity of the minima. Among the methods for keeping a positive definite approximation of the inverse Hessian matrix are Greenstadt's method [28], Marquardt [46], Levenberg [44], and Goldfeld, Quandt and Tratter's alternative [27].

A second problem is that for systems with a small number of variables it might be feasible to find the Hessian matrix, however, for systems with large number of variables and complicated objective functions, it will become a very difficult task. In addition, it will be very hard to write general equations for the evaluation of the elements of the Hessian matrix as done for the elements of the gradient vector $\mathbf{g}$.

Let us, for the sake of argument, say that a Hessian matrix is calculated at every iteration k. Then a still more serious problem occurs. For huge systems it is not practical to take the inverse of the Hessian matrix. Taking this inverse in most cases will require more time than taking more steps using a simpler method such as the steepest descent method.

The problem of retainment of positive definiteness of the inverse Hessian can be solved by the methods noted above. However, problems two and three make using Newton's method quite impractical. These limitations are reasons for looking at the following alternatives which in turn will solve the aforementioned problems and still keep a super-linear rate of convergence.

25.1.3 Quasi-Newton or Large Step Gradient Techniques

From Equation 25.16, we can write the following generalized recursive algorithm to update the state vector such that a minimum E will be approached,

$$\mathbf{x}_{k+1} = \mathbf{x}_k + \eta_k^* \mathbf{H}_k \nabla_\mathbf{x} E \Big|_{\mathbf{x}_k} \tag{25.18}$$

where η_k^* is a weighting factor, and $\mathbf{H}_k$ is a square symmetric matrix. Depending on the choice of $\mathbf{H}_k$, different optimization algorithms will be resulted. Therefore, $\mathbf{H}_k$, multiplied by the gradient of E will provide a direction of descent in the objective function E and η_k^* is the optimal step in that descent direction as provided by some *line search* method. If $\mathbf{H}_k$ in Equation 25.18 is made equivalent to the identity matrix, I, then the method reduces to the steepest descent technique which provides linear convergence. Making $\mathbf{H}_k$ equivalent to the inverse of the Hessian matrix, $\mathbf{G}^{-1}$, of the quadratic approximation of E, as previously defined, the method will reduce to the Newton minimization technique which provides quadratic convergence.

Instead of using the real inverse-Hessian, Quasi-Newton methods use an approximation to the inverse-Hessian provided by an iterative updating scheme. Quasi-Newton methods usually start with an approximation to the inverse-Hessian matrix such as the identity matrix. Different updates for $\mathbf{H}_k$ are then used, leading to different types of Quasi-Newton methods. Updates to matrix $\mathbf{H}_k$ are done recursively in different directions of the inverse-Hessian space, based on the information obtained from the function and gradient behavior in that direction. Depending on whether these updates are done in one or two directions at a time, rank one or rank two methods are generated. Those Quasi-Newton methods which retain a positive definite $\mathbf{H}_k$ are called *variable metric* methods. Not all Quasi-Newton methods use variable metric updates. Newton-like methods in general try to keep Newton's condition 25.19 satisfied.

$$\mathbf{H} \Delta \mathbf{g}_k = \Delta \mathbf{x}_k \tag{25.19}$$

where,

$$\mathbf{H} \overset{\Delta}{=} \mathbf{G}^{-1} \tag{25.20}$$

Condition 25.19 is automatically satisfied for a quadratic function, if $\mathbf{H}$ is the exact inverse Hessian matrix. However, since in Quasi-Newton methods, the inverse Hessian is supposed to be approximated, instead of $\mathbf{H}_k \Delta \mathbf{g}_k = \Delta \mathbf{x}_k$, these methods try to keep the following relation satisfied at each step k,

$$\mathbf{H}_{k+1} \Delta \mathbf{g}_k = \Delta \mathbf{x}_k \tag{25.21}$$

This relationship is referred to as the *Quasi-Newton condition* and it means that the inverse Hessian matrix should be updated such that relation 25.21 is satisfied.

In 1959, Davidon [12] introduced the idea of Quasi-Newton methods. In 1965, Barnes [3] and Broyden [9] independently introduced a method for solving a set of simultaneous linear equations of the same form as Equation 25.19. Barnes' equation is a more general one and includes Broyden's method as a special case. Equation 25.22 gives this update,

$$\Delta \mathbf{H}_k = \frac{(\Delta \mathbf{x}_k - \mathbf{H}_k \Delta \mathbf{g}_k) \mathbf{z}_k^T}{\mathbf{z}_k^T \Delta \mathbf{g}_k} \tag{25.22}$$

where $\mathbf{z}_k$ is a direction in which the update to $\mathbf{H}_k$ is done.

25.1.3.1 Rank One Updates of the Inverse Hessian

A rank-one update to the inverse Hessian matrix $\mathbf{H}_k$ would mean the following,

$$\mathbf{H}_{k+1} = \mathbf{H}_k + \alpha \mathbf{u} \mathbf{u}^T \tag{25.23}$$

where $\mathbf{u}$ is a direction of update. Setting $\mathbf{z}_k$ in Equation 25.22 equal to the error of Equation 25.19, when $\mathbf{H}$ is approximated by $\mathbf{H}_k$, will produce Broyden's rank-one update, Equation 25.24.

$$\Delta \mathbf{H}_k = \frac{(\Delta \mathbf{x}_k - \mathbf{H}_k \Delta \mathbf{g}_k)(\Delta \mathbf{x}_k - \mathbf{H}_k \Delta \mathbf{g}_k)^T}{(\Delta \mathbf{x}_k - \mathbf{H}_k \Delta \mathbf{g}_k)^T \Delta \mathbf{g}_k} \tag{25.24}$$

This update, as applied to Quasi-Newton problems, was introduced by Broyden [10], Davidon [13] and others [19, 49] independently. Let N denote the dimension of the state vector, $\mathbf{x}$. This update has the property that if $\Delta \mathbf{x}_1, \Delta \mathbf{x}_2, \cdots, \Delta \mathbf{x}_N$ are linearly independent, then, at $k = N+1$, $\mathbf{H}_k = \mathbf{G}^{-1}$ for quadratic for any quadratic function, $E(\mathbf{x})$.

Another important feature of Broyden's update is that η_k^* of Equation 25.18 does not necessarily have to minimize E in the $\mathbf{s}_k$ direction. As long as η_k^* is such that $\mathbf{H}_{k+1}$ will not become singular and the denominator of Equation 25.24 is not made zero, any η_k^* could be used in conjunction with the Broyden update. However, some unattractive features also exist for this update. If the objective function is non-quadratic, as in the case of general neural networks, the following less than satisfactory aspects of Broyden update exist,

1. $\mathbf{H}_k$ may not retain its positive definiteness in which case it is necessary to use one of the methods in Section 25.1.2, such as *Greenstadt's method*, to force this matrix to be positive definite.
2. The correction, $\Delta \mathbf{H}_k$ of Equation 25.24 may sometimes become unbounded (sometimes even for quadratic functions, due to *round-off errors*).
3. if $\Delta \mathbf{x}_k$, given by Equation 25.18, is by chance in the same direction as $\Delta \mathbf{x}_{k+1}$, the, $\Delta \mathbf{H}_{k+1}$ becomes singular or undetermined. Therefore, if

$$\mathbf{H}_k \Delta \mathbf{g}_k = \Delta \mathbf{x}_k \tag{25.25}$$

or

$$(\mathbf{H}_k \mathbf{g}_k - \Delta \mathbf{x}_k)^T \Delta \mathbf{g}_k = 0 \tag{25.26}$$

then, $\mathbf{H}_{k+1}$ should be set to H_k, namely, $\Delta \mathbf{H}_k = 0$.

25.1.3.2 Pearson's Updates

Pearson [53] introduced other directions of update for the projection of the error of Equation 25.19. In his *number 2* method, he proposed, $\mathbf{z}_k = \Delta \mathbf{x}_k$ in Equation 25.22, which in turn generated the following update for $\mathbf{H}_k$,

$$\Delta \mathbf{H}_k = \frac{(\Delta \mathbf{x}_k - \mathbf{H}_k \Delta \mathbf{g}_k)\Delta \mathbf{x}_k^T}{\Delta \mathbf{x}_k^T \Delta \mathbf{g}_k} \tag{25.27}$$

His *number 3* method used the complementary possibility of, $\mathbf{z}_k = \mathbf{H}_k \Delta \mathbf{g}_k$. This gives the following update,

$$\Delta \mathbf{H}_k = \frac{(\Delta \mathbf{x}_k - \mathbf{H}_k \Delta \mathbf{g}_k)(\mathbf{H}_k \Delta \mathbf{g}_k)^T}{(\mathbf{H}_k \Delta \mathbf{g}_k)^T \Delta \mathbf{g}_k} \tag{25.28}$$

Pearson's methods do not guarantee positive definiteness of the inverse Hessian Matrix and usually lead to ill-conditioned matrices. Therefore, it is a good idea to reset the inverse Hessian approximation to Identity, every N iterations, i.e., $\mathbf{H}_{Nt} = \mathbf{I} \quad \forall \ t = \{0, 1, 2, \cdots\}$.

25.1.3.3 Rank Two Updates

The rank-one update does not leave much of a freedom for choosing the directions of update. This motivated the formulation of rank-two updates such that the objective is still to satisfy Relation 25.21 at every step k. The general rank two updates are of the following form,

$$\mathbf{H}_{k+1} - \mathbf{H}_k + \alpha \mathbf{u}\mathbf{u}^T + \beta \mathbf{v}\mathbf{v}^T \tag{25.29}$$

where, the directions $\mathbf{u}$ and $\mathbf{v}$ and scaling factors α and β should be chosen. It would be a good idea to update these directions relative to $\Delta \mathbf{x}_k$ and $\mathbf{H}_k \Delta \mathbf{g}_k$. This would give the following general update,

$$\Delta \mathbf{H}_k = \alpha \frac{\Delta \mathbf{x}_k \mathbf{y}^T}{\mathbf{y}^T \Delta \mathbf{g}_k} + \beta \frac{\mathbf{H}_k \Delta \mathbf{g}_k \mathbf{z}^T}{\mathbf{z}^t \Delta \mathbf{g}_k} \tag{25.30}$$

A natural choice to retain symmetry is to pick $\alpha = 1$, $\beta = -1$, $\mathbf{y} = \Delta \mathbf{x}_k$, and $\mathbf{z} = \mathbf{H}_k \Delta \mathbf{g}_k$. This will generate the *Davidon-Fletcher-Powell (DFP)* [23] update given by Equation 25.31.

$$\Delta \mathbf{H}_k = \frac{\Delta \mathbf{x}_k \Delta \mathbf{x}_k^T}{\Delta \mathbf{x}_k^T \Delta \mathbf{g}} - \frac{\mathbf{H}_k \Delta \mathbf{g}_k \Delta \mathbf{g}_k^T \mathbf{H}_k^T}{\Delta \mathbf{g}_k^T \mathbf{H}_k^T \Delta \mathbf{g}_k} \tag{25.31}$$

This algorithm works properly, in general, if $\mathbf{g}_k$ is calculated with minimal error and $\mathbf{H}_k$ does not become ill-conditioned. Define,

$$\mathbf{A}_i = \frac{\Delta \mathbf{x}_i^T \Delta \mathbf{x}_i}{\Delta \mathbf{x}_i^T \Delta \mathbf{g}_i} \tag{25.32}$$

and,

$$\mathbf{B}_i = \frac{\mathbf{H}_i \Delta \mathbf{g}_i \Delta \mathbf{g}_i^T \mathbf{H}_i^T}{\Delta \mathbf{g}_i^T \mathbf{H}_i^T \Delta \mathbf{g}_i} \tag{25.33}$$

Then,

$$\sum_{i=0}^{k-1} \mathbf{A}_i \rightarrow \mathbf{H} \quad as \quad k \rightarrow N \tag{25.34}$$

and

$$\sum_{i=0}^{k-1} \mathbf{B}_i \rightarrow \mathbf{H}_0 \quad as \quad k \rightarrow N \tag{25.35}$$

which results in

$$\mathbf{H}_k \rightarrow \mathbf{H} \quad as \quad k \rightarrow N \tag{25.36}$$

for a quadratic function.

Proof.
Using Equation 25.19, substitute, for $\Delta \mathbf{g}_i$ in Equation 25.32,

$$\mathbf{A}_i = \frac{\Delta \mathbf{x}_i^T \Delta \mathbf{x}_i}{\Delta \mathbf{x}_i^T \mathbf{G} \Delta \mathbf{x}_i} \tag{25.37}$$

Compute the summation of $\mathbf{A}_i$ over N consecutive steps,

$$\sum_{i=0}^{N-1} \mathbf{A}_i = \sum_{i=0}^{N-1} \frac{\Delta \mathbf{x}_i \Delta \mathbf{x}_i^T}{\Delta \mathbf{x}_i^T \mathbf{G} \Delta \mathbf{x}_i} \tag{25.38}$$

$$= \frac{\eta_i^* \mathbf{s}_i \mathbf{s}_i^T \eta_i^*}{\eta_i^* \mathbf{s}_i^T \mathbf{G} \mathbf{s}_i \eta_i^*} \tag{25.39}$$

$$= \frac{\mathbf{s}_i \mathbf{s}_i^T}{\mathbf{s}_i^T \mathbf{G} \mathbf{s}_i} \tag{25.40}$$

Therefore, for quadratic functions when $\mathbf{s}_i$ are conjugate about $\mathbf{G}$ (see Definition 23.12),

$$\sum_{i=0}^{N-1} \mathbf{A}_i = \mathbf{H} \tag{25.41}$$

The DFP method provides $\mathbf{s}_i$ which are conjugate about $\mathbf{G}$ and thus Equation 25.41 holds. Similarly, $\sum_{i=0}^{N-1} \mathbf{B}_i$ can be shown to approach $\mathbf{H}_0$ and to keep $\mathbf{H}_k$ positive definite as $k \to N$.
Q.E.D.

$\square$

An important property of this update is that if $\Delta \mathbf{x}_k^T \Delta \mathbf{g}_k > 0$ for all k, then the approximate inverse Hessian matrix will retain its positive definiteness. This condition can be imposed by using a *line search* method which satisfies the following relation,

$$\mathbf{g}_{k+1}^T \Delta \mathbf{x}_k \geq \sigma \mathbf{g}_k^T \Delta \mathbf{x}_k \tag{25.42}$$

The condition of Equation 25.42 stems from two conditions by *Wolfe* [70] and *Powell* [55] which are called the *Wolfe-Powell conditions* by *Fletcher* [23].

Eventually, condition 25.42 means that the curvature estimate should be positive where the updating is done. In Equation 25.42,

$$\sigma \in [\tau, 1], \tag{25.43}$$

where,

$$\tau \in \left[0, \frac{1}{2}\right], \tag{25.44}$$

These are parameters of the *line search* termination. [23]

In 1970, *Broyden* [11], *Fletcher* [22], *Goldfarb* [26], and *Shanno* [61] suggested the *BFGS* update which is dual with the *DFP* update. This means that if one applies the *DFP* method to updating the Hessian matrix from the following,

$$\mathbf{G}_{k+1} \Delta \mathbf{x}_k = \Delta \mathbf{g}_k \tag{25.45}$$

rather than Equation 25.21, and then apply the Sherman-Morrison inversion formula (see Section 23.4) to obtain an expression for $\Delta\mathbf{H}_k$, them the BFGS formula given by Equation 25.46 will be obtained.

$$\Delta\mathbf{H}_k = \left(1 + \frac{\Delta\mathbf{g}_k^T\mathbf{H}_k\Delta\mathbf{g}_k}{\Delta\mathbf{x}_k^T\Delta\mathbf{g}_k}\right)\frac{\Delta\mathbf{x}_k^T\Delta\mathbf{x}_k}{\Delta\mathbf{x}_k\Delta\mathbf{g}_k} - \frac{\Delta\mathbf{x}_k\Delta\mathbf{g}_k^T\mathbf{H}_k + \mathbf{H}_k\Delta\mathbf{g}_k\Delta\mathbf{x}_k^T}{\Delta\mathbf{x}_k^T\Delta\mathbf{g}_k} \tag{25.46}$$

The *BFGS* update has all the qualities of the *DFP* method plus the fact that it has been noted to work exceptionally well with *inexact line searches* and a global convergence proof exists [23] for the *BFGS* update. No such proof has yet been presented for the convergence of *DFP*.

25.1.3.4 Updates Through Variational Means

Greenstadt [29] developed a general updating scheme using variational means by minimizing the Euclidean norm of the update to the inverse Hessian. This generated the following general updating formula,

$$\Delta H_k = \frac{1}{\nu}\left[\Delta\mathbf{x}_k\Delta\mathbf{g}_k^T\mathbf{M} + \mathbf{M}\Delta\mathbf{g}_k\Delta\mathbf{x}_k^T - \right.$$
$$\left. \mathbf{H}_k\Delta\mathbf{g}_k\Delta\mathbf{g}_k^T\mathbf{H}_k\frac{1}{\nu}\left(\Delta\mathbf{g}_k^T\Delta\mathbf{x}_k - \nu\right)\mathbf{M}\Delta\mathbf{g}_k\Delta\mathbf{g}_k^T\mathbf{M}\right] \tag{25.47}$$

where,

$$\nu \overset{\Delta}{=} \Delta\mathbf{g}_k^T\mathbf{M}\Delta\mathbf{g}_k \tag{25.48}$$

In Equation 25.47, $\mathbf{M}$ is a positive definite matrix. Greenstadt, in his paper, proposed two possible values for $\mathbf{M}$,

$$\mathbf{M} = \mathbf{H}_k \tag{25.49}$$
$$\mathbf{M} = \mathbf{I} \tag{25.50}$$

The updates by setting $\mathbf{M}$ of Equation 25.47 to those prescribed by Equations 25.49 and 25.50 do not retain positive a positive definite $\mathbf{H}_k$ in general. Goldfarb [26] proposed the use of $\mathbf{M} = \mathbf{H}_{k+1}$ which does provide a positive definite approximation to the inverse Hessian matrix. This method is identical to the *BFGS* update discussed in the previous section.

25.1.3.5 Self-Scaling Quasi-Newton Methods

Suppose that the objective function E is scaled by a number c and results in a new objective function,

$$E' = cE \tag{25.51}$$

This objective function has the same minimizer as E and its gradient and inverse Hessian are given in terms of those of E by,

$$\mathbf{g'} = c\mathbf{g} \tag{25.52}$$

$$\mathbf{H'} = \frac{1}{c}\mathbf{H} \tag{25.53}$$

The Newton step for finding the minimizer of a quadratic function, $\mathbf{x}^*$, is $\mathbf{x}^* = \mathbf{x}_k - \mathbf{Hg}_k$. Similarly, the Newton update for E' is,

$$\begin{aligned} \mathbf{x}^* &= \mathbf{x}_k - \mathbf{H'} - \mathbf{g}'_k \\ &= \mathbf{x}_k - \left(\frac{1}{c}\right) c\,\mathbf{H}\,\mathbf{g}_k \\ &= \mathbf{x}_k - \mathbf{Hg}_k \end{aligned} \tag{25.54}$$

Therefore, the Newton step is invariant under scaling while Quasi-Newton methods are generally not invariant under such scaling and will give different results.

Take Broyden's single parameter class of updates described by Equations 25.55 and 25.56, which includes the BFGS and DFP updates as special cases,

$$\mathbf{H}_{k+1} = \mathbf{H}_k - \frac{\mathbf{H}_k \Delta \mathbf{g}_k \Delta \mathbf{g}_k^T \mathbf{H}_k^T}{\Delta \mathbf{g}_k^T \mathbf{H}_k \Delta \mathbf{g}_k} + \theta_k \mathbf{v}_k \mathbf{v}_k^T + \frac{\Delta \mathbf{x}_k \Delta \mathbf{x}_k^T}{\Delta \mathbf{x}_k^T \Delta \mathbf{g}_k} \tag{25.55}$$

where,

$$\mathbf{v}_k = \left(\Delta \mathbf{g}_k^T \mathbf{H}_k \Delta \mathbf{g}_k\right)^{\frac{1}{2}} \left(\frac{\Delta \mathbf{x}_k}{\Delta \mathbf{x}_k^T \Delta \mathbf{g}_k} - \frac{\mathbf{H}_k \Delta \mathbf{g}_k}{\Delta \mathbf{g}_k^T \mathbf{H}_k \Delta \mathbf{g}_k}\right) \tag{25.56}$$

Setting θ_k to 1 in Equation 25.55 produces the *BFGS* update and setting it to zero gives the *DFP* update.

This class of updates is generally not invariant under scaling. This motivated Oren and Spedicato [51, 52] to modify this Broyden's single parameter family (Equations 25.55 and 25.56) by introducing a new parameter mu_k such that by the appropriate choice of μ_k and θ_k, they could have an update which is invariant under scaling of Equation 25.51. This is the general update given by Equation 25.57.

$$\mathbf{H}_{k+1} = \mu_k \left(\mathbf{H}_k - \frac{\mathbf{H}_k \Delta \mathbf{g}_k \Delta \mathbf{g}_k^T \mathbf{H}_k^T}{\Delta \mathbf{g}_k^T \mathbf{H}_k \Delta \mathbf{g}_k} + \theta_k \mathbf{v}_k \mathbf{v}_k^T\right) + \frac{\Delta \mathbf{x}_k \Delta \mathbf{x}_k^T}{\Delta \mathbf{x}_k^T \Delta \mathbf{g}_k} \tag{25.57}$$

where $\mathbf{v}_k$ is given by Equation 25.56.

In addition, Shanno and Phua [63] proposed an initial scaling method which makes Broyden's single parameter class of update self-scaling. The two approaches to making updates invariant under scaling are discussed in more detail in the following two sections.

25.1.3.6 Self-Scaling variable Metric (SSVM) Algorithms

Equation 25.57, when used with exact *line searches*, will become a member of Huang's family [37] where in his notation,

$$\rho_k = \frac{1}{\prod\limits_{i=0}^{k} \mu_i} \tag{25.58}$$

Equation 25.57, leaves a lot of freedom in choosing the parameters, μ_k and θ_k, such that invariance under scaling is achieved. Oren suggested, in [51], that μ_k and θ_k be picked in the following manner,

$$\mu_k = \phi_k \frac{\mathbf{g}_k^T \Delta \mathbf{x}_k}{\mathbf{g}_k^T \mathbf{H}_k \Delta \mathbf{g}_k} + (1 - \phi_k) \frac{\Delta \mathbf{x}_k^T \Delta \mathbf{g}_k}{\Delta \mathbf{g}_k^T \mathbf{H}_k \Delta \mathbf{g}_k} \tag{25.59}$$

where $\phi_k, \theta_k \in [0, 1]$. This choice will provide a set of μ_k such that,

$$\frac{\Delta \mathbf{x}_k^T \Delta \mathbf{g}_k}{\Delta \mathbf{g}_k^T \mathbf{H}_k \Delta \mathbf{g}_k} \le \mu_k \le \frac{\Delta \mathbf{x}_k^T \mathbf{H}_k^{-1} \Delta \mathbf{x}_k}{\Delta \mathbf{x}_k^T \Delta \mathbf{g}_k} \tag{25.60}$$

In another approach, Oren and Spedicato [52] tried picking μ_k and θ_k based on heuristics such that μ_k is as close as possible to unity and θ_k is such that it would offset an estimated estimated bias in the $|\mathbf{H}_k \mathbf{G}|$. In a third approach, Oren and Spedicato [52] picked those μ_k and θ_k that minimize the condition number of $\left(\mathbf{H}_k^{-1} \mathbf{H}_{k+1} \right)$. This choice will put a bound on the condition number of the inverse Hessian approximate and therefore will provide numerical stability. Minimizing this condition number, the following relationship is held between μ_k and θ_k,

$$\theta_k = \frac{b(c - b\mu_k)}{\mu_k(ac - b^2)} \tag{25.61}$$

where,

$$a \overset{\Delta}{=} \Delta \mathbf{g}_k^T \mathbf{H}_k \Delta \mathbf{g}_k \tag{25.62}$$

$$b \overset{\Delta}{=} \Delta \mathbf{x}_k^T \Delta \mathbf{g}_k \tag{25.63}$$

and,

$$c \overset{\Delta}{=} \Delta \mathbf{x}_k^T \mathbf{H}_k^{-1} \Delta \mathbf{x}_k$$

$$= \eta^{*2} \mathbf{g}_k^T \mathbf{H}_k \mathbf{g}_k \tag{25.64}$$

Then, using *Fletcher*'s concept of duality [23], *Oren* and *Spedicato* [52] found those μ_k and θ_k which would make their update *self-dual*. This set is given by,

$$\theta = \frac{1}{1 + \sqrt{\frac{ac}{b^2}}} \tag{25.65}$$

and,

$$\mu_k = \sqrt{\frac{c}{a}} \tag{25.66}$$

Reference [52] gives four sets of switching rules for picking μ_k and θ_k. Algorithms based on Oren and Spedicato's updates are called *Self Scaling Variable Metric (SSVM)* algorithms. *SSVM* methods maintain positive definiteness of the approximation to the inverse Hessian matrix provided that $\Delta \mathbf{x}_k^T \Delta \mathbf{g}_k > 0 \,\forall\, k$. Condition 25.42 is again used in the *line searches* to impose this inequality. For a general nonlinear objective function, the SSVM algorithms provide a set of search directions which are invariant under scaling of the objective function. Also, for a quadratic function, these algorithms have the property that they monotonically reduce the condition number of the inverse Hessian approximate.

A draw-back of the *SSVM* algorithms is that they fail to converge to the inverse Hessian matrix for a quadratic function.[63] This convergence is especially desirable for methods employed for minimizing non-quadratic objective functions.[47]

The SSVM algorithms, in general, perform well with objective functions which depend on lots of variables. This makes them ideal for usage in neural network learning. These algorithms perform exceptionally well (better than all other updates in general) for homogeneous objective functions. A homogeneous objective function $E(\mathbf{x})$ is such that,

$$E(\mathbf{x}) = \tau^{-1}(\mathbf{x} - \mathbf{x}^*)\mathbf{g}(\mathbf{x}) + E(\mathbf{x}^*) \tag{25.67}$$

where τ is the degree of homogeneity and $\mathbf{x}^*$ is the minimizing state.[38] Differentiating Equation 25.67 gives,

$$\mathbf{x}^* - \mathbf{x} - (\tau - 1)\mathbf{H}(\mathbf{x})\mathbf{g}(\mathbf{x}) \tag{25.68}$$

Equation 25.68 suggests that the Newton step should be multiplied by $(\tau - 1)$ in order to get to the minimum. This makes the switch-2 SSVM methods superior to all other Quasi-Newton methods, when used on homogeneous functions.[51, 52, 63]

The lack of convergence of the approximate inverse Hessian to its true value in SSVM updates motivated *Shanno* and *Phua* to investigate methods which would make the Broyden single parameter class self-scaling.[63]

25.1.3.7 Initial Scaling of the Inverse Hessian Approximate

In [65], Spedicato proposes the initialization such that $\mathbf{H}_0$ is the inverse of a diagonal matrix with its diagonal being the diagonal of the true Hessian matrix at $\mathbf{x}_0$. This, however, is not practical since it is very hard to evaluate the Hessian for the objective function of a multi-layer neural network. Shanno and Phua [63] proposed an initial scaling such that,

$$\mathbf{H}_0 = \eta_0^* \mathbf{H}'_0 \tag{25.69}$$

where η_0^* is the initial linear step given by the *line search* algorithm and $\mathbf{H}'_0$ is the initial guess for the inverse Hessian (usually the identity matrix, $\mathbf{I}$). This makes Broyden's one parameter class of updates, given by Equations 25.55 and 25.56, self-scaling and invariant under scaling of the objective function. Initial scaling of Equation 25.69 will therefor give,

$$\mathbf{H}_0 = \eta_0^* \mathbf{I} \tag{25.70}$$

as the new initial guess for the inverse Hessian, if no better estimate of $\mathbf{H}$ is available.

Another initial scaling, proposed by Shanno and Phua [63], uses Oren-Spedicato's SSVM algorithm and finds the μ_0 provided by that algorithm which minimizes the condition number of $(\mathbf{H}_k^{-1} \mathbf{H}_{k+1})$. Then it scales the initial estimate of the inverse Hessian by that value. For example, consider the BFGS update for which $\theta_0 = 1$; μ_0 is given by Equation 25.61 to be,

$$\mu_0 = \frac{b}{a} \tag{25.71}$$

Since μ should be equal to one for the BFGS method, the initial estimate of $\mathbf{H}$ is scaled by μ_0,

$$\mathbf{H}_0 = \frac{b}{a} \mathbf{H}'_0 \tag{25.72}$$

This initial scaling can be evaluated in the same manner for all the members of the Broyden single parameter class of updates using 25.61 and the appropriate θ_0 for that update.

These initial scalings were shown by Shanno and Phua [63] to improve the performance of the BFGS method over the SSVM methods of Oren and Spedicato in all the cases tested but the special case of homogeneous objective functions.

25.1.3.8 Quasi-Newton Methods with Inexact Line Searches

Quasi-Newton methods which work well with inexact *line searches* such as the BFGS and SSVM methods have become very popular due to their reduction of the computational burden associated with *line searches*. Hoshino [35] presented, with his Quasi-Newton algorithm, a correction term which would maintain the orthogonality of the search direction and gradient at the termination point of an inexact *line search*.

Davidon also presented a new algorithm which has drawn a lot of attention in the field of optimization [63, 14] . His algorithm uses no *line searches*, optimally conditions the inverse Hessian approximate, and uses the square root of the inverse Hessian approximate which improves the numerical stability of his algorithm. The following two sections describe the theoretical details of these two approaches to weaken or eliminate *line searches*.

Hoshino's Method

Hoshino [35] presented a new variable metric update which generally works well and has properties similar to those of the BFGS and the DFP methods. However, this method in general has shown to give updates with condition numbers larger than the BFGS and SSVM methods. Hoshino's update is given by Equation 25.73.

$$
\begin{aligned}
\mathbf{H}_{k+1} = \mathbf{H}_k &+ \frac{1}{\Delta \mathbf{g}_k^T \Delta \mathbf{x}_k + \Delta \mathbf{g}_k^T \mathbf{H}_k \Delta \mathbf{g}_k} \left(\left[1 + \frac{2 \Delta \mathbf{g}_k^T \mathbf{H}_k \Delta \mathbf{g}_k}{\Delta \mathbf{g}_k^T \Delta \mathbf{x}_k} \right] \Delta \mathbf{x}_k \Delta \mathbf{x}_k^T - \right. \\
&\left. \Delta \mathbf{x}_k \Delta \mathbf{g}_k^T \mathbf{H}_k - \mathbf{H}_k \Delta \mathbf{g}_k \Delta \mathbf{x}_k^T - \mathbf{H}_k \Delta \mathbf{g}_k \Delta \mathbf{g}_k^T \mathbf{H} \right)
\end{aligned}
\tag{25.73}
$$

An attractive feature of Hoshino's Quasi-Newton minimization is his theoretical approach to the use of his update with inexact *line searches*. *Inexact line searches* are desired to reduce the number of function evaluations. These *inexact line searches* will sometimes give gradients at their termination point which are not perpendicular to the search direction. This will slow down the convergence of the minimization scheme. To evaluate the Quasi-Newton direction of update at any step $k+1$, Hoshino uses a modified gradient which is forced to be perpendicular to the search direction. Consider the *line search* terminating at $\mathbf{x}_{k+1}$. Also, consider $\mathbf{x}'_{k+1}$ to be the true minimum in the direction of search. Furthermore, denote the step from $\mathbf{x}_k$ to the minimum $\mathbf{x}'_{k+1}$ by $\Delta \mathbf{x}'_k$ and define the scalar, ε_k such that,

$$
\Delta \mathbf{x}'_k = \varepsilon_k \Delta \mathbf{x}_k
\tag{25.74}
$$

Therefore, the gradient at the true minimum, $\mathbf{g}'_{k+1}$ will be given by,

$$
\mathbf{g}'_{k+1} = \mathbf{g}_{k+1} + (\mathbf{g}_{k+1} - \mathbf{g}_k)\varepsilon_k
\tag{25.75}
$$

$$
= \mathbf{g}_{k+1} + \varepsilon_k \Delta \mathbf{g}_k
\tag{25.76}
$$

The true minimum would be at the point where the gradient is perpendicular to the direction of search or,

$$\Delta\mathbf{x}_k^T \mathbf{g}'_{k+1} = 0 \tag{25.77}$$

Solving for the ε_k which satisfies this condition gives,

$$\varepsilon_k = -\frac{\Delta\mathbf{x}_k^T \mathbf{g}_{k+1}}{\Delta\mathbf{x}_k^T \Delta\mathbf{g}_k} \tag{25.78}$$

Then, the expression for the modified gradient is given using this scalar factor by,

$$\mathbf{g}'_{k+1} = \mathbf{g}_{k+1} - \frac{\Delta\mathbf{x}_k^T \mathbf{g}_{k+1}}{\Delta\mathbf{x}_k^T \Delta\mathbf{g}_k} \Delta\mathbf{g}_k \tag{25.79}$$

This new gradient gives the Quasi-Newton direction at step $k+1$ to be,

$$\mathbf{x}_{k+1} = -\mathbf{H}_{k+1}\mathbf{g}'_{k+1} \tag{25.80}$$

Hoshino does not use this modified gradient for his update to the inverse Hessian approximate. This gradient is only used to obtain the Quasi-Newton step. Reference [35] provides a stability analysis for this scheme.

Davidon's Optimally Conditioned Quasi-Newton Method with No Line Search
   In 1975, Davidon [14] made an important contribution to the improvement of Quasi-Newton methods by introducing his optimally conditioned method which is free of *line searches*. Schnabel [59] has devoted most of his PhD dissertation to evaluating Davidon's method. The method conducts updates to the inverse Hessian approximate which are optimally conditioned in the same sense as the optimal conditioning of Oren and Spedicato. This conditioning is done by minimizing the condition number of $(\mathbf{H}_k^{-1}\mathbf{H}_{k+1})$ which has been obtained by minimizing $\frac{\lambda_1}{\lambda_N}$ in the Eigenvalue problem,

$$\mathbf{H}_{k+1}\mathbf{u} = \lambda \mathbf{H}_k \mathbf{u} \tag{25.81}$$

Previously, researchers had been trying to minimize the ratio of the condition number $\mathbf{H}_{k+1}$ to the condition number of $\mathbf{H}_k$. However, doing this would generate invariance under orthogonal transformations only, while the optimal conditioning used by Davidon and Oren and Spedicato is invariant under all invertible linear transformations.

   Davidon's update to the inverse Hessian approximate is given by Equation 25.84. This general update includes some updates such as the DFP and BFGS updates as special cases. In Equation 25.84, the value of θ_k is chosen such that the condition number of $(\mathbf{H}_k^{-1}\mathbf{H}_{k+1})$ is minimized. Davidon uses $\Delta\mathbf{x}_0$ as the initial value of w, namely,

$$\mathbf{w}_0 = \Delta \mathbf{x}_0 \tag{25.82}$$

Following $\mathbf{w}$'s are then obtained by Equation 25.83 for quadratic functions. For non-quadratic functions the following is used,

$$\mathbf{w}_k = \Delta \mathbf{x}_k \tag{25.83}$$

This update is part of Davidon's algorithm which does not use a *line search*. However, if sufficient reduction is not experienced by the objective function, a simple inexact *line search* is used to impose sufficient function reduction.

$$
\mathbf{H}_{k+1} = \mathbf{H}_k + \frac{(\Delta \mathbf{x}_k - \mathbf{H}_k \Delta \mathbf{g}_k) w_k^T + w_k (\Delta \mathbf{x}_k - \mathbf{H}_k \Delta \mathbf{g}_k)^T}{\mathbf{w}_k^T \Delta \mathbf{x}} -
$$
$$
\frac{(\Delta \mathbf{x}_k - \mathbf{H}_k \Delta \mathbf{g}_k)^T \Delta \mathbf{g}_k \mathbf{w}_k \mathbf{w}_k^T}{(\mathbf{w}_k^T \Delta \mathbf{g}_k)^2} \theta_k \Upsilon_k \Upsilon_k^T \tag{25.84}
$$

where,

$$\mathbf{w}_{k+1} = w_k (\Delta \mathbf{x}_k - \mathbf{H}_k \Delta \mathbf{g}_k)^T \Delta \mathbf{g}_k - (\Delta \mathbf{x}_k - \mathbf{H}_k \Delta \mathbf{g}_k) \mathbf{w}_k^T \Delta \mathbf{g}_k \tag{25.85}$$

and

$$\Upsilon_k = \frac{\Delta \mathbf{x}_k - \mathbf{H}_k \Delta \mathbf{g}_k}{(\Delta \mathbf{x}_k - \mathbf{H}_k \Delta \mathbf{g}_k)^T \Delta \mathbf{g}_k} - \frac{\mathbf{w}_k}{\mathbf{w}_k^T \Delta \mathbf{g}_k} \tag{25.86}$$

This method has three important features. First, to improve the numerical stability and accuracy of his algorithm, Davidon updates a Jacobian matrix which is the square root of the inverse Hessian approximate.

$$\mathbf{H}_k = \mathbf{J}_k \mathbf{J}_k^T \tag{25.87}$$

By this factorization, the condition number of the Jacobian matrix $\mathbf{J}_k$ is of the order of the square root of the condition number of the matrix $\mathbf{H}_k$. This smaller condition number improves the stability of the method in practical applications. An update in the Jacobian matrix of the following form,

$$\mathbf{J}_{k+1} = (1 + \mathbf{u}\mathbf{v}^T)\mathbf{J}_k \tag{25.88}$$

translates to the following update in the inverse Hessian approximate,

$$\mathbf{H}_{k+1} = (1 + \mathbf{u}\mathbf{v}^T)\mathbf{H}_k(1 + \mathbf{v}\mathbf{u}^T) \tag{25.89}$$

This factorization produces positive definite inverse Hessian approximates. Some rank two updates in the Broyden family such as BFGS, DFP and optimally conditioned updates correspond to rank one updates in the Jacobian matrix. This results in fewer computational operations and less round-off error.

A second feature of this algorithm is motivated by the fact that one could approximate the gradient at the minimum of a quadratic function by a linear interpolation similar to the approach of Hoshino discussed in the previous section.[6] Davidon does not use the actual change in the gradient and the actual step size, he instead uses projections of these changes. This allows him to avoid *line searches*. Shanno and Phua have devoted a paper to the discussion of these projections.[62] Despite the use of these projections, Davidon's method still maintains a positive definite approximation to the inverse Hessian matrix for quadratic and non-quadratic functions. This ensures quadratic convergence for Davidon's algorithm.

In the real implementation of his algorithm, Davidon uses an equivalent form of Equation 25.84 given by Equation 25.90,

$$
\mathbf{H}_{k+1} = \mathbf{H}_k + \frac{\mathbf{e}'_k \Delta \mathbf{x}'^T_k + \Delta \mathbf{x}'_k \mathbf{e}'^T_k}{\Delta \mathbf{x}'^T_k \Delta \mathbf{g}'_k} -
$$
$$
\frac{\mathbf{e}'^T_k \Delta \mathbf{g}'_k \Delta \mathbf{x}'_k \Delta \mathbf{x}'^T_k}{(\Delta \mathbf{x}'^T_k \Delta \mathbf{g}'_k)^2} \theta'_k \Delta \mathbf{g}'^T_k \mathbf{H}_k \Delta \mathbf{g}'_k \boldsymbol{\xi}'_k \boldsymbol{\xi}'^T_k
\tag{25.90}
$$

where

$$
\Delta \mathbf{x}'_k \triangleq \mathbf{Q}_k \Delta \mathbf{x}_k
\tag{25.91}
$$

$$
\Delta \mathbf{g}'_k \triangleq \mathbf{Q}^T_k \Delta \mathbf{g}_k
\tag{25.92}
$$

$$
\mathbf{e}_k \triangleq \Delta \mathbf{x}'_k - \mathbf{H}_k \Delta \mathbf{g}'_k
\tag{25.93}
$$

$$
\boldsymbol{\xi}'_k \triangleq \frac{\Delta \mathbf{x}'_k}{\Delta \mathbf{x}'^T_k \Delta \mathbf{g}'_k} - \frac{\mathbf{H}_k \Delta \mathbf{g}'_k}{\Delta \mathbf{g}'^T_k \mathbf{H}_k \Delta \mathbf{g}'_k}
\tag{25.94}
$$

and $\mathbf{Q}_k$ are the projection matrices which split the space of $\Delta \mathbf{x}_k$ into two parts, $\mathbf{Q}_k \Delta \mathbf{x}_k$ which is a subspace which has already been explored and $(\mathbf{I} - \mathbf{Q}_k) \Delta \mathbf{x}_k$ which is conjugate to the said subspace and which would have been the direction taken if *line searches* were conducted in the previous k steps. See Lemmas 2.2 and 2.3 of Davidon's paper [14] for a more detailed description of $\mathbf{Q}_k$.

The third feature of Davidon's algorithm is that the directions, $\mathbf{w}_k$ of Equation 25.84 need not be orthogonal to the error of the Quasi-Newton step. These directions are chosen such that the updates and new directions satisfy the following conditions,

$$
(\mathbf{H}_{k+1} - \mathbf{H}_k)\mathbf{s} = 0 \ \forall \ \mathbf{s} \in \mathscr{S}_{\mathbf{H}}
\tag{25.95}
$$

and

$$
\mathbf{w}_k \perp \mathscr{S}_{\mathbf{H}} \perp \Delta \mathbf{x}_k - \mathbf{H}_k \Delta \mathbf{g}_k
\tag{25.96}
$$

where,

$$\mathscr{S}_{\mathbf{H}} = \{\Delta\mathbf{g}_0, \cdots, \Delta\mathbf{g}_{k-1}\} \tag{25.97}$$

This ensures approximate inverse Hessian matrices which satisfy the following condition,

$$\mathbf{H}_{k+1}\Delta\mathbf{g}_j = \Delta\mathbf{x}_j \quad \forall \, j \leq k \tag{25.98}$$

This condition ensures convergence to the minimum of a quadratic function after N updates to the inverse Hessian approximate when *inexact line searches* (or even *no line searches*) are used.

Figures 25.3, 25.4, and 25.5 give a flow chart of Davidon's minimization algorithm without any *line searches*. Davidon's update, when no projections are used and the inverse Hessian approximate is updated directly, can be described by Equations 25.55, 25.56, 25.61, 25.62, 25.63, 25.64, and 25.99.

$$\theta = \begin{cases} \frac{b(c-b)}{ac-b^2} & for \; b \leq \frac{2\,a\,c}{a+c} \\ \frac{b}{b-a} & for \; b > \frac{2\,a\,c}{a+c} \end{cases} \tag{25.99}$$

Davidon's update can also be used in accordance with the scaling of the initial guess for the inverse Hessian to make the update invariant under scaling of the objective function. The first method of initial scaling discussed in Section 25.1.3.7, has produced much improvement on the scheme.[63]

25.1.4 Conjugate Gradient Methods

Consider a quadratic function $E(\mathbf{x})$, $\mathbf{x} \in \mathscr{R}^N$,

$$E(\mathbf{x}) = a + \mathbf{b}^T\mathbf{x} + \frac{1}{2}\mathbf{x}^T\mathbf{G}\mathbf{x} \tag{25.100}$$

where $\mathbf{G}$ is positive definite and has full rank, N.

Take a normalized starting direction $\mathbf{s}_0 \in \mathscr{R}^N$ such that $\|\mathbf{s}_0\|_{\mathscr{E}} = 1$ and an initial vector $\mathbf{x}_0$. Find the step size η_0^* such that,

$$\mathbf{x}_1 = \mathbf{x}_0 + \eta_0^*\mathbf{s}_0 \tag{25.101}$$

minimizes E along the direction $\mathbf{s}_0$.

Then,

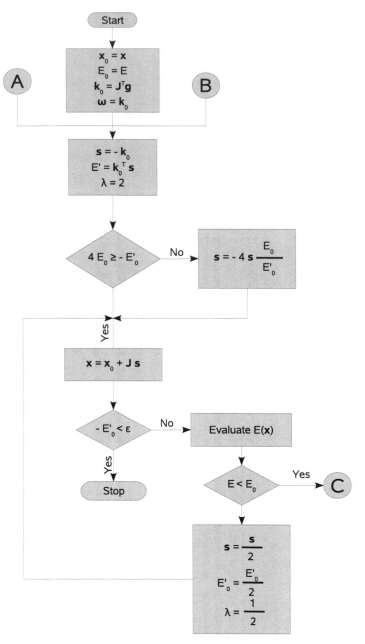

Fig. 25.3: Flowchart of Davidon's Quasi-Newton Method – Part 1

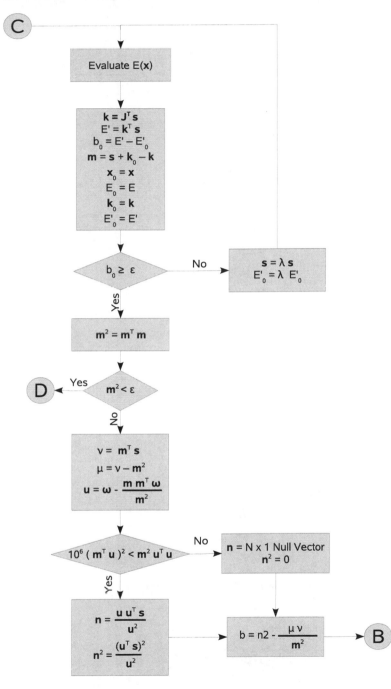

Fig. 25.4: Flowchart of Davidon's Quasi-Newton Method – Part 2

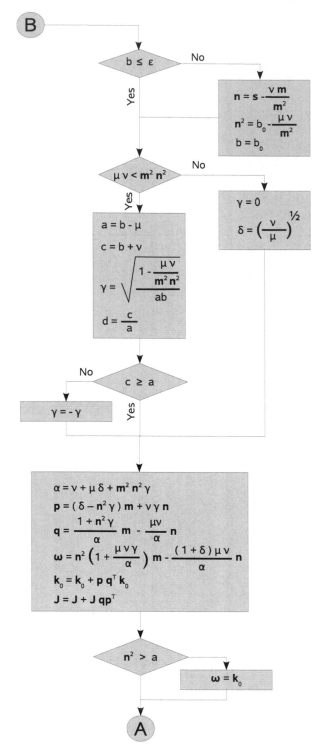

Fig. 25.5: Flowchart of Davidon's Quasi-Newton Method – Part 3

$$E(\mathbf{x}_1) = a + \mathbf{b}^T \left(\mathbf{x}_0 + \eta_0^* \mathbf{s}_0\right) + \frac{1}{2} \left(\mathbf{x}_0 + \eta_0^* \mathbf{s}_0\right)^T \mathbf{G} \left(\mathbf{x}_0 + \eta_0^* \mathbf{s}_0\right) \qquad (25.102)$$

To find the minimum of E in the $\mathbf{s}_0$ direction,

$$\left. \frac{\partial E}{\partial \eta_0} \right|_{\eta_0^*} = 0 \qquad (25.103)$$

or,

$$\nabla^T E(\mathbf{x}_0) \mathbf{s}_0 + \mathbf{s}_0^T \mathbf{G} \eta_0^* \mathbf{s}_0 = 0 \qquad (25.104)$$

Solving for η_0^* in Equation 25.104,

$$\eta_0^* = -\frac{\mathbf{s}_0^T \nabla E(\mathbf{x}_0)}{\mathbf{s}_0^T \mathbf{G} \mathbf{s}_0} \qquad (25.105)$$

If a unidirectional minimization is done for N times in N directions $\mathbf{s}_k$, $k \in \{0, 1, \cdots, N-1\}$ which are mutually conjugate about $\mathbf{G}$, the Hessian (matrix of the second partial derivatives) of E, then,

$$\begin{aligned}
\mathbf{x}_N &= \mathbf{x}_0 + \sum_{k=0}^{N-1} \eta_0^* \mathbf{s}_k \\
&= \mathbf{x}_0 - \sum_{k=0}^{N-1} \frac{\mathbf{s}_k^T \nabla E(\mathbf{x}_k)}{\mathbf{s}_k^T \mathbf{G} \mathbf{s}_k} \mathbf{s}_k
\end{aligned} \qquad (25.106)$$

However,

$$\nabla E(\mathbf{x}_k) = \mathbf{G} \mathbf{x}_k + \mathbf{b} \qquad (25.107)$$

For simplification, define.

$$\mathbf{g}_k \overset{\Delta}{=} \nabla_x E(\mathbf{x}_k) \qquad (25.108)$$

then,

$$\begin{aligned}
\mathbf{s}_k^T \mathbf{g}_k &= \mathbf{s}_k^T \left(\mathbf{G} \mathbf{x}_k + \mathbf{b}\right) \\
&= \mathbf{s}_k^T \left(\mathbf{G} \left(\mathbf{x}_0 + \sum_{i=0}^{k-1} \eta_i^* \mathbf{s}_i\right) + \mathbf{b}\right) \\
&= \mathbf{s}_k^T \left(\mathbf{G} \mathbf{x}_0 + b\right) \quad \text{since} \quad \mathbf{s}_k^T \mathbf{G} \mathbf{s}_i = 0 \ \forall i < k \text{ due to conjugacy}
\end{aligned} \qquad (25.109)$$

The expression for $\mathbf{x}_N$ can be written as,

$$\mathbf{x}_N = \mathbf{x}_0 - \sum_{k=0}^{N-1} \frac{\mathbf{s}_k^T \left(\mathbf{G} \mathbf{x}_0 + \mathbf{b}\right) \mathbf{s}_k}{\mathbf{s}_k^T \mathbf{G} \mathbf{s}_k} \qquad (25.110)$$

By Theorem 23.1,

$$\mathbf{x}_0 = \sum_{k=0}^{N-1} \frac{\mathbf{s}_k^T \mathbf{G} \mathbf{x}_0 \mathbf{s}_k}{\mathbf{s}_k^T \mathbf{G} \mathbf{s}_k} \tag{25.111}$$

Substituting from Equation 25.111 into Equation 25.110,

$$\mathbf{x}_N = -\sum_{k=0}^{N-1} \frac{\mathbf{s}_k^T \mathbf{b} \mathbf{s}_k}{\mathbf{s}_k^T \mathbf{G} \mathbf{s}_k}$$

$$= \sum_{k=0}^{N-1} \frac{\mathbf{s}_k \mathbf{s}_k^T \mathbf{b}}{\mathbf{s}_k^T \mathbf{G} \mathbf{s}_k} \tag{25.112}$$

However, from Theorem 23.2,

$$\mathbf{G}^{-1} = \mathbf{H} = \sum_{k=0}^{N-1} \frac{\mathbf{s}_k \mathbf{s}_k^T}{\mathbf{s}_k^T \mathbf{G} \mathbf{s}_k} \tag{25.113}$$

which, after substitution in Equation 25.112 gives,

$$\mathbf{x}_N = -\mathbf{G}^{-1} \mathbf{b} \tag{25.114}$$

Equation 25.114 is equivalent to the Newton step which minimizes a quadratic function. Therefore, $\mathbf{x}_N$ is the minimizer of E. This shows that the minimum of a quadratic function can be reached in at most N linear minimizations if these minimizations are done along a full set of directions mutually conjugate about the Hessian matrix of the objective function. In general, the method of conjugate directions provides quadratic convergence. A few different conjugate direction methods are discussed in the following sections.

25.1.4.1 Fletcher-Reeves Conjugate Gradient Method

The *Fletcher-Reeves* [24] *conjugate gradient* method generates a set of search directions $\mathbf{s}_i, i \in \{0, 1, \cdots, N-1\}$ such that $\mathbf{s}_k$ is a linear combination of $\mathbf{g}_k$ and all $\mathbf{s}_j, j \in \{0, 1, \cdots, k-1\}$ with the combination of weights picked such that $\mathbf{s}_k$ is conjugate about the Hessian matrix, $\mathbf{G}$, of a quadratic objective function to all $\mathbf{s}_j, j \in \{0, 1, \cdots, k-1\}$. These weights (coefficients) are chosen such that only the two most recent gradients are needed for their evaluation at each iteration. This solution was reached by influences from Hestenes and Steifel [31] and Beckman [4] as noted in [24].

Reference [24] provides a method for finding conjugate directions based on the two most recent gradients. Here, a derivation is given for the calculation of these directions, leading to the method of conjugate gradients due to *Fletcher* and *Reeves* [24].

Write the direction at iteration $k + 1$, $\mathbf{s}_{k+1}$, such that $\mathbf{s}_{k+1}$ is a linear combination of the gradient $\mathbf{g}_{k+1}$ and direction $\mathbf{s}_k$,

$$\mathbf{s}_{k+1} = -\mathbf{g}_{k+1} + \omega_k \mathbf{s}_k \tag{25.115}$$

where ω_k is a weight to be chosen. Doing this for all k, $\mathbf{s}_{k+1}$ becomes a linear combination of $\mathbf{g}_{k+1}$ and $\mathbf{s}_i, i \in \{0, 1, \cdots, k\}$.

In Equation 25.115, ω_k should be chosen such that $\mathbf{s}_{k+1}$ is conjugate to $\mathbf{s}_k$ about the Hessian matrix $\mathbf{G}$ of the objective function, $E(\mathbf{x})$, namely,

$$\mathbf{s}_k^T \mathbf{G} \mathbf{s}_{k+1} = 0 \tag{25.116}$$

From Equation 25.7,

$$\mathbf{s}_k^T = \frac{\Delta \mathbf{s}_k^T}{\eta_k^*} \tag{25.117}$$

and from Equations 25.117, 25.25 and the definition of H (Equation 25.20),

$$\mathbf{s}_k^T = \frac{\Delta \mathbf{g}_k^T \mathbf{G}^{-1}}{\eta_k^*} \tag{25.118}$$

Substituting from Equation 25.118 and 25.115 into Equation 25.116,

$$\frac{\Delta \mathbf{g}_k^T}{\eta_k^*} \mathbf{G}^{-1} \mathbf{G} \left(-\mathbf{g}_{k+1} + \omega_k \mathbf{s}_k \right) = 0 \tag{25.119}$$

or,

$$\frac{\Delta \mathbf{g}_k^T}{\eta_k^*} \left(-\mathbf{g}_{k+1} + \omega_k \mathbf{s}_k \right) = 0 \tag{25.120}$$

and for any arbitrary η_k^*,

$$\Delta \mathbf{g}_k^T \left(-\mathbf{g}_{k+1} + \omega_k \mathbf{s}_k \right) = 0 \tag{25.121}$$

Let us write out all the terms in Equation 25.121 using the definition of $\Delta \mathbf{g}_k$,

$$-\mathbf{g}_{k+1}^T \mathbf{g}_{k+1} + \mathbf{g}_k^T \mathbf{g}_{k+1} + \omega_k \mathbf{g}_{k+1}^T \mathbf{s}_k - \omega_k \mathbf{s}_k^T \mathbf{s}_k = 0 \tag{25.122}$$

The η_k^*s are chosen such that they minimize $E(\mathbf{x})$ in the direction of $\mathbf{s}_k$. Writing Equation 25.105 using 25.107 and 25.108,

$$(\mathbf{G}\mathbf{x}_k + \mathbf{b})^T \mathbf{s}_k + \mathbf{s}_k^T \mathbf{G} \eta_k \mathbf{s}_k = 0 \tag{25.123}$$

Using Equation 25.7 in Equation 25.123 and factoring out $\mathbf{s}_k^T$,

$$\mathbf{s}_k^T \left(\mathbf{Gx}_k + \mathbf{b} + \mathbf{G}\Delta\mathbf{x}_k \right) = 0 \tag{25.124}$$

or,

$$\mathbf{s}_k^T \left(\mathbf{b} + \mathbf{G} \left(\mathbf{x}_k + \Delta\mathbf{x}_k \right) \right) = \mathbf{s}_k^T \left(\mathbf{b} + \mathbf{Gx}_{k+1} \right)$$
$$= 0 \tag{25.125}$$

However, from Equation 25.107,

$$\mathbf{b} + \mathbf{Gx}_{k+1} = \mathbf{g}_{k+1} \tag{25.126}$$

Substituting from Equation 25.126 into Equation 25.126,

$$\mathbf{s}_k^T \mathbf{g}_{k+1} = 0 \tag{25.127}$$

Writing Equation 25.107 for any arbitrary iteration l,

$$\mathbf{g}_l = \mathbf{b} + \mathbf{Gx}_l \tag{25.128}$$

The transition from $\mathbf{x}_k$ to $\mathbf{x}_l$ using the argument in the previous section is given by,

$$\mathbf{x}_l = \mathbf{x}_k + \sum_{j=k}^{l-1} \eta_j^* \mathbf{s}_j \tag{25.129}$$

Substituting from Equation 25.129 into Equation 25.128,

$$\mathbf{g}_l = \mathbf{b} + \mathbf{G} \left(\mathbf{x}_k + \sum_{j=k}^{l-1} \eta_j^* \mathbf{s}_j \right) \tag{25.130}$$

or,

$$\mathbf{g}_l = \mathbf{g}_k + \sum_{j=k}^{l-1} \eta_j^* \mathbf{s}_j \tag{25.131}$$

Multiply Equation 25.131 from the left by $\mathbf{s}_{k-1}^T$,

$$\mathbf{s}_{k-1}^T \mathbf{g}_l = \mathbf{s}_{k-1}^T \mathbf{g}_k + \sum_{j=k}^{l-1} \eta_k^* \mathbf{s}_{k-1}^T \mathbf{Gs}_j \tag{25.132}$$

For conjugate directions and using Equation 25.127 in 25.132,

$$\mathbf{s}_k^T \mathbf{g}_l = 0 \quad \forall \ 0 \le k < l-1 \tag{25.133}$$

and combining Equations 25.127 and 25.133,

$$\mathbf{s}_k^T \mathbf{g}_l = 0 \quad \forall \ 0 \le k \le l-1 \tag{25.134}$$

Furthermore, substituting for $\mathbf{s}_k$ in Equation 25.127 using 25.115,

$$\mathbf{s}_k^T \mathbf{g}_{k+1} = (-\mathbf{g}_k + \omega_{k-1}\mathbf{s}_{k-1})^T \mathbf{g}_{k+1} \tag{25.135}$$

or,

$$\mathbf{s}_k^T \mathbf{g}_{k+1} = -\mathbf{g}_k^T \mathbf{g}_{k+1} + \omega_{k-1}\mathbf{s}_{k-1}^T \mathbf{g}_{k+1} \tag{25.136}$$

Using Equation 25.134 in 25.136,

$$\mathbf{g}_k^T \mathbf{g}_{k+1} = 0 \tag{25.137}$$

Substituting for $\mathbf{s}_k$ from Equation 25.115 into 25.122,

$$\begin{aligned}
-\mathbf{g}_{k+1}^T \mathbf{g}_{k+1} + \mathbf{g}_k^T \mathbf{g}_{k+1} - \omega_k \mathbf{g}_{k+1}^T \mathbf{g}_{k+1} + \omega_k \omega_{k-1} \mathbf{g}_{k+1}^T \mathbf{s}_{k-1} + \\
\omega_k \mathbf{g}_k^T \mathbf{g}_k - \omega_k \omega_{k-1} \mathbf{g}_k^T \mathbf{s}_{k-1} = 0
\end{aligned} \tag{25.138}$$

Using Equation 25.134 and 25.137 in 25.138,

$$-\mathbf{g}_{k+1}^T \mathbf{g}_{k+1} + \omega_k \mathbf{g}_k^T \mathbf{g}_k = 0 \tag{25.139}$$

Solving for ω_k in from Equation 25.139,

$$\omega_k = \frac{\mathbf{g}_{k+1}^T \mathbf{g}_{k+1}}{\mathbf{g}_k^T \mathbf{g}_k} \tag{25.140}$$

Substituting for ω_k from Equation 25.140 into 25.115,

$$\mathbf{s}_{k+1} = -\mathbf{g}_{k+1} + \frac{\mathbf{g}_{k+1}^T \mathbf{g}_{k+1}}{\mathbf{g}_k^T \mathbf{g}_k} \mathbf{s}_k \tag{25.141}$$

Using Equation 25.141, the *Fletcher-Reeves conjugate gradient* method could be summarized by the following steps:

1. Set $k = 0$ and $\mathbf{s}_0 = -\mathbf{g}_0$
2. At the kth iteration, find η_k^* such that $\mathbf{x}_{k+1} = \mathbf{x}_k + \eta_k^* \mathbf{s}_k$ minimizes E in the direction of $\mathbf{s}_k$
3.

$$\mathbf{s}_{k+1} = -\mathbf{g}_{k+1} + \frac{\mathbf{g}_{k+1}^T \mathbf{g}_{k+1}}{\mathbf{g}_k^T \mathbf{g}_k} \mathbf{s}_k \tag{25.142}$$

4. If $\|\mathbf{s}_k\|_{\mathscr{E}} < \varepsilon$, then terminate minimization
5. If $k = N$ then go to step 1, otherwise, go to step 2

25.1.4.2 Partan's Iterative Conjugate Gradient Method

Partan's minimization method [32] is another popular conjugate gradient minimization technique which has been used in developing many commercial minimization software packages. The Iterative Partan method is described by the following procedure.

1. Set $\mathbf{s}_0 = -\mathbf{g}_0$ at the initial point, $\mathbf{x}_0$
2. Find η_0^* such that $\mathbf{x}_1 = \mathbf{x}_0 + \eta_0^* \mathbf{s}_0$ minimizes E along the direction of $\mathbf{s}_0$
3. $\mathbf{s}_1 = -\mathbf{g}_1$
4. Find η_1^* such that $\mathbf{x}_2 = \mathbf{x}_1 + \eta_1^* \mathbf{s}_1$ minimizes E along the direction of $\mathbf{s}_1$
5. Set $\frac{\mathbf{x}_2 - \mathbf{x}_0}{\|\mathbf{x}_2 - \mathbf{x}_0\|_{\mathcal{E}}}$
6. Find η_2^* such that $\mathbf{x}_3 = \mathbf{x}_2 + \eta_2^* \mathbf{s}_2$ minimizes E along the direction of $\mathbf{s}_2$
7. If the termination criteria for the minimization are not met, then set $\mathbf{x}_0 = \mathbf{x}_3$ and go to step 1. Otherwise, terminate minimization.

In practice, the iterative method is not as effective as the following variation which is called the Continuous Partan method.

25.1.4.3 Continuous Partan Minimization Method

The Continuous Partan method [60] has added a few extra steps to the iterative Partan method which enhance the performance of the method in practice. These steps are as follows,

1. Do steps 1 through 6 of the iterative Partan method
2. Set $\mathbf{s}_3 = -\mathbf{g}_3$
3. Find η_3^* such that $\mathbf{x}_4 = \mathbf{x}_3 + \eta_3^* \mathbf{s}_3$ minimizes E along the direction of $\mathbf{s}_3$
4. Set $\frac{\mathbf{x}_4 - \mathbf{x}_1}{\|\mathbf{x}_3 - \mathbf{x}_1\|_{\mathcal{E}}}$
5. Find η_4^* such that $\mathbf{x}_5 = \mathbf{x}_4 + \eta_4^* \mathbf{s}_4$ minimizes E along the direction of $\mathbf{s}_4$
6. If the termination criteria for the minimization are not met, then set $\mathbf{x}_3 = \mathbf{x}_5$ and go to step 2. Otherwise, terminate minimization.

25.1.4.4 The Projected Newton Algorithm

Zoutendijk [74] presented a gradient projection method which is summarized by the following steps:

1. Let $\mathbf{P}_0 = \mathbf{I}$ and start from the initial state, $\mathbf{x}_0$
2. Set

$$\mathbf{P}_k = \mathbf{I} - \mathbf{G}_k \left[\mathbf{G}_k^T \mathbf{G}_k \right]^{-1} \mathbf{G}_k^T$$
$$= \mathbf{P}_{k-1} - \mathbf{P}_{k-1} \Delta \mathbf{g}_k \left[\Delta \mathbf{g}_k^T \mathbf{P}_{k-1} \Delta \mathbf{g}_k \right]^{-1} \Delta \mathbf{g}_k^T \mathbf{P}_{k-1} \tag{25.143}$$

3. If $\mathbf{P}_k\mathbf{g}_k \neq 0$, let $\mathbf{s}_k = -\mathbf{Pg}_k$
4. Minimize $E(\mathbf{x})$ in the direction of $\mathbf{s}_k$
5. If $\mathbf{P}_k\mathbf{g}_k = 0$ and $\mathbf{g}_k = 0$, then terminate
6. If $\mathbf{P}_k\mathbf{g}_k = 0$ and $\mathbf{g}_k \neq 0$ or $k = N$, then set $k = 0$ and $\mathbf{x}_0 = \mathbf{x}_N$ and go to step 1
7. Set $k = k+1$ and go to step 2

For a derivation of the projection matrix of Equation 25.143, see Reference [36]. The above procedure is equivalent to a Quasi-Newton approach with the initial inverse-Hessian approximation picked as $\mathbf{H}_0 = \mathbf{I}$ and then updated by Equation 25.144.

$$\mathbf{H}_{k+1} = \mathbf{H}_k - \frac{\Delta\mathbf{g}_k^T\mathbf{H}_k^T\mathbf{H}_k\Delta\mathbf{g}_k}{\Delta\mathbf{g}_k^T\mathbf{H}_k\Delta\mathbf{g}_k} \tag{25.144}$$

This update is equivalent to the method of Projected Newton-Raphson given by Pearson.[53].

25.2 Gradient-Free Optimization

All the minimization methods listed in Section 25.1 require the evaluation of the gradient vector. For example, in neural network learning, this requirement brings about the need for knowing the exact structure of the network and all of the connections. For large networks, it is very difficult to analytically evaluate these gradients. To evaluate the gradient of the objective function with respect to weights and activation function parameters in the lower (close to input) layers, chain rule requires many derivative evaluations. This difficulty becomes more serious when the number of layers grows. On the other hand, objective function evaluations can easily be performed by the neural network, while gradient evaluations cannot generally be done by the network alone. In neural network learning, this is the main motivation behind the use of optimization methods which require no gradient evaluations. This argument is true with many systems where function evaluations are a lot cheaper than evaluating gradients. The objective is to minimize the need for extra hardware/software for gradient evaluations and to have the system (e.g. the neural network) perform most of the computations. A special advantage of gradient-free methods is that they do not require *regularity* (Definition 24.34) and *continuity* (Definition 24.12) of the objective function. However, they generally require many more function evaluations to converge.

25.2.1 Search Methods

The optimization methods which were described in the previous sections require analytic evaluation of the gradient of the objective function. In a family of optimization methods called search methods, the directions of minimization are evaluated solely based on the objective function values. In general, gradient based minimization methods converge faster than gradient-free methods. However, when many variables are involved, as in the case of neural-network learning problems, the analytical evaluation of the gradient becomes complicated. A popular gradient-free minimization scheme is discussed by the following.

25.2.1.1 Hooke-Jeeves and Wood Direct Search Method

The direct search method as implemented by Hooke-Jeeves [33] and Wood [71] is based on exploratory searches in the directions of independent variables one at a time, while keeping the rest of the variables constant. These methods are known to work poorly when there are terms in the objective function involving the product of a few design variables.[32] This makes the direct search method a poor choice for application to the neural-network learning problem.

25.2.2 Gradient-Free Conjugate Direction Methods

There are a few methods which use only objective function evaluations to predict a search direction which is conjugate to one or more directions about the Hessian of the quadratic approximation to the objective function. Among these methods are Rosenbrock's method, the Davies-Swann-Campey method (a modified version of Rosenbrock's method), *Smith's method* and *Powell's first and second methods*. These *gradient-free minimization techniques* are discussed here, in more detail.

25.2.2.1 Rosenbrock's Method

Rosenbrock's method [57] starts with a full set of orthonormal directions $\mathbf{s}_i, i \in \{0, 1, \cdots, N-1\}$ which could be the directions corresponding to the N independent variables of the objective function. It searches along these directions (i.e. starts out like the Hooke-Jeeves direct search method) for a sufficient reduction in the objective function. After completing N searches, it takes the final value of the state $\mathbf{x}$ and subtracts from it the initial value of the state. Let us denote the value of the state vector after these N searches by $\mathbf{x}_1$ and its initial value by $\mathbf{x}_0$. Following the previous nomenclature,

$$\Delta \mathbf{x}_0 = \mathbf{x}_1 - \mathbf{x}_0 \tag{25.145}$$

$\Delta \mathbf{x}_0$ gives the new direction $\mathbf{s}_0$ upon normalization and then a Gram-Schmidt orthogonalization process as described in Section 23.3 is employed to obtain the rest of the directions, $\mathbf{s}_i, i \in \{1, 2, \cdots, N-1\}$ which are made orthonormal to $\mathbf{s}_0$. The search through all the new N directions is repeated again and a new direction is found. These $\mathbf{s}_0$ directions at every iteration, k, tend to line up with the principal axes of the Hessian of the quadratic approximation to the objective function (Eigenvectors of $\mathbf{G}$). This makes Rosenbrock's method similar to conjugate direction methods in convergence properties when applied to the minimization of a quadratic objective function.

This method has a very serious problem with its applicability to practical problems as the neural network learning problem or Support Vector Machine computations. The search directions generated by the method could sometimes become zero. In that case, the scheme fails. Davies, Swann and Campey made a modification to Rosenbrock's method to reduce the chances of this type of failure.

25.2.2.2 The Davies-Swann-Campey Method

Davies, Swann and *Campey* [66] presented a variation of the *Rosenbrock gradient-free minimization method* which makes it more practical. This method is named after them and abbreviated to the *DSC method*. In the *DSC method*, a *gradient-free linear minimization method* is used to find the minimum of the objective function along the N directions in contrast to *Rosenbrock's method* that makes a mere reduction in the objective function. Another modification done to *Rosenbrock's method* is a reordering of the directions which allows the retainment of nonzero directions separate from the zero directions and minimization is done in those nonzero directions until termination occurs or all the directions become zero.

25.2.2.3 Powell's Method

In most minimization methods, the state vector is updated in a direction ($\mathbf{s}_k$) dictated by the method and a linear search is used to find the optimum step size of the update in that direction (η_k^*). This update will be of the following form at every step of the minimization,

$$\mathbf{x}_{k+1} = \mathbf{x}_k + \eta_k^* \mathbf{s}_k \tag{25.146}$$

The main purpose of the minimization algorithm is to find a sequence of directions in which to perform the linear searches. If the objective function is quadratic, then the best set of directions are in general the set which are mutually conjugate about the Hessian matrix of that function. However, since a gradient-free minimization scheme is intended, the Hessian matrix and even the gradients are not evaluated. This makes the task of finding a mutually conjugate set of directions very difficult. *Powell* [56] states a theorem on conjugacy that helps him develop an algorithm which tends to line up two consecutive new directions of search with conjugate directions.

Powell [56] devised two methods, in evolution from $\widehat{\text{Smith}}$'s method [64], which minimize the objective function $E(\mathbf{x})$ by successive linear searches in directions which are generated by the methods and tend to become conjugate about the Hessian of the quadratic approximation to the objective function E.

These methods are based on two theorems stated by *Powell* [56] in the following manner:

Theorem 25.1 (Powell's First Theorem). *"If $\mathbf{q}_1, \mathbf{q}_1, \cdots, \mathbf{q}_m, m \leq n$ are mutually conjugate directions, then the minimum of the quadratic function, $E(\mathbf{x})$, where $\mathbf{x}$ is a general point in the m-dimensional space containing $\mathbf{x}_0$ and the directions $\mathbf{q}_1, \mathbf{q}_1, \cdots, \mathbf{q}_m$, may be found by searching along each of the directions once only."*

Theorem 25.2 (Powell's Second Theorem). *"If $\mathbf{x}_0$ is the minimum in a space containing the direction $\mathbf{q}$, and $\mathbf{x}_1$ is also the minimum in such a space, then the direction $(\mathbf{x}_1 - \mathbf{x}_0)$ is conjugate to $\mathbf{q}$."*

Powell's paper [56] provides the proofs to these theorems. Using these two theorems, *Powell* presented his first method given by the following steps.

Powell's First Method
Set $\mathbf{s}_k = \hat{\mathbf{e}}_k$, where $\hat{\mathbf{e}}_k$ is the unit vector such that

$$(\hat{\mathbf{e}}_k)_{[j]} \stackrel{\Delta}{=} \begin{cases} 1 \ \forall \ j = k \\ 0 \ \forall j \neq k \end{cases} \tag{25.147}$$

Then,

1. Find η_k^* which minimizes $E(\mathbf{x}_{k-1} + \eta_k \mathbf{s}_k)$
   and set $\mathbf{x}_k = \mathbf{x}_{k-1} + \eta_k^* \mathbf{s}_k, k = \{1, 2, \cdots, N\}$
2. Replace $\mathbf{x}_k$ by $\mathbf{s}_{k+1}$ for $k = \{1, 2, \cdots, N-1\}$
3. Replace $\mathbf{s}_N$ by $\mathbf{x}_N - \mathbf{x}_0$
4. Find η_N^* which minimizes $E(\mathbf{x}_0 + \eta_N \mathbf{s}_N)$
5. Set $\mathbf{x}_0 = \mathbf{x}_0 + \eta_N^* \mathbf{s}_N$ and go to step 1

This procedure finds points $\mathbf{x}_1, \cdots, \mathbf{x}_N$ which minimize the quadratic approximate of the objective function in the $\mathbf{s}_1, \cdots, \mathbf{s}_N$ directions and generates the new direction,

$$s_{N+1} = \frac{x_N - x_0}{\|x_N - x_0\|_{\mathscr{E}}} \tag{25.148}$$

This procedure makes up one iteration of *Powell's first method*. Through renumbering, the procedure is repeated for the new directions $s_1, \cdots, s_N$. *Powell* claims that after N iterations, the minimum of the quadratic function is reached. However, this claim has been shown by *Zangwill* [72] to be false in general. *Zangwill* provides a counter example in [72] for which *Powell's method* will never converge to the minimum. This lack of convergence is due to a mistake in the statement of *Powell's first theorem*. *Powell's methods* could sometimes generate linearly dependent directions which will not span the entire space. To correct for this mistake, *Powell* states that sometimes it is not wise to accept any new direction provided by his method. *Zangwill* gives a correction for *Powell's first theorem* which would solve this problem. He states that in Theorem 25.1, the directions $q_1, \cdots, q_m$, must be such that they span the entire m-dimensional space. This gives the motivation behind his second minimization method. This method as simplified by *Zangwill* [72] is as follows.

Powell's Second Method
Set $s_k^1 = \hat{e}_k$. x_0^1 is chosen randomly and $\varepsilon : 0 < \varepsilon \le 1$ is given as the threshold for accepting a new direction. Furthermore, $\delta^1 = 1$ and $r = 1$. Then,

1. Find η_k^{r*} which minimizes $E(x_{k-1}^r s_k^r)$ and set $x_k^r = x_{k-1}^r + \eta_k^r s_k^r$ where $k = \{1, \cdots, N\}$.
2. Define $\alpha^r \overset{\Delta}{=} \|x_N^r - x_0^r\|_{\mathscr{E}}$ and $s_{N+1}^r = \frac{x_N^r - x_0^r}{\alpha^r}$.
3. Find $\eta_{N+1}^r{}^*$ which minimizes $E(x_N^r + \eta_{N+1}^r s_{N+1}^r)$
4. Set $x_0^{r+1} = x_{N+1}^r = x_N^r + \eta_{N+1}^r{}^* s_{N+1}^r$
5. Set $\eta_s^r = \max \eta_k^r, k = 1, \cdots, N$ where s is the $\arg\max_k \eta_k^r$
6. If $\frac{\eta_s^r \delta^r}{\alpha^r} \ge \varepsilon$, then

$$s_k^{r+1} = s_k^r \ \forall \ k \ne s$$
$$s_s^{r+1} = s_{N+1}^{r+1}$$
$$\delta^{r+1} = \frac{\eta_s^r \delta^r}{\alpha^r}$$

7. If $\frac{\eta_s^r \delta^r}{\alpha^r} < \varepsilon$, then

$$s_k^{r+1} = s_k^r \ for \ k = 1, \cdots, N$$
$$\delta^{r+1} = \delta^r$$

8. Set $r = r + 1$ and go to step 1.

Figure 25.6 represents the convergence Powell's update to form a set of conjugate directions.

In the above algorithm, δ^r is the determinant,

$$\delta^r = det([s_1, s_2, \cdots, s_N]) \tag{25.149}$$

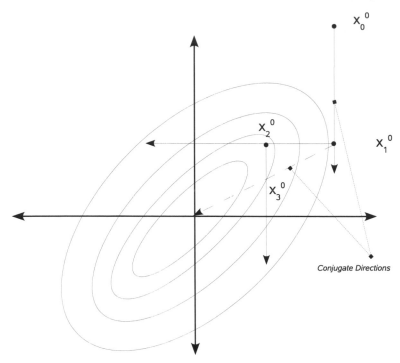

Fig. 25.6: Powell's Convergence to a Set of Conjugate Directions

For $r = 1$, since the directions, $\mathbf{s}_k, k \in \{1, 2, \cdots, N\}$ coincide with the columns of the identity matrix, the determinant, $\delta^1 = 1$. As the method proceeds in iterations, the objective is to find a set of directions which would have the largest determinant, for if the determinant approaches zero, then the set of directions approach a linearly dependent set. When a new direction, $\mathbf{s}_{N+1}^r$ is found through step 3, if it replaces any direction $\mathbf{s}_s^r$, then the new determinant of the direction set will be given by $\delta^{r+1} = \frac{\eta_s^r \delta^r}{\alpha^r}$. Since the size of this determinant is largest if the largest η_s^r is used, then the direction that should be replaced with the new direction should be the one that corresponds to the largest linear step η_s^r. However, if this new determinant is seen to be smaller than some computational tolerance ε, then replacing any direction with this new direction will make the new set of directions linearly dependent. In this case, the new direction is rejected and minimization takes places again with the previous set of directions.

In addition, in [21], *Fletcher* presents a modified version of *Smith's method* which makes it a rival of *Powell's methods*.

25.3 The Line Search Sub-Problem

In the preceding, we have seen many different techniques, each of which at any iteration gives rise to a direction which tends to point toward the minimum of the objective function, $E(\mathbf{x})$. For any such direction, we have been prescribing the use of a *line search* to be able to evaluate the location of the next state of the minimization. This translates to a magnitude η away from the starting point of each of these sub-problems in optimization where an initial point of $\mathbf{x}_k$ is provided along with a direction, $\mathbf{s}_k$, in which the value of $E(\mathbf{x})$ should be minimized. This is a search problem in the $\mathbf{s}_k$ direction and the resulting state will be denoted as $\mathbf{x}_{k+1}$, which is the next state of the minimization path. Therefore, the dependent variable of this unidirectional optimization is some multiplier of the unit length direction, $\mathbf{s}_k$, namely, η. Equation 25.150 shows the relationship between the new state vector $\mathbf{x}_{k+1}$ and the starting point, $\mathbf{x}_k$ with respect to parameter η.

$$\mathbf{x}_{k+1}(\eta_k) = \mathbf{x}_k + \eta_k \mathbf{s}_k \tag{25.150}$$

Since we are interested in that η_k which makes the value of $\mathbf{x}_{k+1}$, the minimizer of $E(\mathbf{x})$ along $\mathbf{s}_k$, we will denote that specific parameter by η_k^*. The line search problem assumes that η_k is the only variable along the direction $\mathbf{s}_k$ and tries to minimize $E(\eta_k)$. This problem is a special case of the larger problem which we have been addressing up to now. It is a minimization problem with one dependent variable, η_k. Therefore, gradient and non-gradient methods also exist for this sub-problem where the gradient is taken with respect to η_k and since it only has one dimension, it coincides with the partial derivative, $\frac{\partial E}{\partial \eta_k}$.

If this derivative is available, the problem becomes simple. One may set the derivative equal to zero and solve for that positive η_k which minimizes $E(\eta_k, \mathbf{x}_k, \mathbf{s}_k)$. The usual second derivative test may be done to make sure the resulting η_k^* is the minimizer of E and the value of the new state may be written in terms of the newly found optimal step size, η_k^*, namely,

$$\mathbf{x}_{k+1} = \mathbf{x}_k + \eta_k^* \mathbf{s}_k \tag{25.151}$$

Line search problems may be categorized into *exact* and *inexact line searches*. *Exact line searches* are generally costly and are provided by the underlying structure of the objective function. However, we usually do not know the exact nature of an objective function. Normally, the objective function is approximated in a locality. Therefore, exact line searches are not very practical and in some cases their solution may not even exist. Modern optimization problems relax the line search and generally require that one of more mild criteria are met. For example, in the minimization case, they would require that with each step, the value of the objective function would decrease. In Section 25.1.3.3 we discussed some such conditions. An example is the *Wolfe-Powell conditions*. We saw that in essence, the *Wolfe condition* makes assumptions on the curvature and therefore defines certain parameters for calculating the *line search* termination. The condition of Equation 25.42 stemmed from the *Wolfe-Powell conditions* [23] and provided us guidelines for the *line search* to be able to satisfy the positive curvature criterion.

25.4 Practical Considerations

In this section we will review different practical considerations when it comes to optimization. First, we will review large-scale optimization techniques which address memory shortage, sparsity of the optimization parameters, and separability considerations. Next, we will examine the numerical stability of different techniques and review techniques for increasing stability by using initial and self-scaling techniques, including the transformation of the problem such that the overall numerical stability is increased.

25.4.1 Large-Scale Optimization

There are large-scale versions of most of the algorithms listed in this chapter. These include modifications to the steepest descent and gradient techniques only using the gradient and most effectively, modifications to Quasi-Newton methods to handle large-scale problems. Among the modifications to the steepest descent technique, one may note the process of training a Feedforward Neural Network introduced by Rumelhart [58]. Please refer to Chapter 14 for the problem definition. In this optimization problem, the objective function is derived from a sum of squares of errors between the desired output of the network and its real output with any set of weights. This sum is over all the output neurons as well as all the different patterns being tested.

In the true sense, the steepest descent algorithm requires that the sum of squares of errors is computed over all the training samples and once a single pass is made through all those samples, the gradient may be updated. Rumelhart, takes the approach that evaluates the gradient of the error (objective function) over a single pattern and makes a correction to the weights. This method, then, presents another pattern and goes through these updates every time a pattern is passed to the network. Theoretically, this is not the same as applying the steepest descent. It is, however, an approximation which tries to deal with the fact that if there are many patterns and that waiting for all the patterns to be presented will delay the optimization process. Although, this method approaches the minimum more slowly compared to processing all the patterns and then computing the gradient, it does achieve faster intermediate convergence, since the system weights are adjusted more frequently.

Also, it has shown some resilience toward over-training since it is not as accurate as the true steepest descent technique. For the same reason, it also has more of a tendency to end up at the global minimum that the more efficient steepest descent technique.

This is akin to what happens in the annealing process for crystal generation in metals, which in the domain of Neural Network training, is called simulated annealing.[40] Simulated annealing uses statistical perturbations to move the state out of a local minimum, sometimes in the wrong direction, to reach the global minimum instead of being trapped at a local one. These types of techniques are generally known as *stochastic relaxation* methods [73] and have been used in variety of situations such as the *modified k-means* algorithm [43]. See Section 11.2.5.2 for more on the *annealing process*.

In the same spirit as the approximations made to the steepest descent technique, but with somewhat different objectives and approaches, the Quasi-Newton methods discussed earlier in this chapter have been modified to handle large-scale optimization problems in a more practical setting. There area number of approaches to reducing the memory and processing intensity of these optimization algorithms. Most of the effort has been in the handling of systems with a large Hessian matrix. There are generally three different approaches depending on the problem at hand.[50]

1. Limited Memory Quasi-Newton Methods
2. Sparse Quasi-Newton Methods
3. Partially Separable Quasi-Newton Methods

25.4.1.1 Limited Memory Quasi-Newton Methods

These approaches limit the amount of history which is kept in a Quasi-Newton update of the Hessian matrix. These methods, such as the Limited-Memory BFGS or L-BFGS, only store the last m pairs of change vectors, $\Delta \mathbf{x}_k$ and $\Delta \mathbf{g}_k$, instead of storing the whole $\mathbf{H}_k$. In these techniques, the said stored set of pairs of vectors are used to evaluate the new direction in which to perform the *line search*. The *line search* will be identical to the normal Quasi-Newton technique. As a new direction is computed the mth oldest pair of change vectors is deleted and the most recently computed pair is stored, keeping the last m change pairs in memory at any time.

The L-BFGS technique operates on the Hessian matrix, $\mathbf{G}_k$, not the inverse Hessian. In this approach, the Hessian matrix approximate, $\mathbf{G}_k$ is obtained from its initial value and the last m differences as follows,

$$
\mathbf{G}_k = \mathbf{G}_0 - \begin{bmatrix} \mathbf{G}_0 \mathbf{U}_k & \mathbf{V}_k \end{bmatrix} \begin{bmatrix} \mathbf{U}_k^T \mathbf{G}_0 \mathbf{U}_k & \mathbf{L}_k \\ \mathbf{L}_k^T & -\mathbf{D}_k \end{bmatrix}^{-1} \begin{bmatrix} \mathbf{U}_k^T \mathbf{G}_0 \\ \mathbf{V}_k^T \end{bmatrix} \tag{25.152}
$$

where, when $k \leq m$,

$$\mathbf{U}_k = [\Delta \mathbf{x}_0, \cdots, \Delta \mathbf{x}_{k-1}]$$
$$\mathbf{V}_k = [\Delta \mathbf{g}_0, \cdots, \Delta \mathbf{g}_{k-1}] \qquad (25.153)$$

$\mathbf{L}_k, \mathbf{D}_k : \mathscr{R}^k \mapsto \mathscr{R}^k$ and,

$$(\mathbf{L}_k)_{i,j} = \begin{cases} \Delta \mathbf{x}_{i-1}^T \Delta \mathbf{g}_{j-1} \; \forall \, i > j \\ 0 \forall i \le j \end{cases} \qquad (25.154)$$

$$\mathbf{D}_k = diag \left[\Delta \mathbf{x}_0^T \Delta \mathbf{g}_0, \cdots, \Delta \mathbf{x}_{k-1}^T \Delta \mathbf{g}_{k-1} \right]$$

When $k > m$,

$$\mathbf{U}_k = [\Delta \mathbf{x}_{k-m}, \cdots, \Delta \mathbf{x}_{k-1}]$$
$$\mathbf{V}_k = [\Delta \mathbf{g}_{k-m}, \cdots, \Delta \mathbf{g}_{k-1}] \qquad (25.155)$$

$\mathbf{L}_k, \mathbf{D}_k : \mathscr{R}^k \mapsto \mathscr{R}^k$ and,

$$(\mathbf{L}_k)_{i,j} = \begin{cases} \Delta \mathbf{x}_{i+k-m-1}^T \Delta \mathbf{g}_{j+k-m-1} \; \forall \, i > j \\ 0 \forall i \le j \end{cases} \qquad (25.156)$$

$$\mathbf{D}_k = diag \left[\Delta \mathbf{x}_{k-m}^T \Delta \mathbf{g}_{k-m}, \cdots, \Delta \mathbf{x}_{k-1}^T \Delta \mathbf{g}_{k-1} \right]$$

Depending on the condition number of the inverse Hessian matrix (see section 25.4.2) and its density (the opposite sense of sparsity), m will vary. It will be larger for larger condition numbers and more dense matrices. [50] prescribes a value between 3 and 20 for most practical purposes, using L-BFGS. The Limited-Memory techniques, are very easy to implement and are usually quite effective. However, they lose the quadratic convergence characteristics of Quasi-Newton updates and lean more toward a linear convergence, but are generally slightly better than linear – will outperform steepest descent and conjugate gradient techniques. Note that the general technique of L-BFGS may be applied to any Quasi-Newton technique. See [50] for a more complete treatment.

25.4.1.2 Sparse Quasi-Newton Methods

The second approach, so called Sparse Quasi-Newton Methods, attracted a lot of attention in the late 1980s and early 1990s. Almost all the researchers involved in the development of Quasi-Newton techniques in the earlier decades introduced a sparse version of theirs and their colleagues' methods. Generally, these techniques tried to apply the knowledge of sparsity of the Hessian and Inverse Hessian matrices to their approximations given through Quasi-Newton techniques. These methods have quite a number of deficiencies including the fact that the approximate Hessian and Inverse Hessian matrices are no longer guaranteed to be positive definite, the methods are not scaling invariant, and normally they require at least as many function evaluations as the Limited-Memory techniques described in Section 25.4.1.1.

25.4.1.3 Partially Separable Quasi-Newton Methods

These methods are quite effective, but are also complex in nature. Using intimate knowledge of the objective function and its relationship with the state vector, they try to decouple the problem into separate problems where,

$$E(\mathbf{x}) = \sum_{i=1}^{M} E_i(\mathbf{x})$$

$$\nabla E(\mathbf{x}) = \sum_{i=1}^{M} \nabla E_i(\mathbf{x})$$

$$\nabla^2 E(\mathbf{x}) = \sum_{i=1}^{M} \nabla^2 E_i(\mathbf{x})$$

If, either the M different problems happen to become decoupled so that each use a disjoint subset of the variables, or otherwise if there is some specific structure or sparsity to the Hessian (inverse Hessian) matrices of the M subproblems in contrast with the original problem, then it makes sense to use the decoupled version. Unfortunately, such techniques are quite problem-specific and are not readily generalized to apply to any large class of problems. Although, It is important to be aware of the possibility in case the occasion arises.

25.4.2 Numerical Stability

when it comes to the implementation of most of the Quasi-Newton techniques, Numerical stability is of utmost importance. Some of the self-scaling and initial-scaling techniques discusses earlier in this chapter try to deal with this specific problem. Theoretically, their results are not very different from non-scaling techniques. However, in practice, the scaling helps establish better numerical stability of the updates.

One very important numerical instability problem that usually comes up, is the case where the condition number ($\frac{\lambda_1}{\lambda_N}$ – λ_i is the ith Eigenvalue of the Inverse Hessian approximation, $\mathbf{H}_k$) is very large. Historically, to deal with related problems in the field of signal processing, methods such as Square-Root filtering have been used. These methods try to represent the ill-conditioned matrix with its square root, in the form of a *Cholesky factorization* [8].

If a matrix $\mathbf{H}_k$ is factored into $\mathbf{L}_k \mathbf{L}_k^T$, the condition number of $\mathbf{L}_k$ will be equal to $\sqrt{\frac{\lambda_1}{\lambda_N}}$. So, if the updates are done to L_k instead of H_k, they will be much more numerically stable. [50] covers some such cases for the *BFGS* method – using the Hessian Matrix, $\mathbf{G}_k$ instead of the inverse Hessian. It reports that usually, the numerical gains are not great. However, they will be substantial if the condition number is very large. Still, the self-scaling techniques, may do better even in those conditions.

25.4.3 Nonsmooth Optimization

Most of the developments described in this chapter started with the assumption that the objective function was a *smooth function* (see Definition 24.20) of the dependent variables. In fact, in the beginning of this chapter, we started with the premise that most smooth functions may be approximated by quadratic functions in small intervals. However, in practice this assumption would fall apart when we deal with problems such as network scheduling and queuing theory [23], radiation therapy [30] and other linear programming or integer programming problems. There are many examples where gradients do not exist and the smoothness criterion is not met.

There are many new techniques which have been developed over the past three or four decades. One very popular technique for handling nonsmooth optimization problems is the use of *subgradients* which are defined and used in the absence of gradients [30]. Since most of the problems in the speaker recognition field seem to be easily modeled using smooth functions and quadratic or higher degree objective functions, we will not treat the nonsmooth case in this book. The interested reader may refer to [23] for an overview of the nonsmooth optimization problem and [30] for a good reference on subgradient techniques.

25.5 Constrained Optimization

Up to this point, we have been generally considering the optimization of smooth functions without any constraints. In a way, it is good to remove complicated factors, such as constraints and nonsmooth behavior to be able to understand complex optimization techniques. However, once we enter the world of practice, these factors start to become important. In Section 25.4.3, we briefly discussed nonsmooth functions. We also noted that in most cases, related to the problems in speaker recognition, optimization functions may be assumed to be smooth.

Constraints, on the other hand, are important in the practice of speaker recognition. As we shall see, constraints are generally imposed in the form of functions which should be considered in the optimization problem simultaneously with the main objective function. Therefore, right at the beginning, we would have to consider the smoothness of the constraints as well. Here, we assume that the functions presenting the constraints on the value of the state vector (dependent variable) as well as the objective function of optimization are smooth (see Definition 24.20).

As we stated in the beginning of this chapter, we may discuss the minimization problem without any loss of generality. We saw that any local optimization problem may be reduced into a minimization problem with a change of variables. Let us call the main objective function of the minimization problem, $E_0(\mathbf{x})$, where $\mathbf{x}^*$ is the minimizer of this objective function in a small neighborhood. Again, without any loss of generality, let us assume that we will always remain in a region where the local minimum is of interest and that $E_0(\mathbf{x})$ is smooth and convex in that region. This simplifies the problem so that we do not have to worry about global versus local minima.

Then, we may define the general constrained minimization problem as follows,

$$\mathbf{x}^* = \arg\min_{\mathbf{x}\in\Omega} E_0(\mathbf{x}) \tag{25.157}$$

where $\mathbf{x} : \mathscr{R}^1 \mapsto \mathscr{R}^N$ and Ω is the *feasibility region* where $\mathbf{x}$ meets all the constraints of the problem and is defined as follows,

$$\Omega = \{\mathbf{x} : (E_{n_e}(\mathbf{x}) = 0 \; \forall \; n_e \in \mathscr{N}_{n_e}) \wedge (E_{n_i}(\mathbf{x}) > 0 \; \forall \; n_i \in \mathscr{N}_{n_i})\} \tag{25.158}$$

The points $\{\boldsymbol{\xi} = \mathbf{x} : \mathbf{x} \in \Omega\}$ are known as *feasibility points* of the constrained optimization problem [23]. $E_{n_e}(\mathbf{x})$ and $E_{n_i}(\mathbf{x})$ are the *equality* and *inequality* constraints, respectively.

Consider a set of indices, $\mathscr{N}_c = \{1, 2, \cdots, N_c\}$, for all the constraint equations, such that $\mathscr{N}_e$ is the set of indices related to equality constraints with the number of elements, $|\mathscr{N}_e| = N_e$, and $\mathscr{N}_i$ is the set of indices related to inequality constraints with $|\mathscr{N}_i| = N_i$. Then, we will have the following relations between the index sets and their sizes, respectively,

$$\mathscr{N}_c = \mathscr{N}_e \cup \mathscr{N}_i \tag{25.159}$$

$$N_c = N_e + N_i \tag{25.160}$$

To simplify the discussion of constraints, it makes sense to define a vector representation of the constraint functions. Let us define the vector functions, $\mathbf{c}^{(e)}$, $\mathbf{c}^{(i)}$, and $\mathbf{c}$, as follows,

$$\mathbf{c}^{(e)}(\mathbf{x}) : \mathscr{R}^N \mapsto \mathscr{R}^{N_e} \; \wedge \; \left(\mathbf{c}^{(e)}\right)_{[n_e]}(\mathbf{x}) = E_{n_e}(\mathbf{x}), n_e \in \mathscr{N}_e \tag{25.161}$$

$$\mathbf{c}^{(i)}(\mathbf{x}) : \mathscr{R}^N \mapsto \mathscr{R}^{N_i} \; \wedge \; \left(\mathbf{c}^{(i)}\right)_{[n_i]}(\mathbf{x}) = E_{n_i}(\mathbf{x}), n_i \in \mathscr{N}_i \tag{25.162}$$

$$\mathbf{c}(\mathbf{x}) : \mathscr{R}^N \mapsto \mathscr{R}^{N_c} \; \wedge \; \mathbf{c}(\mathbf{x}) \overset{\Delta}{=} \{\mathbf{c}^{(e)^T}(\mathbf{x}), \mathbf{c}^{(i)^T}(\mathbf{x})\}^T \tag{25.163}$$

Also, to simplify the notation, we may define the *Jacobian matrix* of the constraint vector functions with respect to **x** as follows,

$$
\begin{aligned}
\{\mathbf{J}_\mathbf{x}^{(e)} : \mathscr{R}^{N_e} \mapsto \mathscr{R}^N\} \ \text{where} \ \left(\mathbf{J}_\mathbf{x}^{(e)}\right)_{[n][n_e]} &\triangleq \frac{\partial E_{n_e}}{\partial(\mathbf{x})_{[n]}} \\
\{\mathbf{J}_\mathbf{x}^{(i)} : \mathscr{R}^{N_i} \mapsto \mathscr{R}^N\} \ \text{where} \ \left(\mathbf{J}_\mathbf{x}^{(i)}\right)_{[n][n_i]} &\triangleq \frac{\partial E_{n_i}}{\partial(\mathbf{x})_{[n]}} \\
\{\mathbf{J}_\mathbf{x} : \mathscr{R}^{N_c} \mapsto \mathscr{R}^N\} \ \ \text{where} \ (\mathbf{J}_\mathbf{x})_{[n][n_c]} &\triangleq \frac{\partial E_{n_c}}{\partial(\mathbf{x})_{[n]}}
\end{aligned}
\tag{25.164}
$$

If the independent variable of the gradient in the definition of the *Jacobian matrix* is **x**, sometimes we will drop the subscript, **x**, for simplicity of notation. i.e., $\mathbf{J}_\mathbf{x}$ will be written as **J**. Furthermore, column n_c of **J** is denoted by $\boldsymbol{\gamma}_{n_c}$. $\boldsymbol{\gamma}_{n_c}$ is also called the *normal vector* [23] of constraint E_{n_c}, since it is usually normal to the surface described by the constraint. For *inequality constraints*, $\boldsymbol{\gamma}_{n_c}$ points toward the *feasibility region* of the constraint. The *normal vectors* for the different types of constraints may be directly defined as follows,

$$
\boldsymbol{\gamma}_{n_e} \triangleq \nabla_\mathbf{x} E_{n_e}(\mathbf{x}) \tag{25.165}
$$

$$
\boldsymbol{\gamma}_{n_i} \triangleq \nabla_\mathbf{x} E_{n_i}(\mathbf{x}) \tag{25.166}
$$

$$
\boldsymbol{\gamma}_{n_c} \triangleq \nabla_\mathbf{x} E_{n_c}(\mathbf{x}) \tag{25.167}
$$

Note that the equality and inequality constraints defined in Equation 25.158 are quite general, since the functions E_{n_e} and E_{n_i} may take on any form as long as they are smooth such that they would allow the use of the nonlinear optimization techniques discussed in the rest of this chapter. Of course, if they are not smooth, the same argument as in Section 25.4.3 may be easily extended to them. For example, if we have a condition where the values of **x** must take on discrete values, then *integer programming* [30] techniques must be utilized. Unfortunately, handling *integer programming* would not be feasible in this book. However, *Fletcher* [23] describes a technique called *branch and bound*, which may be used to reduce the discrete case into a set of smooth optimization problems which may then be solved using the methods discussed in this section.

One of the most practical, popular, and general methods of handling the constrained optimization problem of Equation 25.157 is to use *Lagrange multipliers*. A historic use of *Lagrange multipliers* for handling constraints comes from *Hamilton's principle* of classical dynamics in minimizing the *Hamiltonian*, subject to *holonomic* or *non-holonomic* constraints [20], yielding the *Euler-Lagrange* equations of dynamics. Similar techniques have been used to handle the optimization of nonlinear objective functions with general *equality* and *inequality constraints*.

25.5.1 The Lagrangian and Lagrange Multipliers

We defined the *normal vectors*, $\boldsymbol{\gamma}_{n_c}$, to be the gradient vectors of the constraint functions. Therefore, the *Taylor series expansion* (Definition 24.42) of any constraint function, E_{n_c}, about the minimizer of $E_0(\mathbf{x})$, denoted by $\mathbf{x}^*$, may be written, up to the first order approximation, as follows,

$$E_{n_c}(\mathbf{x}^* + \Delta\mathbf{x}) = E_{n_c}(\mathbf{x}^*) + \boldsymbol{\gamma}_{n_c}^T\Big|_{\mathbf{x}^*} \Delta\mathbf{x} + \mathscr{O}(\|\Delta\mathbf{x}\|_{\mathscr{E}}^2) \tag{25.168}$$

We saw in Equation 25.7 that $\Delta\mathbf{x}$ may be decomposed into a magnitude, $\eta > 0$, and a direction, $\mathbf{s}$,

$$\Delta\mathbf{x} \equiv \eta\mathbf{s} \tag{25.169}$$

In Equation 25.169,

$$\|\Delta\mathbf{x}\|_{\mathscr{E}} = |\eta|\,\|\mathbf{s}\|_{\mathscr{E}} \tag{25.170}$$

Therefore, $\|\mathbf{s}\|_{\mathscr{E}}$ may take on any value, since $|\eta|$ can always correct for the magnitude of $\Delta\mathbf{x}$. However, it is customary to normalize the direction, $\mathbf{s}$, to have unit norm,

$$\|\mathbf{s}\|_{\mathscr{E}} = 1 \tag{25.171}$$

Using the identity in Equation 25.169 to plug in for $\Delta\mathbf{x}$ in Equation 25.168, and using the choice of Equation 25.171, the *Taylor series* expansion may be written in terms of $\mathbf{s}$,

$$E_{n_c}(\mathbf{x}^* + \eta\mathbf{s}) = E_{n_c}(\mathbf{x}^*) + \eta\boldsymbol{\gamma}_{n_c}^T\Big|_{\mathbf{x}^*} \mathbf{s} + \mathscr{O}(|\eta|^2) \tag{25.172}$$

Let us consider a *sequence* (see Section 6.7.2) *of deviations* in $\mathbf{x}$, away from the minimizer, $\mathbf{x}^*$, that approaches $\mathbf{0}$. In other words,

$$\{\Delta\mathbf{x}\}_1^k : \Delta\mathbf{x}_k \to 0 \ \ as \ \ k \to \infty \tag{25.173}$$

where

$$\Delta\mathbf{x}_k \stackrel{\Delta}{=} \mathbf{x}_k - \mathbf{x}^* \tag{25.174}$$

$$= \eta_k\mathbf{s}_k \tag{25.175}$$

We may write the sequence of Equation 25.173 in terms of the definition of Equation 25.175,

$$\{\eta\mathbf{s}\}_1^k : \eta_k \to 0 \wedge \mathbf{s}_k \to \mathbf{s} \ \ as \ \ k \to \infty \tag{25.176}$$

As the sequence in Equation 25.176 converges, $\mathbf{s}_k \to \mathbf{s}$. $\mathbf{s}$ is known as the *feasible direction* evaluated at point $\mathbf{x}^*$, which is the point of convergence of $\mathbf{x}_k$. This sequence is known as a *directional sequence* [23] and may be defined for any point, $\mathbf{x}$. At $\mathbf{x}$, the set of such *feasible directions* may be denoted as $\mathscr{S}_f(\mathbf{x})$. $\mathscr{S}_f(\mathbf{x})$ is known as the *set of feasible directions* at $\mathbf{x}$ and we use a special shorthand for this set, at the minimizer,

$$\mathscr{S}_f^* \triangleq \mathscr{S}_f(\mathbf{x}^*) \tag{25.177}$$

Note that there may be two different types of constraints described by Equation 25.168.

In the first scenario, $n_c \in \mathcal{N}_e$; namely, the constraint of interest is an *equality constraint*. For better readability, let us use the index, n_e, for this case. Since we are only interested in moving along a *feasible direction*, for a small $\mathbf{x}$, the following would have to hold,

$$E_{n_e}(\mathbf{x}^* + \Delta\mathbf{x}) = E_{n_e}(\mathbf{x}^*)$$
$$= 0 \tag{25.178}$$

Plugging Equation 25.178 into Equation 25.168 and choosing $\|\Delta\mathbf{x}\|$ small enough such that the second order and higher order terms vanish, we will have the following relation for the *linearized equality constraints*[2],

$$\boldsymbol{\gamma}_{n_e}^T \Big|_{\mathbf{x}^*} \Delta\mathbf{x} = \boldsymbol{\gamma}_{n_e}^{*T} \Delta\mathbf{x}$$
$$= 0 \tag{25.179}$$

Using Equation 25.169, Equation 25.179 may be rewritten in terms of $\mathbf{s}$, as follows,

$$\boldsymbol{\gamma}_{n_e}^{*T} \mathbf{s} = 0 \tag{25.180}$$

In the second scenario, $n_c \in \mathcal{N}_i$, in other words, the constraint is an *inequality constraint*. Again, we resort to using n_i for this case. Since we would like to take a step in a *feasible direction*, then by the definition of *inequality constraints* given in Equation 25.158, the following relation would have to hold,

$$E_{n_i}(\mathbf{x}^* + \Delta\mathbf{x}) - E_{n_i}(\mathbf{x}^*) \geq 0 \tag{25.181}$$

from which, taking $\|\Delta\mathbf{x}\|$ to be small enough in order to be left with the first order approximation, and plugging the inequality of Equation 25.181 into Equation 25.168, we will have the following, for *linearized inequality constraints*,

$$\boldsymbol{\gamma}_{n_i}^T \Big|_{\mathbf{x}^*} \Delta\mathbf{x} = \boldsymbol{\gamma}_{n_i}^{*T} \Delta\mathbf{x} \tag{25.182}$$
$$\geq 0 \tag{25.183}$$

---

[2] In Section 25.5.1.4, we will treat the case when the original constraints are used without linearization.

As with the case of equality constraints, if we write $\Delta\mathbf{x}$ in terms of a magnitude, η, and a direction, $\mathbf{s}$, Equation 25.183 may be rewritten in terms of $\mathbf{s}$, using the identity in Equation 25.169, as follows,

$$\boldsymbol{\gamma}_{n_i}^{*T}\mathbf{s} \geq 0 \tag{25.184}$$

The set of *feasible directions* for the *linearized constraints* are therefore those directions which conform to Equation 25.180 for the *linearized equality constraints* and to the inequality in Equation 25.184 for the *linearized inequality constraints*. This set of *feasible directions*, as a function of $\mathbf{x}$, may then be designated by $\mathscr{S}_{fl}$, where the l in the subscript stands for *linearized*. A formal definition of this set is as follows,

$$\mathscr{S}_{fl}(\mathbf{x}) \triangleq \{\mathbf{s} : \mathbf{s}^T\boldsymbol{\gamma}_{n_e}(\mathbf{x}) = 0 \ \forall \ n_e \in \mathscr{N}_e \ \wedge \ \mathbf{s}^T\boldsymbol{\gamma}_{n_i}(\mathbf{x}) \geq 0 \ \forall \ n_i \in \mathscr{N}_i\} \tag{25.185}$$

For the set of *feasible directions* associated with the minimizer of Equation 25.157, $\mathbf{x}^*$, we may use the following shorthand notation,

$$\mathscr{S}_{fl}^* \triangleq \mathscr{S}_{fl}(\mathbf{x}^*) \tag{25.186}$$

Sometimes we need to specify the set of *feasible directions* associated with the *linearized equality constraints*, only. Likewise, we may need to specify the set which is only associated with the *linearized inequality constraints*. The following two definitions allow for this distinction,

$$\mathscr{S}_{fl}^{(e)}(\mathbf{x}) \triangleq \{\mathbf{s} : \mathbf{s}^T\boldsymbol{\gamma}_{n_e}(\mathbf{x}) = 0 \ \forall \ n_e \in \mathscr{N}_e\} \tag{25.187}$$

$$\mathscr{S}_{fl}^{(i)}(\mathbf{x}) \triangleq \{\mathbf{s} : \mathbf{s}^T\boldsymbol{\gamma}_{n_i}(\mathbf{x}) \geq 0 \ \forall \ n_i \in \mathscr{N}_i\} \tag{25.188}$$

In the same spirit as in the definition of Equation 25.186, the respective sets for the minimizer, $\mathbf{x}^*$, would be given by the following two definitions,

$$\mathscr{S}_{fl}^{(e)*} \triangleq \mathscr{S}_{fl}^{(e)}(\mathbf{x}^*) \tag{25.189}$$

$$\mathscr{S}_{fl}^{(i)*} \triangleq \mathscr{S}_{fl}^{(i)}(\mathbf{x}^*) \tag{25.190}$$

In Section 25.5.1.2, we will modify the definition of $\mathscr{S}_{fl}(\mathbf{x})$ to include only *active constraints*. This concept will be discussed in detail.

25.5.1.1 Equality Constraints

Note that for the first scenario, where the constraints are *equality constraints*, if direction $\mathbf{s}$ in Equation 25.180 is a direction of descent, then we would have the following relation,

$$\mathbf{s}^T \mathbf{g} < 0 \tag{25.191}$$

where,

$$\mathbf{g} \triangleq \nabla_x E_0(\mathbf{x}) \tag{25.192}$$

The set of *descent directions*, $\mathscr{S}_d$, at any point $\mathbf{x}$ is, therefore, defined as the set of directions, $\mathbf{s}$, that obey the inequality of Equation 25.191. Here is this definition in mathematical notation,

$$\mathscr{S}_d(\mathbf{x}) \triangleq \{\mathbf{s} : \mathbf{s}^T \nabla_x E_0(\mathbf{x}) < 0\} \tag{25.193}$$

If $\mathbf{x} = \mathbf{x}^*$ happens to be the minimizer of the optimization problem of Equation 25.157, the following shorthand notation is used,

$$\mathscr{S}_d^* \triangleq \mathscr{S}_d(\mathbf{x}^*) \tag{25.194}$$

If $\mathbf{x}^*$ is a minimizer of $E_0(\mathbf{x})$, while having a *feasible direction* $\mathbf{s}$, given by Equation 25.180 (i.e. $\mathbf{s} \in \mathscr{S}_{fl}^{(e)*}$), then by definition, Equation 25.191 would never be satisfied. This is because if $\mathbf{x}^*$ is a minimizer of $E_0(\mathbf{x})$, then $\mathbf{s}^T \mathbf{g}^*$ cannot be negative. Therefore, $\mathbf{s}$ may not be a direction of descent at $\mathbf{x}^*$.

According to the above, the only solution to the *constrained minimization problem* with *linearized equality constraints* would exist if the following would be true,

$$\mathbf{x}^* = \mathbf{x} : \mathscr{S}_{fl}^{(e)}(\mathbf{x}) \cap \mathscr{S}_d(\mathbf{x}) = \{\varnothing\} \tag{25.195}$$

We will pursue this idea in its general form in the statement of Lemma 25.1 and explore the connection between linearized constraints and original constraints in Section 25.5.1.4.

Following this argument, the only possible solution would be if $\mathbf{g}^*$ is a linear combination of the *normal vectors*, $\boldsymbol{\gamma}_{n_e}^*$ [23], namely,

$$\mathbf{g}^* = \mathbf{J}_{\mathbf{x}}^{*(e)} \boldsymbol{\lambda}^{*(e)} \tag{25.196}$$

where $\boldsymbol{\lambda}^{*(e)}$ are the linear combination weights and are known as the *Lagrange multipliers* associated with these *equality constraints* and the minimizer $\mathbf{x}^*$. $\mathbf{J}_{\mathbf{x}}^{*(e)}$ is the *Jacobian matrix* associated with the *equality constraints* evaluated at the minimizer, $\mathbf{x}^*$. If $\mathbf{J}_{\mathbf{x}}^{*(e)}$ has *full rank*, the *Lagrange multipliers* may be computed by using the *Moore-Pensrose generalized inverse (pseudo-inverse)* (Definition 23.16) of the *Jacobian matrix* as follows,

$$\boldsymbol{\lambda}^{*(e)} = \left(\mathbf{J}_{\mathbf{x}}^{*(e)}\right)^\dagger \mathbf{g}^* \tag{25.197}$$

Fletcher [23] proves the linear combination statement in Equation 25.196 by using contradiction.

Therefore, for a *feasible point* to be a minimizer of $E_0(\mathbf{x})$, the following set of equations should be valid,

$$\mathbf{g}^* - \mathbf{J}_\mathbf{x}^{*(e)}\boldsymbol{\lambda}^{*(e)} = 0 \tag{25.198}$$

$$E_{n_e}(\mathbf{x}^*) = 0 \tag{25.199}$$

Equations 25.198 and 25.199 constitute $N + N_e$ equations with the same number of unknowns ($\mathbf{x}$ and $\boldsymbol{\lambda}^{(e)}$), and may be solved for $\mathbf{x}^*$ and $\boldsymbol{\lambda}^{*(e)}$.

Another way of writing Equations 25.198 is in its integral domain and using Equations 25.199 to arrive at the following definition,

$$\mathscr{L}(\mathbf{x}, \boldsymbol{\lambda}^{(e)}) \triangleq E_0(\mathbf{x}) - \boldsymbol{\lambda}^{(e)T}\mathbf{c}^{(e)}(\mathbf{x}) \tag{25.200}$$

where $\mathscr{L}(\mathbf{x}, \boldsymbol{\lambda}^{(e)})$ is known as the *Lagrangian function* associated with the constrained minimization problem. Being in the integral form, for the optimal point, $\mathbf{x}^*, \boldsymbol{\lambda}^{*(e)}$, the Lagrangian will have to be constant,

$$\mathscr{L}(\mathbf{x}^*, \boldsymbol{\lambda}^{*(e)}) = Const \tag{25.201}$$

Therefore, an equivalent form for Equations 25.198 and 25.199 would be given by setting the gradient of the *Lagrangian* with respect to its independent variables, $\mathbf{x}$ and $\boldsymbol{\lambda}$, equal to zero,

$$\nabla_{\mathbf{x},\boldsymbol{\lambda}^{(e)}}\mathscr{L}(\mathbf{x}, \boldsymbol{\lambda}^{(e)})\Big|_{\mathbf{x}^*,\boldsymbol{\lambda}^{*(e)}} = \mathbf{0} \tag{25.202}$$

In Equation 25.202, the gradient with respect to $\mathbf{x}$ and $\boldsymbol{\lambda}$ is defined as follows,

$$\nabla_{\mathbf{x},\boldsymbol{\lambda}}\mathscr{L} \triangleq \begin{bmatrix} \nabla_\mathbf{x}\mathscr{L} \\ \nabla_\boldsymbol{\lambda}\mathscr{L} \end{bmatrix} \tag{25.203}$$

As we will see later, $\mathbf{x}^*$ and $\boldsymbol{\lambda}^*$ designate a saddle point in the *Lagrangian*.

An intuitive interpretation of *Lagrange multipliers* is as the measure of sensitivity of the objective function to changes in the constraint equations. *Fletcher* [23] shows this by initially assuming to have only *equality constraints* and then perturbing each of the equality constraint functions, $E_{n_e}, n_e \in \mathcal{N}_e$, by a small amount, ε_{n_e}, such that the equality constraints become,

$$E_{n_e}(\mathbf{x}) = \varepsilon_{n_e} \tag{25.204}$$

In vector form, this perturbation may be denoted by,

$$\boldsymbol{\varepsilon} : \mathscr{R}^1 \mapsto \mathscr{R}^{N_e} \quad \text{where} \quad (\boldsymbol{\varepsilon})_{[n_e]} = \varepsilon_{n_e} \tag{25.205}$$

Then, the *Lagrangian* in Equation 25.200 may be rewritten, using the new constraint equations given by Equation 25.204, as follows.

$$\mathscr{L}(\mathbf{x}, \boldsymbol{\lambda}, \boldsymbol{\varepsilon}) = E_0(\mathbf{x}) - \boldsymbol{\lambda}^{(e)^T} \left(\mathbf{c}^{(e)}(\mathbf{x}) - \boldsymbol{\varepsilon} \right) \tag{25.206}$$

$$= E_0(\mathbf{x}) - \boldsymbol{\lambda}^T \left(\mathbf{c}(\mathbf{x}) - \boldsymbol{\varepsilon} \right) \tag{25.207}$$

Obviously, if $\boldsymbol{\varepsilon} = \mathbf{0}$, Equation 25.207 reduces to Equation 25.200. However, with a non-zero value for $\boldsymbol{\varepsilon}$, the values of $\mathbf{x}^*$ and $\boldsymbol{\lambda}^{*(e)}$ would change as functions of $\boldsymbol{\varepsilon}$, namely, the new values would be $\mathbf{x}^*(\boldsymbol{\varepsilon})$ and $\boldsymbol{\lambda}^{*(e)}(\boldsymbol{\varepsilon})$. According to the new constraints, Equation 25.204, the second term in Equation 25.206 would vanish at the new solution evaluated at perturbation $\boldsymbol{\varepsilon}$. The following equality follows,

$$\mathscr{L}(\mathbf{x}^*(\boldsymbol{\varepsilon}), \boldsymbol{\lambda}^{*(e)}(\boldsymbol{\varepsilon}), \boldsymbol{\varepsilon}) = E_0(\mathbf{x}^*(\boldsymbol{\varepsilon})) \tag{25.208}$$

Therefore, the slope of the objective function, $E_0(\mathbf{x})$, is equivalent to the slope of the *Lagrangian*, $\mathscr{L}$, with respect to the perturbation, $\boldsymbol{\varepsilon}$, and may be computed using the chain rule as follows [23],

$$\frac{dE_0}{d\varepsilon_{n_e}} = \frac{d\mathscr{L}}{d\varepsilon_{n_e}}$$

$$= \nabla_{\mathbf{x}}^T \mathscr{L} \frac{\partial \mathbf{x}}{\partial \varepsilon_{n_e}} + \nabla_{\boldsymbol{\lambda}^{(e)}}^T \mathscr{L} \frac{\partial \boldsymbol{\lambda}^{(e)}}{\partial \varepsilon_{n_e}} + \frac{\partial \mathscr{L}}{\partial \varepsilon_{n_e}} \tag{25.209}$$

Using Equation 25.207, this simply reduces to the value of $(\boldsymbol{\lambda})_{[n_e]}$, i.e.,

$$\frac{dE_0}{d\varepsilon_{n_e}} = \frac{d\mathscr{L}}{d\varepsilon_{n_e}} = (\boldsymbol{\lambda})_{[n_e]} \tag{25.210}$$

Equation 25.210 shows that the *Lagrange multipliers* are measures of the rate of change of the value of the objective function with respect to a perturbation in the *equality constraints*.

25.5.1.2 Active Constraints

Recall the definition for the *feasibility region*, Ω, given in Equation 25.158. In that definition, the constraints were also defined as a set of *equality* and *inequality* constraints. Let us examine these constraints for any point, $\mathbf{x} = \boldsymbol{\xi}$. We may define a new set of constraint indices, as a function of $\mathbf{x}$, that only include those constraints for which $\mathbf{x} = \boldsymbol{\xi}$ is located on the boundary of the the feasibility region associated with the said constraint. In other words, the set of *active constraint indices* is defined as a function of $\mathbf{x}$ as follows,

$$\mathscr{N}_a(\mathbf{x}) : \{ n_a = n_c : E_{n_c}(\mathbf{x}) = 0 \} \tag{25.211}$$

Of course, if $\mathbf{x} = \boldsymbol{\xi}$ is any *feasible point*, $\boldsymbol{\xi} \in \Omega$, then all the *equality constraints* would be considered *active constraints*, namely,

$$\mathcal{N}_e \subset \mathcal{N}_a(\boldsymbol{\xi}) \quad \forall \; \boldsymbol{\xi} \in \Omega \tag{25.212}$$

However, only those *inequality constraints* are considered to be part of the set of active constraints, if the feasible point of interest, $\mathbf{x} = \boldsymbol{\xi}$, is on the boundary of the *feasibility region* associated with the said *inequality constraints*, dictating strict equality, namely, $E_{n_a}(\boldsymbol{\xi}) = 0$. In other words,

$$\mathcal{N}_a^{(i)}(\mathbf{x}) \triangleq \mathcal{N}_a(\mathbf{x}) \cap \mathcal{N}_i \tag{25.213}$$
$$= \{n_a = n_i : E_{n_i}(\mathbf{x}) = 0\} \tag{25.214}$$

Given the definition of Equation 25.213 and the statement of Equation 25.212, the set of active constraints for any *feasible point*, $\mathbf{x} = \boldsymbol{\xi}$, is given by the following union,

$$\mathcal{N}_a(\boldsymbol{\xi}) = \mathcal{N}_a^{(i)}(\boldsymbol{\xi}) \cup \mathcal{N}_e \tag{25.215}$$

It is important to note the significance of *active constraints*. These are the only constraints which are important in determining the solution to the constrained optimization problem. In other words, all inactive inequality constraints will be automatically met, if all active constraints are met for a set of feasible points. In Chapter 15 we discuss the geometric interpretation of the *active constraints*, akin to *support vectors* which are those feasible points, $\mathbf{x}$, representing all the *active constraints* for the optimization problem. In fact,

$$\mathcal{S}_{s.v.} \equiv \mathcal{N}_a(\mathbf{x}^*) \tag{25.216}$$

where $\mathbf{x}^*$ is the solution to the constrained minimization problem – see Equation 15.27 for a definition of $\mathcal{S}_{s.v.}$.

In the next section, we will use the concepts defined in this and the previous section for further treatment of inequality constraints. Since the set of *active constraints* is especially important for the solution of the constrained optimization problem, let us define a shorthand notation for this set as follows,

$$\mathcal{N}_a^* \triangleq \mathcal{N}_a(\mathbf{x}^*) \tag{25.217}$$
$$\mathcal{N}_a^{*(i)} \triangleq \mathcal{N}_a^{(i)}(\mathbf{x}^*) \tag{25.218}$$

We may write Equation 25.215 for the minimizer, $\mathbf{x}^*$, using the new shorthand notation,

$$\mathcal{N}_a^* = \mathcal{N}_a^{*(i)} \cup \mathcal{N}_e \tag{25.219}$$

In Equation 25.188, we defined the *set of feasible directions* associated with the *linearized inequality constraints*. Since only the *active inequality constraints* affect the solution, we may define the set of *feasible directions* associated with the *linearized active inequality constraints* as follows,

$$\mathscr{S}_{fl}^{(ai)}(\mathbf{x}) \triangleq \{\mathbf{s} : \mathbf{s}^T \boldsymbol{\gamma}_{n_a}(\mathbf{x}) \geq 0 \ \forall \ n_a \in \mathscr{N}_a^{(i)}\} \tag{25.220}$$

A new set of feasible directions may now be defined which would include all the feasible directions associated with the active linear constraints,

$$\mathscr{S}_{fl}^{(a)}(\mathbf{x}) \triangleq \mathscr{S}_{fl}^{(e)}(\mathbf{x}) \cap \mathscr{S}_{fl}^{(ai)}(\mathbf{x}) \tag{25.221}$$

In the same manner, the *set of feasible directions* associated with the *original active inequality constraints*[3] may be denoted by $\mathscr{S}_f^{(ai)}(\mathbf{x})$, in which case, the set of feasible directions associated with all *active constraints* would be,

$$\mathscr{S}_f^{(a)}(\mathbf{x}) \triangleq \mathscr{S}_f^{(e)}(\mathbf{x}) \cap \mathscr{S}_f^{(ai)}(\mathbf{x}) \tag{25.222}$$

25.5.1.3 Inequality Constraints

Since the active inequality constraints, $\{E_{n_a}(\mathbf{x}) : n_a \in \mathscr{N}_a^{(i)}(\mathbf{x})\}$, are the only important inequality constraints which govern the solution to the optimization problem, we may write Equation 25.184 for $\{n_a \in \mathscr{N}_a^{*(i)}\}$ as follows,

$$\boldsymbol{\gamma}_{n_a}^T \bigg|_{\mathbf{x}^*} \mathbf{s} \geq 0 \ \forall \ n_a \in \mathscr{N}_a^{*(i)} \tag{25.223}$$

where $\mathbf{s}$ is the feasible direction associated with inequality constraint E_{n_a}. For the sake of simplicity, let us define the following compact notation,

$$\boldsymbol{\gamma}_{n_a}^{*(i)} \triangleq \boldsymbol{\gamma}_{n_a} \bigg|_{\mathbf{x}^*} \ : \ n_a \in \mathscr{N}_a^{*(i)} \tag{25.224}$$

Using this notation, Equation 25.223 may be written as,

$$\boldsymbol{\gamma}_{n_a}^{*(i)T} \mathbf{s} \geq 0 \tag{25.225}$$

It can be shown, quite the same way as we did at the beginning of Section 25.5.1.1, that only the minimizers, $\mathbf{x}^*$, which satisfy the equivalent of Equation 25.198 for $n_a \in \mathscr{N}_{n_a}^{*(i)}$ can be solutions to the constrained optimization problem with the said *linearized inequality constraints*. In this case, we need an added non-negativity condition for the *Lagrange multipliers* associated with the active inequality constraints which will be justified at the end of this section. It means that for $\mathbf{x}^*$ to be a minimizer, there should not be any feasible descent direction at $\mathbf{x}^*$. In mathematical notation,

---

[3] The constraints are not linearized.

$$\mathscr{S}_{fl}^{*(ai)} \cap \mathscr{S}_d^* = \{\varnothing\} \tag{25.226}$$

Equation 25.226 is true *if and only if* a separation hyperplane exists that separates $\mathscr{S}_d^*$ from $\mathscr{S}_{fl}^{*(ai)}$ – see Theorem 23.3.

It is quite easy to see that based on *Farkas' Lemma* (Lemma 23.1), for Equation 25.226 to hold, the following would have to be valid for *active linearized inequality constraints*,

$$\mathbf{g}^* - \mathbf{J_x}^{*(ai)} \boldsymbol{\lambda}^{*(ai)} = \sum_{n_a \in \mathscr{N}_a^{*(i)}} \boldsymbol{\gamma}_{n_a}^{*(i)T} (\boldsymbol{\lambda}^*)_{[n_a]} \tag{25.227}$$

$$= \mathbf{0} \tag{25.228}$$

such that,

$$(\boldsymbol{\lambda}^*)_{[n_a]} \geq 0 \ \forall \ n_a \in \mathscr{N}_a^{*(i)} \tag{25.229}$$

The *Farkas Lemma* (Lemma 23.1) has a number of variations and has been proven in many different ways [17, 25, 23, 7, 45]. A simple approach [23] for justifying the above statement is to consider a small perturbation in the *active inequality constraints*, from 0, associated with the point on the boundary of the *feasibility region* (Section 25.5.1.2) to a small value, $\varepsilon_{n_a}, n_a \in \mathscr{N}_a^{*(i)}$ – see Equation 25.204. Since the value of the change would have to be positive in order for the point to be feasible according to the inequality constraint, the derivative of the objective function with respect to $\boldsymbol{\varepsilon}_{n_a}$ would have to be non-negative, namely,

$$\frac{dE_0(\mathbf{x}^*)}{d\varepsilon_{n_a}} \geq 0 \ \forall \ n_a \in \mathscr{N}_a^{*(i)} \tag{25.230}$$

Therefore, according to Equation 25.209, applied to the *active linearized inequality constraints*, Equation 25.229 would have to hold. Note that since the constraint of Equation 25.229 is valid for all *active inequality constraints*, it must hold for all inequality constraints. In other words,

$$(\boldsymbol{\lambda}^*)_{[n_i]} \geq 0 \ \forall \ n_i \in \mathscr{N}_i \tag{25.231}$$

In fact, it is not hard to see that the *Lagrange multipliers* associated with *inactive constraints* should be 0. Also, at the *active constraints*, as defined in Section 25.5.1.2, the constraint value is 0 for the feasible minimizers. Therefore, by combining these two statements, we see that the following statement must hold for all linearized constraints,

$$(\boldsymbol{\lambda}^*)_{[n_c]} E_{n_c}(\mathbf{x}^*) = 0 \tag{25.232}$$

In other words, either the *Lagrange multiplier* or the value of the constraint function, evaluated at the minimizer would always have to be 0. This condition is called the *complementarity condition* [23] and the optimization problem is said to have *strict complementarity* if for every constraint, either the *Lagrange multiplier* or the *constraint* is nonzero. Note that in order to have *strict complementarity*, the following condition must hold,

$$\left(\boldsymbol{\lambda}^{*(ai)} \right)_{[n_a]} > 0 \ \forall \ n_a \in \mathcal{N}_a^{*(i)} \tag{25.233}$$

25.5.1.4 Regularity Assumption

To recap the past three sections, if a point, $\mathbf{x} = \mathbf{x}^*$, is a local minimizer, then the following would have to hold for *linearized constraints*,

$$\mathcal{S}_{fl}^{*(a)} \cap \mathcal{S}_d^* = \{\varnothing\} \tag{25.234}$$

We will examine the conditions under which this will be true.

For now, let us examine a more general form of the above statement. Consider the original set of constraints, *without linearization*. In Section 25.5.1, we defined the set of *feasible directions* as a function of $\mathbf{x}$, by $\mathcal{S}_f^{(a)}(\mathbf{x})$, which may be written for any feasible point, $\mathbf{x} = \boldsymbol{\xi}$.

Lemma 25.1 (No Feasible Descent Diretions at the Minimizer). *If $\mathbf{x} = \mathbf{x}^*$ is a local minimizer of the objective function, then the following must hold [23],*

$$\mathcal{S}_f^{*(a)} \cap \mathcal{S}_d^* = \{\varnothing\} \tag{25.235}$$

Proof.
The proof of Equation 25.235 is quite simple and may be shown by re-examining the Taylor series expansion of the objective function, $E_0(\mathbf{x})$, about the minimizer, $\mathbf{x}^*$, in a similar manner as was done for the constraint equations, in Equation 25.172. Let us assume that the direction $\mathbf{s}$ is, in the limit, a *feasible direction*, namely,

$$\mathbf{s} \in \mathcal{S}_f^{*(a)} \tag{25.236}$$

This means that Equation 25.176 will hold for the sequence, $\{\eta \mathbf{s}\}_1^k$. The Taylor series expansion of the objective function, about $\mathbf{x}^*$, may then be written as follows,

$$E_0(\mathbf{x}_k) = E_0(\mathbf{x}^*) + \eta_k \mathbf{s}_k^T \mathbf{g}^* + o(\eta_k) \tag{25.237}$$

where $o(\eta_k)$ is the *Landau asymptotic notation* [42], referring to terms which asymptotically vanish relative to η_k. If indeed, $\mathbf{x}^*$ is the local minimizer of $E_0(\mathbf{x})$, then

$$E_0(\mathbf{x}_k) \geq E_0(\mathbf{x}^*) \tag{25.238}$$

Therefore,

$$\eta_k s_k^T g^* + o(\eta_k) \geq 0 \qquad\qquad (25.239)$$

According to the definition of the magnitude, η, given in Equation 25.169, η is always positive,

$$\eta_k > 0 \qquad\qquad (25.240)$$

so we may divide both sides of Equation 25.239 by η_k to produce the following inequality,

$$s_k^T g^* + o(1) \geq 0 \qquad\qquad (25.241)$$

which as $k \to \infty$, it would result in having $s \in \mathscr{S}_d^{*\complement}$, proving the statement of Equation 25.235. Also, see Section 23.7 which shows that there would be a separating hyperplane between the direction of descent and the feasibility region.

□

As we saw in the past three sections, we need Equation 25.234 to hold in order to be able to find a local minimizer that would meet the active constraints of the constrained optimization problem. Therefore, we need to make the following assumption, called the *regularity assumption*, in order to be able to write the necessary conditions for x^* to be a local minimizer of the objective function and to satisfy all the constraints. This statement, as we shall see, will be summarized in two sets of necessary conditions in Section 25.5.1.5. It will follow with a sufficient condition in that section – Theorems 25.3, 25.4, and 25.5, respectively.

Regularity Assumption:
We assume that the following equivalence holds, in order to be able to state the theorems of Section 25.5.1.5,

$$\mathscr{S}_f^{*(a)} \cap \mathscr{S}_d^* = \mathscr{S}_{fl}^{(a)} \cap \mathscr{S}_d^* \qquad\qquad (25.242)$$

$$= \{\varnothing\} \qquad\qquad (25.243)$$

Kühn and Tucker [41] make a stronger assumption for the statement of their theorem on necessary conditions. They assume that the two sets, $\mathscr{S}_{fl}^{(a)}(\xi)$ and $\mathscr{S}_f^{(a)}(\xi)$ are actually identical at a feasible point, $x = \xi$. They calls this the *constraint quali fication* [41] at a feasible point, ξ,

$$\mathscr{S}_{fl}^{(a)}(\xi) = \mathscr{S}_f^{(a)}(\xi) \qquad\qquad (25.244)$$

Equation 25.244 amounts to the statement that if there is any *linearly feasible direction*, $s \in \mathscr{S}_{fl}^{(a)}(\xi)$ (in relation to the linearized constraints), then a feasible direction, $s \in \mathscr{S}_{fl}^{(a)}(\xi)$ (in relation to the original constraints) will also exist.

It is true that if the above *constraint qualification* holds, then the *regularity assumption* will also hold. However, the statement is not reciprocal. In other words, it is not necessary for the *constraint qualification* of Equation 25.244 to hold, in order for the *regularity assumption* to be valid. Since the *regularity assumption* is what will be necessary in the statements of the *first order necessary conditions* (Theorem 25.3), it is important to see when the two above assumptions coincide.

The *constraint qualification* (Equation 25.244) may be rewritten in the following form,

$$\mathscr{S}_{fl}^{(a)}(\boldsymbol{\xi}) = \mathscr{S}_{f}^{(a)}(\boldsymbol{\xi}) \iff \mathscr{S}_{fl}^{(a)}(\boldsymbol{\xi}) \subseteq \mathscr{S}_{f}^{(a)}(\boldsymbol{\xi}) \land \mathscr{S}_{fl}^{(a)}(\boldsymbol{\xi}) \supseteq \mathscr{S}_{f}^{(a)}(\boldsymbol{\xi})$$

(25.245)

The second part of Equation 25.245 is true and may be easily proven. Lemma 25.2 presents this statement and its proof. However, the first part is not necessarily true. In fact, *Kühn and Tucker* [41] provide a simple example when it does not hold. However, in most cases, the *constraint qualification* may be assumed [23]. Lemma 25.3 provides *sufficient conditions* under which the *constraint qualification* would hold.

Lemma 25.2 (Feasibility Directions Subset of Linearized Feasibility Directions).

$$\mathscr{S}_{f}^{(a)}(\boldsymbol{\xi}) \subseteq \mathscr{S}_{fl}^{(a)}(\boldsymbol{\xi})$$

(25.246)

Proof.
The proof is quite similar to the statements at the beginning of Section 25.5.1, with the caveat that the *Taylor series expansion* should be made about any *feasible point*, $\mathbf{x} = \boldsymbol{\xi}$. Since at $\mathbf{x}$, direction $\mathbf{s}$ will be a feasible direction, $\mathbf{s} \in \mathscr{S}_{f}^{(a)}(\boldsymbol{\xi})$, we see that in the limit, as the *directional sequence* of $\mathbf{s}_k$ converges, it converges to $\mathbf{s} \in \mathscr{S}_{fl}^{(a)}(\boldsymbol{\xi})$, proving the statement of this Lemma.

□

Lemma 25.3 (Constraint Qualification – Sufficient Conditions). *If any of the following conditions hold, then* $\mathscr{S}_{f}^{(a)}(\boldsymbol{\xi}) = \mathscr{S}_{fl}^{(a)}(\boldsymbol{\xi})$ *for any feasible point,* $\mathbf{x} = \boldsymbol{\xi}$.

1. *All constraints functions,* $E_{n_a}(x)$ *are linear.*
2. *The normal vectors evaluated at the feasible point,* $\mathbf{x} = \boldsymbol{\xi}$ *(columns of* $\mathbf{J}_{\mathbf{x}}^{(a)}(\boldsymbol{\xi})$*) are linearly independent.*

Proof.
Proof of sufficient condition 1 is trivial, given the definition of Equation 25.185. See Lemma 9.2.2 of *Fletcher* [23] for the proof for the second sufficient conditions.

□

Note that *Farkas' lemma* does not handle equality constraints directly. However, it is possible to write every equality constraint in terms of two inequality constraints which would both have to be satisfied. In other words,

$$E_{n_e} = 0 \tag{25.247}$$

may be rewritten as two the following inequality constraints,

$$E_{n_e} \geq 0 \tag{25.248}$$
$$-E_{n_e} \geq 0 \tag{25.249}$$

This extension may then be used in order to prove that Equation 25.234 would be true *if and only if* Equations 25.228 and 25.229 are valid.

25.5.1.5 Necessary and Sufficient Conditions for an Optimizer

At this point, we have the prerequisite information to quickly review three related theorems, providing the *necessary and sufficient conditions* for the existence of the solution of Equation 25.157.

Theorem 25.3 (Kühn-Tucker Necessary Conditions – First Order Necessary Conditions for a local minimizer). [4] *The necessary condition for $\mathbf{x}^*$ to be a local minimizer of Equation 25.157 is that it obeys the regularity assumption (Equation 25.242) and that there exist Lagrange multipliers, $\boldsymbol{\lambda}^*$, that obey the following Kühn-Tucker conditions,*

$$\nabla_{\mathbf{x}} \mathscr{L}(\mathbf{x}, \boldsymbol{\lambda}) \Big|_{\mathbf{x}^*, \boldsymbol{\lambda}^*} = \mathbf{0} \tag{25.250}$$
$$E_{n_e}(\mathbf{x}^*) = 0 \ \forall \ n_e \in \mathcal{N}_e \tag{25.251}$$
$$E_{n_i}(\mathbf{x}^*) \geq 0 \ \forall \ n_i \in \mathcal{N}_i \tag{25.252}$$
$$(\boldsymbol{\lambda}^*)_{[n_i]} \geq 0 \ \forall \ n_i \in \mathcal{N}_i \tag{25.253}$$
$$(\boldsymbol{\lambda}^*)_{[n_c]} E_{n_c}(\mathbf{x}^*) = 0 \ \forall \ n_c \in \mathcal{N}_c \tag{25.254}$$

Proof.
The proof for each of the statements in the theorem was presented gradually, for the different conditions, in the past few sections.

$\square$

In Theorem 25.3, we pointed out the necessary conditions for the existence of the *local minimizer* of Equation 25.157. The local minimizer, also known as the

---

[4] According to *Nocedal and Wright* [50], the conditions derived by *Kühn and Tucker* [41] and published in 1951 were independently derived by *W. Karush* [39] in his Master's thesis in 1939. For this reason some have called these conditions the *Karush-Kühn-Tucker conditions*, or *KKT* in short.

Kühn-Tucker point, is based on a stationary point of the *Lagrangian* according to the *gradient* of the *Lagrangian* being zero, with respect to the state vector $\mathbf{x}$. This is similar to the *Euler-Lagrange* [48] equations in dynamics. Before examining the *sufficient conditions* for $\mathbf{x}^*$ to be the local minimizer, let us define the *Hessian matrix* of the *Lagrangian* (Equation 25.200) with respect to $\mathbf{x}$ as follows,

$$\mathbf{G_x}(\mathbf{x},\boldsymbol{\lambda}) \triangleq \nabla^2_{\mathbf{x}}\mathscr{L}(\mathbf{x},\boldsymbol{\lambda}) \tag{25.255}$$

Note that for a *quadratic* objective function, the *Hessian matrix*, $\mathbf{G_x}(\boldsymbol{\lambda})$, is a function of $\boldsymbol{\lambda}$ only. Let us define $\mathbf{G_x^*}$ as the *Hessian matrix* with respect to $\mathbf{x}$, of the *Lagrangian*, evaluated at the *Kühn-Tucker point*, $\mathbf{x}^*$, and the vector of *Lagrange multipliers* associated with $\mathbf{x}^*$, $\boldsymbol{\lambda}^*$. Then, the following theorem provides the *second order necessary condition* for $\mathbf{x}^*$ to be the local minimizer of Equation 25.157.

Theorem 25.4 (Second Order Necessary Condition for a local minimizer). *The necessary second order condition for the Kühn-Tucker point, $\mathbf{x}^*$, to be the local minimizer of Equation 25.157 is that the Hessian matrix of the Lagrangian (Equation 25.200), evaluated at the Kühn-Tucker point, $\mathbf{x}^*$, be positive semi-definite,*

$$\mathbf{s}^T \mathbf{G}_x^* \mathbf{s} \geq 0 \ \forall \ \mathbf{s} \tag{25.256}$$

Proof.
See [50, 23].   □

Although the positive semi-definiteness of the Hessian matrix is necessary for a point which satisfies the *KT conditions* of Theorem 25.3, it does not guarantee that point $\mathbf{x}^*$ would be a local minimizer. In fact the point may be a *saddle point* with respect to $\mathbf{x}$. To guarantee that such a *Kühn-Tucker point* is a local minimizer, the Hessian matrix evaluated at the *KT point* must be *positive definite*, hence the following theorem on sufficiency.

Theorem 25.5 (Sufficient Condition for a local minimizer). *The sufficient condition for the Kühn-Tucker point, $\mathbf{x}^*$, to be the local minimizer of Equation 25.157 is that the Hessian matrix of the Lagrangian (Equation 25.200), evaluated at the Kühn-Tucker point, $\mathbf{x}^*$, be positive definite,*

$$\mathbf{s}^T \mathbf{G}_x^* \mathbf{s} > 0 \ \forall \ \mathbf{s} \neq \mathbf{0} \tag{25.257}$$

Proof.
See [50, 23].

□

25.5.2 Duality

The concept of duality is used to represent the optimization problem in a simpler form in the, so called, *dual space*. The dual form of the problem is usually simpler to solve, computationally. Generally, there are many different dual transformations which may be used. The original optimization problem is referred to as the *primal problem* and the transformed problem is called the *dual problem*. The dual problem is usually represented using the vector of *Lagrange multipliers*, $\boldsymbol{\lambda}$ as the independent variable, in contrast to $\mathbf{x}$ which is the independent variable of the *primal problem*.

Depending on the primal problem of interest, there are many different dual forms: *Fenchel's (Conjugate) duality* [68, 18, 67] (using the *Legendre-Fenchel transform* [67]), *geometric duality* [16], *inference duality* [34], *Dorn's duality* [15], *Lagrangian duality* [1], *LP duality* [25], *Wolfe's duality* [69], etc. Here, we are concerned with the *Wolfe duality* [69] which is built on the premise that the primal problem is *convex* – see Section 24.1.4.

In *Wolfe duality*, if the primal problem is not convex (even if it is pseudo-convex), then generally, the dual problem may not have any solution [45, 23]. As a rule, if the primal problem is one of minimization, the dual problem becomes a maximization problem. The *Wolfe dual* problem [69] is represented in terms of the *Lagrangian function*, using the vector of *Lagrange multipliers*, $\boldsymbol{\lambda}$, as the independent variable of the dual problem. The *Wolfe dual* is built upon the first order necessary conditions stated in Theorem 25.3, where the constraints are translated to a maximization condition on the *Lagrangian function*.

The following is the statement of the *Wolfe duality theorem* [69, 23]:

Theorem 25.6 (Wolfe Duality Theorem). *Let $\mathbf{x}^*$ be the solution to the following convex primal optimization problem,*

$$\mathbf{x}^* - \arg\min_{\mathbf{x} \in \Omega} E_0(\mathbf{x}) \tag{25.258}$$

where

$$\Omega = \{\mathbf{x} : (E_{n_i}(\mathbf{x}) \geq 0 \ \forall \ n_i \in \mathcal{N}_{n_i})\} \tag{25.259}$$

Assuming that $E_0(\mathbf{x})$ and $E_{n_i}(\mathbf{x}), n_i \in \mathcal{N}_{n_i}$ are $\mathfrak{C}^1$ continuous (see Definition 24.19) and that the regularity assumption (Equation 25.242) holds, the following Wolfe dual problem is associated with the above primal problem,

$$\mathbf{x}^*, \boldsymbol{\lambda}^* = \arg\max_{\mathbf{x}, \boldsymbol{\lambda} \in \bar{\Omega}} \mathscr{L}(\mathbf{x}, \boldsymbol{\lambda}) \tag{25.260}$$

where

$$\bar{\Omega} = \{\mathbf{x}, \boldsymbol{\lambda} : (\nabla_{\mathbf{x}}\mathscr{L}(\mathbf{x}, \boldsymbol{\lambda}) = \mathbf{0} \ \wedge \ \boldsymbol{\lambda} \succeq \mathbf{0})\} \tag{25.261}$$

and the minimum value of the primal objective function is equal to the maximum value of the dual objective function, namely,

$$E_0(\mathbf{x}^*) = \mathscr{L}(\mathbf{x}^*, \boldsymbol{\lambda}^*) \tag{25.262}$$

Proof. Note that the two conditions in Equation 25.261 follow from the first order Kühn-Tucker necessary conditions given by Equations 25.250 and 25.253. Equation 25.250 is also known as the *dual feasibility* condition. Also, according to another one of the first order Kühn-Tucker necessary conditions, Equation 25.254, the equality of Equation 25.262 follows. In addition, the constraint functions of the primal problem are included in the construction of the Lagrangian function.

Also, since $\boldsymbol{\lambda} \succeq \mathbf{0}$ and $\mathbf{c}(\mathbf{x}) \succeq \mathbf{0}$, then,

$$
\begin{aligned}
\mathscr{L}(\mathbf{x}^*, \boldsymbol{\lambda}^*) &= E_0(\mathbf{x}^*) \\
&\geq E_0(\mathbf{x}^*) - \boldsymbol{\lambda}^T \mathbf{c}(\mathbf{x}^*) \\
&= \mathscr{L}(\mathbf{x}^*, \boldsymbol{\lambda})
\end{aligned}
\tag{25.263}
$$

If we expand the right hand side of Equation 25.263 about $\mathbf{x}$, using a Taylor series expansion, we will have the following,

$$
\begin{aligned}
\mathscr{L}(\mathbf{x}^*, \boldsymbol{\lambda}) &= \mathscr{L}(\mathbf{x}, \boldsymbol{\lambda}) + (\mathbf{x}^* - \mathbf{x})^T \nabla_{\mathbf{x}}\mathscr{L}(\mathbf{x}, \boldsymbol{\lambda}) \\
&\quad + \frac{1}{2}(\mathbf{x}^* - \mathbf{x})^T \nabla_{\mathbf{x}}^2 \mathscr{L}\Big|_{\mathbf{x}, \boldsymbol{\lambda}} (\mathbf{x}^* - \mathbf{x}) + \mathscr{O}(\Delta \mathbf{x}^3)
\end{aligned}
\tag{25.264}
$$

$$
\geq \mathscr{L}(\mathbf{x}, \boldsymbol{\lambda}) + (\mathbf{x}^* - \mathbf{x})^T \nabla_{\mathbf{x}}\mathscr{L}(\mathbf{x}, \boldsymbol{\lambda}) \tag{25.265}
$$

The above inequality stems from the second order necessary conditions, stating that $\nabla_{\mathbf{x}}^2 \mathscr{L}$ is positive semi-definite (convexity of the Lagrangian function with respect to $\mathbf{x}$).

Also note that due to the *dual feasibility* (Equation 25.250),

$$\mathscr{L}(\mathbf{x}, \boldsymbol{\lambda}) + (\mathbf{x}^* - \mathbf{x})^T \nabla_{\mathbf{x}}\mathscr{L}(\mathbf{x}, \boldsymbol{\lambda}) = \mathscr{L}(\mathbf{x}, \boldsymbol{\lambda}) \tag{25.266}$$

Combining Equations 25.263, 25.265 and 25.266, we will have the following,

$$\mathscr{L}(\mathbf{x}^*, \boldsymbol{\lambda}^*) \geq \mathscr{L}(\mathbf{x}, \boldsymbol{\lambda}) \tag{25.267}$$

Equation 25.267 says that $\mathbf{x}^*$ and $\boldsymbol{\lambda}^*$ are the maximizers of the *Wolfe dual problem* (Equation 25.260).

□

Note that Theorem 25.6 does not make any direct provision for handling *equality constraints*. However, it is very simple to include such constraints in the form of the intersection of two inequality constraints. For example, take the following equality constraint,

$$E_{n_e} = 0 \tag{25.268}$$

It may be rewritten as the following two inequality constraints,

$$E_{n_e} \geq 0 \tag{25.269}$$

$$-E_{n_e} \geq 0 \tag{25.270}$$

whose intersection would produce the same result as having the original equality constraint of Equation 25.268, since both constraints of Equations 25.269 and 25.270 would have to hold at the same time.

An interpretation of the maximization of the dual problem is the fact that the optimizers $\mathbf{x}^*$ and $\boldsymbol{\lambda}^*$ actually define a saddle point where the primal variables, $\mathbf{x}^*$, minimize $E_0(\mathbf{x})$. The value of $\mathbf{x}^*$ is then used to find the *dual variables*, $\boldsymbol{\lambda}^*$, which maximize the *Lagrangian* for $\mathbf{x}^*$. The pair, $\{\mathbf{x}^*, \boldsymbol{\lambda}^*\}$, defines a saddle point.

25.5.2.1 Quadratic Objective Function with Linear Inequality Constraints

In most of this chapter, we have approximated the objective function with a quadratic function in a local area. In addition, some of the practical constrained optimization problems in speaker recognition, such as those of support vector machines, work with simple quadratic objective functions. Let us examine a generic quadratic function of the form,

$$\hat{E}_0(\mathbf{x}) = \frac{1}{2}\mathbf{x}^T\mathbf{G}\mathbf{x} + \mathbf{g}^T\mathbf{x} + C \tag{25.271}$$

where, C is a constant term. Also, since we are interested in convex problems, we assume that $\mathbf{G}$ is positive definite,

$$\mathbf{x}^T\mathbf{G}\mathbf{x} > 0 \ \forall \ \mathbf{x} \tag{25.272}$$

For the purpose of minimization, we can drop the constant term, C, which is only an offset. Therefore, we can define,

$$E_0(\mathbf{x}) \triangleq \hat{E}_0(\mathbf{x}) - C \tag{25.273}$$

producing the following objective function,

$$E_0(\mathbf{x}) = \frac{1}{2}\mathbf{x}^T\mathbf{G}\mathbf{x} + \mathbf{g}^T\mathbf{x} \tag{25.274}$$

Furthermore, for practical purposes, let us assume that the minimization problem is subject to *linear inequality constraints*.

Therefore, for a quadratic objective function, the primal minimization problem of Equation 25.258 may be written as follows,

$$\mathbf{x}^* = \arg\min_{\mathbf{x} \in \Omega} E_0(\mathbf{x}) \tag{25.275}$$

$$= \arg\min_{\mathbf{x} \in \Omega} \frac{1}{2}\mathbf{x}^T \mathbf{G}\mathbf{x} + \mathbf{g}^T \mathbf{x} \tag{25.276}$$

where

$$\Omega = \{\mathbf{x} : (E_{n_i}(\mathbf{x}) \geq 0 \ \forall \ n_i \in \mathcal{N}_{n_i})\} \tag{25.277}$$

$$= \{\mathbf{x} : (\mathbf{J}_{\mathbf{x}}^T \mathbf{x} - \mathbf{b} \succeq \mathbf{0})\} \tag{25.278}$$

where $\mathbf{J}_{\mathbf{x}}$ is the *Jacobian matrix* associated with the inequality constraints and is defined by Equation 25.164.

The *Lagrangian function* for the above primal problem will then be,

$$\mathcal{L}(\mathbf{x}, \boldsymbol{\lambda}) = E_0(\mathbf{x}) - \boldsymbol{\lambda}^T \mathbf{c}(\mathbf{x}) \tag{25.279}$$

$$= \frac{1}{2}\mathbf{x}^T \mathbf{G}\mathbf{x} + \mathbf{g}^T \mathbf{x} - \boldsymbol{\lambda}^T \left(\mathbf{J}_{\mathbf{x}}^T \mathbf{x} - \mathbf{b}\right) \tag{25.280}$$

which leads to the following *Wolfe dual* problem, according to Theorem 25.6.

$$\mathbf{x}^*, \boldsymbol{\lambda}^* = \arg\max_{\mathbf{x}, \boldsymbol{\lambda} \in \bar{\Omega}} \mathcal{L}(\mathbf{x}, \boldsymbol{\lambda}) \tag{25.281}$$

$$= \arg\max_{\mathbf{x}, \boldsymbol{\lambda} \in \bar{\Omega}} \frac{1}{2}\mathbf{x}^T \mathbf{G}\mathbf{x} + \mathbf{g}^T \mathbf{x} - \boldsymbol{\lambda}^T \left(\mathbf{J}_{\mathbf{x}}^T \mathbf{x} - \mathbf{b}\right) \tag{25.282}$$

where

$$\bar{\Omega} = \{\mathbf{x}, \boldsymbol{\lambda} : (\nabla_{\mathbf{x}} \mathcal{L}(\mathbf{x}, \boldsymbol{\lambda}) = \mathbf{0} \ \wedge \ \boldsymbol{\lambda} \succeq \mathbf{0})\} \tag{25.283}$$

$$= \{\mathbf{x}, \boldsymbol{\lambda} : (\mathbf{G}\mathbf{x} + \mathbf{g} - \mathbf{J}_{\mathbf{x}}\boldsymbol{\lambda} = \mathbf{0} \ \wedge \ \boldsymbol{\lambda} \succeq \mathbf{0})\} \tag{25.284}$$

The *dual feasibility* relation of Equation 25.284 (the first expression) may be used to solve for $\mathbf{x}$ in terms of the *Lagrange multipliers*, $\boldsymbol{\lambda}$,

$$\mathbf{x} = \mathbf{G}^{-1}(\mathbf{J}_{\mathbf{x}}\boldsymbol{\lambda} - \mathbf{g}) \tag{25.285}$$

The expression of Equation 25.285 may be plugged into Equation 25.282 such that the dual problem is written solely in terms of $\boldsymbol{\lambda}$. Also, since the *dual feasibility* expression has already been used to solve for $\mathbf{x}$, the constraints become much simpler and reduce to the non-negativity constraint on $\boldsymbol{\lambda}$. The Lagrangian may be written in terms of the *Lagrange multiplier* by plugging in for $\mathbf{x}$,

$$\mathcal{L}(\boldsymbol{\lambda}) = \frac{1}{2}(\mathbf{J_x}\boldsymbol{\lambda} - \mathbf{g})^T \mathbf{G}^{-T} \mathbf{G} \mathbf{G}^{-1}(\mathbf{J_x}\boldsymbol{\lambda} - \mathbf{g})$$
$$+ \mathbf{g}^T \mathbf{G}^{-1}(\mathbf{J_x}\boldsymbol{\lambda} - \mathbf{g})$$
$$- \boldsymbol{\lambda}^T \mathbf{J_x}^T \mathbf{G}^{-1}(\mathbf{J_x}\boldsymbol{\lambda} - \mathbf{g})$$
$$+ \boldsymbol{\lambda}^T \mathbf{b} \qquad (25.286)$$

$\mathbf{G}$ is assumed to be positive definite and it is symmetric. Also, we may drop the subscript $\mathbf{x}$ from $\mathbf{J_x}$ for simplicity of notation. Then Equation 25.286 would simplify into,

$$\mathcal{L}(\boldsymbol{\lambda}) = \frac{1}{2}(\mathbf{J}\boldsymbol{\lambda} - \mathbf{g})^T \mathbf{G}^{-1}(\mathbf{J}\boldsymbol{\lambda} - \mathbf{g})$$
$$+ \mathbf{g}^T \mathbf{G}^{-1}(\mathbf{J}\boldsymbol{\lambda} - \mathbf{g})$$
$$- \boldsymbol{\lambda}^T \mathbf{J}^T \mathbf{G}^{-1} \mathbf{J}\boldsymbol{\lambda} + \boldsymbol{\lambda}^T \mathbf{J}^T \mathbf{G}^{-1} \mathbf{g} + \boldsymbol{\lambda}^T \mathbf{b} \qquad (25.287)$$

Multiplying out the compound terms in Equation 25.287 and regrouping in terms of degrees of $\boldsymbol{\lambda}$, we may write the simplified *Wolfe dual* problem as the maximization of a quadratic function of $\boldsymbol{\lambda}$,

$$\boldsymbol{\lambda}^* = \underset{\boldsymbol{\lambda} \succeq \mathbf{0}}{\arg\max} \left[-\frac{1}{2}\boldsymbol{\lambda}^T (\mathbf{J}^T \mathbf{G}^{-1} \mathbf{J})\boldsymbol{\lambda} + (\mathbf{g}^T \mathbf{G}^{-1} \mathbf{J} + \mathbf{b}^T)\boldsymbol{\lambda} - \frac{1}{2}\mathbf{g}^T \mathbf{G}^{-1} \mathbf{g} \right] \qquad (25.288)$$

The maximization problem of Equation 25.288 may easily be solved using the many techniques discussed in this chapter. The maximizer of the *Wolfe dual problem*, $\boldsymbol{\lambda}^*$, may then be used to compute the minimizer of the *primal problem*, $\mathbf{x}^*$, by using Equation 25.285 as follows,

$$\mathbf{x}^* = \mathbf{G}^{-1}(\mathbf{J}\boldsymbol{\lambda}^* - \mathbf{g}) \qquad (25.289)$$

Note that since the constraints of the *primal problem* were chosen to be linear, the *sufficient conditions* for the *constraint qualification* according to Lemma 25.3 are automatically met.

For more on *constrained optimization*, see [23], [41], [69], [50], and [54].

25.6 Global Convergence

The optimization techniques, discussed in this chapter, made no distinction between the types of minimizers they sought. They may end up with any local minimum of the objective function. The true objective of most optimization problems, however, is to find the global optimum. Although there are no true systematic methods for reaching a global minimum, there are some algorithms that help in pursuing that goal.

Kirkpatrick, et al. [40] borrowed ideas from metallurgy, used for optimal crystal growth in steel production. In metallurgy, there is a technique for achieving a more uniform crystalline structure while producing steel. In this process, when the molten steel is cooled, the cooling is done slowly and, once in a while, the material is heated again just slightly, to avoid being stuck in sub-optimal crystalline forms. The occasional heating reshapes the crystal and allows it to cool back down into possibly a more stable form (a form with a lower potential energy). The objective is to create a crystal that has the lowest potential energy, hence is most stable, i.e., stronger.

Simulated annealing [40] uses the same idea and usually works well with slowly converging optimization such as steepest descent. In this technique, once the rate of change of the objective function becomes smaller than a certain threshold, there is a random perturbation added to the state vector which generally increases the value of the objective function. This new state may make it possible for the objective function to take a brand new path toward minimization by possibly taking the system out of a small ditch. This process has proven quite effective for systems with large number of variables. Since it is a statistical technique, it is hard to quantify its effectiveness in general.

Sometimes, depending on the problem at hand, we may have some intrinsic information about the global minimizer or the objective function at the global minimum. In most practical problems this is the case and the information may be utilized to either reach the global minimum or at least to help us know when we have done so. One example of such information is when we are faced with a strictly convex function. In that case, we know that there is only one minimizer and that if a minimizer is reached, it must be the global minimizer. Of course, life is not always that simple. With more complex cases, there may be other information that may be used. Sometimes, we may even be able to modify the objective function to aid us in reaching the global minimizer more readily.

One such example is given by [5] which uses the architecture of neural networks for global convergence. In essence [5] introduces a method that changes the optimization problem to facilitate global convergence. See Section 14.2.5 for more information.

References

1. Arrow, K.J., Hurwicz, L.: Reduction of Constrained Maxima to Saddle-Point Problems. In: Proceedings of the Third Berkeley Symposium on Mathematics, Statistics and Probability, vol. 5, pp. 1–20 (1956). Unoversity of California Press
2. Bachmann, P.: Analytische Zahlentheorie, Bd. 2: Die Analytische Zahlentheorie, 3rd edn. Teubner, Leipzig, Germany (1974)

3. Barnes, J.: An Algorithm for Solving Nonlinear Equations Based on the Secant Method. Computer Journal **8**, 66–72 (1965)
4. Beckman, F.S.: The Solution of Linear Equations by the Conjugate Gradient Method. In: A. Ralston, H. Wolf (eds.) Mathematical Methods for Digital Computers. Wiley, New York (1960)
5. Beigi, H.: Neural Network Learning Through Optimally Conditioned Quadratically Convergent Methods Requiring NO LINE SEARCH. In: IEEE-36th Midwest Symposium on Circuits and Systems, vol. 1, pp. 109–112 (1993)
6. Box, M.J.: A Comparison of Several Current Optimization Method. The Computer Journal **9**, 67–77 (1966)
7. Boyd, S., Vandenberghe, L.: Convex Optimization, 7th printing edn. Cambridge University Press (2009). First Published in 2004. ISBN: 978-0-521-83378-3
8. Brezinski, C.: The Life and Word of André Cholesky. Numer Algor **43**, 279–288 (2006)
9. Broyden, C.: A Class of methods for Solving Nonlinear Simultaneous Equations. Mathematics of Computation **19**, 577–593 (1965)
10. Broyden, C.: Quasi-Newton Methods and Their Application to Function Minimization. Mathematics of Computation **21**, 368–381 (1967)
11. Broyden, C.: The Convergence of a Class of Double Rank Minimization Algorithms, Parts I and II. Journal of the Institute of Mathemats and its Applications **6**, 76–90, 222–231 (1970)
12. Davidon, W.: Variable Metric Method for Minimization. Tech. rep., AEC Research and Development Report, ANL-5990 (1959)
13. Davidon, W.: Variance Algorithms for Minimization. Computer Journal **10**, 406–410 (1968)
14. Davidon, W.C.: Optimally Conditioned Optimization Algorithms Without Line Searches. Mathematical Programming **9**, 1–30 (1975)
15. Dorn, W.: Duality in Quadratic Programming. Quarterly Journal of Applied Mathematics **18**(2), 155–162 (1960)
16. Duffin, R., Peterson, E.: Duality Theory for Geometric Programming. SIAM Journal on Applied Mathematics **14**(6), 1307–1349 (1966)
17. Farkas, J.: Theorie der Einfachen Ungleichungen (Theory of the Simple Inequality). Journal für die Reine und Angewandte Mathematic **124**, 1–27 (1902)
18. Fenchel, W.: Convex Cones, Sets and Functions (1951). Lecture Notes by D. W. Blackett, Sep. 1953
19. Fiacco, A.V., McCormick, G.P.: Nonlinear Programming: Sequential Unconstrained Minimization Techniques. J. Wiley and Sons, New York (1968)
20. Flannery, M.: The Enigma of Nonholonomic Constraints. American Journal of Physics **73**(3), 265–272 (1976)
21. Fletcher, R.: Function Minimization Without Evaluating Derivatives. The Computer Journal **8**, 33–41 (1965)
22. Fletcher, R.: A New Approach to Variable Metric Algorithms. Computer Journal **8**, 317–322 (1970)
23. Fletcher, R.: Practical Methods of Optimization, 2nd edn. J. Wiley and Sons, New york (2000). ISBN: 0-471-49463-1
24. Fletcher, R., Reeves, C.: Function Minimization by Conjugate Gradients. The Computer Journal **7**, 149–154 (1964)
25. Gale, D., Kühn, H.H., Tucker, A.W.: Linear Programming and the Theory of Games. In: e.a. Tjalling C. Koopmans (ed.) Activity Analysis of Production and ALllocation, pp. 317–329 (1951). URL http://cowles.econ.yale.edu/P/cm/m13/
26. Goldfarb, D.: A Family of Variable Metric Methods Derived by Variational Means. Mathematics of Computation **24**, 22–26 (1970)
27. Goldfeld, S., Quandt, R., Trotter, H.: Maximization by Quadratic Hill Climbing. Econometrics **34**, 541–551 (1966)
28. Greenstadt, J.: On the Relative Effectiveness of Gradient Methods. Mathematics of Computation **21**, 360–367 (1967)
29. Greenstadt, J.: Variations on Variable-Metric Methods. Mathematics of Computation **24**, 1–22 (1970)

30. Guta, B.: An Oscillation Theory of Handwriting. Universität Kaiserslautern (2003). PhD Thesis
31. Hestenes, M.R., Steifel, E.: Methods of Conjugate Gradients for Solving Linear Systems. J. Res. N. B. S. **49**, 409 (1952)
32. Himmelblau, D.M.: Applied Nonlinear Programming. McGraw-Hill Book Company, New York (1972)
33. Hooke, R., Jeeves, T.: Direct Search Solution of Numerical and Statistical Problems. Journal of Association of Computer Machinery (ACM) **8**(2), 212–229 (1961)
34. Hooker, J.: Inference Duality as a Basis for Sensitivity Analysis. Constraints: An International Journal **4**, 101–112 (1999)
35. Hoshino, S.: A Formulation of Variable Metric Methods. Journal of the Institue of Mathemtics and its Applicatiions **10**, 394–403 (1972)
36. Householder, A.S.: The Theory of Matrices in Numerical Analysis. Blaisdell Publishing Company, Massachussetts (1964)
37. Huang, H.Y.: Unified Approach to Quadratically Convergent Algorithms for Function minimization. Journal of Optimization Theory and Applications **5**(6), 405–423 (1970)
38. Jacobson, D.H., Oksman, W.: An Algorithm that Minimizes Homogeneous Functions of n Variables in n+2 Iterations and Rapidly Minimizes General Functions. Tech. rep., Technical Report 618, Division of Engineering and Applied Physics, Harvard, University, Cambridge, MA (1970)
39. Karush, W.: Minima of Functions of Several Variables with Inequalities as Side Constraints. University of Chicago, Illinois (1939). Masters Thesis
40. Kirkpatrick, S., Gelatt, C.D., Vecchi, M.P.: Optimization by Simulated Annealing. Science **220**(4598), 671–680 (1983)
41. Kühn, H.W., Tucker, A.W.: Nonlinear Programming. In: j. Neyman (ed.) Proceedings of the Second Berkeley Symposium on mathematical Statistics and Probability, pp. 481–492 (1951)
42. Landau, E.G.H.: Handbuch der Lehre von der Verteilung der Primzahlen, 3rd edn. Chelsea Publishing Company, New York (1974). Both volumes in one. First Edition was published in two volumes, in 1909. ISBN: 0-8284-0096-2
43. Lee, D., Baek, S., Sung, K.: Modified k-Means Algorithm for Vector Quantizer Design. IEEE Signal Processing Letters **4**(1), 2–4 (1997)
44. Levenberg, K.L.: A Method for the Solution of Certain Non-Linear Problems in Least Squares. Quarterly of Applied mathematics **2**, 164–168 (1944)
45. Mangasarian, O.L.: Nonlinear Programming. Society for Industrial and Applied Mathematics (SIAM) (1994). Originally Published: New York: McGraw Hill, 1969; ISBN: 0-89871-341-2
46. Marquardt, D.W.: An Algorithm of Least Squares Estimation of Nonlinear Parameters. SIAM journal on Applied Mathematics **11**, 431–441 (1963)
47. McCormick, G.P., Ritter, K.: Methods of Conjugate Directions Versus Quasi-Newton Methods. Mathematical programming **3**, 101–116 (1972)
48. Meirovitch, L.: Methods of Analytical Dynamics. Advanced Engineering Series. McGraw-Hill, New York (1970). ISBN: 0-704-1455-6
49. Murtagh, B., Sargent, E.: A Constrained Minimization Method with Quadratic Convergence. In: R. Fletcher (ed.) Optimization. Academic Press, London (1969)
50. Nocedal, J., Wright, S.J.: Numerical Optimization, 2nd edn. Springer, New York (2000)
51. Oren, S.S.: On the Selection of Parameteres in Self Scaling Variable Metric Algorithms. Mathematical Programming **7**(1), 351–367 (1974)
52. Oren, S.S., Spedicato, E.: Optimal Conditioning of Self-Scaling Variable Metric Algorithms. Mathematical Programming **10**, 70–90 (1976)
53. Pearson, J.: Variable Metric Methods of Minimization. Computer Journal **12**, 171–178 (1969)
54. Pierre, D.A.: Optimization Theory with Applications. Dover Publications, Inc., New york (1986). ISBN: 0-486-65205-X
55. Powell, M.: Some Global Convergence Properties of a Variable Metric Algorithm for Minimization without Exact Line Searches. SIAM-AMS Proceedings **IX** (1976)
56. Powell, M.J.D.: An Efficient Method for Finding the Minimum of a Function of Several Variables Without Calculating Derivatives. The Computer Journal **7**, 155–162 (1964)

57. Rosenbrock, H.H.: Automatic Method for Finding Greatest or Latest Value of Function. The Computer Journal **3**(3), 175–184 (1960)
58. Rumelhart, D.E., Hinton, G.E., Williams, R.J.: Learning Internal Representations by Error Propagation. In: D.E. Rumelhart, J.L. McClelland (eds.) Parallel Distributed Processing: Explorations in the Microstructure of Cognition, vol. 1, pp. 675–695. MIT Press (1986). ISBN: 0-262-63112-1
59. Schnabel, R.B.: Analyzing and Improving Quasi-Newton Methods for Unconstrained Optimization. Cornell University, New York (1977). Ph.D. Thesis
60. Shah, B.V., Buehler, R.J., Kempthorne, O.: Some Algorithms for Minimizing Functions of Several Variables. SIAM journal on Applied Mathematics **12**, 74–92 (1964)
61. Shanno, D.: Conditioning of Quasi-Newton Method for Function Minimization. Mathematics of Computation **24**, 647–656 (1970)
62. Shanno, D.F., Phua, K.H.: Numerical Comparison of Several Variable Metric Algorithms. Tech. rep., MIS Technical Report 21, University of Arizona (1977)
63. Shanno, D.F., Phua, K.H.: Matrix Conditioning and Nonlinear Optimization. Mathematical Programming **14**, 149–160 (1978)
64. Smith, C.S.: The Automatic Computation of Maximum likelihood Estimates. Tech. rep., N. C. B. Scientific Department Report S.C. 846/MR/40 (1962)
65. Spedicato, E.: Computational Experience with Quasi-Newton Algorithm for Minimization Problems of Moderately Large Size. Tech. rep., Report CISE-N-175, CISE Documentation Service, Segrate, Milano (1975)
66. Swann, W.H.: Report on the Development of a New Direct Search Method of Optimization. Tech. rep., Imperial Chemical Industries, Ltd. Central Instrumentation Laboratory Research Note 6413 (1964)
67. Touchette, H.: Legendre-Fenchel Trfansforms in a Nutshell. Web (2007). URL http://www.maths.qmw.ac.uk/~ht/archive/lfth2.pdf
68. Veinott, A.F.: Conjugate Duality for Convex Programs: A Geometric Development. Linear Algebra and its Applications **114–115**, 663–667 (1989)
69. Wolfe, P.: A Duality Theorem for Nonlinear Programming. Quarterly Journal of Applied Mathematics **19**(3), 239–244 (1961)
70. Wolfe, P.: Convergence Conditions for Ascent Methods. SIAM Review **11**(2), 226–235 (1969)
71. Wood, C.F.: Application of Direct Search to the Solution of Engineering Problems. Tech. rep., Westinghouse Reseach Laboratory Scientific Paper 6-41210-1-p1 (1960)
72. Zangwill, W.: Minimizing a Function Without Calculating Derivatives. The Computer Journal **10**, 293–296 (1967)
73. Zeger, K., Vaisey, K., Gersho, A.: Globally optimal vector quantizer design by stochastic relaxation. IEEE Transactions on Signal Processing **40**(2), 310–322 (1992)
74. Zoutendijk, G.: Methods of Feasible Directions. American Elsevier Publishing Company, New York (1960)

Chapter 26
Standards

In this chapter, we will be reviewing some of the standards and developments of standards, related to speaker recognition. As we will see, there are a few standards bodies which have made certain efforts in developing standard transmission, storage and control schemes for speech-related applications. Most of these standards are not directly developed for speaker recognition. A good portion of them are developments for the telephone communication industry. More recently standards have been developed to be used with audio transmission over the Internet and other telecommunication networks.

Speaker recognition is in essence lucky to have these relevant standards at its disposal. The same may not be said for most other biometrics. Indeed this is what we have seen thorough the coverage of different aspects of speaker recognition. It has the advantage of sharing resources with the telephone communication industry and is well suited for this vast existing infrastructure.

However, as we will see, speaker recognition suffers from the lack of attention in the control standards. It will be made clear that there are several efforts in standardizing such controls for speech recognition and text to speech as well as those for voice over IP and other multimedia transmission. Some early efforts in such standardization have ceased and not many new ones are on the way. Another problem is the difference between speech and other biometric media such as fingerprint, iris, retina, etc. Speech is a sequential time series style data. However, most other biometrics, such as the ones note here, deal with single instances of patterns. This makes the controls inherently different. We will discuss these problems in more detail in the future sections.

In the next section, we will deal with the standardization of audio formats. In the following main section, we will review some encapsulation techniques which encompass these formats. Some of these encapsulations are designed for transmission, while others are more suitable for simple storage of audio.

26.1 Standard Audio Formats

Standardization of audio formats is of utmost importance to the transmission and handling of speech. There have been hundreds of different *coder/decoder* (*codec*) definitions which have been utilized in different speech and other audio applications since the inception of digital audio processing. As we will see, by examining some of these *codecs*, there is no ideal algorithm. Each *codec* has been developed to deal with specific requirements. The ITU-T (*ITU Telecommunication Standardization Sector*) is a permanent organ of the *International Telecommunication Union* (*ITU*) which has been instrumental in developing and maintaining some of the most popular standards. The *ITU-T*, often publishes its recommendations and sometimes makes source code available for the implementation of these standards.

In addition to the ITU-T recommendations, there are many different memoranda published by the *Internet Engineering Task Force* (*IETF*). Each of these publications is in the form of a *Request for Comments* (*RFC*). *RFCs* have unique numbers and maintain the incremental recommendations of the *IETF* for standards relating to the Internet and telecommunication industry. In relation to speech, in most cases, the ITU-T recommendations are used within the different *RFCs*. Some of the *RFCs* will eventually be published as Internet standards by the *IETF*. Others may either be abandoned, incorporated in, or superseded by other Internet standards.

In the following few sections, we will review some of the most relevant audio standards, in relation with the applications of speaker recognition. It is up to the vendor and implementer to pick the proper standard format to maintain, transmit and store data in its raw form. In later sections we will review some of the limited efforts in standardizing other speaker recognition interactions.

26.1.1 Linear PCM (Uniform PCM)

The *linear pulse code modulation* (*linear PCM* or *LPCM*) format is the simplest audio format. It is basically a dump of the audio samples resulting from the *ideal sampling process* discussed in Section 3.6.1 and implemented in the *pulse amplitude modulation* manner discussed in Section 3.6. The *linear PCM* format uses the amplitude quantization technique discussed in Section 3.2. The *linear PCM* is discussed (as *uniform PCM*) in [13] in order to act as a transition format for converting between the μ-law and *a*-law formats which are the main recommendations of the ITU-T G.711 document [13].

The main parameters that need to be specified for the *linear PCM* format are the *sampling frequency* and the number of bits of amplitude quantization. The *linear PCM* or *uniform PCM* format only specifies the computation of the stream of samples, but it does not include any specification for the *quantization level* and *sampling frequency* information. Therefore, it is always necessary to use the raw *linear PCM* data inside an encapsulation format such as *WAV* (Section 26.2.1) or *SAFE* (Section 26.2.3).

26.1.2 μ-Law PCM (PCMU)

The most popular *landline telephony audio* format in the United States is the μ-*law PCM* (PCMU) format which was introduced by the *International Telecommunication Union (ITU-T)* recommendation, ITU-T G.711 [13] in 1972 at Geneva, Switzerland. It is one of the two different logarithmic-style encoding techniques recommended by this document for the amplitudes of the PCM data to be able to adopt an 8-bit representation of the audio signal used in international telephone communication.

The original ITU-T G.711 document [13] also prescribes the standard sampling rate of $8000Hz$ with a tolerance of ±50 *parts per million (ppm)*. This translates to a transmission rate of 64-kbps. In an extension of the recommendation, presented in the ITU-T G.711.1 document [30], higher transmission rates of 80 and 96 kbps were added to the recommendation for allowing a better audio quality.

26.1.3 A-Law (PCMA)

A-law PCM is generally similar to μ-law, with the exception of the logarithmic style look-up table which maps the $13 - 14$ bits per sample to an 8 bit sample. It is important to note that the tables have been adjusted to be slightly suboptimal in order to minimize the $\mu - A - \mu$ and $A - \mu - A$ conversion such that the signal is more or less similar to the original signal before conversion. Of course, these two conversions are not entirely reversible. Slight changes will happen to the signal when the above conversions take place. Most telephony systems allow for a choice between A-law or μ-law PCM encoding.

26.1.4 MP3

MP3 is really the *Moving Picture Experts Group (MPEG)* audio level 3 encoding used in MPEG-1 and MPEG-2 video encapsulation standards. MP3 is a patented lossy compression technique which uses audio perception approximations to reduce the coding requirements of the audio signal – see Section 5.1. Its original MPEG-1 implementation introduced limited bit rates and was standardized in 1993 (MPEG-1 Part 3) [9]. It was later supplemented with additional bit rates through the MPEG-2 implementation, known as MPEG-2 part 3 [10]. In total, MP3 has been defined such that the standard is capable of coding the audio from 8kbps to 320-kbps. It encodes original audio from 16 kHz to 48 kHz. The performance of MP3 is best at over 80 kbps. Other *codecs* with superior quality and compression are available, some of which are royalty-free. However, MP3 has established itself in the music industry, not necessarily due to its quality.

26.1.5 HE-AAC

High Efficiency Advanced Audio Coding (HE-AAC) is a very aggressive, lossy and low-bit-rate audio compression technique used in popular applications with streaming in mind such as the *flash animation* format. It was introduced by the MPEG community as a part of MPEG-4. [20] HE-AAC presents superior quality when compared to MP3 at low to medium bit rates.

The current version of the *HE-AAC*, used in *MPEG-4*, is version 2. Version 1 was standardized in 2003. Version 2 includes the use of *parametric stereo coding* which allows for more efficient coding of stereo signals in low-bandwidth situations. At higher bit rates, there is not much of a difference between version 1 and version 2. Only low bitrate stereo is enhanced by the additional *parametric stereo coding* which is present in version 2.

26.1.6 OGG Vorbis

OGG Vorbis [1] is an open-source, variable bit-rate *codec* which, in most cases, performs as well as or better than *MP3*. In fact at the higher quality levels, it is considered to be similar to *MPEG-4 AAC* [1] (see Section 26.1.5). *Vorbis* is the *codec* and *OGG* [23, 6] is the encapsulating mean for delivering the *Vorbis codec*.[38] There are also many open-source tools available including a library called *LibAO* which is available from the *XIPH Open-Source Community* for free.[39]

Vorbis is optimal for a wide range of applications, from 8 kHz telephony to 192 kHz high-end recordings and with many different channel configurations such as monaural, polyphonic (stereophonic, quadraphonic, 5.1 surround sound, ambisonic, or up to 255 discrete channels) [1]. *Vorbis* usually encapsulated by the OGG format [23], however, it may also be used by itself in an *RTP payload* (Section 26.3.5), for streaming, or by any other encapsulation, such as *SAFE* (see Section 26.2.3).

26.1.7 ADPCM (G.726)

Adaptive differential pulse code modulation ADPCM comes in several different flavors depending on the vendor implementation of interest. Some of these implementation such as the *Dialogic* and *OKI* implementations are not really well documented. They are available in middleware form which is present in hardware such as the *Dialogic telephony cards*. However, if one needs to code or decode specific flavors of the implementation, once would have to rely on reverse engineered source codes which are mostly accurate, but do provide some surprises here and there.

The general idea is to start with a *differential pulse code modulation (DPCM)* representation of the signal. This is in essence a *difference coding* applied to the *LPCM* representation discussed in Section 26.1.1. Difference coding assumes that for high enough sampling rates, the difference between the values of adjacent samples cannot be too great. This idea stems from the fact that the signal is being produced by an inertial system incapable of having large accelerations. Since the difference between adjacent samples in the signal is limited, much less number of bits would be needed to transmit this difference than it would be needed to transmit the original signal. Therefore, keeping the first sample value of the *PCM* representation, the consequent values are transmitted in the form of the difference between the previous sample and current sample. Different *ADPCM* techniques have been developed to produce log-like tables for coding these differences.

The *ITU-T* standard of *G.726* [16] is a combination of a 16-kbps *ADPCM codec* added to the existing *G.721 ADPCM* [14] which operated at 32 kbps and *G.723* [15] which described the 24 and 40 kbps *ADPCM*. Therefore, the combined *codec*, *G.726*, supports 16, 24, 32, and 40-kbps *ADPCM* formats.

26.1.8 GSM

Groupe Spécial Mobile (GSM) is the most popular complete network architecture for mobile (cellular) devices. It includes a few different audio *codecs* [26] which are based on linear predictive coding and depending on their version, operate at 6.5 kbps (*half rate*), 13.2 kbps (*full rate – GSM* 06.10), and most recently, 12.2 kbps (*enhanced full rate – GSM* 06.60) which is used in the third generation *GSM* networks known as *3G*. New *4G* versions of the network are under development, internationally.

The original *GSM codecs* (*full rate* and *half rate*) were based on *residual pulse excitation/long term prediction* [26]. The *full rate codec* would take blocks of 160 samples and encode them into a 33-byte code. Essentially this would translate to the transmission of 8000 single byte samples in a 13,200 bits, hence a 13.2 kbps transmission rate. These standard codes are a part of the *ETS 300 961* standard from the *European Telecommunications Standards Institute* (*ETSI*).

The *GSM enhanced full rate* (*GSM-EFR* or *GSM* 06.60) *codec* has a reduced frame length of 244 bits representing each 160-sample block. This means that 8000 single byte samples would be encoded into a stream of 12,200 bits, producing a 12.2 kbps stream. The *GSM-EFR* is part of the *ETS 300 726* standard from *ETSI*. When *GSM-EFR* is used with *RTP* (see Section 26.3.5), an additional 4 bit header (0xC) is attached to each frame, making the frame 248 bits or 31 bytes long [26].

In Section 26.1.9.3 we will discuss the *QCELP codec* which is used in *CDMAOne* and *CDMA2000* mobile networks. Table 26.1 shows the distribution of the subscribers to the different mobile technologies according to *GSMA*. It makes up more than 80% of the total mobile subscribers around the world (almost 3.5 billion subscribers). Therefore, there is a high probability that any mobile audio data would have been coded using one of the *GSM codec* discussed in this section.

Mobile Technology	No. of Subscribers	Percentage
CDMAOne	2,449,937	0.0568%
CDMA2000 1X	309,907,068	7.1899%
CDMA2000 1xEV-DO	118,688,849	2.7536%
CDMA2000 1xEV-DO Rev. A	12,644,062	0.2933%
GSM	3,450,410,548	80.0504%
WCDMA	255,630,141	5.9307%
WCDMA HSPA	133,286,097	3.0923%
TD-SCDMA	825,044	0.0191%
TDMA	1,480,766	0.0344%
PDC	2,740,320	0.0636%
iDEN	22,172,858	0.5144%
Analog	9,593	0.0002%

Table 26.1: Mobile technology subscribers, worldwide, in the second quarter of 2009 according to GSMA [8]

26.1.9 CELP

Code excited linear prediction is a method which uses *linear predictive coding* (Section 5.4) in conjunction with a fixed codebook and an adaptive one, used with the introduction of a delay, as the excitation input to the *LP* model to generate the approximation to the original signal [32]. It is the basis for a variety of different techniques which will be described in the following few sections.

26.1.9.1 OGG Speex

OGG Speex [32] is another royalty-free audio format, from the XIPH community, which is based on *CELP* (see Section 26.1.9). It is aimed at producing higher performance for lower bit rates when applied to human speech. The *Speex codec* has generally been designed with Voice over IP (VoIP) in mind. It may be defined in a the OGG container and transmitted through protocols such as the RTP (see Section 26.3.5).

Speex has been optimized for 8, 16 and 32 kHz sampling rates. It provides a constant bitrate as well as a variable bitrate coding scenario. The variable bitrate has been designed to handle fricatives (see Section 3.6.5) at low bit rates.

26.1.9.2 G.729

Conjugate structure algebraic code excited linear prediction (CS-ACELP) is a standard low bit rate *codec* which produces an 8-kbps stream of audio. The ITU-T provides its full recommendations [17] for this *codec* as well as *open-source C code* for its implementation. In addition, *RFC-4749* [29] defines the payload for an extension of this *codec* (*G.729.1*) for the *real-time transport protocol* (*RTP*) – see Section 26.3.5. This *codec* is generally used in voice over *IP* systems for its low bit rate and relative high quality.

26.1.9.3 QCELP

Qualcomm code excited linear prediction (*QCELP*) [5] is a variable bit-rate speech *codec* developed by *Qualcomm*[1] in the mid 1990s. It is another *CELP* technique (see Section 26.1.9) in the same class as the *CS-ACELP codec* discussed in Section 26.1.9.2. It is embedded in the *code division multiple access* (*CDMA*) which is

---

[1] Qualcomm is a registered trademark of Qualcomm Incorporated.

a *spread spectrum* method used by the cellular telephone networks (*CDMAOne* and later by *CDMA2000*) as the speech *codec* of choice. *CDMAOne* and *CDMA2000* mobile networks, according to *GSMA*, make up less than 11% of the total mobile network subscriptions. If we include all other flavors of networks derived from *CDMA*, they would still make up less than 19% of the total world subscribers – see Table 26.1. As we saw in Section 26.1.8, more than 80% are *GSM* users which use one of the *GSM audio codecs* of Section 26.1.8.

However, it is important to note that they all use linear predictive coding in some manner. The *QCELP codec* uses a variable transmission rate by adjusting the frame rate with the *pitch* and *loudness* (see Section 5.1) of the audio being transmitted.

Effectively, the transmission rate will be higher for higher pitched audio and lower for low-pitched audio. There is also a built-in silence detection to reduce the transmission of audio on silent portions of the input signal. The *codec* uses a feature extraction technique and transmits the features, instead of the sample points themselves. The features are then used to reconstruct the audio. As the name suggests, a *linear predictive coding* technique (see Section 5.4) is used for feature extraction. The source code for *QCELP* is available from the *Qualcomm FTP site* [25].

26.1.10 DTMF

Dual tone multi-frequency (*DTMF*) is a simple standard for transmitting tones which represent the numbers and special characters on a dial pad of a telephone system. The DTMF is only used for transmitting simple frequencies. It is designed to use two simple tones being transmitted at once to represent each number on a keypad, with the additional support of the asterisk ($*$) and the pound or hash sign (#). These tone combinations have been designed to reduce the chance of occurrence in a natural speech and audio transmission setting so that errors are reduced in the recognition of the intended codes. In Section 24.13.1 we discussed *Goertzel's algorithm* which is used to detect DTMF signals by producing efficient means for computing the *discrete cosine transform* (*DCT*).

26.1.11 Others Audio Formats

As we shall see in Section 26.2.1, the WAV encapsulation format alone, at the time of writing this textbook, supports 104 different *codecs*. Therefore, it is easy to see that there are many different *codec* which may be utilized to encode and decode audio and specifically speech. As we will see in the motivations of the *Standard Audio*

Encapsulation Format (SAFE), discussed in Section 26.2.3, the need existed for trying to consolidate the use of these formats. *SAFE* tries to simplify and standardize the sharing of speech data by trying to focus on a handful of standard *codecs* based on certain criteria, the most important of which are being royalty-free and simple to implement.

26.2 Standard Audio Encapsulation Formats

There are several audio encapsulation formats which are basically wrappers, specifying the information about the actual audio formats they encompass. Of these encapsulations, *WAV* is the most well-known. However, due to its ever-growing supported *coder/decoders* (*codecs*) sometimes it makes sense to choose an encapsulation format with a more limited and controlled set of *codecs*. In the next few sections, in addition to *WAV*, some other specialized encapsulation formats are discussed.

26.2.1 WAV

Waveform audio encapsulation format (*WAV*) was developed jointly by *Microsoft* and *IBM*. [2] Of course *WAV* is really not an audio format, but an audio encapsulation format. At the time of the writing of this textbook, it supports more than 104 audio formats including the *linear pulse code modulation* (*LPCM*) format. The wave header starts with the keyword, *RIFF* which stands for *Resource Interchange File Format*, and continues with information about the size of the raw audio and the *codec* being used, plus some extra information which may be written by the owner of the audio. It allows for the specification of the number of channels. If there is more than one channel, then the data alternates for the different channels along the length of the audio stream. There are other pieces of information about byte alignment as well as sample size in bits.

Some vendors have bastardized the wave encapsulation to fit their own needs. Therefore, it is not that uncommon to see wave files which do not decode properly. In practice, it probably contains the most number of variations by vendors among encapsulation standards.

---

[2] Each of the company names listed is a trademark of its respective corporation.

26.2.2 SPHERE

SPeech HEader REsources (*SPHERE*) is a standard audio header used by the *National Institute of Standards and Technology* (*NIST*), in its speaker, speech and language recognition evaluations. The *SPHERE* audio encapsulation generally starts with a 1024 byte header and follows with raw audio samples. The *SPHERE* 2.6*a* software package may be downloaded from the NIST site [21].

26.2.3 Standard Audio Format Encapsulation (SAFE)

One characteristic that distinguishes speaker recognition (identification, verification, classification, tracking, etc.) from other biometrics is that it is designed to operate with devices and over channels that were created for other technologies and functions. That characteristic supports broad, inexpensive, and speedy deployments.

The explosion of mobile devices has exacerbated the mismatch problem and the challenges for interoperability. Standard Audio Format Encapsulation (SAFE) [2] was proposed to handle interoperability that supports all types of audio interchange operations while, at the same time, limiting the audio formats to a small set of widely-used, open standards.

The *SAFE* proposal has been incorporated into an ANSI standard for audio format of raw data interchange for use in speaker recognition[3]. Also, it has been incorporated in the speaker recognition data interchange draft standards by the *ISO/IEC JTC1/SC37* project 19794-13 (*voice data*) [12].

In this section we will discuss an audio format encapsulation which has been proposed by the author of this book as a part of a draft standard developed by the M1 (biometrics) committee of the computing technology sector (InterNational Committee for Information Technology Standards) of the American National Standards Institute (**ANSI/INCITS**) that is undergoing the public review process.[18] It was also submitted to **ISO/JTC1 SC37** (biometrics) as a U.S. contribution to the Data Interchange Format for Voice.[12]

The main idea of this proposal was to be able to use exciting audio formats by bringing them under one cover so that different needs of the *speaker biometrics community* were met without having to resort to using proprietary formats. Considering the various scenarios for audio interchange, three different goals are most

---

[3] Proposed by Homayoon Beigi and Judith Markowitz [2] and incorporated in whole, by the M1 (biometrics) committee of ANSI/INCITS, as part of INCITS 456: *Speaker Recognition Format for Raw Data Interchange (SIVR-1)*, announced in 2010 [18].

prevalent. Table 26.2 presents these scenarios and the proposed audio format(s) for each case. This section describes the different cases in more detail.

Quality	Format
Lossless	Linear PCM (LPCM)
Amplitude Compression	μ-law (PCMU) and A-law (PCMA)
Aggressive variable bit-rate compression	OGG Vorbis
Streaming	OGG Media Stream

Table 26.2: Audio Interchange Scenarios

Macro	Value
AF_FORMAT_UNKNOWN	0x0000
AF_FORMAT_LINEAR_PCM	0x0001
AF_FORMAT_MULAW	0x0002
AF_FORMAT_ALAW	0x0003
AF_FORMAT_OGG_VORBIS	0x0004
AF_FORMAT_OGG_STREAM	0x1000

Table 26.3: Macros

26.2.3.1 The Uncompressed Non-Streaming Case

Linear Pulse Code Modulation (LPCM) is the method of choice for this kind of audio representation. There is no compression involved in either the amplitude domain or the frequency domain. The bare-minimum information needed in the header for this format is the number of channels, the sampling rate and the sample size (in bits). Table 26.4 includes this header data and some additional information. Microsoft WAV is not included because it is not a format; it is more of an encapsulation. WAV supports Linear PCM plus more than 104 other audio formats, most of which are proprietary *coder/decoders* (*codecs*) and many of which use some method of compression. Supporting *WAV* is tantamount to supporting all the *codecs* which *WAV* supports. That is not in line with the basic goals of the encapsulation proposed here.

26.2.3.2 Amplitude Compression with No Streaming

Logarithmic PCM includes two algorithms which were proposed in the *G*.711 ITU-T Recommendations of 1988 [13] operating at a sampling rate of 8-kHz with 8-bits per sample (64-kbps) with extensions to 80-kbps and 96-kbps as prescribed by the wide-band extension of *G*.711.1 [30]. In this scenario, the amplitude of the signal goes through some logarithmic transformation to increase the dynamic range of the signal. This conserves the number of bits needed to represent a sample. These two algorithms have been very effective techniques and have been used in telephony applications for many years. In the *G*.711 μ-law (PCMU) and *A*-law (PCMA) coding algorithms, each sample is coded to be represented by 8 bits with an 8-kHz sampling rate which amounts to a bit rate of 64 kbps. These two algorithms are known as PCMU and PCMA, respectively. Most telephony products use either PCMU or PCMA for capturing or recording audio. Supporting these algorithms should cover a wide variety of applications.

26.2.3.3 Variable Bit-Rate

These days, the first format that may come to mind is *MP3*. Unfortunately, *MP3* is a proprietary format with many patents attached to it. In contrast, *OGG Vorbis* is an *open-source*, *variable bitrate* format which, in most cases, performs as well as or better than *MP3*, see section 26.1.6.

26.2.3.4 The Streaming Case

The *OGG media stream* [23, 6] may be used to stream audio (and video). It is included here as the streaming encapsulation technique. It is completely open-source and can be used with many *codecs* including *MP3*. It is, however, recommended that *OGG Vorbis* be used in conjunction with the *OGG media stream* to achieve a streaming objective.

26.2.3.5 Audio Encapsulation Header

Table 26.4 contains the fields of the proposed data header. It (in conjunction with Table 26.3) constitutes the core of this proposal. After the proposed header, the data format will follow, either as a whole or in the form of a stream which is handled by the OGG header immediately following the proposed header.

In a typical speaker recognition session there may be different *Instances* of audio which may have common information such as the sampling rate, the sample size, the number of channels, etc. This proposal assumes that any such feature will be set once as a default value and that it may be overridden later on, per instance, as the local instance information may change from the overall SIV session information.

Type	Variable	Description
U16	ByteOrder	Is set to 0xFF00 by the audio file producer
U16	HeaderSize	Size of the header in bytes
Boolean	Streaming	This will 0 for non-streaming and 1 for streaming. This boolean variable is redundant since the AF_FORMAT for streaming audio is greater than 0x0FFF. However, it is used for convenience.
U64	FileLengthInBytes	In Bytes not including the header
U64	FileLengthInSamples	In Number of samples
U16	AudioFormat	See AF_FORMAT macros
U16	NumberOfChannels	Number of channels, *N.B.*, Channel data alternates
U32	SamplingRate	Sampling rate in samples per second – This is the audio sampling rate and not necessarily the sampling rate of the carrier which may be variable.
U64	AudioFullSecondsOf	It is the truncated number of seconds of audio
U32	AudioRemainderSamples	This is the number of samples of audio in the remainder which was truncated by the above variable
U16	BitsPerSample	Number of bits per sample, may be 0 for formats which use variable bits

Table 26.4: Audio Format Header

ByteOrder is a two-byte, binary code which is written at the time of the creation of the data. It is written as 0xFF00. When the data is read, if it is read as 0xFF00, it means that the machine reading the data has the same byte order as the machine writing the data. If it is read as 0x00FF, it means that the machine reading the data has a different byte order than the machine writing the data and that triggers a byte-swap which is applied to all subsequent information over one-byte in length.

FileLengthInSamples is a convenience measure for using LPCM, PCMU and PCMA. For these cases, *FileLengthInSamples* may be deduced from the *FileLength-InBytes*, *NumberOfChannels*, *SamplingRate* and *BitsPerSample*. It is not, however, readily computable for formats with a variable bit-rate compression. In order for it to be independent of the information which may be embedded in the encapsulated headers of OGG Vorbis, OGG Media Stream or any other format which may be added in the future, this value is included in the proposed header. Since *FileLength-InSamples* is designed for convenience, it may be set to 0.

AudioFullSecondsOf and *AudioRemainderSamples* define *FileLengthInSamples* when the number of samples is so large that an overflow may occur. *AudioFullSec-ondsOf* is the total number of seconds (in integer form) where the fractional remainder has been truncated. *AudioRemainderSamples* denotes the number of samples remaining in that truncated remainder. For example, if the total audio is 16.5 seconds long and if the sampling rate is 8-kHz, then *AudioFullSecondsOf* will be 16. The truncated remainder will then be 0.5 seconds which multiplied by 8000-Hz will produce 4000 samples which means the value of *AudioRemainderSamples* is 4000. This method of handling of the total number of seconds of audio avoids the use of floating point numbers which are most problematic in cross-platform interchanges. It also supports very long files where specifying the total number of samples can lead to an overflow.

26.2.3.6 Comments on Format Choices

It is important to note that *Standard Audio Format Encapsulation* (*SAFE*) includes a number of lossless alternatives, including no compression through *LPCM*, amplitude compression through μ-law and *a*-law, and overall lossless compression through *FLAC*. In addition, however, it allows for those who want aggressive compression, to have access to it.

In its recommendations, *SAFE* does not single out *Vorbis* for lossless compression. It uses *OGG Media* which can carry *Vorbis* as well as *Speex*, etc. However, it is true that a recommendation is in place for using *Vorbis* for the following reason: As we saw in Section 26.1.9.1, *Speex* is a *CELP*-based compression algorithm. It basically uses *linear prediction* to achieve its compression, much like many cellular compression algorithms such as *QCELP* and *G.729*. Since these algorithms are designed to utilize *human speech* features picked by a specific coding technique (namely CELP) to compress the audio, the decompressed audio only portrays what these features leave allow. This promotes a bias toward a speech feature. *Vorbis*, on the other hand, has been designed to handle a wider range of audio content (namely *HiFi music*). This allows for less bias toward a specific set of speech models. In addition, although the domain of *SAFE* is *speaker recognition*, we would have to allow for *event classification* which is also a part of the speaker recognition arena. *Speex* would degrade non-speech audio.

With all of the above said, one can still use *Speex* or other audio formats supp supported by *OGG media*. The *SAFE ANSI* standard only specifies *OGG Media* and not *Vorbis* specifically. It does, however, recommend *Vorbis* for the format used within the *OGG Media wrapper* for a greater interoperability standardization and less inherent bias in the compression. The specific details of the *SAFE* standard may be found in a recent paper by Beigi and Markowitz [2].

26.3 APIs and Protocols

Despite the importance of these operations, there is remarkably little work on SIV standards.[19] There have been a handful of *Application Programming Interfaces* (*APIs*) which were either specifically created to apply to speaker recognition or have some connection with the field, such as general biometric APIs. There have also been certain protocols which were mainly developed for speech recognition and *text to speech* (*TTS*) applications, but have added a few anchors to give minimal treatment of speaker recognition. These anchors are unfortunately very incomplete and are mostly directed toward the partial treatment of specific branches of speaker recognition – mainly speaker verification.

26.3.1 SVAPI

Speaker Verification Application Programming Interface (*SVAPI*) was developed by a consortium of companies interested in doing speaker verification. Some of these the same companies which were involved in the core development of another doomed *API*, namely, the *Speaker Recognition API* (*SRAPI*). The work on *SVAPI* started in 1996 and *SVAPI* 1.0 was made available in June, 1997. Version 2.0 which was completed in 1999 was finally abandoned in 2003. The core members of the consortium were *Novell*, which handled the administrative and distribution work, and *IBM*, *Dialogic*, *ITT Industries*, *Motorola*, *Texas Instruments*, and *T-NETIX Inc.* [4] Some U.S. government organizations were also involved.

The idea behind this *API* was to allow *Value Added Resellers* (*VARs*) and *Independent Software Vendors* (*ISVs*) to integrate products with speaker verification technology without being tied to a proprietary API. The main language of the SVAPI was $C++$. It tackled all aspects of speaker verification including the data manipulation and storage. Of course, it allowed enough flexibility for glue codes to be written so that each vendor's proprietary information would still be carried along in the predefined structures of the *API*. The *API* was very well designed, but was quite complicated and the complication may have been what lead to its final demise. It added quite a large baggage to very simple operations and the companies that developed it finally abandoned it. Although *SVAPI* stands for *Speaker Verification API*, but it was quite capable of handling Speaker Identification, Classification and other modalities of Speaker Recognition.

26.3.2 BioAPI

The *ISO/IEC Joint Technical Committee*, *JTC1 SC37*, has been developing another *API* which is aimed at handling a high-level generic biometric authentication model which would be suitable for any form of biometric technology. At the time of the writing of this book, version 1.1 of the *BioAPI* was available. [11] It covers enrollment, identification, and verification for generic biometrics. However, since the con-

---

[4] Each of the company names listed is a trademark of its respective corporation.

sortium was formed mostly by *fingerprint recognition* organizations and since they tried to cover similar biometrics, temporal associations were not considered. So, although *BioAPI* claims to be compatible with any generic biometric, it is really not suitable for Speaker Recognition. For example, it contains no streaming capabilities. Also, it is not apparent if any work is being done to extend it to handle temporal events or not.

At the present, *speaker recognition* is not very high on the list of this consortium. Section 26.2.3 shows that there is some preliminary interest in audio by the same technical committee, however, this interest is quite minimal. *BioAPI* uses *Common Biometric Exchanged Formats Framework* (*CBEFF*) which is going to be extended to handle speaker recognition. Work is on the way in collaboration with the *VoiceXML forum* to make this extension. Section 26.2.3 discusses the audio encapsulation proposal which is being considered as a part of *CBEFF*. There is hope that once *CBEFF* is extended to handle *speaker recognition* specific data at which time *BioAPI* will be closer to handling biometrics.

26.3.3 VoiceXML

Voice Extensive Markup Language (*VoiceXML*) is a *markup language* which was basically conceived around 1999 for creating *ASR* and *TTS* applications. It has been developed and maintained by the *World Wide Web Consortium* (*W3C*) and the *Internet Engineering Task Force* (*IETF*).

Already, myriads of speech applications have been developed using VoiceXML and it has been adopted by a great number of vendors, VARs and SIVs.[33] Since the middle of 2006, a the VoiceXML Forum Speaker Biometrics Committee has been working on the inclusion of SIV in the next version of VoiceXML. VoiceXML version 3.0 has included the recommendations of this committee.[4] The *SIV module* of *VoiceXML 3.0* [37] possesses the three minimal functionalities of *enrollment*, *verification*, and *identification*. Unfortunately, it does not include any more advanced functionalities such as *classification*, *segmentation*, etc. It is hoped that the standards in this arena would be combined with the ANSI, ISO, and BioAPI. If that happens, it will highly increase the spreading of *speaker recognition* in the industry.

There is some collaboration among the *VoiceXML Forum*, *BioAPI*, *ANSI*, and *ISO* activities. At the present, the inclusion of *SIV* capabilities in *CBEFF*, discussed in sections 26.3.2 and 26.2.3, seems to provide the best chance in establishing *speaker recognition* support for the discussed *protocols*, *APIs*, and *languages*.

26.3.4 MRCP

Media resources control protocol (*MRCP*) started as a control protocol for handling text to speech and speech recognition. In fact version 1.0 [28] does not touch upon speaker recognition at all. At the time of writing this book, version 2.0 of *MRCP* was at its final development stages. *MRCP 2.0* has been designed by a public forum. It has capabilities for doing *text to speech* (*TTS*), *automatic speech recognition* (*ASR*), *recording* and a limited capability for doing *speaker identification* and *verification* (*SIV*). It uses information from Section 6 of *RFC-4313* [22], as reference, for developing the *SIV* part of the *protocol*. Since version 2 is not complete, only a draft is available at [3].

MRCP communicates by sending messages using the *transmission control protocol* (*TCP*) or alternatively through *transport layer security* (*TLS*) for secure interactions. In fact, *MRCPv2* borrows functionalities from *HTTP* headers. It supports independent and simultaneous *ASR* and *SIV* sessions. *Text-dependent* and *text-independent* modalities are supported by this version. The verification module has enrollment and verification controls. The speaker identification part of the controls is quite limited. In fact, it is treated as a special case of speaker verification. Unfortunately this is the exact opposite of the true nature of the two functionalities. Indeed speaker verification should be a special case of identification. However, in the perspective of *MRCPv2*, identification is treated as verification where the *ID* to be matched is a group *ID* including all the speakers to be used as potential targets! *MRCPv2* expects cookies to be used for server authentication. It does not handle anything related to the security of the audio. It is considered to be handled by the user. Also, speaker models are never handled by the client or the server.

The protocol is compatible with VoiceXML (see section 26.3.3). VoiceXML may be built on top of the MRCP client to be able to use higher level control of the MRCP resources. There are 4 main categories of resources in MRCP, "synthesizer resource, recognizer resource, recorder resource, and speaker identification and verification resource." The resource for SIV is called *speakverify* and it has the following capabilities, "*SIV* using one or multiple utterances, simultaneous recognition and recording, using live or buffered utterances, voiceprint manipulation: creation, querying, and deletion." *MRCP* also supports the *natural language semantics markup language* (*NLSML*) which is an *extensible markup language* (*XML*) data format capable of storing enrollment information as well as verification results. There are plans to support *extensible multimodal annotation* (*EMMA*) in the future either as an alternative to *NLSML* or in lieu of it.

Unfortunately, the *SIV* feature of *MRCP* is quite limited and does not seem to be a priority for the community, developing the protocol. Still, most of the efforts of the forum is focused on the *TTS* and *ASR* portion of the protocol. For *TTS*, the *speech synthetic markup language* (*SSML*) [36] is used, which is an *XML* based language standardized by the *W3C*. In the same spirit, the *W3C Grammar XML* (*GrXML*) [35] is used to configure the speech recognition grammars. Future versions of *MRCP* will be probably use *VoiceXML 3.0* and higher which have some speaker recognition capabilities built-in. *UniMRCP* [31] is an open source implementations of *MRCP*.

26.3.5 Real-time Transport Protocol (RTP)

RTP was first introduced by *RFC-1889* [7] in 1996 and later superseded by a standard described in *RFC-3550* [27]. It is a transport protocol for real-time applications interested in streaming multimedia. *RFC-3550* [27] describes two parts. The first is the *RTP data transfer protocol* and the second is the *RTP control protocol*. Some applications may use both protocols, while others may only be interested in the data transfer protocol. For example, *MRCP* (Section 26.3.4) replaces the *RTP control protocol* with its own *control protocol* and only uses the *RTP data protocol*.

In defining the *RTP data*, we are concerned with *payload type identification, sequence numbering, timestamping* and *delivery monitoring*. Generally, since *RTP* is concerned with transporting multimedia data such as audio and video, applications run in conjunction with the *user datagram protocol* (*UDP*) [24] to be able to utilize the *checksum* and *multiplexing* features of *UDP*.

RTP is quite powerful and versatile and was designed with multi-destination conferencing applications in mind and it allows for multicasting to several destinations [27]. However, it is important to note that *RTP*, by itself, does not have any *quality of service* (*QOS*) capabilities. It utilizes the services of the lower-level layer on which it is built, such as *UDP*.

26.3.6 Extensible MultiModal Annotation (EMMA)

Extensible multimodal annotation (*EMMA*) [34] is a markup language which has been introduced as a recommendation of the *multimodal interaction working group* of the *World Wide Web Consortium* (*W3C*) to facilitate web access using multimodal interaction. *EMMA* includes hooks in order to be able to use audio transmission mechanisms such as the use of the *session initiation protocol* (*SIP*) voice over *IP protocol* utilizing the *Real-time Transport Protocol* (*RTP*) for audio transmission. It does not define any audio transmission protocol. As it is apparent from the use of *RTP* within *SIP*, *EMMA* only wraps other standards which themselves are usually wrappers for transmitting audio and not the actual audio format itself. However, it does allow for the *TCP* configuration information and the indication of the sampling rate and other data specific information. The work on *EMMA* is ongoing and by no means complete.

References

1. Barbato, L.: RTP Payload Format for Vorbis Encoded Audio. RFC 5215 (Proposed Standard) (2008). URL http://www.ietf.org/rfc/rfc5215.txt
2. Beigi, H., Markowitz, J.: Standard audio format encapsulation (SAFE). Telecommunication Systems **47**(3), 235–242 (2011). URL http://dx.doi.org/10.1007/s11235-010-9315-1. 10.1007/s11235-010-9315-1, Special Issue on Biometric Systems and Applications, Published Online: May 26, 2010
3. Burnett, D., Shanmugham, S.: Media Resource Control Protocol Version 2 (MRCPv2). Website (2010). URL http://tools.ietf.org/html/draft-ietf-speechsc-mrcpv2-21
4. Daboul, C., Shinde, P.: Speaker Identification and Verification Requirements for VoiceXML Applications (2007). URL http://www.voicexml.org/biometrics/SIV_Requirements_20070522.pdf
5. Das, A., Paksoy, E., Gersho, A.: Multimode and Variable-Rate Coding of Speech. In: W. Kleijn, K. Pliawal (eds.) Speech Coding and Syntehsis. Elsevier, New York (1995)
6. Goncalves, I., Pfeiffer, S., Montgomery, C.: Ogg Media Types. RFC 5334 (Proposed Standard) (2008). URL http://www.ietf.org/rfc/rfc5334.txt
7. Group, A.V.T.W., Schulzrinne, H., Casner, S., Frederick, R., Jacobson, V.: RTP: A Transport Protocol for Real-Time Applications. RFC 1889 (Proposed Standard) (1996). URL http://www.ietf.org/rfc/rfc1889.txt. Obsoleted by RFC 3550
8. GSMWorld: Connections by Bearer Technology. World Wide Web (2010). URL http://www.gsmworld.com/about-us/index.htm
9. ISO: ISO/IEC 11172-3:1993 (1993). URL http://www.iso.org/iso/iso_catalogue/catalogue_tc/catalogue_detail.htm?csnumber=22412
10. ISO: ISO/IEC 13818-3:1998 (1998). URL http://www.iso.org/iso/iso_catalogue/catalogue_ics/catalogue_detail_ics.htm?csnumber=26797
11. ISO: ISO/IEC JTC1 SC37, Biometric Advanced Programming Interface. Website (2008). URL http://www.bioapi.org
12. ISO: ISO/JTC1 SC37 WG3, Biometric Data Interchange Format (2009). URL http://www.iso.org/iso/standards_development/technical_committees/list_of_iso_technical_committees/iso_technical_committee.htm?commid=313770
13. ITU-T: G.711: Pulse Code Modulation (PCM) of Voice Frequencies. ITU-T Recommendation (1988). URL http://www.itu.int/rec/T-REC-G.711-198811-I/en
14. ITU-T: G.721: 32 kbit/s adaptive differential pulse code modulation (ADPCM). ITU-T Recommendation (1988). URL http://www.itu.int/rec/T-REC-G.721/e
15. ITU-T: G.723: Extensions of Recommendation G.721 Adaptive Differential Pulse Code Modulation to 24 and 40 kbit/s for Digital Circuit Multiplication Equipment Application. ITU-T Recommendation (1988). URL http://www.itu.int/rec/T-REC-G.723/e
16. ITU-T: G.726: 40, 32, 24, 16 kbit/s Adaptive Differential Pulse Code Modulation (ADPCM). ITU-T Recommendation (1990). URL http://www.itu.int/rec/T-REC-G.726/e
17. ITU-T: G.729: Coding of speech at 8 kbit/s using conjugate-structure algebraic-code-excited linear prediction (CS-ACELP). ITU-T Recommendation (2008). URL http://www.itu.int/rec/T-REC-G.729/e
18. Markowitz, J.A.: Project 1821 - INCITS 456:200x, Information Technology - Speaker Recognition Format for Raw Data Interchange (SIVR-1) (2009). URL http://www.incits.org/scopes/bsr8_1821.htm
19. Markowitz, J.A.: Standards for Speaker Recognition. In: S.Z. Li (ed.) Encyclopedia of Biometrics. Springer, New York (2009). ISBN: 978-0-387-73003-5

20. Meltzer, S., Moser, G.: MPEG-4 HE-AAC v2 – audio coding for today's digital media world. World Wide Web (2005). URL http://www.ebu.ch/fr/technical/trev/trev_305-moser.pdf
21. NIST: Speech File Manipulation Software (SPHERE) Package Version 2.6a. World Wide Web (1996). URL http://www.itl.nist.gov/iad/mig/tools/sphere_26atarZ.htm
22. Oran, D.: Requirements for Distributed Control of Automatic Speech Recognition (ASR), Speaker Identification/Speaker Verification (SI/SV), and Text-to-Speech (TTS) Resources. RFC 4313 (Informational) (2005). URL http://www.ietf.org/rfc/rfc4313.txt
23. Pfeiffer, S.: The Ogg Encapsulation Format Version 0. RFC 3533 (Informational) (2003). URL http://www.ietf.org/rfc/rfc3533.txt
24. Postel, J.: User Datagram Protocol. RFC 768 (Standard) (1980). URL http://www.ietf.org/rfc/rfc768.txt
25. Qualcomm: QCELP Source Code. FTP (1996). URL ftp://ftp.qualcomm.com/pub/vocoder/IS96a
26. Schulzrinne, H., Casner, S.: RTP Profile for Audio and Video Conferences with Minimal Control. RFC 3551 (Standard) (2003). URL http://www.ietf.org/rfc/rfc3551.txt
27. Schulzrinne, H., Casner, S., Frederick, R., Jacobson, V.: RTP: A Transport Protocol for Real-Time Applications. RFC 3550 (Standard) (2003). URL http://www.ietf.org/rfc/rfc3550.txt
28. Shanmugham, S., Monaco, P., Eberman, B.: A Media Resource Control Protocol (MRCP) Developed by Cisco, Nuance, and Speechworks. RFC 4463 (Informational) (2006). URL http://www.ietf.org/rfc/rfc4463.txt
29. Sollaud, A.: RTP Payload Format for the G.729.1 Audio Codec. RFC 4749 (Proposed Standard) (2006). URL http://www.ietf.org/rfc/rfc4749.txt
30. Sollaud, A.: RTP Payload Format for ITU-T Recommendation G.711.1. RFC 5391 (Proposed Standard) (2008). URL http://www.ietf.org/rfc/rfc5391.txt
31. UniMRCP: Open Source MRCP Project. Website (2010). URL http://www.unimrcp.org
32. Valin, J.M.: Introduction to CELP Coding. Manual (2010). URL http://speex.org/docs/manual/speex-manual/manual.html
33. W3C: Voice Extensible Markup Language (VoiceXML. Website (2008). URL http://www.voicexml.org
34. W3C: Extensible MultiModal Annotation (EMMA). Website (2009). URL http://www.w3c.org/TR/emma
35. W3C: Speech Recognition Grammar Specification. Website (2010). URL http://www.w3c.org/TR/speech-grammar
36. W3C: Speech Synthesis Markup Language (SSML). Website (2010). URL http://www.w3c.org/TR/speech-synthesis11
37. W3C: Voice Extensible Markup Language (VoiceXML). Website (2010). URL http://www.w3c.org/TR/voicexml30
38. *0.8* 1.2 Vorbis I Specifications (2004). URL http://xiph.org/ao/doc/
39. LibAO OGG Audio API. The XIPH Open-Source Community (2004). URL http://xiph.org/ao/doc/

Bibliography

1. Abe, S.: Support Vector Machines for Pattern Classification. Advances in Pattern Recognition. Springer-Verlag, London (2005). ISBN: 1-85233-929-2
2. Abu-El-Quran, A., Gammal, J., Goubran, R., Chan, A.a.: Talker Identification Using Reverberation Sensing System. In: Sensors, 2007 IEEE, pp. 970–973 (2007)
3. Adami, A., Kajarekar, S., Hermansky, H.a.: A new speaker change detection method for two-speaker segmentation. In: Acoustics, Speech, and Signal Processing, 2002. Proceedings. (ICASSP '02). IEEE International Conference on, vol. 4, pp. IV–3908–IV–3911 (2002)
4. Ahmad, A.R., Khalia, M., Viard-Gaudin, C., Poisson, E.: Online handwriting recognition using support vector machine. In: IEEE Region 10 Conference, vol. 1, pp. 311–314 (2004)
5. Ahn, S., Kang, S., and, H.K.: Effective speaker adaptations for speaker verification. In: Acoustics, Speech, and Signal Processing, 2000. ICASSP '00. Proceedings. 2000 IEEE International Conference on, vol. 2, pp. II1081–II1084 (2000)
6. Aizerman, A., Braverman, E.M., Rozoner, L.I.: Theoretical foundations of the potential function method in pattern recognition learning. Automation and Remote Control **25**, 821–837 (1964). Translated into English from Russian journal: Avtomatika i Telemekhanika, Vol. 25, No. 6, pp. 917–936, Jun 1964
7. Akaike, H.: Information Theory and an Extension of the Maximum Likelihood Principle. In: Proceedings of the 2nd International Symposium on Information Theory, Supp. to Problems of Control and Informatioin Theory, pp. 267–281 (1972)
8. Akaike, H.: A new look at the statistical model identification. IEEE Transactions on Automatic Control **19**(6), 716–723 (1974)
9. Akkermans, A., Kevenaar, T., Schobben, D.a.: Acoustic ear recognition for person identification. In: Automatic Identification Advanced Technologies, 2005. Fourth IEEE Workshop on, pp. 219–223 (2005)
10. Al Marashli, A., Al Dakkak, O.: Automatic, Text-Independent, Speaker Identification and Verification System Using Mel Cepstrum and GMM. In: 3rd International Conference on Information and Communication Technologies: From Theory to Applications (ICTTA 2008), pp. 1–6 (2008)
11. Ali, S., Silvey, S.: A General Class of Coefficients of Divergence of One Distribution from Another. Journal of the Royal Statistical Society **28**(1), 131–142 (1966)
12. Alkhaldi, W., Fakhr, W., Hamdy, N.a.: Automatic speech/speaker recognition in noisy environments using wavelet transform. In: Circuits and Systems, 2002. MWSCAS-2002. The 2002 45th Midwest Symposium on, vol. 1, pp. I–463–6 (2002)

13. Alsteris, L.D., Paliwal, K.K.: ASR on Speech Reconstructed from Short-Time Fourier Phase Spectra. In: Proceedings of the International Conference on Spoken Language Processing (ICSLP) (2004)
14. Alsteris, L.D., Paliwal, K.K.: Evaluation of the Modified Group Delay Feature for Isolated Word Recognition. In: Proceedings of the International Symposium on Signal Processing and its Applications, vol. 2, pp. 715–718 (2005)
15. Ambikairajah, E.a.: Emerging features for speaker recognition. In: Information, Communications and Signal Processing, 2007 6th International Conference on, pp. 1–7 (2007)
16. and, Y.G.: Noise-robust open-set speaker recognition using noise-dependent Gaussian mixture classifier. In: Acoustics, Speech, and Signal Processing, 2002. Proceedings. (ICASSP '02). IEEE International Conference on, vol. 1, pp. I–133–I–136 (2002)
17. Antezana, J., Massey, P., Stojanoff, D.: Jensen's Inequality and Majorization. Web (2004). URL arXiv.org:math.FA/0411442
18. Aristotle: On Sleep and Sleeplessness. HTTP (350 BCE). URL http://classics.mit.edu/Aristotle/sleep.html. Translated by J. Beare
19. Ariyaeeinia, A., Fortuna, J., Sivakumaran, P., Malegaonkar, A.a.: Verification effectiveness in open-set speaker identification. Vision, Image and Signal Processing, IEE Proceedings - 153(5), 618–624 (2006)
20. Aronowitz, H.: Trainable Speaker Diarization. In: InterSpeech, pp. 1861–1864 (2007)
21. Aronowitz, H., Solewicz, Y.A.: Speaker Recognition in Two-Wire Test Sessions. In: Inter-Speech, pp. 865–868 (2008)
22. ARPA: Proceedings of the DARPA Speech Recognition Workshop (1996)
23. ARPA: Proceedings of the DARPA Speech Recognition Workshop (1997)
24. Arrow, K.J., Hurwicz, L.: Reduction of Constrained Maxima to Saddle-Point Problems. In: Proceedings of the Third Berkeley Symposium on Mathematics, Statistics and Probability, vol. 5, pp. 1–20 (1956). Unoversity of California Press
25. Arthur, D., Vassilvitskii, S.: k-means++: The Advantages of Careful Seeding. Technical Report 2006-13, Stanford InfoLab (2006). URL http://ilpubs.stanford.edu:8090/778/
26. Ash, R.B.: Information Theory. Dover Publications, New York (1990). ISBN: 0-486-6652-16
27. Assaleh, K., Mammone, R.: New LP-derived features for speaker identification. IEEE Transactions on Speech and Audio Processing 2(4), 630–638 (1994)
28. Auckenthaler, R., Carey, M., Lloyd-Thomas, H.: Score Normalization for Text-Independent Speaker Verification Systems. Digital Signal Processing 10(1–3), 42–54 (2000)
29. Auckenthaler, R., Parris, E., Carey, M.a.: Improving a GMM speaker verification system by phonetic weighting. In: Acoustics, Speech, and Signal Processing, 1999. ICASSP '99. Proceedings., 1999 IEEE International Conference on, vol. 1, pp. 313–316 (1999)
30. Avrachenkov, K.E., Beigi, H.S., Longman, R.W.: Operator-Updating Procedures for Quasi-Newton Iterative Learning Control in Hilbert Space. In: IEEE Conference on Decision and Control (CDC99), vol. 1, pp. 276–280 (1999). Invited Paper
31. Avrachenkov, K.E., Beigi, H.S., Longman, R.W.: Updating Procedures for Iterative Learning Control in Hilbert Space. Intelligent Automation and Soft Computing Journal 8(2) (2002). Special Issue on Learning and Repetitive Control
32. Avrachenkov, K.E., Beigi, H.S.M., Longman, R.W.: Updating Procedures for Iterative Learning Control in Hilbert Space. Technical Report (1999)
33. Bachmann, P.: Analytische Zahlentheorie, Bd. 2: Die Analytische Zahlentheorie, 3rd edn. Teubner, Leipzig, Germany (1974)
34. Badri, N., Benlahouar, A., Tadj, C., Gargour, C., Ramachandran, V.a.: On the use of wavelet and Fourier transforms for speaker verification. In: Circuits and Systems, 2002. MWSCAS-2002. The 2002 45th Midwest Symposium on, vol. 3, pp. III–344–7 (2002)
35. Bahl, L., Brown, P., deSouza, P., Mercer, R.: A new algorithm for the estimation of hidden Markov model parameters. In: International Conference on Acoustics, Speech, and Signal Processing (ICASSP), pp. 493–497 (1988)
36. Bahl, L.R., Brown, P.F., deSouza, P.V., Mercer, R.L.: Maximum Mutual Information Estimation of Hidden Markov Model Parameters for Speech Recognition. In: International Conference on Acoustics, Speech, and Signal Processing (ICASSP), vol. 11, pp. 49–52 (1986)

37. Bahl, L.R., Brown, P.F., deSouza, P.V., Mercer, R.L.: Speech Recognition with Continuous-Parameter Hidden Markov Models. Computer Speech and Language **2**(3-4), 219–234 (1987)
38. Bahl, L.R., Brown, P.F., deSouza, P.V., Mercer, R.L., Nahamoo, D.: A fast algorithm for deleted interpolation. In: European Conference on Speech Communication and Technology (Eurospeech), pp. 1209–1212 (1991)
39. Bahl, L.R., Jelinek, F., Mercer, R.L.: A Maximum Likelihood Approach to Continuous Speech Recognition. IEEE Transactions on Pattern Analysis and Machine Intelligence (PAMI) **5**(2), 179–190 (1983)
40. Bakis, R., Chen, S., Gopalakrishnan, P., Gopinath, R., Maes, S., Polymenakos, L.: Transcription of Broadcast News Shows with the IBM Large Vocabulary Speech Recognition System. In: Proceedings of the Speech Recognition Workshop (1997)
41. Banach, S.: Sur les Opérations dans les Ensembles Abstraits et leur Application aux Équations Intégrales. Fundamenta Mathematicae **3**, 133–181 (1922)
42. Banach, S., Tarski, A.: Sur la Décomposition des Ensembles de Points en Parties Respectivement Congruentes. Fundamenta Mathematicae **6**, 244–277 (1924)
43. Barankin, E.W., Mitra, A.P.: Generalization of the Fisher-Darmois-Koopman-Pitman Theorem on Sufficient Statistics. The Indian Journal of Statistics **25**(3), 217–244 (1963)
44. Barbato, L.: RTP Payload Format for Vorbis Encoded Audio. RFC 5215 (Proposed Standard) (2008). URL http://www.ietf.org/rfc/rfc5215.txt
45. Barnes, J.: An Algorithm for Solving Nonlinear Equations Based on the Secant Method. Computer Journal **8**, 66–72 (1965)
46. Barras, C., Gauvain, J.L.a.: Feature and score normalization for speaker verification of cellular data. In: Acoustics, Speech, and Signal Processing, 2003. Proceedings. (ICASSP '03). 2003 IEEE International Conference on, vol. 2, pp. II–49–52 (2003)
47. Baum, L.E., Petrie, T.: Statistical Inference for Probabilistic Functions of Finite State Markov chains. The Annals of Mathematical Statistics **37**(6), 1554–1563 (1966)
48. Baum, L.E., Petrie, T., Soules, G., Weiss, N.: A Maximization Technique Occurring on the Statistical Analysis of Probabilistic Functions of Markov Chains. The Annals of Mathematical Statistics **41**(1), 164–171 (1970)
49. Becker, S., Cunn, Y.L.: Improving the Convergence of Back-Propagation Learning with Second Order Methods. In: 1988 Connectionist Models Summer School, pp. 29–37 (1988)
50. Beckman, F.S.: The Solution of Linear Equations by the Conjugate Gradient Method. In: A. Ralston, H. Wolf (eds.) Mathematical Methods for Digital Computers. Wiley, New York (1960)
51. Beigi, H.: Neural Network Learning Through Optimally Conditioned Quadratically Convergent Methods Requiring NO LINE SEARCH. In: IEEE-36th Midwest Symposium on Circuits and Systems, vol. 1, pp. 109–112 (1993)
52. Beigi, H.: Audio Source Classification using Speaker Recognition Techniques. World Wide Web (2011). URL http://www.recognitiontechnologies.com/~beigi/ps/RTI20110201-01.pdf. Report No. RTI-20110201-01
53. Beigi, H., Markowitz, J.: Standard audio format encapsulation (SAFE). Telecommunication Systems **47**(3), 235–242 (2011). URL http://dx.doi.org/10.1007/s11235-010-9315-1. 10.1007/s11235-010-9315-1, Special Issue on Biometric Systems and Applications, Published Online: May 26, 2010
54. Beigi, H.S.: An Adaptive Control Scheme Using the Generalized Secant Method. In: Proceedings of the Canadian Conference on Electrical and Computer Engineering, vol. II, pp. TA7.21.1–TA7.21.4 (1992)
55. Beigi, H.S.: Character Prediction for On-line Handwriting Recognition. In: Proceedings of the Canadian Conference on Electrical and Computer Engineering, vol. II, pp. TM10.3.1–TM10.3.4 (1992)
56. Beigi, H.S.: A Parallel Network Implementation of The Generalized Secant Learning-Adaptive Controller. In: Proceedings of the Canadian Conference on Electrical and Computer Engineering, vol. II, pp. MM10.1.1–MM10.1.4 (1992)

57. Beigi, H.S.: Adaptive and Learning-Adaptive Control Techniques based on an Extension of the Generalized Secant Method. Intelligent Automation and Soft Computing Journal **3**(2), 171–184 (1997)

58. Beigi, H.S.: Pre-Processing the Dynamics of On-Line Handwriting Data, Feature Extraction and Recognition. In: A. Downton, S. Impedovo (eds.) Progress in Handwriting Recognition, pp. 191–198. World Scientific Publishers, New Jersey (1997). ISBN: 981-02-3084-2

59. Beigi, H.S., Li, C.J.: Neural Network Learning Based on Quasi-Newton Methods with Initial Scaling of Inverse Hessian Approximate. In: The IEEE 1990 Long Island Student Conference on Neural Networks, pp. 49–52 (1990). Recipient of Best Paper Award

60. Beigi, H.S., Li, C.J.: New Neural Network Learning Based on Gradient-Free Optimization Methods. In: The IEEE 1990 Long Island Student Conference on Neural Networks, pp. 9–12 (1990). Recipient of Best Paper Award

61. Beigi, H.S., Li, C.J.: Learning Algorithms for Neural Networks Based on Quasi-Newton Methods with Self Scaling. ASME Transactions, Journal of Dynamic Systems, Measurement, and Control **115**(1), 38–43 (1993)

62. Beigi, H.S., Li, C.J.: Learning Algorithms for Feedforward Neural Networks Based on Classical and Initial-Scaling Quasi-Newton Methods. ISMM Journal of Microcomputer Applications **14**(2), 41–52 (1995)

63. Beigi, H.S., Li, C.J., Longman, R.: Learning Control Based on Generalized Secant Methods and Other Numerical Optimization Methods. In: Sensors, Controls, and Quality Issues in Manufacturing, the ASME Winter Annnual Meeting, vol. 55, pp. 163–175 (1991)

64. Beigi, H.S., Maes, S.H., Chaudhari, U.V., Sorensen, J.S.: A Hierarchical Approach to Large-Scale Speaker Recognition. In: EuroSpeech 1999, vol. 5, pp. 2203–2206 (1999)

65. Beigi, H.S., Maes, S.S.: A Distance Measure Between Collections of Distributions and its Application to Speaker Recognition. In: International Conference on Acoustics, Speech, and Signal Processing (ICASSP98) (1998)

66. Beigi, H.S., Maes, S.S.: Speaker, Channel and Environment Change Detection. In: Proceedings of the World Congress on Automation (WAC1998) (1998)

67. Beigi, H.S.M.: Neural Network Learning and Learning Control Through Optimization Techniques. Columbia University, New York (1991). Doctoral Thesis: School of Engineering and Applied Science

68. Beigi, H.S.M., Maes, S.H., Chaudhari, U.V., Sorensen, J.S.: IBM Model-Based and Frame-By-Frame Speaker Recognition. In: Speaker Recognition and its Commercial and Forensic Appications (1998)

69. Belin, P., Grosbras, M.H.: Before Speech: Cerebral Voice Processing in Infants. Neuron **65**(6), 733–735 (2010)

70. Bellman, R.: Dynamic Programming, dover paperback edition edn. Princeton University Press, Princeton, New Jersey (2010). ISBN: 0-691-1466-83

71. Beltrami, E.: Sulle Funzioni Bilineari. Giornale di Mathematiche di Battaglini **11**, 98–106 (1873)

72. Ben, M., Blouet, R., Bimbot, F.: A Monte-Carlo method for score normalization in Automatic Speaker Verification using Kullback-Leibler distances. In: International Conference on Acoustics, Speech, and Signal Processing (ICASSP), vol. 1, pp. 689–692 (2002)

73. BenAbdelkader, C., Cutler, R., Davis, L.a.: Stride and cadence as a biometric in automatic person identification and verification. In: Automatic Face and Gesture Recognition, 2002. Proceedings. Fifth IEEE International Conference on, pp. 372–377 (2002)

74. Benesty, J., Sondhi, M.M., Huang, Y.: Handbook of Speech Processing. Springer, New york (2008). ISBN: 978-3-540-49125-5

75. Benetos, E., Kotti, M., Kotropoulos, C.: Large Scale Musical Instrument Identification. In: Proceedings of the 4th Sound and Music Computing Conference, pp. 283–286 (2007)

76. Bengherabi, M., Tounsi, B., Bessalah, H., Harizi, F.: Forensic Identification Reporting Using A GMM Based Speaker Recognition System Dedicated to Algerian Arabic Dialect Speakers. In: 3rd International Conference on Information and Communication Technologies: From Theory to Applications (ICTTA 2008), pp. 1–5 (2008)

77. Bengio, S., Keshet, J.: Introduction. Automatic Speech and Speaker Recognition. John Wiley and Sons, Ltd., West Sussex, UK (2009). ISBN: 978-0-470-69683-5

78. Bennani, Y., Gallinari, P.a.: A modular connectionist architecture for text-independent talker identification. In: Neural Networks, 1991., IJCNN-91-Seattle International Joint Conference on, vol. ii, pp. 857–860 (1991)

79. Bennani, Y., Gallinari, P.a.: On the use of TDNN-extracted features information in talker identification. In: Acoustics, Speech, and Signal Processing, 1991. ICASSP-91., 1991 International Conference on, pp. 385–388 (1991)

80. Bentley, J.L.: Multidimensional Binary Search Trees Used for Associative Searching. Communications of the ACM **18**(9), 509–517 (1975)

81. Berg, P.W., McGregor, J.L.: Elementary Partial Differential Equations. Holden-Day Series in Mathematics. Holden-Day, San Francisco (1966)

82. Berger, A., Della-Pietra, S., Della-Pietra, V.: A Maximum Entropy Approach to Natural Language Processing. Computational Linguistics **22**(1) (1996)

83. Berger, G., Gowdy, J.a.: TDNN based speaker identification. In: System Theory, 1993. Proceedings SSST '93., Twenty-Fifth Southeastern Symposium on, pp. 396–399 (1993)

84. Bernstein, D.J.: The Tangent FFT. In: Applied Algebra, Algebraic Algorithms and Error-Correcting Codes, pp. 291–300. Springer, Berlin (2007)

85. Besacier, L., Grassi, S., Dufaux, A., Ansorge, M., Pellandini, F.a.: GSM speech coding and speaker recognition. In: Acoustics, Speech, and Signal Processing, 2000. ICASSP '00. Proceedings. 2000 IEEE International Conference on, vol. 2, pp. II1085–II1088 (2000)

86. Besson, P., Popovici, V., Vesin, J.M., Thiran, J.P., Kunt, M.a.: Extraction of Audio Features Specific to Speech Production for Multimodal Speaker Detection. Multimedia, IEEE Transactions on **10**(1), 63–73 (2008)

87. Bezdek, J.: Pattern Recognition with Fuzzy Objective Function Algorithms. Plenum Press, New York (1981)

88. Bhattacharyya, A.: On a Measure of Divergence between Two Multinomial Populations. Sankhya: The Indian Journal of Statistics **7**(4), 401–406 (1946)

89. Bilmes, J.A.: A Gentle Tutorial of the EM Algorithm and its Applications to Parameter Estimation for Gaussian Mixture and Hidden Markov Models (1998). Technical Report Number TR-97-021

90. Bimbot, F., Magrin-Chagnolleau, I., Mathan, L.: Second-order statistical measures for text-independent speaker identification. Speech Communication **17**(1–2), 177–192 (1995)

91. Bingham, N.H.: Studies in the History of Probability and Statistics XLVI. Measure into Probability: From Lebesgue to Kolmogorov. Biometrika **87**(1), 145–156 (2000)

92. Black, H.S.: Modulation Theory. van Nostrand Publishers, New York (1953)

93. Bocklet, T., Maier, A., Bauer, J.G., Burkhardt, F., Nöth, E.: Age and gender recognition for telephone applications based on GMM supervectors and support vector machines. In: International Conference on Acoustics, Speech, and Signal Processing (ICASSP), pp. 1605–1608 (2008)

94. Bogert, B.P., Healy, M.J.R., Tukey, J.W.: The Quefrency Alanysis of Time Series for Echoes: Cepstrum, Pseudo-Autocovariance, Cross-Cepstrum, and Saphe Cracking. In: M. Rosenblatt (ed.) Time Series Analysis, pp. 209–243 (1963). Ch. 15

95. Boll, S.F.: Suppression of Acoustic Noise in Speech using Spectral Subtraction. IEEE Transactions on Acoustics, Speech, and Signal Processing (ASSP) **27**(2), 113–120 (1979)

96. Boltzmann, L.: Vorlesungen über Gastheorie (Lectures on gas theory). Dover Publications, New York (1896–1898). Translated into English by: Stephen G. Brush, ISBN: 0-486-68455-5

97. Bookheimer, S.: Functional MRI of Language: New Approaches to Understanding the Cortical Organization of Semantic Processing. Annual Reviews of Neuroscience **25**, 151–158 (2002)

98. Borel, E.: Les Probabilités Dénombrables et Leurs Applications Arithmétique. Rend. Circ. Mat. Palermo **2**(27), 247–271 (1909)

99. Borgen, H., Bours, P., Wolthusen, S.: Visible-Spectrum Biometric Retina Recognition. In: International Conference on Intelligent Information Hiding and Multimedia Signal Processing (IIHMSP2008), pp. 1056–1062 (2008)

100. Borman, S.: The Expectation Maximization Algorithm – A Short Tutorial (2004)

101. Boser, B.E., Guyon, I.M., Vapnik, V.N.: A Training Algorithm for Optimal Margin Classifiers. In: Proceedings of the fifth annual workshop on Computational learning theory, pp. 144–152 (1992)

102. Bottou, L., Bengio, Y.: Convergence Properties of the k-Means Algorithm. In: G. Tesauro, D. Touretzky (eds.) Advances in Neural Information Processing Systems, vol. 7. MIT Press, Denver (1995). URL http://leon.bottou.org/papers/bottou-bengio-95

103. Boulard, H., Hermansky, H., Morgan, N.: Towards Increasing Speech Recognition Error Rates. Speech Communications **18**, 205–231 (1996)

104. Bovbel, E., Kheidorov, I., Chaikou, Y.a.: Wavelet-based speaker identification. In: Digital Signal Processing, 2002. DSP 2002. 2002 14th International Conference on, vol. 2, pp. 1005–1008 (2002)

105. Box, M.J.: A Comparison of Several Current Optimization Method. The Computer Journal **9**, 67–77 (1966)

106. Boyd, S., Vandenberghe, L.: Convex Optimization, 7th printing edn. Cambridge University Press (2009). First Published in 2004. ISBN: 978-0-521-83378-3

107. Breiman, L., Fiedman, J., Stone, C.J., Olshen, R.A.: Classification and Regression Trees, 1st edn. Chapman & Hall, New york (2002). ISBN: 0-412-04841-8

108. Brezinski, C.: The Life and Word of André Cholesky. Numer Algor **43**, 279–288 (2006)

109. Briant, P.: History of the Persian Empire (From Cyrus to Alexander). Eisenbrauns, Incorporated, Winona Lake, Indiana (2006). URL http://www.sil.org. Translation into English by: Peter T. Daniels, ISBN: 1-57-506120-1

110. Brodmann, K.: Vergleichende Lokalisationslehre der Grosshirnrinde in ihren Prinzipien dargestellt auf Grund des Zellenbaues. J. A. Barth, Leipzig (1909)

111. Broersen, P.: Accurate ARMA models with Durbin's second method. In: IEEE International Conference on Acoustics, Speech and Signal Processing (ICASSP 1999), vol. 3, pp. 15–19 (1999)

112. Brown, P.F., Pietra, S.A.D., Pietra, V.J.D., Lai, J.C., Mercer, R.L.: An Estimate of an Upper Bound for the Entropy of English. Computational Linguistics **18**(1), 32–40 (1992)

113. Broyden, C.: A Class of methods for Solving Nonlinear Simultaneous Equations. Mathematics of Computation **19**, 577–593 (1965)

114. Broyden, C.: Quasi-Newton Methods and Their Application to Function Minimization. Mathematics of Computation **21**, 368–381 (1967)

115. Broyden, C.: The Convergence of a Class of Double Rank Minimization Algorithms, Parts I and II. Journal of the Institute of Mathemats and its Applications **6**, 76–90, 222–231 (1970)

116. Brunelli, R., Poggio, T.: Face Recognition: Features versus Templates. IEEE Transactions on Pattern Analysis and Machine Intelligence **15**(10), 1042–1052 (1993)

117. Buck, J., Burton, D., Shore, J.a.: Text-dependent speaker recognition using vector quantization. In: Acoustics, Speech, and Signal Processing, IEEE International Conference on ICASSP '85., vol. 10, pp. 391–394 (1985)

118. Burges, C.J.: A Tutorial on Support Vector Machines for Pattern Recognition. Data Mining and Knowledge Discovery **2**, 121–167 (1998)

119. Burnett, D., Shanmugham, S.: Media Resource Control Protocol Version 2 (MRCPv2). Website (2010). URL http://tools.ietf.org/html/draft-ietf-speechsc-mrcpv2-21

120. Burrus, C.S., Gopinath, R.A., Guo, H.: Introduction to Wavelets and Wavelet Transforms: A Primer. Prentice Hall, New york (1997). ISBN: 0-134-89600-9

121. Byrne, W., Gunawardana, A.: Discounted Likelihood Lineaer Regression for Rapid Adaptation. In: Proceedings of the European Conference on Speech Communication and Technology (EUROSPEECH), pp. 203–206 (1999)

122. Byron, G.G.: The Bride of Abydos. A Turkish tale. The British Library, London (1813). Reprint: 2010

123. Cadavid, S., Abdel-Mottaleb, M.a.: Human Identification based on 3D Ear Models. In: Biometrics: Theory, Applications, and Systems, 2007. BTAS 2007. First IEEE International Conference on, pp. 1–6 (2007)

124. Calderon, A.P., Zygmund, A.: On the Existence of Certain Singular Integrals. Acta Math. **88**, 85–129 (1952)

125. Campbell, W., Assaleh, K., Broun, C.a.: Speaker recognition with polynomial classifiers. Speech and Audio Processing, IEEE Transactions on **10**(4), 205–212 (2002)

126. Campbell, W., Gleason, T., Navratil, J., Reynolds, D., Shen, W., Singer, E., Torres-Carrasquillo, P.: Advanced Language Recognition using Cepstra and Phonotactics: MITLL System Performance on the NIST 2005 Language Recognition Evaluation. In: The Speaker and Language Recognition Workshop, 2006. IEEE Odyssey 2006, pp. 1–8 (2006)

127. Campbell, W., Sturim, D., Reynolds, D.a.: Support vector machines using GMM supervectors for speaker verification. Signal Processing Letters, IEEE **13**(5), 308–311 (2006)

128. Campbell, W., Sturim, D., Shen, W., Reynolds, D., Navratil, J.a.: The MIT-LL/IBM 2006 Speaker Recognition System: High-Performance Reduced-Complexity Recognition. In: Acoustics, Speech and Signal Processing, 2007. ICASSP 2007. IEEE International Conference on, vol. 4, pp. IV–217–IV–220 (2007)

129. Campbell, W.a.: Generalized linear discriminant sequence kernels for speaker recognition. In: Acoustics, Speech, and Signal Processing, 2002. Proceedings. (ICASSP '02). IEEE International Conference on, vol. 1, pp. I–161–I–164 (2002)

130. Campbell, W.M., Campbell, J.P., Reynolds, D.A., Jones, D.A., Leek, T.R.: Phonetic Speaker Recognition with Support Vector Machines. In: Advances in Neural Information Processing Systems, pp. 1377–1384 (2003)

131. Campbell, W.M., Sturim, D.E., Reynolds, D.A., Solomonoff, A.: SVM Based Speaker Verification using a GMM Supervector Kernel and NAP Variability Compensation. In: IEEE International Conference on Acoustics, Speech and Signal Processing (ICASSP 2006), vol. 1, pp. 14–19 (2006)

132. Camps-Valls, G., Martin-Guerrero, J., Rojo-Alvarez, J., Soria-Olivas, E.: Fuzzy Sigmoid Kernel for Support Vector Classifiers. Neurocomputing **62**, 501–506 (2004)

133. Canaris, J.: A VLSI architecture for the real time computation of discrete trigonometric transforms. The Journal of VLSI Signal Processing **5**(1), 95–104 (1993)

134. Canonge, E.D.: Voiceless Vowels in Comanche. International Journal of American Linguistics **23**(2), 63–67 (1957). URL http://www.jstor.org/stable/1264055. Published by: The University of Chicago Press

135. Cantelli, F.: Sulla Probabilità Come Limite della Frequenza. Atti Accad. Naz. Lincei **26**(1), 39–45 (1917)

136. Carter, R., Aldridge, S., martyn page, steve parker: The Human Brain Book. Dorling Kindersley Ltd., New York (2009)

137. Castaldo, F., Colibro, D., Dalmasso, E., Laface, P., Vair, C.a.: Stream-based speaker segmentation using speaker factors and eigenvoices. In: Acoustics, Speech and Signal Processing, 2008. ICASSP 2008. IEEE International Conference on, pp. 4133–4136 (2008)

138. Cavalli-Sforza, L.L., Cavalli-Sforza, F., Thorne, S.: The Great Human Diasporas: the History of Diversity and Evolution. Basic Books (1995). Translated into English by: Sarah Thorne, ISBN: 0-20-144231-0

139. Cetingul, H., Erzin, E., Yemez, Y., Tekalp, A.a.: On optimal selection of lip-motion features for speaker identification. In: Multimedia Signal Processing, 2004 IEEE 6th Workshop on, pp. 7–10 (2004)

140. Cetingul, H., Yemez, Y., Erzin, E., Tekalp, A.: Discriminative Analysis of Lip Motion Features for Speaker Identification and Speech-Reading. IEEE Transactions on Image Processing **15**(10), 2879–2891 (2006)

141. Cetingul, H., Yemez, Y., Erzin, E., Tekalp, A.a.: Discriminative lip-motion features for biometric speaker identification. In: Image Processing, 2004. ICIP '04. 2004 International Conference on, vol. 3, pp. 2023–2026 (2004)

142. Chan, A., Sherwani, J., Mosur, R., Rudnicky, A.: Four-Layer Categorization Scheme of Fast
     GMM Computation Techniques in Large Vocabulary Continuous Speech Recognition Sys-
     tems. In: Proceedings of the International Conference on Spoken Language Processing (IC-
     SLP), pp. 689–692 (2004)
143. Chan, W., Lee, T., Zheng, N., and, H.O.: Use of Vocal Source Features in Speaker Segmen-
     tation. In: Acoustics, Speech and Signal Processing, 2006. ICASSP 2006 Proceedings. 2006
     IEEE International Conference on, vol. 1, pp. I–I (2006)
144. Chan, W.N., Zheng, N., and, T.L.: Discrimination Power of Vocal Source and Vocal Tract
     Related Features for Speaker Segmentation. Audio, Speech, and Language Processing, IEEE
     Transactions on 15(6), 1884–1892 (2007)
145. Chao, J., Huang, W., Zhang, Y.: SVM Based Speaker Verification and Gender Dependent
     NAP Variability Compensation. In: Bioinformatics and Biomedical Engineering, 2008.
     ICBBE 2008. The 2nd International Conference on, pp. 710–713 (2008)
146. Chatfield, P.: The Tin Trumpet. Volume I. Whittaker & Co., London (1836). P. 183
147. Chau, C.K., Lai, C.S., Shi, B.E.: Feature vs. Model Based Vocal Tract Length Normalization
     for a Speech Recognition-Based Interactive Toy. In: Active Media Technology, Lecture Notes
     in Computer Science, pp. 134–143. Springer, Berlin/Heidelberg (2001). ISBN: 978-3-540-
     43035-3
148. Chau, L.P., Siu, W.C.: Recursive Algorithm for the Discrete Cosine Transform with General
     Lengths. Electronics Letters 30(3), 197–198 (1994)
149. Che, C., Lin, Q., and, D.S.Y.: An HMM approach to text-prompted speaker verification.
     In: Acoustics, Speech, and Signal Processing, 1996. ICASSP-96. Conference Proceedings.,
     1996 IEEE International Conference on, vol. 2, pp. 673–676 (1996)
150. Chen, C., Chen, C., Cheng, P.: Hybrid KLT/GMM Approach for Robust Speaker Identifica-
     tion. IEE Electronic Letters 39(21), 1552–1554 (2003)
151. Chen, C., Chen, C., Cheng, P.a.: Hybrid KLT/GMM approach for robust speaker identifica-
     tion. Electronics Letters 39(21), 1552–1554 (2003)
152. Chen, C.C., Chen, C.T., and, C.M.T.: Hard-limited Karhunen-Loeve transform for text inde-
     pendent speaker recognition. Electronics Letters 33(24), 2014–2016 (1997)
153. Chen, C.T., Chiang, C.T., and, Y.H.C.: Efficient KLT based on overlapped subframes for
     speaker identification. In: Wireless Communications, 2001. (SPAWC '01). 2001 IEEE Third
     Workshop on Signal Processing Advances in, pp. 376–379 (2001)
154. Chen, H., Bhanu, B.a.: Human Ear Recognition in 3D. Pattern Analysis and Machine Intel-
     ligence, IEEE Transactions on 29(4), 718–737 (2007)
155. Chen, K., Xie, D., Chi, H.: A modified HME architecture for text-dependent speaker identi-
     fication. IEEE Transactions on Neural Networks 7(5), 1309–1313 (1996)
156. Chen, K., Xie, D., Chi, H.: Correction to A modified HME architecture for text-dependent
     speaker identification. IEEE Transactions on Neural Networks 8(2), 455–455 (1997)
157. Chen, K.T., Liau, W.W., , Lee, H.M.W.L.S.: Fast Speaker Adaptation Using Eigenspace-
     based Maximum Likelihood Linear Regression. In: Proceedings of the International Confer-
     ence on Spoken Language Processing (ICSLP), pp. 742–745 (2000)
158. Chen, K.T., Wang, H.M.: Eigenspace-based Maximum A Posteriori Linear Regression for
     Rapid Speaker Adaptation. In: International Conference on Acoustics, Speech, and Signal
     Processing (ICASSP), pp. 317–320 (2001)
159. Chen, S., Gopinath, R.A.: Gaussianization. In: Neural Information Processing Systems
     (NIPS) (2000)
160. Chen, S.S., Gopalakrishnan, P.S.: Speaker, Environment and Channel Change Detection and
     Clustering via the Bayesian Inromation Criterion. In: IBM Techical Report, T.J. Watson
     Research Center (1998)
161. Chen, W.C., Hsieh, C.T., and, C.H.H.: Two-Stage Vector Quantization Based Multi-band
     Models for Speaker Identification. In: Convergence Information Technology, 2007. Interna-
     tional Conference on, pp. 2336–2341 (2007)
162. Cheung, K.F.: A Multidimensional Extension of Papoulis' Generalized Sampling Expansion
     with Application in Minimum Density Sampling. In: R.J.M. II (ed.) Advanced Topics in
     Shannon Sampling and Interpolation Theory. Springer Verlag, New York (1993)

163. Cheung, R., Eisenstein, B.: Feature selection via dynamic programming for text-independent speaker identification. IEEE Transactions on Audio, Speech and Signal Processing **26**(5), 397–403 (1978)

164. Childers, D., Wu, K., Bae, K., Hicks, D.: Automatic recognition of gender by voice. In: International Conference on Acoustics, Speech, and Signal Processing (ICASSP-1988), vol. 1, pp. 603–606 (1988)

165. Childers, D.G., Skinner, D.P., Kemerait, R.C.: The cepstrum: A guide to processing. Proceedings of the IEEE **65**(10), 1428–1443 (1977)

166. Childers, D.G., Skinner, D.P., Kemerait, R.C.: Corrections to "The cepstrum: A guide to processing". Proceedings of the IEEE **66**(10), 1290–1290 (1978)

167. Chou, W.: Maximum A Posteriori Linear Regression (MAPLR) Variance Adaptation for Continuous Density HMMS. In: Proceedings of the European Conference on Speech Communication and Technology (EUROSPEECH-2003), pp. 1513–1516 (2003)

168. Christensen, H., Andersen, O.: Speaker Adaptation of Hidden Markov Models using Maximum Likelihood Linear Regression. (1996)

169. Christopher J, B.: A Syntax of Western Middle Iran. No. 3 in Persian Studies Series. Caravan Books, New York (1977). ISBN: 0-88-206005-8

170. Chunrong, X., Jianhuan, Z., and, L.F.: A Dynamic Feature Extraction Based on Wavelet Transforms for Speaker Recognition. In: Electronic Measurement and Instruments, 2007. ICEMI '07. 8th International Conference on, pp. 1–595–1–598 (2007)

171. Cieri, C., Miller, D., Walker, K.: The Fisher Corpus: a Resource for the Next Generations of Speech-to-Text. In: Fourth International Conference on Language Resources and Evaluation (2004). Available at Linguistic Data Consortium (LDC)

172. Clarkson, T., Christodoulou, C., Guan, Y., Gorse, D., Romano Critchley, D.A., Taylor, J.: Speaker identification for security systems using reinforcement-trained pRAM neural network architectures. IEEE Transactions on Systems, Man and Cybernetics **31**(1), 65–76 (2001)

173. Clarkson, T., Gorse, D., Taylor, J.: Hardware realisable models of neural processing. In: First IEE International Conference on Neural Networks, pp. 242–246 (1989)

174. Collet, M., Charlet, D., Bimbot, F.a.: A Correlation Metric for Speaker Tracking Using Anchor Models. In: Acoustics, Speech, and Signal Processing, 2005. Proceedings. (ICASSP '05). IEEE International Conference on, vol. 1, pp. 713–716 (2005)

175. Colombi, J., Anderson, T., Rogers, S., Ruck, D., Warhola, G.a.: Auditory model representation for speaker recognition. In: Acoustics, Speech, and Signal Processing, 1993. ICASSP-93., 1993 IEEE International Conference on, vol. 2, pp. 700–703 (1993)

176. Colombi, J., Ruck, D., Anderson, T., Rogers, S., Oxley, M.a.: Cohort selection and word grammar effects for speaker recognition. In: Acoustics, Speech, and Signal Processing, 1996. ICASSP-96. Conference Proceedings., 1996 IEEE International Conference on, vol. 1, pp. 85–88 (1996)

177. Cooley, J.W., Tukey, J.W.: An Algorithm for the Machine Calculation of Complex Fourier Series. Mathematics of Computation **19**(2), 297–301 (1965)

178. Cortes, C., Mohri, M., Rastogi, A.: Lp Distance and Equivalence of Probabilistic Automata. International Journal of Foundations of Computer Science (IJFCS) **18**(4), 761–779 (2007)

179. Courant, R., Hibert, D.: Methods of Mathematical Physics, 1st english edn. Volume I. WILEY-VCH Verlag GmbH & Co. KGaA, Weinheim (2009). Second reprint of English version Translated and Revised from German Original, 1937, ISBN: 0-471-50447-4

180. Cover, T.M., Thomas, J.A.: Elements of Information Theory, 2nd edn. John Wiley & Sons, New Jersey (2006). ISBN-13: 978-0-471-24195-9

181. Cramér, H.: Mathematical Methods of Statistics. Princeton University Press (1999). ISBN: 0-691-00547-8

182. Csiszár, I.: Eine Informationstheoretische Ungleichung und ihre Anwendung auf den Beweis der Ergodizität on Markofschen Ketten. Publication of the Mathemtical Institute of Hungarian Academy of Science **A:8**, 84–108 (1963)

183. Daboul, C., Shinde, P.: Speaker Identification and Verification Requirements for VoiceXML Applications (2007). URL http://www.voicexml.org/biometrics/SIV_Requirements_20070522.pdf

184. Danielson, G.C., Lanczos, C.: Some Improvements in Practical Fourier Analysis and Their Application to X-ray Scattering From Liquids – Part I. Journal of the Franklin Institute **233**(4), 365–380 (1942)

185. Danielson, G.C., Lanczos, C.: Some Improvements in Practical Fourier Analysis and Their Application to X-ray Scattering From Liquids – Part II. Journal of the Franklin Institute **233**(5), 435–452 (1942)

186. Darmois, G.: Sur les lois de probabilites a estimation exhaustive. Comptes Rendus de l'Académie des Sciences **200**, 1265–1266 (1935)

187. Das, A., Paksoy, E., Gersho, A.: Multimode and Variable-Rate Coding of Speech. In: W. Kleijn, K. Pliawal (eds.) Speech Coding and Syntehsis. Elsevier, New York (1995)

188. Daugman, J., Downing, C.: Epigenetic Randomness, Complexity and Singularity of Human Iris Patterns. Biological Sciences **268**(1477), 1737–1740 (2001)

189. Davidon, W.: Variable Metric Method for Minimization. Tech. rep., AEC Research and Development Report, ANL-5990 (1959)

190. Davidon, W.: Variance Algorithms for Minimization. Computer Journal **10**, 406–410 (1968)

191. Davidon, W.C.: Optimally Conditioned Optimization Algorithms Without Line Searches. Mathematical Programming **9**, 1–30 (1975)

192. Davis, S., Mermelstein, P.: Comparison of Parametric Representations for Monosyllabic Word Recognition in Continuously Spoken Sentences. IEEE Transactions on Acoustics, Speech and Signal Processing **28**(4), 357–366 (1980)

193. Dehak, N.: Discriminative and Generative Approaches for Long- and Short-Term Speaker Characteristics Modeling: Application to Speaker Verification. École de Technologie Supériure, Montreal (2009). PhD Thesis

194. Dehak, N., Dehak, R., Glass, J., Reynolds, D., Kenny, P.: Cosine Similarity Scoring without Score Normalization Techniques. In: The Speaker and Language Recognition Workshop (Odyssey 2010), pp. 15–19 (2010)

195. Dehak, N., Dehak, R., Kenny, P., Brummer, N., Ouellet, P., Dumouchel, P.: Support Vector Machines versus Fast Scoring in the Low-Dimensional Total Variability Space for Speaker Verication. In: InterSpeech, pp. 1559–1562 (2009)

196. Dehak, N., Dumouchel, P., Kenny, P.a.: Modeling Prosodic Features With Joint Factor Analysis for Speaker Verification. Audio, Speech, and Language Processing, IEEE Transactions on **15**(7), 2095–2103 (2007)

197. Dehak, N., Kenny, P., Dehak, R., Dumouchel, P., Ouellet, P.: Front-End Factor Analysis for Speaker Verification. IEEE Transactions on Audio, Speech and Language Processing **19**(4), 788–798 (2011)

198. Dehak, N., Kenny, P., Dehak, R., Glembek, O., Dumouchel, P., Burget, L., Hubeika, V., Castaldo, F.: Support vector machines and Joint Factor Analysis for speaker verification. In: International Conference on Acoustics, Speech, and Signal Processing (ICASSP), pp. 4237–4240 (2009)

199. Delacretaz, D., Hennebert, J.a.: Text-prompted speaker verification experiments with phoneme specific MLPs. In: Acoustics, Speech and Signal Processing, 1998. Proceedings of the 1998 IEEE International Conference on, vol. 2, pp. 777–780 (1998)

200. Dellaert, F.: The Expectation Maximization Algorithm (2002). Technical Report Number GIT-GVU-02-20

201. Dempster, A.P., Laird, N.M., Rubin, D.B.: Maximum Likelihood from Incomplete Data via the EM Algorithm. Journal of the Royal Statistical Society. Series B (Methodological) **39**(1), 1–38 (1977)

202. Deng, L., O'Shaughnessy, D.: Speech Processing, A Dynamic and Optimization-Oriented Approach. Marcel Dekker, Inc., New york (2003). ISBN: 0-824-74040-8

203. Deshpande, M.S., Holambe, R.S.a.: Text-Independent Speaker Identification Using Hidden Markov Models. In: Emerging Trends in Engineering and Technology, 2008. ICETET '08. First International Conference on, pp. 641–644 (2008)

204. Devore, J.L.: Probability and Statistics for Enineering and the Sciences, 3rd edn. Cole Publishing Company, Pacific Grove, CA, USA (1990). ISBN: 0-534-14352-0

205. Dhanalakshmi, P., Palanivel, S., Ramalingam, V.: Classification of Audio Signals using AANN and GMM. Applied Soft Computing **11**(1), 716 – 723 (2011). DOI 10.1016/j.asoc.2009.12.033. URL http://www.sciencedirect.com/science/article/B6W86-4Y3JY8D-1/2/722d39fe60e735af8ddda0be27d48057

206. Dhillon, I.S., Kulis, Y.G.B.: Weighted Graph Cuts without Eigenvectors: A Multilevel Approach. IEEE Transactions on Pattern Analysis and Machine Intelligence (PAMI) **29**(11), 1944–1957 (2007)

207. Dickinson, E.: The Complete Poems. Back Bay Books, Boston, MA (1976). The Brain: Poem number 632

208. Dietterich, T., Bakiri, G.: Solving Multiclass Learning Problems via Error-Correcting Output Codes. Journal of Artificial Intelligence Research **2**, 263–286 (1995)

209. Ding, I.J.: Improvement of MLLR Speaker Adaptation Using a Novel Method. International Journal of Information Technology **5**(1), 12–17 (2009)

210. Doddington, G., Liggett, W., Martin, A., Przybocki, M., Reynolds, D.: Sheep, Goats, Lambs and Wolves: A Statistical Analysis of Speaker Performance in the NIST 1998 Speaker Recognition Evaluation. In: Proceedings of the International Conference on Spoken Language Processing (ICSLP), pp. 1–5 (1998)

211. Donovan, R.E., Woodland, P.C.: A Hidden Markov-Model-Based Trainable Speech Synthesizer. Computer Speech and Language pp. 1–19 (1999)

212. Dorn, W.: Duality in Quadratic Programming. Quarterly Journal of Applied Mathematics **18**(2), 155–162 (1960)

213. Dragomir, S.S.: Some General Divergence Measures for Probability Distributions. Acta Mathematica Hungarica **109**(4), 331–345 (2005)

214. Drygajlo, A.a.: Forensic Automatic Speaker Recognition [Exploratory DSP]. Signal Processing Magazine, IEEE **24**(2), 132–135 (2007)

215. Duda, R.O., Hart, P.E.: Pattern Classification and Scene Analysis. John Wiley & Sons, New York (1973). ISBN: 0-471-22361-1

216. Duffin, R., Peterson, E.: Duality Theory for Geometric Programming. SIAM Journal on Applied Mathematics **14**(6), 1307–1349 (1966)

217. Dugast, C., Devillers, L., Aubert, X.: Combining TDNN and HMM in a hybrid system for improved continuous-speech recognition. IEEE Transactions on Speech and Audio Processing **2**(1), 217–223 (1994)

218. Dugué, D.: Application des propriétés de la Limite au sens du Calcul des Probabilités à lÉtude de Diverses Questions d'estimation. Journal de lÉcole Polytechnique p. 305 (1937)

219. Duhamel, P., Hollmann, H.: Aplit-Radix FFT Algorithm. Electronics Letters **20**, 14–16 (1984)

220. Dumitrescu, M.E.B.: The Application of the Principle of Minimum Cross-Entropy to the Characterization of the Exponential-Type probability Distributions. Annals of the Institute of Statistical Mathematics **38**(1), 451–457 (1986)

221. Dunn, R.B., Reynolds, D.A., Quatieri, T.F.: Approaches to Speaker Detection and Tracking in Conversational Speech. Digital Signal Processing **10**, 92–112 (2000)

222. Durbin, J.: Efficient Estimation of Parameters in Moving Average Models. Biometrika **46**, 306–316 (1959)

223. Durbin, J.: The Fitting of Time Series Models. Revue Institute International de Statistic **28**, 233–243 (1960)

224. Dutta, T.: Dynamic Time Warping Based Approach to Text-Dependent Speaker Identification Using Spectrograms. In: Congress on Image and Signal Processing (CISP '08), vol. 2, pp. 354–360 (2008)

225. Edie, E., Gish, H.: A Parametric Approach to Vocal Tract Length Normalization. In: IEEE International Conference on Acoustics, Speech and Signal Processing (ICASSP 1996), vol. 1, pp. 346–348 (1996)

226. El Hannani, A., Petrovska-Delacretaz, D.a.: Comparing Data-driven and Phonetic N-gram Systems for Text-Independent Speaker Verification. In: Biometrics: Theory, Applications, and Systems, 2007. BTAS 2007. First IEEE International Conference on, pp. 1–4 (2007)

227. El Hannani, A., Toledano, D., Petrovska-Delacretaz, D., Montero-Asenjo, A., Hennebert, J.a.: Using Data-driven and Phonetic Units for Speaker Verification. In: Speaker and Language Recognition Workshop, 2006. IEEE Odyssey 2006: The, pp. 1–6 (2006)

228. ELRA: European Language Resources Association. Web (2010). URL http://www.elra.info. Speech Data Resource

229. Endres, E., Bambach, W., Flösser, G.: Voice Spectrograms as a Function of Age, Voice Disguise and Voice Imitation. Journal of the Acoustical Society of America (JASA) **49**, 1842–1848 (1971)

230. Eronen, A.: Comparison of Features for Musical Instrument Recognition. In: IEEE Workshop on the Applications of Signal Processing to Audio and Acoustics, pp. 19–22 (2001)

231. Eronen, A., Klapuri, A.: Musical Instrument Recognition Using Cepstral Coefficients and Temporal Features. In: International Conference on Acoustics, Speech, and Signal Processing (ICASSP), vol. 2, pp. 753–756 (2000)

232. Erzin, E., Yemez, Y., Tekalp, A.a.: Multimodal speaker identification using an adaptive classifier cascade based on modality reliability. Multimedia, IEEE Transactions on **7**(5), 840–852 (2005)

233. Eskidere, O., Ertas, F.a.: Impact of Pitch Frequency on Speaker Identification. In: Signal Processing and Communications Applications, 2007. SIU 2007. IEEE 15th, pp. 1–4 (2007)

234. Eskidere, O., Ertas, F.a.: Parameter Settings for Speaker Identification using Gaussian Mixture Model. In: Signal Processing and Communications Applications, 2007. SIU 2007. IEEE 15th, pp. 1–4 (2007)

235. Essen, D.C.V.: Surface-Based Approaches to Spatial Localization and Registration in Primate Cerebral Cortex. Nueroimage **23**(Supplement 1), S97–S107 (2004)

236. Essen, D.C.V., Drury, H.A., Dickson, J., Harwell, J., Hanlon, D., Anderson, C.H.: An Integrated Software Suite for Surface-based Analyses of Cerebral Cortex. Journal of American Medical Informatics Association **8**(5), 443–459 (2001)

237. Fabate, A., Nappi, M., Riccio, D., Ricciardi, S.a.: Ear Recognition by means of a Rotation Invariant Descriptor. In: Pattern Recognition, 2006. ICPR 2006. 18th International Conference on, vol. 4, pp. 437–440 (2006)

238. Fadeev, D.K.: Zum Begriff der Entropie einer endlichen Wahrscheinlichkeitss. Deutscher Verlag der Wissenschaften pp. 85–90 (1957)

239. Fahlman, S.E.: Faster-Learning Variations on Back-Propagation: An Empirical Study. In: 1988 Connectionist Models Summer School, pp. 38–51 (1988)

240. Fant, G.: Acoustic Theory of Speech Production – with Calculations based on X-Ray Studies of Russian Articulations, revised edn. Mouton De Gruyter, The Hague (1970). ISBN: 978-9-027-91600-6

241. Farkas, J.: Theorie der Einfachen Ungleichungen (Theory of the Simple Inequality). Journal für die Reine und Angewandte Mathematic **124**, 1–27 (1902)

242. Feller, W.: The fundamental limit theorems in probability. Bulletin of the American Mathematical Society **51**(11), 800–832 (1945)

243. Feller, W.: An Introduction to Probability Theory and Its Applications, 3rd edn. John Wiley & Sons (1968). Volume I

244. Feller, W.: An Introduction to Probability Theory and Its Applications. John Wiley & Sons (1968). Two Volumes

245. Fenchel, W.: Convex Cones, Sets and Functions (1951). Lecture Notes by D. W. Blackett, Sep. 1953

246. Fermentas Nucleotides Catalog. Website (2009). URL http://www.fermentas.com/catalog/nucleotides

247. Fiacco, A.V., McCormick, G.P.: Nonlinear Programming: Sequential Unconstrained Minimization Techniques. J. Wiley and Sons, New York (1968)

248. Fienberg, S.E.: When Did Bayesian Inference Become Bayesian? Bayesian Analysis **1**(1), 1–40 (2006)

249. Finan, R., Sapeluk, A., Damper, R.a.: Comparison of multilayer and radial basis function neural networks for text-dependent speaker recognition. In: Neural Networks, 1996., IEEE International Conference on, vol. 4, pp. 1992–1997 (1996)

250. Fine, S., Navratil, J., Gopinath, R.a.: A hybrid GMM/SVM approach to speaker identification. In: Acoustics, Speech, and Signal Processing, 2001. Proceedings. (ICASSP '01). 2001 IEEE International Conference on, vol. 1, pp. 417–420 (2001)

251. Fisher, R.A.: Theory of Statistical Estimation. Proceedings of Cambridge Philosophical Society 22, 700–725 (1925)

252. Fisher, R.A.: The Design of Experiments. Oliver and Boyd, Edinburgh (1935). 8th Edition, Hafner, New York, 1966

253. Flanagan, J.L.: Speech Analysis, Synthesis and Perception, 2nd edn. Springer-Verlag, New York (1972). ISBN: 0-387-05561-4

254. Flannery, M.: The Enigma of Nonholonomic Constraints. American Journal of Physics 73(3), 265–272 (1976)

255. Fletcher, R.: Function Minimization Without Evaluating Derivatives. The Computer Journal 8, 33–41 (1965)

256. Fletcher, R.: A New Approach to Variable Metric Algorithms. Computer Journal 8, 317–322 (1970)

257. Fletcher, R.: Practical Methods of Optimization, 2nd edn. J. Wiley and Sons, New york (2000). ISBN: 0-471-49463-1

258. Fletcher, R., Reeves, C.: Function Minimization by Conjugate Gradients. The Computer Journal 7, 149–154 (1964)

259. Fogel, L.: A Note on the Sampling Theorem. The Institute of Radio Engineers Transactions on Information Theory 1(1), 47–48 (1955)

260. Foo, S.W., and, E.G.L.: Speaker recognition using adaptively boosted decision tree classifier. In: Acoustics, Speech, and Signal Processing, 2002. Proceedings. (ICASSP '02). IEEE International Conference on, vol. 1, pp. I–157–I–160 (2002)

261. Foote, J., Silverman, H.a.: A model distance measure for talker clustering and identification. In: Acoustics, Speech, and Signal Processing, 1994. ICASSP-94., 1994 IEEE International Conference on, vol. i, pp. I/317–I/320 (1994)

262. Forehand, C.J.: Integrative Functions of the Nervous System. In: R. Rhoades, D.R. Bell (eds.) Medical Physiology: Principles of Clinical Medicine, 3rd edn. Lippincott Williams and Wilkins, Baltimore, MD (2009). ISBN: 0-7817-6852-8

263. Fortmann, T.E., Hitz, K.L.: An Introduction to Linear Control Systems. Marcel Dekker, Inc., New York (1977). ISBN: 0-824-76512-5

264. Fourier, J.B.J.: Théorie Analytique de la Chaleur. Chez Firmin Didot, Père et Fils, Paris, France (1822). Digitized by Google from an Astor Library, New York, Copy: http://books.google.com/books?id=TDQJAAAAIAAJ&printsec=frontcover&dq=Th%C3%A9orie+analytique+de+la+chaleur&q=&hl=en#v=onepage&q&f=false

265. Fox, R.W.: Introduction to Fluid Mechanics, 2nd edn. Addison-Wesley Publishing Company, New york (1978). ISBN: 0-417-01909-7

266. Fraser, D.A.S.: Nonparametric Methods in Statistics. John Wiley & Sons, New York (1957)

267. Freund, Y., Schapire, R.E.: Experiments with a new boosting algorithm. In: Proceedings of the Thirteenth International Conference on Machine Learning (ICML), pp. 148–156 (1996)

268. Fujisaki, T., Beigi, H., Tappert, C., Ukelson, M., Wolf, C.: Online Recognition of Unconstrained Handprinting: A Stroke-based System and Its Evaluation. In: S. Impedovo, J. Simon (eds.) From Pixels to Features III: Frontiers in Handwriting, pp. 297–312. North Holland, Amsterdam (1992). ISBN: 0-444-89665-1

269. Gaab, N.: The Auditory Cortex: Perception, Memory, Plasticity and the Influence of Musicianship. the University of Zürich, Zürich, Switzerland (2004). PhD Thesis

270. Gale, D., Kühn, H.H., Tucker, A.W.: Linear Programming and the Theory of Games. In: e.a. Tjalling C. Koopmans (ed.) Activity Analysis of Production and ALlocation, pp. 317–329 (1951). URL http://cowles.econ.yale.edu/P/cm/m13/

271. Ganapathiraju, A.: Support Vector Machines for Speech Recognition. Mississippi State University, Mississipi (2002). PhD Thesis

272. Ganapathiraju, A., Hamaker, J., Picone, J.: Support Vector Machines for Speech Recognition. In: International Conference on Spoken Language Processing, pp. 2348–2355 (1998)
273. Ganapathiraju, A., Hamaker, J., Picone, J.: Hybrid SVM/HMM Architectures for Speech Recognition. In: SPeech Transcription Workshop, pp. 504–507 (2000)
274. Gao, X., Wang, P., Qi, Y., Yan, A., Zhang, H., Gong, Y.: Comparison Studies of LS-SVM and SVM on Modeling for Fermentation Processes. In: International Conference on Natural Computation (ICNC), pp. 478–484 (2009)
275. Garcia, V., Nielsen, F., Nock, R.: Hierarchical Gaussian Mixture Model (2010)
276. Garofalo, J., Lamel, L., Fisher, W., Fiscus, J., Pallett, D., Dahlgren, N.: Darpa TIMIT: Acoustic-Phonetic Continuous Speech Corpus. CD-ROM (1993). Linguistic Data Consortium (LDC)
277. Gauci, O., Debono, C., Micallef, P.: A maximum log-likelihood approach to voice activity detection. In: International Symposium on Communications, Control and Signal Processing (ISCCSP 2008), pp. 383–387 (2008)
278. Gauss, C.F.: Nachlass, Theoria Interpolationis Methodo Nova Tractata. In: Carl Friedrich Gauss Werke, Band 3, Königlichen Gesellschaft der Wissenschaften: Göttingen, pp. 265–330 (1866). Note: Volume 3 of the collective works of Gauss.
279. Gauvain, J.L., Lee, C.H.: Maximum a Posteriori Estimation for Multivariate Gaussian Mixture Observation of Markov Chains. IEEE Transactions on Speech and Audio Processing 2(2), 291–298 (1994)
280. George, N., Evangelos, D.a.: Hands-free continuous speech recognition in noise using a speaker beam-former based on spectrum-entropy. In: Acoustics, Speech, and Signal Processing, 2002. Proceedings. (ICASSP '02). IEEE International Conference on, vol. 1, pp. I–889–I–892 (2002)
281. George, S., Dibazar, A., Liaw, J.S., Berger, T.a.: Speaker recognition using dynamic synapse based neural networks with wavelet preprocessing. In: Neural Networks, 2001. Proceedings. IJCNN '01. International Joint Conference on, vol. 2, pp. 1122–1125 (2001)
282. Gerald, C.F., Wheatly, P.O.: Applied Numerical Analysis, 3rd edn. Addison-Wesley Publishing Company, Reading, Massachusetts (1985). ISBN: 0-201-11577-8
283. Gers, F.A., Schmidhuber, J.: LSTN Recurrent Networks Learn Simple Context-Free and Context-Sensitive Languages. IEEE Transactions on Neural Networks 12(6), 1333–1340 (2001)
284. Ghitza, O.: Auditory Models and Human Performance in Tasks Related to Speech Coding and Speech Recognition. IEEE Transactions on Speech and Audio Processing 2(1), 115–132 (1994)
285. Glembek, O., Burget, L., Dehak, N., Brümmer, N., Kenny, P.: Comparison of Scoring Methods used in Speaker Recognition with Joint Factor Analysis. In: International Conference on Acoustics, Speech, and Signal Processing (ICASSP), pp. 4057–4060 (2009)
286. Gnanadesikan, R.: Methods for Statistical Data Analysis of Multivariate Observations, 2nd edn. Wiley-Interscience, New York (1997). ISBN: 0-471-1611-95
287. Goertzel, G.: An Algorithm for the Evaluation of Finite Trigonometric Series. American Mathematical Monthly 65, 34–35 (1958)
288. Goldfarb, D.: A Family of Variable Metric Methods Derived by Variational Means. Mathematics of Computation 24, 22–26 (1970)
289. Goldfeld, S., Quandt, R., Trotter, H.: Maximization by Quadratic Hill Climbing. Econometrics 34, 541–551 (1966)
290. Golgi, C.: The neuron doctrine - theory and facts. Lecture (1906). URL http://nobelprize.org/nobel_prizes/medicine/laureates/1906/golgi-lecture.pdf
291. Goncalves, I., Pfeiffer, S., Montgomery, C.: Ogg Media Types. RFC 5334 (Proposed Standard) (2008). URL http://www.ietf.org/rfc/rfc5334.txt
292. Gonzalez-Rodriguez, J., Fierrez-Aguilar, J., Ortega-Garcia, J.a.: Forensic identification reporting using automatic speaker recognition systems. In: Acoustics, Speech, and Signal Processing, 2003. Proceedings. (ICASSP '03). 2003 IEEE International Conference on, vol. 2, pp. II–93–6 (2003)

293. Good, I.J.: Maximum Entropy for Hypothesis Formulation, Especially for Multidimensional Contingency Tables. The Annals od Mathematical Statistics **34**(3), 911–934 (1963)

294. Gopalan, K., Anderson, T., Cupples, E.: A comparison of speaker identification results using features based on cepstrum and Fourier-Bessel expansion. IEEE Transactions on Speech and Audio Processing **7**(3), 289–294 (1999)

295. Gowdy, J.N., Tufekci, Z.: Mel-scaled discrete wavelet coefficients for speech recognition. pp. 1351–1354 (2000)

296. Grandvalet, Y., Bengio, Y.: Entropy Regularization. In: Semi-Supervised Learning. The MIT Press, Boston (2006). Chapter 9

297. Grashey, S., Geibler, C.a.: Using a Vocal Tract Length Related Parameter for Speaker Recognition. In: Speaker and Language Recognition Workshop, 2006. IEEE Odyssey 2006: The, pp. 1–5 (2006)

298. Grasic, M., Kos, M., Zgank, A., Kacic, Z.a.: Comparison of speaker segmentation methods based on the Bayesian Information Criterion and adapted Gaussian mixture models. In: Systems, Signals and Image Processing, 2008. IWSSIP 2008. 15th International Conference on, pp. 161–164 (2008)

299. Gray, H.: Anatomy of the Human Body, 20th edn. LEA and FEBIGER, Philadelphia (1918). URL http://www.Bartleby.com. Online version, New York (2000)

300. Green, P.E., Kim, J., Carmone, F.J.: A preliminary study of optimal variable weighting in k-means clustering. Journal of Classification **7**(2), 271–285 (1990)

301. Greenstadt, J.: On the Relative Effectiveness of Gradient Methods. Mathematics of Computation **21**, 360–367 (1967)

302. Greenstadt, J.: Variations on Variable-Metric Methods. Mathematics of Computation **24**, 1–22 (1970)

303. Grimaldi, M., Cummins, F.a.: Speaker Identification Using Instantaneous Frequencies. Audio, Speech, and Language Processing, IEEE Transactions on **16**(6), 1097–1111 (2008)

304. Group, A.V.T.W., Schulzrinne, H., Casner, S., Frederick, R., Jacobson, V.: RTP: A Transport Protocol for Real-Time Applications. RFC 1889 (Proposed Standard) (1996). URL http://www.ietf.org/rfc/rfc1889.txt. Obsoleted by RFC 3550

305. Gruber, C., Gruber, T., Krinninger, S., Sick, B.: Online Signature Verification With Support Vector Machines Based on LCSS Kernel Functions. IEEE Transactions on Systems, Man, and Cybernetics, Part B **40**(4), 1088–1100 (2010)

306. GSMWorld: Connections by Bearer Technology. World Wide Web (2010). URL http://www.gsmworld.com/about-us/index.htm

307. Gu, Y., Jongebloed, H., Iskra, D., den Os, E., Boves, L.: Speaker Verification in Operational Environments Monitoring for Improved Service Operation. In: Proceedings of the International Conference on Spoken Language Processing (ICSLP), pp. 450–453 (2000)

308. Guiasu, S., Shenitzer, A.: The Principle of Maximum Entropy. The Mathematical Intelligencer **7**(1), 42–48 (1985)

309. Gupta, V., Kenny, P., Ouellet, P., Boulianne, G., Dumouchel, P.a.: Multiple feature combination to improve speaker diarization of telephone conversations. In: Automatic Speech Recognition and Understanding, 2007. ASRU. IEEE Workshop on, pp. 705–710 (2007)

310. Gurbuz, S., Gowdy, J., Tufekci, Z.: Speech spectrogram based model adaptation for speaker identification. In: Southeastcon 2000. Proceedings of the IEEE, pp. 110–115 (2000)

311. Guta, B.: An Oscillation Theory of Handwriting. Universität Kaiserslautern (2003). PhD Thesis

312. Guyon, I.M., Vapnik, V.N., Boser, B.E., Solla, S.A.: Structural Risk Minimization for Character Recognition. In: T. David S (ed.) Neural Information Processing Systems, vol. 4. Morgan Kaufmann Publishers, San Mateo, CA (1992)

313. Haar, A.: Theorie der Orthogonalen Funktionensysteme (Theory of the Orthohonal System of Functions. Mathematische Annalen (Annals of Mathematics) **69**(3), 331–371 (1910). Original in German

314. Hairer, G., Vellekoop, M., Mansfeld, M., Nohammer, C.: Biochip for DNA Amplification and Label-free DNA Detection. In: IEEE Conference on Sensors, pp. 724–727 (2007)

315. Halmos, P.R.: Elements of Information Theory, 2nd printing edn. Springer-Verlag, New York (1974). ISBN: 0-387-90088-8
316. Halmos, P.R., Savage, L.J.: Application of the Radon-Nikodym Theorem to the Theory of Sufficient Statistics. The Annals of Mathematical Statistics **20**(2), 225–241 (1949)
317. Hamerly, G., Elkan, C.: Alternatives to the k-Means Algorithm that Find Better Clusterings. In: Proceedings of the Eleventh International Conference on Information and Knowledge Management, pp. 600–607 (2002)
318. Hardt, D., Fellbaum, K.a.: Spectral subtraction and RASTA-filtering in text-dependent HMM-based speaker verification. In: Acoustics, Speech, and Signal Processing, 1997. ICASSP-97., 1997 IEEE International Conference on, vol. 2, pp. 867–870 (1997)
319. Harsha, B.: A noise robust speech activity detection algorithm. In: Intelligent Multimedia, Video and Speech Processing, pp. 322–325 (2004)
320. Hartley, H.O.: Maximum Likelihood Estimation from Incomplete Data. Biometrics **14**(2), 174–194 (1958)
321. Hartley, R.V.L.: Transmission of Information. Bell System Technical Journal **7**, 535–563 (1928)
322. Hassab, J.C.: On the Convergence Interval of the Power Cepstrum. IEEE Transactions on Information Theory **1**, 111–112 (1974)
323. Hatch, A., Peskin, B., Stolcke, A.a.: Improved Phonetic Speaker Recognition Using Lattice Decoding. In: Acoustics, Speech, and Signal Processing, 2005. Proceedings. (ICASSP '05). IEEE International Conference on, vol. 1, pp. 169–172 (2005)
324. Hatch, A.O., Kajarekar, S., Stolcke, A.: Within-Class Covariance Normalization for SVM-based Speaker Recognition. In: Interspeech (2006)
325. Hayakawa, S., Itakura, F.a.: Text-dependent speaker recognition using the information in the higher frequency band. In: Acoustics, Speech, and Signal Processing, 1994. ICASSP-94., 1994 IEEE International Conference on, vol. i, pp. I/137–I/140 (1994)
326. Haydar, A., Demirekler, M., Yurtseven, M.: Speaker identification through use of features selected using genetic algorithm. IEE Electronic Letters **34**(1), 39–40 (1998)
327. He, J., Liu, L., Palm, G.a.: A discriminative training algorithm for VQ-based speaker identification. Speech and Audio Processing, IEEE Transactions on **7**(3), 353–356 (1999)
328. He, X., Chou, W.: Minimum Classification Error (MCE) Model Adaptation of Continuous Density HMMS. In: Proceedings of the European Conference on Speech Communication and Technology (EUROSPEECH-2003), pp. 1629–1632 (2003)
329. Healy, M.J., Westmacott, M.: Missing Values in Experiments Analysed on Automatic Computers. Journal of the Royal Statistical Society, Series C (Applied Statistics) **5**(3), 203–206 (1956)
330. Heck, L., Sankar, A.: Acoustic Clustering and Adaptation for Improved Speech Recognition. In: Proceedings of the Speech Recognition Workshop (1997)
331. Heideman, M.T., Johnson, D.H., Burrus, C.S.: Gauss and the History of Fast Fourier Transform. Archive for History of Exact Sciences **34**(3), 265–277 (1985)
332. Hermansky, H.: Perceptual linear predictive (PLP) analysis of speech. The Journal of the Acoustical Society of America (JASA) **87**(4), 1738–1752 (1990)
333. Hermansky, H., Morgan, N.: RASTA Processing of Speech. IEEE Transactions on Speech and Audio Processing **2**(4), 578–589 (1994)
334. Hermansky, H., Morgan, N., Bayya, A., Kohn, P.: RASTA-PLP speech analysis technique. In: IEEE International Conference on Acoustic, Speech, and Signal Processing, vol. 1, pp. 121–124 (1992)
335. Hershey, J.R., Olsen, P.A.: Approximating the Kullback Leibler Divergence Between Gaussian Mixture Models. In: International Conference on Acoustics, Speech, and Signal Processing (ICASSP), vol. 4, pp. 317–320 (2007)
336. Hestenes, M.R., Steifel, E.: Methods of Conjugate Gradients for Solving Linear Systems. J. Res. N. B. S. **49**, 409 (1952)
337. Hilbert, D.: Grundzüge Einer Allgemeinen Theorie der Linearen Integralgleichungen (Outlines of a General Theory of Linear Integral Equations). Fortschritte der Mathematischen Wissenschaften, heft 3 (Progress in Mathematical Sciences, issue 3). B.G. Teubner, Leipzig and Berlin (1912). In German. Originally published in 1904.

338. Himmelblau, D.M.: Applied Nonlinear Programming. McGraw-Hill Book Company, New York (1972)

339. Hochreiter, S., Schmidhuber, J.: Long Short-Term Memory. Neural Computing **8**, 1735–1780 (1997)

340. Hodges, N.D.C.: Census of the Defective Classes. Science **VIII** (1889). URL http://www.census.gov

341. Hogg, R.V., Craig, A.T.: Introduction to Mathematical Statistics, 4th edn. Macmillan Publishing Co., Inc., New York (1979). ISBN: 0-023-55710-9

342. Hollerbach, J.M.: An Oscillation Theory of Handwriting. MIT Press (1980). PhD Thesis

343. Homayoon S. M. Beigi, S.H.M.: Speaker, Channel and Environment Change Detection. Technical Report (1997)

344. Hooke, R., Jeeves, T.: Direct Search Solution of Numerical and Statistical Problems. Journal of Association of Computer Machinery (ACM) **8**(2), 212–229 (1961)

345. Hooker, J.: Inference Duality as a Basis for Sensitivity Analysis. Constraints: An International Journal **4**, 101–112 (1999)

346. Horowitz, J.L.: Semiparametric Models in Econometrics. Springer, New York (1998). ISBN: 0-387-98477-1

347. Hoshino, S.: A Formulation of Variable Metric Methods. Journal of the Institue of Mathemtics and its Applicatiions **10**, 394–403 (1972)

348. Householder, A.S.: The Theory of Matrices in Numerical Analysis. Blaisdell Publishing Company, Massachussetts (1964)

349. Hsieh, C.T., Lai, E., Wang, Y.C.: Robust speech features based on wavelet transform with application to speaker identification. IEE Proceedings - Vision, Image and Signal Processing **149**(2), 108–114 (2002)

350. Hsieh, C.T., Lai, E., Wang, Y.C.a.: Robust speech features based on wavelet transform with application to speaker identification. Vision, Image and Signal Processing, IEE Proceedings - **149**(2), 108–114 (2002)

351. HTK. Web (2011). URL http://htk.eng.cam.ac.uk

352. Huang, H.Y.: Unified Approach to Quadratically Convergent Algorithms for Function minimization. Journal of Optimization Theory and Applications **5**(6), 405–423 (1970)

353. Huang, N.E., Shen, Z., Long, S.R., Wu, M.C., Shih, H.H., Zheng, Q., Yen, N.C., Tung, C.C., Lui, H.H.: The Empirical Mode Decomposition and the Hilbert Spectrum for Nonlinear and Non-Stationary Time Series Analysis. Proceedings of the Royal Society of London **454**(1971), 903–995 (1998)

354. Huang, R., Hansen, J.a.: Advances in unsupervised audio classification and segmentation for the broadcast news and NGSW corpora. Audio, Speech, and Language Processing, IEEE Transactions on **14**(3), 907–919 (2006)

355. Huang, X., Acero, A., Adcock, J., wucn Hon, H., Goldsmith, J., Liu, J., Plumpe, M.: A Trainable Text-to-Speech System. In: Proceedings of the International Conference on Spoken Language Processing (ICSLP), pp. 2387–2390 (1996)

356. Hunt, M., Richardson, S., Bateman, D., Piau, A.: An Investigation of PLP and IMELDA Acoustic Representations and of Their Potential for Combination. In: International Conference on Acoustics, Speech, and Signal Processing (ICASSP), vol. 2, pp. 881–884 (1991)

357. Hurley, D., Nixon, M., Carter, J.a.: A new force field transform for ear and face recognition. In: Image Processing, 2000. Proceedings. 2000 International Conference on, vol. 1, pp. 25–28 (2000)

358. Inal, M., Fatihoglu, Y.a.: Self organizing map and associative memory model hybrid classifier for speaker recognition. In: Neural Network Applications in Electrical Engineering, 2002. NEUREL '02. 2002 6th Seminar on, pp. 71–74 (2002)

359. ISO: ISO/IEC 11172-3:1993 (1993). URL http://www.iso.org/iso/iso_catalogue/catalogue_tc/catalogue_detail.htm?csnumber=22412

360. ISO: ISO/IEC 13818-3:1998 (1998). URL http://www.iso.org/iso/iso_catalogue/catalogue_ics/catalogue_detail_ics.htm?csnumber=26797

361. ISO: ISO/IEC JTC1 SC37, Biometric Advanced Programming Interface. Website (2008). URL http://www.bioapi.org

362. ISO: ISO/JTC1 SC37 WG3, Biometric Data Interchange Format (2009).   URL http://www.iso.org/iso/standards_development/technical_committees/list_of_iso_technical _committees/iso_technical_committee.htm?commid=313770

363. ITU-T: G.711: Pulse Code Modulation (PCM) of Voice Frequencies. ITU-T Recommendation (1988). URL http://www.itu.int/rec/T-REC-G.711-198811-I/en

364. ITU-T: G.721: 32 kbit/s adaptive differential pulse code modulation (ADPCM). ITU-T Recommendation (1988). URL http://www.itu.int/rec/T-REC-G.721/e

365. ITU-T: G.723: Extensions of Recommendation G.721 Adaptive Differential Pulse Code Modulation to 24 and 40 kbit/s for Digital Circuit Multiplication Equipment Application. ITU-T Recommendation (1988). URL http://www.itu.int/rec/T-REC-G.723/e

366. ITU-T: G.726: 40, 32, 24, 16 kbit/s Adaptive Differential Pulse Code Modulation (ADPCM). ITU-T Recommendation (1990). URL http://www.itu.int/rec/T-REC-G.726/e

367. ITU-T: G.114: One-way transmission time.   ITU-T Recommendation (2003).   URL http://www.itu.int/rec/T-REC-G.114/e

368. ITU-T: G.122: Influence of national systems on stability and talker echo in international connections. ITU-T Recommendation (2003). URL http://www.itu.int/rec/T-REC-G.122/e

369. ITU-T: G.131: Talker echo and its control.   ITU-T Recommendation (2003).   URL http://www.itu.int/rec/T-REC-G.131/e

370. ITU-T: P.56: Objective measurement of active speech level. ITU-T Recommendation (2003). URL http://www.itu.int/rec/T-REC-P.56/e

371. ITU-T: G.729: Coding of speech at 8 kbit/s using conjugate-structure algebraic-code-excited linear prediction (CS-ACELP).   ITU-T Recommendation (2008).   URL http://www.itu.int/rec/T-REC-G.729/e

372. J., A.: Effect of age and gender on LP smoothed spectral envelope. In: The IEEE Odyssey Speaker and Language Recognition Workshop, pp. 1–4 (2006)

373. Jaakkola, T., Haussler, D.: Exploiting Generative Models in Discriminative Classifiers. In: Advances in Neural Information Processing Systems, vol. 11, pp. 487–493. MIT Press (1998)

374. Jaakkola, T., Meila, M., Jebara, T.: Maximum Entropy Discrimination. In: Advances in Neural Information Processing Systems, vol. 12, pp. 470–476. MIT Press (1999)

375. Jacobson, D.H., Oksman, W.: An Algorithm that Minimizes Homogeneous Functions of n Variables in n+2 Iterations and Rapidly Minimizes General Functions. Tech. rep., Technical Report 618, Division of Engineering and Applied Physics, Harvard, University, Cambridge, MA (1970)

376. James, D.B.: A Method of Unfolding the Cerebral Cortex or Any Other Folded Surface. Kybernetes **27**(8), 959–961 (1998)

377. Jancey, R.C.: Multidimensional Group Analysis. Australian Journal of Botany **14**, 127–130 (1966)

378. Jancey, R.C.: Algorithm for Detection of Discontinuities in Data Sets. Vegetatio **29**(2), 131–133 (1974)

379. Jankowski C.R., J., Quatieri, T., Reynolds, D.a.: Formant AM-FM for speaker identification. In: Time-Frequency and Time-Scale Analysis, 1994., Proceedings of the IEEE-SP International Symposium on, pp. 608–611 (1994)

380. Jaynes, E.T.: Information Theory and Statistical Mechanics. The Physical Review **106**(4), 620–630 (1957)

381. Jaynes, E.T.: Information Theory and Statistical Mechanics II. The Physical Review **108**(2), 171–190 (1957)

382. Jeffreys, H.: An Invariant Form for the Prior Probability in Estimation Problems. Proceedings of the Royal Society of London **186**(1007), 453–461 (1946)

383. Jerri, A.J.: The Shannon Sampling Theorem – Its Various Extensions and Applications: A Tutorial Review. Proceedings of the IEEE **65**(11), 1565–1596 (1977)

384. Jerri, A.J.: Correction to The Shannon Sampling Theorem – Its Various Extensions and Applications: A Tutorial Review. Proceedings of the IEEE **67**(4), 695–695 (1979)

385. Jin, H., Kubala, F., Schwartz, R.: Automatic Speaker Clustering. In: Proceedings of the Speech Recognition Workshop (1997)

386. Jin, M., Soong, F., Yoo, C.a.: A Syllable Lattice Approach to Speaker Verification. Audio, Speech, and Language Processing, IEEE Transactions on **15**(8), 2476–2484 (2007)

387. Jin, Q., Navratil, J., Reynolds, D., Campbell, J., Andrews, W., Abramson, J.a.: Combining cross-stream and time dimensions in phonetic speaker recognition. In: Acoustics, Speech, and Signal Processing, 2003. Proceedings. (ICASSP '03). 2003 IEEE International Conference on, vol. 4, pp. IV–800–3 (2003)

388. Jin, Q., Pan, Y., Schultz, T.a.: Far-Field Speaker Recognition. In: Acoustics, Speech and Signal Processing, 2006. ICASSP 2006 Proceedings. 2006 IEEE International Conference on, vol. 1, pp. I–I (2006)

389. Johnsen, M.H., Svendsen, T., Harborg, E.:

390. Johnson, R.W., Shore, J.E.: Comments on and Corrections to 'Axiomatic Derivation of the Principle of Maximum Entropy and the Principle of Minimum Cross-Entropy'. IEEE Transactions on Information theory **IT-29**(6), 942–943 (1983)

391. Johnson, S.G., Frigo, M.: A Modified Split-Radix FFT with Fewer Arithmetic Operations. IEEE Transactions on Signal Processing **55**(1), 111–119 (2007)

392. Jolliffe, I.: Principal Component Analysis, 2nd edn. Springer, New york (2002)

393. Jordan, M.C.: Mémoire sur les Formes Bilinéaires. Journal de Math'ematiques Pures et Appliquées **19**, 35–54 (1874)

394. Junqua, J.C., Perronnin, F., Kuhn, R.: Voice Personalization of Speech Synthesizer. U.S. Patent (2005). Patent No. U.S. 6,970,820

395. Kablenet: No Snooping on the Public. World Wide Web (2007). URL http://www.theregister.co.uk/2007/08/03/cctv_audio_recording_consultation

396. Kadri, H., Lachiri, Z., Ellouze, N.a.: Robustness Improvement of Speaker Segmentation techniques Based on the Bayesian Information Criterion. In: Information and Communication Technologies, 2006. ICTTA '06. 2nd, vol. 1, pp. 1300–1301 (2006)

397. Kailath, T.: The Divergence and Bhattacharyya Distance Measure in Signal Selection. IEEE Transactions on Communication Technology **15**(1), 52–60 (1967)

398. Kajarekar, S.S., Ferrer, L., Stolcke, A., Shriberg, E.: Voice-Based Speaker Recognition Combining Acoustic and Stylistic Features. In: N.K. Ratha, V. Govindaraju (eds.) Advances in Biometrics: Sensors, Algorithms and Systems, pp. 183–201. Springer, New York (2008)

399. Kanak, A., Erzin, E., Yemez, Y., Tekalp, A.a.: Joint audio-video processing for biometric speaker identification. In: Acoustics, Speech, and Signal Processing, 2003. Proceedings. (ICASSP '03). 2003 IEEE International Conference on, vol. 2, pp. II–377–80 (2003)

400. Kandel, E., Rose, C.: Charlie Rose Series on the Brain. Television (2009). URL http://www.charlierose.com

401. Kandel, E.R.: Making Your Mind, Molecules, Motion and Memory. Lecture (2008)

402. Kao, Y.H., Rajasekaran, P., Baras, J.a.: Free-text speaker identification over long distance telephone channel using hypothesized phonetic segmentation. In: Acoustics, Speech, and Signal Processing, 1992. ICASSP-92., 1992 IEEE International Conference on, vol. 2, pp. 177–180 (1992)

403. Karman, R.E.: A New Approach to Linear Filtering and Prediction Problems. Transactions of the ASME–Journal of Basic Engineering **82**(Series D), 35–45 (1960)

404. Karush, W.: Minima of Functions of Several Variables with Inequalities as Side Constraints. University of Chicago, Illinois (1939). Masters Thesis

405. Kellogg, O.: On the Existence and Closure of Sets of Characteristic Functions. Mathematische Annalen **86**(1002), 14–17 (1922)

406. Kendall, M.G.: The Estimation of Parameters in Linear Autoregressive Time Series. Econometrica **17**, 44–57 (1949). Supplement: Report of the Washington Meeting

407. Kenny, P.: Joint Factor Analysis of Speaker and Session Varaiability: Theory and Algorithms. Technical report, CRIM (2006). URL http://www.crim.ca/perso/patrick.kenny/FAtheory.pdf

408. Kenny, P., Boulianne, G., Ouellet, P., Dumouchel, P.: Factor Analysis Simplified. In: International Conference on Acoustics, Speech, and Signal Processing (ICASSP), vol. 1, pp. 637–640 (2005)

409. Kenny, P., Ouellet, P., Dehak, N., Gupta, V., Dumouchel, P.a.: A Study of Interspeaker Variability in Speaker Verification. Audio, Speech, and Language Processing, IEEE Transactions on **16**(5), 980–988 (2008)

410. Kent, R.G.: Old Persian Grammar Texts Lexicon. 2nd edn. American Oriental Society, New Haven, Connecticut (1953)

411. Kersta, L.G.: Voiceprint Identification. Nature **196**, 1253–1257 (1962)

412. Khintchine, A.Y.: Fundamental Laws of Probability Theory, russian edn. The University Series in Higher Mathematics. Moscow (1927)

413. Khintchine, A.Y.: Sur la Loi des Grands Nombres. Comptes rendus de l'Acadmie des Sciences **189**, 477–479 (1929)

414. Kiernan, L., Mason, J., Warwick, K.: Robust Initialisation of Gaussian Radial Basis Function Networks using Partitioned k-Means Clustering. IET Electronics Letters **32**(7), 671–673 (1996)

415. Kijima, Y., Nara, Y., Kobayashi, A., Kimura, S.a.: Speaker adaptation in large-vocabulary voice recognition. In: Acoustics, Speech, and Signal Processing, IEEE International Conference on ICASSP '84., vol. 9, pp. 405–408 (1984)

416. Kim, J., Scott, C.: L2 Kernel Classification. IEEE Transactions on Pattern Analysis and Machine Intelligence **32**(10), 1822–1831 (2010)

417. Kim, M.S., Yu, H.J.: A New Feature Transformation Method Based on Rotation for Speaker Identification. pp. 68–73 (2007)

418. Kim, N.S.: Feature domain compensation of nonstationary noise for robust speech recognition. Speech Communication **37**(3–4), 59–73 (2002)

419. Kirby, M., Sirovich, L.: Application of the Karhunen-Loeve Procedure for the Characterization of Human Faces. IEEE Transactions on Pattern Analysis and Machine Intelligence **12**(1), 103–108 (1990)

420. Kirkpatrick, S., Gelatt, C.D., Vecchi, M.P.: Optimization by Simulated Annealing. Science **220**(4598), 671–680 (1983)

421. Kishore, S., Yegnanarayana, B.a.: Speaker verification: minimizing the channel effects using autoassociative neural network models. In: Acoustics, Speech, and Signal Processing, 2000. ICASSP '00. Proceedings. 2000 IEEE International Conference on, vol. 2, pp. II1101–II1104 (2000)

422. Kitahara, T., Goto, M., Komatani, K., Ogata, T., Okuno, H.G.: INSTROGRAM: A New Musical Instrument Recognition Technique Without Using Onset Detection Nor F0 Estimation. In: International Conference on Acoustics, Speech, and Signal Processing (ICASSP), vol. 5, pp. 229–232 (2006)

423. Kohler, M., Andrews, W., Campbell, J., Herndndez-Cordero, J.a.: Phonetic speaker recognition. In: Signals, Systems and Computers, 2001. Conference Record of the Thirty-Fifth Asilomar Conference on, vol. 2, pp. 1557–1561 (2001)

424. Kohonen, T.: Self-Organizination and Associative Memory, 3rd edn. Springer-Verlag, Berlin (1989). ISBN: 0-387-51387-6

425. Kolmogorov, A.N.: Grundbegriffe der Wahrscheinlichkeitsrechnung (1933)

426. Kolmogorov, A.N.: Foundations of the Theory of Probability, 2nd english edn. The University Series in Higher Mathematics. Chelsea Publishing Company, New York (1956)

427. Koolwaaij, J., Boves, L.: A new procedure for classifying speakers in speaker verification systems. In: EUROSPEECH (1997)

428. Koolwaaij, J., Boves, L., Jongebloed, H., den Os, E.: On Model Quality and Evaluation in Speaker Verification. In: International Conference on Acoustics, Speech, and Signal Processing (ICASSP), vol. 6, pp. 3759–3762 (2000)

429. Koopman, B.: On Distribution Admitting a Sufficient Statistic. Transactions of the American Mathematical Society **39**(3), 399–409 (1936)

430. Kotel'nikov, V.A.: On the Transmission apacity of Ether and Wire in Electrocommunications. In: Izd. Red. Upr. Svyazi RKKA (The First All-Union Conference on Questions of Communications) (1933). English Translation by C.C. Bissell and V. E. Katsnelson

431. Kotti, M., Benetos, E., Kotropoulos, C.a.: Computationally Efficient and Robust BIC-Based Speaker Segmentation. Audio, Speech, and Language Processing, IEEE Transactions on **16**(5), 920–933 (2008)

432. Kotti, M., Martins, L., Benetos, E., Cardoso, J., Kotropoulos, C.a.: Automatic Speaker Segmentation using Multiple Features and Distance Measures: A Comparison of Three Approaches. In: Multimedia and Expo, 2006 IEEE International Conference on, pp. 1101–1104 (2006)

433. Kratochvil, P.: Tone in Chinese. In: E.C. Fudge (ed.) Phonolgy, Selected Readings. Penguin Books, Middlesex (1973)

434. Kühn, H.W., Tucker, A.W.: Nonlinear Programming. In: j. Neyman (ed.) Proceedings of the Second Berkeley Symposium on mathematical Statistics and Probability, pp. 481–492 (1951)

435. Kuhn, R., Junqua, J.C.: Rapid Speaker Adaptation in Eigenvoice Space. IEEE Transaction of Speech and Audio Processing **8**(6), 695–707 (2000)

436. Kukula, E., Elliott, S.: Implementation of hand geometry: an analysis of user perspectives and system performance. IEEE Aerospace and Electronic Systems Magazine **21**(3), 3–9 (2006)

437. Kullback, S.: Information Theory and Statistics. Dover Publications, Inc., New York (1997). Unabridged publication of the original text published by John Wiley, New York (1959), ISBN: 0-486-69684-7

438. Kullback, S., Leibler, R.A.: On Information and Sufficiency. The Annals of Mathematical Statistics **22**(1), 79–86 (1951)

439. Kumar, A., Ravikanth, C.: Personal Authentication Using Finger Knuckle Surface. IEEE Transactions on Information Forensics and Security **4**(1), 1–13 (2009)

440. Kumar, P., Wasan, S.K.: Comparative Analysis of k-means Based Algorithms. IJCSNS International Journal of Computer Science and Network Security **10**(4), 1–13 (2010)

441. Kuo, B.C.: Digital Control Systems, 2nd edn. Oxford University Press, New York (1992). ISBN: 0-195-12064-7

442. Kwok, J.T., Mak, B., Ho, S.: Eigenvoice Speaker Adaptation via Composite Kernel PCA. In: Advances in Neural Information Processing Systems 16. MIT Press (2003)

443. Ladefoged, P.: A Course in Phonetics, 5th edn. Wadsworth, Boston (2006). ISBN: 1-413-00688-4

444. Lagrange, M., Martins, L., Teixeira, L., Tzanetakis, G.a.: Speaker Segmentation of Interviews Using Integrated Video and Audio Change Detectors. In: Content-Based Multimedia Indexing, 2007. CBMI '07. International Workshop on, pp. 219–226 (2007)

445. Lamel, L., Gauvain, J.L.a.: Speaker recognition with the Switchboard corpus. In: Acoustics, Speech, and Signal Processing, 1997. ICASSP-97., 1997 IEEE International Conference on, vol. 2, pp. 1067–1070 (1997)

446. Landau, E.G.H.: Handbuch der Lehre von der Verteilung der Primzahlen, 3rd edn. Chelsea Publishing Company, New York (1974). Both volumes in one. First Edition was published in two volumes, in 1909. ISBN: 0-8284-0096-2

447. Latry, C., Panem, C., Dejean, P.: Cloud detection with SVM technique. In: IEEE International Geoscience and Remote Sensing Symposium (IGARSS, pp. 448–451 (2007)

448. Lattner, S., Meyer, M.E., Fiederici, A.D.: Voice Perception: Sex, Pitch and the Right Hemisphere. Human Brain Mapping **24**(1), 11–20 (2005)

449. Laver, J.: Principles of Phonetics. Cambridge Press, New York (1994). ISBN: 0-521-45031-4

450. Lawley, D., Maxwell, A.: Factor Analysis as a Statistical Method. Butterworths Mathematical Texts, London (1971)

451. Laxman, S., Sastry, P.a.: Text-dependent speaker recognition using speaker specific compensation. In: TENCON 2003. Conference on Convergent Technologies for Asia-Pacific Region, vol. 1, pp. 384–387 (2003)

452. LDC: Linguiostic Data Consortium. Web (2010). URL http://www.ldc.upenn.edu/. Speech Data Resource

453. Lebesgue, H.: Intégrale, Longueue, Aire. Université de Paris, Paris, France (1902). PhD Thesis

454. Lee, D., Baek, S., Sung, K.: Modified k-Means Algorithm for Vector Quantizer Design. IEEE Signal Processing Letters **4**(1), 2–4 (1997)
455. Lee, K.H., Chung, K., Chung, J.M., Coggeshall, R.E.: Correlation of Cell Body Size, Axon Size, and Signal Conduction Velocity for Invidually Labelled Dorsal Root Ganglion Cells in the Cat. The Journal of Comparative Neurology **243**(3), 335–346 (1986)
456. Lee, L.S.: Voice Dictation of Mandarin Chinese. IEEE Signal Processing Magazine **14**(4), 63–101 (1997)
457. Leggetter, C.J., Woodland, P.C.: Maximum Likelihood Linear Regression for Speaker Adaptation of Continuous Density Hidden Markov Models. Computer Speech and Language **9**(2), 171–185 (1995)
458. Levenberg, K.L.: A Method for the Solution of Certain Non-Linear Problems in Least Squares. Quarterly of Applied mathematics **2**, 164–168 (1944)
459. Levinson, N.: The Wiener RMS (Root-Mean-Square) Error Criterion in Filter Design and Prediction. Journal of Mathematics and Physics **25**, 261–278 (1947)
460. Lévy, C., Linarès, G., Bnastre, J.F.: Compact Acoustic Models for Embedded Speech Recognition. EURASIP Journal on Audio, Speech, and Music Processing **2009** (2009). Article ID 806186, 12 pages, 2009. doi:10.1155/2009/806186
461. Lewis, M.P. (ed.): Ethnologue, 16th edn. SIL International (2009). URL http://www.sil.org
462. Li, B., Liu, W., and, Q.Z.: Text-dependent speaker identification using Fisher differentiation vector. In: Natural Language Processing and Knowledge Engineering, 2003. Proceedings. 2003 International Conference on, pp. 309–314 (2003)
463. Li, C.J., Beigi, H.S., Li, S., Liang, J.: Nonlinear Piezo-Actuator Control by Learning Self-Tuning Regulator. ASME Transactions, Journal of Dynamic Systems, Measurement, and Control **115**(4), 720–723 (1993)
464. Li, H., Zhang, Y.X.: An algorithm of soft fault diagnosis for analog circuit based on the optimized SVM by GA. In: International Conference on Electronic Measurement and Instruments (ICMEI), vol. 4, pp. 1023–1027 (2009)
465. Li, S.Z., Jain, A.K. (eds.): Handbook of Face Recognition. Springer, New York (2005). ISBN: 978-0-387-40595-7
466. Li, Z., Jiang, W., Meng, H.: Fishervioce: A discriminant subspace framework for speaker recognition. In: International Conference on Acoustics, Speech, and Signal Processing (ICASSP), pp. 4522–4525 (2010)
467. Liese, F., Vajda, I.: On Divergences and Informations in Statistics and Information Theory. IEEE Transactions on Information Theory **52**(10), 4394–4412 (2006)
468. Likas, A.C., Vlassis, N., Verbeek, J.: The Global k-Means Clustering Algorithm. Pattern Recognition **36**(2), 451–461 (2003)
469. Lim, Y.S., Choi, J.S., and, M.K.: Particle Filter Algorithm for Single Speaker Tracking with Audio-Video Data Fusion. In: Robot and Human interactive Communication, 2007. RO-MAN 2007. The 16th IEEE International Symposium on, pp. 363–367 (2007)
470. Lin, C.C., Chen, S.H., Lin, T.C., Truong, T.a.: Feature Comparison among Various Wavelets in Speaker Recognition Using Support Vector Machine. In: Multimedia, 2006. ISM'06. Eighth IEEE International Symposium on, pp. 811–816 (2006)
471. Linde, Y., Buzo, A., Gray, R.M.: An Algorithm for Vector Quantizer Design. IEEE Transactions on Communications **28**(1), 84–95 (1980)
472. Lingras, P., West, C.: Interval Set Clustering of Web Users with Rough K-Means. Journal of Intelligent Information Systems **23**(1), 5–16 (2004)
473. Liou, H.S., Mammone, R.a.: Speaker verification using phoneme-based neural tree networks and phonetic weighting scoring method. In: Neural Networks for Signal Processing [1995] V. Proceedings of the 1995 IEEE Workshop, pp. 213–222 (1995)
474. Lippmann, R.P.: An Introduction to Computing with Neural Nets. The IEEE Acoustic, Speech, and Signal Processing Magazine **4**(2), 4–22 (1987). Part 1
475. Liu, H., and, J.Y.: Multi-view Ear Shape Feature Extraction and Reconstruction. In: Signal-Image Technologies and Internet-Based System, 2007. SITIS '07. Third International IEEE Conference on, pp. 652–658 (2007)

476. Livshin, A.A., Rodet, X.: Musical Instrument Identification in Continuous Recordings. In: Proceedings of the 7th International Conference on Digital Audio Effects (DAFX-04), pp. 1–5 (2004)

477. Lloyd, S.: Least squares quantization in PCM. IEEE Transactions on Information Theory **28**(2), 129–137 (1982)

478. Lodi, A., Toma, M., Guerrieri, R.a.: Very low complexity prompted speaker verification system based on HMM-modeling. In: Acoustics, Speech, and Signal Processing, 2002. Proceedings. (ICASSP '02). IEEE International Conference on, vol. 4, pp. IV–3912–IV–3915 (2002)

479. Loève, M.: Probability Theory, 3rd edn. The University Series in Higher Mathematics. C. Scribnet's Sons, New York (1902)

480. Longman, R.W., Beigi, H.S., Li, C.J.: Learning Control by Numerical Optimization Methods. In: Proceedings of the Modeling and Simulation Conference, Control, Robotics, Systems and Neural Networks, vol. 20, pp. 1877–1882 (1989)

481. Loughlin, P.J., Tacer, B.: On the Amplitude- and Frequency-Modulation Decomposition of Signals. The Journal of the Acoustical Society of America (JASA) **100**(3), 1594–1601 (1996)

482. Lundy, T., Buskirk, J.V.: A new matrix approach to real FFTs and convolutions of length 2k. Computing **80**(1), 23–45 (2007)

483. Lung, S.Y., Chen, C.C.a.: Further reduced form of Karhunen-Loeve transform for text independent speaker recognition. Electronics Letters **34**(14), 1380–1382 (1998)

484. MacDonald, D.K.C.: Information Theory and Its Application to Taxonomy. Journal of Applied Physics **23**(5), 529–531 (1952)

485. MacQueen, J.: Some methods for classification and analysis of multivariate observations. In: Proceedings of the Fifth Berkeley Symposium on Mathematical Statistics and Probability, vol. 1, pp. 281–297 (1967)

486. Maddieson, I.: Patterns of Sounds. Cambridge University Press, Cambridge (1984)

487. Maes, S.H., Beigi, H.S.: Open SESAME! Speech, Password or Key to Secure Your Door? In: Asian Conference on Computer Vision (1998)

488. Mahalanobis, P.C.: On the Generalized Distance in Statistics. Proceedings of the National Institute of Sciences of India **12**, 49–55 (1936)

489. Mak, M., Allen, W., Sexton, G.a.: Speaker identification using radial basis functions. In: Artificial Neural Networks, 1993., Third International Conference on, pp. 138–142 (1993)

490. Makhoul, J.: Spectral Linear Prediction: Properties and Applications. IEEE Transactions on Audio, Speech and Signal Processing **23**(3), 283–296 (1975)

491. Makhoul, J., Cosell, L.: LPCW: An LPC Vocoder with Linear Predictive Spectral Warping. International Conference on Acoustics, Speech, and Signal Processing (ICASSP) **1**, 466–469 (1976)

492. Maltoni, D.: A Tutorial on Fingerprint Recognition. In: Advanced Studies in Biometrics, *Lecture Notes in Computer Science*, vol. 3161, pp. 43–68. Springer, Berlin/Heidelberg (2005). ISBN: 978-3-540-26204-6

493. Mangasarian, O.L.: Nonlinear Programming. Society for Industrial and Applied Mathematics (SIAM) (1994). Originally Published: New York: McGraw Hill, 1969; ISBN: 0-89871-341-2

494. Manning, C.D.: Foundations of Statistical Natural Language Processing. The MIT Press, Boston (1999). ISBN: 0-262-13360-1

495. Marian, V., Shildkrot, Y., Blumenfeld, H.K., Kaushanskaya, M., Faroqi-Shah, Y., Hirsch, J.: Cortical Activation During Word Processing in Late Bilinguals: Similarities and Differences as Revealed by Functional Magnetic Resonance Imaging. Journal of Clinical and Experimental Neuropsychology **29**(3), 247–265 (2007)

496. Mariéthoz, J., Bengio, S.: A Comparative Study of Adaptation Methods for Speaker Verification. In: Proceedings of the International Conference on Spoken Language Processing (ICSLP), pp. 25–28 (2002)

497. Markov, A.A.: Investigation of a Remarkable Case of Dependent Samples. pp. 61–80 (1907)

498. Markov, A.A.: On a Case of Samples Connected in Comples Chains. pp. 171–186 (1911)

499. Markov, A.A.: An Example of Statistical Investigation of the Text *Eugene Onegin* Concerning the Connection of Samples in Chains. In: Izd. Red. Upr. Svyazi RKKA (The First All-Union Conference on Questions of Communications), vol. 7, pp. 153–162 (1913). Lecture – English Translation: Classical Text in Translation, Science in Context, **19**(4), pp. 591–600 (2006)

500. Markowitz, J.A.: Project 1821 - INCITS 456:200x, Information Technology - Speaker Recognition Format for Raw Data Interchange (SIVR-1) (2009). URL http://www.incits.org/scopes/bsr8_1821.htm

501. Markowitz, J.A.: Standards for Speaker Recognition. In: S.Z. Li (ed.) Encyclopedia of Biometrics. Springer, New York (2009). ISBN: 978-0-387-73003-5

502. Marquardt, D.W.: An Algorithm of Least Squares Estimation of Nonlinear Parameters. SIAM journal on Applied Mathematics **11**, 431–441 (1963)

503. Marshall, J.C., Fink, G.R.: Cerebral Localization, Then and Now. NeuroImage **20**(Supplement 1), S2 S7 (2003). Convergence and Divergence of Lesion Studies and Functional Imaging of Cognition

504. Martens, J.B.: Recursive Cyclotomic Factorization – A New Algorithm for Calculating the Discrete Fourier Transform. IEEE Transactions on Acoustic, Speech and Signal Processing **32**(4), 750–761 (1984)

505. Martin, A., Doddington, G., Kamm, T., Ordowski, M., Przybocki, M.: The DET Curve in Assessment of Detection Task Performance. In: Eurospeech 1997, pp. 1–8 (1997)

506. Martin, A., Przybocki, M.: The NIST 1999 Speaker Recognition Evaluation – An Overview. Digital Signal Processing **10**, 1–18 (2000)

507. Martin, K.D.: Sound-Source Recognition: A Theory and Computational Model. Massachusetts Institute of Technology, Cambridge, MA (1999). PhD Thesis

508. Martin, K.D., Kim, Y.E.: Musical instrument identification: A pattern-recognition approach. In: 136th Meeting of the Acoustical Society of America (1998)

509. Matrouf, D., Scheffer, N., Fauve, B., Bonastre, J.F.: A Strainghforward and Efficient Implementation of the Factor Analysis Model for Speaker Verification. In: International Conference on Speech Communication and Technology (2007)

510. Matsui, T., Furui, S.a.: Speaker adaptation of tied-mixture-based phoneme models for text-prompted speaker recognition. In: Acoustics, Speech, and Signal Processing, 1994. ICASSP-94., 1994 IEEE International Conference on, vol. i, pp. I/125–I/128 (1994)

511. Matsushita, K.: Decision Rule, Based on Distance, for the Classification Problem. Annals of the Institute of Statistical Mathematics **8**, 67–77 (1956)

512. Mazarakis, G., Tzevelekos, P., Kouroupetroglou, G.: Musical Instrument Recognition and Classification Using Time Encoded Signal Processing and Fast Artificial Neural Networks. In: Advances in Artificial Intelligence, *Lecture Notes in Computer Science*, vol. 3955, pp. 246–255. Springer, Berlin/Heidelberg (2006). URL http://dx.doi.org/10.1007/11752912_26. 10.1007/11752912_26

513. McCormick, G.P., Ritter, K.: Methods of Conjugate Directions Versus Quasi-Newton Methods. Mathematical programming **3**, 101–116 (1972)

514. McLachlan, G.J., Krishnan, T.: The EM Algorithm and Extensions, 2nd edn. Wiley Series in Probability and Statistics. John Wiley & Sons, New York (2008). ISBN: 0-471-20170-7

515. McLachlan, G.J., Peel, D.: Finite Mixture Models, 2nd edn. Wiley Series in Probability and Statistics. John Wiley & Sons, New York (2000). ISBN: 0-471-00626-2

516. Meirovitch, L.: Methods of Analytical Dynamics. Advanced Engineering Series. McGraw-Hill, New York (1970). ISBN: 0-704-1455-6

517. Meltzer, S., Moser, G.: MPEG-4 HE-AAC v2 – audio coding for today's digital media world. World Wide Web (2005). URL http://www.ebu.ch/fr/technical/trev/trev_305-moser.pdf

518. Mendelovicz, E., Sherman, J.W.: Truncation Error Bounds for Signal Sampling. In: 9th Annual Asilomar Conference on Circuits Systems and Computers, p. 16 (1975)

519. Mercer, J.: Functions of Positive and Negative Type, and their Connection with the Theory of Integral Equations. Philosophical Transactions of the Royal Society of London. Series A, Containing Papers of a Mathematical or Physical Character **209**, 415–446 (1909)

520. Meyer, J.: Acoustic Strategy and Topology of Whistled Langues; Phonetic COmparison and Perceptual Cases of Whistled Vowels. Journal of the International Phonetic Association **38**(1), 64–90 (2008)

521. Michiel Hazewinkel, E.: Encyclopaedia of Mathematics. Kluwer Academic Publishers, Amsterdam (2002). ISBN: 1-4020-0609-8

522. Miller, R.L.: Nature of the Vocal Cord Wave. Journal of the Acoustical Society of America **31**, 667–677 (1959)

523. Minsky, M., Papert, S.: Perceptrons: An Introduction to Computation Geometry. Massachusetts Institute of Technology, Cambridge, MA (1969)

524. Molla, K., Hirose, K.a.: On the effectiveness of MFCCs and their statistical distribution properties in speaker identification. In: Virtual Environments, Human-Computer Interfaces and Measurement Systems, 2004. (VECIMS). 2004 IEEE Symposium on, pp. 136–141 (2004)

525. Molla, M., Hirose, K., Minematsu, N.: Robust speaker identification system using multi-band dominant features with empirical mode decomposition. In: 10th International Conference on Computer and Information technology (ICCIT 2008), pp. 1–5 (2008)

526. Molla, M., Hirose, K., Minematsu, N.a.: Robust speaker identification system using multiband dominant features with empirical mode decomposition. In: Computer and Information Technology, 2008. ICCIT 2008. 10th International Conference on, pp. 1–5 (2007)

527. Moreno, P.J., Ho, P.P., Vasconcelos, N.: A Kullback-Leibler Divergence Based Kernel for SVM Classification in Multimedia Applications. In: Advances in Neural Information Processing Systems 16. MIT Press (2004)

528. Morito, M., Yamada, K., Fujisawa, A., Takeuchi, M.a.: A single-chip speaker independent voice recognition system. In: Acoustics, Speech, and Signal Processing, IEEE International Conference on ICASSP '86., vol. 11, pp. 377 380 (1986)

529. Morlet, J., Arens, G., Fourgeau, I., Giard, D.: Wave Propagation and Sampling Theory. Geophysics **47**, 203–236 (1982)

530. Morris, C.: Whistling Turks. Website (1999). URL http://news.bbc.co.uk/2/hi/programmes/from_our_own_correspondent/506284.stm

531. Moschou, V., Kotti, M., Benetos, E., Kotropoulos, C.a.: Systematic comparison of BIC-based speaker segmentation systems. In: Multimedia Signal Processing, 2007. MMSP 2007. IEEE 9th Workshop on, pp. 66–69 (2007)

532. Muroi, T., Takiguchi, T., Ariki, Y.a.: Speaker Independent Phoneme Recognition Based on Fisher Weight Map. In: Multimedia and Ubiquitous Engineering, 2008. MUE 2008. International Conference on, pp. 253–257 (2008)

533. Murtagh, B., Sargent, E.: A Constrained Minimization Method with Quadratic Convergence. In: R. Fletcher (ed.) Optimization. Academic Press, London (1969)

534. Murthy, H., Beaufays, F., Heck, L., Weintraub, M.a.: Robust text-independent speaker identification over telephone channels. Speech and Audio Processing, IEEE Transactions on **7**(5), 554–568 (1999)

535. Murty, L., Otake, T., Cutler, A.: Perceptual Tests of Rhythmic Similarity: I. Mora Rhythm. Language and Speech **50**(1), 77–99 (2007)

536. Naik, P., Tsai, C.L.: Residual Information Criterion for Single-Index Model Selections. Journal of Nonparametric Statistics **16**(1–2), 187–195 (2004)

537. Naini, A.S., Homayounpour, M.M.: Speaker age interval and sex identification based on Jitters, Shimmers and Mean MFCC using supervised and unsupervised discriminative classification methods. In: The 8th International Conference on Signal Processing, vol. 1 (2006)

538. Nathan, K.S., Beigi, H.S., Clary, G.J., Subrahmonia, J., Maruyama, H.: Real-Time On-Line Unconstrained Handwriting Recognition using Statistical Methods. In: International Conference on Acoustics, Speech, and Signal Processing (ICASSP95), vol. 4, pp. 2619–2622 (1995)

539. Nava, P., Taylor, J.a.: Speaker independent voice recognition with a fuzzy neural network. In: Fuzzy Systems, 1996., Proceedings of the Fifth IEEE International Conference on, vol. 3, pp. 2049–2052 (1996)

540. Navratil, J., Jin, Q., Andrews, W., Campbell, J.a.: Phonetic speaker recognition using maximum-likelihood binary-decision tree models. In: Acoustics, Speech, and Signal Processing, 2003. Proceedings. (ICASSP '03). 2003 IEEE International Conference on, vol. 4, pp. IV–796–9 (2003)

541. Nelwamondo, F., Mahola, U., Marwala, T.a.: Improving Speaker Identification Rate Using Fractals. In: Neural Networks, 2006. IJCNN '06. International Joint Conference on, pp. 3231–3236

542. Nelwamondo, F.V., Marwala., T.: Faults Detection Using Gaussian Mixture Models, Mel-Frequency Cepstral Coefficients and Kurtosis. In: IEEE International Conference on Systems, Man and Cybernetics, (SMC'06), vol. 1, pp. 290–295 (2006)

543. Neyman, J.: Frequentist probability and frequentist statistics. The Synthese Journal **360**(1), 97–131 (2004)

544. Nghia, P.T., Binh, P.V., Thai, N.H., Ha, N.T., Kumsawat, P.a.: A Robust Wavelet-Based Text-Independent Speaker Identification. In: Conference on Computational Intelligence and Multimedia Applications, 2007. International Conference on, vol. 2, pp. 219–223 (2007)

545. Nguyen, P.C., Akagi, M., and, T.B.H.: Temporal decomposition: a promising approach to VQ-based speaker identification. In: Multimedia and Expo, 2003. ICME '03. Proceedings. 2003 International Conference on, vol. 3, pp. III–617–20 (2003)

546. Nilsson, N.J.: Problem Solving Methods in Artificial Intelligence. McGraw-Hill Inc., New York (1971). ISBN: 0-070-46573-2

547. Nishida, M., Kawahara, T.: Speaker indexing and adaptation using speaker clustering based on statistical model selection. In: International Conference on Acoustics, Speech, and Signal Processing (ICASSP), vol. 1, pp. 172–175 (2004)

548. Nishitani, N., Schürmann, M., Amunts, K., Hari, R.: Brocas Region: From Action to Language. Physiology **20**(1), 60–69 (2005)

549. NIST: Speech File Manipulation Software (SPHERE) Package Version 2.6a. World Wide Web (1996). URL http://www.itl.nist.gov/iad/mig/tools/sphere_26atarZ.htm

550. NIST: Speaker Recognition Evaluation Site. Web (2010). URL http://www.nist.gov/itl/iad/mig/sre.cfm

551. Noble, B., Daniel, J.W.: Applied Linear Algebra, 2nd edn. Prentice-Hall, Inc., New Jersey (1977). ISBN: 0-130-41343-7

552. Nocedal, J., Wright, S.J.: Numerical Optimization, 2nd edn. Springer, New York (2000)

553. Nolan, F.: The Phonetic Bases of Speaker Recognition. Cambridge University Press, New York (1983). ISBN: 0-521-24486-2

554. Noll, A.M.: Short-Time Spectrum and 'Cepstrum' Techniques for Vocal-Pitch Detection. The Journal of the Acoustical Society of America (JASA) **36**(2), 296–302 (1964)

555. Noll, A.M.: Cepstrum Pitch Determination. The Journal of the Acoustical Society of America (JASA) **41**(2), 293–309 (1967)

556. Normandin, Y., Cardin, R., DeMori, R.: High-Performance Connected Digit Recognition using Maximum Mutual Information Estimation. IEEE Transactions on Speech and Audio Processing **2**(2), 299–311 (1994)

557. Nosrati, M.S., Faez, K., Faradji, F.a.: Using 2D wavelet and principal component analysis for personal identification based On 2D ear structure. In: Intelligent and Advanced Systems, 2007. ICIAS 2007. International Conference on, pp. 616–620 (2007)

558. Nwe, T.L., Sun, H., Li, H., Rahardja, S.: Speaker Diarization in Meeting Audio. In: International Conference on Acoustics, Speech, and Signal Processing (ICASSP), pp. 4073–4076 (2009)

559. Nyberg, H.S.: A Manual of Pahlavi, vol. I. Otto Harrassowitz, Wiesbaden (1964). Texts, Alphabets, Index, Paradigms, Notes

560. Nyberg, H.S.: A Manual of Pahlavi, vol. II. Otto Harrassowitz, Wiesbaden (1974). Idiograms, Glossary, Abbreviations, Index, Grammatical Survey, Corrigenda

561. Nyquist, H.: Certain Factors Affecting Telegraph Speed. Bell System Technical Journal **3**, 324–346 (1924)

562. Nyquist, H.: Certain Topics in Telegraph Transmission Theory. Transactions of the American Institute of Electrical Engineers (AIEE) **47**, 617–644 (1928). Reprint in Proceedings of the IEEE (2002), Vol. 90, No. 2, pp. 280–305

563. Omar, M.K., Pelecanos, J.: Training Universal Background Models for Speaker Recognition. In: The Speaker and Language Recognition Workshop (Odyssey 2010), pp. 52–57 (2010)

564. Omologo, M., Svaizer, P.: Acoustic event localization using a crosspower-spectrum phase based technique. In: International Conference on Acoustics, Speech, and Signal Processing (ICASSP), vol. II, pp. 273–276 (1994)

565. Oppenheim, A., Schafer, R.: From frequency to quefrency: a history of the cepstrum. IEEE Signal Processing Magazine **21**(5), 95–106 (2004)

566. Oppenheim, A.V.: Superposition in a Class of Nonlinear Systems. Massachussetts Institute of Technology, Cambridge, Massachussetts (1964). Ph.D. Dissertation

567. Oran, D.: Requirements for Distributed Control of Automatic Speech Recognition (ASR), Speaker Identification/Speaker Verification (SI/SV), and Text-to-Speech (TTS) Resources. RFC 4313 (Informational) (2005). URL http://www.ietf.org/rfc/rfc4313.txt

568. Oren, S.S.: On the Selection of Parameteres in Self Scaling Variable Metric Algorithms. Mathematical Programming **7**(1), 351–367 (1974)

569. Oren, S.S., Spedicato, E.: Optimal Conditioning of Self-Scaling Variable Metric Algorithms. Mathematical Programming **10**, 70–90 (1976)

570. Ortega-Garcia, J., Cruz-Llanas, S., Gonzalez-Rodriguez, J.a.: Speech variability in automatic speaker recognition systems for forensic purposes. In: Security Technology, 1999. Proceedings. IEEE 33rd Annual 1999 International Carnahan Conference on, pp. 327–331 (1999)

571. O'Shaughnessy, D.: Speech communications : human and machine, 2nd edn. IEEE Press, New York (2000). ISBN. 978-0-780-33449-6

572. Owens, A.J., Filkin, D.L.: Efficient Training of the Back Propagation Network by Solving a System of Stiff Ordinary Differential Equations. In: IEEE/INNS International Conference on Neural Networks (1989)

573. Paliwal, K., Atal, B.: Frequency-related representation of speech. In: Proceedings of the European Conference on Speech Communication and Technology (EUROSPEECH-03), pp. 65–68 (2003)

574. Papoulis, A.: Signal Analysis. McGraw Hill, New York (1977)

575. Papoulis, A.: Probability, Random Variables and Stochastic Processes, 3rd edn. McGraw-Hill, Inc., New York (1991). ISBN: 0-070-48477-5

576. Park, A., Glass, J.a.: A NOVEL DTW-BASED DISTANCE MEASURE FOR SPEAKER SEGMENTATION. In: Spoken Language Technology Workshop, 2006. IEEE, pp. 22–25 (2006)

577. Parker, D.B.: A Comparison of Algorithms for Neuron-Like Cells. In: Neural Networks for Computing, AIP conference Proceeding 151, pp. 327–332 (1986)

578. Parris, E., Carey, M.a.: Language independent gender identification. In: Acoustics, Speech, and Signal Processing, 1996. ICASSP-96. Conference Proceedings., 1996 IEEE International Conference on, vol. 2, pp. 685–688 (1996)

579. Pawlak, Z.: Rough Sets. International Journal of Computer and Information Sciences **11**, 341–356 (1982)

580. Pawlak, Z.: Rough Sets: Theoretical Aspects of Reasoning About Data. Kluwer Academic Publishers, Norwell, Massachusetts (1991)

581. Pearson, J.: Variable Metric Methods of Minimization. Computer Journal **12**, 171–178 (1969)

582. Pelecanos, J., Slomka, S., Sridharan, S.a.: Enhancing automatic speaker identification using phoneme clustering and frame based parameter and frame size selection. In: Signal Processing and Its Applications, 1999. ISSPA '99. Proceedings of the Fifth International Symposium on, vol. 2, pp. 633–636 (1999)

583. Pelecanos, J., Sridharan, S.: Feature Warping for Robust Speaker Verification. In: A Speaker Odyssey - The Speaker Recognition Workshop, pp. 213–218 (2001)

584. Pelleg, D., Moore, A.: x-Means: Extending k-Means with Efficient Estimation of the Number of Clusters. In: Proceedings of the Seventeenth International Conference on Machine Learning, pp. 261–265 (2000)

585. Pellom, B., Hansen, J.: An efficient scoring algorithm for Gaussian mixture model based speaker identification. IEEE Signal Processing Letters **5**(11), 281–284 (1998)

586. Peterson, G., Barney, H.L.: Control Methods Used in a Study of the Vowels. The Journal of the Acoustical Society of America (JASA) **24**(2), 175–185 (1952)

587. Pfeiffer, S.: The Ogg Encapsulation Format Version 0. RFC 3533 (Informational) (2003). URL http://www.ietf.org/rfc/rfc3533.txt

588. Phan, F., Micheli-Tzanakou, E., Sideman, S.a.: Speaker identification using neural networks and wavelets. Engineering in Medicine and Biology Magazine, IEEE **19**(1), 92–101 (2000)

589. Pierre, D.A.: Optimization Theory with Applications. Dover Publications, Inc., New york (1986). ISBN: 0-486-65205-X

590. Pitman, E.: Sufficient Statistics and Intrinsic Accuracy. Mathematical Proceedings of the Cambridge Philosophical Societ **32**, 567–579 (1936)

591. Poh, N., Bengio, S.: F-Ratio and Client-Dependent Normalisation for Biometric Authentication tasks. In: International Conference on Acoustics, Speech, and Signal Processing (ICASSP), vol. 1, pp. 721–724 (2005)

592. Poh, N., Kittler, J.: A Methodology for Separating Sheep from Goats for Controlled Enrollment and Multimodal Fusion. In: Biometric Symposium, pp. 17–22 (2008)

593. Poh, N., Kittler, J., Rattani, A., Tistarelli, M.: Group-Specific Score Normalization for Biometric Systems. In: IEEE Computer Society Conference on Computer Vision and Pattern Recognition Workshops (CVPRW), pp. 38–45 (2010)

594. Polkowski, L.: Rough Sets, 1st edn. Advances in Soft Computing. Physica-Verlag, A Springer-Verlag Company, Heidelberg (2002). ISBN: 3-7908-1510-1

595. Pollack, I., Pickett, J.M., Sumby, W.: On the Identification of Speakers by Voice. Journal of the Acoustical Society of America **26**, 403–406 (1954)

596. Poritz, A.B.: Hidden Markov models: a guided tour. In: International Conference on Acoustics, Speech, and Signal Processing (ICASSP-1988), vol. 1, pp. 7–13 (1988)

597. Postel, J.: User Datagram Protocol. RFC 768 (Standard) (1980). URL http://www.ietf.org/rfc/rfc768.txt

598. Powell, M.: Some Global Convergence Properties of a Variable Metric Algorithm for Minimization without Exact Line Searches. SIAM-AMS Proceedings **IX** (1976)

599. Powell, M.: Radial Basis Functions for Multivariable Interpolation: a Review. In: Algorithms for Approximation. Clarendon Press, New York (1987)

600. Powell, M.J.D.: An Efficient Method for Finding the Minimum of a Function of Several Variables Without Calculating Derivatives. The Computer Journal **7**, 155–162 (1964)

601. Press, W.H., Teukolsky, S.A., Vetterling, W.T., Flannery, B.P.: Numerical Recipes in C++. Cambridge University Press, New York (2002). ISBN: 0-521-75033-4

602. Qualcomm: QCELP Source Code. FTP (1996). URL ftp://ftp.qualcomm.com/pub/vocoder/IS96a

603. Quatieri, T., Baxter, R.A.: Noise reduction based on spectral change. In: IEEE ASSP Workshop on Applications of Signal Processing to Audio and Acoustics (1997)

604. Quatieri, T.F., Dunn, R.B.: Speech enhancement based on auditory spectral change. In: International Conference on Acoustics, Speech, and Signal Processing (ICASSP), vol. 1, pp. 257–260 (2002)

605. R., T., B., S., L., H.: A Model-Based Transformational Approach to Robust Speaker Recognition. In: International Conference on Spoken Language Processing, vol. 2, pp. 495–498 (2000)

606. Rabiner, L., Juang, B.H.: Fundamentals of Speech Recognition. Prentice Hall Signal Processing Series. PTR Prentice Hall, New Jersey (1990). ISBN: 0-130-15157-2

607. Rabiner, L.R.: A tutorial on hidden Markov models and selected applications in speech recognition. Proceedings of the IEEE **77**(2), 257–286 (1989)

608. Raja, G.S., Dandapat, S.: Speaker Recognition Under Stress Condition. International Journal of Speech Technology **13** (2010). Springerlink Online: DOI 10.1007/s10772-010-9075-z

609. Ramussen, C.E., Williams, C.K.I.: Gaussian Processes for Machine Learning (Adaptive Computation and Machine Learning). The MIT Press, Boston (2006). ISBN: 978-026218253-9

610. Rasmussen, T.B., Milner, B.: The role of early left-brain injury in determining lateralization of cerebral speech functions. Annals of the New York Academy of Science **299**, 355–369 (1977)

611. RCFP: Can We Tape? World Wide Web (2008). URL http://www.rcfp.org/taping

612. Reddy, S., Shevadea, S., Murty, M.: A Fast Quasi-Newton Method for Semi-Supervised Classification. Pattern Recognition **In Press, Corrected Proof** (2010). DOI DOI: 10.1016/j.patcog.2010.09.002. URL http://www.sciencedirect.com/science/article/B6V14-5100HJG-3/2/ddad1fa09bf51c3e3b7754415566b061

613. Reynalds, D.A., Torres-Carrasquillo, P.: Approaches and Applications of Audio Diarization. In: International Conference on Acoustics, Speech, and Signal Processing (ICASSP), vol. 5, pp. 953–956 (2005)

614. Réyni, A.: On Measure of Entropy and Information. In: Proceedings of the 4th Berkeley Symposium on Probability Theory and Mathematical Statistics, pp. 547–561 (1961)

615. Reynolds, D., Kenny, P., Castaldo, F.: A Study of New Approaches to Speaker Diarization. In: InterSpeech (2009)

616. Reynolds, D.A.: Experimental evaluation of features for robust speaker identification. IEEE Transactions on Speech and Audio Processing **2**(4), 639–643 (1994)

617. Reynolds, D.a.: The effects of handset variability on speaker recognition performance: experiments on the Switchboard corpus. In: Acoustics, Speech, and Signal Processing, 1996. ICASSP-96. Conference Proceedings., 1996 IEEE International Conference on, vol. 1, pp. 113–116 (1996)

618. Reynolds, D.A.: Comparison of Background Normalization Methods for Text-Independent Speaker Verification. In: Eurospeech 1997, pp. 963–966 (1997)

619. Reynolds, D.a.: Channel robust speaker verification via feature mapping. In: Acoustics, Speech, and Signal Processing, 2003. Proceedings. (ICASSP '03). 2003 IEEE International Conference on, vol. 2, pp. II–53–6 (2003)

620. Reynolds, D.A., Quatieri, T.F., , Dunn, R.B.: Speaker Verification Using Adapted Gaussian Mixture Models. Digital Signal Processing **10**, 19–41 (2000)

621. Reynolds, D.A., Rose, R.: Robust text-independent speaker identification using Gaussian mixture speaker models. IEEE Transactions on Speech and Audio Processing **3**(1), 72–83 (1995)

622. Rice, B.F., Fechner, R.M., Wilhoyte, M.E.: A new approach to multipath correction of constant modulus signals. In: IEEE, vol. 1, pp. 325–329 (1983)

623. arid Rjjan Rafkin, P.I.M.: Using the Fisher Kernel Method for Web Audio Classification. In: International Conference on Acoustics, Speech, and Signal Processing (ICASSP), vol. 4, pp. 2417 – 2420 (2000)

624. Robinson, A., Fallside, F.: A Recurrent Error Propation Speech Recognition System. Computer Speech and Language **5**, 259–274 (1991)

625. Rosenblatt, F.: Principles of Neurodynamics: Perceptrons and the Theory of Brain Mechanisms. Spartan Books, Washington, D.C. (1962)

626. Rosenbrock, H.H.: Automatic Method for Finding Greatest or Latest Value of Function. The Computer Journal **3**(3), 175–184 (1960)

627. Rossing, T.D.: The Science of Sound, 3rd edn. Addison Wesley (2001). ISBN: 0-80-538565-7

628. Roweis, S., Ghahramani, Z.: A Unifying Review of Linear Gaussian Models. Neural Computation **11**(2), 305–345 (1999)

629. Roy, A., Magimai-Doss, M., Marcel, S.: Boosted Binary Features for Noise-Robust Speaker Verification. In: International Conference on Acoustics, Speech, and Signal Processing (ICASSP), vol. 6, pp. 4442–4445 (2010)

630. Rozgic, V., Busso, C., Georgiou, P., Narayanan, S.a.: Multimodal Meeting Monitoring: Improvements on Speaker Tracking and Segmentation through a Modified Mixture Particle Filter. In: Multimedia Signal Processing, 2007. MMSP 2007. IEEE 9th Workshop on, pp. 60–65 (2007)

631. Rumelhart, D.E., Hinton, G.E., Williams, R.J.: Learning Internal Representations by Error Propagation. In: D.E. Rumelhart, J.L. McClelland (eds.) Parallel Distributed Processing: Explorations in the Microstructure of Cognition, vol. 1, pp. 675–695. MIT Press (1986). ISBN: 0-262-63112-1

632. Sae-Tang, S., Tanprasert, C.a.: Feature windowing-based Thai text-dependent speaker identification using MLP with backpropagation algorithm. In: Circuits and Systems, 2000. Proceedings. ISCAS 2000 Geneva. The 2000 IEEE International Symposium on, vol. 3, pp. 579–582 (2000)

633. Saeed, K., Werdoni, M.: A New Approach for hand-palm recognition. In: Enhanced Methods in Computer Security, Biometric and Artificial Interlligence Systems, Lecture Notes in Computer Science, pp. 185–194. Springer, London (2005). ISBN: 1-4020-7776-9

634. Saeta, J.R., Hernando, J.: Model quality evaluation during enrolment for speaker verification. In: Proceedings of the International Conference on Spoken Language Processing (ICSLP), pp. 1801–1804 (2004)

635. Sahin, F.: A Radial Basis Function Approach to a Color Image Classificatioin Problem in a Real Time Industrial Application. Virginia Tech University, Blacksburg, Virginia (1997). URL http://scholar.lib.vt.edu/theses/available/etd-6197-223641. Masters Thesis

636. Salcedo-Sanz, S., Gallardo-Antolin, A., Leiva-Murillo, J., Bousono-Calzon, C.a.: Offline speaker segmentation using genetic algorithms and mutual information. Evolutionary Computation, IEEE Transactions on 10(2), 175–186 (2006)

637. Sambur, M.: Selection of acoustic features for speaker identification. IEEE Transactions on Audio, Speech and Signal Processing 23(2), 390–392 (1975)

638. Sanchez-Reillo, R., Sanchez-Avila, C., Gonzalez-Marcos, A.: Biometric identification through hand geometry measurements. IEEE Transactions on Pattern Analysis and Machine Intelligence 22(10), 1168–1171 (2000)

639. Savage, L.J.: The Foundations of Statistics. John Wiley & Sons, New York (1954)

640. Scheme, E., Castillo-Guerra, E., Englehart, K., Kizhanatham, A.: Practical Considerations for Real-Time Implementation of Speech-Based Gender Detection. In: Proceesings of the 11th Iberoamerican Congress in Pattern Recognition (CIARP 2006) (2006)

641. Schmidt, E.: Zur Theorie der Linearen und Nichtlinearen Integralgleichungen. I. Teil: Entwicklung Willkürlicher Funktionen nach Systemen Vorgeschriebener. Mathematische Annalen 63, 433–476 (1906)

642. Schmidt, E.: Zur Theorie der linearen und nichtlinearen Integralgleichungen. II. Auflösung der Allgemeinen Linearen Integralgleichung. Mathematische Annalen 64, 161–174 (1907)

643. Schmidt, E.: Zur Theorie der Linearen und Nichtlinearen Integralgleichungen. III. Teil: Über die Auflösung der nichtlinearen Integralgleichung und die Verzweigung ihrer Lösungen. Mathematische Annalen 65, 370–399 (1907)

644. Schnabel, R.B.: Analyzing and Improving Quasi-Newton Methods for Unconstrained Optimization. Cornell University, New York (1977). Ph.D. Thesis

645. Schofield, B.R., Coomes, D.L.: Pathways from Auditory Cortex to the Cochlear Nucleus in Guinea Pigs. Hearing Research 216–217, 81–89 (2006)

646. Schölkopf, B., Smola, A., Müller, K.R.: Nonlinear Component Analysis as a Kernel Eigenvalue Problem (1996). Technical Report No. 44

647. Schölkopf, B., Smola, A.J.: Learning with Kernels: Support Vector Machines, Regularization, Optimization, and Beyond (Adaptive Computation and Machine Learning). MIT Press, Cambridge, MA (2002). ISBN: 978-0-262-19475-4

648. Schölkopf, B., Smola, A.J., Müller, K.R.: Kernel Principal Component Analysis. In: B. Schölkopf, C. Burges, A.J. Smola (eds.) Advances in Kernel Methods. MIT Press, Cambridge, MA (2000)

649. Schroeder, M.R.: Recognition of Complex Acoustic Signals. In: T.H. Bullock (ed.) Life Sciences Research Report; 5, p. 324. Abacon Verbag, Berlin (1977)

650. Schulzrinne, H., Casner, S.: RTP Profile for Audio and Video Conferences with Minimal Control. RFC 3551 (Standard) (2003). URL http://www.ietf.org/rfc/rfc3551.txt

651. Schulzrinne, H., Casner, S., Frederick, R., Jacobson, V.: RTP: A Transport Protocol for Real-Time Applications. RFC 3550 (Standard) (2003). URL http://www.ietf.org/rfc/rfc3550.txt

652. Schür, I.: On power series which are bounded in the interior of the unit circle. In: I. Gohberg (ed.) Methods in Operator Theory and Signal Processing, Operator Theory: Advances and Applications, vol. 18, pp. 31–59. Abacon Verbag (1986). Original in German in J. Reine Angew. Math., 147 (1917), pp. 205–232

653. Schwarz, G.: Estimating the Dimension of a Model. Annals of Statistics **6**(2), 461–464 (1978)

654. Sedgewick, R.: Algorithms in C. Addison-Wesley Publishing Company, New york (1990). ISBN: 0-201-51425-7

655. Selinger, A., Socolinsky, D.A.: Appearance-Based Facial Recognition Using Visible and Thermal Imagery: A Comparative Study. Computer Vision and Image Understanding **91**(1–2), 72–114 (2003)

656. Senapati, S., Chakraborty, S., Saha, G.a.: Log Gabor Wavelet and Maximum a Posteriori Estimator in Speaker Identification. In: Annual India Conference, 2006, pp. 1–6 (2006)

657. Senoussaoui, M., Kenny, P., Dehak, N., Dumouchel, P.: An i-Vector Extractor Suitable for Speaker Recognition with Both Microphone and Telephone Speech. In: The Speaker and Language Recognition Workshop (Odyssey 2010), pp. 28–33 (2010)

658. Seo, C., Lee, K.Y., Lee, J.: GMM based on local PCA for speaker identification. IEE Electronic Letters **37**(24), 1486–1488 (2001)

659. Seo, J., Hong, S., Gu, J., Kim, M., Baek, I., Kwon, Y., Lee, K., and, S.I.Y.: New speaker recognition feature using correlation dimension. In: Industrial Electronics, 2001. Proceedings. ISIE 2001. IEEE International Symposium on, vol. 1, pp. 505–507 (2001)

660. Sessler, G.M., West, J.E.: Self-Biased Condenser Microphone with High Capacitance. The Journal of the Acoustical Society of America (JASA) **34**(11), 293–309 (1962)

661. Shah, B.V., Buehler, R.J., Kempthorne, O.: Some Algorithms for Minimizing Functions of Several Variables. SIAM journal on Applied Mathematics **12**, 74–92 (1964)

662. Shahin, I., Botros, N.a.: Speaker identification using dynamic time warping with stress compensation technique. In: Southeastcon '98. Proceedings. IEEE, pp. 65–68 (1998)

663. Shanmugham, S., Monaco, P., Eberman, B.: A Media Resource Control Protocol (MRCP) Developed by Cisco, Nuance, and Speechworks. RFC 4463 (Informational) (2006). URL http://www.ietf.org/rfc/rfc4463.txt

664. Shanno, D.: Conditioning of Quasi-Newton Method for Function Minimization. Mathematics of Computation **24**, 647–656 (1970)

665. Shanno, D.F., Phua, K.H.: Numerical Comparison of Several Variable Metric Algorithms. Tech. rep., MIS Technical Report 21, University of Arizona (1977)

666. Shanno, D.F., Phua, K.H.: Matrix Conditioning and Nonlinear Optimization. Mathematical Programming **14**, 149–160 (1978)

667. Shannon, C.E.: A Mathematical Theory of Computation. The Bell System Technical Journal **27**, 379–423,623–656 (1948). Reprint with corrections

668. Shannon, C.E.: Communication in the Presence of Noise. Proceedings of the Institute of Radio Engineers **37**(1), 10–21 (1949). Reprint available at: Proceedings of the IEEE, Vol. 86, No. 2, Feb. 1998

669. Sharkas, M., Elenien, M.A.: Eigenfaces vs. fisherfaces vs. ICA for face recognition; a comparative study. In: IEEE 9th International Conference on Signal Procesing (ICSP2008), pp. 914–919 (2008)

670. Shearme, J.N., Holmes, J.N.: An Experiment Concerning the Recognition of Voices. Language and Speech **2**, 123–131 (1959)

671. Sheela, K.A., Prasad, K.S.: Linear Discriminant Analysis F-Ratio for Optimization of TESPAR & MFCC Features for Speaker Recognition. Journal of Multimedia **2**(6), 34–43 (2007)

672. Shore, J.E., Johnson, R.W.: Axiomatic Derivation of the Principle of Maximum Entropy and the Principle of Minimum Cross-Entropy. IEEE Transactions on Information theory **IT-26**(1), 26–37 (1980)

673. Shore, J.E., Johnson, R.W.: Properties of Cross-Entropy Minimzation. IEEE Transactions on Information theory **IT-27**(4), 472–482 (1981)

674. Siafarikas, M., Ganchev, T., Fakotakis, N.a.: Wavelet Packet Bases for Speaker Recognition. In: Tools with Artificial Intelligence, 2007. ICTAI 2007. 19th IEEE International Conference on, vol. 2, pp. 514–517 (2007)

675. Siegler, M.A., Jain, U., Raj, B., Stern, R.M.: Automatic Segmentation, Classification and Clustering of Broadcast News Audio. In: Proceedings of the DARPA Speech Recognition Workshop (1997)

676. SIL International. World Wide Web. URL http://www.sil.org

677. Simon, C., Goldstein, I.: Retinal Method of Identification. New York State Journal of Medicine **15** (1936)

678. Sinanović, S., Johnson, D.H.: Toward a Theory of information processing. Signal Processing **87**(6), 1326–1344 (2007)

679. Slomka, S., Sridharan, S.a.: Automatic gender identification optimised for language independence. In: TENCON '97. IEEE Region 10 Annual Conference. Speech and Image Technologics for Computing and Telecommunications'., Proceedings of IEEE, vol. 1, pp. 145–148 (1997)

680. Smith, C.S.: The Automatic Computation of Maximum likelihood Estimates. Tech. rep., N. C. B. Scientific Department Report S.C. 846/MR/40 (1962)

681. Solewicz, Y.A., Aronowitz, H.: Two-Wire Nuisance Attribute Projection. In: InterSpeech, pp. 928–931 (2009)

682. Solla, S.A., Levin, E., Fleisher, M.: Accelerated Learning in Layered Neural Networks. Complex Systems **2**, 625–640 (1988)

683. Sollaud, A.: RTP Payload Format for the G.729.1 Audio Codec. RFC 4749 (Proposed Standard) (2006). URL http://www.ietf.org/rfc/rfc4749.txt

684. Sollaud, A.: RTP Payload Format for ITU-T Recommendation G.711.1. RFC 5391 (Proposed Standard) (2008). URL http://www.ietf.org/rfc/rfc5391.txt

685. Solomonoff, A., Campbell, W., Quillen, C.: Channel Compensation for SVM Speaker Recognition. In: The Speaker and Language Recognition Workshop Odyssey 2004, vol. 1, pp. 57–62 (2004)

686. Sorensen, H., Heideman, M., Burrus, C.: On computing the split-radix FFT. IEEE Transactions on Acoustics, Speech and Signal Processing **34**(1), 152–156 (1986)

687. Sorensen, H., Jones, D., Heideman, M., Burrus, C.: Real-valued Fast Fourier Transform Algorithms. IEEE Transactions on Acoustics, Speech and Signal Processing **35**(6), 849–863 (1987). See Correction

688. Sorensen, H., Jones, D., Jones, D., Burrus, C.: Correction to Real-valued Fast Fourier Transform Algorithms. IEEE Transactions on Acoustics, Speech and Signal Processing **35**(9), 849–863 (1987). Correction to earlier publication in Jun 1987

689. Soria-Olivas, E., Martin-Guerrero, J.D., Camps-Valls, G., Serrano-López, A.J., Calpe-Maravilla, J., Gómez-Chova, L.: A Low-Complexity Fuzzy Activation Function for Artificial Neural Networks. IEEE Transactions on Neural Networks **14**(6), 1576–1579 (2003)

690. Spedicato, E.: Computational Experience with Quasi-Newton Algorithm for Minimization Problems of Moderately Large Size. Tech. rep., Report CISE-N-175, CISE Documentation Service, Segrate, Milano (1975)

691. Stadtschnitzer, M., Van Pham, T., Chien, T.T.: Reliable voice activity detection algorithms under adverse environments. In: International Conference on Communications and Electronics, pp. 218–223 (2008)

692. Stagni, C., Guiducci, C., Benini, L., Ricco, B., Carrara, S., Paulus, C., Schienle, M., Thewes, R.: A Fully Electronic Label-Free DNA Sensor Chip. IEEE Sensors Journal **7**(4), 577–585 (2007)

693. Steinberg, J.C.: Position of Stimulatioin in Cochlea by Pure Tones. Journal of the Acoustical Society of America **8**(3), 176–180 (1937)

694. Stevens, S.S.: The Relation of Pitch to Intensity. Journal of the Acoustical Society of America **6**(3), 150–154 (1935)

695. Stevens, S.S.: A Scale for the Measurement of the Psychological Magnitude: Loudness. Psychological Review of the American Psychological Association **43**(5), 405–416 (1936)

696. Stevens, S.S.: On the Psychophysical Law. Psychological Review **64**(3), 153–181 (1957)
697. Stevens, S.S., Volkmann, J.E.: The Relation of Pitch to Frequency. Journal of Psychology **53**(3), 329–353 (1940)
698. Stevens, S.S., Volkmann, J.E., Newman, E.B.: A Scale for the Measurement of the Psychological Magnitude Pitch. Journal of the Acoustical Society of America **8**(3), 185–190 (1937)
699. Stewart, J.: Calculus, 6th edn. Brooks Cole, New York (2007). ISBN: 0-495-01160-6
700. Stoll, L., Doddington, G.: Hunting for Wolves in Speaker Recognition. In: The Speaker and Language Recognition Workshop (Odyssey 2010), pp. 159–164 (2010)
701. Strauss, E., Wada, J.A., Goldwater, B.: Sex differences in interhemispheric reorganization of speech. Neuropsychologia **30**, 353–359 (1992)
702. Sturim, D.E., Campbell, W.M., Reynolds, D.A., Dunn, R.B., Quatieri, T.: Robust Speaker Recognition with Cross-Channel Data: MIT-LL Results on the 2006 NIST SRE Auxiliary Microphone Task. In: International Conference on Acoustics, Speech, and Signal Processing (ICASSP), vol. IV, pp. 49–52 (2007)
703. Sugiyama, M., Sawai, H., Waibel, A.: Review of TDNN (time delay neural network) architectures for speech recognition. In: IEEE International Symposium on Circuits and Systems, vol. 1, pp. 582–585 (1991)
704. Sun, H., Ma, B., Huang, C.L., Nguyen, T.H., Li, H.: The IIR NIST SRE 2008 and 2010 Summed Channel Speaker Recognition Systems. In: Interspeech (2010)
705. Suykens, J., Vandewalle, J.: Training Multilayer Perceptron Classification Based on a Modified Support Vector Method. IEEE Transactions on Neural Networks **10**(4), 907–9011 (1999)
706. Swann, W.H.: Report on the Development of a New Direct Search Method of Optimization. Tech. rep., Imperial Chemical Industries, Ltd. Central Instrumentation Laboratory Research Note 6413 (1964)
707. Sykes, P.: A History of Persia, 3rd edn. Macmillan and Co. Ltd., London (1958)
708. Tabatabaee, H., Fard, A., Jafariani, H.: A Novel Human Identifier System using Retinal Image and Fuzzy Clustering. In: International Conference on Intelligent Information Hiding and Multimedia Signal Processing (IIHMSP2008), vol. 1, pp. 1031–1036 (2006)
709. Tafazzoli, A.: Tarikhe Adabiate Iran Pish az Eslam (History of Iranian Literature Before Islam), 3rd edn. National Library of Iran, Tehran (1378 Anno Persico). In Persian, ISBN: 9-64-598814-2
710. Takashima, R., Takiguchi, T., Ariki, Y.: HMM-based separation of acoustic transfer function for single-channel sound source localization. In: International Conference on Acoustics, Speech, and Signal Processing (ICASSP), pp. 2830–2833 (2010)
711. Talkin, D.: A Robust Algorithm for Pitch Tracking (RAPT). In: W.B. Kleijn, K.K. Paliwal (eds.) Speech Coding and Synthesis. Elsevier Publishing Company, New York (1995). ISBN: 0-44-482169-4
712. Tang, H., Chen, Z., Huang, T.a.: Comparison of Algorithms for Speaker Identification under Adverse Far-Field Recording Conditions with Extremely Short Utterances. In: Networking, Sensing and Control, 2008. ICNSC 2008. IEEE International Conference on, pp. 796–801 (2008)
713. Tanprasert, C., Wutiwiwatchai, C., and, S.S.T.: Text-dependent speaker identification using neural network on distinctive Thai tone marks. In: Neural Networks, 1999. IJCNN '99. International Joint Conference on, vol. 5, pp. 2950–2953 (1999)
714. Thiruvaran, T., Ambikairajah, E., Epps, J.a.: Normalization of Modulation Features for Speaker Recognition. In: Digital Signal Processing, 2007 15th International Conference on, pp. 599–602 (2007)
715. Thompson, J., Mason, J.S.: The Pre-detection of Error-prone Class Members at the Enrollment Stage of Speaker Recognition Systems. In: ESCA Workshop on Automatic Speaker Recognition, Identification, and Verification, pp. 127–130 (1994)
716. Eye Prints. Time Magazine (1935)
717. Timoshenko, S., Goodier, J.: Theory of Elasticity. McGraw-Hill Book Company, Inc., New York (1951)

718. Tjalkens, T.: State Dependent Coding: How to find the State? In: 43rd annual Allerton Conference on Communication, Control, and Computing (2005). URL http://cslgreenhouse.csl.illinois.edu/allertonarchives/allerton05/PDFs/Papers/V_A_2.pdf

719. Toh, K.A., and, W.Y.Y.: Fingerprint and speaker verification decisions fusion using a functional link network. Systems, Man, and Cybernetics, Part C: Applications and Reviews, IEEE Transactions on **35**(3), 357–370 (2005)

720. Toh, K.A., Kim, J., Lee, S.: Biometric scores fusion based on total error rate minimization. Pattern Recognition **41**(3), 1066 – 1082 (2008). DOI DOI: 10.1016/j.patcog.2007.07.020. Part Special issue: Feature Generation and Machine Learning for Robust Multimodal Biometrics

721. de la Torre, A., Peinado, A.M., Segura, J.C., Perez-Cordoba, J.L., Benitez, M.C., Rubio, A.J.: Histogram Equalization of Speech Representation for Robust Speech Recognition. IEEE Transaction of Speech and Audio Processing **13**(3), 355–366 (2005)

722. Torres, II., Rufiner, II.a.: Automatic speaker identification by means of Mel cepstrum, wavelets and wavelet packets. In: Engineering in Medicine and Biology Society, 2000. Proceedings of the 22nd Annual International Conference of the IEEE, vol. 2, pp. 978–981 (2000)

723. Tosi, O.I.: Voice Identification: Theory and Legal Applications. University Park Press, Baltimore (1979). ISBN: 978-0-839-11294-5

724. Touchette, H.: Legendre-Fenchel Trfansforms in a Nutshell. Web (2007). URL http://www.maths.qmw.ac.uk/~ht/archive/lfth2.pdf

725. Toussaint, G.T.: Comments on "The Divergence and Bhattacharyya Distance Measure in Signal Selection". IEEE Transactions on Communications **20**(3), 485 (1972)

726. Tranter, S., Reynolds, D.a.: An overview of automatic speaker diarization systems. Audio, Speech, and Language Processing, IEEE Transactions on **14**(5), 1557–1565 (2006)

727. Traunmüller, H.: Analytical Expressions for the Tonotopic Sensory Scale. Journal of the Acoustical Society of America **88**(1), 97–100 (1990)

728. Treichler, J.R., Agee, B.G.: A new approach to multipath correction of constant modulus signals. IEEE Transactions on Audio, Speech and Signal Processing **31**(2), 459–472 (1983)

729. Tsybakov, B.S., Iakoviev, V.P.: On the Accuracy of Restoring a Function with a Finite Number of Terms of Kotel'nikov Series. Radio Engineering and Electronic Physics **4**(3), 274–275 (1959)

730. Tufekci, Z., Gurbuz, S.a.: Noise Robust Speaker Verification Using Mel-Frequency Discrete Wavelet Coefficients and Parallel Model Compensation. In: Acoustics, Speech, and Signal Processing, 2005. Proceedings. (ICASSP '05). IEEE International Conference on, vol. 1, pp. 657–660 (2005)

731. Tveit, A., Engum, H.: Parallelization of the Incremental Proximal Support Vector Machine Classifier using a Heap-Based Tree Topology. In: Workshop on Parallel Distributed Computing for Machine Learning (2003)

732. Tyagi, V., Mccowan, L., Misra, H., Bourlard, H.: Mel-Cepstrum Modulation Spectrum (MCMS) Features for Robust ASR. In: IEEE Workshop on Automatic Speech Recognition and Understanding (2003)

733. Tyagi, V., Wellekens, C.: Fepstrum Representation of Speech Signal. In: IEEE Workshop on Automatic Speech Recognition and Understanding, pp. 11–16 (2005)

734. Tzortzis, G., Likas, A.C.: The Global Kernel k-Means Algorithm for Clustering in Feature Space. IEEE Transactions on Neural Networks **20**(7), 1181–1194 (2009)

735. Umesh, S., Cohen, L., Nelson, D.: Fitting the Mel scale. In: IEEE International Conference on Acoustics, Speech, and Signal Processing (ICASSP99), vol. 1, pp. 217–220 (1999)

736. UniMRCP: Open Source MRCP Project. Website (2010). URL http://www.unimrcp.org

737. Disability Census Results for 2005. World Wide Web (2005). URL http://www.census.gov

738. Valin, J.M.: Introduction to CELP Coding. Manual (2010). URL http://speex.org/docs/manual/speex-manual/manual.html

739. Vapnik, V.N.: Estimation of Dependences Based on Empirical Data, russian edn. Nauka, Moscow (1979). English Translation: Springer-Verlag, New York, 1982

740. Vapnik, V.N.: Statistical learning theory. John Wiley, New York (1998). ISBN: 0-471-03003-1

741. Vargha-Khadem, F., Gadian, D.G., Copp, A., Mishkin, M.: FOXP2 and the Neuroanatomy of Speech and Language. Nature Reviews. Neuroscience **6**(2), 131–137 (2005)

742. Veinott, A.F.: Conjugate Duality for Convex Programs: A Geometric Development. Linear Algebra and its Applications **114–115**, 663–667 (1989)

743. Venayagamoorthy, G., Sundepersadh, N.a.: Comparison of text-dependent speaker identification methods for short distance telephone lines using artificial neural networks. In: Neural Networks, 2000. IJCNN 2000, Proceedings of the IEEE-INNS-ENNS International Joint Conference on, vol. 5, pp. 253–258 (2000)

744. Vetterli, M., Nussbaumer, H.J.: Simple FFT and DCT Algorithms with Reduced Number of Operations. Signal Processing **6**, 262–278 (1984)

745. Viswanathan, M., Beigi, H., Tritschler, A., Maali, F.a.: Information access using speech, speaker and face recognition. In: Multimedia and Expo, 2000. ICME 2000. 2000 IEEE International Conference on, vol. 1, pp. 493–496 (2000)

746. Viswanathan, M., Beigi, H.S., Dharanipragada, S., Maali, F., Tritschler, A.: Multimedia document retrieval using speech and speaker recognition

747. Viswanathan, M., Beigi, H.S., Maali, F.: Information Access Using Speech, Speaker and Face Recognition. In: IEEE International Conference on Multimedia and Expo (ICME2000) (2000)

748. Viterbi, A.J.: Error bounds for convolutional codes and an asymptotically optimum decoding algorithm. IEEE Transactions on Information Theory **13**(2), 260–269 (1967)

749. Vogl, T.P., Mangis, J.K., Rigler, A.K., Zink, W.T., Alkon, D.L.: Accelerating the Convergence of the Back-Propagation Method. Biological Cybernetics **59**, 257–263 (1988)

750. Vogt, B.A., Pandya, D.N., Rosene, D.L.: Cingulate Cortex of the Rhesus Monkey: I. Cytoarchitecture and Thalamic Afferents. The Journal of Comparative Neurology **262**(2), 271–289 (1987). Online version published, Oct. 9, 2004

751. Vogt, R., Baker, B., Sridharan, S.: Modelling Session Variability in Text-Independent Speaker Verification. In: Interspeech, pp. 3117–3120 (2005)

752. Vogt, R., Sridharan, S.: Explicit modelling of session variability for speaker verification. Computer Speech and Language **22**(1), 17–38 (2008)

753. Voice Biometrics. Meeting (2008). URL http://www.voicebiocon.com

754. van Vuuren, S.: Comparison of Text-Independent Speaker Recognition Methods on Telephone Speech with Acoustic Mismatch. In: Proceedings of the International Conference on Spoken Language Processing (ICSLP), pp. 784–787 (1996)

755. W3C: Voice Extensible Markup Language (VoiceXML. Website (2008). URL http://www.voicexml.org

756. W3C: Extensible MultiModal Annotation (EMMA). Website (2009). URL http://www.w3c.org/TR/emma

757. W3C: Speech Recognition Grammar Specification. Website (2010). URL http://www.w3c.org/TR/speech-grammar

758. W3C: Speech Synthesis Markup Language (SSML). Website (2010). URL http://www.w3c.org/TR/speech-synthesis11

759. W3C: Voice Extensible Markup Language (VoiceXML). Website (2010). URL http://www.w3c.org/TR/voicexml30

760. Wada, J.A.: A New Method for the Determination of the Side of Cerebral Speech Dominance. A Preliminary Report of the Intra-Cartoid Injection of Sodium Amytal in Man. Igaju to Seibutsugaki **14**, 221–222 (1949)

761. Wada, J.A., Rasmussen, T.B.: Intracarotid Injection of Sodium Amytal for the Lateralization of Cerebral Speech Dominance: Experimental and Clinical Observations. Journal of Neurosurgery **17**, 266–282 (1960)

762. Waibel, A., Hanazawa, T., Hinton, G., Shikano, K., Lang, K.: Phoneme Recognition Using Time-Delay Neural Networks. IEEE Transactions on Acoustics, Speech and Signal Processing **37**(3), 328–339 (1989)

763. Wang, L., Chen, K., Chi, H.: Capture interspeaker information with a neural network for speaker identification. IEEE Transactions on Neural Networks **13**(2), 436–445 (2002)

764. Wang, X.: Text-Dependent speaker verification using recurrent time delay neural networks for feature extraction. In: IEEE Signal Processing Workshop – Neural Netrowks for Signal Processing – III, pp. 353–361 (1993)

765. Wang, Y.: A Tree-based Multi-class SVM Classifier for Digital Library Document. In: International Conference on MultiMedia and Information Technology (MMIT), pp. 15–18 (2008)

766. Wang, Y., chun Mu, Z., Hui Zeng, a.: Block-based and multi-resolution methods for ear recognition using wavelet transform and uniform local binary patterns. In: Pattern Recognition, 2008. ICPR 2008. 19th International Conference on, pp. 1–4 (2008)

767. Wang, Y., Mu, Z.C., Liu, K., and, J.F.: Multimodal recognition based on pose transformation of ear and face images. In: Wavelet Analysis and Pattern Recognition, 2007. ICWAPR '07. International Conference on, vol. 3, pp. 1350–1355 (2007)

768. Wang, Y., Sereno, J.A., Jongman, A., Hirsch, J.: fMRI Evidence for Cortical Modification During Learning of Mandarin Lexical Tones. Journal of Cognitive Neuroscience **15**(7), 1019–1027 (2003)

769. Wasson, D., Donaldson, R.: Speech amplitude and zero crossings for automated identification of human speakers. IEEE Transactions on Audio, Speech and Signal Processing **23**(4), 390–392 (1975)

770. Webb, J., Rissanen, E.a.: Speaker identification experiments using HMMs. In: Acoustics, Speech, and Signal Processing, 1993. ICASSP-93., 1993 IEEE International Conference on, vol. 2, pp. 387–390 (1993)

771. Weber, F., Peskin, B., Newman, M., Corrada-Emmanuel, A., Gillick, L.: Speaker Recognition on Single and Multispeaker Data. Digital Signal Processing **10**, 75–92 (2000)

772. Weiss: Sampling Theorems Associated with Sturm-Liouville Systems. Bulletin of the Mathmatical Society **63**, 242 (1957)

773. Weiss, L.G.: Wavelets and wideband correlation processing. IEEE Signal Processing Magazine **11**(1), 13–32 (1994)

774. Welch, P.: The use of fast Fourier transform for the estimation of power spectra: A method based on time averaging over short, modified periodograms. IEEE Transactions on Audio and Electroacoustics **15**(2), 70–73 (1967)

775. Welsh, D.: Codes and Cryptography. Oxford University Press, New York (1990). ISBN: 0-198-53287-3

776. Wenndt, S., Shamsunder, S.a.: Bispectrum features for robust speaker identification. In: Acoustics, Speech, and Signal Processing, 1997. ICASSP-97., 1997 IEEE International Conference on, vol. 2, pp. 1095–1098 (1997)

777. Whittaker, E.T.: On the Functions which are Represented by the Expansion of Interpolating Theory. Proceedings of the Royal Society of Edinburgh **35**, 181–194 (1915)

778. Whittaker, J.M.: The Fourier Theory of the Carndinal Functions. Proceedings of the Mathematical Society of Edinburgh **1**, 169–176 (1929)

779. Whittaker, J.M.: Interpolutory Function Theory. No. 33 in Cambridge Tracts in Mathematics and Mathematical Physics. Cambridge University Press, Cambridge, England (1935)

780. Wiener, N.: Cybernetics: or Control and Communication in the Animal and the Machine, 2nd edn. The M.I.T. Press, Cambridge (1976). First Edition was published by John Wiley and Sons, New York (1948), ISBN: 0-262-73009-X

781. Wikipedia: A Taste of Freedom. Website. URL http://en.wikipedia.org/wiki/A_Taste_Of_Freedom

782. Wolf, M., Park, W., Oh, J., Blowers, M.a.: Toward Open-Set Text-Independent Speaker Identification in Tactical Communications. In: Computational Intelligence in Security and Defense Applications, 2007. CISDA 2007. IEEE Symposium on, pp. 7–14 (2007)

783. Wolfe, P.: A Duality Theorem for Nonlinear Programming. Quarterly Journal of Applied Mathematics **19**(3), 239–244 (1961)

784. Wolfe, P.: Convergence Conditions for Ascent Methods. SIAM Review **11**(2), 226–235 (1969)

785. Wood, C.F.: Application of Direct Search to the Solution of Engineering Problems. Tech. rep., Westinghouse Reseach Laboratory Scientific Paper 6-41210-1-p1 (1960)

786. Wu, R., Su, C., Xia, K., Wu, Y.: An approach to WLS-SVM based on QPSO algorithm in anomaly detection. In: World Congress on Intelligent Control and Automation (WCICA), pp. 4468–4472 (2008)

787. Wutiwiwatchai, C., Achariyakulporn, V., Tanprasert, C.a.: Text-dependent speaker identification using LPC and DTW for Thai language. In: TENCON 99. Proceedings of the IEEE Region 10 Conference, vol. 1, pp. 674–677 (1999)

788. Xiang, B., Chaudhari, U., Navratil, J., Ramaswamy, G., Gopinath, R.a.: Short-time Gaussianization for robust speaker verification. In: Acoustics, Speech, and Signal Processing, 2002. Proceedings. (ICASSP '02). IEEE International Conference on, vol. 1, pp. I–681–I–684 (2002)

789. Xie, Y., Dai, B., and, J.S.: Kurtosis Normalization after Short-Time Gaussianization for Robust Speaker Verification. In: Intelligent Control and Automation, 2006. WCICA 2006. The Sixth World Congress on, vol. 2, pp. 9463–9467 (2006)

790. Xie, Z.X., and, Z.C.M.: Improved locally linear embedding and its application on multipose ear recognition. In: Wavelet Analysis and Pattern Recognition, 2007. ICWAPR '07. International Conference on, vol. 3, pp. 1367–1371 (2007)

791. Xiong, Z., Chen, Y., Wang, R., Huang, T.a.: A real time automatic access control system based on face and eye corners detection, face recognition and speaker identification. In: Multimedia and Expo, 2003. ICME '03. Proceedings. 2003 International Conference on, vol. 3, pp. III–233–6 (2003)

792. *0.8* 1.2 Vorbis I Specifications (2004). URL http://xiph.org/ao/doc/

793. LibAO OGG Audio API. The XIPH Open-Source Community (2004). URL http://xiph.org/ao/doc/

794. Xu, L., Jordan, M.I.: On Convergence Properties of the EM Algorithm for Gaussian Mixtures. Neural Computation 8(1), 129–151 (1996)

795. Xu, X., and, Z.M.: Feature Fusion Method Based on KCCA for Ear and Profile Face Based Multimodal Recognition. In: Automation and Logistics, 2007 IEEE International Conference on, pp. 620–623 (2007)

796. Yager, N., Dunstone, T.: Worms, Chameleons, Phantoms and Doves: New Additions to the Biometric Menagerie. In: IEEE Workshop on Automatic Identification Advanced Technologies, pp. 1–6 (2007)

797. Yager, N., Dunstone, T.: The Biometric Menagerie. IEEE Transactions on Pattern Analysis and Machine Intelligence (PAMI) 32(2), 220–230 (2010)

798. Yanguas, L., Quatieri, T.a.: Implications of glottal source for speaker and dialect identification. In: Acoustics, Speech, and Signal Processing, 1999. ICASSP '99. Proceedings., 1999 IEEE International Conference on, vol. 2, pp. 813–816 (1999)

799. Yantorno, R.E., Iyer, A.N., Shah, J.K., Smolenski, B.Y.: Usable speech detection using a context dependent Gaussian mixture model classifier. In: International Symposium on Circuits and Systems (ISCAS), vol. 5, pp. 619–623 (2004)

800. Yavne, R.: An Economical Method for Calculating the Discrete Fourier Transform. In: Proceedings of the AFIPS Fall Joint Computer Conference, vol. 33, pp. 115–125 (1968)

801. Yin, S.C., Rose, R., Kenny, P.a.: A Joint Factor Analysis Approach to Progressive Model Adaptation in Text-Independent Speaker Verification. Audio, Speech, and Language Processing, IEEE Transactions on 15(7), 1999–2010 (2007)

802. Yingle, F., Li, Y., Qinye, T.: Speaker gender identification based on combining linear and nonlinear features. In: 7th World Congress on Intelligent Control and Automation. (WCICA 2008), pp. 6745–6749 (2008)

803. kwong Yiu, K., Mak, M.W., and, S.Y.K.: Speaker verification with a priori threshold determination using kernel-based probabilistic neural networks. In: Neural Information Processing, 2002. ICONIP '02. Proceedings of the 9th International Conference on, vol. 5, pp. 2386–2390 (2002)

804. You, C.H., Lee, K.A., Li, H.: A GMM Supervector Kernel with the Bhattacharyya Distance for SVM Based Speaker Recognition. In: International Conference on Acoustics, Speech, and Signal Processing (ICASSP), vol. 1, pp. 4221–4224 (2009)

805. Young, S.: A Review of Large-Vocabular Continuous Speech Recognition. IEEE Signal Processing Magazine 13(5), 45–57 (1996)

806. Youssif, A., Sarhan, E., El Behaidy, W.a.: Development of automatic speaker identification system. In: Radio Science Conference, 2004. NRSC 2004. Proceedings of the Twenty-First National, pp. C7–1–8 (2004)

807. Yuan, L., and, Z.C.M.: Ear Detection Based on Skin-Color and Contour Information. In: Machine Learning and Cybernetics, 2007 International Conference on, vol. 4, pp. 2213–2217 (2007)

808. Yuan, L., Mu, Z.C., and, X.N.X.: Multimodal recognition based on face and ear. In: Wavelet Analysis and Pattern Recognition, 2007. ICWAPR '07. International Conference on, vol. 3, pp. 1203–1207 (2007)

809. Yuan, Z.X., Xu, B.L., Yu, C.Z.: Binary quantization of feature vectors for robust text-independent speaker identification. IEEE Transactions on Speech and Audio Processing 7(1), 70–78 (1999)

810. Yuo, K.H., Hwang, T.H., Wang, H.C.: Combination of autocorrelation-based features and projection measure technique for speaker identification. IEEE Transactions on Speech and Audio Processing 13(4), 565–574 (2005)

811. Zangwill, W.: Minimizing a Function Without Calculating Derivatives. The Computer Journal 10, 293–296 (1967)

812. Zeger, K., Vaisey, K., Gersho, A.: Globally optimal vector quantizer design by stochastic relaxation. IEEE Transactions on Signal Processing 40(2), 310–322 (1992)

813. Zhang, B.: Generalized k-Harmonic Means – Boosting in Unsupervised Learning (2000). Technical Report Number HPL-2000-137

814. Zhang, B., Hsu, M., U.Dayal: k-Harmonic Means – A Data Clustering Algorithm (1999). Technical Report Number HPL-1999-124

815. Zhang, H., and, Z.M.: Compound Structure Classifier System for Ear Recognition. In: Automation and Logistics, 2008. ICAL 2008. IEEE International Conference on, pp. 2306–2309 (2008)

816. Zhang, H.J., Mu, Z.C., Qu, W., Liu, L.M., and, C.Y.Z.: A novel approach for ear recognition based on ICA and RBF network. In: Machine Learning and Cybernetics, 2005. Proceedings of 2005 International Conference on, vol. 7, pp. 4511–4515 (2005)

817. Zhang, L., Zheng, B., Yang, Z.: Codebook design using genetic algorithm and its application to speaker identification. IEE Electronic Letters 41(10), 619–620 (2005)

818. Zhang, Y.F., Mao, J.L., Xiong, Z.Y.: An efficient clustering algorithm. In: International Conference on Machine Learning and Cybernetics, vol. 1, pp. 261–265 (2003)

819. Zhang, Z., and, H.L.: Multi-view ear recognition based on B-Spline pose manifold construction. In: Intelligent Control and Automation, 2008. WCICA 2008. 7th World Congress on, pp. 2416–2421 (2008)

820. Zhang, Z., Dai, B.T., Tung, A.K.: Estimating Local Optimums in EM Algorithm over Gaussian Mixture Model. In: The 25th International Conference on Machine Learning (ICML), pp. 1240–1247 (2008)

821. long Zhao, H., chun Mu, Z., Zhang, X., jie Dun and, W.: Ear recognition based on wavelet transform and Discriminative Common Vectors. In: Intelligent System and Knowledge Engineering, 2008. ISKE 2008. 3rd International Conference on, vol. 1, pp. 713–716 (2008)

822. Zhen, Y., and, L.C.: A new feature extraction based the reliability of speech in speaker recognition. In: Signal Processing, 2002 6th International Conference on, vol. 1, pp. 536–539 (2002)

823. Zheng, N., Ching, P., Lee, T.: Time Frequency Analysis of Vocal Source Signal for Speaker Recognition. In: Proceedings of the International Conference on Spoken Language Processing, pp. 2333–2336 (2004)

824. Zheng, Y.C., Yuan, B.Z.a.: Text-dependent speaker identification using circular hidden Markov models. In: Acoustics, Speech, and Signal Processing, 1988. ICASSP-88., 1988 International Conference on, pp. 580–582 (1988)

825. Zhou, S.K., Chellappa, R., Zhao, W.: Unconstrained Face Recognition, *International Series on Biometrics*, vol. 5. Springer, New York (2008). ISBN: 978-0-387-26407-3

826. Zhu, X.: Semi-Supervised Learning Literature Survey. Technical Report 1530, Computer Sciences, University of Wisconsin-Madison (2008). URL http://pages.cs.wisc.edu/~jerryzhu/pub/ssl_survey.pdf. Originally written in 2005, but modified in 2008.

827. Zhu, X., Barras, C., Meignier, S., Gauvain, J.L.: Combining Speaker Identification and BIC for Speaker Diarization. In: InterSpeech (2005)

828. Zigel, Y., Wasserblat, M.a.: How to Deal with Multiple-Targets in Speaker Identification Systems? In: Speaker and Language Recognition Workshop, 2006. IEEE Odyssey 2006: The, pp. 1–7 (2006)

829. Zilovic, M., Ramachandran, R., Mammone, R.: Speaker identification based on the use of robust cepstral features obtained from pole-zero transfer functions. IEEE Transactions on Speech and Audio Processing 6(3), 260–267 (1998)

830. Zoutendijk, G.: Methods of Feasible Directions. American Elsevier Publishing Company, New York (1960)

831. Zwicker, E.: Subdivision of the Audible Frequency Range into Critical Bands (Frequenzgruppen). Journal of the Acoustical Society of America 33(2), 248–249 (1961)

832. Zwicker, E., Flottorp, G., Stevens, S.S.: Critical Band Width in Loudness Summation. Journal of the Acoustical Society of America 29(5), 548–557 (1957)

Solutions

Problems of Chapter 7

Problem 7.1
Show that

$$\mathcal{H}(X,Y|Z) = \mathcal{H}(X|Z) + \mathcal{H}(Y|X,Z) \tag{S.1}$$

Solution:
If we apply the *chain rule for conditional entropy*, given by Equation 7.35, to Equation S.1, then the left side of Equation S.1 may be written as,

$$\mathcal{H}(X,Y|Z) = \mathcal{H}(X,Y,Z) - H(Z) \tag{S.2}$$

and the right hand size as,

$$\mathcal{H}(X|Z) + \mathcal{H}(Y|X,Z) = \underline{\mathcal{H}(X,Z)} - \mathcal{H}(Z) \tag{S.3}$$
$$+ \,\mathcal{H}(X,Y,Z) - \underline{\mathcal{H}(X,Z)} \tag{S.4}$$

Equations S.2 and S.4 are equal, confirming the identity in Equation S.1.

Problem 7.2
If $\mathcal{I}(X;Y)$ is the *mutual information* between X and Y, show that,

$$\mathcal{I}(X;Y) = \mathcal{H}(X) - \mathcal{H}(X|Y) \tag{S.5}$$
$$= \mathcal{H}(Y) - \mathcal{H}(Y|X) \tag{S.6}$$

Solution:
Let us start with the definition of *mutual information*, given by Equation 7.108 and repeated here for convenience,

$$\mathcal{I}(X;Y) = \mathcal{I}(Y;X)$$
$$= \sum_{i=1}^{n}\sum_{j=1}^{m} p(X_i, Y_j) \ln \frac{p(X_i, Y_j)}{p(X_i)p(Y_j)} \tag{S.7}$$

Using the identity in Equation 6.51, we may rewrite Equation S.7 as follows,

$$\mathscr{I}(X;Y) = \mathscr{I}(Y;X)$$

$$= \sum_{i=1}^{n}\sum_{j=1}^{m} p(X_i,Y_j)\ln\frac{p(X_i|Y_j)\,p(Y_j)}{p(X_i)\,p(Y_j)} \tag{S.8}$$

$$= \sum_{i=1}^{n}\sum_{j=1}^{m} p(X_i,Y_j)\ln p(X_i|Y_j) - \sum_{i=1}^{n}\left(\sum_{j=1}^{m} p(X_i,Y_j)\ln p(X_i)\right) \tag{S.9}$$

Using the fact that $\sum_{j=1}^{m} p(X_i,Y_j) = p(X_i)$ and rearranging the two main parts of Equation S.9, we have,

$$\mathscr{I}(X;Y) = \mathscr{I}(Y;X)$$

$$= -\sum_{j=1}^{m} p(X_i)\ln p(X_i) - \left(-\sum_{i=1}^{n}\sum_{j=1}^{m} p(X_i,Y_j)\ln p(X_i|Y_j)\right) \tag{S.10}$$

$$= \mathscr{H}(X) - \mathscr{H}(X|Y) \tag{S.11}$$

Similarly, if we use the identity, $p(X_i,Y_j) = p(Y_j|X_i)p(X_i)$ instead of $p(X_i,Y_j) = p(X_i|Y_j)p(Y_j)$, in Equation S.7, we shall have,

$$\mathscr{I}(X;Y) = \mathscr{I}(Y;X)$$

$$= \mathscr{H}(Y) - \mathscr{H}(Y|X) \tag{S.12}$$

Problems of Chapter 14

Problem 14.1
neural network used to produce an *exclusive OR* logic. Table 14.1 shows the input/output relationship for a two-input *exclusive OR* unit for the four possible combinations of patterns. Write the expression for the objective function in terms of true output and the expected output of the system. Also, write the expressions for the state vector and the gradient of the objective function with respect to the state vector.
Solution:
The objective function of the minimization problem is given by,

$$E = \sum_{p=1}^{4}\left(o_{p1}^2 - t_{p1}\right)^2 \tag{S.13}$$

and the state super vector is constructed by the following,

$$\boldsymbol{\phi}^1 = \left[\phi_1^1,\phi_2^1\right]^T \tag{S.14}$$

$$\boldsymbol{\phi}^2 = \left[\phi_1^2\right] \tag{S.15}$$

$$\boldsymbol{\omega}^1 = \left[\omega_{11}^1, \omega_{12}^1, \omega_{21}^1, \omega_{22}^1\right]^T \tag{S.16}$$

$$\boldsymbol{\omega}^2 = \left[\omega_{11}^2, \omega_{12}^2\right]^T \tag{S.17}$$

$$\mathbf{x} = \left[\phi_1^1, \phi_2^1, \omega_{11}^1, \omega_{12}^1, \omega_{21}^1, \omega_{22}^1, \phi_1^2, \omega_{11}^2, \omega_{12}^2\right]^T \tag{S.18}$$

From Equation 14.10,

$$\frac{\partial E_p}{\partial (\mathbf{x})_{[j]}} = 2\left(o_{p1}^2 - t_{p1}\right) \frac{\partial o_{p1}^2}{\partial (\mathbf{x})_{[j]}} \tag{S.19}$$

and from Equations 14.24- 14.26,

$$\frac{\partial o_{p1}^2}{\partial \phi_1^2} = o_{p1}^2\left(1 - o_{p1}^2\right) \tag{S.20}$$

$$\frac{\partial o_{p1}^2}{\partial \omega_{11}^2} = o_{p1}^1\, o_{p1}^2\left(1 - o_{p1}^2\right) \tag{S.21}$$

$$\frac{\partial o_{p1}^2}{\partial \omega_{12}^2} = o_{p2}^1\, o_{p1}^2\left(1 - o_{p1}^2\right) \tag{S.22}$$

$$\frac{\partial o_{p1}^2}{\partial \phi_1^1} = \frac{\partial o_{p1}^2}{\partial o_{p1}^1} \frac{\partial o_{p1}^1}{\partial \phi_1^1}$$
$$= \omega_{11}^2\, o_{p1}^2\left(1 - o_{p1}^2\right) o_{p1}^1\left(1 - o_{p1}^1\right) \tag{S.23}$$

$$\frac{\partial o_{p1}^2}{\partial \phi_2^1} = \frac{\partial o_{p1}^2}{\partial o_{p2}^1} \frac{\partial o_{p2}^1}{\partial \phi_2^1}$$
$$= \omega_{12}^2\, o_{p1}^2\left(1 - o_{p1}^2\right) o_{p2}^1\left(1 - o_{p2}^1\right) \tag{S.24}$$

$$\frac{\partial o_{p1}^2}{\partial \omega_{11}^1} = \frac{\partial o_{p1}^2}{\partial o_{p1}^1} \frac{\partial o_{p1}^1}{\partial \omega_{11}^1}$$
$$= \omega_{11}^2\, o_{p1}^2\left(1 - o_{p1}^2\right) i_{p1}\, o_{p1}^1\left(1 - o_{p1}^1\right) \tag{S.25}$$

$$\frac{\partial o_{p1}^2}{\partial \omega_{12}^1} = \frac{\partial o_{p1}^2}{\partial o_{p1}^1} \frac{\partial o_{p1}^1}{\partial \omega_{12}^1}$$
$$= \omega_{11}^2\, o_{p1}^2\left(1 - o_{p1}^2\right) i_{p2}\, o_{p1}^1\left(1 - o_{p1}^1\right) \tag{S.26}$$

$$\frac{\partial o_{p1}^2}{\partial \omega_{21}^1} = \frac{\partial o_{p1}^2}{\partial o_{p2}^1} \frac{\partial o_{p2}^1}{\partial \omega_{21}^1}$$
$$= \omega_{12}^2\, o_{p1}^2\left(1 - o_{p1}^2\right) i_{p1}\, o_{p2}^1\left(1 - o_{p2}^1\right) \tag{S.27}$$

$$\frac{\partial o_{\mathrm{p}1}^2}{\partial \omega_{22}^1} = \frac{\partial o_{\mathrm{p}1}^2}{\partial o_{\mathrm{p}2}^1}\frac{\partial o_{\mathrm{p}2}^1}{\partial \omega_{22}^1}$$

$$= \omega_{12}^2\, o_{\mathrm{p}1}^2\left(1 - o_{\mathrm{p}1}^2\right)\, i_{\mathrm{p}2}\, o_{\mathrm{p}2}^1\left(1 - o_{\mathrm{p}2}^1\right) \tag{S.28}$$

The above equations may be used to construct $\nabla_x E$.

Problems of Chapter 24

Problem 24.1
Prove Theorem 24.1, namely, show that,

$$|s_1 s_2| = |s_1||s_2| \tag{S.29}$$

Solution:
Write the left side of Equation S.29 in terms of the real and imaginary parts of the variables involved,

$$
\begin{aligned}
|s_1 s_2| &= |(\sigma_1 + i\omega_1)(\sigma_2 + i\omega_2)| \\
&= |\sigma_1 \sigma_2 + i\sigma_1 \omega_2 + i\omega_1 \sigma_2 - \omega_1 \omega_2| \\
&= |(\sigma_1 \sigma_2 - \omega_1 \omega_2) + i(\sigma_1 \omega_2 + \omega_1 \sigma_2)| \\
&= \sqrt{(\sigma_1 \sigma_2 - \omega_1 \omega_2)^2 + (\sigma_1 \omega_2 + \omega_1 \sigma_2)^2} \\
&= \sqrt{\sigma_1^2 \sigma_2^2 + \omega_1^2 \omega_2^2 - 2\sigma_1 \sigma_2 \omega_1 \omega_2 + \sigma_1^2 \omega_2^2 + \omega_1^2 \sigma_2^2 + 2\sigma_1 \omega_2 \omega_1 \sigma_2} \\
&= \sqrt{\sigma_1^2 \sigma_2^2 + \omega_1^2 \omega_2^2 + \sigma_1^2 \omega_2^2 + \omega_1^2 \sigma_2^2}
\end{aligned}
\tag{S.30}
$$

Now do the same for the right hand side of Equation S.29,

$$
\begin{aligned}
|s_1||s_2| &= |\sigma_1 + i\omega_1||\sigma_2 + i\omega_2| \\
&= \sqrt{(\sigma_1^2 + \omega_1^2)}\sqrt{(\sigma_2^2 + \omega_2^2)} \\
&= \sqrt{\sigma_1^2 \sigma_2^2 + \sigma_1^2 \omega_2^2 + \sigma_2^2 \omega_1^2 + \omega_1^2 \omega_2^2} \\
&= \sqrt{\sigma_1^2 \sigma_2^2 + \omega_1^2 \omega_2^2 - 2\sigma_1 \sigma_2 \omega_1 \omega_2 + \sigma_1^2 \omega_2^2 + \omega_1^2 \sigma_2^2 + 2\sigma_1 \omega_2 \omega_1 \sigma_2} \\
&= \sqrt{\sigma_1^2 \sigma_2^2 + \sigma_1^2 \omega_2^2 + \omega_1^2 \sigma_2^2 + \omega_1^2 \omega_2^2}
\end{aligned}
\tag{S.31}
$$

We have arrived at the same expression in Equations S.30 and S.31, proving Equation S.29, hence proving Theorem 24.1.

Problem 24.2
Consider,

$$H(s) = U(\sigma, \omega) + iV(\sigma, \omega) \tag{S.32}$$

where

$$U(\sigma, \omega) = \frac{\sigma^3 - \omega^3}{\sigma^2 + \omega^2}$$

$$V(\sigma, \omega) = \frac{\sigma^3 + \omega^3}{\sigma^2 + \omega^2}$$

and show that meeting the Cauchy-Riemann Conditions is not sufficient for a function, $H(s)$, to be analytic. This problem relates to Theorem 24.5.

Solution:

First, let us write the limit versions of the partial derivatives, $U_\sigma, U_\omega, V_\sigma,$ and V_ω:

$$U_\sigma = \lim_{\Delta\sigma \to 0} \frac{(\sigma + \Delta\sigma)^3 - \omega^3}{(\sigma + \Delta\sigma)^2 + \omega^2} - \frac{(\sigma^3 - \omega^3)}{(\sigma^2 + \omega^2)} \tag{S.33}$$

$$U_\omega = \lim_{\Delta\omega \to 0} \frac{\sigma^3 - (\omega + \Delta\omega)^3}{\sigma^2 + (\omega + \Delta\omega)^2} - \frac{(\sigma^3 - \omega^3)}{(\sigma^2 + \omega^2)} \tag{S.34}$$

$$V_\sigma = \lim_{\Delta\sigma \to 0} \frac{(\sigma + \Delta\sigma)^3 + \omega^3}{(\sigma + \Delta\sigma)^2 + \omega^2} - \frac{(\sigma^3 + \omega^3)}{(\sigma^2 + \omega^2)} \tag{S.35}$$

$$V_\omega = \lim_{\Delta\omega \to 0} \frac{\sigma^3 + (\omega + \Delta\omega)^3}{\sigma^2 + (\omega + \Delta\omega)^2} - \frac{(\sigma^3 + \omega^3)}{(\sigma^2 + \omega^2)} \tag{S.36}$$

First, let us consider the point at the origin of the complex plane, namely, $s = 0$,

$$U_\sigma\big|_{\sigma=0, \omega=0} = \lim_{\Delta\sigma \to 0} \frac{-(\Delta\sigma)^3}{(\Delta\sigma)^3} = 1 \tag{S.37}$$

$$U_\omega\big|_{\sigma=0, \omega=0} = \lim_{\Delta\omega \to 0} \frac{-(\Delta\omega)^3}{(\Delta\omega)^3} = -1 \tag{S.38}$$

$$V_\sigma\big|_{\sigma=0, \omega=0} = \lim_{\Delta\sigma \to 0} \frac{-(\Delta\sigma)^3}{(\Delta\sigma)^3} = 1 \tag{S.39}$$

$$V_\omega\big|_{\sigma=0, \omega=0} = \lim_{\Delta\omega \to 0} \frac{(\Delta\omega)^3}{(\Delta\omega)^3} = 1 \tag{S.40}$$

Therefore, since at $s = 0$, $U_\sigma = U_\omega$ and $U_\omega = -V_\sigma$, the Cauchy-Riemann conditions are satisfied. Now, let us take $s \to 0$ along two different paths in the complex plane,

1. $s \to 0$ along the line, $\omega = \sigma$,

$$s = \sigma + i\omega = \sigma + i\sigma = \sigma(1 + i) \tag{S.41}$$

By setting $\omega = \sigma$ in Equation S.32,

$$H(s)\bigg|_{\omega=\sigma} = i\sigma \tag{S.42}$$

and the derivative,

$$\begin{aligned}
\frac{H(s)}{ds}\bigg|_{s\to 0, \omega=\sigma} &= \lim_{s\to 0} \frac{H(s) - H(0)}{s - 0} \\
&= \lim_{\substack{s\to 0 \\ \omega=\sigma}} \frac{i\sigma - 0}{\sigma(1+i) - 0} \\
&= \frac{i}{1+i} \tag{S.43}
\end{aligned}$$

2. $s \to 0$ along the line, $\omega = 0$, namely, the $\mathbb{R}$-axis,

$$s = \sigma + i\omega = \sigma + i0 = \sigma \tag{S.44}$$

The function,

$$\begin{aligned}
H(s)\bigg|_{\omega=0} &= \sigma + i\sigma \\
&= \sigma(1+i)
\end{aligned}$$

and the derivative,

$$\begin{aligned}
\frac{H(s)}{ds}\bigg|_{s\to 0, \omega=0} &= \lim_{s\to 0} \frac{H(s) - H(0)}{s - 0} \\
&= \lim_{\substack{s\to 0 \\ \omega=0}} \frac{\sigma(1+i) - 0}{\sigma - 0} \\
&= 1 + i \tag{S.45}
\end{aligned}$$

We see from Equations S.43 and S.45 that the $\frac{dH(s)}{ds}\big|_{s=0}$ does not exist while the Cauchy-Riemann conditions are satisfied at $s = 0$. Therefore, satisfying the Cauchy-Riemann conditions is not sufficient for a function to be analytic.

Problem 24.3
Given $U(\sigma, \omega) = e^{\sigma} \cos(\omega)$, find the harmonic conjugate $V(\sigma, \omega)$. This problem is related to Definition 24.37
Solution:

$$\begin{aligned}
U &= e^{\sigma} \cos(\omega) \\
U_{\sigma} &= e^{\sigma} \cos(\omega) \\
U_{\omega} &= -e^{\sigma} \sin(\omega) \\
U_{\sigma\sigma} &= e^{\sigma} \cos(\omega) \\
U_{\omega\omega} &= -e^{\sigma} \cos(\omega)
\end{aligned}$$

$$\nabla^2 U = U_{\sigma\sigma} + U_{\omega\omega}$$
$$= e^\sigma \cos(\omega) - e^\sigma \cos(\omega)$$
$$= 0$$

Therefore, the function may be analytic, see Theorem 24.6.

So the Cauchy-Riemann conditions are satisfied, namely,

$$U_\sigma = V_\omega$$
$$U_\omega = -V_\sigma$$
$$V_\omega = e^\sigma \cos(\omega) \implies V = e^\sigma \sin(\omega) + f(\sigma)$$
$$V_\sigma = e^\sigma \sin(\omega) \implies V = e^\sigma \sin(\omega) + g(\omega)$$

$$e^\sigma \sin(\omega) + f(\sigma) = e^\sigma \sin(\omega) + g(\omega) \implies f(\sigma) = g(\omega) \ \forall \ \sigma, \omega$$
$$\text{(S.46)}$$

$$\therefore f(\sigma) = g(\omega) = C \implies V(\sigma, \omega) = e^\sigma \sin(\omega) + c$$

$$H(s) = e^\sigma \cos(\omega) + ie^\sigma \sin(\omega) + iC$$
$$= e^\sigma (\cos(\omega) + i\sin(\omega)) + iC$$
$$= e^\sigma e^{i\omega} + iC$$
$$= e^{\sigma+i\omega} + iC$$
$$= e^s + iC$$

$$\therefore \boxed{H(s) = e^s = e^\sigma [\cos(\omega) + i\sin(\omega)]}$$

This has the interesting consequence that $\sin(s)$ and $\cos(s)$ where $s \in \mathbb{C}$, could become ∞.

Problem 24.4

What is the period of the exponential function,

$$H(s) = e^s \tag{S.47}$$

Solution:
cos and sin have a period of 2π. Also,

$$\cos(2k\pi) + i\sin(2k\pi) = 1 \tag{S.48}$$

where k in any integer.

Writing Equation S.48 in polar coordinates, we have, $e^{i2k\pi} = 1$. Therefore, since we may multiply 1 by e^s without changing its value, we may write,

$$e^s e^{i2k\pi} = e^s = e^{s+i2k\pi} \tag{S.49}$$

Equation S.49 establishes that the period of e^s is $i2\pi$.

Problem 24.5

Show that the only zeros of complex functions $\sin(s)$ and $\cos(s)$ are the zeros of the real sine and cosine functions.

Solution:

If $\sin(s) = 0$ then $e^{is} - e^{-is} = 0$ or $e^{i2s} = 1$.

So, $2is = 2ik\pi \implies s = k\pi, k = \{0, \pm 1, \pm 2, \cdots\}$.

If $\cos(s) = 0$ then $e^{i2s} = -1$ or $s = \frac{\pi}{2} + k\pi, k = \{0, \pm 1, \pm 2, \cdots\}$.

Problem 24.6

Find the z-transform of,

$$h(t) = \cos(\omega t) \tag{S.50}$$

Solution:

Write the expression for z-transform of the output of an ideal sampler,

$$
\begin{aligned}
H(z) &= \sum_{n=0}^{\infty} \cos(\omega nT) z^{-n} \\
&= \sum_{n=0}^{\infty} \frac{e^{i\omega nT} + e^{-i\omega nT}}{2} z^{-n} \\
&= \frac{1}{2}\left[\frac{z}{z - e^{i\omega T}} + \frac{z}{z - e^{-i\omega T}} \right] \\
&= \frac{q}{2}\left[\frac{z}{z - (\cos(\omega T) + i\sin(\omega T))} + \frac{z}{z - (\cos(\omega T) - i\sin(\omega T))} \right] \\
&= \frac{1}{2}\left[\frac{z^2 - z\cos(\omega T) + zi\sin(\omega T) + z^2 - z\cos(\omega T) - zi\sin(\omega T)}{z^2 - 2z\cos(\omega T) + 1} \right] \\
&= \frac{1}{2}\left[\frac{2z^2 - 2z\cos(\omega T)}{z^2 - 2z\cos(\omega T) + 1} \right] \\
&= \frac{z(z - \cos(\omega T))}{z^2 - 2z\cos(\omega T) + 1} \tag{S.51}
\end{aligned}
$$

Index

Fundamentals of Speaker Recognition